S0-AWR-190

GEOCHIMICA ET COSMOCHIMICA ACTA

SUPPLEMENT I

PROCEEDINGS

OF THE

APOLLO 11 LUNAR SCIENCE CONFERENCE

Houston, Texas, January 5–8, 1970

GEOCHIMICA ET COSMOCHIMICA ACTA
Journal of The Geochemical Society and The Meteoritical Society
SUPPLEMENT I

PROCEEDINGS
OF THE
APOLLO 11 LUNAR SCIENCE CONFERENCE
Houston, Texas, January 5–8, 1970

Edited
by
A. A. LEVINSON
University of Calgary, Calgary, Alberta, Canada

VOLUME 2
CHEMICAL AND ISOTOPE ANALYSES

PERGAMON PRESS
NEW YORK • OXFORD • TORONTO • SYDNEY • BRAUNSCHWEIG

Pergamon Press, Inc., Maxwell House, Fairview Park
New York, N.Y. 10523 USA Teletype 137328

Pergamon Press, Ltd. Headington Hill Hall
Oxford OX3 OBW England

Pergamon of Canada Ltd., 207 Queen's Quay West
Toronto 1 Canada

Vieweg & Sohn Gmbh Burgplatz 1, 33 Braunschweig, Germany

Pergamon House, 19a Boundary Street, Rushcutters Bay NSW2011, Australia

Library of Congress Catalogue No. 72-119485
SBN 08-016392-0

Supplement I
GEOCHIMICA ET COSMOCHIMICA ACTA
Journal of the Geochemical Society and the Meteoritical Society

Type-set in Northern Ireland by the Universities Press, Ltd., Belfast, and printed in U.S.A. by
Book Printers, Mamaroneck, New York.

BOARD OF ASSOCIATE EDITORS

Proceedings of the Apollo 11 Lunar Science Conference, Vol. 2, pp. vii to x.

Volume 2 — Contents

C. S. ANNELL and A. W. HELZ: Emission spectrographic determination of trace elements in lunar samples from Apollo 11 991
F. BEGEMANN, E. VILCSEK, R. RIEDER, W. BORN and H. WÄNKE: Cosmic-ray produced radioisotopes in lunar samples from the Sea of Tranquillity (Apollo 11) 995
W. COMPSTON, B. W. CHAPPELL, P. A. ARRIENS and M. J. VERNON: The chemistry and age of Apollo 11 lunar material 1007
J. D'AMICO, J. DE FELICE and E. L. FIREMAN: The cosmic-ray and solar-flare bombardment of the moon 1029
P. EBERHARDT, J. GEISS, H. GRAF, N. GRÖGLER, U. KRÄHENBÜHL, H. SCHWALLER, J. SCHWARZMÜLLER and A. STETTLER: Trapped solar wind noble gases, exposure age and K/Ar-age in Apollo 11 lunar fine material . . . 1037
W. D. EHMANN and J. W. MORGAN: Oxygen, silicon and aluminum in Apollo 11 rocks and fines by 14 MeV neutron activation 1071
A. E. J. ENGEL and C. G. ENGEL: Lunar rock compositions and some interpretations 1081
S. EPSTEIN and H. P. TAYLOR, JR.: The concentration and isotopic composition of hydrogen, carbon and silicon in Apollo 11 lunar rocks and minerals . 1085
P. R. FIELDS, H. DIAMOND, D. N. METTA, C. M. STEVENS, D. J. ROKOP and P. E. MORELAND: Isotopic abundances of actinide elements in lunar material . 1097
I. FRIEDMAN, J. D. GLEASON and K. HARDCASTLE: Water, hydrogen, deuterium, carbon and ^{13}C content of selected lunar material 1103
J. G. FUNKHOUSER, O. A. SCHAEFFER, D. D. BOGARD and J. ZÄHRINGER: Gas analysis of the lunar surface 1111
R. GANAPATHY, R. R. KEAYS, J. C. LAUL and E. ANDERS: Trace elements in Apollo 11 lunar rocks: Implications for meteorite influx and origin of moon 1117
P. W. GAST, N. J. HUBBARD and H. WIESMANN: Chemical composition and petrogenesis of basalts from Tranquillity Base 1143
G. G. GOLES, K. RANDLE, M. OSAWA, R. A. SCHMITT, H. WAKITA W. D. EHMANN and J. W. MORGAN: Elemental abundances by instrumental activation analyses in chips from 27 lunar rocks 1165
G. G. GOLES, K. RANDLE, M. OSAWA, D. J. LINDSTROM, D. Y. JÉROME, T. L. STEINBORN, R. L. BEYER, M. R. MARTIN and S. M. MCKAY: Interpretations and speculations on elemental abundances in lunar samples . . . 1177
K. GOPALAN, S. KAUSHAL, C. LEE-HU and G. W. WETHERILL: Rb–Sr and U, Th–Pb ages of lunar materials 1195
R. E. HANNEMAN: Thermal and gas evolution behavior of Apollo 11 samples 1207
L. A. HASKIN, R. O. ALLEN, P. A. HELMKE, T. P. PASTER, M. R. ANDERSON, R. L. KOROTEV and KATHLEEN A. ZWEIFEL: Rare-earths and other trace elements in Apollo 11 lunar samples 1213

W. HERR, U. HERPERS, B. HESS, B. SKERRA and R. WOELFLE: Determination of manganese-53 by neutron activation and other miscellaneous studies on lunar dust 1233
G. F. HERZOG and G. F. HERMAN: Na^{22}, Al^{26}, Th and U in Apollo 11 lunar samples 1239
D. HEYMANN and A. YANIV: Inert gases in the fines from the Sea of Tranquillity 1247
D. HEYMANN and A. YANIV: Ar^{40} anomaly in lunar samples from Apollo 11 . 1261
H. HINTENBERGER, H. W. WEBER, H. VOSHAGE, H. WÄNKE, F. BEGEMANN, and F. WLOTZKA: Concentrations and isotopic abundances of the rare gases, hydrogen and nitrogen in lunar matter 1269
C. M. HOHENBERG, P. K. DAVIS, W. A. KAISER, R. S. LEWIS and J. H. REYNOLDS: Trapped and cosmogenic rare gases from stepwise heating of Apollo 11 samples 1283
P. M. HURLEY and W. H. PINSON, JR.: Whole-rock Rb–Sr isotopic age relationships in Apollo 11 lunar samples 1311
I. R. KAPLAN, J. W. SMITH and E. RUTH: Carbon and sulfur concentration and isotopic composition in Apollo 11 lunar samples 1317
T. KIRSTEN, O. MÜLLER, F. STEINBRUNN and J. ZÄHRINGER: Study of distribution and variations of rare gases in lunar material by a microprobe technique 1331
T. P. KOHMAN, L. P. BLACK, H. IHOCHI and J. M. HUEY: Lead and thallium isotopes in Mare Tranquillitatis surface material 1345
J. F. LOVERING and D. BUTTERFIELD: Neutron activation analysis of rhenium and osmium in Apollo 11 lunar material 1351
K. MARTI, G. W. LUGMAIR and H. C. UREY: Solar wind gases, cosmic-ray spallation products and the irradiation history of Apollo 11 samples . . 1357
J. A. MAXWELL, L. C. PECK and H. B. WIIK: Chemical composition of Apollo 11 lunar samples 10017, 10020, 10072 and 10084 1369
C. B. MOORE, E. K. GIBSON, J. W. LARIMER, C. F. LEWIS and W. NICHIPORUK: Total carbon and nitrogen abundances in Apollo 11 lunar samples and selected achondrites and basalts 1375
G. H. MORRISON, J. T. GERARD, A. T. KASHUBA, E. V. GANGADHARAM, ANN M. ROTHENBERG, N. M. POTTER and G. B. MILLER: Elemental abundances of lunar soil and rocks 1383
V. R. MURTHY, N. M. EVENSEN and M. R. COSCIO, JR.: Distribution of K, Rb, Sr and Ba and Rb–Sr isotopic relations in Apollo 11 lunar samples 1393
G. D. O'KELLEY, J. S. ELDRIDGE, E. SCHONFELD and P. R. BELL: Primordial radionuclide abundances, solar proton and cosmic ray effects and ages of Apollo 11 lunar samples by non-destructive gamma-ray spectrometry 1407
J. R. O'NEIL and L. H. ADAMI: Oxygen isotope analyses of selected Apollo 11 materials 1425
N. ONUMA, R. N. CLAYTON and T. K. MAYEDA: Apollo 11 rocks: Oxygen isotope fractionation between minerals, and an estimate of the temperature 1429
R. O. PEPIN, L. E. NYQUIST, D. PHINNEY and D. C. BLACK: Rare gases in Apollo 11 lunar material 1435

R. W. PERKINS, L. A. RANCITELLI, J. A. COOPER, J. H. KAYE and N. A. WOGMAN: Cosmogenic and primordial radionuclide measurements in Apollo 11 lunar samples by nondestructive analysis 1455
J. A. PHILPOTTS and C. C. SCHNETZLER: Apollo 11 lunar samples: K, Rb, Sr, Ba and rare-earth concentrations in some rocks and separated phases . 1471
G. W. REED, JR. and S. JOVANOVIC: Halogens, mercury, lithium and osmium in Apollo 11 samples 1487
H. J. ROSE, JR., F. CUTTITTA, E. J. DWORNIK, M. K. CARRON, R. P. CHRISTIAN, J. R. LINDSAY, D. T. LIGON and R. R. LARSON: Semimicro X-ray fluorescence analysis of lunar samples 1493
J. N. ROSHOLT and M. TATSUMOTO: Isotopic composition of uranium and thorium in Apollo 11 samples 1499
J. P. SHEDLOVSKY, M. HONDA, R. C. REEDY, J. C. EVANS, JR., D. LAL, R. M. LINDSTROM, A. C. DELANY, J. R. ARNOLD, H. H. LOOSLI, J. S. FRUCHTER and R. C. FINKEL: Pattern of bombardment-produced radionuclides in rock 10017 and in lunar soil 1503
L. T. SILVER: Uranium–thorium–lead isotopes in some Tranquillity Base samples and their implications for lunar history 1533
A. A. SMALES, D. MAPPER, M. S. W. WEBB, R. K. WEBSTER and J. D. WILSON: Elemental composition of lunar surface material 1575
R. W. STOENNER, W. J. LYMAN and R. DAVIS, JR.: Cosmic-ray production of rare-gas radioactivities and tritium in lunar material 1583
M. TATSUMOTO: Age of the moon: An isotopic study of U–Th–Pb systematics of Apollo 11 lunar samples—II 1595
H. P. TAYLOR, JR. and S. EPSTEIN: O^{18}/O^{16} ratios of Apollo 11 lunar rocks and minerals 1613
S. R. TAYLOR, P. H. JOHNSON, R. MARTIN, D. BENNETT, J. ALLEN and W. NANCE: Preliminary chemical analyses of Apollo 11 lunar samples 1627
F. TERA, O. EUGSTER, D. S. BURNETT and G. J. WASSERBURG: Comparative study of Li, Na, K, Rb, Cs, Ca, Sr and Ba abundances in achondrites and in Apollo 11 lunar samples 1637
K. K. TUREKIAN and D. P. KHARKAR: Neutron activation analysis of milligram quantities of Apollo 11 lunar rocks and soil 1659
G. TURNER: Argon-40/argon-39 dating of lunar rock samples 1665
H. WAKITA, R. A. SCHMITT and P. REY: Elemental abundances of major, minor and trace elements in Apollo 11 lunar rocks, soil and core samples . 1685
H. WÄNKE, R. RIEDER, H. BADDENHAUSEN, B. SPETTEL, F. TESCHKE, M. QUIJANO-RICO and A. BALACESCU: Major and trace elements in lunar material . 1719
R. K. WANLESS, W. D. LOVERIDGE and R. D. STEVENS: Age determinations and isotopic abundance measurements of lunar samples (Apollo 11) . . 1729
J. T. WASSON and P. A. BAEDECKER: Ga, Ge, Ir and Au in lunar, terrestrial and meteoritic basalts 1741
R. C. WRIGLEY and W. L. QUAIDE: Al^{26} and Na^{22} in lunar surface materials: Implications for depth distribution studies 1751

Organic Geochemistry

P. I. Abell, C. H. Draffan, G. Eglinton, J. M. Hayes, J. R. Maxwell and C. T. Pillinger: Organic analysis of the returned Apollo 11 lunar sample 1757
E. S. Barghoorn: Micropaleontological study of lunar material from Apollo 11 1775
A. L. Burlingame, M. Calvin, J. Han, W. Henderson, W. Reed and B. R. Simoneit: Study of carbon compounds in Apollo 11 lunar samples . . 1779
P. Cloud, S. V. Margolis, Mary Moorman, J. M. Barker, G. Licari, D. Krinsley and V. E. Barnes: Micromorphology and surface characteristics of lunar dust and breccia 1793
P. E. Hare, K. Harada and S. W. Fox: Analyses for amino acids in lunar fines 1799
R. D. Johnson and Catherine C. Davis: Total organic carbon in the Apollo 11 lunar samples 1805
K. A. Kvenvolden, S. Chang, J. W. Smith, J. Flores, K. Pering, C. Saxinger, F. Woeller, K. Keil, I. Breger and C. Ponnamperuma: Carbon compounds in lunar fines from Mare Tranquillitatis—I. Search for molecules of biological significance 1813
G. W. Hodgson, E. Bunnenberg, B. Halpern, Etta Peterson, K. A. Kvenvolden and C. Ponnamperuma: Carbon compounds in lunar fines from Mare Tranquillitatis—II. Search for porphyrins 1829
C. W. Gehrke, R. W. Zumwalt, W. A. Aue, D. L. Stalling, A. Duffield, K. A. Kvenvolden and C. Ponnamperuma: Carbon compounds in lunar fines from Mare Tranquillitatis—III. Organosiloxanes in the hydrochloric acid hydrolysates 1845
S. Chang, J. W. Smith, I. Kaplan, J. Lawless, K. A. Kvenvolden and C. Ponnamperuma: Carbon compounds in lunar fines from Mare Tranquillitatis—IV. Evidence for oxides and carbides 1857
S. R. Lipsky, R. J. Cushley, G. G. Horvath and W. J. McMurray: Analysis of lunar material for organic compounds 1871
W. G. Meinschein, T. J. Jackson, J. M. Mitchell, E. Cordes and V. J. Shiner, Jr.: Search for alkanes of 15–30 carbon atom length in lunar fines . . 1875
Sister Mary E. Murphy, V. E. Modzeleski, B. Nagy, W. M. Scott, M. Young, C. M. Drew, P. B. Hamilton and H. C. Urey: Analysis of Apollo 11 lunar samples by chromatography and mass spectrometry: Pyrolysis products, hydrocarbons, sulfur amino acids 1879
R. C. Murphy, G. Preti, M. M. Nafissi-v and K. Biemann: Search for organic material in lunar fines by mass spectrometry 1891
J. Oró, W. S. Updegrove, J. Gibert, J. McReynolds, E. Gil-av, J. Ibanez, A. Zlatkis, D. A. Flory, R. L. Levy and C. J. Wolf: Organogenic elements and compounds in Type C and D lunar samples from Apollo 11 . . 1901
V. I. Oyama, E. L. Merek and M. P. Silverman: A search for viable organisms in a lunar sample 1921
J. H. Rho, A. J. Bauman, Teh Fu Yen and J. Bonner: Fluorometric examination of the returned lunar fines from Apollo 11 1929
J. W. Schopf: Micropaleontological studies of Apollo 11 lunar samples . 1933
Author Index to Vol. 2 1935

Proceedings of the Apollo 11 Lunar Science Conference, Vol. 2, pp. 991 to 994.

Emission spectrographic determination of trace elements in lunar samples from Apollo 11*

CHARLES S. ANNELL and ARMIN W. HELZ
U.S. Geological Survey, Washington, D.C. 20242

(*Received* 27 *January* 1970; *accepted* 6 *February* 1970)

Abstract—Thirteen lunar samples representing basaltic and gabbroic igneous rocks, as well as breccias and fines, were analyzed by d.c. arc emission spectroscopy. Zn, Cu, Ga, Rb, Li, Mn, Cr, Co, Ni, Ba, Sr, V, Be, Nb, Sc, La, Y and Zr were reported. The breccias had higher concentrations of Ni, Zn and Cu than the igneous rocks. Rb, Li, Sc, La, Y and Ba tend to be more concentrated in the igneous rocks.

INTRODUCTION

THIRTEEN Apollo 11 lunar samples were processed and analyzed spectrographically for 44 elements. The samples reported here can be classified into four groups: (1) four samples of Type A, fine-grained, vesicular igneous rocks; (2) one sample of Type B, medium-grained vuggy igneous rock; (3) seven samples of breccias; and (4) one sample of fines of less than 1 mm grain size from the bulk sample collected by Armstrong. A description of these sample groups has been reported by the Lunar Sample Preliminary Examination Team (LSPET, 1969).

The 12 rock samples consisted of one or more chips and some accompanying fines. Each sample was crushed and pulverized prior to testing. Six of the samples were crushed with a steel diamond mortar prior to grinding to an impalpable powder with an agate mortar and pestle. The remaining six samples were both crushed and ground in the agate mortar. The sample of fines, consisting of fine-grained material of less than 1 mm, was quartered twice and one quarter of the original sample was ground in the agate mortar. A 200 mg portion of each finely ground sample was mixed with 50 mg of graphite powder by grinding together in an agate mortar, and the mixture was stored in polyethylene capsules.

ANALYTICAL METHODS AND RESULTS

Three methods of d.c. arc emission spectroscopy were used:

(1) A 15 A arc in air vaporized a 25 mg sample-graphite mixture to completion. First order spectra from 2300 to 4800 Å were photographed using a 3·4 m Ebert spectrograph. Thirty-eight elements were determined in this procedure, with a coefficient of variation of $\pm$15 per cent of the amount present (BASTRON *et al.*, 1960).

(2) A 25 A arc in argon atmosphere was used to selectively volatilize and determine nine elements: Ag, Au, Bi, Cd, Ge, In, Pb, Tl and Zn. A 25 mg sample–graphite mixture was spectrochemically buffered with a 30 mg Na_2CO_3 admixture. The second order spectra in the 2400–3650 Å region was recorded with a 3 m Eagle spectrograph. A coefficient of variation of $\pm$10 per cent of the amount present obtains with this procedure (ANNELL, 1964).

* Publication authorized by the Director, U.S. Geological Survey.

(3) A 15 A arc in air vaporized a 12·5 mg portion of sample–graphite mixture, buffered with a 20 mg K_2CO_3 admixture, for determination of Cs, Rb and Li. The 3 m Eagle spectrograph was used to record first order spectra in the 6500–9000 Å region. A coefficient of variation of ± 10 per cent of the amount present obtains with this procedure (ANNELL, 1967).

The U.S. Geological Survey standards W-1, G-1, BCR-1 and mixtures thereof were used along with synthetic standards prepared in this laboratory. Based on the report by the Preliminary Examination Team on the Apollo 11 samples (LSPET, 1969), a matrix of high Fe, Ti and Si, along with proportionate amounts of Al, Mg, Ca, Na, Mn and Cr, as oxides or carbonates, was prepared and sintered. This matrix was used for dilution of other standards and mixtures to give spectra and interelement reactions

Table 1. Elements detected

Rock Type	A				B	C							D
Sample No.	57, 23*	69, 24	71, 23	72, 12	03, 23	18, 30	21, 27	59, 23	61, 37	65, 17	68, 24	73,—	84, 62
Zn	†	†	†	†	†	23	24	29	27	23	22	23	19
Cu	5·7	8·7	14	6·7	6·7	12	12	21	16	14	12	19	10
Ga	5·0	4·9	4·8	4·5	4·7	4·4	4·6	4·6	5·2	5·0	4·7	3·7	3·8
Rb	4·7	5·5	5·2	5·0	1·0	3·6	4·0	3·0	3·4	2·8	3·3	2·1	2·7
Li	17	18	17	16	9	12	13	12	11	12	14	11	11
Mn	2150	2390	2230	2230	2580	1660	1770	1990	1820	1925	2020	1880	1960
Cr	2790	2760	3060	2860	1860	2340	2480	2380	2730	2390	2600	2330	1740
Co	30	30	33	30	15	32	33	33	35	30	33	29	24
Ni	6·1	6·7	7·0	6·6	2·7	197	184	222	241	169	205	199	185
Ba†	440	420	470	430	160	220	270	240	270	260	250	240	210
Sr‡	140	130	140	130	150	110	130	120	130	140	130	160	130
V	65	72	78	76	82	60	60	57	60	57	58	66	50
Be	3·3	3·3	3·0	2·7	1·5	1·8	2·0	1·7	2·4	2·2	1·9	2·1	1·6
Nb	29	20	24	23	21	25	28	18	21	25	31	14	18
Sc	99	94	97	96	94	66	72	66	67	69	71	64	56
La	26	27	27	25	15	15	22	19	18	16	21	21	16
Y	165	164	162	155	113	97	113	102	103	103	108	89	81
Zr	635	566	644	530	380	429	424	369	393	390	482	322	273

Element abundances in parts per million, arranged according to rock type and listed in approximate order of volatility in the d.c. arc.

* Last four digits of NASA laboratory number.

† Not detected at limit of 4 ppm.

‡ Line width, <5 per cent transmission, from oscillogram.

comparable to those fort he lunar samples. The high concentration of Fe, Ti, Cr and Mn complicated the analysis, since these elements produce very complex spectra in the u.v. region. The many lines recorded from these elements caused interferences that frequently required the use of analytical lines that were less familiar or had poorer detectability. In an effort to resolve some interference, the photographed spectra were recorded on oscillograms, and the recorded lines were critically examined to determine the qualitative presence of certain elements.

The results of the analyses are listed in Table 1. The most obvious demarcation in the trace element concentrations occurs between the Type A and B igneous rocks, as compared to the Type C and D breccias. Nickel has approximately a 100-fold higher concentrations in breccias compared to the igneous rocks. Zn has over five- to sevenfold increase, depending on how far below the 4 ppm limit of detection the concentration may be in the igneous rocks. In addition, several elements with real but smaller

Table 2. Elements not detected by any of three d.c. arc techniques used

Element	Detection limits (ppm)	Element	Detection limits (ppm)
Ag	0·2	Mo	2
As	4	Nd	100
Au	0·2	P	2000
B	10	Pb	1
Bi	1	Pt	3
Cd	8	Re	30
Ce	100	Sb	100
Cs	1	Sn	10
Ge	1	Ta	100
Hf	20	Te	300
Hg	8	Th	100
In	1	Tl	1
		U	500
		W	200

differences in concentrations between the igneous and brecciated rocks can be noted: (1) Cu tends to be approximately twice as concentrated in the breccias; (2) the trace alkalis, Rb and Li, are half again as concentrated in Type A igneous rocks, as in the breccias; (3) a similar trend as in (2) is shown by Sc, La and Y; (4) Ba tends to be concentrated in Type A igneous rocks, while Sr maintains a fairly even concentration in all samples. Although several other examples can be noted, all these elemental distributions can be explained in conjunction with current petrographic and geologic studies.

Table 2 gives a list of elements not detected, along with their lowest detection limit for the method which is most applicable to their determinations.

Table 3 is a comparison of the trace element analyses of our lunar samples with the concentrations of the same elements compiled for terrestrial basalts and ultramafic rocks by TUREKIAN and WEDEPOHL (1961). The elements are listed in the approximate order of their volatility in the d.c. carbon arc in air (AHRENS and TAYLOR, 1961). The

Table 3. Average elemental concentrations of Type A and Type C lunar rocks compared to terrestrial rocks (concentrations in ppm)

	Type A	Type C	Basalts (T & W)*	Ultramafic (T & W)*
Zn	<4	24	105	50
Cu	9	15	87	10
Ga	5	5	17	1·5
Rb	5	3	30	0·2
Li	17	12	17	0·X
Mn	2250	1870	1500	1620
Cr	2870	2460	170	1600
Co	31	32	48	150
Ni	7	202	130	2000
Ba	440	250	330	0·4
Sr	135	131	465	1
V	73	60	250	40
Be	3	2	1	0·X
Nb	24	23	19	16
Sc	97	68	30	15
La	26	19	15	0·X
Y	162	102	21	0·X
Zr	594	401	140	45

* T & W = TUREKIAN and WEDEPOHL (1961).

four most volatile elements: Zn, Cu, Ga and Rb, are strongly depleted relative to the terrestrial basalts.

Acknowledgments—We wish to thank our colleagues at the U.S. Geological Survey for their help and advice during these analyses, and especially Mr. ANTHONY DORRZAPF and Mr. SOL BERMAN.

REFERENCES

AHRENS L. H. and TAYLOR S. R. (1961) *Spectrochemical Analysis*, 2nd edition, Chap. 7, pp. 80–87. Addison-Wesley.

ANNELL C. S. (1964) A spectrographic method for the determination of cesium, rubidium and lithium in tektites. *U.S. Geol. Surv. Prof. Paper* **501-B,** B148–B151.

ANNELL C. S. (1967) Spectrographic determination of volatile elements in silicates and carbonates of geologic interest, using an argon d-c arc. *U.S. Geol. Surv. Prof. Paper* **575-C,** C132–C136.

BASTRON H., BARNETT P. R. and MURATA K. J. (1960) Method for the quantitative analysis of rocks, minerals, ores and other materials by a powder d-c arc technique. *U.S. Geol. Surv. Bull.* **1084G,** 165–182.

LSPET (LUNAR SAMPLE PRELIMINARY EXAMINATION TEAM) (1969) Preliminary examination of lunar samples from Apollo 11. *Science* **165,** 1211–1227.

TUREKIAN K. K. and WEDEPOHL K. H. (1961) Distribution of the elements in some major units of the earth's crust. *Bull. Geol. Soc. Amer.* **72,** 175–192.

Proceedings of the Apollo 11 Lunar Science Conference, Vol. 2, pp. 995 to 1005.

Cosmic-ray produced radioisotopes in lunar samples from the Sea of Tranquillity (Apollo 11)

F. BEGEMANN, E. VILCSEK, R. RIEDER, W. BORN and H. WÄNKE
Max-Planck-Institut für Chemie (Otto-Hahn-Institut), 65 Mainz, Germany

(*Received* 3 *February* 1970; *accepted in revised form* 23 *February* 1970)

Abstract—^{10}Be, ^{14}C, ^{22}Na, ^{26}Al, ^{36}Cl, ^{37}Ar and ^{39}Ar have been measured in lunar fines (10084) and Type A rock (10057). The latter five isotopes have been studied using a selective dissolution technique to obtain activities (dpm/kg) for the bulk samples as well as their production rates from relevant target elements.

Our conclusions are: (1) During the past 400 years the mean flux of the high-energy component of the cosmic radiation ($E \geq 200$ MeV) has been roughly the same in the vicinity of the moon as that averaged over the orbits of meteorites; perhaps it was slightly less. (2) The lunar surface layers have been exposed to an intense flux of low-energy protons down to energies of at least 50 MeV. (3) Averaged over the last 400 years the flux of fast neutrons with $E > 1{\cdot}5$ MeV has been 25 n/cm² sec. Isotopes with different half-lifes indicate either a peculiar energy spectrum, with an exceedingly large fraction of neutrons below 2 MeV, in which case the flux given is a lower limit, or a strong time-dependence of the neutron flux. (4) An attempt to measure the mean flux of thermal neutrons during the last 300,000 years gave inconclusive results, indicating the flux to have been less than 1 n/cm² sec.

INTRODUCTION

THE EXTENSIVE study of cosmic-ray produced stable and radioactive isotopes in meteorites during the last decade has yielded a wealth of information on the intensity of the cosmic radiation in the past as well as on the history of the meteorites. While it appears to be generally agreed upon that on the average the high-energy component ($E > 200$ MeV) of the cosmic radiation has been constant within a factor of two during the past 10^9 yr (HEYMANN and SCHAEFFER, 1962; HONDA and ARNOLD, 1964; VOSHAGE, 1962, 1967), there are a number of cases where short-lived radioisotopes indicate short-term variations (FIREMAN, 1967; BEGEMANN *et al.*, 1969; FIREMAN and GOEBEL, 1969), especially of the low-energy component. Due to the fact, however, that the orbits of individual meteorites are not known there is always an inherent ambiguity as variations in time cannot be separated from variations in space.

In this respect, lunar surface material can be considered to be superior to meteoritic matter as, although the ideas on the origin of the moon are still as divergent as ever, the consensus is that it has been in its present orbit for a rather long period of time. Furthermore, while meteorites undergo ablation losses during their passage through the atmosphere, which to some extent destroy the record of the primary low-energy cosmic rays, material from the surface of the moon should contain such information as well.

It is for this reason and because of marked differences in the abundance of some important target elements in lunar samples and meteorites that the abundance *pattern* of the radioisotopes must be expected to differ from that found in meteorites. In order to facilitate a comparison with the numerous meteoritic data it is necessary to know the production rates not only for the bulk samples but for individual target elements as well. Only then the difference in the chemical composition can be properly taken into account.

EXPERIMENTAL PROCEDURE

(A) *General*

Two samples of lunar material were available for the present investigation: 100·1 g of bulk fines, <1 mm size (10084-18), and 119·1 g of rock (Type A, 10057-40). No information was supplied regarding the original position of our rock specimen within the rock 10057.

Immediately upon receipt (September 18 and October 10, 1969, respectively) their γ-activity was determined by means of a large, high-resolution Ge(Li)-detector (13·8 per cent relative efficiency and 2·85 keV FWHM for the 1·332 MeV ^{60}Co line) and a 4000-channel analyser. The bulk fines were counted as received; the rock specimen was pulverized to pass a 60 μ sieve.

Next, after addition of a ^{37}Cl spike (20 mg NaCl) and 0·5 cm^3 STP carrier Ar the fines were leached under vacuum for 3 hr with 500 cm^3 of 10^{-4} N HNO_3, washed with acetone and dried overnight at 65°C. Subsequently, both samples were fractionally dissolved in order to fractionate as many relevant target elements as possible.

Preliminary experiments on small aliquots showed that about 20 per cent of *all* mineral components *in the fines* went into solution already with cold HNO_3 (1:3), that hot HNO_3 (1:3) preferentially dissolves feldspar, resulting in a fraction enriched in Ca and Al, and that NH_4HF_2 + H_2SO_4 (conc.) at 350°C completely decomposes the residue. Hence, these were the reagents chosen, the time of treatment was 3–4 hr. All steps were performed under vacuum, after addition of carriers (25 mg Be as $BeSO_4$, 25 mg Sc as $Sc_2(SO_4)_3$, 40 mg NaCl, 0·5 cm^3 Ar). The Cl of the NaCl was enriched in ^{37}Cl (96 per cent), but was free of any detectable radioactivity.

(B) *Chemistry*

In order to minimize any contamination wherever possible we tried to apply procedures requiring only such reagents which can be purified via the gaseous phase.

1. *HNO_3-fractions.* The vacuum-system has been in use for work on meteorites for many years (WÄNKE and VILCSEK, 1959). The procedure to collect and purify the Ar is essentially unchanged.

After removal of the reaction vessel from the vacuum-system the undissolved residue was separated by centrifugation. The residues were washed repeatedly with water and finally acetone, dried at 65°C, and preserved for the next step.

An aliquot (5 per cent) of the supernatant was set aside for a determination of relevant elements. From the main solution the chlorine was precipitated as AgCl and centrifuged. After removal of the excess Ag^+-ions silica was separated by dehydration. Iron was extracted with diethyl ether from 6 N HCl. Subsequently, to the aqueous solution the same volume of diethyl ether was added and the liquids saturated with gaseous HCl at 0°C (FISCHER and SEIDEL, 1941).

The precipitate containing Al and Na was purified by repeating the procedure, and Al separated by two successive precipitations with NH_4OH.

Mg was removed from the combined filtrates with an alcoholic solution of $(NH_4)_2CO_3$. After decomposition of the ammonium salts Na was precipitated again with gaseous HCl. The Al containing hydroxides were dissolved in HCl and the Al for further purification converted to aluminate with NaOH. The alcaline solution was acidified and Al $(OH)_3$ precipitated with NH_4OH.

The solution from the first gaseous HCl-precipitation step was extracted with tri-n-butyl phosphate for the separation of Sc (EBERLE and LERNER, 1955). The aqueous phase was evaporated to dryness, the solids taken up with HNO_3, and the rare earth elements extracted with tri-n-butyl phosphate, too (PEPPARD *et al.*, 1953).

In the further procedure Be was separated by precipitation with NH_4OH in the presence of ammonium salts, the hydroxides dissolved in H_2SO_4, ammonium–peroxy-disulphate added to oxidize chromium to chromate, and Be precipitated as hydroxide.

2. *Residue.* The residue after the hot HNO_3 step was mixed thoroughly with a four-fold amount of

HN_4HF_2 and put into a large Pt-beaker inside a glass vessel, which was then connected to the vacuum system mentioned above. After evacuating the system, carefully degassed concentrated H_2SO_4 (4 cm^3/g residue) was added slowly, and the temperature raised to 350°C. The evolved HCl and SiF_4 were condensed in a trap immersed in boiling nitrogen.

When the reaction was complete the decomposition vessel was taken off the vacuum-system, H_2SO_4 added to the cake, and heated to dryness to ensure the complete removal of fluorides. The cake was then elutriated with H_2O and insoluble sulfates brought into solution with HCl. From this solution the hydroxides of Fe, Ti, Cr, Al, Mn, Sc and Be were precipitated with NH_4OH, and redissolved in HCl. The further treatment was identical to that described for the HNO_3-fraction after the removal of silica.

The filtrate of the hydroxide precipitation was further processed for the separation of Na as described above.

3. *Radiochemical purification.* Cl: AgCl was decomposed with Na_2CO_3 and the chlorine purified in form of HCl gas (VILCSEK and WÄNKE, 1963) and converted to NaCl for counting.

Be: Cu was added to the solution (1 N in HCl) and precipitated as CuS. This was followed by a purification step with $Fe(OH)_3$ in NaOH and a $BaSO_4$ scavenge. The Be-acetylacetonate complex was extracted into CCl_4 (ALIMARIN and GIBALO, 1956). After addition of EDTA Be was precipitated with NH_4OH and ignited to BeO.

(C) *Counting*

Ar was counted in the Geiger-region after addition of Q-gas (98·5 per cent He, 1·5 per cent i-butane); the counter volume was 20 cm^3, the total pressure 0·5 atm, the background 0·35 cpm, and the counting efficiency 80 per cent. The ^{37}Ar and ^{39}Ar contributions were separated by following the decay of ^{37}Ar for periods of time from 50 to 90 days. As a further check, the Ar from the different fractions of both samples was combined and the total ^{39}Ar-activity determined more than six half-lifes of ^{37}Ar after the collection of the samples. As the *rock* specimen was received as late as 80 days after collection the ^{37}Ar could not be determined with any degree of accuracy. Hence, we prefer to give no value at all.

The ^{36}Cl- and ^{10}Be-activities were determined with two Sharp thin-window (800 $\mu g/cm^2$) gas-flow detectors, each shielded by a guard counter in anti-coincidence with the sample counter, again using Q-gas. The samples were mounted on shallow gold planchets to increase the counting efficiency by back-scattering. For both isotopes the efficiencies of the 1 in. and $\frac{1}{2}$ in. detector were around 40 and 20 per cent, respectively, depending slightly on the amount of sample counted. For the sample-weights in question the efficiencies for ^{10}Be were 10 per cent lower than those for ^{36}Cl. The background counting rates were 0·16 and 0·07 cpm, respectively, using dummies of a chemical composition the same as the samples.

^{22}Na was measured by placing the separated NaCl in plastic containers between two 3 $\times$ 2 in. NaJ (Tl)-crystals. The two 511-keV annihilation quanta were counted in coincidence, as well as the combined spectra from both crystals registered with a 512-channel analyser. The latter information was used only to check the absence of any interfering activities. The counting efficiency in the 511-keV coincidence mode was 5·0 per cent.

The ^{26}Al was counted as Al_2O_3 in plastic containers, utilizing the Ge(Li)-detector described above. Both the 511 keV and the 1·81 MeV γ-lines were used, the efficiencies were 2·3 and 0·45 per cent, respectively. The activities found via the two γ-lines always agreed within the statistical limits of error.

^{14}C was determined in a 10 g aliquot of the fines which was intended for determination of the Ar-activity in the bulk fines. This Ar, however, was unfortunately lost. The procedure used to produce CO_2—after the addition of lamp black—was essentially that described by GOEL (1962) except that the conversion temperature was 1100°C. The CO_2 was converted to C_2H_2 with Li (BIEN, 1967); the activity determined by means of a thin-wall proportional counter (Oeschger-type) with a total volume of 2·5 l. and a central volume of 1·5 l., working at a pressure of 200 torr. The counting efficiency, with the lower and upper threshold chosen, was 49 per cent, the background 1·21 cpm.

RESULTS AND DISCUSSION

In Table 1, columns 5 and 9, are listed the concentrations in the bulk samples of those target elements which are most important for the production of the radioisotopes studied. (For a comparison with the results obtained in other laboratories see WÄNKE *et al.*, 1970b.) The amounts dissolved in each step of the fractional dissolution procedure are also given. They were determined by wet chemistry and/or neutron activation of aliquots of the various solutions as well as aliquots of the portions left

Table 1. Results of chemical and neutron activation analyses, and decay rates of the radioisotopes measured

	Lunar fines 10084–18					Lunar rock 10057–40			
	HNO_3 (25°C)	HNO_3 (70°C)	residue	sum	bulk	HNO_3 (90°C)	residue	sum	bulk
Chem. analysis (% by weight)									
O	10·6	4·5	26·4	41·5		9·5	30·9	40·4	
Mg	1·9	0·14	2·8	4·8		0·4	3·8	4·2	
Al	3·9	1·8	1·2	6·9		3·4	1·2	4·6	
Si	3·9	1·8	14·0	19·7		4·0	14·9	18·9	
K	0·045	0·014	0·050	0·109		0·057	0·144	0·201	
Ca	3·6	1·6	3·6	8·8		2·3	5·6	7·9	
Ti	0·5	<0·2	3·8	4·3		0·33	5·35	5·7	
Fe	4·2	0·66	7·1	12·0		1·3	13·0	14·3	
Activity (dpm/kg)									
^{10}Be	5·9±1	—	5·8±1	14±3*	—	—	—	—	—
^{14}C	—	—	—	—	39±4	—	—	—	35±5
^{22}Na	32±5	10±2	19±3	61±6	61±15	9±2	24±6	33±7	24±6
^{26}Al	57±13	37±15	27±14	121±25	86±12	25±10	32±16	57±20	70±10
^{36}Cl	6·5±0·3	1·9±0·3	6·2±0·3	14·6±0·6	—	3·30±0·31	8·85±0·38	12·2±0·5	—
^{37}Ar	14·4±0·9	4·7±4·3	14·1±0·9	33·2±4·5	—	—	—	—	—
^{39}Ar	2·43±0·14	1·18±0·43	3·86±0·20	7·5±0·5	8·1±0·3	1·99±0·17	7·09±0·23	9·0±80·29	8·5±0·3

* The entry of ^{10}Be under 'sum' contains a small contribution from the HNO_3 (70°C)-step, which was not measured but calculated via the amount of oxygen in this fraction. The ^{39}Ar-activities under 'bulk' are not based on measurements of bulk samples, proper, but on the combined Ar from the separate fractions.

undissolved after the individual steps. Oxygen was determined by fast neutron activation in the undissolved portions as well, but the entries in columns 2, 3 and 6 are calculated ones assuming all cations to have been present as oxides. The values obtained from the difference between the oxygen content of the initial samples and their respective residues are naturally rather uncertain ("Elephant method", HOUTERMANS, 1950).

The lower part of Table 1 shows the decay rates of the radioisotopes (in dpm/kg) for the individual fractions or the bulk samples, or both. With the exception of ^{26}Al in the 'fines', the agreement between the sum of the activities and that obtained on the bulk samples is good for a given isotope, showing the data to be internally consistent. Where comparable they, furthermore, agree fairly well with those obtained in other laboratories (LSPET, 1970; SHEDLOVSKY *et al.*, 1970; PERKINS *et al.*, 1970; STOENNER *et al.*, 1970; FIREMAN *et al.*, 1970).

Taking into account that the lunar surface has only a 2π-geometry for the exposure to the cosmic radiation, the decay rates for all isotopes and their abundance *pattern* are distinctly different from what is observed in meteorites (see e.g. HONDA and ARNOLD, 1964), reflecting the different chemical composition as well as the presence of low-energy solar primaries.

(A) *Production rates on individual target elements*

In order to separate the two effects as much as possible we shall now proceed to calculate the decay rates of ^{22}Na, ^{26}Al, ^{36}Cl, ^{37}Ar and ^{39}Ar on individual target elements from the data in Table 1.

^{39}Ar. Proton-induced spallation reactions on Ca can be neglected compared to the contribution from Ti, as only the rare isotopes of Ca with a combined abundance of 3 per cent can yield ^{39}Ar. This is reflected in the 600 MeV cross section data of STOENNER *et al.* (1970) who found the production on Ca to be eight times less than that on Ti.

Similarly, the contribution from the $^{41}K(p, 2pn)^{39}Ar$-reaction is insignificant. This can be seen when comparing the activities possibly produced by analogous reactions on other nuclides, e.g. $^{56}Fe(p, 2pn)^{54}Mn$ and $^{48}Ti(p, 2pn)^{46}Sc$, in lunar matter. The Q-values are $-20{\cdot}4$ MeV and $-22{\cdot}1$ MeV, respectively, compared to $-17{\cdot}7$ MeV for the $^{41}K(p, 2pn)^{39}Ar$-reaction; the cross sections will thus be comparable. Allowing for the much lower abundance of ^{41}K we derive from the observed ^{54}Mn-activities of about 50 dpm/kg (SHEDLOVSKY *et al.*, 1970; PERKINS *et al.*, 1970; O'KELLEY *et al.*, 1970) a contribution of ^{39}Ar from ^{41}K of 0·04 dpm/kg; and from the ^{46}Sc-activities of about 10 dpm/kg (PERKINS *et al.*, 1970; O'KELLEY *et al.*, 1970) a value of 0·02 dpm/kg. These estimates are even upper limits, because both ^{54}Mn and ^{46}Sc are assumed to be solely produced by the (p, 2pn)-reaction. In both cases, however, (n, p)-reactions, on ^{54}Fe and ^{46}Ti, respectively, may very well contribute significantly to the activities observed.

With neutrons as bombarding particles the $^{40}Ca(n, 2p)$-, $^{42}Ca(n, \alpha)$-, $^{43}Ca(n, \alpha n)$-reactions as well as similar ones on the heavier isotopes are possible. Of these, however, only the $^{42}Ca(n, \alpha)$-reaction with a Q-value of $+0{\cdot}34$ MeV is of any importance in spite of the low abundance of ^{42}Ca. For $^{40}Ca(n, 2p)$ the Q-value is $-8{\cdot}1$ MeV and the cross section on neighbouring nuclides at 14·7 MeV is smaller than 1 mb (BRAMLITT and FINK, 1963), while in case of the heavier isotopes the contribution will be negligible because of both the low abundance *and* the increasing Q-value.

Hence, the principal reactions responsible for the total ^{39}Ar-activity will be spallation reactions on Fe and Ti, and the $^{39}K(n, p)$- and $^{42}Ca(n, \alpha)$-reactions induced by neutrons in the MeV energy range. Unfortunately, from the experimental data of Table 1 the latter two cannot be separated, and contrary to chondrites where the Ca/K-ratio is about an order of magnitude smaller the contribution from Ca cannot be neglected entirely.

In order to estimate its importance, however, one can compare the cross sections for the two competing reactions. Since no data are available for the $^{42}Ca(n, \alpha)^{39}Ar$-reaction, we shall assume its cross section to be the same as that of the $^{40}Ca(n, \alpha)^{37}Ar$-reaction, as both are exothermic (n, α)-reactions with similar Q-values. For energies above 4 MeV the ratio $\sigma[^{40}Ca(n, \alpha)^{37}Ar]/\sigma[^{39}K(n, p)^{39}Ar]$ is about 0·6 while for lower energies it gradually decreases (JESSEN *et al.*, 1965; BASS, 1968). Hence, an upper limit for the contribution from Ca is obtained by adding to the amount of potassium in the samples 0·6 times the amount of ^{42}Ca.

The production rates $P(X)$ thus obtained are listed in Table 2. As the contributions from Fe and Ti cannot be calculated independently the yield from Ti was set to

Table 2. Production rates of various isotopes on the major target elements, given in dpm/kg target element

Isotope		Mg	Al	Target Si	K	Ca	Fe
^{22}Na	Fines	300	550	50			
	Rock	180	120	100			
^{26}Al	Fines	—	1400	150			
	Rock	—	540	170			
^{36}Cl	Fines					175	
	Rock					152	
^{37}Ar	Fines					400	
	Rock					—	
^{39}Ar	Fines				3300	—	7·9
	Rock				2700	—	7·5

be four times that from Fe, corresponding to a hardness exponent of $n = 2{\cdot}5$ in the semi-empirical spallation formula as given by GEISS *et al.* (1962). This formula has been found to fit the abundance pattern of cosmic ray produced stable and radioactive nuclides in meteorites over a wide mass range. We wish to emphasize that the production rates on potassium are essentially independent of the value of $P(\text{Ti})/P(\text{Fe})$. A change by a factor of two affects $P(\text{K})$ by less than 10 per cent. They, furthermore, are not very sensitive to the correction for the contribution from the $^{42}Ca(n, \alpha)$-reaction. Neglecting it entirely, the production rates are 4400 and 3100 dpm/kg for the fines and rock, respectively (WÄNKE *et al.*, 1970a); assuming the (n, α)-cross section to be equal to the (n, p)-cross section for the whole range of neutron energies yields 2800 and 2500 dpm/kg, respectively.

^{37}Ar *and* ^{36}Cl. The important elements are Ca, Ti and Fe. From the data available at present, it is not possible to determine the contribution from Ti and Fe except to say that it is very minor: In the cold HNO_3-fraction and the residue the same amounts of Ca were found as well as the same ^{37}Ar- and ^{36}Cl-activities, in spite of the fact that the Ca/Ti-ratio in the two fractions is different by a factor of 8 (Table 1). It should be remembered, however, that the entries in Table 2 are upper limits. For example, if the production of ^{36}Cl on Fe and Ti is set equal to that of ^{39}Ar on these elements, the production rate on Ca would be lower by about 20 per cent.

For ^{37}Ar we believe the predominance of Ca again to be due to fast neutrons, specifically to the $^{40}Ca(n, \alpha)^{37}Ar$-reaction.

^{26}Al. At present, the large statistical errors in the decay rates do not allow to determine the contributions from Al and Si with the desired accuracy. Nevertheless, it is quite clear that in the fines as well as the rock Al is the more important target element. Especially in the fines, although the Si/Al-ratio is about 3, by far the largest percentage of the total ^{26}Al is produced on Al.

^{22}Na. The target elements to be considered are Mg, Al and Si. Here again, the entries in Table 2 are tentative only. What they do show, however, is a distinct difference in relative importance of the three target elements in the fines and in the rock. In the latter the production rates on Al and Si are about the same, while that on Mg is slightly higher. In the fines, on the other hand, the contribution from Si is very small and the yield from Al is distinctly *higher* than that from Mg.

(B) *Implications for the cosmic radiation*

Due to the peculiar chemical composition of the lunar material, especially the absence of major amounts of metallic iron and troilite, the only high-energy product which could be determined with sufficient accuracy is ^{39}Ar produced from iron. As mentioned above, the production rate in Table 2 has been calculated under the assumption that for this isotope $P(\mathrm{Ti})/P(\mathrm{Fe}) = 4$. If one uses a ratio of 2 instead—which appears to be the lowest reasonable choice—P(Fe) would be 11 dpm/kg. In small iron meteorites and the metal phase from stone and stony-iron meteorites, exposed in 4π-geometry, it ranges between 20 and 30 dpm/kg (VILCSEK and WÄNKE, 1963; BEGEMANN and VILCSEK, 1969). Hence, during the last 400 yr the intensity of the high-energy cosmic radiation ($E > 200$ MeV) in the vicinity of the moon has been roughly the same as that averaged over the orbits of the meteorites. If there is any difference at all the flux near the moon has been slightly less, in accordance with direct measurements during a solar minimum (O'GALLAGHER and SIMPSON, 1967), as well as conclusions drawn from measurements on meteorites (FIREMAN, 1967; BEGEMANN and VILCSEK, 1969).

For the low-energy component, there is much more information contained in the data. In the fines, e.g. ^{22}Na is more abundantly produced from Al than from Mg, and the yield from Si is much lower. As found in numerous target experiments with proton energies above 100 MeV (for a recent compilation see KIRSTEN and SCHAEFFER, 1969) that $\sigma(\mathrm{Mg}):\sigma(\mathrm{Al}):\sigma(\mathrm{Si}) \approx 2:1{\cdot}5:1$, we conclude that protons in this energy range cannot be responsible for most of the ^{22}Na in the fines. At lower energies, however, the ^{27}Al(p, x)^{22}Na cross section exhibits a pronounced maximum at about 50 MeV (RAYUDU, 1964; FURUKAWA *et al.*, 1965) which apparently does not exist for the Mg(p, x)^{22}Na-reaction. It is true that older measurements show such a maximum for the latter reaction as well (MEADOWS and HOLT, 1957), but this appears to be an artifact introduced by the use of an erroneous excitation function for the ^{27}Al(n, x)^{24}Na monitor reaction. Using a more recent one (HICKS *et al.*, 1956) the maximum vanishes, in accordance with direct measurements by RAYUDU (1964). Hence, it is our contention that protons of around 50 MeV were the most important nuclear active particles for the production of ^{22}Na in the fines.

At the same time this accounts for the low yield on Si as well (Fig. 1). It, furthermore, explains the difference between fines and rock, if one assumes our specimen of the rock to have been partially shielded from these low energy protons. Not only will the bulk activity be lower in the rock, but the ratio $P(\mathrm{Mg}):P(\mathrm{Al}):P(\mathrm{Si})$ will change towards a more equal production on the three target elements as well, in accordance with our observation.

The same conclusion is reached from the production rates of ^{26}Al on Al and Si.

Here, no cross section data are available, but one can compare the entries in Table 2 with those given by FUSE and ANDERS (1964) for stone meteorites. As these authors found P(Si) = 310 dpm/kg the agreement is excellent, because for the surface of a large body this has to be roughly reduced by a factor of two. Obviously, nuclear reactions on Si leading to ^{26}Al induced by low-energy protons are insignificant. This is supported, e.g. by the fact that ^{54}Mn, produced at least partly by the analogous reaction on ^{56}Fe, is found not to depend on depth in a single lunar rock (SHEDLOVSKY *et al.*, 1970). For the production on Al, on the other hand, the contribution from low-energy protons must play a major role, as is apparent from the continuous decrease of P(Al) in lunar fines [P(Al) = 1400 dpm/kg], lunar rock (540 dpm/kg) and

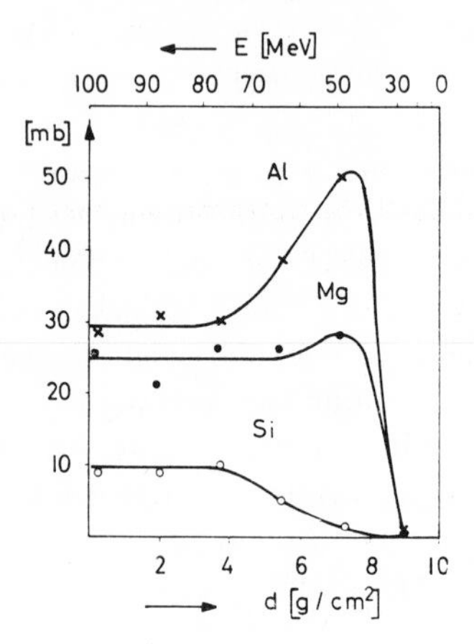

Fig. 1. Cross sections for the production of ^{22}Na from Mg, Al and Si upon bombardment of a thick target with 100 MeV protons (RAYUDU, 1964). The curves drawn through the data points as well as the energy scale were added by us.

meteorites (480/2 = 240 dpm/kg) (FUSE and ANDERS, 1969). Again, this is strikingly confirmed indirectly by the results of SHEDLOVSKY *et al.* (1970), who found a very steep decrease with depth of the activity of ^{55}Fe, an isotope whose position relative to the main target isotope is the same as that of ^{26}Al to Al.

In order to deduce the fast-neutron flux from the ^{39}K(n, p)-produced ^{39}Ar we shall follow the procedure described in detail by BEGEMANN *et al.* (1967). There, it has been shown that for energies above 1·5 MeV the effective cross section for the (n, p)-reaction is 140 mb, rather independent of the shape of the neutron energy spectrum assumed. (The limit of E = 1·5 MeV is solely due to the lack of cross section data for lower neutron energies.) This value combined with a production rate of 3000 dpm/kg potassium, found in both the fines and our rock specimen, yields for the mean flux of fast neutrons ($E >$ 1·5 MeV), averaged over the last 400 yr, ϕ = 25 n/cm^2 sec.

However, the high ratio of the production rates of ^{39}Ar from potassium via the (n, p)-reaction to that of ^{37}Ar from Ca via the (n, α)-reaction indicates that either the neutron energy spectrum contains an extremely large fraction of neutrons below

2 MeV, or the neutron flux is strongly time dependent (Wänke *et al.*, 1970b). (We want to point out that this conclusion is true even when ascribing all the ^{37}Ar to the $^{40}Ca(n, \alpha)$-reaction. If in addition to the contributions from Ti and Fe (see above) any other reactions on Ca and K should be important as well, the argument would be even stronger.) In case there are indeed a large number of neutrons with energies <2 MeV the flux of 25 n/cm^2 sec is a *lower* limit.

Finally, as was mentioned above, the first treatment of the fines consisted of a leach with 10^{-4} N HNO_3 at room temperature. The aim was to leach possibly some chlorine and to deduce the flux of thermal neutrons from the $^{35}Cl(n, \gamma)$-produced ^{36}Cl-activity. The results, however, are not very conclusive. The ^{36}Cl-activity found was $(1{\cdot}2 \pm 1{\cdot}5) \times 10^{-2}$ dpm, the isotope dilution measurement showed an excess of 1·7 mg of chlorine after treatment of the sample, assuming normal isotopic composition for the excess. On the 3σ-significance level this corresponds to a specific activity of less than 3 dpm/kg for a normalized Cl-content of 100 ppm (Eberhardt *et al.*, 1963), or a slow neutron flux of less than 1 n/cm^2 sec, *provided* all the excess chlorine was actually leached from the sample. If part of it was introduced from other sources the specific activity and the neutron flux would be correspondingly higher. For this and other reasons the apparent agreement of this neutron flux with that derived by Albee *et al.* (1970) from measurements of the Gd isotopes is certainly fortuitous.

Acknowledgments—We are grateful to NASA for making available such a generous amount of lunar material for our investigation. We wish to thank the technical staff of our Institute, in particular Mrs. G. Dreibus and Mrs. U. Weidlich for their devoted help.

Finally, the considerable financial support by the Bundesministerium für Wissenschaftliche Forschung is gratefully acknowledged.

References

Albee A. L., Burnett D. S., Chodos A. A., Eugster O. J., Huneke J. C., Papanastassiou D. A., Podosek F. A., Russ G. Price II, Sanz H. G., Tera F. and Wasserburg G. J. (1970) Ages, irradiation history, and chemical composition of lunar rocks from the Sea of Tranquillity. *Science* **167**, 463–466.

Alimarin I. P. and Gibalo I. M. (1956) Die Trennung des Be von Al und anderen Elementen. *Z. anal. Chem.* **156,** 435 (1957). (*Z. Anal. Chim.* **11,** 389–392.)

Bass R. (1968) Private communication.

Begemann F., Vilcsek E. and Wänke H. (1967) The origin of the "excess" argon-39 in stone meteorites. *Earth Planet. Sci. Lett.* **3,** 207–212

Begemann F., Rieder R., Vilcsek E. and Wänke H. (1969) Cosmic ray produced radionuclides in the Barwell and St. Severin meteorites. In *Meteorite Research*, (editor P. Millman), pp. 267–274. D. Reidel.

Begemann F. and Vilcsek E. (1969) Chlorine-36 and argon-39 production rates in the metal of stone and stony-iron meteorites. In *Meteorite Research*, (editor P. Millman), pp. 355–362. D. Reidel.

Bien G. (1967) Personal communication.

Bramlitt E. T. and Fink R. W. (1963) Rare nuclear reactions induced by 14·7 MeV neutrons. *Phys. Rev.* **131,** 2649–2663.

Eberhardt P., Geiss J and Lutz H. (1963) Neutrons in meteorites. In *Earth Science and Meteoritics*, (editors J. Geiss and E. D. Goldberg), pp. 143–168, North-Holland.

EBERLE A. R. and LERNER M. W. (1955) Separation and determination of Sc. *Anal. Chem.* **27,** 1551–1554.

FIREMAN E. L. (1967) Radioactivities in meteorites and cosmic-ray variations. *Geochim. Cosmochim. Acta* **31,** 1691–1700.

FIREMAN E. L. and GOEBEL R. (1969) Argon-37 and argon-39 in recently fallen meteorites and cosmic-ray variations S.A.O. Reprint 906–164.

FIREMAN E. L., D'AMICO J. C. and DEFELICE J. C. (1970) Tritium and argon radioactivities in lunar material. *Science* **167,** 566–568.

FISCHER W. and SEIDEL W. (1941) Die Fällung von $AlCl_3 \cdot 6H_2O$ aus ätherisch-wässriger Salzsäure und ihre Bedeutung als Trennungsoperation. *Z. Anorg. Chem.* **247,** 333–366.

FURUKAWA M., KUME S. and OGAWA M. (1965) Excitation functions for the formation of ^{7}Be and ^{22}Na in proton induced reaction on ^{27}Al. *Nucl. Phys.* **69,** 362–368.

FUSE K. and ANDERS E. (1969) Aluminum-26 in meteorites—VI. Achondrites. *Geochim. Cosmochim. Acta* **33,** 653–670.

GEISS J., OESCHGER H. and SCHWARZ U. (1962) The history of cosmic radiation as revealed by isotopic changes in the meteorites and the earth. *Space Sci. Rev.* **1,** 197–223.

GOEL P. S. (1962) Cosmogenic carbon-14 and chlorine-36 in meteorites. Thesis, Carnegie Inst. Techn.

HEYMANN D. and SCHAEFFER O. A. (1962) Constancy of cosmic rays in time. *Physica* **28,** 1318–1323.

HICKS H. G., STEVENSON P. C. and NERVIK W. E. (1956) Reaction ^{27}Al (p, 3pn)^{24}Na. *Phys. Rev.* **102,** 1390–1392.

HONDA M. and ARNOLD J. R. (1964) Effects of cosmic rays on meteorites. *Science* **143,** 203–212.

HOUTERMANS F. G. (1950) Personal communication.

JESSEN P., BORMANN M., DREYER F. and NEUERT H. (1965) Compilation of experimental excitation functions of some fast neutron reactions up to 20 MeV. I. Inst. f. Experimentalphysik, Universität Hamburg, January 1965.

KIRSTEN T. A. and SCHAEFFER O. A. (1969) High energy interactions in space. In *Interactions of Elementary Particle Research in Science and Technology,* (editor L. C. L. Yuan), Academic Press.

LSPET (LUNAR SAMPLE PRELIMINARY EXAMINATION TEAM) (1969) Preliminary examination of lunar samples from Apollo 11. *Science* **165,** 1211–1227.

MEADOWS J. W. and HOLT R. B. (1951) Some excitation functions for protons on magnesium. *Phys. Rev.* **83,** 1257.

O'GALLAGHER J. J. and SIMPSON J. A. (1967) The heliocentric intensity gradients of cosmic-ray protons and helium during minimum solar modulation. *Astrophys. J.* **147,** 819–827.

O'KELLEY G. D., ELDRIDGE J. S., SCHONFELD E. and BELL P. R. (1970) Elemental compositions and ages of lunar samples by non-destructive gamma-ray spectrometry. *Science* **167,** 580–582.

PEPPARD D. F., FARIS J. P., GRAY P. R. and MASON G. W. (1953) Studies of the solvent extraction behaviour of the transition elements I. Order and degree of fractionation of the trivalent rare earths. *J. Phys. Chem.* **57,** 294–301.

PERKINS R. W., RANCITELLI L. A., COOPER J. A., KAYE J. H. and WOGMAN N. A. (1970) Cosmogenic and primordial radionuclides in lunar samples by non-destructive gamma-ray spectrometry. *Science* **167,** 577–580.

RAYUDU G. S. V. (1964) Nuclear reactions for the study of solar flare protons. Nuclear Chem. and Geochem. Res. at Carnegie Inst. Techn. 1963–64, Progress Report.

SHEDLOVSKY J. P., HONDA M., REEDY R. C., EVANS J. C., LAL D., LINDSTROM R. M., DELANY A. C., ARNOLD J. R., LOOSLI H.-H., FRUCHTER J. S. and FINKEL R. C. (1970) Pattern of bombardment-produced radionuclides in rock 10017 and in lunar soil. *Science* **167,** 574–576.

STOENNER R. W., LYMAN W. J. and DAVIS R., JR. (1970) Cosmic ray production of rare gas radioactivities and tritium in lunar material. *Science* **167,** 553–555.

VILCSEK E. and WÄNKE H. (1963) Cosmic ray exposure ages and terrestrial ages of stone and iron meteorites derived from ^{36}Cl and ^{39}Ar measurements. In *Radioactive Dating,* pp. 381–393. IAEA Vienna.

VOSHAGE H. (1962) Eisenmeteorite als Raumsonden für die Untersuchung des Intensitätsverlaufs der kosmischen Strahlung während der letzten Milliarden Jahre. *Z. Naturforsch.* **17a,** 422–432.

Voshage H. (1967) Bestrahlungsalter und Herkunft der Eisenmeteorite. *Z. Naturforsch.* **22a,** 477–506.
Wänke H. and Vilcsek E. (1959) Argon-39 als Reaktionsprodukt der Höhenstrahlung in Eisenmeteoriten. *Z. Naturforsch.* **14a,** 929–934.
Wänke H., Begemann F., Vilcsek E., Rieder R., Teschke F., Born W., Quijano-Rico M., Voshage H. and Wlotzka F. (1970a) Major and trace elements and cosmic-ray produced radioisotopes in lunar samples. *Science* **167,** 523–525.
Wänke H., Rieder R., Baddenhausen H., Spettel B., Teschke F., Quijano-Rico M. and Balacescu A. (1970b) Major and trace elements in lunar material. *Geochim. Cosmochim. Acta,* Supplement I.

Proceedings of the Apollo 11 Lunar Science Conference, Vol. 2 pp. 1007 to 1027.

The chemistry and age of Apollo 11 lunar material

W. COMPSTON
Department of Geophysics and Geochemistry, Australian National University, Canberra, Australia

B. W. CHAPPELL
Department of Geology, Australian National University, Canberra, Australia

P. A. ARRIENS and M. J. VERNON
Department of Geophysics and Geochemistry, Australian National University, Canberra, Australia

(*Received* 5 *February* 1970; *accepted in revised form* 21 *February* 1970)

Abstract—Major-element, trace-element and Rb–Sr analyses are reported from seven Apollo 11 igneous rocks, two breccias and the fines. The chemistry of these and other Apollo 11 specimens shows that the igneous rocks divide into two groups. Group 1 comprising samples 10017, 10022, 10024, 10049, 10057, 10069, 10071 and 10072, and Group 2, 10003, 10020, 10044, 10045, 10047, 10050, 10058 and 10062. Group 1 is higher in Rb, K, Ba, U and Th by factors of 4 or more; higher in Na, P, S, Zr, Y, REE by factors between 1·5 and 2; slightly higher in TiO_2, and slightly lower in CaO and Al_2O_3. The division does not reflect texture, except that most of Group 1 are LSPET Type A, but probably represents two different lava-flows. Internal mineral isochrons for samples 10017 and 10072 (Group 1) give the age of crystallization as 3·78 ± 0·10 b.y. Incomplete mineral data from samples 10045 and 10047 (Group 2) are consistent with a similar age.

Chemically the fines closely resemble the breccias. Comparisons of Si, Ti, Fe, Al, Ca, Mg and Na show that the regolith at Tranquillity Base could be composed of equal portions of the two lava groups, minus about 5% ilmenite, plus about 15% gabbroic anorthosite. The regolith also requires a component enriched in Sr^{87} which might be the anorthosite, and in Cu, Zn and Ni which might represent meteoritic additions.

U–Pb ages reported for the fines, breccia and igneous rocks must be discordant if the regolith is composed largely of 3·70 b.y. material. We propose that lead was lost from the lavas during extrusion at 3·70 b.y. and concentrated in surface materials which have been incorporated into the present regolith, so that leads from the lavas and regolith lie on 3·70 b.y. secondary isochrons.

INTRODUCTION

THE TRACE and major element compositions of igneous rocks generally show which particular specimens could be cogenetic and hence which might be grouped together validly on a Sr^{87}-evolution diagram. We therefore proposed a study of the chemistry and age of the Apollo 11 samples, first by non-destructively determining nineteen trace elements on each rock specimen, followed by the destructive determination of (a) major elements on one portion of the recovered powder and (b) rubidium and strontium by stable isotope dilution and Sr^{87}/Sr^{86} an another portion. Recent developments of X-ray fluorescence (XRF) techniques permit the accurate and non-destructive determination of many trace elements using samples of 1 to 2 g in weight (NORRISH and CHAPPELL, 1967). In addition, determination of most major elements, to a high precision and with an accuracy probably equal to or better than that obtained with classical analytical methods, is possible by XRF methods using samples as small as 0·3 g (NORRISH and HUTTON, 1969). For this project, we required as many different specimens between 1 g and 5 g weight as possible, to obtain maximum chemical variety and maximum dispersion of Rb/Sr ratios. Seven igneous rocks, two breccias and the <1 mm fines were received from NASA. Contrary to expectations, we found that two different chemical groups were present in the crystalline rocks, signifying two different rock units, and that each of the groups had little internal dispersion in Rb/Sr ratios. In these circumstances, no conclusive interpretation of the total-rock Rb–Sr data could be made, and age-determinations were possible only by separating and analyzing mineral concentrates, which by virtue of the extreme chemical differentiation between crystals and residual glass, had a sufficient range in Rb/Sr to give precise internal Rb–Sr isochrons.

The measurement of rubidium and strontium by the two techniques led to the early discovery of the unreliability of our concentration measurements by isotope dilution for the titanium-rich lunar samples. Although the remedy was simple, the effect could have remained undetected even with duplicate measurements. Reports at the Houston Lunar Science Conference (*Science*, **167**, pp. 417–790) show that there is considerable disagreement between various Rb–Sr analysts on the slope of the line joining total-rock data on the Sr^{87}-evolution diagram. Some of this is certainly due to the titanium interference.

The substance of our results has already been reported (COMPSTON *et al.*, 1970). Although all of our Sr^{87}/Sr^{86} values have been slightly lowered, no essentially new data are given here. However, we now report some further details of the experimental methods and document the precision of our results. More important, we discuss our data and conclusions in the light of the papers given by the other investigators at Houston, and, as far as possible, attempt to give an account of the chemical evolution of the Apollo 11 specimens that we received which is consistent with other data.

ANALYTICAL METHODS

Sample preparation

All rock samples were obtained as chips. All crushing was carried out manually in an agate spherical bowl mortar with an agate pestle, to avoid losses and to restrict the contamination to silica. The powder was sieved through silk screening cloth in Perspex (Lucite) frames. In cases in which the powder was split into fractions for mineral separations, this was done at −60 mesh. Sample 10047 is the only rock in which the mineral components were separated by hand picking, at an average particle size from 170 to 200 μm. Other mineral concentrates were obtained by using heavy liquids and magnetic separation on very fine powder (<40 μm).

X-ray spectrometric measurements

Trace elements were determined by measuring the emission of characteristic radiation from 1 g of powdered sample and correcting for matrix effects by either measuring or calculating mass absorption coefficients (NORRISH and CHAPPELL, 1967). Absorption coefficients for Rb, K_α and Sr K_α radiations were measured directly and these values were extrapolated to all wavelengths shorter than the iron K absorption edge. For longer wavelengths (V, Cr, Co, Ba, La, Ce, Pr and Nd) a direct measurement of absorption coefficient was impractical since this would have involved dilution of the lunar sample with material of low mass absorption coefficient such as cellulose. Thus for these wavelengths, values of the Fe K_α mass absorption coefficient were calculated for each sample using the major element analyses and the absorption coefficient data of HEINRICH (1966), a small correction being applied based on the difference between measured and calculated absorption coefficients on terrestrial basalts. Following the measurement of mass absorption coefficients, the powder was recovered and used for emission measurements. Our sample preparation for emission differed from that normally used

(Norrish and Chappell, 1967) since complete sample recovery was required. The sample was pressed by hand with a tool-steel plunger into a Perspex sleeve fitted with a plastic cover at the bottom to prevent loss of sample during analysis. This method of sample preparation is not as reproducible as our normal method of pelletizing, but tests of Sr K_{α} emission showed a reproducibility of better than 1 per cent. In any case, the measurement of element ratios using similar wavelengths (e.g. rubidium and strontium) is not sensitive to the type of sample preparation, so that although a small

Table 1. Major- and minor-element analyses of ten lunar samples by X-ray fluorescence spectrometry

	Igneous rocks							Breccias		Fines
Group	1	1	1	2	2	2	2			
Type	A	B	B	B	B	B	B	C	C	D
Sample	10072	10017	10024	10003	10045	10047	10062	10018	10061	10084
SiO_2	40·49	40·69	40·25	39·76	39·04	42·16	39·80	41·81	41·87	41·79
TiO_2	11·99	11·92	11·90	10·50	11·32	9·43	10·74	7·99	7·84	7·55
Al_2O_3	7·74	7·78	8·09	10·43	9·51	9·89	10·22	12·34	12·62	13·44
FeO	19·38	19·49	19·46	19·80	19·40	19·11	19·22	16·46	16·45	15·91
MnO	0·24	0·28	0·24	0·30	0·27	0·28	0·30	0·22	0·22	0·21
MgO	7·45	7·51	7·53	6·69	7·73	5·67	7·08	7·79	7·83	7·66
CaO	10·56	10·76	10·66	11·13	11·28	12·15	11·47	12·00	11·96	12·14
Na_2O	0·50	0·51	0·52	0·40	0·36	0·45	0·41	0·46	0·47	0·43
K_2O	0·29	0·30	0·30	0·06	0·05	0·11	0·08	0·17	0·18	0·14
P_2O_5	0·19	0·18	0·20	0·12	0·10	0·11	0·12	0·15	0·14	0·13
S	0·23	0·23	0·22	0·18	0·15	0·18	0·16	0·15	0·15	0·14
T.E. (trace elements as oxides)	0·52		0·56		0·39	0·30		0·45	0·43	0·43
	99·58	99·65	99·93	99·37	99·60	99·84	99·60	99·99	100·16	99·97
O≡S	0·11	0·11	0·11	0·09	0·07	0·09	0·08	0·07	0·07	0·07
	99·47	99·54	99·82	99·28	99·53	99·75	99·52	99·92	100·09	99·90
Trace elements (ppm)										
Ba	300		310		10	88		175	128	134
Rb	5·61	5·72	5·96	0·62	0·62	1·11	0·89	3·60	3·68	2·96
Sr	168·4	163·7	178·3	160·9	137·7	209·2	187·8	158·5	161·6	164·8
Th	3·5	3·7	4·1	1·1	0·4	0·6	0·9	2·4	2·7	2·5
Zr	497	476	375	309	194	334	319	328	342	318
Nb	25		25		14	23		19	19	18
Y	162	159	168	112	73	134	103	106	108	99
La	43		39		16	20		24	23	21
Ce	94		108		32	48		67	37	58
Pr	16		12		6	13		11	15	10
Nd	49		55		17	36		29	20	33
V	22		37		98	13		51	34	36
Cr	2280		2610		2400	1220		1950	1940	1850
Co	34		32		23	16		35	23	34
Ni	<20		<20		<20	<20		200	170	230
Cu	22		16		20	16		32	25	33
Zn	34		14		14	13		54	37	37
Ga	4	4	5	3	3	4	3	4	5	4

error might be introduced into estimation of both rubidium and strontium this should not affect their calculated ratio. Calibrations for rubidium and strontium and other trace elements were made against synthetic standards (Compston *et al.*, 1969).

Sodium was determined on 0·05 g of the recovered sample by flame photometry using a lithium internal standard. A further 0·28 g of powder was used for measuring other major elements on fused samples by the technique of Norrish and Hutton (1969). All spectrometric measurements were made on a Philips PW1220 spectrometer. For the trace elements, individual analyses were recycled where necessary to obtain higher precision (Chappell *et al.*, 1969) so that for elements such as rubidium and thorium, counting times of up to 4 hr were used. The precision of each analysis was calculated from the observed variation in emission during this period. Results of the major and trace element analyses are given in Table 1.

Data for lead included in the Houston report have been omitted as it is clear from other data that our results are too high. The source of this error is not known; it may result from contamination or from some factor associated with the different method of sample preparation for trace element analysis of the lunar samples.

Mass-spectrometric analyses

Initially, the lunar samples were digested in hydrofluoric and perchloric acids and taken up to apparently clear solution in 2·5 N hydrochloric acid. Total-rock samples were processed in 0·05–0·2 g amounts depending on their rubidium content and mineral concentrates from 0·025 g upwards as available. Aliquots were dispensed into beakers containing Rb^{87} and Sr^{84} tracers for separate determination of rubidium and strontium concentrations. Although this procedure is well-documented for terrestrial samples (e.g. COMPSTON *et al.*, 1969), comparison with XRF results shows that it is quite unreliable for the Apollo 11 lunar samples (Table 2). An analysis may be correct for both

Table 2. Comparison of X-ray fluorescence (XRF) with (unreliable) isotope-dilution (ID) results obtained by aliquoting from hydrochloric acid solutions

	Rubidium (ppm)		Strontium (ppm)	
	ID	XRF	ID	XRF
10018 A	4·50	3·60	135·2	158·5
10024 A	7·54	5·96	126·5	178·3
10061 A	3·67	3·68	165·4	161·6
10061 B	3·68		141·3	
10072 A	5·65	5·61	149·8	168·4
10084 A	2·52	2·96	162·1	164·8
10084 B	3·09		153·7	

rubidium and strontium (e.g. 61A) or for either element (61B, 72A) or it may be incorrect for both, usually with rubidium too high and strontium too low (18A, 24A). We consider that rubidium and strontium became absorbed on near-invisible titanium colloids which form and segregate in hydrochloric acid solution. Similar analytical errors have been reproduced using mixtures of ilmenite (containing 300 ppm zirconium) and a standard sample.

The problem was avoided for the Apollo 11 samples by adding, and drying down, a mixed Rb^{87}, Sr^{84} spike solution to the platinum or Teflon dish before weighing in the sample. The rubidium and strontium in the sample apparently equilibrate with the spikes during the initial digestion so that any precipitation from the subsequent hydrochloric acid solution has no effect. Ammonia was added to precipitate the hydroxides, which carried down nearly all of the strontium. Ammonium chloride was fumed away from the separated supernatant in a quartz beaker to leave a concentrate of the alkalis, which was either run directly for rubidium on the mass-spectrometer or run after further rubidium concentration by cation exchange. The hydroxides were washed, redissolved in hydrochloric acid, a few micrograms of common rubidium added, and the solution dried down to equilibrate the trace of spiked rubidium remaining with the added rubidium. This addition of common rubidium minimizes the possible interference of enriched Rb^{87} during Sr^{87}/Sr^{86} measurement. Strontium was concentrated by cation exchange. Total processing blanks were run periodically throughout the analytical programme. They ranged from $0{\cdot}4 \times 10^{-9}$ g to $4{\cdot}3 \times 10^{-9}$ g for rubidium, and from $2{\cdot}3 \times 10^{-9}$ g to $4{\cdot}2 \times 10^{-9}$ g for strontium.

Sr^{87}/Sr^{86} was measured on a single-focussing 12 in. mass-spectrometer using a triple-filament source, Faraday cup collector, automatic magnetic-field peak switching, and digital output. For all samples, runs were obtained to the following specifications: Rb^{85} less than 0·02% of Sr^{87}; tail under Sr^{87} less than 0·01%; Sr^{88} beam stable at 6×10^{-11} A or more; internal precision in ratio per single peak cycle, better than 0·03 per cent. Three independent runs were made per sample mount. Except for sample 10017 mesostasis, no differences in Sr^{87}/Sr^{86} were detected between duplicates or between unspiked and spiked runs, which were accordingly combined. The variations in 95 per cent confidence limits of precision for the mean Sr^{87}/Sr^{86} shown in Table 3 chiefly reflect variation in the number of replicates. Spiked rubidium was observed in some analyses, which were continued until any trend in measured Sr^{87}/Sr^{86} with Rb^{85} disappeared. Corrections were applied to each isotope for non-linearity of the $10^{11}\ \Omega$ input-resistor, as assessed by data obtained with a $10^{10}\ \Omega$ resistor. Recent work has shown that the non-linearity is variable and the limits of precision shown in Table 3 should probably be increased by about $\pm 5 \times 10^{-5}$ to allow for this. Our value for the Eimer and Amend strontium carbonate is 0·70810, which may be used to indicate laboratory bias. In general, our results for the same Apollo 11 samples agree with those of GAST and HUBBARD (1970) and GOPALAN *et al.* (1970), to

Table 3. Isotope-dilution analyses for rubidium and strontium and analyses of Sr^{87}/Sr^{86} of Apollo 11 rocks and minerals

		Rb (ppm)	Sr (ppm)	Rb^{87}/Sr^{86}	Sr^{87}/Sr^{86}
Group 1 lavas					
10017-68	TR A	5·59	167·1	0·0966	0·70448 ± 0·00003
	TR B	5·60	167·3	0·0966	
	PL A	0·92	500·0	0·0053	0·69958 ± 0·00010
	PL B	0·94	492·4	0·0055	
	M A	19·5	282·7	0·1993	0·71022 ± 0·00012
	M B	20·9	292·8	0·2066	0·71046 ± 0·00015
10024-20	TR A	5·75	169·5	0·0965	0·70452 ± 0·00007
	TR B	5·76	169·5	0·0966	
10072-30	TR A	5·60	168·0	0·0959	0·70455 ± 0·00005
	TR B	5·61	169·1	0·0954	
	PL	3·87	531·0	0·0210	0·70051 ± 0·00012
	PX	1·87	85·9	0·0629	0·70294 ± 0·00010
Group 2 lavas					
10003-20	TR A	0·63	159·9	0·0114	0·69984 ± 0·00008
	TR B	0·63	158·6	0·0115	
10045-20	TR *	0·62*	137·7*	0·0130*	0·69993 ± 0·00004
	PL A	1·36	298·3	0·0132	0·69987 ± 0·00007
	PL B	1·31	299·7	0·0126	
	PX A	0·69	85·2	0·0235	0·70024 ± 0·00008
	PX B	0·62	88·2	0·0204	
10047-21	TR *	1·11*	209·2*	0·0153*	0·69997 ± 0·00006
	PL A	0·48	573·8	0·00240	0·69937 ± 0·00010
	PL B	0·59	575·5	0·00296	
10062-36	TR A	0·83	189·8	0·0126	0·69991 ± 0·00005
	TR B	0·94	190·0	0·0142	
Breccias					
10018-20	TR A	3·63	166·1	0·0633	0·70287 ± 0·00006
	TR B	3·66	166·1	0·0638	
10061-25	TR A	3·65	166·2	0·0633	
	TR B	3·77	166·6	0·0652	0·70294 ± 0·00004
	TR C	3·67	165·4	0·0639	
Fines					
10084-44	TR A	2·64	165·0	0·0461	0·70221 ± 0·00005
	TR B	2·67	163·8	0·0470	

* Indicates analysis by X-ray fluorescence spectrometry.
TR denotes total rock, PL plagioclase, PX pyroxene and M mesostasis.
Sr^{87}/Sr^{86} is normalized to 8·3752 for Sr^{88}/Sr^{86}.

within the uncertainty cited by the authors, but not with MURTHY *et al.* (1970) and HURLEY and PINSON (1970). Comparison with the results of ALBEE *et al.* (1970) are made later in this paper.

The reproducibility of our procedures for isotope dilution analysis is shown by the duplicates in Table 3. For the total rock samples, the coefficient of variation in Rb^{87}/Sr^{86} is 1·0 per cent, about double the value normally achieved with rocks having high rubidium contents. For the minerals, the reproducibility was noticeably worse, partly owing to sampling variations as shown by sample 10017 mesostasis, and in part to the general difficulty of determining rubidium in the 50–100 $\times 10^{-9}$ g amounts present in some of the small mineral samples.

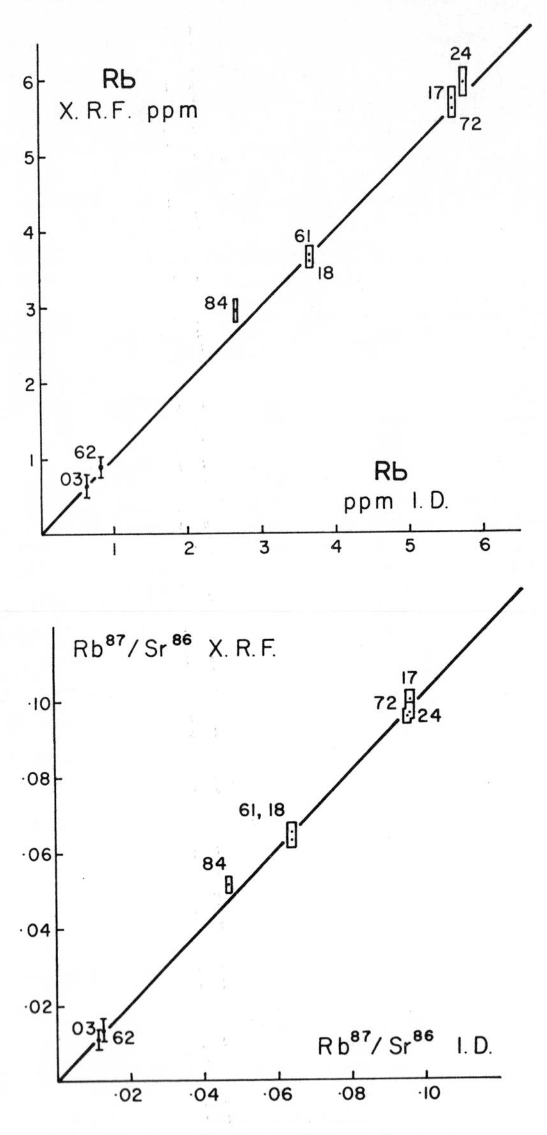

Fig. 1. Comparison of isotope-dilution and X-ray fluorescence measurements for Rb and Rb/Sr in Lunar materials.

Figure 1 shows that the isotope dilution and X-ray fluorescence results agree to within their 95 per cent confidence limits for each sample except 10084. For this sample, the isotope dilution result for rubidium is identified as too low, according to analysis of other splits of 10084 by GAST and HUBBARD (1970), MURTHY *et al.* (1970), PHILPOTTS and SCHNETZLER (1970) and GOPALAN *et al.* (1970). The remaining data show that the limits of bias between the two techniques range from zero to 2·0 per cent for Rb^{87}/Sr^{86}, with the X-ray result averaging 1·0 per cent higher. The results of GOPALAN *et al.* (1970) for other fragments of samples 10017, 10024 and 10072 agree with our own isotope-dilution results for Rb^{87}/Sr^{86} to within +0·5, +0·6 and −2·0 per cent respectively. For sample 10072, HURLEY and PINSON (1970) agree for strontium but are much lower in rubidium, as are

MURTHY *et al.* (1970) for sample 10017. We obtain fair agreement in Rb^{87}/Sr^{86} with GAST and HUBBARD (1970) and with PHILPOTTS and SCHNETZLER (1970), for sample 10017, at $+4{\cdot}0$ and $-3{\cdot}0$ per cent respectively, and similar agreement at ± 3 per cent with PHILPOTTS and SCHNETZLER (1970) for samples 10018 and 10062. These and other comparisons suggest that bias in Rb^{87}/Sr^{86} arising from our spike calibrations is not greater than ± 2 per cent.

CHEMISTRY OF APOLLO 11 SAMPLES

Data for eleven major elements and eighteen minor elements are given in Table 1. These data were presented at the Houston Conference and the crystalline rocks were divided into two groups on a chemical basis (COMPSTON *et al.*, 1970). Many of the Houston papers recognized that the crystalline rocks fell into two chemical groups (e.g. GAST and HUBBARD, 1970) but these were normally identified with the textural subdivision by LSPET into Type A (fine-grained vesicular) and Type B (medium-grained). Since the two chemical groups are not identical with the two textural types, we will discuss the chemical grouping and its implications in more detail.

Chemical subdivision of crystalline rocks

We obtained one Type A sample and six Type B rocks. Table 1 shows that the Type A sample, number 10072, is very similar to the Type B numbers 10017 and 10024, for all major and most minor elements. On the basis of our data, the only discrepant elements are zirconium, vanadium, chromium and zinc, and we have accordingly assigned these three samples to our Group 1, despite the textural differences. The other four crystalline rocks, in Group 2, show more internal variation than the Group 1 rocks but for many elements this variation is well outside the narrow range of Group 1. Thus Group 1 is higher in titanium, magnesium, sodium, potassium, phosphorus, sulphur, barium, rubidium, thorium, zirconium, yttrium and REE, and lower in aluminium and calcium. Our data for other elements do not show significant differences. The most marked differences are for potassium, rubidium and thorium. We have tested the chemical grouping using discriminatory analysis and this gives a significance of difference at better than 99 per cent for most discriminators and better than 99·9 per cent in some cases. Data on other samples presented at Houston support this subdivision, and all rocks studied chemically have been assigned to one of the two groups. While there is a general correspondence between the two chemical groups and the two textural types recognized by LSPET (Group 1 = Type A, Group 2 = Type B), there are some exceptions, as follows: (a) LSPET have assigned 10017 to Type B and AGRELL *et al.* (1970) have described this rock as Type B, but transitional to Type A. Chemically it belongs to Group 1; (b) sample 10024 is stated by the Apollo 11 Sample Information Catalog to have a texture and grainsize typical of Type B but again it has the chemical properties of Group 1; (c) on the other hand, LSPET have assigned 10020 to Type A, but chemically it belongs to Group 2. MORRISON *et al.* (1970) have pointed out that examination of their data at the 1σ level makes it possible to subdivide samples 10020, 10057, 10058 and 10072 into two groups which do not follow the textural types proposed by LSPET but that this subdivision is not possible at the 2_σ level of significance.

The chemical subdivision leads to the following groupings, the textural type being indicated in brackets:

Group 1: 10017(B), 10022(A), 10024(B), 10049(A), 10057(A), 10069(A), 10071(A), 10072(A)

Group 2: 10003(B), 10020(A), 10044(B), 10045(B), 10047(B), 10050(B), 10058(B), 10062(B)

The placing of 10044 is ambiguous; the major element data of AGRELL *et al.* (1970) and ENGEL and ENGEL (1970) and the minor element results of MURTHY *et al.* (1970), SCHMITT *et al.* (1970), and BROWN *et al.* (1970) place this sample clearly in Group 2. However, PHILPOTTS and SCHNETZLER (1970) provide data on this sample that are typical of Group 1 and it appears that their sample was assigned an incorrect number.

Table 4. Averages of analyses of lunar samples

	Crystalline rocks				Breccias		Soil
	Group 1		Group 2				
	This paper (Table 1)	Av. of 6	This paper (Table 1)	Av. of 6	This paper (Table 1)	Av. of 6	Av. of 10084
SiO_2	40·48	40·28	40·19	40·47	41·84	41·69	42·03
TiO_2	11·94	11·88	10·85	10·32	7·92	8·87	7·50
Al_2O_3	7·87	8·95	10·05	10·41	12·48	12·35	13·84
FeO	19·44	19·21	19·47	18·72	16·46	16·39	15·76
MnO	0·25	0·24	0·29	0·27	0·22	0·23	0·21
MgO	7·50	7·60	7·17	6·66	7·81	7·47	7·87
CaO	10·66	10·53	11·29	11·48	11·98	11·81	11·98
Na_2O	0·51	0·64	0·39	0·49	0·47	0·63	0·45
K_2O	0·30	0·31	0·08	0·09	0·18	0·15	0·14
P_2O_5	0·19	0·18	0·11	0·10	0·15	0·12	0·10
S	0·23	0·23	0·16	0·17	0·15	0·15	0·13

Data on additional elements presented at Houston further support our chemical subdivision. Thus uranium, for example, has been recorded at 0·85, 0·87 and 0·86 ppm in samples 10017, 10057 and 10071 by TATSUMOTO and ROSHOLT (1970) and at 0·88 ppm in 10072 by SILVER (1970), all these rocks belonging to Group 1. The same authors record a range from 0·16 to 0·27 ppm for uranium in Group 2 rocks. Because we consider that the chemical subdivision of the crystalline rocks is a fundamental one, we have compiled average major element analyses of the Group 1 and Group 2 rocks from much of the data presented at Houston. These are listed in Table 4, along with our own averages for comparison. The average of six Group 1 rocks refers to analyses of 10017, 10022, 10024, 10049, 10057 and 10072 by the present authors, ENGEL and ENGEL (1970), MAXWELL *et al.* (1970), PECK and SMITH (1970), ROSE *et al.* (1970) and WIIK and OJANPERÄ (1970). The average of eight Group 2 rocks refers to analyses of 10003, 10020, 10044, 10045, 10047, 10050, 10058 and 10062 by the present authors, ENGEL and ENGEL (1970), FRONDEL *et al.* (1970), MAXWELL *et al.* (1970), PECK and SMITH (1970), ROSE *et al.* (1970) and AGRELL *et al.* (1970). Agreement between our own averages and the overall averages is very good. In fact the general agreement for major element analyses on individual samples is surprising when one considers the sampling problems involved in the analysis of small rock chips and the general disagreement found between analysts measuring properly prepared standard samples (FLANAGAN, 1969). For sodium, the general average is higher than our own but this is certainly due to a bias in analyses from one laboratory compared with other analysts who are in good agreement. The only substantial discrepancy appears to be aluminium in Group 1 in which our own average is a little more than 1 per cent lower than the general mean.

It is clear that the Apollo 11 samples are from two distinct chemical units. The fact that within each group there is variation in grain size and the degree of vesicle development, indicates that each group must represent a sampling of rock units large enough to permit textural variation. MASON *et al.* (1970) have pointed out that the range in grainsize of the Apollo 11 rocks suggests that the finest-grained lunar rocks correspond to quenched surfaces of flows, or ejected cinders or bombs, and the coarser-grained samples to flow interiors. We therefore suggest that each group could represent a separate lava flow. A similar suggestion has been made by SCHMITT *et al.* (1970), based on the variation by a factor of four in the REE in their Type B rocks. This variation largely results from inclusion in their Type B samples of the chemically distinct sample 10017 which we assign to Group 1 and thus consider to come from a separate flow. One flow (Group 1) tends to be finer-grained than the other, is marked in particular by higher alkali element, thorium and uranium contents, and is remarkably uniform in chemical composition. The other flow is relatively coarse-grained and apparently more variable in composition. This variation might only be apparent because of the larger sampling error associated with small chips of a coarser-grained rock. However, a comparison of separate potassium determinations by various analysts suggests that the variation in this element (through a factor of two or more) occurs between specimens rather than within specimens. Thus the variation would seem to be real and we suggest that it is associated with slower cooling as indicated by the coarser grainsize relative to the other fine-grained and rapidly cooled flow. The Group 2 rocks would thus represent an initially homogeneous lava in which localized segregation occurred between early formed crystals and the residual liquid during and after eruption. This residual liquid would be enriched in the alkali elements and the sodium, potassium and rubidium contents of the Group 2 rocks are in fact related, increasing in sequence through samples 10045, 10003, 10062 and 10047.

Source of Apollo 11 basalts

In considering the genesis and Rb–Sr chronology of the Apollo 11 lavas it is important to decide, if possible, the place at which the rocks attained their present compositions. RINGWOOD and ESSENE (1970) consider that the lunar basalts have been directly derived from their source regions at depths of 200–400 km, without extensive fractionation en route. On the other hand, O'HARA *et al.* (1970) consider that the rocks probably represent liquid extensively modified by near-surface fractionation. Consideration of REE abundance patterns and in particular the marked negative europium anomaly has led various authors (e.g. GAST and HUBBARD, 1970; SCHMITT *et al.*, 1970) to suggest that plagioclase may have precipitated out of the lunar magmas. RINGWOOD and ESSENE (1970) note that there is little fractionation between rocks so that, for example, the Mg/Fe ratio is remarkably constant both within and between Groups 1 and 2. If the variation in the Apollo 11 basalts were the result of fractional crystallization, it would be necessary to enrich potassium, rubidium and uranium by factors of approximately 3, 7 and 4 respectively, in proceeding from a Group 2 to a Group 1 composition, while not affecting the Mg/Fe ratio. LSPET data on Apollo 12 samples (TAYLOR, personal communication), in fact do show a considerable range in magnesium contents and Mg/Fe ratios while other elements such as potassium and rubidium remain comparatively constant. This is the situation that one would expect if near-surface fractionation had occurred. For example, GAST and HUBBARD (1970) stated that variations in potassium, rubidium, cesium and barium within the otherwise similar Apollo 11 igneous rocks are quite puzzling and that if this were to be explained in terms of simple fractional crystallization processes it would require crystallization of 80–90 per cent of one liquid to produce the other. They note that this is unlikely in view of the near identity of the major element composition of all rocks as stated also by COMPSTON *et al.* (1970). We conclude that the two Apollo 11 groups do not have a derivative relationship, that the distinctive features of the two groups are of primary origin and related to the generation of the two magmas and that the variations within Group 2 are due to localized segregation of a residual liquid phase.

RINGWOOD and ESSENE (1970) have taken the strong enrichment of elements such as uranium, thorium, barium, REE, yttrium, zirconium and titanium, relative to their chondritic abundances, as suggesting that the basalts have been derived by a small degree of partial melting of their source material. The acquisition of the distinctive differences for these elements for the two magmas is also more readily explained by partial melting. Two possibilities exist. The differences could result from heterogeneity in the source material or they could result from differing degrees of partial melting of similar source materials. If the differences were to represent heterogeneity in the source material this would be significant from the point of view of Rb–Sr chronology since a difference in Rb/Sr ratio in the source region might result in differences in the Sr^{87}/Sr^{86} ratios of magmas produced by partial melting.

Comparison of Apollo 11 and Apollo 12 data

TAYLOR (personal communication) has presented LSPET data for the Apollo 12 crystalline rocks. As had been noted in the report, significant differences exist between those data and the Apollo 11 results. We note that although potassium, rubidium, zirconium, yttrium and barium contents are lower than those for Apollo 11, as has been pointed out by LSPET, they are in fact fairly similar in abundance to the Group 2 samples from Apollo 11 and very much lower than the Group 1 contents.

Chemistry and origin of the breccias and fines

Analyses of two breccias (10018 and 10061) and the fines (10084) are given in Table 1. Our average for the two breccias is listed in Table 4 with the average of analyses of samples 10018, 10019, 10048, 10056, 10060 and 10061 by the present authors, ROSE *et al.* (1970) and AGRELL *et al.* (1970) for comparison. The average of six analyses of the fines sample 10084 by the present authors, ENGEL and ENGEL (1970), MAXWELL *et al.* (1970), AGRELL *et al.* (1970), PECK and SMITH (1970) and WIIK and OJANPERÄ (1970) is also given in Table 4. The very close similarity in the two breccias analyzed by us is extended to the other four analyzed so that the six samples belong to a well-defined chemical group. The breccias are themselves very similar in composition to the fines sample 10084. The only apparent discrepancies, from our data, are a lower aluminium content in the breccias (about 1 wt.% Al_2O_3) and a higher rubidium content (about 0·7 ppm). This correspondence supports the suggestion by LSPET that the breccias were formed by shock cementing or lithification of the fine surface material by impact events.

Although the components of the fines are undoubtedly of polygenetic origin, the fact that for most elements studied the fines have contents between those of the Group 1 and Group 2 rocks, indicates that the two rock groups have been the predominant contributors to the fines. We have previously suggested (COMPSTON *et al.*, 1970) that the higher nickel and zinc contents of the fines and breccias could be the result of a contribution from meteorites. KEAYS *et al.* (1970) have studied this in detail and concluded that the fines contain an admixture of 1·5–2% of carbonaceous chondrite-like material. For the major elements, titanium and iron are lower in the fines, and aluminium higher. Various authors at the Houston Conference (ARRHENIUS *et al.*, 1970; CHAO *et al.*, 1970; MASON *et al.*, 1970 and WOOD *et al.*, 1970a) have described a feldspar-rich suite of fragments in the fines, unlike any of the basaltic rocks at Tranquillity Base. Since a feldspar-rich component could contribute to the relatively high aluminium content of the fines, this must be considered in more detail. WOOD *et al.* (1970a, 1970b) have found 3·6% of anorthositic rock fragments in the 1–5 mm size range and suggest that these are derived from the lunar highlands. Their analysis also showed 52·4% of breccia fragments; on a breccia-free basis, the total anorthositic rock fragment component of the 1–5 mm size fraction is close to 7·6 vol.%, equivalent to 6·7 wt.%, assuming densities of 2·9 and 3·3 g/cm^3 for the anorthositic and basaltic rocks. We have listed the average of Group 1 and Group 2 compositions and the composition of the "most common anorthosite type" given by WOOD *et al.* (1970b) in Table 5. The result of combining these in the relative amounts 93·3% and 6·7% is also given in Table 5. The effect of the addition of the anorthositic composition is to raise the aluminium content, and to lower the amounts of titanium and iron, that is, to modify the average basaltic rock composition in precisely the sense required to produce a fines composition. However, the magnitude of the modification is not sufficient to match the composition of the fines. We have noted that the titanium and iron contents of the fines are lower than the average Group 1 composition in close to their stoichiometric proportions in ilmenite (to within 4 per cent). We have suggested (COMPSTON *et al.*, 1970) that ilmenite has been removed from a more primitive soil, perhaps by mechanical sorting of the denser phase during the gardening processes associated with meteorite impact. Our recalculated analysis (basalt plus anorthosite) contains 5·24% excess of ilmenite over the average soil composition. Subtraction of stoichiometric amounts of titanium and iron (ilmenite) to this amount from the basalt plus anorthosite composition, gives the analysis in column 4, Table 5. This can be compared with the fines analysis, column 5. It can be seen that good agreement is obtained, except that aluminium is still too low and, less critically, silica is a little high.

Our calculations have shown that it is possible to transform the average basaltic rock composition to one approaching that of the lunar fines. This supports our earlier

Table 5. Results of recalculation of average crystalline rock composition to incorporate an anorthositic rock component and allow for the removal of ilmenite, and a comparison with the lunar fines

	1	2	3	4	5
SiO_2	40·38	44·9	40·65	43·66	42·03
TiO_2	11·10		10·36	7·60	7·50
Al_2O_3	9·68	26·6	10·79	11·60	13·84
FeO	18·97	6·6	18·14	15·66	15·76
MgO	7·13	7·6	7·16	7·69	7·87
CaO	11·00	14·0	11·19	12·02	11·98
Na_2O	0·45	0·3	0·44	0·47	0·45

1. Mean of Group 1 and Group 2 averages in Table 4.
2. Chemical composition of the "most common anorthosite type" found by WOOD *et al.* (1970b), recalculated from their data.
3. Composition of material containing 93·3 wt.% of mean Group 1 and Group 2 rocks and 6·7 wt.% of anorthosite.
4. Composition of material obtained by removing 5·24% of ilmenite from composition 3.
5. Average of analyses of fines sample 10084.

proposition that the basaltic rocks at Tranquillity Base have provided the dominant component to the fines. If there were any independent basis for making a larger contribution from feldspar-rich rock fragments (about 15 vol.%), a good approximation to the lunar fines composition could be obtained. It is possible that such fragments ejected from the highlands over Tranquillity Base would be laterally sorted so that a larger component would occur in the <1 mm size fraction than is at present found in the 1–5 mm fraction. The time at which the feldspathic component was added could also be critical. If it occurred relatively early in the production of the regolith, the coarser fragments would have been broken up by impact giving a concentration in the finer fractions at the present time. The basaltic rock fragments, on the other hand, are continually being replenished by contributions from bedrock and the boulders present in the regolith, which should result in their concentration in the coarser 1–5 mm fraction approaching a steady state.

AGE OF CRYSTALLIZATION OF APOLLO 11 ROCKS

Rb–Sr mineral ages

The eight-fold enrichment of rubidium in Group 1 over Group 2 rocks, combined with comparatively equal strontium contents, produces two widely separated groups of points on the Sr^{87}-evolution diagram (Fig. 2). No precise linear relationship can be defined within each group as their individual dispersions in Rb^{87}/Sr^{86} are too small Consequently any total-rock age-determinations must be qualified by the assumptions that each group has the same age and that each has the same initial Sr^{87}/Sr^{86}, or that each had a single-stage evolution from some assumed value for initial Sr^{87}/Sr^{86}. As the chemical evidence previously discussed indicates that the groups belong to two separate magmas, such assumptions are unsatisfactory. Only mineral ages can be conclusive.

Minerals were separated from two samples of each group (Table 3). Strontium is enriched and rubidium depleted in the plagioclase. Rubidium is highly enriched in

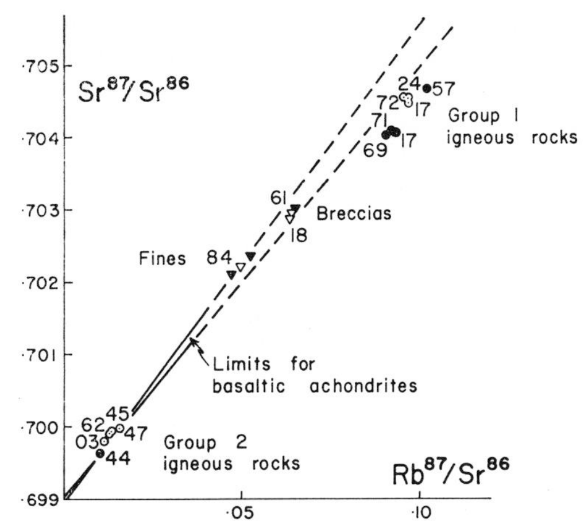

Fig. 2. Sr^{87}-evolution diagram for lunar rocks. Open symbols-this work. Filled symbols from ALBEE *et al.* (1970). Basaltic achondrite limits as given by PAPANASTASSIOU and WASSERBURG (1969).

the mesostasis, as alkali-rich glass and as minute grains of potassium feldspar (e.g. WARE and LOVERING, 1970), which become concentrated in a dense fraction owing to the abundance of fine-grained ilmenite and pyroxene. Figure 3 shows our present results (other mesostasis concentrates will be analyzed in the future). Combining the estimates for samples 10017 and 10072, the Group 1 rocks are dated as 3·78 b.y. with 95 per cent confidence limits of precision of $\pm$0·10 b.y., which may be compared with the more precise figure of 3·65 $\pm$ 0·02 b.y. reported by ALBEE *et al.* (1970), averaged for four Group 1 rocks. These results indicate a small but significant bias between the laboratories, of at least 1 per cent, probably due to calibration of spike solutions. As described previously, it is unlikely that a bias of more than 2 per cent is present in our own data.

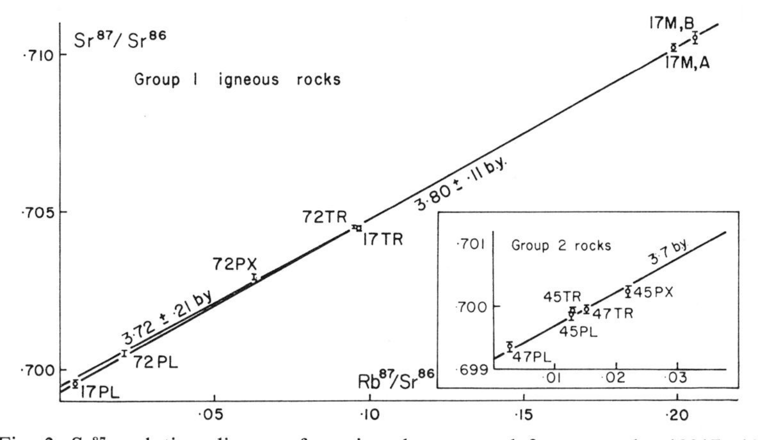

Fig. 3. Sr^{87}-evolution diagram for minerals separated from samples 10017, 10072, 10045 and 10047. TR denotes total-rock, PL plagioclase, PX pyroxene, and M mesostasis.

Our data for minerals from the Group 2 rocks are consistent with a similar age. K–Ar age determinations have been made on rock samples in Group 1 (10017, 10022, 10024 and 10072) and Group 2 (10003, 10044 and 10062) by TURNER (1970). Argon leakage has been detected in rocks 10017 and 10024 and corrected in the remainder by the Ar^{40}–Ar^{39} dating method. TURNER's mean Group 1 age is 3·56 b.y., as compared with 3·83 b.y. for Group 2, but this difference is not significant relative to the observed precision in age within each group. ALBEE *et al.* (1970) report 3·70 $\pm$ 0·08 b.y. for rock 10044 from Group 2, which is indistinguishable from their Group 1 results. Thus both groups of rocks have very similar and probably identical mineral ages close to 3·70 b.y. Our data alone do not reveal any significant difference between the mineral isochrons and the tie-line joining the two rock groups. However, ALBEE *et al.* (1970) detect that rock 10044 (Group 2) is about 0·0002 lower in initial Sr^{87}/Sr^{86} than their four Group 1 rocks. The same difference between groups would be registered by our data if we used 3·70 b.y. as the age of crystallization instead of 3·80 b.y. However, we would obtain values about 0·00015 higher than ALBEE *et al.* (1970), presumably representing another interlaboratory bias as suggested also by comparing the total-rock results in Fig. 2.

The higher initial Sr^{87}/Sr^{86} of the Group 1 rocks has many possible interpretations. For example it may signify that there were two, spatially-separated magma sources which had slightly different Rb/Sr. Or there may have been a single source region containing an interstitial low melting-point fraction rich in trace-elements including Rb, Th, and U, radiogenic Sr and Pb. A low degree of partial melting would selectively sample the low-melting fraction and produce the Group 1 magma, and depending on the Sr^{87} equilibration between the major mineral phases in the source and the interstitial phases, the magma may be enriched in radiogenic Sr. A third possible interpretation is that both magmas come from the same source, but Group 2 was more contaminated by incompatible elements, including radiogenic Sr, located in interstitial phases situated anywhere between the source and the lunar surface.

Figure 2 shows that the Group 2 rocks coincide closely in isochron-coordinates with the basaltic achondrites, particularly so if the interlaboratory bias of 0·00015 is substantiated. This is unlikely to be accidental, and it implies that both the Group 2 rocks and the basaltic achondrites were derived from source regions having the same primary age of 4·39 $\pm$ 0·26 b.y. found by PAPANASTASSIOU and WASSERBURG, (1969). It also implies that the partial melting of the source region at 3·70 b.y. produced little or no fractionation of rubidium from strontium between the final Group 2 magma and the original source, but such a fractionation is necessary in the case of the Group 1 magma.

Comparison with U–Pb dating

Lead isotope data for total-rocks of both groups (TATSUMOTO and ROSHOLT, 1970; SILVER, 1970) do not register 3·70 b.y. in any simple manner. In Pb^{207}/Pb^{204}, Pb^{206}/Pb^{204} coordinates (Fig. 4), all data lie to the right-hand side of the meteorite isochron (MURTHY and PATTERSON, 1962), which shows quite conclusively that two or more stages of lead evolution are involved with every sample. The data are highly correlated, signifying in general a lead isotope mixing line. If the component leads are assumed

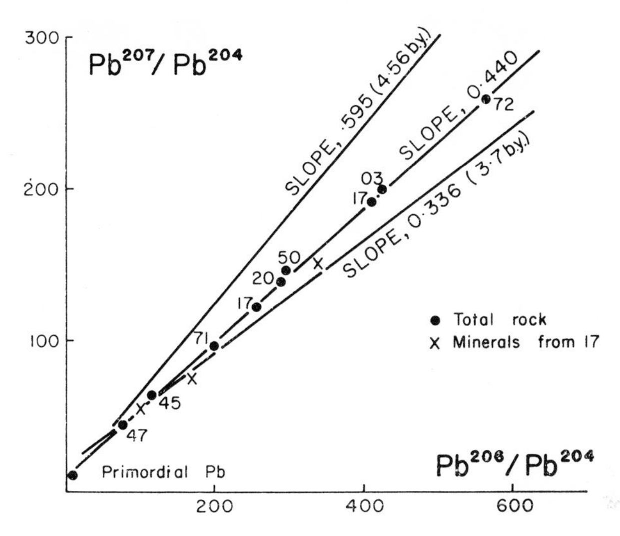

Fig. 4. Lead isochron diagram for lunar samples analyzed by TATSUMOTO and ROSHOLT (1970) and SILVER (1970), showing high correlation of data along mixing-line between 3·70 b.y. and 4·56 b.y. reference isochrons.

to have equal ages and the same primary U–Pb system, the slope of the mixing-line can be given time-significance, and an age of 4·1 b.y. is indicated as the time of generation of the secondary U–Pb systems. As this figure so greatly exceeds the 3·7 b.y. found by Rb–Sr and K–Ar for the time of crystallization on or near the lunar surface, we must conclude either that (a) the mixing-line has no time-significance, or (b) the secondary U–Pb systems were generated some 0·4 b.y. earlier than the lava extrusion, presumably in the lunar mantle.

To consider alternative (a) the excellence of the correlation-line would indicate that mixing is between two particular end-members only. One end member must be highly radiogenic but as a multistage Pb, its age cannot be specified except as 3·7 b.y. or older, from the Rb–Sr ages. The other end-member would lie close to the primary 4·56 b.y. isochron. One simple explanation of the mixing-line is to assume that most of the variation in Pb^{204} represents variable contamination by terrestrial common Pb during handling and analysis. The other end-member, true lunar lead, might be even more radiogenic than sample 10057 (see Fig. 5), or it may be some average value if sample 10057 is overcorrected for processing blank. (Overcorrection by 0·01μ g would increase its isotopic ratios by 25 per cent.) The two data-points for sample 10017 are consistent with common lead contamination but this is inconclusive as the same rock showed a definite sampling variation in Rb–Sr analyses. On the other hand, the trend of lunar fines lead analyses (Fig. 5) strongly suggests common lead contamination.

Alternative (b) would represent the time at which two or more chemical systems of different U/Pb in the moon's interior became closed to lead isotope equilibration. They may be visualized as regions within the moon's mantle which were cool enough by 4·1 b.y. ago to pass below some critical blocking temperature at which effective

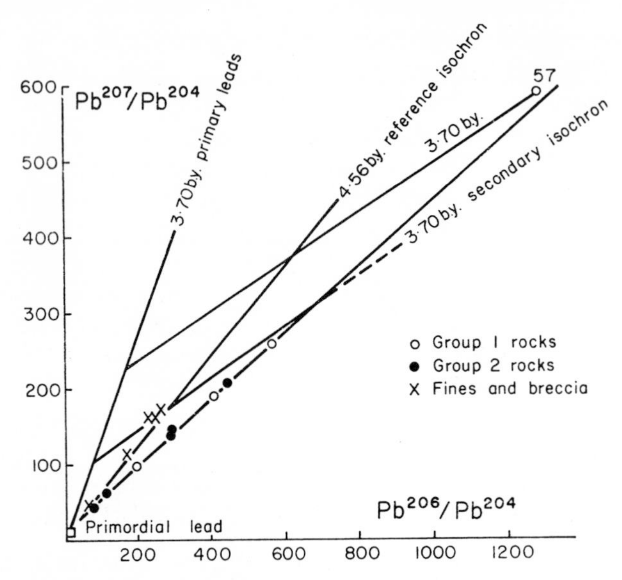

Fig. 5. Lead isochron diagram for lunar samples analyzed by TATSUMOTO and ROSHOLT (1970) and SILVER (1970), showing fines and breccia lead as biased mixture of 3·70 b.y. leads on secondary isochron.

lead isotope diffusion stops. Their separation as different U/Pb systems may have occurred earlier than 4·1 b.y., and from the bias to the right of the meteoritic isochron, it would have involved a nett loss of lead relative to uranium. It would be necessary for magma generation at 3·70 b.y. not to separate U from Pb but to transfer an equal fraction of the uranium and lead of the source, including radiogenic lead, to the lunar surface, without contamination except by uranium and lead from other 4·1 b.y. systems.

THE AGE OF APOLLO 11 FINES

The lunar regolith at Tranquillity Base must be younger than 3·70 b.y. on the reasonable assumption that the Apollo 11 igneous rocks are locally derived, and that they underlie the regolith rather than intrude it. Furthermore, as detailed earlier in this paper, the major and nearly all the minor-element abundances indicate that the regolith at Tranquillity Base is composed mainly of the two local 3·70 b.y. lavas (~85 wt.%) of unknown age. Yet it has an 'age' of 4·65–4·75 b.y. as indicated by apparently concordant U–Pb age-determinations by SILVER (1970), GOPALAN *et al.* (1970), and TATSUMOTO and ROSHOLT (1970), and a 'model' age of about 4·5 b.y. can be calculated from the Rb–Sr data (ALBEE *et al.*, 1970). As pointed out by SILVER (1970), these figures appear to make it very difficult to have more than a small proportion of 3·70 b.y. old material in the local regolith.

In discussing the old regolith 'age,' GOPALAN *et al.* (1970) suggest that the regolith is a sample of many different rocks, possibly from a large area of the lunar surface, while TATSUMOTO and ROSHOLT (1970) infer that it contains only minor amounts of the local lavas and that it is older than the lavas. We wish to indicate instead that the

U–Pb data can be reconciled with a *mainly local* origin for the regolith, by assuming that the (uncontaminated) fines and the (uncontaminated) rock leads lie on 3·70 b.y. secondary isochrons. Figure 5 shows a 4·56 b.y. reference isochron and two 3·70 b.y. secondary isochrons drawn through the fines data of GOPALAN *et al.* (1970) and the fines and breccia of TATSUMOTO and ROSHOLT (1970), and through the most radiogenic rock, 10057. Other analyses of the fines and breccias by SILVER (1970), WANLESS *et al.* (1970) and KOHMAN *et al.* (1970) are assumed to be displaced from these by terrestrial contamination. If the moon accreted at 4·56 b.y., single-stage primary lead evolving from primordial lead in systems of different U/Pb will lie, at 3·70 b.y., along the line shown in Fig. 5. Suppose a mixture of such leads, averaging about 110 in Pb^{207}/Pb^{204} and 80 in Pb^{206}/Pb^{204} coordinates, were available at the lunar surface for incorporation into the regolith. Then the regolith can be viewed as a biased mixture of 3·70 b.y. primary lead and 3·70 b.y. igneous rocks mixing along the lower of the secondary isochrons in Fig. 5.

This hypothesis involves (a) removal of lead relative to uranium during magma generation or during emplacement and cooling of the rocks at the lunar surface, and (b) making this displaced lead available for incorporation into the regolith. It is difficult to imagine a mechanism for (b) if the lead loss occurs at the site of magma generation in the lunar mantle. Loss of lead by 'volatile transfer' or by vacuum evaporation (CHAPMAN and SCHEIBER, 1969) when the lavas were extruded as thin, fluid sheets over the lunar surface (WEILL *et al.*, 1970) seems more feasible. We propose that some such mechanism selectively moved lead from extruding or near-surface lavas onto or into adjacent cooler layers, and that the latter were later transformed into regolith. (It should be noted that such lead 'loss' at 3·70 b.y. would be additional to the deficiency of lead in all lunar samples relative to the earth, as remarked by most authors.)

TATSUMOTO and ROSHOLT (1970) indicate that their U–Pb rock data may lie on a chord on the Concordia diagram which intersects at 3·5–3·6 b.y. due to some metamorphic event. On the above interpretation we would substitute 'lava extrusion' for 'metamorphic event' and we would emphasize that the fines and breccia data must be discordant, displaced above Concordia, along the chord joining the data with the 3·70 b.y. point. All the data on this diagram are subject to displacement up or down along lines joining the plotted points to the origin, due to possible error in U/Pb. Sample 10050, for example, appears to have such error, as it is displaced systematically below the correlation-lines in Pb^{207}/Pb^{204} vs. U^{235}/Pb^{204} and Pb^{206}/Pb^{204} vs. U^{238}/Pb^{204} diagrams, but it plots on the same correlation-line as the other igneous rocks in the Pb^{207}/Pb^{204} vs. Pb^{206}/Pb^{204} diagram. In view of such error, it is difficult to obtain a precise upper intersection between Concordia and the chord from 3·7 b.y. to the data for the fines and breccia, which would yield the age of the (assumed) primary U/Pb system which existed before 3·70 b.y. However, a younger limit of about 4·4 b.y. can be set from the data of GOPALAN *et al.* (1970) and SILVER (1970), and about 4·5 b.y. from TATSUMOTO and ROSHOLT (1970).

KEAYS *et al.* (1970) have suggested that 2 per cent addition of carbonaceous chondrite material to the regolith would produce the increase in copper, zinc and other elements seen in the regolith with respect to the Apollo 11 rocks. Such addition would be an important third component to the mixture of leads proposed here. The enstatite chondrites and carbonaceous chondrites contain a few ppm of non-radiogenic lead (MARSHALL, 1962; REED *et al.*, 1960), so that the addition of 2 per cent to a regolith containing the equivalent of about 0·1 ppm of common lead will shift the fines point in Fig. 5 along the mixing line by 15–20 per cent, towards primordial lead. Such addition could account for some of the displacement of the data from the upper 3·70 b.y. mixing-line.

The maximum Rb–Sr 'age' for the fines, about 4·6 b.y. assuming the basaltic achondrite initial Sr^{87}/Sr^{86}, has no simple meaning. ALBEE *et al.* (1970) have shown that the fines contain a number of discordant age components, and COMPSTON *et al.* (1970) inferred the presence of a component enriched in Sr^{87}, relative to a simple mixture of Group 1 and Group 2 rock constituents in approximately equal amounts

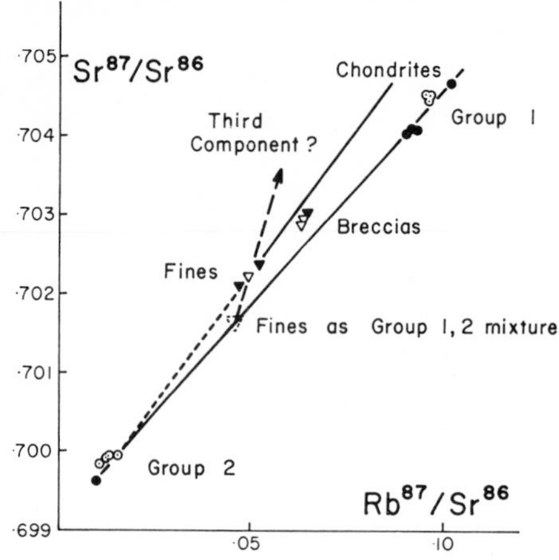

Fig. 6. Sr^{87}-evolution diagram showing fines and breccias as mixtures of Group 1 with Group 2 igneous rocks plus a smaller third component of high Sr^{87}/Sr^{86}.

which the chemistry suggested as a 'soil' model. The identity of this component remains unknown. It is unlikely to be chondritic meteorites, as their mixing-line with fines (Fig. 6) is too little different in slope to the Group 1–Group 2 mixing-line. Far too much would be needed to displace the fines from an original position at the intersection of mixing lines, which latter, moreover, would be too biased towards Group 2. The required component should lie to the left-hand side of the meteorite isochron, as indicated by the broken line in Fig. 6, and it must have at least a two-stage Rb–Sr history. The glass fragment analyzed by ALBEE *et al.* (1970) lies in this field but its potential to displace the fines cannot be assessed until the authors publish their primary data. It is also possible that the unidentified component is part of the feldspathic suite of rocks identified in the regolith (e.g. CHAO *et al.*, 1970; WOOD *et al.*, 1970a), but not yet analyzed by Rb–Sr.

CONCLUSIONS

The igneous rocks collected from Tranquillity Base represent two different rock-units, probably lava flows. The flows are distinguished chemically, not by texture.

Similarities in their major-element chemistry suggest a single source region, but differences in their trace-element and radiogenic isotope chemistry suggest two magmas formed by different degrees of partial melting. Both magmas crystallized after only minor crystal fractionation at 3·70 ± 0·10 b.y.

Major element chemistry shows that the fines and breccias at Tranquillity Base are composed of an equal mixture of the two flows, plus an anorthositic rock component, minus some ilmenite.

The high ages for the regolith by the Rb–Sr and U–Pb methods can be compatible with its formation from mainly 3·70 b.y. lavas. The 3·70 b.y. magmas transported uranium and lead from the source without bias, but at the surface, lead was selectively moved to the top of the sequence. The top lava layers were subsequently transformed into the present regolith, so that lead in the regolith lies on a secondary 3·70 b.y. isochron with the lavas, and contains a slight excess of 3·70 b.y. primary lead. Technical uncertainties in measuring U/Pb obscure the discordant positions of regolith lead (above) and lava lead (below) the Concordia curve. The U–Pb age of the lava source is about 4·5 b.y.

The Rb–Sr model age of the regolith is also about 4·5 b.y. It contains a component enriched in Sr^{87} which had at least a two-stage chemical history.

Acknowledgements—We thank H. Berry and R. Rudowski for sample preparation and mineral separation; J. Pennington, D. Millar, Mrs. A. Brown and Mrs. K. Phillips for assistance with the experimental work; and V. M. Oversby, E. J. Essene, N. G. Ware, A. J. R. White and A. E. Ringwood for helpful discussion.

References

Agrell S. O., Scoon J. H., Muir I. D., Long J. V. P., McConnell J. D. C. and Peckett A. (1970) Mineralogy and petrology of some lunar samples. *Science* **167,** 583–586.

Albee A. L., Burnett D. S., Chodos A. A., Eugster O. J., Huneke J. C., Papanastassiou D. A., Podosek F. A., Russ G. Price II., Sanz H. G., Tera F. and Wasserburg G. J. (1970) Ages, irradiation history and chemical composition of lunar rocks from the Sea of Tranquillity. *Science* **167,** 463–466.

Arrhenius G., Asunmaa S., Drever J. I., Everson J., Fitzgerald R. W., Frazer J. Z., Fujita H., Hanor J. S., Lal D., Liang S. S., Macdougall D., Reid A. M., Sinkankas J. and Wilkening L. (1970) Phase chemistry, structure, and radiation effects in lunar samples. *Science* **167,** 659–661.

Brown G. M., Emeleus C. H., Holland J. G. and Phillips R. (1970) Petrographic, mineralogic and X-ray fluorescence analysis of lunar igneous-type rocks and spherules. *Science* **167,** 599–601.

Chapman D. R. and Scheiber L. C. (1969). Chemical investigations of Australian tektites. *J. Geophys. Res.* **74,** 6737–6776.

Chappell B. W., Compston W., Arriens P. A. and Vernon M. J. (1969) Rubidium and strontium determinations by X-ray fluorescence spectrometry and isotope dilution below the part per million level. *Geochim. Cosmochim. Acta* **33,** 1002–1006.

Chao E. C. T., James O. B., Minkin J. A., Boreman J. A., Jackson E. D. and Raleigh C. B. (1970) Petrology of unshocked crystalline rocks and shock effects in lunar rocks and minerals. *Science* **167,** 644–647.

Compston W., Arriens P. A., Vernon M. J. and Chappell B. W. (1970) Rubidium–strontium chronology and chemistry of lunar material. *Science* **167,** 474–476.

Compston W., Chappell B. W., Arriens P. A. and Vernon M. J. (1969) On the feasibility of NBS 70a K-feldspar as a Rb–Sr age reference sample. *Geochim. Cosmochim. Acta* **33,** 753–757.

Engel A. E. J. and Engel C. G. (1970) Lunar rock compositions and some interpretations. *Science* **167,** 527–528.

Flanagan F. J. (1969) U.S.G.S. standards—II. first compilation of data for the new U.S.G.S. rocks. *Geochim. Cosmochim. Acta* **33,** 81–120.

Frondel C., Klein C. Jr., Ito J. and Drake J. C. (1970) Mineralogy and composition of lunar fines and selected rocks. *Science* **167,** 681–683.

Gast P. W. and Hubbard N. J. (1970) Abundance of alkali metals, alkaline and rare earths and strontium^{87}/strontium^{86} ratios in lunar samples. *Science* **167,** 485–487.

Gopalan K., Kaushal S., Lee-Hu C. and Wetherill G. W. (1970) Rubidium–strontium, uranium and thorium–lead dating of lunar material. *Science* **167,** 471–473.

Heinrich K. F. J. (1966) X-ray absorption uncertainty. In *The Electron Microprobe* (editor T. D. McKinley *et al.*), pp. 296–377. Wiley.

Hurley P. M. and Pinson W. H. Jr. (1970) Rubidium–strontium relations in Tranquillity Base samples. *Science* **167,** 473–474.

Keays R. R., Ganapathy R., Laul J. C., Anders E., Herzog G. F. and Jeffery P. M. (1970) Trace elements and radioactivity in lunar rocks: implications for meteorite infall, solar wind flux and formation conditions of moon. *Science* **167,** 490–493.

Kohman T. P., Black L. P., Ihochi H. and Huey J. M. (1970) Lead and thallium isotopes in Mare Tranquillitatis surface material. *Science* **167,** 481–483.

Marshall R. M. (1962) Mass spectrometric study of the lead in carbonaceous chondrites. *J. Geophys. Res.* **67,** 2005–2015.

Mason B., Fredriksson K., Henderson E. P., Jarosewich E., Melson W. G., Towe K. M. and White J. S., Jr. (1970) Mineralogy and petrography of lunar samples. *Science* **167,** 656–659.

Maxwell J. A., Abbey S. and Champ W. H. (1970) Chemical composition of lunar material. *Science* **167,** 530–531.

Morrison G. H., Gerard J. T., Kashuba A. T., Gangadharam E. V., Rothenberg A. M., Potter N. M. and Miller G. B. (1970) Multielement analysis of lunar soil and rocks. *Science* **167,** 505–507.

Murthy V. R. and Patterson C. C. (1962) Primary isochron of zero age for meteorites and the earth. *J. Geophys. Res.* **67,** 1161–1167.

Murthy V. R., Schmitt R. A. and Rey P. (1970) Rubidium–strontium age and elemental and isotopic abundances of some trace elements in lunar samples. *Science* **167,** 476–479.

Norrish K. and Chappell B. W. (1967) X-ray fluorescence spectrography. In *Physical Methods in Determinative Mineralogy* (editor J. Zussman), pp. 161–214. Academic Press.

Norrish K. and Hutton J. T. (1969) An accurate X-ray spectrographic method for the analysis of a wide range of samples. *Geochim. Cosmochim. Acta* **33,** 431–453.

O'Hara M. J., Biggar G. M. and Richardson S. W. (1970) Experimental petrology of lunar material: the nature of mascons, seas and the lunar interior. *Science* **167,** 605–607.

Papanastassiou D. A. and Wasserburg G. J. (1969) Initial strontium isotopic abundances and the resolution of small time differences in the formation of planetary objects. *Earth Planet. Sci. Lett.* **5,** 361–376.

Peck L. C. and Smith V. C. (1970) Quantitative chemical analysis of lunar samples. *Science* **167,** 532.

Philpotts J. A. and Schnetzler C. C. (1970) Potassium, rubidium, strontium, barium and rare-earth concentrations in lunar rocks and separated phases. *Science* **167,** 493–495.

Reed G. W., Kigoshi K., and Turkevich A. (1960) Determinations of concentrations of heavy elements in meteorites by activation analysis. *Geochim. Cosmochim. Acta* **20,** 122–140.

Ringwood A. E. and Essene E. (1970) Petrogenesis of lunar basalts and the internal constitution and origin of the moon. *Science* **167,** 607–610.

Rose H. J. Jr., Cuttitta F., Dwornik E. J., Carron M. K., Christian R. P., Lindsay J. R., Ligon D. T. and Larson R. R. (1970) Semimicro, chemical and X-ray fluorescence analysis of lunar samples. *Science* **167,** 520–521.

Schmitt R. A., Wakita H. and Rey P. (1970) Abundances of 30 elements in lunar rocks, soil and core samples. *Science* **167,** 512–515.

Silver L. T. (1970) Uranium–thorium–lead isotope relations in lunar materials. *Science* **167,** 468–471.

Tatsumoto M. and Rosholt J. N. (1970) Age of the moon: an isotopic study of uranium–thorium–lead systematics of lunar samples. *Science* **167,** 461–463.

TAYLOR S. R. Personal communication.
TURNER G. (1970) Argon-40/argon-39 dating of lunar rock samples. *Science* **167,** 466–468.
WANLESS R. K., LOVERIDGE W. D. and STEVENS R. D. (1970) Age determinations and isotopic abundance measurements on lunar samples. *Science* **167,** 479–480.
WARE N. G. and LOVERING J. F. (1970) Electron-microprobe analyses of phases in lunar samples. *Science* **167,** 517–520.
WEILL D. F., MCCALLUM I. S., BOTTINGA Y., DRAKE M. J. and MCKAY G. A. (1970) Petrology of a fine-grained igneous rock from the Sea of Tranquillity. *Science* **167,** 635–638.
WIIK H. B. and OJANPERA P. (1970) Chemical analyses of lunar samples 10017, 10072 and 10084. *Science* **167,** 531–532.
WOOD J. A., DICKEY J. S., JR., MARVIN U. B. and POWELL B. N. (1970a) Lunar anorthosites. *Science* **167,** 602–604.
WOOD J. A., MARVIN U. B., POWELL B. N. and DICKEY J. S. JR. (1970b) Mineralogy and petrology of the Apollo 11 lunar sample. Smithsonian Astrophysical Observatory Special Report 307.

Proceedings of the Apollo 11 Lunar Science Conference, Vol. 2, pp. 1029 to 1036.

The cosmic-ray and solar-flare bombardment of the moon

J. D'AMICO, J. DEFELICE and E. L. FIREMAN
Smithsonian Institution Astrophysical Observatory, Cambridge, Massachusetts 02138

(*Received* 2 *February* 1970; *accepted in revised form* 19 *February* 1970)

Abstract—Tritium and argon radioactivities attributable to galactic and solar cosmic-ray interactions were measured in lunar soil and in three lunar rocks. The tritium in the soil was 325 ± 17 dis/min/kg and that in the rocks ranged from 212–250 dis/min/kg. The Ar^{39} in the soil was 12·1 ± 0·7 dis/min/kg, and the Ar^{37}/Ar^{39} ratio was 2·25 ± 0·35. In lunar rocks 10017, 10072 and 10061 the Ar^{39} contents were 16·4 ± 0·9, 15·8 ± 1·0 and 13 ± 3 dis/min/kg, respectively, with Ar^{37}/Ar^{39} ratios of 1·28 ± 0·20, 1·62 ± 0·25 and about 1·5. On the basis of the known galactic cosmic-ray flux and cross sections, at least half the observed radioactivities were produced by solar cosmic rays.

The ($He^3/2H^3$) exposure age of rocks 10017, 10072 and 10061 were 364 ± 40, 202 ± 23 and 372 ± 60 × 10^6 yr, respectively. Their $Ar^{38}/(Ar^{37} + Ar^{39})$ exposure ages were 633 ± 110, 402 ± 72 and 3410 ± 1000 × 10^6 yr, respectively. The older argon exposure ages indicate that there was He^3 loss. The exposure age is simply interpreted as the time since the rocks were thrown out of large impact craters. Rock 10061, which is a breccia, was within 1 m of the lunar surface for approximately 3·4 × 10^9 yr. This indicates that the rate of soil turnover to depths greater than 1 m is slow and that the erosion rate of lunar rocks is small.

INTRODUCTION

THE AMOUNTS of H^3, Ar^{37} and Ar^{39} radioactivities in lunar material are of interest because they give information about galactic and solar-source mechanisms and because, when combined with the stable isotopes He^3 and Ar^{38}, they determine exposure ages (ANDERS, 1962; ARNOLD, 1961; SCHAEFFER, 1962). The properties of the galactic cosmic rays at 1 a.u. are well known (WEBBER, 1967; MEYER, 1969). The properties of solar cosmic rays are less well known. Solar cosmic rays have lower energies than do galactic. The former occur during flares in short bursts whose frequency, although sporadic, depends upon the solar cycle. Both galactic and solar cosmic rays produce radioactive species by interacting with material. Radioactivities present in some solar flares themselves may also be implanted in the material. There is evidence (FIREMAN *et al.*, 1961; TILLES *et al.*, 1963) that H^3 present in the solar flare of November 12, 1960, was implanted in the Discoverer XVII satellite. Since some of the returned lunar material was not shielded either by an atmosphere or by protective outer material, its study should give important information about the solar cosmic rays.

The exposure ages obtained from radioactive and stable isotopes are related to the cratering caused by both small and large impacts and to the movement of the lunar soil. The minute pits on rounded surfaces of lunar rocks indicate that the rocks have been eroded by small-particle bombardment (LSPET, 1969). The craters on the moon are evidence for large meteorite impacts. The lunar soil may be in a state of flux because of this bombardment and also perhaps because of internal stresses. It is therefore of interest to obtain exposure ages for lunar material.

Apparatus and Experimental Procedure

The apparatus used for the argon measurements in lunar materials is similar to that used for argon measurements in meteorites except for the improved proportional counters, whose characteristics are given in a recent article (Fireman and Goebel, 1970). Counter and other improvements made determinations possible in samples as small as 1 g.

Lunar rocks 10072, 10017 and 10061, of petrological types A, B and C, respectively, and a sample of soil fines, which have been described by the LSPET (1969), were analyzed. The samples were vacuum-melted in the presence of argon carrier at 1600°C for 2 hr. The evolved gases were reacted over vanadium foil at approximately 800°C. The noble gases were removed from the vanadium foil at room temperature and their volumes measured. The helium and neon were separated by freezing

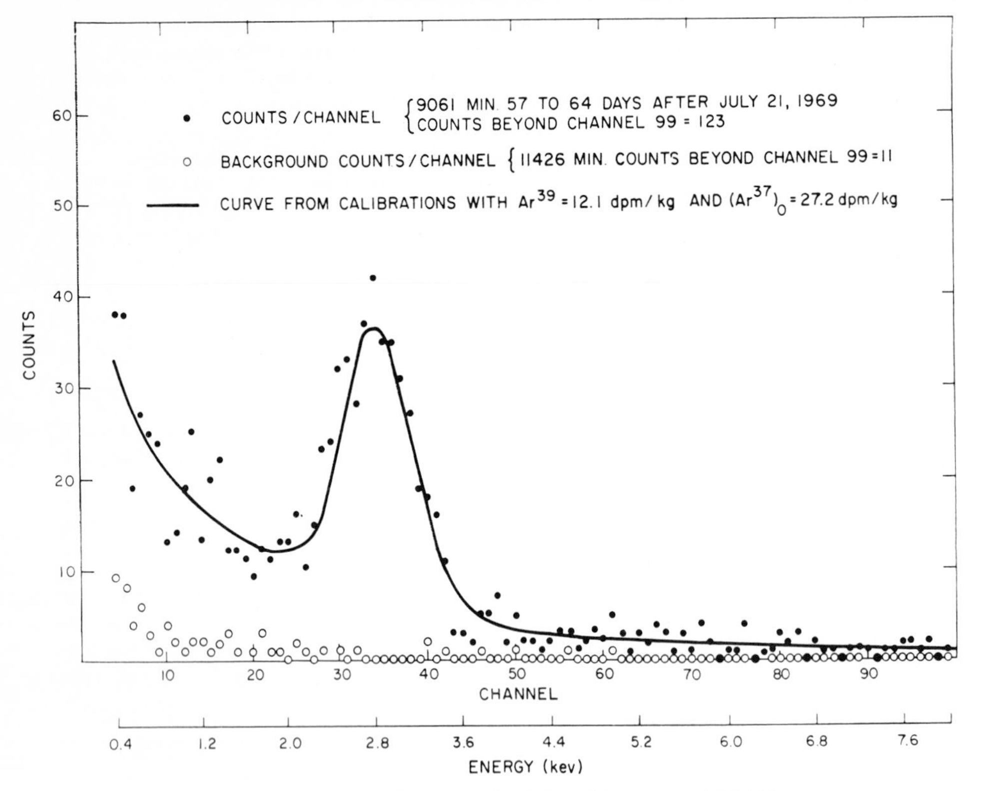

Fig. 1. Argon radioactivity for 9·9 g of lunar soil 10084-24.

the argon and heavier noble gases on charcoal at liquid-nitrogen temperature. Carrier krypton was added to the charcoal. The argon was separated from the krypton at dry-ice temperature and placed in low-level proportional counters of the Davis type (Davis *et al.*, 1968) with volumes between 0·54 and 0·75 cm^3. The argon pressure in the counters was standardized at 1·4 atm. The argon yields ranged between 95 and 99 per cent. Methane (10%) was added to the argon. The counters were removed from the extraction system and counted in a low-level counting system where the backgrounds were 1 c/d or less in 2·8 $\pm$ 0·6-keV channels of a 100-channel analyzer. The 2·2–3·4-keV energies cover the Ar^{37} peak to 0·1 maximum. The backgrounds were less than 4 c/d above the 4·0-keV channel where the Ar^{39} was counted. The counting efficiency for Ar^{37} in different counters varied from 41–45 per cent; and for Ar^{39}, from 14–20 per cent for the channels above 4·0 keV. A thin quartz window at one end of the counter permits the passage of the 5·9-keV X-ray from an Fe^{55} source used to adjust the energy scale. Calibration curves with known amounts of Ar^{37} and Ar^{39} for a typical counter are given in another publication (FIREMAN and GOEBEL, 1970). Figures 1 and 2 show the argon counting data, together with backgrounds, from the 9·9-g lunar soil sample for 9061 min between September 16 and 23, 1969; and from the 10·0-g sample of lunar rock 10017 for 12,936 min between October 3 and 13, 1969. The argon samples were counted for approximately 2 months, and the Ar^{37} activity decayed with a 35-day half-life.

The krypton was removed from the charcoal at 0°C and put into an argon-type counter to which methane had been added. The electronics were adjusted so that the 5·9-keV X-ray from the Fe^{55} source was peaked in channel 30 of the 100-channel analyzer. The 11·9-keV peak from Kr^{81} should then appear in channel 61. The counter efficiency for Kr^{81} was assumed to be the same as for Ar^{37}; and for Kr^{85}, to be 30 per cent above 4·0 keV.

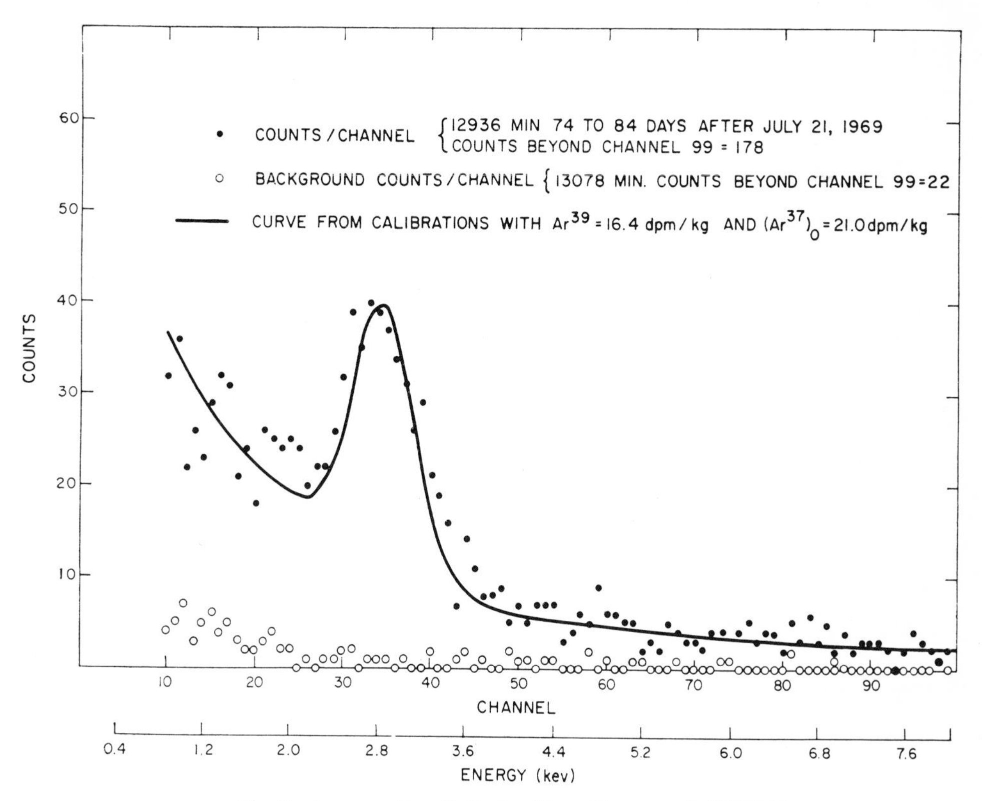

Fig. 2. Argon radioactivity for 10 g of lunar rock 10017-14.

The hydrogen was removed from the vanadium foil while heated, and the volume of the released hydrogen was measured. The hydrogen counter consisted of a copper cathode and a 0·002-in. Therlo center wire in a quartz envelope that had a thin window at one end; its total volume was 42 cm³ (see Fig. 3). The electronics were adjusted so that the 100-channel analyzer scanned energies up to 19·3 keV. With 400-torr pressure of P-10 gas (commercially available counting gas containing 90% argon and 10% methane) and 200-torr pressure of hydrogen, the counting efficiency for tritium between 1 and 19·3 keV was 60 per cent. The resolution for the 5·9-keV source was 20 per cent. The background was 0·14 c/min. The tritium spectrum does not consist of a sharp peak; however, it is different from the background spectrum. Figure 4 shows the tritium and background spectra for the counter of 42 cm³ vol.

Table 1 summarizes the results. For comparison, the radioactivities in the nonmagnetics of the recently fallen Sprucefield meteorite, an L4 chondrite, were measured with the same equipment. The radioactivities in the lunar material are within a factor of 2 of those in the meteorite. There are, however, interesting differences between the lunar samples and also between the lunar samples and meteorites.

Table 1. Radioactivities and gas contents from lunar samples*

Apollo 11 sample	Weight (g)	Tritium (dis/min/kg)	Hydrogen (cm^3/g)	Helium (cm^3/g)	Ar^{39} (dis/min/kg)	Ar^{37} (dis/min/kg)†	Ar^{37}/Ar^{39}
Soil 10084-24	(9·9)	325 ± 17	1·2	0·26	12·1 ± 0·7	27·2 ± 2·2	2·25 ± 0·35
Rock 10017-14	(10·0)	219 ± 7	0·47	$<5 \times 10^{-3}$	16·4 ± 0·9	21·0 ± 2·0	1·28 ± 0·20
Rock 10072-11	(9·4)	234 ± 10	0·76	$<5 \times 10^{-3}$	15·8 ± 1·0	25·7 ± 2·0	1·62 ± 0·25
Rock 10061 surface	(1·49)	235 ± 15	2·5	~0·3	13 ± 3	19 ± 5	~1·5
Rock 10061 interior	(0·96)	231 ± 10	1·4	~0·4	—	—	—
Meteorite							
Sprucefield L4 (non-magnetics)	(21)	290 ± 30	—	$<5 \times 10^{-3}$	7·7 ± 0·4	12·6 ± 1·4	1·64 ± 0·25

* Krypton radioactivities were measured in soil sample 10084-24 and found to be 0·14 ± 0·10 dis/min/kg for Kr^{81} and less than 1 dis/min/kg for Kr^{85}.

† Extrapolated to July 21, 1969.

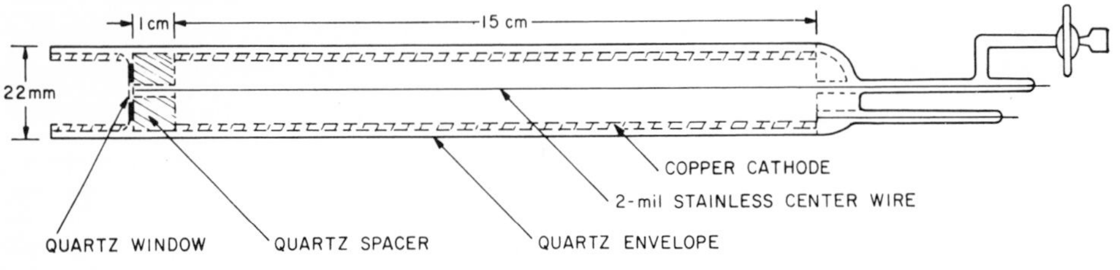

Fig. 3. Hydrogen counter of 42 cm^3 vol.

Tritium and Argon Results

The tritium in the soil was 325 ± 17 dis/min/kg; in rocks 10017 and 10072, 219 ± 7 and 234 ± 10 dis/min/kg. A chip from the surface of rock 10061 had 235 ± 15 dis/min/kg; an interior sample of the same rock had 231 ± 10 dis/min/kg. The higher content in the soil cannot be caused by the small differences in chemical composition between the soil and rocks. The higher H^3 content in the soil is probably caused by a more intense solar bombardment at shallower depths. The soil had 0·26 cm^3/g of helium, in accord with the mass spectrometer results (LSPET, 1969); this content was attributed to the solar wind. The hydrogen in the soil was 1·2 cm^3/g, which may in part be solar-wind hydrogen. The difference between the tritium in the soil and that in the rocks should not be attributed to implanted tritium from solar flares until other possibilities are eliminated. More detailed studies of tritium versus depth are necessary before the excess tritium in the soil can be interpreted.

The tritium contents of rocks 10017, 10072 and 10061 can be combined with their He^3 contents of $(3·1 \pm 0·2) \times 10^{-6}$, $(1·9 \pm 0·1) \times 10^{-6}$ and $(3·5 \pm 0·3) \times 10^{-6}$ cm^3/g

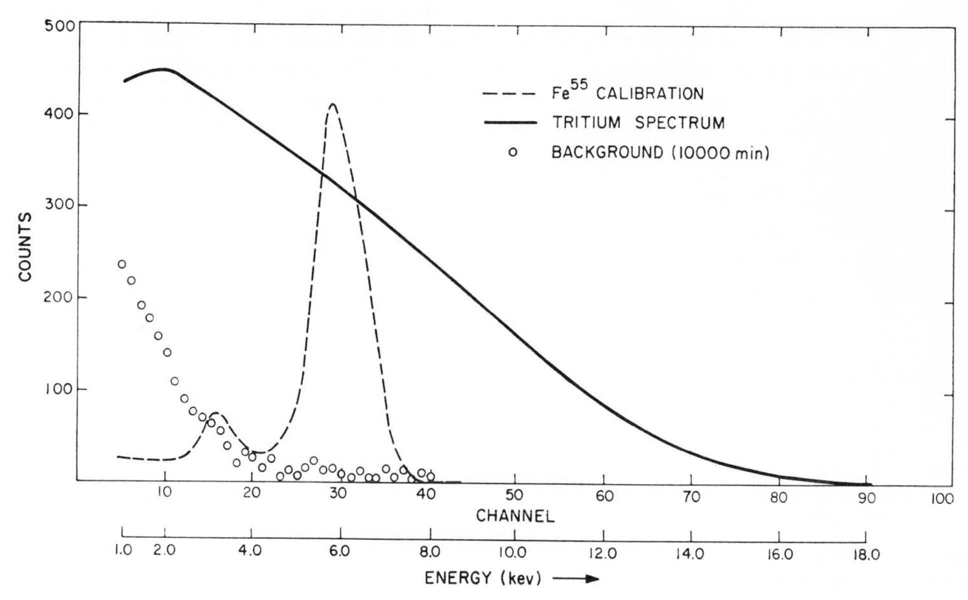

Fig. 4. Tritium and background spectra for hydrogen counter of 42 cm^3 vol.

measured by FUNKHOUSER *et al.* (1970) to obtain cosmic-ray exposure ages. The errors were assigned on the basis of private discussions with D. D. Bogard. The exposure age of rock 10017 is $364 \pm 40 \times 10^6$ yr, of rock 10072 it is $202 \pm 23 \times 10^6$ yr and of rock 10061 it is $372 \pm 60 \times 10^6$ yr, with the He^3 production rate assumed to be twice the present tritium decay rate. On the basis of the large amount of solar Ne^{20} in rock 10061, approximately two-thirds of its He^3 is solar He^3. With this correction the He^3 exposure age is only 120×10^6 yr. These ages are simply interpreted as the time since the rocks were thrown out from large impact craters. There are two possible complications to this simple interpretation: some He^3 may be lost by diffusion, and the production of He^3 may not have been constant because of changes in the amount of surrounding material with time.

The Ar^{39} activity in the soil was $12{\cdot}1 \pm 0{\cdot}7$ dis/min/kg, and those in rocks 10017, 10072 and 10061 were $16{\cdot}4 \pm 0{\cdot}9$, $15{\cdot}8 \pm 1{\cdot}0$ and 13 ± 3 dis/min/kg, respectively. The Ar^{37} activity in the soil was $27{\cdot}2 \pm 2{\cdot}2$ dis/min/kg, and those in rocks 10017, 10072 and 10061 were $21{\cdot}0 \pm 2{\cdot}1$, $25{\cdot}7 \pm 2{\cdot}0$ and 19 ± 5 dis/min/kg, respectively. An exposure age can be obtained from the combination of spallation Ar^{38} (denoted by $Ar_{sp}{}^{38}$ in Table 2) and the argon radioactivities if use is made of the relative production rates obtained by STOENNER, LYMAN and DAVIS (1970). To a first approximation, the Ar^{38} production rate equals that of Ar^{37} plus Ar^{39}. The Ar^{38} contents produced by cosmic rays in rocks 10017, 10072 and 10061 were obtained from the argon measurements by FUNKHOUSER *et al.* (1970) and were corrected for solar argon by assuming an Ar^{36}/Ar^{38} ratio of 5·28 for solar argon and an Ar^{36}/Ar^{38} ratio of 0·70 for cosmic-ray-produced argon. The $Ar^{38}/(Ar^{37} + Ar^{39})$ exposure age of rock 10017 is $633 \pm 110 \times 10^6$ yr, that of rock 10072 is $402 \pm 72 \times 10^6$ yr and that of rock 10061 is $3410 \pm 1000 \times 10^6$ yr. The Ar^{38} exposure age of rock 10061 is remarkably high; the error is caused mainly by the presence of solar argon. Because of the scientific importance of this very old exposure age, an attempt should be made to reduce the error by measuring the rare gases in separated phases. These ages are higher than the He^3 to tritium ages and indicate that He^3 has been lost. The large He^3 loss from rock 10061 is reflected by a low $(He^3/Ne^{21})_c$ ratio; if corrections are made for solar He^3 and Ne^{21}, this ratio is 2·7 compared to a ratio of 10 in rocks in which the He^3 loss is smaller. The old exposure age of rock 10061 indicates that the change in the amount of material surrounding the rocks was small during past aeons. The exposure ages are given in Table 2, together with the cosmic-ray-produced He^3

Table 2. $He^3/2H^3$ and $Ar_{sp}^{38}/(Ar^{37} + Ar^{39})$ cosmic-ray exposure ages

Sample	He^3 (10^{-8} cm/g)*	Exposure age $He^3/2H^3$ (10^6 yr)	Ar_{sp}^{38} (10^{-8} cm^3/g)*	Exposure age $Ar_{sp}^{38}/(Ar^{37} + Ar^{39})$ (10^6 yr)
Rock 10017	310 ± 20	364 ± 40	46 ± 4	633 ± 110
Rock 10072	190 ± 10	202 ± 23	33 ± 3	402 ± 72
Rock 10061†	350 ± 30 110 ± 40‡	372 ± 60 120 ± 60‡	212 ± 48	3410 ± 1000

* Values taken from FUNKHOUSER *et al.* (1970).
† Crystalline lithic fragments from rock 10061.
‡ Corrected for solar He^3, which is taken as $(Ne^{20}–Ne^{22})/50$.

and Ar^{38} contents. EBERHARDT *et al.* (1970) and MARTI *et al.* (1970) measured the Kr/Kr age of rock 10017 to be $510 \pm 50 \times 10^6$ yr, in agreement with the Ar^{38} age. The Kr/Kr ages have not yet been published for rocks 10072 and 10061.

GALACTIC AND SOLAR COSMIC RAYS

The tritium and argon radioactivities in meteorites are attributed to the interactions of galactic and solar cosmic rays with the material. It is reasonable to attribute these radioactivities in lunar material to the same mechanism. Since the flux of galactic cosmic rays at 1 a.u. is known (WEBBER, 1967; MEYER, 1969) and the depth of the lunar samples is small, one can estimate what fraction of the radioactivities are attributable to galactic cosmic rays. For 2π solid angle, the flux of nucleons with energy greater than 1000 MeV is $0{\cdot}8 \pm 0{\cdot}1$ nucleon/cm^2 sec, and that with energies between 400 and 1000 MeV is $0{\cdot}35 \pm 0{\cdot}1$ nucleon/cm^2 sec averaged over the solar cycle. The total tritium-production cross section in oxygen is 38 mbar, independent of the bombardment energy above 400 MeV (FIREMAN and ROWLAND, 1955; CURRIE *et al.*, 1956). For heavier elements,the tritium cross section increases with atomic weight as approximately $A^{2/3}$ for particles above 1000 MeV and is roughly independent of atomic weight for particles between 400 and 1000 MeV (KIRSTEN and SCHAEFFER, 1970). With these fluxes and cross sections, we obtain a tritium production rate of 85 ± 15 dis/min/kg, which is less than half the amount observed. Similar considerations for the Ar^{37} and Ar^{39} production in lunar material on the basis of measured cross sections lead to the conclusion that only a quarter or less of the Ar^{39} observed in the lunar rocks is produced by galactic cosmic rays. It therefore appears that at least half the observed radioactivities are produced by solar cosmic rays.

Relations Between the Exposure Age and Physical Processes on the Lunar Surface

The simplest interpretation of the cosmic-ray exposure age of a lunar rock is the time since the rock was thrown out from a large impact crater. As shown in Table 2, the Ar^{38} exposure ages are older than the He^{3} exposure ages. The differences are the following: a factor of 1·7 for rock 10017, petrological type B; a factor of 2·0 for rock 10072, petrological type A; and a factor of 32 for rock 10061, petrological type C. The rock classification is discussed by the LSPET (1969). Type A is a fine-grained crystalline rock with crystals of less than 200-μ size. Type B is a coarser grained crystalline rock. Type C is a breccia. The explanation for the difference between the Ar^{38} and He^{3} ages is that some He^{3} is lost by diffusion. Eberhardt *et al.* (1970) observed He^{3}

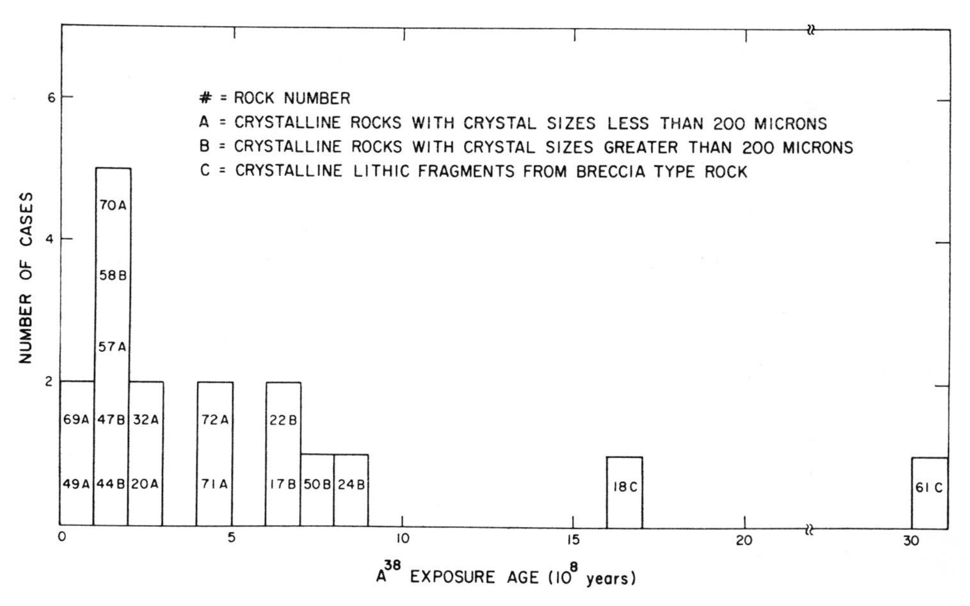

Fig. 5. Distribution of Ar^{38} exposure ages of lunar rocks.

loss from the feldspar fractions of lunar rocks. Figure 5 is a histogram of the Ar^{38} exposure ages obtained from the argon contents measured by Funkhouser *et al.* (1970) and an assumed (Ar^{37} + Ar^{39}) content of 37 dis/min/kg, which is the mean of the values measured for rocks 10017, 10072 and 10061. It appears that the two rocks, 10061 and 10018, of the breccia type, which have Ar^{38} exposure ages of $3{\cdot}1 \times 10^{9}$ and $1{\cdot}7 \times 10^{9}$ yr, have older Ar^{38} exposure ages than do the crystalline rocks of types A and B, which range between 50 and 900 $\times\ 10^{6}$ yr. Rocks 10061 and 10018 are formed by the cementing together of crystalline lithic fragments with fine soil, probably by the shock and heat generated by a large impact. The Ar^{38} in the crystalline lithic fragments at the time of the large impact appears to have been retained, while the He^{3} was lost. The oldest exposure age sets limits on the rate of turnover of lunar soil and the erosional lifetime of lunar rocks.

SHOEMAKER *et al.* (1970), on the basis of crater-distribution observations, estimated the turnover time of lunar soil to 1-m depth to be 1×10^9 yr and the lifetime of a 1-m rock bombarded by small and large particles to be $2{\cdot}5 \times 10^8$ yr. Since 1 m of protective material effectively eliminates the solar and galactic cosmic-ray bombardment, the production rate of Ar^{38} at this depth would be negligible. The exposure age of $3{\cdot}1 \times 10^9$ yr requires that the rate of soil turnover be three times slower and the rate of rock erosion be ten times slower than SHOEMAKER *et al.* estimated.

Acknowledgments—The authors are indebted to F. L. WHIPPLE for his constant encouragement. This research was supported in part by contract NAS 9-8105 from the National Aeronautics and Space Administration.

REFERENCES

ANDERS E. (1962) Meteorite ages. *Rev. Mod. Phys.* **34,** 287–325.

ARNOLD J. R. (1961) Nuclear effects of cosmic rays in meteorites. In *Annual Review of Nuclear Science* (editors E. Segré, G. Friedlander and W. E. Meyerhof), Vol. 11, pp. 349–370. Annual Reviews.

CURRIE L. A., LIBBY W. F. and WOLFGANG R. L. (1956) Tritium production by high energy protons. *Phys. Rev.* **101,** 1552–1563.

DAVIS R. JR., HARMER D. S. and HOFFMAN K. C. (1968) Search for neutrinos from the sun. *Phys. Rev. Lett.* **20,** 1205–1209.

EBERHARDT P., GEISS J., GRAF H., GRÖGLER N., KRÄHENBÜHL U., SCHWALLER H., SCHWARZMÜLLER J. and STETTLER A. (1970) Trapped solar wind noble gases, Kr^{81}/Kr exposure ages and K/Ar ages in Apollo 11 lunar material. *Science* **167,** 558–560.

FIREMAN E. L., DEFELICE J. and TILLES D. (1961) Solar flare tritium in a recovered satellite. *Phys. Rev.* **123,** 1935–1936.

FIREMAN E. L. and GOEBEL R. (1970) Ar^{37} and Ar^{39} in recently fallen meteorites and cosmic ray variations. *J. Geophys. Res.* To be published.

FIREMAN E. L. and ROWLAND F. S. (1955) Tritium and neutron production by 2·2-Bev protons on nitrogen and oxygen. *Phys. Rev.* **97,** 780–782.

FUNKHOUSER J. G., SCHAEFFER O. A., BOGARD D. D. and ZÄHRINGER J. (1970) Gas analysis of the lunar surface. *Science* **167,** 561–563.

KIRSTEN T. A. and SCHAEFFER O. A. (1970) High energy interactions in space. In *Interactions of Elementary Particle Research in Science and Technology* (editor L. C. L. Yuan). Academic Press. In press.

LSPET (LUNAR SAMPLE PRELIMINARY EXAMINATION TEAM) (1969) Preliminary examination of lunar samples from Apollo 11. *Science* **165,** 1211–1227.

MARTI K., LUGMAIR G. W. and UREY H. C. (1970) Solar wind gases, cosmic ray spallation products, and the irradiation history. *Science* **167,** 548–550.

MEYER P. (1969) Cosmic rays in the galaxy. In *Annual Review of Astronomy and Astrophysics* (editors L. Goldberg, D. Layzer and J. G. Phillips), Vol. 7, pp. 1–38. Annual Reviews.

SCHAEFFER O. A. (1962) Radiochemistry of meteorites. In *Ann. Rev. Phys. Chem.* **13,** 151–170.

SHOEMAKER E. M., HAIT M. H., SWANN G. A., SCHLEICHER D. L., DAHLEM D. H., SCHABER G. G. and SUTTON R. L. (1970) Lunar regolith at Tranquillity Base. *Science* **167,** 452–455.

STOENNER R. W., LYMAN W. J. and DAVIS R., JR. (1970) Cosmic ray production of rare gas radioactivities and tritium in lunar material. *Science* **167,** 553–555.

TILLES D., DEFELICE J. and FIREMAN E. L. (1963) Measurements of tritium in satellite and rocket material, 1960–1961. *Icarus* **2,** 258–279.

WEBBER W. R. (1967) The spectrum and charge composition of the primary cosmic radiation. In *Handbuch der Physik* (editors S. Flügge and K. Sitte), Vol. 46/2, pp. 181–264. Springer-Verlag.

Proceedings of the Apollo 11 Lunar Science Conference, Vol. 2, pp. 1037 to 1070.

Trapped solar wind noble gases, exposure age and K/Ar-age in Apollo 11 lunar fine material

P. EBERHARDT, J. GEISS, H. GRAF, N. GRÖGLER, U. KRÄHENBÜHL, H. SCHWALLER, J. SCHWARZMÜLLER and A. STETTLER
Physikalisches Institut, University of Berne, 3000 Berne, Switzerland

(*Received* 17 *February* 1970; *accepted in revised form* 25 *February* 1970)

Abstract—Comprehensive results on the distribution and on the elemental and isotopic abundances of the noble gases He, Ne, Ar, Kr and Xe in the Apollo 11 lunar fine material (sample No. 10084-47) are presented. The high absolute and the relative abundances of these gases suggest that they are primarily implanted solar wind particles. A distinct anti-correlation of trapped He, Ne, Ar, Kr and Xe content and grain size was found in seven bulk grain size fractions (range $1{\cdot}4\ \mu$–$130\ \mu$) and four ilmenite grain size fractions (range $22\ \mu$–$105\ \mu$). The grain size dependence follows the relation $c \sim d^{-n}$ with an average exponent for the five noble gases of $n = 0{\cdot}61$ (bulk grain size fractions) and $n = 1{\cdot}0$ (ilmenite fractions). Etching experiments on a $41\ \mu$ ilmenite fraction prove that the trapped He, Ne, Ar, Kr and Xe are concentrated at the surface of the individual grains, in a layer of typically $0{\cdot}2\ \mu$ depth. These experiments conclusively confirm a solar wind origin of the trapped gases. Solar flare particles are ruled out by the observed grain size dependence down to $1{\cdot}4\ \mu$ grain size. The He^4 surface concentration in the ilmenite is 3×10^{16} atoms/cm^2 corresponding to 50 atm partial pressure. Minimal solar wind irradiation times of 300 yr for each grain are required. Saturation effects occur at these irradiation levels.

Ilmenite shows least diffusion loss of light trapped gases and after correction for *in situ* spallation reactions and radioactive decay the following elemental and isotopic abundances for the solar wind are obtained: $He^4/Ar^{36} \geq 7600$; $Ne^{20}/Ar^{36} \geq 33$; $He^4/He^3 = 2700 \pm 100$; $Ne^{20}/Ne^{22} = 12{\cdot}85 \pm 0{\cdot}1$; $Ne^{20}/Ne^{21} = 400 \pm 11$; $Ar^{36}/Ar^{38} = 5{\cdot}32 \pm 0{\cdot}08$; $Ar^{40}/Ar^{36} < 0{\cdot}7$. Kr and Xe solar wind abundances derived from bulk grain size measurements are: $Ar^{36}/Kr^{86} \geq 7600$: $Ar^{36}/Xe^{132} \geq 16000$; Kr^{78}: Kr^{80}: Kr^{82}: Kr^{83}: Kr^{84}: Kr^{86} = 2·00: 12·90: 66·0: 65·75: 325·2: 100; and Xe^{124}: Xe^{126}: Xe^{128}: Xe^{129}: Xe^{130}: Xe^{131}: Xe^{132}: Xe^{134}: Xe^{136} = 2·99: 2·67: 50·45: 634·5: 100: 495·5: 607: 225·8: 184·2. Terrestrial and solar Ar and Kr are thus isotopically identical except for the radiogenic Ar^{40}. Ne and Xe differ significantly. The Ne difference is explained by mass fractionation. Solar wind Kr and carbonaceous chondrite Kr (AVCC-Kr) are isotopically different. AVCC-Xe agrees with solar wind Xe as far as the light, fission-shielded isotopes are concerned. The presence of a fission Xe component in AVCC-Xe is confirmed, possibly resulting from spontaneous fission of super heavy nuclides. The large difference between solar wind Xe and atmospheric Xe remains unexplained.

In the trapped lunar gas an excess of Ar^{40} is found which cannot be of solar origin. It is argued that this Ar^{40} originated from K^{40} decay in the moon, leaked out into the lunar atmosphere, and was driven back into the lunar surface material by the electro-magnetic fields associated with the solar wind. K and Ba determinations were carried out by isotope dilution with the purpose of determining K/Ar and exposure ages. A K/Ar age of $3{\cdot}5 \pm 0{\cdot}4$ aeons is obtained in the $90\ \mu$ bulk grain size fraction. An average cosmic ray exposure age for the grains and fragments of lunar fine sample 10084 of 520 ± 120 m.y. is obtained. The significance of this exposure age with respect to the history of the regolith in the Apollo 11 landing area is discussed.

INTRODUCTION

THE CONCENTRATION and isotopic composition of the noble gases He, Ne, Ar, Kr and Xe were measured in:

Seven grain size fractions of fine lunar material (sample no. 10084-47) (grain sizes between $1{\cdot}4$ and $130\ \mu$);

Four grain size fractions of ilmenite separated from the fine lunar material (grain sizes between 22 and 105 μ);

Seven etched samples from an ilmenite grain size fraction (grain size 41 μ);

A number of individual crystals, spherules and other structural elements separated from fine lunar material;

Three rock samples from lunar rocks 10017 and 10071;

Two feldspar concentrates from lunar rock 10071. In addition K and Ba were determined in several of these samples.

From these results the elemental and isotopic abundances of the trapped noble gases have been calculated and limits on the elemental and isotopic abundances in the solar wind can be inferred. Exposure ages were obtained from Kr^{81}/Kr and also from Xe^{126} measurements. For one grain size fraction of sample 10084 and for the rock samples and the feldspar concentrates K/Ar ages were derived.

A summary of our results has already been published (EBERHARDT *et al.*, 1970a, referred to as *Science* Publication). In this paper we give a complete representation of our results on the Apollo 11 lunar fine material (sample 10084-47) and a detailed discussion of these data. All our data have been reevaluated and checked, and therefore in a few instances small differences appear between the figures presented here and in the *Science* Publication. These differences are generally insignificant and can be neglected. Results in which the difference is non-negligible will be explicitly mentioned.

The experimental techniques used for the determination of the noble gas concentrations and isotopic compositions are similar to those of EBERHARDT *et al.* (1966), MARTI *et al.* (1966), and EUGSTER *et al.* (1969). Relative to the previous work our Kr and Xe blanks had been lowered considerably and were typically $0{\cdot}1 \times 10^{-12}$ cm^3 STP Kr^{86} and $0{\cdot}3 \times 10^{-12}$ cm^3 STP Xe^{132}. Concentrations were either determined by the peak height method, or by the addition of spike after determination of the isotopic composition, or by the standard isotopic dilution method. Some He, Ne and Ar measurements were made on a new mass spectrometer with on line extraction system and extraction blanks of 2×10^{-10} cm^3 STP He, 2×10^{-10} cm^3 STP Ne and 6×10^{-9} cm^3 STP Ar of atmospheric isotopic composition. The K and Ba concentrations were measured with the isotopic dilution technique. Typical blanks were $0{\cdot}3$ μg K and $0{\cdot}07$ μg Ba. Except when noted all determinations represent single runs.

ORIGIN OF TRAPPED GASES

The preliminary examination of the fine lunar material (LSPET, 1969) already revealed the presence of large amounts of noble gases, presumably of solar wind origin. Other possible sources of these trapped gases* are energetic solar particles, an ambient

---

* As in meteorite research we use the term "trapped gases" for the gases which are not the result of *in situ* nuclear processes. Depending on the origin of the trapped gases also other less volatile elements may be trapped at the same time and the term "trapped elements" might be appropriate. Nuclear processes which have been recognized so far as contributing to the observed noble gases in meteorites and lunar samples are: radioactive decay, fission, spallation reactions, neutron capture and some other specific nuclear reactions. This noble gas component can be called nucleogenic. For a detailed discussion see for example EUGSTER *et al.* (1969).

atmosphere early in the history of lunar material, or gases released during the impact of comet nuclei.

The location of the trapped gases within the sample should allow a distinction between these possible sources. Solar wind emplaced trapped gases should be located in an approximately 0·1 μ thick surface layer of the individual grains. The other mentioned sources for trapped gases could hardly lead to such a strong enhancement of the trapped gas concentration at the surface of the individual grains.

To establish the location of the trapped gases in the lunar fine material we determined the concentration of the noble gases as a function of the grain size in unseparated lunar fine material and in separated ilmenite. The bulk grain size fractions were obtained by sieving (35, 90, 160, 300 and 450 mesh) and by sedimentation in acetone. The ilmenite grain size fractions were separated from the bulk grain size fractions with Clerici-solution (density 4·03 g/cm^3). From X-ray powder diffraction patterns and from microscopic examination we estimate that the ilmenite concentrates contain less than 5–8 per cent other minerals (cf. Fig. 1). The average grain sizes of the individual grain size fractions were determined by measuring the major axis a and the medium axis b of 100–500 grains from each fraction. For the finest fractions electron microscope pictures had to be used and only approximately 50 grains each were measured. The average grain size d was calculated from equation (4) given by EBERHARDT *et al.* (1965a).*

The results of the noble gas measurements are given in Tables 1–3. All stable isotopes of all noble gases were measured in the eleven grain size fractions. Figures 2 to 4 show the observed grain size dependence for the five noble gases. For comparison the grain size dependence of the trapped He^4, Ne^{20} and Ar^{36} content observed in the Khor Temiki aubrite is shown (EBERHARDT *et al.*, 1965a, b). A distinct anticorrelation between grain size and trapped gas content is evident for the bulk lunar fines and the separated ilmenite. With good approximation the anticorrelation can be represented by

$$c \propto d^{-n} \tag{1}$$

where c is the noble gas concentration. For the bulk grain size fractions the exponent n is between 0·58 and 0·65 for all the noble gases. This is slightly higher than the exponents observed in the Khor Temiki aubrite. The ilmenite shows a distinctly steeper grain size dependence. The exponents n are between 1·15 and 0·82. Such a grain size dependence results if the trapped gases are located at the surface of the individual grains. The steepness of the grain size dependence observed in the ilmenite is steeper than for the bulk and indicates an approximately constant concentration of trapped gases per unit surface area, independent of grain size.

In a further study on the location of the trapped gases we etched aliquots of the 41 μ ilmenite grain size fraction with dilute HF in order to remove a thin layer from

---

* For the initial publication of the Apollo 11 lunar sample results in the *Science* Publication the average grain size had been calculated using the simplified relation $\delta = \langle \sqrt{(ab)} \rangle$. For the larger grain sizes d and δ are virtually identical, for the smaller grain sizes d is 10–20 per cent larger than δ. The grain size dependencies given in this paper have thus a slightly different slope.

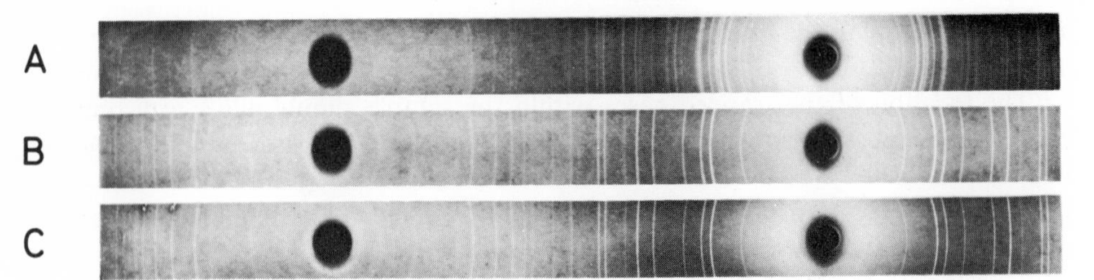

Fig. 1. X-ray diffraction pattern of: (A) 42 μ bulk grain size fraction AA-28 (Apollo 11 lunar fines No. 10084); (B) 41 μ ilmenite grain size fraction AB-1 (Apollo 11 lunar fines No. 10084); and (C) terrestrial ilmenite from Binn Valley, Switzerland. FeK_α radiation, 12 mA, 30 kV, camera dia. 114·83 mm, Straumanis camera arrangement.

Table 1. Results of He, Ne and Ar measurements on bulk grain size fractions, ilmenite grain size fractions, and two series of etched ilmenite grain size fractions. All fractions were separated from Apollo 11 lunar fine material (10084-47)

Sample No.	Grain size (μ)	Surface layer removed (μ)	Weight of analysed sample (mg)	He^4 (10^{-8} cm^3 STP/g)	Ne^{20} (10^{-8} cm^3 STP/g)	Ar^{36} (10^{-8} cm^3 STP/g)	He^4/He^3	Ne^{20}/Ne^{22}	Ne^{22}/Ne^{21}	Ar^{36}/Ar^{38}	Ar^{40}/Ar^{36}
Bulk grain size fractions											
AA-26	130	—	5·7	6,400,000 ±600,000	70,000 ±4,500	11,100 ±700	2,500 ±50	12·42 ±0·3	25·05 ±0·6	5·05 ±0·04	1·36 ±0·02
AA-27	90	—	3·3	8,100,000 ±400,000	87,000 ±4,500	16,500 ±1,300	2,590 ±50	12·40 ±0·1	26·34 ±0·7	5·28 ±0·07	1·21 ±0·02
AA-28	42	—	6·2	10,000,000 ±700,000	120,000 ±8,000	16,800 ±2,000	2,620 ±70	12·40 ±0·1	27·30 ±0·5	5·04 ±0·04	1·16 ±0·02
AA-29	15	—	4·5	22,800,000 ±1,200,000	325,000 ±16,000	31,000* ±2,500	2,570 ±50	12·61 ±0·2	28·65 ±2·0	5·02* ±0·06	1·07* ±0·03
AA-30	3·7	—	3·1	50,000,000 ±5,000,000	480,000 ±25,000	65,000 ±4,500	2,400 ±70	12·63* ±0·25	30·10 ±1·0	5·16 ±0·05	1·03 ±0·02
AA-19	2·0	—	2·2	86,000,000 ±5,000,000	670,000 ±40,000	159,000 ±12,000	2,390 ±60	12·70 ±0·15	30·90 ±1·2	5·39 ±0·06	0·95 ±0·02
AB-4	1·4	—	1·3	124,000,000 ±6,000,000	1,240,000 ±60,000	174,000 ±15,000	2,330 ±50	12·83 ±0·15	30·55 ±0·6	5·32 ±0·09	1·00 ±0·03
Ilmenite grain size fractions											
AB-7	105	—	8·1	10,800,000 ±600,000	52,000 ±2,500	1,970 ±130	2,580 ±50	12·82 ±0·15	29·05 ±0·9	4·96 ±0·04	1·12 ±0·02
AB-5	65	—	9·8	21,300,000 ±1,200,000	97,000 ±5,000	3,900 ±250	2,480 ±90	12·77 ±0·15	29·75 ±1·8	5·12 ±0·05	0·88 ±0·02
AB-1	41	—	2·3	33,200,000 ±1,700,000	154,000 ±8,000	6,300 ±600	2,670 ±50	12·93 ±0·15	29·90 ±1·2	5·15 ±0·08	0·84 ±0·04
AB-11	22	—	6·6	67,500,000 ±3,500,000	300,000 ±15,000	8,900 ±600	2,640 ±80	12·88 ±0·25	30·30 ±1·7	5·28 ±0·06	0·75 ±0·02
Etched ilmenite grain size fraction: series 1											
AB-1	41	—	0·4	38,900,000 ±3,000,000	179,000 ±12,000	6,330 ±450	2,570 ±150	12·90 ±0·25	28·1 ±0·6	5·23 ±0·15	1·15 ±0·35
AB-33	41	0·16	0·3	29,500,000 ±2,500,000	94,100 ±8,000	2,460 ±220	2,650 ±150	12·65 ±0·25	29·1 ±0·7	4·92 ±0·1	—
AB-32	41	0·19	0·6	14,800,000 ±1,000,000	50,700 ±3,000	1,240 ±80	2,440 ±100	12·50 ±0·15	28·6 ±0·6	4·77 ±0·1	—
AB-31	41	0·35	0·3	3,780,000 ±300,000	13,000 ±1,000	360 ±30	2,030 ±100	12·61 ±0·20	23·6 ±0·9	4·05 ±0·25	—
Etched ilmenite grain size fraction: series 2											
AB-1	41	—		33,200,000 ±1,700,000	154,000 ±8,000	6,300 ±600	2,670 ±50	12·93 ±0·15	29·90 ±1·2	5·15 ±0·08	0·84 ±0·04
AB-80	41	0·14	2·8	26,500,000 ±1,300,000	95,000 ±5,000	3,400 ±250	2,900 ±60	12·66 ±0·15	29·90 ±1·3	4·72 ±0·11	0·79 ±0·06
AB-79	41	0·19	2·7	15,500,000 ±8,000,000	50,000 ±3,000	2,000 ±200	2,860 ±60	12·33 ±0·15	29·25 ±1·0	4·62 ±0·08	0·81 ±0·08
AB-78	41	0·35	3·0	5,700,000 ±300,000	20,000 ±1,000	1,080 ±100	2,370 ±50	12·36 ±0·15	25·95 ±0·7	4·10 ±0·06	1·52 ±0·15

* Average of two independent determinations. Weights of second samples analysed: AA-29 6·1mg, AA-30 5·2mg.

the surface of the individual grains. This ilmenite fraction was particularly well separated. The total contamination is estimated from microscopic and X-ray diffraction analysis (cf. Fig. 1) to be less than 5 per cent. In the X-ray diffraction pattern, only the principal line of plagioclase and pyroxene was barely detectable. Comparison with X-ray diffraction pictures taken from lunar glass spherules shows that the glass content of this ilmenite fraction should not be higher than about 5 per cent. The amounts of Fe and Ti dissolved by the HF treatment were measured by atomic absorption. The thickness of the etched layer was calculated from the dissolved Fe and Ti amounts, using the relation given by EBERHARDT *et al.* (1965b). The noble gases remaining in the etched grain size fractions were then measured. The results are given in Tables 1–3. For He, Ne and Ar two independent sets of etching experiments were made, which gave similar results. While the concentration of the spallation produced noble gases was essentially unchanged by the HF treatment, the concentration of all

Table 2. Results of Kr measurements on bulk grain size fractions, ilmenite grain size fractions, and etched ilmenite grain size fraction (series 2). All fractions were separated from Apollo 11 lunar fine material (10084-47). The weights of the samples analysed are given in Table 1

Sample No.	Grain size (μ)	Surface layer removed (μ)	Kr^{86} (10^{-8} cm^3 STP/g)	Kr^{78}/Kr^{86} (×100)	Kr^{80}/Kr^{86} (×100)	Kr^{82}/Kr^{86} (×100)	Kr^{83}/Kr^{86} (×100)	Kr^{84}/Kr^{86} (×100)
Bulk grain size fractions								
AA-26	130	—	2·6 ±0·5	2·395 ±0·05	13·90 ±0·35	67·80 ±0·7	67·85 ±0·5	323·5 ±5
AA-27	90	—	3·3 ±0·6	2·415 ±0·06	14·14 ±0·2	68·10 ±0·8	68·55 ±0·7	328·0 ±4
AA-28	42	—	4·1 ±0·8	2·320 ±0·05	13·77 ±0·2	67·30 ±0·9	67·65 ±0·8	325·0 ±4
AA-29	15	—	5·4 ±1	2·235 ±0·07	13·58 ±0·25	66·75 ±0·8	66·70 ±0·6	325·5 ±2·5
AA-30	3·7	—	14 ±3	2·090 ±0·06	13·03 ±0·25	65·90 ±0·8	65·85 ±0·8	325·0 ±2·5
AA-19	2·0	—	21 ±4	2·170 ±0·05	13·71 ±0·3	66·85 ±0·8	66·25 ±0·9	329·5 ±6
AB-4	1·4	—	40 ±8	2·040 ±0·05	13·14 ±0·2	66·45 ±0·8	66·35 ±0·6	328·0 ±3
Ilmenite grain size fractions								
AB-7	105	—	0·46 ±0·09	4·040 ±0·15	17·58 ±0·5	72·05 ±1·2	73·85 ±0·6	322·0 ±3·5
AB-5	65	—	0·86 ±0·15	3·325 ±0·09	15·90 ±0·35	69·85 ±1·1	71·30 ±0·6	324·5 ±3·0
AB-1	41	—	1·05 ±0·2	3·310 ±0·06	15·71 ±0·3	69·70 ±1·1	70·90 ±1·0	325·5 ±4·0
AB-11	22	—	2·3 ±0·5	2·735 ±0·05	14·66 ±0·25	68·05 ±0·9	69·00 ±1·1	325·0 ±4·5
Etched ilmenite grain size fraction AB-1: series 2								
AB-80	41	0·14	0·51 ±0·1	5·53 ±0·15	21·55 ±0·35	77·8 ±2·0	82·4 ±2·0	321·5 ±5·0
AB-79	41	0·19	0·37 ±0·07	8·10 ±0·5	26·45 ±1·1	85·5 ±2·0	93·4 ±2·0	325·5 ±5·0
AB-78	41	0·35	0·16 ±0·03	12·35 ±0·5	37·7 ±0·9	99·8 ±2·5	111·5 ±3·0	331·0 ±7·0

Table 3. Results of Xe measurements on bulk grain size fractions, ilmenite grain size fractions, and etched ilmenite grain size fraction (series 2). All fractions were separated from Apollo 11 lunar fine material (10084-47). The weights of the samples analysed are given in Table 1

Sample No.	Grain size (μ)	Surface layer removed (μ)	^{132}Xe (10^{-8} cm^3 STP/g)	$\frac{Xe^{124}}{Xe^{132}}$	$\frac{Xe^{126}}{Xe^{132}}$	$\frac{Xe^{128}}{Xe^{132}}$	$\frac{Xe^{129}}{Xe^{132}}$	$\frac{Xe^{130}}{Xe^{132}}$	$\frac{Xe^{131}}{Xe^{132}}$	$\frac{Xe^{134}}{Xe^{132}}$	$\frac{Xe^{136}}{Xe^{132}}$
				$\times 100$							
Bulk grain size fractions											
AA-26	130	—	1·3 ±0·3	0·845 ±0·025	1·086 ±0·04	9·39 ±0·15	106·3 ±0·8	17·19 ±0·25	85·15 ±0·5	37·01 ±0·25	30·32 ±0·25
AA-27	90	—	2·0 ±0·4	0·726 ±0·08	1·090 ±0·05	9·22 ±0·20	104·3 ±0·8	17·00 ±0·5	83·95 ±1·0	37·34 ±0·6	30·91 ±0·4
AA-28	42	—	2·7 ±0·5	0·719 ±0·02	0·941 ±0·055	9·02 ±0·2	103·3 ±1·4	16·79 ±0·15	83·65 ±1·0	37·64 ±0·6	30·94 ±0·5
AA-29	15	—	3·6 ±0·7	0·654 ±0·03	0·760 ±0·02	8·86 ±0·12	104·9 ±1·0	16·73 ±0·35	83·65 ±1·1	37·08 ±0·35	30·39 ±0·4
AA-30	3·7	—	8·9 ±1·8	0·532 ±0·03	0·576 ±0·02	8·44 ±0·07	104·4 ±1·0	16·56 ±0·2	82·45 ±0·7	37·31 ±0·3	30·45 ±0·35
AA-19	2·0	—	13 ±3	0·528 ±0·02	0·529 ±0·02	8·38 ±0·1	104·3 ±1·0	16·50 ±0·2	82·45 ±1·1	36·77 ±1·0	30·34 ±0·35
AB-4	1·4	—	23 ±5	0·530 ±0·03	0·532 ±0·015	8·62 ±0·15	106·4 ±1·0	16·84 ±0·3	83·15 ±0·6	36·75 ±0·25	29·64 ±0·3
Ilmenite grain size fractions											
AB-7	105	—	0·32 ±0·06	1·28 ±0·09	1·70 ±0·06	10·26 ±0·25	103·3 ±2·0	17·40 ±0·6	87·60 ±1·0	37·15 ±1·3	29·9 ±1·0
AB-5	65	—	0·45 ±0·09	0·992 ±0·05	1·32 ±0·09	9·80 ±0·12	106·7 ±2·5	17·44 ±0·4	85·90 ±1·0	36·77 ±0·6	30·02 ±0·5
AB-1	41	—	0·55 ±0·2	0·747 ±0·025	0·907 ±0·03	9·11 ±0·15	105·6 ±1·0	16·81 ±0·4	83·50 ±0·8	37·39 ±0·8	30·51 ±0·3
AB-11	22	—	1·2 ±0·3	0·623 ±0·03	0·639 ±0·025	8·84 ±0·09	105·1 ±1·0	16·81 ±0·2	83·30 ±1·0	36·93 ±0·4	30·02 ±0·4
Etched ilmenite grain size fraction AB-1: series 2											
AB-80	41	0·14	0·30 ±0·06	1·14 ±0·06	1·53 ±0·25	9·46 ±0·15	105·1 ±4·0	16·95 ±1·2	85·5 ±3·0	36·8 ±2·0	30·7 ±1·0
AB-79	41	0·19	0·17 ±0·04	1·26 ±0·15	1·52 ±0·3	9·72 ±0·5	104·1 ±3·0	17·0 ±0·7	85·4 ±1·5	38·0 ±2·5	32·2 ±1·5
AB-78	41	0·35	0·085 ±0·02	1·45 ±0·5	1·73 ±0·6	9·73 ±1·4	104·3 ±5·0	17·4 ±1·8	82·1 ±3·5	38·6 ±3·0	32·2 ±2·5

trapped gases decreased strongly after removing a surface layer with an approximate average thickness of 2000 Å (cf. Fig. 5). We conclude that the trapped noble gases are located within a depth of 2000 Å below the surface of the individual grains. This is in agreement with the observed grain size dependence of the trapped gas content.

The experimental evidence strongly supports a solar wind origin of the trapped gases. The range of ions with an energy of 1 keV per nucleon, as observed in the present day solar wind, is of the order of several hundred to two thousand angstroms. Independent experiments have already established that the solar wind noble gas ions indeed reach the lunar surface (BÜHLER *et al.*, 1969). Our grain size and etching experiments comprise all the noble gases from He to Xe. The light and heavy noble gas species show virtually the same behaviour favouring a solar wind origin not only for trapped He, Ne and Ar but also for Kr and Xe. Therefore, in addition to light elements (BAME *et al.*, 1968) also very heavy ions are accelerated and carried along in the solar wind.

Energetic solar particles, as observed by the nuclear track method in meteorites and also in lunar material, have a typical range of 10–100 μ (PELLAS *et al.*, 1969; LAL and RAJAN, 1969; CROZAZ *et al.*, 1970; FLEISCHER *et al.*, 1970). Hence, trapped noble gases emplaced as energetic solar particles should not show an anticorrelation with grain size down to the one micron range. The concentration of trapped energetic particles would be approximately constant in all grains with a radius smaller than their range. We conclude that energetic solar particles can only be a minor contribution to the trapped gases in lunar material and also in the Khor Temiki aubrite.

For ilmenite the gas concentrations are approximately inversely proportional to the grain size (cf. Fig. 3), i.e. the concentration per cm^2 surface area is independent of grain size. This requires that either the grains were exposed to the solar wind for

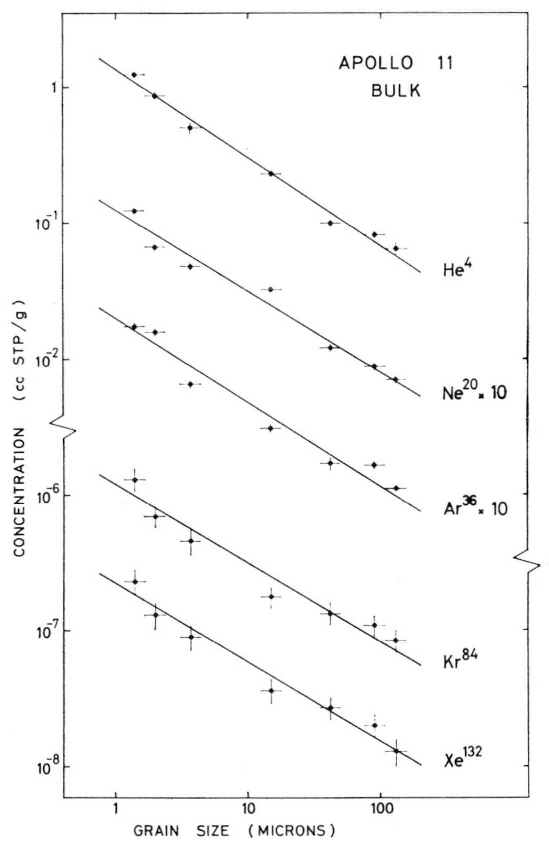

Fig. 2. Grain size dependence of measured He^4, Ne^{20}, Ar^{36}, Kr^{84} and Xe^{132} in unseparated Apollo 11 lunar fine material (No. 10084).

the same time interval, independent of grain size, or the exposure was so long that saturation occurred. The observed surface concentration in the ilmenite grains is 10^{-3} cm^3 He^4/cm^2 (3×10^{16} atoms He^4/cm^2). The present day average He^4/H ratio is 0·04 and the observed He^4 surface concentration would thus correspond to an integrated solar wind bombardment of at least 7×10^{17} atoms/cm^2. Such an irradiation dose would require an exposure time of approximately 300 years at the present solar wind flux level. At these irradiation levels severe saturation effects have to be expected (LORD, 1968), which will most likely influence the relative abundance of the trapped solar wind ions. The observed helium loading of ilmenite surfaces down to a regolith depth of 60 cm would require a minimum irradiation time of a few times 10^7 yr if a trapping efficiency of 100 per cent is assumed. The average exposure age of the Apollo 11 lunar fine material is at least an order of magnitude higher. If we disregard the possibility of gross solar wind flux variations in the past, then diffusion and saturation effects must have lowered the effective trapping efficiency for He to 10 per cent or less.

ELEMENTAL ABUNDANCES OF TRAPPED GASES*

Figure 6 and Table 4 show the observed range of trapped He^4, Ne^{20}, Kr^{86} and Xe^{132} abundances relative to trapped Ar^{36}. No correction for spallation or radiogenic contributions were necessary for these isotopes. The overall abundance pattern is quite similar to the predicted cosmic abundances (ALLER, 1961) and to the trapped

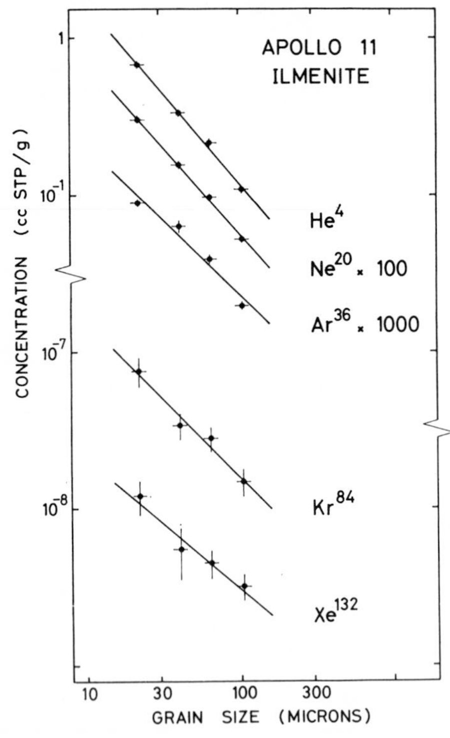

Fig. 3. Grain size dependence of measured He^4, Ne^{20}, Ar^{36}, Kr^{84} and Xe^{132} in ilmenite ($FeTiO_3$) separated from the Apollo 11 lunar fine material (No. 10084).

gas abundances observed in some gas rich aubrites (EBERHARDT *et al.*, 1965a, b; EUGSTER *et al.*, 1969; MARTI, 1969). The Ar–Kr–Xe abundances in the bulk fine material and the separated ilmenite are identical. However, the $(He^4/Ar^{36})_{tr}$ and $(Ne^{20}/Ar^{36})_{tr}$ ratios are considerably higher in the ilmenite than in the bulk material. This is best explained by diffusion loss of He and Ne in the bulk material leading to a

* The indices f, m, r, sp, tr are defined as follows: f: fission; m: measured; r: radiogenic, i.e. decay product of primordial nuclei; sp: spallation; tr: trapped.

relative enrichment of trapped Ar, Kr and Xe. Some individual structural elements of the fine lunar material, such as individual spherules, show even larger diffusion losses than the bulk fine material (Fig. 6).

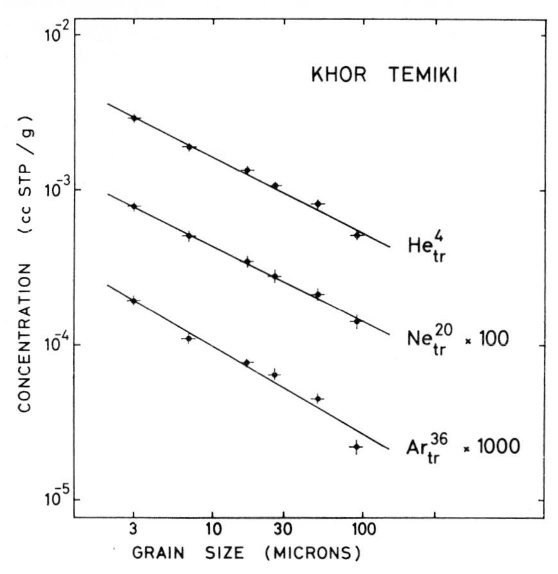

Fig. 4. Grain size dependence of trapped He^4, Ne^{20} and Ar^{36} in the unseparated matrix of the Khor Temiki aubrite (from EBERHARDT *et al.*, 1965a).

Table 4. Elemental and isotopic abundance of trapped gases in the Apollo 11 lunar material (No. 10084) and in separated ilmenite. A range of values is given if our experimental evidence indicates a non-uniform composition

	Bulk grain size fractions	Ilmenite grain size fractions
$(He^4/Ar^{36})_{tr}$	490–790	5300–7600
$(Ne^{20}/Ar^{36})_{tr}$	5·3–10·6	24–33
$(Ar^{36}/Kr^{86})_{tr}$	3500–7600	3900–6000
$(Ar^{36}/Xe^{132})_{tr}$	5300–16000	6000–11400
$(He^4/He^3)_{tr}$	2300–2800	2720 ± 90
$(Ne^{20}/Ne^{22})_{tr}$	12·46–12·83	12·85 ± 0·1
$(Ne^{22}/Ne^{21})_{tr}$	31·0 ± 1·2	31·1 ± 0·8
$(Ar^{36}/Ar^{38})_{tr}$	—	5·32 ± 0·08
$(Ar^{40}/Ar^{36})_{tr}$	0·95 ± 0·06	0·67 ± 0·06

ISOTOPIC COMPOSITION OF TRAPPED GASES

Diffusion losses will affect the isotopic composition of the trapped gases to a much lower degree than the elemental abundances. However, the precision of isotopic ratio measurements is much higher than that of elemental abundance determinations, and a relatively small diffusion loss can produce a detectable change in the isotopic composition. In addition to diffusion, saturation effects may influence isotopic compositions. Trapped gases in samples of different mineralogy, grain size and structural composition could thus have variable isotopic compositions. Ilmenite shows least evidence for diffusion loss and one should expect to obtain from this mineral the most consistent and relevant results on the unaltered isotopic composition of the trapped gases.

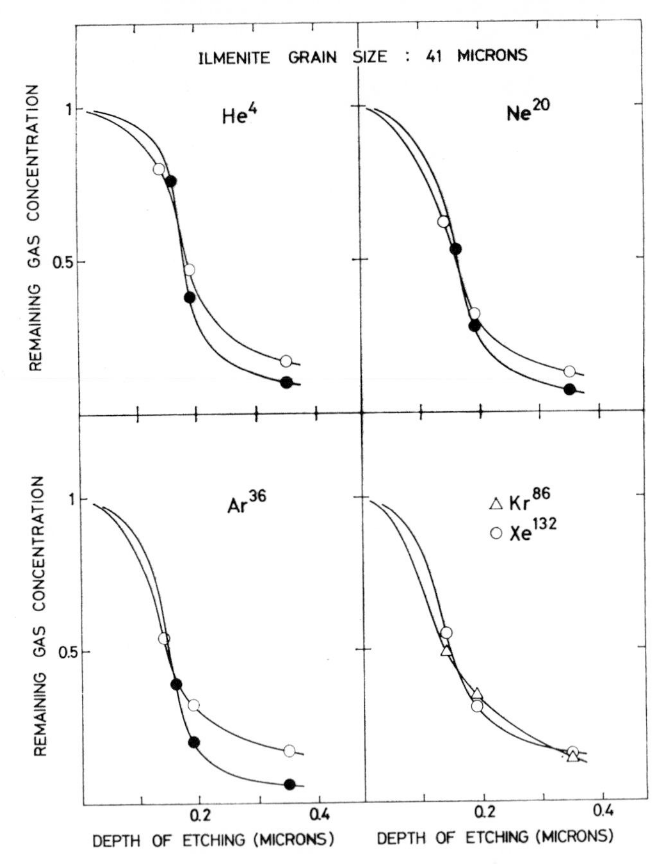

Fig. 5. He^4, Ne^{20}, Ar^{36}, Kr^{84} and Xe^{132} measured in the 41 μ ilmenite grain size fractions separated from Apollo 11 fine material (No. 10084) after removal of a thin surface layer by etching with HF. The horizontal axis is equivalent to the thickness of the layer removed from the surface of the individual grains. The vertical axis gives the concentrations observed in the etched fractions relative to the concentration in the unetched ilmenite. The two curves for He^4, Ne^{20} and Ar^{36} represent the two independent sets of etching experiments performed.

For the less abundant trapped gas isotopes such as He^3, Ne^{21}, $Ar^{38,40}$, $Kr^{78,80}$, and $Xe^{124,126}$ corrections for spallation and radiogenic components are not negligible. These corrections are estimated from correlation diagrams and from abundances in the etched ilmenite fractions, and a detailed discussion is required to deduce the trapped gas isotopic composition. The conclusions of this discussion on the isotopic composition of the trapped gases in Apollo 11 lunar fine material are summarized in Tables 4, 5, 6 and 8.

Helium

The He^3 spallation content in the ilmenite can be estimated from the two samples etched most strongly (AB-31 and AB-78) to be $(380 \pm 100) \times 10^{-8}$ cm³ STP/g. Such a spallation He^3 concentration is in agreement with the average exposure age of the lunar fine material estimated from spallation Xe. After correcting the four ilmenite grain size fractions AB 1, 5, 7 and 11 for spallation He^3 we obtain $(He^4/He^3)_{tr}$ ratios for the trapped gas between 2590 and 2840 with an average of 2720 ± 90 (standard

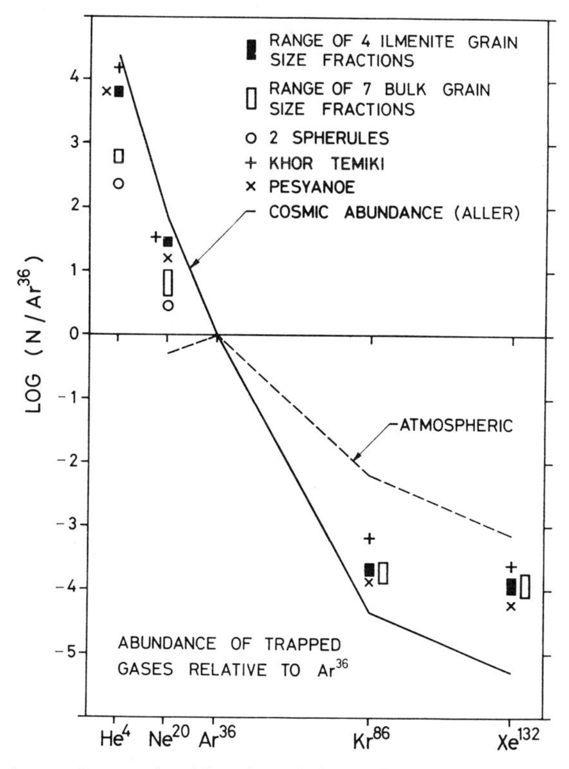

Fig. 6. Abundance of trapped noble gases relative to Ar^{36} observed in Apollo 11 lunar fine material (bulk grain size fractions, ilmenite grain size fractions, separated spherules from No. 10084). The length of the open and shaded bars represent the range of abundance ratios observed in the seven bulk grain size fractions (1·4–130 μ) and the four ilmenite grain size fractions (22–105 μ). The two spherules separated from the fine lunar material had diameters of 0·2 mm. Other data from: ALLER (1961); EBERHARDT *et al.* (1965a); EUGSTER *et al.* (1969); MARTI (1969).

deviation). For the two finer ilmenite grain size fractions the spallation correction is less than 3 per cent and the $(He^4/He^3)_{tr}$ ratios in their trapped component are 2760 and 2680 with an average of again 2720.

The isotopic composition of the trapped He in the bulk grain size fractions is not uniform. The three largest grain size fractions have measured $(He^4/He^3)_m$ ratios higher than 2500. The spallation correction should raise these ratios by approximately 10 per cent, bringing $(He^4/He^3)_{tr}$ in these fractions in agreement with the ilmenite value. In the three finest fractions the measured $(He^4/He^3)_m$ ratios are below 2400 and corrections for spallation He^3 are less than 2 per cent. This difference could either reflect a true grain size effect or it could result from systematic differences in the mineralogical composition of the bulk grain size fractions. Processes which could lead to variations in the $(He^4/He^3)_m$ ratio include diffusion effects, discrimination in trapping, saturation effects, contribution to He^3 from solar flare particles and stripped α particles, and recycling of trapped solar wind particles in a lunar atmosphere. The average U and Th content of the fine lunar material (FIELDS *et al.*, 1970) would contribute less than 10^{-3} cm³ STP He^4/g even in $4{\cdot}5 \times 10^9$ yr. In ilmenite U is depleted (CROZAZ *et al.*, 1970). Thus we can assume that corrections for radiogenic He^4 are negligible in all bulk and ilmenite grain size fractions.

Neon

The spallation Ne concentration in the ilmenite can be calculated from the two ilmenite fractions etched most strongly (AB-31 and AB-78). We obtain 11·9 and 11·2 $\times$ 10^{-8} cm³ STP $Ne^{21}{}_{sp}$/g and use the average of 11·6 $\times$ 10^{-8} cm³ STP $Ne^{21}{}_{sp}$/g ilmenite for the spallation correction. For Ne^{20} and Ne^{22} the spallation correction is always less than 0·03 per cent and can be neglected. For the four ilmenite grain size fractions we obtain $(Ne^{20}/Ne^{22})_{tr} = 12{\cdot}85 \pm 0{\cdot}10$, and after correcting for spallation Ne^{21} we obtain $(Ne^{20}/Ne^{21})_{tr} = 31{\cdot}1 \pm 0.8$. The errors correspond to the range of values derived from the four ilmenite fractions.

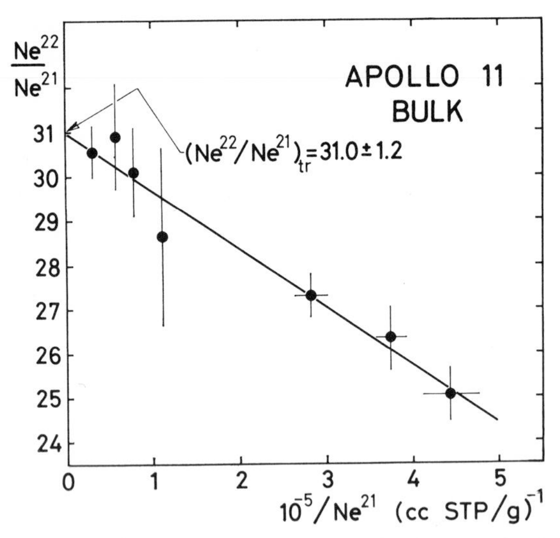

Fig. 7. $(Ne^{22}/Ne^{21})_m$ vs. $1/Ne^{21}{}_m$ correlation diagram for the seven Apollo 11 bulk grain size fractions. In this and the following diagrams the straight lines are least square fits through the measured points. All points were given the same weight independent of the errors. Figures 7–18 show data measured on fractions of No. 10084-47.

The $(Ne^{22}/Ne^{21})_{tr}$ ratio in the bulk material can best be evaluated in a $(Ne^{22}/Ne^{21})_m$ vs. $1/Ne^{21}{}_m$ diagram. The following relation holds:

$$\left(\frac{Ne^{22}}{Ne^{21}}\right)_m = \frac{Ne^{21}{}_{sp}}{Ne^{21}{}_m}\left[\left(\frac{Ne^{22}}{Ne^{21}}\right)_{sp} - \left(\frac{Ne^{22}}{Ne^{21}}\right)_{tr}\right] + \left(\frac{Ne^{22}}{Ne^{21}}\right)_{tr}. \tag{2}$$

If the $Ne^{21}{}_{sp}$ concentration, and the $(Ne^{22}/Ne^{21})_{sp}$ and $(Ne^{22}/Ne^{21})_{tr}$ ratios are the same in all grain size fractions, then the measured $(Ne^{22}/Ne^{21})_m$ ratios should lie on a straight line in this diagram. Within the experimental uncertainties this is indeed the case for the $(Ne^{22}/Ne^{21})_m$ ratios measured in the bulk grain size fractions (see Fig. 7). From the ordinate intercept $(Ne^{22}/Ne^{21})_{tr} = 31{\cdot}0 \pm 1{\cdot}2$ is obtained in the bulk grain size fractions. The error given is the standard deviation of the least square fit plus 50 per cent of the typical analytical error of an individual measurement. The same error definition will be used throughout this paper for other isotope ratios obtained by the ordinate intercept method. The $(Ne^{22}/Ne^{21})_{tr}$ ratio in the bulk is in agreement with the ilmenite ratio. From the slope of the least square line in Fig. 7 the spallation Ne^{21} content in the bulk grain size fractions can be calculated. We obtain $Ne^{21}{}_{sp} = (44 \pm 10) \times 10^{-8}$ cm³ STP/g. The measured $(Ne^{20}/Ne^{22})_m$ ratios can now be corrected for spallation Ne. This spallation correction is always smaller than 1 per cent, and smaller than 0·2 per cent for the finest four fractions. The $(Ne^{20}/Ne^{22})_m$ ratios of the coarsest fractions are consistently smaller than for the ilmenite (cf. Tables 1 and 4). With increasing trapped gas content the $(Ne^{20}/Ne^{22})_{tr}$ ratios have a tendency to increase and for the finest fraction it is virtually identical with the ilmenite ratio.

The $(Ne^{22}/Ne^{21})_{tr}$ ratio in the ilmenite can also be derived from the $(Ne^{22}/Ne^{21})_m$ vs. $1/Ne^{21}{}_m$ plot. We obtain $(Ne^{22}/Ne^{21})_{tr} = 30{\cdot}5 \pm 0{\cdot}8$ and $Ne^{21}{}_{sp} = 7 \times 10^{-8}$ cm³ STP Ne^{21}/g ilmenite. The variation in the measured $(Ne^{22}/Ne^{21})_m$ ratios is rather small and thus the slope of the correlation line somewhat uncertain. Considering this, the spallation Ne^{21} content is in fair agreement with the value derived from the etched ilmenite fractions. The trapped gas ratio agrees with the value calculated above.

The trapped $(Ne^{20}/Ne^{21})_{tr}$ ratio derived in this paper is slightly lower than the value given in the *Science* Paper. This is due to a reevaluation of the measured $(Ne^{22}/Ne^{21})_m$ ratios which lead to a small systematic correction. Other investigators have reported large variations of the trapped $(Ne^{20}/Ne^{22})_{tr}$ ratio in gas released during stepwise heating experiments (REYNOLDS *et al.*, 1970; PEPIN *et al.*, 1970). The reported variations are an order of magnitude larger than the variations we found in the bulk grain size fractions.

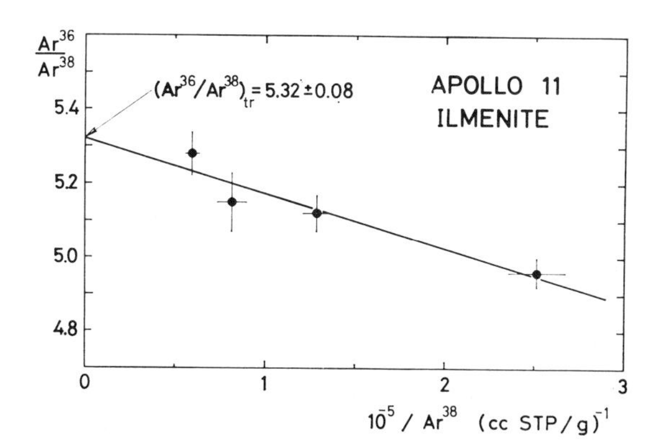

Fig. 8. $(Ar^{36}/Ar^{38})_m$ vs. $1/Ar^{38}{}_m$ correlation diagram for the four ilmenite grain size fractions separated from Apollo 11 fine lunar material.

Argon

The systematic decrease of the $(Ar^{36}/Ar^{38})_m$ ratio in the ilmenite grain size fractions with decreasing trapped gas content is due to spallation argon (cf. Table 1). The following relation holds:

$$\left(\frac{Ar^{36}}{Ar^{38}}\right)_m = \frac{Ar^{38}{}_{sp}}{Ar^{38}{}_m}\left[\left(\frac{Ar^{36}}{Ar^{38}}\right)_{sp} - \left(\frac{Ar^{36}}{Ar^{38}}\right)_{tr}\right] + \left(\frac{Ar^{36}}{Ar^{38}}\right)_{tr}. \tag{3}$$

The high Ti and Fe contents of the ilmenite results in a high spallation yield for Ar. Impurities will contribute only little to the total spallation Ar content. We may thus assume that the $Ar^{38}{}_{sp}$ content and the $(Ar^{36}/Ar^{38})_{sp}$ ratio are the same in all ilmenite grain size fractions. According to equation (3) the $(Ar^{36}/Ar^{38})_m$ ratio and $1/Ar^{38}{}_m$ are then linearly correlated. Figure 8 is such a $(Ar^{36}/Ar^{38})_m$ vs.

$1/Ar^{38}{}_m$ diagram for the ilmenite grain size fractions. The points lie on a straight line, well within the experimental uncertainties. From the intercept of the least square straight line with the ordinate we obtain for the trapped gas in the ilmenite $(Ar^{36}/Ar^{38})_{tr} = 5{\cdot}32 \pm 0{\cdot}08$. From the slope of the correlation line we derive the spallation Ar content in the ilmenite as $Ar^{38}{}_{sp} = (70 \pm 12) \times 10^{-8}$ cm³ STP/g ilmenite. A similar correlation diagram for the two series of etched ilmenite fractions gives for the trapped gas $(Ar^{36}/Ar^{38})_{tr} = 5{\cdot}23 \pm 0{\cdot}16$ and $5{\cdot}32 \pm 0{\cdot}17$ for the etching series 1 and 2 respectively. The larger uncertainties reflect the somewhat greater scattering of the measured ratios along the correlation line, but the agreement with the $(Ar^{36}/Ar^{38})_{tr}$ value derived from the ilmenite grain size fractions is very satisfactory.

The bulk grain size fractions show no clear correlation in an $(Ar^{36}/Ar^{38})_m$ vs. $1/Ar^{38}{}_m$ diagram. This probably reflects a variability in chemical composition leading to variable spallation Ar^{38} contents in these seven grain size fractions. The measured $(Ar^{36}/Ar^{38})_m$ ratios in all these bulk fractions are smaller than or equal to 5·32, with the exception of AA-19. Also in AA-27 the $(Ar^{36}/Ar^{38})_m$

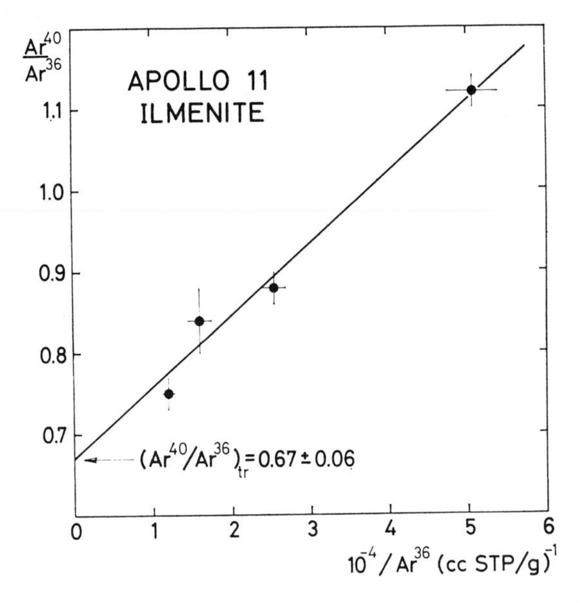

Fig. 9. $(Ar^{40}/Ar^{36})_m$ vs. $1/Ar^{36}{}_m$ correlation diagram for the four ilmenite grain size fractions separated from Apollo 11 fine lunar material. $Ar^{40}{}_{tr}$ is much higher than expected in the Sun. This is discussed in the text.

ratio is rather high, compared to its low trapped gas content. The $(Ar^{36}/Ar^{38})_{tr}$ ratios in all bulk grain size fractions are thus compatible with the ratio found in the ilmenite, with the probable exception of AA-19 and AA-27. The somewhat high $(Ar^{36}/Ar^{38})_m$ ratios measured in these two fractions may be indicative of a trapped gas component in the bulk material with an $(Ar^{36}/Ar^{38})_{tr}$ ratio higher than that in the ilmenite. A high $(Ar^{36}/Ar^{38})_{tr}$ ratio for the trapped gas in lunar fine material has also been reported by PEPIN *et al.* (1970) and by REYNOLDS *et al.* (1970).

The Ar^{40} is a mixture of three components: radiogenic, spallation and trapped Ar^{40}. If the spallation Ar^{40} and radiogenic Ar^{40} concentrations are the same in all fractions, then the measured $(Ar^{40}/Ar^{36})_m$ ratios should fall on a straight line in a $(Ar^{40}/{}^{36})_m$ vs. $1/Ar^{36}{}_m$ diagram. Figure 9 shows this diagram for the ilmenite grain size fractions. Within the experimental uncertainties the measured ratios define a straight line, and from its intercept with the ordinate we obtain for the trapped gas in the ilmenite $(Ar^{40}/Ar^{36})_{tr} = 0{\cdot}67 \pm 0{\cdot}06$. From the corresponding diagram for the bulk grain size fractions (Fig. 10) we derive $(Ar^{40}/Ar^{36})_{tr} = 0.95 \pm 0.06$ in the bulk material. Hence bulk and ilmenite show different $(Ar^{40}/Ar^{36})_{tr}$ ratios. HEYMANN *et al.* (1970) reported in the bulk material $(Ar^{40}/Ar^{36})_{tr} = 1.06$. Their value is derived mostly from samples with larger grain sizes and compares well with our bulk ratio obtained on finer grain sizes.

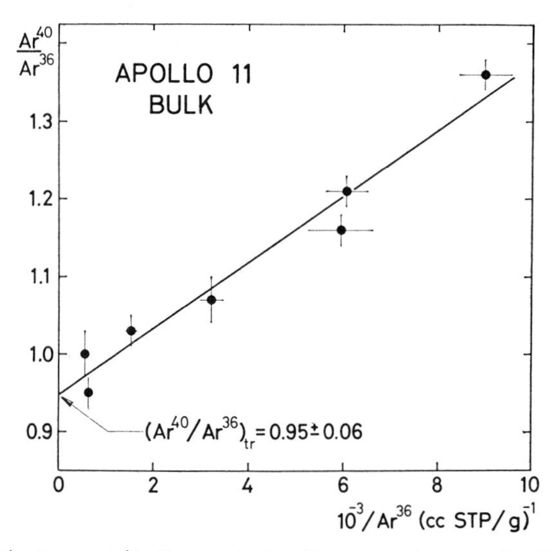

Fig. 10. $(Ar^{40}/Ar^{36})_m$ vs. $1/Ar^{36}{}_m$ correlation diagram for the seven Apollo 11 bulk grain size fractions. $Ar^{40}{}_{tr}$ is much higher than expected in the Sun. This is discussed in the text.

Krypton

All four ilmenite grain size fractions have a distinct Kr-spallation component influencing mainly the abundance of the lighter, low abundance isotopes. Figures 11 and 12 show the corresponding correlation diagrams. The etched ilmenite fractions have been included in these figures to obtain a wider range of data. Within the experimental uncertainties the correlation lines pass through the points representing the isotopic composition of atmospheric Kr and the average isotopic composition of trapped Kr in carbonaceous chondrites (AVCC-Kr).*

All the bulk grain size fractions have lower Kr_{sp}/Kr ratios than the ilmenite fractions, and we shall use these fractions for the more detailed evaluation of the isotopic composition of trapped Kr. Figure 13 shows a $(Kr^{78}/Kr^{86})_m$ vs. $(Kr^{80}/Kr^{86})_m$ correlation diagram for the seven bulk grain size fractions. The least square correlation line passes close to the two points representing AVCC and atmospheric isotopic composition. It is thus impossible to decide from this correlation diagram alone whether the isotopic composition of the trapped gas is closer to AVCC-Kr or to atmospheric Kr. Similar to equations (2) and (3) the following relation holds:

$$\left(\frac{Kr^{78}}{Kr^{86}}\right)_m = \frac{Kr^{86}{}_{sp}}{Kr^{86}{}_m}\left[\left(\frac{Kr^{78}}{Kr^{86}}\right)_{sp} - \left(\frac{Kr^{78}}{Kr^{86}}\right)_{tr}\right] + \left(\frac{Kr^{78}}{Kr^{86}}\right)_{tr}. \quad (4)$$

* The abbreviation AVCC is used to denote the average Kr isotopic composition observed in carbonaceous chondrites. AVCC-Kr is very similar to atmospheric Kr in isotopic composition, showing only a 0·3 per cent per mass unit systematic depletion of the light isotopes relative to atmospheric Kr. In addition, a small fission Kr component may be present in AVCC-Kr. For a detailed discussion see KRUMMENACHER *et al.* (1962); EUGSTER *et al.* (1967b).

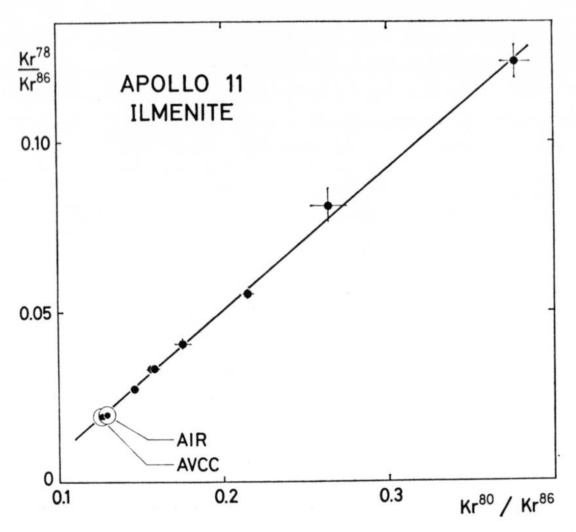

Fig. 11. $(Kr^{78}/Kr^{86})_m$ vs. $(Kr^{80}/Kr^{86})_m$ correlation diagram for the four ilmenite grain size fractions and the three etched ilmenite samples.

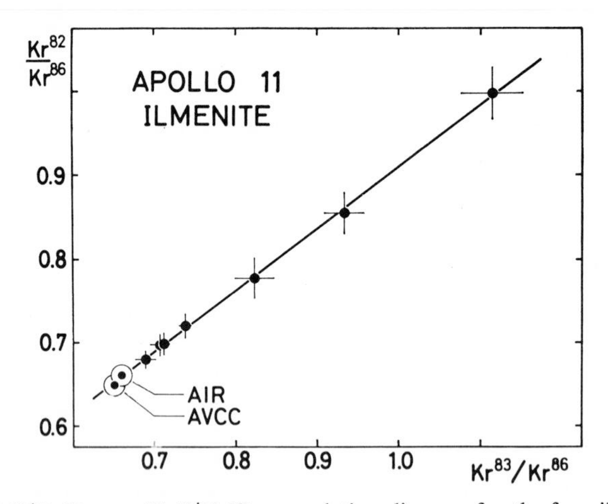

Fig. 12. $(Kr^{82}/Kr^{86})_m$ vs. $(Kr^{83}/Kr^{86})_m$ correlation diagram for the four ilmenite grain size fractions and the three etched ilmenite samples.

Assuming $Kr^{86}{}_{sp} = 0$ relation (4) is reduced to

$$\left(\frac{Kr^{78}}{Kr^{86}}\right)_m = \frac{Kr^{78}{}_{sp}}{Kr^{86}{}_m} + \left(\frac{Kr^{78}}{Kr^{86}}\right)_{tr}. \tag{4a}$$

If the spallation Kr concentration is constant and if the spallation and trapped gas isotope ratios are the same, then the measured $(Kr^{78}/Kr^{86})_m$ ratios should lie on a straight line if plotted as a function of $1/Kr^{86}{}_m$. For five of the seven grain size fractions this is indeed the case (Fig. 14). Distinctly not on the correlation line are AA-26 and AA-19. This is probably due to variations in the spallation Kr content and these samples are omitted from Fig. 14. From the ordinate intercept we obtain $(Kr^{78}/Kr^{86})_{tr} = (2{\cdot}00 \pm 0{\cdot}035) \times 10^{-2}$. In the $(Kr^{80}/Kr^{86})_m$ vs. $1/Kr^{86}{}_m$ correlation diagram again the samples AA-26 and AA-19 fall not on the correlation line and were omitted. We obtain then

$(Kr^{80}/Kr^{86})_{tr} = (12{\cdot}90 \pm 0{\cdot}2) \times 10^{-2}$. These ratios agree with the isotopic composition of atmospheric Kr. They are different from AVCC-Kr (see Table 5).

Table 5. Isotopic composition of solar Kr (lunar trapped Kr). For comparison the isotopic composition of atmospheric and AVCC Kr are given

	Kr^{78}/Kr^{86}	Kr^{80}/Kr^{86}	Kr^{82}/Kr^{86}	Kr^{83}/Kr^{86}	Kr^{84}/Kr^{86}
			× 100		
Solar Kr (lunar trapped Kr) (this paper)	2·00 ±0·035	12·90 ±0·2	66·0 ±0·3	65·75 ±0·3	325·2 ±1·8
Atmospheric Kr (NIER, 1950; EUGSTER *et al.*, 1967a)	1·995 ±0·008	12·96 ±0·04	66·17 ±0·16	66·00 ±0·14	327·3 ±0·7
AVCC Kr (EUGSTER *et al.*, 1967b)	1·927 ±0·014	12·65 ±0·09	65·04 ±0·2	65·11 ±0·2	322·8 ±0·8

The small spallation corrections for the isotopes Kr^{82}, Kr^{83} and Kr^{84} were calculated with the following assumptions:

(1) $(Kr^{78}/Kr^{86})_{tr}$ and $(Kr^{80}/Kr^{86})_{tr}$ are the values given above.

(2) $(Kr^{78}/Kr^{83})_{SP} = 0{\cdot}173$ $(Kr^{80}/Kr^{83})_{SP} = 0{\cdot}492$

$(Kr^{82}/Kr^{83})_{SP} = 0{\cdot}756$ $(Kr^{84}/Kr^{83})_{SP} = 0{\cdot}386$.

These ratios were taken from the spallation Kr spectrum of lunar rock 10017 (EBERHARDT *et al.*, 1970b).

(3) $(Kr^{86}/Kr^{83})_{SP} \equiv 0$.

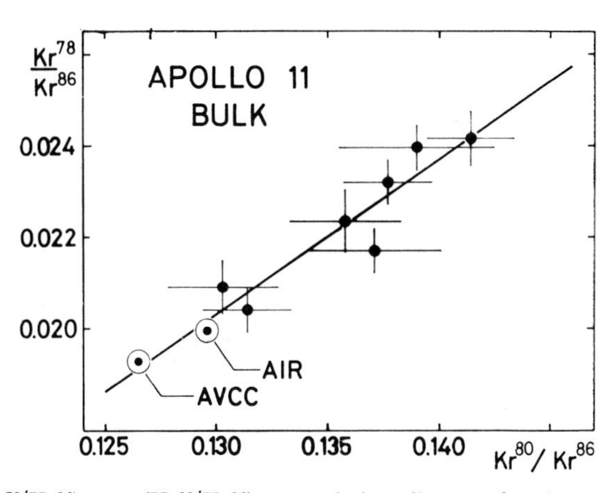

Fig. 13. $(Kr^{78}/Kr^{86})_m$ vs. $(Kr^{80}/Kr^{86})_m$ correlation diagram for the seven Apollo 11 bulk grain size fractions.

Corrections for spallation Kr^{82} and Kr^{83} were all smaller than 3·5 per cent and for spallation Kr^{84} smaller than 1 per cent. The average Kr isotopic compositions obtained from the seven bulk grain size fractions are given in Table 5. Errors for Kr^{82-84} are standard deviations of the ratios obtained for the seven grain size fractions. The overall isotopic composition of lunar trapped Kr is in agreement with atmospheric Kr.

Xenon

All bulk and ilmenite grain size fractions contain sizeable amounts of spallation Xe, affecting mainly the abundances of the lighter isotopes. From $(Xe^{124}/Xe^{130})_m$ vs. $(Xe^{126}/Xe^{130})_m$ and $(Xe^{128}/Xe^{130})_m$ vs. $(Xe^{126}/Xe^{130})_m$ correlation diagrams it is evident that the trapped Xe in the ilmenite and

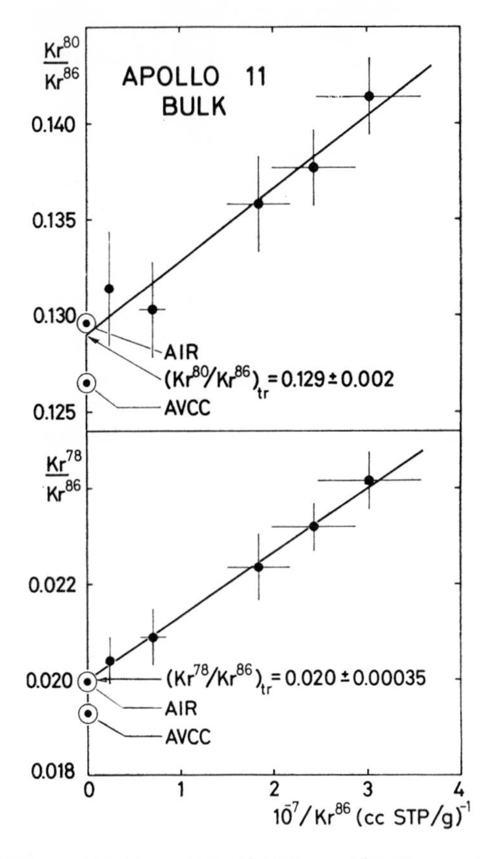

Fig. 14. $(Kr^{78}/Kr^{86})_m$ vs. $1/Kr^{86}{}_m$ and $(Kr^{80}/Kr^{86})_m$ vs. $1/Kr^{86}{}_m$ correlation diagrams for five Apollo 11 bulk grain size fractions. Fractions AA-26 and AA-19 omitted, see text.

in the bulk material cannot have atmospheric isotopic compositions (see Figs. 15 and 16). The correlation lines pass close to the points representing the average isotopic composition of trapped Xe in carbonaceous chondrites (AVCC-Xe).* This suggests that the isotopic composition of trapped Xe in

---

* The abbreviation AVCC is used to denote the average Xe isotopic composition observed in carbonaceous chondrites. AVCC-Xe shows an order of magnitude larger isotopic anomalies than AVCC-Kr. In AVCC-Xe the light isotopes are up to 20 per cent more abundant, the heavy isotopes up to 10 per cent less abundant than in atmospheric Xe (relative to Xe^{130}). For a detailed discussion see REYNOLDS (1960); KRUMMENACHER *et al.* (1962): MARTI (1967): EUGSTER *et al.* (1967b): PEPIN (1968).

ilmenite and bulk material are identical with AVCC-Xe. However, any other type of trapped gas with an isotopic composition represented by a point on the correlation line to the left of the measured ratios could equally well explain the observed data as displayed in Figs. 15 and 16.

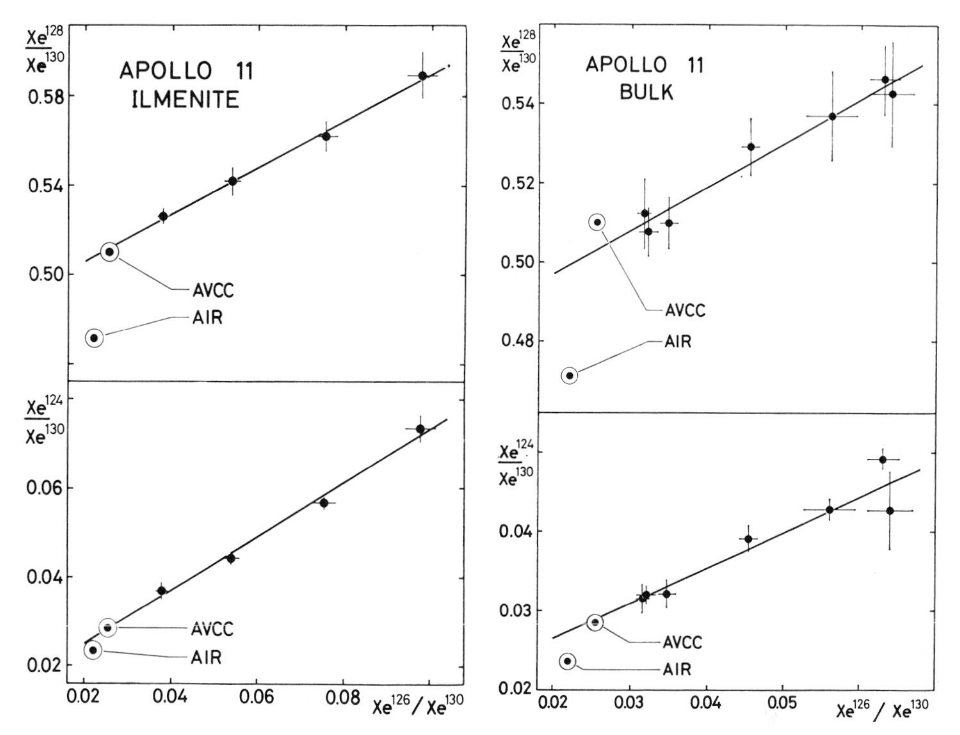

Fig. 15. $(Xe^{124}/Xe^{130})_m$ vs. $(Xe^{126}/Xe^{130})_m$ and $Xe^{128}/Xe^{130})_m$ vs. $(Xe^{126}/Xe^{130})_m$ correlation diagrams for the four ilmenite grain size fractions separated from Apollo 11 fine lunar material.

Fig. 16. $(Xe^{124}/Xe^{130})_m$ vs. $(Xe^{126}/Xe^{130})_m$ and $(Xe^{128}/Xe^{130})_m$ vs. $(Xe^{126}/Xe^{130})_m$ correlation diagrams for the seven Apollo 11 bulk grain size fractions.

More precise conclusions regarding the isotopic composition of the lunar trapped Xe can be obtained from a $(Xe^{128}/Xe^{130})_m$ vs. $1/Xe^{130}{}_m$ correlation diagram. In analogy to equations (2), (3) and (4) the following relation holds:

$$\left(\frac{Xe^{128}}{Xe^{130}}\right)_m = \frac{Xe^{130}{}_{sp}}{Xe^{130}{}_m}\left[\left(\frac{Xe^{128}}{Xe^{130}}\right)_{sp} - \left(\frac{Xe^{128}}{Xe^{130}}\right)_{tr}\right] + \left(\frac{Xe^{128}}{Xe^{130}}\right)_{tr}. \tag{5}$$

The measured $(Xe^{128}/Xe^{130})_m$ ratios should lie on a straight line if the spallation Xe^{130} content is constant. For the bulk grain size fractions this is indeed the case (Fig. 17), except for the largest grain size fraction AA-26 which was consequently omitted. Fraction AA-26 may have a smaller spallation Xe^{130} content than all the other ones. From the ordinate intercept of the least square fitted straight line in Fig. 17 we obtain

$$(Xe^{128}/Xe^{130})_{tr} = (50{\cdot}45 \pm 0{\cdot}8) \times 10^{-2}.$$

This value is in excellent agreement with the isotopic composition of AVCC-Xe (Table 8). Similarly the $(Xe^{124}/Xe^{130})_{tr}$ and $(Xe^{126}/Xe^{130})_{tr}$ ratios in the trapped gas can be derived from the corresponding correlation plots (Fig. 18).

We obtain

$$(Xe^{124}/Xe^{130})_{tr} = (2\cdot99 \pm 0\cdot18) \times 10^{-2}$$

$$(Xe^{126}/Xe^{130})_{tr} = (2\cdot67 \pm 0\cdot2) \times 10^{-2}.$$

These figures again agree with the isotopic composition of AVCC-Xe. Not enough data points are available to allow a similar evaluation of the ilmenite results with sufficient precision. However, in the isotopic composition correlation diagrams (Fig. 15) the correlation lines pass through the AVCC-Xe points and the assumption seems justified that the $(Xe^{124}/Xe^{130})_{tr}$, $(Xe^{126}/Xe^{130})_{tr}$ and $(Xe^{128}/Xe^{130})_{tr}$ ratios of the ilmenite trapped gas are identical with the composition obtained for the trapped gas in the bulk material.

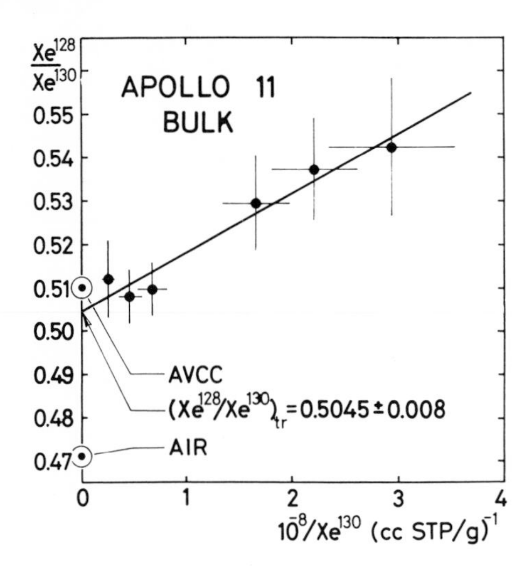

Fig. 17. $(Xe^{128}/Xe^{130})_m$ vs. $1/Xe^{130}{}_m$ correlation diagram for six Apollo 11 bulk grain size fractions. Fraction AA-26 omitted from this diagram, see text.

Small spallation corrections have to be applied for obtaining the isotopic composition of the heavy Xe isotopes in the trapped gas relative to Xe^{130}. They were based on the following well justified assumptions:

(1) $(Xe^{124}/Xe^{130})_{tr}$ and $(Xe^{126}/Xe^{130})_{tr}$ in the trapped gas identical to AVCC-Xe.

(2) $(Xe^{124}/Xe^{126})_{sp} = 0\cdot468$ for bulk material, as derived from a $(Xe^{124}/Xe^{132})_m$ vs. $(Xe^{126}/Xe^{132})_m$ correlation diagram for the bulk samples.

(3) $(Xe^{124}/Xe^{126})_{sp} = 0\cdot617$ in the ilmenite fractions, as derived from a $(Xe^{124}/Xe^{132})_m$ vs. $(Xe^{126}/Xe^{132})_m$ correlation diagram for these fractions.

(4) $(Xe^{129}/Xe^{126})_{sp} = 1\cdot49$; $(Xe^{130}/Xe^{126})_{sp} = 1\cdot00$;
$(Xe^{131}/Xe^{126})_{sp} = 5\cdot46$; $(Xe^{132}/Xe^{126})_{sp} = 0\cdot77$;
$(Xe^{134}/Xe^{126})_{sp} = 0\cdot07$; and $(Xe^{136}/Xe^{126})_{sp} = 0$.

These ratios were taken from the spallation Xe spectrum of lunar rock 10071 (EBERHARDT *et al.*, 1970b).

Corrections for spallation Xe^{130} were smaller than 5 per cent for the bulk grain size fractions and smaller than 8 per cent for the ilmenite fractions. For the heavier isotopes the spallation corrections were correspondingly smaller. The average heavy Xe isotopic compositions obtained for the trapped gas in the nine bulk grain size fractions and in the four ilmenite grain size fractions are given in Table 6. The Xe^{129} to Xe^{136} abundances in the bulk material and in the ilmenite agree extremely well. However, they are distinctly different from either AVCC or atmospheric Xe. The complete isotopic composition of trapped lunar Xe is given in Table 8.

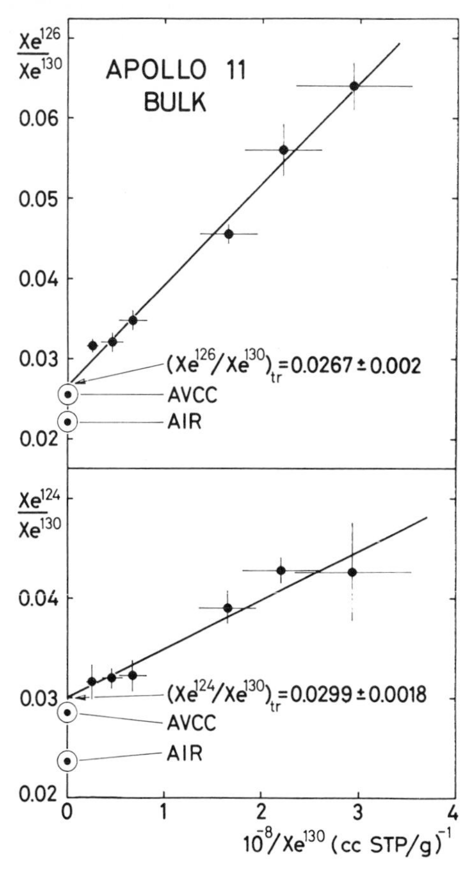

Fig. 18. $(Xe^{124}/Xe^{130})_m$ vs. $1/Xe^{130}{}_m$ and $(Xe^{126}/Xe^{130})_m$ vs. $1/Xe^{130}{}_m$ correlation diagrams for six Apollo 11 bulk grain size fractions. Fraction AA-26 omitted from this diagram, see text.

Elemental and Isotopic Abundances in the Solar Wind

The abundances and isotopic compositions of the trapped gases in the Apollo 11 fine material are not necessarily identical with those of the time integrated solar wind. They may have been altered by diffusion losses, systematic differences in trapping probabilities and by saturation effects. Diffusion coefficients are much larger for the lighter noble gases and diffusion losses should be most severe for the light gases and isotopes. However, the range of the heavier noble gases of the solar wind is smaller and these should be located closer to the surface of the grains than for instance He. This will make the mass dependence of diffusion losses less steep than would be

Table 6. Average heavy Xe isotope abundances of the trapped gas in the seven bulk grain size fractions and in the four ilmenite grain size fractions of sample No. 10084

	$\frac{Xe^{129}}{Xe^{130}}$	$\frac{Xe^{131}}{Xe^{130}}$	$\frac{Xe^{132}}{Xe^{130}}$	$\frac{Xe^{134}}{Xe^{130}}$	$\frac{Xe^{136}}{Xe^{130}}$
			$\times 100$		
Average of seven bulk grain size fractions	634·5 $\pm$3·5	495·9 $\pm$3·0	606·9 $\pm$5·0	225·8 $\pm$3·5	185·2 $\pm$4·0
Average of four ilmenite grain size fractions	635·0 $\pm$5·5	495·0 $\pm$4·5	606·7 $\pm$7·5	225·8 $\pm$4·5	183·6 $\pm$3·0

Corrections for spallation Xe were applied, see text. No corrections for a possible radiogenic Xe^{129} component were made. Errors are standard deviations.

expected from the variation of the diffusion coefficients alone. Systematic differences in the trapping probability will favour the heavy noble gases. At typical solar wind energies, trapping probabilities for all noble gas ions in aluminium and aluminium oxide are 80 per cent or higher (MEISTER, 1969). We expect that for silicates and other lunar material systematic differences in trapping probabilities should be of the same magnitude. Trapping probabilities below 100 per cent for He are in agreement with the observed lunar solar wind albedo (BÜHLER *et al.*, 1969).

The nature and the influence of saturation effects are difficult to assess. The surface concentration of trapped He^4 in the ilmenite grains is 10^{-3} cm^3 He^4/cm^2. It is located in a 0·2 μ thick surface layer. Consequently the He^4 concentration in this layer is 50 cm^3 He^4/cm^3, corresponding to a He partial pressure of 50 atm. The $(H_2/He^4)_{tr}$ ratio in trapped lunar gas is approximately 2·5 (EPSTEIN and TAYLOR, 1970; HINTENBERGER *et al.*, 1970) and the H_2 partial pressure higher than 100 atm, provided the H_2 also shows this pronounced surface enrichment. A pressure of 200 atm corresponds to a density of 6×10^{21} atoms/cm^3, whereas the major lunar minerals have densities of 10^{23} atoms/cm^3. Approximately 6 per cent of the atoms in the surface layer of the individual grains are thus trapped solar wind particles and we expect heavy damage to the original lattice. Gas retention properties and trapping probabilities will be changed by this extensive solar wind bombardment.

Release of trapped gases by the subsequent solar wind bombardment will become important when the gas concentrations are close to saturation levels. In addition, sputtering will slowly erode the gas loaded surfaces and preferentially remove the trapped solar wind ions with the shortest range. No experimental evidence on the combined influence of all these effects occurring at saturation levels on the elemental and isotopic composition is presently available and we have to disregard this effect. A somewhat more refined discussion of the influence of saturation effects will become feasible when the elemental and isotopic abundances of some solar wind noble gas species have been measured by controlled collection experiments such as the SWC experiment (BÜHLER *et al.*, 1969).

The four ilmenite fractions have consistently the highest $(He^4/Ar^{36})_{tr}$ and $(Ne^{20}/Ar^{36})_{tr}$ ratios (Table 4, Fig. 6). The ilmenite fractions have also considerably higher He^4 concentrations than the corresponding bulk fractions (Table 1). Diffusion loss should thus be least in the ilmenite and the $(He/^4Ar^{36})_{tr}$ and $(Ne^{20}/Ar^{36})_{tr}$ ratios of the trapped gas in the ilmenite should represent the best approximation to the true solar wind composition (Table 7). Some bulk grain size

Table 7. Solar wind elemental and isotopic compositions estimated from trapped gases in Apollo 11 lunar surface material

	Solar wind	Terrestrial atmosphere	Cosmic (ALLER, 1961)
He^4/Ar^{36}	$\geq$7600	n.r.	26000
Ne^{20}/Ar^{36}	$\geq$33	0·5	72
Ar^{36}/Kr^{86}	$\geq$7600	160	22000
Ar^{36}/Xe^{132}	$\geq$16000	1500	200000
He^4/He^3	2700 ± 100	n.r.	—
Ne^{20}/Ne^{22}	12·85 ± 0·1	9·80 ± 0·08	—
Ne^{20}/Ne^{21}	400 ± 11	338 ± 2·5	—
Ar^{36}/Ar^{38}	5·32 ± 0·08	5·32 ± 0·01	—
Ar^{40}/Ar^{36}	<0·7	n.r.	—
Kr	similar to terrestrial Kr		—
Xe	see Table 8		—

n.r.: not relevant.
For comparison terrestrial and estimated cosmic abundances are given. Errors given for solar wind isotopic compositions are analytical errors only.

fractions show higher $(Ar^{36}/Kr^{86})_{tr}$ and $(Ar^{36}/Xe^{132})_{tr}$ ratios than the ilmenite and we have chosen them as lower limits for the true solar wind abundances. The solar wind composition derived from the lunar surface material represents some average prevailing during the effective exposure period. It may be different from the recent average solar wind composition.

The mass dependence of the abundances of the noble gases is less steep than would be predicted from cosmic abundances (cf. also Fig. 6). We cannot yet decide whether this is due to saturation and diffusion effects during or after trapping of the solar wind, to mass discrimination in the solar wind or whether this reflects the true chemical composition of the solar wind reservoir. The $(He^4/Ar^{36})_{tr}$ and $(Ne^{20}/Ar^{36})_{tr}$ ratios in ilmenite and bulk material are different by an order of magnitude. Cosmic abundances and ilmenite abundances differ only by a factor 2–3, and this could, in our opinion, be due to moderate diffusion losses in the ilmenite. For Kr and Xe the situation is different. The $(Kr^{86}/Ar^{36})_{tr}$ and $(Xe^{132}/Ar^{36})_{tr}$ ratios are similar in the ilmenite and bulk fractions. If one tries to explain the high Kr and Xe abundances by diffusion then bulk material and ilmenite would have lost Ar, Kr and Xe in the same relative proportions. This seems rather unlikely in view of the large differences observed between the $(He/Ar)_{tr}$ and $(Ne/Ar)_{tr}$ ratios in these two materials. Thus other processes might be responsible for the enrichment of Kr and Xe in the trapped gases (see below).

Discrimination between trapped gas and solar wind should be considerably smaller for isotopic ratios than for elemental abundances. The isotopic compositions of the trapped gas in the ilmenite can thus be considered as reasonable approximations of the true solar wind isotopic composition averaged over the effective exposure period (Table 7).

For the discussion of the solar wind noble gas isotopic compositions we will assume that these ratios represent a good approximation to the true isotopic compositions prevailing at the solar surface. A comparison of solar isotopic compositions with those in other gas reservoirs in the solar system, such as the terrestrial atmosphere and trapped meteoritic gases, might then give clues to their origin and evolution. The lunar $(He^4/He^3)_{tr}$ ratio is of the same magnitude as the ratio observed for meteoritic trapped He (ZÄHRINGER and GENTNER, 1960). In aubrites (Pesyanoe, Khor Temiki) the $(He^4/He^3)_{tr}$ ratio is considerably higher than in the Apollo 11 lunar material and is close to 4000 (ZÄHRINGER, 1962; EBERHARDT *et al.*, 1965a). These two aubrites have very high $(He^4/Ar^{36})_{tr}$ and $(Ne^{20}/Ar^{36})_{tr}$ ratios and diffusion loss can hardly have significantly influenced the trapped $(He^4/He^3)_{tr}$ ratio. Gas rich chondrites show lower $(He^4/He^3)_{tr}$ ratios (HINTENBERGER *et al.*, 1965; PEPIN and SIGNER, 1965) and in carbonaceous chondrites the $(He^4/He^3)_{tr}$ ratio seems highly variable (ANDERS *et al.*, 1970). This variability in the trapped $(He^4/He^3)_{tr}$ ratio of different sources may in part reflect long time variations in the He^4/He^3 ratio at the solar surface. He^3 leaking from the interior of the sun (SCHATZMAN, 1970) could lead to a secular decrease of the He^4/He^3 ratio at the solar surface. Nuclear reactions at the solar surface could also decrease the He^4/He^3 ratio with time. The high $(He^4/He^3)_{tr}$ ratio in aubrites would then correspond to an irradiation early in the history of the solar system.

The low terrestrial Ne^{20}/Ne^{22} and Ne^{20}/Ne^{21} ratios are in good agreement with a mass dependent fractionation process (e.g. gravitational escape). Terrestrial Ne represents the remaining few per mill of the original gas which had initially solar isotopic composition. If we calculate the overall fractionation factor from the difference between the terrestrial and solar Ne^{20}/Ne^{22} ratio, then the terrestrial Ne^{20}/Ne^{21} ratio can be predicted from the solar value. The predicted value depends on the specific model and on the assumed mass dependence of the fractionation factor. A well mixed reservoir (Rayleigh distillation) and a fractionation factor α equal to $\sqrt{(m_1/m_2)}$ would lead to a terrestrial Ne^{21}/Ne^{22} ratio of $(2{\cdot}82 \pm 0{\cdot}08) \times 10^{-2}$; a value in good agreement with the terrestrial Ne^{21}/Ne^{22} ratio of $(2{\cdot}90 \pm 0{\cdot}03) \times 10^{-2}$. In the fractionation process 99·6 per cent of the initial Ne would have to be lost. A comparison of the cosmic and terrestrial Ne^{20}/Ar^{36} ratios shows that at least 99·3 per cent of the Ne must have been lost in a process fractionating the Ne^{20}/Ar^{36} ratio, in good agreement with the loss factor estimated from the fractionation of the Ne isotope ratios. In addition large non-fractionating gas losses must have occurred.

WETHERILL (1954) had pointed out that the terrestrial Ne^{21} abundance might have been changed during the 4·5 aeons life-time of the earth by the addition of Ne^{21} produced in spallation reactions or (α, n) reactions on O^{18}. Our preliminary, slightly higher $(Ne^{20}/Ne^{21})_{tr}$ ratio given in the *Science* Paper for the trapped solar gas had pointed to the presence of such a Ne^{21} component in the terrestrial atmosphere. From the revised ratio we conclude now that such a Ne^{21} component must be smaller than a few per cent.

Similarly, the agreement found between the fractionated solar Ne and terrestrial Ne also limits the high energy particle and neutron flux to which solar matter could have been subjected after the separation of the terrestrial matter from the solar material.

Terrestrial and solar Ar^{36}/Ar^{38} agree within the present experimental uncertainty.

The isotopic compositions of terrestrial and solar Kr are compared in Fig. 19. As customary for the discussion of general Kr and Xe anomalies, δ-values are given. These represent the relative deviation in isotopic composition from atmospheric Kr or Xe. The δ-value representation tends to imply that the terrestrial atmosphere corresponds to the normal, unaltered primary gas reservoir. This is probably not

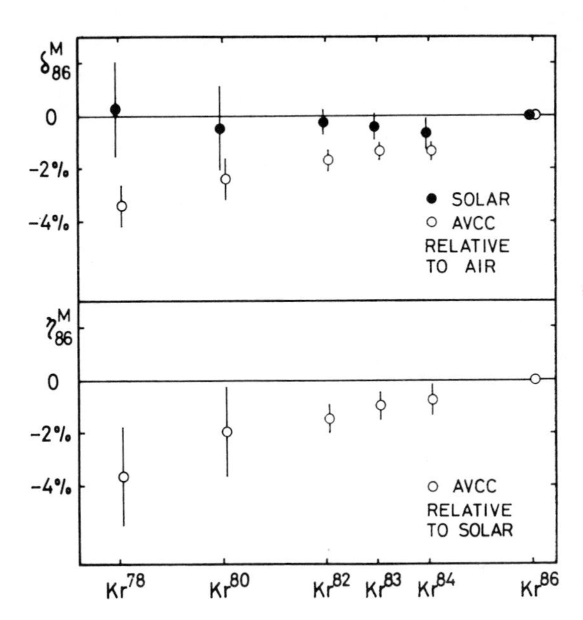

Fig. 19. Comparison of isotopic composition of solar, AVCC and atmospheric Kr.

$$\delta^M{}_{86} = [(Kr^M/Kr^{86})_{sample}/(Kr^M/Kr^{86})_{atm} - 1]$$

is the relative deviation from atmospheric composition;

$$\eta^M{}_{86} = [(Kr^M/Kr^{86})_{sample}/(Kr^M/Kr^{86})_{solar} - 1]$$

is the relative deviation from solar composition. AVCC-Kr from EUGSTER *et al.* (1967b).

true and solar Kr and Xe ought to be a much better approximation to the isotopic compositions prevailing at the time of formation of the solar system. In analogy to the δ-values we therefore define

$$\eta^M{}_R = [(Kr^M/Kr^R)_{sample}/(Kr^M/Kr^R)_{solar} - 1].$$

The η-values give the isotopic composition of a sample relative to the solar gas. Figure 19 shows the η-values for AVCC-Kr. For Kr the δ and η values are very similar, however, for Xe they will be significantly different.

Solar and air Kr agree in isotopic composition. However, this agreement might very well be accidental. AVCC-Kr differs from air Kr only by a systematic mass fractionation of 0·3 per cent per mass unit. It seems conceivable to us that the true solar Kr composition could be more AVCC-like and that a minor mass fractionation in the acceleration (GEISS *et al.*, 1970) and trapping of the solar wind particles changed the isotopic composition sufficiently to bring it close to the composition of air Kr.

Compared with cosmic abundances Kr is heavily enriched relative to Ar^{36} in the terrestrial atmosphere (see Table 7 and Fig. 6). The identity between the isotopic compositions of atmospheric and solar Ar and Kr requires elemental fractionation without changes in isotopic composition. Thus the same process which led to the depletion of atmospheric Ne cannot be solely responsible for the Kr enrichment. Leakage of solar wind Kr into the terrestrial atmosphere (CAMERON, 1962) cannot have contributed appreciably to terrestrial Kr. Otherwise also solar wind Xe would have leaked in and atmospheric Xe should have an isotopic composition similar to trapped lunar Xe. In our opinion, the terrestrial Xe, Kr, Ar fractionation could best be explained either by processes involving the "chemical" affinity (e.g. adsorption) or by processes in which several initially independent reservoirs were individually fractionated and later mixed.

Table 8. Isotopic composition of solar Xe (lunar trapped Xe). For comparison the isotopic compositions of AVCC-Xe, chondritic trapped Xe, Pesyanoe 1000°C Xe, and atmospheric Xe are given

	$\frac{Xe^{124}}{Xe^{130}}$	$\frac{Xe^{126}}{Xe^{130}}$	$\frac{Xe^{128}}{Xe^{130}}$	$\frac{Xe^{129}}{Xe^{130}}$	$\frac{Xe^{131}}{Xe^{130}}$	$\frac{Xe^{132}}{Xe^{130}}$	$\frac{Xe^{134}}{Xe^{130}}$	$\frac{Xe^{136}}{Xe^{130}}$
				× 100				
Solar Xe (lunar trapped Xe) (this paper)	2·99 ±0·18	2·67 ±0·2	50·45 ±0·8	634·5 ±3·5	495·5 ±3	607 ±5	225·8 ±3·5	184·2 ±3
AVCC-Xe (EUGSTER *et al.*, 1967b)	2·854 ±0·065	2·550 ±0·030	51·00 ±0·45	—	508·1 ±3·9	621·9 ±4·2	237·6 ±2·0	199·6 ±1·8
Chondritic trapped Xe (MARTI, 1967)	2·802 ±0·065	2·517 ±0·050	50·15 ±0·5	634·8 ±5·0	505·3 ±3·1	620·0 ±3·1	236·2 ±1·7	198·4 ±1·6
Pesyanoe 1000°C Xe* (MARTI, 1969)	—	—	—	633·1 ±9·5	495·0 ±6·5	608 ±7	222·2 ±3·2	179·6 ±2·8
Atmospheric Xe (NIER, 1950)	2·360 ±0·013	2·199 ±0·012	47·11 ±0·21	649·0 ±3·1	519·9 ±2·4	660·1 ±2·2	256·2 ±1·1	217·7 ±0·9

* Appropriate spallation correction applied.

Table 8 and Fig. 20 give a comparison of the isotopic composition of trapped lunar Xe, with other Xe in the solar system. The fission shielded isotope Xe^{130} is used for reference to facilitate the discussion of possible fission Xe components. The Berne AVCC-Xe values used here are virtually identical with the earlier Berkeley AVCC composition, except for Xe^{124} and Xe^{126} where the Berkeley values have somewhat larger errors. AVCC-Xe and the average trapped chondritic Xe have identical isotopic compositions, with the possible exception of Xe^{128}. Trapped lunar Xe and AVCC-Xe have the same abundances of Xe^{124}, Xe^{126} and Xe^{128}. The heavier isotopes are distinctly less abundant in lunar trapped Xe. The isotopic composition of lunar $Xe^{131-136}$ agrees well with the isotopic composition of the Xe released at 1000°C

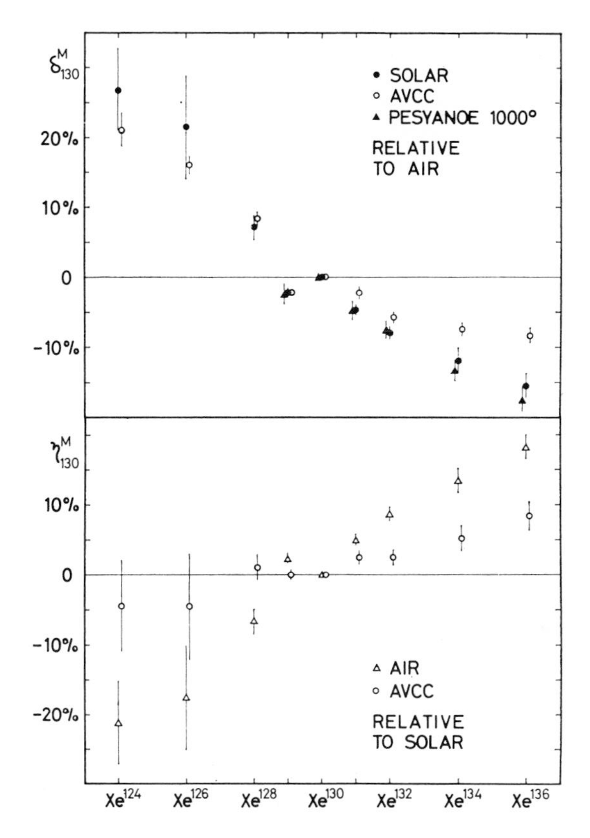

Fig. 20. Comparison of isotopic composition of solar, AVCC-Xe, Pesyanoe 1000°C Xe, and atmospheric Xe.

$$\delta^M{}_{130} = [(Xe^M/Xe^{130})_{sample}/(Xe^M/Xe^{130})_{atm} - 1]$$

is the relative deviation from atmospheric composition;

$$\eta^M{}_{130} = [(Xe^M/Xe^{130})_{sample}/(Xe^M/Xe^{130})_{solar} - 1]$$

is the relative deviation from solar composition. AVCC-Xe from EUGSTER *et al.* (1967b), Pesyanoe 1000°C Xe from MARTI (1969).

from the Pesyanoe aubrite (MARTI, 1969). This result supports Marti's conclusion that, after appropriate spallation corrections, the 1000°C Pesyanoe Xe abundances represent solar Xe.

The difference between the abundances of the heavy isotopes in AVCC-Xe and solar Xe can be attributed to a fission component in AVCC-Xe (cf. PEPIN, 1968). This is especially evident from the η representation of the Xe isotopic anomalies (see Fig. 20). The required fission spectrum would be:

$$(Xe^{131}/Xe^{136})_f = 0{\cdot}80 \pm 0{\cdot}35$$
$$(Xe^{132}/Xe^{136})_f = 0{\cdot}95 \pm 0{\cdot}45$$
$$(Xe^{134}/Xe^{136})_f = 0{\cdot}75 \pm 0{\cdot}30$$

The uncertainties in the fission spectrum are considerable, reflecting the relatively small difference in isotopic composition between AVCC and solar Xe. The differences between the above AVCC fission spectrum and previously derived ones (PEPIN, 1968; EUGSTER *et al.*, 1967b) are the result of dissimilarities in the assumptions. The large experimental errors preclude a detailed comparison with other known fission spectra. However, the high $Xe^{131}{}_f$ yield is striking and could, if confirmed by further measurements, exclude as progenitors such spontaneous fission isotopes as U^{238} and Pu^{244} (HYDE, 1964; EBERHARDT and GEISS, 1966; HOHENBERG *et al.*, 1967; WASSERBURG *et al.*, 1969). The AVCC fission component has been attributed to a possible long-lived isotope from the "island of stability" near $N = 184$ (ANDERS and HEYMANN, 1969; DAKOWSKI, 1969). However, if fission Kr is indeed present in AVCC-Kr (EUGSTER *et al.*, 1967b), we would have in the AVCC fission gas $(Kr^{86}/Xe^{136})_f \sim 0{\cdot}1$, an unexpectedly high ratio for the fission of transuranian or superheavy nuclides.

The Xe from the 1000°C Pesyanoe temperature release experiment shows slightly lower Xe^{134} and Xe^{136} abundances than the trapped lunar gas. The errors overlap (see Fig. 20) and the difference cannot yet be considered as established. It might however be taken as an indication that the lunar trapped gas contains either a very minor fission Xe component or some Xe with air-like Xe isotopic composition. Our Xe isotopic composition results on the etched ilmenite (see Table 1) seem to support this.

The overall isotopic abundance pattern of solar Xe and AVCC-Xe are very similar, except for the fission Xe component in AVCC discussed above. The large difference in isotopic composition observed between solar wind and atmospheric Xe is thus as difficult to explain as the difference between AVCC and atmospheric Xe. Many processes have been discussed (KURODA, 1960; KRUMMENACHER *et al.*, 1962; REYNOLDS, 1963; CAMERON, 1962) for resolving the AVCC-air Xe problem. The systematic trend in the anomaly (see Fig. 20) strongly suggests mass dependent fractionation processes as important (KRUMMENACHER *et al.*, 1962). The fractionation would be very large for a heavy element like Xe and, furthermore, no correspondingly large fractionation has been found between solar, AVCC and atmospheric Ar and Kr. The assumption of a completely different origin for terrestrial Xe might be required to explain these Xe isotope anomalies, with terrestrial Xe originating essentially from a different nucleosyntheses than solar Xe.

THE Ar^{40} PROBLEM

The primary cosmic abundance of Ar^{40} is expected to be very low with an Ar^{40}/Ar^{36} ratio of 10^{-4} or lower (CLAYTON, 1968). The decay of K^{40} in an environment of cosmic chemical abundances could raise the Ar^{40}/Ar^{36} ratio, but it would still remain much smaller than 1 per cent. MARTI (1967a) found in the Novo-Urei achondrite an upper limit of 0·35 for the $(Ar^{40}/Ar^{36})_{tr}$ ratio in the trapped gas. In Novo-Urei the trapped gases are not necessarily solar wind particles, but the low ratio indicates the presence of a primary gas reservoir in the solar system with an $(Ar^{40}/Ar^{36})_{tr}$ ratio lower than 0·35. It is thus unlikely that the high $(Ar^{40}/Ar^{36})_{tr}$ ratio in the lunar trapped gases (see Table 4) is representative for the primary solar wind, and other sources for the large amounts of trapped Ar^{40} have to be considered.

A lunar atmosphere, transient or steady, due to outgassing of the moon could be a source of the excess trapped Ar^{40}. The lunar fine material contains 4×10^{-4} cm³ STP/g trapped Ar^{40}. An average regolith thickness of 10 m would correspond to a total trapped Ar^{40} content of 7×10^{-1} cm³ STP Ar^{40}/cm² lunar surface. With chondritic composition the total radiogenic Ar^{40} content of the moon would be 1·2 $\times 10^4$ cm³ STP Ar^{40}/cm² lunar surface. To explain the trapped Ar^{40} in the fine lunar material as retrapped lunar radiogenic Ar^{40} would require a degassing-trapping process with an overall transfer efficiency of 6×10^{-5}. For a 10 per cent out-gassed moon a retrapping efficiency of 6×10^{-4} or better for Ar^{40} atoms in the lunar atmosphere would be required. The observed correlation between solar wind implanted Ar^{36} and excess Ar^{40} (HEYMANN *et al.*, 1970) implies that the retrapping mechanism must be connected with the solar wind. Elastic collisions with the solar wind particles has been proposed as a possible mechanism for accelerating and trapping Ar^{40} from a lunar atmosphere into the lunar surface (HEYMANN *et al.*, 1970).

Momentum transfer to neutral Ar atoms in the lunar atmosphere by collision with the predominant H^+ and He^{4++} ions of the solar wind is rare, collision times may be of the order of one year. Moreover, it is virtually impossible to impart in such a collision to an Ar atom enough energy for the subsequent trapping in the lunar surface. Heavier ions ($A \geq 12$) could transfer enough energy (~1 keV), but the lifetime of an Ar atom in the lunar atmosphere against collision with such a solar wind ion is a few hundred to a thousand years. Ar atoms in the lunar atmosphere will become ionized by charge exchange and solar UV with a typical time constant of 10 days (cf. MICHEL, 1964). After ionization Ar^{40} is removed from the lunar atmosphere by the electromagnetic field of the solar wind within seconds. Consequently the probability of a collision between an Ar^{40} atom and a heavy ion of the solar wind is only 10^{-4}. Thus collisional retrapping can account for some of the excess Ar^{40}, but in our opinion, cannot be the predominant mechanism.

After the Ar^{40} atoms are ionized they are—as mentioned by T. Gold during the Apollo 11 Lunar Science Conference—accelerated by the moving magnetic fields of the solar wind. For a magnetic field strength of 5 γ the induced electrical field is of the order of only 2 mV/m. Typical roll radii of the resulting cycloidal orbits are 10 lunar diameters. The initial ion velocity is orthogonal to the solar wind velocity and, depending on the direction of the magnetic field, the Ar^{40} ions will be either directed away from the moon and lost, or they will hit the moon with an energy of the order of 10^2 eV to 10^3 eV within seconds. At the lower energies they are neutralized but not trapped and thus recycled into the neutral atmosphere. Ions with trajectories resulting in impact energies of 1 keV or more will become trapped.

The retrapped Ar^{40}, both from this process and from the collision process mentioned above, had a much lower energy of impact than the solar wind Ar. The range and temperature release pattern of the retrapped Ar^{40} atoms should thus be significantly different from the trapped primary solar wind Ar.

Retrapping of evolved lunar gases may not be limited to Ar^{40}. A large fraction of the trapped primary solar wind ions is released by subsequent ion bombardment, by meteorite impacts, or by thermal effects. The He will escape from the moon, but a fraction of the heavier noble gases may become retrapped similar to Ar^{40}. Such a process could be partially responsible for the observed high Kr and Xe abundances in the trapped gases. Cometary gas released during the impact of comet nuclei may also lead to a temporary lunar atmosphere rich in heavy noble gases.

Spallation Produced Noble Gases

The spallation gases in the lunar fine material are generally much less abundant than the trapped gases. However, from the slope of the previously discussed correlation diagrams and from the etched ilmenite samples we can obtain accurate spallation concentrations and isotopic ratios for a number of pertinent spallation isotopes. These results are compiled in Table 9 and will be briefly discussed.

Table 9. Average concentrations and isotopic composition for spallation produced isotopes in Apollo 11 bulk and ilmenite grain size fractions (separated from 10084)

	Bulk fractions	Ilmenite fractions	Units
$He^3{}_{sp}$	—	380 ± 100	10^{-8} cm³ STP/g
$Ne^{21}{}_{sp}$	44 ± 10	9 ± 3	10^{-8} cm³ STP/g
$Ar^{38}{}_{sp}$	—	70 ± 12	10^{-8} cm³ STP/g
$Kr^{78}{}_{sp}$	135 ± 30	130 ± 40	10^{-12} cm³ STP/g
$(Kr^{80}/Kr^{78})_{sp}$	$2{\cdot}9 \pm 0{\cdot}5$	$2{\cdot}38 \pm 0{\cdot}09$	
$(Kr^{82}/Kr^{78})_{sp}$	—	$3{\cdot}31 \pm 0{\cdot}07$	
$(Kr^{83}/Kr^{78})_{sp}$	—	$4{\cdot}50 \pm 0{\cdot}14$	
$Xe^{126}{}_{sp}$	130 ± 30	34 ± 8	10^{-12} cm³ STP/g
$(Xe^{124}/Xe^{126})_{sp}$	$0{\cdot}47 \pm 0{\cdot}08$	$0{\cdot}62 \pm 0{\cdot}04$	
$(Xe^{128}/Xe^{126})_{sp}$	$1{\cdot}5 \pm 0{\cdot}4$	$1{\cdot}44 \pm 0{\cdot}10$	

The averages given for the bulk fractions should be representative for the average Apollo 11 fines with grain size $<130\,\mu$.

Our data do not allow a derivation of $He^3{}_{sp}$ in the bulk material. The $He^3{}_{sp}$ value for ilmenite was deduced from the most strongly etched fractions (AB-31, AB-78). For the bulk material $Ne^{21}{}_{sp}$ was obtained from the slope of the correlation line in Fig. 7. The ilmenite $Ne^{21}{}_{sp}$ value given in Table 9 is the average of the value derived from most strongly etched fractions (AB-31, AB-78) and the value obtained from the appropriate correlation diagram. The low Ne^{21} content in the ilmenite reflects the low Mg, Si and Al concentrations in this mineral phase. For ilmenite we estimate that the Ne^{21} production rate should be approximately 10–20 per cent of the production rate of bulk material [with elemental production rates given by BOCHSLER *et al.* (1968) and cosmic ray hardness $n = 2$]. This is in good agreement with the observed $Ne^{21}{}_{sp}$ content particularly if the 5 per cent contamination of the ilmenite by other minerals is taken into account.

The Ar^{38}_{sp} for ilmenite was calculated from the slope of the correlation line in Fig. 8. Our data seem to indicate a variable Ar^{38}_{sp} content in the bulk grain size fractions and no average is given. The Kr^{78}_{sp} concentration in the bulk material is deduced from the slope of the correlation lines in Fig. 14. For ilmenite the average spallation Kr^{78} concentration was derived by making the appropriate correction for trapped Kr ($Kr^{86}_{sp} = 0$ assumed). The $(Kr^{80}/Kr^{78})_{sp}$, $(Kr^{82}/Kr^{78})_{sp}$, and $(Kr^{83}/Kr^{78})_{sp}$ ratios were derived from the corresponding correlation diagrams (Figs. 11–13 and other similar plots). The spallation Kr content of the bulk material is small and allows only the determination of the $(Kr^{80}/Kr^{78})_{sp}$ ratio. In the ilmenite Kr^{78}_{sp} is high relative to the other Kr spallation isotopes, when compared to the spallation spectrum of the Stannern achondrite (MARTI *et al.*, 1966). This is most likely due to the different target element composition present in ilmenite. Zr is enriched in the ilmenite (ARRHENIUS *et al.*, 1970) and is the major target element for the production of spallation Kr. In Stannern Sr is the major target element.

The Xe^{126}_{sp} content of the bulk is derived from Fig. 18. For the ilmenite it is calculated from the measured ratios by correcting for trapped Xe. The Xe spallation isotope ratios are derived from correlation diagrams. The Xe spallation spectra of the bulk material and the ilmenite are somewhat different.

K/Ar AND EXPOSURE AGES

The K and Ba contents were measured in two bulk grain size fractions (see Table 10). In AA-19 the radiogenic Ar^{40} is masked by the trapped component. However, in the coarser fraction AA-27 the Ar^{40}_{r} content can be calculated with reasonable accuracy and K/Ar age of 3·5 aeons is obtained. The average Ar^{40}_{r} content of the

Table 10. K, Ba, radiogenic Ar^{40}, and spallation Xe^{126} contents; K/Ar and exposure ages of bulk grain size fractions AA-27 and AA-19

	K (ppm)	Ba (ppm)	Ar^{40}_{r} 10^{-8} cm³ STP/g	Xe^{126}_{sp} 10^{-12} cm³ STP/g	K/Ar age (10^9 yr)	Exposure age (10^6 yr)
AA-27	1080 ±30	150 ±6	4300 ±1100	120 ±25	3·5 ±0·4	600 ±140
AA-19	1310 ±40	270 ±10	—	155 ±40	—	440 ±120

seven bulk grain size fractions can be estimated from the slope of Fig. 9. We obtain $Ar^{40}_{r} = 4300 \times 10^{-8}$ cm³ STP/g. With the average K content in 10084 (GAST and HUBBARD, 1970; WÄNKE *et al.*, 1970) of 1145 ppm an age of 3·4 aeons results, confirming the age obtained for AA-27.

The exposure age could be calculated from any one of the spallation isotopes. We prefer Xe^{126}_{sp} because the heavy noble gases are less prone to diffusion loss and because we have measured the concentration of Ba, the main target element, in these two grain size fractions. The Xe^{126}_{sp} production rate is taken from our measurements on lunar rock 10017 and 10071 and is based on the Kr^{81}/Kr radiation age and the measured Ba content of these two rocks. We obtain as an average for the two rocks $Xe^{126}_{sp} = 1{\cdot}3 \times 10^{-15}$ cm³ STP/g 10^6 yr ppm Ba. This rate includes the production from all target nuclei, not just Ba. It was normalized to Ba because the REE/Ba ratios are the same in 10084, 10017 and 10071 within 7 per cent (GAST and HUBBARD, 1970). The production rate given here and the measured Ba concentrations were used for calculating the average exposure ages of the two grain size fractions (Table 10).

The average exposure age of the Apollo 11 fine material is a factor of 7 lower than the Rb/Sr ages (ALBEE *et al.*, 1970; GOPALAN *et al.*, 1970) and the K/Ar ages (TURNER, 1970) of the Apollo 11 crystalline rocks which are thought to be fragments of the local bedrock underlying the regolith (LSAPT, 1970). It is noteworthy that the absorption length of the cosmic ray nuclear active component is smaller than the average thickness of the regolith by about the same factor. This coincidence fits into the general picture that a regolith with a thickness of only several meters is overlying these bedrocks which were formed several billion years ago. In principle this can be explained by two types of models:

1. The regolith has been accumulating continuously since the melting of the bedrock. Some mixing may have occurred, but accumulation dominates mixing.
2. The regolith formed soon after bedrock melting and has been continuously and thoroughly mixed since that time.

The second model would require a certain concentration of nuclei from slow neutrons capture even in material which is now at or near the surface. One can compare the Gd anomalies measured by ALBEE *et al.* (1970) in lunar fine material with anomalies predicted from model calculations of the type given by EBERHARDT *et al.* (1963). Such a comparison indicates that for most of the cosmic ray irradiation period the material was relatively close to the surface ($\lesssim$30 cm). This would support a formation of the regolith along the lines of model 1.

Acknowledgments—We thank Dr. R. GIOVANOLI, Dr. H. VON GUNTEN, Dr. T. PETERS and P. BOCHSLER for their support of this work and for discussions. The assistance of MADELEINE BILL, RUTH KÜPFER, H. HOFSTETTER, V. HORVATH, E. LENGGENHAGER, A. SCHALLER, U. SCHWAB and H. WYNIGER is acknowledged. This work was supported by the Swiss National Science Foundation grants NF 2.73.68, NF 5079 and NF 2.30.68.

REFERENCES

ALBEE A. L., BURNETT D. S., CHODOS A. A., EUGSTER O. J., HUNEKE J. C., PAPANASTASSIOU D. A., PODOSEK F. A., RUSS G. PRICE II, SANZ H. G., TERA F. and WASSERBURG G. J. (1970) Ages, irradiation history, and chemical composition of lunar rocks from the Sea of Tranquillity. *Science* **167**, 463–466.

ALLER L. H. (1961) *The Abundance of the Elements*. Interscience.

ANDERS E. and HEYMANN D. (1969) Elements 112 to 119: were they present in meteorites? *Science* **164**, 821–823.

ANDERS E., HEYMANN D. and MAZOR E. (1970) Isotopic composition of primordial helium in carbonaceous chondrites. *Geochim. Cosmochim. Acta* **34**, 127–131.

ARRHENIUS G., ASUNMAA S., DREVER J. I., EVERSON J., FITZGERALD R. W., FRAZER J. Z., FUJITA H., HANOR J. S., LAL D., LIANG S. S., MCDOUGALL D., REID A. M., SINKANKAS J. and WILKENING L. (1970) Phase chemistry, structure and radiation effects in lunar samples. *Science* **167**, 659–661.

BAME S. J., HUNDHAUSEN A. J., ASBRIDGE J. R. and STRONG I. B. (1968) Solar wind ion composition. *Phys. Rev. Lett.* **20**, 393–395.

BOCHSLER P., EBERHARDT P., GEISS J. and GRÖGLER N. (1968) Rare gas measurements in separate mineral phases of the Otis and Elenovka chondrites. In *Meteorite Research*, (editor P. M. Millman), Vol. 12, Chap. 68, pp. 857–873. D. Reidel.

BÜHLER F., EBERHARDT P., GEISS J., MEISTER J. and SIGNER P. (1969) Apollo 11 solar wind composition experiment: first results. *Science* **166,** 1502–1503.

CAMERON A. G. W. (1962) The formation of the sun and planets. *Icarus* **1,** 13–69.

CLAYTON D. D. (1968) *Principles of Stellar Evolution and Nucleosynthesis.* McGraw-Hill.

CROZAZ G., HAACK U., HAIR M., HOYT H., KARDOS J., MAURETTE M., MIYAJIMA M., SEITZ M., SUN S., WALKER R., WITTELS M. and WOOLUM D. (1970) Solid state studies of the radiation history of the lunar samples. *Science* **167,** 563–566.

DAKOWSKI M. (1969) The possibility of extinct superheavy elements occurring in meteorites. *Earth Planet. Sci. Lett.* **6,** 152–154.

EBERHARDT P., GEISS J. and LUTZ H. (1963) Neutrons in meteorites. In *Earth Science and Meteoritics,* (editors J. Geiss and E. D. Goldberg), Chap. 8, pp. 143–168. North Holland.

EBERHARDT P., GEISS J. and GRÖGLER N. (1965a) Ueber die Verteilung der Uredelgase im Meteoriten Khor Temiki. *Tschermaks Mineral. Petrogr. Mitt.* **10,** 535–551.

EBERHARDT P., GEISS J. and GRÖGLER N. (1965b) Further evidence on the origin of trapped gases in the meteorite Khor Temiki. *J. Geophys. Res.* **70,** 4375–4378.

EBERHARDT P., EUGSTER O., GEISS J. and MARTI K. (1966) Rare gas measurements in 30 stone meteorites. *Z. Naturforsch.* **21a,** 414–426.

EBERHARDT P. and GEISS J. (1966) On the mass spectrum of fission xenon in the Pasamonte meteorite. *Earth Planet. Sci. Lett.* **1,** 99–101.

EBERHARDT P., GEISS J., GRAF H., GRÖGLER N., KRÄHENBÜHL U., SCHWALLER H., SCHWARZMÜLLER J. and STETTLER A. (1970a) Trapped solar wind noble gases, Kr^{81}/Kr exposure ages and K/Ar ages in Apollo 11 lunar material *Science* **167,** 558–560.

EBERHARDT P., GEISS J., GRAF H., GRÖGLER N., KRÄHENBÜHL U., SCHWALLER H., SCHWARZMÜLLER J. and STETTLER A. (1970b) To be published.

EPSTEIN S. and TAYLOR H. P., JR. (1970) $^{18}O/^{16}O$, $^{30}Si/^{28}Si$, D/H and C^{13}/C^{12} studies of lunar rocks and minerals. *Science* **167,** 533–535.

EUGSTER O., EBERHARDT P. and GEISS J. (1967a) ^{81}Kr in meteorites and ^{81}Kr radiation ages. *Earth Planet. Sci. Lett.* **2,** 77–82.

EUGSTER O., EBERHARDT P. and GEISS J. (1967b) Krypton and xenon isotopic composition in three carbonaceous chondrites. *Earth Planet. Sci. Lett.* **3,** 249–257.

EUGSTER O., EBERHARDT P. and GEISS J. (1969) Isotopic analyses of krypton and xenon in fourteen stone meteorites. *J. Geophys. Res.* **74,** 3874–3896.

FIELDS P. R., DIAMOND H., METTA D. N., STEVENS C. M., ROKOP D. J. and MORELAND P. E. (1970) Isotopic abundances of actinide elements in lunar material. *Science* **167,** 499–500.

FLEISCHER R. L., HAINES E. L., HANNEMAN R. E., HART H. R. JR., KASPER J. S., LIFSHIN E., WOODS R. T. and PRICE P. B. (1970) Particle track, X-ray, thermal and mass spectrometric studies of lunar material. *Science* **167,** 568–571.

GAST P. W. and HUBBARD N. J. (1970) Abundance of alkali metals, alkaline and rare earths, and strontium-87/strontium-86 ratios in lunar samples. *Science* **167,** 485–487.

GEISS J., HIRT P. and LEUTWYLER H. (1970) On acceleration and motion of ions in corona and solar wind. *Solar Phys.* in press.

GOPALAN K., KAUSHAL S., LEE-HU C. and WETHERILL G. W. (1970) Rubidium–strontium, uranium and thorium–lead dating of lunar material. *Science* **167,** 471–473.

HEYMANN D., YANIV A., ADAMS J. A. S. and FRYER G. E. (1970) Inert gases in lunar samples. *Science* **167,** 555–558.

HINTENBERGER H., VILCSEK E. and WÄNKE H. (1965) Ueber die Isotopenzusammensetzung und über den Sitz der leichten Uredeglase in Steinmeteoriten. *Z. Naturforsch.* **20a,** 939–945.

HINTENBERGER H., WEBER H. W., VOSHAGE H., WÄNKE H., BEGEMANN F., VILCSEK E. and WLOTZKA F. (1970) Rare gases, hydrogen and nitrogen: concentrations and isotopic composition in lunar material. *Science* **167,** 543–545.

HOHENBERG C. M., MUNK M. N. and REYNOLDS J. H. (1967) Spallation and fissiogenic xenon and krypton from stepwise heating of the Pasamonte achondrite; the case for extinct plutonium 244 in meteorites; relative ages of chondrites and achondrites. *J. Geophys. Res.* **72,** 3139–3177.

HYDE E. K. (1964) The nuclear properties of heavy elements. In *Fission Phenomena*, Vol. 3. Prentice Hall.

KRUMMENACHER D., MERRIHUE C. M., PEPIN R. O. and REYNOLDS J. H. (1962) Meteoritic krypton and barium versus the general isotopic anomalies in meteoritic xenon. *Geochim. Cosmochim. Acta* **26,** 231–249.

KURODA P. K. (1960) Nuclear fission in the early history of the earth. *Nature* **187,** 36–38.

LAL D. and RAJAN R. S. (1969) Observations on space irradiation of individual crystals of gas-rich meteorites. *Nature* **223,** 269–271.

LORD H. C. (1968) Hydrogen and helium ion implantation into olivine and enstatite: retention coefficients, saturation concentrations, and temperature-release profiles. *J. Geophys. Res.* **73,** 5271–5280.

LSAPT (LUNAR SAMPLE ANALYSIS PLANNING TEAM) (1970) Summary of Apollo 11 Lunar Science Conference. *Science* **167,** 449–451.

LSPET (LUNAR SAMPLE PRELIMINARY EXAMINATION TEAM) Preliminary examination of lunar samples from Apollo 11. *Science* **165,** 1211–1227.

MARTI K., EBERHARDT P. and GEISS J. (1966) Spallation, fission, and neutron capture anomalies in meteoritic krypton and xenon. *Z. Naturforsch.* **21a,** 398–413.

MARTI K. (1967) Isotopic composition of trapped krypton and xenon in chondrites. *Earth Planet. Sci. Lett.* **3,** 243–248.

MARTI K. (1967a) Trapped xenon and classification of chondrites. *Earth Planet. Sci. Lett.* **2,** 193–196.

MARTI K. (1969) Solar-type xenon: a new isotopic composition of xenon in the Pesyanoe meteorite. *Science* **166,** 1263–1265.

MEISTER J. (1969) Ein Experiment zur Bestimmung der Zusammensetzung und Isotopenverhältnisse des Sonnenwindes: Einfangverhalten von Aluminium für niederenergetische Edelgasionen. Ph. D. Thesis, University of Berne.

MICHEL F. C. (1964) Interaction between the solar wind and the lunar atmosphere. *Planet. Space Sci.* **12,** 1075–1091.

NIER A. O. (1950) A redetermination of the relative abundances of the isotopes of neon, krypton, rubidium, xenon and mercury. *Phys. Rev.* **79,** 450–454.

PELLAS P., POUPEAU G., LORIN J. C., REEVES H. and AUDOUZE J. (1969) Primitive low-energy particle irradiation of meteoritic crystals. *Nature* **223,** 272–274.

PEPIN R. O. and SIGNER R. (1965) Primordial rare gases in meteorites. *Science* **149,** 253–265.

PEPIN R. O. (1968) Neon and xenon in carbonaceous chondrites. In *Origin and Distribution of the Elements*, (editor L. H. Ahrens), Vol. 30, pp. 379–386. Pergamon.

PEPIN R. O., NYQUIST L. E., PHINNEY D. and BLACK D. C. (1970) Isotopic composition of rare gases in lunar samples. *Science* **167,** 550–553.

REYNOLDS J. H. (1960) Isotopic composition of primordial xenon. *Phys. Rev. Lett.* **4,** 351–354.

REYNOLDS J. H. (1963) Xenology. *J. Geophys. Res.* **68,** 2939–2956.

REYNOLDS J. H., HOHENBERG C. M., LEWIS R. S., DAVIS P. K. and KAISER W. A. (1970) Isotopic analysis of rare gases from stepwise heating of lunar fines and rocks. *Science* **167,** 545–548.

SCHATZMAN E. (1970) Private communication.

TURNER G. (1970) Argon-40/argon-39 dating of lunar rock samples. *Science* **167,** 466–468.

WÄNKE H., BEGEMANN F., VILCSEK E., RIEDER R., TESCHKE F., BORN W., QUIJANO-RICO M., VOSHAGE H. and WLOTZKA F. (1970) Major and trace elements and cosmic-ray produced radioisotopes in lunar samples. *Science* **167,** 523–525.

WASSERBURG G. J., HUNEKE J. C. and BURNETT D. S. (1969) Correlation between fission tracks and fission-type xenon from an extinct radioactivity. *Phys. Rev. Lett.* **22,** 1198–1201.

WETHERILL G. W. (1954) Variations in the isotopic abundances of neon and argon extracted from radioactive minerals. *Phys. Rev.* **96,** 679–683.

ZÄHRINGER J. and GENTNER W. (1960) Uredelgase in einigen Steinmeteoriten. *Z. Naturforsch.* **15a,** 600–602.

ZÄHRINGER J. (1962) Ueber die Uredelgase in den Achondriten Kapoeta und Staroe Pesjanoe. *Geochim. Cosmochim. Acta* **26,** 665–680.

Proceedings of the Apollo 11 Lunar Science Conference, Vol. 2, pp. 1071 to 1079.

Oxygen, silicon and aluminum in Apollo 11 rocks and fines by 14 MeV neutron activation

W. D. EHMANN and J. W. MORGAN
Department of Chemistry, University of Kentucky, Lexington, Kentucky 40506

(*Received* 30 *January* 1970; *accepted in revised form* 25 *February* 1970)

Abstract—Fast neutron activation analysis of Apollo 11 lunar material yielded: Type A (2 rocks) 38·5% O, 18·9% Si, 4·0% Al; Type B (7 rocks) 39·4% O, 18·7% Si, 5·0% Al; Type C (18 rocks) 41·1% O, 19·7% Si, 6·6% Al; Type D (3 aliquants) 40·8% O, 20·2% Si, 7·2% Al and 12·4% Fe. The crystalline rocks (Types A and B) can be put into two distinct chemical groups on the basis of Al contents. The fines and breccias are significantly higher in O, Si, and Al than the crystalline rocks. An apparent deficiency in the fines of about 1% O may be due to reduction by solar wind H.

INTRODUCTION

THE 14 MeV neutron activation analyses for O, Si and Al reported here represent the first stage of a cooperative multielement activation analysis scheme for a large number of major, minor and trace elements (GOLES *et al.*, 1970). The 14 MeV neutron activation method is rapid, and, except for very minor residual radioactivity and radiation damage, is essentially non-destructive. It is probably the only reliable method available for the direct analysis of O in bulk samples, and indications of O depletion in the lunar surface (GAULT *et al.*, 1967), make such measurements of considerable importance. Both Si and Al are routinely reported in wet chemical whole rock analyses; however, there is general agreement (MASON, 1962) that the procedures, at least for Al, leave much to be desired. The rapidity of the activation method enabled the data collection portion of the analyses for these three elements to be carried out on 27 lunar rocks in a period of about three weeks, so that the samples could be transferred to the next laboratory with a minimum of delay.

ANALYTICAL METHOD

Apparatus

14 MeV neutrons were produced by a Kaman Nuclear model A-1250 Cockcroft-Walton generator, and the neutron yield monitored by a low geometry enriched BF_3 detector. The pneumatic single sample transfer system, using dry N_2 as propellant gas, and the sequential programming circuit are essentially those described by VOGT *et al.* (1965). Minor modifications were made to interface with a data acquisition system based on a Nuclear Data ND 2201 4096 channel analyzer. Gamma-ray activity measurements were made using a 10 cm × 10 cm well type NaI (Tl) detector.

Preparation of samples

Samples of 27 lunar rocks were received under a double dry nitrogen seal as irregular-shaped chips, ranging in weight between 0·45 and 2·4 g. The design of our usual polyethylene rabbits (EHMANN and MCKOWN 1968) was modified to take $\frac{3}{8}$ in. i.d. capsules. The lunar rocks fitted this larger capsule except for two (10048-32; 10061-32) which were sealed in polyethylene bags within the outer rabbit container. Before use, all polyethylene parts for the rabbit assembly were immersed in absolute ethanol and agitated ultrasonically for 20 minutes. They were dried in a jet of high purity dry N_2, transferred to a vacuum desiccator, and dried under vacuum. Samples were transferred to inner capsules in a glove box under dry N_2. After sealing the capsule was positioned in the outer container with polyethylene spacers, so that the apparent center-of-mass was centered in the neutron beam position. The outer container was flushed with dry N_2 before it, too, was sealed.

One 5 g sample of lunar fines was received and three aliquants of about 1 g each were prepared under dry N_2 by quartering. Duplicate aliquants of BCR-1, Knippa basalt K89 and vesicular basalts VB-1 and VB-2 were also taken and made up in $\frac{3}{8}$ in. i.d. capsules so that the analytical conditions were closely comparable with those of the rocks.

Preparation of standards

The encapsulation of the standards was identical to that described for the samples. Fused optical quartz L-1 (pieces of a broken quartz lens donated by Dr. W. Blackburn of the University of Kentucky) was used as a standard for Si and O. Potassium dichromate NBS# 136b was used, as received, for an additional standard for the O determinations. In addition to powdered standards, chunks of L-1 quartz, similar in shape to the rock chips, were prepared. Initially, these were used together with powdered standards for Si and O analyses, until it was established that geometry effects introduced no detectable systematic bias. For O, a two way comparison was possible, with powdered quartz and NBS# 136b $K_2Cr_2O_7$. The quartz chunks agreed with both of these to an accuracy of about 0·4 relative per cent.

Opal glass (NBS# 91) and potassium feldspar (NBS# 70a) were dried at 110°C and used for Al standards. The abundances of this element in these materials just bracket those reported in the preliminary analyses of lunar samples (LSPET, 1969). In addition, these standard materials contain Si as a major constituent and this element produces a primary interference in the Al determination. A correction for this interference was made specifically for each analysis using the measured Si content of the appropriate samples. The apparent Al abundance of the standards was also adjusted accordingly. Any residual error in the estimation of this correction will be largely self cancelling for samples and standards possessing similar Al/Si ratios.

Procedure

Detailed procedures for the precise determination of Si and O are given by MORGAN and EHMANN (1970). The chemical composition of the Apollo 11 rocks and fines is such that interferences are insignificant for O and Si determinations. Slight changes were made to increase the speed of analysis. The dwell time used for multi-scaling in the O analyses was increased from 0·4 sec to 0·8 sec to reduce the number of channels used. In the Si determinations, counting time was shortened from 300 sec to 150 sec, after calculation showed that the precision would not be significantly affected.

Aluminum analyses were made using the 0·84 MeV gamma-ray of ^{27}Mg. Appropriate correction was made for the primary interference from the reaction $^{30}Si(n, \alpha)$ ^{27}Mg. Samples and standards were irradiated for 1 min, and, following a delay of 300 sec to allow decay of ^{28}Al, were counted for 300 sec. The significant interference from the 0·85 MeV gamma-ray of ^{56}Mn was empirically determined by recounting after a delay of about 100 min. In principle, Fe analyses can be based on the ^{56}Mn activity, but these were made only for the lunar fines, using the Knippa basalt K89 as a standard. Taking the Fe abundance of this rock as 9·4 per cent, the means of triplicate analyses on each of three aliquants of fines were 12·6 per cent, 12·3 per cent and 12·3 per cent. The mean of these values, 12·4 $\pm$ 0·2% Fe, is in good agreement with the results obtained by other workers.

Results

The results for O, Si and Al in 27 lunar rocks and 3 aliquants of fines are given in Table 1. The O results are the mean of from 5 to 12 replicate runs and for Si 5–8

Table 1. Oxygen, silicon and aluminum abundances in individual Apollo 11 lunar rocks and lunar fines by 14 MeV neutron activation

N.A.S.A. No.	Oxygen (wt. %)*	Silicon (wt. %)*	Aluminum (wt. %)*
Type A			
10022-32	39·3 ± 0·4	19·4 ± 0·1	4·2 ± 0·1
10069-23	37·6 ± 0·1	18·3 ± 0·1	3·7 ± 0·2
Mean	38·5 ± 1·2	18·9 ± 0·8	4·0 ± 0·4
Type B			
10003-36	38·1 ± 0·4	17·9 ± 0·1	5·1 ± 0·1
10024-20	38·9 ± 0·1	18·5 ± 0·1	4·1 ± 0·2
10047-14	40·1 ± 0·1	20·1 ± 0·1	5·5 ± 0·3
10050-29	40·5 ± 0·5	18·4 ± 0·1	5·3 ± 0·2
10058-21	39·9 ± 0·3	19·0 ± 0·2	5·5 ± 0·2
10062-28	38·0 ± 0·2	18·0 ± 0·2	5·3 ± 0·1
10071-21	40·3 ± 0·2	19·1 ± 0·1	4·2 ± 0·1
Mean	39·4 ± 1·0	18·7 ± 0·8	5·0 ± 0·6
Type C			
10018-21	40·1 ± 0·4	20·1 ± 0·1	6·6 ± 0·1
10019-11	39·9 ± 0·2	18·9 ± 0·1	6·5 ± 0·2
10021-22	41·8 ± 0·6	19·6 ± 0·1	6·7 ± 0·1
10048-32	39·8 ± 0·2	18·1 ± 0·1	6·1 ± 0·1
10056-23	41·3 ± 0·3	20·2 ± 0·1	5·7 ± 0·1
10059-30	40·0 ± 0·1	19·1 ± 0·1	6·4 ± 0·1
10060-17	40·3 ± 0·2	20·0 ± 0·1	6·2 ± 0·1
10061-32	41·7 ± 0·2	18·9 ± 0·1	6·7 ± 0·1
10063-05	41·9 ± 0·3	20·3 ± 0·2	6·6 ± 0·2
10064-10	40·5 ± 0·4	19·4 ± 0·1	5·9 ± 0·1
10065-14	41·6 ± 0·7	19·3 ± 0·2	6·6 ± 0·1
10066-05	41·0 ± 0·2	20·2 ± 0·1	6·9 ± 0·1
10067-05	41·6 ± 0·1	20·6 ± 0·1	7·0 ± 0·1
10068-20	40·3 ± 0·3	19·3 ± 0·1	6·3 ± 0·2
10070-05	43·4 ± 0·3	20·6 ± 0·2	7·1 ± 0·1
10073-21	41·4 ± 0·3	20·5 ± 0·1	7·3 ± 0·1
10074-05	42·1 ± 0·5	19·3 ± 0·1	7·0
10075-06	40·4 ± 0·2	19·8 ± 0·3	7·4 ± 0·1
Mean	41·1 ± 1·0	19·7 ± 0·7	6·6 ± 0·5
Fines			
10084-50	42·2 ± 0·5	20·4 ± 0·1	7·2 ± 0·2
10084-50	40·5 ± 0·9	20·3 ± 0·1	7·3 ± 0·1
10084-50	39·8 ± 0·5	20·0 ± 0·2	7·1 ± 0·1
Mean	40·8 ± 1·2	20·2 ± 0·2	7·2 ± 0·1

* Error limits for individual samples are standard deviations of the means calculated generally on the basis of 6 replicate analyses for Si and O and 2 replicate analyses for Al. Those for the means are standard deviations of individual rock abundances within each type.

replicate determinations were made. In the case of Al each value in Table 1 is the mean of duplicate determinations, except for rock 10074-05 which is just a single determination, and for the three aliquants of fines and rocks 10022-32 and 10073-21 where triplicate analyses were performed.

Table 2. Comparison of O, Si and Al analyses of Apollo 11 rocks and fines

N.A.S.A. No.	Oxygen (wt. %)	Silicon (wt. %)	Aluminum (wt. %)	Reference
Type A				
10022		18·7	4·6	a
	39·3	19·4	4·2	This work
Type B				
10003		17·7	5·8	a
		18·6	5·5	b
	38·1	17·9	5·1	This work
10024		18·2	5·0	a
		18·8	4·3	b
	38·9	18·5	4·1	This work
10047		19·3	5·2	a
	40·1	20·1	5·5	This work
10050		19·1	4·7	a
	40·5	18·4	5·3	This work
10058		19·5	6·2	a
		18·4	5·4	c
	39·9	19·0	5·5	This work
10062		18·1	6·4	a
		18·6	5·4	b
	38·0	18·0	5·3	This work
Type C				
10018		19·5	6·5	b
	40·7	19·6	6·1	d
	40·1	20·1	6·6	This work
10019		19·2	7·3	a
	39·9	18·9	6·5	This work
10048		19·7	6·8	a
	39·8	18·1	6·1	This work
10056		19·4	5·7	a
		19·8	5·7	c
	41·3	20·2	5·7	This work
10060		19·4	6·2	a
		18·7	6·0	c
	41·4	19·8	6·2	d
			5·2	e
		19·6	6·3	j
	40·3	20·0	6·2	This work
10061		19·6	6·7	b
	41·7	18·9	6·7	This work
Fines				
10084		19·5	7·1	b
		20·2	7·3	c
	41·5*	19·7	6·9	d
		20·1	7·0	e
		19·8	7·3	f
		19·7	7·4	g
	41·9	19·7	7·3	h
		19·4	7·6	i
		19·7	7·2	j
	40·8	20·2	7·2	This work
Standard Rocks				
BCR–1	45·5	25·5	7·2	k
	45·8	25·7	7·0	This work
K89	41·8†	18·1	5·3	l
	42·6‡			
	42·3	17·8	5·1	This work
VB–1	42·7	21·9	7·9	This work
VB–2	43·3	22·0	8·0	This work

A large number of analyses by several different methods is now available for the lunar rocks and soil, and some of these were made on rocks analyzed by us. Comparisons of analyses for rocks and fines are made in Table 2 and indicate no significant systematic bias. The accuracy of the O, Si and Al analyses of lunar rocks and soil was also checked by simultaneously analyzing BCR-1, NASA Knippa basalt K89, and NASA vesicular basalts VB-1 and VB-2. These results are given in the same table.

Our O results for the lunar rocks and fines appear to be lower than those obtained by WÄNKE *et al.* (1970) by about 0·5–1·0% O. Although this discrepancy is of the same order as our analytical uncertainty it is surprising that in all three cases where comparisons are possible our results are lower. In the past our O results on standard rocks have been marginally high by $\simeq$0·3% O (e.g. MORGAN and EHMANN, 1970) probably due to variable $-H_2O$ (LANGMYHR, 1970). As a result, very special care was taken to handle the lunar rocks and fines under high purity dry N_2 at all times to exclude the possibility of water contamination. It is suggested that our O analyses closely reflect the true abundance of O in the samples under lunar conditions.

DISCUSSION

(1) *Chemical grouping of crystalline rocks*

The crystalline rocks were classified by LSPET (1969) on the basis of crystal texture. COMPSTON *et al.* (1970) have defined two chemical groupings which do not correspond to the LSPET types. Group 1 is characterized by very high Rb, K, Ba and Th relative to Group 2. Of particular relevance to this discussion is the apparent depletion of Al in the Group 1 rocks. Of the specimens analyzed here, COMPSTON *et al.* classified 10022 and 10024 in Group 1 and 10003, 10047 and 10058 in Group 2. Of the remaining crystalline rocks listed in Table 2, 10069 and 10070 are clearly in Group 1 from the high Ba abundances (ANNEL and HELZ, 1970), and 10050 and 10062 can be ascribed to Group 2 on the basis of the low abundances of K (TUREKIAN and KHARKAR, 1970; ROSE *et al.*, 1970). For the crystalline rocks analyzed in this work

---

Refs. to Table 2:

(a) ROSE *et al.* (1970). Si by SP; Al by CG and XRF.
(b) COMPSTON *et al.* (1970). Si and Al by XRF.
(c) MORRISON *et al.* (1970). Si by AA; Al by NA.
(d) WÄNKE *et al.* (1970). O, Si and Al by NA.
(e) SMALES *et al.* (1970). Si by XRF; Al by XRF and NA.
(f) MAXWELL *et al.* (1970). Si and Al by CG and AA.
(g) PECK and SMITH (1970). Si and Al by CG.
(h) WIIK and OJANPERÄ (1970). Si by CG and SP; Al by CG and AA. O calculated by difference.
(i) ENGEL and ENGEL (1970). Si and Al by CG.
(j) AGRELL *et al.* (1970). Si and Al by CG.
(k) FLANAGAN (1969).
(l) Unpublished data provided by NASA.

* Sum of leach fractions yields 40·7%.
† O calculated from oxides scaled to 100%.
‡ O calculated by difference.
AA = atomic absorption; CG = classical gravimetry; NA = neutron activation; SP = spectrophotometry; XRF = X-ray fluorescence.

the Al abundances in each group form a very narrow distribution. The four Group 1 rocks (10022-32; 10024-20; 10069-23; 10071-21) have a mean of $4{\cdot}1 \pm 0{\cdot}2$ and the five Group 2 rocks (10003-36; 10047-14; 10050-29; 10058-21; 10062-28) have a mean of $5{\cdot}3 \pm 0{\cdot}2$. The means of the two groups are very significantly different ($t = 9{\cdot}6$; $p < 0{\cdot}01$) at the 95 per cent confidence level. Clearly the determination of Al by 14 MeV neutron activation as done in this work provides a ready method to characterize the crystalline rocks brought back by Apollo 11. The mean values of $39{\cdot}0 \pm 1{\cdot}1$ and $39{\cdot}3 \pm 1{\cdot}2$% O and $18{\cdot}8 \pm 0{\cdot}5$ and $18{\cdot}7 \pm 0{\cdot}9$% Si for Groups 1 and 2 respectively do not differ significantly ($t = 0{\cdot}4$ and $0{\cdot}3$, respectively).

(2) *Breccias and fines*

The mean values for O, Si and Al in the breccias are significantly higher (95 per cent confidence level) than for either group of crystalline rocks. Again, Al is very useful diagnostically, as the mean abundances in breccias and crystalline rocks differ very significantly, and there is a distinct hiatus between the two groups of values. The abundances of the three elements in the lunar fines closely resemble those found in certain breccias, for example 10073-21. However, because of the very small variance of the Al content of triplicate aliquants of fines, the difference between the mean Al values is significant at the 95 per cent confidence level ($t = 2{\cdot}15$). It appears, that the breccias are not simply impacted soil, a conclusion reached independently from mineralogical evidence (DUKE *et al.*, 1970).

(3) *Apparent oxygen deficiency*

Several groups have reported high summations in total silicate analyses, particularly for the lunar fines (MAXWELL *et al.*, 1970; ROSE *et al.*, 1970; PECK and SMITH, 1970; WIIK and OJANPERÄ, 1970). These apparent errors were ascribed to iron and/or titanium being more reduced than was assumed for the oxide calculations. ROSE *et al.* measured "total reducing capacity" chemically and expressed it as FeO. Iron values measured independently by X-ray fluorescence were then subtracted to give the "excess reducing capacity". This was found in all the samples analyzed, but was greatest in breccias, and particularly in fines. MAXWELL *et al.*, by a more direct method, found no anomaly in rock 10017 and only a small one in rock 10020. The fines, however, showed a large difference, equivalent to 1% FeO. An excess of reducing capacity of this size is too large to be due solely to metallic iron and MAXWELL *et al.* suspect the presence of Ti(III). ROSE *et al.* also checked total Fe in the fines chemically and arrived at $4{\cdot}1$% FeO excess reducing capacity. These discrepancies are difficult to resolve; however, MAXWELL *et al.* ground their samples to 100 mesh before analysis. Unless this was carried out under dry N_2 there is a very real possibility of oxidation (CHAO, 1963). ROSE *et al.* give no details of sample preparation. It should be noted, however, that ROSE *et al.* analyzed a fines sample (311079) different from that studied by MAXWELL *et al.* (10084-132).

Our data indicate that the O depletion of the original sample of fines 10084-50 may be even larger than this. It should be noted that our sample was quartered and packaged in a dry N_2 filled glove box. The mean O values for triplicate aliquants analyzed in our work show a rather large scatter. This is much larger than that found using the same method on well-mixed powders (MORGAN and EHMANN, 1970), and probably reflects a real inhomogeneity. Grinding of the sample under dry N_2 and then splitting by quartering would probably have given more consistent results, however, this procedure was not permitted within the terms of our contract. The direct O mean for the soil is 40·8 per cent with a standard deviation for the mean ($\sigma/\sqrt{n}$) of $\pm 0{\cdot}7\%$ O. The mean of three apparent abundances derived from chemical elemental analyses calculated as oxides (MAXWELL *et al.*, 1970; PECK and SMITH, 1970; WIIK and OJANPERÄ, 1970) is $42{\cdot}4 \pm 0{\cdot}1$ per cent. The mean O depletion can be obtained from the difference between the direct and indirect determinations and is $1{\cdot}6 \pm 0{\cdot}7\%$ O. The high summations of the three total silicate analyses cited above indicate an O depletion of $0{\cdot}6 \pm 0{\cdot}0_3\%$. The difference between these two estimates of O depletion is not significant at the 95 per cent confidence level ($p \sim 0{\cdot}3$).

It is interesting to speculate upon the cause of the apparently smaller O depletion observed in the crystalline rocks when compared to the breccias and particularly the fines. The work of MAXWELL *et al.* (1970) indicates that the O depletion in the crystalline rocks is minimal or possibly even non-existent. The high abundance of solar wind rare gases in the fines and breccias (LSPET, 1969) suggest the possibility that some reduction could take place by solar wind H. Assuming a solar wind origin for the Xe in fines and breccias (LSPET, 1969), it is possible to calculate from the relative solar abundances of H and Xe that more than sufficient solar wind H has impinged upon the soil and breccias to combine with all the O in these materials. On the other hand, a similar calculation shows that solar wind H could *at best* only reduce about 1 or 2 per cent of the O in the crystalline rocks.

(4) *Oxygen in the moon, earth and meteorites*

In order to compare the O abundances found in the lunar rocks and fines with terrestrial and meteoritic abundances, in Fig. 1 we have plotted O against Si abundance (EUGSTER, 1969). Meteorite abundances are taken from VOGT and EHMANN (1965a, b). Because of the problem of terrestrial oxidation, only chondrite analyses for *falls* have been used. In terrestrial rocks (EUGSTER, 1969), Si and O are correlated and can be expressed as:

$$\%\,O = 0{\cdot}415\%\,Si + 35{\cdot}0. \qquad (1)$$

Terrestrial abundances in Fig. 1 are represented by a line based on equation 1. A regression line calculated for lunar rocks can be written as:

$$\%\,O = 1{\cdot}15\%\,Si + 18{\cdot}2. \qquad (2)$$

The lunar fines do not follow this regression, and show a variation of O abundance for almost constant Si. The ordinary chondrites lie on the lunar rock regression, although the O and Si abundances within the H or L group chondrites are not correlated. When the values for 13 H and L group chondrites are introduced into the lunar rock regression they cause no significant change in slope (1·11) or intercept

(18·6). The low O/Si ratio in the ordinary chondrites is due to the presence of FeS and Fe–Ni but in the lunar rocks the presence of significant amounts of ilmenite (31·6% O) and of Ti(III) is largely responsible.

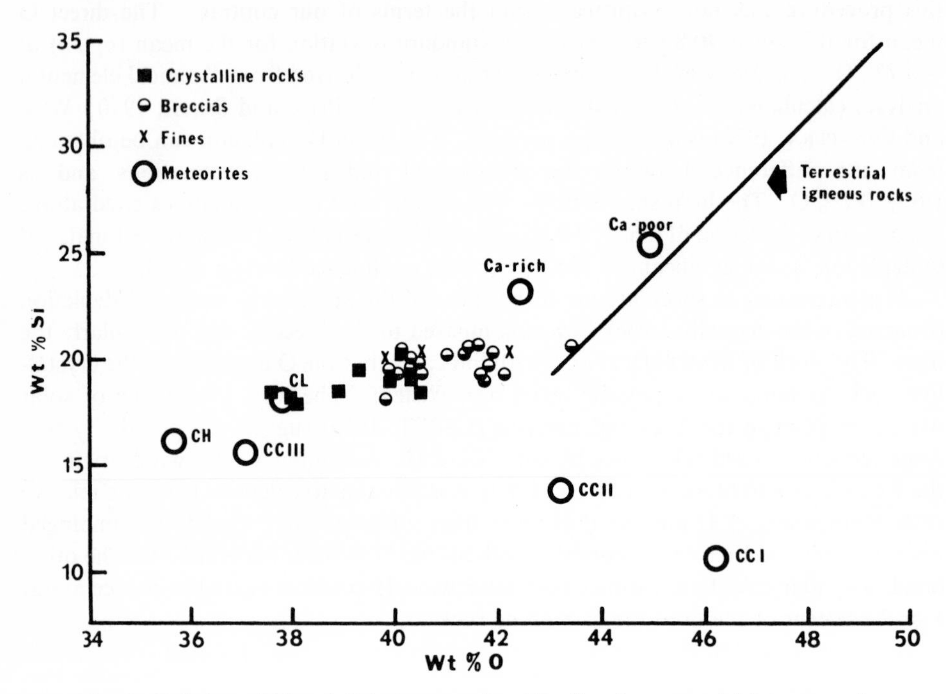

Fig. 1. Relation of Si and O in lunar rocks. The line for terrestrial igneous rocks is based on that derived by EUGSTER (1969). Large circles represent stony meteorite averages, where Ca-rich and Ca-poor refer to the appropriate achondrites, CCI, II and III represent carbonaceous chondrite types and CH and CL are H and L group chondrites respectively.

Acknowledgments—Miss CAROL ANDERSON and Mrs. MARGARET DOTSON assisted with data reduction and manuscript preparation. Financial support was given by the University of Kentucky Research Foundation and by NASA contract NAS 9-8017.

REFERENCES

AGRELL S. O., SCOON J. H., MUIR I. D., LONG J. V. P., MCCONNELL J. D. C. and PECKETT A. (1970) Mineralogy and petrology of some lunar samples. *Science* **167**, 583–586.

ANNEL C. and HELZ A. (1970) Emission spectrographic determination of trace elements in lunar samples. *Science* **167**, 521–523.

CHAO E. C. T. (1963) The petrographic and chemical characteristics of tektites. In *Tektites* (editor J. A. O'Keefe), Chap. 3, p. 79. University of Chicago Press.

COMPSTON W., ARRIENS P. A., VERNON M. J. and CHAPPELL B. W. (1970) Rubidium–strontium chronology and chemistry of lunar material. *Science* **167**, 474–476.

DUKE M. B., CHING CHANG WOO, BIRD M. L., SELLERS G. A. and FINKELMAN R. B. (1970) Lunar soil; size distribution and mineralogical constituents. *Science* **167**, 648–650.

EHMANN W. D. and MCKOWN D. M. (1968) Heat-sealed polyethylene sample containers for neutron activation analysis. *Anal. Chem.* **40**, 1758.

ENGEL A. E. J. and ENGEL C. G. (1970) Lunar rock compositions and some interpretations. *Science* **167,** 527–528.

EUGSTER H. P. (1969) Oxygen, abundance in common igneous rocks. In *Handbook of Geochemistry,* (editor K. H. Wedepohl), Vol. II, part I, Chap. 8, p. E-1. Springer–Verlag.

FLANAGAN F. J. (1969) U.S. Geological Survey Standards—II. First compilation of data for the new U.S.G.S. rocks. *Geochim. Cosmochim. Acta* **33,** 81–120.

GAULT D. E., ADAMS J. B., COLLINS R. J., GREEN J., KUIPER G. P., MAZURSKY H., O'KEEFE J. A., PHINNEY R. A. and SHOEMAKER E. M. (1967) Surveyor V: Discussion of chemical analysis. *Science* **158,** 641–642.

GOLES G. G., RANDLE K., OSAWA M., SCHMITT R. A., WAKITA H., EHMANN W. D. and MORGAN J. W. (1970) Elemental abundances by instrumental activation analysis in chips from 27 lunar rocks. *Geochim. Cosmochim. Acta,* Supplement I.

LANGMYHR F. J. (1970) The reporting of analytical results for reference minerals and rocks. *Geochim. Cosmochim. Acta* **33,** 1561–1562.

LSPET (LUNAR SAMPLE PRELIMINARY EXAMINATION TEAM) (1969) Preliminary examination of lunar samples from Apollo 11. *Science* **165,** 1211–1227.

MASON B. (1962) *Meteorites,* Chap. 10, p. 158. John Wiley.

MAXWELL J. A., ABBEY S. and CHAMP W. H. (1970) Chemical composition of lunar material. *Science* **167,** 530–531.

MORGAN J. W. and EHMANN W. D. (1970) Precise determination of oxygen and silicon in chondritic meteorites by 14 MeV neutron activation using a single transfer system. *Anal. Chim. Acta* **49,** 287–299.

MORRISON G. H., GERARD J. T., KASHUBA A. T., GANGADHARAM E. V., ROTHENBERG A. M., POTTER N. M. and MILLER G. B. (1970) Multielement analysis of lunar soil and rocks. *Science* **167,** 505–507.

PECK L. C. and SMITH V. C. (1970) Quantitative chemical analysis of lunar samples. *Science* **167,** 532.

ROSE H. J., JR., CUTTITTA F., DWORNIK E. J., CARRON M. K., CHRISTIAN R. P., LINDSAY J. R., LIGON D. T. and LARSON R. R. (1970) Semimicro chemical and X-ray fluorescence analysis of lunar samples. *Science* **167,** 520–521.

SMALES A. A., MAPPER D., WEBB M. S. W., WEBSTER R. K. and WILSON J. D. (1970) Elemental composition of lunar surface material. *Science* **167,** 509–512.

TUREKIAN K. K. and KHARKAR D. P., Neutron activation analysis of milligram quantities of lunar rocks and soil. *Science* **167,** 507–509.

VOGT J. R. and EHMANN W. D. (1965a) Silicon abundances in stony meteorites by fast neutron activation analysis. *Geochim. Cosmochim. Acta* **29,** 373–383.

VOGT J. R. and EHMANN W. D. (1965b) An automated procedure for the determination of oxygen using fast neutron activation analysis; Oxygen in stony meteorites. *Radiochim. Acta* **4,** 24–28.

VOGT J. R., EHMANN W. D. and MCELLISTREM M. T. (1965) An automated system for rapid and precise fast neutron activation analysis. *J. Appl. Radiat. Isotop.* **16,** 573–580.

WÄNKE H., BEGEMANN F., VILCSEK E., RIEDER R., TESCHKE F., BORN W., QUIJANO-RICO M., VOSHAGE H. and WLOTZKA F. (1970) Major and trace elements and cosmic-ray produced radioisotopes in lunar samples. *Science* **167,** 523–525.

WIIK H. B. and OJANPERÄ P. (1970) Chemical analysis of lunar samples 10017, 10072 and 10084. *Science* **167,** 531–532.

Proceedings of the Apollo 11 Lunar Science Conference, Vol. 2, pp. 1081 to 1084.

Lunar rock compositions and some interpretations*

A. E. J. ENGEL
Scripps Institution of Oceanography, University of California, San Diego, California 92037

and

CELESTE G. ENGEL
U.S. Geological Survey, Scripps Institution of Oceanography, San Diego, California 92037

(*Received* 12 *February* 1970; *accepted* 12 *February* 1970)

Abstract—Samples of igneous "gabbro," "basalt," and lunar regolith have compositions fundamentally different from all meteorites and terrestrial basalts. The lunar rocks are anhydrous and without ferric iron. Amounts of titanium as high as 7 wt. % suggest either extreme fractionation of lunar rocks or an unexpected solar abundance of titanium. The differences in compositions of the known, more "primitive" rocks in the planetary system indicate the complexities inherent in defining the solar abundances of elements and the initial compositions of the earth and moon.

THIS report discusses some of the results, limitations, and interpretations of studies of the textures, mineralogy, and chemical composition of lunar samples "gabbro" 10044, gabbroic "basalt" 10057, and the dust (regolith) 10084-28 (LSPET, 1969). The chemical data (Table 1) were obtained by gravimetric, photometric, and flame photometric techniques. Amounts of alumina were adjusted on the basis of X-ray fluorescent analyses for Cr and Mn in the R_2O_3 group (Table 1).

Analyses of Apollo 11 samples pose problems as difficult as interpretations of the data. The difficulties in our analytical studies resulted from limitations in size of the samples (4 g), time, and the large amounts of titanium.

Interpretations must be made in ignorance of the interrelations of the lunar rock complexes from which the samples came. Our interpretations are based largely upon, and biased by, our understanding of properties and origins of the more "primitive" terrestrial basalts and gabbros, as well as meteorites, especially the "basaltic" achondritic meteorites (ENGEL *et al.*, 1965).

The three lunar samples consist largely of the minerals pyroxene, ilmenite, and plagioclase. The pyroxene is a pale pink to beige titaniferous augite, and the plagioclase is bytownite. These same minerals with somewhat analogous diabasic textures are common in terrestrial mafic igneous rocks. One major difference is in the proportions of the minerals. The pyroxene is most abundant. It surrounds and interlocks with the plagioclase in diabasic (ophitic) texture. In terrestrial basalts and gabbros, plagioclase commonly equals or exceeds pyroxene in abundance and forms a crystal mesh that envelops the pyroxene. The second striking difference is the abundance of ilmenite in the lunar samples, a reflection of the large concentrations of titanium and iron. Actually TiO_2 is three to five times more abundant in the lunar samples than in the most titaniferous achondritic meteorite, Angra dos Reis (TiO_2, 2·39).

* This article was reprinted from *Science* (Vol. 167, pp. 527–528 (1970)) without further refereeing or revision by the author.

Table 1. Chemical compositions (wt.%) of Apollo 11 regolith and two rocks. Analyst, C. G. Engel. Al_2O_3 values corrected after X-ray fluorescence analyses for Mn and Cr by J. S. Wahlberg (U.S. Geological Survey, Denver, Colorado)

	10044 Gabbro	10057 Vesicular diabase (basalt)	10084-28 Regolith (dust)
SiO_2	42·01	39·79	41·50
TiO_2	8·81	11·44	7·50
Al_2O_3	11·67	10·84	14·31
Fe_2O_3	0·00*	0·00 ?	0·06 ?
FeO	17·98	19·35	15·62
MnO	0·24	0·20	0·22
MgO	6·25	7·65	7·95
CaO	12·18	10·08	11·84
Na_2O	0·48	0·54	0·48
K_2O	0·11	0·32	0·16
H_2O^+	0·00	0·00	0·00
H_2O^-	0·00	0·01	0·01
P_2O_5	0·08	0·17	0·10
Total	99·81	100·39	99·75

* Actual value, —0·08.

The lunar rocks also are anhydrous and without Fe^{+3} (Table 1). The vesicular (scoriaceous) texture of lunar "basalt" 10057 indicates that this rock originally contained volatiles, but if water or hydroxyl were components of the volatiles, they were expelled. This vesiculation and degassing of the rock would be possible only if the basaltic melt cooled at or very near a dry lunar surface, presumably in a near vacuum.

All terrestrial basalts—and many meteorites—contain some ferric iron and combined water and have formed under higher fluid pressures. In terrestrial basalts both ferric iron and combined water commonly approach or exceed 1 weight per cent (Table 2). The most titaniferous terrestrial basalts common to oceanic and many continental volcanoes are also most enriched in K, Na, ferric iron, and combined water

Table 2. Compositions (wt.%) of lunar gabbro, average basaltic achondrite, and common terrestrial basalts. The H_2O^+ content in terrestrial oceanic tholeiitic basalt and alkali-olivine basalt due in part to secondary alteration

	Apollo 11 gabbro 10044	Basaltic achondrite	Oceanic tholeiitic basalt	Alkali-olivine basalt
SiO_2	42·01	48·51	50·01	48·01
TiO_2	8·81	0·48	1·37	2·92
Al_2O_3	11·67	13·04	16·18	15·97
Fe_2O_3	0·00	1·11 ?	2·32	3·87
FeO	17·98	15·90	7·07	7·56
MgO	6·25	7·87	7 71	5·26
CaO	12·18	11·00	11·33	9·04
Na_2O	0·48	0·50 ?	2·79	3·73
K_2O	0·11	0·08 ?	0·22	1·89
H_2O^+	0·00	0·07	0·87	1·33
P_2O_5	0·08	0·19	0·13	0·42

(Table 2 and Fig. 1). These are the alkali-olivine basalts that many petrologists infer are derived from depths of 50–500 km in the earth's mantle. A minority contain from 4 to 7 per cent TiO_2, but invariably other major elements such as Si, Al, and P also exceed lunar concentrations (Table 1). Other compositional differences are even more striking. Many alkali-olivine basalts have appreciable ferric iron, combined water or hydroxyl, high Ba, U, Th, Pb, Zr, Th/U, low K/Rb and, relative to meteorites, more fractionated abundance distribution patterns of rare earths (ENGEL *et al.*, 1965).

The far-more-abundant tholeiitic basalts of the ocean crust, ridges and rises actually appear to be the most "primitive" terrestrial basalts. They show many interesting similarities in composition with basaltic achondrite meteorites and have

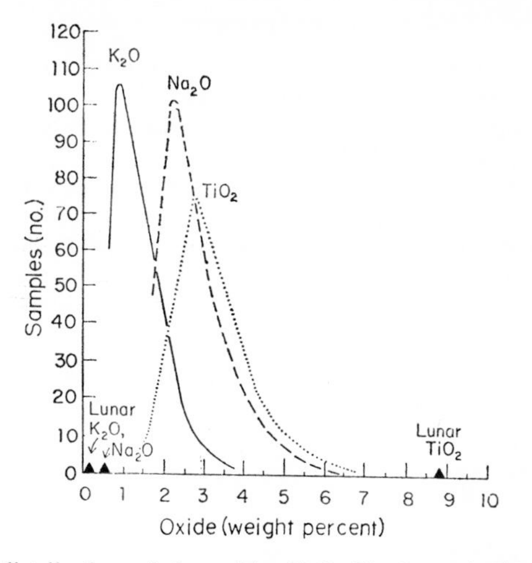

Fig. 1. Frequency distribution of the oxides K_2O, Na_2O, and TiO_2 in alkali-olivine basalts of the oceans. Values for the lunar "gabbro" are plotted as closed triangles on the horizontal coordinate.

abundance distribution patterns of rare earths like those of many meteorites (ENGEL *et al.*, 1965; GAST, 1967; HASKIN *et al.*, 1966). Compared to the lunar samples (Table 2) the oceanic tholeiitic basalts contain approximately the same amounts of Mg, Ca, K, P, U, Th, Y and Yb, but far less Fe, Ti, Ba and Zr, and higher concentrations of Si, Al and Na. Curiously, the one major element ratio common to the most "primitive" terrestrial basalts and meteorites is Al/Ca (0·9–1). But in the lunar "basalt" and "gabbro" this ratio is 0·8 and 0·7, respectively. In many ultramafic lavas extruded from the earth's mantle Al/Ca is as low as 0·6.

Variations in ratios such as Al/Ca, Si/Mg and Na/K and in amounts of Ti and P seem to emphasize the probability that a variety of igneous processes have fractionated all known planetary rocks. The three major igneous processes are (i) partial melting, (ii) fractional crystallization, and (iii) distillation and volatile transfer (ENGEL *et al.*, 1965). The effects of widely varying pressures, temperatures, and gravity also are obvious.

Hence, if the earth and moon (and meteorites?) have a common, closely related source and origin, several generalizations follow: (i) available earth rocks are too highly differentiated (fractionated) from their initial source rocks to offer major insights into initial earth, or earth-moon, compositions and processes; (ii) the moon is a rock complex with quite divergent rock types; (iii) lunar and meteoritic matter are products of intermediate and divergent environments and processes in planetary evolution; and (iv) solar and cosmic abundances of elements are more uncertain than ever. Perhaps the most provoking questions are whether current studies of the moon will provide important insights into early earth history, the origin of meteorites, and tektites.

Acknowledgments—This work was supported by NASA grant NAS 9-7894.

REFERENCES

ENGEL A. E. J., ENGEL C. G. and HAVENS R. G. (1965) Chemical characteristics of oceanic basalts and the upper mantle. *Bull. Geol. Soc. Amer.* **76,** 719–734.

GAST P. W. (1967). Isotope geochemistry of volcanic rocks. In *Basalts* (editor H. H. Hess and A. Poldervaart), Vol. 1, pp. 325–358. Interscience.

HASKIN L. A., FREY F. A. and SCHMITT R. H. (1966). Meteoritic, solar, and terrestrial rare-earth distributions. *Phys. Chem. Earth* **7,** 167–321.

LSPET (LUNAR SAMPLE PRELIMINARY EXAMINATION TEAM) (1969) Preliminary examination of lunar samples from Apollo 11. *Science* **165,** 1211–1227.

Proceedings of the Apollo 11 Lunar Science Conference, Vol. 2, pp. 1085 to 1096.

The concentration and isotopic composition of hydrogen, carbon and silicon in Apollo 11 lunar rocks and minerals

SAMUEL EPSTEIN and HUGH P. TAYLOR, JR.
Division of Geological Sciences, California Institute of Technology, Pasadena, California 91109

(*Received* 16 *February* 1970; *accepted in revised form* 25 *February* 1970)

Abstract—Total hydrogen (as H_2 gas and H_2O) extracted from lunar soil is about 34 μmoles/g and that extracted from lunar breccia about 58 μmoles/g. The lunar H_2 gas is extremely depleted in deuterium relative to mean ocean water; 20 ppm as against 157 ppm. Even this low concentration of deuterium in lunar hydrogen gas is a maximum value and can be due partially or wholly to contamination by terrestrial water. Material balance calculation indicates a still lower deuterium concentration of 10 ppm in the lunar hydrogen; this hydrogen is therefore of solar wind origin.

The H_2O extracted from the lunar samples forms 1/3–1/5 of the total hydrogen, and its deuterium concentration is much higher than in the lunar H_2 gas. H_2O evolved at low temperatures is higher in D/H than the H_2O extractable at higher temperatures. The D/H ratio of both the lunar H_2 gas and the associated H_2O extracted from the lunar samples can be accounted for by the cross-contamination between essentially deuterium-free (less than 10 ppm) solar wind hydrogen and terrestrial water.

The concentrations of carbon in the lunar soil and breccia are 143 and 262 ppm and have δ values of $+18{\cdot}6$ and $+10{\cdot}8$ (relative to PDB standard), respectively. The carbon is thus enriched in C^{13} relative to terrestrial carbon and could have acquired its heterogeneous distribution and isotopic composition by a variety of processes including solar wind bombardment and/or meteoritic impact.

The Si^{30}/Si^{28} ratio of lunar and terrestrial igneous rocks are very similar, with a total variation in the lunar rocks of less than one per mil. The lunar mineral assemblages show a consistent pattern of relative Si^{30} enrichment in the following order: clinopyroxene–plagioclase–cristobalite. The Si^{30} is distinctly enriched in the lunar soil and breccia, relative to the ratio in lunar igneous rocks.

INTRODUCTION

THE RELATIVE abundance of the stable isotopes of the elements hydrogen, carbon, oxygen and silicon were measured for the lunar samples returned from the Sea of Tranquillity by the Apollo 11 expedition. A preliminary report of these results was published elsewhere (EPSTEIN and TAYLOR, 1970). A more detailed report on the oxygen isotope results is presented separately in this issue (TAYLOR and EPSTEIN, 1970). The present paper gives a more extensive discussion of the experimental procedure and of the results dealing with determinations of the concentration of hydrogen and carbon and of the isotopic compositions of hydrogen, carbon and silicon.

EXPERIMENTAL PROCEDURE FOR EXTRACTION OF HYDROGEN AND CARBON

A sketch of the extraction apparatus is shown in Fig. 1. Basically it is a high vacuum glass line that permits the heating of the lunar sample in vacuum by induction with an R. F. generator, and permits the collection and analysis of the various gases emitted by the lunar sample. Where necessary some of the evolved gases can be passed over hot CuO for purposes of converting CO and H_2 to H_2O and CO_2, the CO_2 being the gas used in the mass spectrometer for C^{13}/C^{12} determinations. Passage

* Contribution No. 1719, Division of Geological Sciences, California Institute of Technology, Pasadena, California 91109.

through a U-tube containing hot uranium (700°C) permits the reduction of H_2O to H_2, H_2 being the gas used in the mass spectrometer for D/H determination. In certain cases, the hot uranium was also used for purification of the rare gases.

The quantities of evolved hydrogen gas were in the micromole range and were expected to be of solar wind origin. Since it was expected that the deuterium concentration in the solar wind would be very low or nonexistent (SEGRÈ, 1964), it was particularly important to avoid contamination of the lunar hydrogen with terrestrial hydrogen or water, which have relatively high deuterium concentrations ($\approx$140–160 ppm deuterium). Similarly, it was important to separate the helium from the hydrogen because He^3 would be recorded as HD in the hydrogen mass spectrometer, as this instrument actually measures the HD/H_2 ratio.

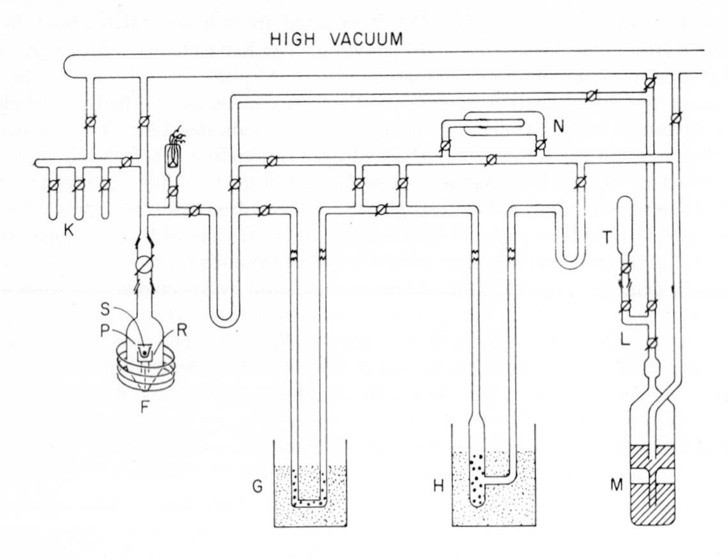

Fig. 1. Sketch of apparatus for extraction of hydrogen and carbon. F—pyrex glass envelope assemblage, containing quartz holder (R), platinum crucible (P) and lunar sample (S); K—glass fingers to store CO_2 and H_2O; G—furnace surrounding U-tube containing CuO; H—furnace surrounding U-tube containing uranium; N—palladium thimble in glass enclosure; M—Toepler pumps; L—calibrated volume; T—sample tube.

With some of these problems in mind, our lunar soil sample (10084-131) was placed in a P_2O_5-dried, nitrogen-filled dry box upon receipt from the Lunar Receiving Laboratory. Also prior to loading of the sample into the platinum crucible, the glass envelope and platinum crucible were preheated in vacuum for several hours. The crucible was heated to over 1400°C and the glass envelope was gently torched to insure that the inner surface of the envelope was thoroughly degassed and subjected to a higher temperature than that reached at any time during the actual extraction of the lunar gas sample. Also the preheating thoroughly degassed the platinum crucible which was found to give as much as 10 μmoles of H_2O even after the crucible was dried at 110°C overnight in the laboratory atmosphere. After the degassing procedures, the reaction vessel was sealed, removed from the vacuum line, and placed in the dry box. By opening the reaction vessel and loading the crucible with lunar sample in the dry-box, the degassed Pt and glass surface were exposed only to a dry nitrogen atmosphere.

Prior to loading of the sample the blanks were run to determine the approximate amounts of CO_2 and H_2 that were evolved from the glass walls and platinum. In all cases, even when the line was heated during the blank runs to temperatures above those expected during the extraction procedure, less than 0·5 μmoles and in most cases less than 0·2 μmoles of H_2 and CO_2 were collected. Exceptions to these results were observed when a glow discharge was allowed to occur in the glass envelope or when a 50 μ or more oxygen pressure was permitted to be in contact with the very hot platinum crucible. Under these conditions 10 or more μmoles of both CO_2 and H_2O were formed in a molar ratio of one, indicating that the stopcock grease is being decomposed or oxidized.

As mentioned above one of our objectives was to eliminate deuterium as a contaminant from the extraction line. Since terrestrial water around Pasadena or Houston contains about 150 ppm deuterium and since water on the surface of the glass lines was the most likely contaminant, the whole line, except for the envelope containing the lunar sample, was flushed with essentially deuterium-free hydrogen (<5 ppm deuterium). This was accomplished by introducing the deuterium-free hydrogen into the line and processing it as if it were a sample extracted from the lunar sample. The logic behind this step was that a deuterium-free water contamination would affect the D/H ratio of the lunar sample to lesser degree than terrestrial water contamination provided that the D/H ratio of the lunar sample was less than one-half the terrestrial ratio.

Once the platinum crucible was loaded with the lunar sample and the system thoroughly evacuated, the temperature of the sample was raised to about 100–150°C and pumped for at least 2 hr. Since the lunar samples were heated to at least these temperatures on the lunar surface during the lunar daytime, there was no danger that pumping at these temperatures would remove any meaningful gas sample. Actually, no appreciable degassing of the lunar sample could be detected until its temperature was raised to about 300°C. After the 100–150°C degassing, the extraction line was isolated from the vacuum and the temperature of the lunar sample raised slowly to about 500°C. The emitted gas was passed through a liquid nitrogen cooled trap while the non-condensable gas was pumped into a calibrated volume for measurement. The condensable gas of interest was primarily H_2O and CO_2 while the non-condensables of interest were hydrogen, the rare gases, CO and possibly N_2. The gases collected during the 150–500°C range are referred to as fraction A in Table 1. Actually fraction A contained very little non-condensable gas and consisted primarily of H_2O and some CO_2. The H_2O and CO_2 were not, at this time, passed through the rest of the extraction system but stored in one of the glass fingers attached to the line. The temperature of the platinum crucible was now slowly raised to about 550°C. The gases collected during the 500–550° temperature interval are referred to as fraction B in Table 1. The condensable water and carbon dioxide of fraction B were stored in a second glass finger. The non-condensable gases of fraction B were cycled over hot CuO (750°C) and converted to CO_2 and H_2O, and then the rare gases were purified by passing over hot uranium. The CO_2 and H_2O were separated and the H_2O converted to H_2 for the mass spectrometer analyses.

Fraction C of the gas sample was extracted when the temperature of the lunar sample was raised from about 550°–600°C. The gases in fraction C were similar in composition to fraction B and treated similarly. Fraction D gas sample was extracted when the lunar sample was heated between 600°C and 1400°C. This gas was also treated similarly to fraction B but fraction D was almost entirely composed of carbon monoxide resulting either from trapped CO in the lunar sample or from reaction between carbon and FeO, as was shown to occur by Clayton and Epstein (1958). By far the largest fraction of the total lunar carbon was found in fraction D.

After the non-condensable gases were separated and measured volumetrically and isotopically, the δ values and concentrations of the CO_2 and H_2O stored in the glass fingers were determined by passing the mixture through CuO and then converting the water to hydrogen. As is noted in Table 1, it was not always possible to measure each gas fraction because of insufficient yield for reliable isotope analyses. In such cases gases from several fractions were combined for a volume and isotope determination.

Since our results show that the hydrogen gas extracted from the lunar sample was unusually depleted in deuterium it was particularly desirable to ascertain that the small concentration of deuterium in the lunar hydrogen was not due to contamination from terrestrial water or hydrocarbons degassing from the glass walls of the line. Such contamination was thought to be particularly likely in the CuO and U furnaces. It was possible to prevent this contamination of the lunar hydrogen by bypassing these furnaces, and by separating the hydrogen directly from the other gases by permitting the hydrogen gas alone to pass through a hot palladium thimble (at 650°C), which is not permeable to any of the other sample gases. This hydrogen could then be measured directly in the mass spectrometer. As our results in Table 1 show, there is very little difference in the δD/H values for the hydrogen extracted by the two different procedures, thus indicating that both methods of extraction are valid.

Table 1. Isotopic and concentration data for hydrogen, carbon, and rare gases extracted from samples 10084, 10087 and 10061 from the Sea of Tranquillity

Sample fractions	Hydrogen gas: Conc. (μ mole/g)	Hydrogen gas: Deuterium conc. δ D/H	Hydrogen gas: Deuterium conc. ppm	Carbon: Conc. (ppm)	Carbon: $\delta^{13}C/^{12}C$	Water: Conc. (μ mole/g)	Water: Deuterium conc. δ D/H	Water: Deuterium conc. ppm	Rare gases: Conc. (cm^3/g)	Rare gases: H_2 gas/rare gases (vol. ratio)
Sample 10084, No. 2 (2·40 g)										
A	None			Trace		6·05	−150	133	None	
B	8·56	−888	18	13·4 (B + C)	+17·8 (B + C)	4·09	−365	100	0·100	4·6
C	12·93	−856	23			0·93 (C + D)	−483 (C + D)	81 (C + D)	0·275	2·5
D	1·45	−615	60	141·7	+16·1				0·140	0·56
Total	22·94	−853	23	155·1	+16·3	11·07	−261	116	0·515	2·4
Sample 10084, No. 1 (1·85 g)										
A	None			85 (A–C)	+18·6 (A–C)					
B	10·30	−849	24			11·58 (B–D)	−275 (B–D)	114 (B–D)		
C	11·34	−828	27							
D	0·72	−609	61	58						
Total	22·36	−830	27	143		11·58	−275	114		
Sample 10061 (0·55 g), vacuum-sealed breccia										
A	None			74 (A–C)	+10·1 (A–C)					
B	17·86	−882	18			8·9 (B–D)	−412 (B–D)	92 (B–D)	0·044	5·0
C	32·06	−868	21						0·190	2·1
D	Trace			188	+10·9				Trace	
Total	49·92	−873	20	262	+10·8	8·9	−412	92	0·234	2·6
Sample 10084*†, No. 3 (1·71 g)										
A				None		7·83 (A + B)	−329 (A + B)	105 (A + B)		
B	9·40	−850	24	Trace						
Sample 10087† (~2 g) vacuum-sealed										
B	4·47	−850	24							
C‡	20·87	−843	25							
Total	25·34	−844	25							

Note: δ values for hydrogen relative to standard mean ocean water (SMOW) in pre mil. Deuterium concentration of SMOW assumed to be 157 ppm. δ values for carbon are relative to PDB standard in per mil. Fraction A gases evolved below 500°C, almost wholly water. Fraction B gases evolved between 500°C and 550°C, a mixture of CO_2, CO, H_2, H_2O and rare gases. Fraction C gases evolved between 550°C and 650°C, a mixture of CO_2, CO, H_2, H_2O and rare gases. Fraction D gases evolved between 650°C and 1500°C, dominantly CO.

* Hydrogen purified through palladium thimble at ~650°C.

† Lunar dust sample heated in quartz tube to release hydrogen (to avoid contact of released H_2 and H_2O with hot platinum.

‡ Lunar dust exposed to water of $\delta \approx -980$ per mil prior to extraction of hydrogen fraction C.

DISCUSSION OF HYDROGEN ISOTOPE DATA

The isotope data for lunar soil (10084) sample 2 are the most complete. Fraction A contained no H_2 nor any appreciable CO_2; it was primarily H_2O which has a δ value of $-150‰$. This value is probably not unlike the δ-value for water vapor found in Pasadena or in Houston. The lunar dust has been subjected to intense bombardment of high-energy particles for millions of years. It would not be surprising if the surfaces of the fine dust particles were chemically highly activated and could adsorb and retain atmospheric moisture very strongly, so that it could not be pumped away until temperatures of over 300°C were reached. Also it is unlikely that lunar water would be isotopically so similar to terrestrial water. It therefore seems almost certain that the water in fraction A is water vapor from Houston or Pasadena, H_2O contamination from the rockets of the lunar module, or water vapor emitted by the astronauts while on the moon.

The δ-values of the water in fraction B are lower than in fraction A. It is difficult to imagine how such a decrease in δD/H could be accomplished by isotope fractionation during the extraction procedure. On the other hand it is unlikely to be so readily extractable if it were "water" intrinsic to the lunar rocks. The most reasonable explanation is that the water in fraction B is also primarily terrestrial in origin but it has become depleted in deuterium either by exchange with, or by oxidation of, some lunar hydrogen gas, the oxidation taking place by the reaction with FeO of the lunar rocks. This interpretation becomes more attractive when one considers the water in fractions C and D and the change of the δ values of the hydrogen gas with progressive degassing of the lunar sample. There is a complementary rise in the δD/H of hydrogen gas suggesting that during the heating isotopic exchange takes place between the terrestrial water and the lunar hydrogen; this possibly occurs with the platinum acting as a catalyst, or perhaps during reaction of Fe and FeO with water and hydrogen respectively. Our results strongly suggest that the solar wind hydrogen has a δD/H less than $-888‰$ and that the water extracted from the lunar samples is largely terrestrial in origin. Considering the possible different sources of isotope fractionations during the escape of hydrogen from the moon and earth, it would be most fortuitous for the water on these two bodies to be isotopically alike. It is very important to prove the presence or absence of primary lunar water in our samples. Very strong evidence for its absence would be the presence of truly deuterium-free hydrogen gas, which conceivably could be obtained if more precautions were taken against sample contamination.

Sample 1 of lunar soil (10084) essentially confirms the results for sample 2. The H_2/H_2O ratio is very similar. The δ values of the hydrogen and water for the two aliquots are very much alike. The δ value of the H_2 is slightly lower in sample 1 but the δ value of the H_2O is slightly higher. It is therefore possible that a little more isotopic exchange between the H_2 and H_2O took place in sample 1. The results for the breccia sample (10061), vacuum packed at the Lunar Receiving Laboratory and opened and loaded in the dry box in our laboratory, show that the H_2/H_2O ratio and the concentration of hydrogen is more than twice as large for the breccia as for the dust. This must be due to an increase in the solar wind component because the hydrogen to rare gas ratio is similar to the ratio for the lunar soil. Since the rare gases are primarily due to solar wind, the accompanying increase in the H_2 concentration must also be due to solar wind. This result is also good evidence that the H_2 we measured is primarily solar wind hydrogen and not reduced H_2O. Also the fact that the δD/H for the breccia hydrogen is lower than for the lunar soil hydrogen is consistent with our own interpretation because the larger the H_2/H_2O ratio the less important should be the contribution from exchange with the H_2O. Conversely, the lower δ value for the associated water in the breccia also shows the expected larger contribution of a low-deuterium component when isotopic exchange occurs with a larger fraction of hydrogen gas. The dependence of the D/H ratio of the hydrogen and water on their relative proportions is certainly suggestive of cross contamination. It is also interesting that in spite of the H_2/H_2O ratio of the breccia being more than double the ratio in the lunar soil, the δ of the H_2 is no lower than -873. This suggests that perhaps the solar wind hydrogen is not entirely free of deuterium because one cannot choose a contaminating H_2O of a single isotopic composition that can eliminate D completely.

The experiments for sample 3 lunar soil and for the vacuum-sealed lunar dust (10087) were special experiments to test the validity of the first three experiments. Fraction B in sample 3 was extracted using a quartz tube as the sample container, and heating with an electrical-resistance furnace. This was done to ascertain whether the platinum crucible facilitated the exchange between the H_2O and H_2. Also this is one of the experiments in which the hydrogen was separated from the other gases by means of the hot palladium thimble. The isotope results are similar to those for sample 2 or sample 1 showing that the changes in procedure did not affect the results. The other extraction on lunar soil sample 10087 shows that the vacuum packed lunar samples gave hydrogen isotopically similar to the previously described samples. Furthermore, after fraction B was extracted the lunar sample was exposed to deuterium-free H_2O vapor to "exchange" away contaminating deuterium-rich water. This effort failed to affect markedly the δ-values of the hydrogen gas. We intend to repeat this experiment, exposing some lunar soil or breccia before any heating and before extraction of gases is done, because it is possible that the preliminary heating deactivated the surfaces of the lunar dust particles.

ESTIMATION OF D/H RATIO OF LUNAR HYDROGEN

By utilizing the relative amounts of total H_2O and total H_2 given for each experiment in Table 1, and by (1) making an assumption regarding the isotopic composition of the lunar hydrogen, it is possible to calculate from material balance considerations the D/H ratio of the terrestrial hydrogen contaminant, or conversely by (2) assuming the D/H ratio of the contaminating hydrogen or water it is possible to estimate the D/H ratio of the lunar hydrogen. This calculation was done for the samples 1 and 2 of the lunar soil and for the vacuum-sealed breccia (10061). Sample 3 of the lunar soil and sample 10087 were not included in our present calculation because in these cases complete extraction of the H_2 and H_2O was not made. Table 2 shows the results of this calculation.

Table 2. Material balance calculation: Partition of D/H ratio between hydrogen gas and water extracted from the lunar samples

Sample	δD/H value of adsorbed H_2O necessary to give deuterium-free hydrogen by material-balance	Calculated δD/H of lunar hydrogen, assuming that δD/H of the adsorbed H_2O is equal to: -150‰	-75‰
10084-2 (2·40 g)	+45	−904 (15 ppm)	−943 (9 ppm)
10084-1 (1·85 g)	+57	−891 (17 ppm)	−930 (11 ppm)
10061 (0·55 g)	+300	−917 (13 ppm)	−936 (10 ppm)

If it is assumed that prior to terrestrial contamination the lunar hydrogen was free of deuterium the contaminant would have δD/H values of +45, +57 and +300 respectively in samples 2 and 1 of the lunar soil (10087) and for the breccia. If it is assumed that the contaminating water had a δD value of −150‰ the original lunar hydrogen would have had δ-values of −904, −891, −917 respectively. If the contaminating H_2O has a δ-value of −75‰ (a value reasonable for Houston water vapor) the lunar hydrogen would have δ-values of −943, −930 and −936‰ respectively. If the first condition prevailed, namely no deuterium in lunar hydrogen, the contaminating water must have been unusually concentrated in deuterium relative to terrestrial H_2O, particularly for the case of the breccia. This is possible only if the contamination took place on the moon by the water from the space suits or from the oxidized fuel where evaporation and deuterium enrichments might have occurred. The third calculation, where it is assumed that the contaminant water has a δ value of −75‰ seems most reasonable. In this calculation one obtains a consistent δ-value for the hydrogen of about -935 ± 5 for all the three samples, suggesting that lunar hydrogen can contain some deuterium (about 10 ppm). The best procedure to determine the actual δ values for lunar hydrogen is to avoid water contamination or if that is impractical, to use water that is essentially free of deuterium in operations where water might come into contact with lunar samples (e.g. in the cooling systems of the astronauts).

Although the possibility that the lunar rocks contain some water has not been entirely eliminated by our experiments there is no doubt that the largest fraction of the hydrogen we obtain is of solar wind origin in that it is as loosely held as solar wind rare gases and comes off with about the same facility. The largest fraction of the extracted water has a δD/H value characteristic of terrestrial water vapor and degasses at a temperature below that at which the solar wind degasses, suggesting that it is probably adsorbed terrestrial water. An insignificant amount of water is extracted at a temperature above 600°C. At present it can be estimated that less than a few micromoles of water per gram of rock exist on the surface of the moon and it is more likely that the surface of the moon is almost devoid of primary water. Any water that may exist can be water due to the oxidation on the moon of solar wind hydrogen.

It is of interest to compare our results with those of other workers who investigated the lunar hydrogen. HINTENBERGER *et al.* (1970) have simply established an upper limit of deuterium concentration as being about 30 per cent of terrestrial abundance. Our results place a much more severe limitation on the deuterium content of lunar hydrogen. Their ratio of H_2/rare gases is given to be about 2 which is somewhat lower than our ratio. Both are probably lower limit values since there still exists the possibility that our rare gases contained some impurity (N_2?).

The results of FRIEDMAN *et al.* (1970) are only in some ways in agreement with ours. For example we obtain a total of 34 μmoles/g of total hydrogen ($H_2 + H_2O$) containing about 48 ppm deuterium for the lunar soil whereas FRIEDMAN *et al.* obtain a higher value of 43 μmoles/g containing 80 ppm deuterium. Although our breccia samples are not the same, their H_2/H_2O ratio is always considerably lower than ours. In spite of the apparent greater amount of H_2O present in their samples our extracted water is isotopically much more similar to terrestrial water. One is led to conclude that some of the lunar H_2 present in the lunar sample was converted to water during the extraction procedure used by FRIEDMAN *et al.* Their criteria of water extractable above 600°C being lunar water is thus invalid since this water could have been oxidized lunar hydrogen. We do not believe in their other criteria that water held by lunar soil at temperatures above 50°C or even above 300°C equates with lunar water. There are numerous terrestrial materials that will absorb water and retain sufficient quantities of it in vacuum at elevated temperatures. They have used the low δD values of water as their most important criteria of its lunar origin. In spite of our higher H_2/H_2O ratio and a lower concentration of water on the whole, the H_2O in our samples is much more enriched in deuterium. We can only conclude that during their extraction oxidation of hydrogen took place to decrease the H_2/H_2O ratio and decrease the δD values of the water they have separated from the hydrogen. Indeed their higher amounts of extractable total hydrogen (and also total deuterium) suggests that their samples suffered a higher degree of contamination with terrestrial water. This is certainly possible since they did not load their samples in a dry nitrogen environment and exposed their apparatus and platinum crucibles to the atmosphere prior to the extraction. We may also note that in other ways the data of FRIEDMAN *et al.* (1970) indicated a more oxidizing environment than we observed, because they obtained mainly CO_2 and SO_2 during extraction, whereas we collected CO with very little CO_2 and we detected a very faint odor of H_2S when our sample vessel was finally opened in air. Our measured δC^{13} values are also much higher than those reported by FRIEDMAN *et al.*

The implications of the presence of even small amounts of water on the moon are so important that it is imperative that valid criteria for its presence on the moon be used. We believe that such criteria are still lacking.

C^{13}/C^{12} RATIOS OF LUNAR CARBON

The carbon data are unusual in that δC^{13} values of both the lunar soil and breccia are appreciably more positive than any common terrestrial carbon. These results are in excellent agreement with those of KAPLAN and SMITH (1970). The agreement shows that our extraction procedure for the carbon is probably valid, namely the simple heating of the lunar samples to over 1400°C. The carbon apparently reacts completely with the ilmenite or other oxides to form CO and CO_2. Our method also allows the extraction of CO trapped in bubbles as observed by BURLINGAME *et al.* (1970). Since the possible contaminants of the lunar carbon (the rocket fuel, commercial greases and oils) have δ values that are very negative (e.g. rocket fuel −35‰, grease $\sim$ −30‰) it might be expected that the more positive the δC^{13} value for the lunar rocks, the more uncontaminated is the extracted lunar carbon.

The similar δ-values of the carbon in fractions B and C of sample 2 of lunar soil indicate that it is unlikely that carbon sources of different composition and different δ values are mixed together in the lunar soil. It would be fortuitous indeed if carbon given off at 600°C would be present as the same carbon compound mixture as that driven off at temperatures between 600° and 1400°C. Similar δ values for carbon from the different fractions suggest a relatively homogeneous source of carbon. There could be several reasons for the unusually high δC^{13} value in lunar soil and for the difference in δC^{13} value between lunar breccia and soil. One of these, as suggested by KAPLAN and SMITH (1970), involves the isotopic fractionation associated with stripping of the carbon and sulfur by the solar wind bombardment of the lunar soil. Solar wind carbon may also constitute an important fraction of the lunar carbon. However, we have no way at present of estimating the C^{13}/C^{12} ratio of the solar wind, so the importance of this process cannot be evaluated.

The different contributions of meteoritic carbon might cause heterogeneity in the concentration and isotopic composition. KEAYS *et al.* (1970) estimate that the lunar soil contains about 2 per cent of carbonaceous chondrite-type material. Types 1 and II carbonaceous chondrites contain about 2 wt % carbon having a δC^{13} of about −4 to −11 (BOATO, 1954). If KEAYS *et al.* (1970) interpretation is correct, the meterorite carbon contribution alone should result in at least a 400 ppm concentration in lunar soil and breccia. The observed carbon concentrations of 150–260 ppm would then indicate that significant loss of carbon has occurred through vaporization of volatile carbon compounds during impact of the meteorites with the lunar surface. Isotopic fractionations might be expected to occur during such a process, and it is reasonable to suppose that C^{12} would be preferentially lost in the volatile phase. This could account for the observed 15–30 per mil enrichment of C^{13} in the lunar dust and breccia relative to that in the carbonaceous chondrites.

VARIATION OF Si^{30}/Si^{28} RATIO IN LUNAR ROCKS

Silicon tetrafluoride is one of the by-products of the extraction of oxygen from silicate rocks by the fluorination method (see TAYLOR and EPSTEIN, 1970). A silicon mass spectrometer has been constructed which can make use of this SiF_4 for the determination of the Si^{30}/Si^{28} variation in terrestrial and lunar silicate minerals and rocks. Thus it was possible to obtain both O^{18}/O^{16} and Si^{30}/Si^{28} analyses for a single sample. The results of these analyses are shown in Table 3 and Fig. 2.

Previous determinations of Si^{30}/Si^{28} in natural materials are those of REYNOLDS and VERHOOGEN (1953) and TILLES (1961). There is as yet insufficient Si^{30}/Si^{28} data for terrestrial silicates to form an adequate basis for the understanding of the overall geochemistry of silicon isotopes. For the common terrestrial minerals and rocks, the variation in the Si^{30}/Si^{28} appears to be about an order of magnitude smaller than the variation of the O^{18}/O^{16} ratio. However, the information that does exist permits some meaningful comparisons between δSi^{30} relationships in lunar and terrestrial rocks and minerals.

Wherever comparisons are available it appears that in a set of coexisting minerals the Si^{30} is enriched in quartz and feldspar relative to pyroxene, biotite, or hornblende. In the lunar igneous samples the cristobalite is richest in Si^{30} (one sample), consistently followed by the plagioclase and then the pyroxene. The δSi^{30} values of the plagioclase and pyroxene from Type A and Type B lunar rocks occupy a very narrow range of about 0·3‰. Within this same range, we also find the δSi^{30} of tektites, of plagioclase in an oceanic basalt and in the Bonsall Tonalite, quartz and muscovite from the Elberton granite, K-feldspar from a carbonatite, and the quartz of a number of pegmatites in Southern California. It seems then that the Si^{30}/Si^{28} in lunar igneous minerals is

Table 3. Si^{30}/Si^{28} ratios of Apollo 11 rocks and minerals

Sample	δSi^{30}(‰)*	Avg. dev.	No. of runs
10068-21, breccia (Type C)			
Whole rock	+0·22		1
Brown glass spherule (4·3 mg)	−0·03		1
10084-130, fines (Type D)			
Whole rock	+0·10	±0·11	3
10085-coarse fines (Type D)			
Brown glass fragment	+0·16		1
10044-31, microgabbro (Type B)			
Cristobalite	+0·34		1
Plagioclase	−0·12	±0·02	4
Clinopyroxene	−0·24	±0·04	3
Whole rock (calc.)	−0·17		
10058-7-1, microgabbro (Type B)			
Plagioclase	0·00	±0·05	4
Clinopyroxene	−0·33		1
Whole rock (calc.)	−0·18		
10003-10, microgabbro (Type B)			
Plagioclase	−0·04		1
Clinopyroxene	−0·13		1
Whole rock (calc.)	−0·09		
10017-32, basalt (Type A)			
Plagioclase	−0·15		1
Clinopyroxene	−0·22		1
Whole rock (calc.)	−0·19		
10057-39, basalt (Type A)			
Plagioclase	−0·01		1
Whole rock	−0·26	±0·06	2

* Relative to a standard quartz from the Rose Quartz Pegmatite, Pala, California.

remarkably similar to a variety of terrestrial igneous rocks and minerals and to tektites. The lunar soils, lunar breccia and some of the achondrite meteorites are, however, significantly different from the Type A and B lunar rocks in their Si^{30}/Si^{28} ratio.

The similarities in the Si^{30}/Si^{28} ratio accentuate what was observed for the O^{18}/O^{16} ratio (TAYLOR and EPSTEIN, 1970), namely that the moon and earth are isotopically very similar with respect to the two most abundant elements composing these bodies. Proposed mechanisms of the formation and history of these two bodies cannot involve any significant isotopic fractionations. The difference in δSi^{30} between the lunar igneous rocks and the breccia and soil can only be explained if either a more Si^{30}-rich component has been added from meteorite infall or from elsewhere on the moon, or if preferential boiling off or stripping of the Si^{28} took place. This enrichment of the soils and breccia in the heavier isotope was also found by us for O^{18} and C^{13}, and by KAPLAN and SMITH (1970) for S^{34}. This indicates that some mechanism has operated on the lunar soil to enrich it in the heavy isotopes of silicon, oxygen, carbon, and sulfur relative to the local basalts and gabbros from which it is largely constituted (DUKE *et al.*, 1970). If this is a single process, one possibility that comes to mind is the aforementioned fractional vaporization, as this may uniformly favor escape of the light isotopes in the vapor phase. This perhaps could have come about either during meteorite impact-melting or possibly in a continuous fashion during solar wind bombardment.

Summary

In dealing with the hydrogen found in lunar materials from the Apollo 11 expedition the main difficulty appears to be the problem of exactly defining the origin of all the extractable hydrogen. There is little doubt that most of the total hydrogen is of solar wind origin and contains little deuterium; less than 20 ppm. The water in the lunar samples is probably nearly all of terrestrial origin but a small fraction of this water can still be lunar water, either as oxidized solar wind or primary lunar water.

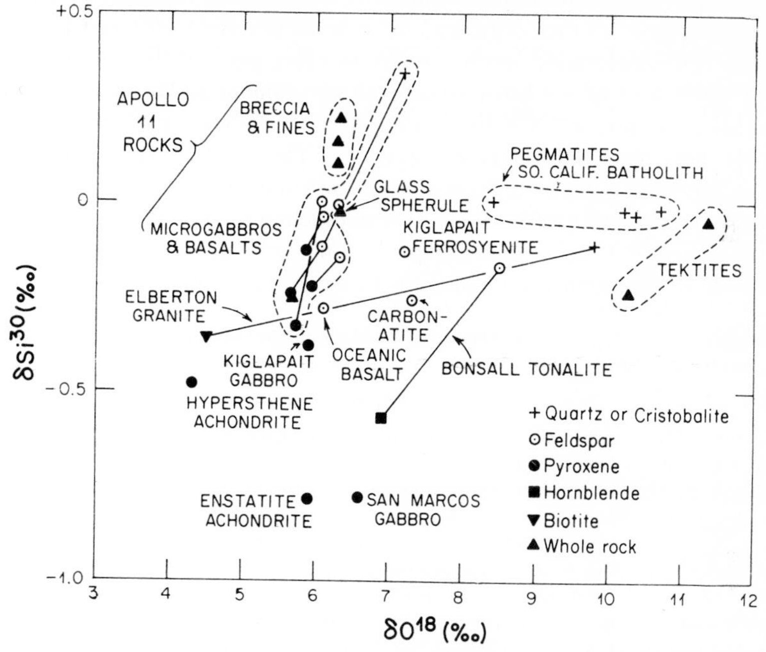

Fig. 2. Plot of δSi^{30} vs. δO^{18} for lunar minerals and rocks. For comparison, data for two tektites and for minerals from several terrestrial igneous rocks and two meteorites (Norton County and Tatahouine) are also plotted. Data-points for coexisting minerals from the same rock are connected by lines. Note that minerals which are O^{18}-rich are also Si^{30}-rich, and that the lunar fines and breccia are richer in Si^{30} than the lunar igneous rocks.

The determination of the D/H ratio of such possible primary lunar water must still await further experimentation. A knowledge of the amount of primary water and its isotopic composition would be of great interest in our evaluation of the fate and role of water in the solar system during the formation of the planets.

The knowledge of the D/H ratio of solar wind hydrogen to better than 10–20 ppm deuterium is also as yet uncertain. If the deuterium concentration in the solar wind is controlled by the equilibrium thermonuclear burning of hydrogen to helium in the sun, the D/H ratio would be $\sim 10^{-17}$ (Fowler *et al.*, 1962). Its presence in solar wind, at concentrations tentatively suggested by our work (≈ 10 ppm), would therefore indicate the possible existence of interesting nuclear reactions in the sun, such as high energy spallation and neutron irradiation on the surface of the sun. Some of these reactions have been used to account for the presence of D, Li, Be and B in the solar system (FOWLER *et al.*, 1962).

The concentration and C^{13}/C^{12} ratio of the carbon extracted from the lunar soil and breccia are different, the concentration being 143 and 260 ppm and the δC^{13} being 18·6 and 10·8‰, respectively. The consistency between the isotope data of KAPLAN and SMITH (1970) and ours adds confidence to the correctness of the high C^{13}/C^{12} ratio for lunar rocks. The reasons for the unusual δ values and for the variable concentrations are as yet not known but possible suggestions include fractional evaporation from meteoritic and solar wind bombardment and presence of a source of C^{13}-rich material in igneous rocks on other parts of the moon.

The Si^{30}/Si^{28} ratio of the lunar silicates is very similar to the ratio for terrestrial rocks and tektites. The Si^{30}/Si^{28} distribution between coexisting minerals is consistent with the few data observed for terrestrial rocks. The Si^{30}/Si^{28} ratio of a hypersthene achondrite, however, is different from that of the lunar rocks, analogous to the differences observed in their O^{18}/O^{16} ratios. Also paralleling the O^{18}/O^{16} ratio, the Si^{30}/Si^{28} ratio of the lunar soil is higher than that of the lunar igneous rocks. The suggestion for explaining the C^{13} enrichments could apply to the explanation of the Si^{30} enrichments.

Acknowledgments—This research was supported by National Aeronautics and Space Administration Contract No. NAS 9-7944.

REFERENCES

BOATO G. (1954) The isotopic composition of hydrogen and carbon in the carbonaceous chondrites. *Geochim. Cosmochim. Acta* **6**, 209–220.

BURLINGAME A. L., CALVIN M., HAN J., HENDERSON W., REED W. and SIMONEIT B. R. (1970) Lunar organic compounds: Search and characterization. *Science* **167**, 751–752.

CLAYTON R. N. and EPSTEIN S. (1958) The relationship between O^{18}/O^{16} ratios in coexisting quartz, carbonate, and iron oxides from various geological deposits. *J. Geol.* **66**, 352–373.

DUKE M. B., WOO C. C., BIRD M. L., SELLERS G. A. and FINKELMAN R. B. (1970) Lunar soil: Size distribution and mineralogical constituents. *Science* **167**, 648–650.

EPSTEIN S. and TAYLOR H. P., JR. (1970) $^{18}O/^{16}O$, $^{30}Si/^{28}Si$, D/H and $^{13}C/^{12}C$ studies of lunar rocks and minerals. *Science* **167**, 533–535.

FOWLER W. A., GREENSTEIN J. L. and HOYLE F. (1962) Nucleosynthesis during the early history of the solar system. *Geophys. J. Roy. Astron. Soc.* **6**, 148–220.

FRIEDMAN I., O'NEIL J. R., ADAMI L. H., GLEASON J. D. and HARDCASTLE K. (1970) Water, hydrogen, deuterium, carbon, carbon-13, and oxygen-18 content of selected lunar material. *Science* **167**, 538–540.

HINTENBERGER H., WEBER H. W., VOSHAGE H., WÄNKE H., BEGEMANN F., VILSCEK E. and WLOTZKA F. (1970) Rare gases, hydrogen, and nitrogen: Concentrations and isotopic composition in lunar material. *Science* **167**, 543–545.

KAPLAN I. R. and SMITH J. W. (1970) Concentration and isotopic composition of carbon and sulfur in Apollo 11 lunar samples. *Science* **167**, 541–535.

KEAYS R. R., GANAPATHY R., LAUL J. C., ANDERS E., HERZOG G. F. and JEFFERY P. M. (1970) Trace elements and radioactivity in lunar rocks: Implications for meteorite infall, solar-wind flux, and formation conditions of moon. *Science* **167**, 490–493.

REYNOLDS J. and VERHOOGEN J. (1953) Natural variations in the isotopic composition of silicon. *Geochim. Cosmochim. Acta* **3**, 224–234.

SEGRÈ E. (1964) *Nuclei and Particles*, p. 564. Benjamin.

TAYLOR H. P. and EPSTEIN S. (1970) O^{18}/O^{16} ratios of Apollo 11 lunar rocks and minerals. *Geochim. Cosmochim. Acta*, Supplement I.

TILLES D. (1961) Natural variations in isotopic abundances of silicon. *J. Geophys. Res.* **66**, 3003–3013.

Proceedings of the Apollo 11 Lunar Science Conference, Vol. 2, pp. 1097 to 1102.

Isotopic abundances of actinide elements in lunar material*

P. R. FIELDS, H. DIAMOND, D. N. METTA, C. M. STEVENS, D. J. ROKOP and P. E. MORELAND
Chemistry Division, Argonne National Laboratory, Argonne, Illinois 60439

(*Received* 6 *February* 1970; *accepted in revised form* 21 *February* 1970)

Abstract—The abundances of uranium and thorium were measured mass spectrometrically as $0{\cdot}59 \pm 0{\cdot}02$ ppm and $2{\cdot}24 \pm 0{\cdot}06$ ppm, respectively, in the fines of Apollo 11 bulk sample. The ratio $^{235}U:^{238}U$ was $0{\cdot}007258 \pm 0{\cdot}000016$, in agreement with terrestrial uranium. The following upper limits were set: $^{236}U:^{238}U \leqslant 3 \times 10^{-9}$, $^{239}Pu \leqslant 1 \times 10^{-9}$ ppm, $^{244}Pu \leqslant 9 \times 10^{-11}$ ppm and $^{247}Cm \leqslant 1{\cdot}25 \times 10^{-10}$ ppm. No unusual alpha-particle or spontaneous-fission activity was found. The search for transactinide nuclides is continuing.

INTRODUCTION

THE PURPOSE of this investigation was to measure the abundances and isotopic composition of thorium, uranium and any transuranium elements that might also be present in lunar material using mass spectrometric and radiometric techniques. Among the transuranium elements we made a specific search for ^{244}Pu and ^{247}Cm, two nuclides which had been produced synthetically in reactors and thermonuclear explosions and whose half-lives had been measured as $(8{\cdot}28 \pm 0{\cdot}10) \times 10^7$ yr (BEMIS *et al.*, 1969; also FIELDS *et al.*, 1966) and $(1{\cdot}64 \pm 0{\cdot}24) \times 10^7$ yr (FIELDS *et al.*, 1963), respectively. The abundances of uranium and thorium not only contribute to the total distribution of the elements in lunar material but are also important in calculating the radioactive heat production on the moon. The ratio U^{235}/U^{236} can be a sensitive measure of the neutron flux on the moon's surface. The possible presence of ^{244}Pu or ^{247}Cm, two extinct or nearly extinct nuclides, was interesting because they would have indicated an infusion from a supernova that occurred quite close to the solar system within the last billion years. Another object of our study was to examine the separated actinide and transactinide element fractions by alpha-particle pulse analysis and spontaneous-fission counting to detect the presence of short-lived nuclides, that might have been produced by bombardment, or might have fallen onto the moon's surface after having been produced elsewhere.

EXPERIMENTAL PROCEDURE AND RESULTS

Earlier work by FIELDS *et al.* (1966) on terrestrial samples showed that it was necessary to work in a laboratory that had never been used for heavy element chemistry; consequently, a new laboratory was equipped for the lunar analysis. All the glassware was carefully leached, and most of the reagents were distilled before use. In some cases special high-purity reagents were used without further treatment, but they

* Work supported by the USAEC and NASA Contract T-76536.

were tested for their thorium and uranium contents before being used in the chemical separation procedure.

An alpha count of an 8-mg sample of the fines gave $0{\cdot}036 \pm 0{\cdot}011$ c/min. The single spontaneous-fission event observed in 13 days of counting could have been due to counter background.

Two one-gram samples of Apollo 11 fines from the bulk sample (sample 10084-75) were dissolved, and known quantities of ^{230}Th (5×10^{-8} g), ^{233}U (7×10^{-8} g), ^{236}Pu (5×10^{-14} g) and ^{242}Cm (2×10^{-14} g) tracers were added. Before use, each of the tracers was carefully purified and then analyzed by mass spectrometry to determine the isotopic distribution of each of the elements. Reagent blanks containing the same tracers were run concurrently with the samples through the same chemical procedure. It was thought that a one gram sample would be adequate for the determination of the uranium and thorium abundances, and we wanted to conserve our sample for further analysis in the light of the preliminary results obtained from the heavier transuranium-element fractions.

After the fines were completely dissolved using HF and $HClO_4$, the solution was converted into 6 N HNO_3, made 0·1 N in $NaNO_2$ to reduce plutonium to the +4 valence state, and extracted several times with 0·4-F "aliquat 336" in xylene (KOCH, 1965; HORWITZ *et al.*, 1966; GERONTOPULOS and RIGALI, 1964). The organic phases were scrubbed with 6 M HNO_3 and then diluted with an equal volume of xylene and 20 per cent by volume of 2-ethyl hexanol. The organic phases were then extracted several times with 0·5 M HCl. The 0·5 M HCl fraction contained the thorium, uranium, neptunium and plutonium. The aqueous 6 N HNO_3 residue from the "aliquat 336" extraction contained the actinides above plutonium ($Z > 94$), and probably, the transactinide elements.

The thorium, uranium, neptunium and plutonium fraction was made 10 M HCl, a trace of HNO_3 added and the solution passed through a Dowex A-1 anion column. The effluent plus 4 column-volumes of wash with 10 M HCl removed the thorium, plutonium was then removed with 1·25% HI in 10 M HCl, the neptunium was next eluted with 4 M HCl, and finally the uranium was eluted with 0·1 M HCl. If the individual actinide elements were not sufficiently pure, additional ion-exchange-column separations were used. These fractions were eventually measured mass spectrometrically or by alpha-pulse analyses.

The 6 M HNO_3 solution containing any heavier actinides and transactinide elements was converted to a 2 M HCl solution, extracted with 0·2 F "aliquat 336", and the organic layer was washed with 2 M HCl several times. The aqueous acid layers were then combined, saturated with HCl gas, and extracted with 0·8 F "aliquat 336". The organic layer was scrubbed with a saturated HCl solution and the HCl layers were combined.

It has been observed (HORWITZ, 1969) that the transition elements in the 6th period of the periodic table exhibit anionic character in HCl solutions. The distribution coefficients for quaternary amine extractions appear to follow those for anion exchange resins (KRAUS and NELSON, 1956); hence the transactinide elements whose atomic number might range from 108 to 114 would be expected to extract into the 0·2 F and 0·8 F "aliquat 336" layers. The combined HCl layers would contain the heavier actinide elements. These latter elements were further purified by additional "aliquat 336" extractions (HORWITZ *et al.*, 1966), HDEHP [di (2-ethylhexyl) orthophosphoric acid] extractions (HORWITZ *et al.*, 1969), and a Dowex-50 cation-exchange column using conc. HCl as the eluant (DIAMOND *et al.*, 1954). The transplutonium actinides, purified from inert materials, could now be separated from each other by a Dowex-50 cation-exchange column using a α-hydroxy isobutyrate as an eluant (CHOPPIN *et al.*, 1956), or they could be mass-spectrometrically analyzed and alpha-pulse analyzed as a group before separating into individual-element fractions.

The isotopic compositions of the separated uranium and thorium fractions were measured in a 30-cm mass spectrometer which offered maximum precision for microgram-sized samples whose isotopic components do not differ by more than 10^4 in abundance. The plutonium and curium fractions were analyzed with a 250-cm mass spectrometer which had been designed to give maximum sensitivity for detection of actinide elements. The results, corrected for the backgrounds introduced by the reagents and by the tracers, are shown in Table 1, columns 2 and 3. The corrections

Table 1. Micrograms (metal weight) of actinide element per gram of lunar fines

Sample number	1	2	3
Uranium	$0{\cdot}600 \pm 0{\cdot}018$	$0{\cdot}582 \pm 0{\cdot}018$	
Thorium	$2{\cdot}27 \pm 0{\cdot}06$	$2{\cdot}21 \pm 0{\cdot}06$	
^{244}Pu	$<3 \times 10^{-10}$	$<1 \times 10^{-10}$	$<9 \times 10^{-11}$*
^{239}Pu	$<1 \times 10^{-9}$		$<1 \times 10^{-9}$*
^{240}Pu	$<3 \times 10^{-10}$		
^{247}Cm	$<1{\cdot}25 \times 10^{-10}$		

* Preliminary results.

were less than 1 per cent for uranium and thorium, but accounted for all of the ^{239}Pu observed. The reported uncertainties are estimated standard deviations, from uncertainties in sampling, counter geometry, alpha counting, and mass spectrometry. No allowance was made for the uncertainties in the half-lives of the ^{233}U ($1{\cdot}62 \times 10^5$ yr) and ^{230}Th ($8{\cdot}0 \times 10^4$ yr) tracers.

It was almost impossible to observe alpha particles of other isotopes of the heavy elements in the various fractions because they were obscured by the larger α-counting rates of the ^{230}Th, ^{233}U, ^{236}Pu and ^{242}Cm tracers that had been added. Consequently, a 10-g sample (3) of the fines was dissolved and only ^{236}Pu tracer added. The plutonium tracer was used to allow increased sensitivity for ^{244}Pu and ^{239}Pu determinations in this larger sample. Sample 3 was also chemically separated into the same element fractions as described earlier. In addition to examining the uranium and plutonium mass spectrometrically, all fractions except plutonium were examined by alpha particle pulse analyses, using a 300 mm^2 solid-state detector (Au–Si) and a 1024-channel pulse-analyzer set to detect all known natural α-activities.

The 30-cm mass spectrometer with an electron multiplier was tested for internal bias by measuring National Bureau of Standards sample U500. A secondary standard, used alternately with the lunar sample was (natural abundance) NBS 949a for which the measured $U^{235}:U^{238}$ was $0{\cdot}007261 \pm 0{\cdot}000032$, in good agreement with the NBS value. Each portion of the uranium was run at 4 or 5 filament temperatures and no measurable fractionation was observed. The $^{235}U:^{238}U$ ratio in lunar fines was 0·007258 ± 0·000016 (Standard deviation) in good agreement with 0·007257 ± 0·0000073 (SEABORG, 1968) found in terrestrial uranium and with the less precisely known meteoritic (HAMAGUCHI *et al.*, 1957) uranium. These results agree with those of other investigators (TATSUMOTO and ROSHOLT, 1970; WANLESS *et al.*, 1970). Another portion of the uranium from sample (3) was analyzed in a special tandem mass spectrometer which can determine neighboring isotopes of widely disparate abundances. This measurement set an upper limit for the $^{236}U:^{238}U$ ratio of 3×10^{-9}. A small ^{239}Pu peak was perceived, but a background for this sample has not yet been measured. A portion of the plutonium fraction was also examined for ^{244}Pu, but no peak was observed. This phase of the analysis is continuing. Preliminary upper limits are reported for ^{239}Pu and ^{244}Pu in Table 1.

Pulse analysis of the uranium fraction (Fig. 1) showed α-particle energies corresponding to ^{238}U, ^{235}U and ^{234}U. The $^{235}U:^{238}U$ ratio was consistent with the ratio observed by the mass measurements. The ratio of the activities $^{238}U:^{234}U = 1·06 \pm 0·09$ in agreement with the expectation of radioactive equilibrium. Similar measurements were made by TATSUMOTO and ROSHOLT (1970).

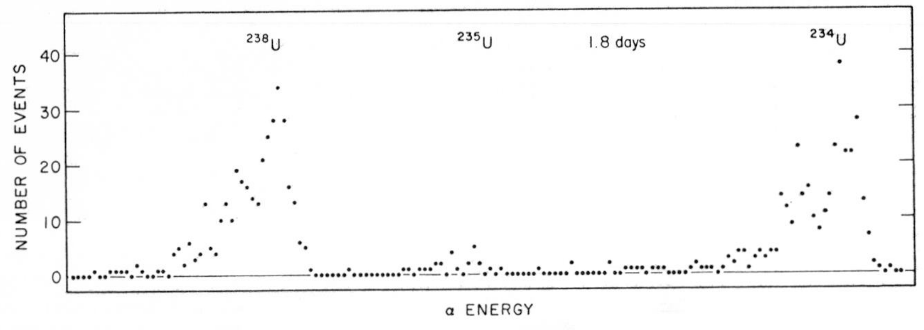

Fig. 1. Alpha-particle energy spectrum of uranium from lunar fines.

The thorium fraction was pulse-analyzed for 3·8 days (Fig. 2). The midpoint of the counting occurred about 4 days after the thorium had been purified from uranium, actinium, radium, lead and virtually all other likely contaminants by ion exchange. The ^{227}Th had been first separated from any ^{227}Ac about 15 days before the pulse analysis. Examination of Fig. 2 discloses that

(1) The $^{232}Th:^{230}Th$ ratio 1·13 ± 0·04, is consistent with the ration 1·21 ± 0·05 predicted by the ^{238}U and ^{232}Th abundances found by mass spectrometer measurements. Thus it can be concluded that there has been no appreciable fractionation of uranium from thorium in the last 2×10^5 yr.

(2) The ^{227}Th and its radioactive daughters are more abundant than the equilibrium concentration expected from the known amount of ^{235}U in the sample; however, a reagent blank accounts for approximately all of the apparent excess. The excess ^{227}Th was traced to an ^{227}Ac impurity in the lanthanum carrier used in the chemical separations. (Although the lanthanum had been purified from thorium before being used in the chemical procedure, enough time elapsed that the ^{227}Th grew back into the solution.)

Inspection of the transplutonium actinide fraction by alpha particle counting and by pulse analysis showed less than 0·001 dis/min per g of lunar fines. There were no spontaneous fissions in this fraction in 10 days of counting.

CONCLUSIONS

The abundances of uranium and thorium are in reasonable agreement with the gamma counting results of O'KELLEY and co-workers (LSPET, 1969). The $^{235}U:^{238}U$

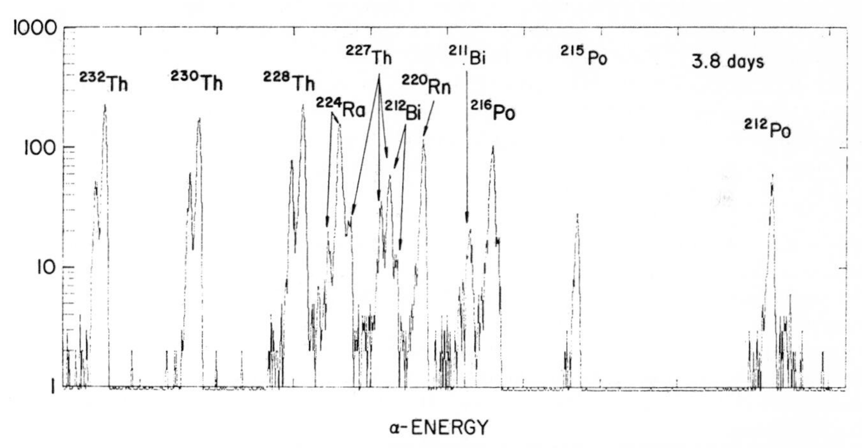

Fig. 2. Alpha particle energy spectrum of thorium fraction from lunar fines. Some of the Th^{227} and its decay products are due to reagent background.

ratio of the lunar sample is in agreement with meteoritic and terrestrial values indicating that the elements of the moon have undergone the same history of nucleogenesis and irradiation as those of the earth and meteorites, and hence, probably the rest of the solar system. The limit set for ^{236}U implies that the product of ^{235}U capture cross section and neutron flux was such that each uranium atom in the lunar fines was exposed to less than 4×10^{-22} neutrons/sec if the irradiation was steady and continuous. Assuming the neutrons are thermal and the appropriate neutron capture cross section of ^{235}U is 100 barns (WESTCOTT *et al.*, 1965), then the uranium was exposed to a neutron flux of less than 4 neutrons/(cm^2/sec). This is consistent with the findings of ALBEE and co-workers (1970) of 1·1 and 0·4 n/(cm^2/sec) from lunar gadolinium isotopic composition, and with the limits set by MURTHY *et al.* (1970). The absence of ^{244}Pu and ^{247}Cm in lunar material argues against nearby (one light-year or less) supernovae explosions within the last billion years. The precise limits depend strongly upon the details of the model used in the calculations.

Acknowledgments—The work was aided by the AEC and by NASA contract T76536. E. P. HORWITZ contributed to the planning and testing of the chemical procedures, and C. H. YOUNGQUIST provided some of the chemical apparatus.

REFERENCES

ALBEE A. L., BURNETT D. S., CHODOS A. A., EUGSTER O. J., HUNEKE J. C., PAPANASTASSIOU D. A., PODOSEK F. A., RUSS G. PRICE II, SANZ H. G., TERA F. and WASSENBURG G. J. (1970) Ages, irradiation history and chemical composition of lunar rock from the Sea of Tranquillity. *Science* **167,** 463–466.

BEMIS C. E., JR., HALPERIN J. and EBY R. (1969) The alpha decay half lives of ^{242}Pu and ^{244}Pu. *J. Inorg. Nucl. Chem.* **31,** 599–604.

CHOPPIN G. R., HARVEY B. G. and THOMPSON S. G. (1956) A new eluant for the separation of the actinide elements. *J. Inorg. Nucl. Chem.* **2,** 66–68.

DIAMOND R. M., STREET K., JR. and SEABORG G. T. (1954) An ion-exchange study of possible hybridized 5f bonding in the actinides. *J. Amer. Chem. Soc.* **76,** 1461–1469.

FIELDS P. R., FRIEDMAN A. M., LERNER J., METTA D. and SJOBLOM R. K. (1963) Possible existence of curium in Nature. II. *Phys. Rev.* **131,** 1249–1250.

FIELDS P. R., FRIEDMAN A. M., MILSTED J., LERNER J., STEVENS C. M., METTA D. and SABINE W. K. (1966) Decay properties of ^{244}Pu, and comments on its existence in nature. *Nature* **212,** 131–134.

GERONTOPULOS P. and RIGALI L. (1964) The extraction of thorium from nitric acid solutions by quaternary ammonium nitrate (Aliquat-336). *Radiochim. Acta* **3,** 122–123.

HAMAGUCHI H., REED G. W. and TURKEVICH A. (1957) Uranium and Barium in Stone Meteorites. *Geochim. Cosmochim. Acta* **12,** 337–347.

HORWITZ E. P., BLOOMQUIST C. A. A., SAURO L. J. and HENDERSON D. J. (1966) The liquid–liquid extraction of certain tripositive transplutonium ions from salted nitrate solutions with a tertiary and quaternary amine. *J. Inorg. Nucl. Chem.* **28,** 2313–2324. E. P. HORWITZ, private communication.

HORWITZ E. P., BLOOMQUIST C. A. A., HENDERSON D. J. and NELSON D. E. (1969) The extraction chromatography of Americium, curium, berkelium and californium with Di (2-ethylhexyl) orthophosphoric acid. *J. Inorg. Nucl. Chem.* **31,** 3255–3271. Also E. P. HORWITZ, private communication.

KRAUS K. A. and NELSON J. (1956) Anion exchange studies of the fission products: *Proc. Int. Conf. Peaceful Uses of Atomic Energy*, Geneva, 1955, Vol. 7, *Nuclear Chemistry and Effects of Irradiation*, p. 113. United Nations, New York.

KOCH G. (1965) Reprocessing by quarternary ammonium nitrates: laboratory studies. In *Solvent Extraction Chemistry of Metals*, (editor H. A. C. MacKay), p. 247. Macmillan.

LSPET (LUNAR SAMPLE PRELIMINARY EXAMINATION TEAM) (1969) Preliminary examination of lunar samples from Apollo 11. *Science* **165,** 1211–1227.

MURTHY V. R., SCHMITT R. A. and REY P. (1970) Rubidium–strontium age and elemental and isotopic abundances of some trace elements in lunar samples. *Science* **167,** 476–479.

SEABORG G. T. (1968) Uranium. *Encyclopedia of the Chemical Elements*, (editor C. A. Hampel), p. 776. Reinhold.

TATSUMOTO M. and ROSHOLT J. N. (1970) Age of the moon: An isotopic study of uranium–thorium–lead systematics of lunar samples. *Science* **167,** 461–463.

WANLESS R. K., LOVERIDGE W. D. and STEVENS R. D. (1970) Age determinations and isotopic abundance measurements on lunar samples. *Science* **167,** 479–480.

WESTCOTT C. H., EKBERG K., HANNA G. C., PATTENDEN N. J., SANATONI S. and ATTREE P. M. (1965) A survey of values of 2200 m/s constants for four fissile nuclides. *Atomic Energy Review* **3,** No. 2, 3–60.

Proceedings of the Apollo 11 Lunar Science Conference, Vol. 2, pp. 1103 to 1109.

Water, hydrogen, deuterium, carbon and C^{13} content of selected lunar material*

IRVING FRIEDMAN, J. D. GLEASON and K. G. HARDCASTLE
U.S. Geological Survey, Denver, Colorado 80225

(*Received* 29 *January* 1970; *accepted in revised form* 20 *February* 1970)

Abstract—The water content of the breccia is 150–455 ppm, with a δD from -580 to -870 per mil. Hydrogen gas content is 40–53 ppm with a δD of -830 to -970 per mil. Contamination by earth water is thought to be low, but formation of water by solar wind protons is postulated. The isotopic equilibration of hydrogen gas with lunar water is possible. On heating in vacuum, 79–140 ppm CO_2 were evolved, with $\delta C^{13} = +2{\cdot}3$ to $+5{\cdot}1$ per mil and $\delta O^{18} = 14{\cdot}2$–$19{\cdot}1$ per mil. Non-CO_2 carbon is 22–100 ppm, $\delta C^{13} = -6{\cdot}4$ to $-23{\cdot}2$ per mil. Lunar dust contains 810 ppm H_2O (D = 80 ppm) and 188 ppm total carbon ($\delta C^{13} = -17{\cdot}6$ per mil). Carbon is present as CO_2 and probably as carbides and elemental carbon. Differences in carbon isotopic composition found by different investigators may be due to inhomogeneous distribution of different forms of carbon with different δC^{13} values.

INTRODUCTION

DURING the past 20 years a large amount of information on the relative abundance of the stable isotopes of hydrogen, carbon, oxygen, and sulfur in terrestrial and meteoritic materials has been accumulated. Consequently, it was of interest to investigate the relative abundance of these isotopes in lunar materials. In addition, determination of the water and hydrogen contents in lunar samples is important because a knowledge of these contents leads to a better understanding of the development and history of the earth's hydrosphere.

EXPERIMENTAL

For the deuterium, the water, and some of the carbon analysis, the weighed samples were placed in a platinum crucible and the crucible was then inserted into a vacuum system. The crucible was heated, step by step, to 1350°C and the H_2, He, H_2O, CO_2 and SO_2 were collected, separated, and measured, and the isotopic abundances were determined on isotope ratio mass spectrometers. After melting, the sample was then mixed with approximately half its weight of previously combusted CuO, and the resulting mixture was combusted at 950°C in oxygen at 1 atm pressure. The collected H_2O was analyzed for D/H and the CO_2 for C^{13}/C^{12} and O^{18}/O^{16}.

The results of the deuterium analyses are expressed as atomic ratios of deuterium to hydrogen, D/H $\times 10^6$, and also as δD per mil SMOW. The C^{13}/C^{12} analyses are given as δC^{13} per mil PDB, whereas the O^{18}/O^{16} ratios are expressed as δO^{18} per mil SMOW.

Geoatmospheric contamination was removed by heating the samples to 50°C in vacuum, after which the temperature was raised and the volatiles were collected. No volatiles were emitted from the samples until a temperature of approximately 300°C was reached. Samples were heated at 550°C from 2 to 5 hr and the volatiles processed. Most of the hydrogen, some helium, more than half the H_2O, and a little CO_2 were released at this stage. The temperature was then raised to 950°C and maintained there for several hours, during which time practically all the remaining H_2O, CO_2 and SO_2 and some He were evolved. Finally the crucible was heated to 1350°C to 1400°C for 15 min to release the remainder of the SO_2, CO_2 and He. The final combustion of the melted sample in oxygen yielded CO_2 and a negligible amount of H_2O. Results of the analysis are given in Table 1.

---

* Publication authorized by the Director, U.S. Geological Survey.

A 9·3-g sample of lunar dust (10084) and a 2·8-g piece of breccia (10060-11), which had been sealed in an aluminum vacuum container in Houston until opened for processing in our laboratory, were combusted directly without being melted in vacuum. Another 3·7-g piece of the vacuum-sealed 10060-11 was processed by heating in a vacuum, as described above, to compare results with those for breccia samples 10046-21 and 10046-22 that had been exposed to laboratory atmosphere for several months.

Table 1. Results of heating and combustion

Sample			Vacuum Heating								
			Temp.	He		H_2			H_2O		
No.	Type	Wt. (g)	°C	cm^3/g	He^4/He^3	ppm	D/H × 10^6	δD	ppm	D/H × 10^6	δD
			300– 550	<0·01	—	45	27	−830	105	70	−560
10046-21	Breccia	1·704	550– 950	0·15	4000 ± 300	5	—	—	350	29	−820
			950–1300	—	—	—	—	—	—	—	—
			Total heating	0·15	4000 ± 300	50	27	−830	455	38	−760
			300– 550	0·05	4000	40	5	−970	190	96	−400
			550– 950	0·003	—	20	—	—	154	36	−770
10046-22	Breccia	3·574	950–1350	0·08	—	—	—	—	28	28	−820
			Total heating	0·13	~ 4000	60	—	—	372	66	−580
			300– 550	0·7	2000	53	22	−870	62	22	−860
10060-11	Breccia	3·712	550– 950	0·1	2000	13	—	—	89	20	−870
			950–1350	—	—	—	—	—	1	—	—
			Total heating	0·8	2000 ± 200	66	~22	−860	152	21	−870
10060-11	Breccia	2·766	950 ←								
10017-16	Crystal	2·028	1350	—	—	~1	—	—	~25	—	—
10084-43†	Dust	9·311	950 ←								

* A small amount of CO_2 released between 550°C and 950°C was lost from this sample.
† Mineral separation on the lunar dust sample (10084) using heavy liquids and a magnetic separator was unsuccessfully attempted. The separated fractions were then recombined, washed in pure acetone, dried and combusted as described. As a result, the deuterium value is probably high due to contamination.

RESULTS

The hydrogen gas emitted by the samples may have been present as molecular hydrogen trapped in the solid material. Alternately, it may have been generated during heating by the reactions of water with carbon, metallic iron, or ferrous iron present in the silicates. The fact that all the hydrogen was liberated below 550°C, whereas appreciable amounts of water continued to be released above that temperature, would tend to rule out reactions with the reduced forms of iron or with carbon. The possibility still exists that the hydrogen equilibrated isotopically with the water vapor. At 550°C the equilibrium constant of the reaction $H_2O + HD \rightleftarrows HDO + H_2$ is 1·3 (SUESS, 1949). The hydrogen evolved at 550°C is depleted in deuterium by a factor of 2 to 20 compared with the water evolved at the same temperature—far in excess of the amount to be expected under equilibrium conditions at the temperature of our experiments.

With the exception of 10060-11, the isotopic compositions of the hydrogen and water are those to be expected for isotopic equilibrium at temperatures from 0° to 200°C, which is well within the lunar temperature range. It is possible that the

of selected Apollo 11 material

			Oxygen Combustion					Total				
CO_2			H_2O			CO_2		H_2			C	
ppm	δC^{13}	δO^{18}	ppm	D/H × 10^6	δD	ppm	δC^{13}	ppm	D/H × 10^6	δD	ppm	δC^{13}
			5	—	—	102	−23·2	100	33	−790	181	−12·0
79	+2·3	+19·1										
			37	—	—	78	−22·7	110	~36	−770	218	−4·8
140	+3·1	+18·8										
			10	—	—	22	−25·7	83	~22	~−860	>101	~−3·2
79*	+3·0	+14·2										
———	———	———→	720	51	−680	113	−10·6	80	51	−680	113	−10·6
—	—	—	27	—	—	69	−20·0	>6	—	—	>69	~−20·0
———	———	———→	810	80	−490	188	−17·6	90	80	−490	188	−17·6

Concentrations = ±20 per cent of amount present.
D/H × 10^6 = ±20 per cent; δC^{13} = ±0·2‰; δO^{18} = ±0·2‰; δD = ±10‰.
δC^{13} values are per mil deviations from PDB.
δO^{18} values are per mil deviations from SMOW.
δD values are per mil deviations from SMOW.
The D/H × 10^6 was calculated from the δD SMOW values, using the relation D/H × 10^6 = 0·158 (1000 + δD per mil).

hydrogen originates as solar wind protons and then slowly reaches isotopic equilibrium with the water already present in the rocks. The isotopic composition of the hydrogen therefore may not be that of the solar wind. The water may be present as liquid or gas inclusions in the solid phases or as water present in the glass structures. It is interesting to note that the water content of the breccias is about equal to that found for freshly erupted Hawaiian basalt, and seems high in view of the much lower oxygen fugacity under which the lunar basalts formed. Some or all of the lunar water may have been formed on the lunar surface by reaction of solar wind protons with oxygen in silicates and oxides. One way of eliminating this source of lunar water would be to examine samples taken from the center of crystalline or glassy rocks (not breccias), well below the depth of penetration of solar protons.

HINTENBERGER *et al.* (1970) report finding approximately 200 ppm of total hydrogen ($H_2O + H_2$) with a D/H of $\sim$44 × 10^6 in breccia 10021, while EPSTEIN and TAYLOR (1970) found approximately 118 ppm of total hydrogen with a D/H of 31 × 10^6 for breccia No. 10061. The results that we report of approximately 100 ppm total hydrogen with a D/H of 20–30 × 10^6 are in essential agreement with these authors.

Although all the samples that we analyzed have had some exposure to terrestrial water vapor, we do not believe that terrestrial water is a serious contaminant in these samples. This opinion is based on several lines of evidence.

(1) The water is not evolved below about 300°C. This would tend to rule out absorbed water.

(2) All the breccia samples, both those kept open to the terrestrial environment and those sealed in vacuum, reacted similarly when heated and gave about the same amounts of hydrogen and water with about the same deuterium content.

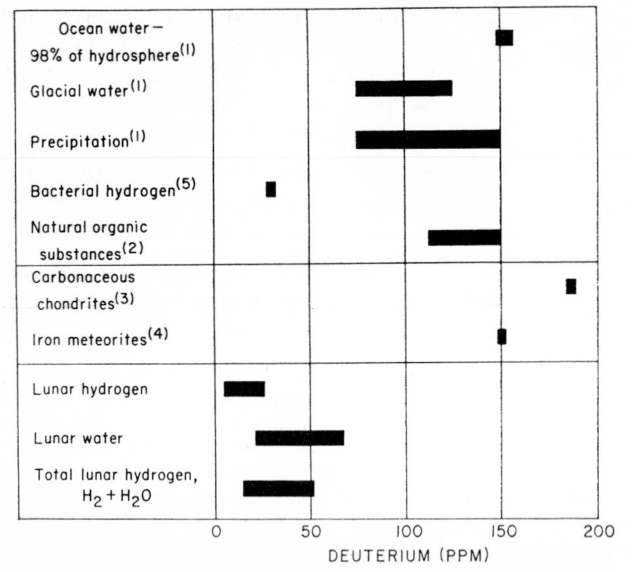

Fig. 1. The deuterium content of hydrogen containing materials from terrestrial, meteoritic and lunar sources. HD calculated using a value of 158 × 10^6 for SMOW. (1) FRIEDMAN *et al.* (1964); (2) BOKHOVEN *et al.* (1956); (3) BOATO (1954); (4) EDWARDS (1955); (5) KRICHEVSKY *et al.* (1961).

(3) The deuterium abundance of the water is far lower than that to be expected from water picked up from the terrestrial atmosphere by the glass phase present in the rocks. As determined from data on volcanic glasses containing secondary water that is derived from meteoric or atmospheric sources, the secondary water should range from about 145 ppm D (hydrated in Houston) to about 125 ppm D (hydrated in Denver). (FRIEDMAN and SMITH, 1958.)

We do not think that the lower water content of the vacuum-sealed sample indicates large-scale contamination by terrestrial water in the other samples because there appears to be great variability in water content between two pieces of the same breccia that had been exposed to the terrestrial atmosphere for several months.

Contamination with unexpended rocket fuel dumped on the lunar surface can be ruled out by the results of FLORY *et al.*, which were published in the report of the LSPET (1969).

Additions of rocket exhaust gases cannot be as easily determined as can those of organic fuel, and these gases still remain a possibility as a contaminant, particularly in the very porous, high-surface-area breccias. The CO_2 results discussed later, however, tend to rule out rocket exhaust contamination.

The 2·8-g piece of vacuum-sealed 10060-11 that was crushed in the laboratory prior to its combustion in oxygen and exposure to the laboratory air for approximately 5 min has approximately 5 ppm more hydrogen than the total hydrogen + water collected by heating the other vacuum-sealed piece in vacuum. In addition, the water from the combusted sample is slightly enriched in deuterium as compared with the water from the other sample. These differences may be due to the uptake of "heavy" (approximately 150–200 ppm D) water from the atmosphere during sample handling, or they may represent inhomogeneity in the breccia. From our experience with the several pieces of breccia 10046, we favor the latter explanation—that of sample inhomogeneity.

Figure 1 is a plot of the deuterium abundances of terrestrial, meteoritic, and lunar material. If the analysis given here is of true lunar water, deuterium on the earth is enriched by a factor of about 3–5 over that in the moon. If the moon is a sample of the primitive earth, the earth will have lost at least two-thirds of its original hydrogen by escape from the exosphere to outer space.

The total carbon content of the lunar samples is reasonably constant at 150 ± 50 ppm, with a δC^{13} of −3 to −20. This is within the range to be expected for average earth carbon (see Fig. 2). Contamination by exhaust gases is again a possibility, but the tenacity with which the CO_2 is held in the samples would tend to rule out this possibility. The CO_2 could have been formed in our laboratory by reduction of the silicates by elemental carbon, but this reaction usually does not begin at temperatures under about 1600°C. The water-gas reaction ($C + H_2O \rightarrow CO + H_2$) is also a high-temperature reaction (approximately 900°C). In addition, the evolution of SO_2 is difficult to explain as an artifact of our experimental procedure or as rocket-exhaust contamination. We believe that the CO_2 is truly lunar and may be present as liquid or gaseous inclusions in the solid phases. The carbon extracted by combustion after sample heating and removal of CO_2 is probably present as either carbides or elemental carbon.

The O^{18}/O^{16} ratio of the CO_2 varies from +14·2 to +19·1. These values are much heavier than the δO^{18} in the silicates and oxides analyzed by FRIEDMAN *et al.* (1970), and this fact is another indication that the CO_2 did not form from the reduction of silicates by carbon. The heavy oxygen may be due to equilibration between the CO_2 and H_2O and the silicates and oxides. It is known that the oxygen of CO_2 in fluid inclusions exchanges with the oxygen present in the host mineral (RYE and O'NEIL, 1968).

BOATO (1954) has analyzed carbon in both carbonaceous chondrites (−3·7 to −18·8‰) and ordinary chondrites (−24·5 ± 0·2‰). These values are in the same range as that determined by our analysis of lunar materials. CLAYTON (1963) has found very heavy carbon (+58·6 to +64·4 per mil) in carbonate minerals from Orgueil, a carbonaceous chondrite. This heavy carbon exceeds the range of the carbon in the lunar samples that we have examined.

Other investigators (BURLINGAME *et al.*, 1970; EPSTEIN and TAYLOR, 1970; KAPLAN and SMITH, 1970) have reported various amounts of CO when samples of dust (10084 or 10086) or breccias were heated. BURLINGAME *et al.* find 160 ppm C, all

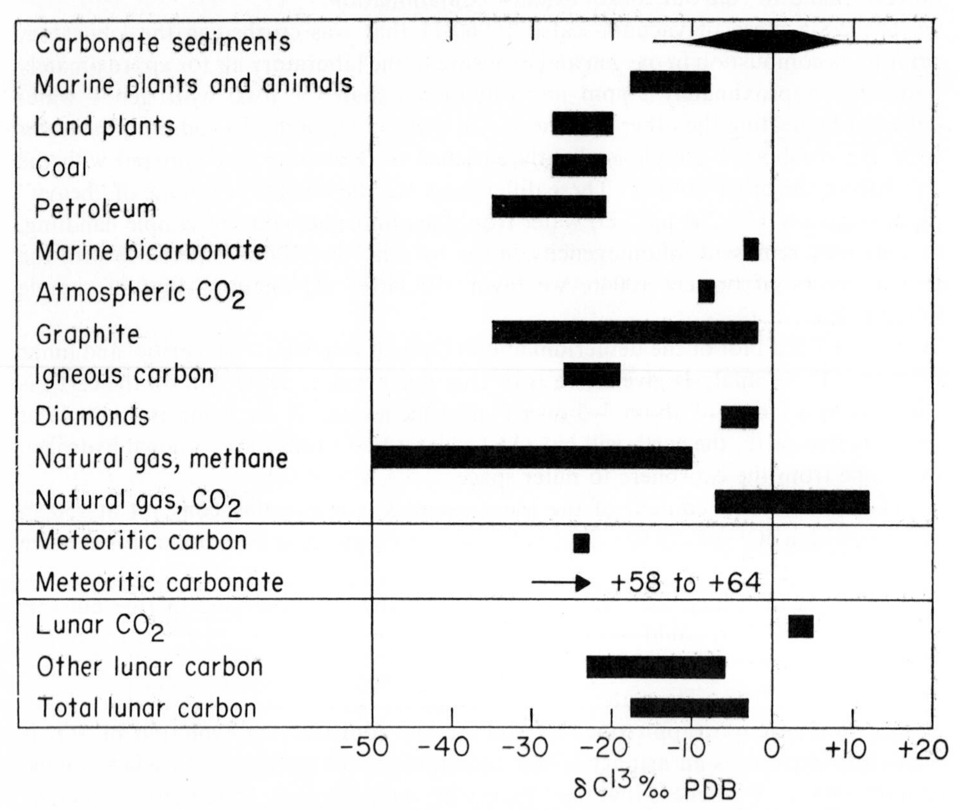

Fig. 2. The relative C^{13} abundance of terrestrial, meteoritic and lunar sources. CRAIG (1953); LLOYD (1964); SILVERMAN and EPSTEIN (1958); ZARTMAN *et al.* (1961).

present as CO. EPSTEIN and TAYLOR report both CO_2 and CO, but observe no difference in δC^{13} between material evolved between 500°–650°C, and 650°–1500°C. KAPLAN and SMITH find that about one-fourth of the carbon in the lunar fines is evolved as CO_2 during vacuum heating to 750°C and about one-fourth as a mixture of possibly CO and hydrocarbon decomposition products (neither positively identified). Vacuum heating of the dust to 750°C gave 86 ppm C with a δC^{13} of approximately 0 per mil. Combustion of the residue yields 59 ppm additional carbon with δC^{13} of +9·4. The total carbon in this sample is 145 ppm with $\delta C^{13} = +4$ per mil. Four dust samples combusted directly by KAPLAN and SMITH gave carbon results from 116 to 170 ppm and δC^{13} from +17·2 to +20·2 per mil.

The conclusion seems obvious. The lunar dust is very inhomogeneous, and contains varying amounts of CO_2, perhaps CO, elemental carbon and carbides, all having very different δC^{13}. If any carbonate from carbonaceous chondrites is present, we may expect to find carbon of +58 to +64 per mil, as well as carbon of −3 to −19 per mil. If ordinary chondritic carbon is present, values of −25 per mil can be expected.

The only lunar sample, other than the dust, that is common to more than one investigator is breccia No. 10060. KAPLAN and SMITH (1970) measured 137 and 132 ppm C with δC^{13} of +1·6 and +2·7, whereas we obtained by direct combustion a value of 113 ppm and δC^{13} of −10·6 per mil. Again, we believe that most of the differences represent inhomogeneities in the breccia.

REFERENCES

BOATO G. (1954) The isotopic composition of hydrogen and carbon in the carbonaceous chondrites. *Geochim. Cosmochim. Acta* **6**, 209–220.

BOKHOVEN C. and THEEUWEN H. H. J. (1956) Deuterium content of some natural substances. *Kon. Ned. Akad. Wetensch., Proc. Ser.* **B59**, 78–83.

BURLINGAME A. L., CALVIN M., HAN J., HENDERSON W., REED W. and SIMONEIT B. R. (1970) Search for and characterization of lunar organic compounds. *Science* **167**, 751–752.

CLAYTON R. N. (1963) Carbon isotope abundance in meteoritic carbonates. *Science* **140**, 192–193.

CRAIG H. (1953) The geochemistry of the stable carbon isotopes. *Geochim. Cosmochim. Acta* **3**, 53–92.

EDWARDS G. (1955) Isotopic composition of meteoritic hydrogen. *Nature* **176**, 109–111.

EPSTEIN S. and TAYLOR H. (1970) O^{18}/O^{16}, Si^{30}/Si^{28}, D/H and C^{13}/C^{12} studies of lunar rocks and minerals returned by the Apollo 11 Expedition. *Science* **167**, 533–535.

FRIEDMAN I., O'NEIL J. R., ADAMI L. H., GLEASON J. D. and HARDCASTLE K. (1970) Water, hydrogen, deuterium, carbon, carbon-13 and oxygen-18 content of selected lunar material. *Science* **167**, 538–540.

FRIEDMAN I., REDFIELD A. C., SCHOEN B. and HARRIS J. (1964) The variation of the deuterium content of natural waters in the hydrologic cycle. *Rev. Geophys.* **2**, 177–224.

FRIEDMAN I. and SMITH R. L. (1958) The deuterium content of water in some volcanic glass. *Geochim. Cosmochim. Acta* **15**, 218–228.

HINTENBERGER H., WEBER H. W., VOSHAGE H., WÄNKE H., BEGEMANN F., VILSCEK E. and WLOTZKA F. (1970) Rare gases, hydrogen and nitrogen: Concentrations and isotopic composition in lunar material. *Science* **167**, 543–545.

KAPLAN I. R. and SMITH J. W. (1970) Concentration and isotopic composition of carbon and sulfur in Apollo 11 lunar samples. *Science* **167**, 541–543.

KRICHEVSKY M. I., SESLER F. D., FRIEDMAN I. and NEWELL M. (1961) Deuterium fractionation in a marine pseudomonad. *J. Biol. Chem.* **236**, 2520–2525.

LSPET (LUNAR SAMPLE PRELIMINARY EXAMINATION TEAM) (1969) Preliminary examination of lunar samples from Apollo 11. *Science* **165**, 1211–1227.

LLOYD R. M. (1964) Variations in the oxygen and carbon isotope ratios of Florida Bay mollusks and their environmental significance. *J. Geol.* **72**, 84–111.

RYE R. O. and O'NEIL J. R. (1968) The O^{18} content of water in primary fluid inclusions from Providencia, North-Central Mexico. *Econ. Geol.* **63**, 232–238.

SILVERMAN S. R. and EPSTEIN S. (1958) Carbon isotopic compositions of petroleums and other sedimentary organic materials. *Bull. Amer. Assoc. Petrol. Geol.* **42**, 998–1012.

SUESS H. (1949) Das Gleichgewicht $H_2 + HDO \leftrightarrows HD + H_2O$ und die weiteren Austrauschgleichgewichte im System H_2, D_2 und H_2O. *Z. Naturforsch.* **5**, 328–332.

ZARTMAN, R., WASSERBURG G. J. and REYNOLDS J. H. (1961) Helium, argon and carbon in some natural gases. *J. Geophys. Res.* **66**, 277–306.

Proceedings of the Apollo 11 Lunar Science Conference, Vol. 2, pp. 1111 to 1116.

Gas analysis of the lunar surface*

J. G. FUNKHOUSER, O. A. SCHAEFFER
Department of Earth and Space Sciences, State University of New York, Stony Brook, L.I., New York 11790

D. D. BOGARD
NASA Manned Spacecraft Center, Houston, Texas 77058

and

J. ZÄHRINGER
Max-Planck-Institut für Kernphysik, Heidelberg, Germany

(*Received* 13 *February* 1970; *accepted* 13 *February* 1970)

Abstract—The rare gas analysis of the lunar surface has lead to important conclusions concerning the moon. The large amounts of rare gases found in the lunar soil and breccia indicate that the solar atmosphere is trapped in the lunar soil as no other source of such large amounts of gas is known. The cosmogenic products indicate that the exposure ages of the seventeen lunar rocks measured vary from 20 to 400 million years with some grouping of the ages. The most striking feature is the old potassium–argon age which for the fourteen rocks analyzed varies from 2·5 to 3·8 billion years. It is concluded that Mare Tranquillitatis crystallized about 4 billion years ago from a molten state produced by a large meteorite impact or volcanic flow.

THE RARE gases offer important clues to the history of the lunar surface layers. Variations in isotopic abundances and absolute amounts of the noble gases indicate that a separation from the primordial gas component took place early in the moon's evolution such that subsequent additions of rare gases are rather easily traced. We now have evidence that the processes that add rare gases to the moon include radioactive decay and solar wind, solar flare and cosmic ray implantation, and nuclear reactions induced by the higher energy particles. Interpretation of rare gas measurements leads to surface exposure ages, the time of crystallization of rocks, and insight into the composition of the solar surface.

Most of the gas analyses were made with a 6 in., 60° magnetic deflection mass spectrometer. A number of the 'gassier' samples were analyzed on a less sensitive 5 in., 180° instrument.

Standard techniques of gas extraction, chemical purification with hot titanium, and separation were employed (LSPET, 1969). All samples were initially degassed at 125°–150°C for 5–24 hr prior to introduction into the vacuum system. Subsequent adsorption of atmospheric rare gases was considered negligible in most samples after we observed the small amount of gases released at 400° to 450°C in several of the samples.

The 6 in. spectrometer was operated at a sensitivity of 2×10^{-10} cm^3/mv for He, Ne, Ar and 3×10^{-13} cm^3/mv for Kr and Xe. The 1650°C-furnace blank comprised

* This article was reprinted from *Science* (Vol 167, pp. 561–563) without further refereeing or revision by the author.

Table 1. Rare gas content of fines and breccia (concentration in units of 10^{-8} cm^3/g).

	Helium		Neon			Argon				
Sample	^4He	4/3	^{20}Ne	20/22	22/21	^{36}Ar	40/36	36/38	^{84}Kr	^{132}Xe
				Fines						
10010	11,000,000	2430	200,000	12·4	31·6	35,000	1·1	5·20	21	10
10010	19,000,000	2540	313,000	12·8	29·4	34,000	1·2	5·17	20	4·1
				Breccia						
10018	24,000,000	2650	480,000	13·4	27·8	72,000	2·1	5·16	35	12
10021	37,000,000	3300	750,000	12·4	30·6	100,000	2·1	5·13	28	10
10023	25,000,000	2570	350,000	12·7	29·3	52,000	2·4	5·15	24	11
10027	15,000,000	2930	320,000	12·2	29·9	67,000	2·2	5·08	41	12
10048	21,000,000	2600	350,000	11·1	28·6	55,000	2·1	5·22	31	12
10061	47,000,000	3400	710,000	12·1	30·0	89,000	1·8	5·14	48	11
10068	25,000,000	2780	330,000	12·2	29·0	42,000	2·7	5·22	20	7·5

most of the background and it normally amounted to $<5 \times 10^{-9}$ cm^3 ^3He, 1×10^{-7} cm^3 ^4He, 9×10^{-9} cm^3 ^{20}Ne, 7×10^{-8} cm^3 ^{40}Ar, 6×10^{-12} cm^3 ^{84}Kr, and 2×10^{-12} cm^3 ^{132}Xe. Long term reproducibility determined from daily standards, was within ± 5 per cent for all five noble gases. Isotopic ratios were corrected for mass spectrometer discrimination and are accurate to ± 2 per cent for the gases He, Ne, and Ar and less than 1 per cent for the gases Kr and Xe. The accuracy in the absolute amount of each gas is estimated to be within ± 20 per cent.

The high gas content of the fines and breccia allowed the use of milligram size samples. Only interior portions of crystalline rocks were analyzed and these were mechanically cleaned of contaminating fines; nevertheless, some degree of solar contamination was unavoidable. Most samples of the crystalline rocks measured ranged from 10 to 100 mg.

Because of the large amounts of solar gases in the fines and breccia (Table 1), surface exposure and K–Ar ages could only be calculated for the crystalline rocks (Table 2) plus two lithic fragments handpicked from two different breccias. Both the radiation and K–Ar ages are depicted (Fig. 1). Corrections for determining spallation

Table 2. Rare gas content of crystalline rocks (concentrations in units of 10^{-8} cm^3/g)

Sample	^3He	^4He	^{20}Ne	^{21}Ne	^{22}Ne	^{36}Ar	^{38}Ar	^{40}Ar	^{84}Kr	^{132}Xe
10017	310	58,000	110	47	56	37	47	4100	0·080	0·040
10018*	420	260,000	5100	76	460	690	230	6100	0·51	0·17
10020	110	20,000	72	14	20	20	19	1800	0·046	0·031
10022	350	63,000	210	52	67	59	50	5700	0·13	0·063
10024	260	50,000	220	32	76	75	68	4100	0·46	0·39
10032	150	15,000	750	16	73	98	32	7400	0·071	0·041
10044	82	34,000	250	8·8	27	36	15	2500	0·094	0·046
10047	96	25,000	210	15	30	43	19	1500	0·090	0·036
10049	25	77,000	73	2·6	8·7	11	5·0	5600	0·059	0·021
10050	440	80,000	900	51	110	150	73	2500	0·35	0·21
10057	44	64,000	110	5·7	14	19	9·7	4800	0·020	0·013
10058	58	21,000	64	7·1	12	15	12	3700	0·015	0·007
10061*	350	95,000	13,000	74	1000	2300	620	8200	1·58	0·46
10069	28	69,000	38	4·0	7·0	6·0	5·4	6500	0·007	0·005
10070	110	98,000	790	11	68	110	32	2400	0·071	
10071	320	73,000	177	37	52	36	34	6!00	0·050	0·029
10072	190	75,000	235	26	44	41	36	7600	0·033	0·012

* Lithic fragments handpicked from breccia.

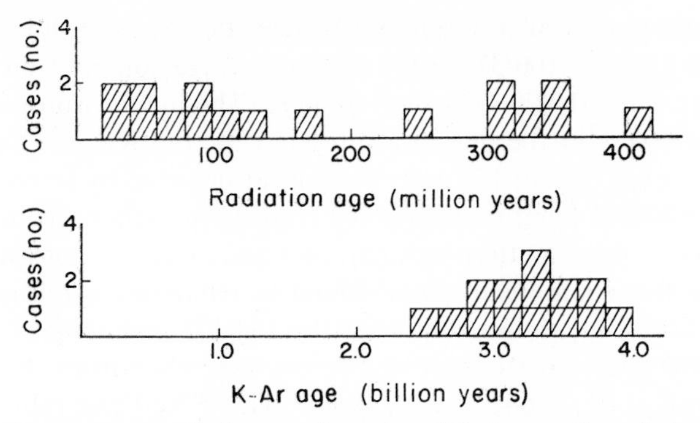

Fig. 1. Histograms of radiation ages (upper) and K–Ar ages (lower) of crystalline rocks.

gases were applied in the normal manner using solar ratios measured in the fines and breccia. No atmospheric ^{40}Ar was assumed present in the calculation of radiogenic Ar, and the solar correction was negligible in most all cases.

The breccia and fines contain extraordinarily large amounts of noble gases which are derived predominantly from the solar wind. Compared to relative cosmic abundances, depletion of the major elemental isotope decreases with increasing atomic weight, as would be expected, invoking loss by diffusion and sputtering-saturation mechanisms. The average relative abundances of ^{4}He, ^{20}Ne, ^{36}Ar, ^{84}Kr, ^{132}Xe in both the fines and breccia samples, normalized to ^{36}Ar (100), are 42,000; 660; 100; 0·054; 0·020, respectively. Both the absolute and relative amount of gas varies by a factor of three, even within a single sample, yet the isotopic abundances vary only slightly. The ratio $^{4}He/^{3}He$ is about 2800, in agreement with estimates of solar abundances (CLAYTON, 1968). The average value for $^{20}Ne/^{22}Ne$ is 12·4 and for $^{22}Ne/^{21}Ne$, about 30·7 (corrected for spallation ^{21}Ne). Both these ratios are similar to those in gas rich meteorites. The high, yet variable, $^{20}Ne/^{22}Ne$ ratio, relative to the terrestrial atmosphere, might be attributed to ^{23}Na (p, α) ^{20}Ne reactions generated by solar flares in the surface layers of the lunar fines. Shock or some similar process apparently tends to homogenize the noble gases throughout the fines and breccia on a small scale as indicated by temperature release patterns. Yet millimeter-size lithic fragments removed from the breccia contain only small amounts of cosmic noble gases.

The breccias returned by Apollo 11 generally contain more gas than the fines. The isotopic ratios are similar with one notable exception: $^{40}Ar/^{36}Ar$ is about 1·1 in the fines, but averages 2·2 in the breccia samples. Theoretical calculations for elemental synthesis predict an $^{40}Ar/^{36}Ar$ ratio of less than 0·01 per cent (CLAYTON, 1968), thus the large amounts of ^{40}Ar observed must be of radiogenic origin, not produced by *in situ* decay of ^{40}K but adsorbed and/or injected into the fines and breccia in characteristically different amounts. Perhaps the breccia were not derived locally, but from fines with a higher $^{40}Ar/^{36}Ar$ ratio in another site. The $^{36}Ar/^{38}Ar$ ratio is a remarkably constant $5·28 \pm 0·06$.

The major Kr and Xe component in the fines and breccia is ascribed to a solar wind origin. The isotopic composition among these samples varied only slightly and showed an average composition for xenon 124/126/128/129/130/131/132/134/136 of 0·57/0·65/8·5/104/16·5/82·5/≡100/37·1/30·4, and an average composition for krypton 78/80/82/83/84/86 or 0·72/4·10/20·4/20·4/≡100/30·8. The lighter isotopes of Kr and Xe show a variable excess in the fines and breccia over the Kr and Xe trapped in carbonaceous chondrites. Xenon-134 and -136 appear depleted by several per cent in the solar xenon component from that observed trapped in carbonaceous chondrites. This particular xenon isotopic pattern in lunar fines and breccia is consistent with the trapped plus fission two-component xenon found in temperature release studies of carbonaceous chondrites but shows values for the 134/132 and 136/132 ratios lower than any temperature release performed to date on these chondrites (REYNOLDS and TURNER, 1964; FUNK *et al.*, 1964; ROWE, 1968). The $^{129}Xe/^{132}Xe$ ratio in the solar component is the same as that found in typical carbonaceous chondrites.

The breccias have many of the characteristics of gas rich meteorites. It is not unreasonable, then, to suggest that gas rich meteorites were formed in a manner similar to the lunar breccia; that is, on a surface of a large body irradiated by solar and cosmic particles, with a certain amount of surface mixing, and then lithified by shock generated by meteoritic impact.

All the crystalline rocks measured contain noble gases produced by spallation reactions. The amount of the spallogenic gases depends on the time the sample was exposed to cosmic rays, whereas the relative isotopic abundance reflects both the chemistry and the degree of radiation shielding of the sample. Elemental isotopic abundances of 15 crystalline rocks show an average value of 9 for $^{3}He/^{21}Ne$ and 0·9 for $^{21}Ne/^{38}Ar$.

Assuming 2π-geometry for the lunar surface rocks and, therefore, a production rate of $1·0 \times 10^{-8}$ cm^3 per million yr for ^{3}He, then absolute radiation ages can be assigned to the different samples. Production rates, relative to ^{3}He, have been derived for ^{21}Ne and ^{38}Ar from excitation functions applied to the average chemical composition of the crystalline rocks. The variation in the measured $^{3}He/^{21}Ne$ ratio appears to be dependent upon the effective energy spectrum rather than chemical composition. The ^{38}Ar is generated primarily from Ca, and the production rate is also a strong function of radiation hardness. The calculated $^{21}Ne/^{38}Ar$ ratio is rather uncertain because of limited cross-section data for Ca. Nevertheless, exposure ages determined from the calculated production rates of ^{21}Ne and ^{38}Ar agree within 20 per cent of the ^{3}He based radiation age in almost all of the 17 samples measured.

Production rates of ^{21}Ne based on the measured ^{22}Na content of three rocks (*l*) agree well with calculated values as does a ^{3}He-T age on a crystalline rock that was analyzed for tritium (DAVIS, 1970). This fact, the general concordancy of the ^{3}He, ^{21}Ne, ^{38}Ar ages, and the concordant K–Ar and U–He ages (discussed later) imply that no widespread diffusive gas loss has occurred for the spallogenic gases in the crystalline rocks.

The radiation ages (Fig. 1) range from 20 to 400 million yr and fall into two rather amorphous groups at about 100 and 350 million yr. Spallation gases in lithic fragments from two breccia samples indicate radiation ages of approximately 350 million yr. Thus, the breccia and the rocks were scattered on the surface of Mare Tranquillitatis by few, but distinct, events, most probably meteoritic impact. The estimated thickness of the regolith in the vicinity of the Apollo 11 landing site (LSPET, 1969) and an assumed age of the basement rock of about 4 billion yr leads to an average rate of 'soil' formation of 1–2 mm per million years. In addition, the fact that even the rocks with the youngest surface ages exhibited rounding and that no rocks were found with radiation ages greater than about 400 million yr again implies a combined erosion and surface coverage rate of 1–2 mm per million yr. Some degree of turnover in the top meter of the regolith must be assumed in order to provide lunar rocks older than a few tens of million years that still remain uncovered.

The lunar crystalline rocks show a wide isotopic variation in krypton and xenon, all characterized by excesses of xenon 124–132 and krypton 78–84 relative to ^{136}Xe and ^{86}Kr, respectively. These excesses are attributed to cosmic-ray spallation reactions in the rocks, mainly on Ba, Sr, Y, and Zr. The spallation spectrum for the various rocks are all similar and are (normalized to $^{126}Xe \equiv 1$ and $^{83}Kr \equiv 1$): 124/126/128/129/130/131/132/134 $\equiv$ 0·55 $\equiv$ 1/1·5/1·7/1·07/6·3/0·9/0·07, and 78/80/82/83/84 $\equiv$ 0·2/0·48/0·76 $\equiv$ 1/0·38. These spectra are quite similar to the spallation spectra found in achondrites except for a much higher ^{131}Xe and a somewhat lower ^{84}Kr. The higher 131 yield is probably an effect of low energy protons as barium is the major target for spallation Xe production, and 730-meV proton spallation of barium does not show this high ^{131}Xe yield (FUNK, 1967). The amounts of ^{126}Xe and ^{83}Kr in the rocks vary by almost an order of magnitude among samples because of the differing exposure histories, yet these amounts correlate reasonably well with the spallation ^{38}Ar. The slight isotopic excesses of the lighter isotopes of Kr and Xe in the lunar fines and breccia over that in carbonaceous chondrites is also probably due to spallation reactions. Not only is the amount of excess ^{126}Xe in different breccia variable, but in amount it is only slightly greater than the spallation ^{126}Xe found in those crystalline rocks with the longest exposure history.

Radiogenic Ar and He are apparent in all crystalline rocks examined. K–Ar ages were obtained for 14 rocks and fragments for which K was known (LSPET, 1969). Potassium was determined by emission spectroscopy but not from aliquots used for noble gas analysis. The K–Ar ages, including two from lithic fragments from two breccia samples, range from 2·5 to 3·8 billion years (Fig. 1). The U–He ages are subject to large errors because of the solar wind correction to the measured ^{4}He and the uncertainty in the U–Th content derived from the K values (LSPET, 1969); nevertheless, most of the ages are concordant within this error with the K–Ar data. Thus, the spread in the K–Ar ages is probably real to some extent, and not just caused by variable leakage of radiogenic Ar.

Mare Tranquillitatis, then, probably first crystallized about 4 billion yr ago from a large molten body produced either by meteoritic impact or volcanism. Subsequent large impacts or volcanic events would further mold the mare to its present form and produce younger rocks. Despite the time spread of approximately 1 billion yr in volcanic events, it appears that the features of the moon at Mare Tranquillitatis were well delineated at least 2 billion years ago.

Acknowledgments—We wish to acknowledge helpful discussions with D. CLAYTON (Rice University) regarding nuclear reactions, especially proton reactions on sodium to produce neon. We thank W. HART, W. HIRSCH, D. MOORE, C. POLO, L. SIMMS, R. WILKIN of Brown & Root–Northrop (MSC) and G. BARBER, R. WARASILA of SUNY, Stony Brook, for their assistance in the analyses. This work was supported by NASA contract NAS 9-8820.

References

Clayton D. D. (1968) *Principles of Stellar Evolution and Nucleosynthesis*. McGraw-Hill.

Davis R. (1970). Private communication.

Funk H. and Rowe M. W. (1967) Spallation yield of xenon from 730 MeV proton irradiation of barium. *Earth Planet. Sci. Lett.* **2,** 215–219.

Funk H., Podosek F. and Rowe M. W. (1967) Fissiogenic xenon in the Renazzo and Murray meteorites. *Geochim. Cosmochim. Acta* **31,** 1721–1732.

LSPET (Lunar Sample Preliminary Examination Team) (1969) Preliminary examination of lunar samples from Apollo 11. *Science* **165,** 1211–1227

Rowe M. W. (1968) On the origin of excess heavy xenon in primitive chondrites. *Geochim. Cosmochim. Acta* **32,** 1317–1326.

Reynolds J. H. and Turner G. (1964) Rare gases in the chondrite Renazzo. *J. Geophys. Res.* **69** 3263–3281.

Proceedings of the Apollo 11 Lunar Science Conference, Vol. 2, pp. 1117 to 1142.

Trace elements in Apollo 11 lunar rocks: Implications for meteorite influx and origin of moon

R. GANAPATHY, REID R. KEAYS, J. C. LAUL and EDWARD ANDERS
Enrico Fermi Institute and Department of Chemistry
University of Chicago, Chicago, Illinois 60637

(*Received* 3 *February* 1970; *accepted in revised form* 27 *February* 1970)

Abstract—Neutron activation analysis of 12 lunar samples showed that Type D fines and Type C breccias are enriched 3- to 100-fold in Ir, Au, Zn, Cd, Ag, Br, Bi, Te and Tl, relative to Type A, B rocks. Smaller enrichments were found for Co, Cu, Ga, Pd, Rb and Cs. The solar wind at present intensity can account for only 2 per cent of this enrichment; an upper limit to the average proton flux during the last $3{\cdot}65 \times 10^9$ yr thus is 9×10^9 cm^{-2} sec^{-1}. The remaining enrichment seems to be due to a $\sim$1·9 per cent admixture of carbonaceous-chondrite-like material, corresponding to an average influx rate of meteoritic and cometary matter of $3{\cdot}8 \times 10^{-9}$ g cm^{-2} yr^{-1} at Tranquillity Base. This result is consistent with the observed crater frequency, and also represents a firm upper limit on the micrometeorite influx rate.

Type A, B rocks are depleted 10- to 100-fold in Ag, Au, Zn, Cd, In, Tl and Bi, relative to terrestrial basalts and a Ti-rich gabbro. Interelement correlations suggest that Tl is lithophile, while Ag, Ir, Au, Br, and Bi are siderophile and Zn, Cd, and Pb are chalcophile.

The 100-fold depletion of the Pb–In group and other volatiles implies either that the moon accreted at a higher mean temperature than did the earth ($\sim$650°K vs. $\sim$550°K), or, more probably, that it formed as an original satellite of the earth, at a distance of 5–20 earth radii. Owing to its orbital motion, the moon would have a much higher encounter velocity and hence lower accretion rate for interplanetary dust than did the earth, particularly in the final stages of accretion when the Pb–In group and water began to condense. This factor may also be responsible for the lower abundance of the alkalis.

The 100-fold depletion of gold relative to terrestrial rocks suggests that removal of siderophile elements from the crust of the moon and earth took place under different physico-chemical conditions, and hence in two separate events. This contradicts the fission hypothesis for the origin of the moon. The low abundance of nickel–iron in the moon may be due to preferential accretion of metal in the central planet. There appear to be no geochemical reasons for postulating an exotic origin of the moon.

INTRODUCTION

FOR OUR measurements, we selected 16 trace elements known to be strongly fractionated in meteorites and planets. They included 4 siderophile elements (Au, Co, Ir, Pd), and 12 volatile ones (Ag, Bi, Br, Cd, Cs, Cu, Ga, In, Rb, Te, Tl, Zn), representing both the "normally depleted" and "strongly depleted" groups of LARIMER and ANDERS (1967). We hoped that these data would provide some clues to chemical processes during and after formation of the moon, and to the influx rate of meteoritic matter.

This particular group of elements did not lend itself to determination by rapid instrumental methods, particularly at the low abundance levels expected. We therefore had no choice but to resort to radiochemical neutron activation analysis, in spite of its time-consuming nature.

A preliminary account of this work was published in *Science* (KEAYS *et al.*, 1970). The present paper contains new data on Te as well as a complete review of other authors' results for the same 16 elements. The discussion has been greatly expanded on the basis of a more thorough analysis of our own and other workers' data.

EXPERIMENTAL

Samples

Seven lunar rocks and two aliquots of fines were analyzed. A third aliquot of fines was separated into 3 sieve fractions, which were analyzed separately: +100 mesh (24%), 100 to 325 mesh (69%), and −325 mesh (7%). One type B sample, 10058, was lost due to failure of a zirconium crucible during fusion.

Four terrestrial samples were also measured: duplicates of Columbia River basalt BCR-1, and one sample each of Knippa basalt and Adirondack pyroxene gabbro. The latter, first described by BUDDINGTON (1939), contains 6·9% TiO_2 and may be the closest terrestrial analogue of the Apollo 11 rocks (OLSEN, 1969).

Procedure

Details of the chemical procedure will be published separately (KEAYS and GANAPATHY, in preparation). Samples of ~0·1 g were prepared in an office area where no chemicals had ever been stored or handled, to reduce the risk of contamination. The samples and group monitors (evaporated on Specpure MgO powder) were sealed in quartz vials, and irradiated for 4 days in the hydraulic facility of the Oak Ridge reactor, at a thermal flux of 2×10^{14} neutrons cm^{-2} sec^{-1}. The irradiation can contained 8 samples and one duplicate set of 4 group monitors. After radiochemical processing, the specific activity of duplicate monitors always agreed to within 5 per cent. Most samples were counted on NaI scintillation counters except Rb, Cs, Co (Ge-Li detector), Ag, Pd (β–γ coincidence counter), Bi and Tl (low level β-proportional counter). Radiochemical purity was checked by γ-spectrum and half-life; where appropriate also by β/γ ratio, β–γ coincidence rate (Bi, Tl), or peak-to-total ratio. A few Tl samples were recycled. Chemical yields were determined gravimetrically, except for Ir, where reactivation was used.

The shortest-lived species measured was 13-hr Cu^{64}. Longer-lived isotopes were used for all other elements. Results for Te had to be based on 117d Te^{123m} rather than 8d I^{131}, because the latter nuclide was also produced in copious amounts from uranium fission. Even the Te^{123m} data could be obtained only after complete decay of fission-produced 78-hr Te^{132}, too late to be included in any of the figures of this paper. Owing to the low sample-to-background ratios, the Te results may be in error by as much as 10–30 per cent.

The Pd-numbers (determined via Ag^{111}) are somewhat uncertain, because Ag^{111} is also made from Ag^{109} by double-neutron capture. Corrections for this effect were rather large, and not too well determined: ~10% for Ag/Pd = 0·5 and ~60% for Ag/Pd = 5.

Owing to the slowness of our procedure and the need to conserve lunar material, we generally performed only one analysis on each sample. The reproducibility of our procedure seems to be fairly good, judging from triplicate analyses of lunar soil 10084-49 (one of the three actually being a weighted mean of three individual analyses on sieve fractions), duplicates of Columbia River basalt BCR-1, and replicates of 12 chondrites (KEAYS and GANAPATHY, in preparation).

Some assurance for the absolute accuracy of our results comes from the generally good agreement of our data with published analyses of BCR-1 and Apollo 11 samples. Moreover, the apparent neutron fluxes derived from the specific activities of the 15 monitors generally agreed within a factor of 2 or 3, and were close to the nominal flux of $(2 \pm 1) \times 10^{14}$ n cm^{-2} sec^{-1} (Table 1). This suggests that errors such as self-shadowing, incomplete exchange with carrier, loss by volatilization or recoil, and cross-contamination of group monitors were small. Much of the inter-element variation in Table 1 must be due to uncertainties in absolute counting efficiencies, which were not well known in most cases.

Since the chemical yield of the bromine monitors in irradiation 2 was suspiciously high, we rechecked all bromine samples and monitors by re-irradiation. Appreciably lower yields were found for two monitors in irradiation 2 and for two samples: 10084 and 10061. The discrepancy was due to coprecipitation of some impurity, perhaps Ag_2SO_3. The bromine values in Table 2 have been corrected. Since it was too late to make any changes in the figures themselves, we have indicated corrections in the figure captions.

Table 1. Apparent neutron flux inferred from monitors ($\times 10^{14}$ cm^{-2} sec^{-1})*

Irradiation†	Ir	Au	Zn	Cd	Ag	Bi	Tl	Br	Pd	Co	Cu	Ga	Rb	Cs	In
1	—	3·0	1·1	2·2	2·3	3·6	2·4	1·5	2·9	2·6	1·9	1·2	0·8	0·8	0·9
2	1·0	2·4	1·3	2·4	2·3	3·7	3·0	3	3·1	2·4	2·3	3	0·8	0·8	1·1
3	1·3	2·3	1·2	3·2	2·2	3·5	2·9	2	2·8	2·3	2·0	2	0·7	0·7	0·9

* Apparent fluxes are based on estimated counting efficiencies, and may therefore have systematic errors from element to element. The quoted neutron flux for the Oak Ridge reactor is 2×10^{14} n cm^{-2} sec^{-1} ($\pm 50\%$).

† The following samples were included in the three irradiations:
1. BCR–1 (2 samples), Knippa basalt.
2. 10072, 10084 (5 samples).
3. 10020, 10057, 10047, 10050, 10048, 10061, APG.

Results

Our results are shown in Table 2. We have also listed data by other investigators, except for Cu, Co, Ga and Rb, where the measurements were numerous and generally concordant.

Doubtful measurements

Italicized measurements are considered doubtful, for reasons listed below.

10047. This rock (powdered at the LRL) gave higher values for 7 elements than did any of our remaining A or B samples. Although these results may be real, we fear that they may reflect contamination during crushing.

Iridium. 10072: much higher than any other A, B rock. Probably laboratory contamination.

Gold. Several results by Wänke *et al.* (1970) were much higher than ours.

Silver. Most of the high results are regarded as questionable, on suspicion of being due to contamination by In–Ag vacuum gaskets on Apollo sample return containers. They include our data on 10072, 10047 and 10061, as well as results by Morrison *et al.* (1970) and Turekian and Kharkar (1970).

Bromine. Data by Morrison *et al.* (1970) were discarded, because they either were higher than ours or were given to only one significant figure. Three values by Reed *et al.* (1970) were discarded, because they were markedly higher than our results on the same or similar samples.

Palladium. Results by Morrison *et al.* (1970) were 1–2 orders of magnitude higher than ours.

Cesium. Results by Morrison *et al.* (1970), though roughly consistent with other measurements, were given to only one significant figure.

Indium. As in the case of silver, the values covered an enormous range, from 1 to 2000 ppb. The contamination problem for In is even more severe than that for Ag. Not only is In the major constituent of the vacuum gasket alloy, but it is also likely to be enriched in the surficial oxide film, owing to its more electropositive character.

Table 2. Abundances in Apollo 11 lunar samples and terrestrial rocks*

Sample No.	Type	Ir (ppb)	Au (ppb)	Zn (ppm)	Cd (ppb)	Ag (ppb)	Bi (ppb)	Tl (ppb)	Br (ppb)	Pd (ppb)	Te (ppb)	Co (ppm)	Cu (ppm)	Ga (ppm)	Rb (ppm)	Cs (ppb)	In (ppb)
10020	A	0·027	0·075	1·29	6·37	2·27	0·15	0·33	16·8	1·5	13	5·65	6·57	1·90	0·74	30·6	*14·6*
				2·1f		*100*f			*100*f	*90*f						*200*f	
				*26*e		*1300*j											
				10d													
10057	A	0·023	0·017	1·71	3·15	0·69	0·27	1·09	25·2	7·3	8	27·2	3·52	>3·5	3·68	159	3·2
			*1·6*k	2·9f	*900*f	*40*f			*100*f	*100*f						*200*f	3·0b
																200k	2·9b
																	2·7k
																	*70*f
10072	A	*4·02*	0·14	1·81	6·47	*17·3*	0·73	0·92	36	3		27·2	4·94	4·73	5·98	159	*179*
				7·0f	*1000*f	*600*f			*100*f	*100*f						*300*f	*49*b
				*24*m					*79*g								*54*b
				*34*c													*2000*f
10017	A/B		*8·7*k	47m	44h				80g							120k	*138*k
				*49*e												155d	1h
				30d													
10049	A			9·3d												166d	
10071	A			11d												170d	
10047†	B	*0·240*	*0·33*	5·76	*255*	*24·7*	*2·15*	0·57	*102*	2	13	14·4	13·3	5·35	1·25	44·6	*109*
				*13*c					*330*g								
10050	B	0·007	0·031	1·75	2·56	1·42	≤0·16	0·33	10·4	1·4	11	15·2	15·2	4·41	0·60	25·9	4·4
10003	B			7·4d												22d	
10024	B			*14*c													
10044	B		*1·9*k			*410*j			*216*g								
10045	B			*14*c													
10058	B			9·3f	*700*f	*70*f			*300*f	*200*f						27d	*600*f
																*300*f	
10062	B			9·9d		*170*j										32d	
10056	—			2·7f	*900*f	*200*f			*60*f	*100*f						*60*f	2·4b
						*160*j											3·5b
																	*60*f
10048	C	6·88	2·66	28·6	78·3	23·6	1·62	2·83	138	13·0	72	34·8	9·19	5·85	4·15	128	*95·5*
																	*56*b
																	*65*b
10061	C	9·18	3·42	29·2	106	*163*	2·79	2·70	246	7	73	34·2	22·0	5·79	3·99	146	*1430*
				27a					150g								
				*37*c					260g								
10018	C			23a													
				*54*c													
10019	C								*380*g								5h
10021	C			24a													*14*b
																	*30*b
10046	C			30f	*800*f	20f			*200*f	*100*f						*200*f	6b
						*1300*j											*25*b
																	*80*f
10059	C			29a													
10060	C			25f	*300*f	10f			*300*f	6i						190i	*911*b
																*200*f	*1150*b
																	4i
10065	C			23a													
10068	C			22a													
10073	C			23a												98d	
				24d													
10084-1‡	D	6·28	2·87	10·6	28·3	4·33	1·43	7·35	67	8·3		26·1	6·99	4·89	3·36	99	*11·6*
10084-2‡	D	6·67	1·99	21·0	28·3	8·53	1·96	1·91	91	9·2		26·7	8·94	5·34	3·17	94	*1950*
10084-3‡	D	10·77	4·41	34·2	46·0	24·1	3·77	6·74	153	8·6		32·5	11·2	5·72	3·27	104	*1690*
10084-W‡	D	6·88	2·38	19·5	29·6	8·67	1·97	3·54	90	8·9		27·0	8·64	5·25	3·22	96	*1470*
10084	D	7·62	4·15	21·1	53·3	8·89	1·55	1·65	87	11·0		26·8	8·07	5·41	3·33	98	*524*
		6·93	2·01	21·0	35·4	8·60	1·37	1·47	77	9·4		28·1	7·75	5·24	3·09	94	*768*
			2·1k	22f	56h	*100*f			*200*f	*40*f						104d	*677*b
				19a	*300*f	*250*j			51g							120k	*681*b
				*36*m					230g							120i	*750*k
				*37*c												*200*f	*580*h
				*47*e													*500*i
				24d													*2000*i
																	*500*f
BCR-1‖	Bas.		0·43	127	99·0	26·4	40·2	274	97	13·0		35·4	17·1	20·1	44	930	95·5
			1·33	126		26·7	44·4	283	152	11·0		36·7	18·8	22·5	48	920	107
Knippa‖	Bas.		2·85	125			88·3	20·3	1300			76·8	15·0	20·2	26	745	115
APG‖	Gab.		3·59	125	160	34·8	5·84	24·5	80		10	67·6	17·6	1·41	3·31	104	112
A, B ave.		0·066	0·041	7·41	12·5	1·46	0·33	0·63	22·1	3·40	16	20·9	9·33	4·48	3·58	121	2·9
C ave.		8·03	3·04	25·6	92	17·87	2·20	2·76	199	8·67	72	31·7	15·3	5·01	3·59	140	5·0
D ave.		7·14	2·66	21·1	43·6	8·72	1·63	2·22	107	9·77		29·9	12·3	4·63	3·10	105	
$\left(\frac{\text{C–AB ave.}}{\text{Cl}}\right)\%$		1·80	2·00	5·51	7·87	4·21	1·64	2·87	3·65	0·95	2·2						
$\left(\frac{\text{D–AB ave.}}{\text{Cl}}\right)\%$		1·60	1·75	4·15	3·08	1·87	1·14	2·15	1·75	1·14							

* Doubtful values are shown in italics.
† Powder prepared at LRL; probably contaminated.
‡ Sieve fractions: 1 = +100 mesh (24%).
2 = 100 to 325 mesh (69%).
3 = −325 mesh (7%).
W = weighted average of individual analyses of fractions 1, 2, 3.
‖ BCR–1 = Columbia River basalt; Knippa = Knippa basalt; APG = titanium-rich Adirondack pyroxene gabbro, 1 mile SSW of Brown Point, Willsboro Quadrangle. We are indebted to A. F. Buddington for this sample. Powdered material is available from G. G. Goles, Department of Geology, University of Oregon, Eugene, Oregon 97403.

a. Annell and Helz (1970); b. Baedecker and Wasson (1970); c. Compston *et al.* (1970); d. Gast and Hubbard (1970); e. Maxwell *et al.* (1970); f. Morrison *et al.* (1970); g. Reed *et al.* (1970); h. Schmitt *et al.* (1970); i. Smales *et al.* (1970); j. Turekian and Kharkar (1970); k. Wänke *et al.* (1970); m. Wiik and Ojanperä (1970).

On geochemical grounds, one would expect In to be about as abundant as Tl, i.e. 0·3 to 8 ppb. Accordingly, we discarded all In values greater than 10 ppb. Even the remaining ones should be regarded as upper limits.

Sampling error

The small size of our samples (~0·1 g) brought about the risk of sampling error, particularly for the coarse-grained (Type B) rocks. In order to check this possibility, R. A. Schmitt kindly agreed to analyze our samples by his non-destructive neutron activation analysis procedure for several major elements (Table 3). Part of the data reduction was performed by H. Wakita at Oregon State University, and the remainder, by the authors. Samples marked with a dagger were later analyzed by us in their entirety; others represent powdered aliquots.

Table 3. Abundance of Na, Al, Ca, and Ti in Apollo 11 lunar samples*

Sample	Type	Na (%)	Al (%)	Ca (%)	Ti (%)
10020-25†	A	0·28	5·53	8·3	7·1
10057-31†	A	0·39	4·54	7·4	6·8
10072-23	A	0·41	4·55	7·6	8·1
10050-26†	B	0·32	5·83	8·9	7·3
10047-32	B	0·32	5·50	7·4	6·2
10048-31†	C	0·37	6·93	7·5	5·6
10061-31†	C	0·38	6·83	7·6	5·6
10084-49 (−325 mesh)	D	0·41	8·49	8·4	5·4

* Instrumental neutron activation analysis by R. A. Schmitt and H. Wakita.

† Analyzed for trace elements in their entirety. Others are aliquots of samples analyzed for trace elements.

The Na, Ca and Al contents agreed to within 5–8 per cent of literature values, and even Ti varied by no more than 20 per cent. This suggests that the feldspar and ilmenite contents of our samples were close to average. Our data thus should be fairly representative, at least for those trace elements that reside mainly in major phases.

Meteoritic Component in C and D Material

Our data are shown in Fig. 1, arranged by decreasing abundance in A and B rocks. Several elements, especially the less abundant ones (Ir, Au), are significantly enriched in C (breccia) and D (soil) over A and B rocks. Inasmuch as breccia and soil probably are largely derived from A,B rock, it seems likely that the increment is due to addition of extraneous material: meteoritic and solar-wind.

The enrichment shows up even better when the results are averaged for each class (Table 2, lines 3–5 from bottom; Fig. 2). *All* 16 elements seem to be enriched in soil and breccia, although the smaller enrichments may in part be due to non-representative sampling. Let us therefore confine ourselves to those 9 elements whose enrichment exceeds a factor of two. (A tenth element, Te, is not plotted.)

We can attempt to characterize the chemical composition of this extraneous component by subtracting the mean abundance in A,B rocks from the observed abundance in C,D material. To facilitate comparison with possible source materials, we shall normalize the data to Cl chondrites (bottom two lines in Table 2; Fig. 3). This approach had previously been used by Mazor and Anders (1967) in a study of gas-rich meteorites.

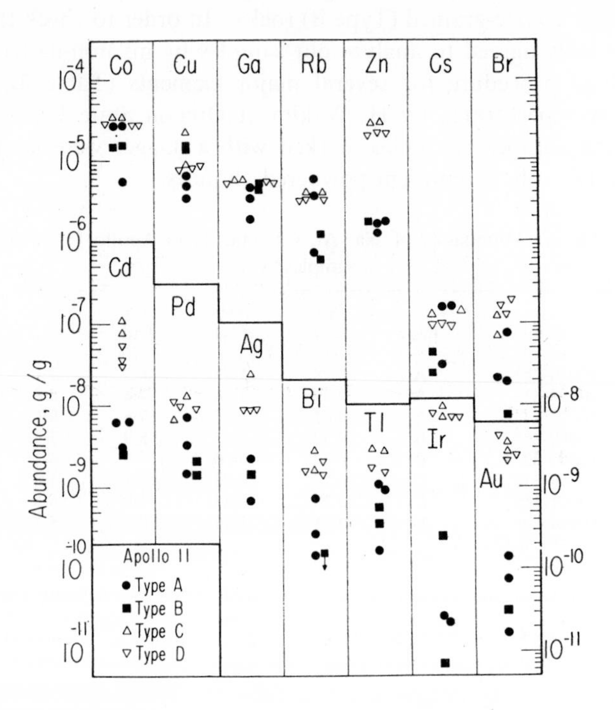

Fig. 1. Abundances of 14 trace elements in lunar samples, arranged by decreasing abundance in Type A and B rocks. Results for Types C and D (open symbols) often lie above those for Types A and B (filled symbols). The highest Br value for Type A and all Br values for Type D are to be lowered by a factor of 2 (see note added in proof at end of "Procedure").

An earlier version of this figure, based on our data alone, was given in our preliminary report on this work (Keays *et al.*, 1970). The present figure includes data from other laboratories, as well as three additional elements: Ni, Ge and Pb (non-radiogenic Pb^{204} only).

If the extraneous component consisted largely of ordinary chondrites, the histogram should follow a staircase pattern, paralleling the ordinary chondrite curve (dashed). If it consisted largely of iron meteorites, it should show a large peak for the first four elements, followed by a steep drop. The observed histogram shows neither of these trends. It is virtually flat, paralleling the carbonaceous chondrite curve except for a hump at Ag, Zn, Cd and Pb.

There is no material known that exactly matches the histogram in Fig. 3. Carbonaceous chondrites, Type I enstatite chondrites, and cosmic matter come close, but are slightly deficient in Ag, Zn, Cd, Pd. The last 3 elements are chalcophile, and we are therefore inclined to believe that the discrepancy is due to sampling. Our A,B "average" is based on relatively few measurements and may be unrepresentative, failing to include some Zn,Cd-rich rock that contributed to the soil. Indeed, data on

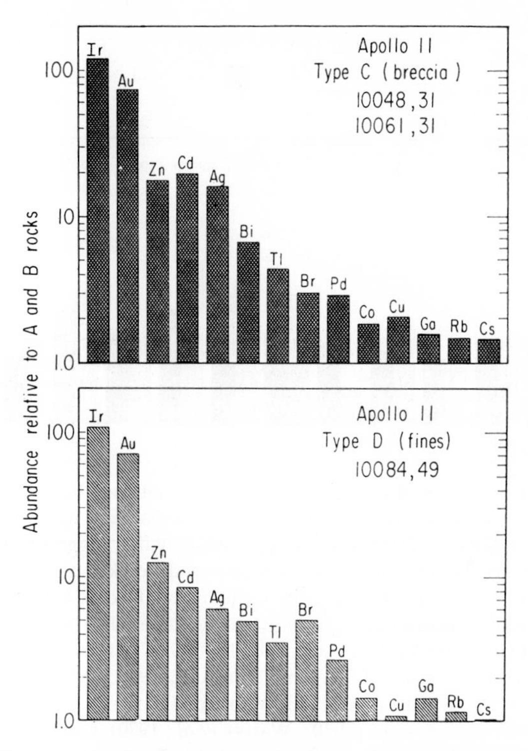

Fig. 2. For the first 9 elements, mean abundances in soil and breccia exceed those in Type A,B rocks, by a factor greater than 2. This suggests the presence of an extraneous (meteoritic?) component. The Br value for Type C should be 9. See note added in proof at end of "Procedure".

major elements, lanthanides, and Hf (SCHMITT *et al.*, 1970; GOLES *et al.*, 1970) suggest that a straight A,B average does not quite match C,D composition; some particular B rocks (e.g. 10044 and 10045) perhaps having contributed to a disproportionate degree (GOLES *et al.*, 1970). In order to bring Zn, Cd and Pb into line, their abundances in A,B rocks or in Cl chondrites would have to be raised twofold. The high Ag abundance, in turn, may be due to contamination by In–Ag vacuum gaskets.

The remaining abundances in Fig. 3 lie mostly in the range 0·01–0·02. If the highest 5 values are excluded (Zn and one each of the Cd, Ag and Pb values), the average is 0·0188. Insofar as these elements are concerned, Types C and D could have been made from Type A,B rocks by adding ~1·9 per cent of carbonaceous-chondrite-like material. ÖPIK (1969) had previously estimated a meteoritic component of 3 per cent in the lunar regolith.

The requirement that the extraneous component be "carbonaceous-chondrite-like" in composition sets a few limits on its nature. Type I and II carbonaceous chondrites and Type I enstatite chondrites are acceptable source materials, as are

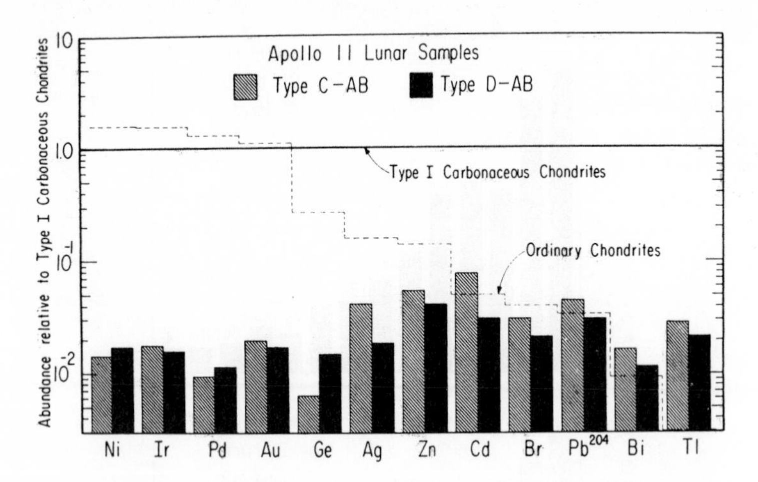

Fig. 3. Abundance pattern of meteoritic component, obtained by subtracting average abundances in A,B rocks from those in C and D material, and normalizing to Type I carbonaceous chondrites. Distribution is approximately flat, suggesting that meteoritic component consists largely of carbonaceous-chondrite-like material, admixed in an amount of ~1·9 per cent. Ordinary chondrite material would have given a staircase pattern parallel to the dashed curve; iron meteorite material, a large peak for the first four elements. Data from Table 1, except for Ni (LSPET, 1969; ANNELL and HELZ, 1970; COMPSTON *et al.*, 1970; GAST and HUBBARD, 1970; MAXWELL *et al.*, 1970; MORRISON *et al.*, 1970; WÄNKE *et al.*, 1970; WIIK and OJANPERÄ, 1970), Ge (BAEDECKER and WASSON, 1970; SMALES *et al.*, 1970; WÄNKE *et al.*, 1970), and Pb (TATSUMOTO and ROSHOLT, 1970). The tellurium data in breccias (Table 2), obtained too late for plotting, fall at 0·024 and hence also fit this trend. Note added in proof at end of "Procedure". The Br values should be changed to 0·0365 for C and 0·0175 for D.

other types of relatively undifferentiated cosmic matter: trapped solar-wind ions and cometary material. (The high water content of some of these materials does not disqualify them, as such water would evaporate during impact and quickly be lost from the moon.)

Solar wind

We can estimate the solar-wind contribution during the last 3·65 AE, the age of the Tranquillity Base region (ALBEE *et al.*, 1970). Let us use the element Ir for illustration. The present-day proton flux is 2×10^8 cm^{-2} sec^{-1} (TILLES, 1965); assuming a cosmic Ir/H ratio of $1{\cdot}65 \times 10^{-11}$ (CAMERON, 1968) we obtain an integrated flux of $1{\cdot}22 \times 10^{-7}$ g Ir cm^{-2} in 3·65 AE. The average thickness of the regolith in southern

Mare Tranquillitatis is 4·6 m (OBERBECK and QUAIDE, 1968). This value agrees with SHOEMAKER *et al.*'s (1970) estimates for the immediate vicinity of Tranquillity Base, 3–6 m. For a mean packed density of 1·8 g/cm^3 (LSPET, 1969), a 10 per cent content of large rock fragments ineffective in trapping solar wind, and complete vertical mixing, one would thus expect a mean Ir content of $1{\cdot}64 \times 10^{-10}$ g Ir/g, only about 2·2 per cent the value in C and D samples.

We cannot completely exclude the possibility that the solar wind intensity was higher in the past. An average proton flux of 9×10^{9} cm^{-2} sec^{-1} would be required to account for the entire Ir in C and D material. But this is surely an unrealistic limit: where then is the Ir from the meteorites that made the lunar craters? Moreover, the observed Xe content in lunar soil, $3{\cdot}6 \times 10^{-7}$ cm^3 STP/g (LSPET, 1969), agrees well with the amount expected for the present-day flux, $3{\cdot}2 \times 10^{-7}$ cm^3 STP/g, for a trapping and retention efficiency of 1. The most plausible interpretation thus is that only some 2 per cent of the extraneous component is derived from the solar wind. The rest must be meteoritic.

Meteorites

If ~98 per cent of the extraneous component is of "meteoritic" (to be exact, "particulate") origin, the average influx rate at Tranquillity Base during the last 3·65 AE is $3{\cdot}8 \times 10^{-9}$ g cm^{-2} yr^{-1}. This figure agrees fairly well with an analogous estimate for the earth, $(1{\cdot}2 \pm 0{\cdot}6) \times 10^{-8}$ g cm^{-2} yr^{-1}, obtained by BARKER and ANDERS (1968) from the Ir and Os content of Pacific and Indian Ocean sediments. At least part of the difference may be due to the disproportionately greater gravitational enhancement of the earth's capture cross section, which amounts to a factor of 4 at a geocentric velocity of 6 km/sec and a factor of 2 at 10 km/sec. Also, the BARKER and ANDERS value was more in the nature of an upper limit than an actual value, owing to ambiguities in interpretation.

Not too much should be made of this agreement, though. The terrestrial value is an average for the last 10^5 yr over some 10^7–10^8 km^2, while the lunar value is an average for $>10^9$ yr at a single location. Both the mechanisms and efficiencies of lateral and vertical mixing are surely different for the two bodies, particularly for large projectiles. Nonetheless, the close agreement between soil and breccias suggests that our results may be representative of at least the southern part of Mare Tranquillitatis. The breccias apparently originated at some depth within the regolith and at some distance (kilometers?) from Tranquillity Base. Yet the extraneous components in breccia and soil generally agree to within 20 per cent.

The trace element enrichment in the −325 mesh fraction of the soil (Fig. 4) probably reflects a larger admixture of meteoritic material, consistent with the higher abundance of glass and highly shocked material in the finest fraction. The major element composition of this fraction (Table 3) is close to that of bulk soil; hence it is unlikely that mineralogical sorting is responsible. Meteorite impact, on the other hand, can probably account for the enrichment. Material closest to the impact center will be comminuted most severely and will receive the greatest admixture of projectile matter. Grain crushing by subsequent impacts in the regolith would be an additional mechanism tending to enrich the finest fraction in meteorite material.

Sources of the meteoritic component

We can try to learn something about the nature of the meteoritic component by comparing the observed content of meteoritic material, 1·88 per cent, with theoretical estimates from cratering theory. Two approaches are available, one relating the content of meteoritic material to projectile velocity, and the other, to crater counts.

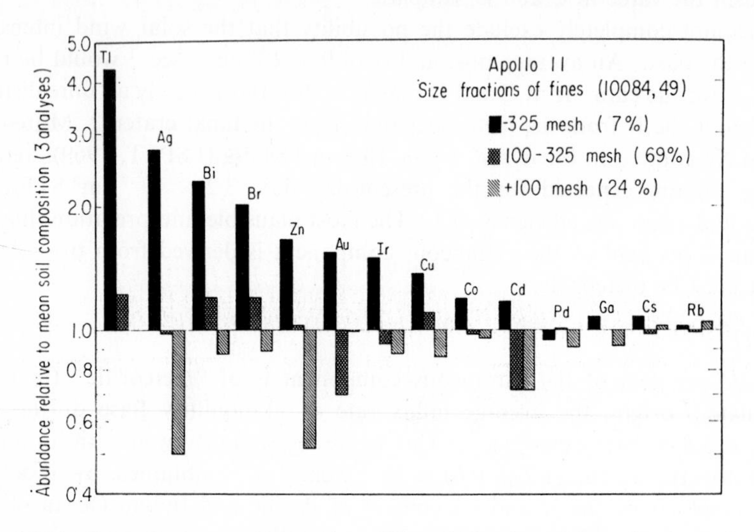

Fig. 4. Trace element enrichment in −325 mesh fraction of the soil apparently reflects increased content of meteoritic material.

Projectile velocity. In ÖPIK's (1958, 1961a) theory of craters, the ratio of eroded mass to projectile mass, M/μ, depends on impact velocity w, the crushing strength s, and density ρ of the target rock, and a dimensionless factor k that allows for the extra momentum of backfiring:

$$M/\mu = kw(\rho/s)^{\frac{1}{2}}.$$

The parameter k is a slowly varying function of impact velocity, projectile density, etc.; it generally falls between 2 and 5 and can be computed exactly from relations given by ÖPIK. For Apollo 11 lunar rocks, $\rho = 3{\cdot}3$ g/cm³. The crushing strength is estimated as 9×10^8 dynes/cm² (ÖPIK, 1969). Setting $M/\mu = 1/0{\cdot}0188 = 53$, and $k = 2{\cdot}0$, we obtain an impact velocity of 4·4 km/sec, corresponding to an encounter velocity of 3·7 km/sec. This value is suspiciously low. The Apollo asteroids, which are probably typical of the crater-forming population, fall into two velocity groups: an "asteroidal" group of mean geocentric velocity 14 km/sec and a "cometary" group of 28 km/sec (ANDERS and ARNOLD, 1965). Long-period or parabolic comets will have still higher velocities, so that a reasonable average for cometary objects is 40 km/sec (ÖPIK, 1969). For asteroidal objects, a round figure of 15 km/sec seems appropriate.

The relative proportions of asteroidal and cometary impacts are not well known.

For the two limiting cases of a purely asteroidal and a purely cometary flux, $M/\mu =$ 252 and 1080, corresponding to meteoritic components of 0·40 per cent and 0·09 per cent in the regolith. This leaves an apparent excess of 1·48 to 1·79 per cent meteoritic material, not accounted for by crater-forming impacts.

Some of this excess may be due to multiple impacts. Below a crater diameter of 186 m, the crater distribution in Mare Tranquillitatis seems to be in a steady state (SHOEMAKER *et al.*, 1969, p. 67). Small impacts that do not penetrate to the base of the regolith add meteoritic material without eroding fresh lunar rock. Thus the ratio M/μ slowly decreases with time.

The major part of the excess is probably due to micrometeorites, which likewise impact the regolith without causing further erosion of bedrock. ÖPIK (1956, 1969) has predicted a micrometeorite influx rate of $1·05 \times 10^{-8}$ g cm^{-2} yr^{-1}, on the basis of zodiacal light measurements. Our data suggest lower values, $\lesssim 3·0 \times 10^{-9}$ to $\lesssim 3·7 \times 10^{-9}$ g cm^{-2} yr^{-1}. Corresponding limits for the earth (corrected for the earth's increased gravitational cross section, at a geocentric velocity of 6 km/sec, as appropriate for dust) are $\lesssim 5·7 \times 10^4$ to $\lesssim 7·0 \times 10^4$ tons/yr. They happen to agree with a currently accepted upper limit of $\lesssim 7 \times 10^4$ tons/yr based on particle collections and satellite-borne impact detectors (PARKIN AND TILLES, 1968).

Crater counts. Let us, for the moment, neglect a possible micrometeorite contribution, and compare the observed influx rate with independent estimates based on crater counts. The density of primary craters in southern Mare Tranquillitatis is given by

$$F = 10^{12·20} C^{-2·58}$$

where F is the cumulative number of craters of diameter greater than C meters, per 10^6 km^2 (TRASK, 1966). Judging from Fig. 3 of TRASK, this relation holds up to diameters of 2·7 km. Above this limit, the data can be represented fairly well by a function of less steep slope:

$$F = 10^{8·92} C^{-1·69}.$$

These equations can be expressed in terms of mass influx, if D, the ratio of crater diameter to projectile diameter, is known. From the relations given by ÖPIK (1961a) we find, in round figures, $D = 10$ for asteroids (15 km/sec) and $D = 20$ for comets (40 km/sec). Assuming a projectile density of 2·6 g/cm^3 (carbonaceous chondrites) we obtain the following expressions for the cumulative mass M (g) per 10^6 km^2 as a function of projectile diameter, x (m):

Asteroids: $M = 3·48 \times 10^{16}\, x^{0·42}$ for $C \leq 2·7$ km
$M = 2·98 \times 10^{13}\, [x_2^{1·31} - x_1^{1·31}]$ for $C \geq 2·7$ km

Comets: $M = 5·8 \times 10^{15}\, x^{0·42}$ for $C \leq 2·7$ km
$M = 9·2 \times 10^{12}\, [x_2^{1·31} - x_1^{1·31}]$ for $C \geq 2·7$ km.

It is not obvious to what limiting crater diameter these distributions should be integrated. The largest post-mare crater near Mare Tranquillitatis is Theophilus, 105 km in dia. But most impact ejecta are deposited near the crater, and hence large, widely-spaced craters contribute less than their share to the major part of the regolith.

ÖPIK (1969) estimates the effective throwout range to be $15 + 0{\cdot}3B_0$ km, where B_0 is the crater diameter in km, and suggests on this basis that only craters of <2·5 km dia. provide anything like continuous blanketing. To make matters worse, there is no assurance that meteoritic and lunar debris are ejected with equal efficiency, in identical patterns.

In view of these uncertainties, we shall attempt a graphical comparison. Figure 5

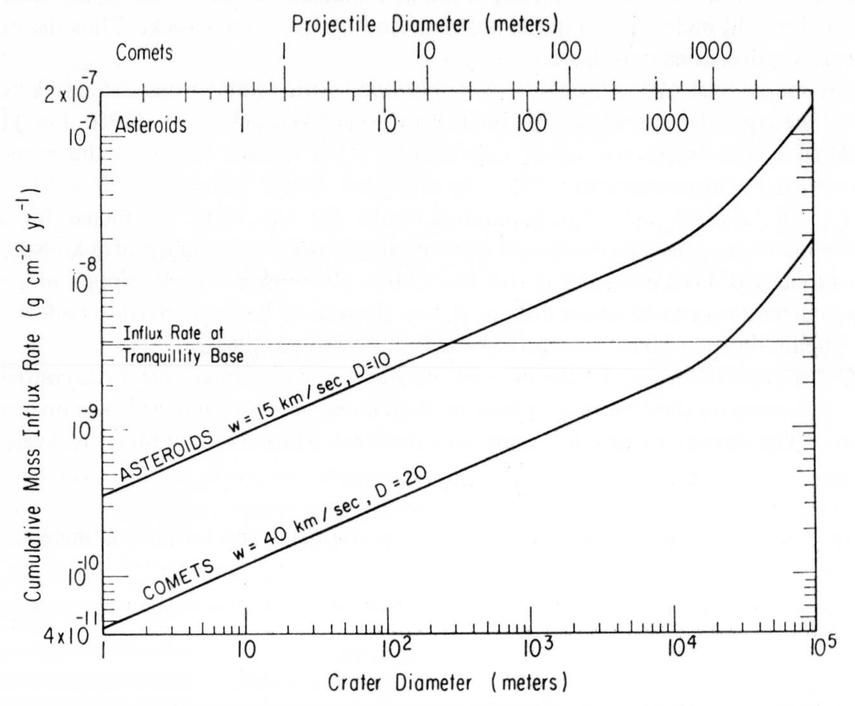

Fig. 5. Mass influx rates predicted from observed crater density in Mare Tranquillitatis (TRASK, 1966), for asteroidal and cometary projectiles (w = impact velocity, D = ratio of crater diameter to projectile diameter). Observed influx rate (horizontal line) can be accounted for by craters of ≤0·3 km or ≤25 km dia., if all projectiles are asteroids or comets, respectively. Larger craters do not broadcast their ejecta far enough to make a uniform contribution to the regolith.

shows cumulative influx rates deduced from crater counts, for both asteroidal and cometary cases. These can be compared with the observed rate.

If asteroids are the dominant projectiles, the observed meteoritic component would seem to come largely from craters less than 0·3 km in diameter. Inasmuch as ≥0·3 km craters are spread, on the average, only 1·2 km apart, the effective fallout range for meteoritic material would have to be on this order, less than the ≥15 km estimated by ÖPIK. Alternatively, the ejection efficiency for meteoritic material may be less than 100 per cent, particularly for larger craters.

For comets, craters up to 25 km dia. would have to contribute, to account for the observed influx rate. The nearest crater in that size range is Sabine, 90 km from

Tranquillity Base. However, if a substantial part of the meteoritic material came from micrometeorites, the size limit would be correspondingly reduced. Data from other lunar sites will be required to estimate the relative importance of asteroidal, cometary, and micrometeoritic material.

Prevalence of carbonaceous-chondrite-like material. The meteoritic component thus is derived from craters between 0 and at least 0·3 km dia., possibly up to 25 km dia. Most of the projectiles in the corresponding mass range appear to be of carbonaceous chondrite composition. This is remarkable, in view of the scarcity of carbonaceous chondrites among recovered meteorites. Several authors have previously suggested that carbonaceous chondrites comprise the major part of the meteoritic influx, but are underrepresented in collections because their friability leads to preferential destruction in the atmosphere (MAZOR and ANDERS, 1967; SHOEMAKER and LOWERY, 1967; MCCROSKY, 1968).

Two possible origins of carbonaceous chondrites have been proposed. HERBIG (1961) was the first to suggest that they might be derived from comets, and this suggestion has been taken up by other authors. If they are cometary, their typically high geocentric velocity would be a further factor causing preferential destruction.

On the other hand, carbonaceous chondrites show evidence of preterrestrial exposure to liquid water (DUFRESNE and ANDERS, 1962), which is easier to reconcile with an asteroidal than with a cometary origin. WHIPPLE (1966) has proposed that small, "half-baked" asteroids might consist of carbonaceous chondrite material throughout, but this is unlikely if accretion of solid bodies began at high temperatures (LARIMER and ANDERS, 1967; ANDERS, 1968; TUREKIAN and CLARK, 1969). More likely, the material approached carbonaceous chondrite composition only toward the end of the accretion process, when temperatures were less than 400°K. The surface layers of all asteroids hence should consist of carbonaceous-chondrite-like material (FISH *et al.*, 1960; ANDERS, 1963). It is interesting in this connection that much of the zodiacal dust (which in turn is the main source of micrometeorites) seems to be of asteroidal origin (SINGER, 1969).

The two origins are not mutually exclusive: both comets and asteroids may contain carbonaceous chondrite material. In any event, it is interesting that so large a part of the interplanetary debris consists of very primitive material.

COMPOSITION OF TYPE A AND B ROCKS

Inter-element correlations

A few correlations show up in our data. Rb, Cs, Tl (and perhaps also Co, Pd, Bi) correlate with K (Fig. 6). This suggests that Tl is mainly lithophile in the lunar crust, as in the earth (AHRENS, 1948; SHAW, 1950) but not in the chondrites (REED *et al.*, 1960) or eucrites (LAUL *et al.*, unpublished work).

Ag, Au, Ir, Br and Bi are fairly well correlated, judging from the fact that their abundances in the 5 rocks decrease in almost identical order (Fig. 7). The highest points are due to rock 10047, though, which may have been contaminated. Presumably these elements are present in a metallic phase, such as Fe or Cu (which shows an abundance pattern similar to that in Fig. 7).

Zn, Cd and Pb (non-radiogenic Pb^{204} only, from TATSUMOTO and ROSHOLT, 1970; SILVER, 1970) correlate with each other (Fig. 8) but not with Au (again with the exception of sample 10047). Presumably these elements are chalcophile, as on earth, whereas Au is siderophile. We cannot test this inference by direct correlation plots against S, because S, like Zn, Cd and Pb, shows little variation in A,B rocks. This lack of variation is, of course, consistent with the postulated similarity in geochemical character.

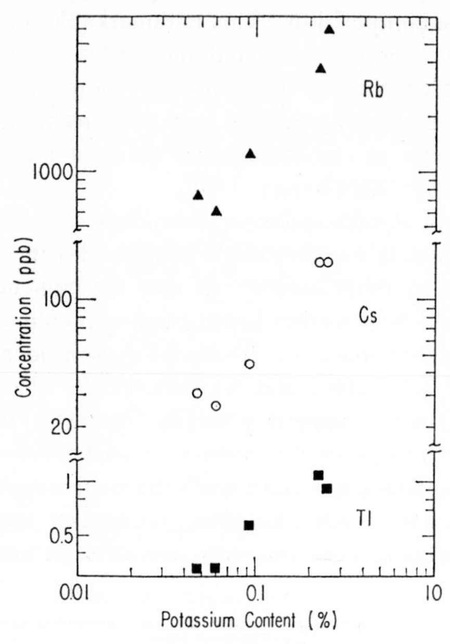

Fig. 6. Thallium correlates with K, Rb and Cs in lunar rocks, which suggests lithophile character.

Gross compositional trends

Some tentative conclusions about the moon may be drawn from a comparison of A,B rocks with other materials. In order to extend this comparison to as many elements as possible, we have averaged the more reliable analyses of Apollo 11 rocks, as published in the January 30, 1970, issue of *Science*, and normalized them to cosmic abundances (Fig. 9) and TAYLOR's (1964) average continental basalts (Fig. 10). (Insufficient data were available for oceanic basalts, which otherwise might have been a more appropriate reference material.) Elements are arranged according to group number in the Periodic Table (long form), with symbol size proportional to period number.

Ideally, we would like to derive whole-moon abundances for each element. By comparing these with cosmic abundances, we can then attempt to reconstruct the cosmochemical processes that led to the formation of the moon. But this is rarely possible for a differentiated body. Only for elements that are almost quantitatively concentrated at the surface (U, Th, Ba, La) or in the interior (siderophile elements) is there any hope of obtaining a reliable estimate. Moreover, the Apollo 11 rocks all came from a single region, whose composition differs measurably from that of other parts of the lunar crust (FRANZGROTE *et al.*, 1970; PATTERSON *et al.*, 1970; LSPET, 1970; WOOD *et al.*, 1970). Thus any of our generalizations from the Apollo 11 rocks must be regarded as tentative.

The *alkalis* (Group IA) are depleted by factors of 0·15 to 0·25, relative to continental basalts (Fig. 10). A whole-moon abundance can be estimated via the K/U ratio, which is 2800 in Apollo 11 rocks (LSPET, 1969; O'KELLEY *et al.*, 1970), and

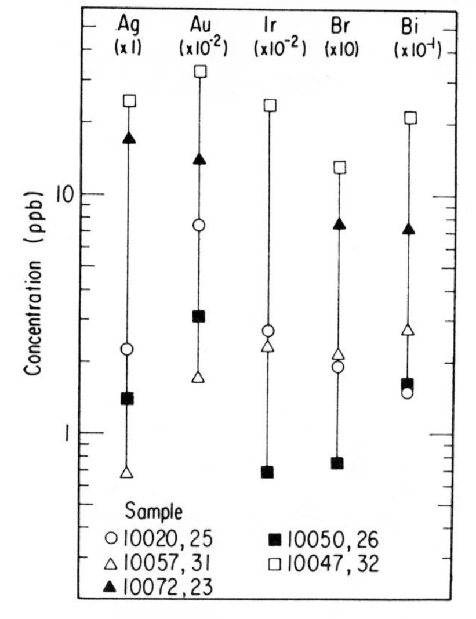

Fig. 7. Abundances of Ag, Au, Ir, Br and Bi in 5 lunar rocks decrease in almost identical order. Presumably all five elements are siderophile, and reside in a metallic phase (Fe or Cu).

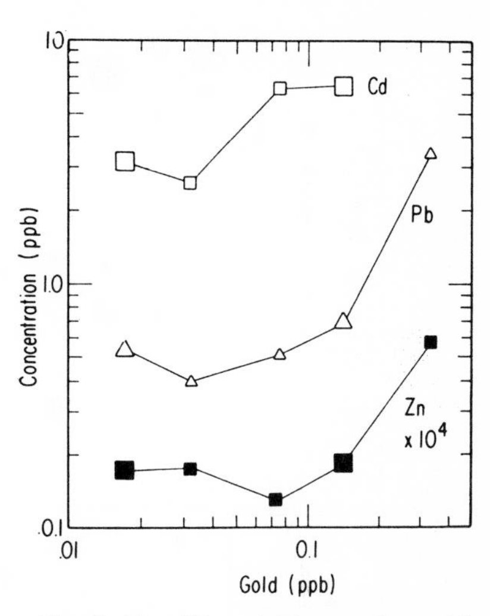

Fig. 8. Zn, Cd and Pb correlate with each other, but not with (siderophile) Au. These three elements may be chalcophile. Alkali-rich rocks are indicated by larger symbols.

1400–2200 in Apollo 12 materials (LSPET, 1970). These values are lower than the terrestrial ratio of 10,000 (WASSERBURG *et al.*, 1964 and references cited therein), the cosmic ratio of 60,000 (CAMERON, 1968), or the Type I carbonaceous chondrite ratio of 55,000 (EDWARDS and UREY, 1955; MASON, 1962; MORGAN and LOVERING, 1968). If the Apollo 11, 12 ratios are assumed to be typical of the moon as a whole, the alkalis are depleted by a factor of $\sim$0·04 relative to the cosmic ratio.

It does not seem likely that the alkalis were volatilized from the lunar surface, as suggested by O'HARA *et al.* (1970), and others. The Rb^{87}/Sr^{87} dating evidence shows that the Rb depletion did not occur during the 3·65 AE melting event, but at a much earlier time: within 11 m.y. of the formation of the meteorite parent bodies 4·7 AE ago (ALBEE *et al.*, 1970; see also HURLEY and PINSON, 1970). Insofar as the chronology is concerned, this loss could have occurred just before or just after the accretion of the moon. The latter alternative faces three objections: (1) volatilization from a melt should have strongly fractionated the alkalis from each other; (2) there was no significant depletion of Rb (or even the more volatile Pb) in the 3·65 AE melting event; (3) this process would not work for the similarly depleted earth, which has a much higher escape velocity.

Probably the fractionation occurred in the solar nebula. In terms of the model of LARIMER and ANDERS (1967), the moon may have accreted from ~96 per cent alkali-depleted high-temperature material and ~4 per cent low-temperature material. For the earth, the corresponding proportions would be 85 per cent and 15 per cent.

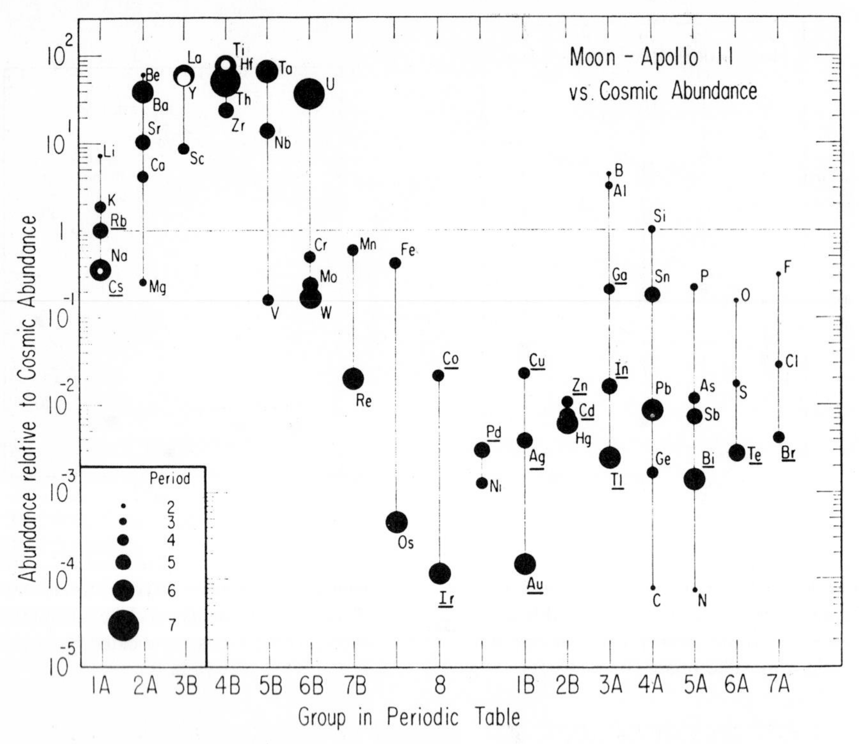

Fig. 9. Abundances in Apollo 11 rocks, normalized to cosmic abundances (CAMERON, 1968). Elements measured by the authors are underlined; all other were taken from *Science* **167**, 461–583 (1970).

These figures assume a normal (= cosmic) U-content in the high-temperature fraction. However, LARIMER and ANDERS (1970) have noted that the first condensate from the solar nebula would be enriched in U, Th, Zr, Ti, etc. (LORD, 1965; LARIMER, 1967). If the earth or moon had accreted mainly from this early condensate of higher-than-cosmic U-content (GAST, 1968), a correspondingly larger amount of low-temperature fraction would be required to account for the observed K/U ratios.

The greater alkali depletion of the moon may actually be a natural consequence of simultaneous formation of earth and moon, as we shall see in our discussion of volatile elements.

The *refractory* Group 3B to 5B elements are notably enriched relative to continental basalts (Fig. 10). Some authors have attributed this enrichment to selective volatilization of all lower-boiling elements. But the enrichment factors relative to cosmic abundances (Fig. 9) are nearly constant at ~60 for Ba, Y, La, Ti, Hf, Ta, Th and U. If the Apollo 11 basalts were a volatilization residue, then the starting material must have had cosmic composition, and have been volatilized to a residue of only 1/60. It seems more plausible to attribute the enrichment to magmatic differentiation. A constant enrichment factor of 60 then implies that these elements, like Ba, U and Th on earth, are almost quantitatively concentrated in the crust.

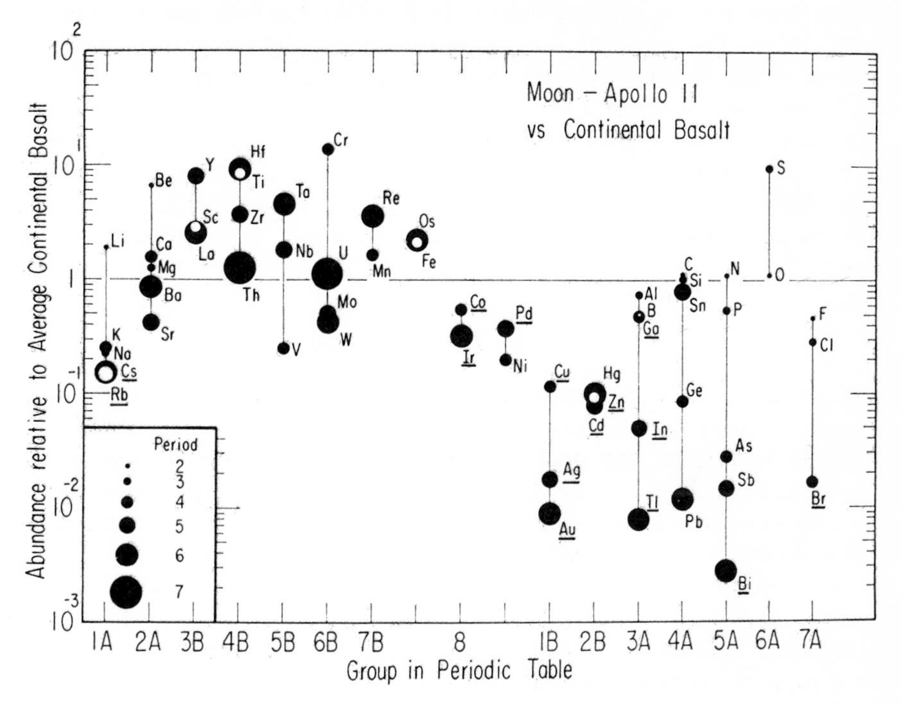

Fig. 10. Abundances in Apollo 11 rocks, normalized to average continental basalt (TAYLOR, 1964)

The *siderophile* elements show an "unearthly" depletion pattern. Re and Os are less strongly depleted than in terrestrial rocks, but Co, Ir, Ni, Pd and especially Au are more strongly depleted (Fig. 10). This contradicts the fission hypothesis for the origin of the moon (WISE, 1963; O'KEEFE, 1969). If the moon had lost its siderophiles while still part of the earth, it should show a terrestrial siderophile pattern. The fact that it has a characteristic pattern of its own suggests that removal of siderophiles from the moon was a separate event, that proceeded under different physico-chemical conditions.

The behavior of gold shows particularly clearly that conditions were different. On earth, gold is much less depleted than are the platinum metals, by only a factor of

$\sim 10^{-2}$ compared to $\sim 10^{-4}$ for Ir. In the Apollo 11 rocks, both Au and Ir show similar depletions, by a factor of $\sim 10^{-4}$ (Fig. 9).

Several factors may be responsible for these differences. Conditions on the moon certainly were more reducing than on earth, while pressure effects must have been less important. Also, the scarcity of water may have depressed the solubility of siderophile elements in the silicate phase.

The *volatile* and *chalcophile* elements in Groups 1B–7A are markedly depleted (Figs. 9, 10). Because of their volatility, these elements have been used as cosmothermometers, to determine the temperature at which various meteorite classes separated from the solar nebula (UREY, 1952, 1954; LARIMER and ANDERS, 1967; ANDERS, 1968). This approach can also be extended to differentiated planets, taking advantage of the fact that some of these elements tend to concentrate in the crust (RINGWOOD, 1966; ANDERS, 1968).

The most interesting group are the relatively most volatile, "strongly depleted" elements: Zn, Cd, Hg, In, Tl, Pb, Bi and Br (LARIMER and ANDERS, 1967). They are grossly depleted in Apollo 11 rocks, relative to both cosmic abundances and continental basalts. Much of our remaining discussion will be concerned with these elements.

DEPLETION OF Pb–In GROUP

It is somewhat risky to estimate whole-moon abundances of these elements from the Apollo 11 rocks alone. Most of the metals in Groups IIA–VA have at least some chalcophile tendencies, and since we concluded in our earlier discussion of the meteoritic component that some Zn- and Cd-rich rocks probably exist on the lunar surface, we must bear in mind the possibility that the true lunar abundance of these elements is higher than indicated by the Apollo 11 rocks. This concern is perhaps least warranted for Tl, which correlates with the alkalis and hence seems to be lithophile (Fig. 6). We shall tentatively assume that the Tl depletion factor from Fig. 10, 1×10^{-2}, is in fact characteristic of the moon as a whole. The Pb depletion factor, as inferred from the isotopic composition of lunar lead, is nearly identical (TATSUMOTO and ROSHOLT, 1970; SILVER, 1970).

Were these elements lost from the moon or were they never accreted? There are three principal possibilities which we shall examine in detail.

(1) *Loss from lunar surface.* We know that Pb, Tl, Bi, etc. were not lost in recent meteorite impacts on the moon, else these elements would have been underabundant in the extraneous component (Fig. 3). Impact and vaporization thus do not cause any loss of Pb, Bi, Tl, though more volatile substances such as water, sulfur and carbon compounds apparently do escape. We also know from Pb^{206}/U^{238} and Pb^{207}/U^{235} ages that the 3·6–4 AE melting event was accompanied by little or no lead loss (TATSUMOTO and ROSHOLT, 1970; SILVER, 1970). Apparently lava flows exposed to the lunar vacuum do not lose appreciable amounts of Pb, Bi, Tl during their normal freezing times. The fact that Hg, an element of far greater volatility, is no more strongly depleted than are Pb, Bi, Tl (REED *et al.*, 1970; Figs. 9, 10) also speaks against loss from a surface melt. The chronometric evidence provides a further

restriction: just as in the case of Rb, the loss must have occurred during or shortly after the moon's formation ~4·6 AE ago.

If this loss happened *after* accretion, conditions must have differed in some singular way from those in more recent times. Perhaps the surface was hot. ÖPIK (1967, 1969) has suggested that the moon accreted so rapidly (at a rate of several km/yr) that its surface was at a temperature of ~900°C during the formation of the pre-mare craters (see also RINGWOOD, 1966; RINGWOOD and ESSENE, 1970). Much of the moon's share of Pb, Bi, Tl, etc. would be accreted only at that late stage, when temperatures in the nebula had fallen low enough for these elements to condense. The gravitational potential energy of accretion amounts to ≥690 cal/g at that stage, enough to heat and vaporize much of the incoming material (UREY, 1952). It is not at all obvious that Pb, Bi, etc. would be lost under these conditions, as they were not lost in recent impacts or in quiescent melting events. Perhaps the combination of impacts *and* a hot surface led to losses, although it seems doubtful that material accreting at a rate of kilometers per year can be outgassed to the required degree (~99%). The efficiency of the process might be higher if other, more abundant volatiles (H_2O, CO) were to escape at the same time (RINGWOOD, 1966). Still, the high abundance of Hg makes it very doubtful that such a surface loss did occur.

(2) *Loss in the solar nebula.* The depletion of Pb, Tl, etc. may have occurred in the solar nebula. If accretion temperatures were high, these elements would have been largely left behind in the gas phase. Because we do not know exactly how temperature and accretion rate varied with time we shall consider two limiting cases. In the first we assume that the planets accreted at a *constant* temperature, from material containing a small but constant amount of volatile metals. In the second, we assume that the bulk of the accretion happened at too high a temperature to allow appreciable condensation of volatile metals. Volatiles were brought in at the very end, when temperatures in the nebula had fallen low enough to convert the remaining dust to volatile-rich, carbonaceous-chondrite-like material.

(2a) *Volatiles acquired throughout accretion.* The abundance of the Pb–In group in the earth appears to be similar to that in ordinary chondrites (Fig. 7 of ANDERS, 1968). In terms of the present model, this implies a similar accretion temperature, ~550°K, for both (Fig. 5 of ANDERS, 1968). The Pb, Tl depletion in the low-temperature fraction of the moon is ~$\frac{1}{25}$ that of the earth, taking into account our earlier estimate that the moon contains only $\frac{1}{4}$ as much low-temperature material. This corresponds to an accretion temperature some 90°K higher.

If we accept this result at face value, we would have to conclude that the moon accreted in a warmer part of the solar nebula, where condensation of the Pb–In group was retarded. But the temperature derived here is somewhat fictitious, valid only in the unlikely case that accretion happened at *constant* temperatures. If, as seems more likely, accretion took place at *falling* temperatures, much of the Pb, Tl, etc. would have been acquired only in the final stages, when the earth and moon had nearly grown to their present sizes (ANDERS, 1968; TUREKIAN and CLARK, 1969). Indeed, KOKUBU *et al.* (1961) had previously inferred from the high deuterium content of terrestrial surface waters that much of the earth's water may have been brought in by carbonaceous chondrites; the type of material expected in the solar nebula below

350–400°K. The earth's entire inventory of Pb, Bi, Tl, In and H_2O could have been supplied by 10^{-2}–10^{-3} earth masses of carbonaceous chondrite material, and hence this process probably happened when accretion of the earth and moon was more than 99 per cent complete. Under these conditions, the accretion rate of volatiles would also depend on the mass of the body, as we shall show below.

(2b) *Volatiles acquired in terminal stage of accretion.* Consider a planet sweeping up dust and planetesimals traveling at velocity U relative to the planet's orbital velocity. From the equations of ÖPIK (1951) the ratio q of the planet's capture cross section to its geometric cross section equals

$$q = 1 + \frac{2m/M}{(r/R)U^2}$$

where m and M are the masses of planet and sun, r is the radius of the planet, and R is the radius of its orbit. The abundance X of a volatile element per gram of planet will be equal to the integrated flux It of volatile-rich material times the ratio of capture cross section to mass:

$$X = 3Itq/r\rho$$

where ρ is the density of the planet. Using subscripts e and m for earth and moon, we write for the abundance ratio of volatiles in the two bodies:

$$\frac{X_e}{X_m} = \frac{[q_e - q_m(r_m/r_e)^2]\rho_m r_m}{q_m \rho_e r_e} = 0{\cdot}164\,\frac{(q_e - 0{\cdot}074\,q_m)}{q_m}.$$

The term $q_m(r_m/r_e)^2$ corrects for the overlap between the moon's and earth's spheres of influence in the case of an earth-bound moon; it is omitted for the case of a sun-bound, independent moon.

In order to evaluate this equation, we need to know U. It is generally assumed that dust and planetesimals in the solar nebula moved in nearly circular orbits, as long as any gas was present to damp radial and out-of-plane velocity components. Encounter velocities with the earth would therefore be small, $U_e \approx 0{\cdot}01$–$0{\cdot}02$. (A typical orbit with $U = 0{\cdot}02$ is $a = 1{\cdot}0$ a.u., $e = 0{\cdot}1$, $i = 5{\cdot}75°$. Most particle orbits are likely to have had smaller e, i and U.)

For the moon, two radically different situations would obtain, depending on whether it originated as an independent planet or as an earth satellite. In the former case, its velocity relative to the earth must have been very small, to allow capture; this implies $U_m \leq 0{\cdot}01$–$0{\cdot}02$ relative to dust in circular orbits. In the latter case, U_m would have been larger, owing to the moon's orbital motion around the earth. At a distance of d earth radii, the moon's orbital velocity will be $7{\cdot}96\, d^{-\frac{1}{2}}$ km/sec, or $U_m \approx 0{\cdot}267\, d^{-\frac{1}{2}}$. An earthbound moon will therefore encounter interplanetary dust at quite high velocities ($U = 0{\cdot}12$ at $d = 5$), and have a proportionately smaller capture cross section.

A few typical values for the expected trace element ratio, X_e/X_m, are given in Table 4. In the first two lines, the moon is an independent planet, orbiting the sun at about 1 a.u. In the next four lines, the moon is an earth satellite.

Table 4. Accretion of volatile elements by earth and moon

Earth–moon distance earth radii	U_e	U_m	$\frac{q_e}{q_m}$	$\frac{X_e}{X_m}$
≫100	0·01	0·01	22	3·6
≫100	0·02	0·02	21	3·4
5	0·01	0·12	982	161
10	0·01	0·08	750	123
5	0·02	0·12	245	40
10	0·02	0·08	187	31

U = velocity relative to orbiting dust particles, in units of earth's orbital velocity (29·77 km/sec.)
q = ratio of capture to geometric cross section.
X = abundance of volatile trace element, per gram of planet.

Clearly, abundance differences by more than a factor of 100 can be produced by having the moon originate as an earth satellite, if the geocentric velocities of the dust were as low as 0·01. There is no longer any need to invoke temperature differences. But is it reasonable to suppose that the moon formed in geocentric orbit?

There are at least two astronomical observations favoring a nearby origin of the earth and moon. ÖPIK (1961b) has noted that the craters in the lunar highlands have small preferential ellipticity components, probably due to tidal deformation. He inferred an earth-moon distance of 5·8 ± 1 earth radii at the time of formation of the highland craters. The distance for the final stages of the moon's accretion (which obviously must have preceded formation of the differentiated highlands) was estimated as 4·5–4·8 earth radii (ÖPIK, 1967).

HARTMANN (1968) observed that the density of pre-mare craters increased twofold between the lunar equator and poles, independently of latitude. The effect was shown by all classes of pre-mare craters, but not by post-mare craters. The only satisfactory explanation seems to be that the moon was close enough to the earth during the formation of pre-mare craters to have its equatorial regions "shielded" by the earth's gravitation.

Thus it appears that accretion in earth orbit has two advantages over other mechanisms for trace element depletion. It is supported by two lines of astronomical evidence, and can quantitatively explain the observed depletion factor without further *ad hoc* assumptions. In fact, the same mechanism may also be responsible for the fourfold alkali depletion in the moon. The temperature for 50 per cent condensation of alkalis is much higher than that of the Pb–In group (~1050° vs. 450–550°K), and accretion of alkalis must therefore have begun much earlier, when accretion rates were not yet as strongly biased in favor of the earth as at later stages. Nonetheless, the mass ratio of earth and moon, 81 : 1, shows that the overall bias was substantial. An average alkali depletion by a factor of 4 thus is not unreasonable.

ORIGIN OF THE MOON

Though the clues found here may not be specific enough to settle the origin of the moon conclusively, they do favor a simultaneous origin, as a planet–satellite system (SCHMIDT, 1950; RUSKOL, 1960, 1962, 1963; LEVIN, 1965; ÖPIK, 1961b, 1967, 1969). A point that needs emphasis, in the light of the recent preference for capture or fission models, is the consistency of simultaneous formation with all available evidence. The low K/U ratio requires that the high-temperature material of earth and moon be depleted in alkalis. This is not a universal characteristic of planetary matter: carbonaceous chondrites show it, but ordinary and enstatite chondrites do not. The low abundance of alkalis and elements of the Pb–In group requires a higher content of low-T material in the earth; which, as we saw, might come about naturally if the moon accreted in earth orbit.

The low iron content of the moon has been widely regarded as a sign of remote origin, for reasons that are less than compelling. True, the density differences among planets and meteorites show that metal–silicate fractionations occurred over distances of some tenths of astronomical units. But there is nothing in these data to preclude such fractionations over much shorter distances, as between a planet and its satellites. Indeed, the smooth density decrease of the Galilean satellites of Jupiter shows that such fractionations can also occur over distances of a few planetary radii (WILDT, 1961):

Satellite	Distance (planetary radii)	Density (g/cm³)
Io	5·9	4·03
Europa	9·4	3·78
Ganymede	15·0	2·35
Callisto	26·4	2·06

Numerous authors, beginning with EUCKEN (1944), have discussed mechanisms for metal–silicate fractionations (LATIMER, 1950; WOOD, 1962; ANDERS, 1964; HARRIS and TOZER, 1967; LARIMER and ANDERS, 1970). There is good reason to believe that metal and silicate particles formed in the solar nebula (WOOD, 1962, 1963; LORD, 1965; LARIMER, 1967). Once temperatures had fallen below the Curie point, ferromagnetism of the metal particles would enlarge their capture cross section by a factor of $\sim 10^4$, causing rapid accretion to larger aggregates (HARRIS and TOZER, 1967). Trace element data do indeed show that the metal–silicate fractionation in chondrites took place below the Curie point (LARIMER and ANDERS, 1970).

The earth, as the larger body, must have begun its growth sooner than did the moon, and hence should have acquired a larger share of the more-rapidly accreting metal (TUREKIAN and CLARK, 1969). The depletion of siderophile elements in the lunar basalts does not necessarily imply a fission origin, as suggested by O'KEEFE (1969). Removal of metal phase to the lunar interior during melting would produce the same effect. Indeed, the severe depletion of gold suggests that the removal took place on the moon, not on the earth. Geophysical data do not preclude a core of 30 per cent the moon's radius, or 6 per cent its mass (NAKAMURA and LATHAM, 1969), but the presence of a core is not a necessary consequence of *in situ* siderophile element depletion. Inward migration of metal to moderate depths would suffice.

The moon's unequal moments of inertia have often been cited as evidence against differentiation (UREY, 1952 and later papers). But LEVIN (1967) has shown that this effect may be due to differences in insolation between polar and equatorial regions.

The principal conclusion thus is that the geochemistry of the moon does not require an exotic mode of origin, such as capture or fission. On the contrary, the data favor a simultaneous origin, as a planet–satellite system.

Acknowledgments—We are indebted to ANNIE PIERCE and FRANK QUINN for technical assistance, and to MICHAEL E. LIPSCHUTZ of Purdue University for the use of counting equipment. SANDRA CROMARTIE and RUDY BANOVICH provided valuable help in the preparation of the manuscript and illustrations. This work was supported in part by NASA Contract NAS 9-7887 and AEC Contract AT(11-1)-382.

REFERENCES

AHRENS L. H. (1948) The unique association of thallium and rubidium in minerals. *J. Geol.* **56,** 578–579.

ALBEE A. L., BURNETT D. S., CHODOS A. A., EUGSTER O. J., HUNEKE J. C., PAPANASTASSIOU D. A., PODOSEK F. A., PRICE R. G., II, SANZ H. G., TERA F. and WASSERBURG G. J. (1970) Ages, irradiation history, and chemical composition of lunar rocks from the Sea of Tranquillity. *Science* **167,** 463–466.

ANDERS E. (1963) On the origin of carbonaceous chondrites. *Ann. N. Y. Acad. Sci.* **108,** 514–533.

ANDERS E. (1964) Origin, age, and composition of meteorites. *Space Sci. Rev.* **3,** 583–714.

ANDERS E. (1968) Chemical processes in the early solar system, as inferred from meteorites. *Acc. Chem. Res.* **1,** 289–298.

ANDERS E. and ARNOLD J. R. (1965) Age of craters on Mars. *Science* **149,** 1494–1496.

ANNELL C. and HELZ A. (1970) Emission spectrographic determination of trace elements in lunar samples. *Science* **167,** 521–523.

BAEDECKER P. A. and WASSON J. T. (1970) Gallium, germanium, indium, and iridium in lunar samples. *Science* **167,** 503–505.

BARKER J. L., JR. and ANDERS E. (1968) Accretion rate of cosmic matter from iridium and osmium contents of deep-sea sediments. *Geochim. Cosmochim. Acta* **32,** 627–645.

BUDDINGTON A. F. (1939) Adirondack igneous rocks and their metamorphism. *Geol. Soc. Amer. Mem.* **7,** 36.

CAMERON A. G. W. (1968) A new table of abundances of the elements in the solar system. In *Origin and Distribution of the Elements*, (editor L. H. Ahrens), pp. 125–143. Pergamon.

COMPSTON W., ARRIENS P. A., VERNON M. J. and CHAPPELL B. W. (1970) Rubidium–strontium chronology and chemistry of lunar material. *Science* **167,** 474–476.

DUFRESNE E. R. and ANDERS E. (1962) On the retention of primordial noble gases in the Pesyanoe meteorite. *Geochim. Cosmochim. Acta* **26,** 251–262.

EDWARDS G. and UREY H. C. (1955) Determination of alkali metals in meteorites by a distillation process. *Geochim. Cosmochim. Acta* **7,** 154–168.

EUCKEN A. (1944) Physikalisch-chemische Betrachtungen über die früheste Entwicklungsgeschichte der Erde. *Nachr. Akad. Wiss. Göttingen, Math.-Phys. Kl.* (1), 1–25.

FISH R. A., GOLES G. G. and ANDERS E. (1960) The record in the meteorites—III. On the development of meteorites in asteroidal bodies. *Astrophys. J.* **132,** 243–258.

FRANZGROTE E. J., PATTERSON J. G., TURKEVICH A. L., ECONOMOU T. E. and SOWINSKI K. P. (1970) Chemical composition of the lunar surface in Sinus Medii. *Science* **167,** 376–379.

GAST P. W. (1968) Implications of the Surveyor V chemical analysis. *Science* **159,** 897.

GAST P. W. and HUBBARD N. J. (1970) Abundance of alkali metals, alkaline and rare earths, and strontium-87/strontium-86 ratios in lunar samples. *Science* **167,** 485–487.

GOLES G. G., OSAWA M., RANDLE K., BEYER R. L., JÉROME D. Y., LINDSTROM D. J., MARTIN M. R., MCKAY S. M. and STEINBORN T. L. (1970) Instrumental neutron activation analyses of lunar specimens. *Science* **167,** 497–499.

HARRIS P. G. and TOZER D. C. (1967) Fractionation of iron in the solar system. *Nature* **215,** 1449–1451.

HARTMANN W. K. (1968) Lunar crater counts, V: Latitude dependence and source of impacting bodies. *Commun. Lunar Planet. Lab., Univ. of Ariz.* **7,** 139–144.

HERBIG G. H. (1961) Comments during discussion. *Proc. Lunar Planet. Colloq.* **2,** (4) 64.

HURLEY P. M. and PINSON W. H., JR. (1970) Rubidium–strontium relations in Tranquillity Base samples. *Science* **167,** 473–474.

KEAYS R. R., GANAPATHY R., LAUL J. C., ANDERS E., HERZOG G. F. and JEFFERY P. M. (1970) Trace elements and radioactivity in lunar rocks: Implications for meteorite infall, solar-wind flux, and formation conditions of moon. *Science* **167**, 490–493.

KOKUBU N., MAYEDA T. and UREY H. C. (1961) Deuterium content of minerals, rocks, and liquid inclusion from rocks. *Geochim. Cosmochim. Acta* **21**, 247–256.

LARIMER J. W. (1967) Chemical fractionations in meteorites—I. Condensation of the elements. *Geochim. Cosmochim. Acta* **31**, 1215–1238.

LARIMER J. W. and ANDERS E. (1967) Chemical fractionations in meteorites—II. Abundance patterns and their interpretation. *Geochim. Cosmochim. Acta* **31**, 1239–1270.

LARIMER J. W. and ANDERS E. (1970) Chemical fractionations in meteorites—III. Major element fractionations in chondrites. *Geochim. Cosmochim. Acta* **34**, 367–388.

LATIMER W. M. (1950) Astrochemical problems in the formation of the earth. *Science* **112**, 101–104.

LEVIN B. J. (1965) The structure of the moon. *Proc. Caltech-JPL Lunar Planet. Conf.* 61–76.

LEVIN B. J. (1967) Thermal effects on the figure of the moon. *Proc. Roy. Soc.* **A296**, 266–269.

LORD H. C. III (1965) Molecular equilibria and condensation in a solar nebula and cool stellar atmospheres. *Icarus* **4**, 279–288.

LSPET (LUNAR SAMPLE PRELIMINARY EXAMINATION TEAM) (1969) Preliminary examination of lunar samples from Apollo 11. *Science* **165**, 1211–1227.

LSPET (LUNAR SAMPLE PRELIMINARY EXAMINATION TEAM) (1970) Preliminary examination of lunar samples from Apollo 12. *Science* **167**, 1325–1339.

MASON B. (1962) The carbonaceous chondrites. *Space Sci. Rev.* **1**, 621–646.

MAXWELL J. A., ABBEY S. and CHAMP W. H. (1970) Chemical composition of lunar material. *Science* **167**, 530–531.

MAZOR E. and ANDERS E. (1967) Primordial gases in the Jodzie howardite and the origin of gas-rich meteorites. *Geochim. Cosomochim. Acta* **31**, 1441–1456.

MCCROSKY R. E. (1968) Distributions of large meteoric bodies. *Smithson. Astrophys. Observ. Spec. Rep.* **280.**

MORGAN J. W. and LOVERING J. F. (1968) Uranium and thorium abundances in chondritic meteorites. *Talanta* **15**, 1079–1095.

MORRISON G. H., GERARD J. T., KASHUBA A. T., GANGADHARAM E. V., ROTHENBERG A. M., POTTER N. M. and MILLER G. B. (1970) Multielement analysis of lunar soil and rocks. *Science* **167**, 505–507

NAKAMURA Y. and LATHAM G. V. (1969) Internal constitution of the moon: Is the lunar interior chemically homogeneous? *J. Geophys. Res.* **74**, 3771–3780.

OBERBECK V. R. and QUAIDE W. L. (1968) Genetic implications of lunar regolith thickness variations. *Icarus* **9**, 446–465.

O'HARA M. J., BIGGAR G. M. and RICHARDSON S. W. (1970) Experimental petrology of lunar material: The nature of mascons, seas, and the lunar interior. *Science* **167**, 605–607

O'KEEFE J. A. (1969) Origin of the moon. *J. Geophys. Res.* **74**, 2758–2767.

O'KELLEY G. D., ELDRIDGE J. S., SCHONFELD E. and BELL P. R. (1970) Elemental compositions and ages of lunar samples by non-destructive gamma-ray spectrometry. *Science* **167**, 580–582.

OLSEN E. (1969) Pyroxene gabbro (anorthosite association): similarity to Surveyor V lunar analysis. *Science* **166**, 401–402.

ÖPIK E. J. (1951) Collision probabilities with the planets and distribution of interplanetary matter. *Proc. Roy. Irish Acad.* **54**, 165–199.

ÖPIK E. J. (1956) Interplanetary dust and terrestrial accretion of meteoritic matter. *Irish Astron. J.* **4**, 84–135.

ÖPIK E. J. (1958) Meteorite impact on solid surface. *Irish Astron. J.* **5**, 14–33.

ÖPIK E. J. (1961a) Notes on the theory of impact craters. *Proc. Geophys. Lab.-Lawerence Radiation Lab. Cratering Symp.*, Washington D.C., March 28–29 Vol. 2, Paper S, 1-28. UCRL Report 6438.

ÖPIK E. J. (1961b) Tidal deformations and the origin of the moon. *Astron. J.* **66**, 60–67.

ÖPIK E. J. (1967) Evolution of the moon's surface. I. *Irish Astron. J.* **8**, 38–52.

ÖPIK E. J. (1969) The moon's surface. *Ann. Rev. Astron. Astrophys.* **7**, 473–526.

PARKIN D. W. and TILLES D. (1968) Influx measurements of extraterrestrial material. *Science* **159**, 936–946.

PATTERSON J. H., TURKEVICH A. L., FRANZGROTE E. J., ECONOMOU T. E. and SOWINSKI K. P. (1970) Chemical composition of the lunar surface in a terra region near the crater Tycho. *Science*, in press.

REED G. W., KIGOSHI K. and TURKEVICH A. (1960) Determinations of concentrations of heavy elements in meteorites by activation analysis. *Geochim. Cosmochim. Acta* **20**, 122–140.

REED G. W., JOVANOVIC S. and FUCHS L. H. (1970) Trace elements and accessory minerals in lunar samples. *Science* **167**, 501–503.

RINGWOOD A. E. (1966) Chemical evolution of the terrestrial planets. *Geochim. Cosmochim. Acta* **30**, 41–104.

RINGWOOD A. E. and ESSENE E. (1970) Petrogenesis of lunar basalts and the internal constitution and origin of the moon. *Science* **167**, 607–610.

RUSKOL E. L. (1960) The origin of the moon. I. Formation of a swarm of bodies around the earth. *Astron. Zh.* **37**, 690–702.

RUSKOL E. L. (1962) The origin of the moon. In *The Moon*, (editors Z. Kopal and Z. K. Mikhailov), pp. 149–155. Academic.

RUSKOL E. L. (1963) On the origin of the moon. II. The growth of the moon in the circumterrestrial swarm satellites. *Astron. Zh.* **40**, 288–296.

SCHMIDT O. J. (1950) Origin of the planets and their satellites. *Izv. Acad. Nauk SSSR, Ser. Fiz.* **14**, 29–45.

SCHMITT R. A., WAKITA H. and RAY P. (1970) Abundances of 30 elements in lunar rocks, soil and core samples. *Science* **167**, 512–515.

SHAW D. M. (1950) The geochemistry of Tl. *Geochim. Cosmochim. Acta* **2**, 118–154.

SHOEMAKER E. M. and LOWERY C. J. (1967) Airwaves associated with large fireballs and the frequency distribution of energy of large meteoroids. *Meteoritics* **3**, 123–124.

SHOEMAKER E. M., MORRIS E. C., BATSON R. M., HOLT H. E., LARSON K. B., MONTGOMERY D. R., RENNILSON J. J. and WHITAKER E. A. (1969) *Surveyor Program Results*, Chap. 3, pp. 19–128. *NASA Spec. Publ. No.* 184.

SHOEMAKER E. M., HAIT M. H., SWANN G. A., SCHLEICHER D. L., DAHLEM D. H., SCHABER G. G. and SUTTON R. L. (1970) Lunar regolith at Tranquillity Base. *Science* **167**, 452–455.

SILVER L. T. (1970) Uranium–thorium–lead isotope relations in lunar materials. *Science* **167**, 468–471.

SINGER S. F. (1969) Interplanetary dust. In *Meteorite Research*, (editor P. M. Millman), Chap. 7, pp. 590–599. D. Reidel.

SMALES A. A., MAPPER D., WEBB M. S. W., WEBSTER R. K. and WILSON J. D. (1970) The elemental composition of lunar surface material. *Science* **167**, 509–512.

TATSUMOTO M. and ROSHOLT J. N. (1970) Age of the moon: An isotopic study of U–Th–Pb systematics of lunar samples. *Science* **167**, 461–463.

TAYLOR S. R. (1964) Abundance of chemical elements in the continental crust: A new table. *Geochim. Cosmochim. Acta* **28**, 1273–1285.

TILLES D. (1965) Atmospheric noble gases: Solar-wind bombardment of extraterrestrial dust as a possible source mechanism. *Science* **148**, 1085–1088.

TRASK N. J. (1966) Size and spatial distribution of craters estimated from the Ranger photographs. In *Ranger VIII and IX. Part II.* Chap. 4, Sec. Bl, pp. 252–263. JPL NASA Tech. Rep. No. 32–800.

TUREKIAN K. K. and CLARK S. P., JR. (1969) Inhomogeneous accumulation of the earth from the primitive solar nebula. *Earth Planet. Sci. Lett.* **6**, 346–348.

TUREKIAN K. K. and KHARKAR D. P. (1970) Neutron activation analysis of milligram quantities of lunar rocks and soils. *Science* **167**, 507–509.

UREY H. C. (1952) *The Planets.* Yale University Press.

UREY H. C. (1954) On the dissipation of gas and volatilized elements from protoplanets. *Astrophys. J. Suppl.* **1**, (6) 147–173.

WÄNKE H., BEGEMANN F., VILCSEK E., RIEDER R., TESCHKE F., BORN W., QUIJANO-RICO M., VOSHAGE H. and WLOTZKA F. (1970) Major and trace elements and cosmic-ray produced radioisotopes in lunar samples. *Science* **167**, 523–525.

WASSERBURG G. J., MACDONALD G. F. J., HOYLE F. and FOWLER W. A. (1964) Relative contributions of uranium, thorium, and potassium to heat production in the earth. *Science* **143**, 465–467.

WHIPPLE F. W. (1966) Before Type I carbonaceous chondrites. Paper presented at 29th Meeting of the Meteoritical Society, Washington, D.C., Nov. 3–5.

WIIK H. B. and OJANPERÄ P. (1970) Chemical analyses of lunar samples 10017, 10072, and 10084. *Science* **167**, 531–532.

WILDT R. (1961) Planetary interiors. In *The Solar System, Part III. Planets and Satellites*, (editors B. M. Middlehurst and G. P. Kuiper), Chap. 5, pp. 159–212. University of Chicago Press.

WISE D. U. (1963) An origin of the moon by rotational fission during formation of the earth's core. *J. Geophys. Res.* **68**, 1547–1554.

WOOD J. A. (1962) Chondrules and the origin of the terrestrial planets. *Nature* **194**, 127–130.

WOOD J. A. (1963) On the origin of chondrules and chondrites. *Icarus* **2**, 152–180.

WOOD J. A., DICKEY J. S., JR., MARVIN U. B. and POWELL B. N. (1970) Lunar anorthosites. *Science* **167**, 602–604.

Proceedings of the Apollo 11 Lunar Science Conference, Vol. 2, pp. 1143 to 1163.

Chemical composition and petrogenesis of basalts from Tranquillity Base*

P. W. GAST, N. J. HUBBARD and H. WIESMANN
Lamont–Doherty Geological Observatory of Columbia University, Palisades, New York 10964

(*Received* 5 *February* 1970; *accepted in revised form* 25 *February* 1970)

Abstract—The chemical composition of rocks from Tranquillity Base show that the igneous rocks can be grouped into high Rb and low Rb groups. Both rock types have REE abundance patterns with very large Eu anomalies and relatively unfractionated trivalent REE abundance patterns. It is suggested that the low Rb groups are mesocumulates. The chemistry of the more primitive high Rb igneous rocks can be explained by extensive fractional crystallization of a liquid with a high normative plagioclase content. Alternatively, partial melting of a plagioclase, pyroxene, spinel mantle could explain the observed composition. Either case requires a lunar crust or mantle richer in aluminum, calcium, rare earth elements and barium than chondritic meteorites.

INTRODUCTION

THE STUDY of lunar samples poses a challenge to combine accuracy, sensitivity and breadth in a single scheme of analysis. In this study we have attempted to exploit the potential of the stable isotope dilution method to achieve this goal. We have determined the concentration of Sr, Ba, K, Rb, Cs and ten REE by stable isotope dilution. In addition, by using quantitative chemical separations we have been able to determine Na, Ca, Zn and Ti by other methods. Furthermore, the extention of this procedure to isotope dilution determinations of several other elements, e.g. Zr, Hf, U, Th, Li and Pb is straight forward using the scheme developed here.

The principle advantages of the stable isotope dilution method are: the great potential for very high accuracy measurements and the possibility of combining concentration measurements with determination of isotopic compositions. With very high sensitivity mass spectrometers the sensitivity of the method for many elements is often comparable to that obtained by neutron activation methods. The major limitation of the method with regard to sensitivity is the requirement for very low sample blanks. That is, very pure reagents and chemical separations made in a controlled environment are required to obtain nominal sensitivity.

ANALYTICAL PROCEDURE USED IN THIS STUDY

Mass Spectrometry

The mass spectrometer used for all isotopic analyses except alkali metals is nearly identical with the instrument described by SHIELDS *et al.* (1966). It will not be further described here. Isotopic analyses of alkali metals were made on a six-inch radius of curvature instrument briefly described by GAST (1962). Surface emission ion sources were used for all elements; Cs, Rb and K were run sequentially on previously outgassed Pt filaments, La was determined as LaO on a single Re filament, all other elements were determined using triple Re filaments. Ce, La, Ba and sometimes Nd were analyzed as separated elements. Nd, Sm, Eu and Gd were analyzed as one mixture. Gd, Dy, Er, Yb and Lu were determined in a second mixture.

* Lamont–Doherty Geological Observatory Contribution No. 1504.

Tracer Calibrations

The enriched isotopes and isotopic purity for tracers used in this study are listed in Table 1. In order to minimize pipetting and weighing errors during addition of these isotope tracers to the sample solutions three combined tracer solutions were employed; they were, Ce and La, Nd, Sm, Eu and Gd and Yb, Er, Dy and Lu.

Table 1. Spike compositions and blank determinations

	Spike composition		Avg. of blank determinations (μg)
La	6·71%	La^{138}	—
Ce	92·77%	Ce^{142}	0·01
Nd	91·06%	Nd^{143}	—
Sm	97·46%	Sm^{149}	—
Eu	98·76%	Eu^{153}	—
Gd	91·77%	Gd^{155}	—
Dy	90·0%	Dy^{161}	—
Er	91·1%	Er^{167}	—
Yb	95·96%	Yb^{171}	—
Lu	70·2%	Lu^{176}	—
Ba	93·6%	Ba^{135}	0·15
Sr	99·35%	Sr^{84}	0·005
K	99·12%	K^{41}	2·5
Rb	99·15%	Rb^{87}	0·025
Cs	37·3%	Cs^{137}	<0·0001

Tracer solutions were calibrated against standard solutions whose concentrations were determined both gravimetrically and titrimetrically. The EDTA standard solution used to calibrate standard REE, Sr, and Ba solutions was calibrated by titrating it against a standard solution of Zn Cl_2 prepared from 99·999% pure metallic Zn. Comparison of concentrations of standard solutions determined gravimetrically and titrimetrically is shown in Fig. 1. Except for the light REE, they are in rather good agreement. The progressive increase in differences between these two methods from Eu to La is ascribed to absorption of CO_2 by the oxides of the REE during drying and weighing of these salts. The titrimetric calibrations were adopted for calibration of the isotope tracer solutions. Thus, with the exception of the alkali metals, all tracer solutions used in this study are calibrated in terms of a single standard solution, made from 99·999% pure Zn metal. The titrimetric calibration of the standard solution has the further advantage of being independent of many cationic and all anionic impurities and non-stoichiometric compositions in the salts used for making the standards.

Chemical separations

Samples ranging in size from 180 to 300 mg were dissolved in HF and $HClO_4$. Tracer solutions usually adding up to ~30 ml were added to the solution of the samples. This solution was evaporated nearly to dryness to assure mixing of tracer and sample. This mush was taken up in 50 ml 0·5 M oxalic acid, heated and quantitatively transferred to a 50 cm long Dowex 50 × 8, 200–400 mesh ion exchange column. This column consisted on a 60 cm length of $\frac{1}{2}$ in. polypropolene tubing fitted with a stopcock on one end and a liquid reservoir on the other end. After adding the sample solution, this column was eluted with 50 ml of 0·5 N oxalic acid, 400 ml 1 N HCl, 250–300 ml 2·0 HCL, 600 ml 2·5 N HNO_3 and 100 ml 4 N HNO_3 in succession. Results of a typical elution are shown in Fig. 2. Rb and Cs were subsequently separated from Mg on a 15 cm × 0·9 cm Dowex 50 × 8 200–400, column by eluting with 0·6 N HNO_3. Aliquots from the elution of the 50 and 15 cm column were used for mass spectrometric analysis. Na, Zn and Ca containing fractions are diluted to standard volumes. Aliquots of these solutions were subsequently used for determination of Na and Zn by atomic absorption spectrometry and Ca by EDTA titration. Tracer experiments using Zr^{95}, Th^{228}, and analysis of individual 20 ml aliquots showed that Ti, Zr, Th and U were quantitatively removed

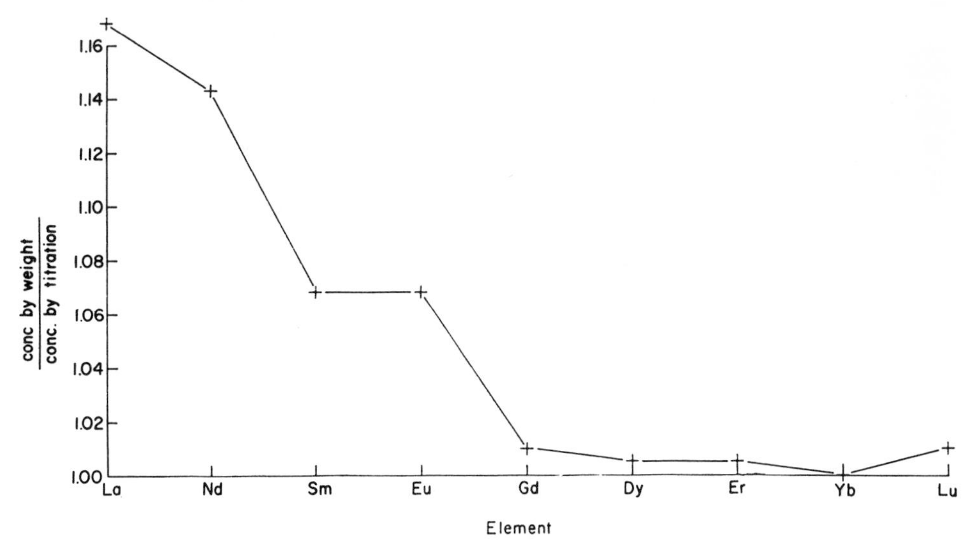

Fig. 1. Comparison of REE standard solution concentrations determined gravimetrically and by EDTA titration.

by the oxalic acid elution. The entire oxalic acid fraction was evaporated with 1 ml of $HClO_4$ to destroy H_2OX. Ti determinations were made colorimetrically on a 2% aliquot of this sample. The remaining aliquot was set aside for subsequent determination of U, Zr and Hf.

During the course of the lunar sample investigations, we determined 3 sets of overall blanks and 2 sets of data on standard rocks. Blank determinations are shown in Table 1. Data obtained on 9 lunar samples and 2 standards are shown in Table 2. Prior to this study, we analyzed a number of meteorite samples using essentially the technique described here. These results are also shown in Table 2. The errors shown for the lunar samples are estimated from replicate measurements made on the same samples.

Sr isotope compositions

The concentration and isotopic composition of Sr were measured during the same mass spectrometer analysis by using a highly enriched Sr^{84} spike (99·35% Sr^{84}). When sufficiently small amounts of this spike are used ($Sr^{84}/Sr^{86} < 0{\cdot}3$) no corrections on the measured $Sr^{87/86}$ ratio are required, aside

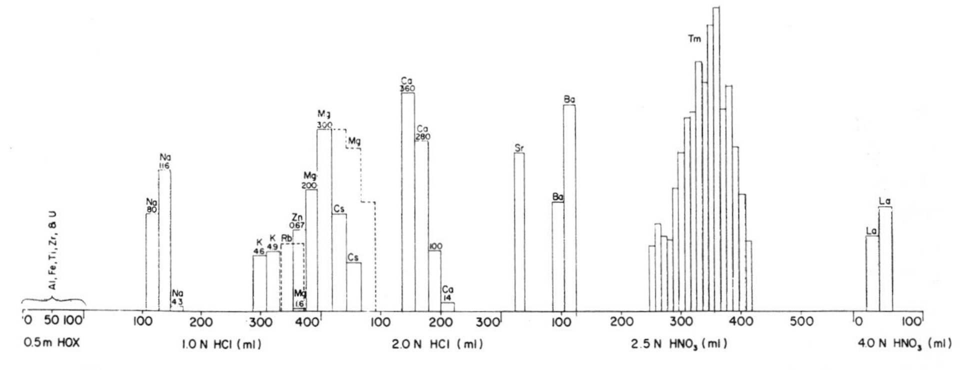

Fig. 2. A typical elution curve for our 50 cm Dowex 50 × 8 ion exchange columns. The vertical axis represents relative concentrations for each individual element, not the relative concentration of two different elements.

Table 2. Concentrations of 16 elements in 9 samples from

	Analytical* errors	Soil		Breccia	High Rb igneous rocks						Low Rb
Sample No.		10084		10073	10049		10017	10071	10062†		10020
Na‡	±*10%*	*0·30*	*0·34*	*0·35*	*0·40*	—	*0·38*	*0·39*	*0·33*	—	*0·28*
K	±3%	1200	1200	1200	2730	2900	2610	2770	628	660	486
Rb	±3%	—	2·79	2·89	6·24	6·20	5·63	5·93	0·90	0·81	0·63
Cs	±10%	0·104	0·102	0·098	0·166	0·177	0·155	0·170	0·032	—	—
Sr	±2·5%	168·8	173·9	167·5	160·8	—	174·8	160·9	141·6	196·4	149·8
Ba	±4%	176	188	175	338	330	309	327	80·5	140	77·1
La	±4%	16·4	16·6	—	29·2	28·8	26·6	28·8	14·2	14·5	8·11
Ce	±4%	47·4	47·7	46·5	84·2	82·8	77·3	83·5	42·7	44·8	25·8
Nd	±4%	36·9	36·3	35·4	64·3	62·8	59·5	64·5	36·7	37·5	23·9
Sm	±4%	13·0	13·1	12·4	22·5	22·3	20·9	22·7	13·3	13·7	9·47
Eu	±4%	1·76	1·70	1·70	2·31	2·29	2·14	2·32	2·02	2·06	1·60
Gd	±4%	16·1	16·8	15·9	29·6	29·3	27·4	29·3	17·7	18·2	12·8
Dy	±4%	19·5	19·2	18·3	34·0	33·4	31·7	33·5	20·0	20·4	15·8
Er	±4%	12·2	11·8	11·4	21·2	20·9	20·0	21·3	12·4	12·8	10·0
Yb	±4%	11·9	11·5	11·1	20·2	20·2	19·2	20·5	12·1	12·3	9·87
Lu	±7%	1·64	1·50	1·56	—	—	2·66	2·87	1·73	—	1·43
$Sr^{87/86}$§		0·7015 ±0·0005		0·7025 ±0·0006	0·7043 ±0·0006		0·7046 ±0·0003	0·7041 ±0·0004	0·7002 ±0·0004		0·7005 ±0·0006

* ~90% confidence limits based on replicate analysis of samples.
† The Sr and Ba concentrations in the first analyses are probably incorrect due to a spiking error.
‡ Atomic absorption analysis of separated fraction.
§ $Sr^{88/86}$ normalized to 8·375.

from normal fractionation corrections. As an interlaboratory comparison, we analyzed Pasamonte for $Sr^{87/86}$ and obtained 0·6998 ± 0·0004 which compares very well with the more accurate ratio of 0·69960 reported by Papanastassiou and Wasserburg (1969). Results for our study are shown on Fig. 3. The isochron drawn in this figure is roughly consistent with the separated mineral data reported by Albee *et al.* (1970), Gopalan *et al.* (1970), and Compston *et al.* (1970).

Discussion of the Results

The results of our studies of REE, Ba and Sr concentrations normalized to the chondrite concentrations of Frey *et al.* (1968), are shown in Fig. 4. In Fig. 5 we have compared our results for these elements with those of Philpotts and Schnetzler (1970) where we have analyzed common samples. A small systematic difference between our data and their data is shown. This difference is similar to that seen for our two spike calibration methods (Fig. 1). We suggest that the different concentrations may in fact be due to different spike calibration procedures used by ourselves and Philpotts, etc. The differences between the NAA results of Haskin *et al.* (1970) are less systematic and often greater than those reported for two mass spectrometric analyses. Other NAA analyses available to us are either incomplete or clearly much inferior to the M.S. data and are, therefore, not considered in our comparison of these elements in lunar rocks.

Tranquillity Base, two terrestrial basalts and 6 meteorites

igneous rocks			Standard samples			Meteorites					
10003		10058	BCR–1		Knippa Basalt	Pasamonte	Pueblito de Allende Ca-rich inclusion	Stannern	Crab Orchard	Bruder-heim	Hol-brook
0·29	—	*0·29*	—	—	—	*0·41*	—	—	—	—	—
470	440	877	15,000	13,400	9,800	429	96·4	—	—	—	—
0·49	0·50	0·98	46·6	46·7	27·9	0·19	3·5	—	—	—	—
0·022	—	0·027	—	—	—	—	—	—	—	—	—
158·6	152·7	218·3	330	326	1,019	56·6	180	—	—	—	—
108	106	117	713	686	672	38·4	47·3	49·2	—	—	—
15·5	14·7	11·5	26·2	26·0	54·7	3·36	4·63	5·57	1·19	0·377	—
47·2	45·5	40·2	54·9	54·9	111·0	8·84	11·5	14·4	1·94	1·00	0·892
40·0	38·3	41·2	28·8	28·8	54·3	6·54	8·40	10·6	1·46	0·715	0·690
14·4	14·0	17·2	6·74	6·74	11·2	2·13	2·82	3·40	0·474	0·219	0·229
1·81	1·76	2·64	1·98	1·95	3·42	0·691	1·30	0·710	0·0279	0·0780	0·0814
19·5	19·0	23·6	—	—	9·30	—	3·87	—	—	0·274	0·303
21·9	21·6	27·0	6·22	6·18	6·18	3·29	4·90	5·05	0·791	0·380	0·402
13·6	13·4	16·3	3·69	3·73	2·83	2·22	3·44	3·32	0·517	0·248	—
13·2	13·0	15·5	3·74	3·63	1·71	2·27	3·96	3·18	0·508	0·238	0·246
1	—	2·14	0·634	0·547	0·216	0·328	—	—	—	—	—
0·7015 ±0·0005		0·6996 ±0·0005	—	—	—	0·6998 ±0·0004	—	—	—	—	—

Stable isotope dilution was used to determine all elements except Na, which was done by atomic absorption. Concentrations are given in ppm, except Na which is in per cent. The concentrations of Ba through Yb in this table are slightly different from those published in GAST and HUBBARD, 1970, due to a computer error in the earlier data.

The data of this study along with similar studies (referenced in Fig. 6) suggest that the collection of igneous rocks from Tranquillity Base fall into two compositional and perhaps two textural groups (the A and B types of LSPET). The textural classification is not always clear. For example, two of the rocks analyzed in this study, 10017 and 10020 are designated as type B and type A respectively by LSPET. The chemistry is at variance with this classification. The compositional groups are clearly distinguished by their Rb and K content. As far as we can establish no intermediate concentrations of Rb or K have been found for the igneous rocks. We will refer to these groups as the "high Rb group" and the "low Rb group." The relative abundance of 20 elements in these two groups is shown in Fig. 7. The high Rb rocks are consistently higher, by more than a factor of 2, in Ba and U. The existence of the two groups is clearly established by this comparison. The high Rb rocks are remarkably similar with regard to these elements. In fact rocks 10049 and 10071 which are texturally easily distinguished are indistinguishable in terms of these elements. The low Rb group is much more variable in terms of REE and Ba abundances. The abundance of light REE in particular is more depleted and Eu anomalies are slightly less pronounced. The average Sm/Eu ratio of low Rb rocks is 6·5 compared to an average of 9·7 for the high Rb rocks.

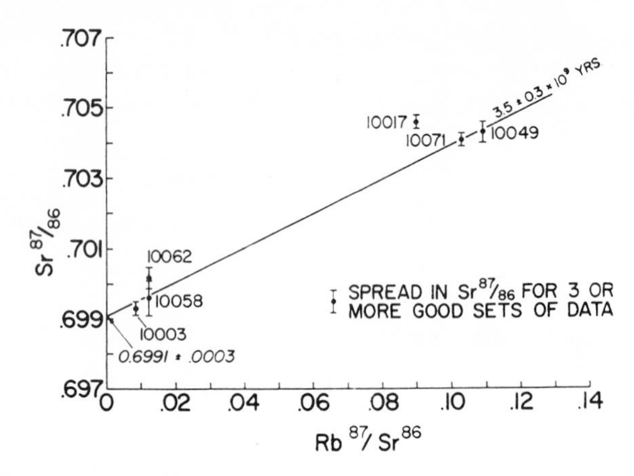

Fig. 3. Isochron diagram for 6 lunar rocks. ($\lambda = 1{\cdot}39 \times 10^{-11}$ yr^{-1}) sample 10017 was ignored in drawing the isochron.

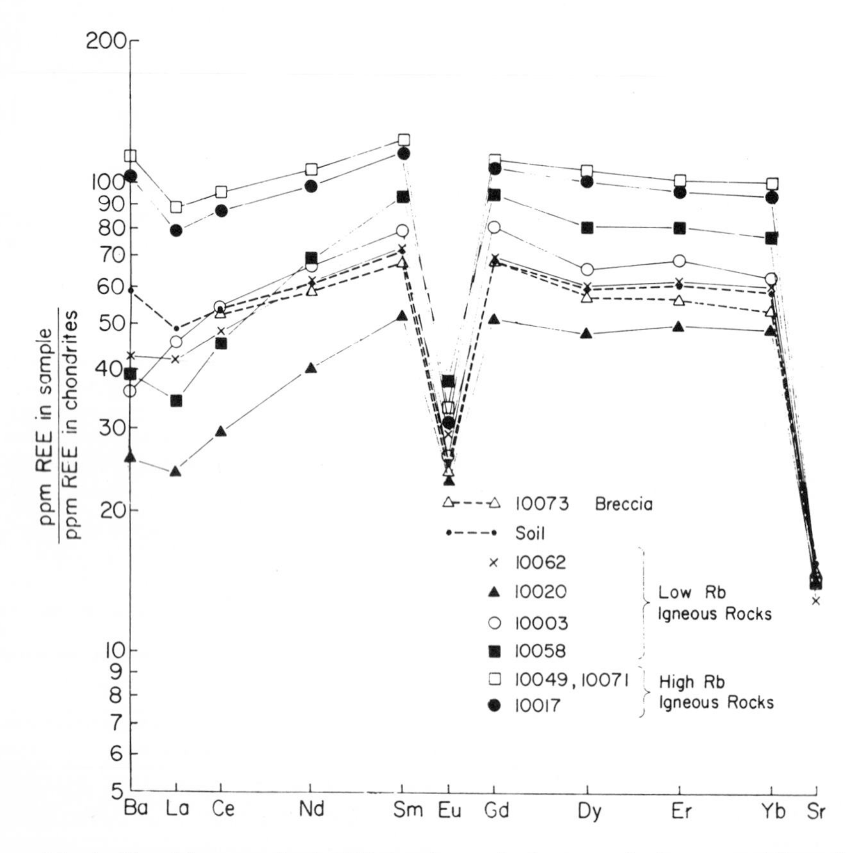

Fig. 4. Chondrite normalized REE abundances for 9 rare earth elements and Ba and Sr. Chondrite REE concentrations are from Frey *et al.* (1968) Dy is interpolated to be 0·335 ppm, Ba is taken to be 3 ppm and Sr 11 ppm.

Soil and breccia samples from Apollo 11 are all very similar to each other. For most elements analyzed here the concentrations are intermediate between the two igneous rock types. Ni, Zn, Ti, Al and Fe concentrations and K/Rb ratios are exceptions to this generalization. The average concentrations for TiO_2, Al_2O_3 and FeO are compared in Fig. 7 relative to the average for the low Rb lavas.

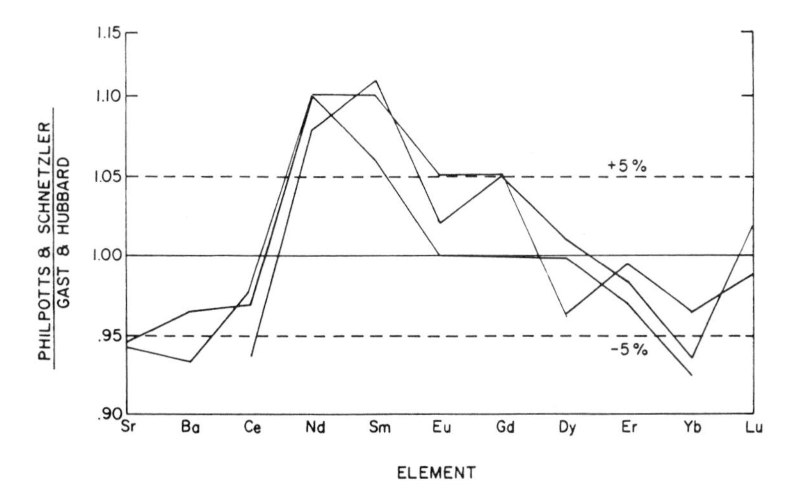

Fig. 5. Comparison of our data with that of PHILPOTTS and SCHNETZLER (1970) for Apollo 11 samples 10084, 10062 and 10017.

INTERPRETATION

Petrogenesis of the igneous rocks

In the absence of geologic observations on the units from which the analyzed rock specimens were derived, we must infer certain geologic relationships from the chemical composition and textures rather than from direct observation. In particular, we inquire which of the two groups, if any, of the igneous rocks may be a primary liquid, i.e. is the direct product of melting or partially melting a portion of the moon which then cooled and crystallized in place without significant fractionation after emplacement. And, if neither, which of the two liquids is a more direct descendent from such a primary liquid. We also ask if the soil and breccias may contain additional components not sampled in the igneous rocks. Once a hypothetical parent liquid is established we may ask how such a liquid is related to the bulk composition of the moon. And finally, somewhat speculatively we ask how the proposed bulk composition was derived from the primitive dispersed matter of the solar system. The compositions of the igneous rocks illustrated in Figs. 4 and 6, show that the compositions of the Rb rich igneous rocks are less variable and vary more regularly than those of the Rb poor rocks. In addition, the relative abundances of the elements in the Rb rich rocks are less fractionated with respect to chondrites, with the exception of Eu and Sr, than those of the Rb poor rocks. The slopes of REE abundance curves in the Rb poor rocks between La and Sm are generally greater and more variable than those for the Rb rich rocks (cf. particularly sample 10058 and 10003, Fig. 4). We infer from this observation that the low Rb rocks are in part the product of a local variable chemical fractionation process.

The interpretation of both the differences within the group of igneous rocks and the overall characteristics of the lunar rocks is facilitated by comparing the abundances

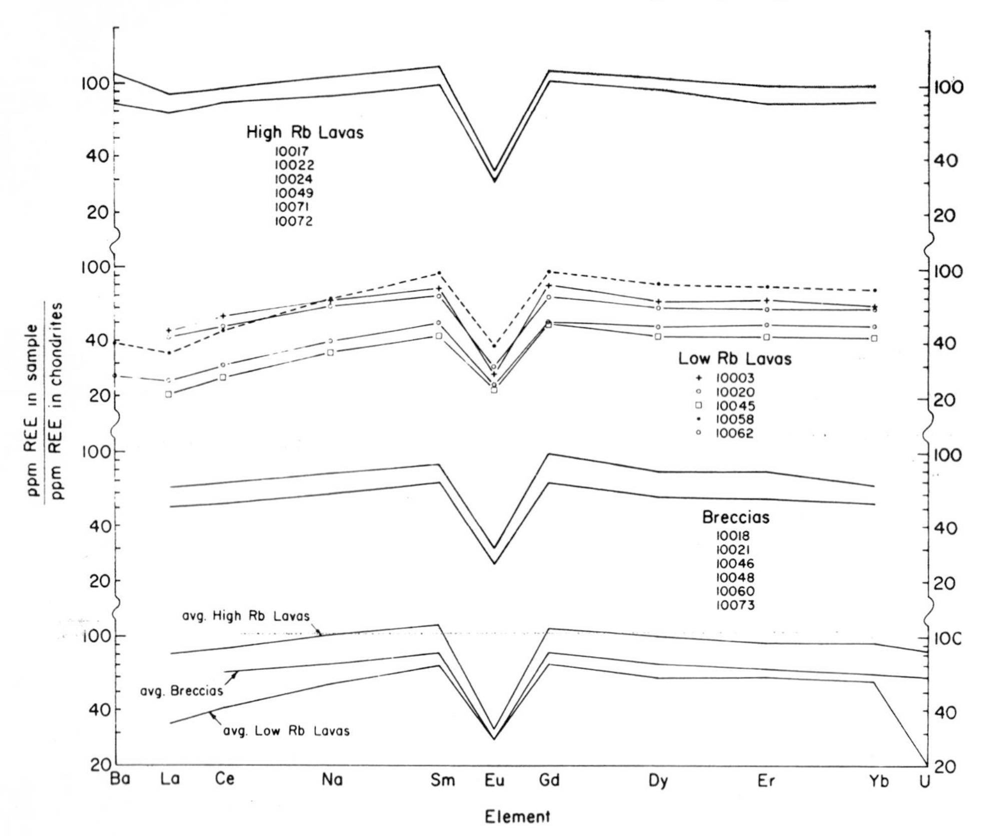

Fig. 6. Chondrite normalized REE and Ba abundances for Apollo 11 samples analyzed by PHILPOTTS and SCHNETZLER (1970) HASKIN *et al.* (1970) and the authors. Normalizing values are as for Fig. 4. Uranium (TATSUMOTO and ROSHOLT, 1970) has been added to the average plots. Sample 10024 is listed as 10044-24 in PHILPOTTS and SCHNETZLER (1970) but cannot be 10044 and is clearly a high Rb rock, probably 10024.

of the REE and Ba in lunar rocks with those observed in a variety of terrestrial rocks. Data for these rocks selected from a number of analyses made by the technique described here are shown in Fig. 8. It is quite clear that the lunar samples are unlike any of the terrestrial samples. Terrestrial rocks most similar to the lunar rocks in REE are the oceanic ridge basalts, in particular, one sample V21-40 which is a glassy rock similar to to a dacite in composition. Gabbroic rocks from oceanic ridge regions, as illustrated by sample V25-6-124, are distinguished by more regular downward slopes than the average extrusive rock from the oceanic ridges. Both the textures, bulk element chemistry and REE abundance patterns, indicate a cumulative origin for these gabbroic rocks (KAY *et al.*, 1970; MIYASHIRO *et al.*, 1970).

In general, the terrestrial rocks show that fractionation of REE and Ba relative to each other is very common in terrestrial rocks. This is quite consistent with the observed distribution coefficients between basaltic liquids and mafic minerals and

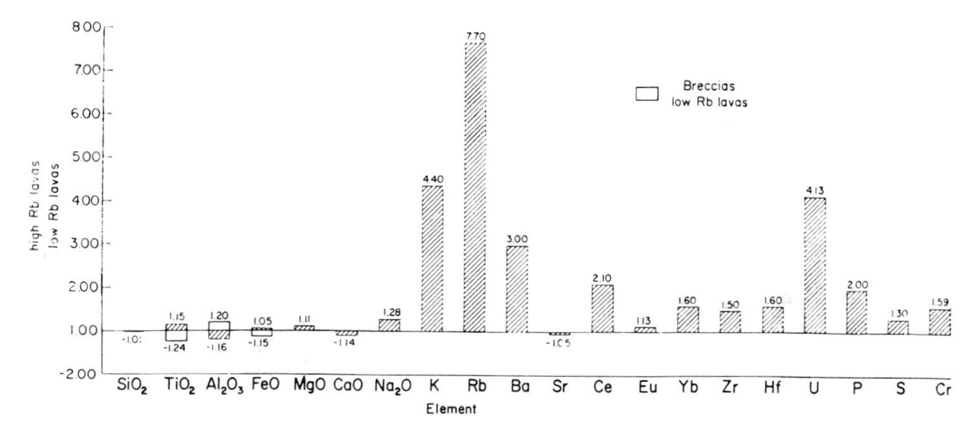

Fig. 7. Bar diagram showing the relative concentrations of 20 elements for average high Rb and average low Rb lavas. Data for SiO_2, TiO_2, Al_2O_3, FeO, CaO, MgO, Na_2O, P and S, are from COMPSTON *et al.* (1970), WIIK and OJANPERÄ (1970), ENGEL and ENGEL (1970), PECK and SMITH (1970), MAXWELL *et al.* (1970) and AGRELL *et al.* (1970). K, Rb, Ba and Sr data are from PHILPOTTS and SCHNETZLER (1970), COMPSTON *et al.* (1970), MURTHY *et al.* (1970), GOPALAN *et al.* (1970), HURLEY and PINSON (1970), and this report. Ce, Eu and Yb data are from PHILPOTTS and SCHNETZLER (1970), HASKIN *et al.* (1970), and this report. Zr, Hf and Cr data are from COMPSTON *et al.* (1970), GOLEŠ *et al.* (1970), MORRISON *et al.* (1970), TUREKIAN and KHARKAR (1970). U data are from TATSUMOTO and ROSHOLT (1970). Although the two groups have been defined by their Rb concentrations other elements K, Ba, Cs and probably U and Th will produce the same distinct grouping. Some elements, SiO_2, Al_2O_3, FeO, MgO, Cr, Zr, Hf, Eu and Sr, have concentrations that overlap between the two groups. The concentrations of other elements, TiO_2, CaO, Na_2O, S and P_2O_5 are consistent with this grouping but show much smaller inter-group differences than do K and Ba.

the production of these liquids by small degrees of partial melting (GAST, 1968a, PHILPOTTS and SCHNETZLER, 1970). The relative lack of fractionation of the REE for the ORB has been ascribed to complete removal of REE from the source of these liquids (GAST, 1968; KAY *et al.*, 1970).

The similarities between the oceanic ridge basalts and lunar basalts and the comparison of cumulates, differentiates and parent liquids from oceanic ridges are both useful in elucidating the origin of the chemical characteristics of the lunar basalts. Many of the differences between the low Rb and high Rb liquids can be explained if the low Rb liquids are in part cumulates from the high Rb liquids. GRIFFIN and MURTHY (1969) and PHILPOTTS and SCHNETZLER (1970) have shown that calcic plagioclase fractionates K and Rb from each other and is at the same time depleted in both elements relative to a basaltic liquid from which it crystallizes. Thus plagioclase enriched cumulates should have higher K/Rb ratios and lower K contents than their parent liquids, in the absence of other K rich phases. PHILPOTTS and SCHNETZLER (1970) have shown that liquidus clinopyroxenes are generally depleted in light REE relative to heavy REE. Similarly their data on elements from Apollo 11 samples suggests that the REE content of ilmenite and pyroxene are quite similar. Thus as long as pyroxene and ilmenite are dominant hosts for REE we infer that cumulates

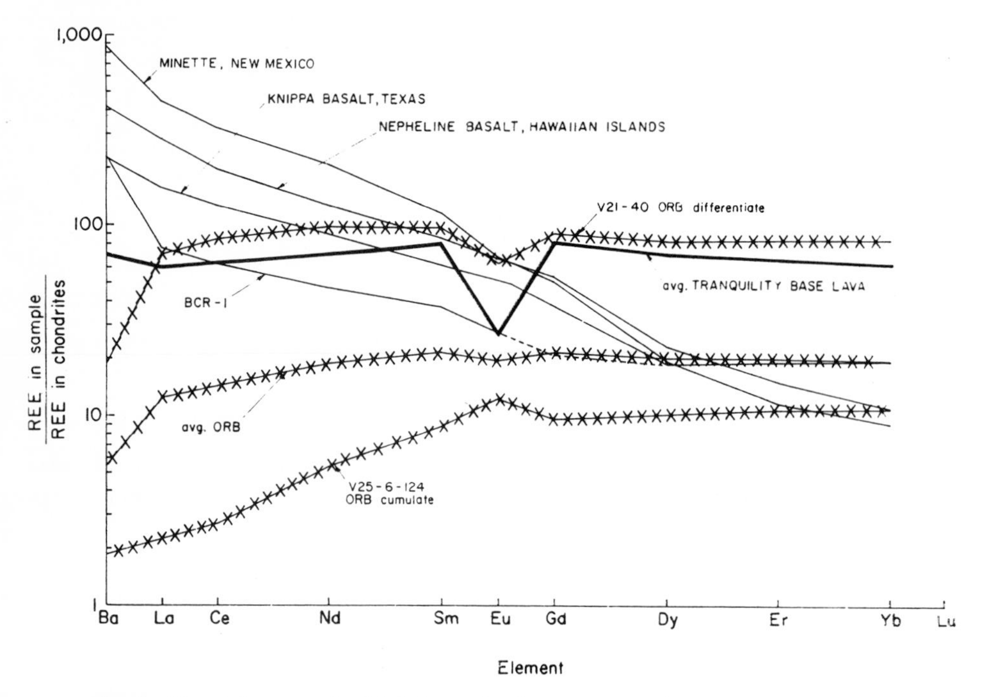

Fig. 8. Chondrite normalized REE and Ba data for selected earth rocks. Data are from KAY *et al.* (1970), KAY (research in progress), and this report. Chondrite values are those given for Fig. 4. The ORB cumulate, V25-6-124, has a bulk composition very similar to the Apollo 11 samples except for much higher Na_2O concentration (MIYASHIRO, 1970).

should be light REE depleted relative to the parent liquid. Depending on the proportions of plagioclase and pyroxene in the cumulate it is also possible that the cumulate is enriched in Eu/Sm relative to the parent liquid. All of these characteristics are illustrated by gabbroic sample V25-6-124. Also KAY *et al.* (1970) show that the K/Rb ratio of one gabbro, sample A150-20-4C, is higher (1380) than that of associated extrusive rocks. By analogy, we thus suggest that as a first approximation the low Rb rocks may be explained as meso-cumulates produced from the liquid similar to the high Rb liquids. Without rather accurate knowledge of the distribution coefficients between crystalline phases and a basaltic liquid for the elements under consideration it is difficult to evaluate whether or not the phases observed in the low Rb rocks represent an early stage of crystallization (10%–20%) or a more advanced crystallization (50%–70%). In the former case the high Rb rocks may represent the liquid residue of this crystallization. In either case, we must conclude that the differentiation inferred from the differences between the high and low Rb lavas cannot explain the dominant chemical characteristics of the liquids from which these rocks were derived. We will assume that the high Rb rocks are closer to the composition of the liquid from which either the high Rb or low Rb liquids were derived rather than vice versa and thus focus our attention on explaining the characteristics of this composition.

The chemical characteristics of these rocks are in many ways contradictory. The high concentrations of Ti, Zr, Hf, Sc, Cr and trivalent REE would normally be associated with extensive fractionation of minor elements between a silicate liquid and co-existing solids. In addition, the pronounced Eu anomaly can only be produced by rather extensive fractionation of the REE between certain solid phrases and a basaltic liquid. In strong contrast with these characteristics is the nearly chondritic relative abundance of the trivalent REE, Ba, U and Th, cf. Fig. 6. These observations seem to require a peculiar type of liquid–crystal fractionation in which large trivalent and quadrivalent ions are rather completely excluded from crystalline phases and thus not fractionated and divalent ions are readily accepted by the crystalline phases. The low Al content and high Ti and Fe content are also very unusual characteristics for liquids in equilibrium with common terrestrial mafic mineral assemblages. Finally, we note that the unusual characteristics of the liquids are not easily explained by non-igneous processes, e.g. vaporization of Eu and Sr from a high temperature liquid or vice versa, i.e. preferential condensation of trivalent REE from a high temperature gas (cf. e.g. LARIMER and ANDERS, 1970). Two possible kinds of explanations emerge from the constraints we have considered: (1) that the liquids we observed have undergone some extensive fractional crystallization before extrusion, or (2) that they were produced from a parent material that is chemically and mineralogically very different from that of most terrestrial basalts.

We have shown (Fig. 8) that in some rather rare situations extensive fractionation of basaltic liquids (ORB) has apparently taken place without fractionation of the REE, except for depletion of Eu relative to other REE. KAY *et al.* (1970) have concluded that the observed effects can be explained by fractional crystallization dominated by plagioclase and olivine. A similar mechanism may be pertinent in the lunar case. A quantitative test of this model requires a knowledge of the distribution coefficients of the relevant elements between a basaltic liquid and plagioclase, in particular, those for Sr, Ba and Eu. Values for Gd and Sm distribution coefficients are also necessary but need not be known as accurately. The phenocryst studies of PHILPOTTS and SCHNETZLER (1969) and SCHNETZLER and PHILPOTTS (1968, 1970) clearly show that Eu and Sr are extensively fractionated from trivalent elements. However, the range of distribution coefficients that can be inferred from their data is still quite wide, e.g. $K_{\mathrm{Eu}}^{\mathrm{liq/plag}}$ ranges from 1 to 5. Moreover, the distribution of Eu between a basaltic liquid and plagioclase may also depend on the p_{O_2} of the liquid. For example, if the Eu^{+3}/Eu^{+2} ratio varies significantly in terrestrial basalts, we infer from the more reduced nature of the lunar liquids (LSPET, 1969) that potential fractionation of Eu from Sm and Gd should be even greater in lunar liquids than in terrestrial liquids (HASKIN *et al.*, 1970 and PHILPOTTS and SCHNETZLER, 1970). The frequency and magnitude of europium anomalies in reduced meteorite systems (Fig. 9) supports this suggestion. Also, the ratio of Eu in plagioclase over Eu in pyroxene in coexisting phases in lunar rocks (PHILPOTTS and SCHNETZLER, 1970), is much greater (×10 or more) than it is in coexisting terrestrial phenocrysts. Thus, in calculating the effects of plagioclase removal we have used $K_{Eu}^{liq/plag}$ values somewhat

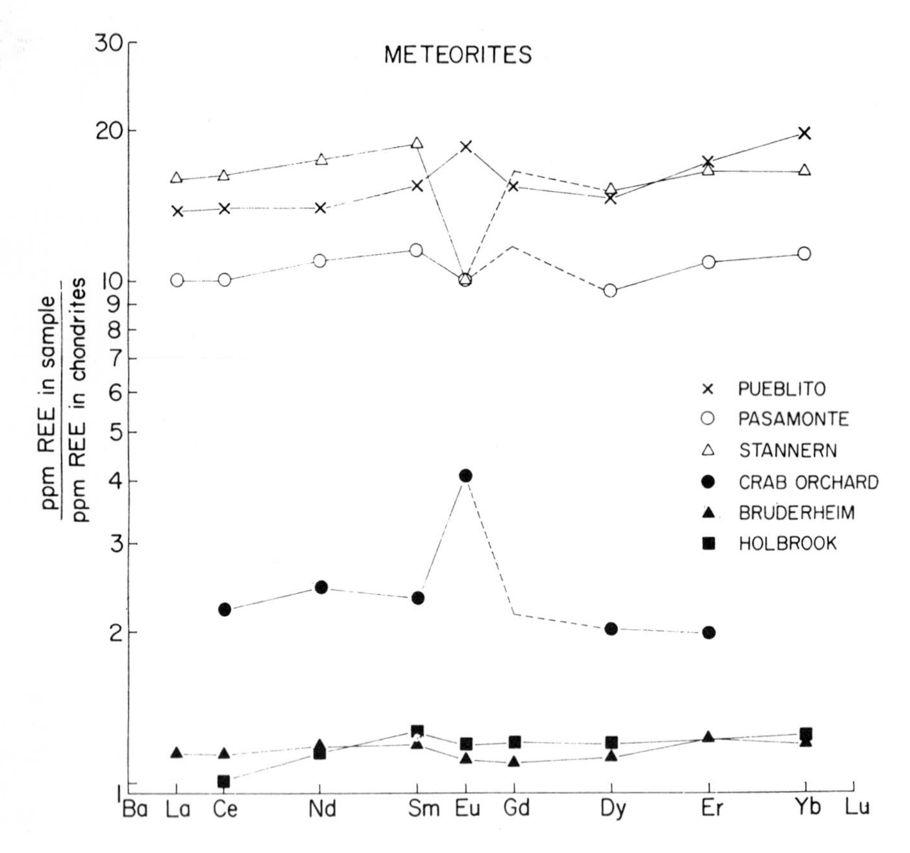

Fig. 9. Chondrite normalized REE data for some meteorites. Data are from this report. Chondrite normalization values are those given for Fig. 4.

below those inferred by SCHNETZLER and PHILPOTTS. The initial REE, Ba and Sr values for a range of distribution coefficients, for the condition that there should be no Eu anomaly in the initial liquid, are shown in Fig. 10. The bulk composition of these several parent liquids for the high Rb liquids are also calculated and shown in Table 3. From this model we infer that the parent liquid for the Tranquillity Base basalts could have been much more like terrestrial basalts. That is a high alumina, (~25% Al_2O_3) liquid with fairly high TiO_2 contents (2·5–3·5%) and an approximately chondritic REE abundance pattern. The MgO content of this hypothetical liquid is

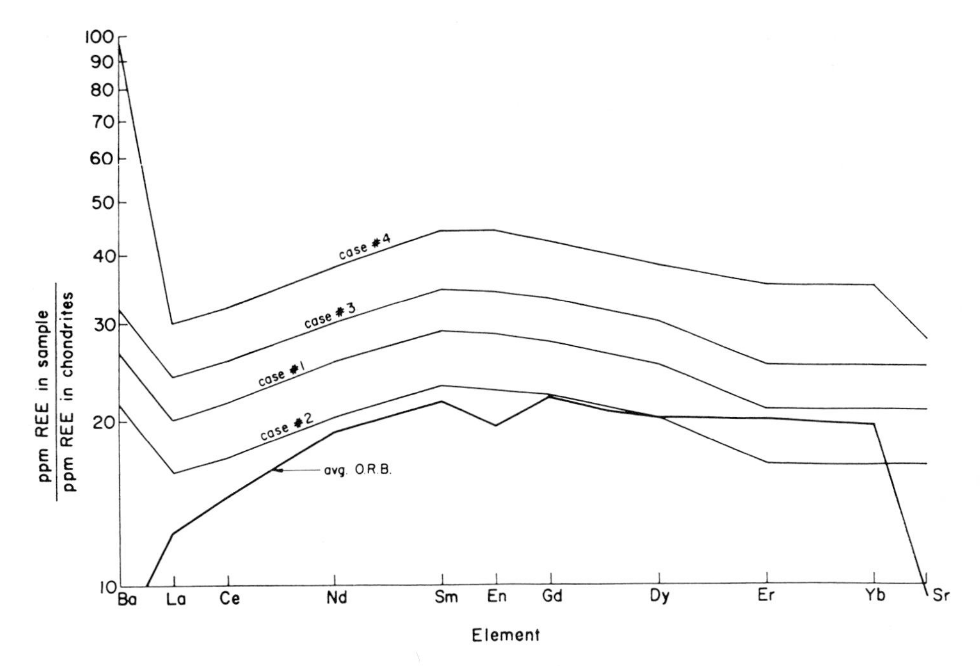

Fig. 10. Chondrite normalized REE, Ba and Sr data for four hypothetical parent liquids of Apollo 11 igneous rocks and average oceanic ridge basalts. See Table 3 for the parameters used in calculating the hypothetical parent liquids.

Table 3. Approximate bulk chemical compositions of the calculated hypothetical parent liquids shown in Fig. 10. The distribution coefficients and phase percentages are shown for each case.

	Case 1	Case 2	Case 3	Case 4*
SiO_2	43·2	43·4	43·0	42·7
TiO_2	2·9	2·3	3·5	4·3
Al_2O_3	25·0	24·6	23·7	21·8
FeO	6·8	6·7	5·8	6·7
MgO	6·3	7·8	6·5	7·0
CaO	14·3	13·8	13·9	13·3
% liq	25	20	30	37
% plag (An 92)	65	65	60	53
% Oliv (Fo 80)	10	15	10	10
$K_{Eu}^{liq/plag}$	1·0	1·5	0·77	0·5
$K_{Sr}^{liq/plag}$	0·56	0·72	0·24	0·4
$K_{Ba}^{liq/plag}$	10·0	10·0	10·0	1·2
$K_{REE}^{liq/plag}$	20·0	20·0	20·0	~20
$K_{REE,Ba,Sr}^{liq/oliv}$	>20·0	>20·0	>20·0	>20

* *K* values for this case on those used in the partial melting calculations, Fig. 11.

essentially unknown until limits can be established for the amount of olivine that has crystallized.

This proposed mechanism can be criticized on both textural and petrochemical grounds. Unlike the case of the oceanic ridge basalts, there is little textural evidence in the Tranquillity basalts that plagioclase is the dominant liquidus mineral (HARGRAVES *et al.*, 1970; BROWN *et al.*, 1970; MASON *et al.*, 1970). Moreover, experimental work with synthetic mixtures (RINGWOOD and ESSENE, 1970; WEILL *et al.*, 1970) clearly indicates that compositions similar to the high rubidium liquids crystallize ilmenite before plagioclase upon cooling. However, O'HARA *et al.* (1970) have shown that plagioclase has a very wide field of occurrence in these compositions. And at slightly lower TiO_2 contents (less than 9 per cent) plagioclase crystallizes before ilmenite. It seems plausible that with a very modest amount of segregation of ilmenite the high rubidium liquids are derived from an alumina rich liquid like that envisioned above. The occurrence of anorthosite gabbro in the soil overlying and intermixed with the igneous rocks is a strong indication that plagioclase rich cumulates (or residues) may not be uncommon in the vicinity of the site where the basalts were collected. The possible inverse chemical relationship of these gabbroic fragments to the basalts provides an interesting test of the hypothesis we have proposed. That is, liquidus plagioclase derived from a liquid having no europium anomaly like that postulated above should have significantly greater Eu/Sm ratios than a liquidus plagioclase derived from a liquid which already has a europium anomaly resulting from a partial melting process.

It is clear that the crystallization of a large volume of liquid is not the only mechanism that can produce the observed chemical characteristics (HASKIN *et al.*, 1970). One can quite easily construct a situation which would produce such a liquid by partial melting. This condition would be obtained if plagioclase remains as a residual phase and if pyroxene and spinel along with some plagioclase are the phases going into the liquid. FREY (1969) has shown that spinels, unlike pyroxenes, are enriched in light REE relative to heavy REE in a lherzolitic assemblage. If this observation can be generalized, it is readily seen that partial melting of a plagioclase, pyroxene, spinel, olivine assemblage would produce a liquid with essentially unfractionated relative REE abundances—except of course for the depletion in europium—even for very small amounts of partial melting. A hypothetical case using rather arbitrary distribution coefficients, shown in Table 4, and the method of calculation proposed by GAST (1968) and SHAW (1970) is shown in Fig. 11. The results of our calculation, Fig. 11, show that liquids with abundance patterns rather similar to those observed in the basalts from Tranquillity Base can be obtained by partial melting using distribution coefficients well within the range of those inferred from studies of coexisting phases, except for europium where we have assumed a distinctly lower value for $K_{Eu}^{liq/plag}$ on the assumption that the Eu^{+3}/Eu^{+2} ratio is significantly lower than that in terrestrial basalts. Also, the distribution of REE between spinel and clinopyroxene is very poorly known and based on two mineral pairs which give quantitatively inconsistent results. However, both samples clearly indicate that the light and heavy REE are extensively fractionated between spinel and clinopyroxene. It is of interest that this combination of minerals not only can produce an europium

Table 4. Distribution coefficients used for partial melting calculations

	$K^{l/\alpha}$ px	$K^{l/\beta}$ plag	$K^{\gamma/\alpha}$ spinel	$K^{l/\rho}$ olivine
La	8	16	24	20
Sm	2·2	25	0·15	20
Eu	2·0	0·5	0·08	20
Gd	1·8	25	0·07	20
Yb	1·4	33	0·05	20
Sr	3	0·4	0·05	14
Ba	20	1·2	13	20

The pyroxene/liquid, plagioclase/liquid, and olivine/liquid values are based on the phenocryst studies of PHILPOTTS and SCHNETZLER (1969) and SCHNETZLER and PHILPOTTS (1968, 1970). The value of $K_{Eu}^{liq/plag}$ was arbitrarily reduced by approximately a factor of 3. The pyroxene/spinel values are based on the data of FREY (1969).

anomaly but produces a broad maximum in abundance around Sm and Gd.* Secondly, it is clear that even if the distribution coefficients (Table 4) were increased by factors of as much as 3 or 4 the observed concentrations cannot be produced from a source with chondritic REE concentrations. This is also true for terrestrial basalts (GAST, 1968b). We thus infer that the liquids which produced the basalts from Tranquillity Base must have been derived from a source that is already strongly enriched in REE, Sr and Ba relative to chondritic meteorites. One of two possible explanations for this observation may be put forth. One, the moon is chemically differentiated and has a crust or mantle enriched in REE, Sr, Ba, U, Th and other large ions, or, two, the moon is a chemically heterogeneous body with a primitive crust or mantle enriched in refractory elements. The postulated plagioclase residue may also explain the unusually low Na content in the liquids in that plagioclase would provide a site in which to retain Na in the residue. The composition of the source of the liquids would be quite similar to eucrites, i.e. high in Fe, Al and Ca with REE, Ba and Sr contents 10–15 times those for chondrites. If the high Rb liquids are produced by this process we can infer that the K and U contents of the "enriched lunar crust" are 200–300 ppm K and 0·06–0·1 ppm U.

In summary the chemical characteristics of the Tranquillity Base basalts do not clearly distinguish which of the two mechanisms noted above are responsible for their origin. However, in either case we infer that plagioclase is a very common phase in the fractionation of the lunar basaltic liquids. It is furthermore clear that these mechanisms are not mutually exclusive. A liquid produced by partial melting, in equilibrium with plagioclase, i.e. saturated with plagioclase, would crystallize plagioclase as a liquidus mineral. Thus some early removal of plagioclase may still be involved in the production of the liquids that produced the observed basalts. Also

* KAY (personal communication) informs us that spinel megacrysts from a single flow (BINNS, 1969) are depleted by more than a factor of 10 relative to their host liquid. If these data represent an equilibrium situation, our model cannot explain the observed pattern. The model presented here could hold if another mineral that is rich in light REE, perhaps $CaTiO_3$ occurs in the lunar crust or mantle.

neither of the above possibilities is consistent with a chondritic major element and refractory element composition for the "lunar crust or mantle."

The fractionation process responsible for the observed chemical compositions is further constrained by the isotopic relationships observed for the igneous rocks and the soil. Both the lead and strontium isotope compositions reported for the igneous rocks suggests that these rocks were not derived from a well mixed homogeneous liquid. In particular, ALBEE *et al.* (1970) have shown that sample 10044 has a lower initial Sr^{87}/Sr^{86} ratio than several other igneous rocks. The initial Pb isotope ratios for the igneous rocks cannot be determined from the lead isotope data alone. However, if we accept the Rb–Sr dates we can infer these ratios from the observations.

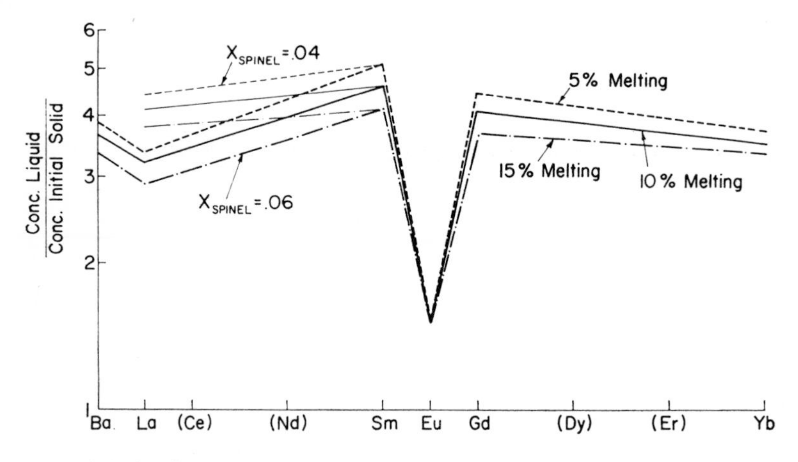

Fig. 11. Calculated concentrations of Ba, La, Sm, Eu, Gd and Yb in a liquid produced by partial melting of a clinopyroxene (32%), plagioclase (22%), spinel (6%) and olivine (40%) assemblage. Liquid consists of 60% pyroxene, 30% plagioclase and 10% spinel. Distribution coefficients used for these calculations are summarized in Table 5. Note that the patterns and relative concentrations of these elements are quite insensitive to the degree of partial melting. The relative abundances can be increased somewhat by increasing the proportion of olivine, without significantly changing the shape of the pattern.

From the observed U^{238}/Pb^{204} and Pb^{206}/Pb^{204}, etc., ratios we can readily find that the isotope ratios 3·65 b.y. were not identical, in particular, samples 10057 and 10017 were derived from sources that already had rather high Pb^{206}/Pb^{204} ratios. If this conclusion is true the simplest interpretation of the observed isotope compositions is a two-stage model (GAST, 1969). The results of such a calculation are shown in Table 5. These data also show that the isotopic compositions and U/Pb ratios of the sources of the igneous rocks produced 3·65 b.y. ago differed for several of the rocks analyzed by TATSUMOTO and ROSHOLT (1970). The times of fractionation of uranium and lead inferred for this model vary significantly from ages of crystallization inferred from Rb–Sr ages. Nevertheless, two important conclusions are derived from this calculation. First, it is unlikely that all the igneous rocks had a common isotopic composition at the time they were formed. In particular, this leads us to the corollary that these

Table 5. Two stage evolution parameters for lead isotope compositions of TATSUMOTO and ROSHOLT (1970)

	$(U^{238}/Pb^{204})_1$	(U^{238}/Pb^{204}) obs.	t_1	$(Pb^{206}/Pb^{204})t_1$		$(Pb^{207}/Pb^{204})t_1$	
10071	83	222·5	3·66	32·3 *32·3**	±5	41·3 *41·3*	±3
10017	217	492·8	3·15	103· *39·5*	±12	119· *71·2*	±6
10057	580	1392	3·78	141· *194·*	±60	200· *249·*	±30
10003	222	491·3	3·40	87· *54·3*	±25	104· *77·6*	±13
10020	166	338·3	3·20	74· *24·6*	±18	86·4 *56·1*	±9
10050†	—	397·2	$<3·0$	—		—	

* All italicized isotope ratios are calculated to 3·65 b.y. Others to time t_1.
† Not calculated because errors in isotope ratios much larger than for other samples.

rocks must be derived from an isotopically and chemically heterogeneous source. In other words, these observations appear to exclude derivation of the igneous rocks from large well-mixed magma chambers. Secondly, we note that the change in the U^{238}/Pb^{204} ratio needed to explain the observed isotope composition is relatively constant, ranging from a factor of 2 to 3.

Soils and breccias

The difference in the isotopic compositions (TATSUMOTO and ROSHOLT, 1970) of the soil and breccias and igneous rocks, in particular, the apparent lack of fractionation of the U/Pb ratio of the soil with respect to the ratio that this material has 4·6 b.y., is a most puzzling feature of the Tranquillity Base soil samples. As we have noted, the igneous rocks clearly show that their formation involved a substantial chemical fractionation, that is, increases in their Rb/Sr and U/Pb ratios. In addition, the creation of a residue or second unit with depleted Rb/Sr and U/Pb ratios must have accompanied their formation. The soil isotopic compositions were either derived from a different unfractionated primitive material or must have recombined both the uranium and rubidium enriched units with the uranium and rubidium depleted units during the 3·6 b.y. history of formation of the local regolith. It is significant that at the same time the soil, aside from a small nickel rich meteorite component (KEAYS *et al.*, 1970) and an Al rich and Ti poor component, retains most of the characteristics of the igneous rocks. This striking similarity is clearly shown in Fig. 6. There are significant differences in the aluminum, iron and titanium contents (Fig. 7) of the breccias in that the average concentrations of these elements, as well as the K/Rb ratios, are outside the combined ranges of the high and low Rb lavas. The average K/Rb ratio of the breccias is 390 and of the high and low Rb lavas 440 and 770. The soil and breccias can be approximated by equal portions of the high and low Rb lavas to which ~1/3 glassy material (WOOD *et al.*, 1970), with ~20·0% Al_2O_3, ~3·0% TiO_2 and ~11·0% FeO, and ~0·5% iron meteorites has been added. The K/Rb ratio of this simulated soil may be too high but could be lowered if the glass is a high K and Rb material with a low K/Rb ratio.

In summary, when we consider all aspects of the chemistry and isotopic relationships of the soil, breccias and igneous rocks together, we are forced to the conclusion that the melting and crystallization of the igneous rocks and the genesis of the europium and strontium anomalies and titanium contents may be different events. That is, it is hard to imagine why the source of the absent europium and strontium is not represented in the soil which has re-combined uranium and lead and rubidium and strontium so that they appear unfractionated relative to the 4·6 b.y. old lunar material. One possible explanation of this paradox may be that these 3·7 b.y. old rocks make up only a small fraction of the underlying bedrock that has been "gardened" in the impact processes that have formed the surface. That is, most of the soil was formed from 4·6 b.y. old mare rocks similar in composition to the 3·7 b.y. rocks.

PLANETOLOGICAL IMPLICATIONS

Even though the materials from Tranquillity Base are probably chemically fractionated relative to a primitive lunar composition some strong inferences regarding this composition can be drawn. Both the isotopic composition of lead and strontium in the basalts place significant constraints on the parent/daughter ratios, i.e. U/Pb and Rb/Sr ratios that may be inferred for the source of these liquids. If the source of these liquids is not the primitive undifferentiated moon it must be derived at a very early stage in the formation of the moon. The U/Pb ratios inferred for this source are very much higher than those found in the earth or in chondritic meteorites. The apparent differences range from a factor of ten to almost a factor of 100. A similar comparison can be made for the ratio of Rb/Sr using the isotopic composition of terrestrial and meteoritic strontium. The average present day Sr^{87}/Sr^{86} ratio in the upper ~200 km of the earth (0·0703–0·705) suggests that this ratio was 0·7001 or higher 3·6 b.y. ago in the upper mantle and crust of the earth. The existing data on very ancient rocks (BROOKS *et al.*, 1969) is consistent with this deduction. The initial ratios for the Tranquillity Base basalts that can be inferred from our measurements and those of other workers, in particular, ALBEE *et al.* (1970) and COMPSTON *et al.* (1970) are significantly lower than this (0·6992–0·6993). We infer from this that the source of the igneous rocks had a Rb/Sr ratio diminished by at least a factor of three relative to the upper part of the earth.

We have already noted that Na is depleted by more than a factor of five relative to terrestrial basalts. Cu and Zn also occur in very low abundance compared to terrestrial basalts. The inference that the source of the lunar rocks is depleted in these elements and other elements relative to the earth's mantle seems plausible.

The depletion of the earth in certain volatile elements in the whole earth relative to primitive solar nebula has been suggested from both isotopic relationships (Rb and Sr) and their abundance in the crust of the earth (GAST, 1968a). The data now available from the moon suggest that this depletion is even more marked for the moon than it is for the earth. This is also shown by the low ratio of K/U for lunar rocks (LSPET, 1969). It is of interest that Li which is much more refractory than the other alkali metals does not appear to be depleted in the lunar materials or achondrites. Furthermore, there is little evidence that the depletion of volatiles is mass dependent. The absence of any evidence for variation in the degree of depletion with mass is strong evidence for the separation of volatiles and refractory elements in the absence of a strong gravitational field. We suggest that the composition of solid material accumulating to form the earth and the moon was separated from the hydrogen rich solar nebula in the temperature range 1000°–1500°K. LARIMER and ANDERS (1970), have shown that at this temperature very extensive element fractionation would take place. It is even likely that there may be developed a significant fraction of material that is strongly enriched in certain elements due to a depletion of major elements such as Mg and Si. This possibility was first suggested to explain the apparent excess of Fe in the earth and mars by UREY (1952). GAST (1968a) has suggested that other refractory elements, e.g. Sr, U and Ba may also be enriched in the whole earth. Such material may be so depleted in silicon that a significant fraction of the metals occur as oxides rather than silicates. In addition this material could have Fe/Mg ratios and TiO_2 contents that greatly exceed those of chondrites. The partial melting products of such a high Fe, low Na assemblage may be quite different from those of a Mg-rich assemblage.

The fractionation of various elements between a vapor phase and a solid phase is very sensitive to temperature. Variations in temperature in both time and space could thus markedly effect the composition of solid matter condensing and aggregating in a primitive nebulae. The composition of small bodies being formed in such a nebulae could vary significantly from place to place due to varying opacities and varying distances from the sun. In situations where the temperature variations are fast compared to rates at which such bodies could equilibrate with their gaseous environment, significant variations in composition of primitive bodies due to these temperature changes can be preserved. It is possible that such variations in composition of the first objects to condense could produce significant heterogeneity in the composition of the moon and may account for the apparent variation in composition of the earth and the moon. Specifically we infer that the average earth has a greater proportion of material accumulated at low temperatures than the portion of the moon represented by the Apollo 11 samples.

Acknowledgments—This research was supported by NASA grant NAS 9-7895. We thank Mr. Robert Kay for operating the computer and providing valuable discussion.

REFERENCES

AGRELL S. O., SCOON J. H., MUIR I. D., LONG J. V. P., MCCONNELL J. D. C. and PECKETT A. (1970) Mineralogy and petrology of some lunar samples. *Science* **167**, 583–586.

ALBEE A. L., BURNETT D. S., CHODOS A. A., EUGSTER O. J., HUNEKE J. C., PAPANASTASSIOU D. A., PODOSEK F. A., RUSS G. PRICE II, SANZ H. G., TERA F. and WASSERBURG G. J. (1970) Ages, irradiation history, and chemical composition of lunar rocks from the Sea of Tranquillity. *Science* **167**, 463–466.

BINNS R. A. (1969) High-pressure megacrysts in basanitic lavas near Armidale, New South Wales. *Amer. J. Sci.* (Schairer Vol.) **267-A**, 33–49.

BROOKS C., HART S. R., KROGH T. E. and DAVIS G. L. (1969) Carbonate contents and $^{87}Sr/^{86}Sr$ ratios of calcites from Archaen meta-volcanics. *Earth Planet. Sci. Lett.* **6**, 35–38.

BROWN G. M., EMELEUS C. H., HOLLAND J. G. and PHILLIPS R. (1970) Petrographic, mineralogic, and X-ray fluorescence analysis of lunar igneous-type rocks and spherules. *Science* **167**, 599–601.

COMPSTON W., ARRIENS P. A., VERNON M. J. and CHAPPELL B. W. (1970) Rubidium-strontium chronology and chemistry of lunar material. *Science* **167**, 474–476.

ENGEL A. E. J. and ENGEL C. G. (1970) Lunar rock compositions and some interpretations. *Science* **167**, 527–528.

FREY F. A. (1969) Rare earth abundances in a high-temperature peridotite intrusion. *Geochim. Cosmochim. Acta* **33**, 1429–1447.

GAST P. W. (1962) The isotopic composition of strontium and the age of stone meteorites—I. *Geochim. Cosmochim. Acta* **26**, 927–943.

GAST P. W. (1968a) Upper mantle chemistry and evolution of the Earth's crust. *The History of the Earth's Crust*, pp. 15–28.

GAST P. W. (1968b) Trace element fractionation and the origin of tholeiitic and alkaline magma types. *Geochim. Cosmochim. Acta* **32**, 1057–1086.

GAST P. W. (1969) The isotopic composition of lead from St. Helena and Ascension Islands. *Earth Planet. Sci. Lett.* **5**, 353–359.

GAST P. W. and HUBBARD N. J. (1970) Abundance of alkali metals, alkaline and rare earths, and strontium-87/strontium-86 ratios in lunar samples. *Science* **167**, 485–487.

GOLEŠ G. G., OSAWA M., RANDLE K., BEYER R. L., JEROME D. Y., LINDSTROM D. J., MARTIN M. R., MCKAY S. M. and STEINBORN T. L. (1970) Instrumental neutron activation analyses of lunar specimens. *Science* **167**, 497–499.

GOPALAN K., KAUSHAL S., LEE-HU C. and WETHERILL G. W. (1970) Rubidium–strontium, uranium and thorium–lead dating of lunar material. *Science* **167**, 471–473.

GRIFFIN W. L. and MURTHY V. RAMA (1969) Distribution of K, Rb, Sr and Ba in some minerals relevant to basalt genesis. *Geochim. Cosmochim. Acta* **33**, 1389–1414.

HARGRAVES R. B., HOLLISTER L. S. and OTALORA G. (1970) Compositional zoning and its significance in pyroxenes from three coarse-grained lunar samples. *Science* **167**, 631–633.

HASKIN L. A., HELMKE P. A. and Allen R. O. (1970) Rare-earth elements in returned lunar samples. *Science* **167**, 487–490.

HURLEY P. M. and PINSON W. H., JR. (1970) Rubidium–strontium relations in Tranquillity Base samples. *Science* **167**, 473–474.

KAY R., HUBBARD N. J. and GAST P. W. (1970) Chemical characteristics and origin of oceanic ridge volcanic rocks. *J. Geophys. Res.* in press.

KEAYS R. R., GANAPATHY R., LAUL J. C., ANDERS E., HERZOG G. F. and JEFFERY P. M. (1970) Trace elements and radioactivity in lunar rocks: Implications for meteorite infall, solar-wind flux, and origin of moon. *Science* **167**, 490–493.

LARIMER J. W. and ANDERS E. (1970) Chemical fractionations in meteorites III. Abundance patterns and their interpretation. *Geochim. Cosmochim. Acta* **34**, 367–388.

LSPET (LUNAR SAMPLE PRELIMINARY EXAMINATION TEAM) (1969) Preliminary examination of lunar samples from Apollo 11. *Science* **165**, 1211–1227.

MASON B., FREDRIKSSON K., HENDERSON E. P., JAROSEWICH E., MELSON W. G., TOWE K. M. and WHITE J. S., JR. (1970) Mineralogy and petrography of lunar samples. *Science* **167**, 656–659.

MAXWELL J. A., ABBEY S. and CHAMP W. H. (1970) Chemical composition of lunar material. *Science* **167**, 530–531.

MIYASHIRO A., SHIDO F. and EWING M. (1970) Crystallization and differentiation in abyssal tholeiites and gabbros from mid-oceanic ridges. *Earth Planet. Sci. Lett.* in press.

MORRISON G. H., GERARD J. T., KASHUBA A. T., GANGADHARAM E. V., ROTHENBURG A. M., POTTER N. M. and MILLER G. B. (1970) Multielement analysis of lunar soil and rocks. *Science* **167**, 505–507.

MURTHY V. RAMA, SCHMITT R. A. and REY P. (1970) Rubidium–strontium age and elemental and isotopic abundances of some trace elements in lunar samples. *Science* **167**, 476–479.

O'HARA M. J., BIGGAR G. M. and RICHARDSON S. W. (1970) Experimental petrology of lunar material: the nature of mascons, seas, and the lunar interior. *Science* **167**, 605–607.

PAPANASTASSIOU D. A. and WASSERBURG G. J. (1969) Initial strontium isotopic abundances and the resolution of small time differences in the formation of planetary objects. *Earth Planet. Sci. Lett.* **5**, 361–376.

PECK L. C. and SMITH V. C. (1970) Quantitative chemical analysis of lunar samples. *Science* **167**, 532.

PHILPOTTS J. A. and SCHNETZLER C. C. (1969) Phenocryst-matrix partition coefficients for K, Rb, Sr and Ba, with applications to anorthosite and basalt genesis. *Geochim. Cosmochim. Acta* **34,** 307–322.

PHILPOTTS J. A. and SCHNETZLER C. C. (1970) Potassium, rubidium, strontium, barium, and rare-earth concentrations in lunar rocks and separated phases. *Science* **167,** 493–495.

RINGWOOD A. E. and ESSENE E. (1970) Petrogenesis of lunar basalts and the internal constitution and origin of the moon. *Science* **167,** 607–610.

SCHNETZLER C. C. and PHILPOTTS J. A. (1968) Partition coefficients of rare-earth elements and barium between igneous matrix material and rock forming mineral phenocrysts—I. In *Origin and Distribution of the Elements* (editor L. H. Ahrens), pp. 929–938. Pergamon.

SCHNETZLER D. C. and PHILPOTTS J. A. (1970) Partition coefficients of rare-earth elements between igneous matrix material and rock-forming mineral phenocrysts—II. *Geochim. Cosmochim. Acta* **34,** 331–340.

SHAW D. M. (1970) Trace element fractionation during anatexis. *Geochim. Cosmochim. Acta* **34,** 237–243.

SHIELDS W. R. (1966) Analytical mass spectrometry section: Instrumentation and procedures for isotopic analysis. NBS Tech. Note 227.

TATSUMOTO M. and ROSHOLT J. N. (1970) Age of the moon: An isotopic study of uranium–thorium–lead systematics of lunar samples. *Science* **167,** 461–463.

TUREKIAN K. K. and KHARKAR D. P. (1970) Neutron activation analysis of milligram quantities of lunar rocks and soils. *Science* **167,** 507–509.

UREY H. C. (1952) *The Planets.* Yale University Press.

WEILL D. F., MCCALLUM I. S., BOTTINGA Y., DRAKE M. J. and MCKAY G. A. (1970) Petrology of a fine grained igneous rock from the Sea of Tranquillity. *Science* **167,** 635–638.

WIIK H. B. and OJANPERÄ P. (1970) Chemical analyses of lunar samples 10017, 10072, and 10084. *Science* **167,** 531–532.

WOOD J. A., DICKEY J. S., JR., MARVIN U. B. and POWELL B. N. (1970) Lunar anorthosites. *Science* **167,** 602–604.

Proceedings of the Apollo 11 Lunar Science Conference, Vol. 2, pp. 1165 to 1176.

Elemental abundances by instrumental activation analyses in chips from 27 lunar rocks

GORDON G. GOLES, KEITH RANDLE and MASUMI OSAWA
Center for Volcanology, University of Oregon, Eugene, Oregon 97403
ROMAN A. SCHMITT and HIROSHI WAKITA
Department of Chemistry and Radiation Center, Oregon State University, Corvallis, Oregon 97331
WILLIAM D. EHMANN and JOHN W. MORGAN
Department of Chemistry, University of Kentucky, Lexington, Kentucky 40506

(*Received* 9 *February* 1970; *accepted in revised form* 25 *February* 1970)

Abstract—Analyses of 27 chips from lunar rocks by instrumental activation techniques provided data on Si, Ti, Al, Fe, Mg, Ca, O, Na, Ba, La, Ce, Sm, Eu, Tb, Ho, Yb, Lu, U, Zr, Hf, Ta, Mn, Cu, Co, Sc, V and Cr. These abundances may be used to comment on the merits of various analytical methods and to survey geochemical characteristics of Apollo 11 rocks. We propose on compositional grounds that rocks 10024 and 10071 should be classified as "basalts" rather than as "microgabbros". The "microgabbro" 10050 is compositionally anomalous, with low Fe (in which it resembles breccias), about 6·5% Mg which is the highest Mg content we have measured in these materials, and low La and Hf contents. It may be a specimen from a geologic unit distinct from that on which Eagle landed. Breccia 10056 is also compositionally anomalous, but it resembles "microgabbros" like 10047 closely enough so that one need not adduce exotic provenance as an explanation.

TWENTY-SEVEN chips from diverse rocks of the Apollo 11 collection were provided for analysis by our instrumental activation techniques. Three purposes were served by this work: (1) A library of specimens for which many elemental abundances are known is made available for further, possibly grossly destructive, investigations. Our work leaves minute amounts of artificially-induced long-lived radioactivities in the specimens and introduces minor radiation damage, principally from fast-neutron bombardment, but the specimens are physically intact and essentially unmodified chemically. (2) We gain experience in comprehensive analyses of gram-sized specimens, in case it ever becomes necessary to undertake elaborate sequential investigations of especially rare materials. (3) We learn much about the geochemistry of a wide variety of Apollo 11 rocks. This paper is mainly concerned with the first and second of these aims; we here discuss briefly some technical innovations, report our data, and compare them with abundances determined by other workers. Geochemical interpretations of the data are in large part reserved for other papers in this issue.

Of the rock chips, four are "basalts", five are "microgabbros" and the remainder are breccias. Two of the "basalts", 10024 and 10071, were initially classified as Type B rocks or "microgabbros" (WARNER, personal communication). We consider them to be "basalts" on compositional grounds, but detailed petrographic studies should be undertaken to clarify their classification.

Data on Si, O and one set of Al abundances were obtained by two of us (W.D.E. and J.W.M.) using 14-MeV neutron activation analysis as described by MORGAN and

EHMANN (1970). Samples and standards were packaged for irradiation and counting under a dry nitrogen atmosphere in the manner of EHMANN and MCKOWN (1968). Silicon and oxygen abundance estimates are based on 5–12 replicate analyses of each specimen, yielding standard deviations from the mean of about $\pm 0{\cdot}12\%$ Si and $\pm 0{\cdot}29\%$ O. Aluminum abundance estimates are based on only two analyses (for the EHMANN–MORGAN data set) of each specimen, which yield an average deviation from the mean of $\pm 0{\cdot}2\%$ Al.

Data on Ti, Al (the second set), Mg, Ca, Na, Mn, Cu and V were obtained by the Oregon State University group (R.A.S. and H.W., with some assistance from their co-workers). A brief account of the activation levels, nuclear reactions and detectors employed is given in this issue (WAKITA *et al.*, 1970). A more detailed description is given by SCHMITT *et al.* (1970). During photonuclear activation (with 23-MeV bremsstrahlung) for determination of Mg, which was done at the electron linear accelerator of Gulf General Atomic, San Diego, the vial containing specimen 10050-29 ruptured. Modifications of the method of EHMANN and MCKOWN have been devised to obviate this problem, which arises because of the large decelerations experienced by the specimen container in the "catcher" of the electron linear accelerator.

Table 1. Major elements in igneous rocks (wt. %)

	"Basalts"				"Microgabbros"					Range in error esti-
	10022-32	10069-23	10024-20	10071-21	10003-36	10047-14	10058-21	10062-28	10050-29	mates
Si	19·4	18·3	18·5	19·1	17·9	20·1	19·0	18·0	18·4	0·1–0·2
Ti	7·1	7·2	7·5	7·0	7·1	6·0	6·1	6·7	6·6	0·3–0·4
Al*	4·2	3·7	4·1	4·2	5·1	5·5	5·5	5·3	5·3	0·2–0·3
	4·7	3·8	4·3	4·5	5·7	5·7	6·3	6·1	5·8	0·1–0·2
Fe	15·7	14·1	15·4	14·9	15·3	15·1	14·9	15·0	13·6	0·2–0·4
Mg	4·1	3·7	4·3	4·4	4·9	3·5	3·8	—	6·5	0·2–0·4
Ca	7·4	7·1	7·1	7·2	8·3	8·7	8·1	9·3	8·2	0·7–0·9
O	39·3	37·6	38·9	40·3	38·1	40·1	39·9	38·0	40·5	0·1–0·5

* First set of values for Al are those determined by W.D.E. and J.W.M. The second set (single determinations only) were done by R.A.S. and H.W.

Table 2. Major elements in breccias (wt. %)

	10018-21	10019-11	10021-22	10048-32	10056-23	10059-30	10060-17	10061-32	10063-05	10064-10	Range in error estimates
Si	20·1	18·9	19·6	18·1	20·2	19·1	20·0	18·9	20·3	19·4	0·1–0·2
Ti	4·9	4·9	4·9	4·8	5·8	4·7	5·1	4·4	5·3	5·6	0·2–0·3
Al*	6·6	6·5	6·7	6·1	5·7	6·4	6·2	6·7	6·6	5·9	0·2
	7·7	7·5	7·2	6·7	6·4	7·0	6·0	7·8	7·7	5·6	0·2–0·3
Fe	13·1	12·7	13·1	13·2	14·8	12·9	13·1	12·63	13·1	12·8	0·18–0·4
Mg	5·1	3·8	5·0	4·1	3·9	5·1	4·9	5·9	4·7	4·3	0·2–0·5
Ca	8·8	9·0	7·7	8·0	9·4	8·1	—	—	9·7	—	0·8–1·0
O	40·1	39·9	41·8	39·8	41·3	40·0	40·3	41·7	41·9	40·5	0·1–0·6

	10065-14	10066-05	10067-05	10068-20	10070-05	10073-21	10074-05	10075-06	Range in error estimates
Si	19·3	20·2	20·6	19·3	20·6	20·5	19·3	19·8	0·1–0·3
Ti	4·7	4·9	5·3	4·7	5·0	4·9	4·7	4·5	0·2–0·3
Al*	6·6	6·9	7·0	6·3	7·1	7·3	7·0	7·4	0·2
	6·6	7·8	8·1	6·8	7·8	7·7	9·3	8·7	0·2–0·3
Fe	13·1	12·8	13·9	12·8	12·6	12·6	11·9	12·1	0·2–0·3
Mg	5·0	4·6	7·2	3·9	5·2	4·7	4·1	4·7	0·2–0·4
Ca	9·4	8·6	8·7	8·7	8·8	8·9	9·3	8·5	0·8–0·9
O	41·6	41·0	41·6	40·3	43·4	41·4	42·1	40·4	0·1–0·7

* First set of values for Al are those determined by W.D.E. and J.W.M. The second set (single determinations only) were done by R.A.S. and H.W.

Data on Fe, Ba, La, Ce, Sm, Eu, Tb, Ho, Yb, Lu, U, Zr, Hf, Ta, Co, Sc and Cr were obtained by the University of Oregon group (G.G.G., K.R. and M.O., with much assistance from their co-workers). These abundances were determined by techniques similar to those of GORDON *et al.* (1968). A noteworthy extension of the previous techniques is the determination of U according to AMIEL's (1962) method, but using an array of fourteen BF_3 neutron detectors.

Elemental abundances and error estimates are presented in Tables 1–4. Errors cited are estimates of single standard deviations; owing to editorial policy, only ranges are given. Major elements in Tables 1 and 2 are presented in the conventional order, with O last; minor and trace elements in Tables 3 and 4 are listed in the order suggested by TAYLOR (1965). The complete Lunar Receiving Laboratory specimen designation is given only in Tables 1 and 2. The igneous rocks which seem in some way anomalous ("basalts" 10024 and 10071, "microgabbro" 10050) are listed out of numerical order to emphasize their peculiar character. Although some of the breccias (notably 10056) may also be compositionally anomalous, we do not feel confident enough of their distinct character to place them in a separate subgroup.

Averages of major element abundances are compared in Table 5. Where necessary, we have reclassified rocks to conform to the scheme used in this paper. It is apparent from Table 5 that the degree of agreement between estimates of bulk composition obtained by different workers is very satisfactory. Almost all differences may be accounted for by sampling rather than analytical problems. Perhaps TUREKIAN and KHARKAR (1970) are systematically too low for Ti, and the Ca values of MORRISON *et al.* (1970) seem to be too high. There is a systematic difference between the two sets of Al abundances reported in this paper, which of course cannot be related to sampling difficulties, but it is not yet clear which set should be considered the more reliable one. In any case, a reasonable inference from the data of Table 5 is that our instrumental techniques are fully competitive with more-conventional rock analysis methods.

Owing to sampling problems and possible marked heterogeneity from one rock to another within a given class, we have not averaged minor- and trace-element abundances for purposes of making comparisons. Rather, Table 6 presents for comparison various sets of determinations of elements in different samples of selected rocks. Rocks were chosen for inclusion in Table 6 on the basis of having extensive data sets available, with a secondary consideration that of providing a representative selection. Agreement between the results in Table 6 is gratifying, especially to one who recalls problems associated with securing agreement among analyses of standard rocks such as G-1 and W-1 in diverse laboratories. Some of the disagreements exhibited in Table 6 may be related to sampling but others seem to arise from serious analytical errors. We comment below on both types of disagreements, on an element-by-element basis.

General agreement among Na abundances is good, although the LSPET (1969) values scatter rather widely and the values reported by ROSE *et al.* (1970) are systematically too high by about a factor two. For rocks of these compositions, the instrumental activation analysis technique for Na is clearly one of the best available (see also SCHMITT *et al.*, 1965; and STUEBER and GOLEŠ, 1967).

In contrast, our method for determination of Ba is of marginal utility. It is not certain whether this element is quite heterogeneously distributed in these rocks or whether the accuracies and precisions implied by other workers are overoptimistic. Two isotope dilution results for Ba in rock 10058 differ by about 6 per cent of their

Table 3. Minor and trace elements in igneous rocks (ppm)

	"Basalts"				"Microgabbros"					Range in error esti- mates
	10022	10069	10024	10071	10003	10047	10058	10062	10050	
Na	3550	3650	3620	3640	2700	3490	3020	3140	2630	50–70
Ba	220	250	170	450	220	—	140	230	—	50–80
La	25·9	23·7	23·0	25·8	13·5	11·3	11·8	13·1	7·2	0·3–0·5
Ce	81	65	76	84	37	46	39	38	34	2–3
Sm	20·3	18·0	19·2	20·0	13·0	18·9	14·0	11·9	11·8	0·2–0·3
Eu	2·15	2·04	—	2·12	1·84	2·71	2·14	2·07	2·0	0·08–0·2
Tb	5·7	4·8	—	5·7	3·5	4·1	3·5	3·3	2·1	0·2–0·6
Ho	8·2	6·9	8·1	9·2	4·0	7·9	5·5	4·4	4·6	0·4–0·9
Yb	21	20·8	19·6	20·8	15·3	18·2	14·0	13·5	11·1	0·5–3
Lu	2·69	2·67	3·20	3·08	2·62	2·88	1·94	1·94	1·96	0·05–0·12
U	0·67	0·78	0·67	0·69	0·31	0·16	0·18	0·27	0·21	0·04–0·10
Zr	130	520	650	210	560	—	190	290	—	80–160
Hf	19·6	17·8	20·0	19·1	11·6	13·2	11·2	11·8	8·6	0·3–1·0
Ta	1·8	2·7	2·4	2·0	—	2·6	1·6	1·0	2·2	0·3–0·5
Mn	1770	1600	1640	1650	1740	2100	1870	1790	1990	100–120
Cu	—	12	—	11	—	—	—	—	—	2
Co	29·8	26·0	28·4	27·1	14·1	12·2	14·4	13·8	13·6	0·4–0·7
Sc	76·6	72·4	76·2	73·2	74·0	92·0	80·8	74·7	88·9	0·8–1·1
V	89	87	84	92	63	63	78	75	117	6–10
Cr	2250	2130	2290	2170	1390	1250	1800	1540	2120	30–100

Table 4. Minor and trace

	10018	10019	10021	10048	10056	10059	10060	10061	10063
Na	3720	3530	3430	3490	3370	3590	3630	3550	3380
Ba	280	—	350	200	240	—	—	260	—
La	16·9	15·5	17·5	17·3	11·0	18·1	17·7	16·8	16·7
Ce	61	54	61	38·1	34	59	61	48·6	—
Sm	14·6	12·7	15·0	13·2	17·8	15·1	15·4	13·2	12·9
Eu	1·82	1·78	1·80	1·91	2·63	1·78	1·84	1·78	1·83
Tb	3·6	—	4·2	3·8	5·0	3·7	3·7	3·4	—
Ho	5·3	5·0	6·9	4·6	6·5	5·5	5·3	3·7	4·7
Yb	15·2	11·7	14·5	15·2	18·0	12·5	13·2	13·1	11·0
Lu	2·14	1·84	2·25	1·90	2·5	1·97	2·30	1·94	1·76
U	0·60	0·49	0·56	0·69	0·18	0·52	0·51	0·59	0·51
Zr	340	580	250	240	340	—	770	240	490
Hf	12·9	10·8	12·2	14·5	13·8	11·5	12·1	13·1	13·1
Ta	1·4	1·7	1·6	1·9	1·6	1·6	2·1	—	—
Mn	1590	1510	1560	1560	1970	1440	1650	1450	1620
Cu	—	—	—	—	—	—	—	—	16
Co	32·7	34·5	30·7	32·2	11·9	34·0	31·6	33·7	35·2
Sc	60·3	60·9	61·8	62·7	91·6	61·1	64·0	59·6	62·2
V	67	63	73	67	47	64	58	80	90
Cr	1880	1870	1950	1950	1280	1900	1880	1930	1940

mean value (GAST and HUBBARD, 1970; and MURTHY *et al.*, 1970), but two isotope dilution results for another "microgabbro", 10062, differ by about 50 per cent of their mean value (GAST and HUBBARD, 1970; and PHILPOTTS and SCHNETZLER, 1970). Some of our values agree well with isotope dilution results, but others seem to lie as much as two standard deviations above them (see data for 10071, 10003, 10062 and 10021).

The degree of agreement among rare earth abundances is extraordinary. Cerium abundances appear to reflect sampling or analytical difficulties, even if one considers only activation analysis or isotope dilution results, but these do not obscure fundamental differences between specimens. Samarium abundances of MORRISON *et al.* (1970) seem to be systematically too high, and those of WÄNKE *et al.* (1970) too low, but these characteristics are discernible only against a background of excellent agreement in other data. Terbium presents analytical difficulties to the activation analyst and much of the scatter in abundances of this element may be related to them. Holmium data are sparse, but perhaps the values of MORRISON *et al.* are too high. LSPET (1969) values for Yb are certainly much too low, as seem to be (to a lesser degree) those of TUREKIAN and KHARKAR (1970). Two out of the three values for Yb by MORRISON *et al.* seem to be too high. Excluding these data, Yb abundances suggest that this element is rather homogeneously distributed in these rocks. Our Lu values may be systematically too high, and those of TUREKIAN and KHARKAR are too low.

Most of the Zr values reported by LSPET (1969) are clearly too high, but even upon excluding these data much scatter remains. ROSE *et al.* (1970) may have overestimated their sensitivity for this element, but in any case they were working close to their detection limit. In view of the significance which abundances of this element are likely to have for questions of lunar petrogenesis (GOLEŠ *et al.*, 1970), it would be

elements in breccias (ppm)

10064	10065	10066	10067	10068	10070	10073	10074	10075	Range in error estimates
3650	3600	3420	3590	3280	3740	3220	3750	3350	60–180
290	220	—	—	150	310	—	280	430	50–100
19·6	17·8	17·4	20·1	16·4	17·3	12·8	13·8	14·9	0·3–0·4
59	63	62	68	60	56	48	55	50	1·7–5
15·5	14·6	15·1	16·7	14·4	13·1	11·5	11·5	11·5	0·2–0·3
1·77	1·73	1·70	2·4	1·80	1·74	1·60	1·73	1·62	0·07–0·2
3·7	4·0	2·8	3·1	3·6	3·1	—	2·8	3·1	0·3–0·5
5·5	6·7	6·5	7·5	6·6	5·8	5·0	5·0	5·4	0·4–0·8
14·8	14·5	11·8	13·8	12·2	14	7·2	12	11·2	0·4–1·5
2·46	2·01	1·90	2·20	2·6	1·80	1·76	1·7	1·89	0·06–0·15
0·65	0·54	0·56	0·54	0·61	0·62	0·45	0·49	0·52	0·05–0·08
520	—	—	—	700	360	—	500	390	70–180
13·9	12·1	10·6	15·4	11·0	12·8	8·9	11·9	8·8	0·3–0·5
1·7	2·1	2·1	2·1	1·8	1·0	1·6	1·0	1·4	0·3–0·4
1600	1540	1590	1820	1470	1520	1580	1420	1540	80–110
—	—	—	—	15	12	14	10	10	2
29·0	31·6	33·8	35·9	31·7	37·3	31·1	30·9	28·7	0·4–0·9
60·5	62·6	60·3	66·0	60·9	57·4	62·0	53·7	56·8	0·6–1·6
73	84	59	71	46	82	82	78	85	6–17
1850	1890	1910	2040	1890	1860	1900	1770	1790	50–80

Table 5A. Averages of major-element abundances in "basalts" (wt. %)

	Si	Ti	Al	Fe	Mg	Ca	O
This work (4 rocks)	18·8 ± 0·5	7·2 ± 0·2	4·1 ± 0·2* 4·3 ± 0·4†	15·0 ± 0·7	4·1 ± 0·3	7·2 ± 0·1	39·0 ± 1·1
COMPSTON *et al.* (1970) (3 rocks)	18·9	7·2	4·2	15·1	4·5	7·6	—
GAST and HUBBARD (1970) (3 rocks)	—	7·6	—	—	—	7·5	—
MORRISON *et al.* (1970) (3 rocks)	20·0	6·4	4·6	15·3	4·3	10·2	—
ROSE *et al.* (1970) (2 rocks)	18·7	7·6	4·8	14·5	4·8	7·4	—
TAYLOR *et al.* (1970) (7 rocks)	19·3	6·5	5·2	14·2	4·6	7·5	—
WÄNKE *et al.* (1970) (2 rocks)	19·3	6·8	4·2	14·3	4·5	8·3	40·6

Table 5B. Averages of major-element abundances in "microgabbros" (wt. %)

	Si	Ti	Al	Fe	Mg	Ca	O
This work (5 rocks)	18·7 ± 0·9	6·5 ± 0·5	5·3 ± 0·2* 5·9 ± 0·9†	14·8 ± 0·7	4·7 ± 1·4	8·5 ± 0·5	39·5 ± 1·1
COMPSTON *et al.* (1970) (4 rocks)	18·8	6·3	5·3	15·1	4·1	8·2	—
GAST and HUBBARD (1970) (4 rocks)	—	5·8	—	—	—	7·9	—
ROSE *et al.* (1970) (6 rocks)	18·8	6·6	5·6	14·5	4·2	8·1	—
TAYLOR *et al.* (1970) (8 rocks)	19·0	5·5	6·2	14·0	4·8	7·6	
TUREKIAN and KHARKAR (1970) (2 rocks)	—	3·8	—	12·5	—	—	—

Table 5C. Averages of major-element abundances in breccias (wt. %)

	Si	Ti	Al	Fe	Mg	Ca	O
This work (18 rocks)	19·7 ± 0·7	5·0 ± 0·4	6·6 ± 0·5* 7·4 ± 0·9†	13·0 ± 0·6	4·6 ± 0·6	8·8 ± 0·6	41·1 ± 1·0
COMPSTON *et al.* (1970) (2 rocks)	19·5	4·7	6·6	12·8	4·7	8·6	—
MORRISON *et al.* (1970) (3 rocks)	19·7	5·2	6·0	13·9	4·3	10·4	—
ROSE *et al.* (1970) (4 rocks)	19·4	5·6	6·5	12·7	4·4	8·4	—
TAYLOR *et al.* (1970) (3 rocks)	19·6	5·3	6·2	13·7	5·4	7·4	—
TUREKIAN and KHARKAR (1970) (2 rocks)	—	3·8	—	12·5	—	—	—
WÄNKE *et al.* (1970) (2 rocks)	19·7	5·1	6·2	11·6	4·8	—	41·1

* Determined by W.D.E. and J.W.M.
† Determined by R.A.S. and H.W. Single determinations only.

worthwhile to undertake a detailed investigation of its distribution by, e.g. isotope dilution or radiochemical activation techniques. An efficient way to attack this problem would be to use some of the chips reported on here, for which accurate Hf data are already available with which to construct Zr/Hf ratios.

Activation analysis determinations of Mn seem to be more reliable than any of the various spectrographic or X-ray fluorescence results. The relatively subtle differences from one rock to another in Mn contents probably are best mapped with our technique.

LSPET (1969) values for Co are too low; otherwise the agreement for this element is good.

Scandium values reported by LSPET tend to scatter in an apparently unrealistic way. Our Sc results are systematically slightly lower than those of other activation analysts or of ANNEL and HELZ (1970); since we used standard rock W-1 as a flux monitor for this element, it is not easy to rationalize this disagreement.

LSPET values for V are too low, but in several cases might represent the lower end of the actual range of abundances in small samples. Without more information, it is not possible to decide whether we are faced with sampling or analytical problems for this element.

Agreement among Cr abundances is unsatisfactory, even if one focusses on activation analysis results alone. For example, four determinations of Cr in breccia 10060 by activation techniques are: 1880 ± 60, 2200, 2800 and 1820 (all in ppm). It seems likely that there are severe sampling problems here, and that all results (with the possible exception of those by LSPET) are equally valid from an analytical standpoint.

We shall conclude by discussing briefly the evidence for geochemical clans among these rocks. As has been pointed out by numerous investigators at the Apollo 11 Conference, there are first-order distinctions between igneous rocks and clastic materials (breccias and "soil"). We can now discern second-order distinctions within these classes, exhibited in abundances of many elements. Consider Tables 1 and 2, especially data for Ti, Al and Ca. Titanium is relatively high in "basalts" and, on the average, intermediate in "microgabbros" and low in breccias. Aluminum is distributed in a mirror-image way (EHMANN and MORGAN, 1970), as is Ca, although less strikingly. In phratry A, the "basalts", there is no chemical distinction between clan A-1, rocks 10022 and 10069 whose classification was never in doubt, and clan A-2, rocks 10024 and 10071. (This statement is valid for the data of Table 3 as well.) In phratry B, "microgabbros" fall into two clans, the populous B-1 (10003, 10047, 10058, 10062) and the apparently rare B-2 clan of rock 10050. Rock 10050 is chemically very unusual, with low Fe (near the upper end of the breccia range), the highest Mg content we have measured, and strikingly-low La and Hf contents. It does not seem likely that it can be petrogenetically related in a simple way to the other "microgabbros".

Among the breccias of phratry C, we may pick out a C-2 clan consisting only of rock 10056. This rock has a Ti content like those of the B-1 clan, rather low Al, and the highest Fe content among the breccias. Its trace-element abundances resemble very closely those of "microgabbro" 10047. Presumably our sample of 10056 contains one or more clasts of rocks of clan B-1 which tend to dominate the geochemical relationships. Perhaps one cannot even exclude mislabeling as an explanation of these observations. Our chip (10056-23) appears on superficial examination to be a

Table 6. Comparisons of minor and trace elemental

	10022	10071	10003	10058	10062
Na	3550 ± 70 3000[f] 6700[l]	3640 ± 70 3900[c]	2700 ± 60 3000[b] 2900[c] 6300[l]	3020 ± 60 2900[c] 4100[f] 3200[g] 5000[l]	3140 ± 60 3000[b] 3300[c] 5100[l] 2990[p]
Ba	220 ± 70 100[f] 277[h]	450 ± 80 470[a] 327[c]	220 ± 60 160[a] 108[c]	140 ± 70 117[c] 85[f] 140[g] 124[h]	230 ± 70 79·9[c] 134[k]
La	25·9 ± 0·5 26·4 ± 0·7[d]	25·8 ± 0·4 27[a] 27·8[c]	13·5 ± 0·3 15[a] 15·1[c] 14·1 ± 0·2[d]	11·8 ± 0·3 11·7[c] 16[g]	13·1 ± 0·3 13·8[c] 12[p]
Ce	81 ± 2 68 ± 2[d]	84 ± 2 83·5[c]	37 ± 2 47·2[c] 41·3 ± 0·7[d]	39 ± 2 40·2[c] 45[g]	38 ± 2 42·7[c] 40·2[k] 48[p]
Sm	20·3 ± 0·3 21·2 ± 0·1[d]	20·0 ± 0·3 22·7[c]	13·0 ± 0·2 14·5[c] 13·1 ± 0·1[d]	14·0 ± 0·2 17·1[c] 22[g]	11·9 ± 0·2 13·3[c] 14·7[k] 10[p]
Eu	2·15 ± 0·10 2·04 ± 0·01[d]	2·12 ± 0·15 2·29[c]	1·84 ± 0·09 1·82[c] 1·80 ± 0·01[d]	2·14 ± 0·11 2·64[c] 3·0[g]	2·07 ± 0·08 2·02[c] 2·07[k] 1·8[p]
Tb	5·7 ± 0·6 4·7 ± 0·1[d]	—	3·5 ± 0·3 3·26 ± 0·07[d]	3·5 ± 0·2 5·4[g]	—
Ho	8·2 ± 0·8 5·5–8·7[d]	—	4·0 ± 0·6 3·7–4·4[d]	5·5 ± 0·5 9[g]	—
Yb	21 ± 3 17·7 ± 0·5[d] 7[f]	20·8 ± 1·6 20·2[c]	15·3 ± 0·7 12·7[c] 11·9 ± 0·3[d]	14·0 ± 1·3 15·2[c] 5[f] 22[g]	13·5 ± 1·3 12·1[c] 11·3[k] 6·3[p]
Lu	2·69 ± 0·09 2·55 ± 0·02[d]	3·08 ± 0·07 2·87[c]	2·62 ± 0·08 1·81[c] 1·69 ± 0·01[d]	1·94 ± 0·08 2·14[c] 2·3[g]	1·94 ± 0·07 1·73[c] 1·76[k] 0·87[p]
U	—	0·69 ± 0·09 0·873[o]	0·31 ± 0·05 0·26 ± 0·03[i] 0·29 ± 0·02[j] 0·268[o]	0·18 ± 0·05 0·20[g]	—
Zr	130 ± 110 1000[f] ~200[l]	210 ± 110 644[a]	560 ± 160 380[a] 309[b] <200[l]	190 ± 100 250[f] 380[g] <200[l]	290 ± 80 319[b] <200[l]
Hf	—	—	—	11·2 ± 0·4 13[g]	11·8 ± 0·3 10[p]

abundances in selected individual rocks (ppm)

10050	10018	10021	10056	10060	10061
2630 ± 50 3800[f] 4900[l]	3720 ± 180 3400[b] 3920[q]	3430 ± 70 1500[f] 3470[p]	3370 ± 60 3200[g]	3630 ± 70 3400[g] 5800[l] 3800[n] 3600[q]	3550 ± 70 3500[b] 3700[f]
—	280 ± 70 220[a] 175[b] 200[k]	350 ± 80 270[a] 105[f] 211[k]	240 ± 80 100[g]	—	260 ± 70 270[a] 128[b] 90[f]
	16·9 ± 0·4 15[a] 24[b] 18[q]	17·5 ± 0·4 22[a]	11·0 ± 0·3 13[g] 12[p]	17·7 ± 0·4 20·8 ± 0·4[d] 25[n] 18[q]	16·8 ± 0·4 18[a] 23[b]
—	61 ± 2 67[b] 52·8[k]	61 ± 2 57·2[k]	34 ± 2 42[g] 75[p]	61 ± 3 58 ± 1[d] 62[g] 59·0[k] 56[n]	48·6 ± 1·9 37[b]
—	14·6 ± 0·3 16·3[k] 8·5[q]	15·0 ± 0·2 17·2[k] 12[p]	17·8 ± 0·3 23[g] 17[p]	15·4 ± 0·3 15·4 ± 0·1[d] 24[g] 17·5[k] 16[n] 8·7[q]	—
—	1·82 ± 0·09 1·84[k] 1·68[q]	1·80 ± 0·09 1·91[k] 1·8[p]	2·63 ± 0·16 2·5[g] 2·9[p]	1·84 ± 0·08 2·06 ± 0·02[d] 2·0[g] 1·98[k] 1·9, 2·6[n] 1·61[q]	—
—	—	—	5·0 ± 0·5 5·4[g]	3·7 ± 0·4 3·6 ± 0·1[d] 5·0[g] 6[n]	—
—	—	—	6·5 ± 0·8 9[g]	5·3 ± 0·6 4·4–5·4[d] 10[g] 7[n]	—
11·1 ± 0·5 2·7[f]	15·2 ± 0·4 11·8[k] 11·1[q]	14·5 ± 0·7 4·5[f] 12·7[k]	18·0 ± 1·5 20[g] 8·3[p]	13·2 ± 0·7 13·2 ± 0·3[d] 22[g] 12·7[k] 14[n] 10·9[q]	13·1 ± 0·5 1·8[f]
—	2·14 ± 0·07 1·87[k] 1·56[q]	2·25 ± 0·07 1·99[k]	2·5 ± 0·3 1·8[g] 1·2[p]	2·30 ± 0·08 2·00 ± 0·02[d] 2·0[g] 1·92[k] 1·8[n] 1·57[q]	—
0·21 ± 0·05 0·156[o]	0·60 ± 0·08 0·60 ± 0·09[i]	0·56 ± 0·09 0·54 ± 0·08[i] 0·39[p]	0·18 ± 0·05 0·21[g] 0·45[p]	0·51 ± 0·07 0·60[g] 0·60[m] 0·4[n]	0·59 ± 0·08 0·674[o]
— 700[f] $<$200[l]	340 ± 100 429[a] 328[b]	250 ± 90 424[a] 1500[f]	340 ± 110 410[g]	770 ± 110 580[g] ~300[l] 300, 320, 340[■]	240 ± 70 393[a] 342[b] 400[f]
—	—	—	13·8 ± 0·5 11[g] 11[p]	13·1 ± 0·3 13[g] 12[n]	—

Table 6 (continued). Comparisons of minor and trace

	10022	10071	10003	10058	10062
Ta	—	—	—	1·6 ± 0·3 1·0[g]	1·0 ± 0·3 1·8[p]
Mn	1770 ± 110 2000[f] 1900[l]	1650 ± 100 2230[a]	1740 ± 100 2580[a] 2300[b] 2200[l]	1870 ± 110 4300[f] 1900[g] 2100[l]	1790 ± 110 2300[b] 2100[l] 1510[p]
Co	29·8 ± 0·7 15[f]	27·1 ± 0·7 33[a]	14·1 ± 0·4 15[a]	14·4 ± 0·4 7[f] 14[g]	13·8 ± 0·4 13[p]
Sc	76·6 ± 1·0 110[f]	73·2 ± 1·0 97[a]	74·0 ± 0·8 94[a]	80·8 ± 1·1 130[f] 87[g]	74·7 ± 0·8 76[p]
V	89 ± 8 36[f]	92 ± 9 78[a]	63 ± 7 82[a]	78 ± 9 32[f] 41[g]	—
Cr	2250 ± 100 2800[f] 2500[l]	2170 ± 80 3060[a]	1390 ± 30 1860[a] 1800[l]	1800 ± 70 3700[f] 1500[g] 1600[l]	1540 ± 40 1700[l] 1310[p]

a. ANNEL and HELZ (1970); emission spectrography.
b. COMPSTON *et al.* (1970); X-ray fluorescence, flame photometry.
c. GAST and HUBBARD (1970); isotope dilution, atomic absorption, X-ray fluorescence, wet chemical.
d. HASKIN *et al.* (1970); radiochemical activation analysis.
e. KEAYS *et al.* (1970); radiochemical activation analysis.
f. LSPET (1969); principally emission spectrography.
g. MORRISON *et al.* (1970); activation analysis and spark-source mass spectrography.
h. MURTHY *et al.* (1970); isotope dilution.
i. O'KELLEY *et al.* (1970); gamma-ray spectrometry.

single clast of an intensely microbrecciated, relatively coarse-grained igneous rock.

We conclude that of the two anomalous chips we have analysed, 10056 seems to be related to the "microgabbro" clan B-1 and thus does not pose any extraordinary problems of provenance or petrogenesis. Rock 10050, however, may well have originated as an impact-mobilized fragment from a geologic unit distinct from that immediately underlying the regolith at the Apollo 11 landing site. If so, a detailed study of this rock would be very valuable.

Acknowledgments—We thank the Oregon State University and Reed College reactor groups for assistance in neutron activations. V. P. GUINN and J. MACKENZIE of Gulf General Atomic, San Diego, assisted with the Mg analyses. W. D. LOVELAND of Oregon State University helped greatly in setting up the technique for determining U used here. We are grateful for assistance from the secretaries of the Department of Geology, University of Oregon, especially J. M. HEER. This work was supported in part by NASA contract NAS 9-7961, NSF grant GA-1383, and NASA grant NGL 38-003-010 (University of Oregon), by NASA grant NGR 38-002-020 and NASA contract 9-8097 (Oregon State University), and by NASA contract NAS 9-8017 and the University of Kentucky Research Foundation (University of Kentucky).

elemental abundances in selected individual rocks (ppm)

10050	10018	10021	10056	10060	10061
—	1·4 ± 0·3 2·1[q]	—	1·6 ± 0·3 2·2[g] 2·6[p]	2·1 ± 0·4 1·7[g]	—
1990 ± 120 3900[f] 2100[l]	1590 ± 100 1660[a] 1700[b] 1050[q]	1560 ± 90 1770[a] 1700[f] 1580[p]	1970 ± 120 2000[g] 2140[p]	1650 ± 100 1600[g] 1800[l] 1620[n] 1340[q]	1450 ± 80 1820[a] 1700[b] 2400[f]
13·6 ± 0·5 15·2[e] 10[f]	32·7 ± 0·8 32[a] 35[b] 24·0[q]	30·7 ± 0·7 33[a] 13[f]	11·9 ± 0·4 15[g] 14[p]	31·6 ± 0·6 32[g] ~30[n] 30·0[q]	33·7 ± 0·6 35[a] 23[b] 34·2[e] 12[f]
88·9 ± 1·0 170[f]	60·3 ± 0·7 66[a] 69[q]	61·8 ± 0·8 72[a] 68[f]	91·6 ± 1·6 97[g] 109[p]	64·0 ± 0·7 70[g] 70[n] 70[q]	59·6 ± 0·7 67[a] 55[f]
117 ± 9 80[f]	67 ± 14 60[a] 51[b]	73 ± 9 60[a] 22[f]	47 ± 9 56[g]	58 ± 8 62[g] ~90[n]	80 ± 6 60[a] 34[b] 32[f]
2120 ± 60 4800[f] 2400[l]	1880 ± 60 2340[a] 1950[b] 1900[q]	1950 ± 50 2480[a] 2500[f]	1950 ± 50 1400[g] 1410[p]	1880 ± 60 2200[g] 2300[l] 2800[n] 1820[q]	1930 ± 50 2730[a] 1940[b] 3000[f]

j. Perkins *et al.* (1970); gamma-ray spectrometry.
k. Philpotts and Schnetzler (1970); isotope dilution.
l. Rose *et al.* (1970); X-ray fluorescence, wet chemical.
m. Silver (1970); isotope dilution.
n. Smales *et al.* (1970); diverse techniques.
o. Tatsumoto and Rosholt (1970); isotope dilution.
p. Turekian and Kharkar (1970); activation analysis.
q. Wänke *et al.* (1970); activation analysis.

References

Amiel S. (1962) Analytical applications of delayed neutron emission in fissionable elements. *Anal. Chem.* **34,** 1683–1692.

Annell C. and Helz A. (1970) Emission spectrographic determination of trace elements in lunar samples. *Science* **167,** 521–523.

Compston W., Arriens P. A., Vernon M. J. and Chappell B. W. (1970) Rubidium–strontium chronology and chemistry of lunar material. *Science* **167,** 474–476.

Ehmann W. D. and McKown D. M. (1968) Heat-sealed polyethylene sample containers for neutron activation analysis. *Anal. Chem.* **40,** 1758.

Ehmann W. D. and Morgan J. W. (1970) Oxygen, silicon and aluminum in lunar samples by 14 MeV neutron activation. *Science* **167,** 528–530.

Gast P. W. and Hubbard N. J. (1970) Abundance of alkali metals, alkaline and rare earths and Strontium-87/Strontium-86 ratios in lunar samples. *Science* **167,** 485–487.

Goleš G. G., Osawa M., Randle K., Beyer R. L., Jerome D. Y., Lindstrom D. J., Martin M. R., McKay S. M. and Steinborn T. L. (1970) Instrumental neutron activation analyses of lunar specimens. *Science* **167,** 497–499.

Gordon G. E., Randle K., Goleš G. G., Corliss J. B., Beeson M. H. and Oxley S. S. (1968) Instrumental activation analysis of standard rocks with high-resolution γ-ray detectors. *Geochim. Cosmochim. Acta* **32,** 369–396.

HASKIN L. A., HELMKE P. A. and ALLEN R. O. (1970) Rare earth elements in returned lunar samples. *Science* **167**, 487–490.

KEAYS R. R., GANAPATHY R., LAUL J. C., ANDERS E., HERZOG G. F. and JEFFERY P. M. (1970) Trace elements and radioactivity in lunar rocks: implications for meteorite infall, solar-wind flux, and formation conditions of Moon. *Science* **167**, 490–493.

LSPET (LUNAR SAMPLE PRELIMINARY EXAMINATION TEAM) (1969) Preliminary examination of lunar samples from Apollo 11. *Science* **165**, 1211–1227.

MORGAN J. W. and EHMANN W. D. (1970) Precise determination of oxygen and silicon in chondritic meteorites by 14-MeV neutron activation with a single transfer system. *Anal. Chim. Acta* **49**, 287–299.

MORRISON G. H., GERARD J. T., KASHUBA A. T., GANGADHARAM E. V., ROTHENBERG A. M., POTTER N. M. and MILLER G. B. (1970) Multielement analysis of lunar soil and rocks. *Science* **167**, 505–507.

MURTHY V. R., SCHMITT R. A. and REY P. (1970) Rubidium–strontium age and elemental and isotopic abundances of some trace elements in lunar samples. *Science* **167**, 476–479.

O'KELLEY G. D., ELDRIDGE J. S., SCHONFELD E. and BELL P. R. (1970) Elemental compositions and ages of lunar samples by nondestructive gamma-ray spectrometry. *Science* **167**, 580–582.

PERKINS R. W., RANCITELLI L. A., COOPER J. A., KAYE J. H. and WOGMAN N. A. (1970) Cosmogenic and primordial radionuclides in lunar samples by nondestructive gamma-ray spectrometry. *Science* **167**, 577–580.

PHILPOTTS J. A. and SCHNETZLER C. C. (1970) Potassium, rubidium, strontium, barium and rare-earth concentrations in lunar rocks and separated phases. *Science* **167**, 493–495.

ROSE, H. J. JR., CUTTITTA F., DWORNIK E. J., CARRON M. K., CHRISTIAN R. P., LINDSAY J. R., LIGON D. T. and LARSON R. R. (1970) Semi-micro chemical and X-ray fluorescence analysis of lunar samples. *Science* **167**, 520–521.

SCHMITT R. A., SMITH R. H. and GOLEŠ G. G. (1965) Abundances of Na, Sc, Cr, Mn, Fe, Co and Cu in 218 individual meteoritic chondrules via activation analysis, 1. *J. Geophys. Res.* **70**, 2419–2444.

SCHMITT R. A., LINN T. A. JR. and WAKITA H. (1970) The determination of fourteen common elements in rocks via sequential instrumental activation analysis. *Radiochim. Acta*. In press.

SILVER L. T. (1970) Uranium–thorium–lead isotope relations in lunar materials. *Science* **167**, 468–471.

SMALES A. A., MAPPER D., WEBB M. S. W., WEBSTER R. K. and WILSON J. D. (1970) Elemental composition of lunar surface material. *Science* **167**, 509–512.

STUEBER A. M. and GOLEŠ G. G. (1967) Abundances of Na, Mn, Cr, Sc and Co in ultramafic rocks. *Geochim. Cosmochim. Acta* **31**, 75–93.

TATSUMOTO M. and ROSHOLT J. N. (1970) Age of the Moon: an isotopic study of uranium–thorium–lead systematics of lunar samples. *Science* **167**, 461–463.

TAYLOR S. R. (1965) Geochemical analysis by spark source mass spectrography. *Geochim. Cosmochim. Acta* **29**, 1243–1261.

TAYLOR S. R., JOHNSON P. H., MARTIN R., BENNETT D., ALLEN J. and NANCE W. (1970) Preliminary chemical analyses of 20 Apollo 11 lunar samples. *Science*. In press.

TUREKIAN K. K. and KHARKAR D. P. (1970) Neutron activation analysis of milligram quantities of lunar rocks and soils. *Science* **167**, 507–509.

WAKITA H., SCHMITT R. A. and REY P. (1970) Elemental abundances of major, minor and trace elements in Apollo 11 lunar rocks, soil and core samples. *Geochim. Cosmochim. Acta* Supplement I.

WÄNKE H., BEGEMANN F., VILCSEK E., RIEDER R., TESCHKE F., BORN W., QUIJANO-RICO M., VORSHAGE H. and WLOTZKA F. (1970) Major and trace elements and cosmic-ray produced radioisotopes in lunar samples. *Science* **167**, 523–525.

Proceedings of the Apollo 11 Lunar Science Conference, Vol.2, pp. 1177 to 1194.

Interpretations and speculations on elemental abundances in lunar samples

GORDON G. GOLES, KEITH RANDLE, MASUMI OSAWA, DAVID J. LINDSTROM, DOMINIQUE Y. JÉROME, TERRY L. STEINBORN, ROBERT L. BEYER, MARILYN R. MARTIN and SHEILA M. MCKAY
Departments of Chemistry and Geology, and Center for Volcanology, University of Oregon, Eugene, Oregon 97403

(*Received* 19 *February* 1970; *accepted in revised form* 2 *March* 1970)

Abstract—Abundances of SiO_2, Ti, Al, Na, Ba, La, Ce, Sm, Eu, Tb, Dy, Ho, Yb, Lu, U, Hf, Zr, Ta, Mn, Co, Fe, Sc, V, Cr and In are reported for Apollo 11 samples. These and companion data are interpreted in the contexts of environmental, paragenetic and petrogenetic problems. We cannot at present set reliable limits on the meteoritic contamination of lunar "soil". A cryptic component, constituting perhaps two-thirds of present "soil" and chemically distinct from known lunar specimens, is characterized by use of a mixing diagram. It contains about 43% SiO_2, 5·8% TiO_2, 16% Al_2O_3, 14% FeO, 8·1% MgO and 12% CaO and has an approximate norm of 45% plagioclase (An_{96}), 40% sub-calcic augite, 4% olivine (Fo_{70}) and 11% ilmenite, all by weight. It is probably very ancient.

Data on fractions of the "microgabbro" 10044 are used to comment on Eu abundance anomalies. Simple one-stage fractionation of plagioclase is not favored as an explanation.

A model proposed by GOLD, in which extant igneous rocks were formed (by impact?) from "soil" which is taken as primary material, is discussed and shown to be implausible.

Relatively low and nearly constant Na values are probably related to petrogenesis rather than to volatilization during paragenesis. Rock clans, noted previously, probably represent samples initially derived from two flow units underlying the Apollo 11 landing site; speculations on stratigraphic and textural relations are presented.

An argument is presented that the moon's surface regions, down to perhaps tens of kilometers depth, have been very dry for most of their history. A speculative early history of intense surficial heating by compression of nebular gases and impact events is outlined, and the chemical and physical structure which might result may account for many known features of the moon.

INTRODUCTION

OUR AIMS in this paper are threefold: to present data which, owing to space limitations, were excluded from our previous publication (GOLEŠ *et al.*, 1970a; hereafter Paper I), to comment more fully on geochemical interpretations of those data and of abundances presented in a companion paper (GOLEŠ *et al.*, 1970b; hereafter Paper II), and to describe some speculations on the moon's history and structure. All of us have contributed to the arduous task of determining the abundances here reported (the senior author least of all). The first four authors are principally responsible for the interpretations put forth, and the senior author is principally responsible for the speculations.

Table 1 contains elemental abundances determined by techniques similar to those of GORDON *et al.* (1968) in the whole-rock samples we analysed. Data determined by the same techniques but in different experiments are given in Paper II, and will be drawn upon to give a broader base to our interpretations. Comparisons with results of other workers as outlined in Paper II demonstrate very satisfactory agreement in all but a few cases.

Table 1. Elemental abundances in whole-rock samples (ppm except where noted)

	"Basalts"				"Microgabbros"		"Breccias"					Range in error estimates
			10071-27		10044-28					10060-14		
	10022-44	10017-27	*A*	*B*	*A*	10045-27	10018-25	10019-25	10021-21	*A*	*B*	
SiO_2 (%)	43·2	43·0	42·2	—	43·3	—	43·1	43·9	44·6	44·4	45·2	1·2–1·3
Ti (%)	7·6	6·7	7·4	—	5·8	—	4·6	5·2	4·8	5·4	5·4	0·04–0·4
Al (%)	4·09	3·92	4·14	—	5·0	—	—	6·29	6·67	6·20	6·11	0·06–0.2
Na (%)	0·28	0·34	0·307	—	0·321	—	0·356	0·38	—	0·33	0·367	0·010–0·06
Ba	270	290	—	250	130	120	—	250	340	—	230	30–120
La	23·8	25·7	23·4	21·9	11·2	6·7	16·9	14·4	17·9	17·8	18·6	0·2–0·5
Ce	80·4	79·1	83	73	42	27	55	48	52	64	55	1·5–8
Sm	19·1	20·0	18	18	16·3	9·3	13·2	12·0	15·4	14·1	15·4	0·2–2
Eu	2·26	2·27	2·03	2·00	2·64	1·51	1·87	1·93	2·00	1·97	1·88	0·04–0·12
Tb	4·52	3·88	4·3	3·8	3·89	1·9	2·62	2·67	3·12	—	2·89	0·11–0·2
Dy	30	27·2	—	31	24·5	15·4	19·0	18·1	22·1	21·7	—	1·0–2
Ho	9·8	6·4	—	8·0	5·6	—	4·8	5·6	6·0	5·4	6·0	0·4–0·8
Yb	17·9	18·7	16·6	14·7	14·6	8·0	11·4	11·1	12·4	12·7	13·1	0·5–1·6
Lu	2·5	2·7	2·3	2·60	1·91	1·4	1·4	1·44	2·0	1·74	1·85	0·10–0·2
U	0·80	0·64	0·62	0·66	0·20	0·17	0·53	0·36	0·53	0·53	0·60	0·04–0·10
Hf	18·1	17·9	16·4	15·1	13·8	6·5	11·0	10·6	12·3	12·9	12·8	0·2–0·4
Zr	590	—	—	—	460	—	330	500	300	170	$<$100	100–130
Ta	1·0	1·2	2·2	2·0	2·3	2·0	1·5	2·0	1·4	1·7	1·8	0·2–0·4
Mn	1740	1700	1730	—	2060	—	1510	1580	1660	1680	1650	30–100
Co	29·0	26·7	27·9	25·0	11·8	18·7	30·9	31·6	30·9	29·1	27·7	0·4–1·9
Fe (%)	14·1	13·5	13·6	12·7	13·2	13·4	11·3	11·3	12·2	12·1	12·07	0·17–0·2
Sc	78·3	75·8	77	67·9	87·7	77·7	58·8	61·2	61·8	58·0	63	0·6–3
V	70	61	89	—	32	—	—	50	59	55	53	8–14
Cr	2320	2170	2270	2010	1270	2380	1880	1890	1950	2020	1950	30–150
In*	—	—	—	—	—	—	—	—	50	1·2	1·0	0·4–20

* Less than 0·1 ppm In except where a value is given.

In Tables 1, 2 and 3, we have acceded to editorial policy and given ranges of error estimates rather than individual values. The complete set of error estimates (single standard deviations, principally from Poisson counting statistics) is available on request. Individual values should be used for any but the crudest computations of averages or comparisions of data. Elements are listed in these tables in a modification of the order suggested by TAYLOR (1965).

"Basalts" 10017 and 10071 in Table 1 belong to the A-2 clan defined in Paper II; i.e., they have at one time or another been classified as "microgabbros" (WARNER, personal communication), presumably on grounds of being rather coarser-grained than is, for instance, 10022. As noted in Paper II, there are no compositional distinctions between rocks of clan A-1 and those of A-2. "Microgabbro" 10045 probably is a member of clan B-2, exemplified in Paper II by rock 10050. That two of the seven "microgabbros" we have studied seem to belong to clan B-2 suggests that these rocks, if from a geologic unit distinct from that immediately underlying the regolith at Eagle's landing site, were not derived from a great distance.

Compositional differences between "basalts" and "microgabbros" and between igneous rocks and clastic materials were mentioned in Papers I and II and will be discussed below.

Abundances determined in the various samples of our "soil" aliquant are presented in Table 2. Samples 10084A and 10084B were taken with a nickel spatula from the original material, 10084-51. Most of the original specimen was then sieved and the intermediate fraction was divided into two samples, 10084D1 and 10084D2. The only possibly-significant compositional differences are found for Ti, Al, Co and Sc. In the fine fraction, 10084E, Ti, Al and Co are enriched and Sc is depleted. Cerium may be enriched in 10084E as well, but this observation is not supported by data on other rare earths. A crude mass balance may be made for Al, which seems to be depleted

Table 2. Elemental abundances in "soil" samples (ppm except where noted)

	Splits of 10084-51		Coarse (+100 mesh) fraction	Splits of 100–325 mesh fraction		Fine (−325 mesh) fraction	Range in error estimates
	10084A	10084B	10084C	10084D1	10084D2	10084E	
SiO_2 (%)	45·2	43·5	44·7	42·7	43·1	44·4	1·2–1·3
Ti (%)	5·1	5·3	4·9	4·2	4·2	6·1	0·2–0·5
Al (%)	7·42	7·24	6·88	7·17	7·14	8·07	0·10–0·16
Na (%)	0·321	0·326	0·28	0·315	0·29	0·342	0·008–0·05
Ba	140	170	500	90	230	170	30–140
La	14·5	14·2	15·6	14·4	14·7	15·6	0·3–0·4
Ce	50·4	52·8	51·2	46·7	47·8	66	1·2–2
Sm	12·6	12·5	13·0	12·7	11·6	13·9	0·2–0·4
Eu	1·76	1·80	1·87	1·79	1·86	1·84	0·04–0·08
Tb	2·47	2·7	2·7	3·1	2·71	3·11	0·13–0·2
Dy	18·8	—	—	—	17·6	—	1·2–1·3
Ho	5·8	6·6	—	—	—	—	0·6–0·8
Yb	10·5	11·3	11·3	10·6	10·5	11·1	0·5–1·2
Lu	1·57	1·58	1·53	1·45	1·6	1·67	0·05–0·4
U	0·45	0·41	0·52	0·36	0·40	0·50	0·06–0·09
Hf	10·2	10·8	10·5	9·78	9·4	10·0	0·19–0·3
Zr	290	—	280	320	—	—	80–100
Ta	1·2	1·4	1·24	1·2	1·3	1·3	0·19–0·2
Mn	1590	1610	1540	—	1570	1430	40–80
Co	30·3	30·6	30·4	29·6	30·0	34·4	0·4–1·0
Fe (%)	11·2	11·4	11·5	10·8	11·1	11·4	0·2–0·3
Sc	58·9	60·0	62·5	56·6	57·8	53·8	0·6–1·3
V	107	81	130	47	68	60	12–20
Cr	1900	1940	1870	1800	1870	1920	40–100
In	0·8	0·9	<0·1	1·7	1·1	1·7	0·3–0·7

in the coarse fraction, but in the other cases we may be dealing with statistical excursions rather than real effects.

Our specimen of "microgabbro" 10044 consisted of coarsely-ground powder of sufficient mass so that it was feasible to split and sieve it. Splitting was done with a small stainless-steel sample splitter. Mineral grains were hand-cobbed from the

Table 3. Elemental abundances in fractions of "microgabbro" 10044 (ppm except where noted)

	Original material	Splits of coarse (+100 mesh) fraction		Splits of fine (−100 mesh) fraction		Plagio-clase–rich separate	Clino-pyroxene-rich separate	Opaque-minerals separate	Range in error estimates
	10044A	10044D	10044H	10044B	10044C	10044E (44·38 mg)	10044F (100·4 mg)	10044G (71·7 mg)	
SiO_2 (%)	43·3	45·3	46·0	46·3	44·0	46·5	51·0	7·5	0·2–1·5
Ti (%)	5·8	5·8	5·6	5·3	6·0	⩽0·9	1·1	29·7	0·3–1·2
Al (%)	5·0	4·35	4·36	6·1	5·96	18·1	1·78	0·4	0·09–0·6
Na (%)	0·321	0·35	0·27	0·388	0·330	1·080	0·61	0·0393	0·0004–0·08
Ba	130	—	130	190	150	160	160	—	30–100
La	11·2	9·2	9·5	12·9	12·6	4·4	2·52	4·85	0·07–0·3
Ce	42	32	37	50	48	14	24	21	2–7
Sm	16·3	15·0	16·7	19·1	18·8	5·75	7·60	7·3	0·09–0·7
Eu	2·64	2·16	2·16	3·33	3·26	6·15	0·57	0·32	0·08–0·18
Tb	3·89	4·04	3·88	4·66	4·84	1·1	—	0·8	0·11–0·3
Dy	24·5	—	—	29·5	—	—	—	—	1·5–1·9
Ho	5·6	—	—	8·0	—	—	—	—	0·6–0·8
Yb	14·6	15·1	15·6	16·4	16·6	2·6	10·1	11·3	0·2–1·3
Lu	1·91	2·2	2·26	2·3	2·1	0·35	1·47	1·60	0·14–0·5
U	0·20	0·17	0·16	0·29	0·22	—	—	—	0·04–0·06
Hf	13·8	12·8	13·7	16·0	16·1	2·22	3·3	16·4	0·12–0·6
Zr	460	—	220	330	—	—	—	—	130–150
Ta	2·3	1·3	1·9	2·1	2·4	—	—	4·1	0·3–0·7
Mn	2060	2100	2200	1970	2030	140	2400	3000	30–200
Co	11·8	10·7	11·2	11·9	13·3	1·6	12·6	18·0	0·2–0·5
Fe (%)	13·2	13·7	13·7	12·9	13·2	0·82	12·4	31·4	0·03–0·5
Sc	87·7	100·6	101·4	81·8	82·5	2·87	152·7	95·0	0·05–1·5
V	32	47	51	44	50	—	100	130	8–30
Cr	1270	1390	1440	1140	1180	18·3	2050	2420	1·5–60

coarse fraction. Masses of mineral separates are small, and their proportions approximate the modal proportions, so splits of the coarse fraction remain representative of +100 mesh material in the original powder to a good approximation. Table 3 embodies our results on these fractions.

INTERPRETATIONS

There are sets of problems related to interpretation of analyses of lunar materials, which may serve as a framework for our discussion. (1) *Environmental* problems treat the interaction of materials exposed at the lunar surface with the particle and radiation fluxes of interplanetary space. These include studies of cosmic-ray and solar wind irradiation effects, erosion by micrometeorite impact and thermal shock, "gardening" of the regolith by impact of large meteoroids, and the input of new material to the regolith from deeper geologic units by still-larger meteoroids. This set of problems often takes the form of using the lunar surface to study some other object or some exotic phenomena. (2) A set of problems which we may term *paragenetic*, are related to what happens to lunar magmas during and after their eruption onto the lunar surface, while they are cooling and crystallizing to become igneous rocks. The way in which this set is defined is based on the assumption that the "basalts" and "microgabbros" from the Apollo 11 collection are truly igneous rocks, on which there seems to be almost universal consensus. The definition would have to be extended in obvious ways if future missions were to discover rocks we would consider of intrusive igneous or metamorphic origins. (3) *Petrogenetic* problems are of special interest to us. How did magmas of the peculiar composition evidenced by our results and those of many other workers arise? What were their parental materials like? Was the heat source internal or, in one sense or another, external?

The three sets mentioned are to some degree independent, and as was demonstrated at the Apollo 11 Lunar Science Conference, it is possible to do excellent research in one area without any knowledge of another. A fourth set, *historical* problems dealing with the origin and evolution of the moon, cannot be treated seriously without at least partial solutions to the other three. While we shall introduce a few speculative ideas on the moon's history in a later section, we do not believe that the time is ripe for serious discussions of this kind.

Among environmental problems is that of how much contribution to the lunar "soil" there has been from meteoroid infall. In Paper I we placed crude limits on the extent of meteoritic contamination of "soil" at the Apollo 11 landing site, based on an argument using Co contents. Assuming that the "soil" consists only of two components, ground-up igneous rocks like those sampled and chondritic meteorites, one may compute contaminations of 1–3 per cent from various models. For example, if we follow KEAYS *et al.* (1970) and assume that the chondritic component is on the average like carbonaceous chondrites, a Co abundance of about 500 ppm in this component would be appropriate (SCHMITT *et al.* 1970a). The Co increment which must be added to igneous rocks to make "soil" depends strongly on how one chooses to weigh "basalts" and "microgabbros". A simple average of Co contents for all igneous rocks in Table 1 here and in Table 3 of Paper II (by rock, not by analysed sample) yields an estimate of chondritic contamination of about 2 per cent.

The two-component assumption is, however, demonstrably false, a fact of which we were not fully aware when writing Paper I. At least three additional components exist. (1) Fragments of iron meteorites have been observed in the "soil" (ADLER *et al.*, 1970; AGRELL *et al.*, 1970; ALBEE *et al.*, 1970; ANDERSON *et al.*, 1970; CARTER and MACGREGOR, 1970; CHAO *et al.*, 1970; DUKE *et al.*, 1970; FREDRIKSSON *et al.*, 1970; FRONDEL *et al.*, 1970; HERZENBERG and RILEY, 1970; KEIL *et al.*, 1970; MCKAY *et al.*, 1970; MASON *et al.*, 1970; MUIR *et al.*, 1970; NAGATA *et al.*, 1970; QUAIDE *et al.*, 1970; RAMDOHR and EL GORESEY, 1970; SCLAR, 1970; STRANGWAY *et al.*, 1970; WÄNKE *et al.*, 1970; WARE and LOVERING, 1970; WOOD *et al.*, 1970). This component, while minute, has a high Co abundance and so should be taken into account. Unfortunately, no reliable estimates of the fraction of "soil" which consists of iron meteoritic debris are known to us. (2) The "soil" at the Apollo 11 landing site contains material resembling anorthositic gabbros or, crudely, Ca-rich achondrites (AGRELL *et al.*, 1970; ALBEE *et al.*, 1970; ANDERSON *et al.*, 1970; ARRHENIUS *et al.*, 1970; BROWN *et al.*, 1970; CHAO *et al.*, 1970; DUKE *et al.*, 1970; FREDRIKSSON *et al.*, 1970; FRONDEL *et al.*, 1970; KEIL *et al.*, 1970; KING *et al.*, 1970; MASON *et al.*, 1970; ONUMA *et al.*, 1970; SHORT, 1970; WOOD *et al.*, 1970). The proportion of this component in the "soil" which is recognizable as glass beads or grains of unusual composition or as lithic fragments is small, but there may well be additional material of this composition present cryptically. A rough guide to the Co contents of this anorthositic component may perhaps be those of Ca-rich achondrites. Cobalt contents of eucrites average about 8 ppm, with an apparent range from 2 to 22 ppm; howardites, which stand in approximately the same textural relationship to eucrites as Apollo 11 breccias do to "microgabbros", average about 19 ppm Co with a range from 3 to 54 ppm (SCHMITT *et al.*, 1970(a). Thus, unless the total proportion of this component is appreciably larger than the 3·6 per cent found among lithic fragments by WOOD *et al.* (1970), it is not likely that it has an important effect on the Co mass balance. KING *et al.* (1970) confirmed the estimate of WOOD *et al.*, and MCKAY *et al.* (1970) found only 0·8 per cent anorthosites in the 4–10-mm chips which they examined.

(3) As we argue below and as other authors have implied (ALBEE *et al.*, 1970; ANNELL and HELZ, 1970; DUKE *et al.*, 1970; Paper I; KEAYS *et al.*, 1970; MOORE *et al.*, 1970; REED *et al.*, 1970; ROSE *et al.*, 1970; SCHMITT *et al.*, 1970b; TUREKIAN and KHARKAR, 1970; VON ENGELHARDT *et al.*, 1970), there is a large fraction of the "soil" (and of most breccias) which is chemically distinct from known rock types and whose Co abundance consequently cannot be predicted. At present, the uncertainties introduced by the existence of this component force us to concede that we cannot estimate meteoritic contamination of Apollo 11 "soil" in a reliable way. These uncertainties, which arise from the limited nature of our data, probably do not seriously affect the conclusions of KEAYS *et al.* (1970) which are based on a much larger set of critical abundance estimates.

Consider now the chemical evolution of Apollo 11 "soil". A major contribution to the "soil" must be from the igneous rocks found embedded in it and which are being eroded at respectable rates (LSPET, 1969; BROWN *et al.*, 1970; CROZAZ *et al.*, 1970; FLEISCHER *et al.*, 1970; FREDRIKSSON *et al.*, 1970; FRONDEL *et al.*, 1970; FUNKHOUSER *et al.*, 1970; KING *et al.*, 1970; MCKAY *et al.*, 1970; MARTI *et al.*,

1970; QUAIDE *et al.*, 1970; SHOEMAKER *et al.*, 1970; SKINNER, 1970; VON ENGELHARDT *et al.*, 1970). Compositionally, breccias generally lie between the "soil" and an appropriate average of igneous rocks and so need not be considered as an independent component. Since the "soil" differs in composition from the average of the igneous rocks, there must be a second major component as mentioned above. A mixing diagram, such as Fig. 1, may be used to estimate its proportion in the "soil" and its

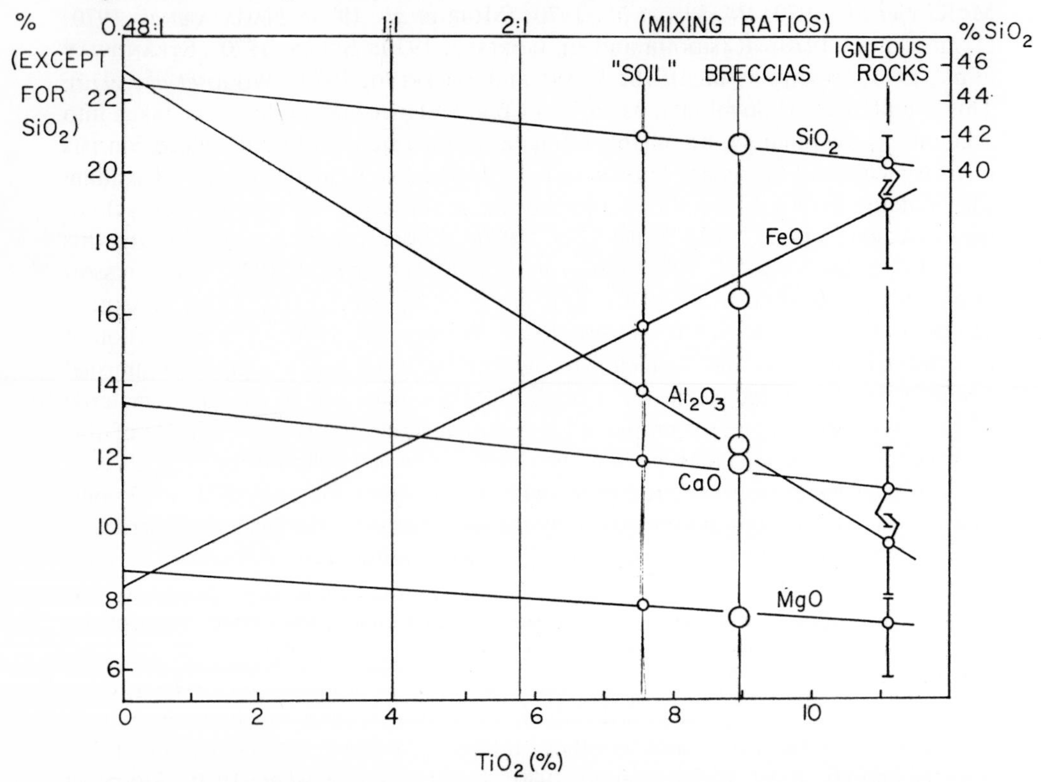

Fig. 1. A mixing diagram for Apollo 11 materials in which TiO_2 is taken as the independent variable. Averages of bulk compositions for "soil", breccias and igneous rocks are given, with ranges of determinations in the igneous rocks indicated by bars. The range in TiO_2 for igneous rocks is about 10 to 12·5 per cent. Although positions of the mixing lines as defined by the extreme points are uncertain, the steep slopes for FeO and Al_2O_3 and shallower ones for SiO_2, CaO and MgO would survive any reasonable modifications of the diagram. These features limit plausible mixing ratios (given at the top of the diagram) to the approximate range 1:1 to 2:1, cryptic component to average igneous rocks.

composition. While it is feasible to deal with models in which three or more components are taken into account, for simplicity we assume that only two major components are present. Our knowledge of the average composition of igneous rocks is not precise enough to warrant elaborate treatment of this problem. One should note that this assumption is equivalent to postulating that the second component is itself a mixture of materials from different sources.

Among major elements, differences in Ti contents between igneous rocks and "soil" seem to be most significant. Consequently, we have used TiO_2 as the independent variable for our mixing diagram, which was constructed after review of all available data on bulk composition of Apollo 11 materials. At the time of making this review, we did not have all final abundance estimates of Paper II and so the mixing lines are defined by averages of abundances reported by other authors in igneous rocks and in "soil". The diagram would not be changed noticeably if it were redrawn to include our data and those of Paper II. Even though the locations of the mixing lines are clearly uncertain (see ranges indicated for data on igneous rocks), the "average breccia" fits well at about 60% "soil" and 40% igneous rock.

If the second major component, whose properties we wish to estimate, had no Ti at all it would still be necessary to add about one part of it to two parts of ground-up

Table 4. Estimates of composition and norm of the cryptic major component of Apollo 11 "soil" (wt.%)

	Mixing Ratio	
	1:1	2:1
SiO_2	43·6	42·8
TiO_2	4·0	5·8
Al_2O_3	18·2	16·0
FeO	12·2	14·0
MgO	8·4	8·1
CaO	12·7	12·3
	99·1	99·0
Plagioclase (An_{96})	51·2	45·0
Ilmenite	7·6	11·0
Ferromagnesian silicate residue (recalculated to 100%)		
SiO_2	51·3	53·0
FeO	21·3	20·5
MgO	20·7	18·8
CaO	6·7	7·7

igneous rock like the average of those analysed, in order to match the composition of the "soil". Not only is the complete absence of Ti implausible, the proportions predicted on that extreme assumption for the other major oxides are unreasonable. Figure 1 could be modified, by weighting "microgabbros" much more heavily than "basalts", so that the predicted bulk compositions resembled the anorthosites of Wood *et al.* (1970) more closely, but we believe this to be an implausible condition. As the diagram stands, we find reasonable bulk compositions in the range of 1:1 or 2:1 mixing ratios (about 4–5·8% TiO_2 in the cryptic component). In Table 4, we give these compositions, the amounts of plagioclase (An_{96}) and ilmenite which would be present if all Al and Ti were in those minerals, and the residue which should be accounted for in ferromagnesian silicates. (We have chosen to use An_{96} in these estimates in order to secure an approximate mass balance for Na.) The ferromagnesian residue compositions may be matched by mixtures of olivine (chosen to be Fo_{70}) and sub-calcic augite of a kind common in "basalts". Ninety percent augite and ten percent olivine would match the residue for the 2:1 mixing ratio, and this set of parameters gives the best overall agreement for all bulk compositional data and with petrological intuition. We have ignored the residual, silica-rich phases which have been observed in Apollo 11 igneous rocks (ADLER *et al.*, 1970; AGRELL *et al.*, 1970; ANDERSON *et al.*, 1970; ARRHENIUS *et al.*, 1970; BAILEY *et al.*, 1970; BROWN *et al.*, 1970; DOUGLAS *et al.*, 1970; DUKE *et al.*, 1970; FRONDEL *et al.*, 1970; KEIL *et al.*, 1970; MASON *et al.*, 1970; ROEDDER and WEIBLEN, 1970; SKINNER, 1970; WARE and LOVERING, 1970; WEILL *et al.*, 1970), but in view of other, grave, sources of uncertainty in our approach this omission is not serious.

To summarize, we suggest that the "soil" is composed of one part igneous rocks like those observed at the sampling site, mixed with about two parts of ground-up rocks with average composition like that shown in the third column of Table 4. The cryptic component so defined would be expected to be a mixture itself of various rock types, but if we assume it to be a single igneous rock type it would have an approximate norm of 45% plagioclase, 40% augite, 4% olivine and 11% ilmenite, all by mass. Surely some material like the anorthositic lithic fragments described by WOOD *et al.* (1970) is present as part of this cryptic component, yet the two have quite different

Table 5. Approximate model abundances of selected trace elements in cryptic component, 2:1 mixing ratio (ppm)

La	13	Eu	1·6	Lu	1·2
Ce	51	Dy	15	Hf	8·6
Sm	10	Yb	8·4	Sc	49

norms. Either the anorthositic fraction of the cryptic component is minor or it is balanced by mafic materials. The cryptic component appears to be unlike the "original liquid" postulated by GAST and HUBBARD (1970) or any of the Ca-rich achondrites, but it resembles the recalculated pyroxene gabbro composition of OLSEN (1969) rather closely. Clasts of this approximate composition should be searched for in Apollo 11 breccias and lithic fragments. That this component is so prominent in a compositional sense but, apparently, is not represented at all among extant igneous rocks suggests that it is ancient.

Perhaps the slight enrichment of Al in the fine fraction of "soil", noted above, may be taken as independent evidence for the existence of the cryptic component. Also, some of the glass beads or individual grains or fragments of unusual composition (e.g., those observed by AGRELL *et al.*, 1970; ARRHENIUS *et al.*, 1970; DUKE *et al.*, 1970) may well be referable to the cryptic component or its mafic fraction rather than to the anorthositic gabbro component of WOOD *et al.* (1970).

Assuming that a mixing ratio of about 2:1 is reasonable, we may attempt to estimate abundances of several trace elements in the cryptic component. Model abundances of seven rare earths, Hf and Sc are given in Table 5, derived from the data of this paper and Paper II. Owing to pronounced differences in abundances of these elements in analysed "basalts" and "microgabbros" (which we weighted equally by rock), the estimates of Table 5 are very uncertain. Nevertheless, it is clear that some of the peculiar trace-element geochemical characteristics of Apollo 11 igneous rocks are shared by the cryptic component. These will be mentioned when we treat petrogenetic problems.

Data on abundances in fractions of the "microgabbro" 10044 (Table 3) lead to several interesting paragenetic questions. In Paper I we presented a brief outline of a model for correcting for contamination of one mineral separate by the others. We assumed that the plagioclase-rich separate is contaminated with 0·5 per cent of each of the other separates, the clinopyroxene-rich separate with 1 per cent of each of the others, and the opaque minerals separate with about 2% plagioclase and 5% clinopyroxenes separates. No way to improve on these assumptions has occurred to us in the interim, and we shall use the same model here. Table 6 embodies corrected

Table 6. Abundances in mineral separates, corrected for cross-contamination (ppm except where noted)

	Plagioclase-rich separate	Clinopyroxene-rich separate	Opaque minerals separate
SiO_2 (%)	46·7	51·5	4·3
Ti (%)	$\leqslant$0·8	0·8	31·0
Al (%)	18·3	1·62	~zero
Na (%)	1·09	0·05	0·016
Ba	~160	≈160	—
La	4·4	2·48	4·99
Ce	14	24	21
Sm	5·68	7·62	7·3
Eu	6·21	0·51	~0·17
Tb	~1·1	—	~0·8
Yb	~2·5	10·2	11·5
Lu	0·34	1·48	1·63
Hf	2·14	~3·2	17·4
Mn	~114	2400	3100
Co	1·5	12·7	19·3
Fe (%)	0·80	12·3	33·2
Sc	~1·8	~160	~90
Cr	~zero	2070	2500

abundance estimates for the three separates, using the stated model.

We cannot emphasize too strongly the formal nature of these computations. The model requires only a knowledge of observed abundances and assumptions as stated; no knowledge of the actual mineralogy of the separates is required (or implied). We believe that the plagioclase-rich separate may be clean enough so that corrected abundances for it are close to the actual abundances in that phase (about An_{87}, according to Table 6). The other separates are contaminated with minor phases; see for example the corrected SiO_2 content of the opaque minerals separate, which would be difficult to reduce to zero by any plausible choices of assumptions and manipulations of errors. As has been shown by various authors (e.g., RAMDOHR and EL GORESEY, 1970; WEILL *et al.*, 1970), several of these minor phases are extraordinarily rich in elements of interest to us. Consequently, their presence in at least two of our mineral separates in significant amounts must be taken into account.

In Paper I we suggested that there is about 50 per cent more plagioclase in the fine fraction of 10044 than in the coarse fraction. We now think that this estimate was biased by considering elements which are concentrated in residual phases, although it would still be reasonable from the Eu abundances of Tables 3 and 6. From Cr contents, plagioclase is over-represented in the fine fraction of 10044 by about 20 per cent. Chromium does not concentrate in residual phases, to our knowledge.

The corrected rare earth abundances of Table 6 were presented graphically in Paper I, and were interpreted without considering the effect of residual phases. While that interpretation might still be valid, an alternative and in some ways simpler hypothesis can be presented. If essentially all rare earths are present in two components, plagioclase and residual phases, we can understand the pronounced similarity between relative abundance patterns of the "clinopyroxene" and "opaque minerals" fractions. Both are simply reflecting the abundances of rare earths in the residual phases. (We do not exclude that some rare earths are sited in clinopyroxenes, or that the minor differences between relative abundances in the two separates are so generated.) Then the strong depletion of Eu relative to other rare earths in those two fractions may be related to the prior crystallization of plagioclase, rather than reflect the reduced state of Eu in the melt.

Similarly, the distribution of Hf exhibited in Table 6 almost certainly does not reflect any preference of that element for sites in opaque minerals, but only its high concentration in residual phases. The surprisingly high content of Sc in the opaque minerals separate may well be related to a similar degree of concentration in residual phases.

These considerations lead us into petrogenetic problems. The Eu abundance anomaly in Apollo 11 materials has attained much notoriety, and many authors have suggested that it reflects large-scale plagioclase crystallization and fractionation. Our alternative interpretation of the rare earths in the two separates discussed above as being representative of residual phases suggests several other possible ways of assembling magmas with striking Eu depletions. For instance, one might extract by some means residual melt from a large intrusive body during its final stages of crystallization. One might partially re-melt a large body at depth in which rare earths (but, to some degree, not Eu) had been concentrated in late phases in a way reminiscent of what we think we see in the residual phases of 10044. Europium-depleted magmas generated by such mechanisms could be further differentiated or mixed with Ti- and Fe-rich magmas in order to avoid the difficulty of postulating a magma which was formed in equilibrium with plagioclase but which does not have plagioclase on or near the liquidus (a criticism which might be leveled against the model of HASKIN *et al.*, 1970). These and similar schemes should be conceptually explored before lunar theorists fixate on simple plagioclase fractionation. Of course, based on present knowledge of rare earth distributions in diverse mineral phases, plagioclase may play an important role in some sense in the evolution of the lunar Eu anomalies. We merely wish to emphasize that this role need not have been played on one stage in a simple manner.

One should note that the rare earth abundances estimated for the cryptic component of the "soil" (Table 5) could represent an earlier stage in the evolution of highly-fractionated lunar magmas. Europium depletion is apparent, but does not seem as pronounced as in extant igneous rocks. Other rare earths, with the probably-spurious exception of Ce, are enriched in comparison to chondrites by factors of 40–50, resembling the least-rare-earth-rich of the known Apollo 11 rocks. The Hf content estimated for the cryptic component also lies near the lower end of the known range in Apollo 11 rocks, and the Sc content is lower than any known in lunar materials at present.

Most of the discussion above is stated in the context of a commonly-accepted model of the origin of lunar surface materials, in which erosional, depositional, and diagenetic processes make clastic materials from igneous rocks derived from the Moon's interior (plus a bit of extra lunar material). The similarity between this model and the geologic processes we have studied on Earth's surface is comforting, but

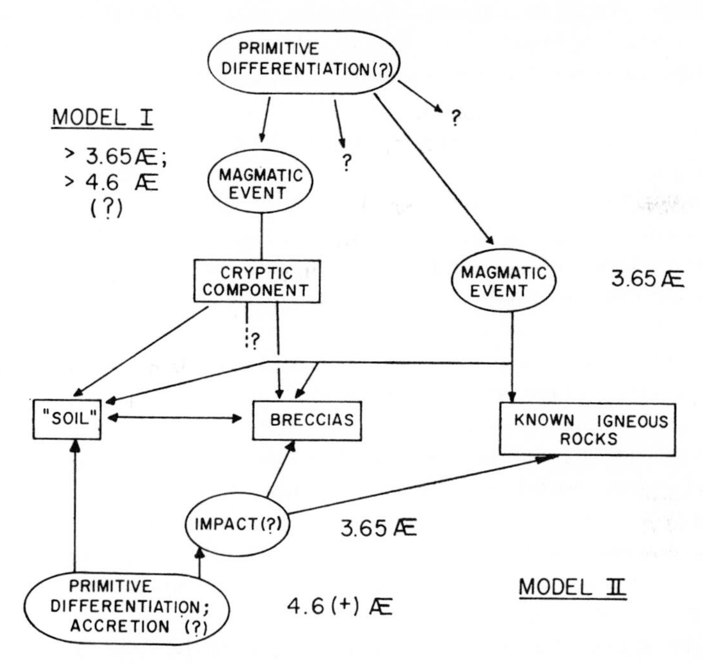

Fig. 2. A schematic representation of two generalized models for the origin of Apollo 11 materials. Present (i.e., within the last aeon—10^9 yr—or so) relationships between "soil", breccias and igneous rocks are indicated near the center of the figure. Owing to the complexity of these relationships, some such as the grinding of igneous rocks to contribute to "soil" are not explicitly shown as on-going processes. Events in Model I become younger downwards, in Model II younger upwards. Results of ALBEE *et al.* (1970) have been drawn upon to set limits on time scales. Model I relies heavily on magmatic events, Model II on a single large-scale magma-generating event which may have been impact of a planetesimal on the lunar surface.

should lead us to be suspicious of too easy a transfer of basic concepts to what is, after all, a very different planet. An alternative model has been proposed by GOLD (personal communication), in which the "soil" is the primary material, with the igneous rocks derived from it by surficial magma-forming events in which appreciable differentiation takes place. In this model ("Model II" in Fig. 2), formation of breccias might well be a by-product of the impacts (?) in which the parental magmas of the igneous rocks were made. As Fig. 2 indicates, Gold's model can be made to appear simpler; only one fundamental event need occur after accretion of the Moon to account for the Apollo 11 materials.

This magma-forming event, however, must do some very complicated chemistry. As our data and the preceding discussion indicate, in order to make the observed igneous rocks from material of "soil" composition there must be a marked loss of Al_2O_3, perhaps of Co, slight losses of SiO_2, CaO and MgO, and perhaps a slight loss of Na. Marked gains of TiO_2, FeO and Sc, and slight gains of Eu and Mn, are required. This would be a very peculiar set of coherence relationships, and we have been unable to think of any process or simple combination of processes which could plausibly account for these chemical differentiations in one event. Consequently we shall continue to favor Model I, in all its diverse permutations, because it possesses inherent complexities which begin to approach those of the materials it is intended to explain.

In Paper I we pointed out the narrowness of the range of Na values we have observed. Data of Paper II, for which Na was determined by Schmitt and Wakita at Corvallis, confirm this characteristic. The senior author had thought that this geochemical feature might be related to paragenesis rather than to petrogenesis, in that volatilization of Na from magmas erupted onto the lunar surface might in part account for the nearly-constant abundances of this element. If such volatilization occurs, it may be limited to an early stage in the paragenesis. WEILL *et al.* (1970) have shown convincingly that plagioclase in rock 10022 begins to form only after 60–70 per cent solidification of the original melt, at which point the Na/Ca ratio of the liquid in equilibrium with plagioclase is roughly consistent with the composition observed for the feldspars (WEILL and MCCALLUM, personal communications). Some loss of Na from the system as a whole at this stage is not excluded by the data, but would seem to be minor at best. If Na is lost early in the cooling of lunar lava flows but not late, perhaps the loss mechanisms are strongly temperature-dependent. Perhaps the formation of plagioclase stabilizes the Na content, or the kinetics of diffusion to the surface of the flow are modified by extensive crystallization.

In any case, if one persists in adducing some form of the volatilization hypothesis it is necessary to explain where the Na (and presumably other elements) goes. We do not have an explanation which we consider wholly satisfactory, but we note that if lunar lava flows are very extensive and very hot, they may sufficiently warm a temporary atmosphere of their own volatiles above them so that on the average the volatiles escape from the Moon's gravitational field before diffusing to colder surfaces (see also O'HARA *et al.*, 1970).

A serious difficulty with this hypothesis lies in the same near-constancy of Na contents which suggested it in the first place. If volatilization of Na had been important, one might expect a strong correlation of Na contents with cooling rates and therefore with textures. Our data do not indicate such a correlation. Consequently, we believe it is more likely that Na contents are a petrogenetic rather than a paragenetic feature. Noting the low Na contents of Apollo 12 specimens (TAYLOR, personal communication) and of other portions of the lunar surface (FRANZGROTE *et al.*, 1970; TURKEVICH, personal communication), we suggest that the moons' surface as a whole may be an alkali-poor province. Why this should be so remains a fundamental problem, and will be speculated upon in a later section.

The clans mentioned above and defined in Paper II are at least in part of petrogenetic origin. For instance, ALBEE *et al.* (1970) find that 10044 (a member of clan B-1) is clearly distinct on the basis of Rb–Sr systematics from 10017, 10057, 10069 and 10071 (all members of phratry A), "indicating that the samples represent at least two different rock bodies." COMPSTON *et al.* (1970) distinguish between our phratries A and B (but not between clans B-1 and B-2) and state that "it seems likely that two separate rock units were sampled, including textural variants of both." Our data suggest that the B rocks are chemically more diverse than the A rocks, but the distinctions between the B-1 and B-2 clans could be reflections of differentiation of minor phases as indeed might be expected from textural evidence. Thus, while there may be three rock units represented in the Apollo 11 igneous specimens, we shall adopt the conservative view that there are two.

The most plausible explanation of these observations is that the specimens were derived from two flow units underlying the regolith at the Apollo 11 landing site. The sub-equal proportions of representatives of the two units suggest that little if any clastic material is intercalated between members of the sequence. This is not an overly restrictive condition on the time spans between eruptive events, since only 4 m of regolith have been generated on the youngest flow during the last 3·65 AE (SHOEMAKER *et al.*, 1970; ALBEE *et al.*, 1970). A possible explanation for the coarser textures of members of phratry B, in comparison to those of most of phratry A, is that the former arise from a lava which flowed over surfaces of relatively high relief and formed magma pools in many places. Such pools would necessarily have been shallow, otherwise appreciable differentiation should have occurred owing to the low viscosity of these magmas (MURASE and MCBIRNEY, 1970; WEILL *et al.*, 1970), but could easily account for the textural variations observed. The younger flow, presumably the source of the A rocks, could then have been very thin and relatively uniform if it were erupted before the underlying unit had been pitted to any significant degree by impacts.

The lowest unit in the column of which we have any knowledge, however vague, is perhaps that which gave rise to much of the mixture which we have termed the cryptic component of the "soil".

FURTHER SPECULATIONS

It is obvious that the beginnings of this section lie somewhere above. Here two topics will be discussed: (1) evidence suggesting that the Moon has been dry, at least within a few tens or hundreds of kilometers of its surface, for most of its history; (2) speculative comments on the history and structure of the Moon.

One of the most striking geochemical features of the Apollo 11 specimens is the contrast of their rare earth abundance patterns with those of terrestrial rocks. This feature was pointed out in Paper I and by numerous other authors; we here merely emphasize the close similarity between rare earth distributions in Apollo 11 materials and in abyssal basalts (see especially HASKIN *et al.*, 1970). We may thus view this feature in light of the contrast between abyssal basalts and continental rocks on Earth. There is growing consensus that most continental rocks are ultimately derived from partial melts formed near a subduction zone along which water is being injected into the upper mantle via engulfed crustal materials. In contrast, although water may play a role in inducing partial melting under mid-oceanic ridges and rises, abyssal basalts are derived from what seem to be much drier regions of Earth's upper mantle (Kay *et al.*, 1970). Perhaps the potentiality of making a systematic geochemical distinction between light and heavy rare earths, such as seen in continental rocks, exists only when appreciable water is present in the source regions of magmas. (Other complexing agents may play the role here suggested for water—see GOLES, 1968, for example—but it can be shown that in most cases these would be closely associated with water anyway.) One may thus speculate that the rare earth patterns of Apollo 11 materials record the formation of their parental magmas under conditions at least as dry as those in Earth's upper mantle under oceanic ridges and rises. Recall that we have tentatively identified two distinct groups of igneous rocks among Apollo 11 specimens, and have discussed the cryptic component of the "soil". The latter is almost certainly very ancient. All of these materials were ultimately derived from magmas formed under very dry conditions, according to the speculation developed above and the data in Tables 1, 2 and 5. It would seem implausible that all could have been derived from some limited subsurface region of the Moon which happened to be very dry. Rather, the line of argument here developed suggests that on a very large scale and for most of its history, the Moon's subsurface regions have been dry.

Weak supporting evidence is provided by Zr/Hf ratios, as mentioned in Paper I. Review of all available data on Zr and Hf abundances in Apollo 11 materials yields a range in this ratio (by mass) of 7–51 with a median value of about 20 or 25. Perhaps it is significant that Hf, with electrons in 4f orbitals available to form covalent bonds with appropriate ligands, is enriched more than is Zr. Note the similar relationship between abundances of lanthanides and those of Y and, to some extent, La itself (SCHMITT *et al.*, 1970b). The low Zr/Hf ratios, relative to those commonly observed in terrestrial rocks, may again be related to differentiation under dry conditions. Some abyssal basalts seem to have Zr/Hf ratios in the range of about 20–40 (CORLISS, personal communication; ENGEL *et al.*, 1965). However, observations of Zr/Hf ratios in this lower range merely suggest that the materials possessing this characteristic at one time in their petrogenetic history were differentiated under dry conditions. The argument concerning rare earth patterns above suggests that *never* during what may have been a long and complex differentiation history were Apollo 11 materials fractionated under wet conditions.

A pragmatic effect of this speculative argument is that one should not count on finding a permafrost layer near the Moon's surface.

Perhaps we must account for an extensively outgassed Moon, at least in its surface regions. Certainly we must account for extensive and unusual (in terrestrial *and meteoritic* terms) differentiation which led to the materials observed in Mare Tranquillitatis. Perhaps we must account for alkali loss on a planetary scale during the early (before 4·5 AE ago) history of the Moon. Certainly we must account for a very short time scale for the development of at least some of these features, and for the occurrence of profound cosmochemical changes at the same time that the Earth was undergoing similar changes (ALBEE *et al.*, 1970). Meteorites were also affected by this major cosmochemical event (TATSUMOTO and ROSHOLT, 1970), which may well have been the dissipation of the solar nebula in the proto-Sun's Hayashi flash. Probably the deep interior of the Moon is chemically different and, especially, less dense than would be predicted from the constitution of the surface materials (RINGWOOD and ESSENE, 1970).

A speculation which seems to bring all of these features under one roof is that Moon's surface was strongly heated during the last stages of accretion (before dissipation of the solar nebula) by a combination of quasi-adiabatic compression of nebular gases and the conversion of gravitational potential energy into heat through large-scale impacts (cf. OSTIC, 1965; UREY, 1969). If this putative stage of rapid accretion and intense heating were short-lived, the deep interior would not be affected and could now have a primitive composition, perhaps like carbonaceous chondrites. Surficial regions, down to a depth of perhaps many tens or a few hundreds of kilometers, would be outgassed (even of alkalies) and strongly differentiated. When the solar nebula is dissipated, the temporary lunar atmosphere would rapidly disperse as well. ÖPIK (1963) has shown that chemical separations would not occur in this process, and alkalies and other constituents volatile at the proposed high temperatures would be lost with the permanent gases. The temporary atmosphere thus assumes significance not only as a means of slowing the radiative cooling of the lunar surface after a large impact event, but also as a carrier for missing constituents at a slightly later stage.

Earth's surface, or those of its predecessor bodies, would presumably have been subjected to the same process, but our planet is so efficient a heat engine that the record would have been almost completely erased by later interactions between crust and mantle. Only very sensitive isotopic investigations suggest that the Earth as a whole has lost constituents volatile at high temperatures (GAST, 1968).

The Moon's present structure would then be a cold, primitive interior overlain by a strongly-differentiated, very dry, alkali-poor exterior. Acoustic coupling between the two might well be weak, giving rise to a wave-guide effect in the surficial materials. (The observations of LATHAM *et al.*, 1970, seem to reflect seismic peculiarities related to smaller-scale structures, but the possibilities of observing a structure such as that speculated on here should be considered.)

A pragmatic effect of the structure suggested by these speculations is that, if one wishes to find indigenous water on the Moon, one should look for surface expressions of the lunar equivalents of kimberlite pipes.

Acknowledgments—We are grateful to our colleagues Y. BOTTINGA, I. S. MCCALLUM, G. A. MCKAY, D. F. WEILL and A. R. MCBIRNEY for stimulating criticism and valuable advice. A penetrating comment by A. L. ALBEE spurred us to reconsider some previous interpretations. Colleagues and reactor crews at Oregon State University and Reed College gave freely of their time and efforts to aid the experimental work. One of us (K.R) wishes to thank M. Franeck for his valuable assistance in the preparation of computer programs. L. J. PURVIS of the Department of Geology, University of Oregon, provided secretarial services under trying conditions. Our families deserve our thanks for their patience. This work was supported in part by NASA contract NAS 9-7961, NSF grant GA-1383, and NASA grant NGL 38-003-010.

REFERENCES

ADLER I., WALTER L. S., LOWMAN P. D., GLASS B. P., FRENCH B. M., PHILPOTTS J. A., HEINRICH K. J. F. and GOLDSTEIN J. I. (1970) Electron microprobe analysis of lunar samples. *Science* **167**, 590–592.

AGRELL S. O., SCOON J. H., MUIR I. D., LONG J. V. P., MCCONNEL J. D. C. and PECKET A. (1970) Mineralogy and petrology of some lunar samples. *Science* **167,** 583–586.

ALBEE A. L., BURNETT D. S., CHODOS A. A., EUGSTER O. J., HUNEKE J. C., PAPANASTASSIOU D. A., PODOSEK F. A., RUSS G. PRICE II, SANZ H. G., TERA F. and WASSERBURG G. J. (1970) Ages, irradiation history, and chemical composition of lunar rocks from the Sea of Tranquillity. *Science* **167,** 463–466.

ANDERSON A. T., JR., CREWE A. V., GOLDSMITH J. R., MOORE P. B., NEWTON J. C., OLSEN E. J., SMITH J. V. and WYLLIE P. J. (1970) Petrologic history of Moon suggested by petrography, mineralogy, and crystallography. *Science* **167,** 587–590.

ANNELL C. and HELZ A. (1970) Emission spectrographic determination of trace elements in lunar samples. *Science* **167,** 521–523.

ARRHENIUS G., ASUNMAA S., DREVER J. I., EVERSON J., FITZGERALD R. W., FRAZER J. Z., FUJITA H., HANOR J. S., LAL D., LIANG S. S., MACDOUGALL D., REID A. M., SINKANKAS J. and WILKENING L. (1970) Phase chemistry, structure, and radiation effects in lunar samples. *Science* **167,** 659–661.

BAILEY J. C., CHAMPNESS P. E., DUNHAM A. C., ESSON J., FYFE W. S., MACKENZIE W. S., STUMPFL E. F. and ZUSSMAN J. (1970) Mineralogical and petrological investigations of lunar samples. *Science* **167,** 592–594.

BROWN G. M., EMELEUS C. H., HOLLAND J. G. and PHILLIPS R. (1970) Petrographic, mineralogic, and X-ray fluorescence analysis of lunar igneous-type rocks and spherules. *Science* **167,** 599–601.

CARTER J. L. and MACGREGOR I. D. (1970) Mineralogy, petrology, and surface features of lunar samples 10062,35, 10067,9, 10069,30 and 10085,16. *Science* **167,** 661–663.

CHAO E. C. T., JAMES O. B., MINKIN J. A., BOREMAN J. A., JACKSON E. D. and RALEIGH C. B. (1970) Petrology of unshocked crystalline rocks and shock effects in lunar rocks and minerals. *Science* **167,** 644–647.

COMPSTON W., ARRIENS P.A., VERNON M. J. and CHAPPELL B. W. (1970) Rb–Sr chronology and chemistry of lunar material. *Science* **167,** 474–476.

CROZAZ G., HAACK U., HAIR M., HOYT H., KARDOS J., MAURETTE M., MIYAJIMA M., SEITZ M., SUN S., WALKER R., WITTELS M. and WOOLUM D. (1970) Solid state studies of the radiation history of the lunar samples. *Science* **167,** 563–566.

DOUGLAS J. A. V., DENCE M. R., PLANT A. G. and TRAILL R. J. (1970) Mineralogy and deformation in some lunar samples. *Science* **167,** 594–597.

DUKE M. B., WOO C. C., BIRD M. L., SELLERS G. A. and FINKELMAN R. B. (1970) Lunar soil: Size distribution and mineralogical constituents. *Science* **167,** 648–650.

ENGEL A. E. J., ENGEL C. G. and HAVENS R. G. (1965) Chemical characteristics of oceanic basalts and the upper mantle. *Geol. Soc. Amer. Bull.* **76,** 719–734.

FLEISCHER R. L., HAINES E. L., HANNEMAN R. E., HART H. R., JR., KASPER J. S., LIFSHIN E., WOODS R. T. and PRICE P. B. (1970) Particle track, X-ray, thermal, and mass spectrometric studies of lunar material. *Science* **167,** 568–571.

FRANZGROTE E. J., PATTERSON J. H., TURKEVICH A. L., ECONOMOU T. E. and SOWINSKI K. P. (1970) Chemical composition of the lunar surface in Sinus Medii. *Science* **167,** 376–379.

FREDRIKSSON K., NELEN J., MELSON W. G., HENDERSON E. P. and ANDERSEN C. A. (1970) Lunar glasses and microbreccias: Properties and origin. *Science* **167,** 664–666.

FRONDEL C., KLEIN C., JR., ITO J. and DRAKE J. C. (1970) Mineralogy and composition of lunar fines and selected rocks. *Science* **167,** 681–683.

FUNKHOUSER J. G., SCHAEFFER O. A., BOGARD D. D. and ZÄHRINGER J. (1970) Gas analysis of the lunar surface. *Science* **167,** 561–563.

GAST P. W. (1968) Upper mantle chemistry and evolution of the Earth's crust. In *The History of the Earth's Crust*, (editor R. A. Phinney), pp. 15–27. Princeton University Press.

GAST P. W. and HUBBARD N. J. (1970) The abundance of alkali metals, alkaline and rare earths, and $^{87}Sr/^{86}Sr$ ratios in lunar samples. *Science* **167,** 485–487.

GOLEŠ G. G. (1968) Rare-earth geochemistry of Pre-cambrian plutonic rocks. In *Proc. 23rd Int. Geol. Congr.*, (editors M. Malkovsky and J. Kantor), Vol. 6, pp. 237–249. Academia.

GOLEŠ G. G., OSAWA M., RANDLE K., BEYER R. L., JEROME D. Y., LINDSTROM D. J., MARTIN M. R.,

McKay S. M. and Steinborn T. L. (1970a) Instrumental neutron activation analyses of lunar specimens. *Science* **167**, 497–499.

Goleš G. G., Schmitt R. A., Ehmann W. D., Randle K., Osawa M., Wakita H. and Morgan J. W. (1970b) Elemental abundances by instrumental activation analyses in chips from 27 lunar rocks. *Geochim. Cosmochim. Acta*, Supplement I.

Gordon G. E., Randle K., Goleš G. G., Corliss J. B., Beeson M. H. and Oxley S. S. (1968) Instrumental activation analysis of standard rocks with high-resolution γ-ray detectors. *Geochim. Cosmochim. Acta* **32**, 369–376.

Haskin L. A., Helmke P. A. and Allen R. O. (1970) Rare-earth elements in returned lunar samples. *Science* **167**, 487–490.

Herzenberg C. L. and Riley D. L. (1970) Mössbauer spectrometry of lunar samples. *Science* **167**, 683–686.

Kay R., Hubbard N. J. and Gast P.W. (1970) Chemical characteristics and origin of oceanic ridge volcanic rocks. Preprint.

Keays R. R., Ganapathy R., Laul J. C., Anders E., Herzog G.F. and Jeffery P. M. (1970) Trace elements in lunar rocks: implications for meteorite infall, solar-wind flux, and formation conditions of Moon. *Science* **167**, 490–493.

Keil K., Prinz M. and Bunch T. E. (1970) Mineral chemistry of lunar samples. *Science* **167**, 597–599.

King E. A., Jr., Carman M. F. and Butler J. C. (1970) Mineralogy and petrology of coarse particulate material from lunar surface at Tranquillity Base. *Science* **167**, 650–652.

Latham G. V., Ewing M., Press F., Sutton G., Dorman J., Nakamura Y., Toksöz N., Wiggins R., Derr J. and Duennebier F. (1970) Passive seismic experiment. *Science* **167**, 455–457.

LSPET (Lunar Sample Preliminary Examination Team) (1969) Preliminary examination of lunar samples from Apollo 11. *Science* **165**, 1211–1227.

Marti K., Lugmair G. W. and Urey H. C. (1970) Solar wind gases, cosmic ray spallation products and the irradiation history. *Science* **167**, 548–550.

Mason B., Fredriksson K., Henderson E. P., Jarosewich E., Melson W. G., Towe K. M. and White J. S., Jr. (1970) Mineralogy and petrography of lunar samples. *Science* **167**, 656–659.

McKay D. S., Greenwood W. R. and Morrison D. A. (1970) Morphology and related chemistry of small lunar particles from Tranquillity Base. *Science* **167**, 654–656.

Moore C. B., Lewis C. F., Gibson E. K. and Nichiporuk W. (1970) Total carbon and nitrogen abundances in lunar samples. *Science* **167**, 495–497.

Morrison G. H., Gerard J. T., Kashuba A. T., Gangadharam E. V., Rothenberg A. M., Potter N. M. and Miller G. B. (1970) Multielement analysis of lunar soil and rocks. *Science* **167**, 505–507.

Muir A. H., Jr., Housley R. M., Grant R. W., Abdel-Gawad M. and Blander M. (1970) Mössbauer spectroscopy of moon samples. *Science* **167**, 688–690.

Murase T. and McBirney A. R. (1970) Viscosity of lunar lavas. *Science* **167**, 1491–1493.

Nagata T., Ishikawa Y., Kinoshita H., Kono M., Syono Y. and Fisher R. M. (1970) Magnetic properties of the lunar crystalline rocks and fines. *Science* **167**, 703–704.

O'Hara M. J., Biggar G. M. and Richardson S. W. (1970) Experimental petrology of lunar material: The nature of mascons, seas, and the lunar interior. *Science* **167**, 605–607.

Olsen E. (1969) Pyroxene gabbro (anorthosite association): similarity to Surveyor V lunar analysis. *Science* **166**, 401–402.

Onuma N., Clayton R. N. and Mayeda T. K. (1970) Oxygen isotope fractionation between minerals and an estimate of the temperature of formation. *Science* **167**, 536–538.

Öpik E. (1963) Dissipation of the Solar Nebula. In *Origin of the Solar System*, (editors R. Jastrow and A. G. W. Cameron), pp. 73–75. Academic Press.

Ostic R. G. (1965) Physical conditions in gaseous spheres. *Mon. Not. Roy. Astron. Soc.* **131**, 191–197.

Quaide W., Bunch T. and Wrigley R. (1970) Impact metamorphism of lunar surface materials. *Science* **167**, 671–672.

Ramdohr P. and El Goresey A. (1970) Opaque minerals of the lunar rocks and dust from Mare Tranquillitatis. *Science* **167**, 615–618.

REED G. W., JOVANOVIC S. and FUCHS L. H. (1970) Trace elements and accessory minerals in lunar samples. *Science* **167,** 501–503.

RINGWOOD A. E. and ESSENE E. (1970) Petrogenesis of lunar basalts and the internal constitution and origin of the Moon. *Science* **167,** 607–610.

ROEDDER E. and WEIBLEN P. W. (1970) Silicate liquid immiscibility in lunar magmas, evidenced by melt inclusions in lunar rocks. *Science* **167,** 641–644.

ROSE H. J., JR., CUTTITTA F., DWORNIK E. J., CARRON M. K., CHRISTIAN R. P., LINDSAY J. R., LIGON D. T. and LARSON R. R. (1970) Semimicro chemical and X-ray fluorescence analysis of lunar samples. *Science* **167,** 520–521.

SCHMITT R. A., GOLEŠ G. G. and SMITH R. H. (1970a) Elemental abundances in stone meteorites. In preparation.

SCHMITT R. A., WAKITA H. and REY P. (1970b) Abundances of 30 elements in lunar rocks, soil, and core samples. *Science* **167,** 512–515.

SCLAR C. B. (1970) Shock-wave damage in minerals of lunar rocks. *Science* **167,** 675–677.

SHOEMAKER E. M., HAIT M. H., SWANN G. A., SCHLEICHER D. L., DAHLEM D. H., SCHABER G. G. and SUTTON R. L. (1970) Lunar regolith at Tranquillity Base. *Science* **167,** 452–455.

SHORT N. M. (1970) Evidence and implications of shock metamorphism in lunar samples. *Science* **167,** 673–675.

SKINNER, B. J. (1970) High crystallization temperatures indicated for igneous rocks from Tranquillity Base. *Science* **167,** 652–654.

STRANGWAY D. W., LARSON E. E. and PEARCE G. W. (1970) Magnetic properties of lunar samples. *Science* **167,** 691–693.

TATSUMOTO M. and ROSHOLT J. N. (1970) The age of the Moon: An isotopic study of U–Th–Pb systematics of lunar samples. *Science* **167,** 461–463.

TAYLOR S. R. (1965) Geochemical analysis by spark source mass spectrography. *Geochim. Cosmochim. Acta* **29,** 1243–1261.

TAYLOR S. R., JOHNSON P. H., MARTIN R., BENNETT D., ALLEN J. and NANCE W. (1970) Preliminary chemical analyses of 20 Apollo 11 lunar samples. *Science,* in press.

TUREKIAN K. K. and KHARKAR D. P. (1970) Neutron activation analysis of milligram quantities of lunar rocks and soils. *Science* **167,** 507–509.

UREY H. C. (1969) Early temperature history of the Moon. *Science* **165,** 1275.

VON ENGELHARDT W., ARNDT J., MÜLLER W. F. and STÖFFLER D. (1970) Shock metamorphism in lunar samples. *Science* **167,** 669–670.

WÄNKE H., BEGEMANN F., VILCSEK E., RIEDER R., TESCHKE F., BORN W., QUIJANO-RICO M., VOSHAGE H. and WLOTZKA F. (1970) Major and trace elements and cosmic-ray produced radioisotopes in lunar samples. *Science* **167,** 523–525.

WARE N. G. and LOVERING J. F. (1970) Electron–microprobe analyses of phases in lunar samples. *Science* **167,** 517–520.

WEILL D. F., MCCALLUM I. S., BOTTINGA Y., DRAKE M. J. and MCKAY G. A. (1970) Petrology of a fine-grained igneous rock from the Sea of Tranquillity. *Science* **167,** 635–638.

WOOD J. A., MARVIN U. B., POWELL B. N. and DICKEY J. S., JR. (1970) Mineralogy and petrology of the Apollo 11 lunar sample. *Geochim. Cosmochim. Acta,* Supplement I.

Proceedings of the Apollo 11 Lunar Science Conference, Vol. 2, pp. 1195 to 1205.

Rb–Sr and U, Th–Pb ages of lunar materials

K. GOPALAN,* S. KAUSHAL,† C. LEE-HU and G. W. WETHERILL
Institute of Geophysics and Department of Planetary and Space Science,
University of California, Los Angeles, California 90024

(*Received* 3 *February* 1970; *accepted in revised form* 25 *February* 1970)

Abstract—K, Rb and Sr concentrations and Sr isotopic compositions have been measured on 6 samples of lunar crystalline rock, one of microbreccia, and a sample of fine lunar dust. Similar measurements have been made on density fractions separated from two samples of crystalline rock. These define isochrons having slopes corresponding to ages of 3575 m.y. ($2\sigma = 215$ m.y.) and 4070 m.y. ($2\sigma = 230$ m.y.) with initial Sr^{87}/Sr^{86} ratios of 0·6996 and 0·6989, respectively, ($2\sigma = 0{\cdot}0004$). These are interpreted as the time of crystallization of these rocks, however it is not clear whether or not these two events occurred at significantly different times.

Lead from the fine surface material is highly radiogenic. Both uranium-lead ages as well as the thorium–lead age are nearly concordant at 4700 m.y. This is interpreted as indicating that the source of this lead evolved in an approximately closed 4700 m.y. old system characterized by the U/Pb and Th/Pb ratios of the surface material, and that the age of this system is essentially equal to the age of the moon.

The lead isotope data are interpreted to exclude a fission origin of the moon at any time significantly later than the time of origin of the solar system, and to present severe difficulties for a lunar origin of tektites.

INTRODUCTION

THIS paper is a somewhat extended version of that published previously (GOPALAN *et al.*, 1970). Since the preparation of this earlier paper, we have extended our measurements to include three more lunar samples, and have studied in more detail the linearity of the mass spectrometric recording system.

The long range goal of these studies is to establish a lunar time scale which will permit the assignment of ages to the times of significant events in lunar history, such as the time at which the volatile elements were lost, that of formation of the oldest highland terranes, the time(s) of maria formation and filling, and the time at which the younger ray craters were formed. This will require combining our results with those of other laboratories, who may be utilizing different methods and techniques, or who may be pursuing somewhat different problems by the same methods. When such a time scale is established, it should prove possible to make use of the photogeological data to assign ages to many lunar features and areas not sampled by Apollo landings, and thereby permit a much greater understanding of lunar history, and its relation to the history of the earth and of other members of the solar system.

It was expected that measurements of these first lunar samples would permit us to orient ourselves with regard to the usefulness of lunar samples for dating purposes,

* Present address: Tata Institute for Fundamental Research, Homi Bhabha Road, Bombay-5, India.

† Deceased, December 2, 1969.

to establish the approximate range of ages with which we will be dealing in the dating of lunar rocks, and to make a start toward the establishment of a lunar chronology. These expectations have in large measure been fulfilled.

RUBIDIUM–STRONTIUM MEASUREMENTS

Our initial Rb–Sr work has emphasized the study of the crystalline basaltic rocks, in the belief that the interpretation of results on such rocks should be more straightforward than that on the microbreccia or dust samples.

Six of these rocks have been analyzed for Rb, Sr, K and Sr isotopic composition by the mass spectrometric isotope dilution method, using Rb^{87}, Sr^{84} and K^{41} as isotopic tracers. The chemical procedures are identical with those described previously in our work on meteorites (GOPALAN and WETHERILL, 1968, 1969, 1970a; KAUSHAL and WETHERILL, 1969, 1970). As before, Sr^{87}/Sr^{86} ratios were normalized to $Sr^{86}/Sr^{88} = 0{\cdot}11940$. The contribution of laboratory contamination was monitored by blank analyses carried out simultaneously with those of the samples. It was found that correction for this contamination would change the reported concentrations by only 0·1 per cent or less, and was consequently negligible.

Table 1. Sr^{87}/Sr^{86} ratios in standard samples

Pasamonte Basaltic achondrite	Sea water	M.I.T. Eimer and Amend $SrCO_3$ Lot 492327
0·6998	0·7089	0·7083
0·6996	0·7088	
	0·7095	

It was found that the Rb/Sr ratios of all the rocks studied were very low, 0·3 or less, and consequently the radiogenic enrichments of the Sr^{87}/Sr^{86} ratios over the primordial value were less than 1 per cent. Accurate measurement of these small differences in isotopic composition was limited by the capability of our recording system, and consequently great care was necessary in order to obtain meaningful data. Stable strontium ion beams were obtained at the same intensity for each analysis, and a procedure was developed for measurement of the recorder base line which avoided as much as possible the introduction of random errors from run to run. The linearity of the vibrating reed electrometer and recorder were carefully studied using a stable 30 V power supply with a voltage divider accurate to 0·001 per cent, and appropriate corrections for non-linearity were applied. Even with these precautions it was not obvious that differences in Sr^{87}/Sr^{86} ratios of less than 0·0007 could be detected. However, preliminary work on isotopic standards (KAUSHAL and WETHERILL, 1969) indicates that with this recording system, differences of less than 0·0006 can be clearly resolved. This work, together with the reproducibility of measurements on interlaboratory standards (Table 1) as well as that of the lunar samples themselves, give some basis for believing that somewhat smaller relative errors, perhaps as small as $\pm$ 0·0002, are obtainable. Conversion to digital recording is currently in progress, and it will then be possible to repeat, with greater precision, the analyses of these samples without further consumption of lunar material.

The concentration measurements involved no unfamiliar problems. Experimental errors of ± 2 per cent in the Rb^{87}/Sr^{86} ratios are assigned, based on previous experience in the measurement of these concentrations in meteorites and in terrestrial rocks. Somewhat larger errors (± 3 per cent) are assigned to the potassium analysis, because of the possibility of isotope fractionation during the mass spectrometric analysis.

The results of the analyses of these rocks are given in Table 2. The strontium concentrations are all rather similar. However the rocks may be divided into two groups on the basis of their rubidium concentrations, or their Rb/Sr ratios. Four of the samples (10017, 10022, 10024, 10072) have Rb^{87}/Sr^{86} ratios ranging from ·096 to ·099, whereas the other two crystalline rocks have much lower Rb^{87}/Sr^{86} ratios, in the

Table 2. K, Rb and Sr analytical results

Sample	K (μg/g)	Rb (μg/g)	Sr (μg/g)	Rb^{87}/Sr^{86}	Sr^{87}/Sr^{86}
10017-41	2500	5·80	173·0	0·0964	0·7045
10022-45	—	5·66	165·9	0·0982	0·7047
10024-24	2814	6·21	184·5	0·0968	0·7045
10072-38	2539	5·72	168·2	0·0978	0·7045
10050-30	665	0·788	188·8	0·01199	0·7002
10047-29	785·8	0·937	193·5	0·01392	0·6998
10019-26 (breccia)	1424	3·31	166·4	0·0572	0·7025
10084-25 (dust)	1100	2·83	163·8	0·0497	0·7019
10017-41 Density fractions:					
$\rho < 2{\cdot}96$	—	9·52	549·6	0·0498	0·7022
$2{\cdot}96 < \rho < 3{\cdot}3$	4090	9·19	256·8	0·1030	0·7047
$3{\cdot}15 < \rho < 3{\cdot}25$	—	7·79	134·4	0·1666	0·7083
$3{\cdot}25 < \rho$	843	2·30	45·8	0·1446	0·7068
$3{\cdot}3 < \rho$ (coarse)	1545	4·01	78·0	0·1479	0·7070
10024-24 Density fractions:					
$\rho < 2{\cdot}96$	—	13·76	597·0	0·0663	0·7028
$2{\cdot}96 < \rho < 3{\cdot}15$	—	19·42	518·3	0·1078	0·7050
$3{\cdot}15 < \rho < 3{\cdot}32$	—	15·43	284·6	0·1560	0·7082
$3{\cdot}3 < \rho$	—	2·82	47·6	0·1704	0·7087
50, 30 Density fraction $\rho < 2{\cdot}96$	682	0·658	222·2	0·0085	0·6995

vicinity of 0·01. Only the fine dust (10084) and the microbreccia sample (10019) were found to have intermediate values of this ratio. Other workers (ALBEE *et al.*, 1970; COMPSTON *et al.*, 1970; GAST and HUBBARD, 1970; HURLEY and PINSON, 1970, MURTHY *et al.*, 1970; WANLESS *et al.*, 1970) have also analyzed these and other Apollo 11 rocks. All of the crystalline rocks, other than fragments in the lunar fines, (ALBEE *et al.*, 1970) clearly fall into one or the other of these Rb/Sr groups.

The whole rock data are plotted on a strontium evolution diagram in Fig. 1. No isochrons have been drawn on this diagram for the following reason. Rb–Sr measurements define the time at which a Rb–Sr fractionation took place in the chemical system being studied. In the case of rock systems, this requires measurement of at least two samples with differing Rb–Sr ratios, which had the same Sr isotopic composition at the time of the event being dated. This is the case for cogenetic materials in which the Sr isotopic composition was initially homogeneous. For whole-rock dating, it is therefore preferable that independent evidence exists for the cogenesis of the rocks. In the absence of such evidence, the entire burden of the argument for cogenesis

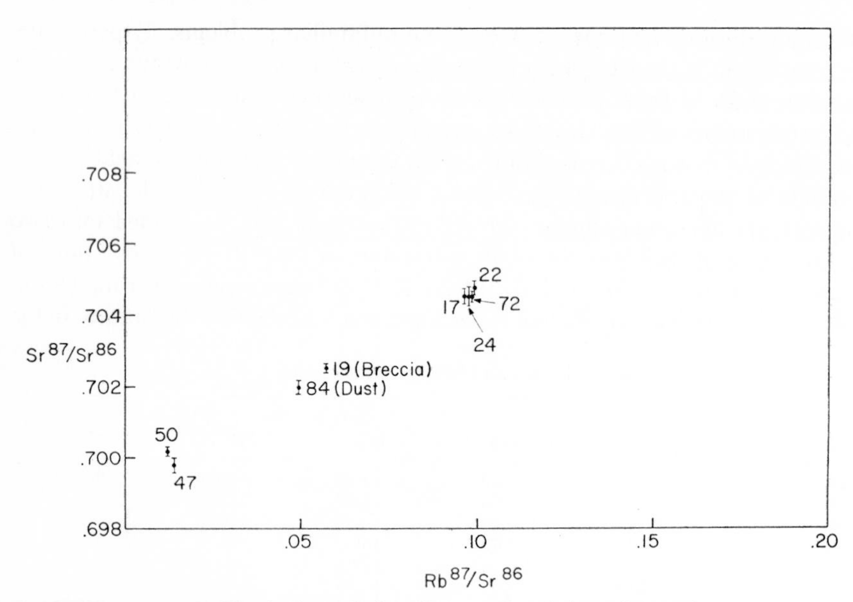

Fig. 1. Sr evolution diagram for whole rock samples.

rests upon the linearity of the data when plotted on a Sr evolution diagram. This independent evidence is lacking in the case of these lunar rocks. Because of the fact that the points fall into two discrete groups, together with the relatively large experimental errors in the Sr^{87}/Sr^{86} ratios, it is not possible to use the linearity of the data points as an argument for cogenesis. Therefore an isochron based on these whole rocks could well be meaningless.

Within the experimental errors of these Rb–Sr measurements, the dust and breccia could be formed from a mixture of the two Rb/Sr groups found at the Apollo 11 site. However some contribution from other sources is required by the U–Pb data, as will be discussed later.

In order to obtain data points to which meaningful isochrons could be fitted, we have directed our effort to the study of the "internal ages" of single rock samples, obtaining varying Rb–Sr ratios by separation of density fractions of rocks fragments which have been finely ground and sieved. These are not pure mineral separates, and were prepared in this simple way in order to conserve lunar material, and to reduce the possibility of laboratory contamination by minimal processing of the material.

The separation procedures are the same used in our work on density fractions from enstatite and shocked hypersthene chondrites (GOPALAN and WETHERILL, 1970a, b). In these earlier investigations it was shown that problems of contamination and leaching by the heavy liquids used (tetrabromoethane and methylene iodide) were negligible.

Density fractions have been studied on two rocks belonging to the high Rb/Sr group (samples 10017-41 and 10024-24), and preliminary work has been done on a sample from the low Rb/Sr group (sample 10050-30). A 2·2 g sample of rock 10017 was first simply crushed in a percussion mortar, and 219 mg of the fragments were separated into three density fractions, $\rho < 2{\cdot}96$ g/cm^3, $2{\cdot}96 < \rho < 3{\cdot}3$ and $\rho > 3{\cdot}3$. These samples were analyzed. The densest fraction consisted largely of coarse ($>150\ \mu$) intergrowths of ilmenite, plagioclase, and pyroxene. In order to separate these phases, a second sample (154 mg) of the crushed rock was mixed with the coarse fraction and then ground in an agate mortar to pass through an 88 μ nylon sieve. Several density fractions were prepared, of which the $3{\cdot}15 < \rho < 3{\cdot}25$ and $\rho > 3{\cdot}25$ fractions were analyzed.

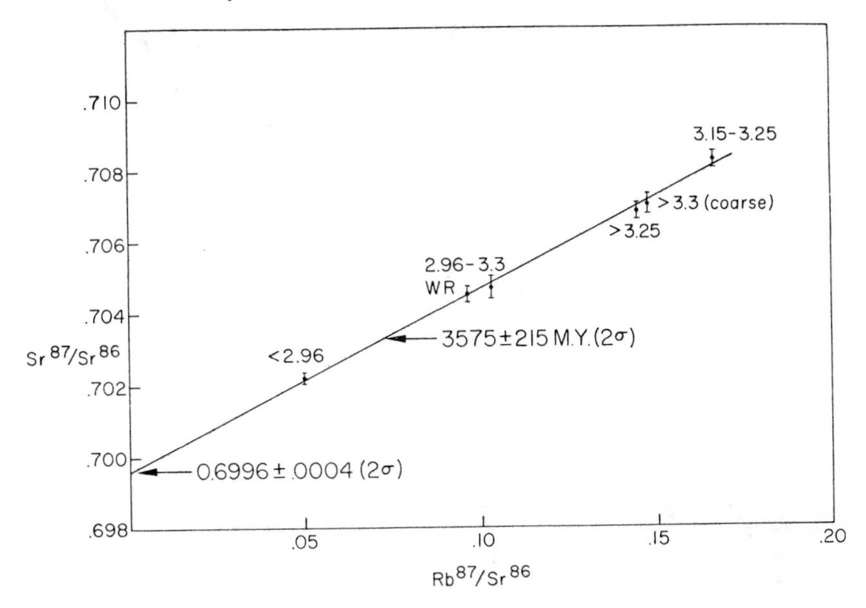

Fig. 2. Sr evolution diagram for density fractions from sample 10017-41. Age calculated using λ (Rb^{87}) $= 1{\cdot}39 \times 10^{-11}$ yr^{-1}.

Approximately 350 mg samples of rocks 10024 and 10050 were first crushed, and then ground to $<88\ \mu$. These were then separated into fractions of density $\rho < 2{\cdot}96$, $2{\cdot}96 < \rho < 3{\cdot}15$, $3{\cdot}15 < \rho < 3{\cdot}32$ and $\rho > 3{\cdot}3$.

The analytical results obtained on these density fractions are given in Table 2, and plotted on Sr evolution diagrams in Figs. 2 and 3.

The data points from the density fractions have been fitted to straight lines by a double regression least squares procedure (YORK, 1966). The slopes of these best-fit straight lines correspond to ages of 3575 and 4070 m.y. for samples 10017 and 10024 respectively. The error brackets shown for the individual data points are the 1σ errors calculated from the mass spectrometric data. It is not claimed that these represent a particularly good measure of the accuracy of the isotopic analyses, and may be subject to systematic as well as additional random error. However, it is believed that these errors represent the most objective available method of weighting the experimental points for the least squares analysis. The 2σ errors calculated by the least squares procedure are somewhat more than 200 m.y. for both of the rocks

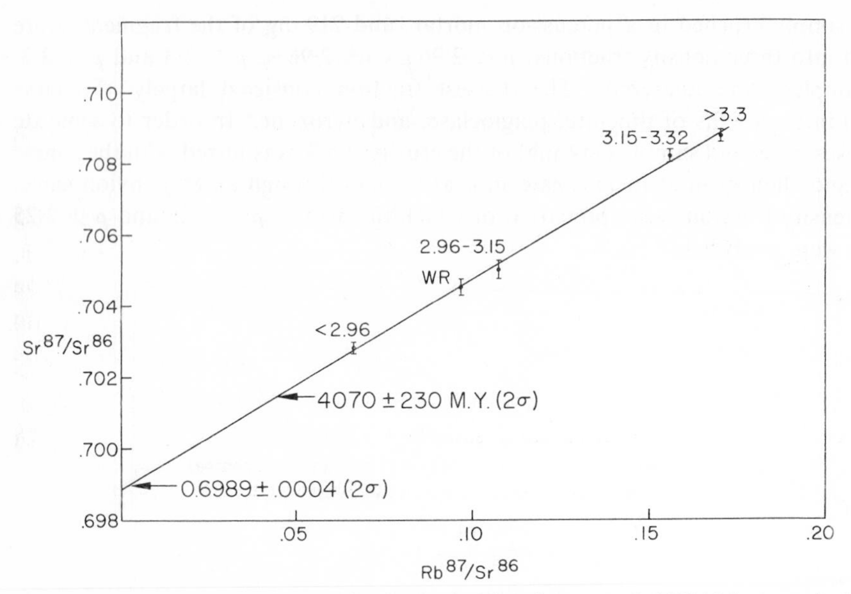

Fig. 3. Sr evolution diagram for density fractions from sample 10024-24. Age calculated using λ (Rb^{87}) = $1 \cdot 39 \times 10^{-11}$ yr^{-1}.

studied. The initial Sr^{87}/Sr^{86} ratios are 0·6996 ($2\sigma = 0 \cdot 0004$) for sample 10017 and 0·6989 ($2\sigma = 0 \cdot 0004$) for sample 10024. From these results it appears that the data define linear arrays quite well, and that the calculated ages indicate the time of formation of these rocks to be approximately 3800 m.y. ago, distinctly more recent than the time of formation of the moon, found to be ~4700 m.y. in age by the U–Pb method. A more precise evaluation of the experimental errors, as well as further investigation of whether or not these two rocks are slightly different in age must await anticipated improvements in instrumental accuracy, as well as further studies with isotope standards of known relative composition.

The 3575 m.y. age found for sample 10017 is in agreement with the more accurate result of 3590 ± 80 m.y. reported by ALBEE *et al.* (1970) and the result of 3800 ± 110 m.y. of COMPSTON *et al.* (1970). The initial Sr^{87}/Sr^{86} ratio also agrees with that of the values 0·69933 ± 0·00007, and 0·6994 ± 0·0002 given by these workers for this rock. There are no other published analyses of separated phases from sample 10024. ALBEE *et al.* (1970) have reported ages for five Apollo 11 rocks ranging from 3590 ± 80 m.y. to 3700 ± 70 m.y. in age, while COMPSTON *et al.* (1970) has reported a mineral isochron corresponding to an age of 3800 ± 300 m.y. for sample 10072. Our result for sample 10024 falls outside the range of ages reported by ALBEE *et al*, and further work will be necessary to determine if this difference is real.

The low initial ratios indicate that the material from which these rocks were derived underwent a differentiation from the higher Rb/Sr ratio of the chondrites and the sun early in lunar history. The Sr^{87}/Sr^{86} ratio of chondritic material at the time of formation of these rocks was about 0·712, much higher than the initial ratio found in these rocks. The initial ratios found are compatible with no more than 100 m.y.

of strontium evolution in a chondritic environment, and indicate that the source of lunar basalts, like those of the earth, was involved in a very early Rb–Sr fractionation. UREY (1952) has presented arguments supporting the conclusion that all the maria were formed at nearly the same time, during the terminal stage of lunar accretion. If this is so, the presence of much younger basalt in Mare Tranquillitatis, indicates that at least some of the mare filling greatly postdates these impact events.

It may also be expected that lunar capture involving a close approach of the moon to the earth more recently than 3600 million years ago, accompanied by extensive lunar volcanism and magma production, would have destroyed these rocks, as well as the evidence, based on crater frequency, for even older rocks in the lunar highlands.

URANIUM, THORIUM, AND LEAD MEASUREMENTS

The concentrations of uranium, thorium, and lead, as well as the lead isotopic composition, have been measured in a sample of fine surface material (sample 10084-25). In making this measurement, it was thought that, of the available material, this sample was the one most likely to indicate the overall lead evolution of the moon,

Table 3. U, Th and Pb results

		Concentrations: (μg/g)		
U	Th	$Pb^{206}_{radiogenic}$	$Pb^{207}_{radiogenic}$	$Pb^{208}_{radiogenic}$
0·549 ± 0·005	2·08 ± 0·02	0·483 ± 0·010	0·325 ± 0·007	0·496 ± 0·025
		Ages (m.y.)		
$\frac{U^{238}}{Pb^{206}}$	$\frac{U^{235}}{Pb^{207}}$	$\frac{Pb^{207}}{Pb^{206}}$	$\frac{Th^{232}}{Pb^{208}}$	
4655 ± 100	4707 ± 100	4740 ± 50	4717 ± 250	

analogous to the mantle lead system sampled in the formation of terrestrial basalts, or the lead system sampled in the formation of terrestrial lead ores. Although considerable uranium–lead fractionation occurs in the earth, terrestrial mixing processes counteracts in large measure the effects of these fractionations, and permit a good understanding of the average isotopic evolution of terrestrial lead. In this way the age of the earth was calculated by HOUTERMANS (1953) to be 4500 m.y., based on the difference in the Pb^{206}/Pb^{204} and Pb^{207}/Pb^{204} ratios between lead in modern terrestrial basalts, and primordial lead as found in the troilite of the iron meteorite Canyon Diablo, as measured by PATTERSON (1953, 1955).

Two samples of the lunar dust weighing 1·7 g and 0·91 g were analyzed by the isotope dilution method, using U^{235} and Th^{230} as isotopic tracers. In the first analysis Pb^{208} tracer was used, in the second Pb^{206} tracer was added. The results of the two lead analyses then permit calculation of the abundance and concentration of all the lead isotopes, after subtraction of the contribution from the tracers. The chemical procedure was that of TATSUMOTO (1966). The results of these analyses are shown in Table 3.

The lead in the lunar surface material is found to be highly radiogenic. Even before correcting for the blank, Pb^{206}/Pb^{204} was equal to 108, and after correction for a total lead blank of about 0·2 μg, the ratio becomes 231. The Pb^{207}/Pb^{204} and Pb^{208}/Pb^{204} ratios are similarly radiogenic. These ratios may well be still higher, since the residual Pb^{204} may represent terrestrial contamination beyond that introduced by the blank. This lead is far more radiogenic than that found in terrestrial basalts, and is approached only by the basaltic achondrite Nuevo Laredo ($Pb^{206}/Pb^{204} = 50·3$, $Pb^{207}/Pb^{204} = 34·9$) as measured by PATTERSON (1955); however the lunar material is distinctly even more radiogenic than this.

If it is assumed that this lead has evolved in a system in which the U^{238}/Pb^{206} ratio changes only because of the decay or uranium, an "age of the moon" analogous to HOUTERMANS' (1953) age of the earth can be calculated. Assuming the initial ratio to be the primordial lead used in the calculation of the age of the earth, the age is found to be 4740 $\pm$ 50 m.y. In addition, the individual ages calculated from the Pb^{206}/U^{238}, Pb^{207}/U^{235} and Pb^{208}/Th^{232} ratios, agree with this result within experimental error. In other words, the concentrations of U, Th and the lead isotopes are consistent with the evolution of lead in a 4700 m.y. old system characterized by the U/Pb and Th/Pb ratios of this surface material. Because the lead is so highly radiogenic, this result is insensitive to the exact isotopic composition of the primordial lead. For example if *modern* terrestrial lead is used instead of the primordial lead found in Canyon Diablo, the calculated Pb^{207}/Pb^{206} age is changed by only 6 m.y.

Similar measurements on lunar dust have been reported by SILVER (1970), TATSUMOTO and ROSHOLT (1970) and WANLESS *et al.* (1970). KOHMAN *et al.* (1970) have reported lead isotopic composition. With the exception of the data of WANLESS *et al.*, there is rather good agreement between our results and those of these other workers. The concentrations of radiogenic lead found by WANLESS *et al.* are considerably higher than those found in the present investigation, while their thorium concentrations are lower. The only discrepancy between our results and those of the remaining investigators is that whereas the ratio of radiogenic Pb^{207}/Pb^{206} found by these workers range from 0·62 to 0·65, we find 0·67. This results in our Pb^{207}/Pb^{206} age being 50–100 m.y. older, and in reversal of the direction of slight discordance in the U^{238}/Pb^{206} and U^{235}/Pb^{207} ages. Our Th^{232}/Pb^{208} age is also somewhat higher than reported by SILVER *et al.* and TATSUMOTO *et al.*, but our lead data from the analysis using Pb^{206} tracer, from which the radiogenic Pb^{208} concentration was calculated, were of poor quality, and the disagreement is probably within experimental error.

The discrepancy in the Pb^{207}/Pb^{206} ratio is harder to explain. The mean deviation of the mass spectrometric data was 0·3 per cent, much less than the discrepancy, and if the Pb^{207} had actually been 3 per cent less, this difference would have been readily measurable. Measurement of NBS isotopic lead standard 982 showed that fractionation effects and systematic errors in our lead isotope measurements are less than 1 per cent, and do not explain the differences. It is possible that a small impurity was hidden under the Pb^{207} peak, causing it to be apparently higher. However, the only peaks seen at other mass positions were at mass 203 and 205, and were presumably due to thallium. The height of these peaks was about 1 per cent that of the Pb^{207}. It is also possible that the lunar dust is somewhat inhomogeneous, although this is not indicated by the concentration data. It is planned to investigate this problem further.

Of course, this age of the moon is dependent on the model used in its calculation. On the other hand, the short half-life of U^{235}, the highly radiogenic nature of the lunar lead, together with the fact that the U^{235}/Pb^{207}, U^{238}/Pb^{206} and Th^{232}/Pb^{208} ages are all concordant, places restrictions on the class of plausible models for which the lunar age is significantly younger. For example, a slightly younger moon (e.g. 4500 m.y. old) may have undergone a U/Pb fractionation in the source material from which the 3650 m.y. old rocks were derived. This would alter the original isotopic composition of the lead in these rocks and ultimately that of the soil derived therefrom. An enrichment of lead with respect to uranium of a factor of about two. over the period prior to the formation of the rocks, would lead to an initial Pb^{207}/Pb^{206} ratio equivalent to that of a 4700 m.y. old moon which had not undergone this fractionation. Subsequent rock formation and erosion, with little net U/Pb fractionation would lead to discordant U–Pb ages requiring about a 5 per cent change in the U/Pb concentration to achieve concordance. This is believed to be beyond the experimental errors of these measurements. Recent loss of lead could remove this discordance. More extreme fractionations of this kind will be required if the moon is less than 4500 m.y. old. There is no independent evidence for these assumed U/Pb fractionations, and these assumptions are therefore of an *ad hoc* nature.

Another possibility is that the soil constitutes an unrepresentative mixing of several sources. It is most reasonable to suppose that the lunar soil at the Apollo 11 site was formed primarily by erosion of local rocks, with a smaller contribution from more distant rocks, subsequent to the formation of the Apollo 11 crystalline rocks, i.e. less than 3650 m.y. ago. In order for the true age of the moon to be significantly younger than the age indicated by the data obtained on the soil, it is necessary that the soil be preferentially enriched in lead of the isotopic composition present in the source of these younger rocks at the time of their formation, relative to the contribution from radiogenic lead generated since the formation of these rocks. Such unrepresentative mixtures could also result in soil lead having an apparent $Pb^{207}Pb^{206}$ age of 4700 m.y., and again the U–Pb lead ages would be slightly discordant. Again, the discordance could be removed by recent lead loss. As discussed below, the Apollo 11 crystalline rocks cannot be the source of the old radiogenic component, and there may be a problem in obtaining the necessary amount of this component, let alone an excess of it.

On the other hand, older lunar ages can be obtained by a model in which the U^{238}/Pb^{204} ratio was very low, e.g. that found in the carbonaceous chondrites, for some arbitrary length of time, whereafter, 4700 m.y. ago it increased to its present value. However, for significantly older lunar ages, models of this kind result in lunar ages older than that of the solar system, based on the Rb–Sr and K–Ar ages of chondrites, together with I–Xe and Pu^{244} formation intervals. Use of a shorter Rb^{87} half-life, as determined by FLYNN and GLENDENIN (1959) would aggravate this discrepancy.

We have taken the point of view that the lunar dust represents a well-mixed U, Th–Pb system in which the effects of U–Pb and Th–Pb fractionations are averaged out, and information reflecting only the major U–Pb fractionation accompanying lunar formation is preserved. From this point of view there is no difficulty understanding the concordant ages obtained, and it is to be expected that the age found in this way is older than that found for the time of crystallization of the lunar rocks. This simply reflects the fact that the moon is older than any of its constituent rocks. However, if the model for the formation of the lunar dust should turn out to be incorrect, then some other explanation of the concordant results may be required. TATSUMOTO and ROSHOLT (1970) and SILVER (1970) have reported U–Pb measurements on crystalline rocks. These measurements indicate that, although these rocks contain excess radiogenic lead, if their age is assumed to be that given by the Rb–Sr data, they nevertheless contain less radiogenic lead than would be expected in a 4700 m.y. old closed system. If this result is common to all the rocks at the Apollo 11 site, then it follows that the lunar dust at this site must not be entirely locally derived. It must also contain a significant contribution from rocks inheriting an excess of original lead at the later time of fractionation, in order to balance the deficiency found in the Apollo 11 rocks. In view of the small quantity of U–Pb data available for the crystalline rocks, the difficulty of knowing the adequacy of the sampling at the Apollo 11 site, and lack of definite knowledge regarding the sources of the lunar dust, it is difficult to say at present whether or not this represents a difficulty requiring explanation.

The extremely high U^{238}/Pb^{204} ratio and the associated highly radiogenic lead found in the lunar dust has some interesting consequences regarding the fission theory of lunar origin, as well as the question of the source of tektites. The U^{238}/Pb^{204} ratio found (≥ 210) is much higher than the terrestrial value of about 9. This is not in itself an insuperable difficulty in a fission theory, because it is possible that lead could be volatilized and lost at the time of formation of the moon. However the isotopic composition of the lead indicates that the U–Pb fractionation resulting from this loss of lead took place 4700 m.y. ago, and that the fission took place during the time of formation of the earth–moon system and the remainder of the solar system. Therefore varieties of the fission theory, in which hundreds of millions of years are required for the earth to be heated by long-lived radioactivities, followed by core formation and rotational instability, are not compatible with the results of this work.

TILTON (1958) has shown that the lead in tektites from several areas is remarkably similar, in fact, identical, in isotopic composition to modern terrestrial lead. This led him to conclude that tektites were of terrestrial origin. It was argued that this conclusion was perhaps invalid, since a moon which fissioned from the earth's mantle might have precisely the U^{238}/Pb^{204} ratio of the source of terrestrial basalts, and would result in modern lunar ejecta containing lead identical to that found on earth. Now, however, we have seen that lunar lead can be grossly dissimilar to terrestrial lead, and furthermore this sample of lunar lead is not from some isolated mineral possibly having a very special history, but from a sample which must represent some sort of an average of rocks spread over a significant area, including possibly the lunar highlands. Therefore, even if the moon did fission from the earth, it is clear that this did not cause lunar lead evolution to be precisely controlled so as to preserve an identity between modern terrestrial lead and that found on the moon.

References

Albee A. L., Burnett D. S., Chodos A. A., Eugster O. J., Huneke J. C., Papanastassiou D. A., Podosek F. A., Russ G. Price II, Sanz H. G., Tera F. and Wasserburg G. J. (1970) Ages, irradiation history and chemical composition of lunar rocks from the Sea of Tranquillity, *Science* **167,** 463–466.

Compston W., Arriens P. A., Vernon M. J. and Chappell B. W. (1970) Rubidium–strontium chronology and chemistry of lunar material. *Science* **167,** 474–476.

Flynn K. and Glendenin L. E. (1959). Half-life and beta spectrum of Rb^{87}. *Phys. Rev.* **116,** 744–748.

Gast P. W. and Hubbard N. J. (1970) Abundance of alkali metals, alkaline and rare earths, and strontium-87/strontium-86 ratios in lunar samples. *Science* **167,** 485–487.

Gopalan K., Kaushal S., Lee-Hu C. and Wetherill G. W. (1970) Rubidium–strontium, uranium, and thorium–lead dating of lunar material. *Science* **167,** 471–473.

Gopalan K. and Wetherill G. W. (1968) Rubidium–strontium age of hypersthene (L) chondrites. *J. Geophys. Res.* **73,** 7133–7136.

Gopalan K. and Wetherill G. W. (1969) Rubidium–strontium age of amphoterite (LL) chondrites. *J. Geophys. Res.* **74,** 4349–4358.

Gopalan K. and Wetherill G. W. (1970a) Rubidium–strontium studies on enstatite chondrite *J. Geophys. Res.* in press.

Gopalan K. and Wetherill G. W. (1970b) Rb^{87}–Sr^{87} studies of shocked hypersthene chondrites. To be submitted to *J. Geophys. Res.*

Houtermans F. G. (1953) Determination of the age of the earth from the isotopic composition of meteoritic lead. *Nuovo Cimento* **10,** 1623–1633.

Hurley P. M. and Pinson W. H., Jr. (1970) Rubidium–strontium relations in Tranquillity Base samples. *Science* **167,** 473–474.

Kaushal S. and Wetherill G. W. (1969) Rb^{87}–Sr^{87} age of bronzite (H group) chondrites. *J. Geophys. Res.* **74,** 2717–2725.

Kaushal S. and Wetherill G. W. (1970) Rb^{87}–Sr^{87} age of carbonaceous chondrites. *J. Geophys. Res.* **75,** 463.

Kohman T. P., Black L. P., Ihochi H. and Huey J. M. (1970) Lead and thallium isotopes in Mare Tranquillitatis surface material. *Science* **167,** 481–483.

Murthy V. R., Schmitt R. A. and Rey P. (1970) Rubidium–strontium age and elemental and isotopic abundances of some trace elements in lunar samples. *Science* **167,** 476–479.

Patterson C. C. (1953) The isotopic composition of meteoritic, basaltic, and oceanic leads and the age of the earth. *Proc. First Conf. Nuclear Process in Geological Settings.* National Research Council 36–40.

Patterson C. C. (1955) The Pb^{207}/Pb^{206} ages of some stone meteorites. *Geochim. Cosmochim. Acta* **7,** 151–153.

Patterson C. C., Tilton G. and Inghram M. G. (1955) Age of the Earth. *Science* **121,** 69–75.

Silver L. T. (1970) Uranium–thorium–lead isotope relations in lunar materials. *Science* **167,** 468–471.

Tatsumoto M. (1966) Isotopic composition of lead in volcanic rocks. *J. Geophys. Res.* **71,** 1721–1733.

Tatsumoto M. and Rosholt J. N. (1970) Age of the Moon: An isotopic study of uranium–thorium–lead systematics of lunar samples. *Science* **167,** 461–463.

Tilton G. R. (1958) Isotopic composition of lead from tektites, *Geochim. Cosmochim. Acta* **14,** 323–330.

Urey H. C. (1952) *The Moon*, p. 39ff. Yale University Press.

Wanless R. K., Loveridge W. D. and Stevens R. D. (1970) Age determinations and isotopic abundance measurements on lunar samples. *Science* **167,** 479–480.

York D. (1966) Least square fitting of a straight line. *Can. J. Phys.* **44,** 1079–1086.

Proceedings of the Apollo 11 Lunar Science Conference, Vol. 2, pp. 1207 to 1211.

Thermal and gas evolution behavior of Apollo 11 samples

R. E. HANNEMAN
General Electric Research and Development Center, Schenectady, New York 12301

(*Received* 6 *February* 1970; *accepted in revised form* 5 *March* 1970)

Abstract—This paper reports on preliminary measurements of the gases evolved on heating Apollo 11 fines and important related physical changes in the samples during thermal treatment. The Ar^{40} anomoly, the origin of a lunar lava-like structure, and the oxidation state of the samples are also discussed.

1. INTRODUCTION

THIS communication reports on preliminary measurements of gas evolution and physical behavior of Apollo 11 samples upon heating. Also included are a treatment of the Ar^{40} anomaly in lunar fines and comments on the thermal origin of some observed microstructures.

2. GAS ANALYSES AND RESULTS

Mass spectrometry studies on the various gases present in lunar fines were carried out on a 180° sector C.E.C.-Bell & Howell 21–104 mass spectrometer using 250 mg samples for each run. The gases were evolved by heating in two different modes: (1) constant linear heating at various rates from room temperature to 1100°C; and (2) isothermal heating at temperatures of 200°C, 500°C, 800°C and 1000°C. Control samples were run to establish background corrections.

Gases identified are H_2, He, Ne, Ar, Kr, N_2 and contaminants from the original sample including CO, CO_2, CH_4, HCl, C_6H_6 and H_2S at low levels. No detectable oxygen is evolved on heating. This result is expected based on our earlier detection (FLEISCHER *et al.*, 1970) of free iron and various partially oxidized minerals in the lunar samples which would be oxidized by any oxygen present.

It is notable that the temperatures required to drive off most of the above gases from these samples are quite high. For a 30 min isothermal heating cycle the hydrogen and helium are essentially depleted from the lunar fines at 750°C whereas nitrogen, neon and other heavier noble gases require at least 900°C for depletion. This increased temperature with larger entrapped atoms is reasonable in terms of expected relative diffusion coefficients and solubilities.

The relative abundances of the principal species in the lunar fines (samples 10084, 4·5 to 4·13) are summarized in Table 1. The overall average concentration of He^4 in the lunar fines is approximately $0{\cdot}26 \pm 0{\cdot}02$ cm³ STP/g but the concentration is quite particle size dependent, with the smallest particles exhibiting the most helium enrichment. This is in accord with the observations of EBERHARDT *et al.* (1970) and HINTENBERGER *et al.* (1970) and indicative of a dominant helium origin from the solar wind.

The measured ratio of H^1 to He^4 in the fines is approximately 8:1 whereas the currently accepted value for the solar wind is 17:1 LAMBERT (1967). The low observed ratio is most likely dominated by the preferential diffusional loss of hydrogen

Table 1. Relative abundances of important gases in the lunar fines (He^4 concentration = 0·26 cm³/g STP)

Isotope	Normalized abundance (by vol.)
H^1	1·0
He^4	$1{\cdot}2 \times 10^{-1}$
Ne^{20}	$1{\cdot}2 \times 10^{-3}$
Ar^{36}	$1{\cdot}8 \times 10^{-4}$
N^{14}	0·6

relative to helium rather than differences in saturation effects or preferential sputtering loss of hydrogen or any change in solar wind isotope ratios with time, or it could be a combination of diffusional losses plus saturation effect (HEYMANN *et al.*, 1970; EBERHARDT *et al.*, 1970) This conclusion is based on our linear heating experiments which show that hydrogen invariably evolves more completely than helium at each temperature. The H^1/He^4 ratio observed for lunar fines in this study compares to 7·7 reported by STOENNER *et al.*, 1970; FIREMAN *et al.*, 1970.

No quantitative measurements of nitrogen level in the lunar fines have been reported previously, although HINTENBERGER *et al.* (1970) did measure it in the breccia. The values found here of 1·3 ± 0·1 cm³/g STP includes adsorbed and trapped nitrogen both. Detectable nitrogen is still evolving at 1000°C after 30 min from the lunar fines.

The Ar^{40} level found in the lunar fines in this study is high and corroborates the Ar^{40} anomoly reported by various workers (HEYMANN *et al.*, 1970; HINTENBERGER *et al.*, 1970; MARTI *et al.*, 1970; PEPIN *et al.*, 1970). Our Ar^{40}/Ar^{36} ratios range from a high value of over 30:1 for large particles to values in the fines of 0·9 to 2·5:1. The other studies cited above have ranged from 40:1 for rocks to 0·7:1 for some fines. The absolute concentrations of Ar^{40} in the fines in this study range from $3{\cdot}5 \times 10^{-4}$ to $9{\cdot}7 \times 10^{-4}$ cm³/g STP. Corresponding Ar^{40} values for large particles (>0·25 mm) are in the range 4 to 6×10^{-5} cm³/g STP.

In the case of large particle samples where Ar^{40}/Ar^{36} ratios of 30:1 or more are found, the corresponding average potassium concentrations are generally 2000 ppm to 2700 ppm compared to 1100 ppm for the fines according to most available data. In the large particle samples the data are quite consistent with radiogenic decay of K^{40} over a few billion years, being totally dominant in supplying the Ar^{40}. There is no Ar^{40} anomoly for that case.

On the other hand the level of Ar^{40} present in the fines greatly exceeds that available from radiogenic decay of the measured K^{40} level. There are several possible mechanisms that have been advanced in an attempt to account for the Ar^{40} anomoly in lunar fines. First of all, the solar wind contribution appears to be quite large since most of the He^4, Ne^{20}, Ar^{36} and other gases are found within approximately 0·3 μ of the surfaces of the fines (EBERHARDT *et al.*, 1970; HINTENBERGER *et al.*, 1970). This does not explain why the K^{40} is depleted, in general, by a factor of up to two in the lunar fines versus rock and coarse material.

We propose here a new explanation for the observed gas distribution. Assume that due to meteoritic bombardment the lunar fines undergo periodic heating to intermediate temperatures where some potassium depletion can occur by volatilization. The solubility and permeability of potassium within the lunar minerals at elevated temperature is expected to be substantially larger than that of argon, based on studies of diffusion of alkali and inert gas atoms in various oxides (DOREMUS, 1962). Thus heating at some prior time in history would have selectively depleted potassium relative to argon. Solar wind contribution to the Ar^{40} and Ar^{36} species would have continued to be entrapped near the surface of the fines since the event or events that preferentially depleted the potassium. The observed depletion of K^{40} in the fines relative to coarse breccia and especially fine rock are consistent with the preferential potassium depletion possibility (see for example, PHILPOTTS and SCHNETZLER, 1970; GAST and HUBBARD, 1970).

Other mechanisms have been proposed in the recent past including: Ar^{40} escape from the lunar depths and recapture on the surfaces of the lunar fines by shock implantation, simple dissolution, or "pumping" by elastic collisions with solar wind or solar cosmic ray ions (HEYMANN *et al.*, 1970).

The preferential depletion process discussed above could operate independently or in parallel with previously proposed mechanisms. Further detailed measurements will be needed to differentiate which process is dominant in causing the Ar^{40} anomaly.

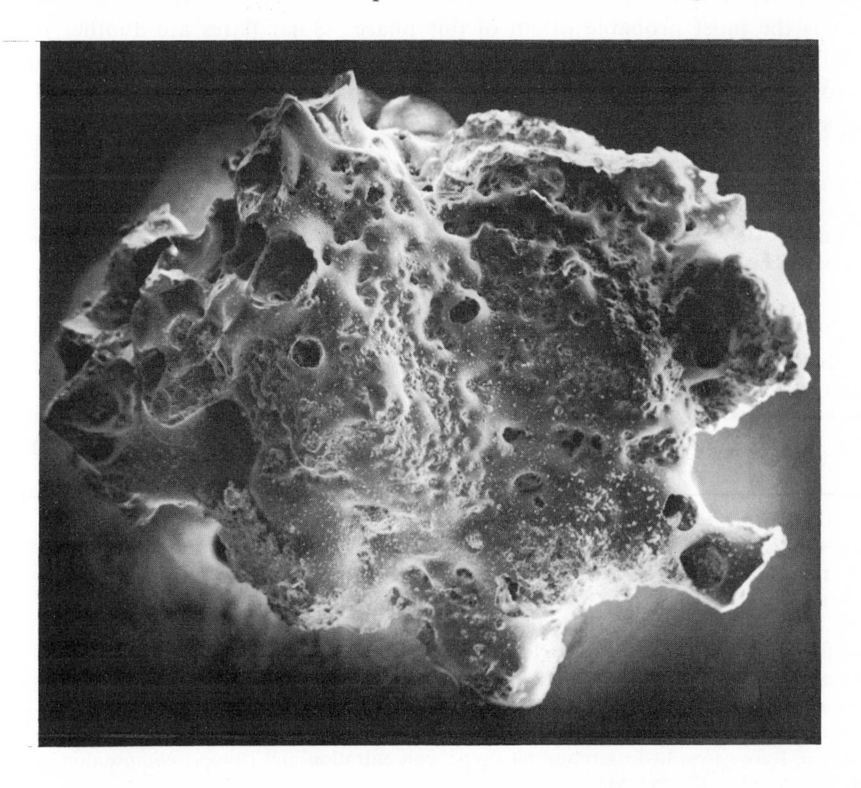

Fig. 1. Scanning electron micrograph of lunar lava-like structure at 50 ×.

3. PHYSICAL BEHAVIOR ON HEATING

Thermogravimetric analyses (TGA) were run on the lunar fines under inert gas and vacuum conditions by standard methods with 10–100 mg samples (10, 084, 4). Linear heating rates of 150°/hr and 350°/hr were used in the TGA runs. It was observed that unless oxygen bearing impurities in the system were held to very low levels (i.e. <5 ppm in He at 1 atm) the specimens exhibited a net weight gain on heating through the range 600–1200°C due to oxidation while the noble gases, hydrogen, nitrogen and other gases were evolving. This result independently demonstrates that the lunar dust is substantially below the fully oxidized state and that its oxidation rate is fairly rapid. The TGA weight losses observed in the oxygen-free atmosphere in a platinum boat on heating are in agreement with total gas evolution reported from the mass spectrometry results above. Melting of the fines in dense alumina does cause substantial chemical reaction with the boat.

Heating of the lunar fines in high purity helium or vacuum reproducibly produced melting accompanied by observable gas evolution at 1165 ± 15°C when carried out in an inert platinum boat. Upon rapid cooling this yielded the porous, dark "lunar lava-like phase" observed abundantly in the lunar fines and shown in Fig. 1. The intense heating and melting of the lunar fines and powder followed by fairly rapid cooling under reducing conditions, such as would be encountered with meteoritic impact is the most probable origin of this phase. Solar flares are another remote possibility for the origin of this structure but this is quite unlikely because of the enormous radiant energy increment needed to flash heat the surface temperature of the moon above 1165°C.

Acknowledgments—The author is grateful to G. SCHACHER, C. VAN BUREN and E. LIFSHIN for their valuable experimental contributions in the mass spectrometry, TGA and scanning electron microscopy studies. He is also indebted to R. L. FLEISCHER, H. HART, and J. S. KASPER for stimulating discussions. This work was supported in part by NASA under contract NAS 9-7898.

REFERENCES

DOREMUS R. (1962), Diffusion in non-crystalline silicates. *Modern Aspects of the Vitreous State*, (editor J. C. Mackenzie), Vol. 2, pp. 1–72. Butterworth.

EBERHARDT P., GEISS J., GRAF H., GRÖGLER N., KRÄHENBÜHL U., SCHWALLER H., SCHWARZMÜLLER J. and STETTLER A. (1970) Trapped solar wind noble gases, Kr^{81}/Kr exposure ages and Kr/Ar ages in Apollo 11 lunar material. *Science* **167**, 558–560.

FIREMAN E., D'AMICO J. C. and DEFELICE J. C. (1970) Tritium and argon radioactivities in lunar material. *Science* **167**, 566–568.

FLEISCHER R. L., HAINES E. L., HANNEMAN R. E., HART H. R. JR., KASPER J. S., LIFSHIN E., WOODS R. T. and PRICE P. B. (1970) Particle track, X-ray, and mass spectrometric studies of lunar material. *Science* **167**, 568–571.

GAST P. W. and HUBBARD N. J. (1970) Abundance of alkali metals, alkaline and rare earths and Strontium-87/Strontium-86 ratios in lunar samples. *Science* **167**, 485–487.

HEYMANN D., YANIV A., ADAMS J. A. S. and FRYER G. E. (1970) Inert gases in lunar samples *Science* **167**, 555–558.

HINTENBERGER H., WEBER H. W., VOSHAGE H., WÄNKE H., BEGEMANN F., VILCSEK E. and WLOTZKA F. (1970), Rare gases, hydrogen and nitrogen: concentration and isotopic composition in lunar material. *Science* **167**, 543–545.

LAMBERT D. L. (1967), Abundance of helium in the Sun. *Nature* **215**, 43–44.

MARTI K., LUGMAIR G. W. and UREY H. C. (1970) Solar-wind gases, cosmic ray spallation products and the irradiation history. *Science* **167**, 548–550.

PEPIN R. O., NYQUIST L. E., PHINNEY D. and BLACK D. C. (1970) Isotopic composition of rare gases in lunar samples. *Science* **167**, 550–553.

PHILPOTTS J. A. and SCHNETZLER C. C. (1970) Potassium, rubidium, strontium, barium and rare-earth concentrations in lunar rocks and separated phases. *Science* **167**, 493–495.

STOENNER R. W., LYMAN W. J. and DAVIS R. JR. (1970) Cosmic ray production of rare gas radioactivities and tritium in lunar material. *Science* **167**, 553–555.

Proceedings of the Apollo 11 Lunar Science Conference, Vol. 2, pp. 1213 to 1231.

Rare earths and other trace elements in Apollo 11 lunar samples

LARRY A. HASKIN, RALPH O. ALLEN, PHILIP A. HELMKE, THEODORE P. PASTER, MICHAEL R. ANDERSON, RANDY L. KOROTEV and KATHLEEN A. ZWEIFEL
Department of Chemistry, University of Wisconsin, Madison, Wisconsin 53706

(*Received* 2 *February* 1970; *accepted in revised form* 25 *February* 1970)

Abstract—The abundances of twelve rare-earth elements (REE) in nine samples and of twenty-four additional elements in four samples of lunar materials from Tranquillity Base have been determined by neutron activation analysis. Compared with their concentrations in the Bruderheim chondritic meteorite, the elements Au, Co, Ni, Cu, As, Sb, Zn, and Se are strongly depleted and Ag, In, Ta, Ba, Sc, Hf, Sr, and the REE strongly enriched. The depleted elements in the lunar rocks are those found predominantly in the metal phase of the Bruderheim chondrite and the enriched elements are found mainly in the troilite and silicate phases of the chondrite. This suggests that the silicate parent of the lunar basalts equilibrated with a metal phase sometime before it gave rise to the highly differentiated lunar basalts.

The relative abundances of the REE in the lunar rocks are similar to those in subalkaline basalts from the ocean ridges, except for Eu, which is severely depleted. Detailed studies of partial melting and fractional crystallization have been made using distribution coefficients for the REE from the literature. An adjustment for the distribution coefficient of Eu was made to account for the influence of the relatively low oxygen fugacities in the lunar magma. Changes in relative REE abundances of the magnitude required to produce the REE distributions found in the lunar rocks from a parent material with a relative REE distribution like that in chondrites can be readily accomplished by partial melting of a solid composed of plagioclase and orthopyroxene in a single process of partial melting. Such changes are more difficult to produce in a single process of fractional crystallization. The concentrations of the REE in the lunar basalts are probably too high to have been generated in a single process from a parent material with REE concentrations equal to those of chondrites.

INTRODUCTION

WE HAVE analyzed six chips of lunar igneous rocks (types A and B), powdered portions of two lunar breccias (type C), and 'fine' material less than 1 mm in diameter (type D) for twelve of the rare-earth elements (REE). We have also analyzed one sample of each type of material for twenty-four additional elements. The samples are from the Sea of Tranquillity and were brought to earth by the Apollo 11 astronauts. They were given to us following their preliminary examination at the Lunar Receiving Laboratory in Houston, Texas.

The classification of the lunar igneous rocks discussed in this paper into type A or B is based on the concentrations of K, Rb and Ba in those rocks. This classification does not in every case match the original ones based on sizes of mineral grains in the chips examined at the Lunar Receiving Laboratory (LSPET, 1969).

EXPERIMENTAL

All elements were determined by neutron activation analysis. Details of the procedures will be published elsewhere, but the following summary includes most of the important features.

The igneous rocks were received as chips weighing approximately 1 g each. Each was pulverized in a Plattner mortar until all portions passed through a fine (100 mesh) nylon sieve, then thoroughly mixed. The breccias and fines were received as cuts of powders and were not further handled. All grinding, mixing, and transferring of each sample was done inside a fresh polyethylene glove bag. Portions of the samples (0·3–0·7 g) were weighed and sealed into tubes made of polyethylene or quartz. Individual standards for 36 elements to be determined were mixed into twelve groups and similarly packaged for neutron irradiation. An iron wire was coiled around that part of each tube containing a sample or a standard. Comparison of specific activities induced in these wires makes it possible to correct for neutron flux gradients in the reactor. Samples analyzed only for REE were irradiated for 1 hr and those analyzed for all thirty-six elements were irradiated for 11 hr in the University of Wisconsin reactor at a neutron flux of $\sim 1 \times 10^{13} n/cm^2/sec$.

After being irradiated, the samples were fused with Na_2O_2 in a zirconium crucible in the presence of carrier for every element to be determined. The sample was then split chemically into twelve groups of elements, or else a single group, the REE, was removed from it. For each element group, a precipitate was formed and mounted on a card for radioassay. Appropriate carriers were added to each of the standards and a precipitate was prepared for each. Samples and standards were assayed by gamma-ray spectrometry on NaI or Ge-Li detectors at appropriate times to observe the decay of elements with short, intermediate, and long half-lives. Following radioassay, a chemical yield was determined for each element by titration or by activation of the carriers present in the mounted samples and standards.

RESULTS

The concentrations of the REE found for each rock, plus those in an andesite from the east Pacific rise (EPR-D-3) (FREY *et al.*, 1968) and in the USGS standard basalt (BCR1) (FLANAGAN, 1969) are given in Table 1. The concentrations for the

Table 1. REE concentrations in ppm for lunar rocks, a submarine andesite, and a basalt (BCR-1)

	Type						
Element	A 10022-38	A 10022-38	A 10057-63	A 10072-25	B 10045-25	B 10045-25	B 10003-46
La	26·4 ± 0·7	26·4 ± 1·3	28·2 ± 0·4	22·7 ± 0·5	6·7 ± 0·3	6·7 ± 1·6	14·1 ± 0·2
Ce	68 ± 2	69 ± 2	75 ± 1	69 ± 1	22·5 ± 0·4	22·5 ± 1·2	41·3 ± 0·7
Nd	66 ± 4	64 ± 8	69 ± 4	51 ± 4	21 ± 2	21·2 ± 4·5	42·5 ± 3
Sm	21·2 ± 0·1	21·2 ± 0·1	20·8 ± 0·1	17·9 ± 0·2	7·94 ± 0·05	8·9 ± 0·2	13·1 ± 0·1
Eu	2·04 ± 0·01	1·98 ± 0·014	2·18 ± 0·02	2·07 ± 0·04	1·52 ± 0·03	1·56 ± 0·02	1·80 ± 0·01
Gd	25 ± 2	22·9 ± 1·1	26 ± 3	26 ± 2	12·3 ± 0·9	14·1 ± 2·5	17 ± 2
Tb	4·7 ± 0·1	4·31 ± 0·05	5·0 ± 0·1	4·3 ± 0·1	2·11 ± 0·06	2·16 ± 0·10	3·26 ± 0·07
Dy	31·2 ± 0·3	29·0 ± 0·7	34·7 ± 0·2	31·2 ± 0·3	14·4 ± 0·1	14·6 ± 0·6	22·4 ± 0·3
Ho	5·5–8·7	5·8–6·7	5·8–7·0	5·2–8·4	2·2–3·5	2·9–3·4	3·7–4·4
Er	16 ± 3	15·6 ± 1·7	19 ± 2	16 ± 3	8·7 ± 2	10·7 ± 1·3	12 ± 1
Yb	17·7 ± 0·5	17·3 ± 0·5	18·8 ± 0·2	16·6 ± 0·5	8·6 ± 0·3	8·46 ± 0·25	11·9 ± 0·3
Lu	2·55 ± 0·02	2·39 ± 0·02	2·66 ± 0·02	2·24 ± 0·01	1·17 ± 0·01	1·17 ± 0·02	1·69 ± 0·01

B 10020-45	C 10048-47	C 10048-47	C 10060-30	D 10084-52	D 10084-52	EPR–D–3	BCR–1
8·4 ± 0·2	20·2 ± 0·3	22·0 ± 0·5	20·8 ± 0·4	16·6 ± 0·4	16·9 ± 0·8	20·4 ± 0·4	25·2 ± 1·0
25·6 ± 0·6	56 ± 1	57·4 ± 1·9	58 ± 1	47 ± 1	47·3 ± 1·7	61 ± 1	54·2 ± 1·2
31 ± 3	41 ± 7	39 ± 4	46 ± 15	43 ± 6	41 ± 18	57 ± 10	30·5 ± 4·3
9·94 ± 0·06	14·8 ± 0·2	15·1 ± 0·2	15·4 ± 0·1	13·7 ± 0·1	13·66 ± 0·06	15·9 ± 0·1	7·23 ± 0·37
1·75 ± 0·01	1·95 ± 0·02	1·95 ± 0·03	2·06 ± 0·02	1·77 ± 0·02	1·74 ± 0·01	4·05 ± 0·03	1·97 ± 0·04
16 ± 2	20 ± 1	19·7 ± 1·0	24 ± 2	15 ± 2	11·3 ± 1·1	22 ± 2	8·02 ± 1·29
2·59 ± 0·06	3·5 ± 0·1	3·44 ± 0·06	3·6 ± 0·1	3·2 ± 0·2	3·03 ± 0·07	3·6 ± 0·1	1·15 ± 0·05
17·8 ± 0·1	25·0 ± 0·3	24·9 ± 0·2	26·3 ± 0·2	20·1 ± 0·1	21·0 ± 0·8	25·1 ± 0·2	6·55 ± 1·41
3·0–3·6	4·2–5·2	4·2–4·7	4·4–5·4	4·2–4·7	4·3–4·6	4·5–5·5	1·34 ± 0·12
9 ± 1	14 ± 4	14 ± 3	16 ± 5	12 ± 2	10·4 ± 1·6	19 ± 6	3·51 ± 0·88
9·8 ± 0·2	12·5 ± 0·3	12·4 ± 0·2	13·2 ± 0·3	11·3 ± 0·3	11·35 ± 0·14	14·8 ± 0·3	3·48 ± 0·12
1·41 ± 0·01	2·10 ± 0·02	2·00 ± 0·05	2·00 ± 0·02	1·66 ± 0·01	1·69 ± 0·01	2·29 ± 0·02	0·526 ± 0·15

Table 2. Concentrations of some elements in lunar rocks and basalt BCR-1

Element	Type A 10022-38	Type B 10045-25	Type C 10048-47	Type D 10084-52	BCR–1
Ag ppb	2·3 ± 0·5	4·9 ± 1·8	16·0 ± 5·6	27·1 ± 6·2	18 ± 5
Cl ppm	19·3 ± 0·8	6·8 ± 0·2	65·4 ± 1·3	24·1 ± 1·5	30·5 ± 1·0
Br ppb	129 ± 2	56 ± 1	125 ± 3	229 ± 6	187 ± 8
Mn ppm	1720 ± 50	2155 ± 43	1700 ± 60	1530 ± 19	1343 ± 40
Ta ppm	1·0 ± 0·4	1·8 ± 0·4	—	1·4 ± 0·3	0·90 ± 0·09
Au ppb	0·10 ± 0·05	<0·2	2·5 ± 0·7	2·8 ± 0·8	0·9 ± 0·9
Ba ppm	228 ± 21	95 ± 5	167 ± 27	168 ± 111	656 ± 4
In ppb	7·8 ± 1·1	14·0 ± 0·8	180 ± 6	860 ± 25	113 ± 2
Cu ppm	5·1 ± 0·6	6·2 ± 0·3	11·1 ± 0·6	10·1 ± 1·6	14·6 ± 0·9
As ppb	63 ± 14	73 ± 7	—	37 ± 9	840 ± 60
Sb ppb	5·6 ± 2·5	7·2 ± 1·7	8·8 ± 4·4	5·1 ± 0·8	620 ± 300
Se ppm	0·7 ± 0·5	0·8 ± 0·4	1·6 ± 0·8	0·8 ± 0·6	0·105 ± 0·005
Sc ppm	76 ± 4	78 ± 7	65·8 ± 1·3	61·7 ± 1·9	31·9 ± 0·6
Hf ppm	21·5 ± 2·6	7·7 ± 1·0	11·7 ± 1·5	7·6 ± 0·5	5·23 ± 0·24
Cr ppm	2254 ± 35	2320 ± 40	2160 ± 200	1824 ± 55	19 ± 5
Ga ppm	2·9 ± 0·8	4·0 ± 0·3	5·9 ± 0·4	5·1 ± 0·3	25·3 ± 2·0
Co ppm	29·0 ± 0·4	16·1 ± 2·0	35·0 ± 4·2	26·8 ± 0·7	35·8 ± 0·7
Zn ppm	2·9 ± 1·3	2·9 ± 2·9	30·2 ± 2·2	22·8 ± 0·8	100 ± 30
Ni ppm	<10	<10	214 ± 4	200 ± 30	12·4 ± 3·2
Fe %	14·5 ± 0·3	14·9 ± 0·4		12·2 ± 0·3	9·3 ± 0·4
K ppm	2200 ± 10	401 ± 28	1410 ± 60	1060 ± 20	9160 ± 170
Rb ppm	5·7 ± 0·6	0·80 ± 0·15	4·16 ± 0·31	3·2 ± 0·6	44·9 ± 2·1
Cs ppm	0·20 ± 0·02	—	0·124 ± 0·007	0·11 ± 0·03	0·91 ± 0·05
Sr ppm	164 ± 23	144 ± 10	190 ± 10	169 ± 13	350 ± 20

other elements in four lunar samples and BCR-1 are given in Table 2. The uncertainties accompanying the values for the REE in BCR-1 are standard deviations based on results of three independent analyses by three separate analysts in our laboratories. All other uncertainties given in Tables 1 and 2 are standard deviations based on counting statistics. The precision is limited principally by the counting statistics. For the REE, there is an additional systematic uncertainty of ±1–2 per cent for most of the samples which is not included in Table 1.

The good agreement between pairs of duplicate analyses for REE in lunar rocks in Table 1 confirms the precision indicated by the triplicate analysis of BCR-1. The absolute accuracy for REE and other elements is believed to equal the precision. Values for BCR-1 and the lunar soil agree well with those obtained by many other analysts, but the spread among literature values for a given element is considerable.

Discussion

Comparisons of element concentrations

We compare below the concentrations of individual elements in the different types of lunar rocks, in terrestrial basalts, and in the Bruderheim chondritic meteorite. We have recently analyzed both whole rock chips and separated phases from that meteorite (Allen, 1970). In making the comparison, we use results of other analysts as well as our own for the rocks from Tranquillity Base. To save space, those results are not specifically referenced as they appear; they are taken from the following papers: Annell and Helz (1970), Baedecker and Wasson (1970), Compston *et al.*

(1970), Ehmann and Morgan (1970), Gast and Hubbard (1970), Gopalan *et al.* (1970), Keays *et al.* (1970), Maxwell *et al.* (1970), Morrison *et al.* (1970), Murthy *et al.* (1970), Philpotts and Schnetzler (1970a), Reed *et al.* (1970), Schmitt *et al.* (1970), Smales *et al.* (1970), Turekian and Kharkar (1970), Wänke *et al.* (1970), Wanless *et al.* (1970), Wiik and Ojanperä (1970).

The following elements are significantly (5–1000 times) less concentrated in the lunar igneous rocks than in the Bruderheim chondrite: Au, Co, Ni, Cu, As, Sb, Zn and Se. Elements that are more concentrated (5–100 times) in the lunar igneous rocks than in the Bruderheim chondrite include Ag, In, Ta, Ba, Sc, Hf, REE, Sr and in type A rocks, K, Rb and Cs. Elements having approximately the same concentrations in the lunar basalts as in the Bruderheim chondrite are Cr, Mn, Ga, Cl, Br and in type B rocks, K, Rb and Cs. Except for Se and Zn, those elements less abundant in the lunar rocks than in the Bruderheim meteorite were found in the meteorite to be predominantly in the metal phase. The principal phase containing Se in the meteorite was not discovered; the silicate phases contained most of the Zn. Those elements that are more abundant in the lunar rocks than in the Bruderheim chondrite were found in the chondrite principally in the troilite (Ag, In) and in the silicate phases. Those elements with comparable concentrations in the meteorite and the lunar igneous rocks were found in the meteorite to be predominantly in the silicate phases, except for Mn and Cr, whose major phase (presumably chromite) was not analyzed. The parent silicate material from which the lunar basalts are derived thus appears to have had extracted from it sometime during its evolution those elements which readily enter a metallic iron phase. Those lithophilic elements (Cl, Br, Zn, K, Rb and Cs) that are not as strongly enriched in the lunar basalts as the other lithophilic elements are relatively volatile and may have distilled away at some time prior to the formation of the lunar basalts.

Ag and In, the elements which are enriched in the lunar basalts and which are the only elements studied that reside predominantly in the sulfide phase of the Bruderheim meteorite are, unfortunately, the components of the gasket used in the pressure seal of the boxes in which the rocks were transported from the moon. It is not clear, therefore, whether their enrichment in the lunar basalts is geochemically significant or due to contamination.

The concentrations of most of the elements in the lunar basalts are similar to those in common terrestrial basalts (Taylor, 1964). These similarities suggest that basalts from both bodies formed by similar processes from similar parent materials. Concentrations of Ag, Au, Cu, Sr, Ba, In, Ga, Ni and the lighter REE in the lunar basalts are near the lowest values found in terrestrial basalts. The heavier REE, Sc, Cr and perhaps Se are more concentrated in the lunar basalts than in common terrestrial basalts. Concentrations of Ni, Sb, As and the alkali metals are much less in the lunar than in the terrestrial basalts and may reflect differences in their parent silicate materials.

The ranges of concentration for most of these elements in type A and type B rocks overlap. Exceptions are K (about 650 to 950 ppm in type B compared to about 2100–2770 ppm in type A), Rb (about 0·5–1·2 ppm in type B, 5·5–6·2 ppm in type A), and Ba (about 65–120 ppm in type B, 250–350 ppm in type A). Other elements whose ranges for the two rock types barely overlap, if at all, are Cl, Br, Hf, Co, Cr, and Cs (more concentrated in type A) and Ag, Mn and Cu (more concentrated in type B).

Several elements are more abundant in types C and D material than in types A and B; these include Ag, Cl, Br, Au, In, Cu and Zn. These may indicate a higher contribution of meteoritic matter to the soils and breccias than is found in the igneous rocks (KEAYS *et al.*, 1970).

Most of the In and Ag may be a result of contamination from the rock boxes but, if so, it is curious that all type C rocks were more severely contaminated than all type A and B rocks examined. The concentrations of Mn and perhaps As and Hf appear to be lower in type C rocks than in type A or B rocks. It has been shown that, on the basis of their different concentrations of the major elements Ti and Al, the type C and D materials do not correspond to simple mixtures of type A and B rocks (SCHMITT *et al.*, 1970).

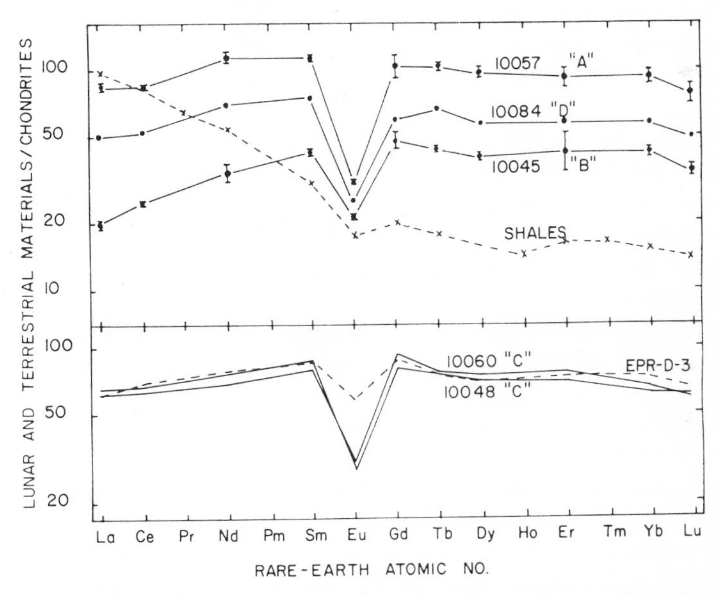

Fig. 1. The REE concentrations in lunar samples (solid lines), terrestrial shales (upper dashed line), and an andesite from the east Pacific rise are compared with those in a composite of nine chondritic meteorites. The REE concentrations in ppm for the chondrites have been divided, element by element, into those for the other samples, and the resulting ratios have been plotted on a logarithmic scale against REE atomic number.

Rare-earth elements

The abundances of the REE in five of the lunar samples we analyzed are compared with those for a composite sample of chondritic meteorites in Fig. 1 (HASKIN and GEHL, 1962; CORYELL *et al.*, 1963; HASKIN *et al.*, 1968).

Samples 10045 and 10057 represent the extremes of REE concentrations encountered. The breccias (10048 and 10060) have REE abundances similar to those in the type D 'fines'. The concentrations in the other igneous rocks vary continuously between those of rocks 10045 and 10057. Type A rocks tend to have higher REE concentrations than type B rocks, according to all the analytical data produced so far (GAST and HUBBARD, 1970; GOLEŠ *et al.*, 1970; HASKIN *et al.*, 1970; MORRISON *et al.*, 1970; PHILPOTTS and SCHNETZLER, 1970a; SCHMITT *et al.*, 1970; TUREKIAN and KHARKAR, 1970; WÄNKE *et al.*, 1970; WIIK and OJANAPERÄ, 1970.)

The *relative* abundances of the REE in the lunar rocks are fairly similar to those which are found in chondritic meteorites and which are presumed to represent the primordial relative abundances of those elements in the solar system. Compared with the chondritic values, the elements Nd through Dy are somewhat enriched in lunar rocks relative to the heaviest and lightest REE. As is frequently observed for terrestrial rocks, the relative abundance of Eu in the lunar rocks is anomalously low (TOWELL *et al.*, 1965; HASKIN *et al.*, 1966; PHILPOTTS and SCHNETZLER, 1968). Most common rocks at the earth's surface, including basalts from the continents and oceanic islands, have relative REE abundances similar to those for the composite of North American shales shown in Fig. 1 (HASKIN *et al.*, 1968). The concentrations of the heavier REE are several times higher in the lunar samples than in common terrestrial materials.

The REE are strongly concentrated in the lunar rocks compared with chondritic meteorites and most other objects that have been suggested as approximately representative of average, non-volatile solar matter from which the terrestrial planets might have formed. The samples of lunar rock that have the highest REE concentrations also have the largest Eu anomalies, but they have ratios of La and Ce to the heavier REE that are the closest to those found in chondrites. The REE contents of the lunar rocks at Tranquillity Base probably are not typical of those for lunar surface rocks, since they come from one of the maria and the maria represent only about one-fifth of the lunar surface. The Sm content of the lunar soil, (a typical REE at about its average concentration at Tranquillity Base) is seventy-five times higher than that of the composite of chondrites. At that level of concentration, the maria would contain, per kilometer of depth, about 15 per cent of all the REE in the moon if the average REE concentrations in the moon were about equal to those in the chondrites. The average depths of the maria most probably exceed 10 km, so the REE contents of the Apollo 11 lunar rocks exceed the average for the maria, or the moon is not similar to the chondrites in REE content, or both.

The Ca-rich achondrites have relative REE abundances similar to those found in chondrites, but their REE concentrations are about seven times higher. (SCHMITT *et al.*, 1963, 1964; SCHNETZLER and PHILPOTTS, 1969; PHILPOTTS *et al.*, 1967). Subalkaline basalts from terrestrial ocean ridges also have relative REE abundances similar to those in chondrites and their REE concentrations are, on the average, about twenty times greater (FREY and HASKIN, 1964; FREY *et al.*, 1968). In addition, they are somewhat depleted in La and Ce as are the lunar basalts. They seldom show any trace of depletion of Eu, however. An andesite from the east Pacific rise (EPR-D-3, FREY *et al.*, 1968) was analyzed along with the lunar breccias 10048 and 10060. The REE distributions for these rocks are compared with those for chondrites in the lower part of Fig. 1. The resemblance is striking; the concentrations of the REE in these samples, except for Eu, are identical.

Besides their similarities in REE contents, the lunar and subalkaline oceanic basalts are both volcanic rocks and are roughly similar in gross composition and texture. Probably the chemical processes which produced both were sufficiently similar to produce their very similar REE distributions. Their REE distributions are almost certainly those of the liquids which chilled to form them.

Fractional crystallization and partial melting

Two processes which produce or act upon silicate melts, which are believed to take part in the formation of basalts, and which would affect REE concentrations are partial melting and fractional crystallization. Mathematical models for the behavior of trace elements during these processes have been developed (see, for example, McINTIRE, 1963; and GAST, 1968). Their use requires a knowledge of the distribution coefficients for the trace elements of interest among the solid and liquid phases involved in the partition. Distribution coefficients for REE for several minerals important in the formation of basalts have been obtained from analyses of phenocrysts and host matrices in basalts (SCHNETZLER and PHILPOTTS, 1968; PHILPOTTS and SCHNETZLER, 1970b; ONUMA *et al.*, 1968) treatment of associated gabbros and ultramafic rock as melt-residue pairs (FREY, 1969) analysis of rocks and minerals from the Skaergaard intrusion (PASTER and HASKIN, to be published), and laboratory studies of synthetic systems (CULLERS *et al.*, 1970). The results from the different techniques do not agree exactly, but the range of variation is small enough that the values of interest can be used with confidence.

Fractional crystallization occurs as a body of molten silicate liquid slowly cools and one or more equilibrium solid (mineral) phases are precipitated from it. Once separated, the crystals do not maintain thermodynamic equilibrium with the liquid as the liquid further cools and its composition changes. In fractional melting, an increment of equilibrium liquid is formed from one or more mineral phases. Once formed, the liquid is removed from the solid phase from which it derived. The liquids from all increments are continuously mixed together to form a single liquid which is not in thermodynamic equilibrium with the solid which gave rise to the last increment that joined the liquid. In either case, a derivative phase is formed (the solid in the case of fractional crystallization; the liquid in the case of partial melting) and a residual parent phase remains until the process has been completed (the liquid in the case of fractional crystallization; the solid in the case of fractional melting).

Whenever the solid phase is a mixture of minerals, the progress of either process can only be accurately calculated by numerical methods which take into account the changing proportions of the minerals with each increment of crystallization or melting (McINTIRE, 1963; GAST, 1968). For the present purpose, it is an adequate approximation to presume that the mineral proportions remain the same and that the distribution coefficient for the trace element in question remains constant throughout the entire process. In such a case, the concentration of a particular REE, e.g. Sm, in the residual phase ($C_{R,Sm}$) relative to the average concentration for the entire system ($C_{A,Sm}$) is given by equation (1) (McINTIRE, 1963) in which x is the fraction of completion of the process (i.e. the fraction of the system that has solidified in the case of fractional crystallization or the fraction of the system that has melted in partial melting).

D_{Sm} is the distribution coefficient for Sm, defined as *the ratio of the equilibrium concentration for* Sm *in the derivative phase to its equilibrium concentration in the*

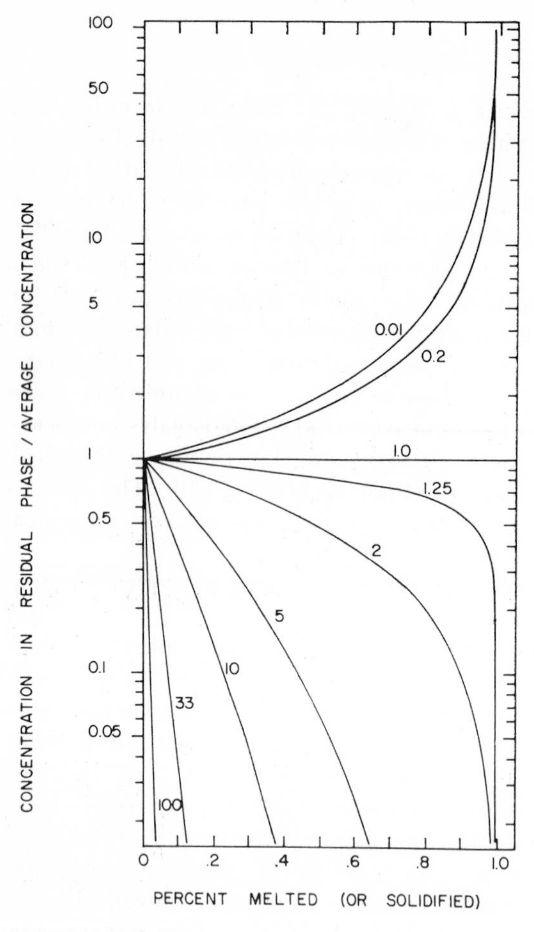

Fig. 2. The ratio of concentration of a trace element in the residual phase to its average concentration for the system has been plotted along the ordinate as a function of the fraction of completion of fractional crystallization or partial melting along the abscissa. The value of the distribution coefficient for each curve is shown beside that curve. See the text and equation (1) for an explanation of the terms.

residual phase (concentration ratio of solid to liquid in the case of fractional crystalli zation, ratio of liquid to solid for partial melting).

$$C_{R,Sm}/C_{A,Sm} = (1 - x)^{D_{Sm}-1}. \qquad (1)$$

Figure 2 shows the changes in concentration for the trace element in the residual phase (i.e. $C_{R,Sm}/C_{A,Sm}$) with fraction of completion of the process (x) for several values of D.

The ratio of the average concentration ($C_{P,Sm}$) of the total amount of the derived phase to the average concentration for the system at any fraction of completion of the process can be obtained from equation (1) and a mass balance for the system.

$$C_{P,Sm}/C_{A,Sm} = [1 - (1 - x)^{D_{Sm}}]/x. \tag{2}$$

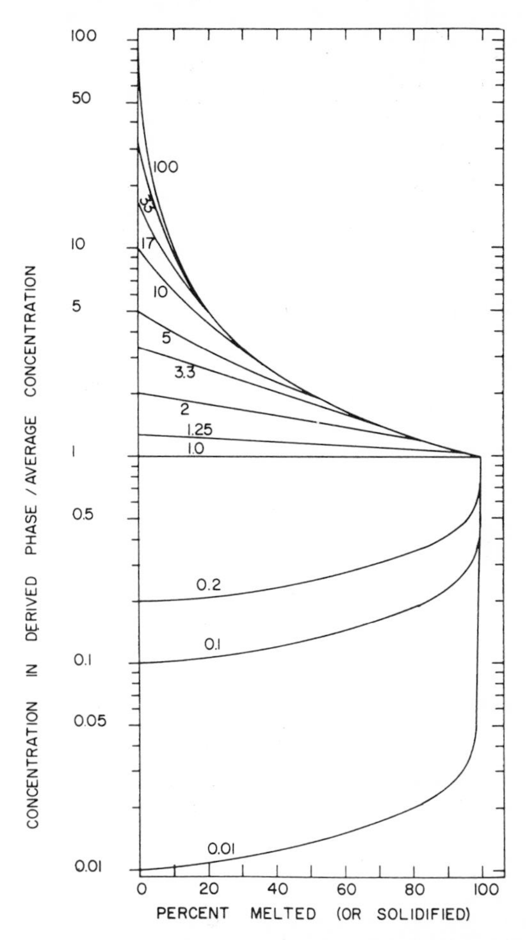

Fig. 3. The ratio of concentration of a trace element in the derivative phase to its average concentration for the system has been plotted along the ordinate as a function of the fraction of completion of the process along the abscissa. See text and equation (2) for explanation.

Figure 3 shows the changes in the concentration of the derived phase ($C_{P,Sm}/C_{A,Sm}$) with the fraction of completion of the process for several values of D.

We now examine the possibility that the *relative* REE abundances found in the lunar rocks can be derived simply by partial melting or by fractional crystallization from a starting material with *relative* REE abundances like those found in the composite of chondrites. If so, could the starting material have had REE *concentrations* similar to those in the chondrites as well? To do this, we consider the changes in the ratio of concentrations for two REE (e.g. Sm and Eu) in the *liquid* phases for the two

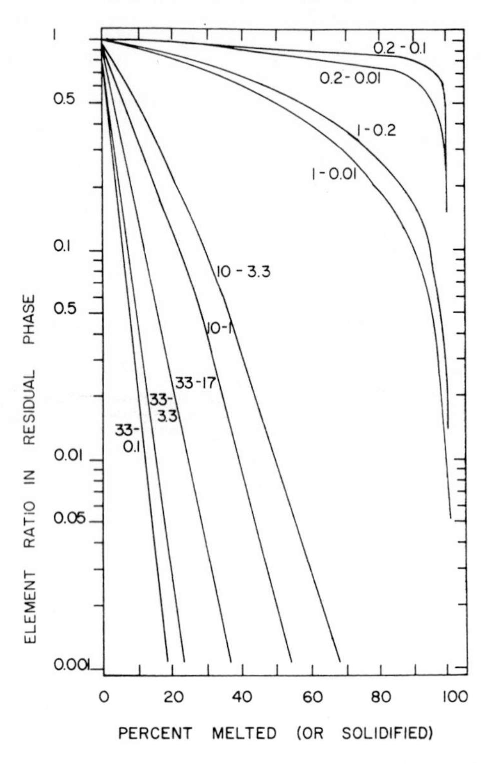

Fig. 4. The ratio of concentrations of one trace element to another in the residual phase, relative to the ratio of their average concentrations for the system, is plotted along the ordinate as a function of the fraction of completion of the process along the abscissa. See text and equation (3) for explanation. The numbers accompanying each curve are the distribution coefficients for the two trace elements used to generate that curve.

processes as compared with the average ratio for the entire system. For a residual phase, the appropriate ratio is obtained from equation (1).

$$\frac{C_{R,Sm}/C_{A,Sm}}{C_{R,Eu}/C_{A,Eu}} = (1 - x)^{D_{Sm} - D_{Eu}}. \tag{3}$$

Note that the rate at which the ratio of concentrations changes in the residual phase depends on the *difference* between the distribution coefficients for the two elements. Figure 4 shows how this ratio changes with the fraction of the system solidified or melted for several values of D.

The ratio of concentrations of Sm and Eu in the derivative phase is given by equation (4), which is derived from equation (2).

$$\frac{C_{\mathrm{P,Sm}}/C_{\mathrm{A,Sm}}}{C_{\mathrm{P,Eu}}/C_{\mathrm{A,Eu}}} = \frac{1 - (1 - x)^{D_{\mathrm{Sm}}}}{1 - (1 - x)^{D_{\mathrm{Eu}}}}. \tag{4}$$

The dependence of this ratio on the values for the partition coefficients is more complex than it was for the residual phase (equation 3). Changes in the ratio with

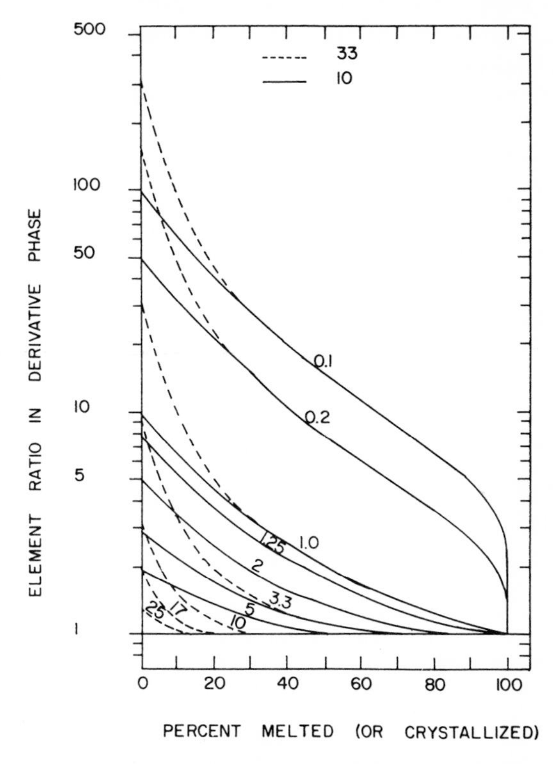

Fig. 5. The ratio of concentrations of one trace element to another in the derivative phase, relative to the ratio of their average concentrations for the system, is plotted along the ordinate as a function of the fraction of completion of the process along the abscissa. See text and equation (4) for explanation. The number accompanying each curve is one of the distribution coefficients used to generate that curve. The other distribution coefficient used was 33 (dashed lines) or 10 (solid lines).

extent of melting or solidification of the system are illustrated in Fig. 5 for various values of D.

Partition coefficients for minerals that might crystallize from or melt to produce a mafic silicate are given in Table 3. These are derived from terrestrial systems and, except for those for Eu, are presumed to apply to lunar silicate systems as well.

The distribution coefficient for Eu is a combination of coefficients for Eu^{2+} and

Eu^{3+}. The only mineral known so far that selectively concentrates Eu^{2+} is feldspar (TOWELL *et al.*, 1965; HASKIN *et al.*, 1966; PHILPOTTS and SCHNETZLER, 1968). This may be caused by some mechanism which depends on the equilibrium distribution of Eu^{3+} between the liquid and the surface of crystallizing feldspar or it may be caused by a mechanism which depends on the equilibrium distribution of Eu^{2+} between the liquid and the crystallizing feldspar. Lunar basalts contain metallic iron, and thus are derived from more reducing liquids than terrestrial basalts. The ratio of free ion activity of Eu^{2+} to that of Eu^{3+} in lunar magmas would then be higher than in the terrestrial magmas from which the values for the distribution coefficients for Eu in feldspar in Table 3 were derived. Those values would be too small for the lunar magmas if the mechanism for selective incorporation of Eu into feldspars involved the distribution of Eu^{2+} between the liquid and the feldspar. They might be too large if the distribution of Eu^{3+} between the two phases is involved. This is not necessarily so. Studies of Gd indicate that only about 0·1 per cent of the Gd^{3+} in silicate liquids is present as the free ion (CULLERS *et al.*, 1970). In the case of Eu, the Eu^{2+} free ion activity could be greatly increased under reducing conditions without an appreciable change in the total amount of Eu in the liquid in the 3+ oxidation state and without an appreciable change in the free ion activity of Eu^{3+}.

The difference in ratios of free ion activities of Eu^{2+} and Eu^{3+} between lunar and terrestrial magmas can be related to any other activity that is dependent on the redox conditions of the silicate liquids, for example, the oxygen fugacities (f oxygen).

$$Eu^{3+} + 1/2O^{=} \rightleftharpoons Eu^{2+} + 1/4O_2. \tag{5}$$

If we assume that the activity of oxide ion is constant in silicate magmas, then the ratio of free ion activity of Eu^{2+} (A_2) to Eu^{3+} (A_3) is given by equation (6) in which

Table 3. Distribution coefficients for REE (concentration in solid/concentration in liquid)

Mineral	La	Sm	Eu	Yb	Eu(L)
Plagioclase*	~0·15	0·026	0·055	—	0·22
Plagioclase*	~0·33	0·166	0·438	0·299	1·33
Plagioclase*	~0·07	0·041	0·194	0·024	0·78
Plagioclase†	0·09	0·023	—	0·008	—
Orthopyroxene*	~0·028	0·047	0·060	0·244	
Orthopyroxene*	~0·030	0·100	0·079	0·670	
Orthopyroxene‡	0·001	0·015	0·024	0·14	
Orthopyroxene§	~0·001	0·014	0·023	0·11	
Pigeonite*	~0·012	0·100	0·068	0·400	
Calcic clinopyroxene*	~0·030	0·090	0·091	0·092	
Calcic clinopyroxene*	~0·070	0·261	0·260	0·227	
Calcic clinopyroxene‡	0·002	0·32	0·41	0·60	
Calcic clinopyroxene§	<0·002	0·43	0·48	0·53	
Olivine*	~0·01	0·011	0·010	~0·023	
Olivine‡	0·012	0·003	0·003	0·013	
Olivine§	~0·01	0·006	0·006	~0·01	

* PHILPOTTS and SCHNETZLER, 1970b.
† PASTER and HASKIN, unpublished.
‡ FREY, 1969.
§ ONUMA *et al.*, 1968.

the constant K is the equilibrium constant multiplied by the square root of the oxide ion activity.

$$A_2/A_3 = K/(f\ oxygen)^{1/4}. \tag{6}$$

The oxygen fugacity in the lunar magma has been estimated at about 10^{-3} times that for the terrestrial basaltic magmas (ANDERSON *et al.*, 1970; SIMPSON and BOWIE, 1970). This would mean that the ratio of free ion activity of Eu^{2+} to Eu^{3+} in the lunar magma was about six times greater than in terrestrial basalt magmas. We do not know the value for that ratio in terrestrial magmas. If the ratios of free ion activity of Eu^{3+} to the total concentration of Eu in terrestrial magmas were appreciably different from the corresponding ratios for Sm and Gd in those magmas, anomalous relative abundances of Eu should be found in minerals other than feldspar which crystallize from the magmas (or in liquids produced by equilibrium partial melting). Most terrestrial systems studied have been so disturbed by the presence of feldspar that independent Eu anomalies for other minerals have not been observed. We therefore *assume* that in terrestrial basaltic magmas the ratio of free ion activity of Eu^{3+} to total Eu is comparable to those for Sm and Gd. On the basis of this assumption, the distribution coefficient for feldspars derived from terrestrial materials can be split into a component due to Eu^{3+}, approximately equal to the average of the distribution coefficients for the adjacent elements Sm and Gd, and a component due to Eu^{2+}, equal to the difference between the total distribution coefficient for Eu and the estimated component due to Eu^{3+}. From the work of PHILPOTTS and SCHNETZLER (1970b), who found that the ability of plagioclase to take up Eu^{2+} appears to vary strongly with the composition of the plagioclase, we estimate that the Eu^{3+} component for plagioclases with anorthite fractions exceeding 0·8 contributes about 40 per cent to the total distribution coefficient and the Eu^{2+} component about 60 per cent. If we increase the component for Eu^{2+} by a factor of six to account for the lower oxygen fugacity in the lunar magma, the partition coefficients for plagioclase designated as Eu(L) in Table 3 result.

Possibly, the component for Eu^{3+} should have been reduced. An ilmenite separate from the lunar rocks has a strong negative Eu anomaly compared with the whole rock (GOLES *et al.*, 1970; PHILPOTTS and SCHNETZLER, 1970a). This might be interpreted to mean that the concentration of Eu^{3+} was much lower than those of Sm^{3+} and Gd^{3+} when the ilmenite crystallized because most of the Eu in the parent liquid was present as Eu^{2+}. However, ilmenite is not known to accept the REE as readily as the other major phases in the rocks (pyroxene and plagioclase). The ilmenite separates were impure and may have been contaminated by traces of a mineral with a high affinity for the REE that crystallized when solidification of the rock was nearly complete. Eu would be strongly depleted in such a mineral because it would have been selectively removed from the liquid by crystallization of feldspar prior to crystallization of the mineral.

Genesis of lunar magmas

Can the ratio of concentrations of Sm to Eu in the lunar rocks be produced from the ratio for chondritic meteorites by fractional crystallization or partial melting that involves an equilibrium with calcic plagioclase? The values for D_{Sm} and the uncorrected values for D_{Eu} in plagioclase in Table 3 are even smaller than those corresponding to the top lines in Fig. 4. This indicates that the change in ratio of concentration of Sm to Eu from the chondritic to the lunar values could only result after a very high percentage of the initial liquid had crystallized; from equation (3) this is calculated to exceed 92 per cent crystallization as plagioclase for the most favorable case (rock 10045 and the second set of partition coefficients) and to exceed 99 per cent for rock 10057 (same partition coefficients). When the 'corrected' values for D_{Eu} are used, 34–97 per cent of the original liquid needs to crystallize to produce rock 10045, and 70 to nearly 100 per cent to produce rock 10057, depending on which set of distribution coefficients is used. Still larger distribution coefficients would generate the desired concentration ratios in a smaller fraction of solidification, but the actual concentrations in the residual liquid would not be as rapidly increased, as seen from Fig. 4.

In partial melting, the liquid is the derivative phase. The appropriate values for D_{Sm} and D_{Eu} are the reciprocals of those in Table 3. The values for D_{Sm} correspond approximately to the curves marked 33 in Fig. 5. Equation (4) cannot readily be solved for the fractions of partial melting required to give the required ratios of concentration of Sm to Eu to rocks 10045 and 10057, so we have estimated them from Fig. 5. With the terrestrial values for D_{Eu} in plagioclase, the required ratio for rock 10045 can be obtained when about 10–30 per cent of the solid has partially melted, and the ratio for rock 10057 at 0 to about 5 per cent partially melted. With the corrected values for D_{Eu}, 15–45 per cent partial melting is required to produce the ratio of Sm to Eu in rock 10045, 5–20 per cent to produce that in rock 10057. This presumes that plagioclase is the only solid phase present.

REE distribution coefficients for plagioclase are such that crystallization of that mineral from a liquid should increase the ratio of the concentration of Sm to La in the liquid. The extent of crystallization required to produce the ratio of concentration of Sm to La in rock 10045 is 80 to nearly 100 per cent; that for rock 10057 is 83 to nearly 100 per cent. These values fall in the same range required to produce the ratios of Sm to Eu in those rocks.

From Fig. 5 we estimate that 15–30 per cent of partial melting of plagioclase would produce the ratio of Sm to La in rock 10045 and 3–15 per cent would produce the ratio in rock 10057. Again, these values overlap with those required to produce the ratios of Sm to Eu.

Crystallization or melting of plagioclase operates in the wrong direction to produce the ratios of Sm to Yb found in the lunar rock. For this, a mineral is required that preferentially incorporates the heavier REE. The most suitable minerals are pyroxenes. Based on the distribution coefficients in Table 3, if the only minerals crystallizing are orthopyroxene or pigeonite about 40–80 per cent of the liquid would have to crystallize to produce the ratios of Sm to Yb found in rock 10045 and 10057, or 60 to nearly 100 per cent would have to crystallize as calcic clinopyroxene to produce the ratios. From 35 to 50 per cent of a solid composed of only orthopyroxene or pigeonite would have to undergo partial melting before the ratio would be produced, and the ratio could not be produced at all by partial melting of calcic clinopyroxene.

Consider a solid consisting of about 80 per cent calcic plagioclase and 20 per cent orthopyroxene. The following set of distribution coefficients, based on those in Table, 3, could be applicable: $D_{Sm} = 0{\cdot}03$, $D_{La} = 0{\cdot}12$, $D_{Eu} = 0{\cdot}25$ and $D_{Yb} = 0{\cdot}076$. Partial melting, in the same mineral proportions, of up to 9 per cent of the solid would produce a liquid with the required ratios of Sm to Eu and Sm to Yb for rock 10057 and up to 15 per cent would produce those ratios in rock 10045. The ratio of Sm to La in rock 10045 is greater than that in rock 10057, just opposite to the trend for Sm to Eu or Sm to Yb. To produce the ratios of Sm to La in rocks 10045 and 10057, about 7 and 17 per cent partial melting is required.

Crystallization of the above solid from a silicate liquid would have to proceed to the extent of 99·9 per cent to produce the ratios of Sm to Eu and Yb in rock 10057, and to the extent of 96 per cent to produce the Sm to Eu ratio in rock 10045. To produce the ratios of Sm to La in rocks 10057 and 10045 would require crystallization of 95 and 99·98 per cent of the original liquid.

The opposite trend for ratios of Sm to La compared with those of Sm to Eu and Yb indicates that the processes that formed the lunar magmas from a starting material with chondritic relative RE abundances was more complex than the simple models used here account for. Nevertheless, it is significant that the ratios described above for the lunar rocks can be readily generated by a reasonable extent of partial melting of common minerals with measured distribution coefficients. It is less likely that these same minerals could have produced the ratios in those rocks in a reasonable amount of fractional crystallization.

The results restrict severely the kinds of minerals that could have taken part in either of the processes. Addition to the solid of several per cent of a mineral such as calcic clinopyroxene whose distribution coefficient for Sm approaches that of plagioclase for Eu narrows the difference between the values of D_{Sm} and D_{Eu} for the solid. This requires a higher percentage of fractional crystallization or a lower percentage of partial melting to produce the required ratio of Sm to Eu in the liquid phase. Even a fraction of a per cent of a mineral which concentrates the REE, for example, apatite, has this same effect (GAST, 1968).

Selective uptake by feldspar is the only postulated mechanism for producing Eu anomalies in REE distributions for which there is any evidence at the present time. Thus, the generation of a large Eu anomaly appears to require the presence of a large proportion of feldspar in the system. This mineral does not appear to be an equilibrium phase under the conditions of temperature, rock density, and distance beneath the lunar surface that have been suggested for the source regions for the lunar basalts (RINGWOOD and ESSENE, 1970; O'HARA *et al.*, 1970). The lunar highlands, however, are postulated to be composed of anorthosite; this implies crystallization of vast amounts of feldspar (WOOD *et al.*, 1970).

The distribution coefficient is presumed in this treatment to be independent of the bulk composition of the liquid and depends only on the composition of the solid. The similarity of values for distribution coefficients for the REE from different types of rocks supports the validity of this assumption.

The concentration of Sm in rock 10045 is 115 times greater than the average for chondritic meteorites. As seen in Fig. 3, the first drop of liquid from partial melting of a solid with $D_{Sm} = 0{\cdot}01$ (of the order of that for Gd) would have a ratio of Sm to that in the starting material ($C_{A,Sm}$) of only 100. Similarly, from Fig. 2, it can be seen that fractional crystallization would have to proceed well beyond 99 per cent for the concentration in the liquid to be raised to 100 times above its initial concentration. Decreasing the values of the partition coefficients beyond 0·01 does not help, because even with a value of 0·01, the liquid, by the time it comprises one per cent of the system, already contains 99 per cent of the Sm. Thus, the rocks from Tranquillity Base probably could not have formed from material with chondritic REE concentrations by any single process of fractional crystallization or partial melting.

The ratios of Sm to Eu and Sm to Yb in the lunar igneous rocks vary smoothly with their concentrations of Sm. This is indicated in Fig. 6, in which it is seen that the rocks of both types A and B appear to belong to a continuous series. The positions of the points in such a graph are changed considerably by possible systematic errors in the values reported for the REE by different investigators. We have arbitrarily 'corrected' the values for Sm and Eu from other analysts by the factor needed to bring their values for those elements in the type D 'fines' into line with ours. This reduces the scatter somewhat. In effect, Fig. 6 is a graph that shows how the ratio of Sm to Eu changes with the extent to which fractional crystallization or partial melting has occurred, with the concentration of Sm serving as the index of the fraction of completion of the process. In fact, equation (1) can be solved for x and the result substituted for x in equation (3) to show how the ratio of Sm to Eu in the residual phase should change with concentration of Sm.

$$\ln \frac{C_{R,Sm}}{C_{R,Eu}} = \frac{D_{Sm} - D_{Eu}}{D_{Sm} - 1} \ln C_{R,Sm} + \ln \left(\frac{C_{A,Sm}}{C_{A,Eu}}\right)^{(D_{Sm}-1)/(D_{Eu}-1)} \tag{7}$$

[Equation (7) can also be obtained by writing equation (1) for Sm, then for Eu, and eliminating x between them.]

A graph of ln (ppmSm/ppmEu) plotted against ln ppm Sm for the lunar samples (not shown) is linear and has a slope of about 0·70. This requires that D_{Eu} exceed 0·70 for a value of zero for D_{Sm}. That value for D_{Eu} is within the range of the corrected values in Table 3.

No simple equation relating the ratio of concentrations of Sm to Eu to the concentrations of Sm in the derivative phase could be obtained. Values obtained for Sm and Eu from use of equation (2), when plotted in the manner of Fig. 6, give straight lines. The slope of the curve in Fig. 6 is about the same as that from data calculated from equation (2) and based on $D_{Sm} = 0{\cdot}03$ and $D_{Eu} = 0{\cdot}3$. These values are also reasonable in comparison with those of Table 3.

In view of the narrow range of scatter of the points from the least-squares line in Fig. 6 and the corresponding graph of ln (ppm Sm/ppm Eu) versus ln (ppm Sm), we conclude that, to a first approximation, the lunar rocks could have been formed by fractional crystallization (or partial melting) of a common source with relative REE abundances nearly like those of chondrites but REE concentrations at least several times greater than chondrites.

Qualitatively, at least, the concentrations of Sr, Ba, K and Rb support this fractionation trend. Like Eu, Sr is strongly depleted relative to Ba and the other REE, and feldspar is the only major mineral known to concentrate that element strongly. Ba, K and Rb are most concentrated in the same rocks in which the REE are the most concentrated (PHILPOTTS and SCHNETZLER, 1970a; GAST and HUBBARD, 1970). Those elements would be expected to concentrate strongly in the first liquid from partial melting or in the final liquid from fractional crystallization (GAST, 1968). The concentrations of K, Rb and Ba do not, however, overlap between the type A and B rocks or appear to form a continuous series between them.

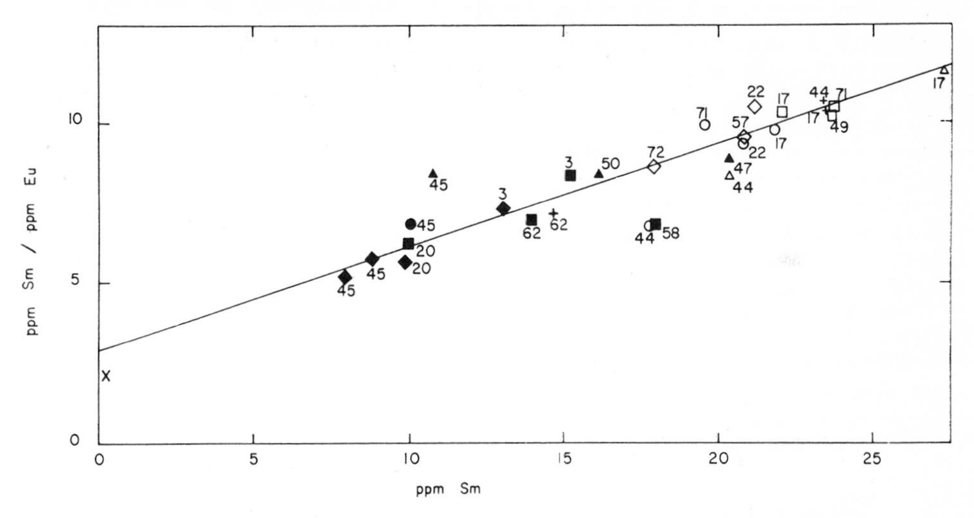

Fig. 6. The ratios of the concentrations of Sm to those of Eu in the lunar rocks have been plotted along the ordinate against the concentrations of Sm along the abscissa. Solid symbols indicate type B lunar rocks, open symbols type A. The various symbols designate the values, 'corrected' as indicated in the text, of GAST and HUBBARD (1970) (squares); GOLES *et al.* (1970) (circles); PHILPOTTS and SCHNETZLER (1970a) (crosses); SCHMITT *et al.* (1970) (triangles); and this work (diamonds). The line is a least squares fit to all the data except the following : 45 triangle, 44 circle, 58 square.

Acknowledgments—We thank the crew of the University of Wisconsin Nuclear Reactor for irradiating the samples with neutrons. We thank the Apollo 11 astronauts and the personnel of NASA whose efforts provided these samples for analysis. We appreciate the assistance of J. BLACKBOURN in part of the radioassay. This work was supported in part by the National Aeronautics and Space Administration under contract NAS 9-7975.

REFERENCES

ALLEN R. O. (1970) Multi-element neutron activation analysis: Development and application to a trace element study of the Bruderheim chondrite. Thesis. To be published.

ANDERSON A. T., CREWE A. V., GOLDSMITH J. R., MOORE P. B., NEWTON J. C., OLSEN E. J., SMITH J. V. and WYLLIE P. J. (1970) Petrologic history of moon suggested by petrography, mineralogy and crystallography. *Science* **167**, 587–590.

ANNELL C. and HELZ A. (1970) Emission spectrographic determination of trace elements in lunar samples. *Science* **167**, 521–532.

BAEDECKER P. and WASSON J. (1970) Gallium, germanium, indium, and iridium in lunar samples. *Science* **167**, 503–505.

COMPSTON W., ARRIENS P., VERNON M. and CHAPPELL B. (1970) Rubidium–strontium chronology and chemistry of lunar material. *Science* **167**, 474–476.

COYRELL C. D., CHASE J. W. and WINCHESTER J. W. (1963) A procedure for geochemical interpretation of terrestrial rare-earth abundance patterns. *J. Geophys. Res.* **68**, 559–566.

CULLERS R., MEDARIS G., and HASKIN L. (1970) Distribution of gadolinium among aqueous and silicate phases. Submitted to *Science*.

EHMANN W. D. and MORGAN J. W. (1970) Oxygen, silicon and aluminum in lunar samples by 14 MeV neutron activation. *Science* **167**, 528–530.

FLANAGAN F. J. (1969) U.S. Geological Survey standards—II. First complication of data for the new U.S.G.S. rocks. *Geochim. Cosmochim. Acta* **33**, 81–120.

FREY F. A. (1969) Rare earth abundances in a high-temperature peridotite intrusion. *Geochim. Cosmochim. Acta* **33**, 1429–1447.

FREY F. A. and HASKIN L. A. (1964) Rare earths in oceanic basalts. *J. Geophys. Res.* **69**, 775–780.

FREY F. A., HASKIN M. A., POETZ J. and HASKIN L. A. (1968) Rare-earth abundances in some basic rocks. *J. Geophys. Res.* **73**, 6085–6098.

GAST P. W. (1968) Trace element fractionation and the origin of tholeiitic and alkaline magma types. *Geochim. Cosmochim. Acta* **32**, 1057–1086.

GAST P. W. and HUBBARD N. J. (1970) Abundance of alkali metals, alkaline and rare earths and strontium-87/strontium-86 ratios in lunar samples. *Science* **167**, 485–487.

GOLES G. G., OSAWA M., RANDLE K., BEYER R., JEROME D., LINDSTROM D., MARTIN M., MCKAY S. and STEINBORN T. (1970) Instrumental neutron activation analysis of lunar specimens. *Science* **167**, 497–499.

GOPLAN K., KAUSHAL S., LEE-HU C. and WETHERILL G. (1970) Rubidium–strontium, uranium, and thorium–lead dating of lunar material. *Science* **167**, 471–473.

HASKIN L. A. and GEHL M. A. (1962) The rare-earth distribution in sediments. *J. Geophys. Res.* **67**, 2537–2541.

HASKIN L. A., FREY F. A., SCHMITT R. A. and SMITH R. H. (1966) Meteoritic, solar, and terrestrial rare-earth distributions. *Phys. Chem. Earth* **7**, 167–321.

HASKIN L. A., HASKIN M. A., FREY F. A. and WILDEMAN T. R. (1968) Relative and absolute terrestrial abundances of the rare earths. In *Origin and Distribution of the Elements* (editor L. H. Ahrens), pp. 889–912. Pergamon.

HASKIN L. A., HELMKE P. A. and ALLEN R. O. (1970) Rare-earth elements in returned lunar samples. *Science* **167**, 487–490.

HURLEY P. M. and PINSON W. (1970) Rubidium–strontium relations in Tranquillity Base samples. *Science* **167**, 473–474.

KEAYS R., GANAPATHY R., LAUL J., ANDERS E., HERZOG G. and JEFFERY P. (1970) Trace elements and radioactivity in lunar rocks: Implications for meteorite infall, solar-wind flux, and formation conditions of moon. *Science* **167**, 490–493.

LSPET (LUNAR SAMPLE PRELIMARY EXAMINATION TEAM) (1969) Preliminary examination of lunar samples from Apollo 11. *Science* **165**, 1211–1227.

MAXWELL J. A., ABBEY S. and CHAMP W. (1970) Chemical composition of lunar material. *Science* **167**, 530–531.

MCINTIRE W. L. (1963) Trace element partition coefficients—a review of theory and applications to geology. *Geochim. Cosmochim. Acta* **27**, 1209–1264.

MORRISON G. H., GERARD J., KASHUBA A. T., GANGADHARAM E., ROTHENBERG A., POTTER N. and MILLER G. (1970) Multielement analysis of lunar soil and rocks. *Science* **167**, 505–507.

MURTHY V. R., SCHMITT R. and REY P. (1970) Rubidium–strontium age and elemental and isotopic abundances of some trace elements in lunar samples. *Science* **167**, 476–479.

O'HARA M. J., BIGGAR G. and RICHARDSON S. (1970) Experimental petrology of lunar material: the nature of mascons, seas and the lunar interior. *Science* **167**, 605–607.

ONUMA N., HIGHUCHI H., WAKITA H. and NAGASAWA H. (1968) Trace element partition between pyroxenes and the host lava. *Earth Planet. Sci. Lett.* **5**, 47–51.

PHILPOTTS J. A., SCHNETZLER C. C. and THOMAS H. H. (1967) Rare-earth and barium abundances in the Bununu howardite. *Earth Planet. Sci. Lett.* **2,** 19–22.

PHILPOTTS J. A. and SCHNETZLER C. C. (1968) Europium anomalies and the genesis of basalt. *Chem. Geol.* **3,** 5–13.

PHILPOTTS J. A. and SCHNETZLER C. C. (1970a) Potassium, rubidium, strontium, barium, and rare-earth concentrations in lunar rocks and separated phases. *Science* **167,** 493–495.

PHILPOTTS J. A. and SCHNETZLER C. C. (1970b) Phenocryst-matrix partition coefficients for K, Rb, Sr and Ba, with applications to anorthosite and basalt genesis. *Geochim. Cosmochim. Acta* **34,** 307–322.

REED G. W., JOVANOVIC S. and FUCHS L. (1970) Trace elements and accessory minerals in lunar samples. *Science* **167,** 501–503.

RINGWOOD A. E. and ESSENE E. (1970) Petrogenesis of lunar basalts and the internal constitution and origin of the moon. *Science* **167,** 607–610.

SCHMITT R. A., SMITH R. H., LASCH J. E., MOSEN A. W., OLEHY D. A. and VASILEVSKIS J. (1963) Abundances of the fourteen rare-earth elements, scandium and yttrium in meteoritic and terrestrial matter. *Geochim. Cosmochim. Acta* **27,** 577–622.

SCHMITT R. A., SMITH R. H. and OLEHY D. A. (1964) Rare-earth, yttrium and scandium abundances in meteoritic and terrestrial matter—II. *Geochim. Cosmochim. Acta* **28,** 67–86.

SCHMITT R. A., WAKITA H. and REY P. (1970) Abundances of 30 elements in lunar rocks, soil, and core samples. *Science* **167,** 512–515.

SCHNETZLER C. C. and PHILPOTTS J. A. (1969) Genesis of calcium-rich achondrites in light of rare-earth and barium concentrations. In *Meteorite Research* (editor P. M. Millman) pp. 206–216. D. Reidel.

SIMPSON P. and BOWIE S. (1970) Quantitative optical and electron-probe studies of the opaque phases. *Science* **167,** 619–621.

SMALES A. A., MAPPER D., WEBB M. S. W., WEBSTER R. K. and WILSON J. D. (1970) Elemental composition of lunar surface material. *Science* **167,** 509–512.

TAYLOR S. R. (1964) Abundance of chemical elements in the continental crust: a new table. *Geochim. Cosmochim. Acta* **28,** 1273–1286.

TOWELL D. G., WOLFOVSKY R. and WINCHESTER J. W. (1965) Rare-earth abundances in the standard granite G-1 and standard diabase W-1. *Geochim. Cosmochim. Acta* **29,** 563–569.

TUREKIAN K. K. and KHARKAR D. P. (1970) Neutron activation analysis of milligram quantities of lunar rocks and soils. *Science* **167,** 507–509.

WÄNKE H., BEGEMANN F., VILCSEK E., RIEDER R., TESCHKE F., BORN W., QUIJANO-RICO M., VOSHAGE H. and WLOTZKA F. (1970) Major and trace elements and cosmic-ray produced radioisotopes in lunar samples. *Science* **167,** 523–525.

WANLESS R., LOVERIDGE W. and STEVENS R. (1970) Age determinations and isotopic abundance measurements on lunar samples. *Science* **167,** 479–480.

WIIK H. B. and OJANPERÄ P. (1970) Chemical analyses of lunar samples 10017, 10072 and 10084. *Science* **167,** 531–532.

WOOD J. A., DICKEY J., MARVIN U. and POWELL B. (1970) Lunar anorthosites. *Science* **167,** 602–604.

Proceedings of the Apollo 11 Lunar Science Conference, Vol. 2, pp. 1233 to 1238.

Determination of manganese-53 by neutron activation and other miscellaneous studies on lunar dust*

W. HERR, U. HERPERS, B. HESS and B. SKERRA
Institut für Kernchemie der Universität Köln, Köln, Germany
and
R. WOELFLE
Institut für Radiochemie der Kernforschungsanlage Jülich, Germany

(*Received* 23 *February* 1970; *accepted* 28 *February* 1970)

Abstract—A highly sensitive determination of spallogenic ^{53}Mn (T = 2 $\times$ 10^{6} yr) was accomplished in 0·99 g of lunar soil. The chemical yield of Mn was determined with "carrier-free" ^{52}Mn tracer. During a 23-day reactor irradiation the ^{53}Mn was transformed into ^{54}Mn(T = 300 days). Appropriate chemical recycling was done by ion exchange and distillation. Interferences of the (n,p) and the (n,2n) nuclear reactions were carefully studied. A ^{53}Mn distintegration rate of 30·3 $\pm$ 5·5 dpm/kg was obtained. This extremely economic method is proposed for further detailed lunar profile measurements. The Re content, which is of possible cosmochemical interest, was determined to be 11 ppb. Appropriate separation techniques were used. The rather weak and complex thermoluminescence properties made a more basic study advisable. Thermogravimetric analysis, mass spectroscopy, and Moessbauer spectroscopy were applied. The presence of ilmenite, metallic Fe etc., and of an unidentified Fe^{2+} containing compound was deduced. Natural thermoluminescence could not be proved with certainty in our surface sample. However, the complexity of the artificial thermoluminescence demands better defined mineral fractions. The fission track method was used to measure U distribution in glass spherules etc.

MANGANESE-53, favoured by its high production cross section and long half-life of about 2 $\times$ 10^{6} yr, is one of the most interesting nuclides in space research. However, detection of ^{53}Mn radioactivity is rather difficult because of its soft x-rays, and only sophisticated "low-level" techniques succeed. This contrasts with ^{26}Al, which can be detected nondestructively by its characteristic γ-cascade with fairly high sensitivity. (Even in our 5-g sample of lunar dust, we were able to detect the stronger γ-emitters.)

Working with meteorites, we developed a very sensitive and therefore economic method in which we convert ^{53}Mn by neutron capture into γ-emitting ^{54}Mn (HERPERS *et al.*, 1967; HERR *et al.*, 1969; HERPERS *et al.*, 1969), as MILLARD (1965) had proposed some years ago.

In 1957 SHELINE and HOOPER (1957) predicted that, in meteorites, the long-lived ^{53}Mn isotope should be produced in relatively high yield. HONDA *et al.* (1961) thereafter confirmed this experimentally. Although the half-life of ^{53}Mn has not yet been measured with sufficient precision, a half value of (1·9 $\pm$ 0·5) $\times$ 10^{6} yr was suggested by KAYE and CRESSY (1965). They compared the ^{53}Mn activity of an iron meteorite of high terrestrial age with the ^{53}Mn activity of a chondrite with short radiation age. However, the method involves a number of uncertainties, which must be allowed for.

---

* This article was reprinted from *Science* (Vol. **167**, pp. 747–749, 1970) without further refereeing but with minor revision by the authors.

MILLARD (1965) showed that the product of the ^{53}Mn (n,γ) activation cross section, σ, and the ^{53}Mn half-life, T, is $(350 \pm 100) \times 10^6$ barns $\times$ years.

From this follows an activation cross section $\sigma = 170$ barns, on the basis of an estimated ^{53}Mn half-life of $T = 2 \times 10^6$ yr. The latter value was accepted as a first approximation in the present study.

The spallation product ^{53}Mn is converted by an (n,γ)-reaction into ^{54}Mn which can be identified conveniently by its characteristic 0·84 MeV γ-line. Because stable ^{55}Mn is present in the lunar fines with an abundance of about 0·2 per cent, we preferred to work without any Mn carrier.

Lunar fines (0·990 g) ($<100\ \mu$m) were fused with NaOH in a Ni crucible. In order to establish the chemical yield of Mn, a practically "carrier-free" ^{52}Mn ($T \simeq 5$ days) tracer was used. Dowex 1 X 8 ion-exchange resin served for purification. For better handling 50 mg of MgO and 5 mg of ZnO (the latter used as internal standard) were added. The chemical yield was 96·0 per cent, as determined by means of a short reactor irradiation and nondestructive γ,γ-coincidence spectroscopy. Our dust sample (grain size $\leq 100\ \mu$) was found to contain 1690 ± 80 ppm normal ^{55}Mn. The sample was then reactor-irradiated for 23 days, together with 10 mg of MnO_2, 5 mg of ZnO, and 2 Fe-standards in a thermal flux of $8 \cdot 10^{12}$ n cm^{-2} sec^{-1} in the FRJ-2 reactor (Dido type) at Juelich. It is important to note that, in the appropriate position, the flux ratio $\phi_{fast}/\phi_{therm} \leq 0{\cdot}001$. Hence, neutron activation results in an activity ratio of

$$\frac{^{53}\text{Mn [dpm]}}{^{54}\text{Mn [dpm]}} \simeq \frac{1}{6000}$$

At the end of the irradiation the relative flux deviations were controlled by counting the ^{65}Zn annihilation radiation in a coincidence counter. Absolute flux calibrations were done with a standardized ^{22}Na sample.

The recycling of sample and standards consisted in taking up in 12N HCl, oxidizing with H_2O_2, adding 47·5 mg Fe (inactive carrier) and treating with a Dowex 1 X 8 ion exchanger. The eluate was brought to dryness, taken up in 2N HCl and loaded on a cation exchanger (Dowex 50). The eluate was digested with strong H_2SO_4, 1 g of $NaIO_4$ was added, and finally the manganese distilled at 40° to 50°C under reduced pressure. The distillation was repeated.

Influence of the (n,p) and the (n,2n) interfering reactions is best demonstrated in the Table 1. While the interference of the ^{54}Fe (n,p) ^{54}Mn reaction can be suppressed largely by radio-chemistry and is only on the order of 2 per cent. That from the other nuclear reaction ^{55}Mn (n,2n) is more serious, contributing up to about 40 per cent. Clearly the latter interference is dependent mainly on the reactor irradiation conditions, and may be reduced by choosing the highest available ratio of ϕ_{therm}/ϕ_{fast}.

Whereas the ^{53}Mn content of most of the investigated smaller meteorites is in the range of 300–600 dpm/kg, the ^{53}Mn content in lunar dust is found to be only 30·3 $\pm$ 5·5 dpm/kg. However, we have to normalize this figure to the amount of $\sim$13 per cent iron present, the main source of spallation nuclides. So, we get a value of ^{53}Mn = 230 dpm/kg Fe of lunar surface material. This still is only about 50 per cent of the amount found and expected in smaller iron meteorites.

On the other hand, from studies of production rates, one knows that ^{53}Mn is mainly formed by secondary reactions and lower energy primaries and that it shows a pronounced depth effect. An increase of ^{53}Mn content up to a 25 cm distance beneath the surface of a spherical iron target has been observed. Evidently we shall have to expect higher ^{53}Mn activities for samples farther below the moon's surface. Moreover, in contrast to meteorites, we must consider a 2π instead of 4π flux of cosmic rays.

Table 1. Determination of spallogenic ^{53}Mn in lunar dust

		Lunar dust (fr. B. < 100 μm) (0·9901 g)	Mn standard	Fe standard
	Weight	1·62 mg Mn	8·03 mg Mn	8·25 mg Fe
Chem. yield	Pre-irrad.	95·9%	—	—
	Recycl.	89·3%	69·0%	65·2%
Carrier after irrad.		47·5 mg Fe	—	10 mg Mn
^{1}spec$_{54}$ (cpm/mg Mn)	(n,γ)	7·64	—	—
	(n,2n)	3·14	3·14*	0
	(n.p)	0·16†	0	(8·92* per mg Fe)
A_0spec$_{54}$ (dpm/mg Mn) at end of irrad.	(n,γ)	110·87	—	—
A_{53} [dpm/kg]		30·3 ± 5·5	—	—

* Neutron flux normalized.
† Originating from Fe-traces in isolated lunar Mn.

Still another point of interest is the relatively long half-life of ^{53}Mn, compared to ^{36}Cl and ^{26}Al. From the high ^{26}Al values (up to 100 dpm/kg) one is inclined to assume that this nuclide is in radioactive equilibrium. If the accretion and mixing rate of lunar dust should be found to be high, then it is conceivable that radioactive equilibrium is not fully established for the longer-lived ^{53}Mn. Further studies on selected samples—for example, cores, should bring the necessary information about spallation profiles. Since the activation method described is so sensitive, manganese-53 at even greater depths (up to several meters) should be measurable.

During the last decade the natural β-activity of ^{187}Re was used by us successfully for geological age determinations (HERR *et al.*, 1967). We showed that the ^{187}Os abundance varies considerably in Re-containing minerals and in iron meteorites. It was pointed out, and D. D. Clayton showed this later in a detailed study, that the major part of the terrestrial (stable) ^{187}Os should be regarded as having resulted from ^{187}Re decay and that most of that decay occurred before the formation of the solar system. These ideas are based on the now accepted s-process theory.

The β-energy of ^{187}Re is extremely weak ($\sim$2 keV) and thus a problem arises about the mode of this decay. Quite recently CLAYTON (1969) outlined the possibility that the radioactivity of ^{187}Re could be dependent on temperature; hence isotopic measurements of solar wind-derived Re from lunar dust could clarify the interesting problem of ^{187}Re cosmochronology. In view of these cosmological questions, we found it necessary to establish the presence of Re in lunar dust. As the abundance in Fe-meteorites is normally below 1 ppm, only radiochemical neutron activation could succeed. Fines (176 mg) were irradiated together with Re standards for 3 days, $\phi = 7 \times 10^{13}$ in the FRJ-2 reactor. Thereafter the sample was fused by melting with NaOH + Na_2O_2 in the presence of 30 mg of Re-carrier. Several $Fe(OH)_3$ precipitations were followed by repeated dry distillation of Re_2O_7. The decay of ^{186}Re was observed. The Re content of soil (sample 10084, grain size $<100\ \mu m$) is calculated to be $11{\cdot}2 \pm 0{\cdot}4$ ppb if one assumes a normal terrestrial isotope abundance.

The next step it is important to check the isotope ratio $^{187}Re/^{185}Re$.

The thermodifferential analysis (TDA) and thermogravimetric analysis (TGA) curves (Fig. 1) show distinct temperature regions where gas-loss and recrystallization

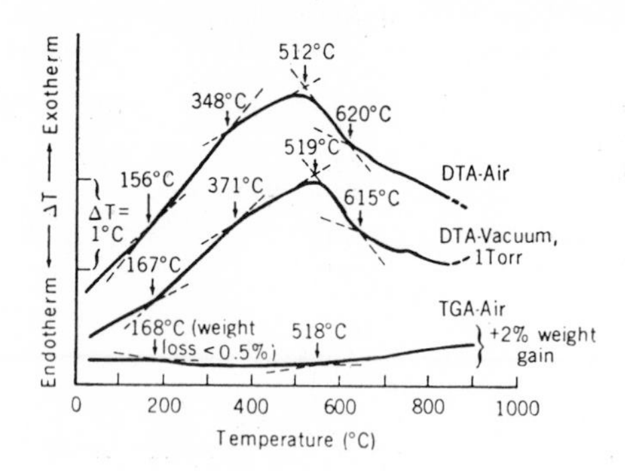

Fig. 1. Differential thermal and thermogravimetric analysis of lunar dust (sample 10084). Heating rate, 20°C/min; reference, Al_2O_3; sample weight, 5–7 mg.

or annealing occurs. A critical temperature is seen at about 510°C. Oxidation at 800°C leads to a gain in weight of about 2 per cent (see the lower TGA curve).

By mass spectroscopy, ratios of $^{4}He/^{3}He = 724 \pm 118$, $^{20}Ne/^{22}Ne = 2{\cdot}34 \pm 0{\cdot}04$, and $^{22}Ne/^{21}Ne = 22{\cdot}6 \pm 6$ were measured in individual grains of dust. Only the $^{20}Ne/^{22}Ne$ ratio was found to be very constant.

Nuclear γ-resonance spectroscopy (with ^{57}Fe) was done on magnetic and non-magnetic fractions of lunar fines. No trace of Fe^{3+} was detected. Troilite was practically not observed in the soil but metallic Fe, ilmenite, pyroxene, and olivine were present. The spectra were compared with those of tektites and the existence of an unidentified Fe^{2+}-containing compound in the lunar soil is proposed (see Table 2).

Table 2. Iron-containing compounds of lunar dust

Source	Fraction of total iron (%)		
	Normal	Magnetic fraction	Nonmagnetic fraction
Olivinic Fe	19·3	18·8	20·1
Pyroxenic Fe	34·0	31·7	43·2
Fe^{2+} in ilmenite	26·0	18·6	24·9
Fe^{3+} in ilmenite	<3·5	<4·6	<3·0
Unidentified compound	10·2	16·8	8·9
Metallic Fe	<7·0	<9·3	—

For the thermoluminescence study different soil fractions were used. However, after 8 weeks of storage time we did not detect any natural THL with certainty. Probably the skin layer of dust was already strongly annealed on the moon. Highest glow intensities occur in the temperature range of 100° to 150°C, if γ-irradiated at

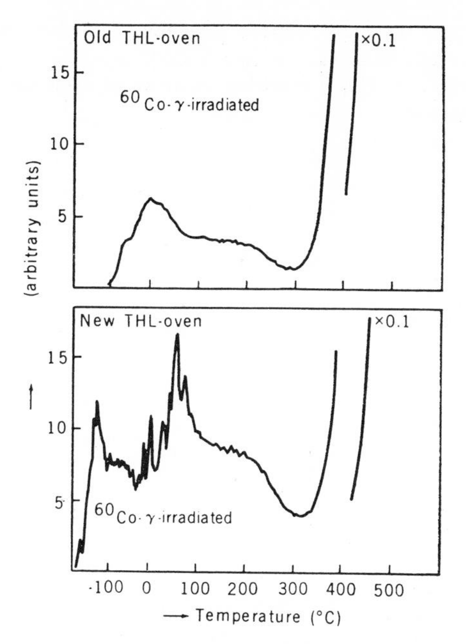

Fig. 2. Thermoluminescence glow curves of lunar fine (sample 10084); (upper curve) 40-mg sample; irradiated for 15 minutes; heating rate, 100°C/min; (lower curve) 47-mg sample; irradiated for 15 minutes; heating rate, 100°C/min. Both measured 5 min after irradiation; both 10^5 rad.

+30°C. At −196°C γ-irradiation the intensity is considerably greater (see Fig. 2). A ^{60}Co-γ-saturation experiment shows that saturation for the soil is reached at a dose of about 10^6 rads.

These preliminary results of thermoluminescence demand better defined and selected samples.

References

Clayton D. D. (1969) Isotopic composition of cosmic importance. *Nature* **224**, 56–57.

Herpers U., Herr W. and Wölfle R. (1967) Determination of cosmic-ray-produced nuclides ^{53}Mn, ^{45}Sc and ^{26}Al in meteorites by neutron activation and gamma coincidence spectroscopy. In *Radioactive Dating and Methods of Low-Level Counting*, pp. 199–205. IAEA, Vienna.

Herpers U., Herr W. and Wölfle R. (1969) Evaluation of ^{53}Mn by (n,γ) activation, ^{26}Al and special trace elements in meteorites by γ-coincidence techniques. In *Meteorite Research*, (editor P. M. Millman), Chap. 32, pp. 387–396. D. Reidel.

Herr W., Wölfle R., Eberhardt P. and Kopp E. (1967) Development and recent applications of the Re/Os dating method. In *Radioactive Dating and Methods of Low-Level Counting*, pp. 499–508. IAEA Vienna.

HERR W., HERPERS U. and WÖLFE R. (1969) Determination of Manganese-53 produced in meteorites by cosmic radiation with the aid of neutron activation. *J. Radioanalyt. Chem.* **2**, 197–203.

HONDA M., SHEDLOVSKY J. P. and ARNOLD J. R. (1961) Radioactive species produced by cosmic rays in iron meteorites. *Geochim. Cosmochim. Acta* **22**, 133–154.

KAYE J. H. and CRESSY P. J. (1965) Half-life of Managanese-53 from meteorite observations. *J. Inorg. Nucl. Chem.* **27**, 1889–1892.

MILLARD H. T., JR. (1965) Thermal neutron activation: Measurement of cross section of Managanese-53. *Science* **147**, 503–504.

SHELINE R. K. and HOOPER J. E. (1957) Probable existence of radioactive manganese-53 in iron meteorites. *Nature* **179**, 85–87.

Proceedings of the Apollo 11 Lunar Science Conference, Vol. 2, pp. 1239 to 1245.

Na^{22}, Al^{26}, Th and U in Apollo 11 lunar samples

GREGORY F. HERZOG and GERALD F. HERMAN
Enrico Fermi Institute and Department of Chemistry, University of Chicago,
Chicago, Illinois 60637

(*Received* 3 *February* 1970; *accepted in revised form* 25 *February* 1970)

Abstract—Positron activities due mainly to Na^{22} and Al^{26} range from 88 to 217 β^+/min-kg in 5 rocks or fragments with weights between 9 and 29 g. The higher activities presumably indicate surface locations. The concentrations of Na^{22}(dpm/kg), Al^{26}(dpm/kg), Th (ppm) and U (ppm) are 73 ± 15, 124 ± 11, 3·05 ± 0·19 and 0·68 ± 0·03 in sample 10072-41, and 33 ± 12, 81 ± 9, 2·52 ± 0·15 and 0·65 ± 0·03 in sample 10061-21. These values are consistent with measurements of other laboratories.

INTRODUCTION

ONE WAY of unravelling the nuclear transformations in lunar surface rocks is to study the distribution of various radionuclides as a function of depth. In the outermost centimeter or so low-energy solar protons and cosmic rays give rise to a large part of the observed activity; naturally occurring U and Th and radionuclides produced by the interactions of high-energy galactic cosmic rays contribute to a sample's gamma spectrum at all depths. Measurements of radioactivity can help to decipher the irradiation history of a sample and may indicate the targets and projectiles which interacted.

We have measured the gamma–gamma coincidence spectra of five small (9–29 g) lunar samples. Measurements on these smaller samples have the advantage of exhibiting in some detail the depth dependence of the nuclear reactions taking place without requiring the destruction of the original material. The measurements might have been even more useful had additional documentation on their original positions been available. We have also attempted to resolve the four principal radionuclides—Na^{22}, Al^{26}, Th^{232} and U in two of five samples, namely 10061-21 and 10072-41.

PROCEDURE

Sample radioactivities were measured by gamma–gamma coincidence spectrometry. One of the spectrometers has been described previously (HEYMANN and ANDERS, 1967); the other with anti-coincidence-shielded 7·6 × 10·2 and 7·6 × 3·8 cm NaI detectors had circuitry for two-parameter analysis and was operated at the bottom of a 16-m well covered by a 1-m layer of lead bricks for maximum background reduction. The resolution of this system for the 662-keV γ-ray of Cs^{137} was 7 per cent for the 'X' detector and 6·7 per cent for the 'Y' detector.

Background measurements were made with three different powdered materials: dunite; Cu cut to a density of 3·3 g/cm^3 with NaCl; and reagent-grade iron. NaCl + Cu and Fe were counted in separate 7·5 cm^3 cylindrical plastic boxes containing 25 and 33 g of material, respectively. 19·5 g of dunite in a 14-cm^3 plastic box served as a third background sample. Counting rates observed for each of these background samples were the same within experimental error. One grand-average background was therefore calculated for use with each lunar sample.

The counting efficiencies of Na^{22}, Al^{26}, Ti^{44}, Sc^{46}, Co^{56}, Co^{58}, Co^{60}, Th^{232} and U were determined with the aid of standard powders made by dispersing known amounts of the radionuclides in a matrix of Fe + MgO or dunite powder. Of the 27 sets of counting efficiencies required (one for each of 9 isotopes for each of 3 samples) 21 were obtained with exact mock-ups filled with the different powders (ROWE *et al.*, 1963) so as to match the lunar samples in bulk density. The others (Ti^{44}, Sc^{46}, Co^{46}, Co^{58}, Co^{60} and U for sample 10072-41) were obtained with approximate mock-ups: cylindrical plastic boxes of about the same size as the sample, filled with the standard powders. Empirical corrections were applied to the counting efficiencies of the approximate mock-ups to compensate for the differences in shape and mass. Cross-comparisons of exact and approximate mock-ups showed that the error remaining after the correction was less than 3 per cent.

The pulses from the NaI detectors were passed to analog-to-digital converters (ADC's) whose output signals constituted our basic data. ADC's X and Y were set to 2048 and 1024 channels and accepted maximum signals corresponding to 4·10 and 2·05 MeV respectively. Sample counting rates were sufficiently low to permit us to store the ADC output event-by-event on a magnetic tape unit. These data, collected in the equivalent of a 2048 × 1024 channel matrix, were used in the subsequent regression analyses. The output signals of the ADC's were also passed, in parallel, to the memory of a two-parameter analyzer where they were stored in the form of a 64 × 64 channel matrix; the 64 × 64 matrix served mainly as a monitor. The electronics were calibrated on the basis of the prominent peaks of the Na^{22} singles spectrum at least once a week and typical drift did not exceed 0·5 per cent.

With the aid of several computer programs, the individual coincidence events were sorted into 38 cells centered about the main photopeaks and Compton ridges of the radionuclides anticipated to be present. The cells typically contained ~7 per cent of the total number of events observed. The cell positions are based on measurements made with actual sources and not on the use of energy calibration curves.

Counting rates in each cell were calculated and, along with the counting efficiency data, were used as input to B34T, a program for regression analysis developed by THORNBER (1966).

RESULTS

The positron activities of our samples are given in Table 1 along with results derived from measurements made by O'KELLEY *et al.* (1970). The gross counting

Table 1. Positron activities (β^+/min-kg)

Sample	Weight (g)	Type	This work	O'KELLEY *et al.* (1970)*
10018-24	25·6	C	107 ± 10	138 ± 20
10026-10	9·3	C	217 ± 20	
10044-36	11·2	B	88 ± 7	
10061-21	28·8	C	103 ± 5	
10072-41	21·7	A	186 ± 13	100 ± 10

* O'KELLEY *et al.* measurements were carried out on the entire rocks.

rates in the 0·511–0·511 MeV coincidence channel were corrected for contributions from Th and U, typically about 5 per cent. Most of the remaining positron activity is due to Na^{22} and Al^{26}; there are also small contributions from Co^{56} and Ti^{44} which we estimate conservatively at 3 per cent.

We analyzed a synthetic sample composed of known amounts of the radioactive species mentioned above in order to check the accuracy of our experimental methods and data reduction procedures. To make the check reasonably rigorous, the ratios of the radionuclides in the synthetic sample were made to approximate those found in the lunar samples. The amounts 'found' in the synthetic sample do indeed agree fairly well with the amounts 'taken' in four cases: Na^{22}, Al^{26}, Th^{232} and U (Table 2).

The other minor constituents were not determined so well, presumably because of inadequate counting time, and are not presented.

Table 2. Analysis of synthetic sample

	Na^{22} (β^+/min-kg)	Al^{26} (dpm/kg)	Th (ppm)	U (ppm)
Taken	2287 ± 50	4926 ± 150	135 ± 4	43 ± 3
Found	1970 ± 550	4580 ± 500	146 ± 7	35 ± 4

In analyzing the data for lunar samples 10061 and 10072 we included only Na^{22}, Al^{26}, Sc^{46}, Co^{56}, Th^{232} and U because Ti^{44}, Co^{58} and Co^{60} were shown to be present in negligible amounts by the LSPET (1969). For the same reason as before, we present only our results for Na^{22}, Al^{26}, Th^{232} and U which may be seen, along with values obtained by other laboratories, in Table 3.

Table 3. Gamma-ray analyses of Apollo 11 lunar samples

Source	Sample No.	Na^{22}* (dpm/kg)	Al^{26} (dpm/kg)	Th (ppm)	U (ppm)
This work	10072-41	73 ± 15	124 ± 11	3·05 ± 0·19	0·68 ± 0·03
O'KELLEY *et al.* (1970)	10072-01	42 ± 9	70 ± 5	2·8 ± 0·17	0·76 ± 0·06
SILVER (1970)	10072-39			3·3 ± 0·1	0·88 ± 0·03
COMPSTON *et al.* (1970)	10072-30			3·5	
MORRISON *et al.* (1970)	10072-24			4·8	0·50
This work	10061-21	33 ± 12	81 ± 9	2·52 ± 0·15	0·65 ± 0·03
TATSUMOTO and ROSHOLT (1970)	10061			2·57 ± 0·02	0·67 ± 0·01
COMPSTON *et al.* (1970)	10061-25			2·7	

* Na^{22} corrected to 7/20/69.

DISCUSSION

The data of Table 4 show that the total amounts of Na^{22} and Al^{26} present near the lunar surface may vary by a factor of 2·6 while the ratio of Na^{22} to Al^{26} remains between 0·45 and 0·65 in most cases. Accordingly, the total positron activity of a sample should usually be a reliable indicator of a sample's original position on the lunar surface. It will fail only for samples of highly unusual composition or exposure age less than a few half-lives of Al^{26}.

O'KELLEY *et al.* (1970) measured 110 β^+/min-kg in the entire rock 10072 whereas we found an 80 per cent higher value for fragment 10072-41. This implies a surface location for the fragment. Conversely, fragment 24 of rock 10018 has a lower positron activity than the whole rock. We therefore inferred that the fragment came from the lower portion of the rock.

Even in the absence of other data we may safely conclude from their positron activities alone that fragment 10044-36 lay buried under the lunar surface while sample 10026 (an entire 9·2 g rock from the contingency sample) was basking more or less directly in the solar wind.

Th and U contents of samples 10061-21 and 10072-41 are in the ratio of $\sim$4 to 1.

Table 4. Na^{22}/Al^{26} ratios

Sample	Na^{22} (dpm/kg)	Al^{26} (dpm/kg)	Na^{22}/Al^{26}	Ref.
10003 (whole)	41 ± 4	74 ± 8	0·55 ± 0·08	a
(-25)	49 ± 2	75 ± 2	0·65 ± 0·03	c
(-25)	56 ± 7	74 ± 7	0·76 ± 0·12	e
10017 (top)	43 ± 6	95 ± 15	0·45 ± 0·09	a
(0–4 mm)	76 ± 10	133 ± 15	0·57 ± 0·10	b
(interior)	33 ± 6	65 ± 10	0·51 ± 0·12	a
(4–12 mm)	43 ± 6	70 ± 10	0·61 ± 0·12	b
(12–30 mm)	37 ± 6	57 ± 9	0·65 ± 0·15	b
(bottom)	30 ± 5	50 ± 7	0·60 ± 0·13	a
(whole)	39 ± 4	73 ± 8	0·53 ± 0·08	a
(-37)	45 ± 2	80 ± 2	0·56 ± 0·03	c
(-37)	47 ± 7	83 ± 7	0·57 ± 0·12	e
10057 (top)	63 ± 7	115 ± 9	0·55 ± 0·08	a
(whole)	41 ± 4	75 ± 8	0·55 ± 0·08	a
(-30)	43 ± 2	84 ± 2	0·51 ± 0·03	c
(-30)	49 ± 8	75 ± 6	0·65 ± 0·13	e
10072 (whole)	46 ± 5	73 ± 8	0·63 ± 0·09	a
(-41)	73 ± 15	124 ± 11	0·59 ± 0·13	f
10018 (whole)	55 ± 8	108 ± 16	0·51 ± 0·10	a
10019 (whole)	47 ± 7	101 ± 15	0·46 ± 0·09	a
10021 (whole)	55 ± 8	110 ± 15	0·50 ± 0·10	a
10061 (-21)	33 ± 12	81 ± 9	0·41 ± 0·13	f
10002 (whole)	51 ± 5	120 ± 12	0·42 ± 0·06	a
(bulk fines)	55 ± 7	108 ± 17	0·51 ± 0·10	b
10084 (-41)	64 ± 3	131 ± 4	0·49 ± 0·02	c
(-113-2)	63 ± 2	137 ± 4	0·46 ± 0·02	c
(-18)	61 ± 6	121 ± 25	0·50 ± 0·15	d
(-113-1)	78 ± 7	138 ± 5	0·57 ± 0·08	e
(-113-2)	73 ± 7	130 ± 5	0·56 ± 0·08	e

a. O'KELLEY *et al.* (1970).
b. SHEDLOVSKY *et al.* (1970).
c. PERKINS *et al.* (1970).
d. WÄNKE *et al.* (1970).
e. WRIGLEY and QUAIDE (1970).
f. Present work.

This seems to be characteristic of lunar rocks and is comparable to the ratio found in many meteorites and some terrestrial basalts; in other terrestrial rocks the ratio sometimes goes as high as 8–10. As may be seen in Table 3 our values are in good agreement with those obtained by other laboratories.

It is interesting to compare the observed Al^{26} activity with that expected from galactic cosmic rays. FUSE and ANDERS (1969) have used the observed Al^{26} contents in meteorites to infer production rates for the principal target elements. These rates presumably reflect production by galactic cosmic rays alone, at a depth of at least a few cm and 4π geometry. The galactic Al^{26} level predicted for rock 10061-21 calculated from these production rates for 2π geometry, is 43 β^+/min-kg, well below the observed value of 81 ± 9 β^+/min-kg.

The meteorite production rates reflect conditions that may have differed in two major respects from lunar conditions. For one, the importance of cosmic-ray secondaries in the meteorite samples may not be comparable to their importance near the lunar surface. Quantitative assessment of this effect is difficult because the average preatmospheric depth of the meteorite samples is not known.

Fortunately we can fall back on the second major difference to account for the discrepancy between calculated and observed Al^{26}, namely the smaller contribution of solar cosmic rays to the production of Al^{26} in meteorites. This difference arises from two sources: (1) The meteorites probably spent most of their time farther from the sun. (2) Most reactions due to solar protons occur near the surfaces of the meteorites; these surfaces have been ablated.

Unfortunately, any quantitative conclusions about the amount of Al^{26} produced by solar protons within a few centimeters of the lunar surface are sensitive to the shape and intensity of the assumed flux (SHEDLOVSKY *et al.*, 1970). Our samples come from the outermost few centimeters and consequently it is difficult to tell how much of the excess Al^{26} is due to solar protons and how much to cosmic-ray secondaries. We would eventually like to measure the Al^{26} activity in a sample that has been shielded from most of the solar particles. For such a sample, the simple calculation based on meteorite cross sections would aid more detailed calculations by isolating the contribution of cosmic-ray secondaries.

Finally we attempt to show that the ratios of Na^{22}:Al^{26} in the rocks and the fines differ and that variations in chemical composition are at least partially responsible. Six measurements with an Al^{26} disintegration rate greater than 115 dpm/kg have been singled out from the data of Table 4. The data of WRIGLEY and QUAIDE (1970) have not been included among the six because those authors have indicated that their values for Na^{22} may be somewhat too high. The value of WÄNKE *et al.* (1970) has been excluded because of its relatively large uncertainty.

For the three Type A or B rocks selected (10017, 10057 and 10072) the average ratio of Na^{22} to Al^{26} is $0{\cdot}57 \pm 0{\cdot}12$. The average of three measurements on the lunar fines 10084 is $0{\cdot}46 \pm 0{\cdot}06$. The probability that these two numbers are different is roughly 0·67. Either documentation or the high Al^{26} disintegration rates (or both) of these samples make it nearly certain that they occupied surface locations at the time of Apollo 11. SHEDLOVSKY *et al* (1970) and PERKINS *et al.* (1970) have suggested that 10017 might have attained the position and orientation in which it was found only within a few half-lives of Al^{26}. The former group has hypothesized that the production of Al^{26} by galactic cosmic rays was greater at this previous location. On the other hand, the contribution of low-energy solar protons would have been reduced by shielding at this depth. The two effects may offset each other so that the observed counting rate for Al^{26} may not be too far from what it would have been had 10017 spent all its time at the surface. It is true that were the observed Al^{26} counting rate 20 per cent *below* saturation, our comparison would be defeated; this seems unlikely in view of the fact that 10017 has the highest Al^{26} activity yet observed in any lunar rock. With the possible exception of 10017, we think it probable that the samples under consideration have had comparable irradiation histories.

Average chemical compositions for the two groups are given in Table 5. Evidently the differences in Al and Na contents point in the right direction.

With the production rates of BEGEMANN *et al.* (1970) we calculate the ratios of Na^{22}/Al^{26} to be 0·58 and 0·49 for the rocks and fines, respectively. The agreement is extraordinarily good and is probably fortuitous. BEGEMANN *et al.* have argued that their 'rock' cross sections may reflect greater shielding from low-energy protons than their 'fines' cross sections. In this case we should compare the fines to rocks with

Al^{26} activity of lower than 85 dpm/kg for example. For this second group of rocks (10003, 10017, 10057, 10072), the average Na^{22}/Al^{26} ratio is $0{\cdot}58 \pm 0{\cdot}11$, again in excellent agreement with the calculation.

Table 5. Average chemical compositions of rocks (10017, 10057 and 10072) and fines (10084) (%)

Sample	Na	Mg	Al	Si	Ref.
10017	0·41	4·8	4·5	19·0	abce
10057	0·37	4·8	5·2	18·1	bde
10072	0·41	4·8	4·5	19·9	ae
Avg.	0·40	4·8	4·7	19·0	
10084	0·33	4·8	7·3*	19·7*	abcd

* EHMANN and MORGAN (1970) also included in these averages.
a. WIIK and OJANPERÄ (1970).
b. WÄNKE *et al.* (1970).
c. MAXWELL *et al.* (1970).
d. ENGEL and ENGEL (1970).
e. LSPET (1969).

Acknowledgments—We are indebted to Drs. G. D. O'KELLEY and J. S. ELDRIDGE of the Oak Ridge National Laboratory for the Al^{26}, Ti^{44}, Sc^{46}, Co^{56} and Co^{58} standards, to Dr. J. A. S. ADAMS of Rice University for the Th and U standards, and to Dr. H. R. HEYDEGGER of the University of Chicago for the Co^{60} standard. We gratefully acknowledge the assistance of Dr. J. C. LAUL in preparation of the standard powders and of J. R. NAPLES in programming and data reduction. Dr. A. TURKEVICH graciously permitted us to use his 15-m counting well. Dr. P. M. JEFFREY made valuable contributions to the experimental phase of this work. Dr. E. ANDERS gave us indispensable aid and advice throughout this project and made many useful editorial suggestions.

REFERENCES

BEGEMANN F., VILCSEK E., RIEDER R., BORN W. and WÄNKE H. (1970b) Cosmic-ray produced radioisotopes in lunar samples. *Geochim. Cosmochim. Acta*, Supplement I.

COMPSTON W., ARRIENS P. A., VERNON M. J. and CHAPPELL B. W. (1970) Rubidium–strontium chronology and chemistry of lunar material. *Science* **167**, 474–476.

EHMANN W. D. and MORGAN J. W. (1970) Oxygen, silicon and aluminum in lunar samples by 14 MeV neutron activation. *Science* **167**, 528–530.

ENGEL A. E. J. and ENGEL C. G. (1970) Lunar rock compositions and some interpretations. *Science* **167**, 527–528.

FUSE K. and ANDERS E. (1969) Al^{26} in meteorites—VI. Achondrites. *Geochim. Cosmochim. Acta* **33,** 653–670.

HEYMANN D. and ANDERS E. (1967) Meteorites with short cosmic-ray exposure ages, as determined from their Al^{26} content. *Geochim. Cosmochim. Acta* **31,** 1793–1809.

LSPET (LUNAR SAMPLE PRELIMINARY EXAMINATION TEAM) (1969) Preliminary examination of lunar samples from Apollo 11. *Science* **165,** 1211–1227.

MAXWELL J. A., ABBEY S. and CHAMP W. H. (1970) Chemical composition of lunar material. *Science* **167,** 530–531.

MORRISON G. H., GERARD J. T., KASHUBA A. T., GANGADHARAM E. V., ROTHENBERG A. M., POTTER N. M. and MILLER G. B. (1970) Multielement analysis of lunar soil and rocks. *Science* **167,** 505–507.

O'KELLEY G. D., ELDRIDGE J. S., SCHONFELD E. and BELL P. R. (1970) Elemental compositions and ages of lunar samples by non-destructive gamma-ray spectrometry. *Science* **167,** 580–582.

PERKINS R. W., RANCITELLI L. A., COOPER J. A., KAYE J. H. and WOGMAN N. A. (1970) Cosmogenic and primordial radionuclides in lunar samples by nondestructive gamma-ray spectrometry. *Science* **167**, 577–580.

ROWE M. W., VAN DILLA M. A. and ANDERSON E. C. (1963) On the radioactivity of stone meteorites. *Geochim. Cosmochim. Acta* **27**, 983–1001.

SHEDLOVSKY J. P., HONDA M., REEDY R. C., EVANS J. C., JR., LAL D., LINDSTROM R. M., DELANEY A. C., ARNOLD J. R., LOOSLI H.-H., FRUCHTER J. S. and FINKEL R. C. (1970) Pattern of bombardment-produced radionuclides in rock 10017 and in lunar soil. *Science* **167**, 574–576.

SILVER L. T. (1970) Uranium–thorium–lead isotope relations in lunar materials. *Science* **167**, 468–471.

TATSUMOTO M. and ROSHOLT J. N. (1970) Age of the moon: An isotopic study of uranium–thorium–lead systematics of lunar samples. *Science* **167**, 461–63.

THORNBER H. (1966) Manual for (B34T, 8Mar66), a stepwise regression program. Univ. of Chicago Center for Math. Studies in Bus. and Econ. Rep. No. 6603.

WÄNKE H., BEGEMANN F., VILCSEK E., RIEDER R., TESCHKE F., BORN W., QUIJANO-RICO M., VOSHAGE H. and WLOTZKA F. (1970) Major and trace elements and cosmic-ray produced radioisotopes in lunar samples. *Science* **167**, 523–525.

WIIK H. B. and OJANPERÄ P. (1970) Chemical analyses of lunar samples 10017, 10072, and 10084. *Science* **167**, 531–532.

WRIGLEY R. C. and QUAIDE W. L. (1970) Al^{26} and Na^{22} in lunar surface materials: implications for depth distribution studies. *Geochim. Cosmochim. Acta*, Supplement I.

Proceedings of the Apollo 11 Lunar Science Conference, Vol. 2, pp. 1247 to 1259.

Inert gases in the fines from the Sea of Tranquillity

DIETER HEYMANN and AKIVA YANIV*
Departments of Geology and Space Science, Rice University, Houston, Texas 77001

(*Received* 5 *February* 1970; *accepted in revised form* 25 *February* 1970)

Abstract—He, Ne, Ar, Kr, and Xe were measured mass-spectrometrically in twenty-two samples of the fines (sample 10084-40) from the Sea of Tranquillity. He^4, Ne^{20}, Ar^{36}, Kr^{84}, and Xe^{132} contents per gram are inversely proportional to fragment size for most of the lithic fragments. Glasses and anorthosite fragments generally contain less trapped gas than lithic fragments of the same size. Radiogenic He_r^4 could not be resolved. Radiogenic Ar_r^{40} is present in the fines. Cosmogenic He_c^3, Ne_c^{21}, and Ar_c^{38} components are present in several of the fragments; apparent 'radiation ages' range from $\sim 100 \simeq 600$ m.y. The principal elemental ratios are below solar-system values. The evidence seems to favor saturation of the surfaces with solar wind such that the observed ratios reflect the abundance ratio in the solar wind and the ratios of diffusion coefficients. However, it is also possible that the lighter inert gases are preferentially removed from the minerals by sputtering. $He^3/He^4 = (4{\cdot}16 \pm 0{\cdot}71) \times 10^{-4}$, in agreement with 'solar' He in carbonaceous chondrites. Ne appears to consist of a trapped and a cosmogenic component. Ar^{36}/Ar^{38} in the trapped gas is nearly identical to the atmospheric value.

INTRODUCTION

THE FIRST sample which we received, 10084-40, consists of 'fines', i.e. fragments smaller than 1 millimeter. From the description of this material by LSPET (1969) and from our own observations under a binocular microscope it became obvious that the fines would make possible several interesting lines of investigation, despite the fact that the precise origin of the materials was not wholly clear. Most importantly, the fines contain fragments ranging in size from less than 1 micron to about 1 mm. A great number of investigators had speculated that the fine materials on the surface of the moon should contain substantial amounts of solar-wind implanted inert gases; and indeed, the first results of LSPET (1969) confirmed this speculation. Earlier, EBERHARDT *et al.* (1965a, 1965b, 1966) had shown that there exists a strong anti-correlation between the trapped gas content and grain size in the Khor Temiki meteorite. The authors suggested that the gases represent most likely trapped low energy *ions such as solar wind particles.*

The fines from Apollo 11 lend themselves easily for a study of gas content vs. fragment size. Accordingly we made splits of sample 10084-40 by picking individual fragments under the microscope, or (below fragment size of 250 μ) by sieving. Also, the fines contain fragments of different mineralogy, chemistry and texture. The most abundant kind we call lithic fragments because they are similar in texture and mineralogy to the crystalline rocks described by LSPET (1969) and other authors. The lithic fragments are probably identical to the soil breccias and basalts discussed by WOOD *et al.* (1970), who found that these constitute nearly 90 per cent by number of all the fragments. In addition the fines contain glassy fragments of various colors and

* On leave of absence from the University of Tel Aviv, Ramat Aviv, Israel.

shapes; most conspicuous among these are the brown or black spheres and spindles. Also the fines contain a small amount of white fragments which we tentatively identified as feldspars, and which are perhaps anorthosite as described by WOOD *et al.* (1970).

Accordingly we prepared samples of these different materials in the hope that the inert-gas study might contribute to the understanding of the origin of the different materials in the fines. Finally, we found that it is quite simple to concentrate the glassy and feldspar fragments by removing strongly magnetic fragments with a hand magnet. The weakly magnetic portion of a sieve fraction 70–105 μ became substantially enriched in glass and feldspar. Thus, a simple magnetic separation made it possible to investigate certain materials in a range of fragment size where the hand-picking of individual fragments became unpractical.

These were the considerations which guided the investigation which we shall describe, and in particular guided the splitting of sample 10084-40 into the sub-samples listed in Table 1.

Table 1. Description of Sub-Samples of 10084-40

Sample No.	Weight (mg)	Size (μ)	Description
1	3·507	all sizes	A scoop from 10084-40; grain size distribution unknown.
2	10·827	all sizes	A scoop from 10084-40; grain size distribution unknown.
3	0·879	870 × 830 × 500	Lithic fragment, grey, rough surface, black glassy coating on much of the surface.
4	0·704	1050 × 700	Lithic fragment, cone-shaped, almost entire surface covered with dark glass.
5	0·644	~500	Eight lithic fragments, like 3, cleaned with brush under microscope.
6	0·637	~250	Twenty-nine lithic fragments, like 3, cleaned like 5.
7	0·574	70–105	Sieve fraction, mostly lithic fragments.
8	0·409	<70	Sieve fraction, mostly lithic fragments.
9	0·499	70–105	Strongly magnetic portion of 7, little if any clear glassy or feldspar fragments; abundant dark glass coating on fragments.
10	0·552	70–105	Weakly magnetic portion of 7, enriched in glassy or feldspar fragments.
11	0·051	~250	One brown sphere and spindle; sphere ~250 μ; spindle ~280 × 130 μ.
12	0·175	~400	Three black spheres; surface generally smooth, but with some pits.
13	0·171	~400	Dark-brown glass, irregularly shaped.
14	0·243	~550	One white fragment, probably feldspar.
15	1·050	~800 × 780	Lithic fragment containing ilmenite, feldspar, and pyroxene.
16	0·519	~350	Four transparent, colorless fragments; probably glass.
17	0·420	~600	One fragment, yellow glass.
18	0·467	~500	One fragment, green glass.
19	0·930	105–250	Acetone-cleaned portion of 105–250 μ fraction.
20	0·743	105–250	Acetone-cleaned portion of 105–250 μ fraction.
21	0·552	105–250	Untreated portion 105–250 μ.
22	0·844	~1000	One white fragment, probably feldspar.

PROCEDURES

After weighing (to ± 0·005 mg) and packaging in aluminum foil the samples were loaded into a side-arm of the melting furnace. Individual samples could then be moved from the side-arm into the furnace proper by means of a hand-magnet.

The furnace, a molybdenum crucible enclosed in a water-cooled pyrex jacket could be induction heated to 1750–1800°C as measured with an optical pyrometer. We melted the samples for 15 min. Following the melting, the gas was cleaned in two steps with Ti–Zr getters.

From previous experience with meteorites we knew the inert gas contents of the packing materials; it turned out that these were in all cases negligibly small. Hence we ran only *furnace blanks*, i.e. we alternated sample runs with runs in which the whole procedure was carried out except for the transfer of a sample into the furnace. All results are corrected for blanks. Normally the corrections amounted to less than 5 per cent of the amounts present; however, in a number of cases (especially for Kr, Xe, and Ar^{40} when these were low in the sample), the correction could be as high as 30 per cent.

Our mass-spectrometer is a Nuclide Analysis—RSS, $4\frac{1}{2}$ in. sector type with Pyrex-glass envelope. All measurements were made in the static mode. The pressures during the measurements varied depending on the amounts of gas; serious peak broadening began at pressures greater than 3×10^{-5} Torr. If this was the case we pumped away part of the sample. This was done by complete evacuation of one part of the system and expanding ('diluting') the remaining gas into this section. The pumping ratio was determined, whenever possible, by the measurement of several peaks (e.g. 3, 20, 36) before and after the dilution. Judging from replicate measurements of pumping ratios, we conclude that the precision of these ratios is $\pm$ 1·5 per cent.

Peaks were obtained by magnetic scanning. Under normal operating conditions we observed the following sensitivities: $He^3 \sim 55$ mV; $He^4 \sim 50$ mV; $Ne^{20} \sim 84$mV; $Ar^{40} \sim 310$mV; $Kr^{84} \sim 470$; and $Xe^{132} \sim 470$ mV, each for 10^{-8} cm³ STP of gas. The unit-mass resolution was 130.

Peaks at masses 1 and 2 were always so small that no corrections for the contribution of H_3^+ at mass 3 was necessary. Corrections were made for the contribution of H_2O^{18} on mass 20; for doubly-ionized Ar^{40} on mass 20 (in our instrument the correction is 0·142 parts of the 40-peak), and for CO_2^{++} on mass 22 (in our case the correction is 0·014 part of the 44-peak).

We have checked from time-to-time for mass discrimination, especially in Ne. In no case was the mass discrimination between 20 and 22 greater than 0·5 per cent (on the basis of the atmospheric ratio 9·80 $\pm$ 0·08 by EBERHARDT *et al.*, 1965c).

Our calibration standards had been previously prepared for meteorite research. The cumulative errors in the calibrations are 1·3 per cent. Prior to the analysis of the lunar samples we checked our Kr- and Xe-standards again with air-standards which were kindly made available by Dr. D. D. Bogard of MSC.

RESULTS AND DISCUSSION

Our results are shown in Tables 2 and 3. The errors in Table 2 were compounded from weighing errors, precision of peak-height measurements, calibration errors, dilution errors, and errors in blank corrections. The errors in Table 3 were compounded from precision of peak-height measurements and from errors assigned to blank corrections.

Radiogenic and cosmogenic gases

The bulk fines contain about 0·4 ppm U and 2·0 ppm Th (LSPET, 1969). If radiogenic He_r^4 was quantitatively retained during an adopted maximum age of 4·6 b.y., He_r^4 would be $\sim$75,000 $\times$ 10^{-8} cm³ STP/g. This is less than 5 per cent of all He^4 now present in samples 1–10 and 19–21; hence negligible. Only in the remaining samples could He_r^4 be a substantial fraction of all He^4 present. But the following evidence suggests that this is not so. The He^3/He^4 ratios in samples 11–18 and 22 are either equal to or greater than those in samples 1 and 2. If samples 11–18 and 22 contained much He_r^4 in addition to solar wind He, the ratios could be *lower* than those in samples 1 and 2. On the other hand, any addition of cosmogenic He *raises* the ratio again, but it is difficult to believe that these samples *always* contain just enough cosmogenic He to compensate for He_r^4, such that the *measured* He^3/He^4 ratios are always at least equal to those in the most gas-rich samples.

Table 2. Inert gas contents of fines, sample 10084-40 (Units 10^{-8} cm³ STP/g)

Sample	He^3	He^4	Ne^{20}	Ne^{21}	Ne^{22}	Ar^{36}	Ar^{38}	Ar^{40}	Kr^{84}	Xe^{132}
1	9230 ±1·9%	24,600,000 2·1%±	240,000 ±2·1%	640 ±3·1%	18,000 ±1·9%	40,300 ±1·8%	7540 ±1·7%	49,400 ±2·0%	18·2 ±4·7%	4·03 ±9·0%
2	8180 ±2·4%	22,500,000 ±2·8%	236,000 ±2·4%	633 ±2·7%	18,000 ±2·5%	39,000 ±2·0%	7260 ±2·0%	48,600 ±2·0%	16·4 ±3·1%	3·43 ±2·6%
3	1010 ±1·9%	1,170,000 ±1·9%	10,100 ±1·7%	150 ±1·9%	855 ±1·7%	1720 ±1·7%	412 ±1·7%	14,300 ±12·0%	2·3 ±10·0%	1·07 ±10·0%
4	746 ±1·7%	2,120,000 ±1·9%	28,000 ±1·7%	93·5 ±3·4%	2180 ±1·7%	8650 ±1·7%	1640 ±1·7%	15,900 ±2·5%	4·66 ±8·4%	1·60 ±17·0%
5	808 ±1·7%	1,910,000 ±1·9%	18,600 ±1·7%	93·4 ±4·3%	1460 ±1·7%	2690 ±1·7%	562 ±1·7%	8720 ±5·1%	1·50 ±20·0%	0·63 ±20·0%
6	1190 ±2·6%	2,610,000 ±1·9%	26,400 ±1·7%	143 ±3·3%	2100 ±1·7%	5000 ±1·7%	1010 ±1·7%	13,600 ±1·9%	2·71 ±15·0%	0·70 ±20·0%
7	2970 ±1·8%	7,500,000 ±1·9%	79,300 ±1·7%	264 ±2·6%	6110 ±1·7%	14,500 ±1·7%	2750 ±1·7%	24,800 ±1·9%	7·4 ±10·0%	1·90 ±20·0%
8	6930 ±1·9%	18,200,000 ±1·9%	202,000 ±1·7%	564 ±2·5%	15,100 ±1·7%	33,300 ±1·7%	6230 ±1·7%	41,200 ±1·9%	13·7 ±7·0%	3·40 ±15·0%
9	3680 ±1·9%	9,780,000 ±1·9%	129,000 ±1·7%	380 ±1·9%	9880 ±1·7%	27,700 ±1·9%	5180 ±1·7%	34,900 ±2·1%	11·9 ±10·0%	3·40 ±15·0%
10	1070 ±2·0%	2,420,000 ±1·9%	29,200 ±1·7%	107 ±6·2%	2230 ±1·9%	3650 ±2·0%	707 ±1·7%	7740 ±2·6%	2·1 ±20·0%	0·86 ±50·0%
11	78 ±16·0%	90,000 ±50·0%	3340 ±5·3%	34 ±50·0%	285 ±8·6%	587 ±5·5%	120 ±6·8%	19,900 ±13·0%		
1A	462 ±4·4%	1,150,000 ±2·2%	26,000 ±2·0%	140 ±5·3%	2080 ±2·7%	7690 ±1·9%	1490 ±2·1%	36,400 ±2·9%	4·8 ±20·0%	0·76 ±50·0%
13	159 ±5·6%	359,000 ±4·4%	13,200 ±2·2%	66 ±11·0%	1040 ±2·8%	1440 ±2·3%	287 ±2·8%	37,600 ±2·8%		
14	114 ±11·0%	195,000 ±3·6%	1760 ±3·7%	13·7 ±15·0%	133 ±4·5%	331 ±3·4%	99 ±3·7%	2440 ±14·0%		
15	91 ±2·6%	224,000 ±1·9%	3020 ±1·9%	7·6 ±12·0%	224 ±2·6%	696 ±1·9%	131 ±1·9%	4190 ±5·3%		
16	80 ±5·2%	203,000 ±1·9%	2160 ±1·9%	33·9 ±6·2%	192 ±4·3%	445 ±4·4%	126 ±2·7%	7180 ±8·1%		
17	235 ±2·7%	285,000 ±2·6%	3330 ±1·9%	46 ±10·0%	302 ±4·8%	202 ±2·8%	38 ±2·8%	3150 ±6·2%		
18	88 ±5·3%	244,000 ±2·6%	7590 ±1·7%	68·9 ±5·2%	646 ±2·6%	919 ±1·7%	190 ±2·6%	6050 ±3·4%		
19	1140 ±1·9%	2,710,000 ±1·9%	36,000 ±2·0%	135 ±3·0%	2820 ±1·7%	8470 ±1·9%	1620 ±1·7%	16,500 ±2·5%	4·39 ±10·0%	1·28 ±15·0%
20	1230 ±1·9%	2,860,000 ±1·9%	35,200 ±2·0%	133 ±3·0%	2710 ±1·7%	*	*	*	*	*
21	3220 ±1·8%	8,540,000 ±1·9%	101,000 ±1·7%	344 ±2·5%	9150 ±1·7%	26,000 ±1·7%	4890 ±1·7%	55,300 ±2·5%	9·95 ±10·0%	2·60 ±15·0%
22	434 ±4·5%	825,000 ±1·9%	7500 ±2·0%	46·7 ±10·0%	594 ±2·7%	1480 ±2·5%	411 ±2·0%	7640 ±5·0%	2·2 ±15·0%	1·2 ±20·0%

The errors listed in this table were compounded from: weighing errors, analytical precision of peak-height determination, dilution (i.e. pumping) errors, calibration errors, and errors in blank corrections.
No entry: the signal was too small for measurement.
* Ar, Xe and Kr pumped away by mistake.

The substantial variation of Ar^{40}/Ar^{36} (Table 3) implies that the fines contain radiogenic Ar^{40} from K^{40}-decay *in situ*. We shall discuss this interesting point more fully in a companion paper.

The presence of cosmogenic components in many of the samples is revealed in Table 3 by concomitant changes in He^3/He^4 (increases), Ne^{20}/Ne^{22} (decreases), Ne^{21}/Ne^{22} (increases), Ar^{36}/Ar^{38} (decreases). Not one of our samples contains the cosmogenic gases in pure or nearly pure form. However, PEPIN *et al.* (1970) have reported that cosmogenic Ne is very similar in composition to that seen in chondrites. It is very unlikely that cosmogenic He could be very different in composition from that in chondrites ($He^4/He^3 \sim 5$). HINTENBERGER *et al.* (1970) report a value of Ar^{36}/Ar^{38} as low as 0·72 inside a crystalline rock. Since the rock may still contain small amounts of trapped Ar with $Ar_T^{36}/Ar_T^{38} = 5{\cdot}35$, we have adopted $Ar_c^{36}/Ar_c^{38} = 0{\cdot}6$.

He_c^3 is calculated with a one-step procedure using $(He^3/He^4)_T = 3{\cdot}73 \times 10^{-4}$, the average of our most gas-rich samples. Corrections for He_c^4 are wholly unnecessary. Likewise Ne_c^{31} can be calculated by a one-step procedure using $Ne_T^{21}/Ne_T^{20} = 2{\cdot}710 \times 10^{-3}$, and neglecting Ne_c^{20}, which is permitted. However Ar_c^{38} must be calculated

Table 3. Element and isotope ratios

Sample	He^4/ Ne^{20}	Ne^{20}/ Ar^{36}	Ar^{36}/ Kr^{84}	Kr^{84}/ Xe^{132}	He^3/He^4 $\times 10^4$	Ne^{20}/Ne^{22}	Ne^{21}/Ne^{22}	Ar^{36}/Ar^{38}	Ar^{40}/ Ar^{36}
1	103	5·95	2200	4·5	3·82	13·36 ±1·0%	0·0356 ±1·5%	5·35 ±1%	1·25
2	95·3	6·05	2400	4·8	3·64	13·13 ±1·0%	0·0352 ±1·5%	5·38 ±1%	1·25
3	116	5·87	750	2·2	8·60	11·79 ±1·0%	0·176 ±1·5%	4·17 ±1%	8·31
4	75·7	3·23	1900	2·9	3·52	12·85 ±1·0%	0·0430 ±2·0%	5·27 ±1%	1·83
5	103	6·90	1800	2·4	4·25	12·68 ±1·0%	0·0637 ±2·0%	4·79 ±1%	3·24
6	98·9	5·29	1800	3·9	4·56	12·58 ±1·0%	0·0681 ±2·0%	4·96 ±1%	2·72
7	94·6	5·45	2000	3·9	3·96	12·98 ±1·0%	0·0431 ±1·5%	5·29 ±1%	1·70
8	90·1	6·08	2400	4·0	3·81	13·37 ±1·0%	0·0373 ±1·5%	5·34 ±1%	1·24
9	75·8	4·68	2300	3·5	3·90	13·09 ±1·0%	0·0384 ±1·5%	5·34 ±1%	1·26
10	82·9	7·99	1700	2·5	4·44	13·10 ±1·0%	0·0481 ±1·5%	5·17 ±1%	2·12
11	26·9	5·68			8·92	11·70 ±6·0%	0·118 ±50%	4·90 ±5%	33·9
12	44·2	3·38	1600	6·3	4·01	12·47 ±2·0%	0·0673 ±3·0%	5·16 ±1·5%	4·73
13	27·2	9·13			4·44	12·65 ±2·0%	0·0635 ±5·0%	5·02 ±2%	26·1
14	111	5·34			5·81	13·29 ±3·5%	0·103 ±7·0%	3·34 ±3%	7·38
15	74·2	4·34			4·06	13·46 ±2·0%	0·0339 ±6·0%	5·33 ±3%	6·02
16	94·0	4·85			3·92	11·21 ±3·0%	0·176 ±4·0%	3·52 ±2%	7·39
17	85·6	16·5			8·04	11·03 ±3·5%	0·151 ±5·0%	5·29 ±3%	15·6
18	32·1	8·25			3·62	11·74 ±2·0%	0·107 ±4·0%	4·83 ±2%	6·58
19	75·3	4·25	1900	3·5	4·21	12·77 ±1·5%	0·0478 ±3·0%	5·24 ±2%	1·94
20	81·3	*	*	*	4·29	12·99 ±1·5%	0·0489 ±3·0%	*	*
21	84·6	3·88	2600	3·8	3·78	13·26 ±1·5%	0·0376 ±3·0%	5·32 ±2%	2·13
22	110	5·08	670	1·8	5·25	12·62 ±2·0%	0·0787 ±4·0%	3·59 ±2%	5·18

The errors in element ratios (e.g. He^4/Ne^{20}) can be calculated from errors given in Table 2. The errors in isotopic ratios (e.g. Ne^{20}/Ne^{22}) were compounded from the precision of peak-height measurements and errors in blank corrections.

with the two end-member compositions mentioned above. The results are given in Table 4.

Let us first consider the last two columns of Table 4 which show He_c^3/Ne_c^{21} and Ne_c^{21}/Ar_c^{38}. It is interesting to see whether these ratios can be accounted for by what is known about the chemistry of our samples. The best approach is to compare the chemistry of the Apollo samples to that of a well-known meteorite class, i.e. the hypersthene chondrites, and calculate expected production ratios in the lunar samples from the average ratios in hypersthene chondrites: $He_c^3/Ne_c^{21} = 5·3$; $He_c^3/Ar_c^{38} = 38$.

Table 4. Cosmogenic gases, radiation ages

Sample	Gas contents, 10^{-8} cm³ STP/g He_c^3	Ne_c^{21}	Ar_c^{38}	Ages, m.y. He^3	Ne^{21}	Ar^{38}	He_c^3/Ne_c^{21}	Ne_c^{21}/Ar_c^{38}
3	570	123	102	570	615	340	4·6	1·2
4	—	18	28	—	90	93	—	0·64
5	98	43	67	98	215	220	2·3	0·64
6	218	72	84	218	360	280	3·0	0·86
7	—	39	38	—	195	127	—	1·0
8	—	15	27	—	75	90	—	0·56
9	—	29	19	—	145	63	—	1·5
10	171	28	28	171	140	94	6·1	1·0
11	45	—	11	45			—	—
12	—	70	62	32	350	200	0·5	1·1
13	25	30	20	25	150	67	0·8	1·5
14	41	9	42	41			4·6	0·21
16	—	28	49	—	140	164	—	0·57
17	129	37	5	129			3·5	7·5
18	—	48	21	—			—	2·3
19	129	37	43	129	185	143	3·5	0·86
20	159	38	—	159	190		4·2	—
21	—	72	39	—	360	130	0·5	1·9
22	126	27	152	126			4·7	0·18

We have used the production ratios of STAUFFER (1962), HINTENBERGER *et al.* (1964), and EBERHARDT *et al.* (1965), but have made allowance for production of Ne_c^{21} and Ar_c^{38} from Ti as follows:

$$He_c^3 = 2{\cdot}00[1{\cdot}179 \times 10^{-2}(O + C) + 0{\cdot}572 \times 10^{-2}]$$

$$Ne_c^{21} = 2{\cdot}2\,Mg + 1{\cdot}35\,Al + Si + 0{\cdot}29S + 0{\cdot}17\,Ca + 0{\cdot}017\,(Fe + Ni + Ti)$$

$$Ar_c^{38} = 16{\cdot}5\,Ca + Fe + Ni + 8\,Ti.$$

The chemical composition of the fines was reported by LSPET (1969); the composition of the hypersthene chondrites was taken from MASON's (1965) review. On the basis of these data, one expects the He_c^3/Ne_c^{21} production ratio in the fines to be substantially greater than in hypersthene chondrites, about 7·6. All of the ratios in Table 4 are below this value. In part this may reflect real compositional differences between individual samples and the bulk of the fines, but we believe that there is a strong indication in the data for substantial He_c^3 losses or greater Ne_c^{21} production than given by the empirical equation for meteorites, or both.

The expected Ne_c^{21}/Ar_c^{38} production ratio in the fines is 1·3. Several ratios in Table 4 are close to this value, but there is also quite a number of cases between 0·6 and 0·9. Two ratios (sample 14 and 22) are about 0·2. In this respect the lunar samples are similar to eucrites, in which Ne^{21}/Ar^{38} ratios as low as 0·21 were observed (HEYMANN, 1968). The low ratios in the eucrites have been ascribed to partial Ne^{21} losses from feldspar. The same explanation may be true in the case of the lunar samples. There is, however, one objection: the He^4/Ne^{20} ratios in samples 14 and 22 are among the highest observed (Table 3), indicating relatively good He retention, and by inference relatively good Ne retention. Either the cosmic-ray produced Ne^{21} is sited less retentively than solar wind Ne^{20}, or the samples with low Ne^{21}/Ar^{38} ratios were irradiated with a proton energy spectrum substantially different from that of chondrites, perhaps solar cosmic rays.

For the calculation of radiation ages we note that the samples fall into a few groups. We have mentioned before that samples 3–9, 19–21 (lithic fragments) should be compositionally similar and this is grossly confirmed by their He_c^3/Ne_c^{21} and Ne_c^{21}/Ar_c^{38} ratios. It is then perhaps not unreasonable to calculate radiation ages for these samples with uniform production rates. We have adopted

$$He_c^3 \; 1.0 = \times 10^{-8} \text{ cm}^3 \text{ STP g}^{-1} \text{ m.y.}^{-1}$$

$$Ne_c^{21} = 0{\cdot}2 \times 10^{-8} \text{ cm}^3 \text{ STP g}^{-1} \text{ m.y.}^{-1}$$

$$Ar_c^{38} = 0{\cdot}3 \times 10^{-8} \text{ cm}^3 \text{ STP g}^{-1} \text{ m.y.}^{-1}$$

For the remaining samples the procedure is much less clearcut. He_c^3 is the cosmogenic isotope least influenced by composition such that a He_c^3 age can be calculated for all the samples that contain resolvable He_c^3. Unfortunately the usefulness of He_c^3 is somewhat impaired by its tendency to diffuse out of certain minerals, especially feldspars (MEGRUE, 1966; HEYMANN *et al.*, 1968). Nevertheless we have calculated He_c^3 ages for all cases using the above listed production rate. Several of the Ne_c^{21}/Ar_c^{38} ratios of samples 11-18 fall in the range of values of samples 1-10. We have assumed that this implies roughly similar composition and we have calculated Ne_c^{21} and Ar_c^{38} ages for these cases as well, using the listed production rates.

The 'ages' in Table 4 can be compared with ages reported for the bulk fines at the Lunar Conference in Houston: 207,215 m.y. (HINTENBERGER *et al.*, 1970); 500 m.y. (EBERHARDT *et al.*, 1970); $\simeq$ 800 m.y. (PEPIN *et al.*, 1970); and 500 m.y. (KIRSTEN *et. al.*, 1970). If we disregard the low He_c^3-ages for samples 11-14, which are probably due to He^3 losses, we find that the apparent ages of sub-samples of the same fines vary from approximately 100–600 m.y. with no systematic trends discernible.

A few comments are in order. Our calculations are based on an analogy with chondrites and no allowance was made for the production by solar cosmic rays. FIREMAN *et al.* (1970) have argued that at least one-half of the observed H^3, Ar^{37} and Ar^{39} were produced by solar cosmic rays. At the present time the effect of solar cosmic rays in our calculations cannot be evaluated because of a lack of cross section data at the proton energies of interest. In general the effect of solar cosmic rays would be to lower the ages listed in Table 4.

The radiation ages can be compared to the Rb–Sr ages for type A and B rocks obtained by ALBEE *et al.* (1970), which indicate a widespread event $\sim$3·6 b.y. ago. This may represent either flooding of the entire mare basin or a local lava flow. The radiation ages of the soil and of individual soil fragments are much younger than this event. Apparently, individual fragments were shielded during a large portion of their lifetime. With a thickness of the regolith of some 3–6 m (SHOEMAKER, 1970) the *average* production rates of regolith material is well below the production rates near the top from which the soil was collected. Good vertical mixing on a time scale of 10^8–10^9 yr is thus commensurate with the observed radiation ages.

Trapped gases vs *fragment size*

Figure 1 shows correlations between trapped gas contents and fragment size for lithic fragments, the most common fragments in the fines.

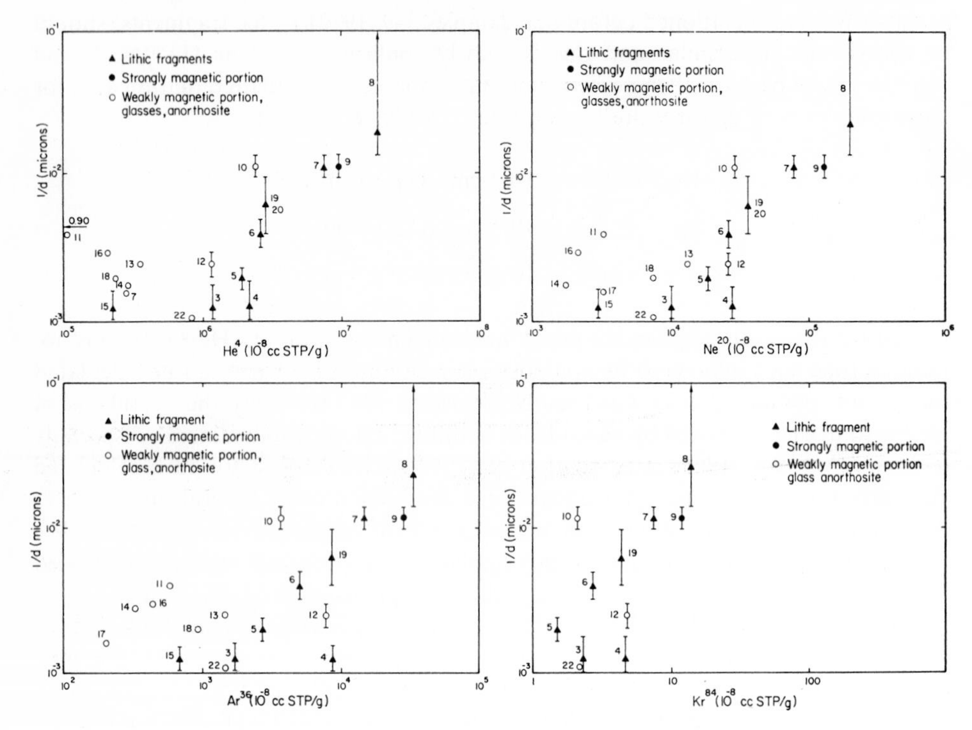

Fig. 1. He^4, Ne^{20}, Ar^{36}, and Kr^{84} vs. $1/d$, the inverse of fragment diameter. For sieve fractions (samples 7, 8, 9, 10 and 19) the upper and lower mesh sizes are used to define the size range. In the remaining cases, the size range indicates roughly the dimensions of the fragment, i.e. greatest and smallest diameter observed under the microscope.

In the range 70–1000 μ the data points fall near 45° lines, which implies a strong anti-correlation between gas content and fragment size. The data can be represented by the following equations:

$$He^4 = [6158/d - 0{\cdot}90] \times 10^{-3}\ \text{cm}^3\ \text{STP/g}$$
$$Ne^{20} = [6890/d - 2{\cdot}5] \times 10^{-5}\ \text{cm}^3\ \text{STP/g}$$
$$Ar^{36} = [11299/d + 3{\cdot}5] \times 10^{-6}\ \text{cm}^3\ \text{STP/g}$$
$$Kr^{84} = [433{\cdot}8/d + 1{\cdot}3] \times 10^{-8}\ \text{cm}^3\ \text{STP/g}$$
$$Xe^{132} = [964{\cdot}4/d + 5{\cdot}9] \times 10^{-9}\ \text{cm}^3\ \text{STP/g}$$

with d = fragment diameter in microns. The results of other investigators, reported at the Lunar Conference, are in good agreement with our data.

The anti-correlation implies that the gases were taken up through the surfaces of the fragments by non-equilibrium processes. The most widely accepted view is that the gases represent solar wind ions, but it is clear that the fines must also contain solar-flare ions and even stopped galactic cosmic rays of relatively low energy. The precise siting of the gases in the fragments cannot be inferred from Fig. 1. However, the ingenious etching studies reported by several investigators (EBERHARDT *et al.*, 1970; HINTENBERGER *et al.*, 1970; KIRSTEN *et al.*, 1970) have shown that the trapped gases are concentrated relatively closely to the outer surfaces of the fragments.

It is somewhat surprising that the correlation does not break down for samples containing only a few fragments such as sample 3 (one fragment) or sample 5 (eight fragments). The amount of gas in any given fragment depends not only on its surface area but also on the duration of its exposure to solar wind. Apparently this average exposure time does not vary greatly from fragment to fragment. Hence there must be continuous and effective 'stirring' at the top of the regolith.

Samples 4, 9 and 12, also shown in Fig. 1, contain more Ar^{36} than lithic fragments of corresponding size. These three samples have in common that they are either wholly made of black glass (sample 12) or that a substantial portion of the surfaces of fragments in these samples are *coated* with the same kind of black glass (Dr. J. A. WOOD has informed us that this glass is Ti and Fe-rich). We believe that the glass was formed by impact melting of gas-rich, very fine dust, followed by rapid quenching such that not too much gas was lost. In such a case one would expect a significant lowering of He^4/Ne^{20} and Ne^{20}/Ar^{36} ratios in the black glass relative to the fines, and this is indeed observed (Table 3).

The remaining samples all have gas contents below those of lithic fragments of the same size. These are mainly clear, or lightly coloured glasses and feldspar fragments (sample 22, though, falls close to the lithic fragments). KIRSTEN *et al.* (1970) have reported that the gas content of clear glass and feldspar is generally below that of ilmenite and Ti-rich glass. Most likely the difference in gas contents reflects different trapping efficiencies of various materials for solar wind ions, but one cannot exclude the possibility that the glassy materials were formed more recently, and hence were not exposed to the solar wind as long as the lithic fragments.

Elemental and isotopic ratios

The principal elemental ratios, He^4/Ne^{20}, Ne^{20}/Ar^{36}, Ar^{36}/Kr^{84} and Kr^{84}/Xe^{132} are shown in Table 3. These are quite different from 'solar-system' values (CAMERON, 1968), and with few exceptions also quite different from values reported for trapped gases in meteorites.

He^4/Ne^{20} is seen to range from 27 to 116 but more than 75 per cent of the values are greater than 70. This is still much lower than most estimates for the cosmic ratio (980, CAMERON, 1968) or the values now seen in carbonaceous chondrites (MAZOR *et al.*, 1970) where He^4/Ne^{20} ranges from an average of 253 ± 76 in the C2's to 354 ± 79 in the C1's. Only the howardites (MAZOR and ANDERS, 1967) show ratios similar to the lunar fines. It is interesting to note that EBERHARDT *et al.* (1970) find a value of 220 in an ilmenite fraction of the fines.

Ne^{20}/Ar^{36} ranges from 3·2 to 16·5, but more than 50 per cent of the cases fall between 4 and 6·5. Again this is much lower than the cosmic value, and lower than the values seen in gas-rich meteorites (generally around 23). EBERHARDT *et al.* (1970) find $Ne^{20}/Ar^{36} = 24 \pm 3$ in an ilmenite fraction of the fines.

The systematic fractionation of the lighter gases and the vast quantities of gas contained in the fines and breccias raise the suspicion that these materials are saturated with solar wind gases, certainly with He, possibly Ne and Ar also. EBERHARDT *et al.* (1970) have strongly advocated this point of view. Exactly what causes the surfaces to become saturated is not clear at this time. The amount of He^4 contained under 1 cm^2 of surface is about $3{\cdot}5 \times 10^{-4}$ cm^3 STP. According to EBERHARDT *et al.* (1970) most of this is sited within 0·2 μ from the surface, hence the average He^4 *concentration* at the surface is $\sim$18 cm^3 STP/cm^2.

Under these conditions it seems quite possible that He^4 was removed from the surface by proton sputtering. Although H^+ sputtering is known to be rather inefficient for the relatively heavy lattice ions, it might be quite effective for the light, interstitial He^4.

A steady state may also be achieved by diffusion of He^4 out of the surface such that the amount lost just compensates the amount implanted from the solar wind. It is clear that one deals with averages over long periods of time. A given fragment may be first exposed to solar wind, then become temporarily shielded, etc. Diffusion losses continue when the fragment was shielded. In the steady state the implantation rate of He^4 is approximately $1{\cdot}9 \times 10^{-13}$ cm^3 STP cm^{-2} sec^{-1}. We have adopted $H^+ = 10^8$ protons cm^{-2} sec^{-1} from SPREITER and ALKSNE, 1969; and $H^+/He^4 = 20$ (cf. BÜHLER *et al.*, 1969). We have made no allowance for shielding, i.e. we assume here that a fragment is always exposed to solar wind. With $He^4 = 3{\cdot}5 \times 10^{-4}$ cm^3 STP cm^{-2}, the fraction of the implanted He^4 that must be lost *on the average* per second is $5{\cdot}4 \times 10^{-10}$. With the equations for diffusion out of a flat plate (JOST, 1960) we calculate $D/h^2 = 5{\cdot}8 \times 10^{-20}$ sec^{-1} (D = diffusion coefficient, h = thickness of the plate). Using $h = 10^{-5}$ cm (0·1 μ) as an average thickness, we find $D \simeq 6 \times 10^{-30}$ cm^2 sec^{-1} as an effective diffusion coefficient needed to maintain He^4 in a steady state at the concentration now seen in the lunar fines, for the case where a fragment was continuously exposed to the solar wind. If the fragment was exposed only part of the time, the value of the diffusion coefficient would become lower.

In the absence of experimentally determined diffusion coefficients in lunar materials it is difficult to test this steady state hypothesis. It is interesting, however, to notice that an observed elemental ratio such as He^4/Ne^{20} would be principally determined by the abundance ratio of the ions in the solar wind and by the square root of the ratio of the diffusion coefficients, i.e. $\sqrt{D_{20}/D_4}$. One important consequence of this is that isotopic ratios such as He^3/He^4 or Ne^{20}/Ne^{22} in the lunar fines cannot be substantially different from the abundance ratios in the solar wind at 1 a.u. He^3 just as He^4 would saturate at a steady state level determined by its diffusion coefficient. Hence the observed He^3/He^4 ratio would differ from the solar wind value by $\sqrt{(D_3/D_4)}$. If isotopic diffusion coefficients vary strictly according to $\sqrt{M}$, then the shift in He^3/He^4 could be approximately -7 per cent; that in Ne^{20}/Ne^{22} only $-2{\cdot}5$ per cent.

ANDERS *et al.* (1970) from a study of trapped He in the carbonaceous chondrites concluded that these meteorites contain two isotopically distinct kinds of He with:

$$(He^3/He^4)_A = (1{\cdot}25 \pm 0{\cdot}76) \times 10^{-4}$$

$$(He^3/He^4)_B = (4{\cdot}20 \pm 0{\cdot}76) \times 10^{-4}.$$

The B-type helium was interpreted as 'solar' helium, most likely trapped solar wind ions. In the lunar fines (Table 3) He^3/He^4 varies because of the admixture of

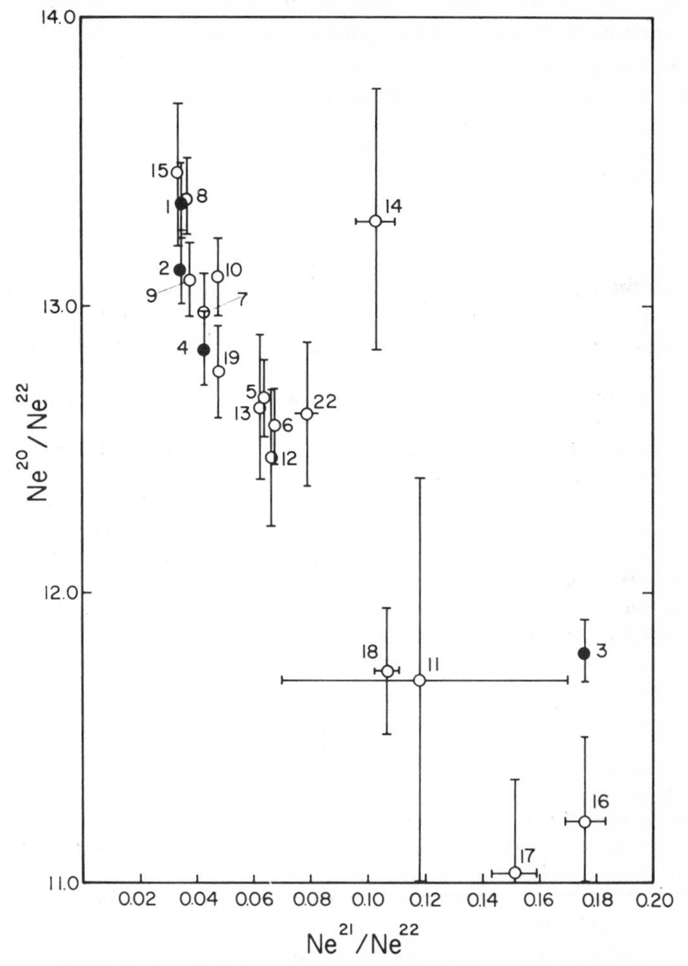

Fig. 2. Ne^{20}/Ne^{22} vs. Ne^{21}/Ne^{22}. A single straight line can be fitted to the points: $(Ne^{20}/Ne^{22}) = 13{\cdot}65–14{\cdot}47\ (Ne^{21}/Ne^{22})$. On the other hand, a near-vertical line could be drawn through samples 1, 2, 4, 7, 8, 9, 10 and 15, which would be the locus of compositions of trapped Ne.

cosmogenic He. However, it is useful to average those He^3/He^4 ratios for samples with $Ne^{20}/Ne^{22} > 13$, because these contain the smallest proportion of cosmogenic gas. The average He^3/He^4 of eight samples is $(4{\cdot}16 \pm 0{\cdot}71) \times 10^{-4}$, in excellent agreement with solar-type He in carbonaceous chondrites.

Our Ne isotopic data are shown in Fig. 2 on a three-isotope plot. The data are consistent with a two component mixture because a single straight line can be fitted to the points:

$$(Ne^{20}/Ne^{22}) = 13{\cdot}65 - 14{\cdot}47\,(Ne^{21}/Ne^{22}).$$

The line would fall close to the curve of PEPIN *et al.* (1970), hence our samples appear to contain only two kinds of neon: trapped solar wind neon and cosmogenic neon. It is thus possible that our samples display the same variation of trapped neon composition, though over a much more restricted range, as was found in the stepwise heating experiments ($Ne^{20}/Ne^{22} \sim 14{\cdot}2$ to $11{\cdot}3$).

Acknowledgments—This work was supported by contract NAS-9-7899 of the National Aeronautics and Space Administration. We thank Dr. J. A. S. ADAMS and Mr. G. E. FRYER for their valuable help.

REFERENCES

ALBEE A. L., BURNETT D. S., CHODOS A. A., EUGSTER O. J., HUNEKE J. C., PAPANASTASSIOU D. A., PODOSEK F. A., RUSS G. PRICE II, SANZ H. G., TERA F. and WASSERBURG G. J. 1970) Ages, irradiation history and chemical composition of lunar rocks from the Sea of Tranquillity. *Science* **167**, 463–466.

ANDERS E., HEYMANN D. and MAZOR E. (1970) Isotopic composition of primordial helium in carbonaceous chondrites. *Geochim. Cosmochim. Acta* **34**, 127–131.

BÜHLER F., EBERHARDT P., GEISS J., MEISTER J., SIGNER P. (1969) Apollo-11 Solar wind composition experiment: first results, *Science* **166**, 1502–1503.

CAMERON A. G. W. (1968) A new table of abundances of the elements in the solar system. In *Origin and Distribution of the Elements*, (editor L. H. Ahrens), pp. 125–143. Pergamon Press.

EBERHARDT P., GEISS J. and GRÖGLER N. (1965a) Über die Verteilung der Uredelgase im Meteoriten Khor Temiki. *Tschermak's Mineral. Petrol. Mit.* **10**, 1–4.

EBERHARDT P., GEISS J. and GRÖGLER N. (1965b) Further evidence on the origin of trapped gases in the meteorite Khor Temiki. *J. Geophys. Res.* **70**, 4375–4378.

EBERHARDT P., EUGSTER O., and MARTI K. (1965c) A redetermination of the isotopic composition of atmospheric neon. *Z. Naturforsch.* **20a**, 623–624.

EBERHARDT P., EUGSTER O. and GEISS J. (1965d) Radiation ages of aubrites. *J. Geophys. Res.* **70**, 4427–4434.

EBERHARDT P., GEISS J. and GRÖGLER N. (1966) Distribution of rare gases in the pyroxene and feldspar of the Khor Temiki meteorite. *Earth Planet. Sci. Lett.* **1**, 7–12.

EBERHARDT P., GEISS J., GRAF H., GRÖGLER N., KRÄHENBÜHL U., SCHWALLER H., SCHWARZMÜLLER J. and STETTLER A. (1970) Trapped solar wind noble gases, Kr^{81}/Kr exposure ages and K/Ar ages in Apollo 11 lunar material. *Science* **167**, 558–560.

FIREMAN E. L., D'AMICIO J. and DEFELICE J. C. (1970) Tritium and argon radioactivities in lunar material. *Science* **167**, 566–568.

HEYMANN D., MAZOR E. and ANDERS E. (1968) Ages of calcium-rich achondrites—I. Eucrites. *Geochim. Cosmochim. Acta* **32**, 1241–1268.

HINTENBERGER H., KÖNIG H., SCHULTZ L., WÄNKE H. and WLOTZKA F. (1964) Die relativen Produktionsquerschnitte für He^3 und Ne^{21} aus Mg, Si, S, und Fe in Steinmeteoriten. *Z. Naturforsch.* **192**, 88–92.

HINTENBERGER H., WEBER H. W., VOSHAGE H., WÄNKE H., BEGEMANN F., VILSCEK E. and WLOTZKA F. (1970) Rare gases, hydrogen and nitrogen: Concentrations and isotopic composition in lunar material. *Science* **167**, 543–545.

JOST W. (1960) *Diffusion in Solids, Liquids and Gases*, p. 41. Academic Press.

KIRSTEN T., STEINBRUNN F. and ZÄHRINGER J. (1970) Rare gases in lunar samples: Study of distribution and variations by a microprobe technique. *Science* **167**, 571–574.

LSPET (LUNAR SAMPLE PRELIMINARY EXAMINATION TEAM). (1969) Preliminary examination of lunar samples from Apollo 11. *Science* **165,** 1211–1227.

MASON B. (1965) The chemical composition of olivine-bronzite and olivine-hypersthene chondrites. *Amer. Mus. Novitates* **2223,** 1–38.

MAZOR E. and ANDERS E. (1967) Primordial gases in the Jodzie howardite and the origin of gas-rich meteorites. *Geochim. Cosmochim. Acta* **31,** 1441–1456.

MAZOR E., HEYMANN D. and ANDERS E. (1970) Noble gases in carbonaceous chondrites. *Geochim. Cosmochim. Acta*. To be published.

MEGRUE G. H. (1966) Rare-gas chronology of calcium-rich achondrites. *J. Geophys. Res.* **71,** 4021–4027.

PEPIN R. O., NYQUIST L. E., PHINNEY D. and BLACK D. C. (1970) Isotopic composition of rare gases in lunar samples. *Science* **167,** 550–553.

REYNOLDS J. H., HOHENBERG C. M., LEWIS R. S., DAVIS P. K. and KAISER W. A. (1970) Isotopic analysis of rare gases from stepwise heating of lunar fines and rocks. *Science* **167,** 545–548.

SPREITER J. R. and ALKSNE A. Y. (1969) Plasma flow around the magnetosphere. *Rev. Geophys.* **7,** 11–50.

SHOEMAKER E. M., HAIT M. H., SWANN G. A., SCHLEICHER D. L., DAHLEM D. H., SCHABER G. G. and SUTTON R. (1970) Lunar regolith at Tranquillity Base. *Science* **167,** 452–455.

STAUFFER H. (1962) On the production ratios of rare gas isotopes in stone meteorites. *J. Geophys. Res.* **67,** 2023–2028.

WOOD J. A., DICKEY J. S., JR., MARVIN U. B. and POWELL B. N. (1970) Lunar anorthosites. *Science* **167,** 602–605.

Proceedings of the Apollo 11 Lunar Science Conference, Vol. 2, pp. 1261 to 1267.

Ar^{40} anomaly in lunar samples from Apollo 11

D. HEYMANN and A. YANIV*
Departments of Geology and Space Science, Rice University, Houston, Texas 77001

(*Received* 5 *February* 1970; *accepted in revised form* 25 *February* 1970)

Abstract—Ar^{40} of up to $\sim 5 \times 10^4$ cm³ STP/g in the fines is correlated to Ar^{36} by:

$$Ar^{40} = 1{\cdot}06\, Ar^{36} + (7{\cdot}12 \pm 1{\cdot}09) \times 10^{-5}.$$

This implies that the "excess" Ar^{40} in the fines is surface correlated. The favored explanation for "excess" Ar^{40} is that it was formed in the moon itself, escaped into the lunar atmosphere, where it was ionized and driven back into the soil by acceleration in the interplanetary electric and magnetic fields. A rough material balance indicates that this would require the release of approximately 0·2 per cent or 1 per cent of all the Ar^{40} produced in the moon, depending on the assumed average K—content. The apparent K–Ar age of the lithic fragments is $4{\cdot}42^{+0{\cdot}24}_{-0{\cdot}28}$ b.y.

INTRODUCTION

WE HAVE seen in the preceding paper that the fines from the Sea of Tranquillity contain substantial amounts of argon, up to nearly $100{,}000 \times 10^{-8}$ cm³ STP/g with Ar^{36}/Ar^{38} quite constant, but Ar^{40}/Ar^{36} varying from 1·25 to 34 (Table 3, preceding paper).

The vast amounts of Ar^{40} pose an intriguing problem, because, if one adopts K = 1000 ppm (LSPET, 1969) as the K-content of the fines, and if one assumes that all of the Ar^{40} in samples 1 and 2 is radiogenic Ar^{40} from K^{40} decay *in situ* one calculates an embarrassingly old K^{40}–Ar^{40} age of greater than 7 b.y. Obviously the fines contain "excess" Ar^{40}, but where did it come from? We favor a lunar origin, in which Ar^{40} diffuses out of the moon and is then implanted into the fines by acceleration in the interplanetary electric and magnetic fields.

Several of the samples were shown to contain cosmogenic Ar_c^{38} (Table 4, preceding paper), hence they contain also cosmogenic Ar_c^{36}. For the purpose of the present paper we shall correct all Ar^{36} values for the cosmogenic component with $Ar_c^{36} = 0{\cdot}6\, Ar_c^{38}$. The largest correction (for sample 22) is nearly 6 per cent, but this is exceptional: in most cases the correction for Ar_c^{36} is less than 1 per cent. Any Ar^{40} correction would be negligible, hence none was made.

CORRELATION BETWEEN Ar^{40} AND Ar^{36} IN THE FINES

Figure 1 contains an important clue to the answer. In this graph we have plotted Ar^{40} vs. Ar^{36} (corrected for Ar_c^{36}) for all of our samples. Let us first consider samples 1, 2, 4, 5, 6, 7, 8, 9 and 19. These consist either of bulk fines or lithic fragments, the most common constituent in the fines. The data points fall on or near a straight line:

$$Ar^{40} = 1{\cdot}06\, Ar^{36} + (7{\cdot}12 \pm 1{\cdot}09) \times 10^{-5}$$

* On leave of absence from the University of Tel Aviv, Ramat Aviv, Israel.

(both Ar^{40} and Ar^{36} in units of 10^{-5} cm³ STP/g). One would expect such a correlation if each sample contained the same amount of radiogenic Ar^{40} from K^{40}-decay *in situ* and another Ar^{40} component which occurs in constant proportion to Ar^{36}, namely $Ar^{40}/Ar^{36} = 1{\cdot}06$.

The first assumption would be satisfied if all the samples contained approximately the same amount of K, and if all the samples had about the same apparent K^{40}–Ar^{40} age. With K = 1000 ppm and $Ar_R{}^{40} = (7{\cdot}12 \pm 1{\cdot}09) \times 10^{-5}$ cm³ STP/g, this age is $4{\cdot}42^{+0{\cdot}24}_{-0{\cdot}28}$ b.y. This result agrees well with the Pb/U, Pb/Pb and Pb/Th ages reported for sample 10084 by TATSUMOTO and ROSHOLT (1970), $4{\cdot}66^{+0{\cdot}07}_{-0{\cdot}16}$ b.y. It also agrees within the limits of error with the Rb–Sr age of magnetic separates of soil as reported by ALBEE *et al.* (1970), $4{\cdot}1 \pm 0{\cdot}1$ b.y. On the other hand, the K–Ar age of the lithic fragments in the soil disagrees with the Rb–Sr ages reported for type A and B rocks, about 3·6 b.y. (ALBEE *et al.*, 1970). The discrepancy might be explained if the soil, in addition to A and B-type material, contains a third component with an age of $\sim$4·5 b.y.

The "excess" Ar^{40}, then, occurs in constant proportion with Ar^{36}, and since we had already concluded in the preceding paper that Ar^{36} is surface correlated, this would imply that Ar^{40} is surface correlated. Does this also imply that Ar^{40}, like Ar^{36} comes to the moon from the Sun via the solar wind? This is very unlikely. Theoretical considerations show that Ar^{40}/Ar^{36} in the Sun should be several orders of magnitude ($\sim 10^{-4}$, CAMERON, 1968) smaller than the ratio of 1·06 seen in Fig. 1. Furthermore, the breccia contain as much as $150{,}000 \times 10^{-8}$ cm³ STP of argon with $Ar^{40}/Ar^{36} = 2{\cdot}2$–$2{\cdot}3$ (LSPET, 1969). Similarly, sample 21, a sieve fraction between 250 and 100 μ has an Ar^{40}/Ar^{36} ratio of 2·13 and falls far above the curve in Fig. 1, whereas the same material after washing with acetone (sample 19) agrees with the linear correlation ($Ar^{40}/Ar^{36} = 1{\cdot}06$). Apparently sample 21 contained a very fine, gas-rich dust with $Ar^{40}/Ar^{36} > 2{\cdot}13$, which was removed by the acetone washing. This variability of Ar^{40}/Ar^{36} speaks strongly against a solar origin for the excess Ar^{40}.

The most probable source for the surface correlated Ar^{40} is the Moon itself. When the moon formed, it may have contained in its interior some primordial argon, low in Ar^{40}. How much, we do not know; the inert gases in the Apollo 11 samples give little if any hint, for the existence in the moon of genuine primordial gas of this kind. From K^{40} decay in the moon, Ar^{40} was accumulated. A gentle heating of the interior of the moon, or large-scale melting processes near the surface, such as seem to have attended the formation of the Sea of Tranquillity would have released a gas phase containing argon enriched in Ar^{40}. This Ar would eventually become available at the surface of the moon, where a mechanism must be found that will firmly implant the Ar in the surfaces of fragments in the "dust", or in the surfaces of rocks.

IMPLANTATION MECHANISMS

There are at least two implantation mechanisms that merit serious consideration. The first is acceleration of ions in the lunar atmosphere by the combined interplanetary electric and magnetic fields, the second is acceleration by elastic collisions of atoms in the lunar atmosphere with solar wind ions. The first mechanism was brought to our attention by T. GOLD (private communication, 1970) at the Lunar Conference in

Houston. On the same subject, there is a detailed paper in preparation by MANKA and MICHEL (R. H. MANKA, private communication, 1970).

When an Ar^{40} atom escapes from beneath the surface of the moon into the lunar atmosphere it may become ionized by charge exchange with solar wind ions and by photoionization. The total lifetime against ionization ($\sim$0·3 yr, MICHEL, 1964) is much shorter than the lifetime against gravitational escape ($\sim 10^8$ yr, cf. BERNSTEIN

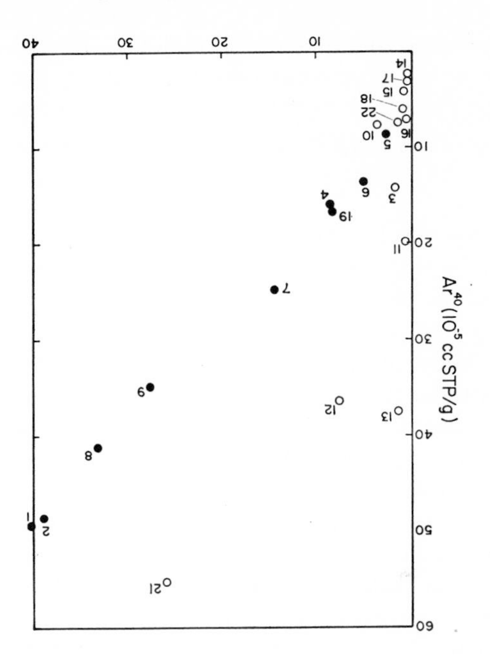

Fig. 1. Ar^{40} vs. Ar^{36} in samples from "fines" (10084). Samples 1, 2, 4, 5, 6, 7, 8, 9 and 19 lie close to a straight line: $Ar^{40} = 1{\cdot}06\ Ar^{36} + (7{\cdot}12 \pm 1{\cdot}09) \times 10^{-5}$. This would imply that the samples contain $(7{\cdot}120 \pm 1{\cdot}086) \times 10^{-5}$ cm^3 STP/g radiogenic Ar^{40} and a surface-correlated component with $Ar^{40}/Ar^{36} = 1{\cdot}06$. The filled symbols were used to calculate the straight line correlation.

et al., 1963). The ion, once formed, will find itself in the interplanetary magnetic field of approximately 10γ at 1 a.u. and in an electric field which is produced by the solar wind moving through a stationary magnetic field. The magnitude and direction of the electric field is given by

$$\mathbf{E} = -\mathbf{V}_{sw} \times \mathbf{B}$$

where $\mathbf{V}_{sw}$ = solar wind velocity.

Near the sunlit surface of the moon the direction of the electric field is roughly perpendicular to the equatorial plane of the moon. The equations of motion of a charged particle, i.c. Ar^{40+}, in the crossed electric and magnetic fields is

$$m\ddot{x} = 0;\ qB\dot{z} = m\ddot{y};\ qE - qB\dot{y} = m\ddot{z}$$

with the x-coordinate along the magnetic field direction, the z-coordinate along the electric field direction; m = mass of ion; q = charge of ion; E = electric field; B = magnetic field.

In terms of the "cyclotron frequency"

$$\omega_c = \frac{qB}{m}$$

the velocity of the ion at any time t is given by:

$$V_x = V_{x,0}$$
$$V_y = (V_{y,0} - qE/m\omega_c) \cos \omega_c t + V_{z,0} \sin \omega_c t + qE/m\omega_c$$
$$V_z = V_{z,0} \cos \omega_c t - (V_{y,0} - qE/m\omega_c) \sin \omega_c t.$$

One may assume that the ion starts at rest in the frame of reference of the moon, hence $V_{x,0} = V_{y,0} = V_{z,0} = 0$. In its initial motion the ion is accelerated along the electrical field, i.e. its motion starts out perpendicular to the direction of the solar wind. Since the radius of gyration of Ar^{40+} under these conditions is on the order of 10^4 km, and since most of the Ar^{40+} ions which will strike the surface of the moon come from within about 100 km above the surface (the scale height for Ar^{40} is approximately 50 km; cf. BERNSTEIN *et al.*, 1963), the ion moves only a short distance along the cycloid before it is stopped. Hence, its motion towards the moon is principally governed by the interplanetary electric field.

The efficiency of this mechanism depends on the trapping probability in the soil, which is a function of the impact energy. The impact energy in turn depends on the height above the lunar surface and the latitude at which the Ar^{40} was ionized. For the trapping probability we use the data by BÜHLER *et al.* (1966), which show that this probability for Ar-ions rises quickly from about 0·2 at 0·5 keV to near 1·0 at about 2·5 keV. Although this trapping efficiency was determined for Al-foil, the authors have pointed out that the results can be used for silicates as well. For the impact energies for Ar ions formed at different altitudes and latitudes we have used the results of MANKA and MICHEL's (1970) calculations. These show that the impact energies range from essentially zero up to a few keV, with a steady increase of energy from the poles to the lunar equator for ions formed at a given height.

Taking into account the density distribution of Ar^{40} as a function of height we estimate that of all the Ar^{40} ions accelerated by this mechanism a few per cent will be trapped upon impact with the surface, roughly one-half will drift away from the moon. The remaining ions, having insufficient energy for implantation, will probably return as neutral Ar^{40} atoms to the lunar atmosphere, where they can go through this cycle again.

The second mechanism calls on impulse transfer by elastic collisions between solar wind ions and neutral Ar^{40} atoms in the lunar atmosphere. It has been proposed by HEYMANN *et al.* (1970). We have now concluded that this mechanism may, in fact, be disregarded. The reason for this is that for low scattering angles, where the differential scattering cross section is relatively high, the Ar atoms gain so little kinetic energy that their trapping probability is very low. At greater scattering angles the kinetic energy of the Ar increases, but the differential scattering cross section decreases substantially (cf. FULS *et al.*, 1957). For example a scattering angle of 49·5° (center of mass system) is required for the energy transfer of 0·45 keV by 25 keV O ions (approximately solar-wind energy with $v = 500$ km/sec) on Ar in an elastic collision. For Ar of 0·45 keV, the trapping probability is $\sim$0·2 (BÜHLER *et al.*, 1966). The differential scattering cross section for O at this angle was derived from that of Ne on Ar as reported by FULS *et al.* (1957), as $\sim 10^{-18}$ cm^2. Hence the product of trapping probability and cross section is $\sim 2 \times 10^{-19}$ cm^2.

This number must now be compared to the charge exchange cross section for H^+ on Ar, which is on the order of 10^{-15} cm^2 (FITE *et al.*, 1960). Charge exchange is one of the processes responsible for producing Ar^+ ions in the first mechanism. But we have seen that the first mechanism has an efficiency of only a few per cent. Hence the product of efficiency and cross section is here $\sim 10^{-17}$ cm^2.

On the basis of these considerations alone, scattering of Ar by O ions in the solar wind is $\sim$50 times less efficient at this scattering angle than the first mechanism. However, the abundance of O in the solar wind is $\sim 10^3$ times less than that of H (cf. BAME *et al.*, 1968). Accordingly the scattering mechanism in this case is much less efficient than the first mechanism.

We have repeated these calculations for H, He, O and Ne at several different scattering angles, and have concluded that all cases the efficiency of the elastic scattering is $\ll 1$ per cent of the first mechanism.

DISCUSSION

We shall now attempt to make a rough material balance for Ar^{40}. Any Ar^{40} released into the lunar atmosphere is removed from the moon by three processes: (I) acceleration of ions in the interplanetary electric and magnetic fields; (II) scattering by elastic collisions, and (III) gravitational escape. The life times of Ar^{40} are $\sim 10^7$ sec, $\sim 10^8$ sec (BERNSTEIN *et al.*, 1963), and $\sim 10^{15}$ sec respectively. In the following we shall therefore use mechanism I as the one that governs both the implantation of Ar^{40} in the moon's surface as well as the removal of Ar^{40} from the moon.

According to BÜHLER *et al.* (1969) the flux of solar-wind He^4 during the Apollo 11 lunar excursion was $(6{\cdot}3 \pm 1{\cdot}2) \times 10^6$ atoms cm^{-2} sec^{-1}. We have adopted $He^4/Ar^{36} \sim 1{\cdot}4 \times 10^4$ following CAMERON (1968). If one assumes that the flux measured by BÜHLER *et al.* represents the average He^4 flux over the total exposure time of the fines to solar wind, then the average Ar^{36} flux was $\sim 4 \times 10^2$ atoms cm^{-2} sec^{-1}.

The ratio $Ar^{40}/Ar^{36} = 1{\cdot}06$ now seen in the fines would seem to imply that the Ar^{40} flux near the equator was also $\sim 4 \times 10^2$ atoms cm^{-2} sec^{-1}. However, this requires that the trapping efficiencies for Ar^{36} and Ar^{40} are the same. This need not be the case, since the terminal velocities of Ar^{40} near the equator ($\sim$1 keV) are much smaller than the kinetic energy of Ar^{36} ($\sim$40 keV). We have seen in the preceding paper that Ar^{36} may be saturated in the fines. Probably Ar^{40} is also saturated, but because of its lower implantation energy it may have saturated at a relatively lower concentration than Ar^{36}. In this sense $Ar^{40}/Ar^{36} \sim 1$ is probably a lower limit for the ratio of the fluxes.

The number of Ar^{40} ions that must be accelerated in the lunar atmosphere is greater than $\sim 4 \times 10^2$ atoms cm^{-2} sec^{-1} for at least two reasons. The first reason is that only a fraction of these ions becomes trapped. We have estimated in the preceding section that this fraction is roughly a few per cent. The second reason is connected with geometrical factors; the Ar^{40} ions arriving at the equator were formed at higher latitudes, both north and south of the equator, and the angle of incidence is about 10° to the surface (MANKA and MICHEL, 1970). Considering all these factors we estimate that the number of Ar^{40} atoms released from the surface should be on the average roughly 2×10^4 atoms cm^{-2} sec^{-1} over the total exposure time of the fines to solar wind. We believe that this estimate is accurate within a factor of ten.

Let us consider two cases for Ar^{40} production in the moon. The first is based on chondritic K content of the moon, 850 ppm. WASSERBURG *et al.* (1964) have pointed out that a K/U ratio $\approx 1 \times 10^4$ is probably characteristic for terrestrial materials, and that for a model earth with this ratio, an average terrestrial K content of $\sim$220 ppm can be concluded. We shall use this number for our second case.

For the chondritic K-content the amount of Ar^{40} produced in the moon since 4–6 b.y. ago is $\sim 6 \times 10^{17}$ moles. This corresponds to an average production of $\sim 8 \times 10^6$ Ar^{40} atoms cm^{-2} sec^{-1}. Comparing this number to required flux of $\sim 2 \times 10^4$ atoms cm^{-2} sec^{-1} we conclude that approximately 0·2 per cent of all the Ar^{40} produced in the moon must have been released into the atmosphere to account for the Ar^{40} excess seen in the fines from Apollo 11. For the second case this fraction is of the order of 1 per cent.

Another possible source for the "excess" Ar^{40} in the fines could be cosmic dust, which on impact is rapidly fused into the surface of fragments in the fines with little if any loss of argon. KEAYS *et al.* (1970) have concluded that the lunar soil contains an admixture of 2 per cent by weight of carbonaceous-chondrite-like material, which may be of meteoritic or cometary origin. If this material, like meteorites, would have a maximum K–Ar age of $\sim$4·5 b.y., it could not account for the "excess" Ar^{40}. In order to explain the "excess", assuming K = 500 ppm, the age must be at least 18 b.y. far greater than the accepted age of the universe.

Acknowledgments—This work was supported by contract NAS-9-7899 of the National Aeronautics and Space Administration. We thank Drs. R. H. MANKA and F. C. MICHEL for making their preliminary results available to us. We also thank Drs. W. D. ARNETT, D. D. CLAYTON, and T. GOLD for helpful discussions.

REFERENCES

ALBEE A. L., BURNETT D. S., CHODOS A. A., EUGSTER O. J., HUNEKE J. C., PAPANASTASSIOU D. A., PODOSEK F. A., RUSS G. PRICE II, SANZ H. G., TERA F. and WASSERBURG G. J. (1970) Ages, irradiation history, and chemical composition of lunar rocks from the Sea of Tranquillity. *Science* **167,** 463–466.

BAME S. J., HUNDHAUSEN A. J., ASBRIDGE J. R. and STRONG I. B. (1968) Solar wind ion composition. *Phys. Rev. Lett.* **20,** 393–395.

BERNSTEIN W., FREDRICKS R. W., VOGL J. L. and FOWLER W. A. (1963) The lunar atmosphere and the solar wind. *Icarus* **2,** 233–248.

BÜHLER F., GEISS J., MEISTER J., EBERHARDT P., HUNEKE J. C. and SIGNER P. (1966). *Earth Planet. Sci. Lett.* **1,** 249–255.

BÜHLER F., EBERHARDT P., GEISS J., MEISTER J. and SIGNER P. (1969) Apollo 11 solar wind composition experiment: first results. *Science* **166**, 1502–1503.

CAMERON A. G. W. (1968) A new table of abundances of the elements in the solar system. In *Origin and Distribution of the Elements* (editor L. H. Ahrens), pp. 125–143. Pergamon.

FITE W. L., STEBBINGS R. F., HUMMER D. G. and BRACKMANN R. T. (1960) Ionization and charge transfer in proton hydrogen atom collisions. *Phys. Rev.* **119**, 663–668.

FULS E. N., JONES P. R., ZIEMBA F. P. and EVERHART E. (1957) Measurements of large-angle single collision between helium, neon, and argon atoms at energies to 100 keV. *Phys. Rev.* **107**, 704–710.

HEYMANN D., YANIV A., ADAMS J. A. S. and FRYER G. E. (1970) Inert gases in lunar samples. *Science* **167**, 555–558.

KEAYS R. R., GANAPATHY R., LAUL J. C., ANDERS E., HERZOG G. F. and JEFFERY P. M. (1970) Trace elements and radioactivity in lunar rocks: Implications for meteorite infall, solar-wind flux, and formation conditions of Moon. *Science* **167**, 490–493.

LSPET (LUNAR SAMPLE PRELIMINARY EXAMINATION TEAM) (1969) Preliminary examination of lunar samples from Apollo 11. *Science* **165**, 1211–1227.

MANKA R. H. and MICHEL F. C. (1970) Lunar atmosphere as a source of Ar^{40} and other lunar surface elements. In preparation.

MICHEL F. H. (1964) Interaction between the solar wind and the lunar atmosphere. *Planet. Space Sci.* **12**, 1075–1091.

TATSUMOTO M. and ROSHOLT J. N. (1970) Age of the Moon: An isotopic study of Uranium–thorium–lead systematics of lunar samples. *Science* **167**, 461–463.

WASSERBURG G. J., MACDONALD G. J. F., HOYLE F. and FOWLER W. (1964) Relative contribution of uranium, thorium and potassium to heat production in the earth. *Science* **143**, 465–467.

Proceedings of the Apollo 11 Lunar Conference, Vol. 2, pp. 1269 to 1282.

Concentrations and isotopic abundances of the rare gases, hydrogen and nitrogen in Apollo 11 lunar matter

H. HINTENBERGER, H. W. WEBER, H. VOSHAGE, H. WÄNKE, F. BEGEMANN and F. WLOTZKA

Max-Planck-Institut für Chemie (Otto-Hahn-Institut), Mainz, Germany

(*Received* 2 *February* 1970; *accepted in revised form* 25 *February* 1970)

Abstract—The concentrations and isotopic abundances of the rare gases have been investigated in fines (10084) and three types of rocks (10003, 10017, 10021, 10044, 10049, 10057). In the fines the results obtained from different grain size fractions and from samples etched with HNO_3 and HF to different degrees demonstrate the strong concentrations of the solar-wind component in the surface layers of the grains. From these data the solar-wind components, the radiogenic ^{40}Ar, and the spallogenic ^{3}He, ^{21}Ne and ^{38}Ar have been determined. The $^{4}He/^{20}Ne$- and $^{4}He/^{36}Ar$-ratios of the solar wind components in separated iron grains from the fines are much higher than in the bulk fines. This we interpret to be due to preferential diffusion loss of helium and neon from the silicates. Krypton and xenon in the rocks show a strong spallation component and compared to calcium-rich achondrites an excess of ^{131}Xe which has been found especially high in the type B rocks 10003 and 10017. As this ^{131}Xe excess correlates with the other spallogenic components a possible explanation is the high abundance of rare earth elements in lunar matter. The investigation of the non-rare gases showed that the breccias 10021 and 10061 contain solar-wind implanted hydrogen which is depleted in deuterium by at least a factor of three as compared to hydrogen of standard mean ocean water. From dissolution experiments some indication was found for solar wind implanted nitrogen in fines 10084.

RARE GASES

THE CONCENTRATIONS and isotopic abundances of the rare gases have been investigated in fines and in all three types of rocks.

The rare gases were released by heat extraction and measured in a double focusing all-metal ultra-high-vacuum mass spectrometer which has been used for several years for meteorite research (SCHULTZ and HINTENBERGER, 1967). The results obtained on the Berkeley Bruderheim Standard, distributed by J. H. Reynolds in 1968 are included in Table 1.

1. *Fines*

In addition to bulk samples two sets of different grain size fractions and a few mg of iron grains have been analysed. The results are compiled in Table 2. The concentrations of all rare gases show a pronounced systematic decrease with increasing grain size.

For grain sizes smaller than 150 μm the rare gas concentrations are almost inversely proportional to the particle size. This is shown in Fig. 1 where the He-content is plotted against the inverse mean grain size. For the sake of simplicity the mean grain size was taken as the arithmetic mean of the limiting mesh widths. Certainly this is an approximation only which will be especially poor for the fraction <30 μm and <50 μm, respectively. Hence it is not surprising that the data point for the <30 μm fraction does not fall on the straight line, which would be the case for a mean grain

Table 1. Rare gas concentrations in fines 10084-18 and in the Bruderheim standard (Errors: <3 per cent for He, Ne and Ar; <10 per cent for Kr and Xe)

Nuclide	Bruderheim standard (150 mg) (10^{-8} cm³ STP/g)	Fines 10084-18 No. 1 (1·32 mg) (cm³ STP/g)	No. 2 (1·34 mg) (cm³ STP/g)	Average (cm³ STP/g)
^{3}He	49·3	$7{\cdot}88 . 10^{-5}$	$7{\cdot}00 . 10^{-5}$	$7{\cdot}44 . 10^{-5}$
^{4}He	531	$2{\cdot}01 . 10^{-1}$	$1{\cdot}79 . 10^{-1}$	$1{\cdot}90 . 10^{-1}$
^{20}Ne	8·74	$2{\cdot}21 . 10^{-3}$	$1{\cdot}98 . 10^{-3}$	$2{\cdot}10 . 10^{-3}$
^{21}Ne	9·62	$5{\cdot}65 . 10^{-6}$	$5{\cdot}45 . 10^{-6}$	$5{\cdot}55 . 10^{-6}$
^{22}Ne	10·6	$1{\cdot}74 . 10^{-4}$	$1{\cdot}63 . 10^{-4}$	$1{\cdot}69 . 10^{-4}$
^{36}Ar	1·48	$3{\cdot}71 . 10^{-4}$	$3{\cdot}80 . 10^{-4}$	$3{\cdot}76 . 10^{-4}$
^{38}Ar	1·48	$7{\cdot}21 . 10^{-5}$	$7{\cdot}39 . 10^{-5}$	$7{\cdot}30 . 10^{-5}$
^{40}Ar	1130	$4{\cdot}09 . 10^{-4}$	$4{\cdot}37 . 10^{-4}$	$4{\cdot}23 . 10^{-4}$
^{84}Kr	0·0077	$2{\cdot}11 . 10^{-7}$	$2{\cdot}65 . 10^{-7}$	$2{\cdot}38 . 10^{-7}$
^{132}Xe	0·011	$2{\cdot}87 . 10^{-8}$	$3{\cdot}08 . 10^{-8}$	$2{\cdot}98 . 10^{-8}$

size of 18 μm instead the arithmetic mean of 15 μm used here. The larger grain-sizes do not show such a regular decrease of rare-gas concentrations with increasing grain size. In some cases even an increase of the concentrations with increasing particle size is observed. This may be due to the occurrence of conglomerates of smaller dust grains which falsify the true grain-size distribution. Further investigations are needed to settle this question.

As can be seen from Table 2 the ratios $^{22}Ne/^{21}Ne$ and $^{40}Ar/^{36}Ar$ vary with grain size as well. This allows to separate the solar wind components from the cosmic ray produced ^{21}Ne and the radiogenic ^{40}Ar. The ^{21}Ne–^{20}Ne-plot in Fig. 2 shows a linear correlation given by

$$^{21}\text{Ne} = (4{\cdot}45 \pm 0{\cdot}48) . 10^{-7}\ \text{cm}^3\ \text{STP/g} + (2{\cdot}45 \pm 0{\cdot}03) . 10^{-3}\ {}^{20}\text{Ne}.$$

The value of $^{21}Ne = (4{\cdot}45 \pm 0{\cdot}48) . 10^{-7}$ cm³/g at $^{20}Ne = {}^{21}Ne$ is thus the spallation component, while the ratio $^{21}Ne/^{20}Ne = (2{\cdot}45 \pm 0{\cdot}03) . 10^{-3}$ is that of the solar wind component. Similarly from the ^{40}Ar–^{36}Ar-plot in Fig. 3 we deduce

$$^{40}\text{Ar} = (7{\cdot}47 \pm 0{\cdot}64) . 10^{-5}\ \text{cm}^3\ \text{STP/g} + (0{\cdot}947 \pm 0{\cdot}019)\ {}^{36}\text{Ar}.$$

The intercept value of $^{40}Ar = (7{\cdot}47 \pm 0{\cdot}64) . 10^{-5}$ cm³ STP/g at $^{36}Ar \approx 0$ is explained as radiogenic argon, while $^{40}Ar/^{36}Ar = 0{\cdot}95 \pm 0{\cdot}02$ represents the isotope ratio of the surface trapped argon in the fines which may be different from the isotope ratio in the solar wind (HEYMANN *et al.*, 1970). The radiogenic ^{40}Ar, together with a potassium content of 1090 ppm, (WÄNKE *et al.*, 1970), yields a gas-retention age of $(4{\cdot}36 \pm 0{\cdot}15) . 10^9$ yr for the fines. This may be compared to $(4{\cdot}42 \pm ^{0{\cdot}24}_{0{\cdot}28}) . 10^9$ yr obtained by HEYMANN *et al.* (1970) and $3{\cdot}5 . 10^9$ yr reported by KIRSTEN *et al.* (1970). A rough value of 1000 ppm has been used in both of these cases for the potassium content.

While there can be little doubt that the surface correlated large amounts of rare gases have been implanted by solar wind, their elemental abundance distribution is almost certainly not the same as that in the solar wind. This is clearly seen when comparing the concentrations in the bulk fines and in the iron grains (Table 2, last column). The isotopic ratios are almost identical, but the ratios of nuclides of different elements differ significantly. The ^{4}He–^{20}Ne-ratio is higher by a factor of 3, the

Table 2. Rare-gas concentrations in cm³ STP/g in the bulk material in various grain-size fractions of fines of Apollo 11 No. 10084-18. Errors in isotope ratios 2 per cent, errors in concentrations for He, Ne and Ar about 3 per cent, for Kr and Xe about 10 per cent. Sample weights between 1 and 6 mg

Nuclides and ratios	Original fines (average)	Grain-size fractions I			
		$<30\ \mu m$	$30–100\ \mu m$	$100–250\ \mu m$	$>250\ \mu m$
^{3}He	$7{\cdot}44 \,.\, 10^{-5}$	$1{\cdot}07 \,.\, 10^{-4}$	$3{\cdot}98 \,.\, 10^{-5}$	$1{\cdot}80 \,.\, 10^{-5}$	$2{\cdot}48 \,.\, 10^{-5}$
^{4}He	$1{\cdot}90 \,.\, 10^{-1}$	$2{\cdot}80 \,.\, 10^{-1}$	$1{\cdot}02 \,.\, 10^{-1}$	$4{\cdot}47 \,.\, 10^{-2}$	$6{\cdot}86 \,.\, 10^{-2}$
^{20}Ne	$2{\cdot}10 \,.\, 10^{-3}$	$3{\cdot}37 \,.\, 10^{-3}$	$1{\cdot}18 \,.\, 10^{-3}$	$5{\cdot}84 \,.\, 10^{-4}$	$1{\cdot}03 \,.\, 10^{-3}$
^{21}Ne	$5{\cdot}55 \,.\, 10^{-6}$	$8.71 \,.\, 10^{-6}$	$3{\cdot}42 \,.\, 10^{-6}$	$1{\cdot}92 \,.\, 10^{-6}$	$3{\cdot}06 \,.\, 10^{-6}$
^{22}Ne	$1{\cdot}69 \,.\, 10^{-4}$	$2{\cdot}68 \,.\, 10^{-4}$	$9{\cdot}44 \,.\, 10^{-5}$	$4{\cdot}74 \,.\, 10^{-5}$	$8{\cdot}35 \,.\, 10^{-5}$
^{36}Ar	$3{\cdot}76 \,.\, 10^{-4}$	$6{\cdot}27 \,.\, 10^{-4}$	$2{\cdot}36 \,.\, 10^{-4}$	$1{\cdot}63 \,.\, 10^{-4}$	$2{\cdot}51 \,.\, 10^{-4}$
^{38}Ar	$7{\cdot}30 \,.\, 10^{-5}$	$1{\cdot}21 \,.\, 10^{-4}$	$4{\cdot}72 \,.\, 10^{-5}$	$3{\cdot}22 \,.\, 10^{-5}$	$5{\cdot}03 \,.\, 10^{-5}$
^{40}Ar	$4{\cdot}23 \,.\, 10^{-4}$	$6{\cdot}60 \,.\, 10^{-4}$	$3{\cdot}04 \,.\, 10^{-4}$	$2{\cdot}36 \,.\, 10^{-4}$	$4{\cdot}17 \,.\, 10^{-4}$
^{84}Kr	$2{\cdot}38 \,.\, 10^{-7}$	$3{\cdot}37 \,.\, 10^{-7}$	$1{\cdot}26 \,.\, 10^{-7}$	$8{\cdot}87 \,.\, 10^{-8}$	$1{\cdot}45 \,.\, 10^{-7}$
^{132}Xe	$2{\cdot}98 \,.\, 10^{-8}$	$4{\cdot}00 \,.\, 10^{-8}$	$1{\cdot}82 \,.\, 10^{-8}$	$1{\cdot}59 \,.\, 10^{-8}$	$2{\cdot}00 \,.\, 10^{-8}$

Nuclides and ratios	Grain-size fractions II					Selected iron grains
	$<50\ \mu m$	$50–100\ \mu m$	$100–150\ \mu m$	$150–200\ \mu m$	$200–300\ \mu m$	
^{3}He	$8{\cdot}88 \,.\, 10^{-5}$	$3{\cdot}62 \,.\, 10^{-5}$	$1{\cdot}92 \,.\, 10^{-5}$	$1{\cdot}96 \,.\, 10^{-5}$	$1{\cdot}89 \,.\, 10^{-5}$	$3{\cdot}87 \,.\, 10^{-5}$
^{4}He	$2{\cdot}14 \,.\, 10^{-1}$	$9{\cdot}25 \,.\, 10^{-2}$	$4{\cdot}84 \,.\, 10^{-2}$	$5{\cdot}01 \,.\, 10^{-2}$	$4{\cdot}81 \,.\, 10^{-2}$	$9{\cdot}80 \,.\, 10^{-2}$
^{20}Ne	$2{\cdot}71 \,.\, 10^{-3}$	$1{\cdot}20 \,.\, 10^{-3}$	$6{\cdot}20 \,.\, 10^{-4}$	$6{\cdot}30 \,.\, 10^{-4}$	$6{\cdot}34 \,.\, 10^{-4}$	$3{\cdot}38 \,.\, 10^{-4}$
^{21}Ne	$7{\cdot}04 \,.\, 10^{-6}$	$3{\cdot}52 \,.\, 10^{-6}$	$1{\cdot}93 \,.\, 10^{-6}$	$1{\cdot}89 \,.\, 10^{-6}$	$1{\cdot}94 \,.\, 10^{-6}$	$9{\cdot}44 \,.\, 10^{-7}$
^{22}Ne	$2{\cdot}18 \,.\, 10^{-4}$	$9{\cdot}64 \,.\, 10^{-5}$	$4{\cdot}96 \,.\, 10^{-5}$	$5{\cdot}02 \,.\, 10^{-5}$	$5{\cdot}10 \,.\, 10^{-5}$	$2{\cdot}73 \,.\, 10^{-5}$
^{36}Ar	$4{\cdot}91 \,.\, 10^{-4}$	$2{\cdot}37 \,.\, 10^{-4}$	$1{\cdot}35 \,.\, 10^{-4}$	$1{\cdot}45 \,.\, 10^{-4}$	$1{\cdot}80 \,.\, 10^{-4}$	$5{\cdot}11 \,.\, 10^{-5}$
^{38}Ar	$9{\cdot}57 \,.\, 10^{-5}$	$4{\cdot}80 \,.\, 10^{-5}$	$2{\cdot}64 \,.\, 10^{-5}$	$2{\cdot}85 \,.\, 10^{-5}$	$3{\cdot}41 \,.\, 10^{-5}$	$1{\cdot}03 \,.\, 10^{-5}$
^{40}Ar	$5{\cdot}57 \,.\, 10^{-4}$	$2{\cdot}90 \,.\, 10^{-4}$	$1{\cdot}97 \,.\, 10^{-4}$	$2{\cdot}12 \,.\, 10^{-4}$	$2{\cdot}46 \,.\, 10^{-4}$	$7{\cdot}68 \,.\, 10^{-5}$
^{84}Kr	$3{\cdot}22 \,.\, 10^{-7}$	$1{\cdot}59 \,.\, 10^{-7}$	$8{\cdot}80 \,.\, 10^{-8}$	$9{\cdot}31 \,.\, 10^{-8}$	$9{\cdot}22 \,.\, 10^{-8}$	$3{\cdot}10 \,.\, 10^{-8}$
^{132}Xe	$4{\cdot}02 \,.\, 10^{-8}$	$2{\cdot}29 \,.\, 10^{-8}$	$1{\cdot}22 \,.\, 10^{-8}$	$1{\cdot}36 \,.\, 10^{-8}$	$1{\cdot}38 \,.\, 10^{-8}$	$5{\cdot}50 \,.\, 10^{-9}$

Nuclides and ratios	Original fines (average)	Grain-size fractions I			
		$<30\ \mu m$	$30–100\ \mu m$	$100–250\ \mu m$	$>250\ \mu m$
$^{4}He/^{3}He$	2550	2620	2560	2480	2770
$^{20}Ne/^{22}Ne$	12·4	12·6	12·5	12·3	12·3
$^{22}Ne/^{21}Ne$	30·5	30·8	27·6	24·7	27·3
$^{36}Ar/^{38}Ar$	5·15	5·12	5·00	5·06	4·99
$^{40}Ar/^{36}Ar$	1·13	1·05	1·29	1·45	1·66
$^{4}He/^{20}Ne$	91	83	86·8	76	67
$^{4}He/^{36}Ar$	505	447	432	274	273
$^{20}Ne/^{36}Ar$	5·59	5·37	5·00	3·58	4·10

Nuclides and ratios	Grain-size fractions II					Selected iron grains
	$<50\ \mu m$	$50–100\ \mu m$	$100–150\ \mu m$	$150–200\ \mu m$	$200–300\ \mu m$	
$^{4}He/^{3}He$	2580	2560	2520	2560	2540	2530
$^{20}Ne/^{22}Ne$	12·4	12·4	12·5	12·5	12·4	12·4
$^{22}Ne/^{21}Ne$	31·0	27·4	25·7	26·6	26·3	28·9
$^{36}Ar/^{38}Ar$	5·16	4·96	5·11	5·09	5·28	4·96
$^{40}Ar/^{36}Ar$	1·13	1·22	1·46	1·46	1·37	1·50
$^{4}He/^{20}Ne$	84	77	78·1	79·5	76	290
$^{4}He/^{36}Ar$	436	390	359	346	267	1920
$^{20}Ne/^{36}Ar$	5·52	5·06	4·59	4·34	3·52	6·61

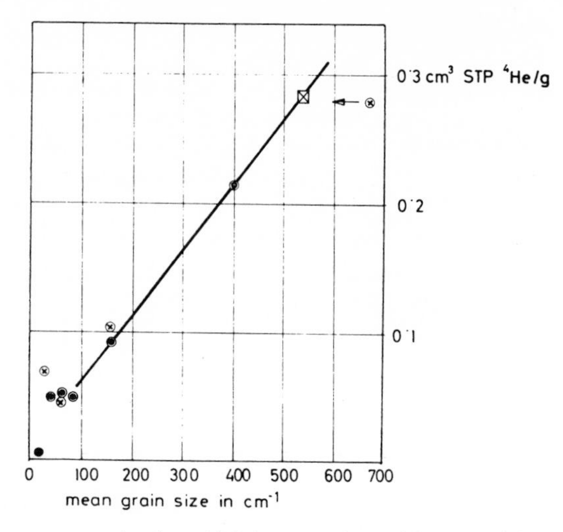

Fig. 1. ^{4}He vs. inverse mean grain size which is approximated by the arithmetic mean of the limiting mesh widths. The data of series I in Table 2 are marked by ⊗, those of series II by ⊙.

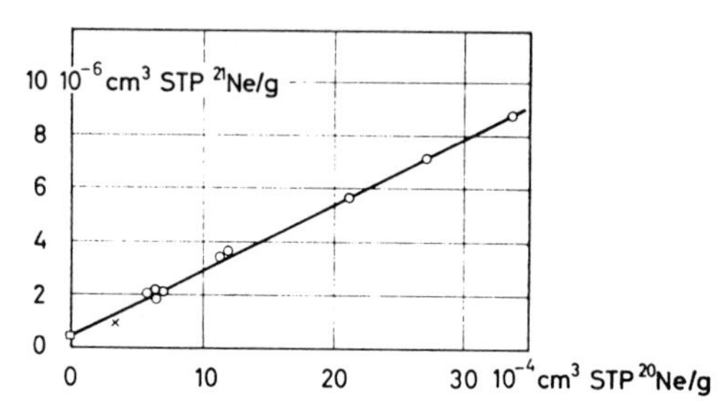

Fig. 2. ^{21}Ne vs. ^{20}Ne for different grain sizes <300 μm in fines 10084. ○ sieved bulk samples, × selected iron grains (not used for the calculation of the regression line), □ intercept.

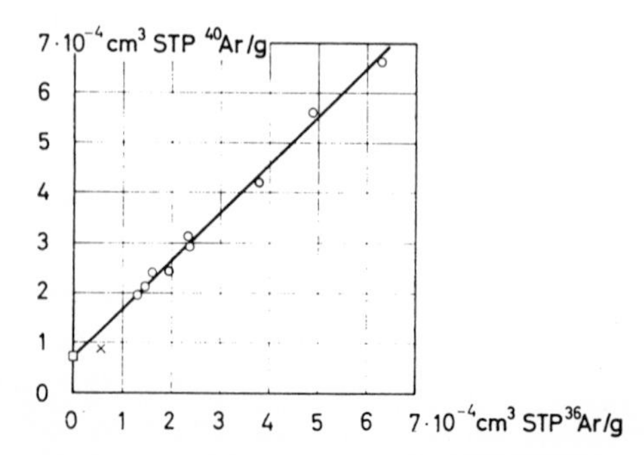

Fig. 3. ^{40}Ar vs. ^{36}Ar for different grain sizes <300 μm in fines 10084. ○ sieved bulk samples, × selected iron grains (not used for the calculation of the regression line), □ intercept.

$^{4}He/^{36}Ar$-ratio higher by a factor of 4 in the iron grains than in the bulk fines. The strong depletion of He and to a lesser extent that of neon we interpret due to preferential diffusion losses of these gases. This is in accordance with our experience (HINTENBERGER *et al.*, 1965) on gas rich stone meteorites, where the metal fraction invariably shows the least or no diffusion losses at all. Although we believe the abundance distribution of the rare gases in lunar metal grains to reflect the solar wind composition better than the bulk fines, even this abundance pattern can have been changed or falsified, e.g. by adhering non-metallic dust. Comparing our results for the metal grains with those for ilmenite of EBERHARDT *et al.* (1970) it appears that most of the silicates have lost their solar wind He almost completely and a considerable fraction of their neon as well. This is in accordance with direct observations by KIRSTEN *et al.* (1970).

Finally small aliquots of bulk fines were etched to different degrees with HNO_3 and HF and the rare gas content in the residues measured. In Table 3 the results are

Table 3. Rare-gas concentrations in cm³ STP/g in the bulk material and in acid-etched fractions of fines of Apollo 11 No. 10084-18. Errors see Table 2. Sample weights used between 1 and 6 mg

Nuclides and ratios	Original fines	Acid-etched fractions (removed wt. %)			Big grains from etched sample
		40	93	99	
^{3}He	$7 \cdot 44 \,.\, 10^{-5}$	$3 \cdot 13 \,.\, 10^{-5}$	$5 \cdot 83 \,.\, 10^{-6}$	$3 \cdot 07 \,.\, 10^{-6}$	$2 \cdot 75 \,.\, 10^{-6}$
^{4}He	$1 \cdot 90 \,.\, 10^{-1}$	$8 \cdot 72 \,.\, 10^{-2}$	$1 \cdot 02 \,.\, 10^{-2}$	$9 \cdot 51 \,.\, 10^{-4}$	$6 \cdot 84 \,.\, 10^{-3}$
^{20}Ne	$2 \cdot 10 \,.\, 10^{-3}$	$5 \cdot 25 \,.\, 10^{-4}$	$4 \cdot 04 \,.\, 10^{-5}$	$3 \cdot 14 \,.\, 10^{-6}$	$3 \cdot 35 \,.\, 10^{-5}$
^{21}Ne	$5 \cdot 55 \,.\, 10^{-6}$	$1 \cdot 78 \,.\, 10^{-6}$	$3 \cdot 57 \,.\, 10^{-7}$	$4 \cdot 37 \,.\, 10^{-7}$	$2 \cdot 04 \,.\, 10^{-7}$
^{22}Ne	$1 \cdot 69 \,.\, 10^{-4}$	$4 \cdot 31 \,.\, 10^{-5}$	$3 \cdot 57 \,.\, 10^{-6}$	$7 \cdot 29 \,.\, 10^{-7}$	$2 \cdot 82 \,.\, 10^{-6}$
^{36}Ar	$3 \cdot 76 \,.\, 10^{-4}$	$3 \cdot 04 \,.\, 10^{-5}$	$1 \cdot 73 \,.\, 10^{-6}$	$9 \cdot 46 \,.\, 10^{-7}$	$1 \cdot 71 \,.\, 10^{-6}$
^{38}Ar	$7 \cdot 30 \,.\, 10^{-5}$	$6 \cdot 44 \,.\, 10^{-6}$	$6 \cdot 65 \,.\, 10^{-7}$	$5 \cdot 69 \,.\, 10^{-7}$	$4 \cdot 46 \,.\, 10^{-7}$
^{40}Ar	$4 \cdot 22 \,.\, 10^{-4}$	$5 \cdot 00 \,.\, 10^{-5}$	$4 \cdot 33 \,.\, 10^{-5}$	$2 \cdot 76 \,.\, 10^{-5}$	$9 \cdot 10 \,.\, 10^{-6}$
^{84}Kr	$2 \cdot 38 \,.\, 10^{-7}$	$2 \cdot 21 \,.\, 10^{-8}$	$2 \cdot 30 \,.\, 10^{-9}$	$4 \cdot 96 \,.\, 10^{-9}$	$1 \cdot 71 \,.\, 10^{-9}$
^{132}Xe	$2 \cdot 98 \,.\, 10^{-8}$	$4 \cdot 60 \,.\, 10^{-9}$	$5 \cdot 70 \,.\, 10^{-10}$	$1 \cdot 10 \,.\, 10^{-9}$	$6 \cdot 80 \,.\, 10^{-10}$
$^{4}He/^{3}He$	2550	2790	1750	310	2490
$^{20}Ne/^{22}Ne$	12·4	12·2	11·3	4·31	11·9
$^{22}Ne/^{21}Ne$	30·5	24·2	10·0	1·67	13·8
$^{36}Ar/^{38}Ar$	5·15	4·71	2·60	1·67	3·83
$^{40}Ar/^{36}Ar$	1·13	1·63	25	29·2	5·32
$^{4}He/^{20}Ne$	90·5	166	252	303	204

given for samples, where 40, 93 and 99 per cent by weight have been removed. Column 6 shows the concentrations in a few mg of large grains (>250 μm) which had been selected from the residuum of an acid-etched sample. In all these cases a systematic decrease of the concentrations of all rare gases was found with increasing amount of material removed by etching. This is clearly demonstrated for ^{4}He, ^{20}Ne and ^{36}Ar by Fig. 4 and provides an additional argument that the rare gases in the fines are concentrated in the outer layers of the grains. Among the light rare gases Ar shows the strongest, He the smallest decrease of concentration with increasing weight of removed material (Fig. 4). This clearly supports the conclusion derived at above that diffusion has drastically altered the distribution of the solar wind rare gases.

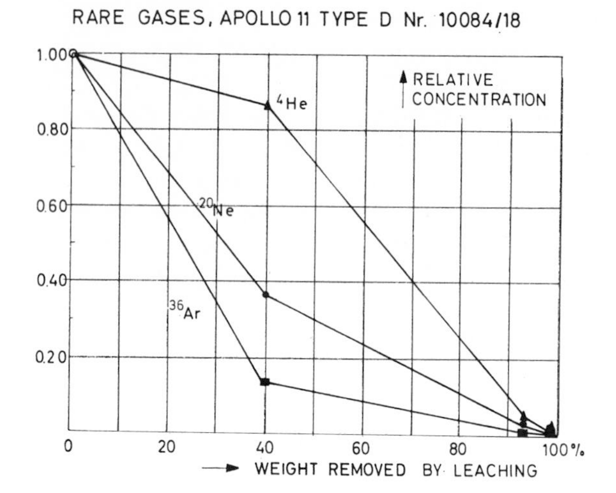

Fig. 4. Relative concentrations of rare gases vs. weight removed by etching.

In the etched samples the isotope ratios with one nuclide abundantly formed by spallation and the other abundantly occurring in the solar wind (as e.g. $^{3}He/^{4}He$, $^{21}Ne/^{20}Ne$ and $^{38}Ar/^{36}Ar$) show an especially strong increase with increasing degree of etching (Fig. 5).

From the data of the original fines and the 40% etched fraction the cosmic ray produced ^{21}Ne component turns out to be $5{\cdot}1 \cdot 10^{-7}$ cm^3 STP/g in reasonable agreement with the value derived from different grain size fractions. As the $^{4}He/^{3}He$ and $^{36}Ar/^{38}Ar$ ratios in the two samples are not very different the cosmic ray produced ^{3}He and ^{38}Ar cannot be calculated with any degree of accuracy. For these two isotopes the

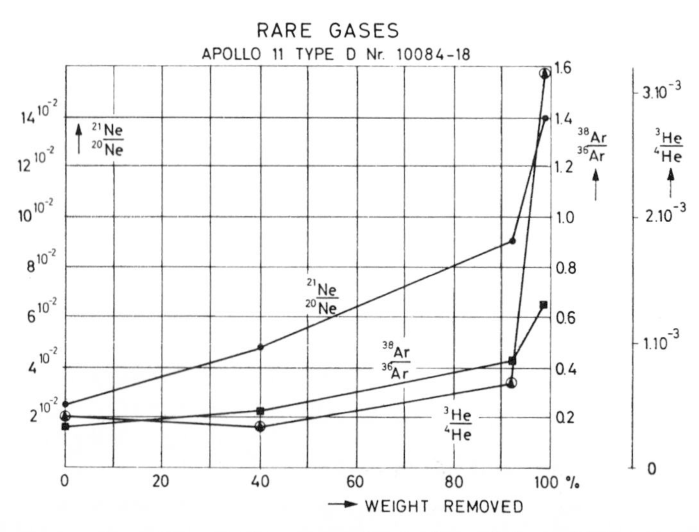

Fig. 5. Ratios of nuclides which are abundantly formed by spallation to nuclides abundantly occurring in solar wind.

spallogenic components can be determined by comparing the data of the original fines with those for the most strongly etched fraction (99 per cent). In this way we find $(^{3}He)_{spall} = 2 \cdot 5 \,.\, 10^{-6}$ and $(^{38}Ar)_{spall} = 4 \cdot 5 \,.\, 10^{-7}$ cm³ STP/g. This value for $(^{38}Ar)_{spall}$ is probably higher than that in the bulk fines, because the 99 per cent sample is enriched in ilmenite which is true for the 93 per cent as well. This change in chemical composition of the latter two fractions is the reason why for the calculation of the $(^{21}Ne)_{spall}$ only the 40 per cent fraction was used, where the concentration of the relevant target elements (Mg + Al + Si) is changed only slightly.

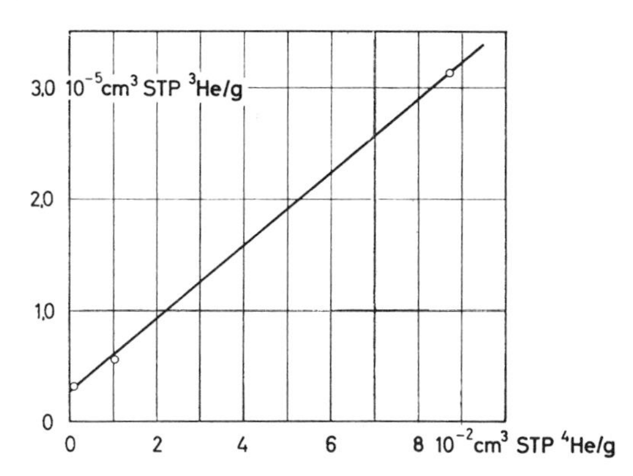

Fig. 6. ³He vs. ⁴He for etched samples of bulk fines 10084.

As the chemical composition has only small influence on the production rate of ³He by the cosmic radiation, this component can also be derived from a ³He–⁴He-plot using the data from all three etched fractions (Fig. 6) which yields

$$^{3}He = (2 \cdot 63 \pm 0 \cdot 15) \,.\, 10^{-6} \text{ cm}^3 \text{ STP/g} + (3 \cdot 29 \pm 0 \cdot 03) \,.\, 10^{-4}\, {}^{4}He.$$

The intercept is to understand as spallation component $(^{3}He)_{spall} = (2 \cdot 63 \pm 0 \cdot 15) \,.\, 10^{-6}$ cm³ STP/g while the figure $(3 \cdot 29 \pm 0 \cdot 03) \,.\, 10^{-4}$ is the (³He/⁴He)-ratio of the solar wind component. If the data of the untreated bulk sample is included the figures change to $(^{3}He)_{spall} = (1 \cdot 5 \pm 1 \cdot 7) \,.\, 10^{-6}$ cm³ STP/g and $(^{3}He/^{4}He) = (3 \cdot 76 \pm 0 \cdot 17) \,.\, 10^{-4}$. We do not have yet an explanation for these differences.

For krypton and xenon in the fines the abundances of all isotopes resemble those found in the terrestrial atmosphere (Figs. 7 and 8).

2. *Crystalline rocks and breccia*

The rare gas concentrations determined in type A, B and C rocks are compiled in Table 4. For type C rock 10021-20 the values are similar to those determined for the fines.

In type A and B rocks all three components solar wind, radiogenic and spallogenic have been detected. It is very probable, however, that for He, Ne and Ar the solar wind component is due to traces of fines attached to the rock chips. A sample from the interior of a chip from rock type A 10057-40 carefully prepared to avoid any such contamination, did not show any component which can be unambiguously

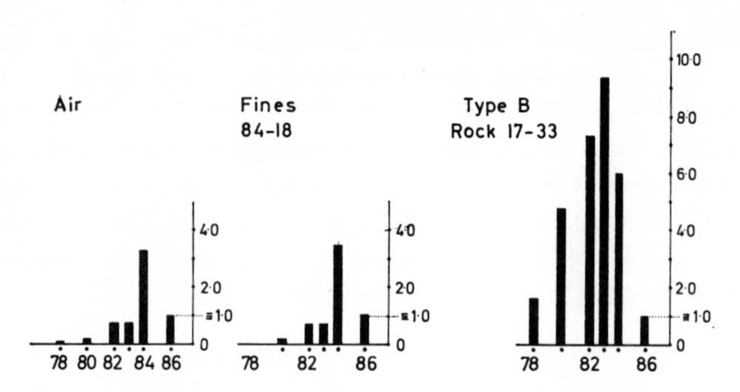

Fig. 7. Isotope abundances of krypton in fines and rocks normalized to ^{86}Kr.

attributed to solar wind (Sample 10057-40-β in Table 4). The results for radiogenic ^{4}He and ^{40}Ar as well as for spallogenic ^{3}He and ^{21}Ne are also included in Table 4. The ages deduced therefrom will later be given in a separate paper.

Of particular interest are the concentrations and isotopic abundances of krypton and xenon in the rocks. The isotopic abundance distribution of Kr and Xe of the most characteristic samples are shown in Figs. 7 and 8; compare also ALBEE *et al.* (1970), REYNOLDS *et al.* (1970), PEPIN *et al.* (1970), MARTI *et al.* (1970).

There is a strong spallogenic component in both gases. In addition in some samples ^{131}Xe shows a peculiar excess compared to meteoritic spallation xenon,

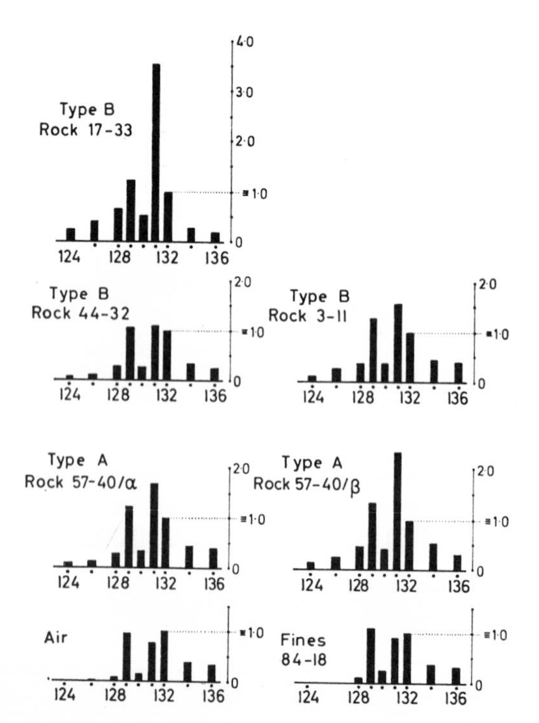

Fig. 8. Isotope abundances of xenon in lunar fines and rocks normalized to ^{132}Xe.

which appears, however, to be correlated with the spallogenic component. In the type B rock 10017-33 the ratio $^{131}Xe/^{132}Xe$ is 3·57 compared with a value of 0·79 in atmospheric xenon and 0·85 in the fines. The measured $^{131}Xe/^{132}Xe$-ratio varies from rock to rock and also from sample to sample within the same rock. In a surface sample of a type A rock (10057-40-α) we observed $^{131}Xe/^{132}Xe$ = 1·68 whereas in a carefully prepared sample from the inner part of the rock (10057-40-β) this ratio is 2·34.

Table 4. Rare-gas concentrations in cm³ STP/g in rocks of type A, B and C of Apollo 11. Errors see Table 2. Sample weights for type A and type B rocks about 100 mg, for type C rocks 3 mg

Nuclides ratios	Type C	Type B			Type A		
	10021-20	10003-11	10017-33	10044-32	10049-20	10057-40α	10057-40β
^{3}He	3·27 . 10^{-5}	1·13 . 10^{-6}	2·92 . 10^{-6}	6·37 . 10^{-7}	2·84 . 10^{-7}	4·39 . 10^{-7}	3·82 . 10^{-7}
^{4}He	9·05 . 10^{-2}	2·24 . 10^{-4}	4·91 . 10^{-4}	1·49 . 10^{-4}	8·89 . 10^{-4}	6·76 . 10^{-4}	5·56 . 10^{-4}
^{20}Ne	1·25 . 10^{-3}	1·05 . 10^{-6}	7·92 . 10^{-7}	1·18 . 10^{-7}	1·64 . 10^{-6}	5·49 . 10^{-7}	6·26 . 10^{-8}
^{21}Ne	3·51 . 10^{-6}	1·29 . 10^{-7}	4·66 . 10^{-7}	6·71 . 10^{-8}	3·15 . 10^{-8}	5·83 . 10^{-8}	5·34 . 10^{-8}
^{22}Ne	9·96 . 10^{-5}	2·13 . 10^{-7}	5·59 . 10^{-7}	8·38 . 10^{-8}	1·61 . 10^{-7}	7·95 . 10^{-8}	6·40 . 10^{-8}
^{36}Ar	1·85 . 10^{-4}	2·86 . 10^{-7}	4·65 . 10^{-7}	9·54 . 10^{-8}	3·03 . 10^{-7}	1·16 . 10^{-7}	4·92 . 10^{-8}
^{38}Ar	3·64 . 10^{-5}	2·20 . 10^{-7}	6·83 . 10^{-7}	1·24 . 10^{-7}	1·02 . 10^{-7}	8·51 . 10^{-8}	8·10 . 10^{-8}
^{40}Ar	5·64 . 10^{-4}	2·63 . 10^{-5}	4·95 . 10^{-5}	4·47 . 10^{-5}	6·47 . 10^{-5}	5·18 . 10^{-5}	5·63 . 10^{-5}
^{84}Kr	8·50 . 10^{-8}	2·80 . 10^{-10}	7·20 . 10^{-10}	2·00 . 10^{-10}	5·40 . 10^{-10}	4·20 . 10^{-10}	1·20 . 10^{-10}
^{132}Xe	2·20 . 10^{-8}	9·90 . 10^{-11}	3·50 . 10^{-10}	1·10 . 10^{-10}	3·10 . 10^{-10}	1·50 . 10^{-10}	8·30 . 10^{-11}
$^{4}He/^{3}He$	2770	198	168	234	3130	1540	1460
$^{20}Ne/^{22}Ne$	12·6	4·93	1·42	1·41	10·2	6·91	0·98
$^{22}Ne/^{21}Ne$	28·4	1·65	1·20	1·25	5·11	1·36	1·20
$^{36}Ar/^{38}Ar$	5·08	1·30	0·68	0·77	2·97	1·36	0·61
$^{40}Ar/^{36}Ar$	3·05	92	106	469	214	447	1140
$(^{21}Ne)_{spall}$		1·3 . 10^{-7}	4·6 . 10^{-7}	6·7 . 10^{-8}	2·7 . 10^{-8}	5·7 . 10^{-8}	5·3 . 10^{-8}
$(^{3}He)_{spall}$		1·1 . 10^{-6}	2·9 . 10^{-6}	6·4 . 10^{-7}	2·3 . 10^{-7}	4·2 . 10^{-7}	3·8 . 10^{-7}
$(^{4}He)_{rad}$		1·4 . 10^{-4}	4·5 . 10^{-4}	1·4 . 10^{-4}	7·4 . 10^{-4}	6·3 . 10^{-4}	5·5 . 10^{-4}
$(^{40}Ar)_{rad}$		2·6 . 10^{-5}	4·9 . 10^{-5}	4·5 . 10^{-5}	6·4 . 10^{-5}	5·2 . 10^{-5}	5·6 . 10^{-5}

A detailed description of the heavy rare gas results will be given later in a separate paper. We do want to point out, however, that the abundance pattern of the spallogenic xenon in these lunar samples cannot be expected to be identical with that observed in calcium-rich achondrites, due to the much higher ratio of the rare earth elements to barium. The contribution from the rare earth elements, in particular from cerium, will be much more important. As the heavier xenon isotopes are shielded this will lead to an increase only for ^{131}Xe, ^{129}Xe and the lighter isotopes, but certainly will be most pronounced for ^{131}Xe.

Hydrogen and Nitrogen

In order to obtain some supplementary information with regard to the solar wind trapped in lunar material the elemental and isotopic abundances of non-rare gases, particularly of hydrogen and nitrogen were measured in a few selected lunar rocks.

Samples of pulverized or crushed rocks 10021-20, 10061-11 and 10049-20, weighing between 240 and 377 mg each, were transferred into a platinum crucible suspended within the quartz tube of the extraction apparatus. The vacuum-packed sample 10061-11 was completely handled in an atmosphere of pure and dry oxygen so that contamination by atmospheric nitrogen and moisture was avoided. The heat extractions were performed in the temperature interval 100–1000°C by heating the quartz tube with a crucible furnace. One sample of 10021-20, after having been treated in this way, was heated again and melted by high frequency.

The gases released were pumped continuously by means of a mercury diffusion pump into a McLeod gauge to collect and determine the amount of gas.

The gas amounts released from various samples of the breccia 10021-20 in the temperature interval 100–1000°C were 2·6–3·0 cm³ STP/g. For the vacuum-packed breccia 10061-11 2·8 cm³ STP/g were obtained, while from one sample of the type A

Table 5. Partial analysis of gases released between 100 and 1000°C from Apollo 11 rocks

Gas	Breccia 10021-20		Breccia 10061-11 vacuum-packed		Blank: degassed breccia 10021-20
	Amount released per gram breccia (cm³ STP/g)	Atomic concentration relative to He	Amount released per gram breccia (cm³ STP/g)	Atomic concentration relative to He	Limits for amount released per gram (cm³ STP/g)
Total	3·0		2·8		⩽0·29
H_2	1·17*	4·33*	1·60	6·50	⩽0·12
He	0·54	1·00	0·49	1·00	
CO	0·39	0·73	0·47	0·96	⩽0·12
N_2	0·105	0·39	0·10	0·41	⩽0·01
^{36}Ar	0·00076	0·0014	0·00067	0·0014	
^{38}Ar	0·00013		0·00013		
^{40}Ar	0·00194		0·00140		
CO_2	0·15	0·28	0·045	0·09	⩽0·05
C†		1·01		1·05	

* These H_2-values may be too low by about 40 per cent due to a variation of the spectrometer sensitivity for H_2.

† Calculated under the assumption that carbon was completely oxidized to CO or CO_2.

rock 10049-20 a gas amount of only 0·65 cm³ STP/g was given off. Less than 0·1 cm³ STP/g were given off when samples of rock 10021-20 were heated subsequently to the melting point. In order to estimate the blank a 238 mg sample of vacuum-molten and crushed rock 10021-20 was heated to 1000°C; an amount of 0·29 cm³ STP/g was collected.

Partial analyses of the gases extracted from the breccias were performed by an 8-in. single focusing mass spectrometer (Varian MAT CH 5). In order to make possible a direct comparison of the rare and non-rare gases a mass resolution of 2500 was chosen. The results are shown in Table 5. Both the elemental and isotopic abundance data confirm that the breccias contain large amounts of the light rare gases which commonly are attributed to trapped solar wind. Obviously, the same holds true for hydrogen, since the hydrogen in the breccia is considerably higher than the total gas amount which is extracted from the type A rock. According to a preliminary analysis, the type A rock contains nitrogen in amounts comparable to those in the breccia in agreement with the results by MOORE *et al.* (1970), while H_2, CO and CO_2 are only slightly higher than in the blank run.

To determine the isotopic composition of the hydrogen released from the breccias the same mass spectrometer was used at a special setting of the ion source potentials.

Under these circumstances, the spectrometer allowed to completely resolve the triplet ^3He–HD–H_3. Hence, no separation of the hydrogen from the other gases was required while the accuracy of the isotopic ratio measurements was still sufficient to confirm our expectation that hydrogen from gas-rich lunar material must be highly depleted in deuterium.

Standard hydrogen samples were prepared from distilled tap water, from standard mean ocean water (SMOW) and from antarctic snow (SNOW) by reduction over uranium at 700°C, and from cylinder hydrogen as well. Their D/H-ratios were determined with an isotope ratio mass spectrometer (Varian MAT GD 150). Using the absolute D/H-ratios as measured by ROTH (1969), namely ≈156 ppm for SMOW and ≈89 ppm for SNOW, values of 144 ppm and 41 ppm were found for distilled tap water and for cylinder hydrogen, respectively.

These standards and the gas samples from two different extractions of 250 mg breccia 10021-20 were alternately run with the high resolution mass spectrometer. Corrections were made to allow for background and molecular flow. The results obtained from the two sets of measurements were

$$(1) \quad D/H = (1{\cdot}08 \pm 0{\cdot}30) \,.\, (D/H)_{\text{Cyl.Hyd.}} = (44 \pm 12) \,.\, 10^{-6}$$

$$(2) \quad D/H = (3{\cdot}5 \pm 0{\cdot}4)^{-1} \,.\, (D/H)_{\text{SMOW}} = (45 \pm 5) \,.\, 10^{-6}.$$

The error limits do only comprise the inaccuracies of the mass spectrometric procedures, not those of the gas extraction. The method was tested by also determining the $^4He/^3He$ ratio in an artificial H_2–^3He–^4He mixture which was investigated independently with the rare gas spectrometer, and in the gas extracted from the breccia as well. The result thus obtained for He from the breccia 10021-20 is $^4He/^3He = 2660 \pm 200$, in reasonable agreement with independent determinations.

The result on the D/H-ratio shows clearly that in the solar wind—and, therefore, in all probability in the solar atmosphere—the deuterium is depleted by at least a factor of three compared to standard mean ocean water. Qualitatively, this is in accordance with spectroscopic observations (KINMAN, 1956) and with theoretical considerations (SALPETER, 1955; FOWLER *et al.*, 1962; BERNAS *et al.*, 1967). EPSTEIN *et al.* (1970) and FRIEDMAN *et al.* (1970) found still lower D/H ratios in their Apollo 11 samples. For several reasons the D/H ratios measured for trapped solar wind hydrogen cannot be expected to reflect the exact solar ratio, and even not the exact solar wind ratio.

A value of about 0·15 (Table 5, rock 10061-11) for the elemental ratio He/H in the breccia will be even less representative for the solar atmosphere or the solar wind. While the abundance ratios of the rare gases have been already altered significantly by diffusion losses alone, here chemical properties will cause additional perturbations. The He/H value of 0·15 found in this study is comparable to the highest values reported for this strongly time dependent ratio in the solar wind (HUNDHAUSEN *et al.*, 1967; HIRSHBERG *et al.*, 1970) and is considerably higher than in the gases released from the gas-rich meteorite Pesyanoe (LORD, 1969).

Our high-resolution mass spectrometer does not resolve the mass doublet $^{15}N^{14}N$–COH. Hence, the $^{15}N/^{14}N$ ratio cannot be determined unless the nitrogen is separated from the gas released from the lunar samples. Therefore, the gas sample extracted from a 280 mg sample of the breccia 10021-20 was run through a gas chromatograph (Varian Aerograph 1533-2B). For the mass spectrometric analysis of the nitrogen fraction again a resolution of about 2500 was chosen in order to detect possible background peaks. Considerable gain variations of the secondary electron multiplier used resulted in a rather moderate precision of this preliminary $^{15}N/^{14}N$ determination. The uncorrected result of 0·319 ± 0·010 per cent for the "lunar" nitrogen has to be compared with a value of 0·322 ± 0·006 per cent obtained for an atmospheric nitrogen sample which was prepared by means of the above-mentioned gas chromatograph.

As the heat extraction experiments do not yet allow to draw definite conclusions with regard to the origin of the nitrogen found a series of dissolution experiments was performed to clarify this question.

Table 6. Nitrogen compounds determined in lunar fines and lunar basalt by various dissolving agents

Lunar fines (10084-18)	Ammonium and nitride nitrogen (ppm)	Nitrate and nitrite nitrogen (ppm)
1. H_2O	11 ± 5	≤ 8
2. H_2SO_4	56 ± 5	≤ 7
3. H_2SO_4 + HF	9 ± 5	≤ 5
Lunar rock (10057-40)		
1. H_2O	≤ 5	17 ± 10
2. H_2SO_4	≤ 5	≤ 5
3. H_2SO_4 + HF	≤ 5	≤ 5

About 100 mg of lunar fines (10084-18) and powdered lunar rock (10057-40) were treated first with 10 ml of water (containing 100 ppm H_2SO_4), thereafter with the same amount of H_2SO_4 1:10 and finally with a mixture of 8 ml H_2SO_4 1:10 and 2 ml 40% HF. Each dissolution step was performed at room temperature for one hour, the last step resulting in a complete dissolution of the sample. The liquids were separated from the residues by centrifugation and after distillation ammonium nitrogen was determined with Nessler's reagens, nitrate and nitrite nitrogen after reduction with Devarda's alloy in the same way (WLOTZKA, 1961). The value obtained for ammonium represents the nitrogen originally present as ammonium or as nitride. Blanks were run for the whole procedure using the same amounts of chemicals and reagents. Table 6 shows the results. Due to the small samples used the amounts of ammonium found in most cases were only slightly higher than those in the blanks. We therefore consider these values to be upper limits only. Significantly higher than the blanks, however, is the value for the sulfuric acid treatment of the lunar fines. Furthermore it is about ten times higher than the upper limit for the lunar type A rock (10057-40). Comparing, then, the results obtained on "fines" and rock it is striking that a pronounced excess of nitrogen is found in the sulfuric acid extract only. In this step the feldspar as well as the surface layers of all other minerals are being dissolved (WÄNKE *et al.*, 1970), the nitrogen must therefore have been located at these sites. As the feldspar from the rock obviously does not yield comparable amounts of nitrogen we consider this to be evidence that the nitrogen has been implanted into the surface layers of the "fines" by solar wind. It appears not to be present as elementary nitrogen, however, but as nitrides or ammonia. This is indicated by the results of a similar experiment where a sample was treated *under vacuum* with H_2SO_4 (1:10), to which was added $HgSO_4$ to avoid the evolution of hydrogen upon dissolution of the traces of metallic iron. During this treatment less than 10 per cent of the total nitrogen—as well as of the hydrogen—was liberated in the gaseous form, which at the same time sets an upper limit to the amount of nitrogen adsorbed on these samples.

Acknowledgements—The co-operation of G. König, Christa Müller, K. F. Roth and S. Specht is gratefully acknowledged. The gas chromatographic separation of nitrogen fractions from atmospheric and lunar samples has been performed by E. Jagoutz and K. Schneider. We thank NASA and the Bundesministerium für Bildung und Wissenschaft for making available Apollo 11 material for this investigation.

References

Albee A. L., Burnett D. S., Chodos A. A., Eugster O. J., Huneke J. C., Papanastassiou D. A., Podosek F. A., Russ Price G., Sanz H. G., Tera F. and Wasserburg G. J. (1970) Ages, irradiation history, and chemical composition of lunar rocks from the Sea of Tranquillity. *Science* **167,** 463–466.

Bernas R., Gradsztajn E., Reeves H. and Schatzman E. (1967) On the nucleosynthesis of lithium, beryllium, and boron. *Ann. Phys.* **44,** 426–478.

Eberhardt P., Geiss J., Graf H., Grögler N., Krähenbühl U., Schwaller H., Schwarzmüller J. and Stettler A. (1970) Trapped solar wind noble gases, Kr^{81}/Kr exposure ages and K/Ar ages in Apollo 11 lunar material. *Science* **167,** 558–560.

Epstein S. and Taylor H. P. (1970) O^{18}/O^{16}, Si^{30}/Si^{28}, D/H, and C^{13}/C^{12} studies of lunar rocks and minerals. *Science* **167,** 533–535.

Fowler W. A., Greenstein J. L. and Hoyle F. (1962) Nucleosynthesis during the early history of the solar system. *Geophys. J.* **6,** 148–220.

Friedman I., O'Neil J. R., Adami L. H., Gleason J. D. and Hardcastle K. (1970) Water, hydrogen, deuterium, carbon, carbon-13, and oxygen-18 content of selected lunar material. *Science* **167,** 538–540.

Heyman D., Yaniv A., Adams J. A. S. and Fryer G. E. (1970) Inert gases in lunar samples. *Science* **167,** 555–558.

Hintenberger H., Vilcsek E. and Wänke H. (1965) Über die Isotopenzusammensetzung und über den Sitz der leichten Uredelgase in Steinmenteoriten. *Z. Naturforsch.* **20a,** 939–945.

Hirshberg J., Alksne A., Colburn D. S., Bame S. J. and Hundhausen A. J. (1970) Observation of a solar flare induced interplanetary shock and helium enriched driver gas. *J. Geophys. Res.* **75,** 1–15.

Hundhausen A. J., Asbridge J. R., Bame S. J., Gilbert H. E. and Strong I. B. (1967) Vela 3 satellite observations of solar wind ions: a preliminary report. *J. Geophys. Res.* **72,** 87–100.

Kinman T. D. (1956) An attempt to detect deuterium in the solar atmosphere. *Mon. Not. Roy. Astron. Soc.* **116,** 77–87.

Kirsten T., Steinbrunn F. and Zähringer J. (1970) Rare gases in lunar samples: study of distribution and variations by a microprobe technique. *Science* **167,** 571–574.

Lord H. C., III (1969) Possible solar primordial hydrogen in the Pesyanoe meteorite. *Earth Planet. Sci. Lett.* **6,** 332–334.

Marti K., Lugmair G. W. and Urey H. C. (1970) Solar wind gases, cosmic-ray spallation products, and the irradiation history. *Science* **167,** 548–550.

Moore C. B., Lewis C. F., Gibson E. K. and Nichiporuk W. (1970) Total carbon and nitrogen abundances in lunar samples. *Science* **167,** 495–497.

Pepin R. O., Nyquist L. E., Phinney D. and Black D. C. (1970) Isotopic composition of rare gases in lunar samples. *Science* **167,** 550–553.

REYNOLDS J. H., HOHENBERG C. M., LEWIS R. S., DAVIS P. K. and KAISER W. A. (1970) Isotopic analysis of rare gases from stepwise heating of lunar fines and rocks. *Science* **167,** 545–548.

ROTH E. (1969) Recent improvements in precise and absolute isotope ratio measurements. *Int. Conf. Mass Spectroscopy, Kyoto,* Sept. 8–12, 1969. Preprint

SALPETER E. E. (1955) Nuclear reactions in stars. II. Protons on light nuclei. *Phys. Rev.* **97,** 1237–1244.

SCHULTZ L. and HINTENBERGER H. (1967) Edelgasmessungen an Eisenmeteoriten. *Z. Naturforsch.* **22a,** 773–779.

WÄNKE H., BEGEMANN F., VILCSEK E., RIEDER R., TESCHKE F., BORN W., QUIJHANO-RICO M., VOSHAGE H. and WLOTZKA F. (1970) Major and trace elements and cosmic-ray produced radioisotopes in lunar samples *Science* **167,** 523–525.

WLOTZKA F. (1961) Untersuchungen zur Geochemie des Stickstoffs. *Geochim. Cosmochim. Acta* **24,** 106–154.

Proceedings of the Apollo 11 Lunar Science Conference, Vol. 2, pp. 1283 to 1309.

Trapped and cosmogenic rare gases from stepwise heating of Apollo 11 samples

C. M. HOHENBERG, P. K. DAVIS, W. A. KAISER, R. S. LEWIS and J. H. REYNOLDS
Department of Physics, University of California, Berkeley, California 94720

(*Received* 2 *February* 1970; *accepted in revised form* 10 *March* 1970)

Abstract—We examined the rare gases from stepwise heating of lunar dust and two crystalline rocks mass spectrometrically. We used isotope correlation diagrams extensively in our analysis of the data in order to identify trapped, cosmogenic, and fissiogenic components and to deduce their release curves. Cosmogenic neon, argon, krypton, and xenon have compositions similar to the cosmogenic gases in calcium-rich achondrites, although some differences occur—notably in ^{131}Xe, which is anomalously abundant in rock 10057 but not in rock 10044. Our sample of rock 10057 contains trapped solar gases, but our sample of rock 10044 does not, suggesting that production of ^{131}Xe may be depth dependent. Fissiogenic xenon in the rocks appears to originate entirely from spontaneous fission of ^{238}U. Isotopes ^{128}Xe, ^{129}Xe, ^{80}Kr and ^{82}Kr are in apparent excess in some of the temperature fractions from the rocks, but except for ^{129}Xe in rock 10044 the excesses are probably "memory" of a previous study of a neutron irradiated meteorite. A genuine excess of ^{129}Xe at two temperatures in rock 10044 is accompanied by excess ^{132}Xe and, possibly, ^{131}Xe and ^{124}Xe. There thus may be more than one cosmogenic component in rock 10044. In none of our lunar work have we seen evidence of any ^{129}Xe from decay of extinct ^{129}I. Rather large variations, with release temperature, in the isotopic composition of trapped helium, neon, and argon can probably be explained by a combination of two effects: (1) deeper burial of heavier isotopes in the solar wind; (2) diffusive separation of isotopes during thermal release. Exposure ages of rocks 10057 and 10044 by the ^{81}Kr–^{83}Kr method are 34 and 70 m.y. respectively. Exposure ages based on ^{83}Kr and ^{130}Xe and the meteoritic production rates for those isotopes, are higher than the Kr–Kr age for rock 10057 as would be expected if the rock spent time at a partially shielded depth. Rock 10044 may have had a simple irradiation history. Our average value for the average exposure age of the lunar fines is 1300 m.y.

INTRODUCTION

WE DESCRIBE the results of stepwise heating of three lunar samples. At each of a series of successively higher temperatures, we examined in a glass mass spectrometer all the rare gases evolved. We continued the programmed heating beyond the melting of the samples until all gases had been extracted. Our objective in the work was to resolve the rare gas mass spectra into the anticipated constituent components—trapped,* cosmogenic, radiogenic and fissiogenic gases—and in the process to search for possible additional components. As in meteorite studies our basic technique has been to plot correlation diagrams for appropriate pairs of isotope ratios (e.g. B/A vs. C/A) in the various temperature fractions from a sample. Two-component mixtures are straight lines on such a diagram, spanning end points corresponding to the two

* We use the term "trapped gases" to designate those gases which were not formed *in situ* by some sort of nuclear transmutation. The "trapped" component can thus include ambient gases dissolved in the minerals at time of formation, solar wind and cosmic-ray gases injected into material at the lunar surface, gases dissolved in existing minerals by diffusion and gases dissolved by shock.

components. If one can somehow infer the end points, one reads the relative concentrations of isotope A in the two components directly from the plot, using the lever rule. Departures from a straight line indicate additional components. In particular it is easy to deduce concentrations of additional monoisotopic components. The method of analysis should become clear upon discussion of examples.

Experimental Methods

We opened the Houston containers at Berkeley in a nitrogen-filled glove-box and loaded the samples into the vacuum systems without exposing them to the atmosphere. One-gram rock chips of samples 10057 and 10044 were broken in half. One half was taken for the work in this paper; the other half was reserved for neutron irradiation. We shall report the analysis of gases in pile-irradiated samples elsewhere. Our ten-gram sample (10084) of lunar fines was split down to working samples of 0·09 g, using a Sepor No. 212 sample-splitter. This method of splitting may not have produced completely representative samples: we noticed that the fine material tended to clump appreciably and it is doubtful that we succeeded in dividing the clumps representatively in the splitting.

We studied half-gram samples of crystalline rocks 10057 and 10044 in "system 1," which is a completely metallic system up to and including the valve which admits the gases to the mass spectrometer, except for a small sample holder and dumper made from a glass of low permeability for helium (Corning type 1720). The "hot-blank" (1500°C heating for one hour) for helium in this system is only $0{\cdot}4 \times 10^{-8}$ cm^3 STP ^{4}He. The samples were heated by radiofrequency induction in a previously outgassed crucible of vapor-deposited tungsten, which is welded to nickel tubing which is in turn welded to the demountable stainless-steel plumbing for the system. Gases released from the outside of the crucible are separately pumped to a waste vacuum. We measured temperatures with a pair of W/W–Re thermocouples attached to the outside of the crucible. In one run (sample 10057) these couples were spotwelded to the crucible, using a platinum tab as an intermediary. Thermocouples so mounted work well, but the platinum slowly sublimes. In the other run (sample 10044) the couples were bound to the crucible with tantalum wire. This scheme was only partially successful: induced currents in the tantalum wire resulted in somewhat high temperature readings. Because of the various problems encountered in the thermometry, the stated temperatures for system 1 may be in error by as much as ± 50°C for rock 10057 and ± 100°C for rock 10044.

We studied the 0·09 g sample of lunar fines (sample 10084) in "system 2," for which the connecting tubulations are Pyrex glass. The crucible is again vapor-deposited tungsten with a separate pumping system for the outside. The "hot blank" for helium for system 2 (1600°C heating for one half hour) is $1{\cdot}0 \times 10^{-7}$ cm^3 STP ^{4}He, which is much larger than for system 1 but completely negligible in comparison with the helium released by sample 10084. In system 2, we measured the higher temperatures directly with an optical pyrometer. We measured the lower temperatures (and checked the pyrometer measurements for the higher temperatures) indirectly: immediately after the run we installed a W/W–Re thermocouple inside the crucible, without making further changes in the system, and reproduced the heating steps in vacuum at exactly the same settings on the induction heater. For the higher temperatures, determinations by the thermocouple and by the optical pyrometer were in excellent agreement. We therefore believed that the stated temperatures for system 2 are accurate within ± 20°C.

In both systems, we purified gases over one or two stages of Zr–Ti alloy, supplemented by a fresh mirror in a titanium bulb. The glass mass spectrometers also carried titanium bulbs in which fresh mirrors were generated before a series of analyses. The rare gases were separated one from another with an activated charcoal trap. The sequence for separation is: liquid nitrogen on charcoal, admit He and Ne; dry ice on charcoal, admit Ar; -39°C on charcoal admit Kr; $+100$°C on charcoal, admit Xe. The He–Ne mixture is looked at in two stages so that each gas can be examined first in one of the two runs. Both systems contain all-metal, gas pipettes filled with pure air and ^{3}He at known pressures. Before and after each sample, several calibrations runs were made with the pipette. These runs furnish both sensitivity factors for the various gases and correction factors for mass discrimination.

We have made no corrections in any of this work for blanks. Instead, blanks were run and recorded separately in the tables. Every fraction reported here thus includes small amounts of atmospheric rare gases. These blank gases are noticeable at the lower temperatures for the runs on the rocks, expecially rock 10044. With one exception (see below) these contributions are very close in size to our hot blanks, showing that the atmospheric gases are *not* coming from the crucible, even in the hot blanks. These results testify to the excellence for rare gas extractions of vapor deposited tungsten crucibles, even though large, so long as they are separately pumped exteriorly by a waste vacuum system. The exception to which we alluded is the neon blank for rock 10057. We ran rock 10057 with a quartz chimney lining the inner surface of the nickel tube connecting with the crucible, so as to catch a fraction of the alkali metals evaporated from the sample. Even after many hours of outgassing near maximum power, small traces of dissolved atmospheric neon were still coming out of the chimney at the higher temperatures. The contribution of blank neon to the *lower* temperatures for rock 10057 can best be estimated from the blank for rock 44, where the quartz chimney was not used.

Background (i.e. "dirt") corrections were made only to ^{78}Kr, ^{80}Kr and ^{81}Kr in the rocks. Entries in the tables where there was such a correction have been flagged. We based the background correction on the 79 peak in the krypton spectrum, a procedure which is somewhat objectionable (except for corrections to ^{81}Kr) but one for which we had no good alternative. In some of the runs a correction had to be applied to $^{20}Ne^{+}$ for $^{40}Ar^{2+}$, although we usually ran neon at electron bombardment voltages which suppressed the doubly charged argon peak to insignificant levels. Again, entries in the tables where such a correction was applied are flagged.

RESULTS FOR LUNAR FINES (SAMPLE 10084)

Complete results for He, Ne, Ar and Xe are set out in Table 1. The krypton data are omitted because they were not precise enough to reveal significant trends, except for an obvious cosmogenic component, which produces correlated changes in the ratios ^{78}Kr/^{82}Kr and ^{80}Kr/^{82}Kr. The summed krypton data appear along with data for the other gases in our summary Table 2.

Xenon in the lunar fines was detected in large concentrations and with an isotopic composition close to that predicted for solar-wind xenon by MARTI (1969) from his study of xenon in the gas-rich achondrite Pesyanoe. For the isotopes of mass 130 and below, the abundance pattern resembles the xenon in average carbonaceous chondrites (AVCC). This fact cannot be established in a total extraction or "melting" run on the sample because of the occurrence of a spallogenic xenon component. But the stepwise heating data enable one to distinguish the components. Figure 1 is a correlation plot of ^{124}Xe/^{130}Xe against ^{126}Xe/^{130}Xe. The points define a line which spans the compositions of AVCC xenon and spallation-type xenon as deduced from the 1200° fraction from rock 10057. The proximity of the most "southwesterly" points on this plot to the AVCC composition, strongly suggests that the solar xenon has the AVCC ratios for the light xenon isotopes. We have used the composition of the 820°C fraction in the fines to represent solar xenon in calculations in this paper, although we recognize that "solar xenon" defined in this way may contain traces of spallation xenon. The isotopes produced abundantly in fission (^{131}Xe, ^{132}Xe, ^{134}Xe and ^{136}Xe.) are all depleted in the lunar fines relative to AVCC xenon, as was noted in the preliminary rare gas studies (LSPET, 1969). Confirmation, in results from the lunar samples, of MARTI's (1969) hypothesis that solar and meteoritic xenon are one and the same—aside from possible fission effects—deepens all the more the mystery about terrestrial xenon. How can it have become so strongly fractionated in mass (almost

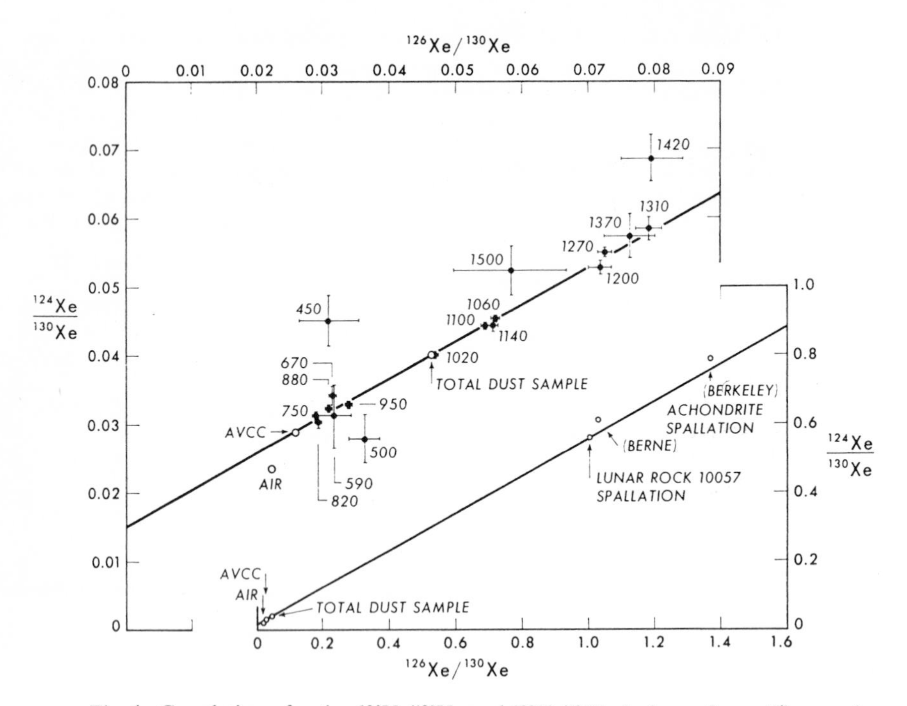

Fig. 1. Correlation of ratios $^{124}Xe/^{130}Xe$ and $^{126}Xe/^{130}Xe$ in lunar fines. The numbers are release temperatures in degrees Celsius for half-hour heatings. These isotopes belong to a two-component system. One component appears to have the AVCC ratios. The other is cosmogenic xenon as deduced from rock 10057.

4 per cent per mass unit!) relative to the xenon elsewhere in the solar system? And without similar effects occurring in argon and krypton?

Helium and neon results are of special interest. We now suspect that separations of isotopes during imbedding of the solar wind gas and in the release by heating are jointly responsible for large variations in isotope ratios. There were errors in some of our data reduction for neon at the time of our report (REYNOLDS *et al.*, 1970) to the Apollo 11 Conference. Because of a very large spread in the ratio $^{20}Ne/^{22}Ne$, without variations, except for spallation effects, in the $^{21}Ne/^{22}Ne$ ratio, we had to suppose that at least two trapped neon components were present. The corrected data no longer support this conclusion. The run of the helium isotopes during outgassing is very instructive. In Fig. 2 we plot the release curves for ^{3}He and ^{4}He for the dust sample, 10084. Both curves are normalized to 100 per cent total release. One notes that the curves are similar but displaced slightly in temperature: ^{3}He release occurs slightly earlier than ^{4}He release. The initial $^{3}He/^{4}He$ ratio of 0·000635 is 36 per cent higher than the total ratio. In a single stage diffusion process the initial enrichment in the ratio should be $(4/3)^{1/4} = 1{\cdot}074$. But we have a two stage process in this instance: ^{4}He was imbedded deeper than the ^{3}He when the solar wind struck the lunar surface, because the streaming of the solar wind is hydrodynamic, with equal velocities for the two helium isotopes. The initial burial depth for isotopes will be proportional to

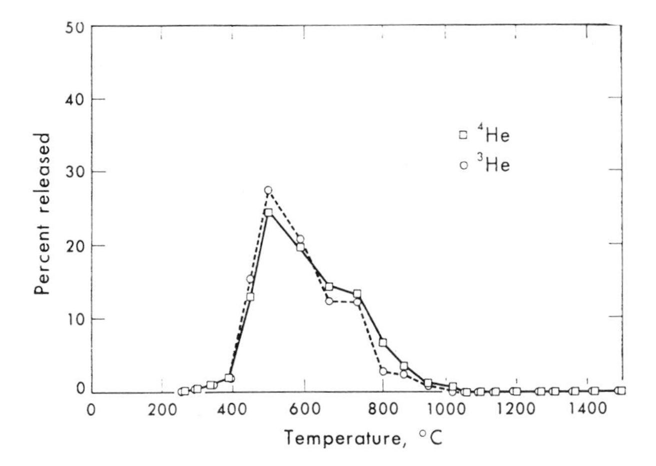

Fig. 2. Release of helium isotopes in stepwise heating of lunar dust. The curves have been normalized to 100 per cent total release.

mass.* And deeper burial of the mass 4 component, enhances the initial separative effect.

Release curves for isotopes ^{20}Ne and ^{22}Ne (not shown) exhibit a similar effect if one looks carefully at the low temperature and high temperature "tails" for the neon release, where the effect is greatest. We see an initial enrichment for ^{20}Ne/^{22}Ne of 11 per cent, which again exceeds the single stage enrichment of $(22/20)^{1/4} = 1{\cdot}024$ and for the same reason. Figure 3 shows the neon ratios for all three of our samples on a standard correlation plot for neon. Rock 10044, which was an interior piece, has practically no trapped gas. Because of ^{40}Ar^{2+} interference no precise ^{20}Ne/^{22}Ne ratios were obtained below 700°C where fractionation of the isotopes in the trapped component would be most visible The ratios thus fall along a line (the "rock line," PEPIN *et al.*, 1970) joining the compositions of trapped and cosmogenic neon. Point C with coordinates ^{20}Ne/^{22}Ne $= 0{\cdot}82$, ^{21}Ne/^{22}Ne $= 0{\cdot}90$ represents cosmogenic neon in a typical stone meteorite. In the dust sample, neon is dominated by the trapped component except at the highest temperatures where the trapped component is almost gone and the bulk of the cosmogenic component is being released. In this sample at low temperatures we thus see diffusive separation of the isotopes, trending roughly along a line of slope

$$d \ln (^{20}\text{Ne}/^{22}\text{Ne})/d \ln (^{21}\text{Ne}/^{22}\text{Ne}) = 2.$$

At later temperatures, where the last of the trapped neon is being released, an appreciable enrichment of the heavier isotopes has built up. We see from the intersections

* It is questionable to what extent the initial distributions of isotopes with depth have been preserved. There is conclusive evidence (EBERHARDT *et al.*, 1970) for saturation effects in trapped helium and neon in the bulk fines and for diffusion of trapped helium into the grains in the bulk fines to depths of about 10 μ (KIRSTEN *et al.*, 1970). Both these effects could redistribute the isotopes. But we find, surprisingly, that the distortion of the isotope ratios in the initial release is just the mass ratio of the isotopes for helium, neon, and argon alike.

Table 1. Helium, neon, argon and xenon from stepwise heating

Temperature (°C)	Ave. 1550° Blank	265	270	290	300	340	390	450	500	590	670	750	820
$^{4}He/^{3}He$‡	—	1575·0‡	1751·0‡	1678·0‡	1666·0‡	1754·0‡	1678·0‡	1754·0‡	1887·0‡	2049·0‡	2487·0‡	2299·0‡	4878·0‡
		±17·0	±34·0	±20·0	±11·0	±9·0	±8·0	±6·0	±11·0	±8·0	±37·0	±5·0	±24·0
[^{3}He], 10^{-8} cm^{3} STP/g	<**0·026***a	**79·4**	**18·1**	**21·3**	**39·1**	**128·0**	**244·0**	**2110·0**	**3690·0**	**2730·0**	**1620·0**	**1690·0**	**385·0**
$^{20}Ne/^{22}Ne$	—	14·27	13·94	14·19	13·59	12·84	12·91	13·49	12·71	13·06	12·87	12·92	12·88
		±0·06	±0·10	±0·08	±0·09	±0·04	±0·10	±0·16	±0·04	±0·10	±0·18	±0·09	±0·09
$^{21}Ne/^{22}Ne$	—	0·0323	0·0300	0·0319	0·0317	0·0313	0·0313	0·0324	0·0309	0·0319	0·0319	0·0324	0·0331
		±0·0030	±0·0010	±0·0005	±0·0003	±0·0003	±0·0003	±0·0004	±0·0002	±0·0003	±0·0006	±0·0003	±0·0004
[^{22}Ne], 10^{-8} cm^{3} STP/g	**0·025***	**26·8**	**4·61**	**5·76**	**13·6**	**111·0**	**244·0**	**4500·0**	**4060·0**	**6190·0**	**7300·0**	**7300·0**	**1890·0**
$^{38}Ar/^{36}Ar$	—	0·1816	0·1784	0·1807	0·1800	0·1787	0·1799	0·1810	0·1835	0·1802	0·1829	0·1897	0·1920
		±0·0007	±0·0016	±0·0011	±0·0008	±0·0013	±0·0010	±0·0010	±0·0006	±0·0010	±0·0011	±0·0009	±0·0006
$^{40}Ar/^{36}Ar$	—	38·25	34·59	30.99	25·35	14·32	11·23	5·60	3·64	2·877	2·128	0·923	0·820
		±0·09	±0·17	±0·09	±0·08	±0·07	±0·04	±0·02	±0·01	±0·010	±0·007	±0·003	±0·002
[^{36}Ar], 10^{-8} cm^{3} STP/g	**0·01***	**3·87**	**0·79**	**0·78**	**1·25**	**4·26**	**9·8**	**9·8**	**270·0**	**886·0**	**3780·0**	**13300·0**	**9000·0**
$^{124}Xe/^{130}Xe$	—	—	—	—	—	—	—	0·045	0·0280	0·0312	0·0342	0·0312	0·0302
								±0·004	±0·003	±0·004	±0·0013	±0·0006	±0·0008
$^{126}Xe/^{130}Xe$	—	—	—	—	—	—	—	0·0309	0·0363	0·0318	0·0316	0·0290	0·0293
								±0·0046	±0·0023	±0·0024	±0·0005	±0·0005	±0·0005
$^{128}Xe/^{130}Xe$	—	0·516	0·508	0·727	0·623	0·656	0·696	0·521	0·540	0·508	0·507	0·509	0·515
		±0·033	±0·054	±0·012	±0·058	±0·053	±0·030	±0·028	±0·010	±0·017	±0·014	±0·005	±0·004
$^{129}Xe/^{130}Xe$	—	6·32	6·92	6·91	6·49	6·74	6·91	6·58	6·57	6·30	6·31	6·31	6·38
		±0·32	±0·44	±0·11	±0·30	±0·37	±0·30	±0·20	±0·08	±0·12	±0·09	±0·03	±0·05
$^{131}Xe/^{130}Xe$	—	5·20	5·47	5·43	5·48	5·60	5·42	5·23	5·09	4·96	4·91	4·93	4·98
		±0·26	±0·35	±0·27	±0·29	±0·30	±0·34	±0·18	±0·07	±0·09	±0·08	±0·03	±0·04
$^{132}Xe/^{130}Xe$	—	6·35	7.01	7·54	6·89	6·90	6·77	6·26	6·21	6·02	5·96	5·97	6·06
		±0·31	±0·41	±0·08	±0·31	±0·37	±0·29	±0·13	±0·05	±0·08	±0·06	±0·03	±0·04
$^{134}Xe/^{130}Xe$	—	2·404	2·553	2·842	2·523	2·385	2·478	2·341	2·270	2·238	2·200	2·174	2·224
		±0·12	±0·18	±0·10	±0·11	±0·15	±0·13	±0·073	±0·032	±0·036	±0·042	±0·013	±0·019
$^{136}Xe/^{130}Xe$	—	2·086	2·374	2·434	2·225	2·138	2·105	1·942	1·840	1·817	1·790	1·763	1·793
		±0·11	±0·17	±0·037	±0·13	±0·13	±0·14	±0·07	±0·024	±0·033	±0·034	±0·013	±0·015
[^{130}Xe], 10^{-12} cm^{3} STP/g	**0·3***	**0·85**	**0·27**	**0·26**	**0·32**	**0·38**	**0·50**	**2·06**	**3·67**	**9·05**	**20·5**	**276·0**	**461·0**

Notes: Boldface entries are gas concentrations. Errors in gas concentrations are 10% for helium, neon, and xenon, 20% for argon. Errors are statistical only. In addition there is a fractional error in the applied mass discrimination factors as follows: Ne 20/22: 0·015; 21/22: 0·05. **Ar** 38/36: 0·05; 40/36: 0·04. **Xe** 124/130: 0·04; 126/130: 0·05; 128/130: 0·01; 129/130: 0·007; 131/130: 0·008; 132/130: 0·006; 134/130: 0·01; 136/130: 0·014.

of 0·089 g of lunar fines (10084–59). Heatings were for ½ hr.

880	950	1020	1060	1100	1140	1200	1270	1310	1370	1420	1500	1420 Blank	Total
2506.0‡	2632·0‡	26740.0‡b	1745·0‡	1416·0‡	1083.0‡	1923·0‡	2283·0‡	3175·0‡	2660·0‡	3270·0‡	3185·0‡	3003·0‡	2138·0‡
±13·0	±14·0	±720·0	±27·0	±14·0	±13·0	±63·0	±63·0	±170·0	±140·0	±300·0	±40·0	±126·0	±6·0
393·0	**143·0**	**5·57**[b]	**13·9**	**4·95**	**3·09**	**0·490**	**0·311**	**0·345**	**0·132**	**0·227**	**0·111**	**0·056**	**13300·0**
12·81	12·50	12·01	11·57	11·21	10·38	9·25	8·94	11·41[b]	9·64	11·67	12·64	14·01	12·85
±0·11	±0·04	±0·09	±0·03	±0·03	±0·02	±0·06	±0·05	±0·10	±0·13	±0·12	±0·09	±0·26	±0·04
0·0339	0·0355	0·0385	0·0408	0·0475	0·0781	0·205	0·211	0·097[b]	0·201	0·092	0·043	0·036	0·0332
±0·0004	±0·0002	±0·0004	±0·0002	±0·0001	±0·0004	±0·002	±0·003	±0·003	±0·004	±0·002	±0·004	±0·011	±0·0001
3050·0	**2480·0**	**1560·0**	**907·0**	**365·0**	**135·0**	**78·3**	**2·31**	**0·53**[b]	**0·33**	**0·35**	**0·13**	**0·043**	**40800·0**
0·1937	0·1950	0·1974	0·1977	0·2009	0·2070	0·2148	0·2279	0·2281	0·2295	0·2316	0·2152	0·2016	0·1925
±0·0004	±0·0009	±0·0002	±0·0003	±0·0005	±0·0008	±0·0005	±0·0005	±0·0008	±0·0005	±0·0007	±0·0008	±0·006	±0·0003
0·857	1·013	1·072	1·208	1·242	1·241	1·491	1·616	2·504	2·167	3·415	4·91	35·4	1·125
±0·001	±0·004	±0.001	±0·001	±0·002	±0·003	±0·002	0·002	±0·007	±0·003	±0·005	±0·01	±0·7	±0·001
6930·0	**3170·0**	**1950·0**	**2790·0**	**2800·0**	**1390·0**	**173·0**	**113·0**	**24·3**	**21·4**	**4·35**	**3·47**	**0·21**	**46700·0**
0.0322	0·0329	0·0401	0·0455	0·0442	0·0442	0.0529	0·0550	0·0583	0·0573	0·0686	0·0523	0·1139	0·0400†
±0·0003	±0·0002	±0·0002	±0·0004	±0·0003	±0·0009	±0·0010	±0·0007	±0·0017	0·0032	±0·0031	±0·004	±0·026	±0·0001
0·0310	0·0340	0·0469	0·0560	0.0545	0·0557	0·0719	0·0725	0·0792	0·0763	0·0796	0·0583	0·0404	0·0464†
±0·0005	±0·0003	±0·0004	±0·0008	±0·0006	±0·0009	±0·0019	±0·0011	±0·0020	±0·0038	±0·0046	±0·008	±0·007	±0·0002
0·518	0·514	0·529	0·541	0·540	0·533	0·556	0·553	0·569	0·561	0·544	0·546	0·589	0.530
±0·003	±0·003	±0·003	±0·004	±0·003	±0·004	±0.005	±0·005	±0·008	±0·015	±0·017	±0·017	±0·088	±0·001
6·33	6·31	6·28	6·26	6.27	6·23	6·18	6·17	6·19	6·18	6·05	5·93	7·10	6·28
±0·03	±0.03	±0·03	±0·04	±0·03	±0·04	±0·05	±0·04	±0·07	±0·08	±0·10	±0·14	±0·85	±0·01
4·95	4·97	4·97	5·00	5·01	5·01	4·96	5·03	5·03	5·06	4·99	4·76	5·53	4·98
±0·02	±0·02	±0·02	±0·03	±0·03	±0·03	±0·04	±0·03	±0·06	±0·06	±0·09	±0·12	±0·68	±0·01
6·01	6·01	5·97	5·94	5·95	5·92	5·82	5·87	5·88	5·87	5·81	5·76	6·30	5·97
±0·03	±0·02	±0·02	±0·03	±0·02	±0·03	±0·04	±0·03	±0·06	±0·07	±0·09	±0·10	±0·76	±0·01
2·209	2·198	2·205	2·187	2·196	2·181	2·136	2·170	2·149	2·177	2·076	2·113	2·233	2·195
±0·013	±0·009	±0·012	±0·013	±0·011	±0·016	±0·018	±0·018	±0·024	±0·032	±0·040	±0·046	±0·29	±0·005
1·785	1·795	1·790	1·776	1·787	1·773	1·741	1·768	1·741	1·773	1·757	1·701	1·707	1·783
±0·008	±0·007	±0·009	±0·010	±0·008	±0·011	±0·015	±0·012	±0·022	±0·028	±0·050	±0·056	±0·22	±0·004
936·0	**1030·0**	**1180·0**	**1700·0**	**1120·0**	**548·0**	**108·0**	**118·0**	**41·2**	**29·4**	**5·19**	**2·82**	**0·25**	**7700·0**

* Absolute blank in cm^3 STP is the tabulated blank times 0·089 g.
† Total of gas fractions where this ratio was obtained.
‡ No discrimination correction available. These ratios may be systematically in error.
a ^3He not measured. ^4He concentration: $\sim 10 \cdot 0 \times 10^{-8}$ cm^3 STP/g.
b Questionable value.

Table 2. Summary of the results for rare gases from lunar samples.

	Lunar fines Sample 10084, 59			Fine-grained Sample 10057, 20		
Isotope	Total	Trapped	Cosmog.	Total	Trapped	Cosmog.
^{4}He/^{3}He	2138·0* ± 6·0	—	—	1809·3*,[b] ± 3·9	—	—
[^{3}He]	**13400·0 ± 1300·0**	—	—	**89·48 ± 7·2**	—	—
^{20}Ne/^{22}Ne	12·85 ± 0·20	12·9[a] ± 0·21	—	12·08 ± 0·15	—	—
^{21}Ne/^{22}Ne	0·0332 ± 0·0017	0·0324[a] ± 0·0016	≡ 0·900	0·120 ± 0·002	≡ 0·032	≡ 0·900
[^{22}Ne]	**40800·0 ± 4080·0**	**40800·0 ± 4080·0**	**40·0 + 78·0–15·0**	**76·6 ± 7·7**	**67·5 ± 6·8**	**9·2 ± 0·9**
^{38}Ar/^{36}Ar	0·193 ± 0·010	0·190[a] ± 0·010	—	0·237 ± 0·002	—	≡ 1·52
^{40}Ar/^{36}Ar	1·125 ± 0·045	—	—	39·19 ± 0·28	—	—
[^{36}Ar]	**46700·0 ± 9300·0**	**46700·0 ± 9300·0**	—	**388·0 ± 39·0**	**372·0 ± 37·0**	**15·1 ± 1·5**
^{78}Kr/^{82}Kr	0·0312† ± 0·0011	0·0307†,[c] ± 0·0013	—	0·0844† ± 0·0016	—	0·224† ± 0·0075
^{80}Kr/^{82}Kr	0·206 ± 0·004	0·201[c] ± 0·004	≡ 0·675	0·4420† ± 0·0066	—	0·670† ± 0·0227
^{81}Kr/^{82}Kr	—	—	—	—	—	0·00722† ± 0·00112
^{83}Kr/^{82}Kr	1·00 ± 0·01	1·00[c] ± 0·01	—	1·06 ± 0·013	—	1·35 ± 0·016
^{84}Kr/^{82}Kr	4·91 ± 0·05	4·91[c] ± 0·05	—	3·57 ± 0·037	—	0·45 ± 0·018
^{86}Kr/^{82}Kr	1·50 ± 0·01	1·50[c] ± 0·01	—	1·02 ± 0·025	—	≡ 0·000
[^{82}Kr]	**7·6 ± 1·5**	**7·5 ± 1·5**	**0·093 + 0·064–0·024**	**0·0337 ± 0·0061**	**0·0237 ± 0·0043**	**0·01 ± 0·0018**
^{124}Xe/^{130}Xe	0·0400 ± 0·0016	0·0302[c] ± 0·0015	—	0·265 ± 0·0037	—	0·57 ± 0·0083
^{126}Xe/^{130}Xe	0·0464 ± 0·0023	0·0293[c] ± 0·0016	≡ 1·03	0·465 ± 0·0097	—	1·03 ± 0·022
^{128}Xe/^{130}Xe	0·530 ± 0·005	0·515[c] ± 0·007	—	1·048 ± 0·053	—	1·61 ± 0·070
^{129}Xe/^{130}Xe	6·28 ± 0·04	6·38[c] ± 0·07	—	4·11 ± 0·072	—	1·03 ± 0·066
^{131}Xe/^{130}Xe	4·98 ± 0·04	4·98[c] ± 0·06	—	6·05 ± 0·040	—	7·50 ± 0·081
^{132}Xe/^{130}Xe	5·97 ± 0·04	6·06[c] ± 0·06	—	3·58 ± 0·023	—	0·32 ± 0·042
^{134}Xe/^{130}Xe	2·20 ± 0·02	2·22[c] ± 0·03	—	1·283 ± 0·0096	—	0·020 ± 0·0166
^{136}Xe/^{130}Xe	1·78 ± 0·03	1·79[c] ± 0·03	—	1·040 ± 0·0083	—	≡ 0·000
[^{130}Xe]	**0·77 ± 0·08**	**0·76 ± 0·08**	**0·013 ± 0·002**	**0·00517 ± 0·00031**	**0·00291 ± 0·00019**	**0·00226 ± 0·00016**

Notes: Boldface entries are gas concentrations in units of 10^{-8} cm^3 STP/g. Errors in this table are total errors.

* No discrimination correction was available.

† Correction has been made for background. Error includes uncertainty in background as well as discrimination corrections.

[a] Based on 750° fraction.

The gas fractions for the stepwise heatings have been summed

Crystalline rock		Medium-grained crystalline rock Sample 10044, 20				
Fission	Excess	Total	Trapped	Cosmog.	Fission	Excess
—	—	213·69*	—	—	—	—
		± 0·15				
—	—	**102·4**	—	—	—	—
		± 13·3				
—	—	3·311	—	—	—	—
		± 0·043				
—	—	0·719	≡ 0·029	≡ 0·905	—	—
		± 0·014				
—	—	**9·5**	**2·0**	**7·5**	—	—
		± 1·0	**± 0·2**	**± 0·8**		
—	—	0·9843	—	≡ 1·52	—	—
		± 0·0088				
—	—	282·8	—	—	—	—
		± 2·0				
—	—	**41·9**	**16·8**	**25·1**	—	—
		± 4·2	**± 1·7**	**± 2·5**		
—	—	0·174†	—	0·233†	—	—
		± 0·0042		± 0·0122		
—	**0·0039†[d,e]**	0·520†	—	0·617†	—	**0·0013†[d,e]**
		± 0·0103		± 0·0265		
—	—	—	—	0·00334†	—	—
				± 0·000806		
—	—	1·09	—	1·27	—	—
		± 0·0074		± 0·025		
—	—	2·89	—	0·51	—	—
		± 0.031		± 0·042		
—	—	0·78	—	≡ 0·000	—	—
		± 0·009				
—	**0·0010[e]**	**0·0311**	**0·0167**	**0·0144**	—	**0·00035[e]**
		± 0·0093	**± 0·0050**	**± 0·0043**		
—	—	0·63	—	0·69	—	—
		± 0·01		± 0·014		
—	—	1·001	—	1·106	—	—
		± 0·022		± 0·028		
—	**0·00028[d,e]**	1·47	—	1·625	—	**0·000047[d,e]**
		± 0·075		± 0·087		
—	**0·00040[d,e]**	2·85	—	1·71	—	**0·00052[d]**
		± 0·052		± 0·057		
—	—	4·55	—	4·28	—	—
		± 0·041		± 0·067		
—	—	1·97	—	0·77	—	—
		± 0·014		± 0·021		
—	—	0·558	—	0·041	—	—
		± 0·0048		± 0·0076		
0·00013[d]	—	0·439	—	≡ 0·000	**0·00015[d]**	—
± 0·00006		± 0·0028			**± 0·00005**	
—	—	**0·00135**	**0·00021**	**0·0014**	—	—
		± 0·000081	**± 0·000029**	**± 0·000080**		

[b] Includes estimate for one gas fraction which was lost.
[c] Based on 820° fraction.
[d] Value is a concentration (not a ratio) for the isotope with numerator for the row.
[e] Questionable effect. See text.

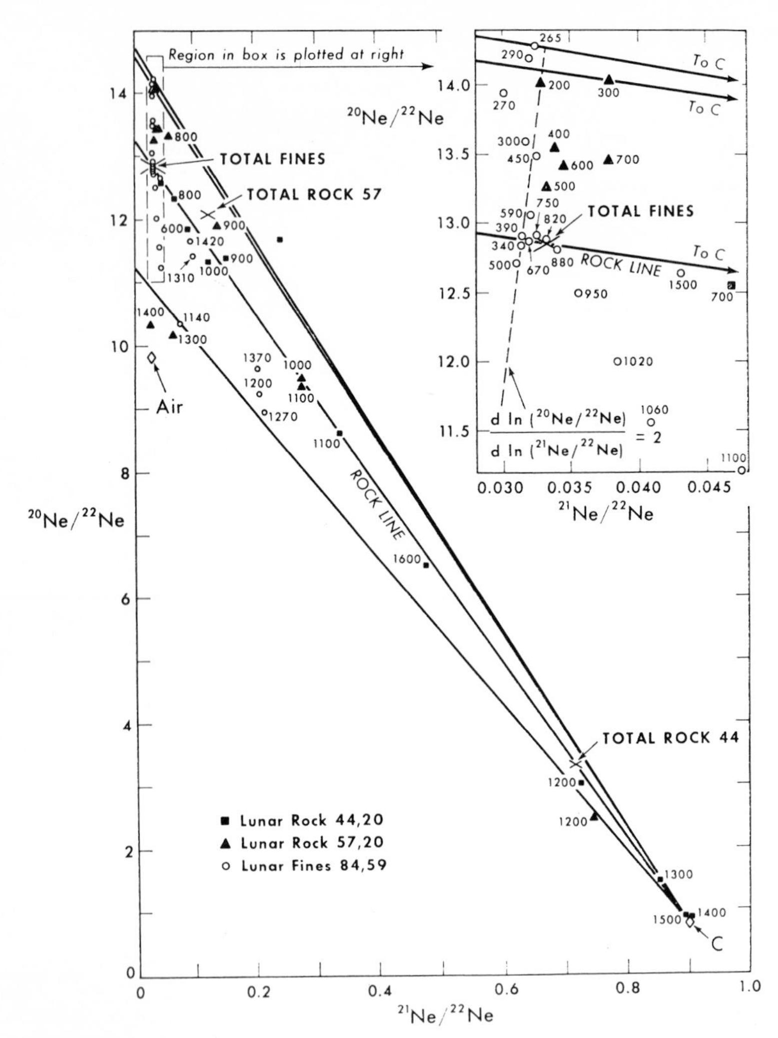

Fig. 3. Isotopic composition of neon fractions released in stepwise heating of lunar samples. The numbers are release temperatures in degrees Celsius. Mass-dependent fractionation should produce compositions on a line with d ln (^{20}Ne/^{22}Ne)/d ln (^{21}Ne/^{22}Ne) = 2. Mixing with cosmogenic neon further displaces the points toward point C. Extreme mixing lines are shown for the fines and for rock 10057. The lower mixing line for rock 10057 was omitted because of high-temperature contamination with atmospheric neon from a quartz chimney. The "rock line" should be the locus of points representing total neon from lunar samples. Rock 10044, in the temperature range plotted, essentially follows the rock line (see text).

of mixing lines through C with the diffusive trend line that in the trapped component the ratio ^{20}Ne/^{22}Ne varies from 14·3 to 10·9 in the course of the release. Rock 10057 is an intermediate case: it contains enough trapped gas and enough cosmogenic gas to show diffusive separation at low temperatures and strong cosmogenic features at high temperatures. At 1200°C and above, atmospheric neon from the quartz chimney

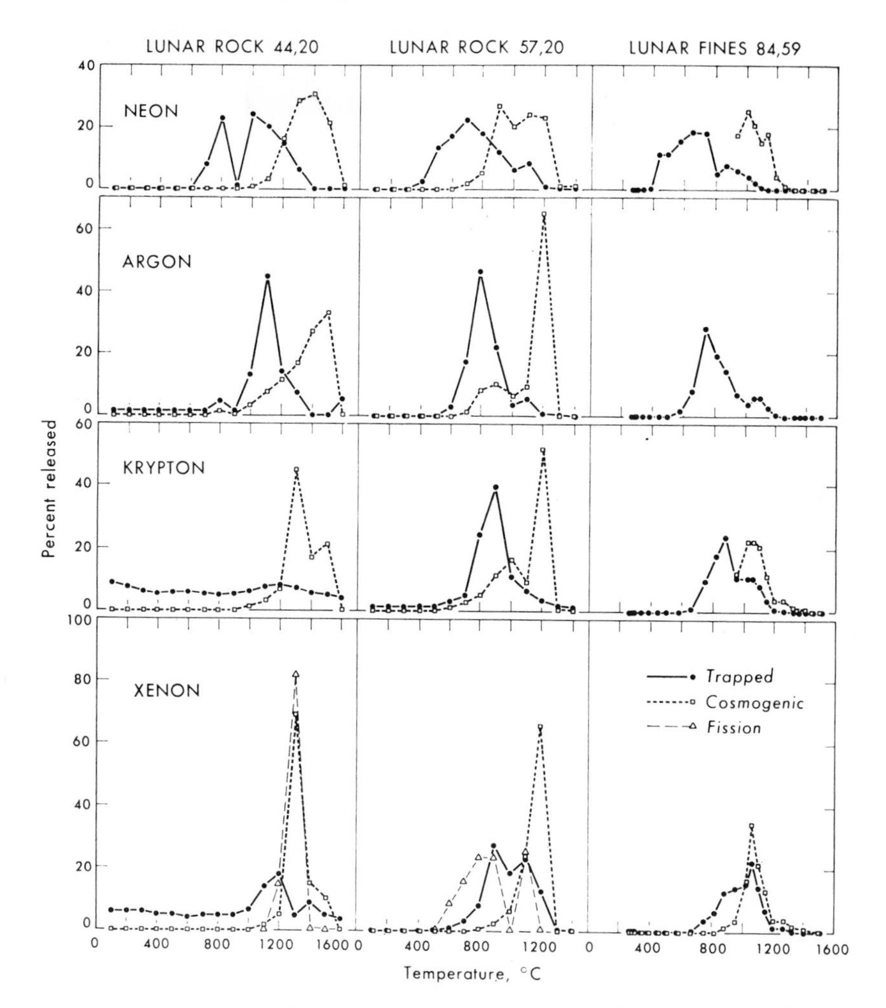

Fig. 4. Release curves for trapped, cosmogenic, and fission components from the samples. Curves are normalized to 100 per cent total release. Partial release curves are normalized to 100 per cent total for the fractions plotted. In the 900°C heating for rock 10044 there was a thermocouple malfunction; it is questionable that the temperature was actually reached.

is dominating the trapped component so that the points fall near a line joining point C with air neon.

Table 2 contains summary data on the lunar fines. For neon, krypton, and xenon we have resolved the total gas released into cosmogenic and trapped components, giving the amounts of each. For argon our estimate of the cosmogenic component would be too inaccurate (see below) to be useful and it was not included in the table. In the calculations we used the cosmogenic mass spectra for the light isotopes of Kr and Xe deduced from the highly cosmogenic gas in the 1200°C fraction from rock 10057. For neon we assumed the cosmogenic ratios corresponding to point C in Fig. 3.

Table 3. Rare gases from stepwise heating of lunar

Temperature (°C)	Ave. 1550° blank	100	200	300	400	500	600	700
^{4}He/^{3}He‡	—	—	1924·0‡	2157·0‡	1878·0‡	2090·0‡	1581·0‡	—[d]
	—	—	± 20·0	± 10·0	± 3·0	± 14·0	± 5·0	—
[^{3}He], 10^{-8} cm^{3} STP/g	a	b	**0·149**	**0·308**	**2·594**	**12·53**	**32·74**	—[d]
^{20}Ne/^{22}Ne	—	11·36§	14·05§	14·04§	13·554§	13·258	13·404	13·456
	—	± 2·48	± 0·38	± 0·12	± 0·046	± 0·055	± 0·008	± 0·061
^{21}Ne/^{22}Ne	—	0·0274	0·0329	0·0378	0·0337	0·0331	0·0344	0·0378
	—	± 0·0025	± 0·0014	± 0·0007	± 0·0003	± 0·0003	± 0·0003	± 0·0002
[^{22}Ne], 10^{-8} cm^{3} STP/g	**0·26*[f]**	**0·0081**	**0·0839**	**0·2071**	**1·672**	**9·183**	**11·52**	**15·18**
^{38}Ar/^{36}Ar	—	0·1846	0·1802	0·1821	0·1852	0·1923	0·1924	0·1887
	—	± 0·0009	± 0·0009	± 0·0007	± 0·0009	± 0·0008	± 0·0006	± 0·0013
^{40}Ar/^{36}Ar	—	267·7	183·30	149·0	84·60	77·70	78·49	36·56
	—	± 0·5	± 0·40	± 0·36	± 0·09	± 0·16	± 0·07	± 0·17
[^{36}Ar], 10^{-8} cm^{3} STP/g	**0·069***	**0·0806**	**0·1158**	**0·2084**	**0·5077**	**2·897**	11·75	**65·46**
^{78}Kr/^{82}Kr	—	—	—	—	—	—	0·0486‖	0·0453‖
	—	—	—	—	—	—	± 0·0083	± 0·0066
^{80}Kr/^{82}Kr	—	—	—	—	—	—	0·4253‖	0·4925‖
	—	—	—	—	—	—	± 0·0245	± 0·0225
^{81}Kr/^{82}Kr	—	—	—	—	—	—	—	—
	—	—	—	—	—	—	—	—
^{83}Kr/^{82}Kr	—	0·980	0·969	0·982	0·972	0·982	0·960	0·958
	—	± 0·019	± 0·018	± 0·022	± 0·036	± 0·039	± 0·017	± 0·012
^{84}Kr/^{82}Kr	—	4·642	4·541	4·627	4·649	4·291	4·198	4·198
	—	± 0·065	± 0·047	± 0·066	± 0·151	± 0·070	± 0·051	± 0·033
^{86}Kr/^{82}Kr	—	1·372	1·400	1·367	1·413	1·312	1·232	1·222
	—	± 0·022	± 0·019	± 0·028	± 0·047	± 0·039	± 0·023	± 0·015
[^{82}Kr], 10^{-12} cm^{3} STP/g	**3·5***	**2·70**	**2·72**	**2·68**	**2·38**	**2·85**	**6·62**	**13·76**
^{124}Xe/^{130}Xe	—	—	—	—	—	—	0·0564	0·0546
	—	—	—	—	—	—	± 0·0051	± 0·0048
^{126}Xe/^{130}Xe	—	—	—	0·0985	—	—	0·0715	0·0795
	—	—	—	± 0·0088	—	—	± 0·0065	± 0·0038
^{128}Xe/^{130}Xe	—	1·257	1·110	0·912	0·839	0·976	1·099	1·142
	—	± 0·072	± 0·059	± 0·047	± 0·061	± 0·037	± 0·032	± 0·025
^{129}Xe/^{130}Xe	—	7·15	6·15	6·70	6·35	6·48	7·02	6·80
	—	± 0·34	± 0·27	± 0·20	± 0·27	± 0·19	± 0·22	± 0·12
^{131}Xe/^{130}Xe	—	5·47	4·88	5·11	4·89	5·03	5·19	5·09
	—	± 0·29	± 0·20	± 0·16	± 0·19	± 0·16	± 0·17	± 0·08
^{132}Xe/^{130}Xe	—	6·78	5·90	6·44	5·95	6·09	6·16	5·90
	—	± 0·28	± 0·23	± 0·19	± 0·22	± 0·17	± 0·17	± 0·087
^{134}Xe/^{130}Xe	—	2·75	2·38	2·431	2·467	2·363	2·393	2·258
	—	± 0·13	± 0·11	± 0·084	± 0·122	± 0·081	± 0·080	± 0·041
^{136}Xe/^{130}Xe	—	2·40	1·914	2·064	2·054	1·996	2·043	1·915
	—	± 0·13	± 0·082	± 0·075	± 0·092	± 0·064	± 0·062	± 0·036
[^{130}Xe], 10^{-12} cm^{3} STP/g	**0·081***	**0·036**	**0·072**	**0·089**	**0·077**	**0·177**	**0·380**	**0·914**

Notes: Boldface entries are gas concentrations. Errors in gas concentrations are: He ± 8%; Ne ± 11% Ar ± 8%; Kr ± 18%; Xe ± 6%.

Unless otherwise specified errors are statistical only. In addition there is a fractional error in the applied mass discrimination factors as follows:

Ne 20/22: 0·01; 21/22: 0·016. **Ar** 38/36: 0·007; 40/36: 0·0065. **Kr** 78/82: 0·03; 80/82: 0·03; 83/82: 0·005; 84/82: 0·01; 86/82: 0·01. **Xe** 124/130: 0·013; 126/130: 0·02; 128/130: 0·05; 129/130: 0·017; 131/130: 0·005; 132/130: 0·005; 134/130: 0·005; 136/130: 0·005.

* Absolute blank in ccSTP is tabulated blank times 0·557 g.

† Total of gas fractions where this ratio was obtained.

‡ No discrimination correction available. These ratios may be systematically in error.

§ Correction has been applied for $^{40}Ar^{2+}$. Error includes uncertainty in this correction.

‖ Correction has been made for background. Error includes uncertainty in background and in discrimination correction.

rock 10057-20. Heatings were for one hour

800	900	1000	1100	1200	1300	1400	Total
2072·0‡	2138·0‡	2224·0‡	1695·0‡	1212·0‡	—	—	1809·3[e]
± 3·0	± 6·0	± 13·0	± 11·0	± 7·0	—	—	± 3·9
6·595	**3·433**	**1·002**	**0·326**	**0·402**	—	—[c]	**89·48**[e]
13·333	11·893	9·492	9·372	2·4972§	10·146§	10·350§	12·078
± 0·016	± 0·028	± 0·014	± 0·008	± 0·0088	± 0·113	± 0·088	± 0·015
0·0596	0·1362	0·2755	0·2713	0·7482	0·0666	0·0296	0·1200
± 0·0002	± 0·0005	± 0·0008	± 0·0006	± 0·0021	± 0·0011	± 0·0003	± 0·0006
12·59	**8·595**	**6·517**	**7·973**	**2·612**	**0·2211**	**0·2431**	**76·60**
0·1941	0·2086	0·2669	0·2792	1·4000	0·2405	0·1888	0·2369
± 0·0005	± 0·0006	± 0·0010	± 0·0007	± 0·0028	± 0·0017	± 0·0012	± 0·0008
18·370	34·72	146·30	43·40	212·40	154·80	235·20	39·19
± 0·037	± 0·05	± 0·28	± 0·04	± 0·33	± 0·84	± 1·00	± 0·12
176·1	**84·97**	**14·68**	**19·89**	**10·76**	**0·2153**	**0·0928**	**387·8**
0·0455‖	0·0451‖	0·0849‖	0·1084‖	0·1984‖	0·0633‖	0·0653‖	0·0844†‖
± 0·0019	± 0·0022	± 0·0036	± 0·0048	± 0·0065	± 0·0125	± 0·0155	± 0·0016
0·3285‖	0·3994‖	0·5323‖	0·3624‖	0·6086‖	0·2819‖	0·3188‖	0·4420†‖
± 0·0111	± 0·0130	± 0·0171	± 0·0129	± 0·0197	± 0·0564	± 0·0603	± 0·0066
—	—	—	—	0·0072‖	—	—	—
—	—	—	—	± 0·0012	—	—	—
0·998	0·981	1·052	1·161	1·300	0·964	0·941	1·062
± 0·011	± 0·010	± 0·010	± 0·010	± 0·012	± 0·018	± 0·030	± 0·012
4·482	4·399	3·215	3·184	1·054	4·347	4·436	3·567
± 0·026	± 0·031	± 0·016	± 0·023	± 0·007	± 0·051	± 0·118	± 0·032
1·310	1·299	0·9025	0·9200	0·1960	1·274	1·236	1·022
± 0·009	± 0·036	± 0·0083	± 0·0093	± 0·0018	± 0·025	± 0·037	± 0·023
63·05	**106·1**	**43·17**	**25·47**	**60·38**	**3·27**	**2·57**	**337·8**
0·0555	0·0632	0·1325	0·2668	0·4557	—	0·2046	0·2650†
± 0·0019	± 0·0014	± 0·0021	± 0·0035	± 0·0027	—	± 0·0137	± 0·0028
0·0792	0·0901	0·2168	0·4604	0·8235	—	0·0897	0·4647†
± 0·0023	± 0·0018	± 0·0045	± 0·0047	± 0·0064	—	± 0·0174	± 0·0050
0·800	0·7659	0·7590	0·9591	1·3770	1·457	0·878	1·048
± 0·017	± 0·0098	± 0·0075	± 0·0118	± 0·0095	± 0·135	± 0·075	± 0·013
6·381	6·288	5·289	4·033	2·152	6·53	6·39	4·107
± 0·078	± 0·071	± 0·055	± 0·037	± 0·018	± 0·49	± 0·26	± 0·058
5·099	5·176	5·456	5·919	6·976	5·91	5·38	6·048
± 0·065	± 0·066	± 0·056	± 0·058	± 0·048	± 0·36	± 0·26	± 0·062
5·780	5·763	4·906	3·614	1·523	5·56	6·13	3·579
± 0·063	± 0·060	± 0·043	± 0·029	± 0·007	± 0·30	± 0·19	± 0·046
2·164	2·110	1·780	1·303	0·4819	2·11	2·48	1·282
± 0·036	± 0·028	± 0·022	± 0·015	± 0·0050	± 0·13	± 0·11	± 0·023
1·789	1·719	1·454	1·046	0·3758	1·609	1·88	1·039
± 0·029	± 0·031	± 0·015	± 0·010	± 0·0049	± 0·093	± 0·55	± 0·027
2·506	**8·567**	**7·118**	**12·96**	**18·78**	**0·035**	**0·061**	**51·78**

[a] ^{3}He not measured. ^{4}He concentration: **0·72** × 10^{-8} cm^3 STP/g.
[b] ^{3}He not measured. ^{4}He concentration: **11·1** × 10^{-8} cm^3 STP/g.
[c] ^{3}He not measured. ^{4}He concentration: **7·2** × 10^{-8} cm^3 STP/g.
[d] Gas lost.
[e] Includes estimate for the one significant gas fraction lost.
[f] High blank (high temperature only) because of quartz chimney (see text).

In addition to amounts, Table 2 contains corrected isotopic compositions for the trapped components. These can safely be identified as the solar component for neon, krypton, and xenon, except for possible fission effects in xenon.

Argon from the fines is a complex system where a number of effects operate together in the stepwise release. The initial $^{38}Ar/^{36}Ar$ ratio of 0·18 rises to 0·19 at 750°C where the release of trapped argon reaches its peak. Judging from the release curves (see Fig. 4) for cosmogenic neon and krypton from the fines, cosmogenic argon has not yet begun to be released at this temperature so that we must again ascribe the change in the ratio to diffusive separation of the isotopes. The initial $^{38}Ar/^{36}Ar$ ratio is 5·3 per cent lower than that for the total. Again, as with helium and neon, the initial distortion in the ratio seems to be just the mass ratio for the isotopes involved. Above 750°C the cosmogenic argon begins to be released, increasing the $^{38}Ar/^{36}Ar$ ratio further. The interplay of the two independent effects on the ratio makes it difficult to compute the amount of cosmogenic argon in the fines with any precision and the computation was not attempted. It was possible, however, to ascertain that the $^{38}Ar/^{36}Ar$ ratio in the trapped component is 0·19 as reported in Table 2.

The sample certainly contains radiogenic ^{40}Ar. The difficulty in knowing how much of the ^{40}Ar is due to potassium decay in the fines themselves has been described very clearly by HEYMANN *et al.* (1970) who find that ^{40}Ar in the lunar fines is predominantly surface correlated with solar ^{36}Ar, but in proportions ($^{40}Ar/^{36}Ar = 1·056$) which greatly exceed what can reasonably be expected from the sun (BERNSTEIN *et al.*, 1963). Our concentrations for ^{36}Ar and ^{40}Ar plot near, but not precisely on, the correlation line in the paper by HEYMANN *et al.* (1970).

Our total rare gas concentrations for sample 10084 are higher than values by HEYMANN *et al.* (1970), HINTENBERGER *et al.* (1970) and MARTI *et al.* (1970) among whom there is rough consensus on the rare gas contents of this sample. For Ne, Kr and Xe our concentrations are about 100 per cent higher than those reported by MARTI *et al.*; for He and Ar, about 50 per cent higher. A possible difference in average grain size between our samples could easily produce differences in gas concentration of this magnitude, since the concentrations depend strongly on grain size.

Figure 4 includes fractional release curves against temperature for the trapped component of Ne, Ar, Kr and Xe in the dust. Cosmogenic release curves are also included, except for argon. As expected, the cosmogenic component is released at higher temperatures than the trapped component. And release of the trapped component is shifted toward higher temperatures as one progresses from helium (Fig. 2) to xenon.

RESULTS FROM FINE GRAINED CRYSTALLINE ROCK 10057

Rock 10057 is a good example of a system where a stepwise heating analysis permits us to sort out the various rare gas components with good success. The stone was heated in 100° steps. All the data are set out in Table 3. The neon results have already been discussed in the section on lunar fines.

The xenon analysis is the most complex. We know just from the uranium content and age of the rock that there is a significant xenon component from spontaneous fission of ^{238}U in addition to the cosmogenic and trapped components. We thus are dealing with at least a three component system for many of the isotopes. HOHENBERG *et al.* (1967) showed in their work on stepwise heating of the Pasamonte calcium-rich achondrite that such a three component system can be unraveled without foreknowledge of the isotopic composition of the cosmogenic and fissiogenic components. An isotope ratio which was pivotal, on one correlation diagram, for subtracting a component was treated as a free parameter. "Success" of the subtraction based on an assigned value for this parameter could be judged by the fit of the residual gas, after

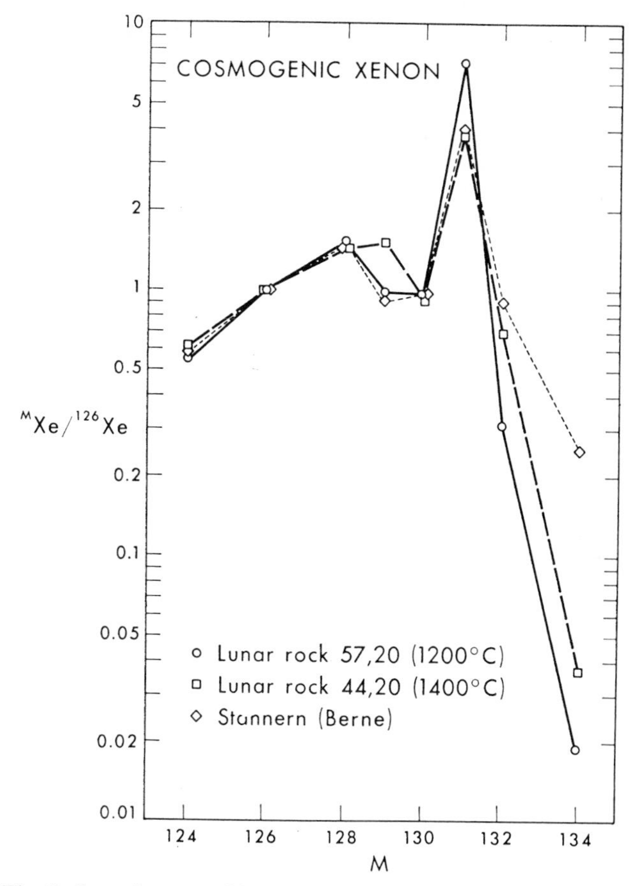

Fig. 5. Isotopic composition of cosmogenic xenon from lunar rocks. The comparison spectrum is cosmogenic xenon from the Stannern calcium-rich achondrite (MARTI *et al.*, 1966). The point for ^{129}Xe in the comparison spectrum is a value based on achondrites by ROWE (1967). Except at mass 131 the errors in the comparison spectrum and the lunar spectra overlap.

the subtraction, to a straight line on a second correlation diagram. With an electronic computer, the computational cycle was repeated rapidly and a best value for the free parameter quickly obtained.

In the present work we have made some short cuts instead of a complete analysis, but we have been able thereby to complete an adequate (if not the best possible) analysis quite quickly "by hand." In rock 10057 the successful short cut was to assume sample, 10084) from the 1200°C fraction then immediately gave us the isotopic composition of the cosmogenic component. The end points were now known on the two-component correlation diagram (trapped and cosmogenic) for $^{124}Xe/^{130}Xe$ vs. $^{126}Xe/^{130}Xe$. This diagram could then be used to partition ^{130}Xe between trapped and cosmogenic components in the various temperature fractions. Trapped xenon was subtracted off by this technique and the residual xenon tested as a two-component system (cosmogenic and fissiogenic xenon) on the correlation diagram for $^{134}Xe/^{132}Xe$ vs. $^{136}Xe/^{132}Xe$. A straight line spanning the compositions of the cosmogenic xenon

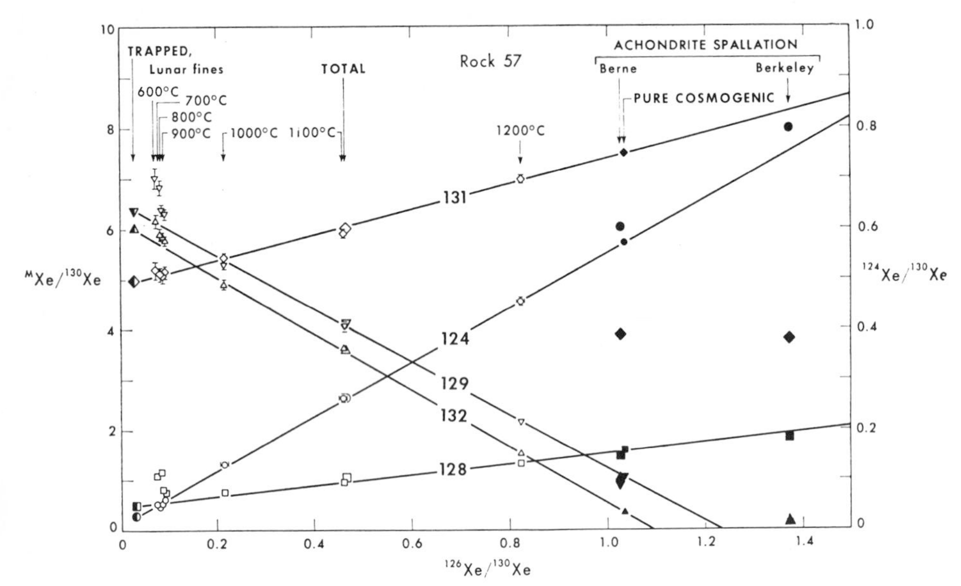

Fig. 6. Graphical analysis of the xenon released in stepwise heating of lunar rock 10057. Small corrections for fission have been made at masses 131 and 132. Isotopes 124, 126, 130 and 132 appear to belong to a simple two-component system (trapped and cosmogenic xenon). Isotope 131 fits a two-component model but with a highly anomalous composition for the cosmogenic component. Isotopes 128 and 129 display excess amounts in the temperature range 600 to 900°C inclusive but these excesses are now thought to be almost certainly due to memory of gas from a neutron-irradiated meteorite previously studied in the mass spectrometer. The errors shown are statistical only. Total errors are larger.

and xenon from spontaneous fission of ^{238}U was found. This last diagram could be used, in turn, for subtracting off fissiogenic xenon in the various temperature fractions where fission corrections were needed.

Some of the results of this analysis will now be presented. In Fig. 5 and Table 2 that the isotope ^{136}Xe in the highly cosmogenic 1200°C fraction was all trapped xenon (no fission or cosmogenic xenon). Subtraction of trapped xenon (as seen in the dust we exhibit the isotopic composition so obtained for the cosmogenic xenon. The spectrum for rock 10057 is compared with the cosmogenic spectrum for the Stannern calcium-rich achondrite. We have plotted the values obtained for this last spectrum by Marti *et al.* (1966) at Berne simply because their values fit the lunar sample better than other spectra do. Their spectrum has been supplemented by a meteoritic value for $(^{129}Xe/^{126}Xe)_{cosmogenic}$ obtained by ROWE (1967). Agreement between the meteoritic and lunar spectra are quite good, except at mass 131 where there is a large and conspicuous excess in the lunar sample. The discrepancies at masses 132 and 134, which are not so large in terms of absolute gas concentration, are not significant when errors are considered.

The correlation diagram for $^{124}Xe/^{130}Xe$ vs. $^{126}Xe/^{130}Xe$ is shown in Fig. 6. The

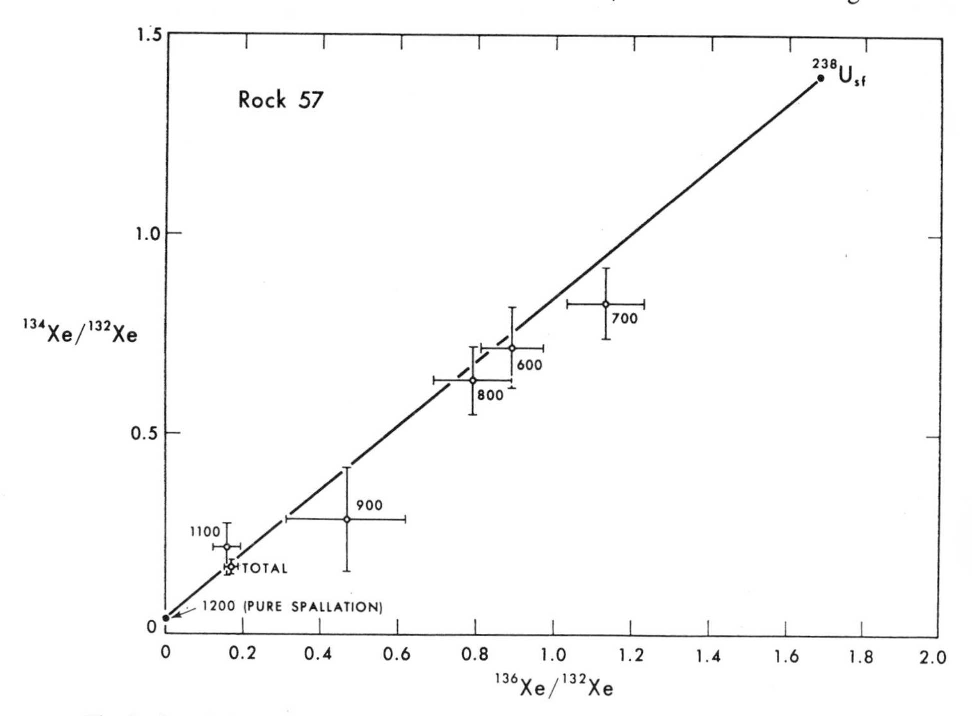

Fig. 7. Correlation of ratios $^{134}Xe/^{132}Xe$ and $^{136}Xe/^{132}Xe$ in rock 10057. Trapped xenon has been subtracted off before computing the ratios. The diagram is consistent with a two-component system of cosmogenic xenon and fissiogenic xenon from spontaneous fission of ^{238}U. The errors shown are statistical only. Total errors are larger.

experimental points fit the line very closely. For each point the amounts of $(^{130}Xe)_{cosmogenic}$ and $(^{130}Xe)_{trapped}$ are in inverse proportion to the line segments between the measured point and the cosmogenic and trapped end points. The release curves for trapped and cosmogenic xenon based on this partitioning are shown in Fig. 4. After subtraction of the trapped component, the correlation line for the heavy isotopes $^{134}Xe/^{132}Xe$ vs. $^{136}Xe/^{132}Xe$ should be a straight line. It is displayed in Fig. 7 where we see that within experimental error the points define a line between the spallation composition and the composition of xenon from the spontaneous fission of ^{238}U (WETHERILL, 1953). The release pattern for the fission component is also shown in Fig. 4. The shape is somewhat unlikely (of course the zero at 1200°C is an artifact: it was assumed so) and one wonders how, if at all, the shape would be altered by a more complete, computerized analysis. But the total amount of fission ^{136}Xe found is in agreement with that expected after $3{\cdot}7 \times 10^9$ yr (ALBEE *et al.*, 1970) decay by spontaneous fission of 800 ppb (LSPET, 1969) uranium: $1{\cdot}3 \times 10^{-12}$ cm^3 STP/g found; $1{\cdot}5 \times 10^{-12}$ cm^3 STP/g expected. In this calculation we used $1{\cdot}0 \times 10^{16}$ yr as the half-life for spontaneous fission of ^{238}U (FLEISCHER and PRICE, 1964). The fission results were used to correct the ratios ^{131}Xe/^{130}Xe and ^{132}Xe/^{130}Xe shown in Fig. 6. The corrections are not large. Figure 6, where ratios ^{M}Xe/^{130}Xe are plotted against ^{126}Xe/^{130}Xe, provides a comprehensive display of the xenon isotope variations in rock 10057. For values of M equal to 124 (already discussed), 131, and 132, the points fall reasonably well on lines joining the 1200°C fraction and the fraction seen at 820°C in the lunar fines. These isotopes, with ^{126}Xe and ^{130}Xe, appear after fission correction to belong to a two-component system of trapped and cosmogenic xenon. For values of M equal to 128 and 129, the points between 600° and 900°C fall above the lines containing the other points, as would occur if there were an excess of ^{128}Xe and ^{129}Xe being released at these temperatures. The amounts seen in excess are given in our summary Table 2. We now believe, for reasons discussed below, that the apparent excesses in ^{128}Xe and ^{129}Xe are due to memory of gas samples from a neutron irradiated meteorite studied earlier in the same mass spectrometer. One should note in Fig. 6 that the correlation lines pass close to the compositions for cosmogenic xenon in calcium-rich achondrites, except at mass 131 where the correlation line observed has the opposite slope for that observed in the Stannern meteorite.

The analysis for krypton from rock 10057 is much simpler than that for xenon because there is no appreciable fission krypton component. The assumption that krypton is the 1200°C fraction contains only trapped krypton at mass 86 should be nearly true. The cosmogenic spectrum for krypton so derived is given in Table 2 and shown in Fig. 8 along with the achondritic spectrum for Stannern as computed by MARTI *et al.* (1966). The correlations of ^{80}Kr/^{84}Kr and ^{82}Kr/^{84}Kr with ^{83}Kr/^{84}Kr revealed an apparent excess of ^{80}Kr and ^{82}Kr for temperatures below 1100°C. Otherwise the correlation lines for krypton were those for a two component system, leading to the release curves shown in Fig. 4. Radioactive ^{81}Kr was observed (especially in the 1200°C fraction), permitting calculation of a ^{81}Kr/^{83}Kr exposure age for the rock (see the section on exposure ages, below).

The ratio of the total ^{80}Kr excess to the total ^{82}Kr excess is 3·8, within 15 per cent or so error. For krypton produced by pile neutrons on bromine, the ratio is $3{\cdot}94 \pm 0{\cdot}05$ (REYNOLDS, 1950). The ratios of excess ^{80}Kr to excess ^{82}Kr in the individual temperature fractions are in good agreement, with few exceptions, with the ratio for the total excess krypton. It thus seems almost certain that we are seeing excess krypton which originated in a neutron irradiation of bromine. It is equally certain that this krypton did not originate in the sample. The highest bromine content (0·38 ppm) reported in lunar material (REED *et al.*, 1970) combined with the highest neutron exposure (17×10^{15} cm^{-2}) reported for lunar material (ALBEE *et al.*, 1970) generates only 10^{-11} cm^3 STP/g of excess ^{80}Kr. We find $3{\cdot}9 \times 10^{-11}$ cm^3 STP/g in rock 10057. The argument is even stronger in rock 10044 where REED *et al.* measured the bromine (0·22 ppm) and ALBEE *et al.* have put an upper limit of about 10^{15} cm^{-2} on the integrated neutron flux. There will be less than $0{\cdot}3 \times 10^{-12}$ cm^3 STP/g excess ^{80}Kr produced; we find

13×10^{-12} cm³ STP/g. We did not have to look far to find a source of the excess krypton and xenon. Ten months prior to the lunar work, the mass spectrometer (but not the sample system) was used to study xenon from a heavily neutron-irradiated sample of the enstatite chondrite Abee (HOHENBERG and REYNOLDS, 1969). Krypton from the meteorite was admitted along with the xenon. The amount of excess ^{80}Kr admitted to the spectrometer from the larger of two samples studied totaled about 6×10^{-8} cm³ STP or about 2800 times the amount of excess ^{80}Kr found in rock 10057. Such an amount of "memory" appears (in hindsight!) to be quite reasonable.

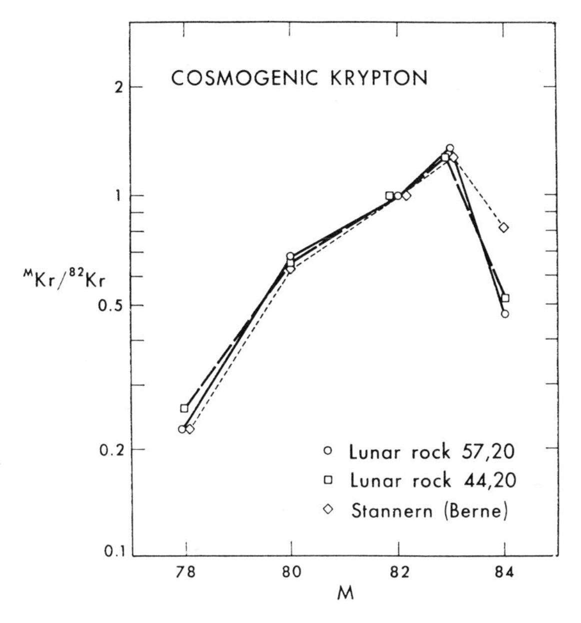

Fig. 8. Isotopic composition of cosmogenic krypton from lunar rocks. The comparison spectrum is cosmogenic krypton from the Stannern calcium-rich achondrite (MARTI *et al.*, 1966). At all masses the errors in the comparison spectrum and the lunar spectra overlap.

The excesses of ^{128}Xe and ^{129}Xe can be ascribed to the same source. In the irradiated meteorite the ratio of excess ^{129}Xe to ^{128}Xe was 1·47. In rock 10057, the ratio was the same, 1·43 (and close to that in all temperature fractions where excesses were observed). The amount of excess ^{128}Xe admitted to the mass spectrometer from the larger run with the meteorite was $0{\cdot}33 \times 10^{-8}$ cm³ STP or 2100 times the amount of excess ^{128}Xe in rock 10057. We need say no more about excess Kr and Xe isotopes in rock 10057!

The release curves for argon and neon, also shown in Fig. 4, were computed in the same way as for the fines (see above). Looking at all the release curves for rock 10057 together, we note that the cosmogenic fraction was released similarly for all the gases, at temperatures near the melting point for the rock. Release of the trapped component takes place at lower temperatures and at temperatures which are shifted upward progressively in going from neon to xenon. Note the agreement in temperature for each gas between the release of the trapped component in rock 10057 and in the fines. The chemical ratios (e.g. He/Ne etc.) are also very similar in rock 10057 and the fines. Our sample of 10057 is either contaminated with lunar dust or contains a surface patch in which solar wind gases have been imbedded. Other samples of rock 10057 which have been run (FUNKHOUSER *et al.*, 1970; WANLESS *et al.*, 1970; HINTENBERGER *et al.*, 1970) appear to contain much less trapped gas.

The $^{40}Ar/^{36}Ar$ ratio in rock 10057 greatly exceeds that in the fines. The excess ^{40}Ar has to be attributed to potassium decay in the rock. Curiously, our value of ^{40}Ar is more than 3 times higher than values reported by other investigators. We cannot explain this discrepancy. In rock 10044 (see below) we measured *less* ^{40}Ar than other investigators. Air argon cannot contribute appreciably to ^{40}Ar in rock 10057, judging from the release curve.

RESULTS FROM MEDIUM GRAINED CRYSTALLINE ROCK 10044

Our calculations for rock 10044 were very similar to those for rock 10057, and the same general sorts of results were obtained. The Figures and Tables contain a parallel display of results for the two rocks. In this section we mainly stress the differences encountered with rock 10044. The complete stepwise heating data are set out in Table 4.

The essential short cut for the xenon calculations in rock 10044 was the assumption that the 1400°C fraction contains only trapped ^{136}Xe. This will not be strictly true, we know, but quite consistent results were obtained with the assumption. The cosmogenic xenon spectrum so obtained is set out in Table 2 and Fig. 5. Cosmogenic xenon in rock 10044 differs markedly from rock 10057 at mass 131. The relative abundance of cosmogenic ^{131}Xe in rock 10044 is more nearly that in calcium-rich achondrites. Variability in the relative production of cosmogenic ^{131}Xe is probably some sort of depth effect. Since our sample of rock 10057 may include some lunar surface (as judged from the concentration of trapped gases) whereas rock 10044 appears (from the lack of trapped gas) to be an interior piece, we tentatively assign high production of ^{131}Xe to surface locations. Cosmogenic ^{129}Xe also differs in relative strength between rocks 10044 and 10057. As we shall see below, the amount of ^{129}Xe in the various temperature fractions is somewhat erratic in this rock, which makes it difficult to be sure about the abundance of ^{129}Xe in the cosmogenic spectrum for rock 10044.

The correlation diagram (Fig. 9) for $^{124}Xe/^{130}Xe$ vs. $^{126}Xe/^{130}Xe$ is well behaved for rock 10044 and permits us to subtract off trapped xenon and to prepare a two component correlation diagram for the heavy xenon isotopes (Fig. 10). The concentrations of trapped xenon in the various temperature fractions of rock 10044 are never greatly above the level of the xenon "hot blank," so that atmospheric values have been used for the isotopic composition of the trapped xenon. The actual trapped component is probably a blend of atmospheric and solar xenon. Figure 10 shows that the heavy isotope data are consistent with the fission xenon component originating from spontaneous fission of ^{238}U. In rock 10044 we find too much ^{136}Xe from fission. Expected from a $3{\cdot}7 \times 10^9$ yr old rock (ALBEE *et al.*, 1970; TURNER, 1970) with 280 ppb

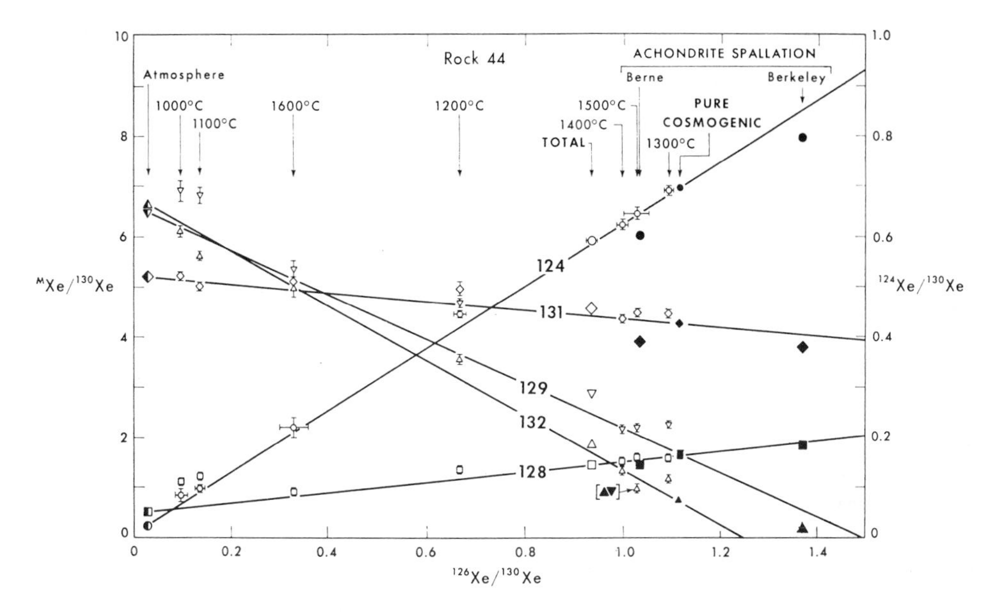

Fig. 9. Graphical analysis of the xenon released in stepwise heating of lunar rock 10044. Small corrections for fission have been made at masses 131 and 132. Isotopes 124, 126, 130, and 131 appear to belong to a simple two-component system (trapped and cosmogenic xenon). Excess ^{128}Xe and ^{129}Xe at 1000, 1100 and 1600°C is thought to be almost certainly memory of gas from an irradiated meteorite previously studied in the mass spectrometer. Isotopes 129 and 132 are in excess at 1200 and 1300°C, where the effects are *not* instrumental. Error shown are statistical only. Total errors are larger.

uranium (Wänke *et al.*, 1970) is $0{\cdot}5 \times 10^{-12}$ cm^3 STP/g; whereas we found $1{\cdot}5 \times 10^{-12}$ cm^3 STP/g. Possibly our sample contained more uranium than 280 ppb.

The comprehensive display of isotope abundance variations in xenon appears in Fig. 9. The isotopes ^{131}Xe and ^{132}Xe have been fission corrected. The most noteworthy feature of the diagram is the erratic behavior of ^{129}Xe. A correlation line for ^{129}Xe/^{130}Xe vs. ^{126}Xe/^{130}Xe has been drawn between trapped xenon and xenon in the 1400°C fraction. The choice leads to positive excesses of ^{129}Xe at most other temperatures, but no negative ones. Use of the cosmogenic ^{129}Xe abundance for rock 10057 would produce an even more steeply dipping correlation line and would lead to even greater excesses of ^{129}Xe in the various temperature fractions. The total excess ^{129}Xe calculated from the construction shown in Fig. 9 is $5{\cdot}2 \times 10^{-12}$ cm^3 STP/g. Only part of this is due to memory of the gas from the irradiated meteorite: we attribute the excesses at 1000, 1100 and 1600°C and part of the excess at 1200°C (totaling perhaps $0{\cdot}7 \times 10^{-12}$ cm^3 STP/g) to this source. The remainder at 1200 and the whole at 1300°C (totaling 5×10^{-12} cm^3 STP/g) is not an instrumental effect. It is accompanied at just those two temperatures by significant amounts of excess ^{132}Xe and possibly some ^{124}Xe and ^{131}Xe. It seems possible that there may be more than one cosmogenic component in rock 10044. It would be interesting to work with mineral separates of this rock as a means of exploring the question further.

Table 4. Rare gases from stepwise heating of lunar

Temperature (°C)	Ave. 1500° Blank	100	200	300	400	500	600	700	800
$^4He/^3He$	—	3817·0‡	1733·0‡	6536·0‡	2320·0‡	2041·0‡	1497·0‡	672·6‡	372·5‡
	—	±190·0	±33·0	±726·0	±151·0	±46·0	±27·0	±2·3	±0·4
[^3He], 10^{-8} cm^3 STP/g	*a*	**0·0004**	**0·0039**	**0·0002**	**0·0020**	**0·0029**	**0·0435**	**4·365**	**24·28**
$^{20}Ne/^{22}Ne$	—	9·1§	10·6§	8·4§	10·5§	11·5§	11·44§	12·55§	12·32
	—	±9·4	±6·8	±11·4	±7·0	±6·0	±0·38	±0·08	±0·05
$^{21}Ne/^{22}Ne$	—	0·0294	0·0514	0·0488	0·0374	0·0444	0·0889	0·0468	0·0678
		±0·0055	±0·0029	±0·0042	±0·0056	±0·0018	±0·0062	±0·0010	±0·0009
[^{22}Ne], 10^{-8} cm^3 STP/g	**0·0042***	**0·0021**	**0·0023**	**0·0017**	**0·0023**	**0·0025**	**0·0077**	**0·176**	**0·482**
$^{38}Ar/^{36}Ar$	—	0·1826	0·1841	0·1849	0·1879	0·1848	0·1895	0·3010	0·6101
	—	±0·0010	±0·0010	±0·0008	±0·0010	±0·0009	±0·0007	±0·0013	±0·0019
$^{40}Ar/^{36}Ar$	—	296·8	300·2	299·4	298·6	298·7	295·6	298·8	363·5
	—	±0·3	±0·7	±0·6	±0·8	±0·5	±0·4	±0·6	±0·8
[^{36}Ar], 10^{-8} cm^3 STP/g	**0·24***	**0·256**	**0·221**	**0·221**	**0·239**	**0·251**	**0·221**	**0·312**	**1·100**
$^{78}Kr/^{82}Kr$	—	—	—	—	—	—	—	—	—
	—	—	—	—	—	—	—	—	—
$^{80}Kr/^{82}Kr$	—	—	—	—	—	—	—	—	—
	—	—	—	—	—	—	—	—	—
$^{81}Kr/^{82}Kr$	—	—	—	—	—	—	—	—	—
	—	—	—	—	—	—	—	—	—
$^{83}Kr/^{82}Kr$	—	0·956	0·950	0·929	0·955	0·975	0·944	0·957	0·941
	—	±0·016	±0·016	±0·021	±0·014	±0·024	±0·020	±0·025	±0·021
$^{84}Kr/^{82}Kr$	—	4·583	4·745	4·635	4·681	4·715	4·635	4·61	4·573
	—	±0·068	±0·059	±0·091	±0·051	±0·099	±0·087	±0·10	±0·078
$^{86}Kr/^{82}Kr$	—	1·316	1·393	1·334	1·364	1·370	1·365	1·367	1·333
	—	±0·023	±0·024	±0·028	±0·020	±0·032	±0·028	±0·033	±0·025
[^{82}Kr], 10^{-12} cm^3 STP/g	**16·3***	**16·43**	**13·46**	**10·85**	**9·77**	**10·20**	**10·19**	**9·78**	**8·50**
$^{124}Xe/^{130}Xe$	—	—	—	—	—	—	—	—	—
	—	—	—	—	—	—	—	—	—
$^{126}Xe/^{130}Xe$	—	—	—	—	—	—	—	—	—
	—	—	—	—	—	—	—	—	—
$^{128}Xe/^{130}Xe$	—	0·735	0·648	0·646	0·694	0·623	0·656	0·668	0·705
	—	±0·034	±0·052	±0·038	±0·048	±0·054	±0·045	±0·049	±0·046
$^{129}Xe/^{130}Xe$	—	6·84	6·17	6·54	6·56	6·43	6·88	6·50	6·02
	—	±0·18	±0·25	±0·31	±0·22	±0·31	±0·30	±0·35	±0·26
$^{131}Xe/^{130}Xe$	—	5·28	4·98	5·18	5·23	5·04	5·35	5·05	4·95
	—	±0·14	±0·18	±0·23	±0·18	±0·23	±0·23	±0·28	±0·18
$^{132}Xe/^{130}Xe$	—	6·56	6·12	6·31	6·51	6·35	6·76	6·36	5·97
	—	±0·12	±0·22	±0·27	±0·19	±0·27	±0·23	±0·33	±0·21
$^{134}Xe/^{130}Xe$	—	2·596	2·334	2·42	2·482	2·45	2·61	2·44	2·297
	—	±0·061	±0·098	±0·12	±0·088	±0·12	±0·10	±0·13	±0·090
$^{136}Xe/^{130}Xe$	—	2·202	1·976	2·125	2·056	1·996	2·24	2·03	1·971
	—	±0·053	±0·078	±0·097	±0·072	±0·091	±0·10	0·11	±0·073
[^{130}Xe], 10^{-12} cm^3 STP/g	**0·124***	**0·116**	**0·106**	**0·101**	**0·093**	**0·103**	**0·082**	**0·094**	**0·097**

Notes: Boldface entries are gas concentrations. Errors in gas concentrations are: He ± 13%; Ne ± 10%; Ar ± 10%; Kr ± 30%; Xe ± 6%.

Unless otherwise specified errors are statistical only. In addition there is a fractional error in the applied mass discrimination factors as follows:

Ne 20/22: 0·01; 21/22: 0·016. **Ar** 38/36: 0·007; 40/36: 0·0065. **Kr** 78/82: 0·03; 80/82: 0·03; 83/82: 0·005; 84/82: 0·01; 86/82: 0·01. **Xe** 124/130: 0·013; 126/130: 0·02; 128/130: 0·05; 129/130: 0·017; 131/130: 0·005; 132/130: 0·005; 134/130: 0·005; 136/130: 0·005.

rock 10044-20. Heatings were for one hour

900[b]	1000	1100	1200	1300	1400	1500	1600	Total
152·72‡	100·60‡	111·00‡	210·8‡	207·0‡	92·06‡	58·46‡	2513·0‡	213·7‡
± 0·28	± 0·10	± 0·15	± 0·2	± 0·5	± 0·27	± 0·19	± 133·0	± 0·15
2·889[b]	**24·43**	**23·80**	**12·19**	**6·79**	**2·468**	**1·132**	**0·00297**	**102·4**
11·39	11·35	8·610	3·062	1·490	0·8779	0·9113§	6·53§	3·311
± 0·08	± 0·04	± 0·034	± 0·010	± 0·012	± 0·0041	± 0·0045	± 0·23	± 0·027
0·1499	0·1239	0·3364	0·7264	0·8515	0·9046	0·8955	0·477	0·7193
± 0·0034	± 0·0011	± 0·0024	± 0·0030	± 0·0087	± 0·0056	± 0·0045	± 0·021	± 0·0072
0·0273[b]	**0·522**	**0·630**	**1·503**	**2·265**	**2·267**	**1·612**	**0·0076**	**9·510**
0·3757	0·4978	0·4551	0·9091	1·191	1·516	1·521	0·4233	0·9843
± 0·0017	± 0·0010	± 0·0058	± 0·0034	± 0·003	± 0·002	± 0·003	± 0·0016	± 0·0036
396·7	352·7	316·6	720·4	346·9	103·5	51·82	249·0	282·8
± 1·4	± 0·5	± 0·4	± 1·8	± 0·7	± 0·9	± 0·07	± 0·5	± 0·8
0·241[b]	**2·842**	**9·385**	**5·100**	**5·111**	**6·907**	**8·404**	**1·099**	**41·91**
—	—	0·050‖	0·0962‖	0·2215‖	0·1739‖	0·1822‖	—	0·1740‖†
—	—	± 0·018	± 0·0098	± 0·0075	± 0·0086	± 0·0086	—	± 0·0043
—	0·286‖	0·448‖	0·513‖	0·619‖	0·493‖	0·536‖	0·137‖	0·522‖†
—	± 0·057	± 0·036	± 0·024	± 0·020	± 0·019	± 0·020	± 0·040	± 0·010
—	—	—	—	0·00236‖	—	—	—	—
—	—	—	—	± 0·00057	—	—	—	—
0·926	0·948	0·944	1·062	1·235	1·192	1·233	0·983	1·0928
± 0·016	± 0·024	± 0·015	± 0·018	± 0·012	± 0·016	± 0·017	± 0·027	± 0·0049
4·547	4·376	3·860	2·967	1·176	1·800	1·560	4·74	2·886
± 0·071	± 0·099	± 0·051	± 0·023	± 0·005	± 0·014	± 0·012	± 0·12	± 0·010
1·314	1·257	1·108	0·808	0·2359	0·4261	0·3275	1·353	0·7802
± 0·022	± 0·032	± 0·018	± 0·010	± 0·0045	± 0·0060	± 0·0036	± 0·038	± 0·0037
10·03	**11·81**	**17·05**	**24·34**	**77·48**	**35·52**	**39·26**	**7·07**	**311·8**
—	0·0845	0·0973	0·444	0·6896	0·6206	0·643	0·222	0·6307†
—	± 0·0091	± 0·0057	± 0·014	± 0·0074	± 0·0093	± 0·016	± 0·019	± 0·0053
—	0·097	0·1379	0·669	1·096	0·998	1·029	0·331	1·001†
—	± 0·012	± 0·0076	± 0·014	± 0·017	± 0·015	± 0·025	± 0·029	± 0·011
0·655	1·102	1·226	1·355	1·579	1·510	1·516	0·914	1·471
± 0·050	± 0·036	± 0·042	± 0·025	± 0·022	± 0·021	± 0·028	± 0·039	± 0·014
6·21	6·91	6·83	4·688	2·252	2·185	2·174	5·34	2·848
± 0·33	± 0·19	± 0·12	± 0·074	± 0·027	± 0·035	± 0·056	± 0·21	± 0·020
4·84	5·24	5·040	4·964	4·463	4·369	4·430	5·10	4·548
± 0·26	± 0·15	± 0·095	± 0·093	± 0·052	± 0·056	± 0·083	± 0·21	± 0·034
6·05	6·37	5·720	3·795	1·265	1·347	1·340	5·06	1·965
± 0·30	± 0·10	± 0·075	± 0·056	± 0·012	± 0·014	± 0·021	± 0·19	± 0·010
2·21	2·452	2·240	1·325	0·2611	0·2929	0·2742	1·790	0·5583
± 0·12	± 0·071	± 0·040	± 0·026	± 0·0041	± 0·0054	± 0·0064	± 0·078	± 0·0041
1·98	2·116	1·840	1·123	0·1870	0·1959	0·1753	1·499	0·4392
± 0·11	± 0·070	± 0·032	± 0·021	± 0·0027	± 0·0031	± 0·0072	± 0·067	± 0·0031
0·071	**0·143**	**0·314**	**0·901**	**7·965**	**1·907**	**1·328**	**0·117**	**13·53**

* Absolute blank in cm^3 STP is tabulated blank times 0·6121 g.
† Total of gas fractions where this ratio was obtained.
‡ No discrimination correction available. These ratios may be systematically in error.
§ Correction has been applied for $^{40}Ar^{2+}$. Error includes uncertainty in this correction.
‖ Correction has been made for background. Error includes uncertainty in background and in discrimination correction.
[a] ^3He not measured. Concentration of ^4He is **0·68** $\times 10^{-8}$ cm^3 STP/g.
[b] Thermocouple problem: doubtful that 900°C was actually reached in this heating.

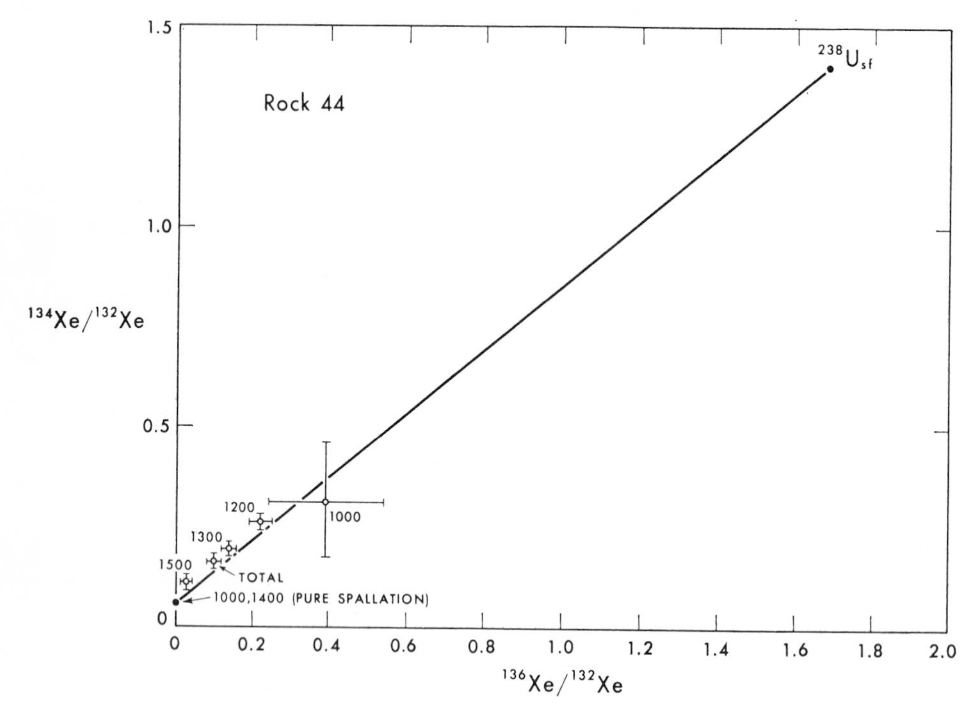

Fig. 10. Correlation of ratios $^{134}Xe/^{132}Xe$ and $^{136}Xe/^{132}Xe$ in rock 44. Trapped xenon has been subtracted off before computing the ratios. The diagram is consistent with a two-component system of cosmogenic xenon and xenon from spontaneous fission of ^{238}U. The errors shown are statistical only.

COSMIC RAY EXPOSURE AGES OF THE SAMPLES

The data obtained in this paper permit calculation in a number of ways of cosmic-ray exposure ages for the samples. We conclude this report by giving results of some of these calculations.

The most precise exposure ages which we can calculate come from our observation in both rocks of cosmogenic ^{81}Kr, a radioactive isotope with a half-life of 0·21 million years. Since decay of this isotope will be in equilibrium, at the time the rock was collected, with its production, measurement of its concentration gives us a production rate for ^{81}Kr. One of the stable cosmogenic Kr isotopes gives us a measure of the time-integrated production. If the relative production rates of the two isotopes can be specified, an exposure age which depends only on mass spectrometer krypton measurements is obtained. Errors in the sensitivity of the mass spectrometer for krypton do not affect the result, since only abundance ratios for krypton isotopes are involved.

We have carried out these calculations using the same assumptions as MARTI *et al.* (1970) who are among the originators of the method. Our results appear in Table 5. We are in fair agreement with MARTI *et al.* (1970) on rock 10057. They obtain 47 m.y. in comparison with our value of 34 m.y.

Table 5. Exposure ages of lunar samples

	$(^{83}Kr/^{81}Kr)_{cosm.}$	P_{81}^{a}	P_{83}^{a}	^{81}Kr–^{83}Kr age in 10^6 yr	$^{83}Kr_{cosm.}^{b}$	Sr (ppm)	Y (ppm)	Zr (ppm)	$^{83}Kr_{cosm.}$ age in 10^6 yr	$^{130}Xe_{cosm.}^{b}$	Ba (ppm)	Ce (ppm)	$^{130}Xe_{cosm.}$ age in 10^6 yr
Nuevo Laredo[c]	—	—	—	—	13 ± 0·39	84·4[d]	—	67[e]	≡ **16**[f]	1·5 ± 0·45	46[g]	10·7[h]	≡**16**[f]
Lunar fines 10084-59	—	—	—	—	1255 + 864 − 324	200[i]	150[i]	390[i]	**1130**[j] **+ 840 − 450**	130 ± 20	220[i]	50[i]	**1450**[j] **± 540**
Lunar rock 10057-20	187·6 ± 29·3	79·3	135	**34 ± 5**	135 ± 24·3	130[i]	210[l]	560[i]	**118**[j] **± 41**	22·6 ± 1·4	280[i]	88[i]	**76**[j] **± 23**
Lunar rock 10044-20	380·2 ± 92·4	76·8	127	**70 ± 17**	183 ± 55	167[k]	160[l]	410[l]	**176**[j] **± 75**	11·4 ± 0·7	285[k]	76[k]	**100**[j] **± 30**

[a] Production rates (P) computed after MARTI *et al.* (1970) $P_{81} = 0{\cdot}95\,[(P_{80} + P_{82})/2]$; $P_{82} \equiv 100$.
[b] Gas concentration in units 10^{-12} cm^3 STP/g.
[c] MUNK (1967).
[d] GAST (1962).
[e] SCHMITT *et al.* (1964).
[f] MUNK (1967); HEYMANN *et al.* (1968).
[h] HAMAGUCHI *et al.* (1957).
[h] SCHMITT *et al.* (1963).
[i] MORRISON *et al.* (1970).
[j] 2π irradiation geometry assumed.
[k] PHILPOTTS and SCHNETZLER (1970).
[l] MAXWELL *et al.* (1970); Value is for Type B Rock 10017.

An ^{81}Kr–^{83}Kr age is very meaningful if the sample has had a simple irradiation history: i.e. if it has been abruptly transferred from a completely shielded environment to its final position on the surface of the moon. If the rock came to its final position in stages, the ^{81}Kr–^{83}Kr method will overestimate its time in the final position but underestimate the integrated period when it was within a meter or so of the surface.

For this reason, it is interesting to compare the ^{81}Kr–^{83}Kr ages with exposure ages determined from various stable cosmogenic nuclides and the production rates inferred for those nuclides from stone meteorites. We have elected to compute exposure ages based on ^{83}Kr and ^{130}Xe, using the calcium-rich achondrite Nuevo Laredo as a meteorite standard. The exposure age for Nuevo Laredo has been found by more than one author to be 16 million years. MUNK (1967) has conveniently measured the ^{83}Kr and ^{130}Xe concentrations from spallation in the Nuevo Laredo stone. We can correct for chemical differences between Nuevo Laredo and the lunar rocks, using the Rudstam calculations carried out by HOHENBERG *et al.* (1967) for Kr and Xe production in the relevant target elements. In other words we take the Kr and Xe production rates from the 16 m.y. exposure in Nuevo Laredo, but we adjust these production rates (using relative yields given by the Rudstam calculations) to allow for differences in chemical composition between Nuevo Laredo and the lunar sample in question. The details, including the chemical compositions we used, appear in Table 5. The production rates found for Nuevo Laredo were halved to take into account the 2π instead of 4π irradiation geometry for samples near the lunar surface. We believe that the ^{130}Xe ages are to be preferred to the ^{83}Kr ages because we have encountered occasional severe variations in our krypton sensitivity which make Kr concentrations somewhat suspect. But in all cases the two methods agree within the errors. In comparing the ^{81}Kr–^{83}Kr ages with the ^{130}Xe ages, we find agreement, within the errors, for rock 10044, suggesting that this rock may have had a simple irradiation history: sudden emplacement on the lunar surface from shielded depth. For rock 10057, the ^{81}Kr–^{83}Kr age of 34 m.y. is significantly less than the ^{130}Xe age of 76 m.y., as if it had spent some time at a partially shielded location. Our average value for the average exposure age of the fines is 1300 m.y. plus or minus about 500 m.y.

Acknowledgments—We thank G. McCRORY for much help in the experimental work. This work received partial support from NASA and from the AEC and bears AEC code number UCB-34P32-72.

Note added in proof: All krypton ratios in Table 4 (of the form ^{M}Kr/^{82}Kr) should be multiplied by the factor 1·008. None of the conclusions reached in the paper will be changed thereby.

REFERENCES

ALBEE A. L., BURNETT D. S., CHODOS A. A., EUGSTER O. J., HUNEKE J. C., PAPANASTASSIOU D. A., PODOSEK F. A., RUSS G. P., SANZ H. G. and WASSERBURG G. J. (1970) Ages, irradiation history, and chemical composition of lunar rocks from the Sea of Tranquillity. *Science* **167,** 463–466.

BERNSTEIN W., FREDRICKS R. W., VOGL J. L. and FOWLER W. A. (1963) The lunar atmosphere and the solar wind. *Icarus* **2,** 233–248.

EBERHARDT P., GEISS J., GRAF H., GRÖGLER N., KRÄHENBÜHL U., SCHWALLER H., SCHWARZMÜLLER J. and STETTLER A. (1970) Trapped solar wind noble gases, Kr^{81}/Kr exposure ages and K/Ar ages in Apollo 11 lunar materal. *Science* **167,** 558–560.

FLEISCHER R. L. and PRICE P. B. (1964) Decay constant for spontaneous fission of U^{238}. *Phys. Rev.* **133,** B63–64.

FUNKHOUSER J. G., SCHAEFFER O. A., BOGARD D. D. and ZÄHRINGER J. (1970) Gas analysis of the lunar surface. *Science* **167,** 561–563.

GAST P. W. (1962) The isotopic composition of strontium and the age of stone meteorites—I. *Geochim. Cosmochim. Acta* **26,** 927–943.

HAMAGUCHI H., REED G. W. and TURKEVICH A. (1957) Uranium and barium in stone meteorites. *Geochim. Cosmochim. Acta* **12,** 337–347.

HEYMANN D., MAZOR E. and ANDERS E. (1968) Ages of calcium-rich achondrites—I. Eucrites. *Geochim. Cosmochim. Acta* **32,** 1241–1268.

HEYMANN D., YANIV A., ADAMS J. A. S. and FRYER G. E. (1970) Inert gases in lunar samples. *Science* **167,** 555–558.

HINTENBERGER H., WEBER H. W., VOSHAGE H., WÄNKE H., BEGEMANN F., VILCSEK E. and WLOTZKA F. (1970) Rare gases, hydrogen, and nitrogen: concentrations and isotopic composition in lunar material. *Science* **167,** 543–545.

HOHENBERG C. M., MUNK M. N. and REYNOLDS J. H. (1967) Spallation and fissiogenic xenon and krypton from stepwise heating of the Pasamonte achondrite; The case for extinct plutonium 244 in meteorites; Relative ages of chondrites and achondrites. *J. Geophys. Res.* **72,** 3139–3177.

HOHENBERG C. M. and REYNOLDS J. H. (1969) Preservation of the iodine-xenon record in meteorites. *J. Geophys. Res.* **74,** 6679–6683.

KIRSTEN T., STEINBRUNN F. and ZÄHRINGER J. (1970) Rare gases in lunar samples: study of distribution and variations by a microprobe technique. *Science* **167,** 571–574.

LSPET (LUNAR SAMPLE PRELIMINARY EXAMINATION TEAM) (1969) Preliminary examination of lunar samples from Apollo 11. *Science* **165,** 1211–1227.

MARTI K. (1969) Solar-type xenon: a new isotopic composition of xenon in the Pesyanoe meteorite. *Science* **166,** 1263–1265.

MARTI K., EBERHARDT P. and GEISS J. (1966) Spallation, fission, and neutron capture anomalies in meteoritic krypton and xenon. *Z. Naturforsch.* **21a,** 398–413.

MARTI K., LUGMAIR G. W. and UREY H. C. (1970) Solar wind gases, cosmic ray spallation products, and the irradiation history. *Science* **167,** 548–550.

MAXWELL J. A., ABBEY S. and CHAMP W. H. (1970) Chemical composition of lunar material. *Science* **167,** 530–531.

MORRISON G. H., GERARD J. T., KASHUBA A. T., GANGADHARAM E. V., ROTHENBERG A. M., POTTER N. M. and MILLER G. B. (1970) Multielement analysis of lunar soil and rocks. *Science* **167,** 505–507.

MUNK M. N. (1967) Argon, krypton, and xenon in Angra dos Reis, Nuevo Laredo, and Norton County achondrites: the case for two types of fission xenon in achondrites. *Earth Planet. Sci. Lett.* **3,** 457–465.

PHILPOTTS J. A. and SCHNETZLER C. C. (1970) Potassium, rubidium, strontium, barium, and rare-earth concentrations in lunar rocks and separated phases. *Science* **167,** 493–495.

REED G. W., JR., JOVANOVIC S. and FUCHS L. H. (1970) Trace elements and accessory minerals in lunar samples. *Science* **167,** 501–503.

REYNOLDS J. H. (1950) A mass spectrometric investigation of branching in Cu^{64}, Br^{80}, Br^{82} and I^{128}. *Phys. Rev.* **79,** 789–794.

REYNOLDS J. H., HOHENBERG C. M., LEWIS R. S., DAVIS P. K. and KAISER W. A. (1970) Isotopic analysis of rare gases from stepwise heating of lunar fines and rocks. *Science* **167,** 545–548.

ROWE M. W. (1967) Cosmic ray spallation and the special anomaly in achondrites. *Earth Planet. Sci. Lett.* **2,** 92–98.

SCHMITT R. A., BINGHAM E. and CHODOS A. A. (1964) Zirconium abundances in meteorites and implications to nucleosynthesis. *Geochim. Cosmochim. Acta* **28,** 1961–1979.

SCHMITT R. A., SMITH R. H., LASCH J. E., MOSEN A. W., OLEHY D. A. and VASILEVSKIS J. (1963) Abundances of the fourteen rare-earth elements, scandium, and yttrium in meteoritic and terrestrial matter. *Geochim. Cosmochim. Acta* **27,** 577–622.

TURNER G. (1970) Argon-40/argon-39 dating of lunar rock samples. *Science* **167,** 466–468.

WÄNKE H., BEGEMANN F., VILCSEK E., RIEDER R., TESCHKE F., BORN W., QUIJANO-RICO M., VOSHAGE H. and WLOTZKA F. (1970) Major and trace elements and cosmic-ray produced radio-isotopes in lunar samples. *Science* **167,** 523–525.

WANLESS R. K., LOVERIDGE W. D. and STEVENS R. D. (1970) Age determinations and isotopic abundance measurements on lunar samples. *Science* **167,** 479–480.

WETHERILL G. W. (1953) Spontaneous fission yields from uranium and thorium. *Phys. Rev.* **92,** 907–912.

Proceedings of the Apollo 11 Lunar Science Conference, Vol. 2, pp. 1311 to 1315.

Whole-rock Rb–Sr isotopic age relationships in Apollo 11 lunar samples

P. M. HURLEY and W. H. PINSON, JR.
Department of Earth and Planetary Sciences, Massachusetts Institute of Technology, Cambridge, Massachusetts 02139

(*Received* 5 *February* 1970; *accepted in revised form* 28 *February* 1970)

Abstract—Whole-rock fragments of lunar samples 10072, 10022, 10058, 10047 and 10020 have ratios of Rb^{87}/Sr^{86} and Sr^{87}/Sr^{86} that lie on an apparent isochron of 3850 ± 200 m.y., and $(Sr^{87}/Sr^{86})_I =$ 0·6991 ± 0·00015, relative to standard sea water $Sr^{87}/Sr^{86} = 0{\cdot}7090$. This would be a valid age only if the samples all had the same initial Sr isotopic abundance.

The fact that the points for samples 10058, 10047 and 10022 lie close to the meteorite reference isochron indicates that these magmas were not greatly fractionated in Rb/Sr relative to source materials from which they were derived. This leads to a ratio of Rb/Sr in the source materials of these magmas of not over 0·01 that must have existed since early in the history of the solar system. This value is lower than average earth or meteorites and indicates either a very early magmatic fractionation or a strong depletion in Rb relative to Sr in some of the primordial material of the moon.

INTRODUCTION

FIVE rock samples weighing 0·82 g (10022-43), 1·27 g (10020-28), 1·14 g (10047-25), 2·36 g (10072-25), and 1·2 g (58·36) and a 5·16 g sample of fines from the lunar soil (10084-39) were available for analysis. This progress report gives analyses on total rock samples only; work on separated minerals and the fine-grained soil material has not been completed. Rubidium, strontium, and the isotopic composition of Sr were determined by mass spectrometric techniques in separate whole pieces of rock ranging upward in size from a few tens of milligrams. These samples were not necessarily assumed to be representative of the rock chips selected by the Lunar Receiving Laboratory, but, by relating the isotopic data from each sample separately, and using different whole fragments in repeated analyses, we hoped to obtain additional information on the homogeneity of the rocks, test the assumption of a lack of short range migration of components, and derive meaningful age evaluations and average compositions.

ANALYTICAL PROCEDURES

Analytical data presented in the preliminary report (HURLEY and PINSON, 1970) have been revised.* Reagent and total contamination levels are being reduced and controlled, and now are negligible in

* Owing to a late start on the lunar samples (November, 1969) and the early scheduling of the Preliminary Report (December, 1969) we did not have sufficient time to investigate several possible sources of instrumental bias that could be significant at the new level of precision required by the low Rb/Sr ratios in the lunar materials. Combinations of switch positions in both the electrometer and expanded scale instruments, and two different sources, were not initially kept identical between unknown and standard analyses of Sr^{87}/Sr^{86}. Only by extending this investigation until the latest time permitted for this present report have we been able to bring all analyses together into a single fixed-bias comparison against the sea water standard. Only the original analyses on samples 10072 and 10022 had to be revised by amounts greater than the stated errors. In each instance the internal consistency of the measurement was satisfactory, which masked the error due to bias.

the cases of Sr, and Rb in Type A basalts. Corrections up to 4% have been made in Rb analyses on materials which are in the range of 1 ppm.

The sample fragments were gently broken in a boron carbide mortar, weighed, and digested in a Parr Industries Teflon-lined bomb at 120–140°C. Before acid digestion the samples were spiked with Rb^{87}. Minimum amounts of reagents† were used: 10 ml of HCl carrying the spike, 3 ml of HF and 1 ml $HClO_4$ per 0·15 g of sample were the approximate amounts. The solution was washed into a Pt dish, covered with a Teflon lid, and heated with a hot plate below and a heat lamp above. After achieving total solution (no visible mineral or chemical residues) the solution was evaporated to get rid of excess HF and SiF_4 and diluted with 0·5 N HCl to 50 ml in a volumetric flask, and aliquots were taken for Sr spiking and for Rb and Sr isotope dilution and Sr isotope ratio analysis. Element separations were achieved on cation exchange columns using Dowex 50, 8 per cent cross-linked, 350 mesh.

The mass spectrometer used for this work consisted of a single focussing 30·5 cm radius, 60° sector analyzer section (Consolidated Electrodynamics Corporation) with detachable source, in-line valve, and cold finger in source housing using single Ta filament surface ionization, Faraday cup collector, Cary vibrating reed electrometer, and expanded scale system of recording residual output voltages on a modified Leeds-Northrop Speedomax G recorder. All scale settings on vibrating reed electrometer and expanded scale system were kept the same for all runs, including the calibration and standardization analyses, so that instrumental fixed bias was removed.

In analyses of Sr isotopic ratios the internal consistency of repeated ratio measurements in a single analysis was made to reach a standard deviation of the mean of less than 0·03 per cent. Allowing for possible unforseen effects, we believe that the overall error will not exceed a 2σ value (two standard deviations) of 0·0002–0·0004 in Sr^{87}/Sr^{86} ratio measurements, depending on the analytical data. This estimate is substantiated by replicate analyses of samples and standards in which all operating settings are kept constant.

Mass spectrometer errors in analyses by isotope dilution are generally not more than 1 per cent (standard deviation) for Sr and 2 per cent for Rb, but in these cases of low Rb content the error due to uncertain contamination may outweight the measurement error. Repeated analyses on separate small rock fragments show also the inhomogeneity of the rocks. Because the Rb, Sr and Sr isotope analyses are performed on each piece separately, the inhomogeneity does not affect the age relationship. All measurements are relative to the Rb and Sr Normal Solutions distributed by the Lunar Receiving Laboratory (prepared by D. S. Burnett, California Institute of Technology).

Our average value for Sr^{87}/Sr^{86} in Atlantic seawater (Sargasso Sea) is 0·7090. This is compared with 0·7091 reported by PAPANASTASSIOU and WASSERBURG (1969) for Pacific seawater. Thus, if the homogeneity of Sr in seawater is assumed, all of our Sr^{87}/Sr^{86} values would have to be increased by 0·0001 to be directly comparable with those reported by the California Institute of Technology group. In Fig. 1 the initial ratio of Sr^{87}/Sr^{86} in basaltic achondrites found by PAPANASTASSIOU and WASSERBURG (1969) have been plotted as 0·6989 instead of their reported value of 0·6990, in order to present a direct comparison of the data against the 4600 m.y. meteorite isochron.

ANALYTICAL RESULTS AND DISCUSSION

For convenience in this report we refer to Type A and Type B rocks as those having approximately 6 ppm and 1 ppm Rb respectively. Thus samples 10072 and 10022 are grouped as Type A, and samples 10058, 10047 and 10020 as Type B. Analyses are listed in Table 1 and plotted in Fig. 1.

In Fig. 1 the line drawn through the sample points is not considered to be an isochron because there is no evidence that the rocks are cogenetic, with an identical

† All reagents used were purified. Demineralized water was twice distilled in quartz. Hydrochloric acid and nitric acid were made by double distillation in vycor glass stills. Hydrofluoric acid was made by bubbling HF gas into quartz-distilled water, and the perchloric acid had been twice distilled in vycor.

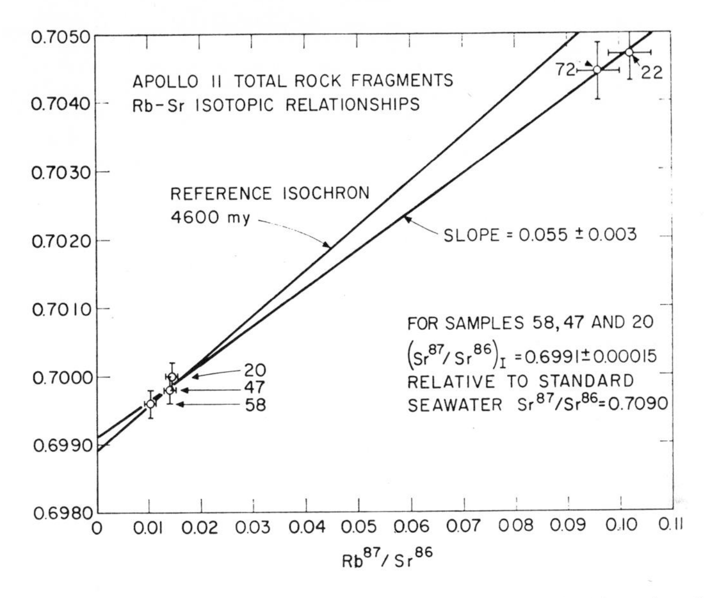

Fig. 1. Rb and Sr isotopic relationships in fragments of Apollo 11 total-rock samples 10072, 10022, 10058, 10047 and 10020. Estimated errors are given as two standard deviations. The slope of the apparent isochron of 0·055 ± 0·003 is equivalent to an age value of 3850 ± 200 m.y. using $Rb^{87}\lambda = 1{\cdot}39 \times 10^{-11}\ yr^{-1}$. However this is not a valid age unless the value of $(Sr^{87}/Sr^{86})_I = 0{\cdot}6991$ applies to all points.

initial ratio of Sr^{87}/Sr^{86}. If they were demonstrated to be cogenetic by other evidence the whole-rock age of the assembly would be 3850 ± 200 m.y., with $(Sr^{87}/Sr^{86})_I = 0{\cdot}6991 \pm 0{\cdot}00015$ relative to standard sea water equal to 0·7090, and for $Rb^{87}\lambda = 1{\cdot}39 \times 10^{-11}\ yr^{-1}$, and with Sr^{86}/Sr^{88} normalized to 0·1194.

Probably the best indication of the true time of volcanism and crystallization of the magmas is given by the mineral isochrons of ALBEE *et al.* (1970), GOPALEN *et al.* (1970) and COMPSTON *et al.* (1970). These range from 3400 to 4050 m.y. with a concentration

Table 1. Elemental and isotopic abundances of Rb and Sr in lunar rock fragments.

Sample	Rb ppm	Rb^{87} ppm	Sr ppm	Sr^{86} ppm	Rb^{87}/Sr^{86} Atom	Sr^{87*}/Sr^{86} N
Type A						
10072	5·44	1·54	167	16·2	0·094	0·7044
10022	6·00	1·70	173	16·7	0·101	0·7047
Type B						
10058	0·61	0·173	172	16·6	0·0103	0·6996
10047	0·93	0·263	194	18·8	0·0138	0·6998
10020	0·74	0·210	152	14·7	0·0141	0·7000

Rb and Sr analyses are relative to Rb and Sr Normal Solutions distributed by Lunar Receiving Laboratory. Errors (2σ) are estimated to be 0·0004 and 0·0002 for Sr^{87}/Sr^{86} for Type A and Type B rocks respectively, and 4 per cent for elemental abundance.

* Normalized to Sr^{86} Sr^{88} = 0·1194.

of values around 3650 m.y. in the analyses of ALBEE *et al.* If the value of 3650 m.y. is assumed to be correct the initial ratio for the Type A rocks in Fig. 1 would be 0·6993, and for the Type B rocks, 0·6991. These are compatible with the initial ratios found by ALBEE *et al.* Therefore we can see no reason to believe that any of these rocks differ in age or initial ratio from those Type A and Type B rocks measured by others; and that although probably coeval, the types are not cogenetic.

These inferred initial ratios are surprisingly low. If rocks 10058 and 10047 are indeed 3650 m.y. in age their initial ratios of Sr^{87}/Sr^{86} are between 0·6990 and 0·6991, or if rocks 10058, 10047 and 10020 are grouped together, the average would be 0·6991, with a 2σ error for the mean of the grouping being 0·00015. This value of Sr^{87}/Sr^{86} for the source region from which these magmas came would imply a growth of radiogenic Sr^{87}/Sr^{86} of $0{\cdot}0002 \pm 0{\cdot}00015$ in 900 m.y., if the age of the moon is assumed to be 4600 m.y. and the primordial value of $Sr^{87}/Sr^{86} = 0{\cdot}6990$ as found from basaltic achondrites by PAPANASTASSIOU and WASSERBURG (1969), after allowing for the difference of 0·0001 between the two laboratories for the seawater standard. The ratio of Rb/Sr in the source region for Type B rocks during this first stage of 900 m.y. therefore must have been $0{\cdot}006 \pm 0{\cdot}004$, which is less than one-fifth of the ratio of Rb/Sr for average earth estimated by HURLEY (1968), and a fortieth of average chondrites. The fact that the points in Fig. 1 for the Type B rocks fall close to the 4600 m.y. meteorite isochron indicates that there was no major difference between the Rb/Sr ratio in the magmas and that in the original source material. On the other hand the Type A points in Fig. 1 clearly lie to the right of the 4600 m.y. isochron. This means a fractionation that increased Rb relative to Sr that must have occurred toward, or at, the end of the 900 m.y. first stage, and probably during the extraction of the magmas at about 3700 m.y.

CONCLUSIONS

(1) A Sr^{87}/Sr^{86} vs. Rb^{87}/Sr^{86} plot of the whole-rock samples 10072, 10022, 10058, 10047 and 10020 shows the points falling on a line with a slope of $0{\cdot}055 \pm 0{\cdot}003$. If cogenetic the samples would give a whole-rock isochron age of 3850 ± 200 m.y. On the other hand if the mineral isochron age of 3650 m.y. found by others is true, these five samples would be compatible with this age value only if the initial ratio of the Type A rocks were 0·6993, and of the Type B rocks, 0·6991.

(2) If the age value of these rocks is either that given by the plot in Fig. 1, or the mineral isochron value found by others, the initial Sr^{87}/Sr^{86} in Type B rocks is unusually low. The source materials from which the Type B magmas were derived had Rb/Sr ratios of not greater than $0{\cdot}006 \pm 0{\cdot}004$, about a fifth that of average earth and a fortieth that of average chondrites. These ratios came into existence at or close to 4600 m.y., and not at the time of volcanism. The ratio of Sr^{87}/Sr^{86} in these materials has not matched that in average earth or chondrites since early in the history of planetary bodies.

Acknowledgments—This research is supported by the U.S. Atomic Energy Commission as part of a project under Contract AT(30-1)01381. The samples, and expenses connected with delivery and reporting, were supplied by the National Aeronautics and Space Administration. NATHANIEL CORWIN of the Woods Hole Oceanographic Institute supplied the Sargasso Sea ocean water sample. We appreciate the help of V. G. POSADAS in the chemical preparations.

REFERENCES

ALBEE A. L., BURNETT D. S., CHODOS A. A., EUGSTER O. J., HUNEKE J. C., PAPANASTASSIOU D. A., PODOSEK F. A., RUSS G. PRICE II., SANZ H. G., TERA F. and WASSERBURG G. J. (1970) Ages, irradiation history, and chemical composition of lunar rocks from the Sea of Tranquillity. *Science* **167**, 463–466.

COMPSTON W., ARRIENS P. A., VERNON M. J. and CHAPPELL B.W. (1970) Rubidium–strontium chronology and chemistry of lunar material. *Science* **167**, 474–476.

GOPALEN K., KAUSHAL S., LEE-HU C. and WETHERILL G. W. (1970) Rubidium–strontium, uranium and thorium–lead dating of lunar material. *Science* **167**, 471–473.

HURLEY P. M. (1968) Correction to: Absolute abundance and distribution of Rb, K and Sr in the earth. *Geochim. Cosmochim. Acta* **32**, 1025–1030.

HURLEY P. M. and PINSON W. H., JR. (1970) Rubidium–strontium relations in Tranquillity Base samples. *Science* **167**, 473–474.

PAPANASTASSIOU D. A. and WASSERBURG G. J. (1969) Initial strontium isotopic abundances and the resolution of small time differences in the formation of planetary objects. *Earth Planet. Sci. Lett.* **5**, 361–376.

Proceedings of the Apollo 11 Lunar Science Conference, Vol. 2, pp. 1317 to 1329.

Carbon and sulfur concentration and isotopic composition in Apollo 11 lunar samples*

I. R. KAPLAN, J. W. SMITH† and E. RUTH
Institute of Geophysics and Planetary Physics, University of California, Los Angeles 90024

(*Received* 5 *February* 1970; *accepted* 20 *February* 1970)

Abstract—The concentration of carbon and sulfur in six lunar samples ranged between 20–200 ppm and 650–2300 ppm, respectively. Carbon was present in gaseous, volatilizable and non-volatile forms. Among the forms of carbon, terrestrial contaminants were recognized. Sulfur appeared to exist only as acid-volatile sulfide. The bulk fines contain the highest concentration of carbon and the lowest concentration of sulfur. They are always enriched in the heavier isotope C^{13} or S^{34}. The fine-grained basaltic rocks show the reverse relationship; lowest carbon, highest sulfide concentrations and no apparent enrichment in heavy isotopes. The breccias are of intermediate composition. A A grain-size separation of the bulk fines indicated that the highest carbon content and lowest sulfur content is present in material with the smallest particle size.

INTRODUCTION

THE PURPOSE of this study was to differentiate between the various forms of carbon and sulfur present in the lunar material and to measure the δC^{13} and δS^{34} ratios of these components. The results presented here relate to six samples, five that were submitted directly to us at UCLA (10084; 10002-54; 10060-22; 10049 and 10057-40), one bulk fine submitted through Ames Research Center (10086) and one obtained from a sample sent to Professor J. William Schopf, U.C.L.A. (10086A).

The study was considered important for several reasons: (i) to determine if organic matter resembling that found in carbonaceous chondrites is present, (ii) to identify primordial carbon compounds which may lead to an understanding of early biochemical evolution on earth, (iii) to search for any evidence for a possible origin of the moon by fission from the earth and (iv) to obtain a better understanding of cosmic isotope abundances.

Parts of this investigation, dealing with identification of carbon compounds, are presented in greater detail by CHANG *et al.* (1970), and by KVENVOLDEN *et al.* (1970) in this volume. The results of these studies will, therefore, only be summarized here insofar as to enable a discussion and interpretation of the data.

EXPERIMENTAL

Carbon

Carbon compounds were analyzed by undertaking the following steps: (1) extraction with organic solvents and water, (2) hydrolysis with hydrochloric acid and search for both soluble and volatile components, (3) degradation of the silicate minerals by repeated treatment with hydrochloric and with hydrofluoric acids, (4) pyrolysis at temperatures from 150 to 750°C, and (5) complete

* Publication No. 825 Institute of Geophysics and Planetary Physics.

† Permanent address: Division of Mineral Chemistry, C.S.I.R.O., P.O. Box 175, Chatswood, Sydney, Australia.

combustion in an oxygen atmosphere. Approximately 1 g of each sample studied was separately combusted.

Combustion of the intact lunar sample was undertaken in a small scale glass apparatus shown in Fig. 1. Each sample of lunar material was placed in a porcelain boat (previously "fired at 1100°C" for one hour in an air furnace) which was put into the combustion tube and evacuated for 48 hr at 150°C and 10^{-3} torr pressure. The rock samples had previously been broken up in a stainless steel diamond mortar, and ground rapidly in an alumina mortar to <1 mm grain size. Prolonged grinding and handling were not undertaken in order to prevent contamination. At the beginning of each combustion step the furnaces were turned on and all the apparatus evacuated to 10^{-3} torr; the catalyst furnace containing platinum foil, copper oxide, and silver wire was heated to 700°C. When the combustion furnace reached a temperature of 1050°C, oxygen (700 torr, scrubbed through

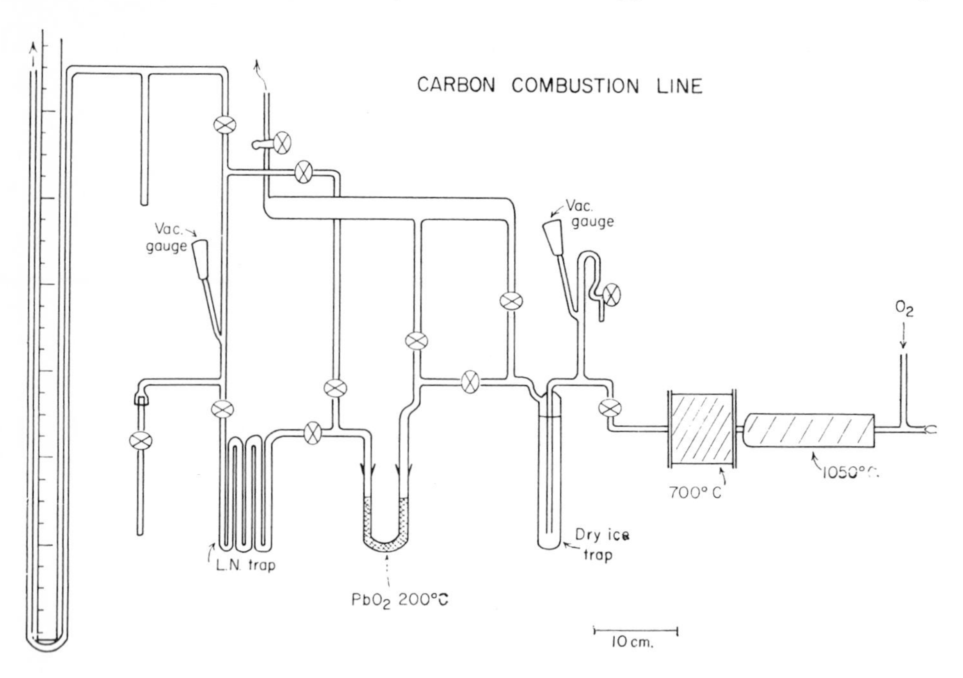

Fig. 1. Carbon combustion line. This line was also modified for combustion of Ag_2S to SO_2.

liquid nitrogen) was introduced into the apparatus and the porcelain boat containing the sample was pushed down the combustion tube (with a bar magnet) into the hot zone of the furnace. Combustion was performed for one hour. At the end of this period, the gases were slowly passed through the Dry Ice trap (to remove water), the lead oxide trap held at 200°C (to remove oxides of nitrogen and sulfur), and into the liquid nitrogen multiple trap. After 10 min, excess oxygen was slowly pumped away and the system finally pumped down to 10^{-3} torr. The gas (CO_2) was transferred by distillation into the "cold finger" of the calibrated manometer, and the pressure was measured to obtain the gas volume. Again, the sample was transferred by distillation into the gas collection tube shown at the bottom of Fig. 1, and from there transported to the mass spectrometer. The results of these combustions are shown in Table 1 and Fig. 2.

In two experiments, samples were pyrolyzed. In one, sample 10084 was pyrolyzed in an apparatus described by CHANG *et al.* (1970); in the second, involving sample 10086a, pyrolysis was carried out in the above apparatus in the absence of oxygen (under 10^{-3} torr pressure), with the catalyst furnace turned off, and the traps isolated. After one hour pyrolysis at the desired

Table 1. Concentration and isotopic composition of carbon and sulfur

Sample		Carbon		Sulfur	
Description	Reference No.	Conc. (ppm)	δC^{13}* (per mil)	Conc. (ppm)	δS^{34}† (per mil)
Bulk Fines	10086	143	+20·2	680	+5·4
		170	+17·2	640‡	+8·2‡
Bulk Fines	10084	147	+18·8	770	+4·7
		116	+19·5		+4·4
Breccia	10002-54	198	+ 8·4	1070	+3·5
		181	+ 9·2		+3·4
Breccia	10060-22	137	+ 1·6	1120	+3·6
		132	+ 2·7		+3·3
Fine-grained rock	10049	63	−18·8	2200	+1·2
		77	−21·4		+1·3
Fine-grained rock	10057-40	21	−25·6	2280	+1·2
		11	−29·8		+1·2

* Relative to PDB standard.
† Relative to Canyon Diablo standard.
‡ Total sulfur by aqua regia oxidation.

temperature, the sample-containing boat was pushed out of the furnace with a bar magnet, the furnaces were brought to the required temperature, oxygen introduced, and combustion allowed to proceed as before. The pyrolysis products were therefore captured as carbon dioxide gas.

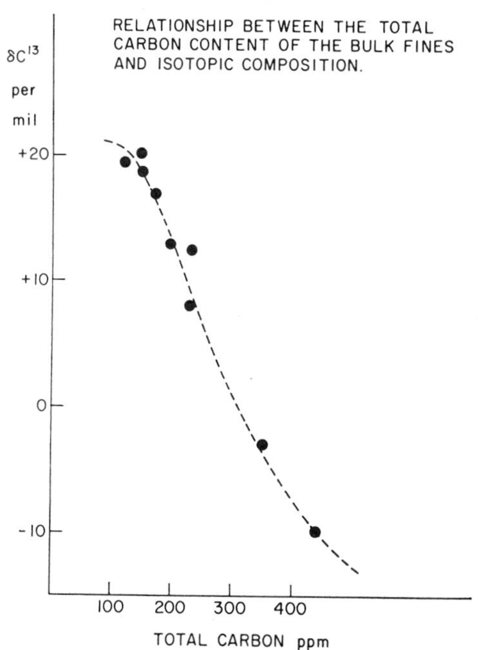

Fig. 2. Relationship between the total carbon content of the bulk fines (10086 and 10084) and δC^{13} of the product CO_2.

In one experiment, 3·543 g of sample 10084 was subdivided by sieving through 140- and 300-mesh 3 in. brass sieves previously cleaned by washing in water and a benzene–methanol solvent mixture. The three fractions obtained were separately combusted as described above after being heated to 150°C at 10^{-3} torr pressure for over 12 hr.

Sulfur

Sulfur was analyzed in the lunar material by several steps carried out in series. First a benzene–methanol (90–10%) extract of 54·6 g lunar fines (10086) was shaken with freshly cleaned strips of copper (KVENOLDEN *et al.*, 1970). This method has been shown to convert copper to copper sulfide when elemental sulfur is present in amounts of 0·5 mg/10 ml solvent.

Sulfide was released by treatment of the lunar material with 85% phosphoric acid under vacuum. The reaction was carried out by mild heating to the boiling point or just below. The released hydrogen sulfide was trapped in 5 ml of a 10% silver nitrate solution kept in a side arm of the reaction flask. The reaction proceeded for two days before termination.

The residue was washed and the separated acid solution treated with a 5% barium chloride solution to precipitate sulfate. The residual solid was oxidized with aqua regia and the solution obtained again treated with a 5% barium chloride solution. A separate intact sample of fines (10086) was oxidized directly with aqua regia.

Barium sulfate was reduced with graphite and converted to silver sulfide by the method originally described by RAFTER (1957). Silver sulfide obtained from the various treatments was combusted using

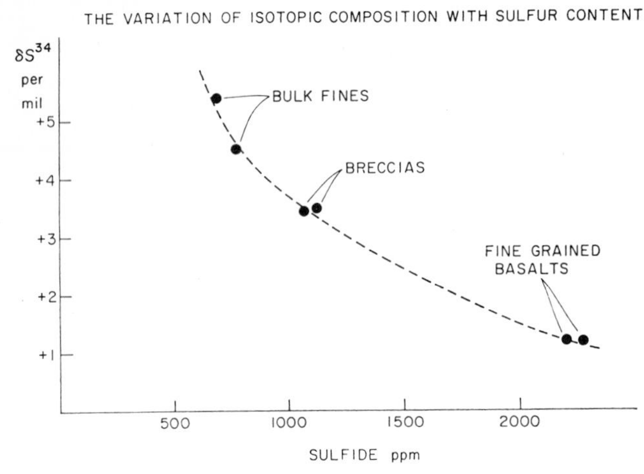

Fig. 3. Variation of δS^{34} with total sulfur content in fines, breccias and fine-grained basalts.

the vacuum system shown in Fig. 1 with the following modifications: (i) the combustion tube was removed and replaced by another tube in which no second furnace or catalysts were present, (ii) the lead oxide trap was removed and replaced with an empty U tube, (iii) silver sulfide were mixed with pretreated cuprous oxide (CuO, which has been baked out at 800°C in partial vacuum) and introduced into the furnace with a bar magnet when the furnace temperature reached 800°C. Combustion proceeded for 10 min without addition of oxygen gas. The resultant sulfur dioxide was purified in the vacuum system in a similar manner to that described for carbon dioxide, and captured in the gas collection tube for mass spectrometric studies.

Mass spectrometry

CO_2 and SO_2 were each measured in two separate 6-in. radius, dual-collecting Nuclide Corp. mass spectrometers. Instrumental resolution in duplicate measurements was able to produce readings of δC^{13} or $\delta S^{34} \pm 0{\cdot}1‰$. The chemical treatment of the samples could produce greater errors, which may amount to δC^{13} or $\delta S^{34} = \pm 0{\cdot}2‰$, especially in small samples.

By definition:

$$\delta C^{13}‰ = \frac{C^{13}/C^{12}\ \text{sample} - C^{13}/C^{12}\ \text{standard}}{C^{13}/C^{12}\ \text{standard}} \times 1000.$$

The standard used is PDB limestone.

$$\delta S^{34}‰ = \frac{S^{34}/S^{32}\text{ sample} - S^{34}/S^{32}\text{ standard}}{S^{34}/S^{32}\text{ standard}} \times 1000.$$

The standard used is Canyon Diablo meteorite troilite (FeS).

Blanks

Several blank runs on the combustion system, in which either an empty boat was "combusted", or sand (previously fired at 1100°C in the atmosphere) was combusted, yielded $<1\ \mu$ mole CO_2 (<10 ppmC). In a single experiment carried out to test possible contamination which may have occurred during combustion of pyrolyzed products, a "blank" value of 3 μ mole CO_2 was obtained. This volume of gas could constitute a significant proportion of that produced in the low temperature pyrolysis (<500°C, Table 3).

In order to test possible contamination during sample manipulation, a 1 g sample of No. 10086 lunar fines which had been previously combusted, was removed from the combustion boat, re-ground loaded back into a new boat and re-combusted. These manipulations yielded a total carbon "blank" equal to 15 ppm carbon. A similar test on fine sand, which was also then sieved, yielded a "blank" of 8 ppmC.

Results

Carbon

Carbon was found to be present mainly as oxides (CO_2 or CO), carbide and probably elemental carbon (Chang *et al.*, 1970). The presence of a significant amount of C—H bond material could not be confirmed, since only small quantities of methane were released during pyrolysis.

The total carbon content was highest in the fines and in the breccia (Table 1). The highest value measured was in breccia No. 10002-54. The lowest values measured were in the fine-grained basaltic rocks. Of the two basaltic rocks measured, No. 10057-40 yielded values only slightly higher than those of blank runs. That carbon is present in the basaltic rocks cannot be stated with confidence.

A sample (3·543 g) of bulk fines, No. 10084, was separated into three grain-size fractions (<300 mesh $=$ 0·7376 g, 300–140 mesh $=$ 1·5847 g and >140 mesh $=$ 1·2208 g). Each was separately analyzed for total carbon (Table 2). The highest value of 261 ppm was found in the finest material, 183 ppm in the intermediate size and 92 ppm in coarsest material. The weighted mean of 168 ppm is a little higher than the value for total carbon in the intact sample (147 and 116 ppm in two analyses).

Table 2. Carbon and sulfur concentration, δC^{13} and δS^{34} from different size fractions of the fines

		Sample			
Mesh size	10084 C Content (ppm)	$\delta C^{13}‰$	Mesh size	10086A S Content (ppm)	$\delta S^{34}‰$
<300	261	+15·6	<140	546	+5·3
140–300	183	+20·3	>140	663	+2·9
>140	92	+15·0			
Weighted Mean	168	+16·8	Weighted Mean	596	+4·0

Slight contamination may have occurred, therefore, during the separation and handling process.

Contamination is a serious problem. It could have resulted anywhere during several operational steps. First, retro engine exhaust products may have been absorbed on the lunar surface. Second, contamination could have occurred during sterilization of the sample boxes, during documentation and during distribution in the Lunar Receiving Laboratory (LRL). Third, contamination could occur during sample handling and preparation for analysis. At the initiation of the study when small samples (200–300 mg) were analyzed, the carbon content was always greatest. When, however, large samples ($\geqslant$1·0 g) were used, the values were less, and if the sample was baked out (under conditions approaching those on the lunar surface) at 150°C under a vacuum of 10^{-3} torr, the values obtained were the lowest. Apparently some volatile impurities can be removed from the samples. The most reliable results are presented in Table 1.

A test for contamination can be measured by the change in δC^{13} with increase in carbon of the lunar fines. Because normal terrestrial contaminants generally have δC^{13} values in the range of -20 to $-30‰$, contaminating carbon would be expected to move the isotopic ratio toward this range of values. Figure 2 shows that this effect appears to be occurring. The most reliable δC^{13} values for the lunar fines, therefore, lie in the range of $+17$ to $+20‰$. The basaltic rock No. 10057-40 may contain contaminant carbon only, whereas No. 10049 is probably a mixture of terrestrial and lunar carbon.

The three size fractions of the fines do not show a widely different isotopic distribution (Table 2); they range between $+15$ and $+20‰$. The finest material may possibly be contaminated by 10–15 per cent of isotopically lighter material, but this is uncertain.

The brecciated rocks Nos. 10002-54 and 10060-22 have isotopic ratios which are intermediate between those of the fines and those of the fine-grained basaltic rocks. Because in this case the sample with the highest carbon content (10002-54, Table 1) also contained the highest C^{13} content, it is difficult to ascribe contamination as an interpretation of the results from these rocks. One must assume that heterogeneity occurs which accounts not only for the spread in total carbon among the different rock groups, but also for the range of δC^{13} values.

Heterogeneity was confirmed by the pyrolysis experiments represented in Table 3 and also described by CHANG *et al.* (1970). Whereas in one pyrolysis experiment from 150°C to 750°C, on sample No. 10086, CO_2 appeared in significant amounts, in a second (duplicate) experiment no carbon dioxide was detected at levels $>$7 ppm. Further, only traces of gas (containing carbon) were measured in the products released below 500°C. Two experiments carried out on sample Nos. 10084 and 10086-A, in which the pyrolysis products were combusted did show that a measurable amount of gas was released below 500°C. The results show that the gases released at the lower temperatures are enriched in the lighter isotope relative to the gases released at temperatures between 500–750°C. As only very small amounts of gas (as CO_2) were collected, the blank error may have been a serious problem in the results at the lower temperatures. Therefore, they are probably not

Table 3. δC^{13} of carbon fractions resulting from pyrolysis and hydrolysis

Experiment	Sample	Carbon content (ppm)	δC^{13}‰
Combustion of intact sample	10086	157*	+18·7*
	10084	132*	+19·1*
Combustion of residue from HCl/HF treatment	10086	30	− 7·9
	10086	42	− 4·3
Pyrolysis at 150–250°C†	10084	27	−25·1
Pyrolysis at 250–500°C	10084	19	−12·0
Pyrolysis at 500–750°C	10084	40	+23·4
Combustion of residue	10084	59	+ 9·4
		Total = 145	Mean = + 4·1
Pyrolysis at 150–250°C	10086-A‡	30	−41·2
Pyrolysis at 250–500°C	10086-A	42	−31·7
Pyrolysis at 500–750°C	10086-A	64	+16·4
Pyrolysis at 750–1050°C	10086-A	160	+13·0
Combustion of residue	10086-A	<1	—
		Total = 296	Mean = + 1·9

* Average of values shown in Table 1.
† Pyrolysis refers to pyrolysis at the temperature given for one hour followed by combustion of the products in oxygen at 1050°C.
‡ This sample has been sieved and represents a grain size <140 mesh.

representative of the indigenous carbon. Sample 10086-A had previously been transferred to a new container and had later been sieved. The gas released at the lower temperature, may therefore also represent some handling contamination. The fact, however, that the gas released at temperatures between 750 and 1050°C is isotopically lighter than that released during pyrolysis between 500 and 750°C argues strongly for the presence of carbon with more than one isotopic range of values.

One possible source of contamination was the retro engine exhaust fuel. To examine this possibility, a sample from trap "A" of a land-based experimental test (FLORY *et al.*, 1969) was obtained through Dr. Al Burlingame, University of California, Berkeley, and its isotopic composition determined. Combustion of the sample yielded δC^{13} value of −34·7‰. Apparently the isotopically heavy values do not result from this source.

Sulfur

Neither elemental sulfur nor sulfate could be positively identified in the lunar material. All the sulfur appeared to be evolved as hydrogen sulfide on treatment with acid; this chemical behavior suggests the presence of a sulfide mineral such as troilite. This suggestion is confirmed by numerous petrographic studies in which troilite has been reported (MASON *et al.*, 1970; SKINNER *et al.*, 1970; EVANS, 1970) as the only significant sulfur mineral. Elemental sulfur may be present in trace amounts (NAGY *et al.*, 1970). It may result, however, from surficial oxidation of sulfide minerals on exposure to the earth's atmosphere. Hydrogen sulfide gas (or even sulfur dioxide gas) may also be trapped in small amounts in the glass spherules, but this has not been confirmed.

The distribution of sulfur follows a pattern different from that of carbon. Quantitatively, the two elements are inversely related, sulfur being in highest concentration in the fine-grained basaltic rock and in lowest concentration in the fines; the brecciated rock once again displaying intermediate values (Table 1). The isotopic distribution, however, appears to show the same distributional pattern as does carbon; the fines are most enriched in the heavy isotope and the basaltic rocks are least enriched (Fig. 3). The results are consistent on repeated analyses, at least for the two breccias and for the two basalts. The fines appear to fall in the range of $\delta S^{34} = +4$ to $+8‰$ (KVENVOLDEN *et al.*, 1970); sample No. 10086 consistently showed more enrichment in δS^{34} than did sample No. 10084 (Table 1).

Sample No. 10086-A, separated into two fractions on a No. 140 mesh sieve showed that the smaller grain size material contains less sulfur, but is enriched in δS^{34} relative to the coarser material >140 mesh (Table 2). This heterogeneity probably accounts for the observed distribution range in the δS^{34} results from analyses on the fines (Table 1).

Contamination in the case of sulfur could occur, but is probably less likely than in the case of carbon. Molybdenum sulfide was used as a lubricant for the collection cases and a sulfur-containing compound was evidently used as an adhesive in the LRL, but neither of these possible contaminants have so far been available for isotopic analysis. Because the sulfur content is considerably higher than carbon content, and the opportunity for contamination during handling and analysis is small, significant contamination of sulfur compounds is unlikely.

DISCUSSION

Results for total carbon content presented here generally agree most closely with those of MOORE *et al.* (1970) and EPSTEIN and TAYLOR (1970). Two- to threefold variation between results of single analyses of the former may have been due to slight contamination of the type discussed above. Contamination may have been particularly evident in their results of analysis on sample 10086-A, where the bulk sample yielded a mean of 142 ppm carbon, whereas the fine-grained dust passing through a 300 mesh sieve was measured as containing 500 ppm carbon. The problem of contamination may also have been present in some of the experiments of FRIEDMAN *et al.* (1970), where all the δC^{13} values for combusted carbon (and ΣC by averaging) were lighter than 0‰ PDB. The results they obtained for δC^{13} of CO_2 released by pyrolysis (vacuum heating) of two breccias, however, are very close to those found by us for breccia sample No. 10060-22 (Table 1) obtained by total combustion. There may be some problem in their combustion technique. The second breccia we analyzed (No. 10002-54), yielded a δC^{13} mean value of $+8{\cdot}8‰$, which is similar to the value of $+10{\cdot}8‰$ for breccia No. 10061 given by EPSTEIN and TAYLOR (1970). Carbon apparently displays a range of isotopic values, possibly representing different carbon forms or different sources of origin.

Heterogeneity in the samples may explain the different results obtained by different workers. BURLINGAME *et al.* (1970) found CO as the major carbon gas released on pyrolysis, EPSTEIN and TAYLOR found a mixture of CO and CO_2 as did ORO *et al.* (1970); FRIEDMAN *et al.* report only CO_2. Others noted significant quantities of methane released during pyrolysis (ABELL *et al.*, 1970; NAGY *et al.*, 1970). KAPLAN and SMITH (1970) and CHANG *et al.* (1970) give evidence for the presence of carbide as well as elemental carbon from chemical experiments. Petrographic evidence for carbide has been given by ANDERSON *et al.* (1970) whereas ARRHENIUS *et al.* (1970) report the detection of a macroscopic (1 mm^3) grain of graphite.

The only quantitative results reported for sulfur are those of ORO *et al.* (1970) for a sample of No. 10086-6 fines and a sample of No. 10002-54 breccia. A wide range of values is noted from 590 to 4200 ppm in sample No. 10086 and 1000 to 2800 ppm in sample No. 10002. Because only small sample sizes were used, and no interpretation of the results is given, we can only conclude that the large spread is the result of heterogeneity in sulfide distribution, and not representative of the bulk sample.

The four questions raised in the Introduction may now be examined in light of the data presented above:

(i) Carbonaceous chondrites contain carbon compounds in several different forms ranging from carbonate to highly reduced extractable and non-extractable carbon-hydrogen compounds. The carbon content of carbonaceous chondrites ranges from 0·3 per cent in type IV, to 4·0 per cent in type I (BELSKY and KAPLAN, 1970). None of these compounds have been confirmed in lunar fines with any confidence. Furthermore, the δC^{13} values of carbonaceous meteorites fall in the range of $+40$ to $+70‰$ for the carbonate fraction and -15 to $-17‰$ for the insoluble organic residue (Fig. 4) whereas the δC^{13} values for the total meteorite are in the range of -5 to $-18‰$ (SMITH and KAPLAN, 1970). Thus there is no evidence from the carbon data for any significant carbon contribution to the lunar fines from carbonaceous chondrites. On the other hand, the values of $-4·3‰$ and $-7·9‰$ for residual carbon remaining after treating sample No. 10086 with a hydrochloric-hydrofluoric acid mixture (Table 3) fall in the range of graphite from iron meteorites (Fig. 4).

Evidence from sulfur data also argue against a simple addition of sulfur from carbonaceous chondrites or other meteorites to the lunar fines. Total sulfur in all meteorites analyzed so far (KAPLAN and HULSTON, 1966) have a δS^{34} range of $\pm 1‰$. The δS^{34} values of troilite in ordinary chondrites and iron meteorites also fall in this range, whereas δS^{34} values of troilite in carbonaceous chondrites have a somewhat wider range of ± 3 per cent. In addition, carbonaceous chondrites type I and II contain sulfate as an important form of sulfur (in type I, the most abundant form of sulfur). δS^{34} values of meteorites resemble most closely the values of the fine-grained basaltic rock, but not the lunar fines.

(ii) In the absence of reliable data on endogenous, reduced carbon-hydrogen compounds in the lunar material, no information can be gained about early biochemical evolution on earth. There is obviously no comparison between the complex compounds that must have been present on the primordial earth and those isolated from the present lunar samples. Given water, however, the carbides in the lunar material could react to form low-molecular-weight hydrocarbons which may then constitute starting material for more complex syntheses. From the information so far obtained, there is no evidence for primitive, terrestrial-like biological or organic chemical processes.

(iii) Based on the data obtained from this study, there is also no strong evidence for an earth-moon relationship, at least during the last 2–3 $\times$ 10^9 yr when sediments began to form on earth. The sulfur content in the lunar basaltic rock is higher than generally found in terrestrial basalt, although the δS^{34} values are similar. Troilite is not a common mineral of the lithosphere, but may, of course, be in the earth's mantle. The δC^{13} value of terrestrial igneous rocks range from -20 to $-25‰$ (CRAIG, 1953). An analysis of a recent basalt flow from Mt. Kilauea, Hawaii (Aloi Crater) yielded 150 ppm C and $\delta C^{13} = -23{\cdot}5‰$. Although this material contained very little extractable organic compounds, its carbon may have been derived in part from graphitization of biogenic carbon as well as terrestrial contamination. Because detailed

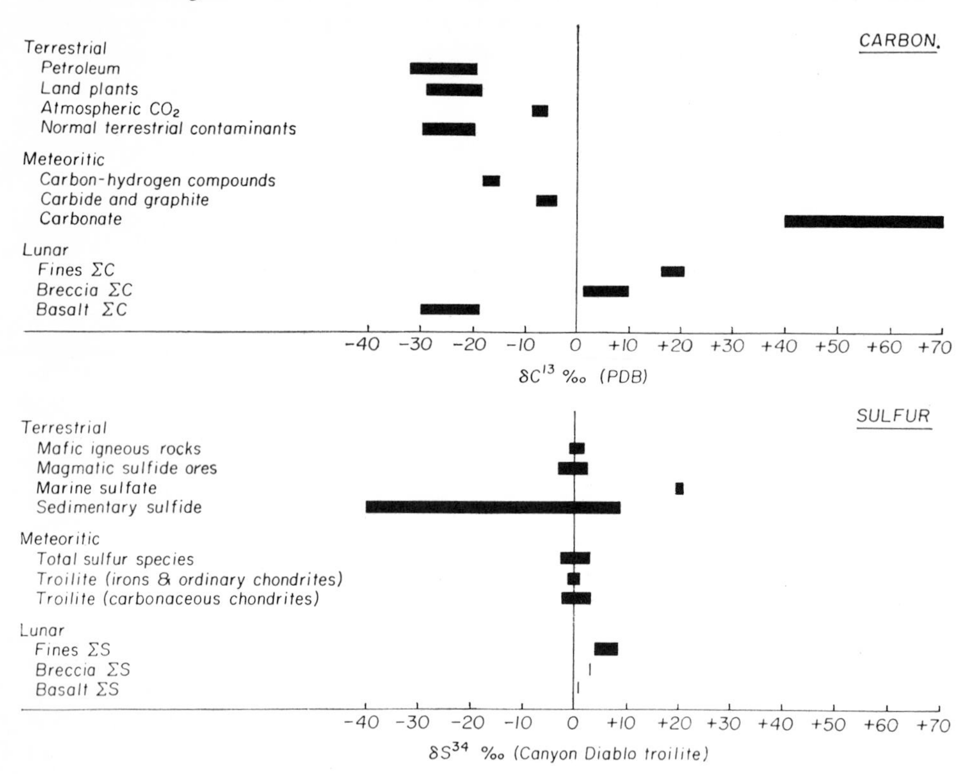

Fig. 4. Distribution diagram of δC^{13} and δS^{34} among various phases of terrestrial, meteoritic and lunar components.

study on the various forms of carbon and their isotopic values in terrestrial basalt has not yet been undertaken, and because the carbon data on the lunar basaltic rocks need to be confirmed, comparison of data is premature.

(iv) Although MOORE *et al.* (1970) and EPSTEIN and TAYLOR (1970) suggest a solar wind origin for the carbon, we believe there is yet insufficient evidence to support or refute this assumption. Several difficulties appear to confront the hypothesis. First, hydrogen gas and rare gases were mainly removed at a low temperature ($<650°C$) in the experiments of EPSTEIN and TAYLOR, whereas CO is evolved mainly at temperatures $>650°C$. It therefore appears that the carbon is bound much more firmly to the solid matrix, and implies an endogenous origin for this element. Second, if the dust is largely derived from erosion of the rocks, and if carbon, nitrogen, sulfur and other light elements are contributed by the solar wind, why is the sulfur content lower in the fines than in the original rock? Third, the carbon content of the brecciated rocks is equal to or higher than that of the fines from which they are purported to have formed. Yet carbon in the breccia is isotopically lighter than in the fines. If carbon was preferentially lost in the "welding" process of breccia formation, one might expect the isotopically heavier carbon to remain in the breccia. On the other hand, the general similarity of δC^{13} values in the various grain sizes of the lunar fines (Table 2) argues for a common source. The same, however, is apparently not the case for sulfur.

A surprising result of the analyses is the presence of carbon predominantly as oxides of carbon, but also as carbide and probably graphite but not as methane. This observation is more surprising in view of the presence of molecular hydrogen in the fines and breccia, and may therefore argue against significant extralunar origin. Because ample petrological evidence exists for at least local melting and possibly crustal differentiation by melting (WOOD *et al.*, 1970), a large part of the carbon (as well as water and other volatiles) may have been lost from the lunar surface as CO_2 and CH_4. Non-volatile graphite and carbides, isolated from silicate or oxide phases, may have therefore been enriched. Subsequent re-melting, either by meteorite impact or sub-crustal magma flows, may have caused oxidation of this carbon to CO_2 and CO. Other reactions, such as the water-gas reaction, $C + H_2O \rightleftharpoons H_2 + CO$, could also account for the deuterium content of the entrapped hydrogen which appears to be greater than would be expected from contribution of solar wind protons (EPSTEIN and TAYLOR, 1970; FRIEDMAN *et al.*, 1970). Such high temperature reactions would, however, not be expected to cause significant isotopic fractionation of the carbon.

The enrichment of carbon and depletion of sulfur in the dust and breccia, relative to the fine-grained basaltic material, shows that some selective process is operating. One possible mechanism is a density separation in the following order: (1) Transfer of material from impact areas immediately adjacent to the site of collection as well as from mineralogically distinct areas, possibly lunar highlands, which permits preferential distribution of fine-grained low density (possibly gas-containing) glass spherules, and (2) "stirring" of the overburden dust profile which allows a settling of the denser sulfide ($\sim$5 g/cm^3) relative to the less dense glass and silicate phases ($<$3 g/cm^3). Cores from Apollo 12 and future missions may indicate if such a density stratification occurs.

Interpretation of the isotopic effect, where both C^{13} and S^{34} are concentrated in the lunar fines is still problematic. High energy processes generally result in relatively small isotopic fractionation, either during equilibration of phases or as a rapid, unidirectional, kinetic separation. In the case of carbon, an isotope fractionation effect may be established by preferential diffusion of the light isotope following Graham's law as a first order approximation:

$$\frac{k_1}{k_2} = \left(\frac{M_2}{M_1}\right)^{\frac{1}{2}}$$

$$\text{For } CO_2\text{: } \frac{k_1(C^{12})}{k_2(C^{13})} = 1{\cdot}011, \qquad \text{For CO: } \frac{k_1(C^{12})}{k_2(C^{13})} = 1{\cdot}018.$$

Thus, slow diffusion from shock-damaged grains may be one mechanism for conserving the heavier isotope. This, however, does not appear to be a valid mechanism for enriching S^{34}.

KAPLAN and SMITH (1970) speculated that enrichment of heavy isotopes in the fine material over that in the breccia and basaltic rock was due to exposure to solar wind protons. LIBBY (1969) has suggested that these protons reduce the lunar surface by reacting with oxides through high energy (450 eV) impingement. Thus, reactions of the following type may be expected:

$$CO + 6H^1 \rightarrow CH_4 + H_2O,$$
$$CO_2 + 8H^1 \rightarrow CH_4 + 2H_2O,$$
$$FeS + 2H^1 \rightarrow Fe + H_2S.$$

The absence of significant quantities of methane and probable absence of trapped hydrogen sulfide, suggest that if these gases formed by the above reactions, they have diffused away or have been swept off the lunar surface. LIBBY estimates 10^8 hydrogen atoms/cm^2/sec are impinging on the lunar surface. At this rate, assuming 100 m.y. exposure of any one surface, and a 0·1% efficiency in reaction of a carbon or sulfur species, amounts of sulfur or carbon in excess of 10 μ/g could be reduced. The equilibrium fractionation factor for the CO_2–CH_4 couple is large, and at the surface temperature of the lunar day (<150°C) it would still be 1·04 (BOTTINGA, 1969).

The calculations are of course highly tentative in view of the absence of data for such effects, but they are given here to indicate that a "hydrogen stripping process" may be a mechanism to be considered. This process coupled with density stratification mentioned earlier, may account for the observed features of heavy isotope and carbon concentration and for sulfur depletion in the surface samples. If this explanation is correct, deeper samples should show an increase in sulfur, less isotopic differentiation and perhaps a different carbon distribution.

Acknowledgments—We wish to thank Drs. CHANG, KVENVOLDEN and PONNAMPERUMA, NASA, Ames Research Center and Dr. J. W. SCHOPF, UCLA, for sharing samples and discussion of results. This study was supported through NASA contract NAS9-8843.

REFERENCES

ABELL P. I., DRAFFAN G. H., EGLINTON G., HAYES J. M., MAXWELL J. R. and PILLINGER C. T. (1970) Organic analysis of the returned lunar sample. *Science* **167,** 757–759.

ANDERSON A. T., JR., CREWE A. V., GOLDSMITH J. R., MOORE P. B., NEWTON J. C., OLSEN E. J., SMITH J. V. and WYLLIE P. J. (1970) Petrologic history of the moon suggested by petrography, mineralogy, and crystallography. *Science* **167,** 587–590.

ARRHENIUS G., ASUNMAA S. K., DREVER J. I., EVERSON J., FITZGERALD R. W., FRAZER J. Z., FUJITA H., HANOR J. S., LAL D., LIANG S. K., MACDOUGALL D., REID A. M., SINKANKAS J. and WILKENING L. (1970) Phase chemistry, structure, and radiation effects in lunar samples. *Science* **167,** 659–661.

BELSKY T. and KAPLAN I. R. (1970) Light hydrocarbon gases, C^{13}, and origin of organic matter in carbonaceous chondrites. *Geochim. Cosmochim. Acta* **34,** 257–278.

BOTTINGA Y. (1969) Calculated fractionation factors for carbon and hydrogen isotope exchange in the system calcite–carbon dioxide–graphite–methane–hydrogen–water vapour. *Geochim. Cosmochim. Acta* **33,** 49–64.

BURLINGAME A. L., CALVIN M., HAN J., HENDERSON W., REED W. and SIMONEIT B. R. (1970) Lunar organic compounds: Search and characterization. *Science* **167,** 751–752.

CHANG S., SMITH J., KAPLAN I., LAWLESS J., KVENVOLDEN K. and PONNAMPERUMA C. (1970) Carbon compounds in lunar fines from Mare Tranquillitatis—IV. Evidence for oxides and carbides. *Geochim. Cosmochim. Acta*, Supplement I.

GRAIG H. (1953) The geochemistry of the stable carbon isotopes. *Geochim. Cosmochim. Acta* **3,** 53–92.

EPSTEIN S. and TAYLOR H. R., JR. (1970) O^{18}/O^{16}, Si^{30}/Si^{28}, D/H, and C^{13}/C^{12} studies of lunar rocks and minerals. *Science* **167,** 533–535.

EVANS H. T., JR. (1970) Lunar troilite: Crystallography. *Science* **167,** 621–623.

FRIEDMAN, I., O'NEIL J. R., ADAMI L. H., GLEASON J. D. and HARDCASTLE K. (1970) The water, hydrogen, deuterium, carbon, carbon-13 and oxygen-18 content of selected lunar material. *Science* **167,** 538–540.

KAPLAN I. R. and HULSTON J. R. (1966) The isotopic abundance and content of sulfur in meteorites. *Geochim. Cosmochim. Acta* **30,** 479–496.

KAPLAN I. R. and SMITH J. W. (1970) Concentration and isotopic composition of carbon and sulfur in Apollo 11 lunar samples. *Science* **167,** 541–543.

KVENOLDEN K., CHANG S., FLORES J., SAXINGER C., PERING K., WOELLER F., SMITH J., BREGÈR I. and PONNAMPERUMA C. (1970) Carbon compounds in lunar fines from Mare Tranquillitatis: I. The general scheme of analysis. *Geochim. Cosmochim. Acta* Supplement I.

LIBBY W. F. (1969) Why is the moon gray? *Science* **166,** 1437–1438.

MASON, B. FREDRIKSSON K., HENDERSON E. P., JAROSEWICH E., MELSON W. G., TOWE K. M. and WHITE J. S., JR. (1970) Mineralogy and petrography of lunar samples. *Science* **167,** 656–659.

MOORE C. B., LEWIS C. F., GIBSON E. K. and NICHIPORUK W. (1970) Total carbon and nitrogen abundances in lunar samples. *Science* **167,** 495–497.

NAGY B., DREW C. M., HAMILTON P. B., MODZELESKI V. E., MURPHY M. E., SCOTT W. M., UREY H. C. and YOUNG M. (1970) Organic compounds in lunar samples: Pyrolysis products, hydrocarbons, amino acids. *Science* **167,** 770–773.

ORÓ J., UPDEGROVE W. S., GIBERT J., MCREYNOLDS J., GIL-AV E., IBANEZ J., ZLATKIS A., FLORY D. A., LEVY R. L. and WOLF C. (1970) Organic elements and compounds in surface samples from the Sea of Tranquillity. *Science* **167,** 765–767.

RAFTER T. A. (1957) Sulphur isotopic variations in nature. I: The preparation of sulphur dioxide for mass spectrometer examination. *N.Z.J. Sci. Tech.* **38B,** 849–957.

SKINNER B. J. (1970) High crystallization temperatures indicated for igneous rocks from Tranquillity Base. *Science* **167,** 652–654.

SMITH J. W. and KAPLAN, I. R. (1970) Endogenous carbon in carbonaceous chondrites. *Science* **167,** 1367–1370.

WOOD J. A., DICKEY J. S., JR., MARVIN U. B. and POWELL B. N. (1970) Lunar Anorthosites. *Science* **167,** 602–604.

Proceedings of the Apollo 11 Lunar Science Conference, Vol. 2, pp. 1331 to 1343.

Study of distribution and variations of rare gases in lunar material by a microprobe technique*

T. KIRSTEN, O. MÜLLER, F. STEINBRUNN and J. ZÄHRINGER
Max-Planck-Institut für Kernphysik, Heidelberg, Germany

(*Received* 2 *February* 1970; *accepted in revised form* 5 *March* 1970)

Abstract—The rare gas distribution in lunar soil 10084-31, breccias 10019-13, 10046-19, 10061-26 and rocks 10057-22, 10057-80, 10085-41, was studied with a micro-helium-probe. Gases are concentrated in grain surfaces and originate from solar wind. ^{4}He concentrations of different mineral components vary by more than a factor of 10 apart from individual fluctuations for each type. Also grains with no detectable ^{4}He exist. Titanium-rich components have the highest, Ca-rich minerals the lowest concentrations. The solar wind was redistributed by diffusion. Lithic fragments in breccias contain no solar gases. Glass pitted surfaces of crystalline rocks contain about 10^{-2} cm^3 ^{4}He/cm^2. Etched dust grains clearly show spallogenic and radiogenic components. The apparent mean exposure age of dust is approximately 500 m.y., its K–Ar age lies between 3·4 and 4·3 b.y. Cavities of crystalline rocks contain ^{4}He, radiogenic argon, hydrogen and nitrogen.

INTRODUCTION

ONE OF the first important observations of the LSPET (1969) was the occurrence of large amounts of He, Ne and Ar in lunar fines and breccias. The purpose of this paper is to gain additional and specific informations about these gases. Since their relative abundances and the isotopic composition resemble the solar abundances as well as the abundances of light primordial rare gases in gas rich meteorites, they could be due to trapped solar wind, to trapped primordial gases, or to stopped solar flare particles. We have studied the distribution of these gases on a microscale since this provides the clue to their origin. Enrichment at the surface of grains would point to solar wind, whereas a more homogeneous distribution would be expected for solar flare particles. Diffusive fractionation of light and heavy rare gases may yield information about the thermal history of lunar samples and about turnover rates. A split of the components could yield cosmic ray exposure ages and radiogenic ages. Also, the original orientation of rock surfaces could be inferred from gas concentration profiles.

EXPERIMENTAL PROCEDURE

The combinations of a microprobe with a He-mass spectrometer (ZÄHRINGER, 1966), and of a Laser beam with a mass spectrometer (MEGRUE, 1967) have previously been applied to gas rich meteorites. With the microprobe technique, samples are bombarded by a 35 keV electron beam of 1 μA with a beam width of $\sim$10 μ. ^{4}He in situ degassed by the electron beam was registered by a 60° mass spectrometer simultaneously with elements like Ti, Fe, Ca, Cr and Mn (Fig. 1). Helium profiles related to the variation of these elements were obtained. The sample holder was provided with a hat to yield efficient gas collection with a resulting sensitivity of 10^{-8} cm^3 ^{4}He/mV sec. Calibration was made by means of a standard leak and by comparison with statically measured test

* This paper is essentially the same as published previously by Kirsten *et al.* (1970). Some Figures and Tables were added for completeness and some changes have been made. Nevertheless, the paper is still not intended to discuss in any detail the excellent contributions of other authors on the same or on related topics presented at the Apollo 11 Conference in Houston, Texas.

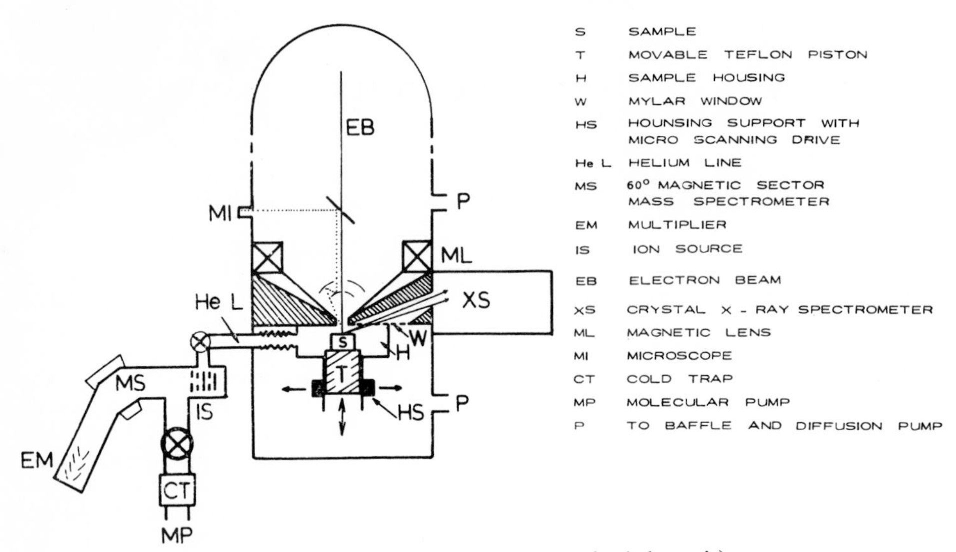

Fig. 1. Outline of the helium microprobe (schematic).

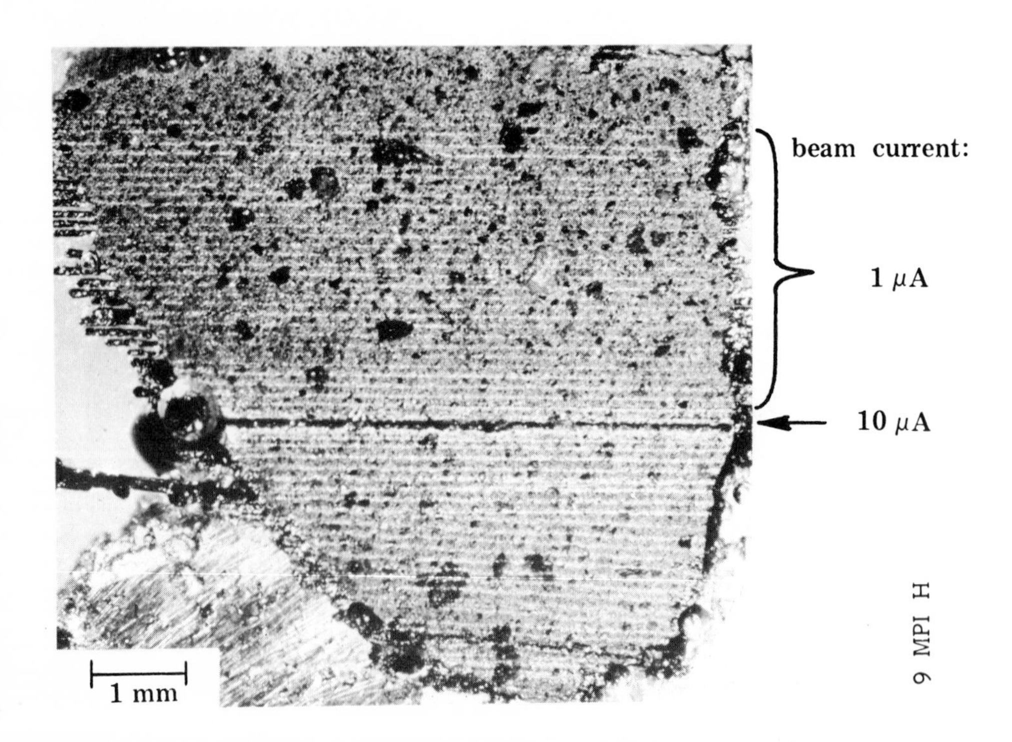

9 MPI H

Fig. 2. Traces of electron beam in polished section of breccia 10061-26.

samples. The technique has been applied in a scanning mode (16 μ/sec) to polished breccias and rocks (Fig. 2) as well as to surfaces of unpolished rocks. A special mechanism kept the beam in focus even for irregularly shaped samples. Second scans along outgassed traces liberated no additional He. The high gas retentivity of the minerals required a larger beam, which in turn caused a larger beam diameter. Therefore the local resolution was sometimes less than the average grain diameter. Hence, to supplement the helium profiles with data about well defined minerals, 800 handpicked grains of five major constituents of lunar dust and of different grain sizes were singly analyzed. Applying a

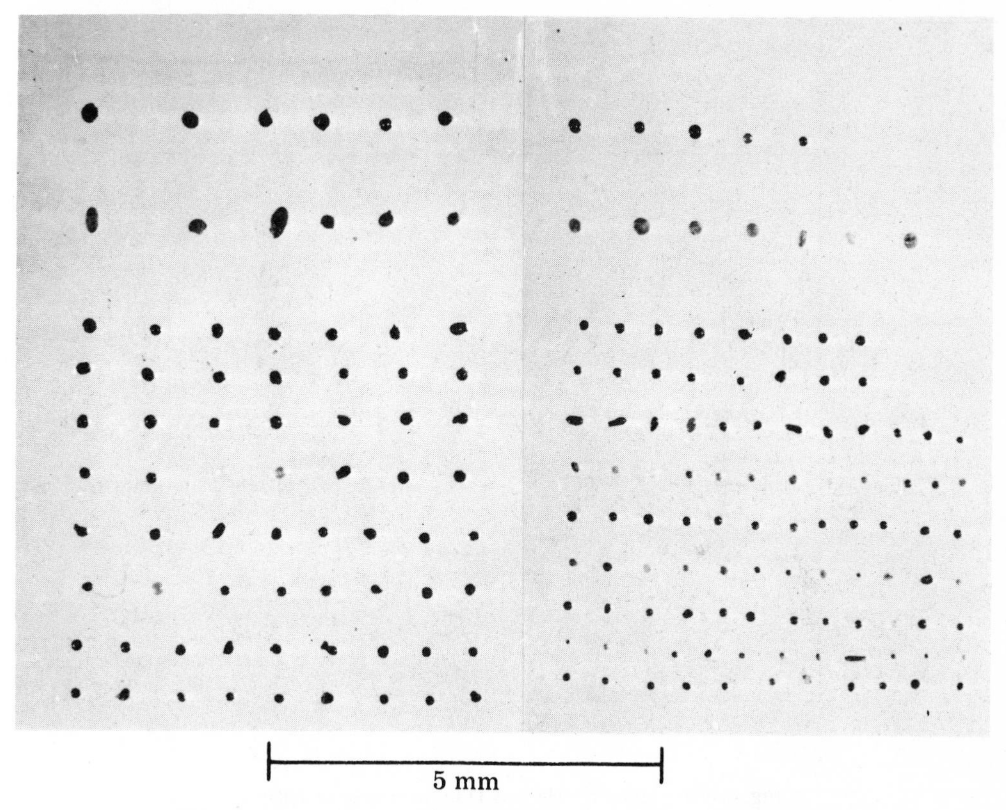

Fig. 3. Glass spherules singly mounted for microprobe bombardment.

method as used by EBERHARDT *et al.* (1965), half of the 800 grains were etched with a HF/H_2SO_4 mixture to remove the outer 2 μ of each grain. Etching velocities were determined under the microscope. Etching times varied between 8 sec (glassy fragments) and 2 min (ilmenite). All grains were singly mounted (Fig. 3) and bombarded with the electron beam, and ^{4}He was analyzed together with Ti or Fe. Each grain was degassed by a single shot. Per hour, 50 grains could be analyzed. During bombardment, the irregularly shaped crystals were melted to spherules. From their diameter, volumes were determined under the microscope. The diameter reduction due to vaporization was found to be 10 per cent for glass spherules. This correction was applied to all types of grains. This causes some uncertainty, however, it should not change the general pattern. Outgassing was complete since second shots gave no additional He-signals.

To supplement the results and to check the calibration of the dynamic measurements, seven grain size fractions of unseparated lunar dust were analyzed for He, Ne and Ar in a static mass spectrometer. Also, an etched portion (adjusted to etch 2 μ of the pyroxene) was measured.

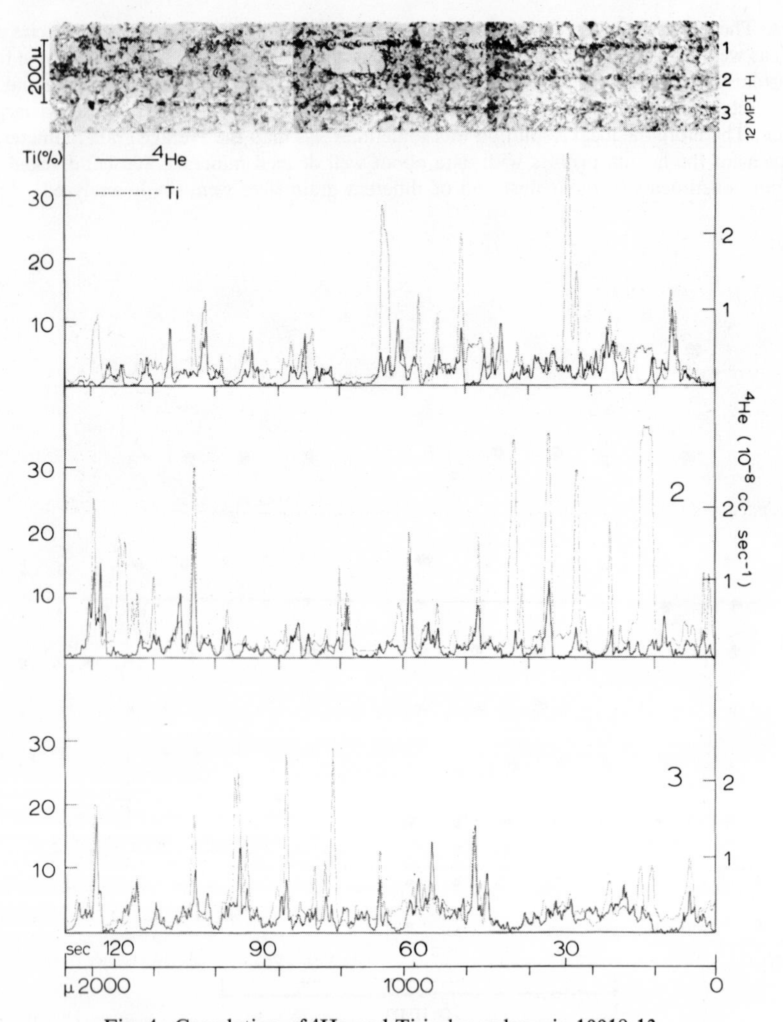

Fig. 4. Correlation of ^{4}He and Ti in lunar breccia 10019-13.

To analyze gases included in cavities of crystalline rocks, chips were broken in the high vacuum system without heating. Active gases were rapidly scanned with a mass filter, rare gases were measured mass spectrometrically.

RESULTS

Fines and breccias

A ^{4}He–Ti scan typical for breccias is shown in Fig. 4. It reveals a very inhomogeneous He-distribution. Helium is concentrated in spots of some ten microns diameter, mostly in very fine grained areas. Larger grains and especially crystalline inclusions are poor in ^{4}He. A rather strong correlation between Ti content and He is found. Many of the larger He-peaks occur together with Ti-enrichments; however, not all Ti-rich spots contain He. The content of Fe, Cr and Mn is also correlated with He, but to a lesser extent and for the latter two elements mostly together with Fe.

On the other hand, Ca and He are inversely related (Fig. 5). These findings occur in hundreds of scans across breccias 10019-13, 10046-19 and 10061-26. The peak height for the He-richest spots corresponds to gas concentrations of the order of 3 cm^3 $^4He/g$. (Scan 16 μ/sec; volumen degassed $\sim 2 \times 10^{-9}$ cm^3/sec).

The averaged ^4He concentrations for up to eight grain size fractions and five mineral components of handpicked dust grains measured by the microprobe technique are shown in Fig. 6. Clearly, the He content decreases in the progression ilmenite,

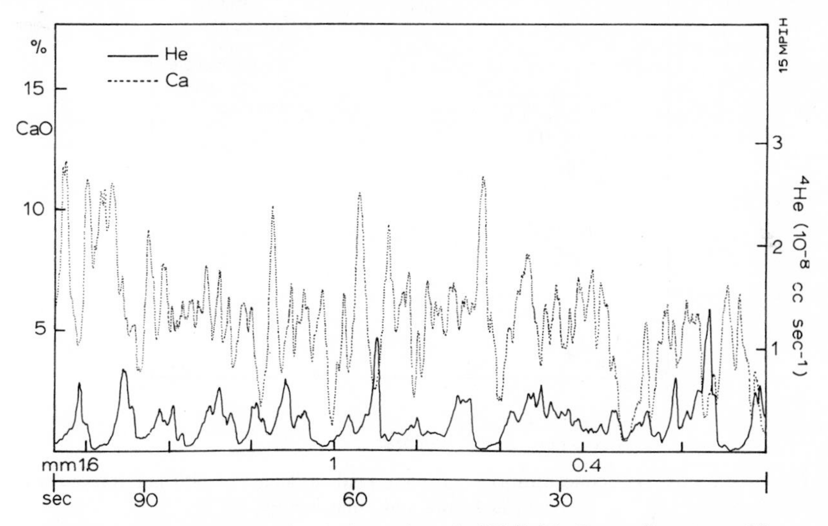

Fig. 5. ^4He and Ca distribution in lunar breccia 10061-26. Beam diameter $\sim$30 μ.

Ti-rich glass fragments, glass spherules, pyroxene, plagioclase, in total by a factor of about 10. It is also seen that He concentrations increase toward smaller grain sizes nearly linearly with the reciprocal diameter. Figure 6 also shows the He reduction due to etching of the outer 2 μ for the sieve fractions of the same five components. Amounts of He are generally lowered by factors of 2–10; nevertheless, even for etched minerals the relation between He and reciprocal diameter essentially remains.

The statistical distribution of He content in single grains of the same size is shown for each component in Fig. 7. The histogram applies to the 120 $\pm$ 20 μ fraction; similar histograms were obtained for all other sieve fractions. Variations are large for each type but largest for Ti-rich components.

Table 1 shows the results of the static measurements on sieve fractions of unseparated dust. The linearity between gas concentration and reciprocal diameter is valid for He as well as for Ne and Ar (Fig. 8). From 180 μ to $<$25 μ diameter, the $^4He/^{20}Ne$ ratio increases slightly. The same is true for the ratios $^{20}Ne/^{36}Ar$ and $^4He/^{36}Ar$ (Fig. 9). After etching, these ratios increase, pronouncedly for smaller grain sizes (Fig. 9). The isotopic ratios $^4He/^3He$, $^{22}Ne/^{21}Ne$, $^{20}Ne/^{22}Ne$, $^{36}Ar/^{38}Ar$ and $^{36}Ar/^{40}Ar$ decrease only very little with increasing grain size (Table 2, Fig. 10). Etching reduces $^4He/^3He$ by about 8 per cent, $^{20}Ne/^{22}Ne$ by about 15 per cent, and $^{22}Ne/^{21}Ne$ by a factor of about 2 for all grain sizes (Fig. 10). The $^{36}Ar/^{38}Ar$ ratio in

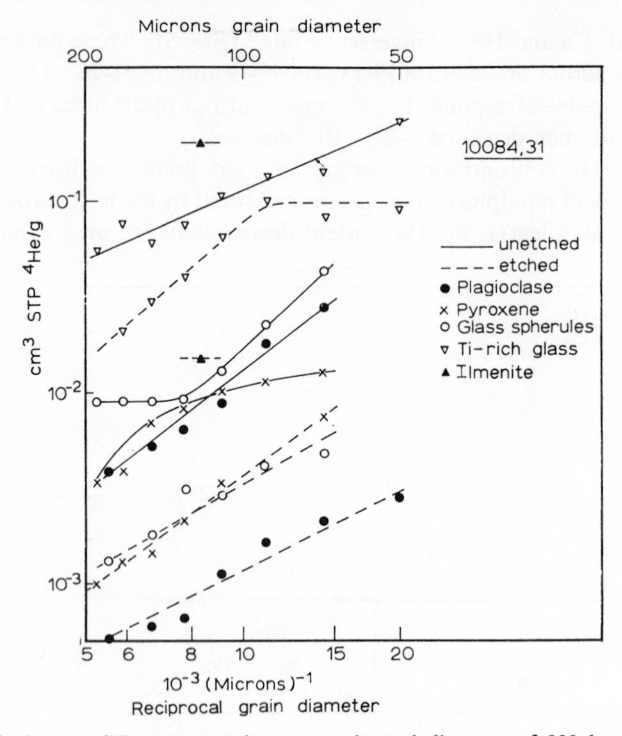

Fig. 6. Average ^{4}He concentrations vs. reciprocal diameter of 800 handpicked dust particles before and after etching.

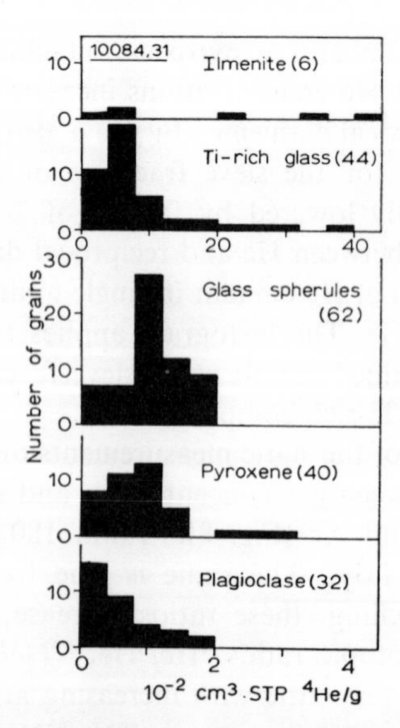

Fig. 7. Fluctuation of ^{4}He concentration in individual grains of separated dust components, $120 \pm 20\ \mu$ diameter.

Table 1. Absolute rare gas concentrations in seven grain size fractions of lunar dust

Grain size diameter	Rare gas concentration, 10^{-4} cm³ STP/g					
	^{4}He		^{20}Ne		^{36}Ar	
	Unetched	Etched	Unetched	Etched	Unetched	Etched
<25 μ	3000	—	39	—	6·8	—
25–50 μ	1020	365	15	1·69	2·6	0·075
50–80 μ	648	278	10	1·58	1·8	0·157
80–100 μ	483	218	7·2	1·60	1·5	0·250
100–125 μ	388	242	6·4	2·10	1·48	0·382
125–160 μ	316	211	5·5	1·80	1·21	0·352
160–200 μ	281	148	4·8	1·69	1·22	0·435

Averages of two independent measurements. Error ±5 per cent.

etched samples decreases with decreasing grain size down to 3·7 for the 25–50 μ fraction. The ^{40}Ar/^{36}Ar ratio increases after etching, especially for small grain sizes, where it reaches a maximum value of 6·6 (Fig. 10).

The elemental isotopic abundances fall in the same range as results obtained on lunar fines by other authors. It may be interesting to note that dust samples 10010 (FUNKHOUSER *et al.*, 1970a), 10084-31 (this work) and 10084-59 (REYNOLDS *et al.*, 1970) have about 40 per cent lower ^{4}He/^{20}Ne ratios than 10084-18 (HINTENBERGER *et al.*, 1970), 10084-29 (MARTI *et al.*, 1970), 10084-40 (HEYMANN *et al.*, 1970), 10084-48 (PEPIN *et al.*, 1970), and 10084-47 (EBERHARDT *et al.*, 1970). Not only absolute, but also relative elemental abundances may vary within the same sample. This was also noted by FUNKHOUSER *et al.* (1970a).

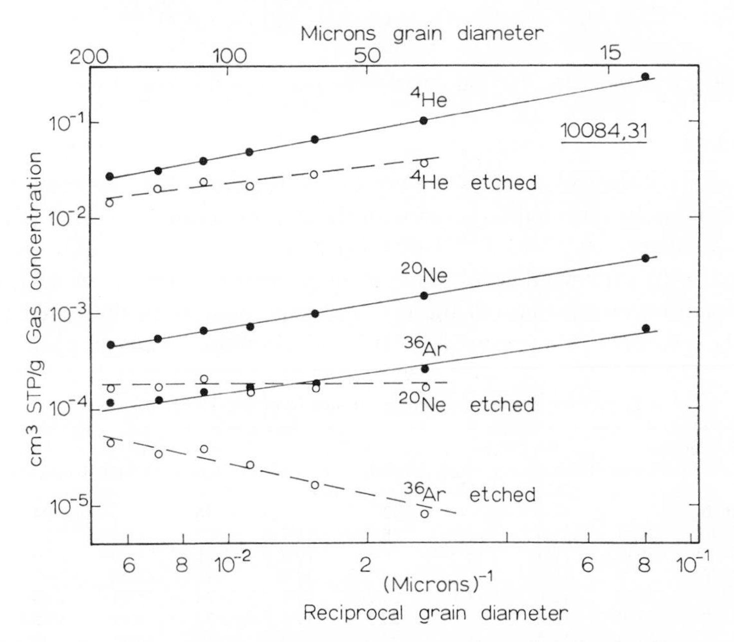

Fig. 8. ^{4}He, ^{20}Ne and ^{36}Ar concentrations vs. reciprocal grain size diameter of unseparated lunar dust before and after etching.

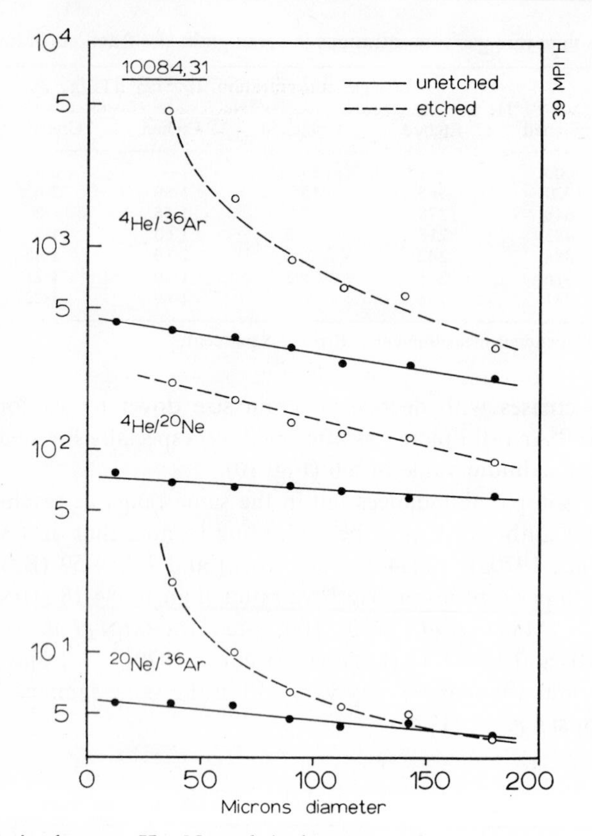

Fig. 9. Ratios between He, Ne and Ar in seven grain size fractions of lunar dust.

Crystalline rocks

Scans across polished sections of the crystalline rock 10057-22 liberated less He than was detectable by our method, even with the beam intensity increased to 25 μA. Two other specimens (10057-80 and 10085-41) had pitted surfaces and were most likely exposed to the solar wind also. Many scans going from a freshly broken surface to the pitted surface clearly showed the rare gas implantation on the pitted surface (Fig. 11). The quantity of He per cm^2 is $\sim 10^{-2}$ cm^3. Repeated scans perpendicular to

Table 2. Isotope ratios in seven grain size fractions of lunar dust

Grain size diameter	$^4He/^3He$ Unetched	$^4He/^3He$ Etched	$^{20}Ne/^{22}Ne$ Unetched	$^{20}Ne/^{22}Ne$ Etched	$^{22}Ne/^{21}Ne$ Unetched	$^{22}Ne/^{21}Ne$ Etched	$^{36}Ar/^{38}Ar$ Unetched	$^{36}Ar/^{38}Ar$ Etched	$^{40}Ar/^{36}Ar$ Unetched	$^{40}Ar/^{36}Ar$ Etched
<25 μ	2620	—	13·1	—	32	—	5·36	—	1·04	—
25–50 μ	2700	2690	13·0	11·2	28·7	13·7	5·25	3·67	1·17	6·6
50–80 μ	2980	2630	12·3	10·9	26·6	10·2	5·26	4·19	1·26	5·8
80–100 μ	2930	2730	12·8	11·5	24·5	12·9	5·19	4·54	1·47	3·1
100–125 μ	2730	2520	12·6	11·1	24·2	13·7	5·24	5·07	1·45	2·93
125–160 μ	2600	2430	12·8	11·1	22·5	12·8	5·22	4·68	1·65	2·86
160–200 μ	2350	2450	12·5	11·3	22·5	13·3	5·19	4·95	1·70	2·29

Error of ratios ± 3 per cent.

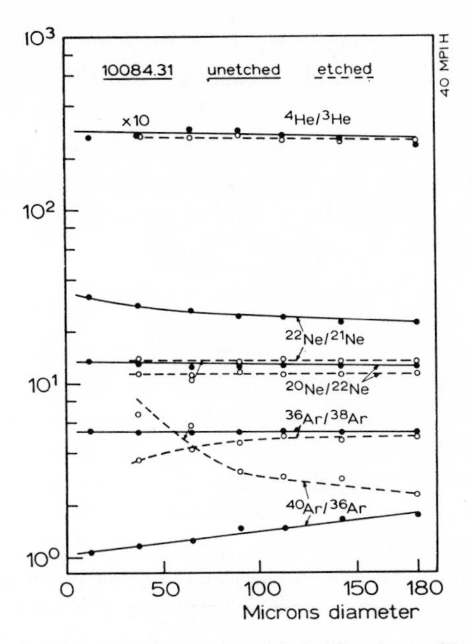

Fig. 10. Isotope ratios in seven grain size fractions of lunar dust.

the pitted surface revealed an average thickness of ~10 μ for the He loaded layer (Fig. 11). Again, if ilmenite grains form the surface, the He concentrations are highest. For penetration depths of about 10 μ, an average ^{4}He concentration of 3 cm^3/g results.

The analysis of gases included in cavities of the crystalline rock 57-22 gave the following results (in 10^{-8} cm^3): In a 69 mg piece: N_2, 700; H_2, 5; ^{4}He, 3; ^{40}Ar, 1·4;

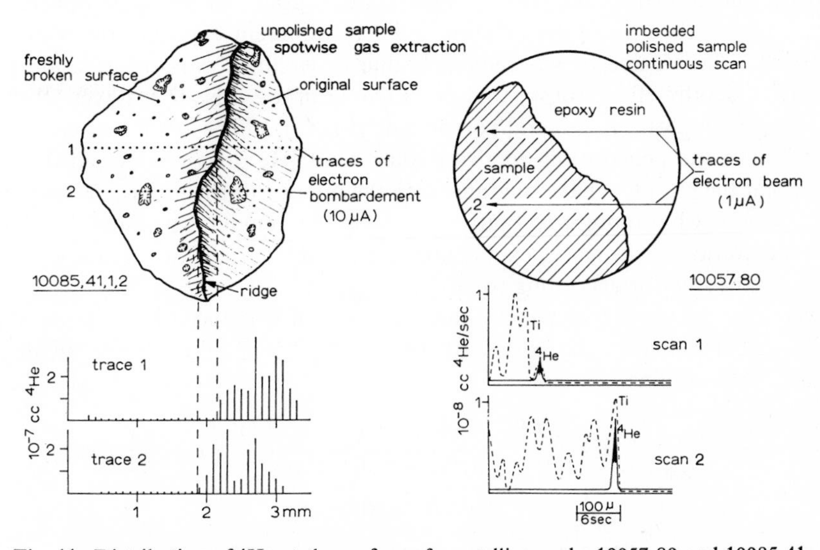

Fig. 11. Distribution of ^{4}He at the surface of crystalline rocks 10057-80 and 10085-41.

$^{40}Ar/^{36}Ar = 700$. In a 50 mg piece: ^{4}He, 0·7; ^{40}Ar, 0·63; H_2 and N_2 not detected. The Ar is clearly radiogenic, probably because of degassing after rock crystallization.

DISCUSSION

The foregoing results show that the rare gases consist of three components. The main contribution seems to come from trapped solar wind particles. There are also radiogenic and spallogenic components.

Solar wind

A solar wind origin is suggested by the proportionality of the gas concentrations with the reciprocal diameter of the grains, both in mineral components and in unseparated sieve fractions. The same was found by EBERHARDT *et al.* (1970); HEYMANN *et al.* (1970) and by HINTENBERGER *et al.* (1970). Since the concentrations increase with the surface to volume ratio, the gases must be concentrated in the outermost surface layers. This is supported by the composition of the gases. The overall pattern agrees with the so called solar abundances, despite of a slight depletion of the lighter gases. Gas bearing grains must have been at the uppermost surface at one time or another. In a similar way, the surfaces of glass pitted crystalline rocks were loaded with gases while they were exposed to the sun, but the interior was always shielded. Breccias are composed of dust and crystalline fragments. It is natural that the matrix contains solar gases while they are undetectable in lithic fragments. These fragments as well as mineral grains without detectable gas content most likely were never exposed to the surface or were degassed at a later stage. The correlation of He with Ti and to some extent with Fe, Cr, and Mn, and the anticorrelation between He and Ca may be explained by different diffusion constants or gas trapping properties of different minerals. The enrichment of trapped gases in Ti-rich components was also found by EBERHARDT *et al.*, (1970.) The fact that grains of the same mineral differ substantially in their gas concentration indicates different exposure times or partial outgassing. The degree of gas reduction due to etching bears information on the penetration depths of the trapped gases. Etching of a 2 μ-layer removes 50–90% of the He depending on the mineral component (Fig. 6). Since 10–50 per cent are still left and since the penetration depth of solar wind ^{4}He does not exceed 0·05 μ, we assume that the gas must have migrated into the interior of the grains. To express the migration somewhat more quantitatively, a simplified model is assumed in which the gas concentration G decreases exponentially from the surface of a grain with radius R towards the interior according to $G(r) = G_{\text{surf.}} \cdot e^{(r-R)/b}$.

The mean thickness b at which the concentration is reduced by a factor of e can then be calculated from the measured degree of gas reduction due to etching of c microns. One obtains

$$\frac{G_{\text{unetched}}}{G_{\text{etched}}} = V = \frac{R^2 - 2Rb + 2b^2(1 - e^{-R/b})}{R'^2 - 2R'b + 2b^2(1 - e^{-R'/b})}\left(1 - \frac{c}{R}\right)^3 e^{c/b}.*$$

$$(R' = R - c).$$

* The factor $(1 - c/R)^3$ has been neglected in KIRSTEN *et al.* (1970).

From the microprobe measurements made on single grains we obtain mean thicknesses of $\sim$0·7 μ for ilmenite, $\sim$0·9 μ for plagioclase, $\sim$1·2 μ for glass spherules, $\sim$1·4 μ for pyroxene and $\sim$2·7 μ for Ti-rich glass fragments, more or less independent from the grain size. For ilmenite, EBERHARDT *et al.* (1970) obtained a mean thickness of 0·2 μ.

If one applies the same formalism to the data on unseparated dust, a mean value for the thickness c of the removed layer must be used. From the weight loss due to etching, $\bar{c} = 5{\cdot}2\ \mu$ was found. The apparent mean thicknesses obtained for the 160–200 μ fraction are $\bar{b}_{He} \sim 8\ \mu$; $\bar{b}_{Ne} \approx \bar{b}_{Ar} \sim 4{\cdot}8\ \mu$. For the 25–50 μ fraction, these values are 4·3 μ; 2·1 μ and 1·4 μ, respectively. If the data by HINTENBERGER *et al.* (1970) are combined with the soil size distribution given by DUKE *et al.* (1970), similar values result. The fact that $\bar{b}_{He}$ for unseparated dust is larger than b_{He} for any of the separated components leads us to believe that the abundant conglomerates consisting of intergrown minerals and glass fragments, which have not been measured separately, are more equilibrated than undisturbed constituents with well defined lattices.

The diffusion lengths decrease with decreasing grain sizes, more pronounced for the heavier gases. Most likely this effect is due to less heating because of a higher turnover rate for the smaller grains. From the He-data, we obtain an average surface concentration of $2{\cdot}5 \times 10^{-4}$ cm^3 ^{4}He/cm^2 for lunar dust; about 3×10^{-3} cm^3 ^{4}He/cm^2 for ilmenite and $\sim 10^{-2}$ cm^3 ^{4}He/cm^2 for pitted rock surfaces.

If the ^{4}He concentration in fines is divided by the solar wind flux [$6{\cdot}3 \times 10^6$ ^{4}He-atoms/cm^2 sec, BÜHLER *et al.* (1969)], one obtains an apparent surface residence time of $\sim$35 yr. The ten times higher value for ilmenite is a direct proof for diffusion loss in bulk fines. Hence, the calculated residence times are lower limits because of shielding, surface saturation, diffusion and sputtering. EBERHARDT *et al.* (1970) arrived at similar conclusions.

Assuming a continuous turnover, the average exposure age of dust ($\sim$500 m.y., penetration $\sim$1 m) yields an irradiation time of $\sim$25000 yr for a 50 μ-layer, which corresponds to the average grain size of lunar dust. Therefore, it is more likely that the actual irradiation time at the surface is some 10000 yr. This time would be sufficient to allow appreciable migration of ^{4}He into the grains at a surface temperature of 97°C for half the time and could account for the occurrence of ^{4}He even 5 μ below the grain surfaces. A rough estimate for the time required to fill the interior of a 100 μ-grain yields $\sim 10^{3\pm1}$ yr. This is based on stepwise heating experiments (FUNKHOUSER *et al.*, 1970b) which gave an apparent average diffusion constant of ^{4}He for lunar dust of $D/a^2 \sim 3 \times 10^{-12}$ sec^{-1}. More detailed diffusion experiments should be undertaken.

EXPOSURE AGE OF LUNAR DUST

The decrease of the isotopic ratios ^{4}He/^{3}He, ^{22}Ne/^{21}Ne, ^{20}Ne/^{22}Ne, ^{36}Ar/^{38}Ar and ^{36}Ar/^{40}Ar with increasing grain size, respectively, after etching is due to the enrichment of the solar wind at the surface, while radiogenic and spallogenic gases are homogeneously distributed. By etching the solar wind is preferentially removed and spallogenic and radiogenic components become relatively enriched in the remaining sample. The spallogenic component can clearly be distinguished from the $^{22}Ne/^{21}Ne$ and $^{36}Ar/^{38}Ar$ ratios in the etched dust fractions. For the solar and spallogenic components, the following values have been used: $(^{22}Ne/^{21}Ne)_{solar} = 30{\cdot}7$ (FUNKHOUSER *et al.*, 1970a); $(^{22}Ne/^{21}Ne)_{spall.} = 1{\cdot}05$; $(^{36}Ar/^{38}Ar)_{solar} = 5{\cdot}3$; $(^{38}Ar/^{36}Ar)_{spall.} = 1{\cdot}6$. The highest reported value for the $^{4}He/^{3}He$ ratio in lunar breccias is 3400 (FUNKHOUSER *et al.*, 1970a), however, in many cases this ratio is ~15 per cent lower. We have used a $(^{4}He/^{3}He)_{solar}$ ratio of 3200 for estimates of ^{3}He exposure ages.

The production rate of ^{3}He in lunar material has been assumed to be half of that in ordinary chondrites because of the 2π geometry at the lunar surface. This is substantiated by the tritium results of STOENNER *et al.* (1970) and FIREMAN *et al.* (1970). For the ratio P_3/P_{21} we have used the average value from crystalline rocks (FUNKHOUSER *et al.*, 1970a). This agrees with ^{22}Na measurements by O'KELLEY *et al.* (1970). The ratio P_{21}/P_{38} is rather sensitive to the Ca- and Ti-content and to the cosmic ray spectra; therefore we adopted the average value found in lunar rocks (FUNKHOUSER *et al.*, 1970a) instead of ^{38}Ar production rates based on ^{37}Ar and ^{39}Ar measurements, which are lower by about a factor of two (FIREMAN *et al.*, 1970; STOENNER *et al.*, 1970). The production rates used to calculate exposure ages are thus:

$$P_3 = 1 \times 10^{-8}\ \text{cm}^3\ {}^{3}\text{He/g, m.y.,} \qquad P_{21} = P_{38} = 0{\cdot}15 \times 10^{-8}\ \text{cm}^3\text{/g, m.y.}$$

All sieve fractions give ^{21}Ne exposure ages of 500 ± 80 m.y. HINTENBERGER *et al.* (1970) found 215 m.y. with $P_{21} = 0{\cdot}2 \times 10^{-8}$ cm^3 ^{21}Ne/g, m.y. A reliable ^{38}Ar exposure age could be estimated only for the 25–50 μ fraction. That age is also about 500 m.y. The apparent ^{3}He exposure age (200 m.y.) is uncertain because of high ^{4}He content and possible diffusion losses. The ^{21}Ne and ^{38}Ar exposure ages are considered as upper limits since spallogenic gases could be partially inherited and since solar flare induced reactions could also have contributed. As in this work, EBERHARDT *et al.* (1970) found an exposure age of 500 m.y., based on ^{126}Xe. MARTI *et al.* (1970) obtained a ^{81}Kr–Kr age of 330 m.y. The high exposure ages found by REYNOLDS *et al.* (1970) do not fit in this picture. It is remarkable that the exposure age of lunar dust exceeds the values obtained for crystalline rocks (FUNKHOUSER *et al.*, 1970a).

RADIOGENIC AGE OF LUNAR DUST

The etched small grain size fractions show a pronounced enrichment of ^{40}Ar. Assuming an $^{40}Ar/^{36}Ar$ ratio of 1 in the trapped component, the radiogenic ^{40}Ar content varies between 4·2 and $7{\cdot}5 \times 10^{-5}$ cm^3/g for the different grain size fractions. The measured average K-concentration in sample 10084-31 is 1150 ppm. (With the same analytical procedure we obtained for Na 0·32%, Cs 93 ppb and U 0·54 ppm. The errors do not exceed 5 per cent.) K–Ar-ages between 3·4 and 4·3 b.y. result. Values between 4·2 and 4·9 b.y. were reported by ALBEE *et al.* (1970); HEYMANN *et al.* (1970) and by MARTI *et al.* (1970). Thus, the K–Ar ages of lunar dust exceed the K–Ar ages of crystalline rocks (FUNKHOUSER *et al.*, 1970), but they are similar to Rb–Sr and Pb–U ages for dust (ALBEE *et al.*, 1970, TATSUMOTO and ROSHOLT, 1970). A similar age difference between fines and crystalline rocks exists for Rb–Sr and Pb–U ages (ALBEE *et al.*, 1970, TATSUMOTO and ROSHOLT, 1970).

Acknowledgements—We thank W. GENTNER, S. KALBITZER and D. STORZER for stimulating discussions and A. HAIDMANN, P. HORN, E. JESSBERGER, D. KAETHER, H. W. MÜLLER and H. RICHTER for their assistance in this work.

REFERENCES

ALBEE A., BURNETT D., CHODOS A., EUGSTER O., HUNEKE J., PAPANASTASSIOU D., PODOSEK F., RUSS G. PRICE II, SANZ H., TERA F. and WASSERBURG G. (1970) Ages, irradiation history, and chemical composition of lunar rocks from the Sea of Tranquillity. *Science* **167,** 463–466.

BÜHLER F., EBERHARDT P., GEISS J., MEISTER J. and SIGNER P. (1969) Apollo 11 solar wind composition experiment: first results. *Science* **166,** 1502–1503.

DUKE M., WOO C., BIRD M., SELLERS G. and FINKELMAN R. (1970) Lunar soil: size distribution and mineralogical constituents. *Science* **167,** 648–650.

EBERHARDT P., GEISS J. and GRÖGLER N. (1965) Further evidence on the origin of trapped gases in the meteorite Khor Temiki. *J. Geophys. Res.* **70,** 4375–4378.

EBERHARDT P., GEISS J., GRAF H., GRÖGLER N., KRÄHENBÜHL U., SCHWALLER H., SCHWARZMÜLLER J. and STETTLER A. (1970) Trapped solar wind noble gases, Kr^{81}/Kr exposure ages and K/Ar ages in Apollo 11 lunar material. *Science* **167,** 558–560.

FIREMAN E., D'AMICO J. and DEFELICE J. C. (1970) Tritium and argon radioactivities in lunar material. *Science* **167,** 566–568.

FUNKHOUSER J. G., SCHAEFFER O. A., BOGARD D. D. and ZÄHRINGER J. (1970a) Gas analysis of the lunar surface. *Science* **167,** 561–563.

FUNKHOUSER J. G., SCHAEFFER O. A., BOGARD D. D. and ZÄHRINGER J. (1970b) LRL. Unpublished results.

HEYMANN D., YANIV A., ADAMS J. A. and FRYER G. (1970) Inert gases in lunar samples. *Science* **167,** 555–558.

HINTENBERGER H., WEBER H. W., VOSHAGE H., WÄNKE H., BEGEMANN F., VILCSEK E. and WLOTZKA F. (1970) Rare gases, hydrogen and nitrogen: concentrations and isotopic composition in lunar material. *Science* **167,** 543–545.

KIRSTEN T., STEINBRUNN F. and ZÄHRINGER J. (1970) Rare gases in lunar samples: study of distribution and variations by a microprobe technique. *Science* **167,** 571–574.

LSPET (LUNAR SAMPLE PRELIMINARY EXAMINATION TEAM) (1969) Preliminary examination of lunar samples from Apollo 11. *Science* **165,** 1211–1227.

MARTI K., LUGMAIR G. and UREY H. C. (1970) Solar wind gases, cosmic ray spallation products, and the irradiation history. *Science* **167,** 548–550.

MEGRUE G. H. (1967) Isotopic analysis of rare gases with a laser microprobe. *Science* **157,** 1555–1556.

O'KELLEY G., ELDRIDGE J., SCHONFELD E. and BELL P. (1970) Elemental compositions and ages of lunar samples by non-destructive gamma-ray spectrometry. *Science* **167,** 580–582.

PEPIN R. O., NYQUIST L., PHINNEY D. and BLACK D. C. (1970) Isotopic composition of rare gases in lunar samples. *Science* **167,** 550–553.

REYNOLDS J. H., HOHENBERG C., LEWIS R., DAVIS P. and KAISER W. A. (1970) Isotopic analysis of rare gases from stepwise heating of lunar fines and rocks. *Science* **167,** 545–548.

STOENNER R. W., LYMAN W. and DAVIS R. (1970) Cosmic ray production of rare gas radioactivities and tritium in lunar material. *Science* **167,** 553–555.

TATSUMOTO M. and ROSHOLT J. (1970) Age of the moon: an isotopic study of uranium–thorium–lead systematics of lunar samples. *Science* **167,** 461–463.

ZÄHRINGER J. (1966) Primordial helium detection by microprobe technique. *Earth Planet. Sci. Lett.* **1,** 20–22.

Proceedings of the Apollo 11 Lunar Science Conference, Vol. 2, pp. 1345 to 1350

Lead and thallium isotopes in Mare Tranquillitatis surface material*

TRUMAN P. KOHMAN, LANCE P. BLACK, HARUHIKO IHOCHI and JAMES M. HUEY
Department of Chemistry, Carnegie-Mellon University, Pittsburgh, Pennsylvania 15213

(*Received* 2 *March* 1970; *accepted* 2 *March* 1970)

Abstract—Lead from Apollo 11 fines is more radiogenic than any meteoritic lead reported and older than any terrestrial radiogenic lead: $^{204}Pb/^{206}Pb/^{207}Pb/^{208}Pb = 1/99._6/69._0/117._1$. Comparison with primordial lead from meteoritic troilite yields a $^{207}Pb/^{206}Pb$ age of $4{\cdot}7 \pm 0{\cdot}1 \times 10^9$ yr. The $^{238}U/^{204}Pb$ ratio is $\gtrsim 90$ and the $^{232}Th/^{238}U$ ratio is $3{\cdot}9 \pm 0{\cdot}1$. The lead content is $\gtrsim 1{\cdot}7 \times 10^{-6}$. Evidently Pb was strongly depleted and Th and U strongly enriched in the formation of this material. Thallium was too low ($\lesssim 5 \times 10^{-9}$) to yield mass spectra, but indications are favorable for eventual observation of extinct natural radioactivity of ^{205}Pb.

WE HAVE made an isotopic analysis of lead from a sample of fines from the Apollo 11 site, from which an age of the lunar surface can be derived, and have initiated a search for isotopic anomalies due to certain extinct natural radionuclides, from which fine time-resolution of the earliest lunar history might be derived. In particular, we have undertaken to determine the abundance and isotopic composition of thallium, which might be altered by the electron-capture decay of ^{205}Pb (half-life $\sim 24 \times 10^6$ yr) to ^{205}Tl. Previous searches for variations in the $^{205}Tl/^{203}Tl$ ratio in meteoritic thallium have given negative results (ANDERS and STEVENS, 1960; OSTIC *et al.*, 1969).

Lead and thallium are isolated by volatilization under vacuum at temperatures close to the melting point of the bulk rock. Anion exchange and electrodeposition are used for further purification. A single-filament mass-spectrometric technique modified from CAMERON *et al.* (1969) is used. Sub-microgram quantities of silica gel, the purified Tl and Pb in 1-M HNO_3, and H_3PO_4 are successively loaded onto the filament; drying is effected by a combination of resistance and radiant heating in air. The mass spectrometer is a Nuclide Corporation Model SU 2·4 (90° deflection, 30-cm radius). Filament temperatures are monitored by an optical pyrometer. The Pb and Tl data are collected by repeated magnet-current switching between peak pairs; approximately 12 cycles are used for each ratio. The Tl is measured at ~740°C and Pb at ~1170°C.

Elemental abundances and yields of various parts of the procedure are determined by isotopic dilution with the use of enriched ^{203}Tl and ^{206}Pb. The ion-exchange purification yields were 95–100 per cent for both Tl and Pb. The electrodeposition yields were 60–70 per cent for Tl and 90–95 per cent for Pb. Because isotopic mixing of spikes and elements in natural solids cannot be achieved, in actual runs with such materials the spikes are added to the solutions of the vapor condensates. Volatilization yields were estimated by comparison of recovered Tl and Pb from standard rocks

* Reprinted from *Science* (Vol. 167, pp. 481–483, 1970) without further refereeing.

AGV-1 and BCR-1 with published determinations (FLANAGAN, 1969; TATSUMOTO, 1968; WAHLER, 1968) as 70–100 per cent for Tl and 60–90 per cent for Pb in heatings of 24–48 hr. Yields seem to be higher at temperatures just below the melting point than just above. The highest temperatures tried have been 1050°C.

The first lunar material used was a 3-g sample (A) of less than 1-mm fines from the Apollo 11 site (sample 10084, 45) ground in a motor-shaken tungsten-carbide capsule-pestle to pass through a 200-mesh nylon screen. It was heated successively at 800°, 850°, and 900°C for 20 hr each. The vapor condensate solution was made up to 10 ml. Two 4-ml aliquots (**1** and **3**) were processed for isotope-abundance measurements and a 2-ml aliquot (**2**) was spiked for abundance determinations. In neither of the unspiked samples could Tl be seen in the mass spectrometer. In run **1**, Pb isotope ratios were obtained only with a chart recorder with ~2 per cent accuracy; the signal appeared at a relatively low temperature (820°C) and disappeared before digital-voltmeter measurements could be made. In run **3** moderately stable ion beams of $\sim 10^{-11}$ A were obtained at 1170°C, and data were collected with the digital voltmeter. Digital-voltmeter data were obtained for both Tl and Pb in run **2**.

Table 1. Mass-spectrometric data and calculations

Quantity	Run A-**1** (unspiked)	Run A-**3** (unspiked)		Run A-**2** (spiked)
		Data	Corrected	
Effective sample mass	1·2 g	1·2 g		0·6 g
Mass Pb spike*				1·094 μg
Mass Tl spike†				221·0 ng
Measured $^{205}Tl/^{203}Tl$	not seen	not seen		0·093
Measured $^{206}Pb/^{204}Pb$	$71\cdot0_2$‡	90·2 ± 0·3‡	$99\cdot_6$	
Measured $^{207}Pb/^{206}Pb$	$0\cdot70_6$	0·6965 ± 0·0005	$0\cdot69_4$	
Measured $^{208}Pb/^{206}Pb$	$1\cdot25_6$	1·1978 ± 0·0006	$1\cdot17_8$	0·333
Calculated $^{207}Pb/^{204}Pb$	$50\cdot_3$‡	62·8 ± 0·2‡	69·0	
Calculated $^{208}Pb/^{204}Pb$	$89\cdot_6$‡	108·0 ± 0·3‡	$117\cdot_1$	
Mass natural Tl				3·3 ng§
Mass natural Pb				$1\cdot01_7$ μg
Tl abundance in sample				$<5\cdot_5 \times 10^{-9}$‖
Pb abundance in sample				$>1\cdot7_0 \times 10^{-6}$**

* $^{204}Pb/^{206}Pb/^{207}Pb/^{208}Pb$ = 1/2688/189·9/36·34.
† $^{250}Tl/^{230}Tl = 0\cdot081_6$
‡ Probably lower limits because of possibility of lead contamination. Not corrected for possible isotope fractionation.
§ Calculated assuming for natural thallium $^{205}Tl/^{203}Tl$ = 2·389 as observed in this instrument (OSTIC *et al.*, 1969).
‖ Probably an upper limit because of nonzero blank.
** Probably a lower limit because of incomplete volatilization.

The results are given in Table 1. The results of run **3** were corrected for a typical blank of 0·07 μg of modern terrestrial Pb (MURTHY and PATTERSON, 1962). Additional contamination during collection, handling, and grinding of the sample would not have been corrected for.

The $^{206}Pb/^{204}Pb$ and $^{207}Pb/^{204}Pb$ ratios are plotted along with similar data on meteoritic and common terrestrial lead in Fig. 1a. The troilite point represents the mean of several measurements selected by MURTHY and PATTERSON (1962), who adopt $^{204}Pb/^{206}Pb/^{207}Pb/^{208}Pb$ = 1/9·56/10·42/29·71 for this and for primordial lead of

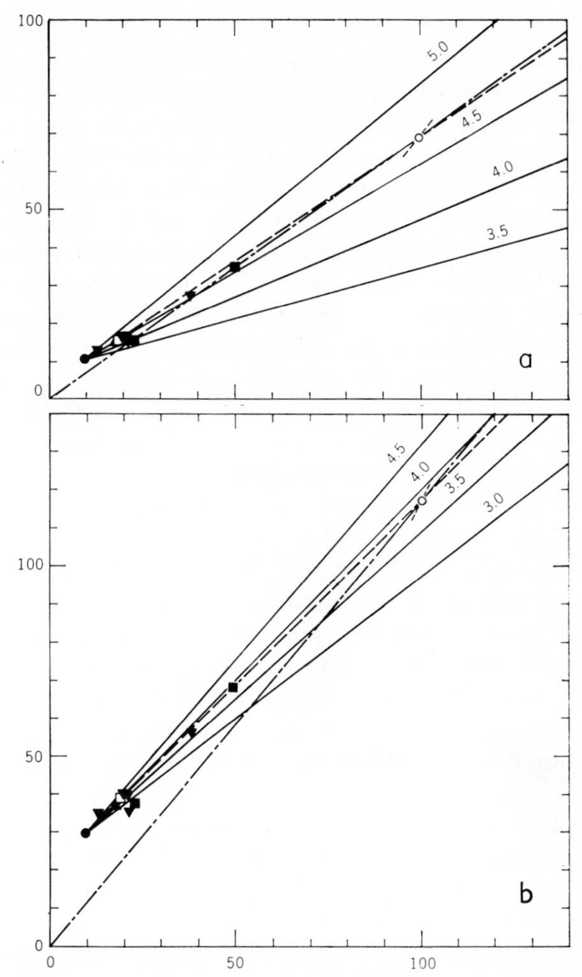

Fig. 1. Comparison of isotopic composition of lead from Apollo 11 lunar surface fines with that of meteorites and the earth. (a) $^{207}Pb/^{204}Pb$ vs. $^{206}Pb/^{204}$. ——— Calculated isochrons for single-stage ages indicated in units of 10^9 years. (b) $^{208}Pb/^{204}Pb$ vs. $^{208}Pb/^{204}Pb$. ——— Calculated locus lines for indicated values of $^{232}Th/^{238}U$ at the present time following $4{\cdot}7 \times 10^9$ yr of single-stage evolution. In both parts: (○) lunar fines sample 10084, 45, A-3; (●) meteorite troilite; (▼) chondrite meteorite; (■) achondrite meteorite; (▲) iron meteorite; (□) typical modern terrestrial common lead; —— —— common-lead contamination line; —— - —— ^{204}Pb error line; - - - - - isotope fractionation line.

the solar system. Subtraction of these ratios from the corresponding A-3 lunar values yields a radiogenic $^{207}Pb/^{206}Pb$ ratio of $0{\cdot}65_1$. With disintegration constants $\lambda(^{238}U) = 1{\cdot}537 \times 10^{-10}$ yr^{-1} and $\lambda(^{235}U) = 9{\cdot}722 \times 10^{-10}$ yr^{-1}, and $(^{238}U/^{235}U)_{now} = 137{\cdot}8$ (Kanasewich, 1968) the age of the lunar surface sample is calculated to be $T = 4{\cdot}6^9 \times 10^9$ yr. Isochrons corresponding to several ages are shown on the figure. The effects of possible corrections for error in the ^{204}Pb peak measurements, for contamination by modern terrestrial lead, and for isotope fractionation in the mass spectrometer are shown by dashed lines. Since the contamination line through the lunar point is almost identical with the isochron which it defines, and the other two corrections are not likely to be appreciable, the error in the age due to these effects should be small. For example, an uncertainty in the $^{207}Pb/^{206}Pb$ ratio of as much as 2 per cent corresponds to an uncertainty of only $\sim 0.03 \times 10^9$ yr in T. Considering all uncertainties we express our result as $4.7 \pm 0.1 \times 10^9$ yr. This age is not substantially different from those derived from the meteorite Pb isochron, or for the earth by a similar calculation with mean oceanic Pb, 4.55×10^9 yr (MURTHY and PATTERSON, 1962; KANASEWICH, 1968), or from meteoritic ^{87}Rb–^{87}Sr isochron ages. However, our radiogenic $^{207}Pb/^{206}Pb$ ratio, 0·65, seems significantly greater than that of chondrite and achondrite meteorites, 0·59 or 0·60, suggesting that the lunar surface may actually be the oldest sample of solar-system matter yet available. On the other hand, a growth model involving two or more stages could be consistent with a more recent formation of the Mare Tranquillitatis surface material.

The $^{208}Pb/^{204}Pb$ and $^{206}Pb/^{204}Pb$ ratios are similarly plotted (Fig. 1b) along with iso-compositional lines corresponding to various $^{232}Th/^{238}U$ ratios, calculated for $T = 4.69 \times 10^9$ yr with $\lambda(^{232}Th) = 4.99 \times 10^{-11}$ yr^{-1} (KANASEWICH, 1968). The above-mentioned subtraction yields a radiogenic $^{208}Pb/^{206}Pb$ ratio of 0.97_1, which corresponds to an atomic $^{232}Th/^{238}U$ ratio 3.8_9 and an elemental Th/U ratio of 3.7_6. Any uncertainty in the $^{208}Pb/^{206}Pb$ ratio corresponds to the same fractional uncertainty in these parent ratios. The effects of ^{204}Pb error, contamination, and fractionation are likewise shown in the figure; again these effects are very small. Considering all uncertainties, including that in the age, we express our result as $^{232}Th/^{238}U = 3.9 \pm 0.1$ and $Th/U = 3.8 \pm 0.1$.

Our lead isotopic composition and the assumption of a closed system aged 4.69×10^9 yr corresponds to a present-day ratio $^{238}U/^{204}Pb$ (μ) ≈ 86. This is undoubtedly a minimum value, because a small amount of common-lead contamination could substantially increase the ^{204}Pb in the isolated sample. We conclude that $\mu \gtrsim 90$. The preliminary γ-ray measurement of similar Apollo 11 fines (LSPET, 1969) indicated a U content of $\sim 0.46 \times 10^{-6}$. Combining this with $\mu = 86$ and our Pb isotopic composition yields a total Pb abundance of $\sim 1.3_3 \times 10^{-6}$, relatively insensitive to common-lead contamination. This is close to the value calculated from the isotope-dilution analysis of the vapor-condensate, $1.7_0 \times 10^{-6}$, which we regard as a lower limit because of probably incomplete volatilization. The γ-ray measurements (LSPET, 1969) gave $\sim 1.6 \times 10^{-6}$ for the Th content, or ~ 3.5 for the Th/U ratio, consistent with our result.

Comparison of the ^{238}U and ^{204}Pb contents of the Apollo 11 fines, which, because of the natural mixing processes by which they were produced, may be fairly representative of the lunar surface, with the same values in type I carbonaceous chondrites (EHMANN, 1968) ($\mu \sim 0.4$), widely regarded as good samples of non-volatile solar-system matter, is made in Fig. 2. Evidently U and Th were highly enriched and Pb highly depleted in the formation of the lunar surface. This process must have been distinctly different from that producing the characteristic elemental assemblages of the ordinary chondrites (EHMANN, 1968) ($\mu \sim 4$) and the earth ($\mu \approx 9$) (Fig. 2).

Unfortunately, too little Tl was present in the 1·2-g sample aliquots to yield mass spectra. The $3._3$ ng of Tl calculated for the 0·6-g aliquot by isotope dilution is lower than that observed in any previous blank and probably still includes a residual blank. The Tl content of this composite lunar surface material, $<5 \times 10^{-9}$, is at least tenfold lower than in type I carbonaceous chondrites, $\sim 74 \times 10^{-9}$ (LAUL *et al.*, 1969), and

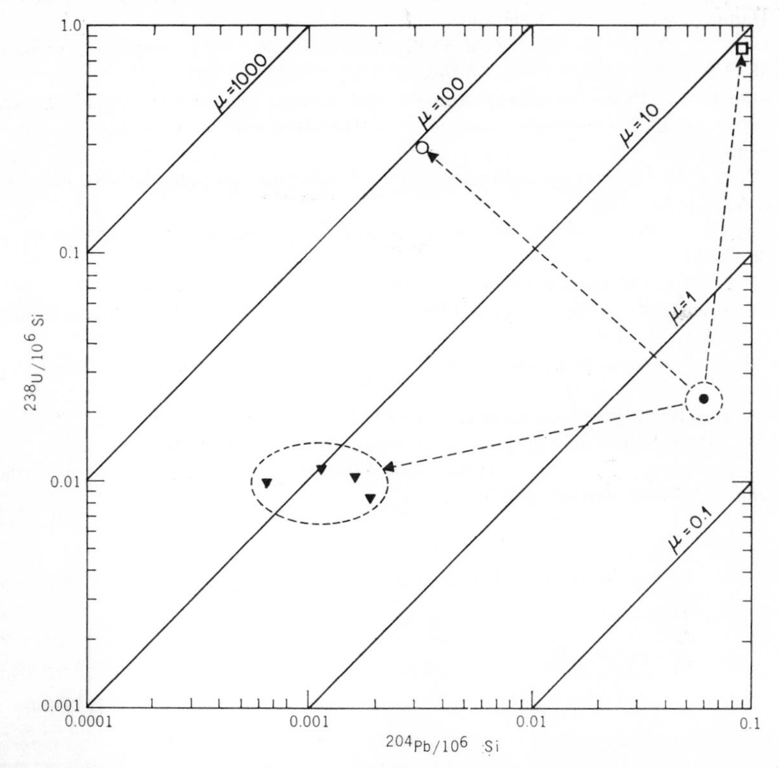

Fig. 2. Comparison of ^{238}U and ^{204}Pb contents of Apollo 11 lunar surface fines (○) with those of type 1 carbonaceous chondrites (●), ordinary chondrites (▼), and the earth (□). μ = present atomic ratio $^{238}U/^{204}Pb$.

may be similar to that observed in some highly equilibrated ordinary chondrites, $\lesssim 1 \times 10^{-9}$ (LAUL *et al.*, 1969). The depletion of Tl seems to be greater than that of ^{204}Pb. This, combined with the extreme antiquity indicated for the lunar surface, indicates favorable conditions for the eventual detection of radiogenic ^{205}Tl from extinct natural radioactivity of ^{205}Pb (KOHMAN, 1956, 1961; ANDERS and STEVENS, 1960; OSTIC *et al.*, 1969).

Acknowledgments—We thank R. G. OSTIC, J. T. TANNER, M. J. TOIA and M. W. HARAMIC of this University, and W. R. SHIELDS and co-workers of the National Bureau of Standards, for assistance in various portions of this investigation. Supported by NASA through contract NAS-9-8073 and by the U.S. Atomic Energy Commission through contract AT(30-1)-844. Acquisition of the mass spectrometer was assisted by grant GA-349 from the National Science Foundation. J.M.H. is a NASA graduate trainee.

REFERENCES

ANDERS E. and STEVENS C. M. (1960) Search for extinct lead 205 in meteorites. *Earth Planet. Sci. Lett.* **7**, 3043–3047.

CAMERON A. E., SMITH D. H. and WALKER R. L. (1969) Mass spectrometry of nanogram-size samples of lead. *Anal. Chem.* **41**, 525–526.

EHMANN W. D. (1968) Prevalence of the elements. In *Encyclopedia of the Chemical Elements* (editor C. A. Hampel), pp. 567–576. Reinhold.

FLANAGAN F. J. (1969) U.S. Geological Survey standards—II. First compilation of data for the new U.S.G.S. rocks. *Geochim. Cosmochim. Acta* **33**, 81–120.

KANASEWICH E. R. (1968) The interpretation of lead isotopes and their geological significance. In *Radiometric Dating for Geologists*, (editors E. I. Hamilton and R. M. Farquhar), pp. 147–223. Interscience.

KOHMAN T. P. (1956) Extinct natural radioactivity: Possibilities and potentialities. *Ann. N.Y. Acad. Sci.* **62,** Art. 21, 503–542.

KOHMAN T. P. (1961) Chronology of nucleosynthesis and extinct natural radioactivity. *J. Chem. Educ.* **38,** 73–82.

LAUL J. C., PELLY I. and LIPSCHUTZ M. E. (1970) Submitted to *Geochim. Cosmochim. Acta*.

MURTHY V. R. and PATTERSON C. C. (1962) Primary isochron of zero age for meteorites and the earth. *J. Geophys. Res.* **67,** 1161–1167.

OSTIC R. G., EL-BADRY H. M. and KOHMAN T. P. (1969) Isotopic composition of meteoritic thallium. *Earth Planet. Sci. Lett.* **7,** 72–76.

TATSUMOTO M. (1968) Private communication to F. S. FLANAGAN (1969).

WAHLER W. (1968) Pulse-Polarographische Bestimmung der Spurenelemente Zn, Cd, In, Tl, Pb, und Bi in 37 geochemischen Referenzproben nach Voranreicherung durch selektive Verdampfung. *Neues Jahrb. Mineral. Abhand.* **108,** 36–51.

Proceedings of the Apollo 11 Lunar Science Conference, Vol. 2, pp. 1351 to 1355.

Neutron activation analysis of rhenium and osmium in Apollo 11 lunar material

J. F. LOVERING* and D. BUTTERFIELD†
Department of Geophysics and Geochemistry, Australian National University, Canberra, A.C.T. 2600, Australia

(*Received* 3 *February* 1970; *accepted in revised form* 23 *February* 1970)

Abstract—Neutron activation analyses for Re and Os in lunar volcanic rocks indicate abundances of 0·0016–1·7 ppb Re and 0·02–0·9 ppb Os which are generally within the range of concentrations observed in terrestrial rocks. These levels are strongly depleted relative to abundance levels in chondritic meteorites and imply some metal phase separation and segregation within the moon if the moon has an overall chondritic composition.

Re and particularly Os levels observed in the secondary lunar rocks (i.e. the breccias) and the surficial fine material indicate possible chondritic contamination levels of up to 2 wt.% or iron meteorite contamination levels <1 per cent.

INTRODUCTION

Re and Os are important elements since their strongly siderophilic properties make them very sensitive indicators of metal–silicate fractionation processes. For example MORGAN and LOVERING (1967b) have shown that terrestrial igneous and deep-seated metamorphic rocks are generally strongly depleted in both Re and Os most likely as a result of overall mantle depletion following the segregation of the earth's metallic core. A similar depletion of Re and Os in the primary lunar rocks would also imply that metal–silicate fractionation processes may well have operated during the evolution of the moon.

In the event that the primary lunar surface rocks are depleted in Re and Os, then relatively high Re and Os abundances in the secondary lunar rocks (i.e. the breccias) and the fine surficial material would imply contamination by either chondritic and/or iron meteorite material, since chondrites and particularly iron meteorites show relatively high Re and Os contents.

EXPERIMENTAL PROCEDURES

Fragments of about 2 g each of two fine-grained volcanic rocks (A) and one medium-grained volcanic rock (B) and two breccias (C) were available for analysis. The fragments were gently broken in a specially cleaned hardened steel mortar and then crushed to −200 mesh (silk sieve) in a new agate mortar. The 5·3 g sample of lunar fines (D) studied had been previously sieved through a 1 mm stainless steel sieve at the LRL. In our laboratory it was homogenised by rolling on weighing paper and then split into two 2·65 g portions one of which was used for the neutron activation analytical study. For analysis duplicate samples of about 0·1 g of each rock and fines powder were split off and sealed in cleaned silica ampoules. Comparator standards were prepared by spiking *Specpure* silicon dioxide with known amounts of rhenium and osmium and these were also sealed in silica ampoules.

---

* Present Address: School of Geology, University of Melbourne, Parkville, Victoria 3052, Australia.

† Present Address: Department of Chemistry, Australian National University, Canberra, A.C.T. 2600.

Batches of six samples and two standards were irradiated for one week in the HIFAR reactor at the Australian Atomic Energy Commission Research Establishment in a predominantly thermal flux of 9×10^{12} neutrons/cm² per sec.

The Re and Os analytical method developed by MORGAN (1965a) was used in a modified form. In particular the primary separation method was changed to solvent extraction in alkaline solution because of the presence of large amounts of titanium in the lunar rocks. Other changes were concerned with the addition of ruthenium carrier and subsequent ruthenium separation to prevent contamination of the Os by fission and normal ruthenium, and the total dissolution of the silica vial container along with the sample it contained to prevent loss of Os on the vial walls. Separations were checked by γ-ray spectrometry and determinations made by β-counting.

Table 1. Rhenium and osmium abundances in Apollo 11 lunar material.

Rock Type	Rock No.	Rhenium (ppb)	Osmium (ppb)	Os/Re
Fine-grained volcanic (A)	10057-21	1·35 ± 0·04	0·026 ± 0·001	0·019
		1·73 ± 0·04	0·019 ± 0·0005	0·011
	10069-20	1·18 ± 0·03	0·897 ± 0·03	0·76
		—	0·750 ± 0·02	—
Medium-grained volcanic (B)	10047-20	0·0016 ± 0·0006	0·352 ± 0·012	220
		0·030 ± 0·01	0·166 ± 0·001	5·6
Breccia (C)	10019-12	4·26 ± 0·16	1·21 ± 0·031	0·28
		5·20 ± 0·12	1·22 ± 0·034	0·23
	10046-20	0·216 ± 0·006	2·67 ± 0·055	12·4
		0·218 ± 0·006	2·22 ± 0·048	10·2
Fines (D)	10084-53	8·68 ± 0·13	3·81 ± 0·06	0·44
		10·62 ± 0·19	13·8 ± 0·3	1·31
Silica ampoule	—	<0·0002	0·027 ± 0·002	—
Average 28 iron meteorites (HIRT *et al.*, 1963)	—	594	7060	11·9
Average 32 chondrites (MORGAN and LOVERING, 1967a)	—	56·9	657	11·5
Terrestrial igneous and deep-seated metamorphic rocks (MORGAN and LOVERING, 1967b)	—	0·01–2·4	≤0·02–5·9	≤0·1–130

DATA AND DISCUSSION

The results of the study are given in Table 1 with 1σ errors based on counting statistics. Since the analytical method called for the dissolution of the silica ampoules along with the irradiated samples, an analysis was made of one of these silica ampoules and the results (Table 1) indicate no measurable Re content and a low Os content. All Os abundances in Table 1 have been corrected for the small Os blank from the silica ampoule.

Despite the extremely low levels (0·001–10 ppb) analytical agreement between duplicates is usually good although there are some major discrepancies in duplicate Re abundances which suggest sampling problems.

Lunar volcanic rock

The two fine-grained (A) volcanic rocks analysed have rather similar Re abundances but the Os abundances differ by about a factor of 40 so that Os/Re ratios range from ~0·02–~0·8. For the one medium-grained (B) volcanic rock available for analysis the Re duplicates varied by a factor of about 30 although Os duplicates were in reasonable agreement. As a result Os/Re ratios on the duplicates were very different (5·6 and 220) suggesting that Os and Re are present in separate phases and also that the sample is inhomogeneous with regard to the Re-bearing phases. However both Re and Os abundance levels and Os/Re ratios for the lunar volcanic rocks are within the observed variation for terrestrial rocks (Table 1).

Since it has been suggested by DUKE and SILVER (1967) that the basaltic achondrites may represent lunar surface rock it is of interest to consider the Re and Os abundances in these meteorites. The only data available at present (Table 2) are analyses of

Table 2. Rhenium and osmium abundances in achondrites (after MORGAN, 1965b)

Achondrite Type	Achondrite Name	Rhenium (ppb)	Osmium (ppb)
Eucrite	Moore County	0·029 ± 0·004	0·20 ± 0·21
		0·091 ± 0·010	0·59 ± 0·24
Howardite	Binda	0·076 ± 0·035	≤0·17
		≤0·072	≤0·17
Nakhlite	Nakhla	≤0·062	0·87 ± 0·14
		0·093 ± 0·006	0·51 ± 0·10
Angrite	Angra dos Reis	0·060 ± 0·010	0·99 ± 0·17
		0·078 ± 0·007	0·56 ± 0·15
Diogenites	Ellemeet	0·056 ± 0·008	0·47 ± 0·12
(Hypersthene		0·133 ± 0·013	0·74 ± 0·11
achondrites)	Johnstown	1·28 ± 0·02	16·8 ± 0·3
		0·26 ± 0·02	—
Aubrite	Bishopville	0·223 ± 0·008	4·77 ± 1·06
(Enstatite		0·281 ± 0·010	5·10 ± 0·71
achondrite)			

relatively poor precision determined by MORGAN (1965b) using an end-window GM tube counter with a relatively high background. However the data are sufficient to conclude that Re and Os levels in basaltic achondrites (i.e. eucrites and howardites), nakhlites, angrites, diogenites and aubrites are all sufficiently low to be generally consistent with levels in lunar volcanic rocks.

Bearing in mind the limited data available at present it still may be concluded that the lunar volcanic rocks show considerable depletion in Re and Os with regard to average chondritic material (see Table 1). If an overall chondritic composition is to be proposed for the moon by analogy with the earth then evolution of lunar volcanic rocks from chondritic material must have involved at some stage separation and segregation of metal phase.

On the other hand REED *et al.* (1970) have reported values of 17 and 755 ppb Os in two type A lunar volcanic rocks. These values are very much higher than our results for these rocks (Table 1) and, if substantiated, would imply either very high meteorite contamination levels in the Tranquillity lava or unfractionated chondrite-like Os abundances in lunar volcanic rocks. However REED *et al.* (1970) have themselves described their Os values as "preliminary data" and our own data would imply hat their values may well be too high.

Breccias and fine surficial material

If it is assumed that the breccias and the fine surficial material are both derived from the break-up of the volcanic rocks at the Apollo 11 site, then calculations can be made of the possible contamination of the lunar surface by meteorite infall. Chondrites and iron meteorites contain relatively high Re and Os abundances (100–10,000 ppb) and high Os/Re ratios (~11) so that the Re and particularly the Os content of the secondary lunar rocks should be sensitive indicators of simple meteorite contamination.

However such calculations are complicated by a least two factors. First the very large range of Re and Os abundances measured in the three lunar volcanic rocks studied here makes it extremely difficult to assume a realistic figure for the base Re and Os abundance levels in the secondary breccias and fine surficial material before any supposed meteorite contamination. Secondly there is the possibility that selective removal of one element with respect to the other during the process of contamination would radically alter the added/Os/Re ratio during incorporation of meteoritic material in the lunar secondary rocks and fines.

The breccia 10019-12 shows an Os abundance (Table 1) which is only marginally higher than the highest levels observed so far by us in three lunar volcanic rocks. Given the observed wide dispersion of Os abundances in the three lunar volcanic rocks it is meaningless to attempt a rigorous calculation of the possible Os contamination from meteoritic infall in this case. On the other hand the Re content is 2–3 times higher than the highest value observed in the lunar volcanic rocks. Simple meteorite infall would lead to about 10 times more Os than Re being added by way of contamination but the Os data alone imply only a limited contribution of meteoritic infall. If the source of the high Re in 10019-12 is due to meteorite infall then the meteorite Os/Re ratio must have been drastically reduced during the contamination process. Selective volatilisation of Os relative to Re from a vaporising impacting meteorite might explain the anomaly under certain conditions but a more rigorous discussion is not possible at the present time.

On the other hand the breccia 10046-20 (Table 1) shows a Re content well within the range observed in the lunar volcanic rocks but the Os content is about three times the highest value measured by us in a lunar volcanic rock. Assuming contamination by simple incorporation of average chondrite material (Table 1), the observed Os levels are consistent with <1 wt.% of such contamination. Since Os abundances in iron meteorites are an order of magnitude greater than chrondrite abundance (Table 1), then any contamination by iron meteorites must be at a much lower level than 1 per cent.

The lunar fines (10084-53) show very variable Re and Os abundances between duplicates indicating severe sampling problems. The highest Os abundance measured (13·8 ppb) is consistent with a 2 per cent contamination by chondritic material while the lower Os abundance of 3·8 ppb indicates a chondrite contamination level of ~1 per cent. If the contaminating material is iron meteorite material then even at the highest Os abundance the level of such contamination would be <1 per cent.

These conclusions are consistent with observations by LOVERING and WARE (1970) that although small particles of meteoritic metal phase are present in both breccias and fines, the amount of this contamination is small and very much less than 1 per cent. It would appear that at the Apollo 11 site the breccias and surficial fines indicate levels of chondritic contamination up to about 2 per cent or iron meteorite contamination less than 1 per cent. On the basis of other data KEAYS *et al.* (1970) have suggested that Type C and D lunar material could have been made from Type A, B rocks by adding 1–2 per cent carbonaceous chondrite-like material.

Acknowledgments—We wish to thank Professor I. Ross for his co-operation in the course of this work and the Australian Atomic Energy Commission for special provision of neutron irradiation facilities. W. BERRY and M. COWAN prepared the crushed samples and the work was supported in part by grants from the Australian Institute of Science and Engineering, Australian Research Grants Committee and the Department of Education and Science.

REFERENCES

DUKE M. B. and SILVER L. T. (1967) Petrology of eucrites, howardites and mesosiderites. *Geochim. Cosmochim. Acta* **31,** 1637–1665.

HIRT G., HERR W. and HOFFMEISTER W. (1963) Age determination by the rhenium-osmium method. In *Radioactive Dating*, pp. 35–43. IAEA, Vienna.

KEAYS R. R., GANAPATHY R., LAUL J. C., ANDERS E., HERZOG G. F. and JEFFERY P. M. (1970) Trace elements and radioactivity in lunar rocks: Implications for meteorite infall, solar–wind flux, and formation conditions of moon. *Science* **167,** 490–493.

LOVERING J. F. and WARE N. G. (1970) Electron probe microanalysis of phases in Apollo 11 lunar samples. *Geochim. Cosmochim. Acta*, Supplement I.

MORGAN J. W. (1965a) The simultaneous determination of rhenium and osmium in rocks by neutron activation analysis. *Anal. Chim. Acta* **32,** 8–16.

MORGAN J. W. (1965b) The application of activation analysis to some geochemical problems. Unpublished Ph.D. Thesis, Australian National University (two volumes).

MORGAN J. W. and LOVERING J. F. (1967a) Rhenium and Osmium abundances in chondritic meteorites. *Geochim. Cosmochim Acta* **31,** 1893–1909.

MORGAN J. W. and LOVERING J. F. (1967b) Rhenium and osmium abundances in some igneous and metamorphic rocks. *Earth Planet. Sci. Lett.* **3,** 219–224.

REED G. W., JR., JOVANOVIC S. and FUCHS L. H. (1970) Trace elements and accessory minerals in lunar samples. *Science* **167,** 501–503.

Proceedings of the Apollo 11 Lunar Science Conference, Vol. 2, pp. 1357 to 1367.

Solar wind gases, cosmic-ray spallation products and the irradiation history of Apollo 11 samples

K. MARTI, G. W. LUGMAIR and H. C. UREY
Chemistry Department, University of California at San Diego,
La Jolla, California 92037

(*Received* 4 *February* 1970; *accepted in revised form* 25 *February* 1970)

Abstract—The isotopic abundances of the rare gases in the fines are found to be similar to those previously reported for gas-rich meteorites. Relative to the heavy gases, Ne and He are depleted by factors of 2·5 and 10 respectively. Kr–Kr^{81} ages of rocks 10017, 10047, 10057, 10071 and of the soil cover a range of from 47 to 509 m.y. K–Ar ages of type A rocks lie between 2·3 and 2·9 billion years, that of type B rock 10047 is 3·5 b.y. It is suggested that varying relative production rates of the Kr and Xe isotopes in the rocks are due to differences in chemical composition as well as different irradiation conditions.

INTRODUCTION

THIS is a revised and expanded version of the initial report on Apollo 11 samples (MARTI *et al.*, 1970). The stable isotopes of the noble gases He to Xe and $2·1 \times 10^5$ yr Kr^{81} have been measured in the soil (fines $\leq$1 mm) and in four rocks (10017, 10047, 10057, 10071) from the Apollo 11 mission. The results show that we are dealing with gases of different origin. The soil contains large amounts of trapped solar wind gases which almost completely mask other components. In the crystalline rocks, on the other hand, cosmic-ray induced spallation and radiogenic gases are predominant. Solar type noble gases have previously been studied in gas-rich meteorites by several authors (see PEPIN and SIGNER, 1965 for references; also EBERHARDT *et al.*, 1965; MARTI, 1969). Furthermore, solar wind gases have been collected on aluminum foil on the lunar surface (BÜHLER *et al.*, 1969). Comparison of all these gases gives information on the elemental and isotopic abundances in the Sun and in the solar wind as well as on nuclear transformations or fractionation mechanisms. Cosmic-ray exposure ages as obtained by the Kr–Kr method (MARTI, 1967), coupled with the relative mass yields of cosmic-ray produced heavy noble gases and the observed neutron effects, give information on the irradiation history of the Apollo 11 material. The locations of the samples within the rocks in general are not known. From rock 10017, however, we have an outside and an interior piece and also documented small chips from the samples analyzed by SHEDLOVSKY *et al.* (1970). The latter samples give information on the production rates at different depths below the top surface. These results will be reported elsewhere together with results from He, Ne and Ar in the rock samples.

EXPERIMENTAL TECHNIQUE

Our soil sample 10084-29 consists of fines <1 mm, and the rock samples used in this study were crushed and also passed through a 1 mm sieve. Aliquots were then used for the analyses of noble gases, of potassium and gadolinium.

The noble gases were extracted at a crucible temperature of about 1700°C for 20 min after the samples were stored in vacuum at 80°C for three days to remove adsorbed terrestrial gases. Average extraction blanks for He^4, Ne^{20}, Ar^{40}, Kr^{84} and Xe^{132} were 11×10^{-8}, $0·15 \times 10^{-8}$, $0·9 \times 10^{-8}$, $0·5 \times 10^{-12}$ and $0·3 \times 10^{-12}$ cc STP respectively. The gases were cleaned by means of four Ti getters and analyzed in four fractions, Xe, Kr, Ar and He + Ne, previously separated from each other by adsorption on charcoal at temperatures of −75°, −120° and −196°C respectively, which were kept constant within 0·1°C. Standards of atmospheric gases were used to determine Kr and Xe sensitivities and the mass discrimination of the mass spectrometer, except for He where a known mixture of He^3 and He^4 was used. Isotopic dilution technique was used to determine the contents of He, Ne and Ar. The detection limit of the mass spectrometer was at 5×10^{-15} cm^3 STP per Kr isotope, and the resolving power allowed a satisfactory resolution of a low hydrocarbon background in the Kr mass region. Extraction system and mass spectrometer were operated at about 50°C to minimize Xe adsorption losses.

K and Ar^{40} were determined in aliquots. Potassium analyses were done on an atomic absorption spectrometer. The samples which ranged in size from 10 to 50 mg were dissolved in a Teflon beaker using HF, HCl and H_3BO_3, and were diluted to volume. Ultrapure KCl was used as the standard. Standards were also made which contained in proper percentages all the major elements of the rocks studied, but no potassium. All solutions were in a matrix of 1:1 dilution of constant boiling (20·24%) HCl. The results were obtained by using two independent methods: (a) A comparison of the sample with standards of both pure KCl and of the made-up rock; (b) A standard addition technique where 3 standards were stepwise added to the sample. The reproducibility was carefully checked on test samples and found to be better than 2 per cent. The accuracy of the K contents is 5 per cent. The chemical separation of Gd is similar to that reported by EUGSTER *et al.* (1969), and will be discussed elsewhere. A chemical yield of about 70 per cent has been obtained. Gd was loaded as the nitrate on a single Re-filament and analyzed in the mass spectrometer as GdO^+. Two detection modes, ion counting and current integration, were used.

RESULTS AND DISCUSSION

The isotopic and relative elemental abundances of the solar wind gases in the soil are given in Table 1. The heavier rare gases Ar to Xe show abundance ratios similar

Table 1. Concentrations (in cm^3 STP/g), isotopic composition and relative elemental abundances in lunar soil 10084-29

	He^4	Ne^{20} $\times 10^{-2}$	Ar^{36} $\times 10^{-4}$	Kr^{84} $\times 10^{-8}$	Xe^{132} $\times 10^{-8}$	He^4/He^3	Ne^{20}/Ne^{22}	Ne^{22}/Ne^{21}	Ar^{36}/Ar^{38}	Ar^{40}/Ar^{36}
Lunar soil 10084-29	0·23 ± 0·02	0·22 ± 0·02	3·3 ±0·3	16 ±2	2·1 ± 0·2	2540 ± 100	12·80 ± 0·10	28·3 ± 0·2	5·25 ± 0·06	1·09 ± 0·02

	He^4/Ne^{20}	Ne^{20}/Ar^{36}	Ar^{36}/Kr^{84}	Kr^{84}/Xe^{132}
Lunar soil 10084-29	104	6·63	2040	7·8
Pesyanoe (MARTI, 1969)	388	16·4	2270	7·5
Atmospheric	0·318	0·522	48·5	27·8
Av. Chondritic	—	—	~200	1·3
Cosmic: ALLER (1961)	355	72	6760	30
SUESS and UREY (1956)	400	61	4300	27
CAMERON (1965)	1000	13	18,000	13

to those found in the Pesyanoe meteorite (MARTI, 1969), and agree within factors of two to three with estimates from cosmic abundance calculations. Relative to the heavy gases, Ne and He in the fines are depleted by factors of about 2·5 and 10 respectively as compared to Pesyanoe, a typical gas-rich meteorite. The concentration levels of the heavy gases in Pesyanoe are about 300 times lower. The agreement between the abundance pattern of Ar, Kr and Xe in lunar soil and in Pesyanoe do indicate that although the light noble gases seem to have reached saturation levels, this is not true for Kr and Xe. If we accept trapping and gas retention efficiencies close to unity, a directional He^4 solar wind flux of $6 \cdot 3 \times 10^6$ atoms/cm^2 sec as reported by BÜHLER *et al.* (1969) and the He/Xe abundance ratio of Pesyanoe, we calculate an average irradiation time of about 10^6 yr to obtain in a dust layer of 1 g/cm^2 the observed Xe concentration levels. The isotopic composition of the solar wind gases (Table 1) is well within the range of values previously reported for some gas-rich meteorites, except for Ar^{40}. All gases He to Xe contain a small but clear-cut cosmic-ray spallation component and therefore do not represent the pure solar wind composition. The measured Ne^{20}/Ne^{22} ratio of 12·80 is little affected by the spallation component, but in the bulk sample the measured ratios $He^4/He^3 = 2540$, $Ne^{22}/Ne^{21} = 28 \cdot 3$ and $Ar^{36}/Ar^{38} = 5 \cdot 25$ are all slightly too low. The concentration of $Ar^{40} = 3 \cdot 6 \times 10^{-4}$ cm^3 STP/g in the dust is unusually high. From nuclear abundance systematics one would expect a much lower relative abundance of the Ar^{40} isotope in the Sun. The measured low Ar^{40}/Ar^{36} ratio of 0·35 in the meteorite Novo Urei reported by MARTI (1967b) also indicates a much smaller value for primordial argon. Less than 30 per cent of the Ar^{40} can be accounted for on the basis of the measured K and a K–Ar age of $4 \cdot 5 \times 10^9$ yr. A stepwise heating experiment using temperature increments of about 200°C has revealed a variable release of the Ar isotopes (Fig. 2). The Ar^{38}/Ar^{36} ratio sharply increases above 1200°C when a spallation component becomes apparent. The lowest ratio $Ar^{40}/Ar^{36} = 0 \cdot 78$ was found in the 800°C fraction, but gives only an upper limit for the occurrence of Ar^{40} in the solar wind. The true value is probably much lower. The excess Ar^{40} over this upper limit combined with the measured potassium yields a K–Ar age of $4 \cdot 65 \times 10^9$ yr (Table 4). These results suggest the presence of a large radiogenic Ar component which is not related to the K in the dust. This is supported by the fact that the release of excess Ar^{40} shows two maxima at $\leq$600° and $\geq$1200°C. HEYMANN *et al.* (1970) and PEPIN *et al.* (1970) have arrived at very similar conclusions. An implantation of ionized Ar^{40} from a temporary lunar atmosphere, together with the solar wind, is a possible origin (HEYMANN *et al.*, 1970; GOLD, 1970).

Inspection of the xenon isotopic composition in the soil (Table 2) reveals large excesses of the light isotopes up to Xe^{131} which can be attributed to spallation, but the neutron-rich isotopes Xe^{134} and Xe^{136} are deficient as compared to both trapped chondritic and atmospheric xenon. The relative abundances of these heavy isotopes are lower than those reported for bulk analyses of the Pesyanoe meteorite, but are rather similar to those reported for the 1000°C fraction of Pesyanoe (MARTI, 1969). From the stepwise heating experiment (Table 2), more information on the composition of solar wind Xe is obtained. The relative abundances of the light isotopes in the low temperature fractions are only little higher than those in trapped chondritic xenon. Isotopic ratios of Xe in the different temperature fractions exhibit linear correlations (Fig. 1) demonstrating a two component mixture of spallation with solar wind xenon. The data points of the neutron poor isotopes of trapped chondritic xenon always fall on or close to the correlation lines. The composition of the Xe isotopes Xe^{124} to Xe^{130} in the solar wind therefore has to be very similar to that found in trapped

Table 2. Concentrations and isotopic composition of xenon in the lunar soil (10084-29), normalized to Xe^{132} = 100%

	Xe^{132} ($\times 10^{-8}$ cm^3 STP/g)	124	126	128	129	130	131	132	134	136
Bulk samples average	2·05 ±0·25	0·632 ±0.012	0·725 ±0·018	8·75 ±0·10	104·5 ± 0·5	16·66 ± 0·08	83·5 ± 0·4	100	36·93 ± 0·22	29·90 ± 0·20
Temperature fractions:										
600°	0·038	0·484 ±0·018	0·474 ±0·015	8·29 ±0·10	104·9 ± 0·6	16·33 ± 0·12	82·4 ± 0·4	100	36·90 ± 0·30	29·95 ± 0·30
800°	0·454	0·495 ±0·015	0·476 ±0·012	8·47 ±0·08	105·4 ± 0·5	16·53 ± 0·10	82·6 ± 0·3	100	36·81 ± 0·20	29·85 ± 0·20
1000°	0·647	0·545 ±0·012	0·595 ±0·012	8·54 ±0·08	104·4 ± 0·4	16·55 ± 0·10	82·7 ± 0·3	100	37·02 ± 0·22	30·19 ± 0·20
1200°	1·168	0·736 ±0·012	0·924 ±0·013	9·00 ±0·09	104·5 ± 0·4	16·74 ± 0·09	84·0 ± 0·3	100	37·11 ± 0·20	30·22 ± 0·20
1400°	0·017	0·910 ±0·018	1·245 ±0·018	9·64 ±0·11	107·5 ± 0·6	17·27 ± 0·13	88·6 ± 0·7	100	36·44 ± 0·30	29·39 ± 0·30
Total	2·32 ±0·30	0·634 ±0·015	0·741 ±0·014	8·76 ±0·09	104·7 ± 0·5	16·64 ± 0·10	83·4 ± 0·4	100	37·02 ± 0·25	30·12 ± 0·25
Pesyanoe (MARTI, 1969)	0·0092	0·488	0·504	8·28	103·3	16·40	82·0	100	37·3	30·7
Atmosphere (NIER, 1950)		0·357	0·335	7·14	98·3	15·17	78·8	100	38·8	33·0

chondritic but clearly different from terrestrial atmospheric Xe. This has also been concluded by EBERHARDT *et al.* (1970), REYNOLDS *et al.* (1970), PEPIN *et al.* (1970), and from a study of solar type xenon in the Pesyanoe meterorite by MARTI (1969). The relative contents of the heavy isotopes are similar in all temperature fractions, although after subtraction of the spallation component slightly higher values are observed in the 1000° and 1200° fractions. Fission Xe from spontaneous fission of U^{238} can account for less than 10 per cent of the effect. It is noted that both Xe^{134} and Xe^{136} in all temperature fractions indicate slightly larger relative abundances than in the 1000°C—fraction of Pesyanoe. Beside a possible mass fractionating process during implantation or release of solar xenon, alternate possible explanations are: (a) Implantation of Xe ions from a temporary lunar atmosphere of chondritic Xe composition; (b) infalling Xe-rich meteoritic matter such as suggested by KEAYS *et al.* (1970); (c) Fission excesses of transuranium elements. From the results given in Tables 2 and 5, there is, however, no evidence in solar wind Kr and Xe for a large scale neutron irradiation of solar matter such as proposed by CAMERON (1962).

As mentioned above, the gases in the rocks are predominantly spallogenic and radiogenic. We have determined cosmic-ray exposure ages for four rocks (Table 3), using the Kr–Kr dating method, which is based on the ratio of a stable Kr isotope and the radioisotope Kr^{81}. This dating method is found to be both accurate and best suited for lunar material for the following reasons: (a) the age determination is based on the isotopic composition of one element only, and, therefore, eliminates

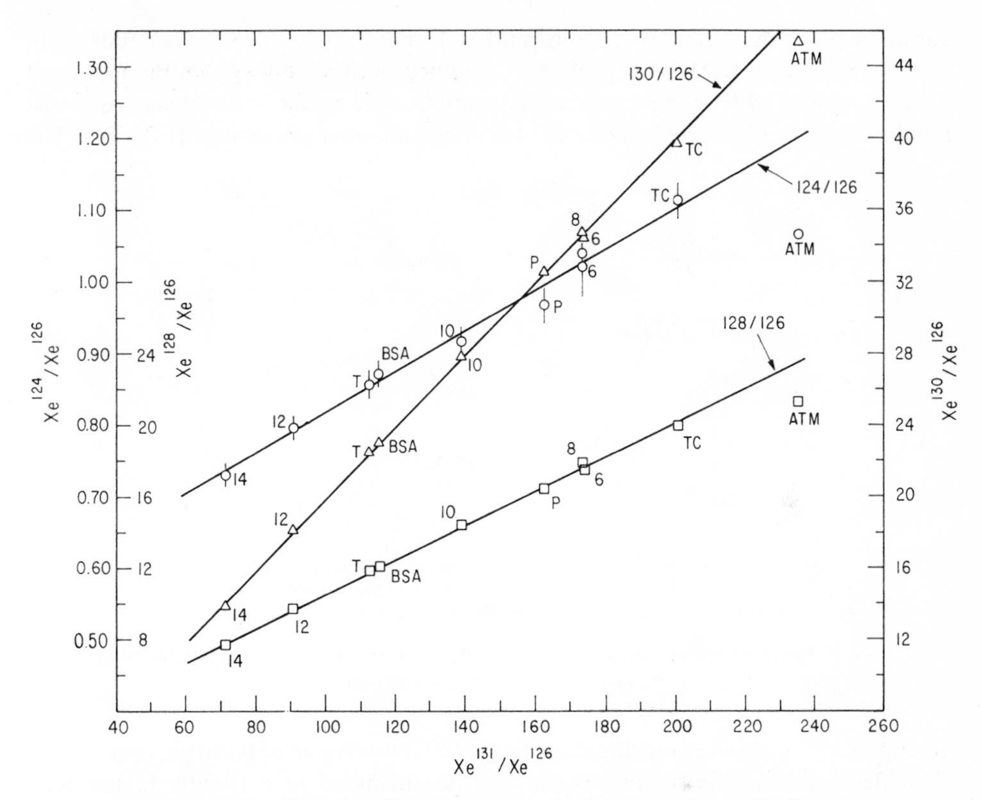

Fig. 1. Correlation plots of the light Xe isotopes obtained in a release of Xe from lunar fines (10084-29) at stepwise increased temperatures. The used symbols are: BSA = Bulk Samples, Average; T = Total of temperature run; TC = Trapped chondritic xenon (MARTI, 1967c; EUGSTER *et al.*, 1967); P = Pesyanoe (MARTI, 1969); ATM = Terrestrial Atmospheric. The temperature points are given in hundreds of °C. Errors exceeding the symbol size are indicated by bars.

errors in concentration measurements; (b) Sr, Y and Zr, the main target elements for the production of Kr, are strongly enriched in lunar material, making it possible to measure Kr^{81} in samples as small as 0·1 g; (c) because of both unusual chemical composition and unusual irradiation conditions, the production rates of those stable isotopes generally used for age determinations in meteorites are not well known. Kr–Kr ages are obtained from (MARTI, 1967)

$$T_r = \frac{Kr^{83}\,\text{spall}}{Kr^{81}}\,\frac{P_{81}}{P_{83}}\,\frac{1}{\lambda}$$

where $\lambda = 3{\cdot}3 \times 10^{-6}$/yr and $P_{81} = (0{\cdot}95 \pm 0{\cdot}05)\,\dfrac{P_{80} + P_{82}}{2}$.

The relative production rates of the Kr isotopes are similar for all the investigated rocks, which is indicated by rather constant P_{81}/P_{83} ratios (Table 3). The largest variation in the Kr^{78}/Kr^{83} spallation ratio is 15 per cent (between rocks 10047 and 10057). Spallation Kr^{83} in the rocks was obtained by subtracting from the measured values a solar wind Kr component corresponding to the excess of Kr^{86} over the lowest relative Kr^{86} yield of rock-specimen 10017-55, an interior sample (Fig. 3). The

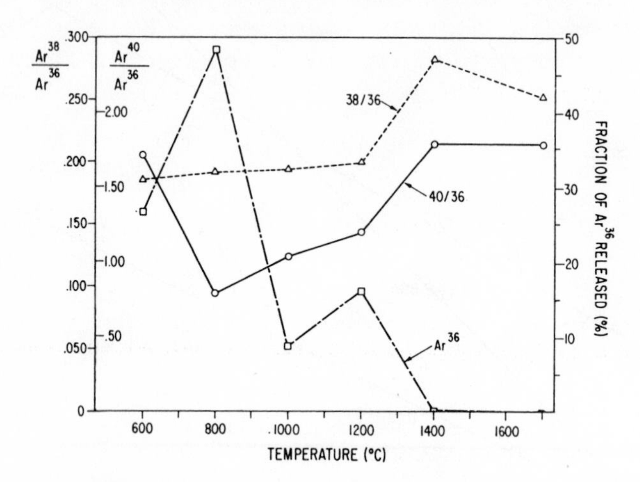

Fig. 2. Isotopic composition and fractional release of Ar from lunar fines (10084-29) at stepwise increased temperatures.

Kr^{83}spall/Kr^{81} ratio of an outside chip of rock 10017 was found to be 10 per cent smaller than that of an inside sample which may be attributed to a slightly higher Kr^{81} content. This has been confirmed for another outside sample and will be discussed in detail elsewhere.

Table 4 summarizes the K–Ar ages obtained for four rocks and a sample of soil. The K contents of the fine grained igneous rocks (type A) are similar, with an average of 0·261 per cent. Rock 10047, the only analyzed coarse grained igneous rock (type B), has a three times lower K content, but also the highest K–Ar age. K in type B

Table 3. Kr–Kr^{81} ages of Apollo 11 rocks and fines

Sample	Kr^{83}/Kr^{81}	Kr^{83}spall/Kr^{81}	P^{81}/P^{83}	Kr^{81}–Kr^{83} age m.y.	$Kr^{81} \times 10^{-12}$ cm³ STP/g
Rocks:					
10017-55	2840 ± 160	2840	0·591	509 ± 29	0·35
10017-56	2597 ± 125	2520	0·588	449 ± 22	0·37
10047-40	495 ± 22	470	0·604	86 ± 4	0·32
10057-58	424 ± 18	266	0·580	47 ± 2	0·37
10071-32	2058 ± 120	2028	0·605	372 ± 22	0·36
Fines:					
10084-29	62000 ± 9300	1800	0·60	330 ± 90	0·40

The errors assigned to the Kr^{81}–Kr^{83} age do not include the uncertainties in the Kr^{81} half life and in the P^{81}/P^{83} production ratios.

rocks seems to be generally lower by factors of 3 to 5 as compared to type A rocks (GAST and HUBBARD, 1970). Furthermore, we note that the K–Ar ages of other type B rocks (10003, 10044, 10062) as reported by TURNER (1970) and ALBEE *et al.* (1970) are systematically higher, both the "total Ar" age and also the age obtained from the high temperature Ar fractions. On the other hand, according to TURNER (1970), rock 10017 has lost more than 48 per cent of its radiogenic Ar^{40}, yielding an apparent age of only 2·30 b.y., in good agreement with our results. There is no apparent correlation between K–Ar ages and cosmic-ray exposure ages.

Table 4. K–Ar ages of Apollo 11 rocks and fines

Sample	Rock type	$Ar^{40} \times 10^{-5}$ cm³ STP/g	K (wt. %)	Age* (b.y.)
Rocks:				
10017-55	A	5·47 ± 0·33	0·274 ± 0·014	2·38 ± 0·11
10017-56	A	4·85 ± 0·30	0·265 ± 0·013	2·26 ± 0·10
10047-40	B	3·70 ± 0·22	0·088 ± 0·004	3·46 ± 0·12
10057-58	A	4·72 ± 0·30	0·259 ± 0·013	2·26 ± 0·11
10071-32	A	7·26 ± 0·50	0·246 ± 0·012	2·93 ± 0·12
Fines:				
10084-29		35·9 ± 3·2	0·117 ± 0·006	6·84 ± 0·18
Excess Ar^{40} only†		≥10·2	0·117 ± 0·006	≥4·65

* $\frac{\lambda_K}{\lambda_{\beta^-}} = 0{\cdot}124$, $\lambda = 5{\cdot}46 \times 10^{-10}$ and $K^{40} = 0{\cdot}0118\%$ were used; errors were added quadratically.

† Excess Ar^{40} over $\frac{Ar^{40}}{Ar^{36}} = 0{\cdot}78$, the lowest ratio observed in the 800°C temperature fraction of the soil sample.

The isotopic abundances of Kr and Xe in the different rocks are given in Tables 5 and 6. Spallation Kr and Xe are clearly the dominant components. In rock specimen 10017-55 no evidence for solar wind gases was found. The isotopic composition of Ne is $Ne^{20}:Ne^{21}:Ne^{22} = 0{\cdot}92:1{\cdot}00:1{\cdot}11$, and $Ar^{38}/Ar^{36} = 1{\cdot}55$, which is compatible with a pure spallation component. Therefore, the obtained mass spectra may represent pure spallation Kr and Xe with a small fission component from U superimposed thereon. Figures 3 and 4 show the mass spectra of Kr and Xe in rock 10017-55 after correcting for spontaneous fission products from U^{238}. A U content of 0·854 ppm (TATSUMOTO and ROSHOLT, 1970), WETHERILL's (1953) data for spontaneous fission of U, a half life for spontaneous fission of $9{\cdot}0 \times 10^{15}$ [average of values reported by SEGRÈ (1952) and FLEISCHER and PRICE (1964)] and a rock age of $3{\cdot}65 \times 10^9$ yr (ALBEE *et al.*, 1970) were used. This fission correction amounts to 9·4 per cent of the measured Xe^{136} and 1·2 per cent of the measured Kr^{86}. The remaining fractions of these two isotopes are considered to represent a mixture of trapped and spallation gases of unknown proportions. The two limiting cases, which assume the remaining Kr^{86} and Xe^{132} to be either pure spallation or pure trapped gases, are shown in Figs. 3 and 4.

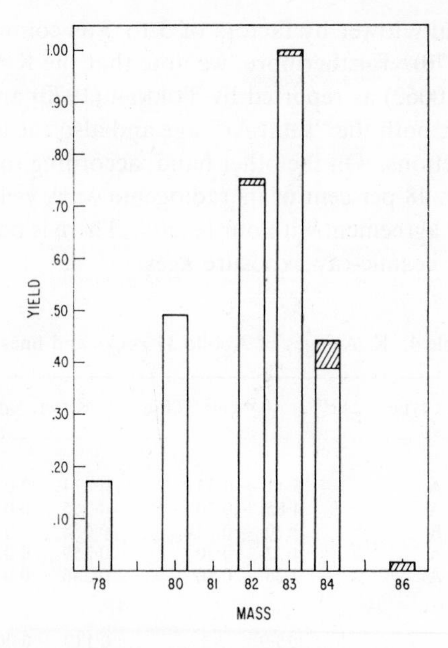

Fig. 3. Mass yields of Kr in rock 10017-55, normalized to Kr^{83} = 1·00. A small amount of fission krypton due to U^{238} (see text) has been subtracted. If all Kr^{86} is assumed to belong to the spallation component, then the upper limits of the bars represent the relative Kr spallation mass yields. If all the Kr^{86} is assumed to belong to a trapped component, the relative spallation yields are obtained by subtracting this trapped component (hatched area).

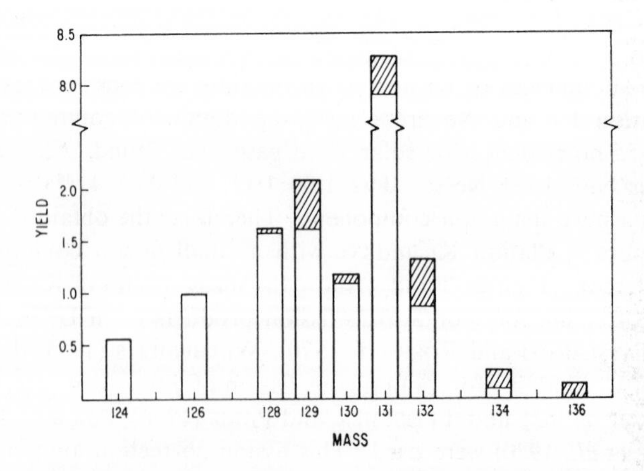

Fig. 4. Mass yields of Xe in rock 10017-55 normalized to Xe^{126} = 1·00. Fission xenon due to U^{238} has been subtracted (see text). If all Xe^{136} is assumed to belong to the spallation component, then the upper limits of the bars do represent the relative Xe spallation mass yields. If all the Xe^{136} is assumed to belong to a trapped component, the relative spallation yields are obtained by subtracting this trapped component (hatched area).

Table 5. Concentrations and isotopic composition of Kr, normalized to $Kr^{83} = 100$, in Apollo 11 rocks, and fines 10084-29

Rock	Type	Kr^{83} ($\times 10^{-12}$ cm^3 STP/g)	78	80	81	82	84	86	$\left(\frac{Kr^{78}}{Kr^{83}}\right)_{spall}$
10017-55	A	990 ± 130	17·22 ± 0·20	49·0 ± 0·3	0·0352 ±0·0020	75·3 ± 0·3	44·1 ± 0·3	1·68 ± 0·03	0·172
10017-56	A	950 ± 140	16·64 ± 0·21	47·8 ± 0·5	0·0385 ±0·0019	75·8 ± 0·4	58·2 ± 0·5	6·38 ± 0·10	0·171
10057-58	A	158 ± 20	11·67 ± 0·20	37·6 ± 0·3	0·236 ±0·010	83·8 ± 0·5	208·1 ± 2·0	57·1 ± 0·9	0·167
10071-32	A	730 ± 95	18·92 ± 0·22	50·5 ± 0·3	0·0486 ±0·0028	76·9 ± 0·3	49·4 ± 0·3	3·79 ± 0·05	0·191
10047-40	B	160 ± 25	18·42 ± 0·22	49·5 ± 0·4	0·202 ±0·009	77·4 ± 0·4	67·2 ± 0·5	9·27 ± 0·20	0·192
Fines 10084-29		33000 ±4300	3·32 ± 0·05	20·15 ± 0·25	—	99·4 ± 0·6	488 ± 2	149·3 ± 1·0	—

We discuss now the large scale variation in the relative Xe spallation mass yields which is paralleled by much smaller variations in the mass yields of Kr. The ratios $\left(\frac{Kr^{78}}{Kr^{83}}\right)_{spall}$ and $\left(\frac{Xe^{131}}{Xe^{126}}\right)_{spall}$ were obtained after correcting for solar wind and fission components by assuming Kr^{86} and Xe^{136} in the rocks to be mixtures of solar wind, U fission and spallation gases. The $\frac{Kr^{86}}{Kr^{83}}$ and $\frac{Xe^{136}}{Xe^{131}}$ spallation ratios are adopted from sample 10017-55, for which we assume that all Kr^{86} and Xe^{136} belong to the spallation component. U contents were taken from TATSUMOTO and ROSHOLT (1970) and SILVER (1970). It is first noted that the lowest Xe^{131}/Xe^{126} ratio is found in rock 10047, our only type B rock. Other results reported on type B rocks (ALBEE *et al.*, 1970; PEPIN *et al.*, 1970, REYNOLDS *et al.*, 1970) indicate similar values for this isotope ratio which also is comparable to that found in the Stannern achondrite (MARTI *et al.*, 1966). Type A rocks, on the other hand, show systematically higher but variable Xe^{131}/Xe^{126} ratios. Type A rocks are not only characterized by a very high Ba content but also by higher abundance ratios Ba/REE and Ba/Sr than type B rocks (GAST and HUBBARD, 1970). The Sr abundance is similar in both types, but the Sr/Zr

Table 6. Concentrations and isotopic composition of Xe, normalized to $Xe^{131} = 100$, in Apollo 11 rocks

Rock	Type	Xe^{131} $\times$ 10^{-12}cc STP/g	124	126	128	129	130	132	134	136	$\frac{Xe^{131}}{Xe^{126}}$ spall
10017-55	A	1100 ± 120	6·61 ±0·06	12·06 ±0·10	19·59 ±0·17	25·27 ±0·20	14·14 ±0·10	16·03 ±0·15	3·15 ±0·04	1·63 ±0·03	8·29
10057-58	A	150 ± 18	5·45 ±0·08	9·80 ±0·12	17·95 ±0·16	49·65 ±0·52	15·31 ±0·14	41·37 ±0·35	14·16 ±0·20	11·55 ±0·20	7·76
10071-32	A	590 ± 65	9·55 ±0·10	16·80 ±0·18	26·26 ±0·20	35·84 ±0·25	18·35 ±0·11	22·36 ±0·20	4·53 ±0·07	2·66 ±0·06	5·83
10047-40	B	30 ± 6	13·20 ±0·14	21·70 ±0·20	33·11 ±0·25	59·0 ±0·5	21·83 ±0·16	39·55 ±0·50	10·33 ±0·13	8·29 ±0·20	3·98

ratios differ also (MORRISON *et al.*, 1970). It seems obvious, therefore, that chemical differences could at least in part explain varying relative production rates.

Given a simple cosmic-ray energy spectrum such as

$$f(E)\,dE = f_0 E^{-\alpha}\,dE,$$

the production rate of an isobar is obtained (GEISS *et al.*, 1962) as

$$P(\Delta A, A_0) = B(A_0, n)(\Delta A)^{-n} \text{ with } n = \tfrac{1}{2}(3\alpha - 1),$$

where A_0 is the mass of the target nucleus and ΔA the total mass loss. A value of $n \approx 3{\cdot}5$ is required to produce Xe^{131}/Xe^{126} spallation ratios of about 8 from Ba as target element. This value is much higher than those found in meteorites ($n \leq 2{\cdot}7$) and clearly requires heavy shielding for most of the time of irradiation. We have to remember, however, that the irradiation geometry is quite different. Close to the surface of the moon a 2π-irradiation of an infinite target does occur. In a deeper position, the effective irradiation angle decreases as the depth increases, and the modulation of primary radiation becomes more effective. Fast secondary neutrons will not escape as readily as in the case of a rather small meteoritic body. Because low-energy spallation Xe is almost exclusively produced from Ba only, the relative production rates of the Xe isotopes will also be affected by variations in the relative abundances of Ba and the rare earth elements (REE).

As an alternative, the evidence of very large relative Xe^{131} yields suggests that special reactions may partially be responsible for the excess Xe^{131} production. Because of the high capture cross section, Gd is a sensitive indicator for thermal neutrons. We have measured the isotopic compositions of Gd in rock 10017 and in our soil sample. The depletion of Gd^{157} is between 0·1 and 0·4 per cent for rock 10017 and $\leq$0·2 per cent in the soil sample corresponding to integrated thermal neutron fluxes of between 4×10^{15} and 16×10^{15} n/cm^2 and $\leq 8 \times 10^{15}$ n/cm^2, respectively. Because of some background interferences, specifically $BaCl^+$, we could not yet take advantage of high analytical precision. This result shows that thermal neutron capture can only contribute 3 per cent to the total Xe^{131} production. No neutron resonances are known for Ba^{130}, and it seems unlikely that the neutron resonance integral will contribute much to the total cross section.

Acknowledgments—We thank M. MEYERS, K. R. GOLDMAN and M. HILLIER for their assistance in the experimental work. We have benefited from discussions with J. R. ARNOLD. Support for this research from NASA contract NAS 9-8107 is acknowledged.

REFERENCES

ALBEE A. L., BURNETT D. S., CHODOS A. A., EUGSTER O. J., HUNEKE J. C., PAPANASTASSIOU D. A., PODOSEK F. A., RUSS G. PRICE II, SANZ H. G., TERA F. and WASSERBURG G. J. (1970) Ages, irradiation history, and chemical composition of lunar rocks from the Sea of Tranquillity. *Science* **167**, 463–466.

ALLER L. H. (1961) *The Abundance of the Elements*. Interscience.

BÜHLER F., EBERHARDT P., GEISS J., MEISTER J. and SIGNER P. (1969) Apollo 11 solar wind composition experiment: First results. *Science* **166**, 1502–1503.

CAMERON A. G. W. (1962) The formation of the Sun and planets. *Icarus* **1**, 13–69.

CAMERON A. G. W. (1965) In *Handbook of Geophysics and Space Environments*. U.S. Air Force Cambridge Research Laboratories, Washington, D.C.

EBERHARDT P., GEISS J. and GRÖGLER N. (1965) Further evidence on the origin of trapped gases in the meteorite Khor Temiki. *J. Geophys. Res.* **70,** 4375–4378.

EBERHARDT P., GEISS J., GRAF H., GRÖGLER N., KRÄHENBÜHL U., SCHWALLER H., SCHWARZMÜLLER J. and STETTLER A. (1970) Trapped solar wind noble gases, Kr^{81}/Kr exposure ages and K/Ar ages in Apollo 11 lunar material. *Science* **167,** 558–560.

EUGSTER O., EBERHARDT P. and GEISS J. (1967) Krypton and xenon isotopic composition in three carbonaceous chondrites. *Earth Planet. Sci. Lett.* **3,** 249–257.

EUGSTER O., TERA F., BURNETT D. S. and WASSERBURG G. J. (1969) The isotopic composition of Gd and neutron capture effects in some meteorites, Preprint, December, 1969.

FLEISCHER R. L. and PRICE P. B. (1964) Decay constant for spontaneous Fission of U^{238}. *Phys. Rev.* **133,** B63.

GAST P. W. and HUBBARD N. J. (1970) The abundance of alkali metals, alkaline and rare earths and Sr^{87}/Sr^{86} ratios in lunar samples. *Science* **167,** 485–487.

GEISS J., OESCHGER H. and SCHWARZ U. (1962) The history of cosmic radiation as revealed by isotopic changes in the meteorites and on the earth. *Space Sci. Rev.* **1,** 197–223.

GOLD T. (1970) Private Communication.

HEYMANN D., YANIV A., ADAMS J. A. S. and FRYER G. E. (1970) Inert gases in Lunar samples. *Science* **167,** 555–558.

KEAYS R. R., GANAPATHY R., LAUL J. C., ANDERS E., HERZOG G. F. and JEFFERY P. M. (1970) Trace elements and radioactivity in lunar rocks: implications for meteorite infall, solar-wind flux, formation condition of the Moon. *Science* **167,** 490–493.

MARTI K., EBERHARDT P. and GEISS J. (1966) Spallation, fission and neutron capture anomalies in meteoritic krypton and xenon. *Z. Naturforsch.* **21a,** 398–413.

MARTI K. (1967a) Mass-spectrometric detection of cosmic-ray-produced K^{81} in meteorites and the possibility of Kr–Kr dating. *Phys. Rev. Lett.* **18,** 264–266.

MARTI K. (1967b) Trapped xenon and the classification of chondrites. *Earth Planet. Sci. Lett.* **2,** 193–196.

MARTI K. (1967c) Isotopic composition of trapped krypton and xenon in chondrites. *Earth Planet. Sci. Lett.* **3,** 243–248.

MARTI K. (1969) Solar-type xenon: A new Isotopic composition of xenon in the Pesyanoe meteorite. *Science* **166,** 1263–1265.

MARTI K., LUGMAIR G. W. and UREY H. C. (1970) Solar wind gases, cosmic-ray spallation products, and the irradiation history. *Science* **167,** 548–550.

MORRISON G. H., GERARD J. T., KASHUBA A. T., GANGADHARAM E. V., ROTHENBERG A. M., POTTER N. M. and MILLER G. B. (1970) Multielement analysis of lunar soil and rocks. *Science* **167,** 505–507.

NIER A. O. C. (1950) A redetermination of the relative abundances of the isotopes of neon, rubidium, xenon and mercury. *Phys. Rev.* **79,** 450–454.

PEPIN R. O. and SIGNER P. (1965) Primordial rare gases in meteorites. *Science* **149,** 253–265.

PEPIN R. O., NYQUIST L. E., PHINNEY D. and BLACK D. C. (1970) Isotopic composition of rare gases in lunar samples. *Science* **167,** 550–553.

REYNOLDS J. H., HOHENBERG C. M., LEWIS R. S., DAVIS P. K. and KAISER W. A. (1970) Isotopic analysis of rare gases from stepwise heating of lunar fines and rocks. *Science* **167,** 545–548.

SEGRÈ E. (1952) Spontaneous Fission. *Phys. Rev.* **86,** 21.

SHEDLOVSKY J. P., HONDA M., REEDY R. C., EVANS J. C., JR., LAL D., LINDSTROM R. M., DELANY A. C., ARNOLD J. R., LOOSLI H-H., FRUCHTER J. S. and FINKEL R. C. (1970) Pattern of bombardment-produced radionuclides in rock 10017 and in lunar soil. *Science* **167,** 574–576.

SILVER L. T. (1970) Uranium–thorium–lead isotope relations in lunar materials. *Science* **167,** 468–471.

SUESS H. E. and UREY H. C. (1956) Abundances of the elements. *Rev. Mod. Phys.* **28,** 53.

TATSUMOTO M. and ROSHOLT J. N. (1970) Age of the moon: An isotopic study of uranium–thorium–lead systematics of lunar samples. *Science* **167,** 461–463.

TURNER G. (1970) Argon-40/argon-39 dating of lunar rock samples. *Science* **167,** 466–468.

WETHERILL G. W. (1953) Spontaneous fission yields from uranium and thorium. *Phys. Rev.* **92,** 907.

Proceedings of the Apollo 11 Lunar Science Conference, Vol. 2, pp. 1369 to 1374.

Chemical composition of Apollo 11 lunar samples 10017, 10020, 10072 and 10084

J. A. Maxwell
Geological Survey of Canada, Ottawa, 4, Ontario
L. C. Peck
United States Geological Survey, Denver, Colorado 80225
and
H. B. Wiik
Finnish Research Council for Sciences, Helsinki, Finland

(*Received* 29 *January* 1970; *accepted* 29 *January* 1970)

Abstract—Major, minor and trace element analyses, by three separate laboratories, are reported for the lunar rocks 10017, 10020 and 10072, and for the fine surficial material, 10084. Brief details of the analytical procedures used are also given. The analyses confirm in general the results previously reported for lunar material, especially the high titanium content; water, carbon dioxide, fluorine, chlorine and Fe(III) are either absent or present in negligible amounts. It was not possible to determine metallic Fe but the anomalous reducing capacity of some lunar material has been demonstrated. The composition of the samples differs markedly from that of known rocks and meteorites.

Introduction

Three samples of Apollo 11 rocks, and one of the lunar surficial material, were split and distributed among three Principal Investigators concerned with making bulk chemical analyses, as follows: Maxwell received 10017-29, 10020-30 and 10084-132, Peck was given 10020-23, 10072-32 and 10084-131, and Wiik received 10017-20, 10072-20 and 10084-102. Individual reports of the resulting analyses have already appeared (Maxwell *et al.*, 1970; Peck and Smith, 1970; Wiik and Ojanperä, 1970) but it has seemed advantageous to combine the data and analytical details into one paper, to facilitate comparison and use of the results obtained in three widely-separated laboratories.

Sample 10084 is an aliquot of the <1 mm fraction of the lunar surficial material, of which 20 g was split into three portions by Wiik in the Lunar Receiving Laboratory. The three rock specimens were broken into smaller pieces in the LRL and approximately equal portions of each (10–20 g) were distributed to each of the Principal Investigators selected to receive them. Further preparation of the samples for analysis was done in the individual laboratories.

Analytical Techniques

A. *At the Geological Survey of Canada* (*JAM*)

The conventional rock analysis methods used to determine most of the constituents are those described by Maxwell (1968), with the exceptions of those for total carbon, carbon dioxide, fluorine, chlorine and total sulfur. The non-distillation, spectrophotometric method of Huang and Johns (1967), as modified by Sen Gupta (1968), was employed for the simultaneous determination of chlorine and fluorine. Total carbon and sulfur, liberated as CO_2 and SO_2 respectively by induction

Table 1. Chemical analyses (wt.%) of lunar material

Constituent	10017 GSC*	GSF*	10020 GSC	USGS*	10072 GSF	USGS	10084 GSC	GSF	USGS
SiO_2	40·78	40·77	39·92	39·95	40·20	40·53	42·28	42·25	42·20
TiO_2	11·71	11·82	10·72	10·52	12·28	11·74	7·35	7·54	7·32
Al_2O_3	8·12	7·92	10·04	10·19	7·78	8·52	13·76	13·83	14·07
Fe_2O_3	0·00	0·0	0·00	0·03	0·0	0·00†	0·00	0·0	0·00‡
FeO	19·82	19·79	19·35	19·14	19·77	19·76	16·02	15·80	15·81
Cr_2O_3	0·36	0·33	0·40	0·38	0·36	0·35	0·33	0·28	0·28
MgO	7·65	7·74	7·81	7·87	8·06	7·68	7·93	7·97	7·93
CaO	10·55	10·58	11·24	11·31	10·27	10·42	12·00	11·96	12·01
Na_2O	0·51	0·51	0·37	0·39	0·52	0·54	0·42	0·43	0·46
K_2O	0·30	0·29	0·05	0·05	0·29	0·27	0·13	0·13	0·12
H_2O^+	0·00		0·00	0·00		0·00	0·00		0·05
		0·00			0·00			0·00	
H_2O^-	0·01		0·01	0·00		0·00	0·01		0·00
P_2O_5	0·13	0·18	0·08	0·07	0·18	0·14	0·11	0·14	0·08
MnO	0·22	0·22	0·24	0·27	0·22	0·24	0·20	0·20	0·21
S	0·22	ND§	0·15	0·18	ND	0·24	0·13	ND	0·14
CO_2	0·00	0·00	0·00	ND	0·00	ND	0·03	0·00	ND
F	0·00	ND	0·00	ND	ND	ND	0·00	ND	ND
Cl	0·00	ND	0·00	ND	ND	ND	0·00	ND	ND
C	0·01	ND	0·01	ND	ND	ND	0·00	ND	ND
Fe^0	ND	0·0	ND	ND	0·0	ND	ND	0·0	ND
Sub total	100·39	100·15	100·39	100·35	99·93	100·43	100·70	100·53	100·68
Less O≡S	0·11	—	0·08	0·09	—	0·12	0·07	—	0·07
Total	100·28	100·15	100·31	100·26	99·93	100·31	100·63	100·53	100·61

* GSC—Geological Survey of Canada: analysts, J. A. Maxwell, S. Abbey (Cr_2O_3, Na_2O, K_2O) and J. G. Sen Gupta (total C and S).

GSF—Geological Survey of Finland: analysts, H. B. Wiik and P. Ojanperä; P. Kauranen, P. Puumalainen, E. Häsänen and R. Rosenberg (Mn, Na, K).

USGS—United States Geological Survey: analysts, L. C. Peck and Vertie C. Smith.

† Actual value obtained, −0·14.

‡ Actual value obtained, −0·12.

§ ND—not determined.

furnace heating (SEN GUPTA, 1970), were determined titrimetrically, the CO_2 by non-aqueous titration (SEN GUPTA, 1969). The CO_2 released by heating the samples with acid was similarly determined. The results are given in Table 1; a reference basalt sample, USGS BCR-1, was run simultaneously with the lunar samples, to provide a means for evaluating the precision and accuracy of the data (Table 2).

The optical emission spectrographic determination of minor and trace elements was made with an air-jet controlled d.c. arc and a Jarrell-Ash 3·4 m grating spectrograph; 30-mg portions of the

Table 2. Chemical analysis (wt.%) of reference basalt (USGS BCR-1)*

SiO_2	— 54·48	H_2O^+	— 0·66
TiO_2	— 2·26	H_2O^-	— 0·86
Al_2O_3	— 13·62	P_2O_5	— 0·347
Fe_2O_3	— 3·77	MnO	— 0·165
FeO	— 8·65	S	— 0·033
MgO	— 3·46	CO_2	— 0·055
CaO	— 6·91	F	— 0·039
Na_2O	— 3·36	Cl	— 0·002
K_2O	— 1·76	C	— 0·00
	Sub total	— 100·43	
	Less O≡S	— 0·02	
	Total	— 100·41	

* Analysts, J. A. Maxwell and S. Abbey (Na_2O and K_2O), Geological Survey of Canada.

Table 3. Minor and trace element content (ppm) of lunar material

Constituent	10017 GSC*	10017 GSF*	10020 GSC	10072 GSF	10084 GSC	10084 GSF	BCR-1 GSC
Cr	2200	2260	3000	2460	2300	1880	<20
Zr	410	430	210	460	260	380	<200
Ba	230		67		150		700
Sr	100		130		140		220
V	49	70	120	82	67	71	420
Ni	<20	<5	<20	<5	190	200	<20
Co	<20	30	<20	30	<20	32	<20
Cu	<5	20	<5	21	<5	13	10
Zn	49	47	26	24	47	36	120
Li	23	ND	12	ND	15	ND	15
Sc	52	77	78	77	51	59	23
Y	160	160	120	175	120	120	34
Yb	<10	19	<10	19	<10	12	<10
Sm	ND	25	ND	21	ND	18	ND
Eu	ND	2·1	ND	2·0	ND	1·8	ND
Tb	ND	4	ND	3	ND	3	ND
Lu	ND	5	ND	5	ND	3	ND
Hf	ND	12	ND	12	ND	8	ND
Ta	ND	5	ND	5	ND	3	ND
Th	ND	4·6	ND	4	ND	3	ND

* GSC—Geological Survey of Canada: all determinations by emission spectroscopy (analysts, W. H Champ, K. A. Church, D. A. Brown, Joanne Crook, G. A. Bender), except Zn and Li (atomic absorption spectroscopy, S. Abbey, analyst); < = less than; ND = not determined.

GSF—Geological Survey of Finland: analysts, H. B. Wiik and P. Ojanperä (atomic absorption spectroscopy, *AAS*); A. Löfgren and A. Savola (emission spectroscopy, *ES*); P. Kauranen, P. Puumalainen, E. Häsänen and R. Rosenberg (neutron activation, *NA*); Sc, V, Cu, Y and Zr by *ES*; Sm, Eu, Tb, Lu, Yb, Hf, Ta and Th by *NA*; Zn by *AAS*; Co by *ES* and *NA*; Ni by *ES*, *NA* and *AAS*.

samples were mixed with a buffer at a dilution factor of 7·5, and 45-mg charges of this mixture taken. No internal standard was used, but intensity ratios of trace element lines were related to selected iron lines in an "external standard" (Fe_2O_3) exposure on the same photographic plate. Standard samples having compositions generally similar to the samples were run simultaneously and used to correct the working curves. The data are given in Table 3, together with the results obtained for the USGS basalt, BCR-1.

During preparation of the samples for analysis, a very small amount of a metallic phase (metallic Fe?) was noted only in sample 10084-132. Although it could not be determined by known methods for the determination of metallic Fe, it would, however, contribute to the value obtained for Fe(II) by the modified Pratt method (MAXWELL, 1968). The possibility of the presence of some Ti(III) has also been suggested and, in order to determine the magnitude of the error likely to be introduced by these constituents into the determination of Fe(II), the procedure was further modified by the addition of a known quantity of an oxidant, Fe(III), to each sample before its decomposition with acid. Following the potentiometric determination of the "Fe(II)", the total Fe was similarly determined in the same solution. The values obtained are given in Table 4 and it is seen that small anom-

Table 4. Reducing capacity (wt.%) of lunar material*

Sample no.	(1) "Fe(II)" as FeO	(2) Total Fe as FeO	Difference (1 − 2)
10017-29	19·82	19·81	+0·01
10020-30	19·43	19·28	+0·15
10084-132	17·05	16·02	+1·03

* Analyst, J. A. Maxwell, Geological Survey of Canada.

alies were found for the two rock samples, 10017-29 and 10020-30, and a significantly large one for the fines, 10084-132. If it is assumed that the difference in Fe(II) was caused only by reduction of the Fe(III) added, then the metallic Fe content of 10084-132 is about 0·3 per cent; similarly, the difference could be caused by the presence of about 0·7% Ti(III). While the source(s) cannot be stated with-certainty, it is suggested (JAM) that metallic iron is the more likely candidate.

The use of atomic absorption spectroscopy and other techniques requiring small sample weights was given early consideration in case the quantity of sample available for analysis should prove to be limited. Although the latter situation did not arise, it was decided to apply a new composite scheme, developed by Sydney Abbey at the Geological Survey of Canada, to the determination of some major and minor constituents of the lunar samples, as a test of its future applicability to samples of this nature; two sample portions were used for the analysis, and the results are given in Table 5.

Table 5. Chemical analyses (wt.%) of lunar material by new composite scheme*

Constituent	10017-29	10020-30	10084-132
SiO_2	40·14	41·00	42·44
TiO_2	11·16	10·28	7·19
Al_2O_3	8·17	9·83	13·66
Total Fe, expressed as FeO	19·38	19·03	15·67
MgO	7·72	7·77	7·87
CaO	10·99	11.96	12·26
Na_2O	0·51	0·37	0·42
K_2O	0·30	0·05	0·13
P_2O_5	0·15	0·07	0·10
MnO	0·25	0·27	0·21
Cr_2O_3	0·36	0·40	0·33
Total	99·13	101·30	100·28

* Analysts, S. Abbey (atomic absorption spectroscopy) and J. L. Bouvier (TiO_2 and P_2O_5, by colorimetry, on solutions prepared for AAS).

A 200-mg sample was fused with a fivefold excess of lithium metaborate and the fusion dissolved in dilute hydrofluoric acid; sufficient boric acid was added to provide a solution which was stable towards glass for at least 2 hr (BERNAS, 1968). Two dilutions were prepared, one containing 50 mg of sample and 300 mg Sr, and the other 10 mg and 150 mg, in 100 ml. The more concentrated solution was used for the determination by atomic absorption spectroscopy of Si and Al, using a nitrous oxide-acetylene flame, and K, Mn and Cr with the air-acetylene flame. Fe, Mg, Ca and Na were determined in the second solution, using an air-acetylene flame. Details of the method will be published elsewhere. Two more aliquots of the original sample solution were used for the photometric determination of phosphorus and titanium.

An additional 500 mg sample was decomposed by acid treatment and the resulting solution, containing free sulfuric acid and some added hydrochloric acid, was analyzed for Li and Zn by atomic absorption spectroscopy, using the method of "standard addition" (ABBEY, 1967).

All determinations were based upon a comparison of sample solutions with solutions similarly prepared from international reference rock samples, using assigned values based upon data prepared by ABBEY (1970) and GOVINDARAJU (1969). All atomic absorption measurements were made with a modified Techtron Model AA-3 instrument.

B. *At the United States Geological Survey (LCP)*

Except where otherwise noted, the classical procedures used are those described by PECK (1964), and the page numbers given in the following brief discussion refer to this publication. The results obtained are given in Table 1.

The procedure for the determination of SiO_2 (p. 62) was modified to eliminate complications caused by the high titanium content of the samples; the separated silica was fused with Na_2CO_3

and the dehydration and filtration repeated. The two filtrates were combined for the precipitation of the ammonium hydroxide group. The determination of titanium was done colorimetrically (p. 70) with a probable precision of $\pm 0{\cdot}1$ per cent absolute.

The known presence of troilite (FeS) and metallic Fe makes the results obtained for Fe(II) uncertain (pp. 39–42); metallic Fe is counted as FeO in the summation. Because Fe(III) is determined by difference after the separate determinations of total Fe and Fe(II), the cumulative errors in both of these determinations are therefore reflected in the value found for Fe(III). The small negative and positive values obtained for Fe(III) are probably the result of such errors, and, within the accuracy limits of the method, it is likely that the samples do not contain any Fe(III).

The value for H_2O^+ reported for sample 10084-131 is probably for water which was adsorbed after the sample was exposed to the terrestrial atmosphere; strongly adsorbed water is not expelled when a silicate is heated at 105°C (H_2O^-).

It is likely that most of the chromium present is in the mafic minerals. A small amount of black material that might have been chromite was observed in sample 10084-131 but there was no evidence of its presence in the other samples. The procedures for the determination of chromium, and of sulfur, are those described by KOLTHOFF and SANDELL (1952).

Metallic iron was not determined because of doubts about the reliability of published methods, and it might be more useful to use sample material for an examination of a separated magnetic phase by other techniques such as Mössbauer spectroscopy. The presence of metallic iron, which is reported as FeO, could account for the unacceptably high summation given in Table 1 for sample 10084-131 (similarly high summations are given for the other splits by both the GSC and GSF). There would have to be about 3 per cent of it, however, and data from other sources suggest that 1 per cent or less is probably present. Other factors, such as powder segregation in samples containing grains having a high density, may be contributing as well to the high summation.

C. *At the Geological Survey of Finland* (*HBW*)

The constituents given in Table 1 were determined by the following methods: MgO, CaO and total H_2O, gravimetric; TiO_2 and P_2O_5, colorimetric; SiO_2, gravimetric and colorimetric; Al_2O_3, gravimetric and atomic absorption spectroscopic; FeO and Fe_2O_3, titrimetric; Cr_2O_3, atomic absorption spectroscopic; MnO, colorimetric and neutron activation; Na_2O and K_2O, flame emission spectroscopic and neutron activation. The determination of CO_2 and metallic Fe are qualitative. The methods used to determine other trace constitutents are given, together with the results, in Table 3.

SUMMARY AND CONCLUSIONS

The compositional data given for the four lunar samples thus confirm, in general, the data previously presented by LSPET (1969), with the exception of the values for Cr, Zr and Ba. It is significant that the best agreement between the replicate results is found in the three analyses for sample 10084, reflecting the greater degree of homogeneity of this fraction of the fine surficial material. The occasional differences which are observed in the duplicate values for the three rocks may in part be caused by inhomogeneity in the distributed portions.

The composition of these four samples do not resemble any known terrestrial rock, nor any known meteorite. The very low nickel content is especially peculiar, and would make a common origin for lunar, terrestrial and meteoritic material very doubtful.

Acknowledgments—We thank the following for their cooperation and assistance: at the Geological Survey of Canada, SYDNEY ABBEY for the atomic absorption spectroscopic determinations and for suggesting the addition of Fe(III) in the Fe(II) determination; J. L. BOUVIER for the determination of phosphorus and titanium in the fluoborate solutions; J. G. SEN GUPTA for the determination of total sulfur and carbon; and W. H. CHAMP, K. A. CHURCH, D. A. BROWN, JOANNE CROOK and G. A. BENDER for the emission spectrographic analyses; at the Geological Survey of Finland, P. OJANPERÄ for the flame photometric and atomic absorption spectroscopic determinations; A. LÖFGREN and A. SAVOLA (emission spectroscopy) and P. KAURANEN, P. PUUMALAINEN, E. HÄSÄNEN and R. ROSENBERG (neutron activation); at the United States Geological Survey, VERTIE C. SMITH.

REFERENCES

ABBEY S. (1967) Analysis of rocks and minerals by atomic absorption spectroscopy. Part I. Determination of magnesium, lithium, zinc and iron. *Geol. Surv. Can. Paper* **67-37**.

ABBEY S. (1970) U.S. Geological Survey standards—a critical study of published analytical data. *Can. Spectrosc.* **15**, 10–16.

BERNAS B. (1968) A new method for decomposition and comprehensive analysis of silicates by atomic absorption spectrometry. *Anal. Chem.* **40**, 1682–1686.

GOVINDARAJU K. (1969) Private communication, Centre de Recherches Pétrographiques et Géochimiques, Nancy, France.

HUANG W. H. and JOHNS W. D. (1967) Simultaneous determination of fluorine and chlorine in silicate rocks by a rapid spectrophotometric method. *Anal. Chim. Acta* **37**, 508–515.

KOLTHOFF I. M. and SANDELL E. B. (1952) *Textbook of Quantitative Inorganic Analysis*, 3rd edition. Macmillan.

LSPET (LUNAR SAMPLE PRELIMINARY EXAMINATION TEAM) (1969) Preliminary examination of lunar samples from Apollo 11. *Science* **165**, 1211–1227.

MAXWELL J. A. (1968) *Rock and Mineral Analysis*. Interscience.

MAXWELL J. A., ABBEY S. and CHAMP W. H. (1970) Chemical composition of lunar material. *Science* **167**, 530–531.

PECK L. C. (1964) Systematic analysis of silicates. *U.S. Geol. Surv. Bull.* **1170**.

PECK L. C. and SMITH V. C. (1970) Quantative chemical analysis of lunar samples *Science* **167**, 532.

SEN GUPTA J. G. (1968) Determination of fluorine in silicate and phosphate rocks, micas and stony meteorites. *Anal. Chim. Acta* **42**, 119–125.

SEN GUPTA J. G. (1969) Private communication, Geological Survey of Canada, Ottawa.

SEN GUPTA J. G. (1970) Rapid combustion methods for determining sulfur in rocks, ores and stony meteorites. A comparative study of the usefulness of resistance-type and induction furnaces. *Anal. Chim. Acta* in press.

WIIK H. B. and OJANPERA P. (1970) Chemical analyses of lunar samples 10017, 10072 and 10048. *Science* **167**, 531–532.

Proceedings of the Apollo 11 Lunar Science Conference, Vol. 2, pp. 1375 to 1382.

Total carbon and nitrogen abundances in Apollo 11 lunar samples and selected achondrites and basalts*

C. B. MOORE, E. K. GIBSON,† J. W. LARIMER,
C. F. LEWIS and W. NICHIPORUK
Arizona State University, Tempe, Arizona 85281

(*Received* 2 *February* 1970; *accepted in revised form* 16 *February* 1970)

Abstract—Total carbon abundances were determined in lunar samples and related rocks by combusting the samples at 1800°C in an oxygen atmosphere and detecting carbon dioxide produced with a gas chromatograph. Weighted mean analyses for bulk lunar fines were 225 and 140 ppm total carbon. The range was from 110 to 350 ppm total carbon. The weighted mean total carbon in fine breccia ("C") was 230 ppm; coarse breccia 100 ppm; fine-grained rock ("A") 70 ppm and medium-grained rock ("B") 64 ppm. Carbon is concentrated in the finest-grained sieved fraction. Achondrites and terrestrial basalts appear to be higher in carbon than lunar rocks.

Nitrogen was determined after opening the lunar samples at 2400°C in a graphite crucible to produce reducing conditions. Molecular nitrogen was determined in a gas chromatograph. Weighted mean analyses were 150 and 100 ppm for bulk fines. The weighted mean total nitrogen for fine breccia ("C") was 125 ppm; for coarse breccia 100 ppm; for fine-grained rock ("A") 115 ppm; and for medium grained rock ("B") 30 ppm. Achondrites and terrestrial basalts contain from 30 to 50 ppm nitrogen.

The total carbon and nitrogen in fines appear to be mixture of indigenous lunar material together with solar wind components.

INTRODUCTION

TOTAL carbon and nitrogen contents of Apollo 11 lunar samples, basaltic achondrites and terrestrial basalts, were determined utilizing analytical techniques developed for these elements in meteorites (MOORE *et al.*, 1965, 1966, 1967, 1969a, b; GIBSON and MOORE, 1970).

Carbon and nitrogen are important because of their crucial role and organogenic elements. Also, their abundances give an indication of the oxidation-reduction conditions and due to their volatility, thermal conditions near the lunar surface.

EXPERIMENTAL

For the determination of total carbon, samples were burned in a flowing oxygen atmosphere at over 1600°C to form CO_2. After necessary purification and trapping of the effluent gases, the CO_2 was detected utilizing a LECO 589–400 gas chromatographic low-carbon analyzer. Samples were heated to about 2400°C in a graphite crucible in a helium atmosphere to reduce all nitrogen compounds to N_2 which was detected in a LECO Nitrox-6 gas chromatographic analyzer. Differential thermal conductivity is utilized as the detection method in both systems. National Bureau of Standards low carbon (101e) and nitrogen (33d) steel standards were used to construct standard analytical curves for both determinations. The combustion–gas chromatographic detection technique determines total carbon and nitrogen but does not discriminate their chemical state in the analyzed samples.

---

* Contribution No. 44 from Center for Meteorite Studies.

† Present Position: NAS-NRC Resident Research Associate, NASA Manned Spacecraft Center, Houston, Texas.

Table 1. Total carbon in Apollo 11 lunar rocks

Samples		Sample wt. (g)	Total carbon (μg/g)	Weighted mean (μg/g)
10086-A	Fines	0·3433	130 ± 4	—
		0·3094	162 ± 6	—
		0·0782 Δ	160 ± 25 Δ	—
		0·1555 Δ	122 ± 13 Δ	—
		(0·8864)	—	142 ± 10
10086-B	Fines	0·1080	190 ± 20	—
		0·2586	181 ± 8	—
		0·2565	109 ± 8	—
		0·3482	353 ± 6	—
		(0·9713)	—	226 ± 10
10002-54	"C" breccia	0·2831	228 ± 7	—
		0·2874	231 ± 7	—
		(0·5705)	—	230 ± 7
10044-37	"C–B" coarse breccia	0·2093	108 ± 9	—
		0·3677	76 ± 6	—
		0·1058	180 ± 18	—
		(0·6828)	—	102 ± 10
10049-23	"A" basalt	0·2268	68 ± 9	—
		0·3464	71 ± 6	—
		(0·5732)	—	70 ± 8
10050-33	"B" basalt	0·2394	152 ± 8	—
		0·3234	15 ± 7	—
		0·3007	47 ± 6	—
		(0·8635)	—	64 ± 8

Table 2. Total nitrogen in Apollo 11 lunar rocks

Sample		Sample weight (g)	Total nitrogen (μg/g)	Weighted mean (μg/g)
10086-A	Fines	0·0971	142 ± 5	—
		0·1822	87 ± 2	—
		0·1239	100 ± 3	—
		0·0749 Δ	107 ± 5 Δ	—
		0·1192 Δ	92 ± 4 Δ	—
		(0·5973)	—	102 ± 4
10086-B	Fines	0·1969	150 ± 2	—
		0·1299	164 ± 5	—
		0·1024	144 ± 4	—
		(0·4292)	—	153 ± 4
10002-54	"C" breccia	0·0896	113 ± 5	—
		0·1588	135 ± 4	—
		0·1146	119 ± 4	—
		(0·3630)	—	125 ± 4
10044-37	"C–B"	0·1666	135 ± 4	—
		0·1266	95 ± 4	—
		0·1993	69 ± 3	—
		(0·4925)	—	98 ± 4
10049-23	"A" basalt	0·1510	85 ± 3	—
		0·1293	35 ± 4	—
		0·1159	247 ± 5	—
		(0·3962)	—	116 ± 4
10050-33	"B" basalt	0·1888	32 ± 2	—
		0·2067	28 ± 2	—
		(0·3955)	—	30 ± 2

Table 3. Total carbon and nitrogen in basaltic achondrites, basalts and LRL control samples

	C	N
Pasamonte	730 ± 20	33 ± 3
Eucrite	780 ± 20	34 ± 3
	710 ± 20	
	610 ± 20	44
	580 ± 20	45
	700 ± 20	
Sioux Co.	650 ± 20	24 ± 3
Eucrite	690 ± 20	30 ± 3
	690 ± 20	
	640 ± 20	33 ± 3
	580 ± 20	
	440 ± 20	
Haraiya	4250 ± 10	31 ± 3
Eucrite	4460 ± 10	42 ± 3
	4740 ± 10	
	5020 ± 10	43 ± 3
	4620 ± 10	43 ± 3
	4980 ± 10	
Yurtuk	1050 ± 20	66 ± 3
Achondrite	1090 ± 20	57 ± 3
	1190 ± 20	
	790 ± 20	51 ± 3
	830 ± 20	50 ± 3
	830 ± 20	
Columbia River basalt	320 ± 20	31 ± 3
BCR–1	230 ± 20	28 ± 3
	360 ± 20	30 ± 3
	370 ± 20	30 ± 3
	370 ± 20	26 ± 3
Vesicular basalt		
Carbon K–71 (LRL)	1890 ± 20	49 ± 3
Nitrogen K–73 (LRL)	1980 ± 20	58 ± 3
Scoriaceous basalt	360 ± 20	38 ± 3
Mokuopuhi flow	360 ± 20	41 ± 3
Hawaii	420 ± 20	
Dunite	278 ± 20	54 ± 3
LEM exhaust test	323 ± 20	55 ± 3
		48 ± 3
		54 ± 3
Fused quartz	30 ± 10	54 ± 3
F–201 control (F–12)	44 ± 10	56 ± 3
Fused quartz	93 ± 7	57 ± 3
F–201 sample (311010)	36 ± 10	
Fused quartz	38 ± 10	54 ± 3
Control sample (C–7)	21 ± 10	
Fused quartz	90 ± 7	60 ± 3
Crushed sample (311008)	35 ± 10	58 ± 3

The results of the analyses in lunar samples are listed in Table 1 for total carbon and in Table 2 for total nitrogen. Table 3 contains carbon and nitrogen contents of selected basaltic achondrites, terrestrial basalts, and Lunar Receiving Laboratory (LRL) control standards. For each lunar sample the weights of material used for individual analyses are listed together with the concentration of carbon or nitrogen detected. No attempt was made to homogenize individual lunar samples. This minimized the possibility of contamination and also provided a test for sample inhomogeneity. Rock samples were crushed by a single stroke in a clean diamond mortar. All of the rock specimens were crushed quite easily to a moderately fine material. To guard against contamination, they were not sieved or run through a mechanical splitter.

The precision indicated as $\pm$ is taken from the 90 per cent confidence level on the line of regression for the reference standards. Analytical precision of the nitrogen analyses is superior to that for carbon because the difference between the thermal conductivities of nitrogen and helium is greater than that for carbon dioxide and helium.

RESULTS

The carbon and nitrogen abundances in the lunar samples show some similarities. Both are highest in the fines samples 10086-A and 10086-B and the fine-grained breccia 10002-45. The coarse-grained basalt 10050-33 contained the lowest concentrations of both carbon and nitrogen. With respect to carbon, the fine-grained basalt 10049-23 is similar to the coarse-grained basalt but its nitrogen content is relatively high. The sample of rock 10044-37 provided for our analysis had the macroscopic characteristics of a coarse-grained basalt but was cataloged as a coarse-grained breccia. In terms of both carbon and nitrogen, it is intermediate to the fine-grained breccia 10002-54 and the coarse-grained basalt 10050-33. Both the basaltic achondrites and the terrestrial basalts have higher carbon contents than the lunar samples. The achondrites were carefully selected and prepared and there is little possibility of contamination by our handling. The terrestrial basalts may, of course, have had secondary carbonates added to them from deuteric or groundwater solutions. The unexpected high carbon abundance in the Haraiya eucrite may be from cometary impact during its preterrestrial history or even possibly from immersion in an organic liquid for density determination by an unknown individual. Their nitrogen contents are similar to the lunar rocks in absolute abundance but their nitrogen–carbon ratios are lower.

The lunar rocks like the enstatite chondrites appear to have been formed under high temperature reducing conditions. Unlike the lunar material, the enstatite achondrites have high carbon contents (0·056–0·56 wt. % carbon) indicating a different evolution or initial composition (MOORE and LEWIS, 1966). MOORE *et al.* (1969) showed that enstatite chondrites which do not contain the mineral sinoite, Si_2N_2O, have five to ten times the nitrogen concentration present in ordinary chondrites. This relative enrichment in nitrogen in both the lunar rocks and enstatite chondrites may be due to their reducing character.

The carbon and nitrogen in Apollo 11 lunar samples may be from four sources: (1) indigenous, (2) meteorite or comet impact, (3) solar wind, (4) contamination. Contamination may have resulted from the LEM landing, astronaut activity, or subsequent laboratory handling.

The possibilities for contamination from nitrogen in returned lunar samples appear to be greater than for carbon. Possible sources are from the LEM exhaust gases and from the nitrogen atmosphere to which the fresh lunar rocks were exposed at the Lunar Receiving Laboratory. The control samples show no evidence of extensive carbon or nitrogen contamination in the LRL at the level of significance determined by our analytical technique.

In order to see if nitrogen or carbon could be easily desorbed from lunar fines, a split of sample 10086-A was heated for 24 hr at 300°C in a flowing helium atmosphere. The results of the analyses of this material are identified in Tables 1 and 2 by the symbol Δ. No appreciable carbon or nitrogen appeared to be removed by this treatment. If contamination by atmospheric nitrogen has taken place, it is strongly bonded to the fines material.

The relatively high nitrogen content of fine-grained basalt 10049-23 would then also appear to be indigenous. This supports the idea that the nitrogen has been retained because of the highly reducing state of the lunar igneous environment. Evidence for chemically reduced nitrogen has been given by HINTENBERGER *et al.* (1970) who reported the detection of a major fraction of NH_4^+ in a sulfuric acid leached sample of lunar fines.

Carbon and nitrogen may be indigenous to the moon as elemental carbon, as interstitial atoms of carbon or nitrogen, as organic material, or as inorganic compounds (carbonate, carbides, ammoniacal nitrogen, molecular nitrogen). Meteoritic material apparently arrives with hypervelocity impacts, causing volatilization and atomization of the projectile and part of the target. Solar wind particles presumably arrive as single atoms or ions.

The total carbon trends seem to support the idea that the values of 70 and 64 μg/g (ppm) found in the lunar basaltic rocks are primarily due to indigenous carbon. The close agreement for the two splits of the fine-grained basalt 10049-23 indicates some degree of homogeneity with respect to the analyzed sample, whereas, the variation between the splits of the coarse-grained basalt 10050-33 indicates a more sporadic distribution of carbon among or between the large crystalline phases.

The two fines samples 10086-A and 10086-B and the fine dark breccia 10002-54 are significantly higher in total carbon than are the basalts. Sample 10086-A received less handling than sample 10086-B in the Lunar Receiving Laboratory and its lower carbon values may be attributed to this. There is no evidence to support this assumption, and the overlap in the range of values in 10086-A and 10086-B together with a constant carbon–nitrogen ratio, suggests a normal sample variation. In order to investigate the distribution with respect to particle size, a split of sample 10086-A was sieved and analyzed for carbon. Material greater than 60 mesh had 70 μg/g $\pm$ 15 total carbon; from 60 to 140 mesh had 115 $\pm$ 20 μg/g; from 140 to 300 mesh, 210 $\pm$ 10 μg/g; and minus 300 mesh, 500 $\pm$ 20 μg/g. Evidently the carbon is concentrated in the finest sized material. Microscopic examination of the material larger than 60 mesh and the 60–140 mesh fraction indicated that it is made up primarily of fine-grained basalt, glassy impactite material, and some crystal fragments. Samples 10086-A and 10086-B may contain different proportions of coarse and fine fragments and hence, have different total carbon abundances. The fine-grained breccia appears to consist primarily of compacted fines and apparently lost no carbon during lithification.

DISCUSSION

In our initial report (MOORE *et al.*, 1970) we noted that the rare-gas analyses reported in the preliminary examination of lunar samples from Apollo 11 (LSPET, 1969) also showed enrichment in the fines and breccia. Assuming that all of the rare-gas content is due to solar wind, some rough calculations were made to indicate how much of the carbon might be attributed to the solar wind. The major unknown in the calculation was a "sticking" or retention factor relating the relative retention on the lunar fines to each of the rare gases. A comparison of Apollo 11 rare-gas abundances with CAMERON's (1968) table of abundances in the solar system, indicated a fractionation in the lunar materials favoring the retention of higher atomic weight gases. If the solar carbon/neon is normalized to the neon in the lunar samples, the solar wind contribution of carbon to the lunar fines is 6 μg/g or about 4 per cent of the total carbon. If the carbon abundance is calculated from the krypton abundance, the solar wind contribution is about 50 μg/g, and if it is calculated from xenon, the solar wind contribution would be 300 μg/g. Assuming similar "sticking" factors, the solar wind contribution would be about five times lower for nitrogen than for carbon.

A general calculation was also made to estimate the possible meteoritic and cometary contribution to the lunar fines. The major unknown in this case was the composition of the meteoritic influx. Carbonaceous chondrites were selected as representative of impact material. This choice was supported by comparing the increase in nickel from the lunar igneous rocks to the lunar breccia and fines as reported in the preliminary examination of samples from Apollo 11 (LSPET, 1970). Nickel and carbon both showed significant increases, and nickel, like carbon, has a much higher concentration in carbonaceous chondrites than in lunar rocks. KEAYS *et al.* (1970) work on trace elements in Apollo 11 rocks also indicated that carbonaceous chondrite-like material is being added to the lunar surface.

MASON's (1963) review of carbonaceous chondrites gives typical concentrations of 2% carbon and 1% nickel in these meteorites. Using these values to calculate the approximate contribution of meteoritic extralunar material indicated that about 0·5–2 per cent has been mixed with lunar material. The nitrogen contribution of meteorites is about 10 per cent that of carbon and again this element did not give an apparent satisfactory material balance. Based on our simplifying assumptions, both solar wind and meteorite influx failed to account for the relatively high nitrogen abundances. They were explained by the possibility that indigenous nitrogen is high in the lunar rocks, as indicated by the fine-grained basalt 10049-23, or that solar wind or meteoritic nitrogen is retained to a greater degree than carbon.

On the basis of the sulfur abundances in the lunar samples by KAPLAN and SMITH (1970) it now appears that carbonaceous chondrite and/or cometary material may not be a major source of carbon and nitrogen in the lunar fines. Sulfur, like carbon and nickel, is much more abundant in the carbonaceous chondrites than in the lunar basalts. An influx of carbonaceous chondrite-like material should then increase the sulfur content in the lunar fines with respect to the lunar basalts. KAPLAN and SMITH (1970) noted the opposite effect. Sulfur in the fine-grained rocks was about 2200 μg/g and in the fines about 700 μg/g. In order for an element to be depleted in the lunar soil, it must be either (1) volatile and lost from the surface by some mechanism or

(2) diluted by the addition of a material depleted in that element. This material may be either extra-lunar or from a different area on the moon. The mixture of anorthosite with Apollo 11 site bedrock as reported by WOOD *et al.* (1970) would be an example of such a mixing effect. It appears that this specific mixing has added aluminum to the Apollo 11 fines. In order to cut the sulfur abundance by more than one-half, greater than a one to one mixture of anorthosite and basalt would be required. WOOD *et al.*'s (1970) data on the distribution of rock types in the soil do not support such a model. Impact vaporization is a more likely mechanism. That meteorite impact is catastrophic enough to destroy the impacting material is supported by the fact that no complex carbonaceous compounds or water have been found in the lunar soil (LSPET, 1969) and that no magnetite has been identified (JEDWAB *et al.*, 1970). Water, organic compounds, and magnetite are abundant in the carbonaceous chondrites. It appears that most of the meteoritic impact material is atomized, reconstituted or severely shocked with the resultant loss of volatile species. The target material of either bedrock and/or regolith must also undergo a similar process. The preliminary examination of the lunar fines (LSPET, 1969) indicated that about half of it was glassy. This estimate together with the fact that the fines have half the sulfur content of the lunar rocks has led us to estimate that about half of the material making up the lunar fines has been heated enough to expel its volatile elements.

Utilizing a model based upon the assumptions that (1) the average proton flux on the lunar surface has been 2×10^8 cm^{-2} sec^{-1} (KEAYS *et al.*, 1970); (2) there is a 100 per cent retention factor for carbon, nitrogen, and sulfur accumulated; (3) half the indigenous volatiles from lunar rocks and solar wind volatiles in the lunar regolith were lost by impact; (4) all meteoritic volatiles are lost on impact; the expected total concentration of carbon, nitrogen, and sulfur in lunar fines may be estimated. The results of such a set of calculations are given in Table 4. The solar wind is assumed to have an elemental distribution similar to CAMERON's (1968) solar abundance table.

Table 4. Calculated and experimental concentrations of volatile elements in lunar fines

Element	Maximum solar wind ($\times \frac{1}{2}$) (μg/g)	Lunar basalt (type A) ($\times \frac{1}{2}$) (μg/g)	Theoretical fines (μg/g)	Reported fines	Reported breccia	Reference
C	150	35	185	142–226	230	a
N	32	55	87	102–153	125	a
S	15	1100	1100	640–770	1070–1120	b
				1400		c
				1200	1500	d

References:
a. This paper.
b. KAPLAN and SMITH (1970).
c. PECK and SMITH (1970).
d. AGRELL *et al.* (1970).

The agreement may be considered satisfactory in view of the poor sampling and gross assumptions. According to this model a major amount of the volatile elements carbon and nitrogen in the lunar fines appears to have been derived from solar wind. Solar wind contributions of carbon and nitrogen would not be concentrated in separate mineral phases but would most likely attach themselves to a nearest neighbor on a particle surface. Since the most abundant element available is oxygen, carbon—oxygen and nitrogen—oxygen bonds would be expected to be common. Also solar wind protons would contribute to carbon—hydrogen bonds. Evidence for carbon compounds of both types have been presented by BURLINGAME *et al.* (1970) and ABELL *et al.* (1970) who have reported that the species CO, CO_2 and a minor amount of CH_4 have been released from lunar fines by pyrolysis.

Acknowledgments—This research was supported in part by grants from the National Science Foundation (NSF-GA 909) and the National Aeronautics and Space Administration (NGL-03-001-001) and by a National Aeronautics and Space Administration contract (NAS9-9989).

REFERENCES

ABELL P. I., DRAFFAN G. H., EGLINTON G., HAYES J. M., MAXWELL J. R. and PILLINGER C. T. (1970) Organic analysis of the returned lunar sample. *Science* **167,** 757–759.

AGRELL S. O., SCOON J. H., MUIR I. D., LONG J. V. P., MCCONNELL J. D. C. and PECKETT A. (1970) Mineralogy and petrology of some lunar samples. *Science* **167,** 583–586.

BURLINGAME A. L., CALVIN M., HAN J., HENDERSON W., REED W. and SIMONEIT B. R. (1970) Lunar organic compounds: Search and characterization. *Science* **167,** 751–752.

CAMERON A. G. W. (1968) A new table of abundances of the elements in the solar system. In *Origin and Distribution of the Elements,* (editor L. H. Ahrens), Vol. 30, Sec. 1, pp. 125–143. Pergamon.

GIBSON E. K. and MOORE C. B. (1970) Inert carrier-gas fusion determination of total nitrogen in rocks and meteorites. *Anal. Chem.* in press.

HINTENBERGER H., WEBER H. W., VOSHAGE H., WANKE H., BEGEMANN F., VILSCEK E. and WLOTZKA F. (1970) Rare gases, hydrogen, and nitrogen: concentrations and isotopic composition in lunar material. *Science* **167,** 543–545.

JEDWAB J., HERBOSCH A., WALLAST R., NAESSENS G. and VAN GEEN-PEERS N. (1970) Search for magnetite in lunar rocks and fines. *Science* **167,** 618–619.

KAPLAN I. R. and SMITH J. W. (1970) Concentration and isotopic composition of carbon and sulfur in Apollo 11 lunar samples. *Science* **167,** 541–543.

KEAYS R. R., GANAPATHY R., LAUL J. C., ANDERS E., HERZOG G. F. and JEFFERY P. M. (1970) Trace elements and radioactivity in lunar rocks: Implications for meteorite infall, solar-wind flux and formation conditions of moon. *Science* **167,** 490–493.

LSPET (LUNAR SAMPLE PRELIMINARY EXAMINATION TEAM) (1969) Preliminary examination of lunar samples from Apollo 11. *Science* **165,** 1211–1227.

MASON B. (1963) The carbonaceous chondrites. *Space Sci. Rev.* **1,** 621–646.

MOORE C. B. and LEWIS C. F. (1965) Carbon abundances in chondritic meteorites. *Science* **149,** 317–318.

MOORE C. B. and LEWIS C. F. (1966) The distribution of total carbon content in enstatite chondrites. *Earth Planet. Sci. Lett.* **1,** 376–378.

MOORE C. B. and LEWIS C. F. (1967) Total carbon content of ordinary chondrites. *J. Geophys. Res.* **72,** 6289–6292.

MOORE C. B. and GIBSON E. K. (1969a) Nitrogen in chondritic meteorites. *Science* **163,** 174–176.

MOORE C. B., GIBSON E. K. and KEIL K. (1969b) Nitrogen abundances in enstatite chondrites. *Earth Planet. Sci. Lett.* **6,** 457–460.

MOORE C. B., LEWIS C. F., GIBSON E. K. and NICHIPORUK W. (1970) Total carbon and nitrogen abundances in lunar samples. *Science* **167,** 495–497.

PECK L. C. and SMITH V. C. (1970) Analysis of moon samples Apollo 11 mission. *Science* **167,** 532.

WOOD J. A., DICKEY J. S., JR., MARVIN U. B. and POWELL B. N. (1970) Lunar anorthosites. *Science* **167,** 602–604.

Proceedings of the Apollo 11 Lunar Science Conference, Vol. 2, pp. 1383 to 1392.

Elemental abundances of lunar soil and rocks

GEORGE H. MORRISON, JESSE T. GERARD, A. THOMAS KASHUBA,*
ESWARA V. GANGADHARAM, ANN M. ROTHENBERG, NOEL M. POTTER
and GARY B. MILLER
Department of Chemistry, Cornell University, Ithaca, New York 14850

(*Received* 30 *January* 1970; *accepted in revised form* 17 *February* 1970)

Abstract—Results are presented for multielement analysis of lunar soil and seven rocks returned by Apollo 11. Sixty-seven elements were determined using spark source mass spectrography and neutron activation. U.S.G.S. standard W-1 was used as comparative standard. Results indicate an apparent uniformity of composition amongst the samples. Comparison with solar, meteoritic and terrestrial abundances reveals depletion of volatile elements and enrichment of the rare earths, Ti, Zr, Y and Hf. Although there is an overall similarity of the lunar material to basaltic achondrites and basalts, the differences suggest detailed geochemical processes special to the history of this material.

INTRODUCTION

AMONG the most important types of information desired from the lunar material returned to earth by the Apollo 11 mission is the characterization of its chemical composition. By comparing the elemental abundance patterns of the lunar material with those of solar, meteoritic, and terrestrial materials, some insight into its cosmological history can be obtained.

Of the various multielement techniques for the survey analysis of solids, spark source mass spectrography (MS) and neutron activation analysis (NAA) provide the greatest sensitivity and scope. Because of the complex nature of the lunar materials with the attendant serious interference problems, the use of either MS or NAA alone would have precluded the determination of many elements. By using both techniques in a complementary fashion on a given limited sized sample, it has been possible to determine 67 elements.

This approach has been used to determine the elemental abundances in the lunar soil and seven rocks representing the three textural classes proposed by the LSPET (1969), i.e. fine-grained vesicular (type A), coarse-grained crystalline (type B), and the breccia (type C). The sample of soil received (10084-55) was the fine fraction (<1 mm) of type D sample 10002.

In addition to the determination of elemental abundances, Auger electron spectroscopy was used to ascertain the chemical nature of the lunar soil surface.

EXPERIMENTAL

Samples

Samples analyzed include three type A rocks (10020-26, 10057-32 and 10072-24), two type B (10046-23 and 10058-26), two type C (10056-20 and 10060-13) and one type D (10084-55). Detailed petrographic descriptions of these rocks may be found in the LSPET Report (1969).

---

* Present address: Jones and Laughlin Steel Corp., 900 Agnew Road, Pittsburgh, Pa.

Sample preparation

One gram of each sample received was crushed and homogenized in an agate electromagnetic micromill to pass 400 mesh. Two hundred mg were removed for NAA and 500 mg for MS.

Procedures

For MS the 500 mg portions were mixed with equal weights of graphite and pressed into electrodes. The compressed electrodes were sparked in a double-focusing instrument according to the procedure of Morrison and Kashuba (1969). Various magnet settings and charge collections were used to reduce the interference from halation. Analysis parameters were optimized for maximum sensitivity of representative metallic elements. Sensitivity factors used in calculating the quantitative results were obtained from parallel experiments run with the U.S.G.S. standard diabase W-1. Accuracy, determined from past analyses of terrestrial and meteoritic materials, averages 14 per cent for trace elements. Precision of the lunar analyses is 5–25 per cent, as indicated by the relative standard deviation of each element's results about the mean.

For NAA, 200 mg of sample were irradiated in cleaned polyethylene vials at progressively increased neutron fluxes for short periods of time in the Cornell TRIGA Mark II reactor. To determine Al and V, a flux of 6×10^{10} n/cm^2 sec was used for 45 sec; for Mg, 2×10^{11} for 40 sec; and for Mn and Ni, 2×10^{12} for 90 sec. After these non-destructive irradiations, the samples were encapsulated in high purity quartz vials and irradiated at a flux of $3{\cdot}5 \times 10^{12}$ n/cm^2 sec for 8·5 hr. The U.S.G.S. standard diabase W-1 was used as a multielement standard and 300 mg were irradiated simultaneously with each sample. Counting was performed on a high resolution 30 cm^3 coaxial lithium drifted germanium detector using a 400 channel analyzer. After allowing decay for 15 hr, both the irradiated samples and W-1 standard were mixed with sufficient unirradiated W-1 to result in total sample-carrier weights of 1 g. In addition to serving as a multielement carrier, the added W-1 insured similar chemistry between the samples and standards when processed according to the radiochemical separation scheme previously described (Morrison *et al.*, 1969).

To complete the values for the major elements, silicon was determined by atomic absorption spectrophotometry (AA). Samples were dissolved in an acid digestion bomb and the resulting solutions diluted with water were aspirated into a 5-cm burner using an acetylene–nitrous oxide flame. A hollow cathode Si lamp was used and the absorption of the 2516 Å line was measured. The U.S.G.S. standard basalt BCR-1 was used as an elemental standard.

Auger electron spectroscopy was performed in order to establish the surface composition of the lunar soil and several spherules recovered from it (Rhodin and Demuth, 1969). One sample of lunar soil and 4 spherules were mounted on individual gold flags using AR grade sodium silicate as a binding agent. A blank was run using sodium silicate. The flags were then placed on a LEED crystal manipulator in an ultra-high vacuum system equipped for grazing incident angle Auger excitation. The samples were outgassed prior to Auger analysis at 400°C for 15 sec.

Results and Discussion

The results of the analyses of the eight lunar samples received are given in Table 1. Results from the method having the better accuracy, as determined by past analyses of terrestrial and meteoritic materials and having the better analytical conditions (higher sensitivity, lower background, etc.) were used. When these criteria were judged equivalent an arithmetic average of the results of the two methods was reported.

The results of both MS and NAA determinations depend on the assumption that terrestrial isotope ratios are the same as in the lunar samples. In the determination of Pb by MS, the isotope ratios did not correspond to published terrestrial values. Of the four stable Pb isotopes, ^{204}Pb was obscured. A summation of the intensities of the remaining three isotopes was used to compute the elemental concentration. Extremely low intensities precluded exact work, but the ^{206}Pb : ^{207}Pb : ^{208}Pb ratios were found to be

Table 1. Elemental abundances in the lunar samples studied

Element	Method*	Type A rocks			Type B		Type C		Soil
		10020-26†	10057-32	10072-24	10058-26	10046-23	10056-20	10060-13	10084-55
				Major elements, wt. %					
Si	AA	19·4	21·5	19·2	18·4	20·6	19·8	18·7	20·2
Al	NAA	5·9	4·0	4·0	5·4	6·2	5·7	6·0	7·3
Ti	NAA	6·1	6·5	6·7	5·8	5·0	5·6	5·1	4·1
Fe	NAA	14·7	15·7	15·4	15·2	13·2	14·8	13·8	12·5
Mg	NAA	5·0	3·7	4·3	3·4	5·5	2·8	4·6	4·6
Ca	NAA	10·3	10·1	10·3	11·0	9·8	11·0	10·4	9·6
Na	NAA	0·28	0·35	0·32	0·32	0·35	0·32	0·34	0·33
K	NAA	0·052	0·22	0·29	0·093	0·17	0·094	0·16	0·11
Mn	NAA	0·20	0·17	0·17	0·19	0·16	0·20	0·16	0·16
Cr	NAA	0·22	0·21	0·24	0·15	0·21	0·14	0·22	0·20
Zr	NAA	0·036	0·056	0·072	0·038	0·062	0·041	0·058	0·039
Ni‡	NAA	0·006	0·004	0·003	0·008	0·007	0·005	0·007	0·017
V	NAA-MS	0·0059	0·0040	0·0062	0·0041	0·0068	0·0056	0·0062	0·0078
P	MS	0·070	0·041	0·070	0·024	0·10	0·033	0·060	0·14
				Trace elements, ppm					
Li	MS	5	8	14	6	16	16	7	6
Be	MS	2	2·5	4	1·5	6	3	3	4
B	MS	1	4	4	2	9	2	3	2
N§	MS	40	70	110	40	260	70	20	110
F§	MS	85	70	100	50	220	30	80	66
Cl§	MS	150	50	60	50	520	16	150	350
Sc	NAA	85	84	86	87	64	97	70	60
Co	NAA-MS	20	24	28	14	42	15	32	40
Cu	NAA	3·7	5·5	18	7·1	9·7	3·8	11	9·9
Zn	NAA	2·1	2·9	7·0	9·3	30	2·7	25	22
Ga	NAA	3·5	4·7	4·3	4·3	4·9	4·3	5·1	4·6
Ge‖	MS	—	1·3	1·1	1·2	—	1·2	1·4	0·7
As¶	NAA	0·03	0·04	0·05	0·07	0·05	0·03	0·09	0·07
Se¶	MS	0·4	—	—	—	0·4	—	0·9	0·2
Br¶	NAA	0·1	0·1	0·2	0·3	0·2	0·06	0·3	0·2
Rb	NAA-MS	2·8	4·8	5·7	1·2	3·6	2·0	4·0	4·4
Sr	NAA-MS	170	130	140	180	170	160	180	200
Y	MS	130	210	250	150	190	180	210	150
Nb	MS	36	42	45	47	38	37	45	33
Mo	NAA-MS	0·4	0·4	0·4	0·4	0·7	0·4	0·7	0·7
Pd¶	MS	0·09	0·1	0·1	0·2	0·1	0·1	—	0·04
Ag§	MS	0·1	0·04	0·6	0·07	0·02	0·2	0·01	0·1
Cd¶	MS	—	0·9	1	0·7	0·8	0·9	0·3	0·3
In§	MS	—	0·07	2	0·6	0·08	0·06	—	0·5
Sn¶	MS	—	0·6	0·4	1·2	—	0·3	—	0·7
Sb	NAA	0·01	0·005	0·01	0·01	0·005	0·005	0·005	0·005
Cs	NAA-MS	0·2	0·2	0·3	0·3	0·2	0·06	0·2	0·2
Ba	NAA-MS	96	280	300	140	280	100	250	220
La	NAA-MS	11	31	35	16	23	13	24	22
Ce	NAA-MS	34	83	96	45	67	42	62	50
Pr	MS	8·7	22	20	13	20	12	13	9
Nd	NAA-MS	40	66	88	72	60	57	82	46
Sm	NAA-MS	14	24	28	22	20	23	24	18
Eu	NAA-MS	1·6	2·1	2·2	3·0	2·0	2·5	2·0	1·9
Gd	NAA-MS	17	26	31	22	20	24	28	20
Tb	NAA-MS	3·5	5·6	6·8	5·4	4·5	5·4	5·0	3·8
Dy	MS	30	42	45	39	30	40	41	25
Ho	NAA-MS	7	8	10	9	9	9	10	6
Er	MS	19	32	35	36	23	27	30	15
Tm	NAA-MS	1·2	2·3	2·8	2·0	1·6	2·1	1·8	1·2
Yb	NAA-MS	15	26	28	22	20	20	22	12
Lu	NAA-MS	1·5	2·2	2·6	2·3	1·8	1·8	2·0	1·4
Hf	NAA-MS	11	15	18	13	11	11	13	9
Ta	NAA	1·3	1·2	1·8	1·0	1·7	2·2	1·7	1·3
W	NAA	0·13	0·42	0·42	0·36	0·35	0·15	0·35	0·25
Pb¶	MS	—	3	3	3	2	1·2	3	6
Th	NAA-MS	1·5	4·5	4·8	1·1	2·8	1·4	3·0	2·3
U	NAA	0·14	0·56	0·50	0·20	0·58	0·21	0·60	0·48

* NAA = Neutron Activation Analysis, MS = Mass Spectrometry, AA = Atomic Absorption.
† NASA laboratory number. ‡ Third decimal place approximate.
§ Possible contamination. ‖ Upper limit. ¶ Approximate.

approximately 2:1:2, and are in excellent agreement with those of TATSUMOTO (1970). The Pb values themselves are imprecise, due to work close to the detection limit.

For the elements Ru, Rh, Te and I, MS determinations were limited by the lack of values for W–1 necessary to compute sensitivity factors. A comparison of intensities shows that Ru and Te are approximately of the same concentrations in the lunar samples as in terrestrial basalt, exemplified by the U.S.G.S. standard BCR–1; Rh is enhanced by a factor of 2–3: and I is depleted by a factor of 2. The elements Re, Os and Ir were not observed; their detection limits are approximately 0·1 ppm for MS in these samples. Hg was not detected by NAA at the 0·02 ppm level.

The accuracy of elemental determinations is dependent on the level of possible contaminants. The values for N, F, Cl, Ag and In may have been influenced by contamination before being received. The samples were received under a dry nitrogen blanket and prepared for analysis in dry air, resulting in the possibility of surface N_2 adsorption. The samples were sealed on the lunar surface in Teflon bags and the sample boxes were sealed using In–Ag alloy O-rings. In addition, a microscopic examination of the samples as received from NASA revealed foreign material in the form of transluscent fibers, possibly from the plastic packaging or handling materials.

The MS values for Ge, Pd and Cd are of limited accuracy, since these were determined at the detection limits. In addition, Ge^{+} ions had Nd^{2+} and Sm^{2+} interferences, for which corrections were attempted. The Ge values should be regarded as upper limits only.

Comparison of our data with the results of other investigators presented at the Lunar Science Conference shows good agreement. For the major elements the agreement is quite good except for Ca which appears to be consistently about 20% high. Because of the large number of trace elements determined in this study, it was difficult to find sufficient data on the same samples to make meaningful comparisons for all elements. In the case of the rare earth elements where comparative data was more available, the agreement is generally good with the exception of Er which appears to be high. The remaining trace elements are in good agreement except for N, F, Cl, Ge, Pd, Ag, Cd, In and Pb which have special analytical problems as discussed above.

Examination of the data in Table 1 shows that, at the 2σ level of significance, there is neither general sample-to-sample variance, not is there a sample type-to-type variation of elemental abundances. Several investigators (GOLEŠ *et al.*, 1970; ANNELL and HELZ, 1970; KEAYS *et al.*, 1970) have suggested that there are compositional differences between type A and B versus C and D materials. On the other hand, GAST and HUBBARD (1970) found that A and B were different and C and D were intermediate to A and B. We do not find such relationships based on a comparison involving 67 elements for the samples studied. The observations of the other studies were based on comparisons involving a small number of elements.

When the variations at the 1σ level are examined, division of the samples into two groups is possible, based on the behavior of only 13 elements (Table 2). For example, if the samples are arranged in order of increasing K content, rocks 10020, 10056 and 10058 fall into one group and rocks 10046, 10057, 10060 and 10072 comprise a second group. On this basis, the lunar fine material 10084 is intermediate between the groups, with values for these elements approaching the second group. These groups do not follow the textural classification suggested by the LSPET (1969). However, it should be emphasized that such a compositional classification is possible only if the 1σ variation of values is regarded as significant. COMPSTON *et al.* (1970) also recognized two groups transgressing the LSPET types A and B. We find breccias included in our two groups, whereas they find the breccias and soil intermediate to their two groups. The rock samples analyzed by us and COMPSTON were different except for one.

Table 2. Grouping at the 1σ level

Element	Average of rocks 10020, 10056, 10058 (ppm)	Soil 10084 (ppm)	Average of rocks 10046, 10057, 10060, 10072 (ppm)
K	800	1100	2100
Th	1·3	2·3	3·8
U	0·18	0·48	0·56
Ba	112	220	278
Rb	2·0	4·4	4·6
Y	150	150	220
La	13	22	28
Ce	40	50	77
Pr	11	9	19
Zr	380	390	620
W	0·21	0·25	0·39
Co	16	40	32
Ga	4·0	4·6	4·8

Of the elements studied, there is greatest variation amongst the eight samples for P, K, Zn and Cu. This variation cannot be attributed to analytical imprecision and must be regarded as reflecting the variations in the samples received by this laboratory. In the case of Zr and Yb, our values are significantly different from the data reported for the same samples by the LSPET. At present we are unable to assess the source of this discrepancy apart from commenting on the possible sampling difficulties in the distribution of the samples by NASA. Similar observations were made by other workers at the recent Lunar Science Conference.

Results from the Auger experiment show that titanium and calcium are present as major constituents in the outer 2–5 atomic layers in the lunar soil sample. The sodium silicate binding material gives rise to oxygen, silicon, chlorine, and potassium peaks. The spherules give rise to small but discernible calcium and titanium peaks. There was no iron detected in the surface layers of any of the samples.

Comparison of the average lunar elemental abundance data with solar abundances (CAMERON, 1968; recalculated on an H, He-free basis), meteorites and terrestrial basalts has been made. In the lunar material, volatile elements such as Sb, As, Br, Cu, Zn and Hg appear depleted. Sodium and potassium are depleted when compared to terrestrial basalts, and enriched when compared to the solar abundances. The rare earths and refractory elements such as Ti, Zr, Y and Hf appear enriched. Comparison of elemental abundances in the lunar material with that of solar abundances indicates that most of the non-volatiles including the rare earths are enriched (Fig. 1).

The lunar REE abundances are higher when compared to solar, meteoritic and basaltic values. When normalized to La, the lunar REE abundances follow a similar pattern to normalized solar and achondritic REE abundances, with the exception of

Eu which is depleted in the lunar material. When the absolute abundances are compared, the REE are substantially enriched in lunar materials relative to continental basalts (10 times) and chondritic meteorites (130 times). Figure 2 shows the average

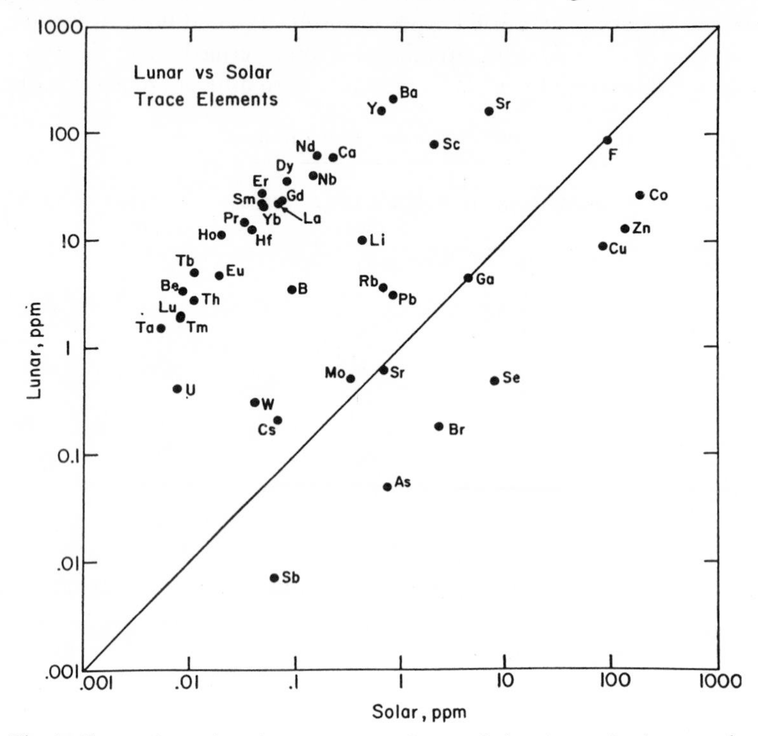

Fig. 1. Comparison plot of average trace elemental abundances for lunar rocks and soil against solar abundances.

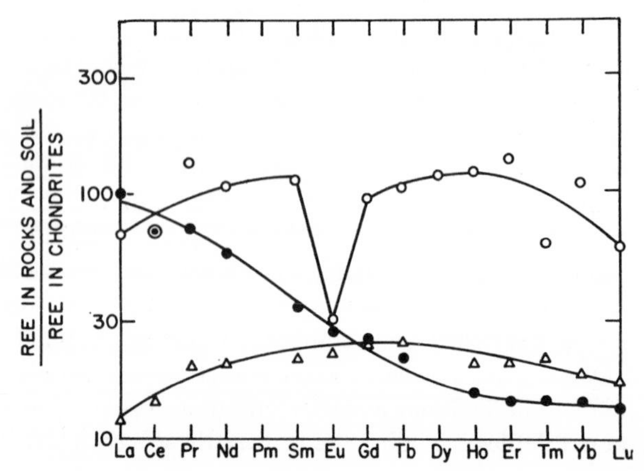

Fig. 2. Relative rare earth abundances. Open circles represent average values for 8 lunar samples, closed circles represent values for composite basalt and triangles represent average values for 12 oceanic basalts.

lunar REE abundances normalized to those of chondrites (SCHMITT *et al.*, 1970) plotted against atomic number. Also shown for comparison are plots for continental and oceanic basalts (FREY *et al.*, 1968). From the figure it can be seen that the degree of enrichment is highest in lunar material. The pattern obtained for lunar material

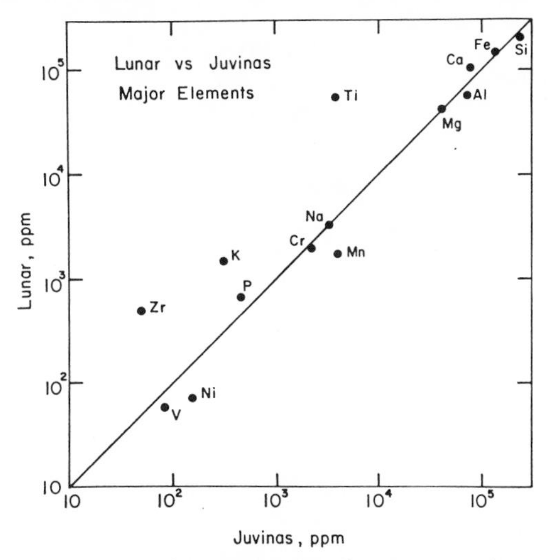

Fig. 3. Comparison plot of average major elemental abundances for lunar rocks and soil vs. Juvinas.

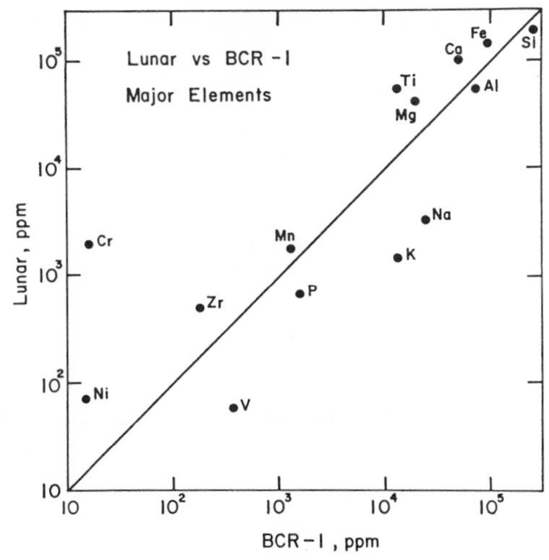

Fig. 4. Comparison plot of average major elemental abundances for lunar rocks and soil vs. BCR-1.

and oceanic basalts is generally similar. There is a tendency for medium-heavy REE (Nd to Ho) to be enriched to a greater degree (about 100 times in lunar materials and about 24 times in oceanic basalts), than the lighter REE (about 70 times in lunar material and 17 times in oceanic basalts). In the case of the lunar REE pattern we also

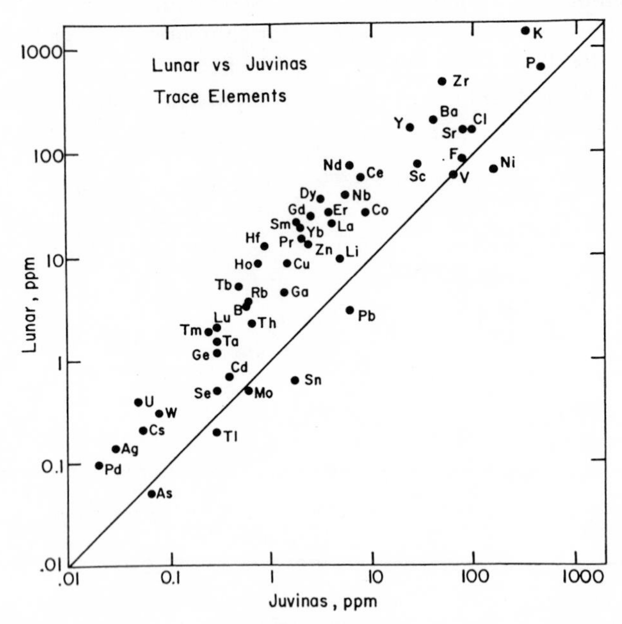

Fig. 5. Comparison plot of average trace elemental abundances for lunar rocks and soil vs. Juvinas.

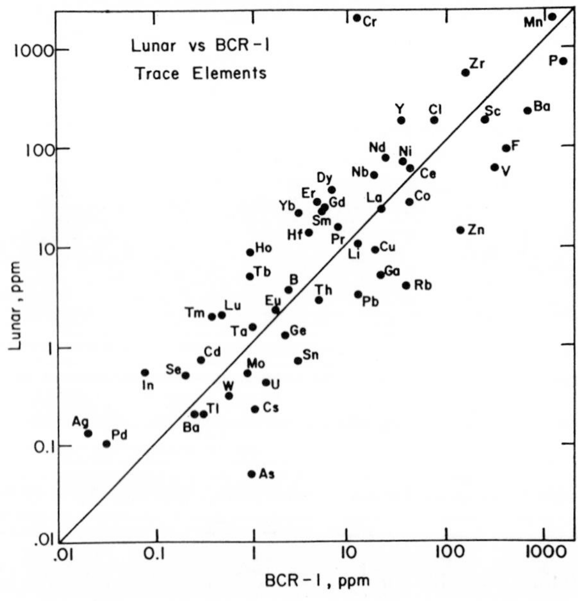

Fig. 6. Comparison plot of average trace elemental abundances for lunar rocks and soil vs. BCR-1.

find exceptional depletion of Eu (up to 30%) as do other workers. In contrast, the distribution pattern for continental basalts indicates that the factor of enrichment of REE decreases smoothly from the lightest REE (100 for La) through Eu (27) to heaviest REE (30 for Lu).

A notable feature of the patterns in Fig. 2 is the striking constancy of the degree of enrichment of Eu in lunar materials (30 times), continental basalts (27 times) and oceanic basalts (22 times) over its chondritic abundance. This may mean that the conditions of fractionation of this element might have been the same in the three kinds of materials. Whether this has any relationship with the fact that Eu is the only rare earth which could exist in the +2 state is not understood.

With regard to major elements, the lunar material closely resembles basaltic achondrites (eucrites), and to a lesser extent, terrestrial basalts, with the notable exception of Ti and Zr which are substantially enriched in the lunar material (Figs. 3 and 4). The trace element distribution in the lunar material is also comparable to that in achondrites and basalts (Figs. 5 and 6). Comparison with basalts shows that the trace elements, though dispersed, are evenly distributed about a 1:1 trend line with a correlation coefficient of 0·90 and a slope of 43°. Comparison with achondrites shows that most trace elements are uniformly enriched in the lunar materials. A least squares fit for this comparison gives a slope of 43° and a correlation coefficient of 0·94. Because of the paucity of sufficient trace element data in the literature for basalts and basaltic achondrites for comparison with the lunar data, our laboratory performed extensive analyses of BCR–1 and Juvinas as representatives of these two groups.

The lunar samples are compositionally different from all materials with which they are compared. Although there is closer similarity to basaltic achondrites and basalts, the lunar material does not match them in all respects. This uniqueness implies detailed geochemical processes special to the history of lunar materials.

Acknowledgments—The authors gratefully acknowledge the support of this work by the National Aeronautic and Space Administration under grant NAS 9-9986, the National Science Foundation under grant GP-6471X, and the Advanced Research Projects Agency through the Cornell Materials Science Center.

REFERENCES

ANNELL C. and HELZ A. (1970) Emission spectrographic determination of trace elements in lunar samples. *Science* **167**, 521–523.

CAMERON A. G. W. (1968) A new table of abundances of the elements in the solar system. In *Origin and Distribution of the Elements*, (editor L. H. Ahrens), pp. 125–143. Pergamon.

COMPSTON W., ARRIENS P. A., VERNON M. J. and CHAPPELL B. W. (1970) Rubidium–strontium chronology and chemistry of lunar material. *Science* **167**, 474–476.

FREY F. A., HASKIN M. A., POETZ J. A. and HASKIN L. A. (1968) Rare earth abundances in some basic rocks. *J. Geophys. Res.* **73**, 6085–6098.

GAST P. W. and HUBBARD N. J. (1970) Abundance of alkali metals, alkaline and rare earths, and strontium-87/strontium-86 ratios in lunar samples. *Science* **167**, 485–487.

GOLES G. G., OSAWA M., RANDLE K., BEYER R. L., JEROME D. Y., LINDSTROM D. J., MARTIN M. R., MCKAY S. M. and STEINBORN T. L. (1970) Instrumental neutron activation analysis of lunar specimens. *Science* **167**, 497–499.

KEAYS R. R., GANAPATHY R., LAUL J. C., ANDERS E., HERZOG G. F. and JEFFERY P. M. (1970) Trace elements and radioactivity in lunar rocks: implications for meteorite infall, solar-wind flux, and formation conditions of the moon. *Science* **167**, 490–493.

LSPET (LUNAR SAMPLE PRELIMINARY EXAMINATION TEAM) (1969) Preliminary examination of lunar samples from Apollo 11. *Science* **165**, 1211–1227.

MORRISON G. H., GERARD J. T., TRAVESI A., CURRIE R. L., PETERSON S. F. and POTTER N. M. (1969) Multielement neutron activation analysis of rock using chemical group separations and high resolution gamma spectrometry. *Anal. Chem.* **41**, 1633–1637.

MORRISON G. H. and KASHUBA A. T. (1969) Multielement analysis of basaltic rock using spark source mass spectrometry. *Anal. Chem.* **41**, 1842–1846.

RHODIN T. N. and DEMUTH J. (1969) Private communication, Cornell University.

SCHMITT ROMAN A., WAKITA HIROSHI and REY PLINIO (1970) Abundances of 30 elements in lunar rocks, soil, and core samples. *Science* **167**, 512–515.

TATSUMOTO MITSUNOBU and ROSHOLT JOHN N. (1970) The age of the moon. An isotopic study of U–Th–Pb system of lunar samples from Apollo 11. *Science* **167**, 461–463.

Proceedings of the Apollo 11 Lunar Science Conference, Vol. 2, pp. 1393 to 1406.

Distribution of K, Rb, Sr and Ba and Rb–Sr isotopic relations in Apollo 11 lunar samples

V. RAMA MURTHY, N. M. EVENSEN and M. R. COSCIO, JR.
Department of Geology and Geophysics, University of Minnesota,
Minneapolis, Minnesota 55455

(*Received* 2 *February* 1970; *accepted in revised form* 25 *February* 1970)

Abstract—Abundances of K, Rb, Sr and Ba in Apollo 11 crystalline rocks show a grouping into two classes corresponding to Type A and B rocks described by LSPET (1969). If these rocks are derived by partial melting, the trace element ratios are consistent with a clinopyroxene-rich source containing about 5 per cent plagioclase. Olivine up to 20 per cent may be present, but larger amounts are unlikely from density considerations. The abundance levels of the four elements in these rocks are consistent with derivation from a source material of achondrite-like composition with K depleted by a factor of 1·5–2·0 and small enrichments of Ba and Sr. Up to 30 per cent plagioclase can be separated from these partial melts without significantly affecting the trace element abundances and can provide a mechanism for explaining the observed europium depletion anomalies and the low Al contents of these rocks.

The Rb–Sr isotopic data for the crystalline rocks cluster into two close groupings on a Sr-evolution diagram. A regression line for the total rock data yields an apparent age of $4{\cdot}42 \pm 0{\cdot}12$ AE and an initial Sr^{87}/Sr^{86} ratio of $0{\cdot}6987 \pm 0{\cdot}0001$. The significance of the apparent age cannot be assessed in the absence of information on whether the rocks are cogenetic and coeval, and have remained closed with respect to Rb–Sr since the time of their formation. Data from the Rb-rich phases of these rocks (ilmenite, mesostasis etc.) indicate that a possible disturbance is such that the apparent ages may be equal to or less than the true ages. We suggest that melting processes on the lunar surface date back to early history of accretion of solid bodies in the solar system and that the inferred chemical characteristics of the source material reflects solar system fractionation processes.

INTRODUCTION

THIS paper describes the distribution of the trace elements K, Rb, Sr and Ba and the isotopic relations of the Rb–Sr system in Apollo11 lunar samples. For our study, we received six crystalline rocks of Types A and B, a breccia and the fine soil sample (LSPET, 1969). Preliminary studies by LSPET (1969) and numerous other studies have shown that the rocks are produced by crystallization of melts. The breccia and the soil are complex mixtures of lithic fragments and material produced by lunar erosional processes.

EXPERIMENTAL PROCEDURES

The first phase of our studies consisted of analyses on total rock fragments of approximately 0·5 g each. The location sites of the trace-elements in these rocks and the behavior of these rocks when subjected to mineral separation procedures are unknown. Our previous experience with ultramafic rocks (GRIFFIN and MURTHY, 1968; 1969) has shown that these elements can be highly labile and we therefore felt that a choice of total rock samples is to be preferred during the first phase. Subsequently we have studied mineral separates and density fractions from powdered lunar rocks. It will be shown later that these separates show some deviant behavior.

All elemental analyses were done by stable isotope dilution techniques. Chemical dissolutions employed specially purified reagents and all operations were performed in a dry filtered N_2 atmosphere in a newly constructed, ultra-clean laboratory. From a single sample, ion-exchange procedures and other microchemical techniques were employed to extract K, Rb, Sr, Ba, V, Cr and the rare-earth elements for elemental and isotopic studies. This paper discusses only the first four elements; other studies will be reported in a later publication. Prior to and during the analyses of the lunar samples, five contamination blank levels were determined. The contamination levels during the period of analysis were, K = 0·16 μg; Rb = 0·8 ng, Sr = 0·3 ng, and Ba = 0·07 μg per total chemical procedure. The abundances of these elements in the lunar samples and the size of the samples chosen made the contamination blank corrections negligibly low. Isotopic tracers were added to aliquots of the initial volumetric solution prepared in a quartz volumetric flask. The isotope tracer solutions were calibrated against gravimetric solutions of Johnson–Matheson spec-pure reagents. For comparative purposes, we have also measured the gravimetric solutions for Rb and Sr supplied by Professor Burnett of the California Institute of Technology. Our results on these shelf standards are lower by 1·1 per cent for Rb and 1·4 per cent for Sr, than the quoted values.

The Sr-isotopic compositions were determined on a 12 in. radius 60° sector solid source mass-spectrometer. Because of the low Rb/Sr ratios in the lunar samples, a reasonable study of the evolution of Sr^{87} in these samples requires an assured precision and reproducibility of the Sr^{87}/Sr^{86} ratios of about 0·01 per cent. This precision and reproducibility was achieved in the present study by the addition of a programmable magnetic field switching system and digital output and analysis of data on-line.

The magnetic field in our system is controlled by a commercial gaussmeter-controller (Varian FR 40) which was modified to accept remote-programmed commands corresponding to desired magnetic field values. Field values can be selected in 0·1 G increments and after an initial warm-up period of about 10 min., the reproducibility and long-term stability are on the same order. The field is sensed by a Hall-effect probe located in the magnet gap. Magnetic field values corresponding to peak tops and backgrounds on either side of the peaks of isotopes to be measured are coded in any desired sequence onto a punched paper tape loop. Through a paper tape reader, and a custom built interface, the program from the tape loop activates the field controller to step through the field values in sequence. In addition to the magnetic field values, the paper tape also contains an isotope reference number and the corresponding range factor for optimum measurement of the ion currents by a digital voltmeter of 0·01 per cent accuracy. These range factors were kept constant for all the Sr-isotopic determinations. A high precision attenuator between the output of the vibrating reed electrometer and the DVM, operable through the tape loop command ensures that readings on the DVM will always be near full scale.

Data analysis and the overall control of the system are done by a Wang Model 370 programmable calculating system with auxiliary storage, typewriter output of results, interfacing to the DVM and the magnetic field programmer. This system functions as a small scale computer which commands the magnet programmer to jump to a field value, wait for a preset period to stabilize the electronics and the field (1–3 sec) and then provide a read command to the DVM, average a preset number of DVM readings and then repeat the cycle. The peak intensities and isotope ratios are printed out after each scan of an isotope spectrum. After a "set" of data consisting of ten scans of the mass spectrum has been taken, the data system prints out the means, deviations and other desired statistical parameters for the particular set.

For Sr-isotopic measurements, 8–12 sets of 10 ratios each were collected. For the final analysis of a given run, the grand mean of all the sets was computed and a 95 per cent confidence error limit calculated. Replicate analysis of Eimer and Amend reagent standard and duplicate analyses of some of the lunar sample indicates that the Sr^{87}/Sr^{86} ratio can be measured to a precision and reproducibility of one part in 10^4.

DATA AND DISCUSSION

1. *Trace-element abundances*

Analytical data on the abundances of K, Rb, Sr and Ba in the lunar samples, separated mineral phases and density fractions are shown in Table 1 along with some relevant elemental ratios.

Table 1. Elemental abundances of K, Rb, Sr and Ba in lunar samples and mineral separates

Lunar sample	K (μg/g)	Rb (μg/g)	Sr (μg/g)	Ba (μg/g)	K/Rb	K/Ba	K/Sr	Ba/Sr	Rb/Sr
Type A									
10017-11 t.r.	2207	5·33	169	290	414	7·6	13·1	1·67	0·0315
10017-11 $\rho > 3{\cdot}32$	2512	6·95	129		361		19·47		0·0539
10017-11 $\rho < 3{\cdot}32$	2999	5·34	456		562		6·6		0·0117
10022-30 t.r.	2289	5·57	163	277	411	8·3	14·0	1·67	0·0342
10069-22 t.r.	2295	5·60	165	277	410	8·3	13·9	1·67	0·0339
Type B									
10045-23 t.r.	424	0·823	109	69·4	515	6·1	3·9	0·63	0·0076
10044-21 t.r.	816	1·15	224	128	710	6·4	3·6	0·59	0·0051
10058-29 t.r.	853	1·23	207	124	693	6·9	4·1	0·60	0·0059
10058-29 ilmenite	523	1·04	55·9	81·2	503	6·4	9·4	1·45	0·0186
10058-29 plagioclase	1112	1·08	514		1030		2·2		0·0021
10058-29 residual fraction	461	0·808	86·2		571		5·3		0·0094
Breccia									
10065-19	1406	3·69	168	200	381	7·0	8·4	1·25	0·0220
Soil									
10084-38	1020	2·88	162	162	354	6·3	6·3	1·00	0·0178

The elemental abundances of K, Rb, Sr and Ba for the crystalline rocks are distinctly grouped corresponding approximately to the fine grained vesicular Type A rocks and the coarser grained Type B rocks (COMPSTON *et al.*, 1970; GAST and HUBBARD, 1970.) The values for the soil sample and the breccia are intermediate and reflect mixing of material largely derived from the crystalline rocks examined. Comparison of the lunar sample data with chondritic and achondritic abundances (Table 2) shows several interesting features. Ba and Sr are strongly enriched in the lunar samples compared to chondrites, with a pronounced enrichment of Ba relative to Sr. K and Rb in Type A rocks are enriched but in Type B rocks are depleted relative to chondrites. The abundances of all four elements show moderate enrichments relative to achondritic values, with Type B rocks approaching achondritic levels.

A striking feature of the elemental data for the rocks is the excellent coherence between K, Ba, Sr and to a lesser extent Rb. The ratios K/Ba, K/Sr and Ba/Sr increase with increase in K content. The K/Rb ratios range from 350–700; although no pronounced trends are seen, there is a suggestion of decrease in K/Rb ratio with increase in K content.

Table 2. Comparative abundances in chondrites, achondrites and lunar samples

	Lunar	Chondrites (Urey, 1964)	Lunar Chondrites	Achondrites (Gast, 1965)	Lunar Achondrites
K	420–2300	800	0·5–2·9	330	1·3–7·0
Rb	0·8–5·6	2·7	0·3–2·1	0·20	4–28
Sr	109–224	11	10–20	80	1·4–2·8
Ba	70–280	4·0	18–70	30	2·2–8·8
K/Rb	354–710	330	1·1–2·1	1700	0·22–0·43
K/Sr	3·9–13·9	80	0·05–0·17	4·1	0·9–3·4
K/Ba	6·1–8·3	230	0·03–0·04	10·3	0·6–0·8
Ba/Sr	0·63–1·67	0·36	1·8–4·6	0·40	1·6–4·2

The trace element ratio trends and the enrichment patterns of the elements in the crystalline lunar rocks must reflect (a) compositional characteristics of the source materials from which these lavas were produced, (b) fractionation processes involving liquid–solid equilibria and (c) processes that may be peculiar to lunar surface, such as evaporation and loss of volatile elements. The extreme enrichments in Ba and Sr almost certainly are related to source characteristics, whereas the K and Rb abundances may have been affected by loss due to volatilization on the lunar surface.

We wish to explore here some of the processes and possibilities mentioned above by a comparison of the trace-element characteristics of partial melts derived from a variety of compositions appropriate for the lunar interior, with the characteristics observed in the lunar crystalline rocks. The limited number of samples available for study, the grouping of the data into essentially two types of rocks and the inadequate knowledge of the geologic setting from which the samples originated place severe limitations on any interpretation. Our efforts here are solely aimed at constructing a reasonable framework of models against which we can compare the presently available samples and those to be obtained in future missions to the moon.

The systematics of partial melting and fractional crystallization processes related to the origin of basaltic rocks have been given by GAST (1968) and have been used to evaluate the K, Rb, Sr and Ba patterns of terrestrial basalts by several authors (GAST, 1968; GRIFFIN and MURTHY, 1969; PHILPOTTS and SCHNETZLER, 1970). To evaluate the characteristics of lunar partial melts, the relevant variables to be considered are: (a) modal composition of the source material, (b) distribution coefficients for the elements between liquid and solid phases and (c) degree of partial melting and fractional crystallization.

O'HARA *et al.* (1970) and RINGWOOD and ESSENE (1970) have both concluded from high pressure experiments on the Apollo 11 samples that the major part of the moon cannot be composed of material similar to Tranquillity Base basalts as these would transform to eclogite at relatively shallow depths corresponding to a pressure of 10 kbar and would produce a moon of considerably greater density than is observed. They suggest that partial melting of a source area composed dominantly of pyroxene would satisfactorily explain the petrology observed in these rocks. Accordingly, we have made a number of calculations to determine the fractionation trends of trace elements and elemental ratios with varying degrees of partial melting of a pyroxene-rich source.

The equations derived by GAST (1968) have been slightly modified to allow for complete melting of a phase, as when large partial melts are made from an initial bulk solid containing only a small amount of a low-melting phase. The distribution coefficients used in all calculations are those derived by GRIFFIN and MURTHY (1968, Table 9A) from terrestrial ultramafic and basaltic rocks.

The four phases chiefly considered were clinopyroxene, orthopyroxene, plagioclase and olivine. Melting models were constructed for initial solids with pyroxene ranging from pure clinopyroxene to pure orthopyroxene and plagioclase contents from 0 to 20 per cent. A number of additional model calculations were made to investigate particular properties of the system, some of which will be discussed later in the paper. All models were examined up to 30 per cent partial melting.

In the pure pyroxene compositions, K, Rb and Ba become rather strongly enriched in the liquids; Sr is only moderately enriched unless orthopyroxene forms 70 per cent or more of the source. If 20 per cent plagioclase is introduced into the initial solid, Ba enrichments are decreased considerably and K enrichments decreased by a smaller factor. Sr enrichments become somewhat lower, and are much more constant with varying orthopyroxene content.

Up to 20 per cent olivine can be added to the source area without affecting the trends of trace element or elemental ratio fractionation, and with only slight effect on the actual numerical values. Both the petrologic studies earlier cited (O'HARA *et al.*, 1970; RINGWOOD and ESSENE, 1970) and density considerations make it unlikely that larger amounts of olivine are present in the lunar interior. For this reason, most of our calculations were made for an olivine-free source.

We have also examined a number of models in which partial melting is followed by fractional crystallization of plagioclase. A number of authors (e.g. MURTHY *et al.*, 1970; GAST and HUBBARD, 1970) have suggested that the negative europium anomalies observed in the Apollo 11 samples may be due to separation of plagioclase cumulates from the parent melts. This hypothesis is especially attractive in view of the very low partial pressures of oxygen indicated by the mineral assemblages in these rocks (LSPET, 1969) which would tend to produce a higher Eu^{+2}/Eu^{+3} ratio than is found in terrestrial systems and thus enhance the incorporation of Eu by plagioclase. Our calculations show that up to 30 per cent plagioclase can be separated from the melts by fractional crystallization without seriously affecting abundances of the elements considered. Whether this is adequate to account for the observed Eu anomalies depends upon the effect of pO_2 on the Eu distribution coefficient (HASKIN *et al.*, 1970). Distribution coefficients for K, Sr and Ba between liquid and plagioclase (GRIFFIN and MURTHY, 1969; PHILPOTTS and SCHNETZLER, 1970) show that plagioclase concentrates Sr relative to K and to a lesser extent, relative to Ba. Thus plagioclase separation will only enhance the K/Sr and Ba/Sr trends produced by partial melting. Based on these trace-element ratio trends and the Eu depletion anomaly, we have earlier suggested (MURTHY *et al.*, 1970) that gabbroic-anorthosite, and anorthositic rocks may also be found on the moon. Detailed petrological examination of the Apollo-11 materials have shown the presence of such rocks, very likely derived from the highlands (WOOD *et al.*, 1970).

In comparing the trace element enrichments produced by the partial melting models with the chondrite-normalized lunar abundances as shown in Table 2, it becomes obvious that the lunar rocks are very difficult to produce by simple fractionation from a source having chondritic abundances of these trace elements. K and Rb are actually depleted relative to chondritic abundances, while the partial melting models are always characterized by fairly high K and Rb enrichments. Ba and Sr on the other hand, are enriched in the lunar samples more than can be accounted for by any but extremely small partial melting (1 per cent) which would also strongly enrich K and Rb.

Normalization of the lunar samples to achondritic abundances, however, produces a much more satisfactory picture. Here the abundance pattern seen is one of moderate to high enrichment in K, Rb, and Ba and moderate enrichment in Sr; this type of fractionation is quite typical of many of the models examined. This suggests that the achondritic abundances of these four elements represent at least a first approximation to the abundances in the source area from which the lunar samples were formed. It then becomes reasonable to investigate the achondritic model more closely to determine how the formation may have occurred and what constraints we must impose on the system in the process.

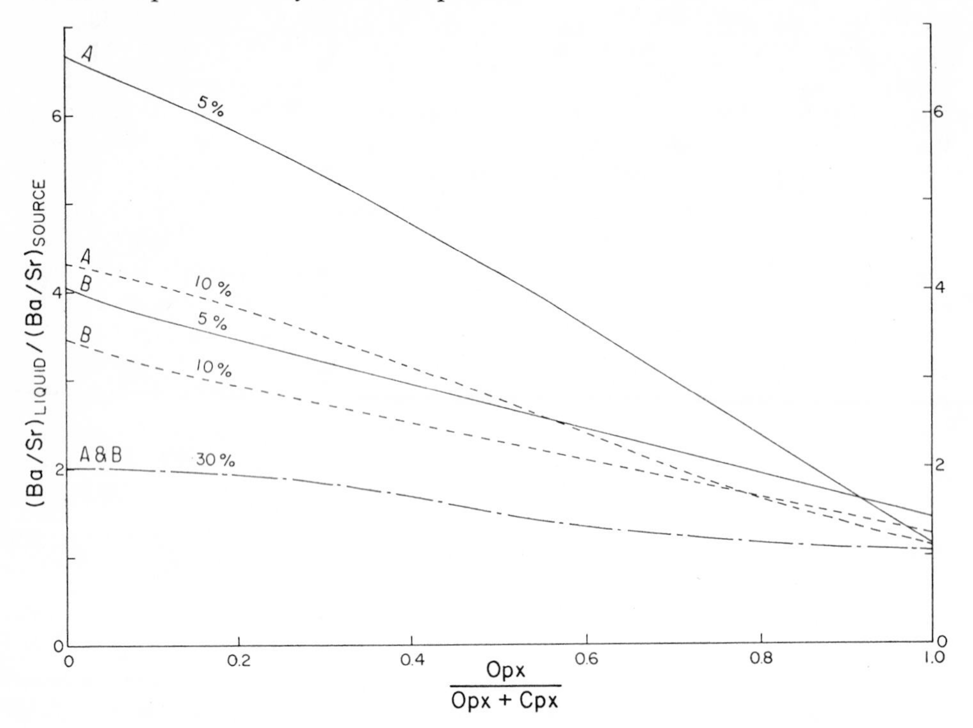

Fig. 1. Fractionation trends of Ba/Sr ratio as a function of orthopyroxene–clinopyroxene ratio of source material. A—100% pyroxene; B—10% plagioclase, 90% pyroxene. Percentage figures indicate degree of partial melting.

It is interesting to note that the Ba/Sr ratios in chondrites and achondrites are virtually identical. This suggests that the definite Ba/Sr fractionation observed in the lunar samples may be quite significant if the source material is at all similar to either of these compositions. Figure 1 demonstrates that if partial melting exceeds 5 per cent, the source material must contain at least half of its pyroxene as clinopyroxene in order to produce the observed range and maximum values of Ba/Sr fractionation. This is particularly true if the source area contains plagioclase in addition to pyroxene, and is in accord with the suggestion of O'HARA *et al.* (1970) that clinopyroxene is dominant over orthopyroxene in the lunar interior. In addition, the maximum Ba/Sr ratios observed in the lunar samples preclude the existence of more than about 10 per cent plagioclase in the source area if they are to be derived by partial melting from either a chondrite or achondrite-like trace element pattern in the lunar mantle.

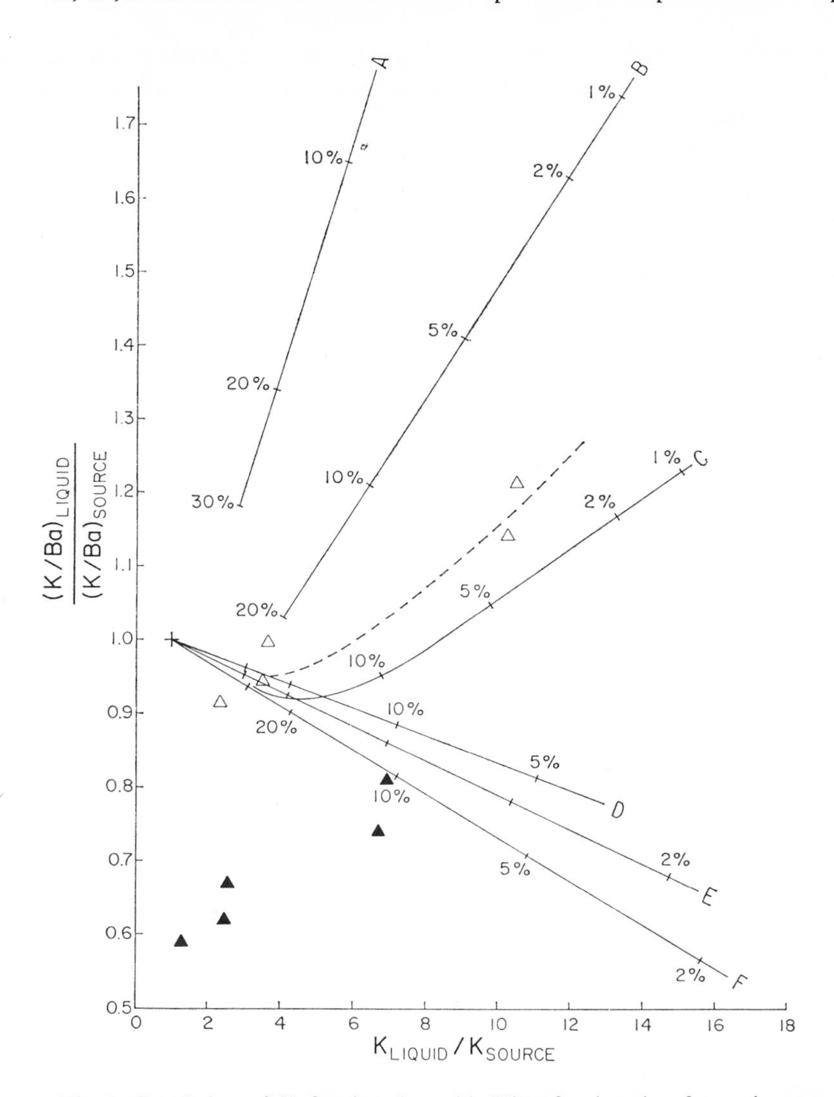

Fig. 2. Correlation of K fractionation with K/Ba fractionation for various source compositions. A—20% plagioclase, 64% clinopyroxene, 16% orthopyroxene; B—10% plagioclase, 72% clinopyroxene, 18% orthopyroxene; C—5% plagioclase, 76% clinopyroxene, 19% orthopyroxene; D—64% clinopyroxene, 16% orthopyroxene 20% olivine; E—80% clinopyroxene, 20% orthopyroxene; and F—20% clinopyroxene, 80% orthopyroxene. Note that for 100 per cent melting, all curves must arrive at the point (1, 1), marked by a +, representing the composition of the starting material. Solid triangles represent lunar crystalline rocks normalized to achondritic abundances. Percentage figures indicate degree of partial melting. Other symbols are explained in the text.

The K/Ba fractionation trends for a wide variety of starting materials are shown in Fig. 2. It is immediately apparent that the lunar rocks, particularly the group B rocks, cannot be derived from a source of achondritic K and Ba abundance by ordinary fractionation processes, since none of the partial melting curves pass through the lunar sample points. Furthermore, if we assume the lunar samples to be cogenetic, at least to the extent of being produced by similar processes from similar starting materials, only partial melting processes involving plagioclase can reproduce the observed trend of increasing K/Ba ratio with increasing K. Although curve C,

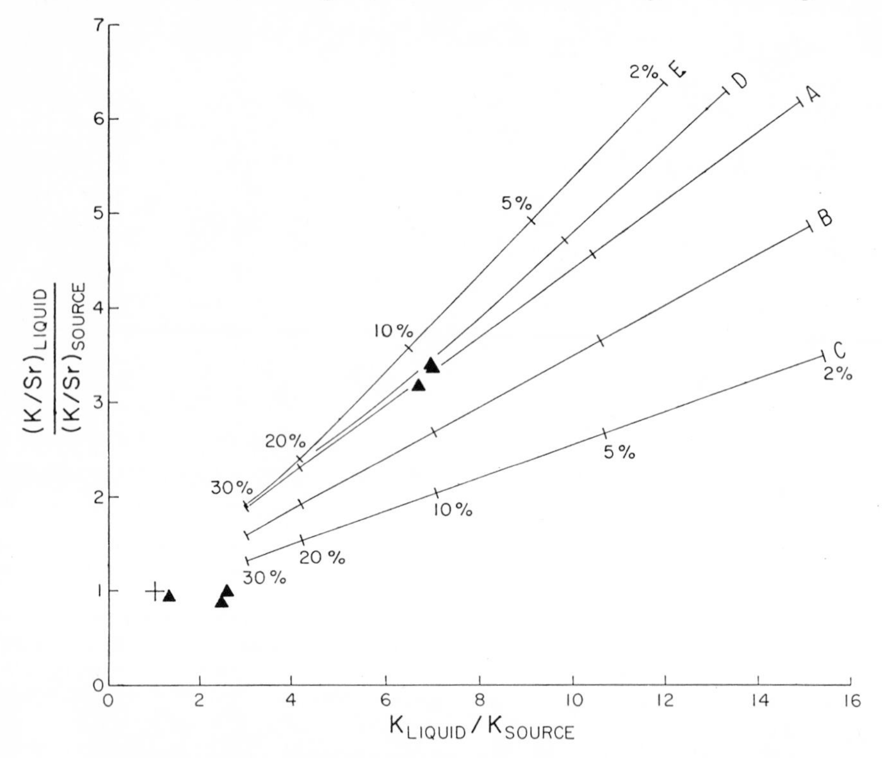

Fig. 3. Correlation of K fractionation with K/Sr fractionation for various source compositions. A—80% clinopyroxene, 20% orthopyroxene; B—60% clinopyroxene, 40% orthopyroxene; C—40% clinopyroxene, 60% othopyroxene; D—5% plagioclase, 76% clinopyroxene, 19% orthopyroxene and E—10% plagioclase, 72% clinopyroxene, 18% orthopyroxene. Solid triangles represent lunar crystalline rocks normalized to achondritic abundances. Percentage figures indicate degrees of partial melting.

representing source material containing 5 per cent plagioclase, has the proper slope, it does not pass through the points representing the achondrite—normalized lunar samples. If, however, the samples are normalized to a source depleted in potassium with respect to achondrites, the corresponding points on the diagram will be moved upward along a series of tie lines from the origin (0, 0) passing through the points. The distance moved along the tie lines will be proportional to the K depletion in the source relative to achondrites, and to the distance from the origin. Alternatively, if the partial melts derived from achondritic source material are subsequently depleted in K, perhaps by a process of volatilization near the lunar surface, the compositions along the partial melting curves will be moved downward along the same tie lines. The net effect of the two processes is the same, and cannot be distinguished on this type of diagram. The open triangles in Fig. 2 represent the effect of K depletion by a factor of $\sim$1·5 upon the lunar sample points; they then lie upon the dashed curve representing partial melting from a source with slightly more than 5 per cent plagioclase and an orthopyroxene/clinopyroxene ratio of 1:4. The group A rocks in this case represent 2–5 per cent partial melts and the group B rocks 20–30 per cent partial melts from this composition. This model is quite sensitive to both K depletion and plagioclase content required to reproduce the slope of the lunar sample data, although changes in the pyroxene ratio or addition of olivine will not greatly affect it. If the potassium depletion is in the source material, it would correspond to a K abundance of $\sim$200 ppm. Potassium depletion by either mechanism could be accompanied by lesser fractionations of Rb, Sr and Ba; this would change the figures quoted slightly but would not affect the overall conclusions.

Figure 3 shows a similar plot for K/Sr fractionation. Again none of the partial melting curves can satisfactorily account for the lunar sample points, although the mismatch is not as bad as in the previous diagram. Although this system is much less sensitive to variations in the plagioclase content of the source material, it can be seen that a high clinopyroxene/orthopyroxene ratio is required to fit the slope of the lunar data. If potassium depletion by a factor of $\sim$1·5 is again applied to the sample points, they provide a reasonable match to curve D (corresponding to curve C in Fig. 2), again at 2–5 per cent melting for group A and 30 per cent melting for group B rocks. The fit is somewhat better if the K depletion is accompanied by a slight Sr enrichment, suggesting that the lunar rocks may be derived from a source material more removed from chondritic compositions than the achondrites already are.

Because the dispersion of the samples within groups A and B does not in general follow these partial melting trends, it is likely that this dispersion is due to later fractionation by other means, following separation of these two groups by partial melting processes.

In summary, the trace element abundances in the Apollo 11 lunar samples are consistent with the derivation of these rocks by varying degrees of partial melting of a dominantly clinopyroxene source area, as suggested by O'HARA *et al.* (1970), containing $\sim$5 per cent plagioclase. Up to 20 per cent olivine is permissible. Source abundances of K, Rb, Sr and Ba are probably similar to those in achondrites, with the potassium depleted either in the source or in the partial melts, perhaps by vaporization. Study of samples from subsequent lunar missions should help to verify or disprove this model, and may aid in determining the lateral homogeneity of the lunar interior and of the processes which have produced the crystalline rocks on the surface.

2. Rb–Sr *isotopic relations*

The Rb–Sr isotopic data for the lunar samples, and mineral and density separates from a Type A rock (10017) and a Type B rock (10058) are shown in Table 3. The Sr^{87}/Sr^{86} values are normalized to Sr^{86}/Sr^{88} ratios of 0·1194. The isotopic ratios

Table 3. Rb–Sr isotopic data of lunar rocks and mineral fractions

Lunar sample	$^{87}Rb/^{86}Sr$ ($\times 10^2$)	$^{87}Sr/^{86}Sr$*	E† ($\times 10^4$)
Group A			
10017-11 t.r.	9·18	0·70473	1·6
10017-11 Heavy fraction ($\rho > 3·32$)	15·57	0·70701	0·30
10017-11 Light fraction ($\rho < 3·32$)	3·39	0·70025	1·8
1022-30 t.r.	9·90	0·70490	0·65
10069-22 t.r.	9·84	0·70483	2·6
		0·70478	1·7
Group B			
10045-23 t.r.	2·17	0·69991	0·92
10044-21 t.r.	1·48	0·69962	0·56
10058-29 t.r.	1·72	0·69997	0·69
		0·69990	0·86
10058-29 ilmenite	5·35	0·70153	1·5
10058-29 plagioclase	0·610	0·69924	1·4
10058-29 residual fraction	2·71	0·70057	0·8
Breccia			
10065-19	6·39	0·70298	0·77
Soil			
10084-38	5·13	0·70179	0·85
Nuevo Laredo achondrite	1·49	0·69968	1·8
Eimer and Amend reagent $SrCO_3$	—	0·70820	2·1

* Normalized to $^{86}Sr/^{88}Sr = 0·1194$. The lunar sample isotopic ratios are relative to the Eimer and Amend reagent standard value of 0·7082.

† E is the 95 per cent confidence limit and is equal to

$$2\left[\frac{\sum_{i=1}^{n}(\bar{X}_i - m)^2}{n(n-1)}\right]^{1/2}$$

where n is the number of sets of 10 scans each and $(\bar{x}_i - m)$ is the difference between the mean value of the ith set and the mean of all sets.

reported for the lunar samples are relative to the Eimer and Amend Reagent $SrCO_3$ value of 0·70820 obtained by applying a measured systematic bias correction factor to our previously reported value of 0·70734 for this standard (MURTHY *et al.*, 1970).

As with the trace element abundances, the isotopic data for the lunar rocks cluster into two groups corresponding to Types A and B, with Type A rocks showing higher Rb/Sr and Sr^{87}/Sr^{86} ratios than Type B rocks, as shown in the Sr-evolution diagram in Fig. 4. The coordinates for the soil and breccia samples are intermediate between Type A and B points showing that to a large extent the soil and breccia are mixtures of Type A and B components. The total range in Sr^{87}/Sr^{86} ratio observed in the rocks is 0·8 per cent, and the dispersion in each group is inadequate to derive any chronologic information by making assumptions of isochronism, i.e. uniform initial Sr^{87}/Sr^{86} ratio and chemical closure for the same period, in each group. In our previous report (MURTHY *et al.*, 1970) we presented isochron ages for Type A and B rocks separately but this was due to an incorrect assignment by type for one of the rocks.

If it can be assumed that both types of crystalline rocks had formed from the same source at one time, it is possible to combine the total rock data to obtain both an

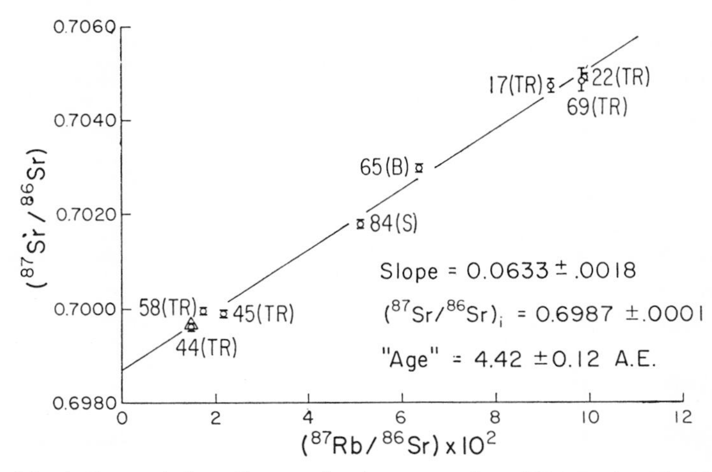

Fig. 4. Sr—evolution diagram for lunar samples. TR—total rock, S—soil, B—breccia $(Sr^{87}/Sr^{86})_i$ = 0·69871 ± 0·00012; slope of the regression line for crystalline rocks is 0·0633 ± 0·0018 corresponding to an apparent age of 4·42 ± 0·12 AE. Coordinates for the soil and breccia fall close to the regression line and represent mixing of Type A and B crystalline rock components. Triangle on top of data point for 44 (TR) indicates the point described by the achondrite, Nuevo Lavedo.

apparent age and an initial Sr^{87}/Sr^{86} ratio for the system. A regression line joining the crystalline rock data on the Sr evolution diagram is shown in Fig. 4. The slope of the line as 0·0633 ± 0·0018 and corresponds to an "age" of 4·42 ± 0·12 AE using a Rb^{87} decay constant of $1{\cdot}39 \times 10^{-11}$ yr^{-1}, and the intercept corresponds to a $(Sr^{87}/Sr^{86})_i$ = 0·69871 ± 0·00012. During the period of lunar sample analysis, we also analyzed the basaltic achondrite, Nuevo Laredo. The $(Sr^{87}/Sr^{86})_i$ for this meteorite, obtained by correction for radiogenic growth for 4·5 AE, is 0·69872 ± 0·00018 compared to 0·699047 ± 0·000118 obtained by Papanastassiou and Wasserburg (1969). Our value is in excellent agreement with that deduced for the lunar samples in the above model. The point described by Nuevo Laredo in the Sr evolution diagram is located on the lunar rock best fit line. Despite this agreement, the significance of this age must rest upon independent arguments that call for a coeval genesis of Type A and B rocks from the same source. Compston *et al.* (1970) have found that both $(Sr^{87}/Sr^{86})_i$ and isochron ages are approximately equal for whole rock and mineral separates, and have therefore suggested that the rocks are coeval and derived from the same source material. The total rock age we have obtained is significantly higher than the combined mineral and total rock age of 3·81 ± 0·07 AE given by Compston *et al.* (1970) or the internal isochron age of 3·65 ± 0·05 by Albee *et al.* (1970).

Because of the above discrepancies and the difficulties with the interpretation of the total rock ages, an effort was made to determine internal isochrons for some individual rocks by using hand-picked mineral fractions and density separates. Figure 5 shows the data for 10017 (Type A) and 10058 (Type B) rocks and their separated phases. The total rock regression line from Fig. 4 is shown for comparison. For 10058, data points of plagioclase, total rock, and the residual fraction (R) from which both plagioclase and "ilmenite" have been hand picked fall on a straight line

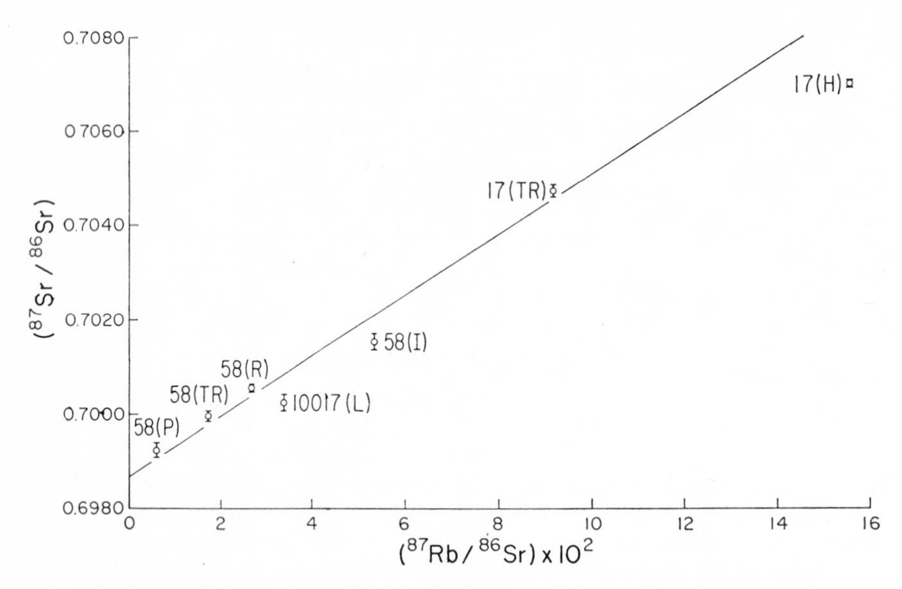

Fig. 5. Sr—evolution diagram for total rock and mineral and density fractions of rocks 10017 and 10058. TR—total rock, P—Plagioclase, I—Ilmenite, H—density fraction $\rho > 3{\cdot}32$, L—density fraction $\rho < 3{\cdot}32$, R—Residual fraction from which both plagioclase and ilmenite have been removed by hand picking. The regression line from the total rock data of Fig. 4 is shown for comparison.

corresponding to an age of $4{\cdot}42 \pm 0{\cdot}1$ and an initial Sr^{87}/Sr^{86} ratio of 0·69886. Both the age and initial ratio agree well with the parameters obtained from the total rock data of Fig. 4. However, the data point for the hand-picked "ilmenite" fraction deviates significantly from the linear relationship exhibited by the other three points for this rock. Similar deviations are shown by density fractions of 10017 containing ilmenite (17 H) and plagioclase, glass, mesostasis fraction (17 L) as shown in Fig. 5. In both rocks, the direction of deviation of these high Rb phases indicates loss of Sr^{87} relative to Rb. We do not know at present whether the disturbance to these systems was induced by the laboratory processes of grinding, mineral separation etc. or by natural causes on the Moon. Rb and Sr are most likely tenuously bound in the "ilmenite" fractions and the mesostasis. We are currently investigating the lability of these elements in these and other phases in order to evaluate the nature of the disturbance of the Rb–Sr isotopic system in mineral phases from lunar crystalline rocks. It is clear however that internal isochron relationships obtained primarily by the use of these Rb-enriched phases (ilmenite, mesostasis) may tend to give minimal ages, and higher $(Sr^{87}/Sr^{86})_i$ ratios, if these systems have been disturbed in the manner found in the present work.

It is instructive to note that using the highly sensitive U–Th–Pb system, TATSUMOTO and ROSHOLT (1970) suggest younger mineral isochron ages compared to toal rocks, indicating possible parent–daughter fractionations in the mineral phases subsequent to their formation. Also, the locations of the crystalline rock data on the Concordia diagram (SILVER, 1970; TATSUMOTO and ROSHOLT, 1970) indicate disturbances in the

U–Th–Pb system, even for the total rocks. In view of the lack of adequate documentation of the behavior of the Rb–Sr isotopic system in the lunar rocks and mineral fractions, the significance of the difference among the numerous Rb–Sr "age" values for the lunar samples cannot be evaluated at present.

From our total rock data and from the mineral fraction data for 10058 (excluding the "ilmenite" for reasons mentioned earlier) we suggest that Rb–Sr fractionation in these rocks occurred at 4·4 b.y. ago. If these fractionation processes can be equated to melting processes that led to the formation of the crystalline rocks, our data suggest that melting processes on the lunar surface date back to the formation of solid bodies in the solar system. By whatever mechanism the earth acquired the moon, whether by fission, volatilization or capture, this process must have taken place in the earliest history of the earth, probably contemporaneously with the terminal stages of accretion of the earth.

The equality of the initial Sr^{87}/Sr^{86} for the lunar rocks and that for the Nuevo Laredo achondrite permits an evaluation of the *time interval* between the formation of the lunar crystalline rocks and the achondritic meteorites, under certain assumptions regarding the Rb/Sr ratio of the reservoir from which these objects were derived (Papanastassiou and Wasserburg, 1969). For example, for a reservoir having a chondritic Rb/Sr ratio (0·25) we infer a time interval of less than 15 m.y. If the achondrites originate from the moon, as suggested by Duke and Silver (1967) and the Rb/Sr ratio is $\sim$0·02, the time interval between the formation of achondrites and the lunar rocks corresponds to about 180 m.y.

Acknowledgments—We are grateful to Mr. Bor-ming Jahn for help in mass spectrometric analyses and general assistance in the project. We are greatly indebted to Mr. Gerald Peterson, and Mr. Stuart Nelson of the Electronic Shop of our Physics Department for the design and construction of much of the equipment related to the automatic data collection system. Financial support for this work through NASA contract NAS 9-7884 is acknowledged.

References

Albee A. L., Burnett D. S., Chodos A. A., Eugster O. J., Huneke J. C., Papanastassiou D. A., Podosek F. A., Russ G. Price II, Sanz H. G., Tera F. and Wasserburg G. J. (1970) Ages, irradiation history, and chemical composition of lunar rocks from the Sea of Tranquillity. *Science* **167**, 463–466.

Compston W., Arriens P. A., Vernon M. J. and Chappel B. W. (1970) Rubidium–strontium chronology and chemistry of lunar material. *Science* **167**, 474–476.

Duke M. B. and Silver L. T. (1967) Petrology of eucrites, howardites and mesosiderites. *Geochim. Cosmochim. Acta* **31**, 1637–1665.

Gast P. W. (1965) Terrestrial ratio of potassium to rubidium and the composition of the earth's mantle. *Science* **147**, 858–860.

Gast P. W. (1968) Trace element fractionation and the origin of tholeiitic and alkaline magma types. *Geochim. Cosmochim. Acta* **32**, 1057–1086.

Gast P. W. and Hubbard N. J. (1970) Abundance of alkali metals, alkaline and rare earths, and strontium-87/strontium-86 ratios in lunar samples. *Science* **167**, 485–487.

Griffin W. L. and Murthy V. Rama (1968) Abundances of K, Rb, Sr and Ba in some ultramafic rocks and minerals. *Earth Planet. Sci. Lett.* **4**, 497–501.

Griffin W. L. and Murthy V. Rama (1969) Distribution of K, Rb, Sr and Ba in some minerals relevant to basalt genesis. *Geochim. Cosmochim. Acta* **33**, 1389–1414.

Haskin L. A., Helmke P. A. and Allen R. O. (1970) Rare Earth Elements in returned lunar samples. *Science* **167**, 487–490.

LSPET (LUNAR SAMPLE PRELIMINARY EXAMINATION TEAM) (1969) Preliminary examination of lunar samples from Apollo 11. *Science* **165,** 1211–1227.

MURTHY V. RAMA, SCHMITT R. A. and REY P. (1970) Rubidium–strontium age and elemental and isotopic abundances of some trace elements in lunar samples. *Science* **167,** 476–479.

O'HARA M. J., BIGGAR G. M. and RICHARDSON S. W. (1970) Experimental petrology of lunar material: the nature of mascons, seas and the lunar interior. *Science* **167,** 605–607.

PAPANASTASSIOU D. A. and WASSERBURG G. J. (1969) Initial strontium isotopic abundances and the resolution of small time differences in the formation of planetary objects. *Earth Planet. Sci. Lett.* **5,** 361–376.

PHILPOTTS J. A. and SCHNETZLER C. C. (1970) Phenocrysts matrix partition coefficients for K, Rb, Sr and Ba, with application to anorthosite and basalt genesis. *Geochim. Cosmochim. Acta* **34,** 307–322.

RINGWOOD A. E. and ESSENE E. (1970) Petrogenesis of lunar basalts and the internal constitution and origin of the moon. *Science* **167,** 607–610.

SILVER L. T. (1970) Uranium–thorium–lead isotope relations in lunar materials. *Science* **167,** 468–471.

TATSUMOTO M. and ROSHOLT J. N. (1970) Age of the moon: an isotopic study of uranium–thorium–lead systematics of lunar samples. *Science* **167,** 461–463.

UREY H. C. (1964) A review of atomic abundances in chondrites and the origin of meteorites. *Rev. Geophys.* **2,** 1–34.

WOOD J. A., DICKEY J. S. JR., MARVIN U. B. and POWELL B. N. (1970) Lunar anorthosites. *Science* **167,** 602–604.

Proceedings of the Apollo 11 Lunar Science Conference, Vol. 2, pp. 1407 to 1423.

Primordial radionuclide abundances, solar proton and cosmic-ray effects and ages of Apollo 11 lunar samples by non-destructive gamma ray spectrometry*

G. DAVIS O'KELLEY and JAMES S. ELDRIDGE
Oak Ridge National Laboratory, Oak Ridge, Tennessee 37830

and

ERNEST SCHONFELD and P. R. BELL†
Manned Spacecraft Center, Houston, Texas 77058

(*Received* 2 *February* 1970; *accepted in revised form* 25 *February* 1970)

Abstract—A gamma-ray spectrometer with low background was used to determine the radioactivity of crystalline rocks, breccias and fine material from the Apollo 11 mission. Measurements were made in special facilities at the Lunar Receiving Laboratory. Nuclides measured were ^{40}K, ^{232}Th, ^{238}U, ^{22}Na, ^{26}Al, ^{44}Ti, ^{46}Sc, ^{48}V, ^{52}Mn, ^{56}Co and ^{60}Co. Upper limits were assigned for the concentrations of ^{7}Be. Concentrations of K, Th and U ranged between 480 and 2550, 1·01 and 3·30 and 0·26 and 0·83 ppm, respectively. Potassium concentrations were near those of chondrites or some oceanic gabbros and basalts. The Th and U contents are near those of ordinary terrestrial basalts and the concentration ratio of Th to U is constant at $3·9 \pm 0·2$, in agreement with the terrestrial value. Products of low-energy nuclear reactions such as ^{56}Co showed pronounced concentration gradients at rock surfaces, which suggest that solar proton activation is the principal source of such nuclides. Concentrations of K, Th, U, ^{22}Na and ^{26}Al determined here were combined with concentrations of rare gases to estimate gas-retention ages and cosmic-ray exposure ages with ranges of 2200–3970 and 38–340 m.y., respectively, for four of the crystalline rocks. Bulk densities were determined for four crystalline rocks, two breccias and a sample of fines.

INTRODUCTION

GAMMA-RAY spectrometry is a useful method for radiochemical analysis of many types of geological samples. Because gamma rays are very penetrating, gamma radioactivity of samples from a few grams to several kilograms can be determined after appropriate calibration. Modern techniques are capable of resolving multiple-component spectra, and so the analysis often is also non-destructive. Gamma-ray analysis is an excellent method for obtaining average concentrations in whole rocks or other massive samples, which may be of value when such large samples contain small-scale inhomogeneities. The power of this method of analysis was demonstrated in early experiments by VAN DILLA et al. (1960) and ROWE (1963), who determined radionuclides in meteorites from singles spectra recorded with large NaI(T1) scintillation spectrometers of low background. The high sensitivity of the coincidence method for detection of ^{26}Al was demonstrated by ANDERS (1960).

Just as for terrestrial samples and meteorites, gamma-ray spectrometry can yield information on the primordial radionuclide content of lunar samples. Radioactive

* Research carried out under Union Carbide's contract with the U.S. Atomic Energy Commission through interagency agreements with the National Aeronautics and Space Administration.

† Present address: Oak Ridge National Laboratory, Oak Ridge, Tennessee 37830.

species of crucial importance in this connection are ^{40}K, together with U, Th and their respective decay products.

Because the moon has neither an atmosphere nor a significant magnetic field to protect it from bombardment by galactic cosmic rays and heavy charged particles from the sun, a variety of radioactive nuclei are constantly being formed by nuclear reactions in the lunar surface. Further, since the lunar samples have been irradiated in known orientations in space and are free from atmospheric ablation, they are in some respects more suitable objects for study than meteorites. A careful study of the induced radionuclides was expected to yield important information concerning the bombardment history of the material and the history of the incident particle fluxes responsible for production of the observed species. It was also hoped that this study would yield information which would prove useful in estimating the rate of turnover of the lunar surface due to impact.

As expected, effects attributable to bombardment by particles from recent solar flares were detected in this study. Because of the short range of the solar particles and the high yields of some nuclei compared to production by galactic cosmic rays, gamma-ray spectrometry offers a technique for determination of the most recent orientation of rocks on the lunar surface before return, in the absence of other documentation.

Many of the radionuclides of interest have short half-lives; hence, measurements must begin as soon as possible after the arrival of the lunar samples on earth. Analyses of samples from Apollo 11 were carried out at the Lunar Receiving Laboratory, Houston, Texas, during and shortly after the sample quarantine period. A preliminary account of this work was reported by LSPET (1969). The preliminary data have now been refined through more detailed calibrations and data analyses (O'Kelley *et al.*, 1970). The new results and a discussion of our experimental techniques which we have not previously published are the subjects of the present report.

Experimental Methods

Gamma-ray spectrometer system

The gamma-ray spectrometer used in these studies has a very low background and is located in the Radiation Counting Laboratory (RCL) of the Lunar Receiving Laboratory (LRL). The location and general features of the RCL were described by McLane *et al.* (1967). Briefly, the heart of the RCL is a vault supplied with radon-free air, enclosed by walls of compacted, crushed dunite 0·9 m thick, inside a welded steel liner. The vault is entered through a maze which also functions as an airlock, and all interior materials have been verified for low radioactive contamination. The entire facility is 15 m below ground and covered by an overburden of earth and rock. With the exception of preamplifiers, all ancillary electronic instrumentation is located in an adjacent control room on the same underground level.

The detector system used for the Apollo 11 mission consisted of two large scintillation detectors at 180° with the sample included between them. Each detector consisted of a crystal of NaI(Tl) 23 cm in dia. and 13 cm long, coupled optically to a pure NaI light guide 23 cm in dia. and 10 cm long. Gamma-ray absorption in the light guide attenuated the gamma radiation from the four RCA type 4521 photomultiplier tubes used to collect the light.

The principal detectors are surrounded by a mantle of NE-102 plastic scintillator (Nuclear Enterprises, Ltd., Winnipeg, Manitoba, Canada) 30 cm thick, viewed by 22 RCA type 4521 photomultiplier tubes. The mantle is connected in anticoincidence with the NaI(Tl) detectors both to suppress background events and to improve the peak-to-total ratio for gamma spectra by rejecting those events for which a gamma ray transfers only a portion of its energy to the NaI(Tl), accompanied by the escape of a degraded photon.

Background is further reduced by enclosing the detectors and anticoincidence mantle inside a lead shield with walls 20 cm thick. The shield weighs about 24 metric tons and was constructed by casting lead "concrete" inside a reinforced steel liner. The "concrete" consisted of 83% lead shot, 16% Chemtree-82, (product of Chemtree Corporation, Central Valley, N.Y. 10917) and 1% LiF. These constituents were mixed with distilled water and cast in place. All materials used in the construction of the shield were carefully selected for low levels of radioactive contamination.

Because the sample is viewed by two detectors, spectra may be recorded either as non-coincident, or "singles," data if an event occurs in one detector only, or as a gamma-gamma coincidence event if a signal is produced in both detectors simultaneously. The data acquisition system includes a coincidence-anticoincidence logic circuit and dual 12-bit analog-to-digital converters interfaced to a Digital Equipment Corporation PDP-9 computer with 16,384-word memory. The computer was programmed* for simultaneous recording of singles data from each detector at a resolution of 255 channels; gamma–gamma coincidence data in a folded matrix with a resolution of 127 × 127 channels but requiring only 8192 storage locations; and a sum-coincidence spectrum of 255 channels. Energy scales were 20 keV/channel for 255-channel spectra and 40 keV/channel for 127-channel spectra. All of these spectra were used in the detailed analysis of each lunar sample, as described below.

Although some methods for storage of coincidence data from two detectors are well established through discussions in the instrumentation literature [e.g. see BECKURTS *et al.* (1964)], the two parameter matrix used for the present work is unusual and warrants a brief description. If the two NaI(Tl) principal detectors are adjusted to identical gains and responses, then the energy coordinates in the matrix $(E_x, E_y) \equiv (E_y, E_x)$. As shown in Fig. 1, the coincidences between the full-energy peaks of a cascade of two gamma rays can be stored at a single location in an energy–energy correlation diagram of triangular shape. For example, in Fig. 1 the coincidences from the 1·17–1·33 MeV gamma cascade in the decay of ^{60}Co are stored in a single memory location which includes both events of the type (E_x = 1·17 MeV, E_y = 1·33 MeV) and (E_x = 1·33 MeV, E_y = 1·17 MeV). It is clear that only half the memory capacity is required for this method of storage compared with the normal two-parameter memory configuration.

The extent to which the design objectives (MCLANE *et al.*, 1967) have been met can be seen from the background performance. In the energy interval 0·1–2·0 MeV, the background of one detector (5200 cm^3 active volume) in the singles mode is 476 counts min^{-1} without anticoincidence and 79 counts min^{-1} with anticoincidence. These values are at least comparable to the best systems described in the literature to date (c.f. WOGMAN *et al.*, 1967).

Sample preparation and system calibration

The proposed operation of the gamma-ray spectrometer system and the associated data reduction techniques required the collection of spectra of all expected radionuclides in a variety of geometrical configurations. To accomplish the desired calibrations before measurement of lunar samples, radionuclide solutions of twelve expected species were purchased from commercial suppliers and calibrated by absolute gamma-ray spectrometry to an overall uncertainty $\leq$5 per cent. Uranium and thorium certified standards were obtained from the U.S. Atomic Energy Commission, New Brunswick Laboratory. Five other radionuclides were prepared by reactor or cyclotron irradiations at Oak Ridge National Laboratory. Potassium standards were prepared from reagent grade potassium chloride, and for some sources from potassium enriched in ^{40}K. Table 1 lists the radionuclide standards measured prior to the Apollo 11 samples.

The accuracy of quantitative radionuclide analysis by the least-squares technique is strongly dependent on the similarity of the standard response function of each radionuclide to the corresponding response in the unknown sample. Therefore, it was necessary to collect the standard

* The PDP-9 programs were written by W. W. BLACK, W. RICHARDSON and R. L. HEATH, Idaho Nuclear Corporation.

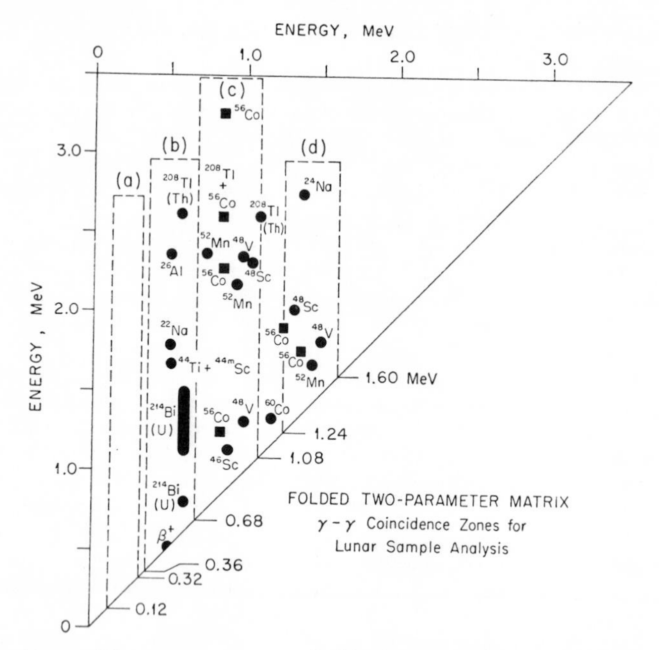

Fig.1. A map of the gamma–gamma coincidence matrix, showing principal coincidences for some nuclides of interest (black regions). Dashed lines enclose four zones (a), (b), (c) and (d), whose component planes were summed for least-squares analysis. Upper energy bounds of each zone are denoted by the horizontal dashed lines.

responses in a variety of configurations to cover a range of lunar sample sizes. In addition, the electronic densities of the standards must simulate those of the lunar samples so that scattering effects will be comparable. Electrolytically reduced iron powder was found to be a suitable dispersing medium in that it is sufficiently free of radioactive contamination and has a bulk density of 3·4–3·5 g/cm^3. Accordingly, small assayed portions (10^4–10^5 dis/min) of each radionuclide were mixed thoroughly with the appropriate amount of iron powder to ensure homogeneity. Then, standard 25-, 50-, 100-, 200-, 300-, 600-, 1260- and 4100-g samples were prepared and measured for all of the radionuclides listed in Table 1. Figure 2 illustrates the extremes in response obtained with a point-source mount and with a 1263-g sample of the mixture that contained 16-d ^{48}V. These two spectra are

Table 1. Radionuclides compiled in approximate libraries for preliminary examination of Apollo 11 samples

Nuclide	Half-life	Source*	Nuclide	Half-life	Source*
^{7}Be	53 d	A	^{52}Mn	5·7 d	C
^{22}Na	2·60 yr	A	^{54}Mn	313 d	A
^{24}Na	15·0 hr	B	^{55}Co	18 hr	C
^{26}Al	7·4 × 10^5 yr	A	^{56}Co	77·3 d	A
^{43}K	22 hr	B	^{58}Co	71 d	A
^{44}Ti	47 yr	A	^{60}Co	5·26 yr	A
^{46}Sc	84 d	A	^{40}K	1·28 × 10^9 yr	D
^{48}Sc	1·83 d	C	U	4·51 × 10^9 yr	E
^{48}V	16·1 d	C	Th	1·39 × 10^{10} yr	E
^{51}Cr	27·8 d	A			

* Source: A, commercial radionuclide preparation; B, produced at ORNL by neutron irradiations; C, produced at ORNL by cyclotron bombardments; D, natural radiochemical; E, certified standard from New Brunswick Laboratory.

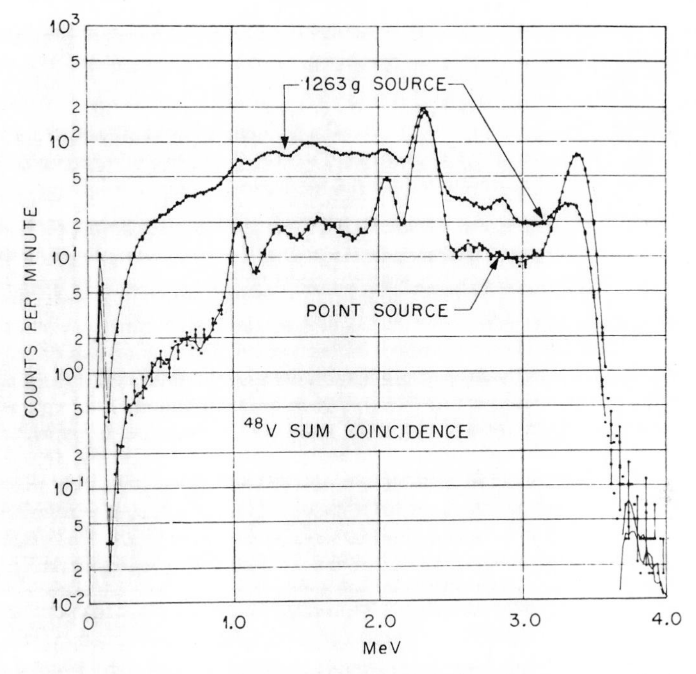

Fig. 2. Sum-coincidence spectra of ^{48}V recorded in the LRL system. The lower curve shows the response to a point source, while the upper curve shows the spectrum of a source distributed in 1263 g of iron powder.

from the sum-coincidence mode of analysis in the analyzer system. It will be noted that much of the structural detail present in the point-source mount (lower curve) is lost through attenuation and scattering by the iron matrix in the extended source (upper curve). From such response differences, it is readily apparent that extensive calibration information is necessary for accurate analysis with techniques that utilize the entire spectrum for least-squares analysis.

Sample sizes for Apollo 11 materials ranged from 157–971 g, and bulk densities ranged from 1·55–3·00 g/cm³. Analyses for the preliminary report (LSPET, 1969) were made by using the approximate libraries collected in advance of the Apollo 11 mission. Due to the need for extrapolations to account for density and mass differences between the approximate library samples and the actual lunar samples, it was necessary to assign rather large error statements to all of the minor constituents in the preliminary report. Following the 50-day quarantine period, exact replicated standards were prepared for the four crystalline rocks and the sample of fines. The replicas of rocks consisted of 0·08 mm aluminum foil shells, either hand-formed around each lunar sample or constructed from contour photography models, reinforced with epoxy cement and filled with a standard radionuclide–iron powder mixture. Electronic densities were calculated from the gross chemical composition (LSPET, 1969) and matched in the replicated standards by use of a mass of iron powder ~5 per cent in excess of the lunar sample mass. The bulk densities of rocks were reproduced by the addition of 3-mm polyurethane spheres (0·016 g/cm³) to the appropriate quantity of iron powder in order to fill completely the exact replica. Lunar fines were simulated with mixtures of powdered iron and dunite. Nine such replicas of each lunar sample were filled with a standardized aliquot of one of each radionuclide found in the preliminary tests. A collection of responses for each set of replicas was made into an "exact library" for final analysis of the lunar samples in the least-squares program.

To check the experimental method a test sample was prepared in a replica of sample 10017 at concentrations similar to those found in the preliminary examination. The test sample was counted for 1035 min, a time similar to that for the lunar sample itself. Table 2 illustrates the accuracy with which the test mixture was analyzed using the least-squares program and the exact library for sample 10017. Note that the errors associated with the experimental determinations in Table 2 are those due to counting statistics only, which ranged from 2·3 to 8 per cent while the absolute differences ranged from 0·8 to 5·0 per cent. The counting statistics, of course, could be improved by measuring the sample longer than the 17-hr interval shown for this test. For a massive one kilogram sample and such a relatively short counting time, this test illustrates the accuracy obtainable with this spectrometer system and data analysis schemes.

Data analysis

Programming and interfacing of the data acquisition system permitted the collection of spectral data as singles (non-coincident) events from each of the two detectors, as gamma–gamma coincident events in the 127 × 127 matrix, and as sum-coincidence spectra. The four sets of data were all collected simultaneously and permitted the determination of most radionuclides with all four spectra. Monoenergetic gamma-ray emitters such as ^{40}K and ^{54}Mn are best determined from the two independent singles spectra.

Table 2. Analysis of sample 10017 test mixture by use of exact library (1035-minute counting interval)

Nuclide	Added	Found*	Difference (%)
K	2270 ppm	2220 ± 50	−2·2
Th	2·85 ppm	2·82 ± 0·07	−1·1
U	0·70 ppm	0·71 ± 0·03	+1·4
^{22}Na	101 dpm/kg	103 ± 6	+2·0
^{26}Al	125 dpm/kg	126 ± 6	+0·8
^{46}Sc	115 dpm/kg	118 ± 5	+2·6
^{54}Mn	106 dpm/kg	111 ± 6	+4·7
^{56}Co	101 dpm/kg	106 ± 8	+5·0

* Errors stated are those due to counting statistics only.

ALPHA-M, a computer program for quantitative radionuclide determination by least-squares resolution of gamma-ray spectra (Schonfeld, 1967) was used exclusively for data analysis of all lunar samples in this study. Each of the four types of spectra were analyzed separately using the NASA Univac 1108 computer complex. The 255-channel singles and sum-coincidence spectra were resolved in the normal mode of analysis with program ALPHA-M using the appropriate spectral libraries.

The 8192-channel gamma–gamma coincidence matrices contain many zero-value channels and are too large to analyze conveniently by least-squares techniques. Statistical consideration concerning the large number of zero-value channels and the desire to condense the matrix so that the gain-shift option of ALPHA-M could be employed, made it desirable to divide the coincidence matrix into several zones. Figure 1 shows the contour map of the matrix with the selected energy zones indicated by dotted lines. Spots indicating transitions for some of the more prominent coincidences are shown in black. Note that ^{56}Co has six major spots as indicated by the black squares.

The PDP-9 computer was programmed to allow a direct addition of any of the 127 planes in the matrix. Accordingly, a scheme was developed whereby the matrix was divided into four segments of 0·12–0·32, 0·36–0·68, 0·68–1·08 and 1·24–1·60 MeV, as shown in Fig. 1. Each of these segments consists of 127-channel spectra at 40 keV per channel. (The matrix is folded, but the computer programming reflects the energy segment horizontally when the diagonal is reached.) Each energy segment of the four 127-channel zones was further restricted by eliminating some of the upper channels which contained no useful data, and some of the lower channels which lay in regions of scattering. The spectra within each zone were summed so that computer analysis of coincidence data was finally performed on four single-parameter spectra. It should be emphasized that the zone summation method just described reduces the total number of events in the coincidence matrix very little. In addition, the least squares program utilizes most of the coincidence data from complex decays such as that of ^{56}Co. This is a distinct advantage over the technique of "spot" analysis where the integrated area at a gamma–gamma intersection of the matrix is directly compared to that of a calibrated standard.

Figure 1 does not show all the possible areas and interferences in the matrix. The energy segment 0·12–0·32 MeV was left blank due to the numerous low-energy transitions occurring in that region. In addition, that segment contains many peaks due to backscatter from high-energy transitions. Accordingly, the computer program contains the option of excluding this region from analysis. Other interferences, such as that of the 1·83-MeV transition from ^{26}Al which affects the 1·28 + 0·511 MeV sum line of ^{22}Na are not shown in the diagram, but are suitably recognized and resolved in the least-squares analysis.

Cobalt-60 appears to be left out in the way the matrix is segmented. This is not the case, however, due to the reflection of zone (d) horizontally at the diagonal (Fig. 1). Hence the 1·17 MeV ^{60}Co appears in coincidence with the 1·24–1·60 MeV segment.

The combination of two computer techniques, namely, that of plane summation in the PDP-9 coupled with the least squares analysis in the Univac 1108, has provided a very powerful technique for resolving a large matrix in the multidimensional gamma-ray spectrometer. We have demonstrated the ability to determine quantitatively mixtures containing 12–15 gamma-emitting radionuclides at the 10–100 dpm/kg level, even in the presence of intense interferences from Th at the 3–4 ppm level. All data from the analysis of Apollo 11 samples and standard libraries is stored on magnetic tape and may be immediately recalled by the PDP-9 for further data analysis should the need arise. In addition, the data analysis techniques are sufficiently flexible to allow a change in the choice of energy segments in the gamma–gamma coincidence matrix if future samples should contain different radionuclides.

Results and Discussion

Analysis of whole rocks and fines

Because of the complex operations involved in handling lunar material, preparation of samples for gamma-ray spectrometry was slow during the quarantine period, and analysis of the first sample could not begin until late on July 29, 1969, four days after samples arrived at the LRL. Thus, radioactive species with half-lives of less than a few days were undetectable.

Another practical difficulty arose because it became necessary to use a variety of sample containers (three types of stainless steel and two of polypropylene) with different gamma-ray attenuation and scattering properties. Small uncertainties in the background corrections arose during the preliminary examination because of variations in radioactive contamination levels in some container types. Since it was not known in advance that some of these containers were to be used, spectrum libraries in some cases had to be acquired during the sample quarantine period, which reduced the time available for lunar sample measurement.

The precision of the measurements could have been improved by use of longer counting times; however, because these analyses were part of the preliminary examination at the LRL, time was limited, and measurement of a large number and variety of samples was emphasized. All samples were measured in at least two separate experiments, and total counting times ranged from about 1000 to 3000 min per sample.

Evidence for the high concentrations of Th and U in the samples was noted from the beginning of our measurements. Not only was this Th and U content important in itself, but it also made detection of weak gamma-ray components difficult because of the intense interferences from the gamma rays of the Th and U decay series.

The results are summarized in Table 3. Radioactive concentrations were obtained for 12 species, and upper limits were estimated for ^{7}Be. The nuclides of shortest half-lives identified were ^{52}Mn (5·7 days) and ^{48}V (16·1 days). The ^{52}Mn was identified by its gamma radiations and half-life, and was detected only in the first two samples received in the RCL. More complete information on radionuclide concentrations was obtained for rock 10017 than for any other sample listed in Table 3. At present, we have not completed our computer analyses on counting data from some of the other samples to obtain concentration estimates for ^{60}Co, ^{44}Ti and ^{7}Be.

The errors listed in Table 3 include, in addition to the statistical errors of counting, estimates of possible systematic errors due to uncertainties in the detector efficiency calibrations. Somewhat larger errors are shown for the breccias in Table 3, because no replicas were used for calibration.

Table 3. Gamma-ray analyses of whole rocks and fines from Apollo 11. Values for short-lived nuclides have been corrected for decay to 0000 hours, CDT, 21 July 1969

	Sample No. (Type)*							
	10057-1 (A)	10072-1 (A)	10003-0 (B)	10017-0 (B)	10018-1 (C)	10019-1 (C)	10021-1 (C)	10002-6 (D)
Weight (g)	897	399	213	971	211·5	234	157	301·5
Bulk density (g/cc)	2·73 ± 0·14	2·37 ± 0·24	2·88 ± 0·35	3·00 ± 0·15	2·0 ± 0·2	2·02 ± 0·15		1·55 ± 0·05
K (ppm)†	2550 ± 130	2300 ± 120	480 ± 25	2430 ± 120	1420 ± 70	1200 ± 60	1600 ± 80	1100 ± 60
Th (ppm)†	3·30 ± 0·20	2·80 ± 0·17	1·01 ± 0·06	3·25 ± 0·18	2·30 ± 0·20	1·90 ± 0·19	2·50 ± 0·25	1·92 ± 0·10
U (ppm)†	0·79 ± 0·06	0·76 ± 0·06	0·26 ± 0·03	0·83 ± 0·07	0·60 ± 0·09	0·43 ± 0·06	0·54 ± 0·08	0·49 ± 0·04
^{7}Be (dpm/kg)	<70		<100	<60				<80
^{22}Na (dpm/kg)	41 ± 4	46 ± 5	41 ± 4	39 ± 4	55 ± 8	47 ± 7	55 ± 8	51 ± 5
^{26}Al (dpm/kg)	75 ± 8	73 ± 8	74 ± 8	73 ± 8	108 ± 16	101 ± 15	110 ± 15	120 ± 12
^{44}Ti (dpm/kg)	<2·5	<2·5		2·1 ± 1·3				<2·5
^{46}Sc (dpm/kg)	10 ± 2	8 ± 2	13 ± 3	13 ± 3	13 ± 4	10 ± 3	13 ± 4	8 ± 2
^{48}V (dpm/kg)			12 ± 9	11 ± 7				
^{52}Mn (dpm/kg)			35 ± 20				33 ± 21	
^{54}Mn (dpm/kg)	32 ± 6	20 ± 4	35 ± 7	33 ± 7	38 ± 10	28 ± 9	21 ± 6	28 ± 7
^{56}Co (dpm/kg)	31 ± 8	40 ± 10	43 ± 10	26 ± 7	33 ± 10	35 ± 10	50 ± 15	40 ± 7
^{60}Co (dpm/kg)				1·1 ± 0·8				

* Classification according to LSPET (1969). Upper limits are 2σ evaluated from least squares analysis.
† Standardization for assay of K, Th, and U with reference to terrestrial isotopic abundances. Equilibrium of Th and U decay series also assumed.

Concentrations listed in Table 3 represent averages taken over the whole sample, usually a rock. Thus, if the sample is inhomogeneous, data obtained on a small sample may not be directly comparable to the average concentrations in Table 3. As shown later, pronounced concentration gradients are found for some induced nuclides. Also, WOOD *et al.* (1970) have reported small potassium-rich areas in crystalline lunar rocks of basaltic composition which would render suspect potassium analyses on small samples, unless large masses were homogenized before aliquots were taken. GAST and HUBBARD (1970) also found that concentrations of some elements in small chips of crystalline lunar rocks may not be representative of the concentrations in larger samples. Gamma-ray spectrometry offers a useful technique for determining average sample concentrations without the sampling errors to which small samples are vulnerable.

Bulk densities for seven samples were obtained from the sample weights and the volumes of the replica shells. It will be noted that the crystalline rocks, breccias and fines show a distinct correlation between bulk density and sample type.

The results in Table 3 generally agree within experimental error with other radioactivity studies (e.g. SHEDLOVSKY *et al.*, 1970; PERKINS *et al.*, 1970; KEAYS *et al.*, 1970; WÄNKE *et al.*, 1970) and with precision chemical analyses (e.g. GAST and HUBBARD, 1970; PHILPOTTS and SCHNETZLER, 1970; TATSUMOTO and ROSHOLT, 1970; WANLESS *et al.*, 1970) on the same samples.

The breccias and fines are very similar in their chemical compositions, as evidenced by their similar concentrations of K, Th and U. This substantiates the hypothesis that the breccias were compacted mostly from the fine material.

With the exception of sample 10003, the crystalline rocks from the Apollo 11 landing site form a distinct group with K, Th and U contents significantly higher than those of the breccias and fines. Sample 10003 is a coarsely crystalline rock which resembles in texture the terrestrial gabbro and differs from the other crystalline rocks in its lower concentrations of K, Th and U. This suggests that rock 10003 may have come from another region of the moon; that it may be the product of an igneous process different from that which formed the other crystalline rocks examined; or that it may have originated in a stratigraphically different location in the same general region. A remote origin for rock 10003 may be possible, since the Apollo 11 landing site is crossed by distinct rays associated with distant craters (LSPET, 1969). Further, the crystalline rocks of the Apollo 12 sample which correspond to the petrographic types A and B as defined by LSPET (1969) closely resemble rock 10003 in their concentrations of K, Th, U, ^{26}Al and ^{22}Na, as discussed in LSPET (1970a, b). On the other hand, the fines and the rare breccias from the Apollo 12 site have much higher concentrations of K, Th and U than do the type A or B rocks from the same site or the fines and breccias from Apollo 11. Clearly, the chemistry of the lunar soil is not uniform in the different maria.

The concentrations of the naturally radioactive elements are of great importance to the thermal history of the moon. Potassium concentrations range from 480 ppm for rock 10003 to 2550 ppm for rock 10057. The lowest values are near the average K concentration of $\sim$850 ppm for chondrites (UREY, 1964) but are distinctly above the average K concentration of 360 ppm (HEYMANN *et al.*, 1968) for eucrites. The range of lunar K concentrations also includes values determined for some oceanic gabbros and basalts (ENGEL and FISHER, 1969). The Th and U contents are near those of terrestrial basalts. The concentration of radioactive elements in the lunar surface materials is much greater than that inferred from thermal models for the mean radioactive element content of the moon (ANDERSON and PHINNEY, 1967). It has been proposed by RINGWOOD and ESSENE (1970) that the required degree of differentiation could be achieved through a small degree of partial melting of a pyroxene-rich assemblage at depths of 200–400 km. Such partial melting might have been initiated either through impact or through internal heating. ALBEE *et al.* (1970) propose that, if the event which formed the mare materials arose through internal heating, the moon would be partially molten today. However, it should be stressed that, although melting occurred to considerable depths below the lunar surface, the source of heat is at present very much open to question.

It was noted by WASSERBURG *et al.* (1964) that systematic differences occur between the K/U ratios for terrestrial materials and meteorites. Further, the terrestrial

K/U ratios for common rock types are remarkably constant over a factor of 40 variation in K concentration (Gast, 1968). Ratios of K/U and Th/U for our lunar samples are listed in Table 4. The errors shown were based on the errors listed for the element concentrations in Table 3. The K/U ratios are very low, as observed in LSPET (1969), and our most recent data show variations from 1850–3230 with a mean of 2680. Again, sample 10003 with a ratio K/U of 1850 shows a slight difference from the other Apollo 11 samples, but approaches the ratio K/U of about 2200 found by LSPET (1970a, b) for the Apollo 12 materials.

The lunar values of the ratio K/U are very low by comparison with chondrites (70,000) and with terrestrial rocks (10,000). Duke and Silver (1967) and, more recently, Ringwood and Essene (1970) have pointed out the possibility that rocks of composition similar to the eucrites might have formed on the moon. Detailed chemical

Table 4. Primordial element concentration ratios in lunar rocks and fines

Sample (type)*	K/U	Th/U
10057-1 (A)	3230 ± 295	4·2 ± 0·4
10072-1 (A)	3030 ± 286	3·7 ± 0·4
10003-0 (B)	1850 ± 144	3·9 ± 0·4
10017-0 (B)	2930 ± 288	3·9 ± 0·4
10018-1 (C)	2370 ± 374	3·8 ± 0·7
10019-1 (C)	2790 ± 414	4·4 ± 0·8
10021-1 (C)	2960 ± 464	4·6 ± 0·8
10002-6 (D)	2250 ± 212	3·9 ± 0·4
Averages:		
All samples	2680	4·0 ± 0·2
Fines and crystalline rocks		3·9 ± 0·2

* Petrologic type according to LSPET (1969).

properties serve to distinguish lunar materials from eucrites. In addition, the average K/U ratio of about 5000 (Goles, 1969) for eucrites lies intermediate between the terrestrial and lunar values and distinct from both. The identification of unbrecciated eucrites has been cited by Heymann *et al.* (1968) as evidence against a lunar origin for eucrites.

Since the K/U ratio is not readily changed by terrestrial igneous processes, we would not expect to see dramatic variations at different locations on the lunar surface. Indeed, the findings of LSPET (1970a, b) at the Apollo 12 site showed that the chemistry at the maria sites is clearly related and that the lower limit for the K/U ratio is reduced to about 1400 at the Apollo 12 site. Thus, it appears that the moon, like the earth, is depleted in potassium and enriched in uranium relative to chondrites (Gast, 1968). The contribution by potassium to radiogenic heating of the moon is small relative to that of thorium and uranium.

The concentration ratios Th/U are also listed in Table 4. Our best average value for this ratio is 3·9 ± 0·2, from which we have excluded measurements on breccias, because calibrations for breccias did not employ replicas and so are less accurate. This value agrees with the determination of 3·8 ± 0·2 by Perkins *et al.* (1970). The lunar value may be compared with ~3·6 for the terrestrial crustal average computed from data in the critical review by TAYLOR (1964), whose model for crustal composition was a combination of averages for granites and basalts in the combination of 1:1. Such crustal averages are highly model-dependent; however, the general agreement between Th/U ratios for earth and moon is not surprising in view of the chemical similarity of these elements. Heavy element analyses of meteorites show considerable spread, and Th/U concentration ratios for chondrites range between 3·4 and 3·7 (MORGAN and LOVERING, 1968).

Because of the interest in ^{22}Na and ^{26}Al due to their roles in dating and as indicators of the fluxes which have irradiated the lunar surface, it should be noted that the concentrations of these species show dramatic differences from concentrations in meteorites. The ratio ^{26}Al/^{22}Na is two to three times greater than that in meteorites. This difference can be understood in terms of the chemical composition of the lunar material and the bombarding fluxes of galactic and solar protons (SHEDLOVSKY *et al.*, 1970).

The concentrations of ^{26}Al and ^{22}Na are very similar in the fines and breccia; in turn, these concentrations are significantly higher than those of the crystalline rocks. For cases which can be compared, our results on ^{22}Na and ^{26}Al agree with those of SHEDLOVSKY *et al.* (1970). We are also in agreement with the results of PERKINS *et al.* (1970) except for the ^{22}Na concentration of fine material. It may be significant that our fines sample 10002-6 was one of the components of sample 10084 and so may show minor differences in radiation exposure from that of the final mixture.

Our data on ^{54}Mn and ^{56}Co concentrations do not show any significant variations with sample type. We are unable to detect the wide variations in ^{54}Mn contents reported by PERKINS *et al.* (1970).

Solar protons are chiefly responsible for production of ^{48}V and ^{56}Co. Studies of solar proton effects will be discussed below.

To date our analyses of the counting data have been applied in greatest detail to the study of rock 10017. Thus, it is only on this sample that a value is available for the ^{60}Co concentration. This nuclide is especially interesting because of the possibility that it may be produced by neutron capture. Such a possibility may be examined by use of the integrated thermal neutron flux of $1{\cdot}7 \times 10^{16}$ neutrons/cm^2 determined for rock 10017 by ALBEE *et al.* (1970) for abundance ratios of Gd isotopes. An average thermal neutron flux of 1·0 neutron/cm^2 sec is obtained from their exposure age of $5{\cdot}5 \times 10^8$ yr. Our exposure age of $3{\cdot}4 \times 10^8$ yr (see Table 6) leads to a higher value of 1·6 neutrons/cm^2 sec. An estimate of the ^{60}Co concentration to be expected based on our sample weight of 971 g and a ^{59}Co concentration of 30 ppm (WIIK and OJANPERA, 1970) yields 0·7 dpm/kg for the lower flux value and 1·1 dpm/kg for the higher flux value. These values of 0·7–1·1 dpm/kg agree with the experimentally determined ^{60}Co concentration of $1{\cdot}1 \pm 0{\cdot}8$ dpm/kg. Although the estimated yield of ^{60}Co is subject to considerable uncertainty at present, it appears that neutron capture may be the principal source of the ^{60}Co determined in our study of rock 10017.

Solar proton effects

In addition to analyses of whole rocks, attempts were made to study radionuclide concentration gradients, since it was expected that solar flare protons could produce high concentrations of some nuclides at the upper surfaces of lunar materials (Ebeoglu and Wanio, 1966; Lal *et al.*, 1967). When a nuclide decays by cascades of coincident radiations and the nuclide is concentrated on the surface of a massive sample ($\frac{1}{4}$ to 1 kg), attenuation of part of the cascade gamma rays by the bulk sample as well as changes in solid angle subtended at the detectors will cause differences in the detection efficiencies of the two detectors. The 0·511–(0·511 + 1·28) MeV coincidence in ^{22}Na and the 0·511–(0·511 + 1·83) MeV coincidence in ^{26}Al are especially sensitive to such positioning effects. These differences can be determined after appropriate calibration if the two parameter storage matrix is operated in the conventional manner, so that events with energy coordinates (E_x, E_y) and (E_y, E_x) are stored in separate locations. The difference in the intensities of the two symmetrical "spots" in the gamma–gamma coincidence matrix serves as the basis for an instrumental method for deriving concentrations at upper and lower surfaces. A replica containing a steep surface concentration gradient of ^{26}Al was used to derive a calibration factor. The method requires centering the sample between the detectors. The centering may be easily accomplished by using as a guide the area of the full energy peak of the ^{40}K gamma ray, since the ^{40}K is uniformly distributed throughout the sample volume.

Surface concentration gradients of nuclides without coincident gamma rays, such as ^{54}Mn, may be derived from the ratio of activities in the two singles spectra from the upper and lower detectors.

Many lunar sample experiments require knowledge concerning the orientation of the given sample and identification of the most recently exposed surface. Optical properties, gas retention age determinations, solar wind effects and magnetics studies are all examples of experiments which require information concerning the sample orientation. The orientation of rock 10017 on the lunar surface was uncertain. Prior to distribution of pieces of 10017, it was decided to check for surface concentration gradients by counting a small piece chipped from the suspected top. The sample, ARA, was part of the segment labelled ARB in Fig. 3. It was approximately oriented on the underside of ARB opposite the segment labelled T4. When measured, it was found to contain radionuclide concentrations *less* than the whole rock average, indicating that the sample was from the true underside of the rock. Results of these measurements are shown in the column labelled "bottom" in Table 5. Sample ARB (with ARA removed from the bottom so that little of the bottom remained) was then tested using the instrumental method. This measurement yielded the results labelled "top" and "interior" for sample 10017 in Table 5. Note that the upper and lower surfaces of sample ARB are actually the "top" and "interior" of rock 10017, since sample ARA had been removed from the bottom of the original rock. Therefore, the top and bottom measurements of the sort described above gave "top" and "interior" values for sample ARB.

A small fragment from the top surface of rock 10057 was also measured to detect concentration gradients. The fragment weighed 7·6 g and was only 4 mm thick. Results from this sample are also shown in Table 5 labelled 10057, "top."

Although errors were large, concentration gradients were observed for ^{56}Co, ^{26}Al, ^{22}Na and ^{54}Mn (Table 5). Because of the low threshold energy for the production of ^{56}Co the concentration gradients for this nuclide are particularly steep. This

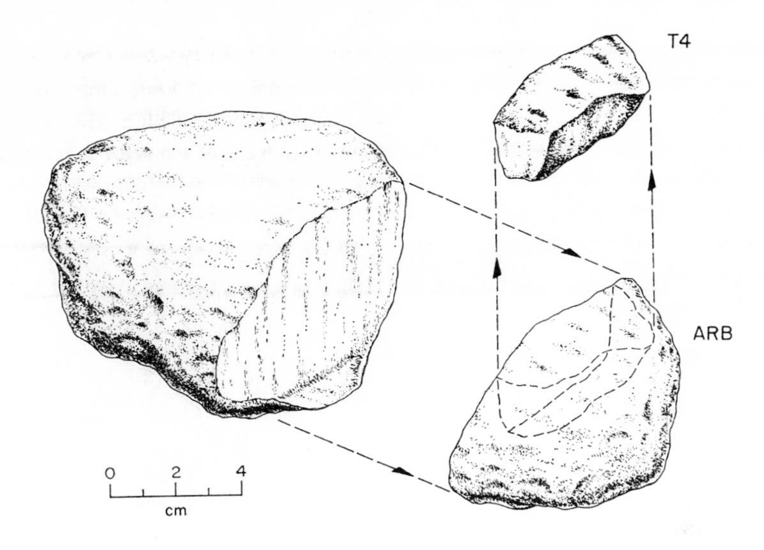

Fig. 3. Exploded view of rock 10017-0 which shows the origin of fragment ARB whose top surface was identified by gamma-ray spectrometry as described in the text. Surface sample T4 was analyzed by SHEDLOVSKY *et al.* (1970).

suggests that solar proton activation is the principal source of the ^{56}Co, especially the flare of 12 April 1969 (SHEDLOVSKY *et al.*, 1970). The data of Table 5 suggest that ^{26}Al, ^{22}Na and ^{54}Mn are produced both by solar and cosmic-ray protons. The presence of such concentration gradients makes difficult a precise comparison between some data in Table 3 and data from fragments of the same rocks.

The concentration of ^{26}Al produced at saturation by galactic cosmic-ray bombardment of lunar surface material may be estimated by use of the empirical production rates derived from data on stony meteorites by FUSE and ANDERS (1969). If the major element composition is known then the production of ^{26}Al in a $2\ \pi$ geometry can be estimated by use of half the production rate of Fuse and Anders. For rock 10017 and the fine material the concentrations of ^{26}Al estimated in this way are 40 and 48 dpm/kg,

Table 5. Radionuclide concentrations in lunar samples as a function of depth

	Sample 10017				Sample 10057	
Nuclide	Top*	Interior*	Bottom	Whole Rock†	Top	Whole rock†
^{26}Al (dpm/kg)	95 ± 15	65 ± 10	50 ± 7	73 ± 8	115 ± 9	75 ± 8
^{22}Na (dpm/kg)	43 ± 6	33 ± 6	30 ± 5	38 ± 4	63 ± 7	41 ± 4
^{54}Mn (dpm/kg)	38 ± 10	28 ± 7	20 ± 7	33 ± 7	60 ± 18	32 ± 6
^{56}Co (dpm/kg)	44 ± 18	—	10 ± 11	26 ± 7	90 ± 25	31 ± 8
K (ppm)	2350 ± 50	2400 ± 50	2400 ± 50	2430 ± 120	2580 ± 100	2550 ± 130

* Concentrations for "Top" and "Interior" of sample 10017 were determined by an instrumental method and are only approximate. See text.

† Values for whole rocks are from Table 3.

respectively. A more sophisticated calculation by SHEDLOVSKY *et al.* (1970) also shows that the production of ^{26}Al by galactic cosmic rays is much lower than the concentrations found on the surfaces of rocks 10017 and 10057. The high concentrations of ^{26}Al in the fines suggest that these samples were taken from a thin (about one cm) layer on the lunar surface.

In a qualitative way the surface concentration gradient for ^{26}Al can be used to estimate an upper limit to the erosion rate of rocks by micrometeorites. After a small

Table 6. Estimation of gas retention ages and ^{22}Na–^{21}Ne exposure ages for crystalline rocks

Quantity	Sample no. 03*	17†	57†	72†
		Gas concentrations (10^{-8} cm^3/g)		
$^{4}He_{r}$‡	1700	54,500	61,200	62,800
^{40}Ar	2600	4100	4800	7600
^{21}Ne	12·9	47	5·7	26
		Gas retention ages (10^6 years)§		
U,Th–^{4}He	2300	2450 ± 150	2700 ± 150	2750 ± 150
K–^{40}Ar	3970	2220 ± 150	2400 ± 150	3200 ± 230
		Exposure ages (10^6 years)§		
^{22}Na–^{21}Ne	90	340 ± 40	38 ± 5	160 ± 20
^{3}He	110	310	44	190
^{3}H–^{3}He	—	375 ± 40‖	45 ± 4¶	205 ± 25‖
^{38}Ar–(^{37}Ar + ^{39}Ar)	—	640 ± 160‖	110¶	410 ± 80‖
^{81}Kr–Kr	—	510 ± 50**	47††	—
		480 avg.††		

* Gas contents from HINTENBERGER *et al.* (1970).
† Gas contents from FUNKHOUSER *et al.* (1970a).
‡ A slight correction of 60(^{20}Ne–1·1 ^{21}Ne), due to FIREMAN (1970), is applied to ^{4}He concentrations to obtain $^{4}He_{r}$.
§ Errors quoted for K–^{40}Ar and ^{22}Na–^{21}Ne ages are combined analytical errors of gas concentrations and our radioactivity determinations of K, Th, U or ^{22}Na. Errors for gas concentrations of samples 10017, 10057 and 10072 were taken as 5 per cent (FUNKHOUSER *et al.*, 1970b).
‖ FIREMAN *et al.* (1970).
¶ STOENNER *et al.* (1970).
** EBERHARDT *et al.* (1970).
†† MARTI *et al.* (1970).

correction for chemical composition, the ^{26}Al concentrations in the fines and the surfaces of the crystalline rocks 10017 and 10057 are similar. The ^{26}Al half-life is $0{\cdot}74 \times 10^6$ yr; therefore, if the erosion rates are higher than a few mm/10^6 yr, the steep surface concentration gradient for ^{26}Al on the rocks would not be observed.

Gas-retention and exposure ages

Some of the information obtained in this study may be combined with gas analysis data to provide estimates of crystallization and cosmic-ray exposure ages. In Table 6 we show gas retention ages for four rocks, derived from elemental analyses for K, Th and U reported in Table 3 above, combined with concentrations of radiogenic ^{4}He and ^{40}Ar from the literature. Where results on the same rocks can be compared, these ages agree with other determinations by the same methods (e.g. ALBEE *et al.*, 1970; EBERHARDT *et al.*, 1970; FUNKHOUSER *et al.*, 1970a; HINTENBERGER *et al.*, 1970).

Except for rock 10003, for which the U, Th–^{4}He age is only about half the K–^{40}Ar age, both gas retention ages are concordant. These results suggest that radiogenic ^{4}He is located in retentive sites, but that rock 10003 may have undergone some significant heating not experienced by the other rocks. A similar exception to the agreement between the two types of retention ages was noted by HINTENBERGER *et al.* (1970) for rock 10044.

Gas retention methods are generally found to give somewhat shorter ages than other methods, due to loss of gas from interstitial phases. This effect is exemplified by the work of ALBEE *et al.* (1970), who found K–^{40}Ar ages for rocks 10017 and 10057 similar to those we report in Table 6, but obtained longer ages of $3{\cdot}2$–$3{\cdot}6 \times 10^{9}$ yr on mineral grains by the same method. The ^{87}Rb–^{87}Sr internal isochrons for these rocks were found by ALBEE *et al.* (1970) to yield an age of $3{\cdot}65 \times 10^{9}$ yr, which the authors believe to be the age of formation of the Sea of Tranquillity. It may be noted that, if the event which formed the mare materials occurred $3{\cdot}65 \times 10^{9}$ yr ago, the K–^{40}Ar age of $4{\cdot}0 \times 10^{9}$ yr determined here for rock 10003 is quite remarkable. This result suggests that the origin of rock 10003 may be different from that of the other rocks examined.

Our estimates of cosmic-ray exposure ages were made by the ^{22}Na–^{21}Ne method. From similar work on meteorites (SPANNAGEL and SONNTAG, 1967) it was assumed that the effective cross sections for production of ^{22}Na and ^{22}Ne were equal, i.e. a ratio of production rates of 0·50. The ratio of these production rates would not be expected to vary greatly with chemical composition, since both spallation products have the same mass and differ only by one atomic number. Specific activities of ^{22}Na from the present work were used in combination with Ne concentrations obtained by FUNKHOUSER *et al.* (1970a) or HINTENBERGER *et al.* (1970). The spallogenic ^{22}Ne was estimated to be 10 per cent greater than the concentration of spallogenic ^{21}Ne. The ages shown in Table 6 are in good agreement with ^{3}He exposure ages determined from the data of FUNKHOUSER *et al.* (1970) and HINTENBERGER *et al.* (1970). The ^{3}He exposure ages were calculated on the basis of 10^{-8} cm^{3} of ^{3}He/g per m.y. exposure. However, it should be mentioned that our estimate of ^{22}Na–^{21}Ne ages are subject to considerable uncertainties in the production rates chosen and the use of ^{22}Na values which include production by solar protons. Further, our radioactivity values in Table 3 are for whole rocks, while the gas analyses were performed on small chips. Thus, the locations from which the chips were taken could be crucial. Fortunately, most chips for gas analysis were from the interior portions of rocks, which leads to data more consistent with the requirements of our radioactivity assays.

The ^{22}Na–^{21}Ne exposure ages in Table 6 and the corresponding ^{3}He exposure ages were used to estimate the production rates for the ^{26}Al–^{21}Ne method. A value for the ratio of production rates $(P_{26}/P_{21}) = 1{\cdot}0 \pm 0{\cdot}1$ was obtained from the average of 6 values. This ratio is quite different from the value for chondrites ($\sim$0·3) and probably reflects the differences in chemical composition of the lunar target material.

The ^{22}Na–^{21}Ne ages agree with the ^{3}H–^{3}He ages but are younger than the ^{38}Ar ages. This suggests that some of the ^{3}He and ^{21}Ne is not retained.

The variations in cosmic-ray exposure ages of a factor of ten seen here are consistent with the concept of a continual series of disturbances of the lunar surface due to impact.

Acknowledgments—The authors gratefully acknowledge contributions to the work reported here by V. A. MCKAY, R. T. ROSEBERRY, T. F. SLISKI, R. E. WINTENBERG and K. J. NORTHCUTT of Oak Ridge National Laboratory; M. K. ROBBINS and W. PORTENIER of Brown and Root-Northrop; J. MORGAN of Lockheed Corp.; R. S. CLARK and J. KEITH, NASA. We thank J. R. ARNOLD, P. W. GAST and O. A. SCHAEFFER for helpful discussions; E. ANDERS and E. L. FIREMAN for suggestions concerning this manuscript; and the management and staff of the Lunar Receiving Laboratory for their hospitality.

REFERENCES

ALBEE A. L., BURNETT D. S., CHODOS A. A., EUGSTER O. J., HUNEKE J. C., PAPANASTASSIOU D. A., PODOSEK F. A., RUSS G. PRICE II., SANZ H. G., TERA F. and WASSERBURG G. J. (1970) Ages, irradiation history, and chemical composition of lunar rocks from the Sea of Tranquillity. *Science* **167,** 463–466.

ANDERS E. (1960) The record in the meteorites—II. On the presence of aluminum-26 in meteorites and tektites. *Geochim. Cosmochim. Acta* **19,** 53–62.

ANDERSON D. L. and PHINNEY R. A. (1967) Early thermal history of the terrestrial planets. In *Mantles of the Earth and Terrestrial Planets,* (editor S. K. Runcorn), pp. 113–126. Interscience.

BECKURTS K. H., GLÄSER W. and KRÜGER G. (editors) (1964) *Automatic Acquisition and Reduction of Nuclear Data.* Gesellschaft für Kernforschung m. b. H. Karlsruhe.

DUKE M. B. and SILVER L. T. (1967) Petrology of eucrites, howardites, and mesosiderites. *Geochim. Cosmochim. Acta* **31,** 1637–1665.

EBEOGLU D. B. and WAINIO K. M. (1966) Solar proton activation of the lunar surface. *J. Geophys. Res.* **71,** 5863–5872.

EBERHARDT P., GEISS J., GRAF H., GRÖGLER N., KRÄHENBUHL U., SCHWALLER H., SCHWARTZMÜLLER J. and STETTLER A. (1970) Trapped solar wind noble gases, Kr^{81}/Kr exposure ages and K/Ar ages in Apollo 11 lunar material. *Science* **167,** 558–560.

ENGEL C. G. and FISHER R. L. (1969) Lherzolite, anorthosite, gabbro, and basalt dredged from the mid-Indian Ocean ridge. *Science* **166,** 1136–1141.

FIREMAN E. L. (1970) Private communication.

FIREMAN E. L., D'AMICO J. and DEFELICE J. C. (1970) Tritium and argon radioactivities in lunar material. *Science* **167,** 566–568.

FUNKHOUSER J. G., SCHAEFFER O. A., BOGARD D. D. and ZÄHRINGER J. (1970a) Gas analysis of the lunar surface. *Science* **167,** 561–563.

FUNKHOUSER J. G., SCHAEFFER O. A., BOGARD D. D. and ZÄHRINGER J. (1970b) Private communication.

FUSE K. and ANDERS E. (1969) Aluminum-26 in meteorites—VI. Achondrites. *Geochim. Cosmochim. Acta* **33,** 653–670.

GAST P. (1968) Upper mantle chemistry and evolution of the earth's crust. In *The history of the earth's crust,* (editor R. A. Finney), pp. 15–27. Princeton University Press.

GAST P. W. and HUBBARD N. J. (1970) Abundance of alkali metals, alkaline and rare earths, and strontium-87/strontium-86 ratios in lunar samples. *Science* **167,** 485–487.

GOLES G. G. (1969) Cosmic abundances. In *Handbook of Geochemistry,* (editor K. H. Wedepohl), Vol. I, pp. 116–133. Springer-Verlag.

HEYMANN D., MAZOR E. and ANDERS E. (1968) Ages of calcium-rich achondrites—I. Eucrites. *Geochim. Cosmochim. Acta* **32,** 1241–1268.

HINTENBERGER H., WEBER H. W., VOSHAGE H., WÄNKE H., BEGEMANN F., VILCSEK E. and WLOTZKA F. (1970) Rare gases, hydrogen, and nitrogen: Concentrations and isotopic composition in lunar material. *Science* **167,** 543–545.

KEAYS R. R., GANAPATHY R., LAUL J. C., ANDERS E., HERZOG G. F. and JEFFERY P. M. (1970) Trace elements and radioactivity of lunar rocks: Implications for meteorite infall, solar-wind flux, and formation conditions of moon. *Science* **167,** 490–493.

LAL D., RAJAN R. S. and VENKATAVARADAN V. S. (1967) Nuclear effects of "solar" and "galactic" cosmic-ray particles in near-surface regions of meteorites. *Geochim. Cosmochim. Acta* **31,** 1859–1869.

LSPET (LUNAR SAMPLE PRELIMINARY EXAMINATION TEAM) (1969) Preliminary examination of lunar samples from Apollo 11. *Science* **165,** 1211–1227.

LSPET (LUNAR SAMPLE PRELIMINARY EXAMINATION TEAM) (1970a) *Lunar Sample Catalog-Apollo 12.* NASA Manned Spacecraft Center Rep. MSC-01512.

LSPET (LUNAR SAMPLE PRELIMINARY EXAMINATION TEAM) (1970b) Preliminary examination of lunar samples from Apollo 12. *Science* **167,** 1325–1339.

MCLANE J. C., KING E. A., FLORY D. A., RICHARDSON K. A., DAWSON J. P., KEMMERER W. W. and WOOLEY B. C. (1967) Lunar Receiving Laboratory. *Science* **155,** 525–529.

MARTI K., LUGMAIR G. W. and UREY H. C. (1970) Solar wind gases, cosmic ray spallation products, and the irradiation history. *Science* **167,** 548–550.

MORGAN J. W. and LOVERING J. F. (1968) Uranium and thorium abundances in chondritic meteorites. *Talanta* **15,** 1079–1095.

O'KELLEY G. D., ELDRIDGE J. S., SCHONFELD E. and BELL P. R. (1970) Elemental compositions and ages of lunar samples by nondestructive gamma-ray spectrometry. *Science* **167,** 580–582.

PERKINS R. W., RANCITELLI L. A., COOPER J. A., KAYE J. H. and WOGMAN N. A. (1970) Cosmogenic and primordial radionuclides in lunar samples by nondestructive gamma-ray spectrometry. *Science* **167,** 577–580.

PHILPOTTS J. A. and SCHNETZLER C. C. (1970) Potassium, rubidium, strontium, barium, and rare-earth concentrations in lunar rocks and separated phases. *Science* **167,** 493–495.

RINGWOOD A. E. and ESSENE E. (1970) Petrogenesis of lunar basalts and the internal constitution and origin of the moon. *Science* **167,** 607–610.

ROWE M. W. (1963) Quantitative measurement of gamma-ray emitting radionuclides in meteorites. *Los Alamos Sci. Lab. Rep.* **LA-2765.**

SCHONFELD E. (1967) ALPHA M—an improved computer program for determining radioisotopes by least-squares resolution of the gamma-ray spectra. *Nucl. Instrum. Methods* **52,** 177–178.

SHEDLOVSKY J. P., HONDA M., REEDY R. C., EVANS J. C., LAL D., LINDSTROM R. M., DELANY A. C., ARNOLD J. R., LOOSLI H.-H., FRUCHTER J. S. and FINKEL R. C. (1970) Pattern of bombardment-produced radionuclides in rock 10017 and in lunar soil. *Science* **167,** 574–576.

SPANNAGEL G. and SONNTAG C. (1967) Cosmic-ray produced activities produced in chondrites. In *Radioactive Dating and Methods of Low-Level Counting*, pp. 231–238. IAEA, Vienna.

STOENNER R. W., LYMAN W. J. and DAVIS R. (1970) Cosmic ray production of rare gas radioactivities and tritium in lunar material. *Science* **167,** 553–555.

TATSUMOTO M. and ROSHOLT J. N. (1970) Age of the moon: An isotopic study of uranium–thorium–lead systematics of lunar samples. *Science* **167,** 461–463.

TAYLOR S. R. (1964) Abundance of chemical elements in the continental crust: A new table. *Geochim. Cosmochim. Acta* **28,** 1273–1285.

UREY H. C. (1964) A review of atomic abundances in chondrites and the origin of meteorites. *Rev. Geophys.* **2,** 1–34.

VAN DILLA M. A., ARNOLD J. R. and ANDERSON E. C. (1960) Spectrometric measurement of natural and cosmic-ray induced radioactivity in meteorites. *Geochim. Cosmochim. Acta* **20,** 115–121.

WÄNKE H., BEGEMANN F., VILCSEK E., RIEDER R., TESCHKE F., BORN W., QUIJANO-RICO M., VOSHAGE H. and WLOTZKA F. (1970) Major and trace elements and cosmic-ray produced radioisotopes in lunar samples. *Science* **167,** 523–525.

WANLESS R. K., LOVERIDGE W. D. and STEVENS R. D. (1970) Age determinations and isotopic abundance measurements on lunar samples. *Science* **167,** 479–480.

WASSERBURG G. J., MACDONALD G. J. F., HOYLE F. and FOWLER W. A. (1964) Relative contributions of uranium, thorium, and potassium to heat production in the earth. *Science* **143,** 465–467.

WIIK H. B. and OJANPERA P. (1970) Chemical analysis of lunar samples 10017, 10072, and 10084. *Science* **167,** 531–532.

WOGMAN N. A., ROBERTSON D. E. and PERKINS R. W. (1967) A large detector, anticoincidence shielded multidimensional gamma-ray spectrometer. *Nucl. Instrum. Methods* **50,** 1–10.

WOOD J. A., MARVIN U. B., POWELL B. N. and DICKEY J. S. (1970) Mineralogy and Petrology of the Apollo 11 lunar sample. *Smithson. Astrophys. Observ. Spec. Rep.* **307.**

Proceedings of the Apollo 11 Lunar Science Conference, Vol. 2, pp. 1425 to 1427.

Oxygen isotope analyses of selected Apollo 11 materials*

JAMES R. O'NEIL and LANFORD H. ADAMI
U.S. Geological Survey, Menlo Park, California 94025

(*Received* 29 *January* 1970; *accepted* 20 *February* 1970)

Abstract—Oxygen isotope ratios of lunar rocks are within the range found for unaltered terrestrial basalts suggesting a common source of planetary material for the moon and earth. Basaltic achondrites and mesosiderites do not appear to be genetically related to lunar material on the basis of their pyroxene isotope ratios. From the oxygen isotope fractionation between plagioclase and ilmenite, the temperature of crystallization of Type B rocks is estimated to be 1100–1300°C.

INTRODUCTION

OXYGEN isotope ratios were measured on several Apollo 11 samples: one Type A rock (10057), two Type B rocks (10017 and 10050), two Type C rocks (10046 and 10060), and the lunar fines (10084). Ratios were also measured on separated minerals from the Type B rocks and various separates made on the lunar fines. It is instructive to compare the oxygen isotope ratios of the lunar materials with the ratios in terrestrial and meteoritic materials, and in addition, it is possible to infer a temperature of crystallization of the rocks from the values of the fractionations between coexisting mineral pairs.

EXPERIMENTAL

Oxygen was liberated from the samples by reaction with BrF_5 at 550° (CLAYTON and MAYEDA, 1963). The oxygen was converted to CO_2 and analyzed on an isotope ratio mass spectrometer. The oxygen isotope ratios of the Apollo 11 samples are expressed as deviations in parts per thousand (δO^{18} values) from the ratio in the SMOW standard. These values are listed in Table 1. The whole-rock analysis of Type B rock 10017 (and all others) was actually made on pulverized rock and not calculated from the modes and isotopic compositions of the constituent minerals. Measurements by ONUMA *et al.* (1970) and by EPSTEIN and TAYLOR (1970) indicate that the breccias and lunar fines are quite similar to one another in oxygen isotopic composition and are a few tenths per mil enriched in O^{18} relative to the Types A and B rocks. Our analysis of breccia 10046 is the one exception to this observation. The δO^{18} value of this sample is 5·7‰, which is 0·5‰ lighter than lunar fines and the other breccias. The isotope ratio of the one other breccia analyzed by us (10060) is identical to that of the fines.

RESULTS

The δO^{18} values of lunar rocks lie between 5·7‰ and 6·2‰ within the 5·5‰–7·0‰ range found for most unaltered terrestrial basalts (TAYLOR, 1968). This striking similarity implies, at least, a common source of pre-planetary material for the moon and the earth.

The δO^{18} value of 5·7‰ for lunar pyroxene is in the 5·3–6·3‰ range found by TAYLOR *et al.* (1965) to be characteristic of pyroxene from ordinary chondrites and enstatite achondrites. Basaltic achondrites and mesosiderites which are mineralogically and chemically similar to lunar materials have δO^{18} values for pyroxene which

* Publication authorized by the Director, U.S. Geological Survey.

Table 1. Oxygen isotope compositions (δO^{18}‰)

Sample	Plagioclase	Pyroxene	Ilmenite	Other	Whole rock
Fine grained					
10057-20	—	—	—	—	5·8 ± 0·1 (2)
Crystalline					
10017-16	6·3 ± 0·1 (2)	5·7 ± 0·1 (2)	3·8 ± 0·1 (2)	—	5·9 ± 0·1 (2)
10050-34	6·2 ± ? (4)	5·6 ± 0·1 (2)	4·0 ± 0·2 (2)	—	—
Breccia					
10046-21	—	—	—	—	5·7 ± 0·2 (2)
10060-11	—	—	—	—	6·2 ± 0·2 (2)
Dust					
10084-143					
Whole rock	—	—	—	—	6·2 ± 0·1 (3)
Dark glass	—	—	—	6·7 (1)	—
Plag (>90%)	6·1 (1)	—	—	—	—
Magnetic separate	—	—	—	5·6 ± 0·1 (2)	—
Ilmenite (~50%)	—	—	5·4 (1)	—	—

δ values are per mil deviations from the SMOW standard.
Numbers in parentheses refer to number of analyses made.

are 1–2 per mil lower and thus do not appear to be genetically related to lunar material.

Because mineral separation on the lunar fines was difficult, the analyses of these separates are not particularly meaningful. Since the whole rock is 6·2 and most separates from the fines and other rock types yield lower values, there must be an isotopically heavy component in the fines (and breccia). Our analysis of a glass separate indicates this is the predominant heavy species (δO^{18} = 6·7‰) but cristobalite is also present and is isotopically heavy (δO^{18} = 7·1‰) (Onuma *et al.*, 1970; Epstein and Taylor, 1970).

Discussion

The temperature of crystallization of Type B rocks can be estimated from the oxygen isotope fractionations between plagioclase and ilmenite and from the laboratory calibrations of the systems plagioclase–water (O'Neil and Taylor, 1967) and magnetite–water (O'Neil and Clayton, unpublished data). A correction must be made for the small positive fractionation (0·2‰) between ilmenite and magnetite at high temperatures. Oxygen yields of 106–114 per cent (over stoichiometric $FeTiO_3$) were obtained for the ilmenite samples, and the measured δO^{18} values of 4·6‰ (10017) and 4·3‰ (10050) were corrected to the values shown in the table on the assumption that the excess yields were due to pyroxene impurities. This uncertainty, the uncertainty in the laboratory calibrations, and the moderate insensitivity of this isotopic thermometer at very high temperatures place severe limits on the precision of the isotopic temperature. From our data we estimate the plagioclase–ilmenite fractionation to be between 1·8 and 2·2 per mil, implying a crystallization temperature of 1100–1300°C for Type B rocks. Perhaps the best argument for the significance of the isotopic temperatures of the lunar rocks at this time is the good correlation of the lunar data with data on rapidly quenched terrestrial basalts and on anorthosites (Anderson, 1966).

The minerals in the Type B rocks have obviously not undergone significant post-solidification exchange which normally would occur in terrestrial rocks that are this coarse grained. This lack of retrograde isotopic exchange and the high melting temperature argues for the fact that little, if any, water was present when the lunar rocks crystallized.

REFERENCES

ANDERSON A. T. (1966) Mineralogy of the Labrieville anorthosite, Quebec. *Amer. Mineral.* **51,** 1671–1711.

CLAYTON R. N. and MAYEDA T. K. (1963) The use of bromine pentafluoride in the extraction of oxygen from oxides and silicates for isotopic analysis. *Geochim. Cosmochim. Acta* **27,** 43–52.

EPSTEIN S. and TAYLOR H. P., JR. (1970) $^{18}O/^{16}O$, $^{30}Si/^{28}Si$, D/H and $^{13}C/^{12}C$ studies of lunar rocks and minerals. *Science* **167,** 533–535.

O'NEIL J. R. and TAYLOR H. P., JR. (1967) The oxygen isotope and cation exchange chemistry of feldspar. *Amer. Mineral.* **52,** 1414–1437.

ONUMA N., CLAYTON R. N. and MAYEDA T. K. (1970) Oxygen isotope fractionation between minerals and an estimate of the temperature of formation. *Science* **167,** 536–538.

TAYLOR H. P., JR., DUKE M. B., SILVER L. T. and EPSTEIN S. (1965) Oxygen isotope studies of minerals in stony meteorites. *Geochim. Cosmochim. Acta* **29,** 489–512.

TAYLOR H. P., JR. (1968) The oxygen isotope geochemistry of igneous rocks. *Contrib. Mineral. Petrol.* **19,** 1–71.

Proceedings of the Apollo 11 Lunar Science Conference, Vol. 2, pp. 1429 to 1434.

Apollo 11 rocks: Oxygen isotope fractionation between minerals, and an estimate of the temperature of formation

NAOKI ONUMA, ROBERT N. CLAYTON* and TOSHIKO K. MAYEDA
Enrico Fermi Institute, University of Chicago, Chicago, Illinois 60637

(*Received* 7 *February* 1970; *accepted* 9 *February* 1970)

Abstract—Oxygen isotopic compositions of separated minerals from three Type A and four Type B rocks are very uniform. $\delta^{18}O$ values are: plagioclase 6·20, clinopyroxene 5·75, ilmenite 4·45 (parts per thousand relative to Standard Mean Ocean Water). The isotopic distribution corresponds to equilibrium at 1120°C. The isotopic composition of lunar pyroxenes falls within the range for pyroxenes of terrestrial mafic and ultramafic rocks, ordinary chondrites, enstatite chondrites, and enstatite achondrites, but above the range for basaltic achondrites, hypersthene achondrites, and mesosiderites. Glass isolated from the lunar soil has a $\delta^{18}O$ of 6·2, significantly richer in ^{18}O than the crystalline rock fragments in the soil.

OXYGEN isotope analyses have been done on the separated major mineral phases of seven lunar rocks of types A and B, on whole rock samples of types C and D, and on glass spherules and rock fragments separated from type D.

One-gram samples of the crystalline rocks were crushed in a steel mortar, ground in an agate mortar under acetone, and sieved through stainless-steel sieves. For the coarser-grained type B rocks, the size fraction between 325- and 500-mesh (25–43 μm) was used for mineral separation. For the finer-grained type A rocks, the fraction passing through the 500-mesh sieve (<25 μm) was used. Mineral separation was accomplished in a few cases by hand-picking grains from the coarser fractions, but usually by centrifugation in heavy liquids (bromoform, methylene iodide, and Clerici solutions). Densities of the liquids used are given for each sample in Table 1. Mineral purity was evaluated by X-ray diffraction and optical microscopy, and was >95 per cent in all samples, and usually >98 per cent.

Oxygen was extracted from 10-mg portions of the separated minerals by the bromine pentafluoride procedure (CLAYTON and MAYEDA, 1963). Manometric measurement of the oxygen yield gave the quantitative determinations of oxygen content (Tables 1 and 2). Experimental error is estimated at ± 1 per cent of the amount present. Oxygen was converted to carbon dioxide and the oxygen isotope ratio was measured with a 60°, 15-cm, double-collecting mass spectrometer. Oxygen extraction was carried out in duplicate for all samples, and each gas sample was further analyzed in duplicate. Results of isotope analyses are given in Tables 1 and 2. Standard error of the mean of the duplicate measurements is estimated to be $\pm 0{\cdot}07$ per mil.

Oxygen contents of all phases analyzed are in agreement with values calculated from the estimated major-element compositions. No evidence of nonstoichiometry with respect to oxygen has been seen.

---

* Also Departments of Chemistry and Geophysical Sciences.

Table 1. Oxygen contents and isotopic compositions of minerals from Type A and Type B rocks

	Type A Rocks			Type B Rocks			
	10020-24	10022-36	10071-22	10017-21	10044-22	10047-15	10058-35
(a) Specific gravity*							
Cr†							<2·38
Pc	2·70–2·82	2·71–2·81	2·71–2·82	2·70–2·77	2·69–2·77	2·70–2·75	2·71–2·82
Cpx	3·36–3·46	3·36–3·44	3·36–3·47	3·33–3·56	hand-pick	hand-pick	3·36–3·52
Ol	3·54–3·65						
Il	>4·2	>4·2	>4·2	>4·3	>4·2	>4·3	>4·2
(b) Oxygen content (wt.%)							
Cr							53·4
Pc	45·8	46·3	47·2	47·3	46·4	46·5	46·7
Cpx	42·0	42·6	42·1	42·9	42·5	42·0	42·6
Ol	40·5						
Il	33·7	33·9	33·1	33·3	32·0	31·9	32·4
(c) $\delta^{18}O$‡ (rel. to SMOW)							
Cr							7·09
Pc	6·19	6·31	6·17	6·16	6·18	6·20	6·00
Cpx	5·74	5·71	5·72	5·81	5·77	5·76	5·53
Ol	5·14						
Il	4·41	4·55	4·47	4·51	4·35	4·29	4·24
(d) $\Delta^{18}O$§							
Pc–Cpx	0·45	0·60	0·45	0·35	0·41	0·44	0·47
Cpx–Il	1·32	1·15	1·24	1·29	1·41	1·46	1·28
Pc–Il	1·77	1·75	1·69	1·64	1·82	1·90	1·75

* Values quoted are specific gravities of liquids which bracket the value for the separated mineral.
† Abbreviations for mineral names:
Cr–cristobalite, Pc—plagioclase, Cpx—clinopyroxene, Ol—olivine, Il—ilmenite.
‡ ^{18}O content in per mil terminology: deviation of $^{18}O/^{16}O$ ratio in parts per thousand from ratio in Standard Mean Ocean Water (SMOW).
§ Oxygen isotope fractionation, $\Delta^{18}O$, defined as 1000 ln α, where α is a fractionation factor for two phases, e.g. $\frac{(^{18}O/^{16}O)Pc}{(^{18}O/^{16}O)Il}$.

Table 2. Oxygen contents and isotopic compositions of various fractions of Type C and Type D rocks

Sample description	O Content (wt.%)	$\delta^{18}O$ (rel. to SMOW)
Type C		
10060-12 whole rock	41·3	6·03
Type D		
10084-46 whole rock	42·6	6·18
10084-46 glass spherules	42·2	6·22
10084-46 glass fragments	42·3	5·88
10084-46 Type A fragments	41·7	5·74
10084-46 Type B fragments	40·6	5·64
10084-46 "Anorthosite" fragments	44·8	5·89

The uniformity of isotopic composition from rock to rock, for a given mineral, is very striking. Of the type A and type B rocks, only sample 10058-35 shows any systematic departure from the mean, each of its minerals having an ^{18}O content 0·2 per mil lower than the means of the other rocks.

The soil and microbreccia have ^{18}O contents similar to one another, and about 0·5 per mil greater than that of the crystalline rocks. From the approximate modal compositions of the crystalline rocks given in the preliminary reports, their whole-rock $\delta^{18}O$ is estimated to be $5{\cdot}7 \pm 0{\cdot}1$ per mil. Fragments of type A and type B rocks were hand-picked from the 50–100 mesh (150–300 μm) fraction of the lunar soil and were found to have isotopic compositions in this range (Table 2). Irregular brown glass fragments (refractive index 1·68) were found to have almost the same composition, and may have been derived by fusion of crystalline rocks.

Glassy spherules, mean dia. 300 μm, were also separated from the soil, and have an ^{18}O-content significantly greater than that of the rock fragments, implying addition of an isotopically heavier component or an isotopic fractionation in the process of formation of the spherules. An addition to the soil of 2 per cent carbonaceous-chondrite-like meteorites, as suggested by KEAYS *et al.* (1970), would cause an ^{18}O enrichment of about 0·2 per mil. Addition of other known types of meteoritic material would have a negligible effect because of the similarity of their isotopic compositions to that of the soil (TAYLOR *et al.*, 1965). Enrichment of ^{18}O in glass resulting from a loss of volatiles in very-high-temperature fusion has been shown experimentally (WALTER and CLAYTON, 1967), and may play some part on the moon.

The mean value of the $^{18}O/^{16}O$ ratio in objects of the solar system may be determined in part by the region in the early solar nebula within which the solid particles accreted. There may have been gradients in isotopic composition which are reflected today in differences among the planets, their satellites, and the meteorites and asteroids. Such differences have been observed for various classes of meteorites by TAYLOR *et al.* (1965). Lacking data for the average oxygen-isotope composition of the moon, perhaps the best we can do at this stage is to compare the isotopic composition of a particular major mineral with that of similar minerals in terrestrial rocks and meteorites. The ^{18}O-content of the Apollo 11 pyroxenes is identical with that of pyroxenes from some mantle-derived rocks: garnet peridotites and pyroxenites from kimberlite pipes (GARLICK, 1966; ANDERSON *et al.*, 1970). The lunar pyroxenes lie at the high end of the distribution for oceanic basalts and andesites (GARLICK, 1966; TAYLOR, 1968; ANDERSON *et al.*, 1970).

The lunar pyroxenes fall into group II of TAYLOR *et al.* (1965), which contains the ordinary chondrites, enstatite chondrites, and enstatite achondrites. They are distinctly richer in ^{18}O than the pyroxenes of group I meteorites: basaltic achondrites, hypersthene achondrites, and mesosiderites.

If we speculate that oxygen isotope gradients in the early solar system were in the direction of decreasing ^{18}O outward, as suggested by the data on meteorites, then the oxygen isotope results on Apollo 11 rocks indicate an initial condensation either at the same position as terrestrial material, or somewhat nearer the sun.

Isotopic fractionations between pairs of minerals can be used to determine the temperature of last equilibration of the minerals. In favorable cases, this may simply be the temperature of their initial crystallization. Isotopic fractionations between pairs of major phases are shown in Table 1. For a given mineral pair the observed fractionations are identical, within experimental error, for all seven rocks analyzed: plagioclase–clinopyroxene, 0·45 ± 0·04; clinopyroxene–ilmenite, 1·31 ± 0·08; plagioclase–ilmenite, 1·76 ± 0·06.

In principle, any two of these fractionations provide two independent measures of "isotopic temperature." A minimal test of the existence of equilibrium is that these

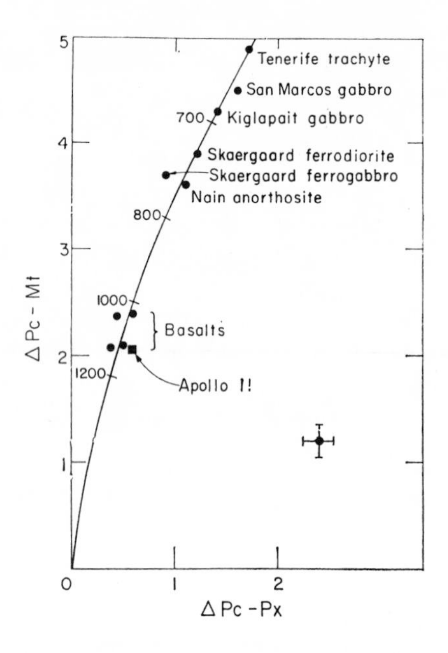

Fig. 1. Plagioclase–clinopyroxene–magnetite "concordancy diagram". Data on terrestrial rocks are from TAYLOR (1968), GARLICK (1966) and ANDERSON *et al.* (1970). Temperatures indicated along concordancy line are determined from the plagioclase–magnetite isotopic thermometer (O'NEIL and TAYLOR, 1967; O'NEIL and CLAYTON, 1964).

two temperatures agree. Unfortunately, laboratory calibration of isotopic thermometers involving pyroxene and ilmenite has not been done, so that this test cannot yet be carried out rigorously. It is, however, possible to compare the relative magnitudes of two isotopic fractionations with those found within the same mineral assemblage in various terrestrial rocks. This is done in Fig. 1, for the assemblage: plagioclase–clinopyroxene–magnetite using analytical data for terrestrial mafic igneous rocks, both intrusive and extrusive (GARLICK, 1966; TAYLOR, 1968; ANDERSON *et al.*, 1970). Because of the dependence of fractionations involving plagioclase on the chemical composition of the plagioclase (O'NEIL and TAYLOR, 1967), all data have been normalized to a feldspar composition of An60. This amounts to an effective increase in the δ-value of lunar plagioclase by 0·16 per mil. Furthermore, the isotopic fractionation between ilmenite and magnetite is very small, but not negligible, at igneous temperatures, so that the ilmenite δ-value has been reduced by 0·20 per mil to determine a hypothetical magnetite δ-value. The resulting point, representing the average fractionation for the seven Apollo 11 samples analyzed has been plotted on the plagioclase–pyroxene–magnetite diagram, and falls on the previously determined line through the data points for terrestrial rocks. It is virtually indistinguishable from the results for recent Hawaiian basalts. Hence the evidence for isotopic equilibrium among the major phases is fairly strong.

A similar "concordancy" test for the system plagioclase–clinopyroxene–olivine indicates that the olivine of sample 10020-24 was in isotopic equilibrium with the other minerals. Data for terrestrial samples are not available to make this test for cristobalite in sample 10058-35, but the magnitude of the cristobalite–plagioclase fractionation of 1·08 is within 0·1 of the value estimated by extrapolation of laboratory calibrations of the quartz–feldspar system to the isotopic temperature inferred for these rocks.

The best estimate of an isotopic temperature can be made from the fractionation between plagioclase and ilmenite. This is possible through the experimental calibration of the plagioclase–magnetite pair (O'Neil and Taylor, 1967; O'Neil and Clayton, 1964), along with the small ilmenite–magnetite fractionation, estimated on the basis of measurements on terrestrial basalts to be about 0·20 at the temperatures involved. The resulting isotopic temperature is 1120°C, with an analytical uncertainty of ±30°C. (Somewhat larger systematic errors may exist in the calibration of this isotopic thermometer, particularly since this temperature lies outside the range of the laboratory calibration.)

The isotopic temperature is very similar to those of terrestrial mafic extrusive rocks, which in turn, are in good agreement with other estimates of extrusion temperatures. Hence it appears that the isotopic temperatures are very close to the initial temperatures of crystallization from a silicate melt. On earth, large differences in isotopic temperatures are found between fine-grained, rapidly-quenched volcanic rocks, and their coarse-grained plutonic equivalents (Garlick, 1966; Taylor, 1968). This has been interpreted as resulting from postcrystallization retrograde isotopic exchange, on a scale comparable to the grain size, during the slow cooling of the plutonic rocks. No terrestrial rock as coarse-grained as the Apollo 11 type B samples has been found with a high isotopic temperature corresponding to its initial crystallization. Apparently the cooling conditions of the lunar samples were such as to prevent significant subsolidus exchange. The absence of water may have been an important factor.

The indistinguishability between type A and B rocks in terms of isotopic compositions or isotopic fractionations suggests that they were derived from magmas of the same isotopic composition, and that in both types the initial isotopic distribution at the time of crystallization has been quenched in, and has subsequently remained unchanged.

Acknowledgment—This research was supported in part by NASA grant NAS-9-7888.

REFERENCES

ANDERSON A. T., CLAYTON R. N. and MAYEDA T. K. (1970) Oxygen isotope studies in basalts. In preparation.

CLAYTON R. N. and MAYEDA T. K. (1963) The use of bromine pentafluoride in the extraction of oxygen from oxides and silicates for isotopic analysis. *Geochim. Cosmochim. Acta* **27,** 43–52.

GARLICK G. D. (1966) Oxygen isotope fractionation in igneous rocks. *Earth Planet. Sci. Lett.* **1,** 361–368.

KEAYS R. R., GANAPATHY R. G., LAUL J. C., ANDERS E., HERZOG G. and JEFFERY P. M. (1970) Trace elements and radioactivity in lunar rocks: implications for meteorite infall, solar wind flux, and formation conditions of the moon. *Science* **167,** 490–493.

O'NEIL J. R. and CLAYTON R. N. (1964) Oxygen isotope geothermometry. In *Isotopic and Cosmic Chemistry*, (editors H. Craig, S. L. Miller and G. J. Wasserburg), pp. 157–168. North-Holland.

O'NEIL J. R. and TAYLOR H. P. (1967) The oxygen isotope and cation exchange chemistry of feldspars. *Amer. Mineral.* **52,** 1414–1437.

TAYLOR H. P., DUKE M. B., SILVER L. T. and EPSTEIN S. (1965) Oxygen isotope studies of minerals in stony meteorites. *Geochim. Cosmochim. Acta* **29,** 489–512.

TAYLOR H. P. (1968) The oxygen isotope geochemistry of igneous rocks. *Contrib. Mineral. Petrol.* **19,** 1–71.

WALTER L. S. and CLAYTON R. N. (1967) Oxygen isotopes: experimental vapor fractionation and variations in tektites. *Science* **156,** 1357–1358.

Proceedings of the Apollo 11 Lunar Science Conference, Vol. 2, pp. 1435 to 1454.

Rare gases in Apollo 11 lunar material

R. O. PEPIN, L. E. NYQUIST, DOUGLAS PHINNEY and
DAVID C. BLACK
School of Physics and Astronomy, University of Minnesota,
Minneapolis, Minnesota 55455

(*Received* 13 *February* 1970; *accepted in revised form* 28 *February* 1970)

Abstract—Stepwise heating and total rare gas data are reported for a variety of lunar samples. Discussion is focussed primarily on the isotopic compositions of trapped and spallation neon, krypton and xenon components in the lunar fines, breccia and crystalline rocks. The total Ne isotopic composition in the fines and breccia, and its variation with outgassing temperature in stepwise heating, resembles isotopic patterns found in Ne from gas-rich meteorites. Trapped Xe and Kr components in the lunar fines, implanted by the solar wind, are isotopically very similar to carbonaceous chondrite Xe with a strong heavy isotope depletion and to terrestrial Kr. It seems probable, at least in some cases, that most of the trapped gases present at low levels in crystalline rocks are due to dust contamination. Spallation effects dominate all three gases in the rocks; spallation yield spectra are generally similar to those deduced in the past from studies of spallation gases produced in various classes of meteorites by the galactic cosmic radiation. There is strong evidence for a continuum of spallation Kr and Xe compositions, with the forms of the spallation yield distributions correlating from Kr to Xe in individual rocks, and from rock to rock for the same type of material. Those rocks which contain a large excess of ^{131}Xe show characteristic spallation yields compatible with production by a softer irradiation spectrum, and also show a distinct Xe target element abundance pattern which does not appear to be reflected in Kr target elements.

THIS paper is an expanded, updated version of the initial report from this laboratory on elemental and isotopic abundances of rare gases in the Apollo 11 samples (PEPIN *et al.*, 1970). Mass spectrometric analyses have been carried out for all rare gas isotopes in six lunar crystalline rocks, one breccia, crystalline rock fragments sieved from the lunar soil, and the fine (<1 mm) fraction of the soil (fines). Nineteen samples were analyzed; of these, fifteen were one-step, high temperature gas extractions and four—one sample of fines, two samples of breccia and a crystalline rock—were stepwise heating experiments. Heavy rare gas data from the full stepwise heating of breccia 10061-38(4)II have not yet been analyzed and will be reported in a later publication. Listings of sample numbers, types and weights are given in Tables A1 and A2.

ANALYTIC PROCEDURE

Two mass spectrometer systems were used in these studies: (I) one 6 in. double-focusing instrument in which all gases were analyzed, and (II) a pair of 6 in. single-focusing instruments, one for He, Ne, Kr and Xe and the second for Ar. The Ar instrument was operated at low resolution, the other two at resolutions sufficiently high to separate out all interfering hydrocarbons. Both spectrometer systems were operated on-line with sample processing systems consisting of high-temperature, high vacuum sample furnaces, Ti–Zr getters, and charcoal traps for gas separation. Samples were heated for one hour at temperatures in both single-step (usually 1650°C) and stepwise heating experiments. He and Ne were analyzed together; Ar, Kr and Xe were separated into >90 per cent pure fractions by selective desorption from charcoal at -135°C, -82°C, and 100°C respectively, and analyzed separately. One hour, 1650°C extraction blanks for both systems are given in Table A1.

Absolute gas concentrations were calculated from ion beam intensities, calibrated by analyzing pipetted samples from a standard gas mixture containing ^{3}He, ^{4}He and atmospheric Ne, Ar, Kr and Xe in known amounts. Calibrations were carried out before and after single-step extractions, and on the average once every two temperature fractions in stepwise heating experiments. Isotopic mass discriminations were determined in the same calibrations, based on current values for the isotopic compositions of atmospheric Ne (EBERHARDT *et al.*, 1965), Ar and Xe (NIER, 1950a, b), and Kr (NIEF, 1960; EUGSTER *et al.*, 1967a).

The lunar materials were received under dry N_2, and were sampled in a dry N_2 hood. The samples to be analyzed were then exposed to air during examination, weighing, packaging and loading in the vacuum oven. No clear evidence of adsorption of atmospheric rare gases was seen in the analyses. Since rare gas concentration levels in the lunar fine material are up to several thousand times higher than in the crystalline rocks, dust contamination is a serious problem in rock studies. Attempts were made to clean the surfaces of rock samples with a high velocity N_2 needle jet; it is not clear from the results that we were entirely successful.

RESULTS

Data are given in Tables A1–A5 in the Data Appendix. Numbers in parentheses, (1)–(5), following the sample designations identify multiple samples from the same specimen, and the Roman numerals I and II the system on which the samples were analyzed. Listed errors in isotopic compositions are 1σ statistical errors, summed quadratically from all contributing sources; uncertainties covering variations in mass discrimination factors between calibrations are included. Errors in absolute concentrations include systematic components—uncertainties in the composition of the standard calibration gas and in the volumes of the gas pipettes—in addition to statistical errors. Blank corrections were relatively unimportant except for Kr and Xe in the crystalline rocks, particularly those analyzed in System II (Table A1). Here, although blanks were monitored carefully and were relatively constant in amount and isotopic composition, it is difficult to guarantee that the corrections for a particular sample were accurately represented by interpolation between adjacent blanks. Therefore those isotopes most sensitive to blank subtractions—$^{84-86}Kr$, ^{129}Xe and $^{132-136}Xe$—could be subject to uncertainties beyond the estimated error limits in Table 1. Memory effects tended to degrade the measurement of the same isotopes; here it was decidedly advantageous to carry out the analyses with ^{82}Kr and ^{130}Xe as reference isotopes.

DISCUSSION

Our primary purpose in this paper is to present the data and briefly discuss certain salient features of the patterns in these data. Extremely interesting effects such as variations in $^{4}He/^{3}He$ and $^{38}Ar/^{36}Ar$ ratios in the various temperature fractions of the fine material, and the bearing of the Ar stepwise heating data on the problem of excess ^{40}Ar in the fines (HEYMANN *et al.*, 1970), probably require confirming experiments and certainly require more time for interpretation.

Helium, neon and argon

All gases in the crystalline rocks are dominated by spallation isotopes, except for the radiogenic daughters ^{4}He and ^{40}Ar. Neon isotopic compositions in the rocks and rock fragments from the soil are shown on a three-isotope correlation diagram in Fig. 1. The linear pattern indicates a highly variable mixture of a single spallation

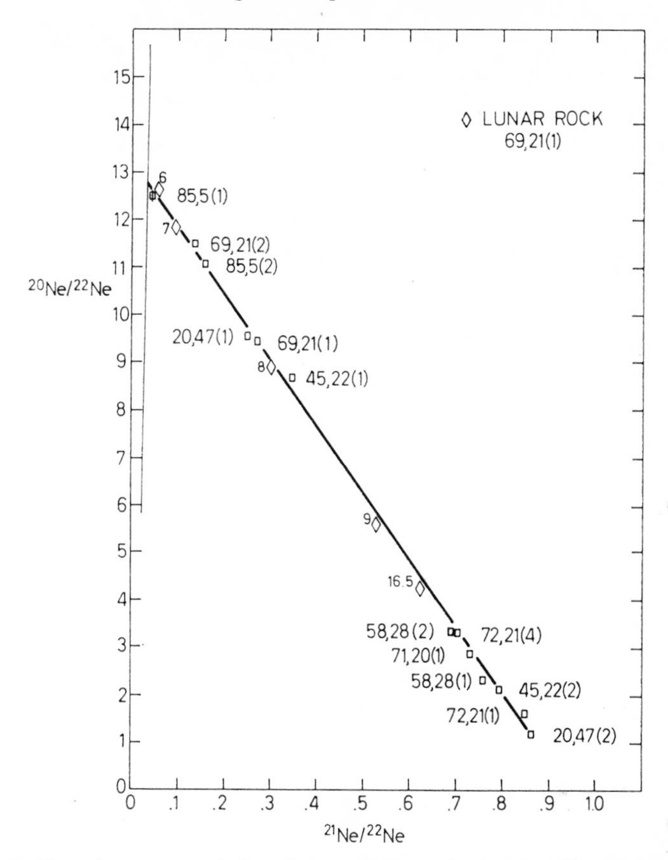

Fig. 1. Three-isotope correlation diagram for neon in the crystalline rocks and coarse fines. Data from single-step gas extractions and a stepwise heating experiment on rock 10069-21(1); numbers on the latter points are outgassing temperatures in 100°C units. The near-vertical line to the left of the diagram is the meteoritic "trend line" (BLACK and PEPIN, 1969). The strong correlation indicates a mixture of spallation neon with a "trapped" component characterized by $^{20}Ne/^{22}Ne \simeq 12{\cdot}6$.

composition, perhaps represented in virtually pure form in 10020-47(2), and a trapped neon component with $^{20}Ne/^{22}Ne = 12{\cdot}6 \pm 0{\cdot}2$. Neon in 10020-47(2) is essentially identical to that in the C3 and LL3 meteorites Felix and Chainpur, which terminate similar correlation lines for the carbonaceous chondrites (PEPIN, 1967). The striking variability in the proportion of trapped to spallation neon in repeat runs on certain samples (e.g. 10020-47 and 10045-22) could be due to minute and variable contamination by lunar fine material at a level $\leqslant 0{\cdot}01$ per cent by mass, despite care in handling and cleaning the rock chips. Direct solar wind irradiation on an exterior surface of a sample chip is another possibility; unfortunately, we have no information on locations of any of these samples in the original rocks. The relatively high trapped Ne concentrations in 10085-5(1-2), 1–2 mm fragments of fine and coarse-grained crystalline rock hand-picked from the 1–10 mm fraction of the soil, probably are due to direct exposure to the solar wind.

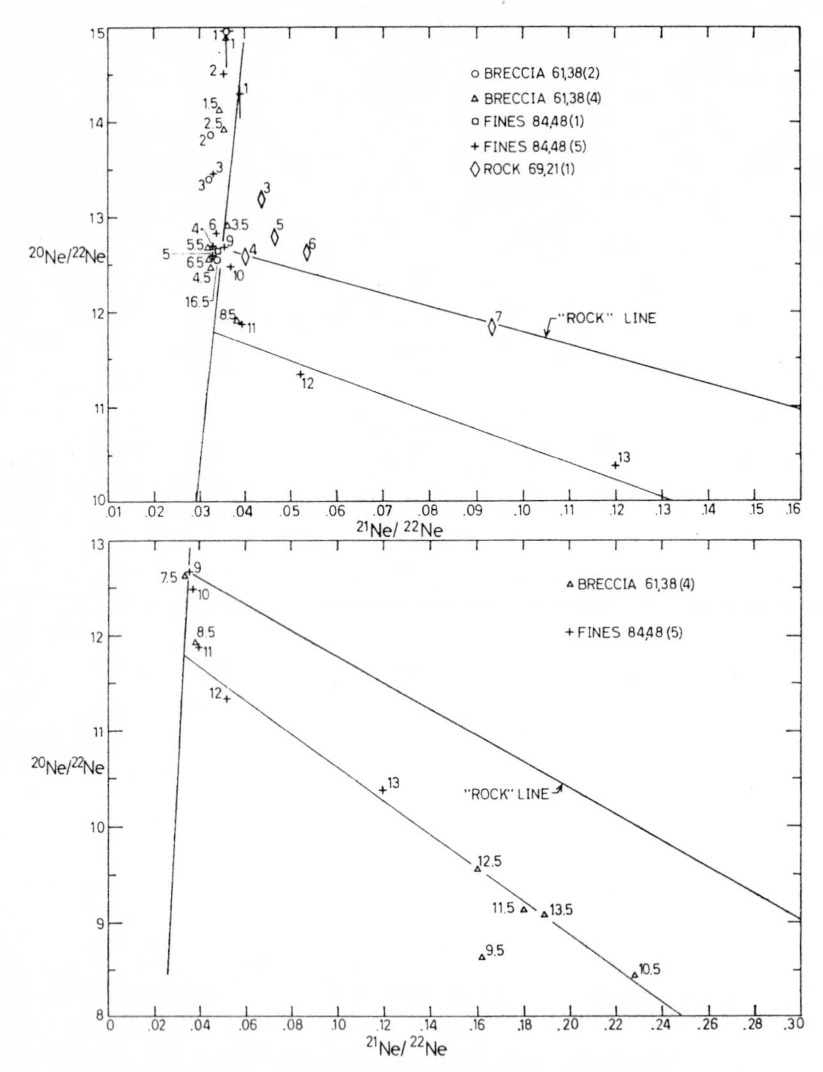

Fig. 2. Isotopic compositions of neon from stepwise heating of fines and breccia, and low temperature fractions from rock 10069-21(1), on three-isotope plots. The rock line and trend line are from Fig. 1; the third line is the apparent correlation defined by high temperature points from the fines and breccia. Almost all Ne in the fines and breccia is released in the steps clustering around $^{20}Ne/^{22}Ne = 12{\cdot}6$. To avoid crowding, not all data points are presented on the graphs. Numbers are outgassing temperatures in 100°C units.

High temperature points from stepwise heating of the type A rock 10069-21 are also plotted in Fig. 1. We carried out this experiment to check the possibility of multiple trapped neon components in the rocks; clearly, in this sample at least, trapped neon released at $T \geqq 600$°C other than the ~12·6-component does not exist at an appreciable level. [REYNOLDS *et al.* (1970) reported trapped neon $^{20}Ne/^{22}Ne$ as low as ~8 in rock 10057-20. We understand that this was the result of an error in the data reduction (HOHENBERG, personal communication).]

Variations in neon isotopic composition in the fines, breccia and rock 10069-21(1) (low temperature) in the temperature fractions of stepwise heating experiments are shown in Fig. 2. Neon in the breccia and fines, with very high $^{20}Ne/^{22}Ne$ ratios in the low temperature fractions, is strikingly similar to neon from the gas-rich achondrite Kapoeta, except that $^{21}Ne/^{22}Ne$ ratios here are generally low; points trending downward with increasing temperature lie to the left of the "trend-line" shown in the figure, instead of moving along it as they typically do in gas-rich meteorite neon (BLACK and PEPIN, 1969). It is clear from the data that $^{21}Ne/^{22}Ne$ at low and intermediate temperatures is not absolutely constant; the variations, although smaller and closer to error limits than those in $^{20}Ne/^{22}Ne$, are real. A similar low-temperature behavior—displaced to a more cosmogenic regime—occurs in rock 10069 (Fig. 2) and is compatible with trace soil contamination, as are the relative abundances of trapped gases which in this rock parallel those in the fines within $\sim$ a factor two.

The principal neon component in the lunar fines and breccia is one with $^{20}Ne/^{22}Ne = 12{\cdot}55$–$12{\cdot}65$, within error of the composition of Neon-B ($^{20}Ne/^{22}Ne = 12{\cdot}5 \pm 0{\cdot}2$) which dominates in the gas-rich meteorites. At high temperatures the neon compositions shown in the bottom part of Fig. 2 trend toward an increasing content of spallation neon. In both the fines and breccia there is strong evidence for the existence of a considerably lower trapped $^{20}Ne/^{22}Ne$, as low as 10 in the 950°C fraction of the breccia. The four highest temperature fractions of the breccia and the two highest temperature fractions of the dust form a linear array which does *not* pass through the rock spallation neon composition.

We have in the past interpreted such patterns in meteorites in terms of mixtures of spallation neon and two trapped neon components with different isotopic compositions (see BLACK and PEPIN, 1969). However, HOHENBERG *et al.* (1970) suggest that both the high $^{20}Ne/^{22}Ne$ ratios at low temperature and the low $^{20}Ne/^{22}Ne$ at high temperature shown in Fig. 2 are consequences of mass fractionation of a single, isotopically uniform trapped component in diffusive release from mass-dependent trapping sites during laboratory heating; in other words, that the apparent isotopic variations in trapped neon are artifacts of the stepwise heating technique. [The key argument is that heavier solar wind isotopes are implanted at greater depth and are more retentively held. Simple diffusive fractionation, with $D \propto 1/\sqrt{M}$, cannot account for the magnitude of the variations in Fig. 2 (see BLACK and PEPIN, 1969).] The idea is attractive, since it is difficult to account for the presence of rather large abundances of trapped neon other than solar wind neon in the lunar soil and breccia. HOHENBERG *et al.* point out that their data on the lunar fines, which are in reasonable agreement with ours, appear to support this interpretation, at least to first order. A more detailed analysis and additional experiments are clearly needed in order to establish whether this mechanism is actually responsible for the patterns in Fig. 2.

Gas release curves from the lunar fines as functions of temperature are given in Fig. 3, at the moment without discussion except to note that they are *not* consistent with a simple volume diffusion model with diffusion constants $D_M \propto 1/\sqrt{M}$. It is on the basis of these outgassing curves, compared with those for rock 10069 (see Fig. 4) and total gas abundance data given in Tables A1 and A2 that we have concluded that the so-called trapped gases in at least some of the crystalline rocks are compatible with dust contamination of the rock samples.

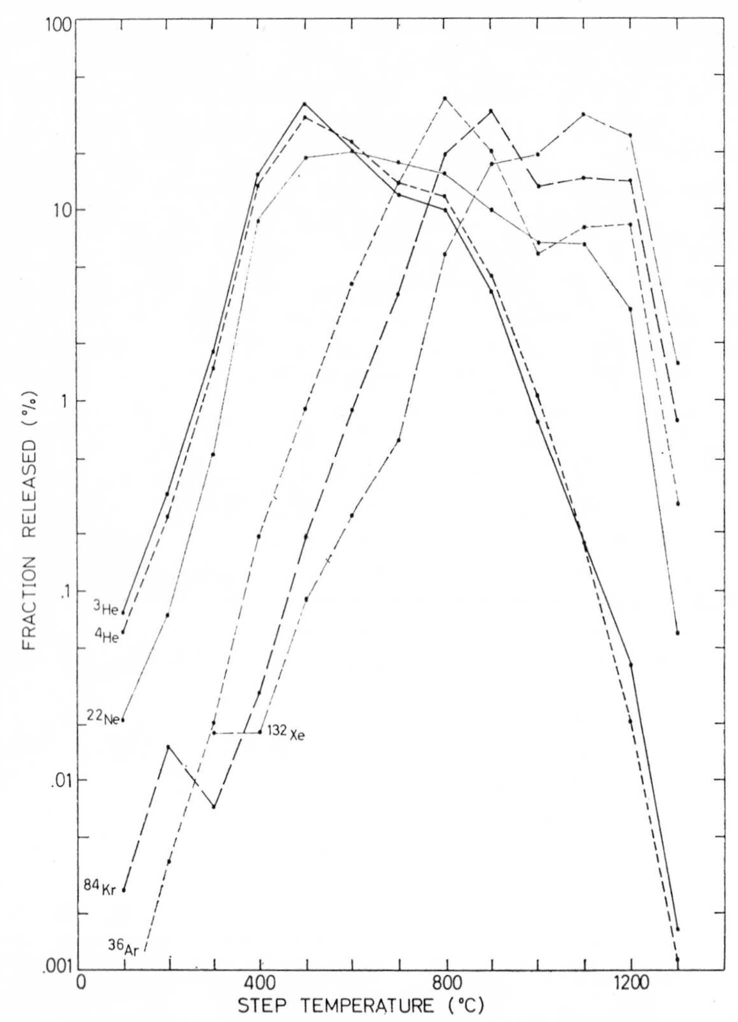

Fig. 3. Thermal release of rare gases during stepwise heating of lunar fines 10084-48(5). Gas amounts in each temperature step are expressed as percentages of the integrated total over all temperatures given in Table A3.

Values of $^{4}He/^{3}He \simeq 2500$, $^{4}He/^{20}Ne = 30–105$, $^{20}Ne/^{22}Ne = 12·6$, $^{36}Ar/^{38}Ar = 5·25$ and $^{40}Ar/^{36}Ar = 1·08–2·23$ are characteristic of bulk analyses of the soil and breccia, but the He and Ar isotopic ratios, as well as the neon, show considerable variation during stepwise heating. In the fines (Table A3), $^{4}He/^{3}He$ increases from ~1700 at 100°C to 3050 at 1000°C, then decreases. $^{36}Ar/^{38}Ar$ falls monotonically from 5·56 at 200°C to 4·54 at 1300°C. Breccia patterns are similar, both for He (Table A4) and for Ar, for which we have preliminary data. Some of these variations are clearly due to variable admixtures of spallation gases, but it is not at all clear that the entire range of isotopic shifts can be so attributed. Here again, HOHENBERG *et al.* (1970) suggest fractionation in the stepwise heating experiment as a contributing mechanism.

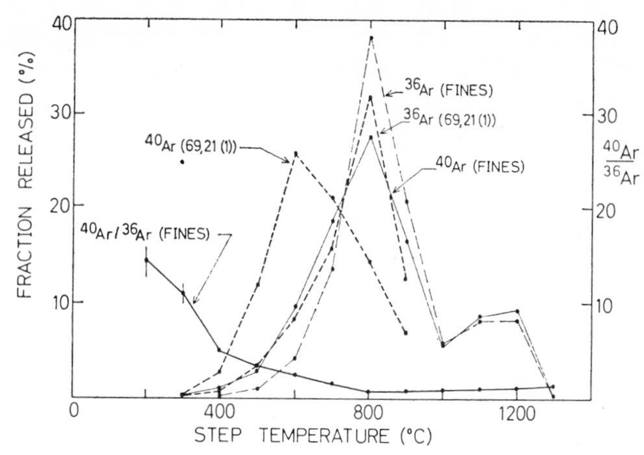

Fig. 4. Thermal release of argon during stepwise heating of lunar fines 10084-48(5) and crystalline rock 10069-21(1). All amounts expressed as percentages of totals given in Tables A3 and A5.

$^{40}Ar/^{36}Ar$ is of particular interest since the lunar fines contain about six times more ^{40}Ar than could have derived from in situ decay of ^{40}K over $4{\cdot}5 \times 10^9$ yr, suggesting that ^{40}Ar must be abundant in the trapped gases [see HEYMANN *et al.* (1970), HEYMANN and YANIV (1970a, b) for a detailed discussion]. A component which is probably *in-situ*-produced radiogenic ^{40}Ar outgasses preferentially at low temperature in the stepwise heating: $^{40}Ar/^{36}Ar = 14{\cdot}4$ in the 200°C fraction and falls to a minimum of 0·78 at 800°C, the point of maximum ^{36}Ar release and therefore the point at which the "trapped" gas composition is best represented. There is evidence that the actual $^{40}Ar/^{36}Ar$ ratio in primordial Ar, without radiogenic contributions from any source, is far lower, not only from theoretical considerations but from measurements in certain meteorites. MARTI (1967a) found trapped $^{40}Ar/^{36}Ar < 0{\cdot}35$ in Novo Urei, and BLACK (1970) sets a limit of $<0{\cdot}05$ from stepwise heating experiments on gas-rich meteorites.

Fractional release outgassing curves for ^{36}Ar and ^{40}Ar, and for the $^{40}Ar/^{36}Ar$ ratio in the fines vs. temperature, are given in Fig. 4. For comparison, the release of ^{36}Ar and ^{40}Ar from rock 10069 is also plotted; here, ^{36}Ar is dominantly trapped Ar and ^{40}Ar primarily *in-situ*-produced radiogenic Ar. Both the ratio plot and the release curves for the fines show a low-temperature preferential outgassing of ^{40}Ar. Above ~700°C, the ^{40}Ar and ^{36}Ar releases correlate very convincingly, suggesting that the major abundance of both isotopes must be sited in much the same way in the dust grains; there is evidence that the siting is surficial, consistent with low-energy ion implantation for *both* isotopes (HEYMANN *et al.*, 1970). The presence of some "excess ^{40}Ar" in the high temperature release is shown by the approximate doubling of the $^{40}Ar/^{36}Ar$ ratio from the "trapped" value of 0·78 at 800°C to 1·5 at 1300°C. Heymann and his associates (HEYMANN *et al.*, 1970; HEYMANN and YANIV, 1970a) find that the bulk of ^{40}Ar in the lunar fines is surface-correlated; assuming that the residual, non-surface-correlated component is ^{40}Ar from *in-situ* ^{40}K decay, they calculate a K–Ar age of $\sim 4{\cdot}4 \times 10^9$ yr. MARTI *et al.* (1970) have calculated a K–Ar age for the soil of $4{\cdot}65 \times 10^9$ yr on the assumption that ^{40}Ar in excess of the "trapped" ratio of

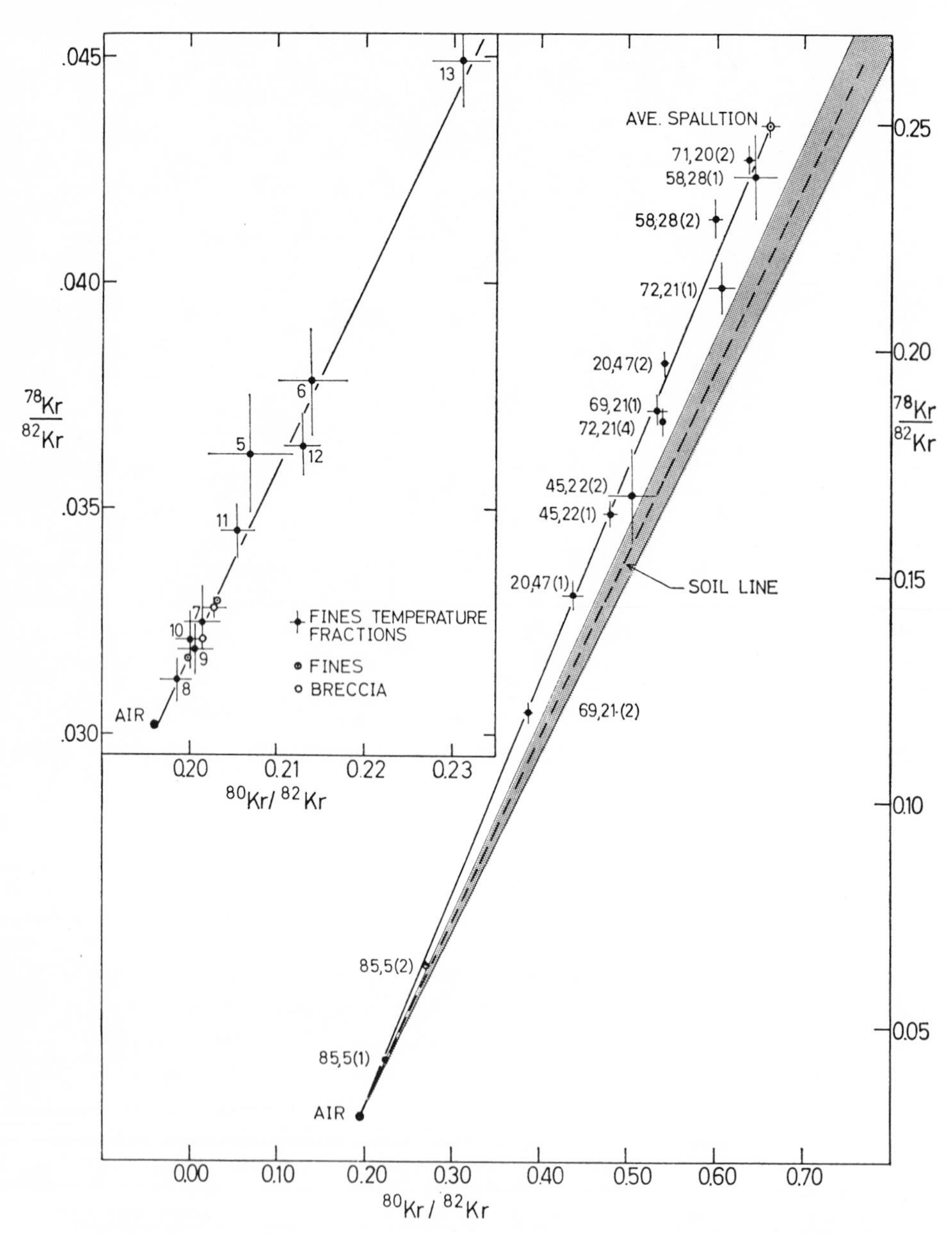

Fig. 5. Correlation diagram for the light Kr isotopes in lunar material. Crystalline rock data are given in the body of the Figure, with data from fines and breccia in the insert on a grossly expanded scale. The average spallation point is taken from Table 1.

0·78 is *in-situ*-produced radiogenic Ar. The same technique using our abundance data yields a somewhat longer age.

Krypton and xenon

Light isotope compositions of Kr and Xe in thirteen samples from six lunar crystalline rocks and crystalline rock fragments from the lunar soil are shown on

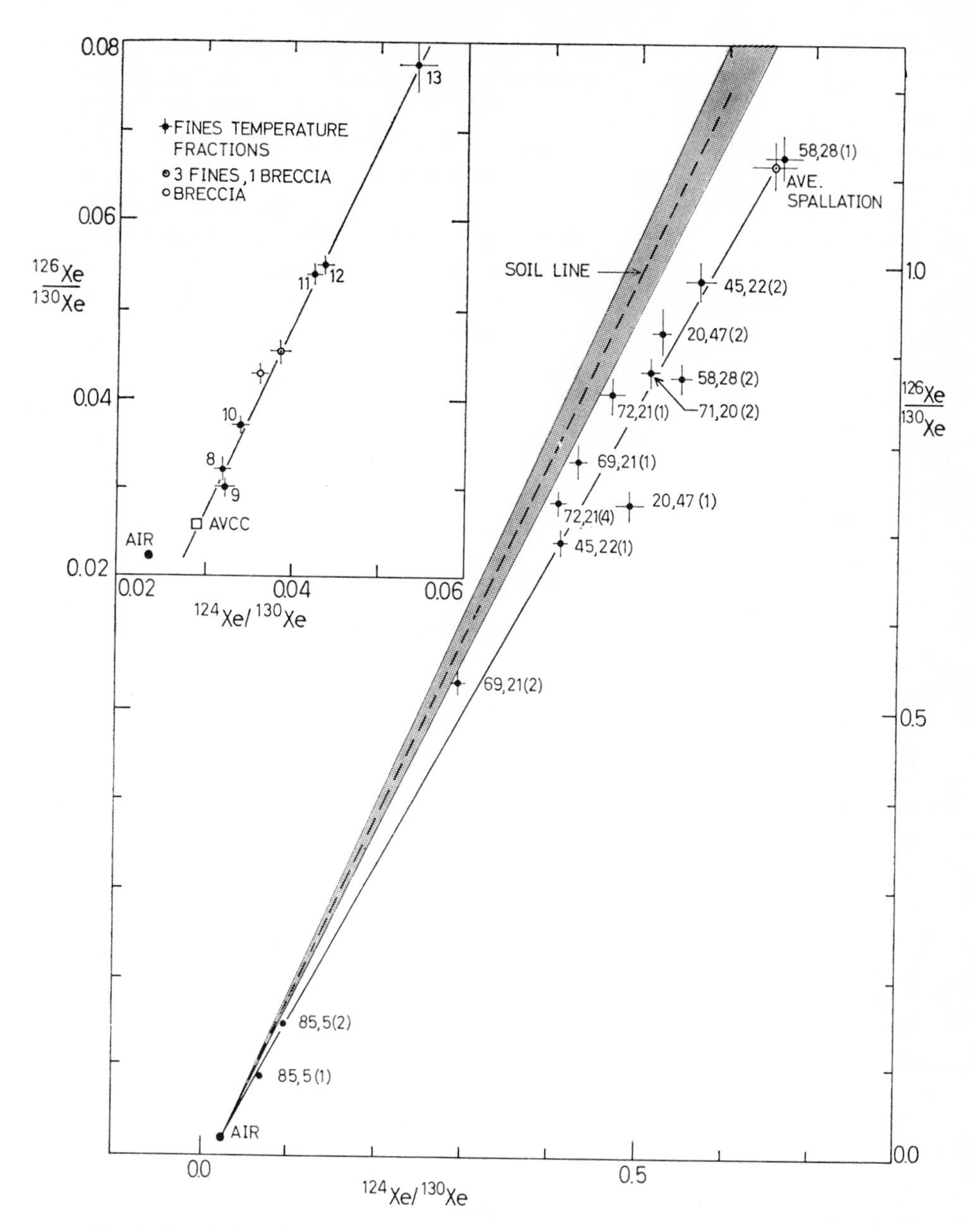

Fig. 6. Correlation diagram for three light Xe isotopes in lunar material. Fines and breccia data are in the insert, rock data in the main figure. Note that data points for rocks 10069-21 and 10072-21 lie to the left of the correlation line, while rocks 10020-47 and 10058-28 tend to lie to the right of the line, representing differences in the spallation yield spectra in these rocks: see text and Table 1.

three-isotope diagrams in Figs. 5 and 6. Practically all the Kr points fall within error on a single linear correlation extending from the air point to the point labeled "Ave. Spallation." This point represents the average composition of spallogenic Kr in these samples, calculated by assuming spallogenic ^{86}Kr $\equiv 0$ and subtracting off a trapped component of atmospheric composition in amount corresponding to the measured ^{86}Kr. The composition of the average spallation component is given in Table 1. Some of the point scatter in Fig. 5 is certainly real, paralleling the larger

Table 1. Spallation yield spectra of Kr and Xe in lunar soil and crystalline rocks

	^{78}Kr	^{80}Kr	^{82}Kr	^{83}Kr	^{84}Kr	^{86}Kr
Average: All rocks	0·249	0·660		1·34	0·56	
	± 0·006	± 0·009	≡ 1	± 0·02	± 0·03	≡ 0
Average: Rocks	0·264	0·683		1·34	0·57	
10020 and 10058	± 0·007	± 0·019	≡ 1	± 0·04	± 0·04	≡ 0
Average: Rocks	0·233	0·639		1·36	0·54	
10069 and 10072	± 0·002	± 0·002	≡ 1	± 0·01	± 0·05	≡ 0
1300°C Fraction	0·28	0·84		1·7	1·2	
Fines 10084-48(5)	± 0·04	± 0·08	≡ 1	± 0·4	± 0·6	≡ 0

	^{124}Xe	^{126}Xe	^{128}Xe	^{129}Xe	^{130}Xe	^{131}Xe	^{132}Xe	^{134}Xe	^{136}Xe
Average: Rocks	0·609		1·424	1·24	0·796	3·71	0·60	0·02	
10020 and 10058	± 0·021	≡ 1	± 0·017	± 0·08	± 0·008	± 0·12	± 0·04	± 0·02	≡ 0
Average: Rocks	0·542		1·586	1·23	0·957	7·18	0·62	0·05	
10069 and 10072	± 0·003	≡ 1	± 0·026	± 0·09	± 0·038	± 0·12	± 0·05	± 0·01	≡ 0
1300°C Fraction	0·49		1·8	3·5	1·2	8·7	2·0		
Fines 10084-48(5)	± 0·05	≡ 1	± 0·4	± 2·0	± 0·5	± 3·3	± 2·0	≤ 1	≡ 0

variations in spallation Xe discussed below, but to an excellent first approximation these samples contain variable mixtures of a spallation component which in the light isotopes is very nearly isotopically uniform from sample to sample, and a trapped component which may well reflect contribution from surface dust and which resembles atmospheric Kr in isotopic composition.

Variations in the composition of spallation Xe in these rocks are much more prominent. A correlation line between the average spallation and air [or average carbonaceous chondrite (AVCC)] composition points in Fig. 6 fits the data about as well as any other single line, but there is considerable scatter in the data points away from a unique linear correlation, and the scatter is in general real. No single, isotopically uniform spallation component can fit these rock data. An extreme case of this is found in one of the heavier Xe isotopes, ^{131}Xe: the variation of spallogenic ^{131}Xe/^{126}Xe ratios by about a factor 2 from rock to rock was noted by a number of rare gas investigators in the initial reports on Apollo 11 samples (REYNOLDS *et al.*, 1970; MARTI *et al.*, 1970; PEPIN *et al.*, 1970; FUNKHOUSER *et al.*, 1970; ALBEE *et al.*, 1970).

Isotopic compositions of spallation Xe in these samples were calculated by first correcting the heavy Xe isotope abundances in Table A1 for contributions from ^{238}U spontaneous fission, assuming a Xe-retention age equal to the average high-temperature Ar-retention age of lunar rocks, $3{\cdot}7 \times 10^9$ yr, found by TURNER (1970), and using uranium concentrations measured by TATSUMOTO and ROSHOLT (1970), SILVER (1970), and MORRISON *et al.* (1970) or estimated from average measured values for fine and coarse-grained rocks. Fission Xe corrections on total ^{136}Xe range from 2·8 per cent (10020-47(1)) to 24 per cent (10069-21(1)). Spallation yield spectra were then found by assuming spallogenic $^{136}Xe \equiv 0$ and subtracting from the fission-corrected rock Xe a component, normalized to ^{136}Xe, with the average isotopic composition of Xe in the lunar fines (Table A2). The average spallation Xe composition for all rocks is plotted in Fig. 6. This is not a particularly meaningful spectrum, since it averages over real variations. To demonstrate this, we selected two pairs of crystalline rocks with quite different $^{131}Xe/^{126}Xe$ ratios in the spallation component: Samples 10020-47(1, 2), and 10058-28(1, 2) with $(^{131}Xe/^{126}Xe)_{spall} = 3{\cdot}71 \pm 0{\cdot}12$, and Samples 10069-21(1, 2) and 10072-21(1, 4) with $(^{131}Xe/^{126}Xe)_{spall} = 7{\cdot}18 \pm 0{\cdot}12$. All other samples have intermediate values for this ratio. The calculated spallation Xe spectra for these pairs of samples are tabulated in Table 1; errors are 1σ deviations from the means. It is completely clear that the dramatic increase in ^{131}Xe yield is accompanied by smaller but still significant changes in the overall relative yield pattern, in just the direction expected for a softer irradiation spectrum. This effect is *not* sensitive to the assumed composition of trapped Xe in (or on) the rocks; subtraction of atmospheric-type Xe leads to yield spectra in which the numbers are slightly changed but the differences between the two pairs of rocks just as pronounced. Spallation yields at ^{129}Xe are curiously low, as if the trapped gas composition we have used overestimates the correction at this isotope. The variations in isotopic composition between the rock pairs at ^{129}Xe and $^{132-134}$Xe are smaller than one might expect from the behavior of the other isotopes, but here variations in the spallation spectra of individual rocks are large; this scatter is probably due to the uncertainty, discussed above, in correcting the heavy isotopes for blanks and memory effects.

If atmospheric rather than lunar fines Xe is used as the trapped component in deriving these spallation spectra, the average yields at every isotope for the rock pair 10069–10072 are almost identical to those obtained by ALBEE *et al.* (1970) for another type A rock, 10017, on the assumption of trapped Xe of atmospheric composition. With this choice the relative ^{129}Xe yields increase from the values in Table 1 to 1·65 for 10069–10072 and 1·54 for 10020–10058. These yields are slightly higher than the ^{128}Xe yields, a pattern which seems to be more in accord with spallation theory than the lower values in Table 1. This raises the possibility, suggested by ALBEE *et al.* that trapped Xe in the lunar rocks is in fact atmospheric in composition (and perhaps in origin, through atmospheric contamination). However, since the trapped gas abundance pattern from Ne through Xe in 10069 agrees fairly well with that in the soil, our tentative conclusion is that we are dealing with solar wind Xe, and we have used this choice in Table 1.

In view of the Xe spallation yield variations between type A and B rocks shown in Table 1, it is instructive to reexamine the Kr data for the same effect. Results for the same rock pairs (Table 1) clearly show the same kind of variations in the light spallogenic Kr isotopes, correlating with the Xe patterns, in agreement with the correlations of $^{78}Kr/^{83}Kr$ vs. $^{126}Xe/^{131}Xe$ noted by MARTI *et al.* (1970). The striking difference in the two sets of spallation spectra is the absence in Kr of variations on the scale of those in ^{131}Xe. The more rapid fall-off in spallation yields toward the lighter isotopes in the type A rocks indicates a softer effective irradiation spectrum, so it seems reasonable to look to low energy nuclear effects in attempting to account for the ^{131}Xe anomaly. A modest decrease from the average galactic cosmic ray energy is not effective; HOHENBERG and ROWE (1969) have measured the Xe spallation yields from 730 MeV protons on Ba, Cs and the rare earths, and the production of ^{131}Xe is not unusually high. ALBEE *et al.* (1970) and MARTI *et al.* (1970) have pointed out that the total measured slow neutron flux in the lunar rocks is inadequate to account for the observed excess ^{131}Xe. ALBEE *et al.* also note that very low energy protons should produce not only ^{131}Xe but also ^{129}Xe and $^{83-84}Kr$. We see no evidence for a specific excess of ^{129}Xe and, in agreement with their results, no pronounced anomalies in the Kr spallation yield distributions.

Variations in chemical composition as well as differences in the irradiation energy spectrum undoubtedly play a role in shaping the spallation yield patterns in Kr and Xe. While it is quite clear that anomalies as striking as the ^{131}Xe excess—and the absence of the analogous effect in Kr—cannot arise from chemical effects alone, there are chemical variations in these rocks which favor Xe production from closely neighboring target elements in type A rocks relative to type B, without any clear associated effect in Kr. Both Ba and rare earth element abundances are markedly higher in type A material, but more significantly the Ba/REE ratio is also about a factor two higher in type A (GAST and HUBBARD, 1970; MURTHY *et al.*, 1970). For target elements contributing to Kr, although in some cases rock-to-rock variations are quite large, on the average the Sr, Y and Zr abundances and the Sr/Y and Sr/Zr ratios are very similar in both A and B material (MORRISON *et al.*, 1970).

Heavy rare gases in the lunar fines. Light isotope variations in Kr and Xe in stepwise heating temperature fractions from the lunar fines, shown in the insets of Figs. 5 and 6, form linear arrays on these correlation diagrams. Bulk measurements on several samples of the fine material and the breccia fall closely along these lines. The correlated variations clearly arise from the nonuniform mixing of spallation and trapped gas components in the various temperature fractions.

There are two noteworthy features of these correlations: they pass essentially through the isotopic compositions of atmospheric Kr and AVCC Xe respectively, and they differ significantly in slope from either the average or individual correlation lines for the crystalline rocks. Both features are characteristic of all three-isotope correlations involving $^{124-130}Xe$ and all Kr isotopes.

One cannot determine uniquely the isotopic compositions of the trapped or spallogenic Kr and Xe in the soil from stepwise heating data alone. Figures 5 and 6, and other three-isotope diagrams constructed from data in Tables 2 and 3, suggest very strongly that the light trapped Xe isotopes follow the carbonaceous chondrite pattern, and that the trapped Kr isotopes are close to atmospheric in composition. This assumption can be tested quantitatively for consistency by assuming a value for one trapped isotopic ratio in each gas and deriving the full isotopic composition from this ratio and the correlation line equations. We have carried out these calculations, assuming $(^{126}Xe/^{130}Xe)_{trapped} = 0{\cdot}0255$, the value for AVCC Xe from EUGSTER *et al.* (1967b) and PEPIN (in preparation), and $(^{82}Kr/^{80}Kr)_{trapped} \equiv (^{82}Kr/^{80}Kr)_{atmosphere} = 5{\cdot}11$ (EUGSTER *et al.*, 1967a). The isotopic compositions of trapped lunar Xe and Kr are then $^{124}Xe/^{126}Xe/^{128}Xe/^{129}Xe/^{130}Xe/^{131}Xe/^{132}Xe/^{134}Xe/^{136}Xe = 0{\cdot}294 \pm 0{\cdot}007/ \equiv$

0·255/5·08 ± 0·11/64·0 ± 0·8/ ≡ 10/49·6 ± 0·6/60·8 ± 0·7/22·42 ± 0·24/18·26 ± 0·25 and $^{78}Kr/^{80}Kr/^{82}Kr/^{83}Kr/^{84}Kr/^{86}Kr$ = 0·1535 ± 0·0015/ ≡ 1/ = 5·11/5·077 ± 0·036/25·23 ± 0·15/7·681 ± 0·054. For comparison, the AVCC Xe composition (PEPIN, in preparation; see also MARTI, 1967b; EUGSTER *et al.*, 1967b) is $^{124}Xe/^{126}Xe/^{128}Xe/^{129}Xe/^{130}Xe/^{131}Xe/^{132}Xe/^{134}Xe/^{136}Xe$ = 0·288 ± 0·004/0·255 ± 0·003/ 5·06 ± 0·03/ . . . / ≡ 10/50·24 ± 0·26/61·37 ± 0·26/23·58 ± 0·22/19·91 ± 0·22. The relative abundances of the light Xe isotopes in lunar trapped Xe are clearly completely consistent with the AVCC pattern. It is also clear that the heavy, unshielded Xe isotopes are significantly depleted relative to AVCC Xe, as noted earlier by EBERHARDT *et al.* (1970), MARTI *et al.* (1970), PEPIN *et al.* (1970) and REYNOLDS *et al.* (1970); the most obvious explanation is a depletion in lunar Xe of the fission Xe component known to be present in the carbonaceous chondrites (see PEPIN *et al.*, 1970). The nominal isotopic composition of trapped lunar Kr is compatible with that of terrestrial Kr within 0·4 per cent at every isotope.

The soil correlation lines, with shaded error regions, are extended in Figs. 5 and 6 into the highly spallogenic regimes occupied by the crystalline rocks. These lines deviate from the rock data in a way consistent with a more rapid drop-off in spallation yield with increasing mass difference from the spallation target elements; i.e. consistent with a lower effective irradiation energy for spallation in the soil as compared to the rocks. Since soil particles spend some fraction of their lifetimes at the very surface of the lunar regolith, there could be a detectable spallation component in the soil arising from solar proton bombardment. We have not yet attempted to specify this possibility quantitatively. Assuming the trapped Kr and Xe compositions given above, one can calculate the spallation yield distributions for Kr and Xe in the lunar soil; this is most accurately done for the spallation-rich 1300°C fraction (Table A3). Results are given in Table 1; while not very precise, they do suggest low-energy production. In Ba/REE ratio, the soil and breccia resemble type A rocks, while Sr/Y/Zr ratios are similar to both A and B rocks (GAST and HUBBARD, 1970; MORRISON *et al.*, 1970). It would seem, therefore, that departures from Type A spallation yield distributions (rocks 10069 and 10072, Table 1) would reflect an energy rather than a chemical effect, provided that the spallation gases in the 1300°C fraction of the soil derive from material which is chemically representative of the soil as a whole.

Finally, it is interesting to note that isotopic compositions of neon, krypton and xenon, and in particular the proportion of spallation to trapped gases, are practically identical within the accuracy of our data in both the soil and the brecciated rock, despite a 2- to 5-fold excess in gas concentrations in the breccia. It seems reasonable on this evidence to accept the view that the breccia formed by shock-lithification of soil, perhaps accompanied by shock-induced concentration of gases in dust particles. But also on this evidence it is unreasonable to argue that the exposure ages of the breccias themselves as calculated from spallation gas contents are meaningful, since this would require that exposure ages of soil and breccia differ by about the same factors that the trapped gas contents differ. In fact, assuming that trapped gas compositions and spallation production rates are the same in the soil and breccia, a simple calculation using the trapped gas composition given above and Xe data from Table A2 suggests that the exposure age of the breccia as an *independent* object does not exceed ≈15 per cent of the total exposure age of the soil.

Data Appendix

Table A1. Rare gases in lunar crystalline rocks

Sample	Weight (mg)	Gas content ($\times$ 10^{-8} cm^3STP/g) ^{3}He	^{4}He	^{22}Ne	^{36}Ar	^{40}Ar	^{20}Ne/^{22}Ne	^{21}Ne/^{22}Ne	^{38}Ar/^{36}Ar
10020-47(1)I	194·0	129·7	54,390	56·28	87·6	1784	9·55	0·2495	0·3675
Type B		± 2·3	± 490	± 0·90	± 1·3	± 22	± 0·03	± 0·0022	± 0·0038
10020-47(2)II	132·1	104·8	10,660	12·11	11·71	1694	1·209	0·8583	1·425
Type B		± 3·9	± 370	± 0·45	± 0·41	± 60	± 0·005	± 0·0034	± 0·014
10045-22(1)II	230·4	85·6	34,200	35·0	—	—	8·68	0·3425	—
Type B		± 3·2	± 120	± 1·3			± 0·03	± 0·0014	
10045-22(2)I	265·7	112·3	12,940	16·57	14·72	1582	2·125	0·7929	1·128
Type B		± 2·0	± 120	± 0·25	± 0·29	± 21	± 0·006	± 0·0032	± 0·012
10058-28(1)I	371·6	69·1	12,090	8·63	8·86	3285	2·345	0·7577	1·101
Type B		± 1·2	± 110	± 0·13	± 0·18	± 36	± 0·007	± .0·0030	± 0·011
10058-28(2)II	180·8	54·2	16,460	8·12	10·26	4800	3·375	0·6882	1·032
Type B		± 2·2	± 690	± 0·31	± 0·36	± 170	± 0·010	± 0·0028	± 0·010
*10069-21(1)I	132·3	38·74	89,110	14·49	18·34	6575	9·40	0·2708	0·4253
Type A		± 0·77	± 890	± 0·22	± 0·13	± 37	± 0·07	± 0·0029	± 0·0042
10069-21(2)II	125·2	38·2	82,300	27·7	47·0	6120	11·49	0·1347	0·2656
Type A		± 1·4	± 2900	± 1·0	± 1·8	± 230	± 0·03	± 0·0005	± 0·0027
10071-20(1)II	207·8	215·0	43,100	41·6	45·3	7020	2·901	0·7298	1·030
Type A		± 8·6	± 1700	± 1·6	± 1·6	± 240	± 0·009	± 0·0029	± 0·010
10072-21(1)I	111·9	183·2	48,270	29·30	24·5	8010	1·639	0·8456	1·306
Type A		± 3·3	± 440	± 0·44	± 0·8	± 280	± 0·005	± 0·0085	± 0·013
10072-21(4)II	98·8	181	53,100	22·8	30·8	7300	3·350	0·0741	1·180
Type A		± 13	± 1900	± 1·1	± 1·2	± 290	± 0·010	± 0·0042	± 0·012
Blanks									
System I		<0·001	0·6	0·009	0·007	1			
System II		0·006	2·6	0·004	0·080	26			

Table A1. (continued)

Sample	Weight (mg)	^{132}Xe (10^{-10} cm^3STP/g)	^{124}Xe	^{126}Xe	^{128}Xe	^{129}Xe	^{130}Xe	^{131}Xe	^{132}Xe	^{134}Xe	^{136}Xe
10020-47(1)I	194·0	0·69	15·33	23·05	38·42	118·5	31·44	147·2	100	30·3	26·0
Type B		± 0.03	± 0·44	± 0·46	± 0·46	± 1·2	± 0·50	± 1·5		± 0·3	± 0·3
10020-47(2)II	132·1	0·35	24·76	43·57	67·81	125·5	47·13	233·3	100	25·4	24·2
Type B		± 0·04	± 0·25	± 1·13	± 1·15	± 3·3	± 0·94	± 4·2		± 1·0	± 1·0
10045-22(1)II	230·4	0·50	14·32	24·01	43·25	124·5	34·82	182·8	100	31·0	23·6
Type B		± 0·04	± 0·30	± 0·50	± 0·61	± 1·6	± 0·52	± 1·8		± 0·8	± 0·6
10045-22(2)I	265·7	0·27	27·22	47·10	74·87	142·5	47·87	263·5	100	29·0	28·4
Type B		± 0·02	± 0·84	± 0·99	± 0·67	± 1·9	± 0·57	± 5·3		± 1·2	± 0·8
10058-28(1)I	371·6	0·15	51·51	87·26	128·1	168·4	77·70	342·1	100	24·9	16·4
Type B		± 0·01	± 1·65	± 1·83	± 1·3	± 5·6	± 1·17	± 8·2		± 0·7	± 0·4
10058-28(2)II	180·8	0·52	23·90	38·18	62·09	128·1	43·63	206·5	100	31·2	25·0
Type B		± 0·04	± 0·41	± 0·65	± 0·75	± 2·7	+ 0·48	± 3·5		± 0·8	± 0·6
*10069-21(1)I	332·3	0·32	23·39	42·79	70·33	126·3	54·62	343·7	100	33·2	27·6
Type A		± 0·03	± 0·63	± 0·98	± 1·27	± 2·0	± 0·87	± 4·1		± 0·5	± 0·6
10069-21(2)II	125·2	0·86	7·89	14·34	29·59	110·2	26·85	175·5	100	36·1	29·5
Type A		± 0·07	± 0·21	± 0·29	± 0·47	± 1·4	± 0·40	± 1·4		± 1·1	± 0·9
10071-20(2)II	207·8	1·98	42·85	74·07	119·3	163·3	83·96	448·0	100	21·1	12·1
Type A		± 0.13	± 0·73	± 1·26	± 1·2	± 0·7	± 0·34	± 1·8		± 0·2	± 0·1
10072-21(1)I	111·9	1·27	34·21	62·56	105·6	152·3	73·09	482·5	100	26·6	18·8
Type A		± 0·08	± 1·13	± 1·50	± 3·4	± 1·8	± 1·10	± 3·9		± 0·6	± 0·4
10072-21(4)II	98·8	2·19	18·26	33·02	60·17	126·1	44·90	297·4	100	32·3	26·2
Type A		± 0·15	± 0·35	± 0·63	± 0·48	± 1·3	± 0·36	± 1·2		± 0·5	± 0·4
Blanks											
System I		0·01									
System II		0·07									

Table A1. (continued)

Sample	Weight (mg)	^{84}Kr (10^{-10} cm^{3}STP/g)	^{78}Kr	^{80}Kr	^{82}Kr	^{83}Kr	^{84}Kr	^{86}Kr
10020-47(1)I	194·0	5·33 ± 0·71	5·25	15·75	36·0	42·4	100	27·8
Type B			± 0·13	± 0·33	± 0·5	± 0·7		± 0·3
10020-47(2)II	132·1	3·48 ± 0·40	11·85	32·50	60·0	73·2	100	21·8
Type B			± 0·20	± 0·23	± 0·6	± 0·7		± 0·5
10045-22(1)II	230·4	4·56 ± 0·55	7·34	21·39	44·7	53·0	100	25·8
Type B			± 0·13	± 0·26	± 0·5	± 0·6		± 0·4
10045-22(2)I	265·7	1·45 ± 0·19	7·87	23·59	46·8	58·0	100	23·8
Type B			± 0·42	± 1·01	± 1·7	± 2·5		± 0·6
10058-28(1)I	371·6	1·01 ± 0·13	17·37	46·67	72·6	97·9	100	18·6
Type B			± 0·63	± 1·63	± 1·3	± 2·7		± 0·7
10058-28(2)II	180·8	1·89 ± 0·21	21·86	57·07	95·2	117·4	100	17·6
Type B			± 0·37	+ 0·46	± 1·0	± 1·2		± 0·4
*10069-21(1)I	332·3	1·00 ± 0·15	12·49	35·61	66·8	85·6	100	23·4
Type A			± 0·22	± 0·43	± 0·7	± 0·8		± 0·4
10069-21(2)II	125·2	4·79 ± 0·57	3·85	12·44	32·1	36·3	100	27·7
Type A			± 0·07	± 0·12	± 0·5	± 0·4		± 0·8
10071-20(2)II	207·8	6·63 ± 0·74	30·84	80·86	127·0	162·7	100	12·7
Type A			± 0·43	± 0·32	± 1·0	± 1·3		± 0·1
10072-21(1)I	111·9	3·31 ± 0·37	24·74	70·03	115·6	156·0	100	14·7
Type A			± 0·62	± 1·26	± 1·6	± 1·9		± 0·5
10072-21(4)II	98·8	7·37 ± 0·83	12·27	35·75	66·4	83·6	100	22·4
Type A			± 0·21	± 0·29	± 0·7	± 0·9		± 0·4
Blanks								
System I		0·01						
System II		0·30						

* Integrated stepwise heating experiment.

Table A2. Rare gases in lunar breccia, fines and rock fragments from the soil

Sample	Weight (mg)	Gas content (× 10^{-8} cm^{3}STP/g)			^{4}He/^{3}He	^{20}Ne/^{22}Ne	^{21}Ne/^{22}Ne	^{38}Ar/^{36}Ar	^{40}Ar/^{36}Ar
		^{3}He	^{22}Ne	^{36}Ar					
10061-38(1)I	12·4	13,580	40,290	63,880	2523	12·65	0·0337	0·1897	2·23
Breccia		± 370	± 930	± 890	± 25	± 0·03	± 0·0003	± 0·0019	± 0·02
*10061-38(2)II	10·3	10,120	59,900	82,900	2618	12·57	0·0331	0·1884	2·19
Breccia		± 400	± 2300	± 3300	± 26	± 0·04	± 0·0002	± 0·0019	± 0·02
*10061-38(4)II	89·8	11,590	87,200	—	2778	12·56	0·0330	—	—
Breccia		± 260	± 2100		± 33	± 0·11	± 0·0004		
10084-48(1)I	13·2	8100	18,030	38,240	2418	12·66	0·0346	0·1898	1·09
Fines		± 900	± 810	± 760	± 24	± 0·04	± 0·0004	± 0·0028	± 0·03
10084-48(2)I	11·6	9320	16,970	36,850	2375	12·63	0·0337	0·1927	1·10
Fines		± 250	± 510	± 740	± 24	± 0·04	± 0·0004	± 0·0020	± 0·01
*10084-48(5)I	57·3	9410	16,910	37,800	2252	12·58	0·0348	0·1912	1·08
Fines		± 110	± 150	± 470	± 27	± 0·14	± 0·0005	± 0·0020	± 0·01
10085-5(1)I	145·1	227·9	624	2210	1653	12·52	0·0425	0·1994	2·69
Fragments		± 4·1	± 10	± 26	± 16	± 0·14	± 0·0003	± 0·0020	± 0·03
10085-5(2)II	40·2	150·6	100·2	166·8	1021	11·10	0·1542	0·2834	54·89
Fragments		± 5·9	± 3·9	± 6·2	± 10	± 0·03	± 0·0006	± 0·0028	± 0·55

Table A2. (continued)

Sample	Weight (mg)	^{84}Kr (10^{-10} cm^{3}STP/g)	^{78}Kr	^{80}Kr	^{82}Kr	^{83}Kr	^{84}Kr	^{86}Kr
10061-38(1)I Breccia	12·4	3250 ± 340	0·677 ± 0·006	4·187 ± 0·021	20·64 ± 0·04	20·75 ± 0·06	100	30·44 ± 0·09
10061-38(2)II Breccia	10·3	4920 ± 590	0·655 ± 0·009	4·115 ± 0·029	20·42 ± 0·08	20·40 ± 0·08	100	30·46 ± 0·12
10084-48(1)I Fines	13·2	2220 ± 310	0·648 ± 0·008	4·084 ± 0·033	20·44 ± 0·06	20·44 ± 0·06	100	30·57 ± 0·12
10084-48(2)I Fines	11·6	2130 ± 220	0·637 ± 0·006	—	20·42 ± 0·04	20·51 ± 0·08	100	30·41 ± 0·06
*10084-48(5)I Fines	57·3	2230 ± 260	0·673 ± 0·013	4·142 ± 0·033	20·37 ± 0·08	20·43 ± 0·10	100	30·43 ± 0·12
10085-5(1)I Fragments	145·1	128 ± 13	0·890 ± 0·014	4·702 ± 0·024	21·11 ± 0·06	21·64 ± 0·09	100	30·31 ± 0·09
10085-5(2)II Fragments	40·2	86 ± 10	1·499 ± 0·018	6·361 ± 0·045	23·49 ± 0·12	24·74 ± 0·10	100	29·80 ± 0·07

Table A2. (continued)

Sample	Wt. (mg)	^{132}Xe (10^{-10} cm^3 STP/g)	^{124}Xe	^{126}Xe	^{128}Xe	^{129}Xe	^{130}Xe	^{131}Xe	^{132}Xe	^{134}Xe	^{136}Xe
10061-38(1)I Breccia	12·4	584 ± 20	0·638 ± 0·013	0·736 ± 0·012	8·82 ± 0·08	104·9 ± 0·6	16·44 ± 0·13	82·88 ± 0·58	100	36·94 ± 0·22	30·14 ± 0·21
10061-38(2)II Breccia	10·3	993 ± 63	0·602 ± 0·010	0·709 ± 0·018	8·78 ± 0·05	105·2 ± 0·4	16·65 ± 0·10	82·94 ± 0·33	100	37·01 ± 0·17	30·18 ± 0·18
10084-48(1)I Fines	13·2	347 ± 15	0·651 ± 0·021	0·753 ± 0·023	8·78 ± 0·10	104·7 ± 0·3	16·73 ± 0·09	82·64 ± 0·41	100	37·22 ± 0·15	30·27 ± 0·12
10084-48(2)I Fines	11·6	348 ± 11	0·655 ± 0·020	0·755 ± 0·011	8·72 ± 0·06	104·5 ± 0·3	16·71 ± 0·07	82·59 ± 0·33	100	36·85 ± 0·15	29·96 ± 0·12
*10084-48(5)I Fines	57·3	325 ± 14	0·647 ± 0·016	0·755 ± 0·022	8·86 ± 0·07	105·1 ± 0·4	16·74 ± 0·10	83·14 ± 0·42	100	36·86 ± 0·18	30·00 ± 0·18
10085-5(1)I Fragments	145·1	19·5 ± 0·6	1·166 ± 0·025	1·526 ± 0·038	9·93 ± 0·08	105·4 ± 0·4	17·15 ± 0·14	86·08 ± 0·43	100	36·85 ± 0·15	30·09 ± 0·12
10085-5(2)II Fragments	40·2	31·2 ± 2·1	1·642 ± 0·036	2·538 ± 0·056	10·25 ± 0·08	100·9 ± 0·4	17·26 ± 0·16	90·64 ± 0·64	100	38·36 ± 0·23	32·40 ± 0·18

* Integrated stepwise heating experiment.

Table A3. Rare gases from stepwise heating of lunar fines; 10084-48(5)I

Temp (°C)	Gas content (× 10^{-8} cm^{3}STP/g) ^{3}He	^{22}Ne	^{36}Ar	^{4}He/^{3}He	^{20}Ne/^{22}Ne	^{21}Ne/^{22}Ne	^{38}Ar/^{36}Ar	^{40}Ar/^{36}Ar
100	7·27 ± 0·17	3·32 ± 0·27	—	1785 ± 18	14·28 ± 0·24	0·0390 ± 0·0009	—	—
200	30·16 ± 0·75	11·67 ± 0·23	0·94 ± 0·06	1730 ± 17	14·51 ± 0·06	0·0352 ± 0·0007	0·1798 ± 0·0018	14·4 ± 0·1
300	163·9 ± 3·1	81·4 ± 1·6	6·63 ± 0·19	1885 ± 19	13·45 ± 0·03	0·0330 ± 0·0002	0·1812 ± 0·0018	11·0 ± 0·1
400	1455 ± 39	1366 ± 33	64·0 ± 1·7	1928 ± 19	12·69 ± 0·03	0·0329 ± 0·0005	0·1805 ± 0·0018	5·05 ± 0·05
500	3364 ± 81	2947 ± 71	339·7 ± 9·2	1986 ± 20	12·59 ± 0·03	0·0329 ± 0·0003	0·1815 ± 0·0018	3·42 ± 0·03
600	1927 ± 52	3200 ± 74	1596 ± 46	2494 ± 25	12·83 ± 0·05	0·0337 ± 0·0003	0·1812 ± 0·0018	2·48 ± 0·02
700	1098 ± 27	2809 ± 65	5170 ± 140	2583 ± 26	12·66 ± 0·04	0·0334 ± 0·0004	0·1835 ± 0·0018	1·47 ± 0·01
800	925 ± 24	2407 ± 55	14,400 ± 360	2712 ± 27	12·62 ± 0·03	0·0336 ± 0·0005	0·1901 ± 0·0019	0·78 ± 0·01
900	352 ± 12	1551 ± 37	7750 ± 210	2642 ± 26	12·70 ± 0·04	0·0352 ± 0·0004	0·1930 ± 0·0019	0·86 ± 0·01
1000	71·8 ± 1·4	1015 ± 28	2180 ± 67	3064 ± 31	12·49 ± 0·02	0·0369 ± 0·0003	0·1945 ± 0·0019	1·02 ± 0·01
1100	16·80 ± 0·39	1036 ± 30	3070 ± 100	2177 ± 22	11·88 ± 0·04	0·0393 ± 0·0003	0·1973 ± 0·0020	1·15 ± 0·01
1200	3·738 ± 0·093	472 ± 14	3130 ± 100	1159 ± 12	11·34 ± 0·03	0·0520 ± 0·0004	0·2036 ± 0·0020	1·22 ± 0·01
1300	0·152 ± 0·012	9·03 ± 0·26	106·4 ± 3·3	1561 ± 16	10.38 ± 0·02	0·1199 ± 0·0013	0·2203 ± 0·0022	1·46 ± 0·01
Total	9410 ± 110	16,910 ± 150	37,800 ± 470	2252 ± 27	12·58 ± 0·14	0·0348 ± 0·0005	0·1912 ± 0·0020	1·08 ± 0·01

Table A3. (continued)

Temperature (°C)	^{84}Kr (10^{-10} cm^{3}STP/g)	^{78}Kr	^{80}Kr	^{82}Kr	^{83}Kr	^{84}Kr	^{86}Kr
100–400	1·21 ± 0·25	—	—	—	—	—	—
500	4·23 ± 0·60	0·769 ± 0·036	4·383 ± 0·083	21·22 ± 0·23	20·76 ± 0·25	100	29·76 ± 0·42
600	19·6 ± 2·2	0·767 ± 0·025	4·345 ± 0·056	20·30 ± 0·30	20·42 ± 0·22	100	29·60 ± 0·30
700	81·2 ± 9·2	0·666 ± 0·017	4·131 ± 0·037	20·50 ± 0·12	20·35 ± 0·16	100	29·80 ± 0·15
800	436·0 ± 51·0	0·636 ± 0·010	4·050 ± 0·032	20·39 ± 0·08	20·24 ± 0·08	100	30·34 ± 0·12
900	732·0 ± 85·0	0·641 ± 0·012	4·036 ± 0·032	20·11 ± 0·08	20·10 ± 0·08	100	30·51 ± 0·12
1000	291·0 ± 33·0	0·651 ± 0·013	4·062 ± 0·032	20·30 ± 0·08	20·46 ± 0·09	100	30·54 ± 0·12
1100	332·0 ± 39·0	0·712 ± 0·014	4·250 ± 0·034	20·61 ± 0·10	20·76 ± 0·10	100	30·50 ± 0·15
1200	314·0 ± 37·0	0·756 ± 0·013	4·425 ± 0·035	20·72 ± 0·08	21·01 ± 0·13	100	30·39 ± 0·12
1300	17·1 ± 2·1	0·970 ± 0·030	4·995 ± 0·070	21·22 ± 0·30	22·06 ± 0·29	100	29·95 ± 0·18
Total	2230 ± 260	0·673 ± 0·013	4·142 ± 0·033	20·37 ± 0·09	20·43 ± 0·10	100	30·43 ± 0·13

Table A3. (continued)

T (°C)	^{132}Xe (10^{-10} cm^3 STP/g)	^{124}Xe	^{126}Xe	^{128}Xe	^{129}Xe	^{130}Xe	^{131}Xe	^{132}Xe	^{134}Xe	^{136}Xe
100–500	0·43 ± 0·06	—	—	—	—	—	—	—	—	—
600	0·81 ± 0·04	—	—	8·53 ± 0·26	105·2 ± 3·5	17·73 ± 1·19	81·84 ± 1·80	100	40·27 ± 0·85	32·65 ± 0·82
700	2·00 ± 0·10	—	—	8·65 ± 0·29	104·8 ± 1·8	16·05 ± 0·63	81·62 ± 1·14	100	36·97 ± 0·92	31·04 ± 0·34
800	19·1 ± 0·60	0·531 ± 0·016	0·528 ± 0·032	8·53 ± 0·15	106·3 ± 0·5	16·67 ± 0·15	82·57 ± 0·66	100	37·23 ± 0·30	30·03 ± 0·24
900	57·0 ± 1·7	0·533 ± 0·021	0·495 ± 0·017	8·52 ± 0·06	105·3 ± 0·4	16·57 ± 0·10	82·26 ± 0·41	100	36·89 ± 0·22	29·91 ± 0·18
1000	63·6 ± 3·0	0·562 ± 0·017	0·610 ± 0·017	8·62 ± 0·05	104·6 ± 0·4	16·54 ± 0·10	82·39 ± 0·33	100	36·85 ± 0·15	29·94 ± 0·18
1100	98·6 ± 4·4	0·719 ± 0·012	0·908 ± 0·021	9·05 ± 0·05	105·0 ± 0·4	16·91 ± 0·07	83·75 ± 0·25	100	36·85 ± 0·15	30·01 ± 0·15
1200	78·4 ± 3·8	0·737 ± 0·012	0·924 ± 0·018	9·07 ± 0·07	105·1 ± 0·4	16·88 ± 0·08	83·48 ± 0·42	100	36·74 ± 0·15	30·06 ± 0·18
1300	5·07 ± 0·38	0·933 ± 0·037	1·338 ± 0·054	9·85 ± 0·20	106·9 ± 1·0	17·26 ± 0·22	88·40 ± 0·80	100	36·58 ± 0·80	29·57 ± 0·60
Total	325 ± 14	0·647 ± 0·016	0·755 ± 0·022	8·86 ± 0·07	105·1 ± 0·4	16·74 ± 0·10	83·14 ± 0·42	100	36·86 ± 0·18	30·00 ± 0·18

Table A4. Rare gases from stepwise heating of lunar breccia 10061-38(2)II (upper table) and 10061-38(4)II (lower table)

Temp (°C)	Gas content (× 10^{-8} cm^{3}STP/g) ^{3}He	^{22}Ne	^{36}Ar	^{4}He/^{3}He	^{20}Ne/^{22}Ne	^{21}Ne/^{22}Ne	^{38}Ar/^{36}Ar	^{40}Ar/^{36}Ar
100	3·72 ± 0·15	1·80 ± 0·09	0·752 ± 0·029	1661 ± 17	14·96 ± 0·39	0·0360 ± 0·0014	0·3262 ± 0·0095	6·0 ± 2·6
200	43·6 ± 1·7	23·3 ± 0·9	23·6 ± 0·8	2132 ± 21	13·86 ± 0·06	0·0326 ± 0·0002	0·1785 ± 0·0018	229·0 ± 2·3
300	102·8 ± 4·1	63·2 ± 2·5	14·90 ± 0·54	2116 ± 21	13·41 ± 0·05	0·0321 ± 0·0002	0·1826 ± 0·0018	32·40 ± 0·32
1650	9960 ± 400	59,800 ± 2300	82,900 ± 3300	2629 ± 26	12·57 ± 0·04	0·0331 ± 0·0002	0·1884 ± 0·0019	2·185 ± 0·022
Total	10,120 ± 400	59,900 ± 2300	82,900 ± 3300	2618 ± 26	12·57 ± 0·04	0·0331 ± 0·0002	0·1884 ± 0·0019	2·185 ± 0·022
100	1·93 ± 0·07	1·30 ± 0·07	—	1980 ± 20	14·91 ± 0·10	0·0361 ± 0·0005	—	—
150	11·9 ± 0·5	8·11 ± 0·41	—	1920 ± 19	14·12 ± 0·09	0·0347 ± 0·0003	—	—
250	51·7 ± 2·0	25·7 ± 1·2	—	1930 ± 19	13·91 ± 0·07	0·0357 ± 0·0004	—	—
350	292 ± 12	256 ± 10	—	3460 ± 35	12·93 ± 0·06	0·0362 ± 0·0004	—	—
450	1610 ± 60	6240 ± 260	—	2690 ± 27	12·47 ± 0·06	0·0327 ± 0·0003	—	—
550	4490 ± 220	31,500 ± 1300	—	2540 ± 25	12·67 ± 0·06	0·0326 ± 0·0003	—	—
650	2500 ± 90	34,800 ± 1500	—	2740 ± 27	12·54 ± 0·06	0·0326 ± 0·0003	—	—
750	2220 ± 80	10,400 ± 500	—	2900 ± 29	12·62 ± 0·06	0·0337 ± 0·0003	—	—
*850	673 ± 26	3970 ± 180	—	3030 ± 30	11·91 ± 0·06	0·0382 ± 0·0004	—	—
950	1·16 ± 0·04	3·81 + 0·17	—	412 ± 4	8·62 ± 0·05	0·1619 ± 0·0019	—	—

Table 4. (continued)

1050	2·46 ± 0·09	5·69 ± 0·23	—	318 ± 3	8·44 ± 0·04	0·228 ± 0·002	—	—
1150	2·41 ± 0·09	4·26 ± 0·18	—	261 ± 5	9·12 ± 0·05	0·180 ± 0·002	—	—
1250	0·0837 ± 0·0036	1·015 ± 0·046	—	2820 ± 28	9·86 ± 0·10	0·160 ± 0·002	—	—
1350	0·146 ± 0·008	5·51 ± 0·25	—	1290 ± 13	9·08 ± 0·05	0·189 ± 0·002	—	—
Total	11,590 ± 260	87,200 ± 2100	—	2780 ± 33	12·56 ± 0·11	0·0330 ± 0·0004		

* Temperature overrun to ~ 1100°C for 5 min in the 850°C fraction.

Table A5. Rare gases from stepwise heating of lunar rock 10069-21(1)I

	Gas content (× 10^{-8} cm^{3}STP/g)							
Temp (°C)	^{3}He	^{4}He	^{22}Ne	^{36}Ar	^{40}Ar	^{20}Ne/^{22}Ne	^{21}Ne/^{22}Ne	^{38}Ar/^{36}Ar
100	—	12·2 ± 1·2	—	—	—	—	—	—
200	0·103 ± 0·005	183·4 ± 2·9	0·022 ± 0·003	0·006 ± 0·002	4·29 ± 0·77	14·1 ± 1·7	—	—
300	1·322 ± 0·033	2913 ± 26	0·46 ± 0·01	0·023 ± 0·004	22·5 ± 0·5	13·19 ± 0·07	0·0434 ± 0·0004	0·256 ± 0·012
400	10·08 ± 0·20	20,540 ± 250	1·72 ± 0·05	0·18 ± 0·01	185·8 ± 2·2	12·60 ± 0·05	0·0403 ± 0·0003	0·2534 ± 0·0062
500	14·31 ± 0·29	33,280 ± 400	2·06 ± 0·04	0·63 ± 0·01	783 ± 10	12·80 ± 0·03	0·0464 ± 0·0004	0·2706 ± 0·0040
600	7·29 ± 0·15	17,250 ± 170	1·94 ± 0·03	1·54 ± 0·03	1706 ± 22	12·64 ± 0·05	0·0537 ± 0·0004	0·2641 ± 0·0026
700	2·326 ± 0·046	5762 ± 58	1·52 ± 0·03	2·91 ± 0·05	1394 ± 18	11·82 ± 0·02	0·0936 ± 0·0007	0·2343 ± 0·0023
800	1·779 ± 0·041	4787 ± 48	1·72 ± 0·04	5·84 ± 0·08	956 ± 13	8·89 ± 0·03	0·3008 ± 0·0021	0·2368 ± 0·0024
900	0·858 ± 0·028	2386 ± 24	1·77 ± 0·03	2·33 ± 0·04	457 ± 7	5·60 ± 0·02	0·5296 ± 0·0037	0·2940 ± 0·0029
1650	0·674 ± 0·019	1993 ± 18	3·25 ± 0·06	4·88 ± 0·07	1067 ± 15	4·26 ± 0·02	0·6254 ± 0·0038	0·9062 ± 0·0091
Total	38·74 ± 0·77	89,110 ± 890	14·49 ± 0·22	18·34 ± 0·13	6575 ± 37	9·40 ± 0·07	0·2708 ± 0·0029	0·4253 ± 0·0042

Acknowledgments—Principal financial support for this investigation was provided by NASA contract NAS 9-8093, with supplementary support by ONR Contract Nonr-710(58) and NSF Grant GA-905. We found the suggestions of the referees for this paper, Dr. Edward Anders, Dr. Charles Hohenberg and Dr. Kurt Marti, very helpful.

References

Albee A. L., Burnett D. S., Chodos A. A., Eugster O. J., Huneke J. C., Papanastassiou D. A., Podosek F. A., Russ G. Price II, Sanz H. G., Tera F. and Wasserburg G. J. (1970) Ages, irradiation history, and chemical composition of lunar rocks from the Sea of Tranquillity. *Science* **167,** 463–466.

Black D. C. and Pepin R. O. (1969) Trapped neon in meteorites—II. *Earth Planet. Sci. Lett.* **6,** 395–405.

Black D. C. (1970) Trapped helium, neon and argon in meteorites: Boundary conditions on the formation and evolution of the solar system. In preparation.

Eberhardt P., Eugster O. and Marti K. (1965) A redetermination of the isotopic composition of atmospheric neon. *Z. Naturforsch.* **21a,** 623–624.

EBERHARDT P., GEISS J., GRAF H., GRÖGLER N., KRÄHENBÜHL U., SCHWALLER H., SCHWARZMÜLLER J. and STETTLER A. (1970) Trapped solar wind noble gases, Kr^{81}/Kr exposure ages and K/Ar ages in Apollo 11 lunar material. *Science* **167,** 558–560.

EUGSTER O., EBERHARDT P. and GEISS J. (1967a) The isotopic composition of krypton in unequilibrated and gas rich chondrites. *Earth Planet. Sci. Lett.* **2,** 385–393.

EUGSTER O., EBERHARDT P. and GEISS J. (1967b) Krypton and xenon isotopic compositions in three carbonaceous chondrites. *Earth Planet. Sci. Lett.* **3,** 249–257.

FUNKHOUSER J. G., SCHAEFFER O. A., BOGARD D. D. and ZÄHRINGER J. (1970) Gas analysis of the lunar surface. *Science* **167,** 561–563.

GAST P. W. and HUBBARD N. J. (1970) Abundance of alkali metals, alkaline and rare earths, and strontium-87/strontium-86 ratios in lunar samples. *Science* **167,** 485–487.

HEYMANN D., YANIV A., ADAMS J. A. S. and FRYER G. E. (1970) Inert gases in lunar samples. *Science* **167,** 555–558.

HEYMANN D. and YANIV A. (1970a) Inert gases in the fines from the sea of tranquillity. *Geochim. Cosmochim. Acta* Supplement I.

HEYMANN D. and YANIV A. (1970b) Ar^{40} anomaly in samples from tranquillity base. *Geochim. Cosmochim. Acta*, Supplement I.

HOHENBERG C. M. and ROWE M. W. (1969) Spallation yields of xenon from irradiation of Cs, Ce, Nd, Dy and a rare earth mixture with 730 MeV protons. *32nd Annual Meeting of the Meteoritical Society*, Houston, Texas, Oct. 29–31, 1969.

HOHENBERG C. M., DAVIS P. K., KAISER W. A., LEWIS R. S. and REYNOLDS J. H. (1970) Trapped and cosmogenic rare gases from stepwise heating of Apollo 11 samples. *Geochim. Cosmochim. Acta*, Supplement I.

MARTI K. (1967a) Trapped xenon and the classification of chondrites. *Earth Planet. Sci. Lett.* **2,** 193–196.

MARTI K. (1967b) Isotopic composition of trapped krypton and xenon in chondrites. *Earth Planet. Sci. Lett.* **3,** 243–248.

MARTI K., LUGMAIR G. W. and UREY H. C. (1970) Solar wind gases, cosmic ray spallation products, and the irradiation history. *Science* **167,** 548–550.

MORRISON G. H., GERARD J. T., KASHUBA A. T., GANGADHARAM E. V., ROTHENBERG A. M., POTTER N. M. and MILLER G. B. (1970) Multielement analysis of lunar soil and rocks. *Science* **167,** 505–507.

MURTHY V. R., SCHMITT R. A. and REY P. (1970) Rubidium–strontium age and elemental and isotopic abundances of some trace elements in lunar samples. *Science* **167,** 476–479.

NIEF G. (1960) As reported in isotopic abundance ratios reported for reference samples stocked by the National Bureau of Standards, (editor F. Mohler) NBS Tech. Note 51.

NIER A. O. (1950a) A redetermination of the relative abundances of the isotopes of carbon, nitrogen, oxygen, argon, and potassium. *Phys. Rev.* **77,** 789–793.

NIER A. O. (1950b) A redetermination of the relative abundances of the isotopes of neon, krypton, rubidium, xenon, and mercury. *Phys. Rev.* **77,** 450–454.

PEPIN R. O. (1967) Trapped neon in meteorites. *Earth Planet. Sci. Lett.* **2,** 13–18.

PEPIN R. O., NYQUIST L. E., PHINNEY D. and BLACK D. C. (1970) Isotopic composition of rare gases in lunar samples. *Science* **167,** 550–553.

REYNOLDS J. H., HOHENBERG C. M., LEWIS R. S., DAVIS P. K. and KAISER W. A. (1970) Isotopic analysis of rare gases from stepwise heating of lunar fines and rocks. *Science* **167,** 545–548.

SILVER L. T. (1970) Uranium–thorium–lead isotope relations in lunar materials. *Science* **167,** 468–471.

TATSUMOTO M. and ROSHOLT J. N. (1970) Age of the moon: An isotopic study of uranium–thorium–lead systematics of lunar samples. *Science* **167,** 461–463.

TURNER G. (1970) Argon-40/argon-39 dating of lunar rock samples. *Science* **167,** 466–468.

Proceedings of the Apollo 11 Lunar Science Conference, Vol. 2, pp. 1455 to 1469.

Cosmogenic and primordial radionuclide measurements in Apollo 11 lunar samples by nondestructive analysis*

R. W. PERKINS, L. A. RANCITELLI, J. A. COOPER,
J. H. KAYE and N. A. WOGMAN
Battelle Memorial Institute, Pacific Northwest Laboratories, P.O. Box 999, Richland
Washington 99352

(*Received* 9 *February* 1970; *accepted in revised form* 28 *February* 1970)

Abstract—The cosmogenic radionuclides ^{7}Be, ^{22}Na, ^{26}Al, ^{44}Ti, ^{46}Sc, ^{48}V, ^{51}Cr, ^{54}Mn, ^{56}Co ^{57}Co and ^{60}Co and the primordial radionuclides ^{40}K, ^{238}U and ^{232}Th were measured by nondestructive gamma-ray spectrometry in 3 aliquots of lunar fines sample 10084, in a type A lunar rock specimen 10057-30 and in two type B rock specimens 10003-25, and 10017-37. The ^{7}Be, ^{44}Ti, ^{48}V and ^{51}Cr were present in very small amounts and only upper concentration limits are reported. The concentrations of the more abundant isotopes ^{22}Na and ^{26}Al were measured to absolute accuracies of 2–4 per cent. From galactic cosmic-ray radionuclide production rates which were derived from meteorite data it was established that in lunar fines the ^{22}Na, ^{26}Al, ^{46}Sc and ^{56}Co were produced mainly from solar cosmic rays. The ^{56}Co lunar fines concentration, which was 50-fold above galactic production rates, is accounted for by the April 12, 1969 solar flare. The high ^{46}Sc concentrations, 8 to 13 dpm/kg, resulted from the high Ti content of the lunar materials. A strong concentration gradient for ^{22}Na but not ^{26}Al in the rock 10017 specimen indicated that the original surface of the rock had been recently exposed for only a short time compared to the $7 \cdot 4 \times 10^{5}$ yr ^{26}Al half-life. The Th:U atomic ratios are constant at about 3·8, in agreement with a common nucleosynthesis for lunar, earth and meteorite materials. Gas retention ages determined by combining reported He and Ar measurements with the primordial radionuclide measurements of the lunar rocks indicate ages of 1·9–2·4 b.y., except for K–Ar age of 3·9 b.y., for rock 10003.

INTRODUCTION

MEASUREMENTS of the primordial and cosmogenic radionuclides in lunar samples from the Apollo 11 flight have provided our first precise information on lunar surface radioactivity. The relative and absolute cosmogenic radionuclide concentrations of lunar samples are far different from those observed in meteorites, reflecting both substantial differences in the chemical composition of the lunar surface and in the incident cosmic-ray energy spectrum. While meteorites are exposed to a somewhat similar but less intense cosmic-ray spectrum, the isotopes produced by solar cosmic rays are largely lost by ablation during entry of the meteorite through the earth's atmosphere. In addition, meteorite orbits are believed to extend to about 3 astronomical units; therefore, the actual solar cosmic-ray exposure is much less. Thus, while the cosmogenic isotopes observed in meteorites result mainly from the high-energy galactic component, those at the lunar surface are produced to a substanitial degree from the solar component of the cosmic-ray spectrum. The concentrations of the cosmogenic radionuclides are directly related to the exposure time of rocks and soil on the lunar surface and to the flux spectrum and intensity. The radionuclide measurement, therefore, provides a basis for estimating these parameters as well as temporal variations in the intensity and energy of the flux, the erosion rates of the rocks,

* This paper is based on work supported by the National Aeronautics and Space Administration–Manned Spacecraft Center, Houston, Texas, under Contract NAS 9-7881.

accretion and mixing of the fines, and any rotation of the rocks on the lunar surface. This type of analysis is performed in a manner similar to that employed in determining meteorite exposures and history from their cosmogenic radionuclide constituents (HONDA and ARNOLD, 1967; KIRSTEN and SCHAEFFER, 1969). The nondestructive primordial radionuclide measurements provide a direct means for differentiating between rocks and soil of different origins, for determining the degree of magmatic differentiation of the Th, U and K, and for comparing whole rock radionuclide concentrations with those determined on rock chips or mineral fractions by destructive analysis. Since the longest-lived cosmogenic radionuclide measured in the work was ^{26}Al, with a half-life of $7{\cdot}4 \times 10^5$ yr, bombardment histories and surface processes of only a few million years can be inferred; however, the primordial radionuclides are indicative of other lunar processes from the very origin of the moon.

Preliminary gamma-ray spectrometric measurements during sample quarantine at the Lunar Receiving Laboratory provided approximate radionuclide concentrations of lunar samples (LSPET, 1969). These measurements at the Lunar Receiving Laboratory were further refined by subsequent calibration work (O'KELLEY *et al.*, 1970). Following release of the lunar samples from quarantine, a large group of radionuclides were measured by radiochemical separation (SHEDLOVSKY *et al.*, 1970) in rock 10017 and in lunar fines in an effort to establish concentration gradients as well as other parameters.

Fig. 1. Lunar samples in counting containers.

The objectives of the present work, which was reported in a very abbreviated and somewhat preliminary form (PERKINS *et al.*, 1970), were to perform very accurate measurements of the radionuclide content of lunar surface rocks and fines and interpret these observations in terms of cosmic-ray exposure and the history of the specimens. For example, the ^{26}Al and ^{22}Na concentrations of lunar fines and rock specimens have been measured to an absolute accuracy of 2–4 per cent, comparable to those previously reported in our meteorite studies (RANCITELLI *et al.*, 1969).

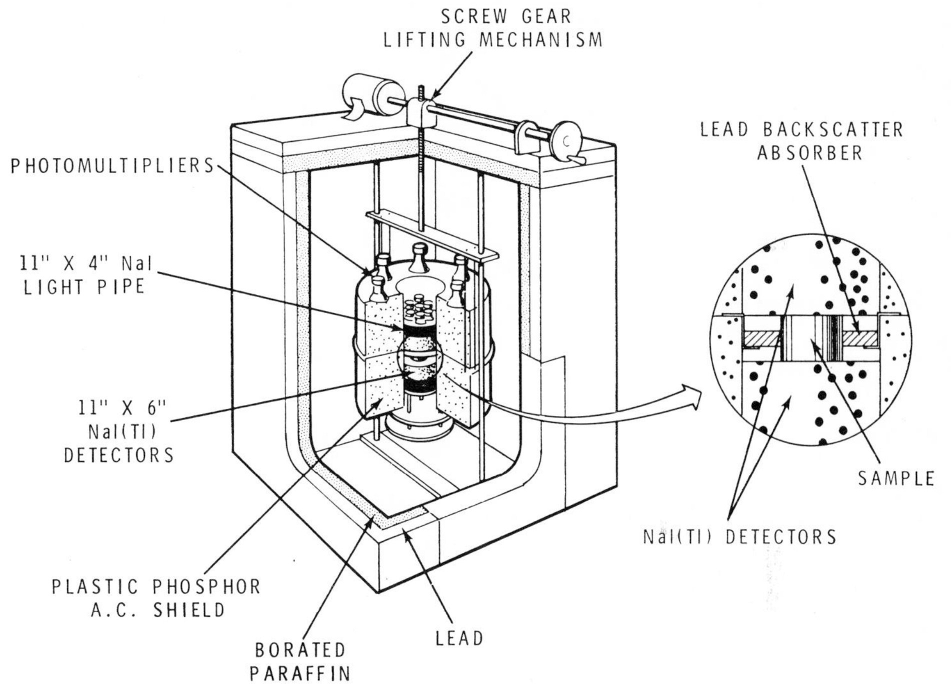

Fig. 2. Large multidimensional gamma-ray spectrometer.

PROCEDURE

All of the measurements were made by nondestructive gamma-ray spectrometric techniques. For these measurements, the lunar samples were maintained in a dry nitrogen atmosphere in their thin containers of stainless steel or aluminum-lined lucite. Figure 1 shows a photograph of the containers with the lunar samples in place. The fines were contained in two types of containers. One type consisted of a 10-cm-dia. by 2·5-cm-deep stainless steel cylinder with 0·5 mm thick top and bottom faces (upper left in Fig. 1). This as well as all other containers, was hermetically sealed with a Viton "O" ring. The other lunar fines container was a 11·3 cm i.d. by 7·5 cm deep, thin-windowed lucite cylinder with a 0·4 mm aluminum liner covering its bottom and sides. The rock samples were wrapped in aluminum foil and bundled in crumpled aluminum foil in 11·3 cm i.d. by 5 cm deep, thin walled lucite containers. A sample of rock 10017 was contained in a similar geometry but was retained in position with stainless steel springs. This sample was not wrapped in aluminum foil and it was possible to view it and determine the location of the original surfaces. The only original surface was facing one of the lucite windows and thus permitted an optimum geometry for studying depth gradients of the radionuclide distribution. The counting was performed on anticoincidence shielded multidimensional gamma-ray spectrometers (MDGRS), employing large principal NaI(Tl) detectors (WOGMAN *et al.*, 1967, 1969) (Fig. 2) and on an anticoincidence shielded Ge(Li) spectrometer (COOPER, 1970). The

MDGRS, which employed either two 28-cm-dia. by 15 cm thick NaI(Tl) crystals or two 23-cm-dia. by 20 cm thick NaI(Tl) crystals, offer a very high sensitivity for the measurement of those radionuclides which decay by emission of several gamma-rays or by emission of a positron and a gamma ray [^{22}Na, ^{26}Al, ^{44}Ti, ^{46}Sc, ^{48}V, ^{56}Co, ^{232}Th (^{208}Tl) and ^{238}U (^{214}Bi)]. The anticoincidence shielded Ge(Li) spectrometer which viewed the sample with two 70 cm^3 detectors, provided an improved sensitivity for the measurement of radionuclides which emit only one photon per disintegration (^{7}Be, ^{51}Cr, ^{54}Mn and ^{57}Co) and served to verify the measurements of the more abundant cascade photon emitters (^{22}Na and ^{26}Al). Two spectra were obtained from each diode. One was a coincidence spectrum containing events where photons interacted simultaneously in one diode and in the large plastic phosphor anticoincidence shield. This spectrum contained most of the events from radionuclides which emit two or more photons per disintegration. The second was an anticoincidence

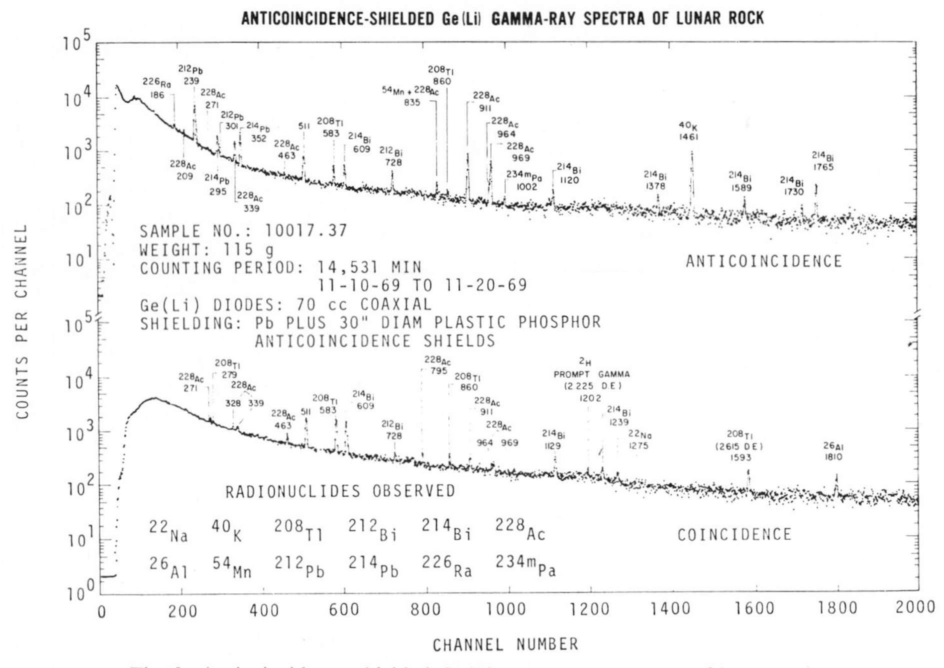

Fig. 3. Anticoincidence-shielded Ge(Li) gamma-ray spectra of lunar rock.

spectrum containing photon events not in coincidence with the anticoincidence shield and thus emphasizing the single photon emitters. The gamma-ray spectra of the 115·5-g sample of rock 10017 are shown in Fig. 3 and illustrate the type of spectral information which is obtained.

The radionuclides which were measured in the lunar samples, together with their half-lives, and the energy region used for the measurement by multidimensional gamma-ray spectrometry are shown in Table 1. The background counting rates normally produced a negligible interference compared with Compton interference. The analysis of multidimensional gamma ray spectra is similar to that of normal spectra and has been covered in earlier work (PERKINS, 1965). The concentrations of the radionuclides in the lunar samples were based on 10,000 to 20,000 min counting intervals on both types of analyzers. Standard mock-ups with the same size, shape, physical density, and approximately the same electron density as each lunar rock and containing known amounts of each radionuclide of interest were prepared from a mixture of 16·3% plaster of paris, 65·6% iron powder and 18·1% water. Mock-ups of the lunar fines were prepared from a mixture of 5% Al_2O_3 and 95% dunite to provide the same density as the fines. The standard radioisotopes were pipetted into the Al_2O_3, dried, and after mixing the labeled Al_2O_3 was then homogeneously mixed with the dunite to form the mock-up.

Table 1. Radionuclides in lunar surface materials and the energy regions used in their measurements by multidimensional gamma-ray spectrometry

Radionuclide	Half-life	Gamma-ray Energy regions (MeV)
^{7}Be	53 d	0·4774
^{22}Na	2·58 yr	(1·2746 + 0·511) and 0·511
^{26}Al	$7·4 \times 10^5$ yr	(1·81 + 0·511) and 0·511
^{44}Ti (^{44}Sc)	47 yr	(1·156 + 0·511) and 0·511
^{46}Sc	84 d	1·1205 and 0·8894
^{48}V	16·1 d	0·9833 and 1·312
^{51}Cr	27·8 d	0·3198
^{54}Mn	303 d	0·8353
^{56}Co	77·3 d	3·26 and 0·8469
^{57}Co	267 d	0·1219 + 0·1363
^{60}Co	5·26 yr	1·173 and 1·332
^{40}K	$1·3 \times 10^9$ yr	1·460
^{226}Ra (^{214}Bi)	1620 yr	0·6094 and 1·120
^{232}Th (^{208}Tl)	$1·4 \times 10^{10}$ yr	0·583 and 2·615

Other types of mock-ups prepared in powdered basalt, and dunite with iron powder added to provide 25% total Fe were shown to have essentially identical counting efficiencies to those in the dunite, thus substantiating the adequacy of the mock-ups (see Table 2). As indicated in Table 2, a very significant change in counting efficiency resulted when a Pb $(NO_3)_2$-Ag NO_3 solution of lunar fines density was used as a mock-up.

Inhomogeneous mock-ups where the activity was all located in the top 25 per cent of 4 cm by 4 cm right circular cylinders produced only a 5 per cent change in the total coincidence counting efficiency regardless of orientation. However, any significant inhomogeneities could easily be determined by differences in the ratios of the coincidence photons from ^{22}Na and ^{26}Al which were deposited in different ratios in the two NaI(Tl) crystals.

The standard radionuclide solutions used in preparing mock-ups were calibrated by counting them in a precisely controlled geometry on a Ge(Li) diode for which counting efficiency vs. photon energy had been very accurately established (PERKINS *et al.*, 1969).

For the very accurate control of the diode calibration an efficiency versus energy calibration curve was prepared using a group of 8 International Atomic Energy standard radionuclide sources for which the disintegration rates were known to 1 per cent or better. (HOUTERMANS, 1970.) Our secondary calibrations against this curve allowed a similar degree of accuracy in the preparation of standard radionuclide solutions and allowed the preparation of standard mock-ups with an accuracy of ±1 to 2 per cent. With the exception of ^{226}Ra, all of the radionuclides were standardized by counting in precisely controlled geometries on the calibrated Ge(Li) diode. The ^{226}Ra standard solution was obtained from the U.S. Bureau of Standards and had been originally prepared at their laboratory by weighing and subsequently dissolving pure radium metal. In using ^{226}Ra as a standard for determining the uranium content of the samples, it was of course assumed that the ^{238}U:^{235}U ratios were identical to terrestrial ratios and that complete secular equilibrium existed in the ^{238}U chain. The mock-ups containing ^{226}Ra were prepared in pyrex containers which were formed in the approximate

Table 2. Coincidence counting efficiency of ^{22}Na gamma rays from a 1 kg sample (counting efficiencies normalized to 1 for the 0·511 and 0·511 + 1·275 MeV coincidence peak)

Material	0·511 and 0·511	0·511 and 1·275	0·511 and (0·511 + 1·275)
Dunite (6% Fe)	1·16	0·67	1
Basalt	1·16	0·68	0·98
Dunite (25% Fe)	1·17	0·68	0·99
$Pb(NO_3)_2$–$AgNO_3$ Solution	1·14	0·79	0·83

shape of the lunar samples. These pyrex containers were sealed to prevent radon loss and allowed to stand for a sufficient time for radon and its daughters to reach secular equilibrium with the ^{226}Ra. Radium-226 standards which had not been sealed in pyrex showed large radon losses and would have produced erroneous results.

The potassium measurements were also made on the assumption that lunar and terrestrial potassium had the same isotopic composition. The instruments used for the lunar sample measurements were calibrated by counting duplicates or triplicates of the mock-ups to a precision of ± 1 to 2 per cent. The radionuclide concentrations in the lunar samples were calculated by direct comparison of their counting rates with those of mock-up standards after application of appropriate background and Compton corrections. A least squares computer program analysis gave radionuclide concentrations and error limits in agreement with normally calculated values.

A serious problem in the cosmogenic radionuclide measurements which is not encountered in meteorite measurements resulted from the high Th and U concentrations in the lunar material. The Th and U, which were 10- to 50-fold higher in the lunar material than in the meteorites, produced high Comptom scatter "background" over the entire coincidence spectrum.

Table 3. Cosmogenic and primordial radionuclides in lunar materials*

Radionuclides	10057-30 (A)	10017-37 (B)	10003-25 (B)	10084-41	10084-113-2	10084-113-1
^{7}Be (dpm/kg)				<112		
^{22}Na (dpm/kg)	43 ± 2	45 ± 2	49 ± 2	64 ± 3	63 ± 2	
^{26}Al (dpm/kg)	84 ± 2	80 ± 2	75 ± 2	131 ± 4	137 ± 4	
^{44}Ti (dpm/kg)	<2					<1·5
^{46}Sc (dpm/kg)	11 ± 2	13 ± 2	8 ± 2	11 ± 1		
^{48}V (dpm/kg)				<10		
^{51}Cr (dpm/kg)	<62			<63		
^{54}Mn (dpm/kg)	41 ± 8	46 ± 17	60 ± 7	24 ± 3		
^{56}Co (dpm/kg)	19 ± 9					53 ± 10
^{57}Co (dpm/kg)	<3·9	<5·0	<4·3	1·3 ± 1·6		
^{60}Co (dpm/kg)	<1	0·8 ± 0·6	<1	0·3 ± 0·8		
K (wt.%)	0·23 ± 0·02	0·26 ± 0·01	0·046 ± 0·006	0·11 ± 0·01	0·11 ± 0·01	
Th (ppm)	3·6 ± 0·1	3·2 ± 0·1	1·08 ± 0·05	2·2 ± 0·1	2·3 ± 0·1	
U (ppm)	0·95 ± 0·05	0·86 ± 0·05	0·29 ± 0·02	0·54 ± 0·03	0·56 ± 0·04	
Wt. of sample (g)	230	115·5	117·2	999	327·0	270·8
Wt. of rock (g)	897	971	213			

*(<) Less than values represent 2σ of the gross counting rate in the photopeak area.

The accuracy of this nondestructive counting technique was evaluated by preparing a mock-up containing a mixture of known amounts of the major radionuclides in the proportions present in lunar material. Analysis of this mock-up by the counting and calculation procedures employed for the lunar samples established the fact that individual radionuclides could be measured with the predicted accuracy from the combined errors of individual mock-ups.

Results and Discussion

Radionuclide concentrations and production rates

The radionuclide concentrations observed in samples of three lunar rocks and in three aliquots of the lunar fines sample 10084 are summarized in Table 3. The standard deviations were 2–4 per cent for ^{22}Na and ^{26}Al and represent the uncertainties resulting from all counting statistics, including the differences in the counting rates of the duplicate or triplicate mock-ups. This high degree of accuracy provides a firm basis for evaluating small changes in the various factors responsible for the observed radionuclide concentrations.

To evaluate the radionuclide concentrations in terms of lunar surface processes, it is helpful to know or estimate radionuclide production rates from the incident galactic and solar cosmic rays. While the galactic-cosmic ray flux is a relatively steady irradiation source being modulated by solar activity, the solar cosmic rays are almost entirely associated with solar flares. Various methods have been employed for calculating the cosmogenic radionuclide production in meteorites and in the lunar surface from galactic cosmic rays. Unfortunately, all of these must be based on rather poorly known excitation functions, cosmic-ray energy spectra, secondary flux build-up, and limited information on production rates in meteorites. Production rates from galactic cosmic rays for some radionuclides in the lunar surface have been estimated by SHEDLOVSKY *et al.* (1970) using available data and appear to be in

Table 4. Calculated production rates of ^{26}Al from galactic cosmic rays in lunar material and meteorites* (dpm/kg)

Material	Al	Si	Fe + Ni	Ca	Ti	S	Total 4π	Total 2π	Observed	Calculated, % Observed
				Lunar material						
Fines (10084)†	32·8	62·0	0·3	0·6	0·1	0·1	95·9	48·0	134	36
Rock A (10057)	19·0	58·6	0·3	0·6	0·2	0·1	78·8	39·4	84	47
Rock B (10017)	20·5	58·9	0·3	0·5	0·2	0·1	80·5	40·3	80	50
Rock B (10003)	27·6	54·6	0·3	0·6	0·2	0·1	83·4	41·7	75	56
				Meteorites						
Saint Séverin‡	5·7	58·0	0·5	0·1	—	0·8	65·1		36	
Denver§	6·2	58·8	0·5	0·1	—	0·8	66·4		48	

* Calculations based on production rates by FUSE and ANDERS (1969).
† Chemical analysis used in the calculations were those of LSPET (1969) for the fines; MAXWELL *et al.* (1970) for rock 10017, ROSE *et al.* (1970) for rock 10003; and WANKE *et al.* (1970) for rock 10057. Sulfur concentrations of 0·22% for all of the rocks, and 0·13% for the fines were based on S measurements in rock 10017 and lunar fines by MAXWELL *et al* (1970).
‡ Chemical analysis from ORCEL *et al.* (1967).
§ Chemical analysis from MASON and JAROSEWICH (1968).

reasonable agreement with observed concentrations. From an evaluation of this and other calculation methods for estimating galactic cosmogenic radionuclide concentrations, it seemed prudent to use a more direct approach which did not require the experience and intuition which must be added to available quantitative information. FUSE and ANDERS (1969) have developed a method for estimating the ^{26}Al production in stony meteorites which is based on a best fit of the observed ^{26}Al and target element concentrations. Using this approach, the expected production rates based on a 2π isotropic galactic flux have been calculated and are presented in Table 4 along with calculations for 2 meteorite samples.

Since the calculated concentrations listed for the 2π isotropic flux include the contribution from secondary buildup, they would be more typical for lunar surface depths of 15–30 cm, where secondary flux buildup results in maximum concentrations, and are probably about twofold higher than the surface concentrations. Extensive measurements in near surface samples from the meteorite Saint Séverin (MARTI *et al.*, 1969) provided the first direct proof that ^{22}Na and ^{26}Al concentrations in near-surface samples are almost twofold lower than in deeper regions of a large meteorite. Thus, the galactic contributions from cosmic-ray primaries alone may be as little as one-half that listed in Table 4. The concentrations of ^{22}Na and ^{26}Al in the near-surface sample of Saint Séverin and in the Denver meteorite (RANCITELLI, 1970), a single 230 g stone, are compared with lunar concentrations in Table 4. The low ^{22}Na and ^{26}Al concentrations in the Denver meteorite are apparently due to a lack of secondary flux buildup because of its small size. Based on the rather extensive radionuclide measurements in Saint Séverin (MARTI *et al.*, 1969), Allende (RANCITELLI *et al.*, 1969), Denver (RANCITELLI, 1970) and Bruderheim (HONDA and ARNOLD, 1964) meteorites, the lunar surface galactic cosmic-ray production rates of the 10 cosmogenic radionuclides of particular interest to this study have been estimated (Table 5). For this estimation it was simply assumed that the isotropic

Table 5. Estimated galactic cosmic-ray radionuclide production in lunar surface fines from meteorite data

	Concentration dpm/kg				
Isotope	Meteorites	2π and surface effects	Change due to compositional effect	Lunar fines (10084)	Galactic production (%)
^{22}Na	81*	20	18	64	28
^{26}Al	61*	15	19	134	14
^{44}Ti§	2†	1	0·7	<1·5	>47
^{46}Sc§	6*	3	3	11	27
^{48}V§	21*	11	7	<10	>70
^{51}Cr§	54*	27	14	<63	>22
^{54}Mn	129*	32	17	24	71
^{56}Co	4‡	2	1	53	2
^{57}Co	32*	8	1	1·3	80
^{60}Co	11‡	6	>0·5	~0·3	—

* Allende meteorite concentrations at greater than 25 cm depth (RANCITELLI *et al.*, 1969).
† Bruderheim meteorite concentration (HONDA and ARNOLD, 1964).
‡ Saint Séverin meteorite near surface concentrations (MARTI *et al.*, 1969).
§ Produced by high energy reactions in meteorites.

flux incident on the lunar surface was half of that impinging on meteorites. Lunar surface production rates (in the top few cm of lunar soil) for radionuclides which result from high ΔA reactions such as ^{44}Ti, ^{46}Sc, ^{48}V and ^{51}Cr would thus be equal to one-half of those in meteorites. Production rates for radionuclides which are produced by low ΔA reactions such as ^{22}Na, ^{26}Al, ^{54}Mn, ^{56}Co, ^{57}Co and ^{60}Co were assumed to be equal to one half of those near the surface of meteorites or one-fourth of those at more substantial depths where production from secondary particles is near a maximum. Since target element concentrations are quite different in the lunar surface than in the meteorites used for comparison, correction for this difference based on available cross section information was applied. From these estimated galactic cosmic-ray production rates and the lunar fines radionuclide content, the percentage of each radionuclide resulting from galactic cosmic-ray interactions was calculated (Table 5). The radionuclide content not accounted for by the galactic component may be attributed to solar proton production. Although this is a very elementary approach, further sophistication without improvements in the accuracy of basic parameters may not produce significantly more reliable values. These results indicate a high solar contribution to the production of many of the radionuclides in the lunar fines.

Since ^{22}Na (2·6 yr) is produced mainly by solar protons, its concentration at any given time will depend markedly on solar flare activity during time periods comparable to its half-life. Meteorite measurements indicate that the galactic cosmic-ray flux has been relatively constant for the past several million years (HONDA and ARNOLD, 1967). Therefore, the fact that ^{26}Al, with a half life of $7 \cdot 4 \times 10^5$ yr is also produced mainly by solar cosmic-rays suggests that the solar flux must have been comparable to its present level for the past million years.

An analysis of the galactic vs. the solar proton production of radionuclides in the lunar rock samples indicates a generally lower solar proton contribution than in the fines. The much lower ^{22}Na and ^{26}Al concentrations in lunar rocks than in the fines do not seem to be accounted for by differences in chemical composition and are apparently due to differences in shielding. Unfortunately, documentation of the rock orientations on the lunar surface and the relationship of our specimens to the original rocks and of the sampling depth of the lunar fines is not presently available to help explain the observed concentration differences. The ^{26}Al concentrations in the three rock samples range from 56 to 63 per cent of that in the lunar fines, while the ^{22}Na concentrations in the three rocks are also low, ranging from 67 to 77 per cent of that in the fines. Since both of these radionuclides are produced by low energy reactions, a closer average distance to the surface which the fines apparently occupied could explain their higher concentrations.

Table 6. Comparison of lunar sample cosmogenic radionuclides with meteorites (dpm/kg)

	Lunar samples				Meteorites			
Isotope	Rock (10057)	Rock (10017)	Rock (10003)	Fines (10084)	A_1*	$A*_2$	B*	C*
^{22}Na	43	45	49	66	69	81	61	56
^{26}Al	84	80	75	134	57	61	43	36
^{54}Mn	41	46	60	24	99	129		48
				RATIOS				
$^{26}Al:^{22}Na$	1·95	1·78	1·53	2·03	0·83	0·75	0·72	0·64
$^{54}Mn:^{22}Na$	1·0	1·0	1·2	0·36	1·4	1·6	1·2	0·86

* Meteorite A_1 is 11 cm deep sample of Allende, A_2 is $>$25 cm sample, B is Denver, C is near-surface sample of Saint Séverin.

The $^{26}Al:^{22}Na$ ratio is two- to threefold higher than that normally present in meteorites (Table 6). This high ratio is mainly explained by the target elements available for production of the nuclides by solar protons. The threefold higher aluminum content of the lunar samples than meteorites would favor ^{26}Al production relative to ^{22}Na from solar protons. Also, Mg, which is a major target element for ^{22}Na production, is about threefold lower in the lunar samples and would further limit ^{22}Na production. The $^{26}Al:^{22}Na$ ratios in lunar samples bear a direct relationship to exposure age and could be employed for estimating the surface exposure age in a manner similar to that used in the age dating of meteorites. However, since the relative production rates of ^{22}Na and ^{26}Al for the low energy solar component are not well known, caution should be exercised in applying this dating technique either with this isotope pair or for other cosmogenic pairs. Regardless of any variations in the ^{22}Na and ^{26}Al production rates with energy, depth, solar cycle, etc., it is obvious that the lunar rocks and fines have surface exposure ages comparable to or much longer than the half-life of ^{26}Al. If one assumed that the production ratios of $^{26}Al:^{22}Na$ were the same in the lunar samples and that the $^{26}Al:^{22}Na$ ratio of about 2 which was observed in both the lunar fines and in the specimen from rock 10057 (see Table 3) represents a saturation value, then the specimens of rocks 10017 and 10003 would clearly indicate undersaturation. In view of known differences in $^{22}Na:^{26}Al$ production rates with energy, it appears that these ratio differences reflect variations in shielding of the rock specimens, or rock movement on the lunar surface, rather than different surface exposures.

The ^{54}Mn concentrations in the lunar samples show an even wider range than ^{22}Na and ^{26}Al and do not appear to show any correlation with them (Tables 3 and 6). The main target element for ^{54}Mn production is iron and the much lower concentrations of iron in the lunar materials than in chondrites help explain the generally lower ^{54}Mn concentrations. The efficient production of ^{54}Mn in lunar materials requires more energetic protons than the production of either ^{22}Na or ^{26}Al, and it appears that galactic protons do account for most of the ^{54}Mn production (Table 5). The lunar fines show a much lower ^{54}Mn concentration than the rocks which is not explained by their somewhat lower iron content. It appears that while the solar protons greatly enhance the production of ^{22}Na and ^{26}Al, they contribute relatively little to ^{54}Mn production; thus, one observes the low ^{54}Mn in lunar fines and the higher concentrations in rocks where a significant secondary flux buildup may occur.

The titanium content of the lunar samples is about 50 times that in meteorites and one would therefore expect cosmogenic isotopes from this target element to be more prevalent. The ^{46}Sc, which is produced from titanium with a cross section of 80 mb at 45 MeV (BRODZINSKI, 1970) was present at concentrations of 8–13 dpm/kg in the lunar rocks and fines. These values are about fourfold higher than the expected galactic production rates (Table 5) based on meteorite observations. This higher ^{46}Sc concentration is attributed mainly to the high titanium content of the samples and to the presence of solar protons in a sufficiently high energy range for its effective production. The ^{44}Ti concentrations in rock 10057 and in the lunar fines sample were <2 and $<1{\cdot}5$ dpm/kg. The cross section for production of ^{44}Ti from Ti is low, (BRODZINSKI, 1970), comparable to that from iron, while the threshold is rather high. Therefore, relatively little solar proton contribution would be expected and ^{44}Ti concentrations comparable to or lower than the observed 1·4 and 2·0 dpm/kg in the Harleton and Bruderheim meteorites (HONDA and ARNOLD, 1964) would be expected (Table 5).

The isotopes ^{48}V and ^{56}Co are both produced by low energy (p, n) reactions on Ti and Fe, respectively. They have almost identical excitation functions (BRODZINSKI, 1970) and are thus excellent indicators of the solar particle flux intensity. Cobalt-56 concentrations of 53 and 19 dpm/kg in the lunar fines and rock 10057, respectively, were observed. The concentration in lunar fines is about fifty times what one would predict from observed meteorite concentrations (Table 5), indicating that about 98 per cent of the ^{56}Co originates from solar proton reactions. Solar flare intensity estimates based on satellite observations (HSIEH and SIMPSON, 1970) indicate that most of the solar proton flux which could have contributed to the observed ^{56}Co concentrations occurred on April 12, 1969. The proton flux of this flare is estimated to have resulted in an integrated lunar surface exposure of about 10^9 proton/cm^2 for proton energies between 10 and 100 MeV (HEYDEGGER, 1970). Based on ^{56}Co production rate measurements in proton irradiated basalt (RANCITELLI, 1970), this could account for the observed ^{56}Co concentration. This same observation has been made by SHEDLOVSKY *et al.* (1970), O'KELLEY *et al.* (1970) and by HEYDEGGER (1970). Because of the short half-life of ^{48}V (16·1 d) and the two-to three-month delay in sample receipt, only an upper limit of 10 dpm/kg could be measured in lunar fines. Based on the nearly identical nature of the excitation functions for ^{48}V and ^{56}Co formation, the ^{48}V concentration from the April 12, 1969, solar event is estimated to be about 3 dpm/kg at the time of sampling or less than 10 dpm/kg when the galactic component is included.

The ^{57}Co concentrations in the lunar samples are more than tenfold lower than in meteorites (Table 5). The principal target element Ni is about tenfold lower in lunar material and this is evidently responsible for the very low ^{57}Co values.

The very low ^{60}Co concentrations (see Table 3) in the lunar materials are to be expected from the low nickel and cobalt concentrations of the lunar surface. Based on integral neutron flux estimates of 10^{15}–10^{16} (ALBEE *et al.*, 1970), neutron capture by stable cobalt could not have contributed significantly to the observed ^{60}Co concentrations. Only upper limits were obtained for ^7Be (53 d) and ^{51}Cr (28 d). Beryllium-7 is a high-energy reaction product and has not yet been detected in meteorites. Chromium-51 is also a high-energy reaction product and its estimated concentration is only about 14 dpm/kg (Table 5). It had decayed by about fivefold at the time of sample receipt and was thus well below its original and measurable concentration.

Concentration gradients

Since solar protons and alpha particles have relatively low energies, their penetration depths are small; and a steeply declining concentration gradient with depth could be expected. A very obvious depth gradient for several radionuclides was observed in rock 10017 (SHEDLOVSKY, *et al.*, 1970) by destructive chemical analysis and also by direct counting of rock specimens (O'KELLEY *et al.*, 1970). The sample of rock 10017 (10017-37), which was analyzed in the present work, was also shown to have a strong concentration gradient for ^{22}Na but essentially none for ^{26}Al. The rock sample, which was mounted in a 7·5 cm cylindrical lucite container similar to those shown in Fig. 1, had its only original surface facing a flat face of the container and was thus well suited for depth gradient measurements. The triple coincidence ^{26}Al photopeaks, which result from the simultaneous capture of a 0·511 MeV annihilation photon by each crystal of a multidimensional gamma-ray spectrometer while the 1·810 MeV gamma ray is seen in only one of the crystals, are very sensitive to source location. In the case of ^{22}Na, a similar combination of photons is emitted, 0·511, 0·511 and 1·275 MeV. From the coincidence counting rate measurements of the triple coincidence peaks compared with those of homogeneous mock-ups and mock-ups with known concentration gradients, it was possible to establish rather precisely the magnitude of the concentration gradient in the lunar rock specimen. While the

^{26}Al indicated essentially no concentration gradient, the ^{22}Na showed a gradient that could be simulated from a mock-up which had a concentration of 90 dpm/kg to a depth of 6 mm and a concentration of 36 dpm/kg for the remaining 30 mm. The average ^{22}Na concentrations of the rock specimen was 45 dpm/kg. Since the work of Shedlovsky *et al.* (1970) and O'Kelley *et al.* (1970) on a specimen of this same rock showed steep concentration gradients for both ^{22}Na and ^{26}Al, some fairly recent shifting of lunar surface orientation of rock 10017 must have occurred. This could be explained if during the past 10^5 yr its orientation had shifted such that an

Table 7. Uranium, thorium and potassium in Apollo 11 lunar samples

U (ppm)				Th (ppm)				
Fines 10084	Rock 10003	Rock 10017	Rock 10057	Fines 10084	Rock 10003	Rock 10017	Rock 10057	Ref. Methods[k]
0·55	0·29	0·86	0·95	2·25	1·08	3·2	3·6	(a), γ
0·49[i]	0·26	0·83	0·79	1·92	1·01	3·25	3·30	(b), γ
0·544	0·268	0·854	0·865	2·09	1·03	3·36	3·42	(c), MS
0·56		0·73		2·17		2·72		(d), MS
0·59				2·24				(e), MS
		0·69	0·80			3·05	3·05	(f), MS
0·55				2·08				(g), MS
Th/U[j]				**K (%)**				
Fines 10084	Rock 10003	Rock 10017	Rock 10057	Fines 10084	Rock 10003	Rock 10017	Rock 10057	
4·2	3·8	3·8	3·9	0·11	0·046	0·26	0·23	(a), γ
4·0	4·0	4·0	4·3	0·11	0·048	0·243	0·255	(b), γ
4·0	3·9	4·1	4·0					(c), MS
3·9		3·7						(d), MS
3·9								(e), MS
		4·5	3·9			0·206	0·201	(f), MS
3·9				0·11		0·250		(g), MS
				0·11		0·239		(h), MS

(a) This work.
(b) O'Kelley *et al.* (1970).
(c) Tatsumoto and Rosholt (1970).
(d) Silver (1970).
(e) Fields *et al.* (1970).
(f) Hintenberger *et al.* (1970).
(g) Gopalon *et al.* (1970).
(h) Philpotts and Schnetzler (1970).
(i) Sample 10002, A component of 10084.
(j) Atomic abundances.
(k) γ = Gamma ray spectrometry.
MS = Mass spectrometry.

originally unexposed surface was moved to a surface position. The rather uniform distribution of micrometeorite pits on all surfaces of some of the lunar rocks suggest that such turnover takes place and with an appreciable probability during the lifetime of a rock.

Primordial radionuclides

The absolute and relative concentrations of the primordial radioactive elements K, U and Th provide important information bearing on the geochronology of the moon, its relationship to the earth, meteorites, and tektites, as well as the process of nucleosynthesis. The observed concentrations of these elements in the lunar fines and rock samples from this study (Table 7) are compared with measurements of other investigators where both larger and much smaller aliquots of the samples were analyzed. Most of the results are in reasonably good agreement if one considers the error limits assigned by the different investigators. The fines are expected to be nearly homogeneous as a result of the presumed manner of formation and the sieving conducted at the Lunar Receiving Laboratory. Since they are also the most widely studied of the lunar samples, they would serve to unveil any bias in the various analytical techniques. For example, the U concentrations cluster narrowly around a value of 0·54 ppm, and very close to the value of 0·55 ppm obtained in this work. This is remarkable agreement when one considers that the sample sizes studied ranged from a few hundred milligrams in the case of mass spectrometric techniques to the kilogram sample employed in this work and indicates homogeneity over a mass range of at least 10^4. Homogeneity is also indicated by the close grouping of reported K and Th concentrations (Table 7). The K content of the fines was reported to be 0·11 per cent by each of these laboratories which measured it, while the six reported Th concentrations grouped closely about the average Th content of 2·12 ppm. Assuming that the laboratories which were in agreement on the fines measurements obtained comparable precision with the rock measurements, it would appear that the rocks are also reasonably homogeneous in their primordial radionuclide content over a sample size range of a few hundred milligrams to a kilogram. This is particularly evident where the very precise measurements of TATSUMOTO and ROSHOLT (1970) on milligram quantities are compared with our measurements on hundred gram quantities.

The atomic abundances of Th relative to U observed in this work are 3·8–3·9 for the three rocks and 4·2 for the fines. The latter value may possibly be somewhat high. Our U measurements are based on one of the U decay products, ^{214}Bi, which is a daughter of radon and some radon loss to the container atmosphere above the fines specimen may have occurred. Since O'KELLEY *et al.* (1970) also based their U measurements on radon daughter counting their low U value for the fines of 0·49 ppm may be due to radon loss prior to their receipt and counting of the sample. The lunar sample Th:U ratios obtained in this work, as well as others summarized in Table 7, are in good agreement with the calculated $^{232}Th:^{238}U$ value of $3{\cdot}8 \pm 0{\cdot}3$ for the average present day solar system (FOWLER and HOYLE, 1960) and with the ratios in chondrites (MORGAN and LOVERING, 1968) and terrestrial crustal rocks (MASON, 1966; KRAUSKOPF, 1967). These constant Th:U ratios, which would be expected from their similar ionic radii and identical outer electron configuration, lend support for the cessation of nucleosynthesis of lunar matter at a time similar to that of the earth and meteorites.

The rare measurement of He and Ar by FUNKHOUSER *et al.* (1970) and HINTENBERGER *et al.* (1970) has been used in conjunction with the K, U and Th concentrations observed in this work to calculate gas retention ages of these samples. These ages, which are presented in Table 8, are with the exception of the K–Ar age of rock 10003, considerably shorter than the solidification ages of these samples as measured by the Sr–Rb (ALBEE *et al.*, 1970), and by Pb–Pb, Pb–Th and Pb–U (TATSUMOTO and ROSHOLT, 1970) techniques. This difference, which was also observed by ALBEE *et al.* (1970), is apparently due to diffusion loss of these radiogenic rare gases. The fact that the K–Ar age of rock 10003 compares favorably with the Pb–Pb ages of TATSUMOTO and ROSHOLT (1970) while the U–Th–He age is only one-half of the K–Ar age and comparable to those of rocks 10017 and 10057 emphasizes the need for an accurate evaluation of diffusion loss, as in the technique of TURNER (1970).

Table 8. Gas-retention ages of Apollo 11 lunar samples (b.y.)

Sample No.	K/Ar	U–Th–He	$^{207}Pb/^{206}Pb$‡
10084	3·3*	—	4·7
10003	3·9†	1·9†	4·0
10017	2·3†	2·0†	4·0
10057	2·3†	2·4†	4·2

* Rare gas data of FUNKHOUSER *et al.* (1970).
† Rare gas data of HINTENBERGER *et al.* (1970).
‡ TATSUMOTO and ROSHOLT (1970).

The uncertainties in the sample orientation on the lunar surface, position of rock specimens within the original rocks, and sampling depths of the lunar fines greatly limit one's ability to reconstruct the past history of lunar and solar processes from even very accurate radionuclide measurements. Radionuclide measurements on samples from a carefully collected and preserved core sample would be extremely valuable in defining precisely both the irradiation history and the mixing, accretion and erosion processes on the lunar surface.

Acknowledgments—We wish to thank D. R. EDWARDS, J. G. PRATT and J. H. REEVES of this Laboratory for their aid in standards preparation and in data acquisition. The unique and sensitive instrumentation which made this work possible was developed during the past decade under sponsorship of the United States Atomic Energy Commission, Division of Biology and Medicine.

REFERENCES

ALBEE A. L., BURNETT D. S., CHODOS A. A., EUGSTER O. J., HUNEKE J. C., PAPANASTASSIOU D. A., PODOSEK F. A., RUSS G. PRICE II., SANZ H. G., TERA F. and WASSERBURG G. J. (1970) Ages, irradiation history, and chemical composition of lunar rocks from the Sea of Tranquillity. *Science* **167**, 463–466.

BRODZINSKI R. L. (1970) Personal communication. Battelle Memorial Institute, Pacific Northwest Laboratories, Richland, Washington.

COOPER J. A. (1970) Unpublished data. Battelle Memorial Institute, Pacific Northwest Laboratories, Richland, Washington.

FIELDS P. R., DIAMOND H., METTA D. N., STEVENS C. M., ROKOP D. J. and MORELAND P. E. (1970) Isotopic abundances of actinide elements in lunar material. *Science* **167**, 499–501.

FOWLER W. A. and HOYLE F. (1960) Nuclear cosmochronology. *Ann. Phys.* **10**, 280.

FUNKHOUSER J., SCHAEFFER O., BOGARD D. and ZÄHRINGER J. (1970) Gas analysis of the lunar surface. *Science* **167**, 561–563.

FUSE K. and ANDERS E. (1969) Aluminum-26 in meteorites—VI. Achondrites. *Geochim. Cosmochim. Acta* **33**, 653–670.

GOPALAN K., KAUSHAL S., LEE-HU C. and WETHERILL G. W. (1970) Rubidium–strontium and uranium, thorium–lead dating of lunar material. *Science* **167**, 471–473.

HEYDEGGER E. R. (1970) Personal communication. University of Chicago, Chicago, Illinois.

HINTENBERGER H., WEBER H. W., VOSHAGE H., WÄNKE H., BEGEMANN F., VILCSEK E. and WLOTZKA F. (1970) Rare gases, hydrogen and nitrogen: concentrations and isotopic compositions in lunar material. *Science* **167**, 543–545.

HONDA M. and ARNOLD J. R. (1964) Effects of cosmic rays on meteorites. *Science* **143**, 203.

HONDA M. and ARNOLD J. R. (1967) Effects of cosmic rays on meteorites. In *Handbuch der Physik*, (editor K. Sitte), Vol. 46, No. 2, pp. 613–632. Springer Verlag.

HOUTERMANS H. (1968) Personal communication. International Atomic Energy Agency, Vienna, Austria.

HOUTERMANS H. (1970) Personal communication. International Atomic Energy Agency, Vienna, Austria.

HSIEH J. and SIMPSON J. (1970) Personal communication. University of Chicago, Chicago, Illinois.

KIRSTEN T. A. and SCHAEFFER O. A. (1969) In *Interactions of Elementary Particle Research in Science and Technology*, (editor L. C. L. Yuan). Academic Press.

KRAUSKOPF K. B. (1967) *Introduction to Geochemistry*, pp. 639–640. McGraw-Hill.

LSPET (LUNAR SAMPLE PRELIMINARY EXAMINATION TEAM) (1969) Preliminary examination of lunar samples from Apollo 11. *Science* **165**, 1211–1227.

LSPET (LUNAR SAMPLE PRELIMINARY EXAMINATION TEAM) (1970) Preliminary examinations of lunar samples from Apollo 12. *Science*, **167**, 1325–1339.

MARTI K., SHEDLOVSKY J. P., LINDSTROM R. M., ARNOLD J. R. and BHANDARI N. G. (1969) Cosmic-ray produced radionuclides and rare gases near the surface of Saint-Séverin meteorite. In *Meteorite Research*, (editor P. M. Millman), p. 22. Springer-Verlag.

MASON B. (1966) *Principles of Geochemistry*, 3rd edition, pp. 45–46. John Wiley.

MASON B. and JAROSEWICH E. (1968) Denver meteorite: a new fall. *Science* **160**, 878–879.

MAXWELL J. A., ABBEY S. and CHAMP W. H. (1970) Chemical composition of lunar material. *Science* **167**, 530–531.

MORGAN J. W. and LOVERING J. F. (1968) Uranium and thorium abundances in chondritic meteorites. *Talanta* **15**, 1079–1095.

O'KELLEY G. D., ELDRIDGE J. S., SCHONFELD E. and BELL P. R. (1970) Elemental compositions and ages of lunar samples by nondestructive gamma-ray spectrometry. *Science* **167**, 580–582.

ORCEL J., DAVID B., KRAUT F., NORDEMANN D. and TOBAILEM J. (1967) Sur la météorite de Saint-Séverin (Charente). Chute du 27 juin 1966. *Compt. Rend.* **D264**, 1556–1564.

PERKINS R. W. (1965) An anticoincidence shielded—multidimensional gamma-ray spectrometer. *Nucl. Instrum. Methods* **33**, 71–76.

PERKINS R. W., RANCITELLI L. A., COOPER J. A., KAYE J. H. and WOGMAN N. A. (1970) Cosmogenic and primordial radionuclides in lunar samples by nondestructive gamma-ray spectrometry. *Science* **167**, 577–580.

PERKINS R. W., RANCITELLI L. A., WOGMAN N. A., COOPER J. A. and KAYE J. H. (1969) Measurement of primordial and cosmogenic radionuclides in lunar surface materials. Semiannual Technical Progress Report to NASA Manned Spacecraft Center, December 1, 1968, through June 1, 1969.

PHILPOTTS J. A. and SCHNETZLER C. C. (1970) Potassium, rubidium, strontium, barium and rare-earth concentrations in lunar rocks and separated phases. *Science* **167**, 493–495.

RANCITELLI L. A. (1970) Unpublished data. Battelle Memorial Institute, Pacific Northwest Laboratories, Richland, Washington.

RANCITELLI L. A., PERKINS R. W., COOPER J. A., KAYE J. H. and WOGMAN N. A. (1969) Radionuclide composition of the Allende meteorite from nondestructive gamma-ray spectrometric analysis. *Science* **166**, 1269–1272.

ROSE H. J., JR., CUTTITTA F., DWORNIK E. J., CARRON M. K., CHRISTIAN R. P., LINDSAY J. R., LIGON D. T. and LARSON R. R. (1970) Semimicro chemical-X-ray fluorescence analysis of lunar samples. *Science* **167**, 520–521.

SHEDLOVSKY J. P., HONDA M., REEDY R. C., EVANS J. C., JR., LAL D., LINDSTROM R. M., DELANY A. C., ARNOLD J. R., LOOSLI H., FRUCHTER J. S. and FINKEL R. C. (1970) Pattern of bombardment-produced radionuclides in rock-10017 and in lunar soil. *Science* **167**, 574–576.

SILVER L. T. (1970) Uranium–thorium–lead isotope relations in lunar materials. *Science* **167**, 468–471.

TATSUMOTO M. and ROSHOLT J. N. (1970) Age of the moon: an isotopic study of uranium–thorium–lead systematics of lunar samples. *Science* **167**, 461–463.

TILTON G. R. (1958) Isotopic composition of lead from tektites. *Geochim. Cosmochim. Acta* **14**, 323–330.

TURNER G. (1970) Argon-40/argon-39 dating of lunar rock samples. *Science* **167**, 466–468.

WÄNKE H., BEGEMANN F., VILCSEK E., RIEDER R., TESCHKE F., BORN W., QUIJANO-RICO M., VOSHAGE H. and WLOTZKA F. (1970) Major and trace elements and cosmic-ray produced radioisotopes in lunar samples. *Science* **167**, 523–525.

WOGMAN N. A., ROBERTSON D. E. and PERKINS R. W. (1967) A large detector, anticoincidence shielded multidimensional gamma-ray spectrometer. *Nucl. Instrum. Methods* **50**, 1–10.

WOGMAN N. A., PERKINS R. W. and KAYE J. H. (1969) An all sodium iodide anticoincidence shielded multidimensional gamma-ray spectrometer for low-activity samples. *Nucl. Instrum. Methods* **74**, 197–212.

Proceedings of the Apollo 11 Lunar Science Conference, Vol. 2, pp. 1471 to 1486.

Apollo 11 lunar samples: K, Rb, Sr, Ba and rare-earth concentrations in some rocks and separated phases

JOHN A. PHILPOTTS and C. C. SCHNETZLER
Planetology Branch, Goddard Space Flight Center, Greenbelt, Maryland 20771

(*Received* 2 *February* 1970; *accepted in revised form* 21 *February* 1970)

Abstract—K, Rb, Sr, Ba and rare-earth element concentrations have been determined by mass spectrometric isotope dilution for eight Apollo 11 lunar samples and for some separated phases. K and Rb are at chondritic levels, Sr at fifteen times, and Ba and rare-earths at 30–100 times chondritic levels. There are trace element similarities between the lunar samples and terrestrial dredge basalts, the bulk earth and basaltic achondrites. Apollo 11 rocks have higher Eu^{2+}/Eu^{3+} than terrestrial basalts, indicating more reducing conditions. The large Eu deficiencies in the Apollo 11 rocks indicate loss of feldspar during extensive fractional crystallization or limited partial fusion of feldspathic source rock at relatively shallow depths. The Apollo 11 rocks are not primary solar system materials. The trace element data distinguish two types of igneous rock and it is proposed that these may represent liquid and partial cumulate. The low K and Rb contents of the lunar igneous rocks are characteristic of their source material. A community of origin for the earth, the moon and the basaltic achondrites is indicated.

INTRODUCTION

THIS paper reports determinations of K, Rb, Sr, Ba and rare-earth concentrations in eight Apollo 11 lunar samples, in a fine fraction from the soil, and in mineral fractions separated from two of the igneous rocks. It is an expanded version of our first publication on the returned lunar samples (PHILPOTTS and SCHNETZLER, 1970a). We have now had the opportunity, however, of drawing on the data and ideas put forward in the first reports of other investigators.

The elements we have analyzed constitute a particularly powerful group for the deciphering of geochemical processes. The concentrations of the volatile alkali metals K and Rb can be useful in considering volatilization or condensation processes. Various element ratios such as K/Rb, Rb/Sr, K/Ba and rare-earth/rare-earth are good indicators of the type and extent of igneous differentiation processes. Rb/Sr is also of importance in the dating of samples. Eu concentrations can serve as indicators of redox conditions.

The major purpose of the present study was to characterize the Apollo 11 samples and to place them in the spectrum of solar-system materials. The long range goal of our work on these and future returned lunar rocks is to elucidate the evolution of the moon. This paper represents our second step in this direction.

RESULTS

The results of our analyses of lunar samples are given in Table 1. All of these results are for single weighings. The analytical technique used was mass-spectrometric stable-isotope dilution (SCHNETZLER *et al.*, 1967a, b). The precision of our whole-rock data is, with a few random exceptions, expected to be better than about ± 2 per cent, as shown by our replicate analyses of the rock standards BCR-1 and W-1 given in

Table 1. Sr, Rb, K, Ba and rare-earth abundances,

	10017-44 A	10044-24? B	10062-29 B	10018-22 C	10021-24 C	10046-18 C	10060-19 C	10084-130 D
Sr	165	167	194	164	165	165	167	162
Rb	5·70	5·64	0·832	3·79	4·03	4·22	4·13	2·78
K	2390	2430	601	1490	1550	1620	1570	1120
Ba	287	285	134	200	211	219	224	170
Ce	75·5	76·6	40·2	52·8	57·2	58·0	59·0	46·1
Nd	66·1	66·1	39·9	45·4	48·9	50·2	50·0	40·5
Sm	23·4	23·4	14·7	16·3	17·2	17·7	17·5	13·9
Eu	2·26	2·21	2·07	1·84	1·91	1·94	1·98	1·77
Gd	28·4	28·6	18·1	20·5	—	21·5	—	—
Dy	32·8	33·6	20·6	21·8	25·2	24·9	26·2	19·5
Er	19·1	19·3	11·8	12·8	13·0	14·8	14·5	11·7
Yb	18·1	16·6	11·3	11·8	12·7	13·0	12·7	10·6
Lu	2·63	—	1·76	1·87	1·99	—	1·92	—
Wt. (g)	0·125	0·286	0·122	0·145	0·137	0·311	0·218	0·311

Table 2. Comparison of our data on lunar rocks, on BCR-1, and on W-1 with those of other investigators suggests tıat the accuracy of our analyses is, in general, better than ±5 per cent, with the *possible* exceptions of Nd and Sm; our results for these elements appear to be 5–10 per cent higher in many (but not all) cases where comparison with other laboratories could be made. The quality of our mineral separate data is somewhat poorer than that of the whole-rock data because of significant blank corrections, which exceed 10 per cent in three cases, and because of less than optimum spiking. Sample weights are included in Table 1 and average blanks in Table 2 so that the interested reader can compute approximately the blank corrections which have been applied to the data if they exceeded 1 per cent of the concentration of any particular element. It should be noted that the order in which the elements are presented in the tables and figures was selected to permit ready estimation of important elemental ratios.

Table 2. Sr, Rb, K, Ba and rare-earth concentrations in rock standards BCR-1 and W-1, and blanks

	BCR-1 (Avg. of 2)* (ppm)	% Dev. from avg.	W-1 (Avg. of 4)* (ppm)	Max. % Dev. from Avg.	Avg. Blanks (μg)
Sr	325	0·3	186 (1)	—	0·05 (5)
Rb	46·6	0·2	21·0 (1)	—	0·003 (5)
K	14,300	0·3	5200 (1)	—	2·5 (5)
Ba	646	0·15	157	1·0	0·15 (2)
Ce	53·9	1·3	23·4	1·3	≳0·0x (2)
Nd	32·1	2·0	15·1	0·6	≳0·00x (2) (Nd–Lu)
Sm	7·44	0·9	3·76 (3)	0·5	
Eu	1·942	0·4	1·112	3·0	
Gd	6·47 (1)	—	4·03	1·2	
Dy	6·36	0·3	3·95	2·3	
Er	3·58	0·8	2·30	3·5	
Yb	3·38	0·6	2·08	1·9	
Lu	0·536	0·8	0·031 (1)	—	

* Unless otherwise specified by number in parentheses.

in ppm by weight, in Apollo 11 samples

10084-130 (<400 mesh)	10044-24?			10062-29			Normalizing values	
	Pyx.	Plag.	Opaque	Pyx.	Plag.	Opaque		
173	62·6	541	29·6	64·6	396	41·3	11	Sr
2·92	1·96	1·67	2·95	0·248	0·230	0·571	3	Rb
1100	811	2160	987	173	413	303	1000	K
180	93·7	271	104	30·9	70·1	62·1	3·6	Ba
46·8	33·3	18·5	33·3	13·0	9·19	20·2	0·787	Ce
40·9	29·9	15·0	27·2	17·3	6·75	17·9	0·652	Nd
14·0	12·6	4·34	8·84	7·71	2·0	6·42	0·208	Sm
1·84	0·93	4·79	0·345	0·768	2·75	0·539	0·071	Eu
16·8	16·6	—	—	9·7	—	—	0·256	Gd
18·9	21·3	4·43	13·6	13·8	2·56	9·33	0·303	Dy
11·1	12·7	3·04	9·64	8·14	1·32	6·04	0·182	Er
9·91	11·6	2·54	11·8	7·57	1·29	7·02	0·188	Yb
1·54	1·85	0·401	2·06	1·19	0·20	1·24	0·034	Lu
0·123	0·0569	0·0188	0·0570	0·0742	0·0520	0·0952		Wt. (g)

It will be noted that in Table 1 one of our samples is labeled "10044-24?". This labeling is consistent with that used in our first report (PHILPOTTS and SCHNETZLER, 1970a) and was the designation for this particular sample when received from the Lunar Receiving Laboratory. The question-mark has been added, however, inasmuch as we now think that the sample that we received is *not* a split from sample 10044. The major evidence for this is the fact, as discussed below, that the lunar igneous rocks fall into two distinct trace-element groups, and our sample "10044?" falls into one group whereas data from numerous other investigators place sample 10044 into the other group. Complicating the matter is the circumstance that we were originally scheduled to receive sample 10024, but instead received this sample labeled 10044-24. RUSSEL HARMON, Assistant Curator of the Lunar Receiving Laboratory agrees (personal communication, January 1970) that a mix-up is likely. We are currently analyzing additional samples 10044 and 10024 in an attempt to resolve this mix-up.

DISCUSSION OF WHOLE-ROCK DATA

To a first approximation, all our Apollo 11 samples are similar to each other for the trace elements considered (Table 1). The similarity is apparent in Fig. 1, in which are plotted trace-element concentrations, normalized to average chondritic abundances (Table 1), for representative samples of each type that we analyzed. This uniformity of all the Apollo 11 samples indicates that these trace-element chemical properties are characteristic for, at least, the near surface of all of Mare Tranquillitatis; if this is correct, the conclusions reached about the petrogenesis of these rocks will have broader applicability. Closer inspection of our own data and those of other investigators reveals that these trace elements serve to distinguish three groupings within the Apollo 11 samples: (a) igneous rocks high in trace elements, (b) igneous rocks low in trace elements, (c) breccias and soil. The differences (which are discussed below) between the two types of igneous rocks are well brought out in Fig. 5 which gives the ranges encountered in a number of studies (primarily GAST and HUBBARD, 1970; HASKIN *et al.*, 1970; MURTHY *et al.*, 1970; PHILPOTTS and SCHNETZLER, 1970a;

SCHMITT *et al.*, 1970; GOPALAN *et al.*, 1970; COMPSTON *et al.*, 1970). For the elements we have studied, the breccias and soil appear to be intermediate between the two types of igneous rocks. The relative concentrations of the trace elements in the breccias and soil are closer to those in the enriched type of igneous rock; trace element concentrations in a roughly equal mixture of high and low type igneous rocks would, of course, be dominated by the contribution from the high type. The breccias and soil, however, are not simply mixtures of the two types of igneous rocks. There is a suggestion of this in the trace-elements we have studied; K/Rb (Table 1, Fig. 1), for example, appears to be lower for the breccias and soil than for either type of igneous rock. The case against simple mixing is stronger for some other elements as GAST and

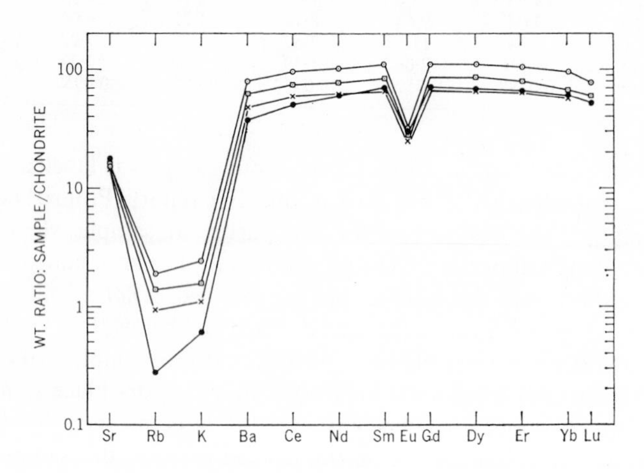

Fig. 1. Chondrite-normalized Sr, Rb, K, Ba and rare-earth concentrations in crystalline rocks 10017-44 (○) and 10062-29 (●), breccia 10060-19 (⊡), and soil 10084-130 (×).

HUBBARD (1970) and COMPSTON *et al.* (1970), for example, have pointed out on the basis of higher Al, Ni and Zn and lower Ti in the breccias and soil than in either type of igneous rock. Finally, the apparent age of the breccias and soil seems to be significantly greater than that of the igneous rocks (e.g. ALBEE *et al.*, 1970; TATSUMOTO and ROSHOLT, 1970).

It has been suggested previously (PHILPOTTS and SCHNETZLER, 1970a) that a limited number of provenance sites may be indicated for the Apollo 11 samples. Data from other investigators (e.g. GAST and HUBBARD, 1970; HASKIN *et al.*, 1970; MURTHY *et al.*, 1970) appear to support this suggestion. For the trace elements considered in this study, the enriched igneous rocks fall into a very tight group with little spread in trace element concentrations relative to the amounts present (Fig. 5). A similar situation holds for the breccias. These data suggest the possibility that the enriched igneous rocks and the breccias among the Apollo 11 samples may have all come from just two sites.

We turn now to a comparison of the Apollo 11 samples and other solar-system rocks. In Fig. 2 are plotted normalized trace element concentrations for an enriched lunar igneous rock (10017, Table 1), for the average terrestrial continental crust (HURLEY, 1968; HASKIN *et al.*, 1966), and for a dredge andesite and a dredge basalt (PHILPOTTS *et al.*, 1969). Trace element concentrations in the terrestrial continental crust are quite different from those in the lunar samples. K and Rb are about an order of magnitude higher, Sr about three times higher, and the heavier rare-earths a factor of 3–4 lower than in the lunar samples. Tektite trace-element contents resemble those of the continental crust rather than those of the lunar samples. The lunar samples do show a marked trace element resemblance to lavas dredged from the floor of the earth's oceans, particularly to the andesite (PD3). The resemblance might arise from their both possibly representing low pressure differentiation products from

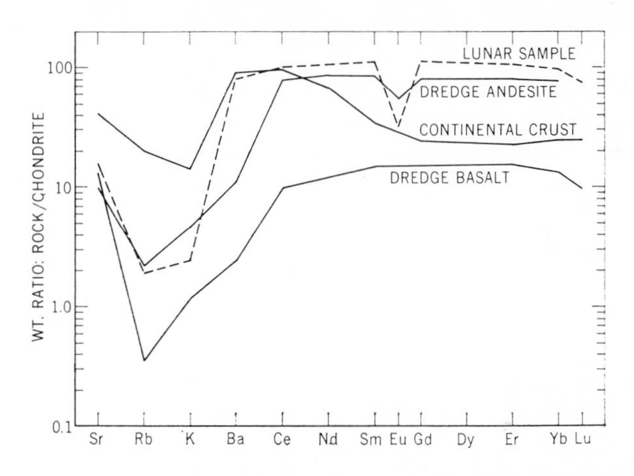

Fig. 2. Chondrite-normalized Sr, Rb, K, Ba and rare-earth concentrations in lunar sample 10017-44, in the average terrestrial continental crust (HURLEY, 1968; HASKIN *et al.*, 1966), and in terrestrial dredge andesite and basalt (PHILPOTTS *et al.*, 1969).

similar source materials. A major difference between the dredge lavas and the lunar samples appears to be the low relative concentration of Ba in the former. It is of interest in this connection that we have analyzed a number of orogenic basalts and andesites which have trace element characteristics similar to those of the dredge lavas except for much higher relative concentrations of Ba—higher, in fact, than those of the lunar samples. There seems to be a distinct possibility, therefore, that there may be terrestrial igneous rocks with relative Ba concentrations similar to those in the lunar samples. Finally it is worth noting that although there may be small differences, possibly a somewhat lower K/Ba, for example, the lunar sample trace-element relative concentrations are similar in many respects to those expected for the bulk earth (e.g. HURLEY, 1968; HASKIN *et al.*, 1966).

The Apollo 11 samples do not have relative or absolute trace element concentrations similar to those of chondritic meteorites. This is apparent from Fig. 1 in which the lunar data are normalized to average chondritic values. The low relative abundance of the volatile alkali elements Rb and K when compared to chondritic values is the

Table 3. Sr, Rb and K concentrations of some calcium-rich achondrites, in ppm by weight

Achondrite	Sr	Rb	K
1. Stannern	87·0	0·83	686
2. Juvinas (Whole rock)	79·6	0·33	298
Juvinas Plagioclase	186	0·42	731
Juvinas Pyroxene	10·8	0·27	65
3. Jonzac	71·7	0·33	300
4. Bununu	44·2	0·25	186
5. Zmenj	28·3	0·34	164
6. Moore County (Whole rock)	77·2	0·087	193
Moore County Plagioclase	168	0·14	415
7. Angra dos Reis	143	0·135	46
8. Serra de Magé	50·8	0·063	61
Shergotty	50·7	0·13	1500

most notable difference; as discussed below, we consider the low alkali concentrations to be characteristic of a significant portion, if not all, of the moon.

The Apollo 11 samples have trace element characteristics suggestively similar in a relative sense to those of the basaltic achondrite meteorites (SCHMITT *et al.*, 1964;

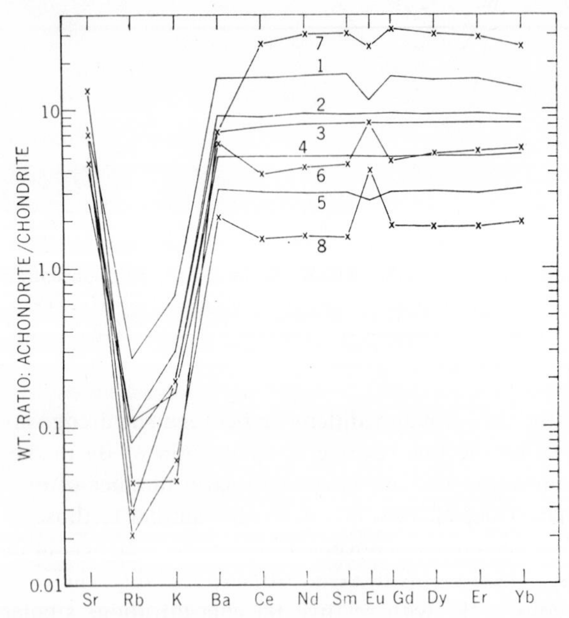

Fig. 3. Chondrite-normalized Sr, Rb, K, Ba and rare-earth concentrations in some Ca-rich achondrites. Numbers correspond with samples presented in Table 3. Ba and rare-earth data from SCHNETZLER and PHILPOTTS (1969).

GAST, 1965), however there is some difference in the absolute level of the trace element concentrations. This is apparent in Fig. 3, in which are plotted chondrite-normalized Ba and rare-earth concentrations (SCHNETZLER and PHILPOTTS, 1969) and K, Rb and Sr concentrations (Table 3) for a number of basaltic achondrites. The level in Stannern (No. 1 in Fig. 3) is the highest reported for the basaltic achondrites; it is of interest that Stannern shows a definite negative Eu anomaly. The similarity in K/Ba for the lunar samples and the basaltic achondrites may be particularly significant inasmuch as this ratio tends not to change much during igneous differentiation (PHILPOTTS and SCHNETZLER, 1970b). Also plotted in Fig. 3 are data for Angra dos Reis (No. 7), a unique titaniferous–augite achondrite with rare-earth concentrations (SCHNETZLER and PHILPOTTS, 1969), including a negative Eu anomaly, that are distinctly similar to those in the lunar samples. However, Ba is lower, and K and Rb (Table 3) much lower, in Angra dos Reis. We consider Angra dos Reis, and also the unbrecciated achondrites Moore County (No. 6) and Serra de Mage (No. 8), to be of cumulative origin (SCHNETZLER and PHILPOTTS, 1969). The fact that these meteorites show tendencies (e.g. lower K and Rb contents) relative to the brecciated basaltic achondrites (which we believe to represent liquids, for the most part) similar to those shown by the low trace-element type of lunar igneous rock relative to the high trace-element type (Fig. 5) is of interest and will be considered below.

DISCUSSION OF MINERAL DATA

Trace element data for plagioclase, pyroxene and opaque fractions separated from the igneous rocks "10044?" and 10062, are presented in Fig. 4, normalized to the concentrations in the respective whole-rocks. The mineral separates were obtained

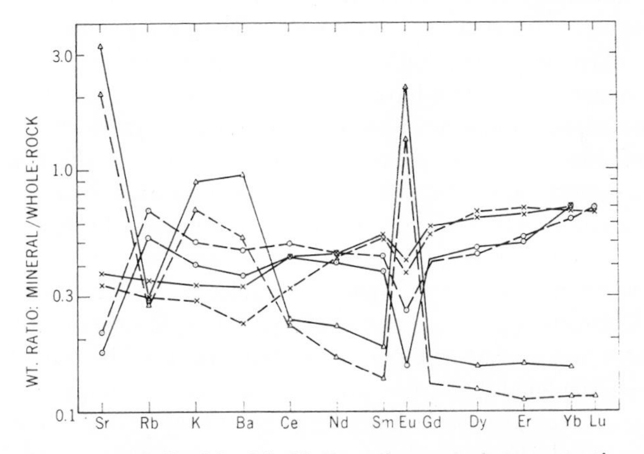

Fig. 4. Whole-rock normalized Sr, Rb, K, Ba and rare-earth concentrations in plagioclases (△), pyroxenes (x) and opaques (⊙) separated from crystalline rocks "10044-24?" (solid lines) and 10062-29 (dashed lines).

using a combination of electromagnetic, heavy-liquid and hand-picking techniques; they were estimated to be more than 97 per cent free of contaminating grains, but some foreign material included within grains may have gone unnoticed, particularly in the case of the opaque separates. Because of the fine grain-size of the rocks and the stringent purification, the mineral yields were very small. Hence the separated phases were probably not typical of these phases in the whole-rock—the procedures favor separation of coarse grains, presumably grains that started to crystallize earlier. This might partly explain the fact, apparent in Fig. 4, that the mineral separates do not contain the various trace elements in concentrations sufficient, in general, to yield the whole rock values; that is, other pyroxene, feldspar, or opaque fractions in the whole rock, presumably forming later than the phases separated, may well have higher trace element concentrations, in general, than those measured. At the same time it is also probable that the differences in absolute trace-element concentration between the whole-rocks and the separated phases are in part due to the presence of interstitial, accessory phases, such as phosphates, alkali feldspar or glass, that are rich in trace-elements. Rare-earth bearing apatite (ADLER *et al.*, 1970) and dysanalyte (RAMDOHR and EL GORESY, 1970), for example, have been found in the Apollo 11 samples. That our separated pyroxene and feldspar may represent early-formed material is supported to some extent by the fact that the trace-element ratios between these minerals and their respective whole rocks are close to those expected on the basis of measured trace-element partition coefficients (PHILPOTTS and SCHNETZLER, 1970b; SCHNETZLER and PHILPOTTS, 1968, 1970) between such minerals and the liquids from which they are crystallizing. The consistency of the mineral to whole-rock trace-element ratios, in spite of the differences between the whole-rock concentrations themselves, tends to indicate that the separates were fairly pure. There is also fairly good agreement with the mineral data presented by GOLES *et al.* (1970) except for Ba in the pyroxene and the size of the Eu anomalies; it is again worth noting that sample 10044 of GOLES *et al.* and our sample "10044?" are almost certainly not from the same rock. In the case of the opaque separates we did not expect such high trace-element concentrations in this phase, particularly the high concentrations of the large Rb cation. Other investigators (ALBEE *et al.*, 1970) have also reported high opaque Rb values. Electron microprobe analyses of fractions of our opaque separates revealed only ilmenite, with very minor amounts of pyroxene and plagioclase. Possibly the high trace element concentrations reflect late crystallization of the separated opaques. Perhaps the consistency is fortuitous and we are seeing contamination by late phases. That the separated opaques or their possible contaminants are late is favored to some extent by the complementary nature of the opaque and feldspar Sr, Rb and K concentrations, for example. Similarly, we believe that the large Eu depletions observed in one of our opaque separates and in that of GOLES *et al.* (1970) can be best attributed in part to Eu having been depleted by some prior plagioclase crystallization. The opaque separates, however, remain somewhat of a puzzle at the present time.

A matter on which the mineral data do throw some light is the redox state of the lunar samples. The partitioning of Eu (an element that differs from the other rare-earths in showing significant natural stability in the divalent state) between two phases can be considered in terms of the partitioning of Eu^{2+} and the partitioning of Eu^{3+}. Feldspars tend to accept considerable Eu^{2+}; the Eu^{2+} phenocryst-liquid partition coefficient for feldspar is apparently larger, often much larger, than the Eu^{3+} partition coefficient (SCHNETZLER and PHILPOTTS, 1970). For other phases, however, it appears that the Eu^{2+} partition coefficient is somewhat smaller than the Eu^{3+} partition coefficient. This results in relative Eu excesses in feldspars and, generally, small deficiencies in other phases (e.g. SCHNETZLER and PHILPOTTS, 1970). The size of these excesses or deficiencies is a function of a number of parameters, including the system's Eu^{2+}/Eu^{3+}, which is related to the operative redox conditions (TOWELL *et al.*,

1965; PHILPOTTS and SCHNETZLER, 1968; TOWELL *et al.*, 1969; PHILPOTTS and SCHNETZLER, 1969). Hence, the partitioning of total Eu between associated phases is expected to be an approximate measure of redox conditions in comparable systems. The ratio of Eu between plagioclase and pyroxene in the lunar rocks is higher than between comparable terrestrial feldspar and pyroxene (PHILPOTTS and SCHNETZLER, 1968; SCHNETZLER and PHILPOTTS, 1968, 1970), and this, assuming equilibration, indicates higher Eu^{2+}/Eu^{3+} than in these terrestrial rocks which in turn implies more reducing conditions. This is in accord with the conclusions of LSPET (1969) and the low ferric iron contents of the Apollo 11 samples (e.g. AGRELL *et al.*, 1970).

PETROGENESIS OF THE IGNEOUS ROCKS

In deciphering the evolution of the Apollo 11 igneous rocks, there are three major sets of data upon which to draw. These are data for (1) the high trace-element igneous rocks, (2) the low trace element igneous rocks, and (3) the bulk moon. For the first two we have concrete data, whereas bulk moon data derives from speculation. Nevertheless there are some characteristics that can be assigned to the bulk moon with a fair degree of confidence. One of these is the probability that the bulk moon has chondritic relative rare-earth concentrations. Part of the evidence for this assumption is the fact that all chondrites (SCHMITT *et al.*, 1964) and many achondrites (Fig. 3) have essentially the same relative rare-earth concentrations. Further, the range of terrestrial rock relative concentrations includes the chondritic values (HASKIN *et al.*, 1966); it appears that the bulk earth may well have chondritic relative rare-earth concentrations. Finally the Apollo 11 rocks (Fig. 1), particularly the breccias, soil and the high trace-element type of igneous rock (Fig. 5), have relative rare-earth concentrations close to those of chondrites.

A notable deviation from chondritic abundances in the Apollo 11 samples is the low relative value of Eu. Possibly the bulk-moon shows a negative Eu anomaly. This is doubtful. Again, the chondrites and many achondrites show no Eu anomaly. Many terrestrial basalts have normal Eu concentrations (HASKIN *et al.*, 1966) and by induction normal Eu can be assigned to the bulk earth. By analogy with these other solar system materials it appears likely that Eu is normal in the bulk moon. Assuming this to be true, then the large negative Eu anomalies in the Apollo 11 samples become an important clue to the evolution of these rocks.

The occurrence of Eu anomalies in feldspar bearing rocks can be attributed for the most part to a feldspar effect (TOWELL *et al.*, 1965; PHILPOTTS and SCHNETZLER, 1968). As discussed earlier, feldspars accept considerably more Eu^{2+}, in general, than trivalent rare-earths in a relative sense. Therefore, given some Eu^{2+} stability, feldspars show positive Eu anomalies relative say to the rare-earth concentrations in an equilibrium liquid. Hence feldspar cumulates possess excess Eu; conversely a liquid that has lost feldspar by crystallization, or a liquid derived by partial fusion in which a feldspathic residuum has been left behind, will possess a relative deficiency of Eu.

The Apollo 11 rocks contain feldspar and it seems reasonable to attribute their negative Eu anomalies to a feldspar effect. This concept has been questioned, however. GOLES *et al.* (1970) have recognized that the Eu depletions may be related to plagioclase fractionation but they believe that they may more plausibly be attributed to incoherence of Eu^{2+} and trivalent rare-earths during magma formation. We find this confusing inasmuch as the feldspar effect is almost certainly due to this incoherence. It would seem that GOLES *et al.* are suggesting an Eu^{2+} effect not involving feldspar. This might fit in with the genetic model for Apollo 11 rocks put forward by RINGWOOD and ESSENE (1970) which did not involve feldspar. Feldspar is the only mineral known at present to take excess Eu. There may well be others but none have been clearly established in the considerable body of data on rock-forming terrestrial minerals

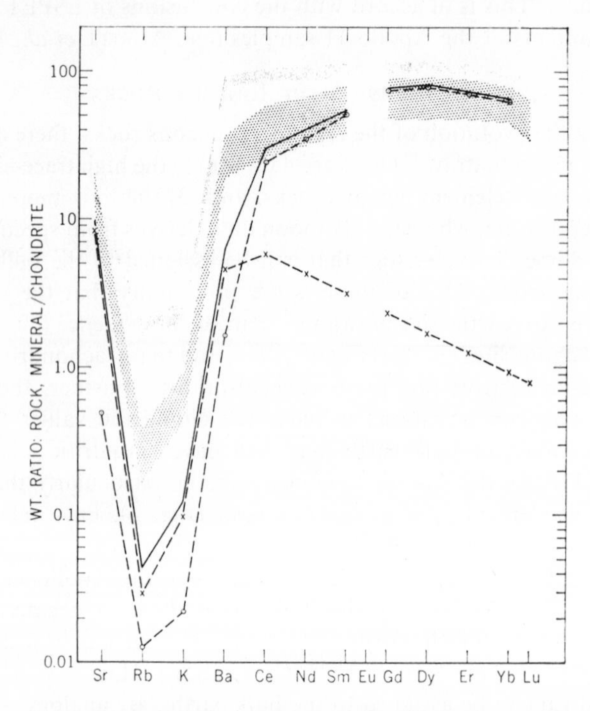

Fig. 5. Chondrite-normalized Sr, Rb, K, Ba and rare-earth concentration range for trace-element rich Apollo 11 igneous rocks (fine stipple), and for trace-element poor Apollo 11 igneous rocks (coarse stipple), and concentrations calculated for an incomplete mixture (solid line) of 30 per cent plagioclase and 50 per cent clinopyroxene in equilibrium with the high trace element igneous rocks and for the contributions to the mixture of the plagioclase (x) and the pyroxene (⨀).

(e.g. SCHNETZLER and PHILPOTTS, 1968, 1970), in studies of achondrite minerals (SCHNETZLER and PHILPOTTS, 1969), nor in the first data on known lunar minerals (above; GOLES *et al.*, 1970). Phenocrysts other than feldspar tend to show small negative Eu anomalies relative to their matrix materials in terrestrial rocks (SCHNETZLER and PHILPOTTS, 1970). As pointed out in the mineral discussion this effect will be functions of Eu^{2+} and Eu^{3+} partition coefficients and the liquids Eu^{2+}/Eu^{3+} and is more exaggerated for Apollo 11 minerals than for comparable terrestrial phases. Crystallization of phases other than feldspar from a lunar melt would therefore produce positive Eu anomalies in the liquid, the opposite situation to that observed in the Apollo 11 rocks. It is expected that non-feldspathic lunar cumulates will have Eu deficiencies relative to the parental melt. It is conceivable that the Apollo 11 rocks could represent cumulates in which the negative Eu anomaly merely reflects the deficiency in the mafic phases. But there are several problems with this interpretation, one of them being that if the high trace-element type igneous rock represents a cumulate then the parental liquid would have to contain seemingly prohibitively high concentrations of some trace-elements. It is concluded then, that, at our present state of knowledge, the negative Eu anomalies in the Apollo 11 rocks are best ascribed to a feldspar effect.

HASKIN *et al.* (1970) have considered the feldspar effect in the problem of generating, by fractional crystallization or partial melting, the variations in abundances of Eu relative to Sm in the Apollo 11 rocks from initial material with chondritic relative abundances of these elements. For a given amount of fractional crystallization, the extent of development of the Eu anomaly depends on the difference between the actual Eu partition coefficient and the partition coefficient for Eu^{3+} (PHILPOTTS and SCHNETZLER, 1968)—a value that can be obtained by interpolation from the other trivalent rare-earth partition coefficients or may be approximated, as HASKIN *et al.* (1970) have done, by the partition coefficient for Sm (or Gd). The larger the difference between the partition coefficients for Eu and Eu^{3+} (Sm), then the larger will be the anomaly for any given extent of fractional crystallization. HASKIN *et al.* used a value of 0·32 for this difference. Measured phenocryst-matrix partition coefficients, however, indicate that larger differences might be possible, even for calcic plagioclase (SCHNETZLER and PHILPOTTS, 1970). Thus the values for the partition coefficients used by GAST and HUBBARD (1970) for Eu and trivalent rare-earths of 0·77 and 0·05, respectively, do not seem unreasonable for terrestrial feldspars. We are not dealing with terrestrial rocks, however, but rather, as discussed in the sections on mineral data, with a system having higher Eu^{2+}/Eu^{3+}. The Eu concentrations in the lunar minerals, considered in light of the other rare-earth concentrations and the measured trivalent partition coefficients, indicate that the Eu partition coefficient for lunar plagioclase is probably between one and two, and that the difference between the partition coefficients for Eu and Eu^{3+} (Sm) is close to (probably somewhat larger than) unity. Hence, whereas HASKIN *et al.* (1970) estimated that between 88 per cent and 99 per cent fractional crystallization of plagioclase would be needed to explain the range of Eu anomalies encountered in the Apollo 11 rocks starting from material with normal Eu, our calculations yield roughly from 50 per cent to 70 per cent crystallization. Addition of other crystallizing phases to the model would, of course, increase the degree of crystallization required, particularly if, as we expect, these other phases tend to reverse the feldspar effect on Eu. Assuming that during much of an extensive fractional crystallization process, the liquid will be at an effective cotectic for the major phases, then we calculate that somewhat in excess of 99 per cent crystallization would be required to generate the Eu anomaly exhibited by the high trace-element type of Apollo 11 igneous rock. Thus we agree with HASKIN *et al.* (1970) that very extensive fractional crystallization would be required to produce the observed Eu depletion. However, the choice of different partition coefficients would seem to place fractional crystallization within the realm of possibility. Further, if the maria represent gigantic lava lakes then considerable fractional crystallization may have taken place. Nevertheless, we tend to agree with HASKIN *et al.* that limited partial fusion appears to be a more likely mode of origin for the high trace-element type of igneous rock. At the same time, however, the feldspathic rock indicated by the Surveyor VII analyses (PATTERSON *et al.*, 1969), considered along with the large Eu anomaly in the Apollo 11 rocks and the expectation that feldspar would float in such titaniferous basalts, tempt the speculation that the lunar highlands represent feldspathic scum produced by flotation during fractional crystallization of basalt, as has been suggested by ANDERSON *et al.* (1970) and WOOD *et al.* (1970).

Experimental studies impose restrictions on the differentiation of the lunar igneous rocks. Garnet is stable at moderate pressures (O'HARA *et al.*, 1970; RINGWOOD and ESSENE, 1970) and this restricts much of the differentiation (i.e. that seemingly involving feldspar) to depths less than about a couple of hundred kilometers. This differentiation could therefore take place within the constraints that UREY (1968, 1969) believes the mascons place on the thermal history of the moon. Also relevant to the feldspar differentiation scheme is the fact pointed out by O'Hara *et al.* (1970), that the Apollo 11 rocks fall close in composition to low pressure cotectic liquids and this implies advanced crystal fractionation (or limited partial fusion) under low pressures. The low pressure cotectic composition would seem to have to be fortuitous in the scheme presented by RINGWOOD and ESSENE (1970) which involves partial fusion at greater than 200 km without extensive fractionation en route and which does not seem to satisfactorily account for the observed Eu anomalies. It should be noted, however, that although feldspar is close to the cotectic, it was not on the liquidus in experiments on Apollo 11 igneous rock composition melts (ANDERSON *et al.*, 1970; KUSHIRO *et al.*, 1970; O'HARA *et al.*, 1970; RINGWOOD and ESSENE, 1970). This does not seem to be consistent with the simple hypotheses of plagioclase fractional crystallization or partial fusion of a feldspathic rock which the Eu data strongly support. Perhaps the experimental studies do not exactly reproduce the natural lunar liquids—perhaps the lunar rocks have excess accessory cumulative olivine or ilmenite, or perhaps there are differences in the volatile contents of the natural and experimental melts. If this is not the case then a more complex genetic history involving at least two fusion events would seem to be required.

We turn now to a consideration of the relationship between the high and low trace-element types of Apollo 11 igneous rocks. If both represent related liquids then, as GAST and HUBBARD (1970) have pointed out, considerable fractional crystallization is required to go from the low trace-element type to the high. The greatest difference between the two types in terms of relative concentration exists for Rb (Fig. 5). More than 90 per cent fractional crystallization is required to cover the range in Rb contents even assuming that the crystallizing solid takes no Rb at all. Such extensive crystallization seems inconsistent with the near identity of the major element composition of the two types (GAST and HUBBARD, 1970; COMPSTON *et al.*, 1970); one might expect, for example, a considerable difference in the Na concentrations. However, at the present time it does not seem possible to absolutely rule out this possibility. GAST and HUBBARD (1970) put forward an alternate explanation involving a third liquid rich in alkalis and Ba which when added to the low-type liquid would give rise to the high-type, allowing for some fractional crystallization. This model, while possible, is aesthetically unappealing. It would seem to mean that the near-chondritic relative rare-earth concentrations of the high trace-element Apollo 11 igneous rock arise fortuitously. Further, we can see no reason, other than unusually perverse sampling, why this model would yield mixed liquids showing considerably less relative range in trace-element concentrations than found in one of the component liquids. An alternate explanation that we have offered (PHILPOTTS and SCHNETZLER, 1970a) is that the low trace-element type of Apollo 11 igneous rock may be partly cumulative in origin. In support of this model we have plotted in Fig. 5 the trace-element concentrations, based on measured phenocryst-matrix partition coefficients (GSFC No. 271, PHILPOTTS and SCHNETZLER, 1970b; SCHNETZLER and PHILPOTTS, 1970), that associated bytownite (times 0·3) and iron-rich augite (times 0·5) would have were they in equilibrium with the trace-element enriched type of lunar rock. The sum of the trace-element concentrations in the incomplete mixture of 30% bytownite and 50% augite (essentially the concentrations in the Apollo 11 samples) have also been plotted. Eu has not been included in the plots inasmuch as the terrestrial partition coefficients, as discussed earlier, are not appropriate. The tendency for the concentrations in the low trace-element igneous rocks to approach those in the hypothetical plagioclase–augite cumulate is apparent. In such cumulates the rare-earths would be dominated by the pyroxene contribution; this may explain why the enriched igneous rock type has more nearly chondritic relative rare-earth concentrations than the low trace-element type of igneous rock. Sr and to a lesser extent K and Ba are contributed mostly by the feldspar. This is also illustrated by the Moore Country achondrite (No. 6, Fig. 3) which we believe to be a cumulate from a liquid of normal brecciated achondrite composition. If the low trace-element type of igneous rock represents cumulates initially having various proportions of minerals and liquid, then this might explain the range in the trace-element content of this type compared to the homogeneity of the enriched type (liquid). A similar genetic model would be derivation of the trace-element rich rock by lesser partial fusion, and the trace-element poor rock by greater partial fusion, of a preexisting rock, probably a cumulate.

Finally, although it is not possible to make firm *a priori* assumptions concerning the characteristics of the source material for trace elements other than the rare-earths, it is of interest to consider how differentiation involving feldspar might effect the concentrations of these other elements so that potential source materials can be identified. Partition coefficients of K, Rb, Sr and Ba for plagioclase and clinopyroxene have been published (PHILPOTTS and SCHNETZLER, 1970b); there is no data available on ilmenite coefficients but approximate values for these may be obtained from the mineral data on Apollo 11 rocks. Also there are examples of other rocks, such as dredge lavas (PHILPOTTS *et al.*, 1970) and the basaltic achondrites (Fig. 3; SCHNETZLER and PHILPOTTS, 1969), whose trace-element concentrations have been considered in terms of plagioclase crystallization. These data indicate that igneous differentiation, especially essentially anhydrous differentiation, is almost certainly not responsible for the low relative concentrations of K and Rb in the lunar samples. The low alkali contents must be characteristic of the source material for the igneous rocks and probably also of the bulk moon. A likely candidate for this source material is the basaltic achondrites, or their source materials, or markedly similar materials. It is of interest that a lunar origin for the basaltic achondrite meteorites has been proposed (DUKE and SILVER, 1967) and the major element chemistry of lunar rocks and these meteorites have many similarities (DUKE, 1969). Also experimental evidence (RINGWOOD and ESSENE, 1970) is not inconsistent with this hypothesis. If the basaltic achondrites are not lunar, and if the moon and the earth did not originate together, then it appears that remarkably similar processes have occurred in several places within the solar system.

SUMMARY AND CONCLUSIONS

(1) Apollo 11 igneous rocks, breccias, and soil all have roughly similar concentrations of K, Rb, Sr, Ba and rare-earth elements.

(2) The trace-element characteristics of the lunar samples are very different from those of chondrites, tektites and the terrestrial continental crust; they are similar in many ways to the characteristics of terrestrial dredge lavas, the bulk earth, and basaltic achondrite meteorites.

(3) The Apollo 11 samples fall into three chemically distinguishable groups: (a) igneous rocks rich in some trace elements such as K and Rb, (b) igneous rocks low in these trace elements, (c) the breccias and soil.

(4) Mineral data for Apollo 11 rocks indicate higher Eu^{2+}/Eu^{3+} than in terrestrial basalts and this implies more reducing conditions.

(5) The large Eu deficiencies in the Apollo 11 rocks indicate loss of feldspar during extensive fractional crystallization or limited partial fusion of feldspathic source rock at relatively shallow depths. The Apollo 11 rocks cannot be considered primary material.

(6) It is suggested that the trace-element rich igneous rocks may represent a liquid whereas the trace-element poor igneous rock may represent a partial cumulate. Alternatively these rocks might result from lesser and greater amounts of partial fusion, respectively, of a preexisting rock, probably a cumulate.

(7) The low K and Rb contents of the igneous rocks are characteristic of their source material.

(8) A community of origin for the earth, the moon and the basaltic achondrites is indicated.

Acknowledgments—We thank NASA for providing the samples, C. W. KOUNS for the phase separations, S. SCHUHMANN, P. SHADID and F. WOOD for proficient analytical assistance, and all those who have suffered our company or lack thereof during the past several months.

Note added in proof:

N. J. HUBBARD (personal communication, February, 1970) has suggested that our Nd and Sm standard powders might contain significant CO_2. Dr. S. R. HART, Carnegie Institution, Washington, D.C., has kindly performed gas chromatographic analyses of these powders and finds 5–15% of CO_2 plus H_2O (110–1050°C) in general agreement with observed weight loss (110°–1000°C). Thus it appears that our Nd and Sm values in ppm are about 10% high. Normalized values are not affected.

References

Adler I., Walter L. S., Lowman P. D., Glass B. P., French B. M., Philpotts J. A., Heinrich K. J. F. and Goldstein J. I. (1970) Electron microprobe analysis of lunar samples. *Science* **167,** 590–592.

Agrell S. O., Scoon J. H., Muir I. D., Long J. V. P., McConnell J. D. C. and Peckett A. (1970) Mineralogy and petrology of some lunar samples. *Science* **167,** 583–586.

Albee A. L., Burnett D. S., Chodos A. A., Eugster O. J., Huneke J. C., Papanastassiou D. A., Podosek F. A., Russ G. Price II, Sanz H. G., Tera F. and Wasserburg G. J. (1970) Ages, irradiation history and chemical composition of lunar rocks from the Sea of Tranquillity. *Science* **167,** 463–466.

Anderson A. T., Jr., Crewe A. V., Goldsmith J. R., Moore P. B., Newton J. C., Olsen E. J., Smith J. V. and Wyllie P. J. (1970) Petrologic history of moon suggested by petrography, mineralogy, crystallography. *Science* **167,** 587–590.

Compston W., Arriens P. A., Vernon M. J. and Chappell B. W. (1970) Rubidium–strontium chronology and chemistry of lunar material. *Science* **167,** 474–476.

Duke M. B. and Silver L. T. (1967) Petrology of eucrites, howardites and mesosiderites. *Geochim. Cosmochim. Acta* **31,** 1637–1665.

Duke M. B. (1969) Surveyor alpha-scattering data: consistency with lunar origin of eucrites and howardites. *Science* **165,** 515–517.

Gast P. W. (1965) Terrestrial ratio of potassium to rubidium and the composition of the Earth's mantle. *Science* **147,** 858–860.

Gast P. W. and Hubbard N. J. (1970) The abundance of alkali metals alkaline and rare-earths and strontium-87/strontium-86 ratios in lunar samples. *Science* **167,** 485–487.

Goles G. G., Osawa M., Randle K., Beyer R. L., Jerome D. Y., Lindstrom D. J., Martin M. R., McKay S. M. and Steinborn T. L. (1970) Instrumental neutron activation analyses of lunar specimens. *Science* **167,** 497–499.

Gopalan K., Kaushal S., Lee-Hu C. and Wetherill G. W. (1970) Rubidium–strontium, uranium, and thorium–lead dating of lunar material. *Science* **167,** 471–473.

Haskin L. A., Frey F. A., Schmitt R. A. and Smith R. H. (1966) Meteoritic, solar and terrestrial rare-earth distributions. In *Physics and Chemistry of the Earth,* (editors L. H. Ahrens, F. Press, K. Rankama and S. K. Runcorn), Vol. 7, pp. 167–321. Pergamon.

Haskin L. A., Helmke P. A. and Allen R. O. (1970) Rare-earth elements in returned lunar samples. *Science* **167,** 487–489.

Hurley P. M. (1968) Absolute abundance and distribution of Rb, K and Sr in the earth. *Geochim. Cosmochim. Acta* **32,** 273–283.

Kushiro I., Nakamura Y., Haramura H. and Akimoto S. (1970) Crystallization of some lunar mafic magmas and generation of rhyolitic liquids. *Science* **167,** 610–612.

LSPET (Lunar Sample Preliminary Examination Team) (1969) Preliminary examination of lunar samples from Apollo 11. *Science* **165,** 1211–1227.

Murthy V. R., Schmitt R. A. and Rey P. (1970) The Rb–Sr age and the elemental and isotopic abundances of some trace elements in lunar samples. *Science* **167,** 476–479.

O'Hara M. J., Biggar G. M. and Richardson S. W. (1970) Experimental petrology of lunar material; the nature of mascons, seas and the lunar interior. *Science* **167,** 605–607.

Patterson J. H., Franzgrote E. J., Turkevich A. L., Anderson W. A., Economou T. E., Griffin H. E., Grotch S. L. and Sowinski K. P. (1969) Alpha-scattering experiment on Surveyor 7: Comparison with Surveyors 5 and 6. *J. Geophys. Res.* **74,** 6120–6148.

Philpotts J. A. and Schnetzler C. C. (1968) Europium anomalies and the genesis of basalt. *Chem. Geol.* **3,** 5–13.

Philpotts J. A. and Schnetzler C. C. (1969) Europium anomalies and the genesis of basalt: a reply. *Chem. Geol.* **4,** 464–465.

Philpotts J. A. and Schnetzler C. C. (1970a) Potassium, rubidium, strontium, barium, and rare-earth concentrations in lunar rocks and separated phases. *Science* **167,** 493–495.

PHILPOTTS J. A. and SCHNETZLER C. C. (1970b) Phenocryst-matrix partition coefficients for K, Rb, Sr and Ba, with applications to anorthosite and basalt genesis. *Geochim. Cosmochim. Acta* **34,** 307–322.

PHILPOTTS J. A., SCHNETZLER C. C. and HART S. R. (1969) Submarine basalts: some K, Rb, Sr, Ba rare-earth, H_2O and CO_2 data bearing on their alteration, modification by plagioclase, and possible source materials. *Earth Planet. Sci. Lett.* **7,** 293–299.

RAMDOHR P. and EL GORESY A. (1970) The opaque minerals of the lunar rocks and dust from Mare Tranquillitatis. *Science* **167,** 615–618.

RINGWOOD A. E. and ESSENE E. (1970) Petrogenesis of lunar basalts, internal constitution and origin of the moon. *Science* **167,** 607–609.

SCHMITT R. A., SMITH R. H. and OLEHY D. A. (1964) Rare-earth, yttrium and scandium abundances in meteoritic and terrestrial matter II. *Geochim. Cosmochim. Acta* **28,** 67–86.

SCHMITT R. A., WAKITA H. and REY P. (1970) Abundances of 30 elements in lunar rocks, soil and core samples. *Science* **167,** 512–514.

SCHNETZLER C. C., THOMAS H. H. and PHILPOTTS J. A. (1967a) The determination of barium in G-1 and W-1 by isotope dilution. *Geochim. Cosmochim. Acta* **31,** 95–96.

SCHNETZLER C. C., THOMAS H. H. and PHILPOTTS J. A. (1967b) Determination of rare-earth elements in rocks and minerals by mass spectrometric, stable isotope dilution technique. *Anal. Chem.* **39,** 1888–1890.

SCHNETZLER C. C. and PHILPOTTS J. A. (1968) Partition coefficients of rare-earth elements and barium between igneous matrix material and rock forming mineral phenocrysts—I. In *Origin and Distributions of the Elements*, (editor L. H. Ahrens), pp. 929–938. Pergamon.

SCHNETZLER C. C. and PHILPOTTS J. A. (1969) Genesis of the calcium-rich achondrites in light of rare-earth and barium concentrations. In *Meteorite Research*, (editor P. Millman), pp. 206–216. D. Reidel.

SCHNETZLER C. C. and PHILPOTTS J. A. (1970) Partition coefficients of rare-earth elements between igneous matrix material and rock forming mineral phenocrysts—II. *Geochim. Cosmochim. Acta* **34,** 331–340.

TATSUMOTO M. and ROSHOLT J. N. (1970) Age of the moon: An isotopic study of uranium–thorium–lead systematics of lunar samples. *Science* **167,** 461–463.

TOWELL D. G., WINCHESTER J. W. and SPIRN R. V. (1965). Rare-earth distributions in some rocks and associated minerals of the batholith of southern California. *J. Geophys. Res.* **70,** 3485–3496.

TOWELL D. G., SPIRN R. V. and WINCHESTER J. W. (1969) Europium anomalies and the genesis of basalt: a discussion. *Chem. Geol.* **4,** 461–464.

UREY H. C. (1968) Mascons and the history of the moon. *Science* **162,** 1408–1410.

UREY H. C. (1969) Early temperature history of the moon. *Science* **165,** 1275.

WOOD J. A., DICKEY J. S., JR., MARVIN U. B. and POWELL B. N. (1970) Lunar anorthosites. *Science* **167,** 602–604.

Proceedings of the Apollo 11 Lunar Science Conference, Vol. 2, pp. 1487 to 1492.

Halogens, mercury, lithium and osmium in Apollo 11 samples*

G. W. REED, JR. and S. JOVANOVIC
Argonne National Laboratory, Argonne, Illinois 60439

(*Received* 2 *February* 1970; *accepted in revised form* 21 *February* 1970)

Abstract—Fluorine, Cl, Br and I in Apollo 11 samples range from 140 to 350, 2·9 to 16, 0·04 to 0·4 and 0·016 to 0·62 ppm, respectively. F/Cl, Cl/Br and Br/I ratios cluster about 22, 51 and 3, respectively, with a few notable exceptions. Some of the fluorine is associated with apatite. Similar concentrations of Cl, Br or I in crystalline rocks and soil and the low Cl/Br ratios relative to cosmic (~50 vs. 400) make it unlikely that these elements are derived from the solar wind. Hg volatilized from samples by stepwise heating exhibited a low-temperature fraction unlike that seen in any meteoritic or terrestrial material. This appears to be surface adsorbed Hg. This component constitutes 25–45 per cent of the total Hg (10^{-9} g/g) in soil and crystalline rocks alike; thus even the rocks must be sufficiently porous so that their interiors are accessible to volatiles on the lunar surface. Lithium concentrations are uniform at about 11 ppm. Osmium varies from 0·4 to 300 ppb. Te is estimated at a few tenths ppm.

INTRODUCTION

OUR OBJECTIVES in this paper are to report those data we have been able to reevaluate and to present the few new results not available earlier. For many trace elements a detailed comparison of the data presented at the Apollo 11 Lunar Science Conference, January 1970, is not warranted at this time but areas of agreement and disagreement will be pointed out.

MEASUREMENTS AND DISCUSSION

The samples measured consisted of soil, LRL crushed and homogenized medium grained rocks, vesicular rock, vuggy rock with ilmenite platelets and microbreccia. The last three were interior rock chips from which we removed and crushed small fragments. All sample handling was confined to a glove box with a nitrogen atmosphere.

Measurements of the amounts of F, Cl, Br, I, Hg, Os, Li and Te in several types of lunar material are presented. All elements were determined by neutron activation except F which was measured separately by photon activation. In contrast to our previous F measurements (REED and JOVANOVIC, 1969), the latter irradiations were monitored by polyvinylfluoride foil sandwiches surrounding samples sealed in fused silica vials and rotated in the photon beam (WILKNISS and LINNENBOOM, 1968). Chemical procedures are described in REED (1964) and REED and ALLEN (1966).

Cl, Br, I, U, Te and Li were measured in one 100–200 mg portion of a sample; Hg, Os, Ru, Cr, Sc and F were measured in another aliquot. However, most F measurements reported were obtained on separate samples. U and Te were by-products of the I measurements. The production of F^{18} from neutron irradiation

* Work supported by USAEC and by NASA contract T-76356.

of Li in silicates (QUIJANO-RICO and WÄNKE, 1969) was the basis of our Li determinations. Hg was volatilized from samples in a closed system. Complete release of Hg is assumed because all the samples were found to be fused after heating to 1200°C (REED and JOVANOVIC, 1967). U, Sc, Cr and Ru results are not included in this report.

Standard counting errors are given in the tables, those from other sources are considered negligible. In the case of the I determination larger errors were estimated (see below).

The data from the halogen measurements are summarized in Table 1. In some experiments the samples were leached for 10 min with boiling water, and the separate results are indicated. Up to 80 per cent of the Cl and Br may be in water soluble

Table 1. Halogen and lithium concentrations in Apollo 11 samples*

Sample	F_{ppm}	Cl_{ppm}	Br_{ppm}	I_{ppb}	Li_{ppm}
Soil 10084-2	144 ± 7	2·9 ± 0·4	0·038 ± 0·011		1·9 ± 0·2
10084-2		$1{\cdot}8_l \pm 0{\cdot}3$	$0{\cdot}079_l \pm 0{\cdot}004$	$19_l \pm 1$	7·2 ± 1·2†
		$9{\cdot}9_s \pm 0{\cdot}7$	$0{\cdot}15_s \pm 0{\cdot}02$	$608_s \pm 190$	
Microbreccia 10061-22	342 ± 60	9·6 ± 1·5	0·15 ± 0·02		4·0 ± 0·5
10061-22		$0{\cdot}74_l \pm 0{\cdot}07$	$0{\cdot}01_l \pm 0{\cdot}01$		
		$6{\cdot}8_s \pm 0{\cdot}7$	$0{\cdot}25_s \pm 0{\cdot}03$		
10019-29		$6{\cdot}4_l \pm 0{\cdot}6$	$0{\cdot}14_l \pm 0{\cdot}03$	$2{\cdot}7_l\ ^{+0{\cdot}3}_{-0{\cdot}9}$	$0{\cdot}14_l \pm 0{\cdot}03$
		$9{\cdot}4_s \pm 0{\cdot}9$	$0{\cdot}24_s \pm 0{\cdot}05$	$71_s \pm 14$	$13_s \pm 2$
Medium grained rock 10044-48	202 ± 7	$12{\cdot}1_l \pm 0{\cdot}1$	$0{\cdot}12_l \pm 0{\cdot}01$	$10_l \pm 2$	9·5 ± 2†
		$2{\cdot}6_s \pm 0{\cdot}2$	$0{\cdot}070_s \pm 0{\cdot}014$	$470_s \pm 157$	
Medium grained rock 10047-48	$28_l \pm 1$	$4{\cdot}8_l \pm 0{\cdot}5$	$0{\cdot}22_l \pm 0{\cdot}04$	$11{\cdot}6_l \pm 1{\cdot}2$	$0{\cdot}31_l \pm 0{\cdot}07$
	$165_s \pm 3$	$9{\cdot}6_s \pm 0{\cdot}1$	$0{\cdot}11_s \pm 0{\cdot}02$	$4{\cdot}9_s \pm 4{\cdot}9$	$16_s \pm 2$
Vesicular rock 10017-22	251 ± 42	15·0 ± 0·1	0·077 ± 0·007	480 ± 140	17 ± 3
Vuggy rock 10072-24	271 ± 62	14·0 ± 0·1	0·070 ± 0·006	370 ± 110	

* Except for F, all data on a line are for a single aliquot; subscripts *l* and *s* refer to water extractable and nonextractable portions, respectively.

† The leach solutions in these experiments were not measured but little of the Li appears to be water soluble.

phases. Duplicate measurements were made in a few cases. Cl and Br in the microbreccia sample are in reasonable agreement; Cl, Br and Li in two soil aliquots disagree with each other by an almost constant factor. Sample inhomogeneity would seem to be a reasonable explanation.

WÄNKE *et al.* (1970) report Cl concentrations similar to those given here but MORRISON *et al.* (1970) obtained results up to an order of magnitude greater. KEAYS' *et al.* (1970) values and MORRISON'S *et al.* (1970) limits are consistent with the Br concentrations reported here.

The average F/Cl ratio is about 22. The vesicular and vuggy rocks have Cl/Br ratios 3·5 times greater than the ratio of about 51 for the other samples.

The measurement of I is complicated because the beta decay curves contain contributions from the many fission product I radioactivities. These dominate the decay because of the high U content of Apollo 11 samples. We have attempted to test the validity of our resolution of the decay curves by comparing fission I activities measured in samples and in U monitors. In all the samples, several hundred counts of a 25-min activity attributable to I^{128} were observed.

An average Br/I ratio of about 2·8 is estimated for four of the six samples measured; samples 10019-29 and 10047-48 are exceptions. Too few measurements have been made to apply a criterion for reliability based on interhalogen ratios.

Fluorine concentrations of 140–340 ppm were found; only one sample, 10047-48, was leached with hot H_2O and 15 per cent of the F was soluble. Our F results are not in agreement with the single soil value of 74 ppm of SMALES *et al.* (1970) or with most of the concentrations reported by MORRISON *et al.* (1970). The latter reports 66 ppm for the soil in agreement with SMALES. A possible explanation for our higher results may be contamination from the teflon bags the samples were returned in. Our interior rock fragments should be relatively free of such contamination and yet they are at least as high as the soil.

The mineral fluorapatite is a common site for F in terrestrial rocks but has never been found in meteorites. It has been observed in several moon rocks, but the bulk of the fluorine is not in fluorapatite. A rough estimate can be made of the amount of F present as fluorapatite in the medium-grained rock 10044-48 in which apatite has been reported most frequently. Apatite constitutes approximately 5 per cent of the volume of the pyroxferroite grains, thus only about 15 ppm F can be accounted for if pyroxferroite constitutes 1 per cent of the rock (FUCHS, 1970). ALBEE *et al.* (1970), ADLER *et al.* (1970) and KEIL *et al.* (1970) report chlorapatite in several samples and this is the site of some of the chlorine observed in this work.

Chlorine and Br contents are so low that they might be accounted for as being derived from the solar wind over the exposure age ($\sim 10^8$ yr) of the surface material. The presence of similar concentrations of Cl and Br in the soil, breccia, and crystalline rocks and the fact that the ratios of leachable (and unleachable) Cl and Br in these materials are much lower than the cosmic ratio ($\sim$50 vs. 400) make it difficult to attribute a significant amount of the halogen contents to a solar wind source.

Hg data are summarized in Table 2 and Fig. 1. The Hg concentrations range from 0·6 to 13 ppb, the largest being associated with the vesicular rock. These concentrations are significantly lower than those in basaltic achondrites (18–9000 ppb) (REED and JOVANOVIC, unpublished) or in chondrites and approach more nearly values observed in some terrestrial rocks (JOVANOVIC and REED, 1968). We were

Table 2. Mercury and osmium concentrations (ppb) in Apollo 11 samples

Sample	Hg_{Total}	$Hg_{>450°C}$	Os
Soil 10084-2	5·3 ± 0·1	0·06 ± 0·01	283 ± 63
10084-1	0·95 ± 0·04	0·22 ± 0·02	0·41 ± 0·26
Microbreccia			
10061-22	1·2 ± 0·1	0·042 ± 0·002	
Medium grained rock			
10044-48	0·60 ± 0·03	0·028 ± 0·003	
Vuggy rock			
10072-24	5·5 ± 0·1	1·1 ± 0·1	4 ± 0·7
Vesicular rock			
10017-22	13 ± 2	0·09 ± 0·01	220 ± 66

assured by members of LRL that their systems were not exposed to Hg vapor. The Hg release patterns from the lunar samples were significantly different from patterns observed in meteorites and terrestrial rocks (JOVANOVIC and REED, 1968): a large release of 25–45 per cent of the total Hg is observed at low temperature (<110°C) and the maximum in the release is shifted from 150°C to 250°C. During isothermal heating at 250°C an initial burst occurs containing 50–80 per cent of the Hg eventually evolved as noted previously for meteorites and terrestrial samples. The large 110°C

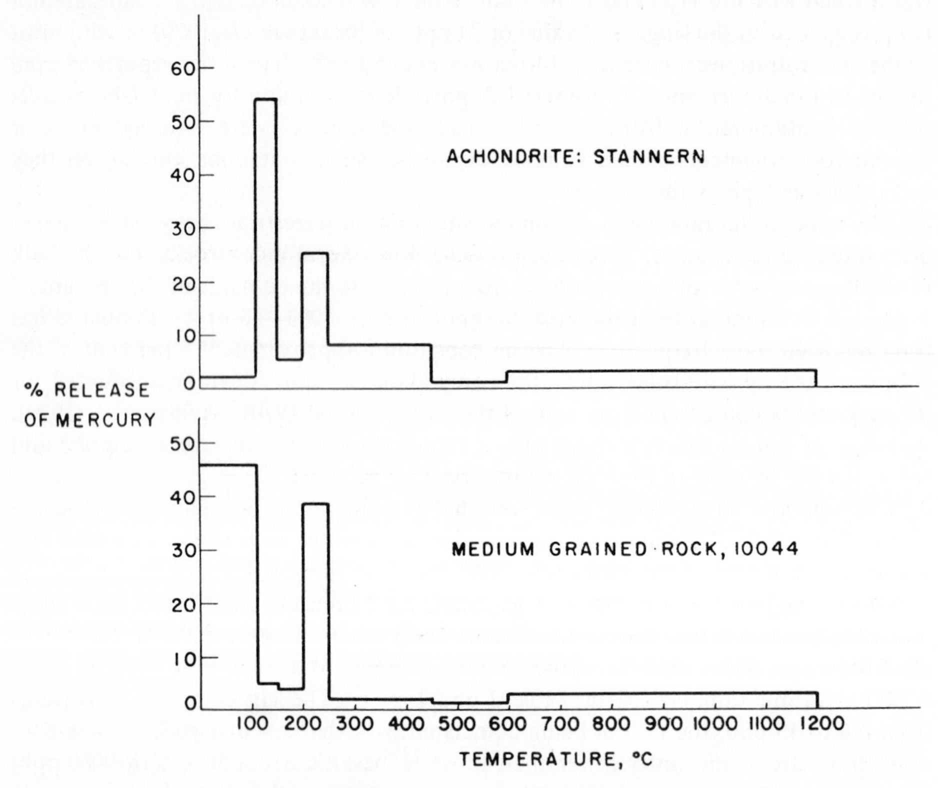

Fig. 1. Hg release in stepwise heating of lunar and meteoritic rocks.

fraction might be explained as free Hg adsorbed on surfaces. It may be Hg released from low temperature (<150°C) traps during thermal cycling on the lunar surface or Hg released during the degradation of surface material. The only deviation from the trend noted is in the soil sample, 10084-1, which yielded a release pattern similar to meteorites and terrestrial rocks and which also gave the highest relative amount of Hg released above 450°C. Some of the low temperature Hg must constitute a tenuous atmosphere on the sunlight side of the moon. The fraction volatilized will depend on the absorbency of the surfaces.

The release trends are consistent with the following interpretation. The most degraded material, the soil 10084-2, has the largest low temperature fraction, followed by the microbreccia. The fine grain rock, 10072-24, has a greater amount of its Hg in high temperature traps. The medium grained igneous rock occupies an intermediate position possibly because of slower cooling to lower temperatures. The relatively large low temperature release for the vesicular rock (10017-22) can be rationalized as resulting from Hg associated with the vapor phase that formed the vesicules and trapped in the glass-like lining of the voids. Subtraction of the amount of Hg released below 110°C does not alter the trends noted.

The Os results reported previously have been revised because of a monitor mix-up. The corrected concentrations are listed in Table 2. Concentrations range from 0·4 to 300 ppb. Samples 10072-24, 10017-22 and 10084-1 were measured in the same irradiation and the data should reliably reflect the relative amounts of Os present. At the Apollo 11 Lunar Science Conference Butterfield and Lovering reported Os contents of 0·022, 0·88 and 1·2 ppb for two type A volcanic rocks and a breccia, respectively.

The Os concentrations are lower than most values reported for chondritic and iron meteorites [0·3– $>$ 1 ppm, HIRT *et al.* (1963) and MORGAN and LOVERING (1967)]. They are higher than a Hawaiian basalt or W-1 and tend to be higher than the ~7·9 ppb reported for the achondrite Nuevo Laredo (BATE and HUIZENGA, 1963). The Os results could be pertinent to attempts to monitor the infall of meteoritic material on the moon by way of the platinum metals (BARKER and ANDERS, 1968). If these preliminary data are verified, the similar value in an igneous rock and a soil may imply that the Os is of lunar origin.

Tellerium was measured in the I samples via β–γ coincidence counting of the I^{131} 0·36 MeV γ-ray. The contribution from fission produced I^{131} was large and could be corrected for in only one irradiation. The Te contents of 10084-2 and 10044-48 are calculated to be 0·8 and 0·3 ppm, respectively. The leach solutions from these samples contained too little I^{131} to be resolved, in marked contrast to the case for meteorites. Since other samples gave negative results the above numbers are considered to be estimates only. The Li contents range from 2 to 17 ppm. A value of 13·1 ppm in W-1, not listed in Table 1, was obtained in this work and is in good agreement with literature values (QUIJANO-RICO and WÄNKE, 1969). The lunar sample Li contents appear to be similar to or lower than those in achondrites (QUIJANO-RICO and WÄNKE, 1969) and gabbroic rocks. Similar Li contents have been reported by LSPET (1969), MORRISON *et al.* (1970), ANNELL and HELZ (1970) and WÄNKE *et al.* (1970).

Acknowledgments—We would like to thank the Chemistry Department and the operating staff of the High Flux Reactor, BNL, for making their facilities available. We are especially grateful to J. HUDIS and Mrs. E. ROWLAND, BNL, for their assistance. Carrier and monitor analyses performed by K. JENSEN, R. BANE and Miss I. FOX, ANL, are gratefully acknowledged. The cooperation of the ANL Linac staff is greatly appreciated.

REFERENCES

ADLER I., WALTER L. S., LOWMAN P. D., GLASS B. P., FRENCH B. M., PHILPOTTS J. A., HEINRICH K. J. F. and GOLDSTEIN J. I. (1970) Electron microprobe analysis of lunar samples. *Science* **167,** 590–592.

ALBEE A. L., BURNETT D. S., CHODOS A. A., EUGSTER O. J., HUNEKE J. C., PAPANASTASSIOU D. A., PODOSEK F. A., RUSS G. PRICE II, SANZ H. G., TERA F. and WASSERBURG G. J. (1970) Ages, irradiation history, and chemical composition of lunar rocks from the Sea of Tranquillity. *Science* **167,** 463–466.

ANNELL C. and HELZ A. (1970) Emission spectrographic determination of trace elements in lunar samples. *Science* **167,** 521–523.

BARKER J. L. and ANDERS E. (1968) Accretion rate of cosmic matter from iridium and osmium contents of deep-sea sediments. *Geochim. Cosmochim. Acta* **32,** 627–645.

BATE G. L. and HUIZENGA J. R. (1963) Abundances of ruthenium, osmium and uranium in some cosmic and terrestrial sources. *Geochim. Cosmochim. Acta* **27,** 345–360.

FUCHS L. H. (1970) Fluorapatite and other accessory minerals in Apollo 11 rocks. *Geochim. Cosmochim. Acta*, Supplement I.

HIRT B., HERR W. and HOFFMEISTER W. (1963) Age determinations by the rhenium-osmium method. Radioactive Dating, pp. 35–43. IAEA, Vienna.

JOVANOVIC S. and REED G. W., JR. (1968) Hg in metamorphic rocks. *Geochim. Cosmochim. Acta* **32,** 341–346.

KEAYS R. R., GANAPATHY R., LAUL J. C., ANDERS E., HERZOG G. F. and JEFFERY P. M. (1970) Trace elements and radioactivity in lunar rocks: implications for meteorite infall, solar-wind flux and formation conditions of moon. *Science* **167,** 490–493.

KEIL K., PRINZ M. and BUNCH T. E. (1970) Mineral chemistry of lunar samples. *Science* **167,** 597–599.

LSPET (LUNAR SAMPLE PRELIMINARY EXAMINATION TEAM) (1969) Preliminary examination of lunar samples from Apollo 11. *Science* **165,** 1211–1227.

MORGAN J. W. and LOVERING J. F. (1967) Rhenium and osmium abundances in chondritic meteorites. *Geochim. Cosmochim. Acta* **31,** 1893–1909.

MORRISON G. H., GERARD J. T., KASHUBA A. T., GANGADHARAM E. V., ROTHENBERG A. M., POTTER N. M. and MILLER G. B. (1970) Multielement analysis of lunar soil and rocks. *Science* **167,** 505–507.

QUIJANO-RICO M. and WÄNKE H. (1969) Determination of boron, lithium and chlorine in meteorites. In *Meteorite Research*, (editor P. M. Millman), p. 132. D. Reisel.

REED G. W., JR. (1964) Fluorine in stone meteorites. *Geochim. Cosmochim. Acta* **28,** 1729.

REED G. W., JR. and ALLEN R. O. (1966) Halogens in chondrites. *Geochim. Cosmochim. Acta* **30,** 779.

REED G. W., JR. and JOVANOVIC S. (1969) Some halogen measurements on achondrities. *Earth Planet. Sci. Lett.* **6,** 316–320.

REED G. W., JR. and JOVANOVIC S. (1967) Hg in chondrites. *J. Geophys. Res.* **72,** 2219–2228.

SMALES A. A., MAPPER D., WEBB M. S. W., WEBSTER R. K. and WILSON J. D. (1970) Elemental composition of lunar surface material. *Science* **167,** 509–512.

WÄNKE H., BEGEMANN F., VILCSEK E., RIEDER R., TESCHKE F., BORN W., QUIJANO-RICO M., VOSHAGE H. and WLOTZKA F. (1970) Major and trace elements and cosmic-ray produced radioisotopes in lunar samples. *Science* **167,** 523–525.

WILKNISS P. and LINNENBOOM V. J. (1968) Use of the Naval Research Laboratory 60-Mev linac for photon activation analysis. *Proc. 2nd Conf. on Practical Aspects of Activation Analysis with Charged Particles*, (editor H. G. Ebert), p. 147–160. EURATOM.

Proceedings of the Apollo 11 Lunar Science Conference, Vol. 2, pp. 1493 to 1497.

Semimicro X-ray fluorescence analysis of lunar samples*

HARRY J. ROSE, JR., FRANK CUTTITTA, EDWARD J. DWORNIK, M. K. CARRON, R. P. CHRISTIAN, J. R. LINDSAY, D. T. LIGON and R. R. LARSON
U.S. Geological Survey, Washington, D.C. 20242

(*Received* 19 *January* 1970; *accepted* 23 *January* 1970)

Abstract—Major and selected minor elements were determined in seven whole rock fragments, five portions of pulverized rock and the lunar soil collected at the Apollo 11 Tranquillity Base. Three different rock types were represented: vesicular, fine-grained basalt; medium to coarse-grained, vuggy gabbroic rocks; and breccia. The ranges (in wt.%) for the major constituents are SiO_2 (38–42), Al_2O_3 (8–14), total iron as FeO (15–20), MgO (6–8), CaO (10–12), Na_2O (0·5–0·9), K_2O (0·05–0·4), total titanium as TiO_2 (8–13), MnO (0·2–0·3); and Cr_2O_3 (0·2–0·4). The high reducing capacity of the samples strongly suggests the presence of Ti (III).

INTRODUCTION

THE RETURNED lunar materials were divided by LSPET (1969) into four groups—type A, fine grained vesicular crystalline igneous rock; type B, medium grained crystalline igneous rock; type C, breccia; type D, fines less than 1 cm in size. All types were represented among the thirteen received for analysis, although the lunar fines were less than 1 mm in size. The rocks are characterized by similar mineral assemblages in varying proportion—clinopyroxene, plagioclase and ilmenite are common to all types. An abundance of glasses is found in the fines and some glass spherules are noted in the breccia. Preliminary semi-quantitative examination LSPET (1969) showed an overall similarity in chemical composition of samples. The quantitative results reported here indicate some variations between rock types.

ANALYTICAL METHODS

The samples were weighed, photographed and subjected to thorough mineralogical examination under clean room conditions. Each whole rock sample was carefully disaggregated in a boron carbide mortar in preparation for X-ray fluorescence and chemical analysis.

A. *Semimicro X-ray fluorescence analysis*

A simple fusion, ROSE *et al.* (1963), with no heavy absorber was made with 60 mg of the sample in 940 mg of $Li_2B_4O_7$ at 1100°C for 15 min. After cooling, the fused bead was brought to 1200 mg with powdered cellulose and then was ground to less than 350 mesh.

Three hundred milligrams of the ground mixture was pressed into a one-half inch dia. pellet backed with cellulose powder. Na, Mg, Al and Si were determined with a soft X-ray spectrometer and P, K, Ca, Ti, Mn, Fe, Cr, Zr and Ni with an X-ray milliprobe (ROSE *et al.*, 1969)

The remainder of the ground mixture was pressed into pellets one-inch in dia. with cellulose powder backing. Al, Si, P, K, Ca, Ti, Cr, Mn, Fe, Ni and Zr were determined with conventional air-path and vacuum spectrometers.

B. *Semimicro chemical analysis*

Chemical determinations of H_2O^- were made by dehydration of the sample at 110°C for 1 hr. Na_2O and MgO were determined by atomic absorption. SiO_2 was determined spectrophotometrically

* Publication authorized by the Director, U.S. Geological Survey.

Table 1. Chemical composition of lunar igneous rocks (in wt.%)

Constituent / Rock No.	10003	10022	10024	10047	10049	10050	10058*	10058†	10062
SiO_2	37·8	40·1	39·0	41·3	41·0	40·9	41·4	41·7	38·8
Al_2O_3	11·0	8·60	9·50	9·80	9·50	8·90	10·7	11·8	12·1
Fe_2O_3	0·00	0·00	0·00	0·00	0·00	0·00	0·00	0·00	0·00
FeO	19·8	18·9	18·5	19·0	18·7	17·3	17·3	18·2	18·3
MgO	7·20	7·74	8·11	6·10	7·03	8·03	6·25	6·30	7·21
CaO	11·0	10·7	10·0	12·2	11·0	11·3	12·1	11·0	12·0
Na_2O	0·85	0·91	0·80	0·65	0·71	0·66	0·79	0·68	0·69
K_2O	0·05	0·30	0·28	0·11	0·36	0·05	0·07	0·09	0·07
H_2O^-	0·00	0·00	0·00	0·00	0·00	0·00	0·00	0·00	0·00
Total Ti as TiO_2	12·0	12·2	13·2	10·2	11·3	12·6	11·1	9·55	10·3
P_2O_5	<0·2	<0·2	<0·2	<0·2	<0·2	<0·2	<0·2	<0·2	<0·2
MnO	0·29	0·25	0·24	0·29	0·25	0·27	0·27	0·27	0·27
Cr_2O_3	0·26	0·37	0·40	0·22	0·32	0·35	0·21	0·24	0·25
ZrO_2	<0·03	0·03	<0·03	<0·03	0·03	<0·03	<0·03	<0·03	<0·03
NiO	<0·001	<0·001	<0·001	<0·001	<0·001	<0·001	<0·001	<0·001	<0·001
Total	100·25	100·10	100·03	99·87	100·20	100·36	100·19	99·83	99·99
Total Fe as Fe_2O_3	22·0	21·0	20·5	21·1	20·8	19·2	19·2	20·2	20·3
R_2O_3	45·8	42·6	44·1	41·7	42·0	41·4	41·4	43·1	43·6

* Whole rock.
† Pulverized.

using molybdenum blue. R_2O_3 determinations were made by triple precipitation with NH_4OH; Al_2O_3 was obtained by difference.

The total reducing capacity of the samples was determined by dissolution in the presence of excess NH_4VO_3 and subsequent back titration of excess vanadium (V). FeO was determined on the lunar fines (sample 311079) after catalytic oxidation.

RESULTS AND DISCUSSION

The results of the analysis are given in Tables 1 and 2. The values are the averages of all determinations. The samples are grouped in numerical sequence according to rock type.

Table 2. Chemical composition of lunar breccia and soil fines (in wt.%)

Constituent / Rock No.	10019	10048	10060	311079
SiO_2	41·1	42·2	41·5	42·2
Al_2O_3	13·7	12·9	11·8	14·1
Fe_2O_3	0·00	0·00	0·00	0·00
FeO	15·7	15·7	17·0	15·3
MgO	7·86	7·54	7·52	7·94
CaO	11·9	11·4	11·6	12·1
Na_2O	0·93	0·52	0·78	0·54
K_2O	0·14	0·17	0·18	0·14
H_2O^-	0·00	0·00	0·00	0·00
Total Ti as TiO_2	8·25	8·95	9·15	7·60
P_2O_5	<0·2	<0·2	<0·2	<0·2
MnO	0·22	0·22	0·23	0·21
Cr_2O_3	0·32	0·31	0·33	0·31
ZrO_2	0·03	<0·03	0·04	0·03
NiO	0·02	0·02	0·02	0·02
Total	100·15	99·91	100·13	100·47
Total Fe as Fe_2O_3	17·4	17·4	18·9	17·0
R_2O_3	40·3	40·0	40·5	39·3

The igneous rock types, basalt (10022) and gabbro, appear in Table 1. Samples 10047 and 10049 were pulverized at the Lunar Receiving Laboratory, while the remaining samples were whole rock fragments. Sample 10058 was received both as a whole rock fragment and as pulverized material. The heterogeneity of the original specimen is apparent when comparing their chemical compositions.

The three breccia samples and the lunar fines (311079) are given in Table 2. Sample 10019 was a whole rock fragment and 10048 and 10060 were pulverized.

The data show an overall similarity of the major constituents and is in agreement with the semiquantitative results reported in the preliminary investigations at the Lunar Receiving Laboratory (LSPET, 1969).

Table 3. Comparison of range and average values* (wt.%) of igneous type rocks with breccias and lunar fines

Element	Basalt and gabbro		Breccia and lunar fines	
oxide	Range	Average	Range	Average
SiO_2	37·8–41·7	40·2	41·1–42.2	41·8
Al_2O_3	8·6–12·1	10·2	11·8–14·1	13·1
FeO	17·3–19·8	18·4	15·3–17·0	15·9
MgO	6·1–8·1	7·1	7·5–7·9	7·7
CaO	10·0–12·2	11·2	11·4–12·1	11·8
Na_2O	0·65–0·91	0·75	0·52–0·93	0·69
K_2O	0·05–0·36	0·15	0·14–0·18	0·16
TiO_2	9·55–13·2	11·4	7·60–9·15	8·49
MnO	0·24–0·29	0·27	0·21–0·23	0·22
Cr_2O_3	0·21–0·40	0·29	0·31–0·33	0·32
ZrO_2		$<0·03$		0·03
NiO		$<0·001$		0·02
Total		99·99		100·23
R_2O_3	41·1–45·8	42·8	39·3–40·5	40·0

* P_2O_5 and S $<0·2\%$, H_2O^- 0·00%.

Closer examination of the quantitative data bear out a number of similarities and discernible differences when comparing the igneous basalt and gabbro rocks with the breccia and lunar fines. Table 3 compares the two groups, giving the ranges of composition and the average composition for all the samples analyzed.

Comparison of the data (Table 3) shows that the SiO_2 and Al_2O_3 ranges for the two groups have some overlap but the average in the igneous rocks is much lower than in the breccia and soil. The average FeO and TiO_2 content is considerably higher in the igneous group and the entire range of composition is characteristically higher than the breccia and fines. MgO and CaO averages are higher in the breccia and fines but are of wider range of composition in the igneous rocks. Na_2O has the same range and average in either group and K_2O has the same average value although a much wider range in the igneous group. MnO has a narrow range but is slightly higher in the igneous rocks than in the breccia and fines. Although averaging nearly the same, Cr_2O_3 has a wider range of composition in the igneous rocks. The relatively high NiO and ZrO_2 content characterize the breccia and soil and are of much lower concentration in the igneous rocks. R_2O_3 values are considerably higher in the igneous rocks while the average summation of all constituents is higher in the breccia and fines.

The high summations were at first considered due to the presence of metallic iron because it had been identified in the samples. A determination of the total reducing capacity (TRC) of the samples as shown in Table 4 modified our thinking. The TRC values were considerably higher than the determined values for FeO. The difference could not be explained on the basis of metallic iron alone because the X-ray fluorescence determination is unaffected by the state of the element and the FeO values should approach the TRC values if no other element in a reduced state is present. The rare occurrence of troilite in the samples tends to discount sulfide minerals as a major source of sulfur.

The excessive reducing capacity of the lunar samples may be explained by the presence of an element other than Fe or S that is present as a major constituent capable of existing in a lower valence state and perhaps existing in several states of oxidation.

Table 4. Comparison of total Fe (as FeO) determined by X-ray spectroscopy with total reducing capacity (as FeO) of the lunar materials (in wt.%)

Sample	Rock-type	Total reducing capacity (TRC)	FeO value	TRC-FeO* (ΔTRC)
10003	Gabbro	22·4	19·8	+2·6
10019	Breccia	19·4	15·7	3·7
10022	Basalt	21·0	18·9	2·1
10024	Gabbro	22·3	18·5	3·8
10047	Gabbro	19·8	19·0	0·8
10048	Breccia	20·3	15·7	4·6
10049	Gabbro	21·8	18·7	3·1
10050	Gabbro	19·2	17·3	1·9
10058†	Gabbro	19·3	17·3	2·0
10058‡	Gabbro	20·3	18·2	2·1
10060	Breccia	20·8	17·0	3·8
10062	Gabbro	20·5	18·3	2·2
311079	Fines	19·4	15·3	4·1

* Difference as FeO.
† Whole rock.
‡ Pulverized sample.

The high vaccum, and absence of water on the lunar surface create an ideal reducing environment so that one might postulate the presence of Ti (III) in order to account for the excessive reducing capacity.

Because of the possibility of the presence of Ti (III), the total reducing capacity experiment described in the section on chemical analysis was modified. The experiment was designed to oxidize catalytically any Ti (III), while leaving the ferrous iron unaffected. Two separate determinations of FeO on the soil fines (sample 311079) that gave a total reducing capacity of 19·4 per cent, gave values of 15·26 and 15·35 per cent after catalytic oxidation, essentially identical to the X-ray value of 15·3 per cent. The figures present a strong case for the presence of Ti (III).

The difference between the total reducing capacity and the ferrous iron values (ΔTRC) can be correlated with the rock type. The ΔTRC values for the igneous type rocks have a range of 0·8–3·8 per cent and an average of 2·3 per cent. For the breccia and lunar fines, the range is 3·7–4·6 per cent with an average of 4·0 per cent.

Studies are currently in progress to give credence to the existence of Ti (III). Examination is being made of the L spectra of titanium in the lunar samples and it will be compared with the spectra of titanium in all states of oxidation. Because Ti (III) in the lunar samples can not exceed several per cent, a severe restriction on the intensity of the fine structure may hamper interpretation. Separation of those minerals containing Ti (III) would significantly improve intensity and interpretation. It has been suggested that some of the clinopyroxenes may contain Ti (III), but the Cr–Ti spinel referred to by a number of investigators appears the most likely host mineral. The small size of our samples preclude its concentration in sufficient quantity to permit study of its L spectra.

References

LSPET (Lunar Sample Preliminary Examination Team) (1969) Preliminary examination of lunar samples from Apollo 11. *Science* **165,** 1211–1227.

Rose H. J., Jr., Adler I. and Flanagan F. J. (1963) X-ray fluorescence analysis of the light elements in rocks and minerals. *Appl. Spectrosc.* **17,** 81–85.

Rose H. J., Jr., Christian R. P., Lindsay J. R. and Larson R. R. (1969) Microanalysis with the X-ray milliprobe. *U.S. Geol. Surv. Prof. Paper* **650–B,** 128–135.

Proceedings of the Apollo 11 Lunar Science Conference, Vol. 2, pp. 1499 to 1502.

Isotopic composition of uranium and thorium in Apollo 11 samples*

JOHN N. ROSHOLT and MITSUNOBU TATSUMOTO
U.S. Geological Survey, Denver, Colorado 80225

(*Received* 12 *February* 1970; *accepted in revised form* 24 *February* 1970)

Abstract—The isotopic composition of uranium and thorium was determined by mass spectrometry and alpha spectrometry in eight lunar samples from Apollo 11. The U^{238}/U^{235} ratio in all samples is the same as that for terrestrial uranium within experimental error (137·8 ± 0·3). The U^{234} daughter is in radioactive equilibrium with parent U^{238} in the samples; however, it could not be demonstrated that Th^{230} is in equilibrium with U^{238} in some rock samples as measured by alpha spectrometry. The reasons for the variations in the isotopic composition of thorium are still being sought.

INTRODUCTION

ACCORDING to the nucleosynthesis theory (BURBIDGE *et al.*, 1957), the heavy elements thorium, uranium and plutonium were formed by successive neutron captures in the *r* (rapid) process of synthesis where neutron capture was more rapid than beta decay. Recent measurements of fission tracks (CANTELAUBE *et al.*, 1967) and concentrations of neutron-rich xenon isotopes (WASSERBURG *et al.*, 1969a), in whitlockite from the St. Severin meteorite, provided further evidence that significant quantities of Pu^{244} and I^{129} existed in the early stages of the solar system. From evidence for existence of these radioisotopes, with 82 million year and 17 million year half lives, respectively, WASSERBURG *et al.* (1969b) and HOHENBURG (1969) concluded that complex models are required for nucleosynthesis of solar system material. In both models it is postulated that the *r*-process products occurred in three modes: a large amount of initial production early in the history of the galaxy, followed by a relatively quiescent period of 3–4 billion years, terminated by a 'last-minute' synthesis about 4·8 billion years ago that possibly initiated the separation of the solar system. TATSUMOTO and ROSHOLT (1970) found that lunar fines and breccia from Apollo 11 sites are just slightly younger (4·66 billion years) than the model time of the terminating nucleosynthetic event. Thus, the lunar environment may provide a unique area for sampling the earliest phases of solar system material available that has enough uranium and thorium for routine isotopic analyses.

URANIUM AND THORIUM ANALYSES

Apollo 11 samples used to determine the isotopic compositions of uranium and thorium are described in more detail in an accompanying paper (TATSUMOTO, 1970).

The isotopic ratios of uranium (U^{238}, U^{235}, U^{234}) were determined by mass spectrometry, and the radioactivity ratios of U^{234}/U^{238} were determined by α-particle spectrometry. The radioactivity ratios of thorium isotopes (Th^{232}, Th^{230}, Th^{228}) were determined by α-particle spectrometry on splits of the same sample used to determine concentrations of uranium and thorium. Uncertainties in radioactivity ratios are standard deviations based on counting statistics.

* Publication authorized by the Director, U.S. Geological Survey.

After the separation of lead for determination of the isotopic abundances of lead isotopes (Tatsumoto and Rosholt, 1970), uranium and thorium were separated by anion exchange from 6 N HCl. Uranium was further purified by hexone extraction, and thorium was further purified by anion exchange from 7 N HNO_3 media. Uranium and thorium were electrodeposited, separately, on platinum disks (Rosholt *et al.*, 1966) for radioactivity measurements. Blanks for uranium and thorium composition were 0·001–0·002 μg.

Radioactivity ratios of U^{234} to U^{238} were determined first, then uranium was removed from the platinum disk by acid dissolution and it was loaded on rhenium side-filaments for solid source mass spectrometry measurements using triple filaments. Two different quantities of uranium—1 μg and 0·3 μg—were used. Approximately 1 μg quantities of uranium were used for measurement of both U^{238}/U^{235} and U^{235}/U^{234}. All U^{238}/U^{235} measurements (Table 1) were made using a Faraday cup for

Table 1. Isotopic composition of uranium in Apollo 11 samples

Sample number	Rock type	U* (ppm)	U^{234}/U^{238} (Activity ratio)	U^{235}/U^{234} Reference† / U^{235}/U^{234} Sample	U^{238}/U^{235} Reference† / U^{238}/U^{235} Sample ~1 μg U	~0·3 μg U
10003	crystalline	0·268	0·98 ± 0·03			1·000 ± 0·002
10017	crystalline	0·854	1·00 ± 0·02	0·996 ± 0·007	0·997 ± 0·003	0·999 ± 0·002
10020	vesicular	0·202	0·98 ± 0·03			1·001 ± 0·002
10050	crystalline	0·156	1·01 ± 0·04			0·998 ± 0·002
10057	vesicular	0·865	1·01 ± 0·02	1·002 ± 0·005	0·998 ± 0·003	0·999 ± 0·002
10071	vesicular	0·873	0·99 ± 0·02	1·002 ± 0·006	1·000 ± 0·003	1·000 ± 0·002
10061	breccia	0·674	1·03 ± 0·03	0·997 ± 0·007	1·001 ± 0·003	0·999 ± 0·002
10084	fines	0·544	1·00 ± 0·03	1·000 ± 0·005	0·999 ± 0·003	1·000 ± 0·002

* The values are from Tatsumoto and Rosholt (1970).
† Terrestrial references used were Republic of Congo Pitchblende and #3633 granite reference (Rosholt *et al.*, 1970).

ion collection. U^{235}/U^{234} ratios were measured using an electron multiplier. Approximately 0·3 μg quantities of uranium from eight samples were measured for U^{238}/U^{235}. Better precision was obtained on the smaller quantities because less isotopic fractionation with variation of filament temperature occurred (Rosholt, unpublished data).

The results in Table 1 indicate that U^{234} is in radioactive equilibrium with U^{238}, and the U^{238}/U^{235} ratio in all samples is the same as terrestrial uranium (Hyde *et al.*, 1964), within experimental error (137·8 ± 0·3).

Thorium-232, with a 4·0 MeV α-particle emission, has been used as a natural yield tracer for the determination of the state of radioactive equilibrium between Th^{230} and parent U^{238} in crustal silicate rocks from earth (Rosholt *et al.*, 1967). The 4·0 MeV α-particle emitted from Th^{232} was measured to determine its radioactivity. If Th^{230} is in radioactive equilibrium with U^{238}, the activity ratio expected for thorium isotopes can be calculated using the atomic ratio of Th^{232}/U^{238} as determined by isotope dilution (Tatsumoto and Rosholt, 1970) and using the decay constants for Th^{232} and U^{238}. The expected activity ratio is

$$(Th^{232}/Th^{230})_{Ex.} = (Th^{232}/U^{238})_{atom} \times (\lambda_{232}/\lambda_{238}),$$

where λ_{232} and λ_{238} are the decay constants for Th^{232} and U^{238}. Values for the measured activity ratios of thorium isotopes and the expected activity ratios of Th^{232}/Th^{230} are shown in Table 2. Radioactive equilibrium could not be demonstrated in four of the samples (10020, 10050, 10057, 10071). A Th^{228}/Th^{232} activity ratio of unity would indicate that Th^{228} is in radioactive equilibrium with parent Th^{232}. In three of the rocks (10020, 10050, 10057) the Th^{228} alpha activity was measured to be 10–20 per cent greater than the 4·0 MeV α-activity of Th^{232}. The reasons for these variations are not yet understood. We do not interpret the variations as representing excess Th^{230} activity, inasmuch as the correlation of Th^{228}/Th^{232} with Th^{230}/Th^{232} ratios indicate the radioactivity of Th^{232} to be less than expected from the Th^{232} concentration. Possibly the daughter product thorium isotopes were not equilibrated with Th^{232} in laboratory processing. Or possibly, in some lunar rocks, a small amount of

Table 2. Radioactivity ratios of thorium isotopes

Sample	Atomic ratio Th^{232}/U^{238}*	$(Th^{232}/U^{238})_{atom} \times (\lambda_{232}/\lambda_{238})$† Expected Th^{232}/Th^{230} activity ratio	Measured activity ratio Th^{232}/Th^{230}	Th^{228}/Th^{232}
10003	3·96	1·26 ± 0.02	1·30 ± 0·04	0·98 ± 0·03
10017	4·07	1·29 ± 0·02	1·26 ± 0·04	1·02 ± 0·03
10020	3·55	1·13 ± 0·02	1·06 ± 0·04	1·09 ± 0·04
10050	3·53	1·12 ± 0·02	0·91 ± 0·07	1·20 ± 0·09
10057	4·08	1·29 ± 0·02	1·17 ± 0·03	1·10 ± 0·03
10071	4·06	1·29 ± 0·02	1·16 ± 0·03	1·03 ± 0·03
10061	3·94	1·25 ± 0·02	1·27 ± 0·04	1·01 ± 0·04
10084	3·97	1·26 ± 0·02	1·25 ± 0·04	1·02 ± 0·04

* The values are from TATSUMOTO and ROSHOLT (1970).
† $\lambda_{232} = 4{\cdot}88 \times 10^{-11}\,yr^{-1}$; $\lambda_{238} = 1{\cdot}537 \times 10^{-10}\,yr^{-1}$.

an isomer of Th^{232} occurs with a mode of decay other than 4·0 MeV alpha particle emission. Isomers of even-even nuclei in some heavy elements are known to exist such as Pb^{202m}, Pb^{204m}, Pb^{206m} and Po^{212m} (HYDE *et al.*, 1964).

DISCUSSION

In their theory for the nucleosynthesis of *r*-process isotopes, BURBIDGE *et al.* (1957) assigned a production ratio for U^{235}/U^{238} of 1·65. WASSERBURG *et al.* (1969b), in the two-spike model for nuclear chronology in the galaxy, predicted the time of cessation of the initial element production to be no later than 8·6 billion years ago. Within that time, differential radioactive decay of U^{235} and U^{238} would have produced a change from the assumed initial U^{238}/U^{235} of 0·61 to a present-day ratio of 700. Uranium in the terminating spike at about 4·8 billion years would have decayed from the same assumed initial U^{238}/U^{235} ratio to a present-day ratio of thirty-one. In this model, only complete mixing of stellar material during the relatively short interval between the predicted terminal spike and formation of the solar system could have produced a constant U^{238}/U^{235} ratio for all stellar matter. There is no evidence, from this investigation, that solar system matter was not completely mixed: indeed, the measured U^{238}/U^{235} in lunar material of different types is nearly identical to that of terrestrial uranium.

Primordial Th^{232}, in *r*-process nucleosynthesis, would have been produced primarily by beta decay of neutron-rich nuclides and to slight extent through neutron capture by Th^{231}. A significant amount of radiogenic Th^{232} would have been produced, up to 400 million years after the 'last-minute' synthesis and after the formation of the solar system, by decay of Pu^{244} progenitor. Some of the radiogenic Th^{232} may have had a different cosmogenic history than the primordial Th^{232}. Therefore, the isotopic composition of many varied lunar rocks should be investigated to search for clues to (1) the variable cosmogenic history of thorium and (2) the reason that concentrations of uranium and thorium are much greater in lunar surface material than in stony meteorites (LOVERING and MORGAN, 1964). Results on the thorium isotopes in the lunar samples are inconclusive, and a preliminary search for an isomer of Th^{232}, as an explanation for low Th^{232} α-activity, has not been fruitful.

Acknowledgments—We thank R. J. KNIGHT for laboratory assistance, and J. S. STACEY and N. V. CARPENTER for development of the electron multiplier used on the mass spectrometer. This study was supported by NASA Contract T–75445.

References

Burbidge E. M., Burbidge G. R., Fowler W. A. and Hoyle F. (1957) Synthesis of the elements in stars. *Rev. Mod. Phys.* **29**, 547–650.

Cantelaube Y., Maurette M. and Pellas P. (1967) Traces d'ions lourds dans les mineraux de la chondrite de Saint Severin. In *Radioactive dating and methods of low-level counting*, p. 215. International Atomic Energy Agency.

Hohenberg C. M. (1969) Radioisotopes and the history of nucleosynthesis in the galaxy. *Science* **166,** 212–215.

Hyde E. K., Perlman I. and Seaborg G. T. (1964) *The nuclear properties of the heavy elements. Detailed radioactivity properties*, Vol. II, p. 437. Prentice-Hall.

Lovering J. F. and Morgan J. W. (1964) Uranium and thorium abundances in stony meteorites. 1, The chondritic meteorites. *J. Geophys. Res.* **69,** 1979–1988.

Rosholt J. N., Doe B. R. and Tatsumoto M. (1966) Evolution of the isotopic composition of uranium and thorium in soil profiles. *Bull. Geol. Soc. Amer.* **77,** 987–1003.

Rosholt J. N., Jr., Peterman Z. E. and Bartel A. J. (1967) Reference sample for determining the isotopic composition of thorium in crustal rocks. In *Geological Survey Research 1967*, pp. 133–136. U.S. Geol. Survey Prof. Paper 575-B.

Rosholt J. N., Peterman Z. E. and Bartel A. J. (1970) U–Th–Pb and Rb–Sr ages in granite reference sample from southwestern Saskatchewan. *Can. J. Earth Sci.* **7,** 184–187.

Tatsumoto M. (1970) An isotopic study of U–Th–Pb systematics of Apollo 11 lunar samples—II. *Geochim. Cosmochim. Acta*, Supplement I.

Tatsumoto M. and Rosholt J. N. (1970) Age of the moon: An isotopic study of uranium–thorium–lead systematics of lunar samples. *Science* **167,** 461–463.

Wasserburg G. J., Huneke J. C. and Burnett D. S. (1969a) Correlation between fission tracks and fission type xenon in meteoritic whitlockite. *J. Geophys. Res.* **74,** 4221–4232

Wasserburg G. J., Schramm D. N. and Huneke J. C. (1969b) Nuclear chronologies for the galaxy. *Astrophys. J.* **157,** L91–L96.

Proceedings of the Apollo 11 Lunar Science Conference, Vol. 2, pp. 1503 to 1532.

Pattern of bombardment-produced radionuclides in rock 10017 and in lunar soil

S.H.R.E.L.L.D.A.L.F.F.*
University of California, San Diego, Department of Chemistry, La Jolla
California 92037

(*Received* 2 *February* 1970; *accepted in revised form* 19 *February* 1970)

Abstract—A large number of radionuclides have been measured in bulk fines (<1 mm) and as a function of depth in lunar rock 10017. Data are reported on Be^{10}, Na^{22}, Al^{26}, Cl^{36}, V^{49}, Mn^{53}, Mn^{54}, Fe^{55}, Co^{56}, Co^{57} and Ni^{59} and upper limits given for Sc^{46}, V^{48}, Cr^{51} and Co^{60}. New chemical procedures and detector systems are summarized.

The results for several nuclides show striking evidence of excess surface production attributable to solar flare particles. Data for short-lived species, Co^{56}, Co^{57}, Mn^{54}, Fe^{55} and Na^{22}, appear consistent with fluxes from known recent events. Long-lived species demonstrate the existence of solar flare protons and alphas at least for the last 10^5–10^6 yr.

Two models have been compared to observation. The simplest, model A, assumes that the rock has been fixed at the surface for $\geqslant 10^7$ yr. In this case the mean energy or rigidity of the solar protons has been higher than presently observed. Model B, consistent with the complex history of rock 10017 as recorded by rare gases and nuclear tracks, assumes a moderately deep bombardment and ejection to the surface 10^5–10^6 yr ago. This permits a flux and spectrum of protons similar to that now observed. Both models have difficulties but model B is now preferred. Ni^{59} is uniquely produced by alpha particles, and requires a flux of about 10^1 α/cm^2 sec averaged over 10^5 yr.

Introduction

The fossil record of bombardment by cosmic ray particles has been studied in great detail in meteorites (Honda and Arnold, 1967; Kirsten and Schaeffer, 1969). This record is however incomplete in at least two respects. First, there is no accurate information on the orbits of meteorites (with two exceptions), which must frequently extend out to the asteroid belt or inside the orbit of Venus. Hence we are not sure of the place to which the observations relate. Second, all meteorites suffer ablation in passing through the atmosphere. Many are broken up as well. The depth of samples below the pre-atmospheric surface is of the order of 3–30 cm, and poorly known. Only recently some meteorites have been found with low ablation (Marti *et al.*, 1969; Amin *et al.*, 1969).

Lunar rock samples are free of both these uncertainties. The surface layers should display not only the record of galactic cosmic rays, but also that of the lower energy (~10–100 MeV) solar flare particles (Fan *et al.*, 1968; McDonald, 1963). A comparison with deeper layers, which are accessible only to higher energy bombardment, can give basic information on the history of solar activity and of the galactic

* A group of colleagues including (in acronymic order) Julian P. Shedlovsky (National Center for Atmospheric Research, Boulder, Colorado 80301), Masatake Honda (University of Tokyo, Tokyo, Japan), Robert C. Reedy, John C. Evans, Jr., Devendra Lal, Richard M. Lindstrom, Anthony C. Delany, James R. Arnold, Heinz-Hugo Loosli, Jonathan S. Fruchter and Robert C. Finkel.

radiation at a point 1 astronomical unit from the sun but generally not influenced by the earth's magnetic field.

Of course the moon's surface has not always been a stationary target. The vertical and horizontal movement of lunar soil and rocks in the process of repeated cratering has been expressively termed 'gardening' (PARKIN, 1965). The mean rate of gardening, as a function of depth, at any site can in principle be determined by study of soil and rock samples.

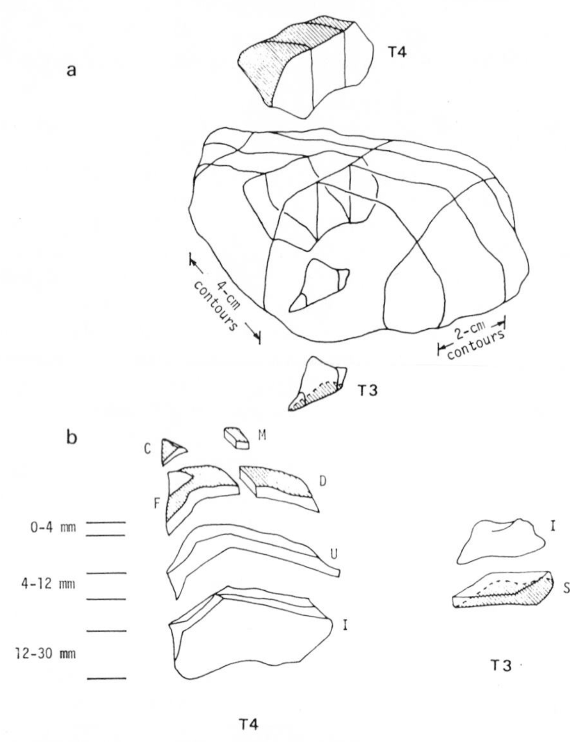

Fig. 1. (a) Schematic of rock 10017, showing location of samples. (b) Subdivision of samples.

Each of the radioactive nuclides studied gives historical information for a time of the order of its mean life. The other major indicators, rare gases and nuclear tracks, integrate over the total exposure time. Nondestructive counting can sample a wider variety of materials. The data are complementary, and the discussion below will include a first attempt to consider them all together.

DESCRIPTION OF SAMPLES

The Apollo 11 samples studied consisted of 100·21 g of 10084-16 bulk fines ($<$1 mm) and two pieces (of mass 19·18 g and 110·26 g) of Type B igneous coarse-grained rock 10017 (total mass = 980 g). We designated the dust as T1 and the two rock pieces as T3 and T4. Figure 1(a) shows a drawing of rock 10017 with an outline of the approximate position of our specimens. Figure 1(b) represents the subdivisions of the pieces T3 and T4, which were carried out in our laboratory before chemical dissolution.

In order to carry out our studies, it was necessary to know which side of rock 10017 was up on the lunar surface. We are indebted to Ernest Schonfeld of the Radiation Counting Laboratory in Houston for the initial suggestion as to the proper orientation, which was opposite to that which had been inferred from the shape of the rock. Schonfeld's orientation, which was made on the basis of non-destructive counting, was later confirmed by our counting data for all nuclides. Attempts to further confirm the surface by particle track counting and non-destructive counting of chips in our laboratory were not successful. The track method was not applicable because of the presence of tracks on all sides. Since track measurements are integral in nature, the information could not be used to determine the most recent orientation of the rock.

Preparation of Samples

Before dissolution the samples were examined and some mineral grains were removed for particle track studies. Each sample was then crushed in a diamond mortar, ground under ethanol in an alumina mortar to less than 120 μm and then extracted with a series of hand magnets of strength from 700 to 1400 G to separate magnetic minerals. The dust yielded 0·8 per cent of a crude magnetic fraction, the rock only 0·05 per cent.

The soil sample (T1) was also leached with acetic acid (0·1 M) (200 ml) for 60 min and with nitric acid (0·1 M) (200 ml) for 60 min in order to extract easily soluble minerals, and to extract Pb^{210} which was expected to be concentrated on the surfaces of the grains. Ten mg of lead carrier were added to each solution during the leach. The Pb^{210} results will be discussed in a separate publication.

Hydrogen sulfide was evolved in moderate quantity by acetic acid and in much greater quantity by nitric acid.

Due to precipitation by this evolved H_2S, the leaches showed no lead on analysis; any small amounts of AgCl obtained on addition of $AgNO_3$ were also masked by Ag_2S. (See the chlorine section below.) Corrections were made in chemical yield data for the amounts of the various elements which were leached out during this procedure.

Still Chemistry

(a) *Description of still*

A special still was used for the destructive distillation of up to 100 g of powdered rock in a mixture of HF and HNO_3. The still is pictured in Fig. 2. The body of the still was a 600 ml teflon beaker (I) which was wrapped in asbestos tape and fitted with a hose clamp (M). The teflon still head (S) was fitted with a viewing port (H), a dropping funnel (R), a nitrogen inlet (E), and a header block (L). The header block (L) had a screw plug which could be removed in the event of blockage in the header pipe (D). The nitrogen outlet pipe (A) led from the traps (F, F, F) to the gas collection apparatus.

(b) *Operation of still*

The crushed ($<$120 μ) sample was added with water (15 ml) to the teflon beaker which served as the still pot. The still was assembled, checked for leaks and was flushed with N_2. Aliquots of HNO_3 (35%) and HF (48%) were added in the following order. HNO_3, HF, HNO_3 + carrier, HF $\times$ 3, HNO_3 + carrier, HF $\times$ 3, HNO_3 + carrier, HF. To ensure complete reaction the sizes of the aliquots were adjusted to the mass of the sample. Carriers for Be, Cl, Sc, V, Co, Ni and Pb were added in amounts calculated to yield convenient weights of final counting samples. The reagent grade chemicals and high purity metals used for the carriers were radiochemically purified and checked for activity by high sensitivity β counting. Except for Cl, which was added as NH_4Cl solution, all carriers were converted to nitrate solutions before addition.

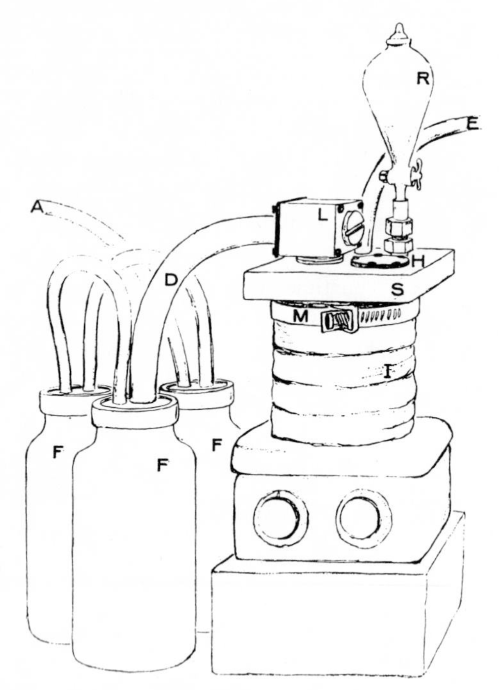

Fig. 2. The lunar still; letter designations in text.

Gas collection was started with the addition of the first aliquot of acid. Five ml of Ar carrier was added by syringe to the gas stream for each liter of N_2 collected. Addition of acid was cautious at first but when the initial reaction had subsided the temperature was increased until distillation commenced. Further acid was added and distillation continued until all the acid had distilled, leaving the sample as a cake. The Si and Cl were collected in the water-filled traps, and the Ar was collected with the N_2 in evacuated steel cylinders. The Ar sample has not yet been purified or counted.

MAIN CHEMISTRY

(a) *Experimental procedures*

The separation procedures outlined below were designed to separate as many as sixteen elements from the samples (see Fig. 3.). Prior to processing the actual lunar samples, a series of eight simulations of the chemistry served to locate problem areas and, in particular, to provide counting blanks.

Procedures for separation and purification were drawn from many sources. Especially useful guides to planning were HILLEBRAND *et al.* (1953) for classical methods, HONDA *et al.* (1961), and the series of booklets on the radiochemistry of the elements compiled by the National Academy of Sciences.

The progress of most metallic elements through the procedure was followed by frequent measurements by atomic absorption spectrometry. This technique was particularly valuable for determining the effectiveness of ion-exchange separations and in recoveries of elements from incomplete separations. Final yields were usually determined by gravimetry and were checked in some cases by atomic absorption spectrometry.

In order to determine the final percentage chemical yields, it was necessary to know the composition of the lunar rock and dust. We measured this composition for the relevant elements by atomic absorption spectrometry. The values we measured are given (Table 1) along with the adopted best values which were determined by consideration of analyses of other workers in addition to our own.

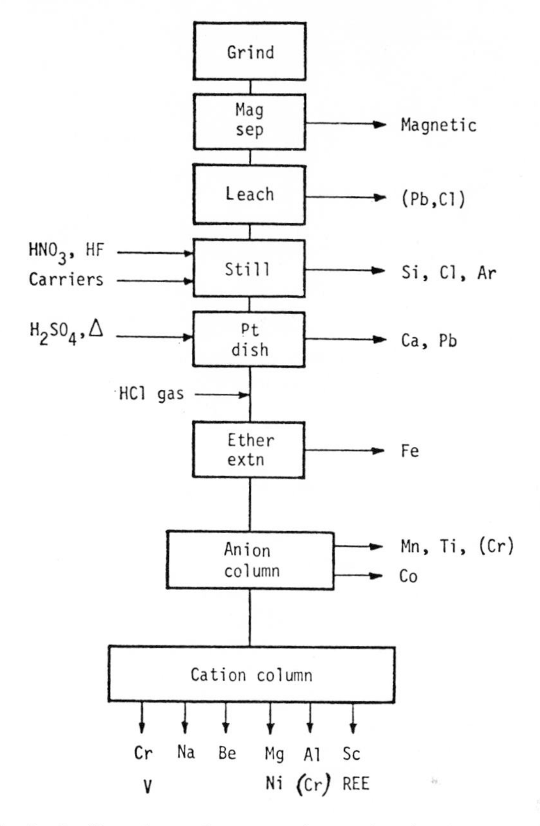

Fig. 3. Outline of sample preparation and main chemistry.

(b) *Solution*

The main chemistry, described below for a 50g. sample, was repeated with minor variations to take account of sample size and improvements in the procedure for eight separate samples of lunar rock and dust. Throughout the procedure teflon, platinum, or Vycor ware was used except in ion exchange experiments. The greenish-white cake which remained in the beaker after the still operation was treated with H_2SO_4 (1:1) to remove excess HNO_3 and fluorides. Initial heating with H_2SO_4 (1:1)

Table 1. Chemical composition

	Bulk lunar fines 10084-76		Rock 10017	
	This work	Adopted	This work	Adopted
Be (ppm)	—	3	—	—
Na (%)	0·40	0·32	0·37	0·37
Al (%)	6·95	7·1	4·0	4·3
Cl (ppm)	—	14	—	14
Sc (ppm)	89	60	85	81
Ti (%)	7·3	4·4	6·3	7·1
V (ppm)	85	72	100	70
Cr (ppm)	1950	1950	—	2200
Mn (ppm)	1600	1600	1950	1900
Fe (%)	12·3	12·3	15	15·3
Co (ppm)	38	35	45	30
Ni (ppm)	237	272	60	60

(40 ml) was carried out in the teflon still beaker. The slurry was then transferred to a 300 ml platinum dish, H_2SO_4 (50 ml) was added, and the mixture taken to fumes of SO_3 on a hot plate. Heating was continued until the mixture of sulfates was converted to a dry cake. This fuming process was repeated using another 30 ml of H_2SO_4 (1:1).

The dry cake was mixed with water (200 ml) and heated to effect dissolution. An excess of boric acid (10 g) was added at this time to insure the complexing of any residual fluoride. A white precipitate (approx. 12 ml) composed of gypsum, $PbSO_4$ and other coprecipitated metallic elements was separated by centrifugation. Only a very small amount of unattacked mineral was observed in the residue.

(c) *Gypsum*

The white sulfate residue was first converted to the carbonate form by adding ~15 g of solid $(NH_4)_2CO_3$ to an aqueous slurry of the residue and gently heating for about 30 min. The residue was separated from the supernatant $(NH_4)_2SO_4$ solution by centrifugation. This procedure was repeated three times and the completeness of the reaction checked by confirming the absence of sulfate in the final supernatant using $BaCl_2$ plus HCl.

All supernatants were combined and heated to decompose excess ammonium carbonate. Then the solution was boiled with aqua regia to decompose ammonium ion, and the residual H_2SO_4 removed by further heating to SO_3 fumes. Dilute HCl was added, and the solution combined with the charge solution of the main chemistry cation exchange column as described below.

The residue from the $(NH_4)_2CO_3$ treatment, now mostly $CaCO_3$ and $PbCO_3$, was dissolved in dilute HCl with the evolution of CO_2. There was very little residue. The pH of the solution was adjusted to 3 with NH_4OAc buffer and hydrogen sulfide was passed through the hot sample solution to precipitate PbS. A small amount of metallic sulfides, mainly iron, were then recovered from the filtrate by precipitation with $(NH_4)_2S$ at higher pH. These sulfides were dissolved in aqua regia and combined with the charge solution of the main chemistry cation exchange column. The recoveries of metals coprecipitated with the gypsum were found to be: Na (10 mg, ~6%), Mn (4 mg, ~4%) and Pb (~80%).

(d) *Iron extraction*

HCl gas was bubbled through the filtrate from the original separation of gypsum and $PbSO_4$. The exothermic solvation reaction caused the temperature of the solution to rise to approximately 80°C. The solution was saturated with HCl at this temperature, cooled in ice, and the Fe(III) extracted with isopropyl ether. About 500 ml of ether was used in two steps. The iron was then back extracted into water.

(e) *Anion exchange*

The bluish-green sample solution remaining after the extraction of iron was heated to expel most of the ether. The solution was then chilled in ice and made 9 M in HCl by passing through HCl gas. The amount of HCl dissolved was monitored by following the weight increase of the solution. The final solution was adjusted to be about a third the volume of the anion exchange column.

A 1-l. anion exchange column (Dowex 1, X-10, 200–400 mesh) was conditioned with 9 M HCl. Just before charging a small amount of precipitate was removed from the sample solution by centrifugation. The precipitate (~7 ml) was composed mainly of gypsum, boric acid and a small amount of Ti, Na and Al salts. This precipitate was dissolved and added to the charge solution of the main chemistry cation exchange column along with the recoveries from the gypsum fraction. The dense 9 M HCl sample solution (360 ml) was charged to the top of the anion exchange column in a slow stepwise fashion to avoid disruption of the resin bed.

The filtrate of the 9 M HCl sample solution and the following 9 M HCl wash were collected. Most of the metallic elements were recovered from less than one column volume of effluent, whereas Mn was collected in 1·2–4 column volumes of effluent (a column volume is in all cases defined as the volume of the wet resin bed). For example about 99 per cent of the total Be and 1 per cent of the total Mn were found in the first fraction; and Mn, most of the Ti, ~10% of the total Cr and a very small amount of the other metallic elements were collected in the second fraction. Co was eluted quantitatively with 4 M HCl. The column was then washed with 0·5 M HCl for iron recovery and then with water to collect the other minor constituents.

(f) *Cation exchange*

The first fraction of the anion exchange column (0–1·2 column volumes) was evaporated, and excess H_2SO_4 removed by fuming in a platinum dish. The moist cake was dissolved in 200 ml of water with heating. At this time the recoveries from the gypsum fraction and from the anion column were added to the solution. A small final residue (~2 ml) composed mainly of gypsum was removed by filtration and the solution diluted to 400 ml with water. The solution was made 0·1% in H_2O_2. The color changed from green to deep wine-red because of the formation of Ti and V peroxide complexes.

A 1·5 l. cation exchange volume (Dowex 50W, X-8, 100–200 mesh) was conditioned with 6 M HCl and washed with water. The 400 ml sample solution was charged and the sample filtrate and eluate (1 M HCl, 0·1% H_2O_2) were collected in the following fractions:

V (plus ~75% of total Cr)	0·8–2·7 column volumes
Na (plus some Ti)	2·7–3·7 column volumes
Be (plus some Mg and Ti)	4·5–5·5 column volumes
Ni (plus some Mn and Mg)	6–10 column volumes

The column was then washed with 3·3 column volumes of 2·5 M HCl, and the effluent collected as the Al (plus ~15 per cent of total Cr) fraction. 0·5 column volumes of water were then passed through the column followed by 2·3 column volumes of a solution 2 M in NH_4OAc and 2 M in HOAc. During the first 1·8 column volumes of the acetate eluate the pH remained at 2 due to elution of 4 M HOAc. The pH then changed to the buffer value of 4·5. The Sc (plus rare earths) eluted very sharply just as the pH changed. After the collection of Sc the column was washed with 0·7 column volumes of water, converted to the H^+ form with 0·7 column volumes of 2 M HCl, and stripped of any remaining metallic ions with 2 column volumes of 6 M HCl.

Specific Element Purifications

(a) *Aluminum*

Al was recovered from the main chemistry cation column. The solution was evaporated to 200 ml of constant boiling HCl. One ml of (30%) H_2O_2 was added to oxidize any Cr to CrO_4^{2-} and $Al(OH)_3$ was precipitated from boiling solution with NH_4OH at pH 6–7. The precipitate was centrifuged, washed with H_2O and dissolved in boiling HCl. An $Fe(OH)_3$ scavenge was next carried out at pH 10 using NaOH, followed by reprecipitation of $Al(OH)_3$ with HCl + NH_4Cl at pH 7. After a PbS scavenge, the Al was loaded onto a 275 cm³ cation column (Dowex 50W, X-8, 100–200 mesh) in H_2O. The column was eluted with 15 column volumes of 1·0 M HCl followed by 5 column volumes of 2·0 M HCl to elute the Al. The cation column effluent was evaporated to minimum volume, made 9 M with conc. HCl, covered with an equal volume of diethyl ether, cooled to less than 4°C, and saturated with HCl gas until precipitation of $AlCl_3$ was complete (Gooch–Havens procedure, Hillebrand *et al.*, 1953). After centrifuging, the precipitate was dissolved and converted completely to the NO_3^- form, after which it was passed through a 25 ml nitrate anion column (Dowex 1, X-8, 100–200 mesh) with 2 column volumes of HNO_3 (1:1). After a second acid PbS scavenge, $Al(OH)_3$ was precipitated with NH_4OH from boiling pH 7 solution. The final precipitate was then ignited for 12 hr at 1200°C, and weighed as Al_2O_3. Overall yields on the aliquot portions of Al purified ranged from 80 to 90 per cent.

(b) *Beryllium*

The Be fraction from the 'main chemistry' cation column was collected and evaporated to a small volume. $Be(OH)_2$ was precipitated with NH_4OH at pH 7, digested on a hot plate, centrifuged and washed with 1% NH_4Cl. The $Be(OH)_2$ was dissolved in a few drops of warm 6 M HCl and placed on a 10 ml cation column (Dowex 50W, X-8, 100–200 mesh) which had been washed with 100 ml of

6 M HCl and then H_2O. The column was eluted with 1 M HCl and 5 ml fractions collected. The Be fraction was collected in column volumes 3–6 of the eluent. $Be(OH)_2$ was reprecipitated as above, washed and dissolved in hot 0·5 M HCl. An acid CuS scavenge was performed followed by a $Fe(OH)_3$ scavenge in 10% NaOH. The Be was precipitated with ammonia twice, then dissolved in a small amount of 6 M HCl and passed through a 5 ml anion column to remove any residual Fe. The final purification was achieved using an acetylacetonate extraction of the Be complex into $CHCl_3$ at pH 8 (MORRISON and FREISER, 1962). The organic material was destroyed by adding 5 ml aqua regia and boiling to near dryness several times. The Be was taken up in dilute HCl, precipitated as $Be(OH)_2$, filtered through Whatman 541, washed with 1% NH_4Cl and ignited to BeO at 1200°C for several hours in a platinum crucible. The yields ranged from 71 to 83 per cent.

(c) *Chlorine*

Twenty mg of NH_4Cl carrier was added during operation of the teflon still. Cl was recovered from the traps as AgCl, nearly all in the first trap. After digesting overnight, the AgCl precipitate was centrifuged, washed with HNO_3(0·1 M), dried and weighed. The precipitate was dissolved in NH_3 (1:1) and passed through a 5 ml cation column (Dowex 50W, X-4, 100–200 mesh) in the ammonium form. The Cl^- was eluted with NH_4OH (1:1) (20 ml) followed by the Ag^+ in HNO_3 (1:1) (20 ml). The Cl^- solution was evaporated to 5 ml and transferred with a dropping pipette into a sublimer where it was carefully taken to dryness. The sample deposition technique is described below. Chemical yields were determined gravimetrically and ranged from 76 to 90 per cent.

(d) *Chromium*

The Cr was recovered from the Ni–Mg, Al and V fractions of the main chemistry. It was separated from the large amount of remaining SO_4^{2-} by collection with 200 mg of $Fe(OH)_3$. $NH_3 + H_2O_2$ were then added to oxidize the Cr to CrO_4^{2-} and precipitate the Fe. Cu carrier was added and precipitated as CuS. Additional NH_3 and H_2O_2 were added to reoxidize the Cr to CrO_4^{2-} after boiling to remove H_2S. The solution was then buffered with NH_4OAc and $Ba(NO_3)_2$ was added to precipitate $BaCrO_4$. The precipitate was dissolved in hot H_2SO_4 (1:20) and the $BaSO_4$ filtered out. The pH was adjusted to 1·7. The solution was cooled to 5°C. H_2O_2 was added and blue perchromic acid was extracted with cold ethyl acetate (MORRISON and FREISER, 1962). After washing the Cr was back extracted into $NH_3 + H_2O$, reduced with $HCl + H_2O_2$ and precipitation as hydroxide was attempted. Decomposition products of ethyl acetate somewhat hindered the precipitation of $Cr(OH)_3$. Perhaps this could have been avoided by extracting into ethyl ether. In this case the Cr acetate complex was finally evaporated into a crucible and ignited directly to Cr_2O_3. The chemical yield, determined gravimetrically, was 63 per cent.

(e) *Cobalt*

The Co was recovered from the 4 M HCl fraction of the main chemistry anion column. After evaporation, the sample was dissolved in HCl (0·1 M) and a CuS scavenge was performed. The filtrate was then boiled to remove H_2S, Fe (3 mg) was added and $Fe(OH)_3$ precipitated with NH_3 at pH 5·5. The Fe precipitate was dissolved in HCl and reprecipitated. The filtrates were subsequently combined. The Co was precipitated with $(NH_4)_2S$ in ammoniacal solution, Millipore filtered and ignited for 12 hr at 800°C to Co_3O_4. After the first counting the sample was redissolved in HNO_3 (1:1), and passed through a 10 ml anion column (Dowex 1, X-8, 100–200 mesh) which had been conditioned with HNO_3 (1:1). The Co was eluted with an additional 5 column volumes of HNO_3 (1:1). After evaporation, the sample was dissolved in HCl (0·01 M) and another $Fe(OH)_3$ scavenge was performed. The Co was reprecipitated, filtered, ignited, and counted again. Chemical yields were determined gravimetrically, and yields were checked by dissolving the counting sample and measuring the amount of Co by atomic absorption spectrometry. Yields ranged from 76 to 89 per cent.

(f) *Iron*

An aliquot equivalent to 1·5 g of Fe was taken for purification from the Fe extracted from the main chemistry. A CuS scavenge was carried out at pH 1·5. The solution was boiled to remove

H_2S, and then the Fe^{2+} resulting from S^{2-} reduction was reoxidized to Fe^{3+} with HNO_3. The solution was next evaporated, taken up in 6 M HCl and loaded onto a 100 ml anion column (Dowex 1, X-8, 100–200 mesh). The column was rinsed with 10 ml of 6 M HCl followed by 100 ml of 3 M HCl and Fe eluted with 200 ml of 0·1 M HCl. Deposition on a Cu holder for counting was accomplished by electroplating from an ascorbic acid—sodium citrate—ammonia solution (HAHN, 1945). The actual electroplating of the sample was preceded by a 5-hr scavenge electroplate at 6 V to remove less electropositive impurities. The electroplating was terminated before the last 100 mg of Fe was plated in order to avoid more electropositive impurities. Because of the ample amount of Fe in most samples, chemical yields were adjusted, both to increase purity and give convenient amounts of counting sample, to about 50 per cent on the aliquots taken.

(g) *Manganese*

The Mn was recovered from two main chemistry fractions, the 9 M HCl anion column effluent and the 1 M HCl cation column effluent. The Mn was separated from the major contaminant, Ti, by means of a mixed solvent anion column (250 ml, Dowex 1, X-10, 200–400 mesh) which was conditioned with a mixture of conc. HCl:isopropanol:H_2O (2:2:1). Ti was eluted with 2 column volumes of the same 2:2:1 mixture. The major portion of the Ti in the sample was in this fraction. Manganese was then eluted with 2 column volumes of conc. HCl:isopropanol:H_2O (2:1:2). The Mn was then passed through a small cation column with 1·0 M HCl followed by a CuS scavenge. Iron carrier was added and $Fe(OH)_3$ was precipitated using a dilute HCl–pyridine system to buffer at approximately pH 7. Finally MnO_2 was precipitated using NH_3 and H_2O_2, and the precipitate was ignited at 1050°C to Mn_3O_4. The chemical yields were determined gravimetrically and checked by dissolving the precipitate and determining Mn by atomic absorption. The final yields ranged from 60 to 87 per cent.

(h) *Nickel*

Ni was recovered from the main chemistry cation column. The solution was rendered slightly acid with HCl and a 1% dimethylglyoxime solution in ethanol (0·5 ml DMG solution for each mg of Ni) added. Ni DMG was then precipitated with dilute NH_4OH. The copious red precipitate was extracted into $CHCl_3$ (30 ml/mgNi). In some cases, due to incomplete separation on the anion exchange column, Mn was recovered from the aqueous layer. The brown $CHCl_3$ solution was next covered with a 50 ml layer of HCl (1:1) and evaporated to dryness. The DMG was destroyed by wet ashing with HNO_3 (2×100 ml) followed by $HClO_4$ (50 ml) and finally conc. H_2SO_4 (25 ml). Fe carrier was then added and precipitated as $Fe(OH)_3$ at pH 5·5. After filtration the Ni was plated onto two small sample holders from an NH_4OH–NH_4SO_4 solution (KIRBY, 1961). The chemical yields were 51 and 86 per cent as determined by atomic absorption spectrometry.

(i) *Scandium*

The Sc fraction was recovered from the main chemistry cation column by elution with NH_4OAc. The NH_4OAc was decomposed by repeated evaporation to fumes with HNO_3. The sample was next dissolved in HCl (dil.) and centrifuged to remove the remaining residue. $Sc(OH)_3$ was precipitated using $NH_4OH + H_2O_2$ and the precipitate was centrifuged and washed. The final purification was achieved by NH_4SCN extraction into ethyl ether (MORRISON and FREISER, 1962). The ether fraction was evaporated, the NH_4SCN was destroyed by careful addition of HNO_3, the Sc was taken up in 6 M HCl and passed through a 5 ml anion column (Dowex 1, X-8, 200–400 mesh) to remove any remaining Fe. The final Sc fraction was precipitated with NH_4OH and ignited at 1000°C to Sc_2O_3. The Sc was later recycled by repeating the NH_4SCN extraction and then passing it through a 5 ml NO_3^- anion column (8 M HNO_3) (Dowex 1, X-8, 200–400 mesh). The final yield after two recycles was 30 per cent determined gravimetrically.

(j) *Sodium*

Na was recovered from the main chemistry cation column. Iron carrier was added and precipitated as $Fe(OH)_3$ with NH_4OH. After evaporating the supernatant, it was dissolved in H_2O and CuS was precipitated with H_2S. The solution was next loaded onto a 25 ml cation column (Dowex 50W,

X-8, 100–200 mesh) with H_2O and the column was eluted with 5 column volumes of 1·0 M HCl. The Na usually came off in column volumes 3–5 inclusive. The Na elution was checked by a simple flame test. The appropriate cation exchange column effluent was evaporated to 10 ml and diethyl ether (10 ml) was added. After the temperature was lowered to 4°C, NaCl was precipitated by bubbling HCl gas into the solution (Gooch–Havens procedure, HILLEBRAND, 1953). After centrifuging the supernatant was then decanted, the NaCl washed into a crucible with acetone and heated at 500°C to constant weight. Chemical yields, determined gravimetrically, ranged from 82% to 93%.

(k) *Vanadium*

The vanadium fraction from the main chemistry cation exchange column also contained some Ti and Cr. The solution was made 10 M in HCl. It was then placed on an anion column (Dowex 1, X-10, 100–200 mesh) and the column eluted with 10 M HCl. V immediately preceded Ti and appeared after approximately 0·5 column volumes of eluent had been passed through the column. The separation of V and Ti was about 95 per cent complete with some of the V remaining in the Ti.

A small amount of conc. HNO_3 was added to oxidize V(IV)–V(V). H_2SO_4 was added and HCl removed by heating. The soluton was diluted with water to be 10% in H_2SO_4 and chilled to $\leqslant$ 5°C. A few milliliters of 6% aqueous cupferron solution were added and the dark brown vanadium cupferrate extracted with chloroform, leaving the Cr in the aqueous phase. The organic phase containing the V was dried into a platinum crucible and ignited at 600°C to V_2O_5, the final counting form. The final chemical yield was 60 per cent as determined gravimetrically and checked by atomic absorption spectrometry on the redissolved sample.

(l) *Atomic absorption*

Analysis of the lunar samples for chemical yield purposes was performed using a Techtron AA-4 atomic absorption spectrometer with a Honeywell Elektronik 19 strip chart recorder. Most of the abundance values obtained were rather close to those obtained by other analysts although in several cases it seemed necessary to change our chemical yield values on the basis of data obtained by other investigators. Good agreement was obtained for Al, Cr, Fe, Mn and Sc. More difficulty was experienced with Co, Ni and V. Major errors occurred with Ti and Na. The problems with Na and Ni were probably due to contamination. The spectrometer also proved invaluable as a semiquantitative tool for following the various elements through their respective chemistries.

The sample for analysis, of the order of 0·5 g, was dissolved in HF–$HClO_4$ (1:3). After dissolution had proceeded to the point that only a white CaF_2–AlF_3 residue remained 500 mg of H_3BO_3/g of sample was added to complete dissolution of the Ca and Al. This first solution was made up to 25 ml in 1:1 HCl and labeled solution 1. A 5 ml aliquot of this solution was then diluted to 50 ml and labeled solution 2. Finally, 5 ml of solution 2 was evaporated to dryness, dissolved in 50 ml of H_2O and labeled solution 3a. Sc, V and Co were determined in solution 1, Cr, Mn, Ni and Ti were determined in solution 2 and Al, Fe and Na were determined in solution 3a.

COUNTING

(a) *Mounting techniques*

A uniform counting sample deposit is highly desirable because it provides reproducible geometry and allows the accurate calculation of a self-absorption correction. Three deposition techniques were found useful.

(1) *Millipore technique*. The apparatus consisted of a rectangular plexiglass cell connected to a vacuum line. A rectangular 0·45 μ cellulose acetate Millipore filter was placed on a polyethylene frit at the bottom of the cell. The oxide was next ground very fine in a solution of 0·1% agar and 1 drop 'Photoflo' dispersant/liter and poured into the cell. The solvent was H_2O for Cr, Mn and Co oxides, and EtOH for Sc and V oxides. The oxide slurry was well mixed with a jet of water or EtOH to insure uniformity, an N_2 tent was placed over the cell to avoid radon pickup, and the vacuum was applied. The Millipore filter was then removed from the cell, trimmed to fit the appropriate Cu sample holder, glued into place and covered with aluminized mylar. After counting, the amount of sample often was measured again to confirm that mounting losses were small.

(2) *Electroplating*. For details see the Fe purification section.

(3) *Ammonium chloride sublimation*. The final Cl solution was evaporated to dryness under N_2 in the Pyrex cup of a sublimation assembly. The cup was capped with a gold-plated copper sample holder for the Tanaka counter and heated while cooling the holder. Sublimation yields were typically 95 per cent.

(b) *Counting systems*

Five high-sensitivity counting systems were used depending on the decay scheme of the nuclide to be measured and the amount of sample. Some characteristics of the counters are given in Table 2, and others are described in the following. A fuller description will be given in EVANS *et al.* (1970).

Table 2. Basic counter characteristics

Counter	Mode	Sample area (cm^2)	Sample holder	Isotope	Counting form	Energy (keV)	Efficiency (%)	Typical back-ground (cpm)
Evans	β–γ (coinc)	10	Cu	Na^{22}	NaCl	511	7·0†	0·005†
			Cu	Al^{26}	Al_2O_3	511	8·1†	0·005†
	β–γ (anti-coinc)		Cu			Total	30–40	0·08
Loosli	X- (anti-coinc)	8·8	Cu	V^{49}	V_2O_5	4·5	3·8†	0·006†
		8·8	Cu	Mn^{53}	Mn_3O_4	5·4	4·3†	0·010†
		10	Cu	Ni^{59}	Ni	6·9	3·8†	0·006†
	X–γ (coinc)	8·8	Cu	Mn^{54}	Mn_3O_4	835	1·1†	0·002†
	X–β–γ-(coinc)	8·8	Cu	Co^{56}	Co_3O_4	511	1·0†	0·007†
		8·8	Cu	Co^{56}	Co_3O_4	847	1·0†	0·004†
		8·8	Cu	Co^{56}	Co_3O_4	>2000	2·3†	0·02†
		8·8	Cu	Co^{57}	Co_3O_4	122	6·4†	0·003†
		8·8	Cu	Co^{60}	Co_3O_4	2500	1·3†	0·002†
Delany 1	X-(anticoinc)	57	Cu	Cr^{51}	Cr_2O_3	4·95	3·7†	0·045†
			Cu	Mn^{53+54}	Mn_3O_4	5·4	3·40†	0·057†
Delany 2	X-(anticoinc)	153	Cu	Fe^{55}	Fe	5·9	6·9†	0·090†
Tanaka	Geiger (anti-coinc)		Lucite	Be^{10}	BeO	Total	21	0·24
	Scint (coinc)	4				0–600	11	0·04
	Geiger		Cu	Cl^{36}	NH_4Cl	Total	39	0·24
	Scint					70–800	21	0·05
Lal	2 El	4	Lucite	Be^{10}	BeO	Total	35	0·56
	β^-		Cu	Cl^{36}	NH_4Cl	Total	49	0·56
Solid state	γ (anticoinc)	4	Lucite	Be	BeO	477	1·2†	0·025†
			Lucite	Mn^{54}	Mn_3O_4	835	0·4†	0·013†

† Efficiency and typical background values are for the area under the full width of the peak at half the maximum height.

(1) *Beta-gamma counting* (*Evans.*). All Na^{22} and Al^{26} samples and one Sc^{46} sample were counted by β–γ coincidence counting on a 10 cm^2 Geiger counter inside a 3 in. × 3 in. NaI(Tl) crystal with a 1 in. × 2 in. well. The Geiger counter was of a flat design, made of lucite covered with gold-coated Mylar. It was operated as a flow counter with standard Q gas. An ND-1100 256 channel analyzer was used to accumulate the gamma spectrum in coincidence with a gate pulse from the Geiger counter. The gross μ meson rate was measured with a scaler to determine counter stability and in addition the Geiger rate in anti-coincidence with the gamma well was measured to check the radiochemical purity of the sample. The counter used two sample holders, stamped out of 0·005-in. OFHC Cu with a shallow depression of 5 cm^2 each.

All Na^{22} and Al^{26} activities were calculated using the 511 KeV β^+ annihilation peak only. Integral intervals at both higher and lower energy were examined to check for spectral purity. No significant excesses were found above that expected from the nuclide being counted. A number of the Al samples were contaminated with a pure β^- activity which was most probably Pb^{210}–Bi^{210}. However, this caused no interference in the coincidence spectrum.

(2) *X-ray and X–γ-coincidence system* (*Loosli*). V^{49}, Mn^{53}, Mn^{54}, Co^{56}, Co^{57}, Co^{60} and Ni^{59} were counted in a system consisting of a 10 cm^2 proportional counter operated inside the well of a 4 in. × 4 in. NaI γ-ray assembly as described by BHANDARI (1969). In order to get better peak resolution for the gas counter as well as better counting stability, a number of modifications were made in the original design, including (1) reduction of the length and width of the active area of the sample holders, (2) use of guard electrodes to minimize end effects, (3) addition of O-ring fittings, and (4) provision of a spring for the center wire. The counter was used as a flow counter with an overpressure of 10 psi of P-10. Solid samples plated or mounted on Millipore filter papers were deposited on two Cu sample holders and covered with aluminium-coated Mylar. These sample holders were bent to semicylinders and used as the counter cathode. The counter was operated in coincidence or anti-coincidence with the gamma-ray assembly or in both modes simultaneously. Data were collected on a Nuclear Data 2200 pulse height analyzer. Gross counting rates were also collected on scalers.

In order to insure reproducibility, calibrations with external sources were performed frequently. A collimated I^{125} source was used to excite Cu-X-rays (8 keV) in the sample holder as well as X-rays from the mounted sample.

From the results of counting Mn and Ni blanks and different background samples, there was no indication in this counting system that excitation of the sample holder or sample by gamma radiation in the shield contributed significantly to the energy region under consideration.

Mn^{53} and Mn^{54} were measured simultaneously in X-anti-coincidence and X–γ mode respectively. The X-AC peak was corrected for Mn^{54}. The Co isotopes present the most complex decay schemes. A so-called $X\beta\gamma$ mode was used for Co^{56}, Co^{57} and Co^{60}. X- and β-radiation are measured together with the small counter pressurized with P-10. All γ-ray peaks of Co^{56} were compared, as well as important continuum ranges. If all agreed, including the region >2 MeV, the activity could be ascribed mainly to Co^{56}. For Tl Co we derived an upper limit (2σ) for the ratio Co^{58}/Co^{56} of 0·2. The limit for Co^{60} was derived from the 2·5 MeV sum peak, since Co^{56} interferes with the two single peaks. The Co^{57} region was corrected for the Co^{56} Compton spectrum.

Apparent weak β^- contamination appeared in most samples of Ni, Mn and Co. By using a base-line subtraction procedure for background subtraction, we believe we have avoided errors from this source.

(3) *X-ray system* (*Delany*). The two large X-ray counters used for the determination of the activities of Fe^{55}, $Mn^{53} + Mn^{54}$, and Cr^{51} were of similar design, differing mainly in size. One had an active area of 57 cm^2, the other 153 cm^2. They were both single-wire internal counters with thick (3 cm) lucite walls. The upper halves of the cylindrical counters were removable and the rectangular (125 μm thick) copper sample holders with the samples deposited on their central active areas were pushed down and around the inside of the cavities. The counters were operated as flow counters using P-10 at 1 atm. and were run in anti-coincidence with the surrounding Geiger ring in the iron shield. The data were accumulated in a ND-110 128 channel analyzer. Scalers were used to record the guard rate and both the gross and the net counting rates above 1 keV. The counters were calibrated using the external I^{125} γ-excitation technique described previously. The net excess activity was determined by baseline subtraction with the shape of the baseline derived from that of a blank of similar mass. The background of the larger X-ray counter contained structure which was attributed to absorption of minimum ionizing particles and to excitation of the sample and the sample holder. This structure increased the uncertainty in the value of the background. Allowance was made for this in assignment of error.

(4) *β^- counting*. Two counting systems were used for the counting of Be^{10} and Cl^{36}. This allowed the opportunity to cross-check all results on these nuclides. Good agreement was obtained. The samples were first counted on a coincidence type β-ray spectrometer based on the design of TANAKA *et al.* (1967) with certain modifications. The system consisted of a 4 cm^2 Geiger counter operated in coincidence with a plastic scintillator block, and in anti-coincidence with a μ-meson guard tray.

The beta spectrum of the plastic scintillator was accumulated in a 256-channel time-shared portion of a 2048-channel ND 2200. The gross beta rate in the Geiger counter and the net Gegier rate in anti-coincidence with the guard tray were also measured. This permitted two semi-independent measurements of the count rate to be made simultaneously. The end point energy as determined from the scintillation spectrum gave an indication of sample purity. The agreement between net Geiger rate and net scintillation rate could be used as a further check. The correct energies for Be^{10} and Cl^{36} were always obtained within the limits of error.

The second system was a two element Geiger coincidence counter of the type described by LAL *et al.* (1967). This counter consisted of two Geiger counters separated by a plastic absorber. The counters were operated in coincidence with each other and in anti-coincidence with a guard tray. A β^- half thickness value was obtained from the ratio of counts occurring in the top counter only to that occurring in both counters. The energy values and count rates obtained in this manner were in general in good agreement with those obtained by the scintillation spectrometer.

(5) *Solid-state counting.* A 40 cm^3 Ge(Li) solid state gamma detector was used in this work for counting of Mn^{54}, Be^7 and cross standardization of counting standards. It will be used for determination of Mn^{53} by neutron activation to Mn^{54}, when a suitable reactor again becomes available. The detector was constructed of low background materials surrounded by a 4 cm thick NaI(Tl) anti-coincidence mantle, and 10 cm of Pb shielding. This system had lower sensitivity for small samples than those previously described. Hence it was used only for the lunar fines, which was the largest single sample.

(c) *Measurements*

Background values, efficiencies and other counter characteristics are given in Table 2. Background counts were taken where possible with samples of the same chemical species, carried through the same chemistry as the samples. Standard samples were prepared from calibrated solutions in forms and thicknesses comparable to those of the samples. Self absorption corrections were calculated by the method of LIBBY (1956) for β emitters, and for X-ray emitters the Gold integral for extended sheet sources was used (ROSSI, 1952).

Our results and upper limits for the Apollo 11 samples are listed in Tables 3 and 4. The errors quoted include all known sources: counting statistics, calibration error, chemical yield and others, but ratios, especially of the same nuclide in different samples, may be more accurate. When upper limits are given they correspond to 2σ added to a possible small signal.

The upper limit quoted in Table 4 for T4-DFCU Ni^{59} is based on the total weight of both the surface samples T4-DFC and the interior sample T4-U. If the assumption is made that all Ni^{59} is produced in the surface layer only, an upper limit of 7 dis/min/kg for T4-DFC Ni^{59} is obtained.

The data are in reasonable agreement with those of other workers, where comparisons are possible. Generally, a lack of orientation information makes comparisons difficult, but soil data should be comparable. The most direct comparisons are with O'KELLEY *et al.* (1970). Here the agreement is satisfactory in all cases. PERKINS *et al.* (1970) and WÄNKE *et al.* (1970) give results which parallel ours but seem higher for some nuclides in the soil. This may be due to poor mixing or sorting of the sample, since a small difference in the ratio of surface to deeper material could account for the difference. HERR *et al.* (1970) obtain a lower value for Mn^{53} by activation.

INTERPRETATION

(a) *Observed effects*

The data summarized in Table 4 must be compared with our expectations from known sources. While radioactive nuclides in lunar surface rocks may arise from many sources, two are certainly present. The first is bombardment by galactic cosmic rays, which has been observed in terrestrial samples and in meteorites for many years (LAL and PETERS, 1967; HONDA and ARNOLD, 1967). Typical energies of the primary particles are in the region of 1–10 BeV/nucleon. Nuclide production from

Table 3

Sample	Nuclide	Counter	Mode	Energy (keV)	Net cpm	dis/min/kg†	Results by other workers
T-1	Be^7	Solid state	γ	477	−0·003 ± 4	<253	<112 PERKINS (1970)
T-1	Be^{10}	Tanaka	Geiger	Total	0·208 ± 10	16·3 ± 1·6	
		Tanaka	scint.	0–600	0·115 ± 6	16·7 ± 1·7	
		Lal	2 El	Total	0·35 ± 2	16·2 ± 2·4	
T4-DFCU	Be^{10}	Tanaka	Geiger	Total	0·108 ± 8	15·1 ± 1·8	
		Lal	2 El	Total	0·145 ± 18	12·2 ± 2·6	
T4-I	Be^{10}	Tanaka	Geiger	Total	0·118 ± 11	16·1 ± 2·0	
		Lal	2 El	Total	0·20 ± 2	15·5 ± 2·6	
T-1	Na^{22}	Evans	β–γ	511	0·060 ± 4	51·3 ± 7·3	51 ± 5 O'KELLEY (1970)
		Lal	2 El	Total	0·41 ± 3	54 ± 9	63 ± 2 PERKINS (1970); 61 ± 6 WÄNKE (1970)
T4-D	Na^{22}	Evans	β–γ	511	0·0190 ± 8	83·1 ± 12·5	39 ± 4 whole rock 10017 O'KELLEY (1970) (T4-D to T3-SI)
T4-FC	Na^{22}	Evans	β–γ	511	0·0266 ± 31	80·4 ± 11·3	
T4-U	Na^{22}	Evans	β–γ	511	0·0413 ± 33	42·9 ± 6·2	
T4-I	Na^{22}	Evans	β–γ	511	0·042 ± 4	36·9 ± 5·5	
T3-SI	Na^{22}	Evans	β–γ	511	0·0181 ± 18	30·9 ± 4·1	
T-1	Al^{26}	Evans	β–γ	511	0·0226 ± 22	107 ± 13	120 ± 12 O'KELLEY (1970); 137 ± 4 PERKINS (1970); 121 ± 25 WÄNKE (1970)
T4-D	Al^{26}	Evans	β–γ	511	0·0300 ± 25	129 ± 16	95 ± 15 top (T4-D, T4-FC)
T4-FC	Al^{26}	Evans	β–γ	511	0·051 ± 6	142 ± 16	
T4-U-1	Al^{26}	Evans	β–γ	511	0·0283 ± 17	79·3 ± 12	65 ± 10 middle (T4-U-1, T4-U-2, T4-I)
T4-U-2	Al^{26}	Evans	β–γ	511	0·0240 ± 21	65·2 ± 11	whole rock 10017 O'KELLEY (1970) (T4-D to T-3S)
T4-I	Al^{26}	Evans	β–γ	511	0·0223 ± 24	66·4 ± 10·3	
T-3S	Al^{26}	Evans	β–γ	511	0·0260 ± 22	65·0 ± 10·3	50 ± 7 bottom
T-1	Cl^{36}	Tanaka	Geiger	Total	0·510 ± 14	16·1 ± 1·5	
		Tanaka	scint.	70–800	0·269 ± 10	15·6 ± 1·5	14·6 ± 0·8 WÄNKE (1970)
		Lal	2 El	Total	0·65	16·3 ± 1·8	
T4-F	Cl^{36}	Tanaka	Geiger	Total	0·075 ± 9	<39‡	
		Tanaka	scint.	70–800	0·034 ± 4	<33‡	
		Lal	2 El	Total	0·061 ± 16	<33‡	
T4-U	Cl^{36}	Tanaka	Geiger	Total	0·197 ± 11	15·9 ± 1·6	
		Tanaka	scint.	70–800	0·105 ± 6	16·0 ± 1·6	
		Lal	2 El	Total	0·264 ± 34	18·1 ± 3·2	

Table 3 (continued)

Sample and	Nuclide	Counter	Mode	Energy (keV)	Net cpm	dis/min/kg†	Results by other workers
T4-I	Cl^{36}	Tanaka Tanaka Lal	Geiger scint. 2 El	Total 70–800 Total	0·258 ± 16 0·139 ± 6 0·32 ± 2	17·3 ± 1·9 17·0 ± 1·9 17·9 ± 2·5	
T-1	Sc^{46}	Evans	β–γ	890 + 1120	0·007 ± 3	<9	8 ± 2 O'Kelley (1970) 11 ± 1 Perkins (1970)
T-1	V^{49}	Loosli	X–AC	4·5	0·009 ± 1_8	8·9 ± 2·4	
T-1	Cr^{51}	Delany-1	X–AC	4·95	0·004 ± 10	<45	<63 Perkins (1970)
T-1	Mn^{53} + Mn^{54}	Delany-1	X–AC	5·4	0·076 ± 10	87 ± 15	
T-1	Mn^{53}	Loosli	X–AC	5·4	0·018 ± 3	46 ± 11	30 ± 6 Herr (1970)
T-1	Mn^{54}	Loosli Solid state	X–γ γ	835 835	0·0026 ± 8 0·0034 ± 21	39 ± 14 18 ± 10	28 ± 7 O'Kelley (1970) 24 ± 3 Perkins (1970)
T4-DFC	Mn^{53}	Loosli	X–AC	5·4	0·025 ± 4	94 ± 24	
T4-DFC	Mn^{54}	Loosli	X–γ	835	0·0015 ± 9	36 ± 20	
T4-U	Mn^{53} + Mn^{54}	Delany-1	X–AC	5·4	0·050 ± 8	93 ± 17	
T4-U	Mn^{53}	Loosli	X–AC	5·4	0·012 ± 5	43 ± 16	
T4-U	Mn^{54}	Loosli	X–γ	835	0·0024 ± 15	39 ± 26	
T4-I	Mn^{53} + Mn^{54}	Delany-1	X–AC	5·4	0·077 ± 9	100 ± 18	
T4-I	Mn^{53}	Loosli	X–AC	5·4	0·018 ± 3	50 ± 12	
T4-I	Mn^{54}	Loosli	X–γ	835	0·0006 ± 7	10 ± 11	
T3-S	Mn^{53}	Loosli	X–AC	5·4	0·005 ± 2_4	48 ± 21	
T3-S	Mn^{54}	Loosli	X–γ	835	0·0009 ± 6	36 ± 30	
T-1	Fe^{55}	Delany-2	X–AC	5·9	0·073 ± 8	216 ± 40	
T4-F	Fe^{55}	Delany-2	X–AC	5·9	0·095 ± 10	455 ± 49	
T4-D	Fe^{55}	Delany-2	X–AC	5·9	0·082 ± 10	436 ± 86	
T4-C	Fe^{55}	Delany-2	X–AC	5·9	0·0099 ± 66	248 ± 171	
T4-U	Fe^{55}	Delany-2	X–AC	5·9	0·021 ± 6	96 ± 25	
T4-I	Fe^{55}	Delany-2	X–AC	5·9	0·022 ± 7	94 ± 33	
T3-S	Fe^{55}	Delany-2	X–AC	5·9	0·016 ± 9	77 ± 49	
T3-I	Fe^{55}	Delany-2	X–AC	5·9	0·023 ± 7	98 ± 34	
T-1	Co^{56}	Loosli	X–β–γ	511 847 >2000	0·012 ± 2_5 0·015 ± 2_3 0·030 ± 4	37 ± 6	40 ± 7 O'Kelley (1970) 53 ± 10 Perkins (1970)
T-1 recycled	Co^{56}	Loosli	X–β–γ	511 847 >2000	0·012 ± 2_6 0·014 ± 2_4 0·030 ± 4_3	44 ± 8	

Table 3 (continued)

Sample and Nuclide		Counter	Mode	Energy (keV)	Net cpm	dis/min/kg†	Results by other workers
T4-D	Co^{56}	Loosli	X–β–γ	511 847 >2000	0·001 ± 1_9 0·004 ± 1_6 0·006 ± 3	135 ± 37	26 ± 7 whole rock 10017 O'KELLEY (1970)
T4-DFC recycled	Co^{56}	Loosli	X–β–γ	511 847 >2000	0·003 ± 1_5 0·003 ± 1_3 0·009 ± 2_4	115 ± 20	
T4-U	Co^{56}	Loosli	X–β–γ	511 847 >2000	0 ± 0·0014 0·0007 ± 12 0 ± 0·0023	<16	
T4-I	Co^{56}	Loosli	X–β–γ	511 847 >2000	−0·0007 ± 15 0·002 ± 1 −0·0005 ± 23	<8	
T-1	Co^{57}	Loosli	X–β–γ	122	0·002 ± 1_5	<1·6	1·3 ± 1·6 PERKINS (1970)
T4-D	Co^{57}	Loosli	X–β–γ	122	−0·0003 ± 11	<7	
T4-DFC	Co^{57}	Loosli	X–β–γ	122	0·002 ± 1	5·8 ± 2·9	
T4-U	Co^{57}	Loosli	X–β–γ	122	0 ± 0·0009	<1·5	
T4-I	Co^{57}	Loosli	X–β–γ	122	0·002 ± 1_2	<2·3	
T-1	Ni^{59}	Loosli	X–AC	6·9	0·005 ± 1_3	3·3 ± 1·0	
T4-DFCU	Ni^{59}	Loosli	X–AC	6·9	0 ± 0·0016	<3	
T-1	Co^{60}	Loosli	X–β–γ	2500	0·0014 ± 12	<5	0·3 ± 0·6 PERKINS (1970)
T4-DFC	Co^{60}	Loosli	X–β–γ	2500	0 ± 0·0008	<16	0·8 ± 0·6 PERKINS (1970)
T4-U	Co^{60}	Loosli	X–β–γ	2500	−0·0009 ± 7	<1·5	
T4-I	Co^{60}	Loosli	X–β–γ	2500	0·0008 ± 7_4	<4·2	

† Corrected for decay to July 21, 1969.
‡ Sample apparently contaminated—correct β^- energy was not observed.

this source changes slowly with depth near the surface, generally increasing with depth as the number of secondaries increases.

The second certain source is bombardment by solar flare protons and alpha particles. These particles, with energies typically in the 10–100 MeV range, are much less penetrating, and are usually brought to rest by ionization rather than disappearing by nuclear interaction. Their ranges in rock are of the order of millimeters.

Thus the signature of solar flare effects is a rapid decrease with depth below the top surface of the rock. An examination of Table 4 shows several strong examples of this. The difference between 0–4 mm and deeper samples is especially striking for Co^{56} and Fe^{55}, and clearly also present for Na^{22}, Al^{26} and Mn^{53}. Since the last two nuclides are long-lived, we have the first clear evidence for solar flare protons in the past. The long-lived products which do not show a significant depth gradient,

Table 4. Summary of counting results† ‡

Nuclide	$t_{1/2}$	Rock 10017				Bulk fines
		T4-DFC 0–4 mm	T4-U 4–12 mm	T4-I 12–30 mm	T3-SI 60 mm	10084-16 T-1
Be^{10}	$2·5 \times 10^6$ yr	14·0 ± 1·8		15·8 ± 2		16·4 ± 1·6
Na^{22}	2·6 yr	81 ± 11	43 ± 6	37 ± 6	31 ± 4	51 ± 7
Al^{26}	$7·4 \times 10^5$ yr	133 ± 16	74 ± 8	66 ± 10	65 ± 10	107 ± 13
Cl^{36}	3×10^5 yr	$<$33	16·0 ± 1·6	17·4 ± 1·9		16·0 ± 1·5
V^{49}	330 d					8·9 ± 2·4
Mn^{53}	2×10^6 yr	94 ± 24	43 ± 16	50 ± 12	48 ± 21	46 ± 11
Mn^{54}	312 d	36 ± 20	39 ± 26	10 ± 11	36 ± 30	29 ± 10
Mn^{53+54}			93 ± 17	100 ± 18		87 ± 15
Fe^{55}	2·5 yr	445 ± 85	96 ± 29	94 ± 33	98 ± 34	216 ± 40
Co^{56}	77 d	125 ± 20	$<$16	$<$8		40 ± 6
Co^{57}	270 d	5·8 ± 2·9	$<$1·5	$<$2·3		$<$1·6
Ni^{59}	8×10^4 yr	$<$3				3·3 ± 1·0

† Corrected for decay to July 21, 1969
‡ Specific activities in dis/min/kg

Be^{10} and Cl^{36}, are produced in higher-energy reactions. Radionuclide production by solar protons in lunar material is also seen and discussed by O'Kelley *et al.* (1970), Perkins *et al.* (1970), Fireman *et al.* (1970), Keays *et al.* (1970) and Marti *et al.* (1970). Heavy nuclei from the sun produce nuclear tracks (Crozaz *et al.*, 1970; Fleischer *et al.*, Lal *et al.*, 1970).

In the following sections we discuss these two modes of production in as quantitative a way as possible. The analysis will begin with calculation of nuclide production due to galactic cosmic rays. Meteorite data are used to check the model. With this source subtracted, the data for short-lived nuclides in the surface layer are compared with the production calculated for known recent solar particle events. When this has been done, the long-lived nuclides are considered in the inverse way, using the observed residual gradient to calculate the flux and rigidity of solar proton spectra in the past. There seems no need at present to consider other mechanisms of formation.

(b) *Galactic cosmic ray production*

The production of radionuclides by galactic cosmic-rays was calculated in a manner similar to that used by Arnold *et al.* (1961) (hereafter referred to as AHL), with one flux spectrum used to represent the total strong-interacting particle energy-distribution, including both the primary and the secondary particles. The basic shape of the flux spectrum used in the calculations, in units of particles cm^{-2} sec^{-1} MeV^{-1}, was

$$\Psi(E) = \frac{dJ}{dE} = K(\alpha + E)^{-2·5}, \tag{b1}$$

where α determines the shape of the spectrum at low energies and K is a normalization constant. The value of α is 1 BeV for galactic primaries in free space and as secondaries are formed with increasing penetration into matter, α decreases. The value of α that best describes the total particle flux spectrum as a function of depth in the moon is shown in Fig. 4 as determined by LAL and VENKATAVARADAN (1970). The surface has a value of α less than that of the primary cosmic-rays because many of the secondaries produced below the surface travel back to the surface.

As discussed in AHL, the above energy distribution agrees quite well with the observed distribution for energies above several hundred MeV, but at lower energies, this shape is not very accurate for the total flux energy-spectrum in dense matter. A shape that includes more low-energy secondaries is more realistic. A flux distribution that includes more low-energy particles than equation (b1) was obtained by using a separate shape for energies below 100 MeV and using equation (b1) for all energies above 100 MeV. The shape of the distribution used below 100 MeV was derived from the neutron energy spectrum observed in the earth's atmosphere by HESS *et al.* (1959)

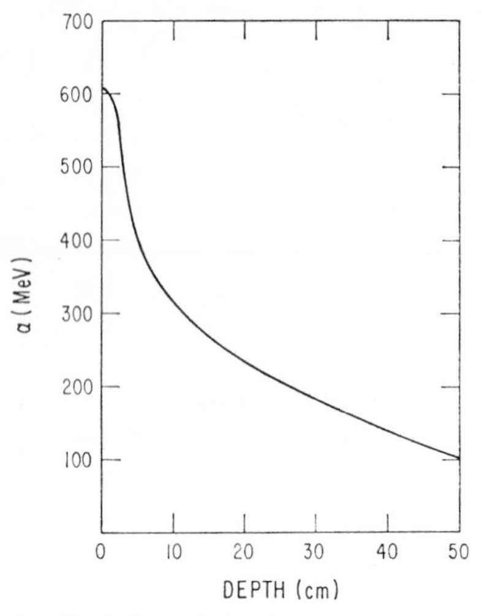

Fig. 4. Variation of the shape parameter α with depth

for neutron energies from 2 to 100 MeV. This shape, denoted H, is very well described by the relation

$$H(2 \leq E \leq 100 \text{ MeV}) = k(E^{-1} + 10E^{-2} + 11E^{-3}) \qquad \text{(b2)}$$

where E is the kinetic energy of the particle in MeV, and is normalized to equation (b1) at 100 MeV. This low energy shape is the same as that used in AHL. The distribution with shape H below 100 MeV and equation (b1) with a value of $\alpha =$ 400 MeV about 100 MeV, denoted $H + (\alpha = 400)$, has the same shape as their $S(10, E)$ distribution and the $H + (\alpha = 200)$ distribution is similar to their $S(100, E)$ distribution.

The normalization constant K was determined by comparing equation (b1) integrated over all energies above 1 BeV to the total flux above this energy for the depth of the sample for which the calculations were being made. This integral flux above 1 BeV was determined assuming an effective absorption mean-free-path for nucleons of 155 g cm^{-2}. This value was used instead of the interaction mean-free-path (100 g cm^{-2}) since many interactions produce fast secondaries that are well collimated in the forward direction and which are above this energy, thus increasing the effective mean-free-path for the total number of particles above 1 BeV. The value used for the free-space integrated flux of protons above 1 BeV averaged over one solar cycle was 1·9 protons cm^{-2} sec^{-1} for 4π sterad (SIMPSON, 1960).

Since the ratio of α-particles to protons above 1 BeV/nucleon is 1/7 (WEBBER, 1962), the contribution of α-particles and the products from the break-up of α-particles to the total flux should be included. Alpha-particles with energies greater than 1 BeV/nucleon were assumed to produce an average of 3·6 particles with a mean interaction length of 90 g cm^{-2}. The contributions from heavier primaries to the total flux were neglected.

Excitation functions (cross-sections as a function of energy) for all the important reactions that produce the radionuclides measured above were determined from experimental results, by analogies to similar reactions, or by calculations using the spallation formulae of RUDSTAM (1966), in a manner similar to that used by AHL, and LAL and VENKATAVARADAN (1967). For high energies, protons and neutrons were considered to have the same cross-sections for forming a given product from a given target. For low energies, neutron-induced reactions were used whenever the difference in proton-induced and neutron-induced reactions for the formation of a specific product from a given target would be significant, e.g. (n, p) reactions (which have no proton-induced reaction equivalent) or (n, α) reactions (which are much more favorable than the corresponding $[p, 3pn]$ reactions). Otherwise, results from proton-induced reactions were used. The excitation functions used for the formation of V^{49} and heavier nuclides from Fe were taken from AHL, with some recent experimental results included for the production of Mn^{54} and Mn^{53} by low-energy neutrons on Fe^{54}.

The cross-sections for the formation of Al^{26} were not well known. These cross-sections were adjusted to reflect the production rate ratios observed by FUSE and ANDERS (1969) for the formation of Al^{26} from Al and Si fractions and to give the observed production rates in meteorites (see below).

For the model as given here, the data used were the flux energy distribution, the chemical composition of the target, and excitation functions for all the important reactions. The flux variation over the sample range was ignored and the energy spectrum of the flux averaged over the entire sample was given by $\Psi(E)$, which usually was taken from equation (b1), or the flux distributions described above that used both equations (b1) and (b2). The flux distributions were normalized as described above. Representing the excitation function by $\sigma(E)$, and the number of target atoms per kilogram of sample by N, the production rate, P, for each reaction in atoms min^{-1} kg^{-1} was calculated from the expression

$$P = \int N\sigma(E)\Psi(E)\,dE. \tag{b3}$$

This model for the calculation of the production rates for formation of radionuclides by galactic cosmic-rays was tested and refined by calculating the production rates for various nuclides that have been measured in meteorites. The L-chondrites Bruderheim and Harleton were used to compare the calculations made using this model to the experimentally determined disintegration rates. Data for Bruderheim are given by HONDA *et al.* (1961) and for Harleton by HONDA and ARNOLD (1967). The pre-fall sample depths are estimated to be 40 and 10 g cm^{-2} for Bruderheim and Harleton, respectively. The attenuation of protons and the attenuation of α-particles and the build-up of secondaries by α-particle break-up were obtained by integration over all solid angles for the assumed geometry of the sample using the techniques described above. The number of particles per cm^2 sec with energies greater than 1 BeV was calculated to be 1·52 for Bruderheim and 2·03 for Harleton. These values, as can be seen by comparing them, are rather insensitive to depth and radius.

A series of calculations were made for each of these meteorites using equation (b3) with the shape parameter of equation (b1), α, varied over a wide range of values for both distribution shapes. Since the spectral shape was only approximately known,

Table 5. Comparison of observed and calculated activities in Bruderheim and Harleton meteorites and in rock 10017 (units of dis/min/kg)

Nuclide	Bruderheim (1·52 particles/cm² sec)† Observed	Bruderheim Calculated $\alpha = 400$ MeV	Bruderheim Calculated $H+$ ($\alpha = 400$)	Harleton (2·03)† Observed	Harleton Calculated $\alpha = 600$ MeV	Harleton Calculated $H+$ ($\alpha = 600$)	Lunar Rock 10017 T4-I (12–30 mm) (1·06)† Observed	Lunar Rock Calculated $\alpha = 600$ MeV	Lunar Rock Calculated $H+$ ($\alpha = 600$)
Short lived									
Fe^{55}	340	135	292	$\leq$180	104	219	94 $\pm$ 33	36	76
Mn^{54}	100	56	97	38	46	76	10 $\pm$ 11‡	16	27
V^{49}	34	25	26	20	25	25	(9)§	9	9
Na^{22}	90	87	99	64	75	84	37 $\pm$ 6	28	32
Long lived									
Mn^{53}	85	56	71	45	47	59	50 $\pm$ 12	16	20
Cl^{36}	7·5	6·5	7·5	7·0	6·7	7·4	17 + 2	8	11
Al^{26}	60	53	67	45	45	56	66 $\pm$ 10	32	43
Be^{10}	19	25	25	21	26	26	16 $\pm$ 2	15	15

† Total flux of particles with energies of greater than 1 BeV used as normalization.
‡ Activities in T4-U and T3-SI are around 35 dis/min/kg.
§ Activity of lunar fines adjusted for chemical composition of rock 10017.

the calculations were used to see which spectral shape gave the best correlation between calculated and experimental results.

The experimentally measured disintegration rates for various radionuclides from these meteorites together with the calculated production rates for several spectral shapes are given in Table 5. The agreement between the calculated and experimental values is very good, especially when a low-energy rise in the energy distribution such as $H+$ ($\alpha = 400$) is used to improve the calculated results for low-energy products such as Fe^{55} and Mn^{54}. The shapes are consistent with what would be expected for the flux distribution at such depths, that is, the deeper sample from the larger object requires more low-energy particles (a lower value for parameter α), reflecting the build-up of secondary particles with depth.

The expected production rates of radionuclides in rock 10017 by galactic cosmic-rays were next calculated using the same model. The outer surface of the rock was taken to have the curvature of a sphere of radius 10 cm and to be exposed to galactic irradiations for a total solid angle of 2π sterad. The calculations were made for the T4-I fragment using a mean depth of 2 cm (7 g cm^{-2}). Using the same normalization techniques as described above, the total flux of particles with energies above 1 BeV was calculated to be 1·06 particles cm^{-2} sec^{-1} (roughly half the Harleton value, reflecting 2π bombardment).

From the experience with the meteorite calculations, the choice of the shape parameter α was fixed within narrow limits. The depth is similar to that of Harleton, for which $\alpha = 600$ MeV gives the best fit (for the calculated variation, see Fig. 4). Since many low-energy secondaries will be coming up from below (both from deeper in the rock and from the lunar regolith), the distribution $H + (\alpha = 600)$ should be a good estimate of the total flux distribution.

While the procedure used to calculate galactic production rates in rock 10017 was consistent with available physical information and theories, and was tested for production rates in meteorites, it should be emphasized that the extrapolation of this method to the moon might introduce some uncertainties in the calculations due to chemical and geometric differences. However, we believe that the results for the calculation of the galactic cosmic-ray production of radionuclides in the deep sample T4-I are, on average, not more than 30 per cent in error. The calculated galactic production rates for T4-I using the $\alpha = 600$ MeV and $H + (\alpha = 600)$ flux distributions are given in Table 5 along with the experimentally determined disintegration rates for the radionuclides measured.

For the short-lived nuclides (Fe^{55}, Mn^{54}, V^{49} and Na^{22}, which have half-lives of 2·6 yr or less), the agreement between the calculated and observed results is generally within the errors of the measured activities. The calculations described below for the known solar flares for the past 10 yr show that the solar contributions for these short-lived isotopes in T4-I are negligible and thus the production of these nuclides should be due almost entirely to galactic cosmic-rays. The comparisons between the results for the $\alpha = 600$ MeV and the $H + (\alpha = 600)$ flux distributions and the experimental results indicate that some additional low-energy flux similar to that in $H + (\alpha = 600)$ probably is needed, but the experimental results are not good enough to allow a quantitative estimate for this low-energy flux.

For the long-lived isotopes (Mn^{53}, Cl^{36} and Al^{26}, which have half-lives of greater than $\sim 3 \times 10^5$ yr) the experimental results tend to be higher than the calculations. Even in the case of the high-energy product Be^{10} there seems to be an excess, since the meteorite values are lower than theoretical, while those for lunar samples are higher. Although, as mentioned above, the different chemistry of the sample introduces some uncertainties in the Al^{26} and Cl^{36} production rate calculations (from the elements Al and Ca, respectively), this is not true for Mn^{53}, which is only produced from Fe and which was fairly well predicted for the two meteorites. A detailed discussion of the observed activities of these long-lived radionuclides in rock 10017 will be given after the contributions of solar particles to the production of radionuclides have been calculated and discussed.

The models of Kohman and Bender (1967) and Honda (1962) represent different approaches to this problem. It would be useful to compare them, but we have not done so.

(c) *Solar particle contributions*

The flux of solar particles has a steep energy distribution. Most of the particles have energies considerably less than 100 MeV. Since these low-energy particles are stopped in several millimeters to a few centimeters of material, the energy spectrum of the flux varies significantly over the range of millimeters, making calculations with the production rate formula used above, equation (b2), very difficult. Hence a new model must be used.

Solar particle energy distributions are usually described by either a rigidity distribution or a power law distribution. The rigidity distribution is

$$\frac{dJ}{dR} = K \exp(-R/R_0), \tag{c1}$$

where R is the rigidity of the particle (rigidity being the momentum of the particle per unit charge, $R = pc/ze$) in units of megavolts (MV), R_0 is a shape parameter, and K is a normalization constant. (A 6 MeV proton has a rigidity of about 100 MV.) Typical R_0's have been about 50–150 MV for recent solar flares for the energy range 10–100 MeV. The power law distribution is

$$\frac{dJ}{dE} = kE^{-\gamma}, \tag{c2}$$

where γ determines the shape and k is a normalization constant. Typical values for γ have been about 2·5–3·5 for the same energy range.

Almost all of the particles emitted from the sun come in short bursts, or flares, whose duration is short compared to the half-lives of observable nuclides. Thus the contribution of each flare to the production of short-lived nuclides can be calculated separately, taking into consideration the integral flux and spectrum of the flare as well as the decay of the various nuclides since the flare.

The average shape and intensity of the solar particles over past solar cycles is not known. LAL and VENKATAVARADAN (1967) made an estimate using the observed flares for the last solar cycle that the average flux above 10 MeV was about 8 protons cm^{-2} sec^{-1} $sterad^{-1}$ with a mean rigidity (R_0) of about 100 MV. This average flux may be too high since the last cycle appears to have been more active than others over the past few hundred years as observed from sun-spot numbers. In the actual calculations, the activities were calculated for flux distributions using a series of shapes (both R_0's and γ's) and an arbitrary normalization. The observed results were then used to determine the flux and spectrum that best fit, as discussed below.

Any calculation for the production rates of radionuclides by solar particles must take into account the rapid slowing and stopping of these low energy particles. This can be seen in the fact that a proton of 50 MeV energy has a range in a lunar rock of only 8 mm. To calculate production rates for these solar particles, the total production of nuclides along their path lengths should be considered.

The calculations were made for successive layers in rock 10017, with the results given in units of atoms formed per minute per cm^2. The production rate expression used for these low-energy solar particles in a layer from the surface to a depth d

involved the product of two terms—one giving the ratio of the number of atoms of a specific nuclide formed to the number of incident particles at an energy E for unit energy loss of the particle as it slows down, $f(E)$, and the other being the total number of particles per unit area and time with energy E somewhere in the layer, $\Phi(d, E)$.

For a differential energy interval $\mathrm{d}E$ at an energy E, the differential range of a particle, $(\mathrm{d}x/\mathrm{d}E)$, was determined in units of mg cm^{-2} MeV^{-1}. If the number of atoms of target element per unit mass of material is N and the cross-section for a given product from this target at an energy E is $\sigma(E)$, the production ratio is

$$f(E) = N \frac{\mathrm{d}x}{\mathrm{d}E}(E)\sigma(E). \tag{c3}$$

For a layer of depth d, the number of particles that have energy E somewhere in the layer, $\Phi(d, E)$, was calculated for a given incident flux on the surface of the layer. For each solid angle Ω, the number of particles that have an energy E somewhere inside the layer is the integral flux from energy E to the energy $E^*(d_\Omega)$. $E^*(d_\Omega)$ is the energy of a particle that, after having traversed a distance d_Ω in the layer, has energy E, where d_Ω is the distance that a particle must travel to reach the depth d for the incident solid angle Ω. The total number of particles, in units of particles cm^{-2} sec^{-1}, with energy E within this layer of depth d is obtained by integrating this integral flux over all solid angles:

$$\Phi(d, E) = \int_0^{4\pi} \mathrm{d}\Omega \int_E^{E^*(d_\Omega)} \Psi(E', \Omega)\, \mathrm{d}E'. \tag{c4}$$

For the moon, the incident differential flux of particles, $\Psi(E', \Omega)$, was taken to be the same for all angles of incidence of less than 90° to the normal of the moon's surface and zero for all other angles. The distance d_Ω includes geometric factors for the shape of the rock as well as the angle of incidence and the depth of the layer. The attenuation of the incident solar-particles by nuclear interactions is negligible for the short distances considered here and was ignored in the calculations.

The production rate in a layer of depth d was calculated using the expression

$$P(d) = \int_0^\infty f(E)\Phi(d, E)\, \mathrm{d}E, \tag{c5}$$

and the results were in units of atoms formed per minute per unit area. This was readily converted to a production rate per unit mass by dividing by the thickness of the layer in units of mass per unit area (g/cm^2). The outer surface of the rock was assumed to have the same curvature as a sphere of radius 10 cm. The range-energy calculations were made using the stopping-power $(\mathrm{d}E/\mathrm{d}x)$ expression as given by WILLIAMSON *et al.* (1966), assuming a density of 3·5 g cm^{-3}.

The excitation functions used for the solar-proton calculations were taken from thin-target proton cross-sections. Many cross-sections were taken directly from the literature and the rest were derived by analogy to similar reactions, in a manner similar to that described by LAL and VENKATAVARADAN (1967). Experimental cross-sections were available for most of the short-lived nuclides from all major targets, but cross-sections for most of the long-lived nuclides have not been measured, although the Mn^{53} excitation function used for protons on Fe was derived at low energies from measured (p, α) reactions on Fe. Contributions from secondary particles formed by the incident solar-particles were neglected since, in the short distances traversed by the solar particles, only a small fraction of the incident particles will have reacted to produce secondaries. This assumption may introduce an appreciable error for harder flux distributions. Cross-sections for the reactions of α-particles on Fe were also compiled.

(d) *Results*

The excess activities of short-lived nuclides in surface layers of the lunar sample can be calculated from data in Tables 4 and 5. The galactic production changes slowly with depth, so that the values calculated for the deep sample, T4-I, are applicable also at the surface. A second method available for short-lived (but not long-lived) nuclides is to subtract the experimental T4-I values from those for the surface sample T4-DF. This is equivalent because with the spectra observed since 1959 very little production from solar bombardment takes place at deeper levels. The net results, representing the top 1·4 g cm^{-2}, are divided by 700 cm^2 kg^{-1} to convert dis/min kg^{-1} to dis/min cm^{-2}.

HSIEH and SIMPSON (1970) and KINSEY and MCDONALD (1970) have kindly provided us with observed particle fluxes and spectra for the period January 1967 to May 1969, mainly from data obtained on the IMP IV satellite. Data from each event are ascribed to a single point in time, since the duration of a solar flare is at most a few days, short compared to the half-lives considered. Solar protons only are considered, except in the case of Co^{57} where alpha particles make an appreciable contribution.

Data for flares of the last cycle (1952–1963) are taken from tables of Webber (in MCDONALD, 1963). Since several of these flares were much larger than anything seen in the present cycle (cycle 20), they contribute strongly to the production of Fe^{55} and Na^{22}.

The integral fluxes are then corrected for decay, using the half-life of each nuclide considered. To aid in visualizing the results, we have found it useful next to convert the total number of protons to an equivalent steady state flux by dividing by the mean life of the nuclide.

As an example, the April 12, 1969 event is reported by HSIEH and SIMPSON (1970) to have produced a total of $1{\cdot}4 \times 10^9$ protons cm^{-2} (4π flux) of 10 MeV or more, with a differential spectrum from 10–100 MeV close to $E^{-3{\cdot}1}$. For Co^{56} this is corrected for decay to the date of collection, July 21, 1969, to $5{\cdot}7 \times 10^8$ protons cm^{-2}. This is equivalent in production to a steady state flux of 60 protons cm^{-2} sec^{-1}. The next most important event for this nuclide, that of November 18, 1968, contributes an equivalent of 8 protons cm^{-2} sec^{-1}. The total for all events is an equivalent of 77 protons cm^{-2} sec^{-1}. The data of KINSEY and MCDONALD (1970) give 57 protons cm^{-2} sec^{-1} on the same basis.

This can now be compared with the calculations made as described above, which are normalized to a 4π steady flux of 100 protons cm^{-2} sec^{-1}. The cross-section data in this case are experimental up to 60 MeV (WILLIAMS and FULMER, 1967).

The results of these calculations are given in Table 6, along with the experimental values. Note changes in both since SHEDLOVSKY *et al.* (1970), due to improved data and calibrations.

The agreement is good enough to leave little doubt that solar particles are an adequate explanation for the effects observed. There is a tendency for experiment to be lower than calculation, especially for Co^{56} for which both are most accurate. This is to be expected. The most obvious sources of error are (1) deviation from the horizontal of our surface sample, (2) loss of fragile surface material from rock 10017 in handling and transit, and (3) directionality of arrival of the solar particles at the beginning of the April 12 event, when Tranquillity Base was in darkness. All these effects tend to lower the observed value.

In particular the cases of Fe^{55} and Na^{22} should be noted. About 50 per cent of their production is calculated to be caused by the flares of cycle 19, not only because

Table 6. Short-lived nuclides in 0–4 mm layer of rock 10017 from solar proton data: Experiment vs. Calculation

Nuclide	$t_{1/2}$	Expt. (dis/min/cm²)	Calc. (dis/min/cm²)
Co^{56}	77 d	0·18	0·22†, 0·30‡
Co^{57}	270 d	0·01	0·007†, 0·008‡
Mn^{54}	312 d	$\leq$0·04	0·02
Fe^{55}	2·6 yr	0·5	0·29
Na^{22}	2·6 yr	0·06	0·06

† Preliminary data of KINSEY and MCDONALD.
‡ Preliminary data of HSIEH and SIMPSON.

of their high integral fluxes but also because of their higher values of R_0 which favor production of these nuclides. We find it notable that data derived from much less direct measurements are as concordant with our experiments as those from recent satellite data.

The long-lived nuclides show a different pattern. Observed activity in deep samples strikingly exceeds the calculated galactic production for Mn^{53}, Al^{26} and Cl^{36}. Even for Be^{10}, where the agreement appears good, normalization to the meteorite results suggests a real excess of roughly 4 dis/min kg^{-1} in T4-I. The depth trends are discussed above.

We have not yet been able to include 3×10^5 yr Kr^{81}, as measured by MARTI *et al.* (1970), in our calculations. However, the observed surface excess is in line with our data.

In our earlier paper (SHEDLOVSKY *et al.*, 1970) we dealt briefly with two models which might explain the data. We present them again here because they figure in the more detailed discussion which follows. Model A ascribes the excess production at depth to a hard solar proton spectrum (high R_0) averaged over a long period of time. The higher energy protons penetrate and activate deeper layers. This model, if correct, yields values for mean flux and rigidity over the mean lives of each nuclide. In Model B the rock is considered to have been buried at a depth of $\sim$100 g cm^{-2} below the surface until a time of the order of 10^5–10^6 years ago. The higher production in this case results from the larger number of low energy secondaries produced by galactic bombardment at these depths. The excess surface activity of Mn^{53} and Al^{26} was produced after the rock came to the surface.

We now give a fuller discussion of our data in terms of each model.

(1) *Model A*. This model makes fewer assumptions about the history of rock 10017. It can be judged by its consistency or inconsistency with the data for the several nuclides. Since Mn^{53} and Al^{26} show the best determined depth profiles, we begin with them.

The excess of Mn^{53} is about 74 dis/min kg^{-1} in the 0–4 mm layer, and about 25 dis/min kg^{-1} in the deeper layers. This profile can be crudely matched with several fluxes and R_0's: an average flux of 25 protons cm^{-2} sec^{-1} and the very high value $R_0 = 400$ MV, a flux of 50 protons cm^{-2} sec^{-1} and $R_0 = 150$ MV, and of course intermediate R_0's and fluxes. The match is better for higher R_0's. Below $R_0 = 150$ the profile becomes unacceptably steep.

For Al^{26} the results are similar. The range of R_0 is 400–125 MV. The required flux ranges from 10 to 25 protons cm^{-2} sec^{-1}, lower than for Mn^{53}. The uncertainty in the calculation is greater here, because of the dependence on target chemistry.

For Be^{10}, a comparatively high-energy product whose half-life is similar to that of Mn^{53}, the results are in poor agreement, especially for high R_0's, since even an excess of 4 dis/min kg^{-1} is produced by a very small flux. The least disagreement is for $R_0 = 125$–150 MV, which narrows the possible range if the data are accepted at face value. Here as for Mn^{53} target chemistry cannot cause serious errors.

For Cl^{36}, the required average flux is high, for example 300 protons cm^{-1} sec^{-1} for $R_0 = 150$ MV. Since the half-life of Al^{26} is $7{\cdot}4 \times 10^5$ yr, and of Cl^{36} 3×10^5 yr, the two fluxes are inconsistent even for the extreme case in which all the proton flux was received recently, and no decay has occurred. The equivalent mean fluxes can be at most in the ratio of the half-lives (Cl^{36}/Al^{26}) or about 2·5, instead of the ratio of about fifteen suggested by this model. However, the production of Cl^{36} by protons is particularly sensitive to target chemistry, and the true excess may be much smaller than we calculate.

WÄNKE *et al.* (1970) have studied the target chemistry of Al^{26} and Cl^{36} by leaching experiments in lunar soil, which may be compared with their earlier work in meteorites. The data suggest that we may have underestimated galactic production of Cl^{36} from Ca. However, their data and ours are not sufficiently precise at the present time to resolve these issues. Hence the Cl^{36} arguments against Model A are not conclusive.

Some support for Model A is provided by FLEISCHER *et al.* (1970) who found a rather hard spectrum of bombarding Fe nuclei from track measurements. Since at present rigidities for protons and heavy nuclei have been found to be similar (FAN *et al.*, 1968), this may indicate that the proton spectrum has been harder in the past.

In summary, this model has the attraction of simplicity and lack of free parameters. It does not appear to account well for the data, but it cannot be excluded.

(2) *Model B*. This model is consistent with a number of observed facts:

1. The rare gas data for rock 10017 show a long bombardment age of order 5×10^8 yr. For the stable rare gases no depth gradients are seen, while Kr^{81} (3×10^5 yr) is higher at the surface (MARTI *et al.*, 1970). This absence of gradients in the stable

nuclides demonstrates that the rock was bombarded by a uniform flux through nearly all its history. Unless solar protons are a recent phenomenon, the long bombardment must have occurred while the rock was buried (not too deeply because of the large integral production) beneath the surface. The very high integral neutron flux observed by ALBEE *et al.* (1970) also requires a moderately deep bombardment.

2. Nuclear track studies of several groups (CROZAZ *et al.*, 1970; FLEISCHER *et al.*, 1970; LAL *et al.*, 1970) show ages calculated from the flux of Fe nuclei of the order of 10^7 yr. Since this age is much shorter than the rare gas age, there is further evidence of burial over long periods. The track age is much longer than that required for the exposure in Model B, but this can be accounted for by repeated exposure.

3. Galactic production is higher for Mn^{53} and Al^{26} at depth than it is at the surface, as can be seen from the work of HONDA (1962), or the comparison of the data for Bruderheim and Harleton (Table 5). The estimated values of 35 dis/min kg^{-1} Mn^{53} and 55 dis/min kg^{-1} Al^{26} at a depth of 100 g cm^{-2} are close to those observed in T4-I.

In Model B, the time of last ejection of rock 10017 to the surface is an arbitrary parameter. The difficulty is to fix it recently enough so that the bulk of the activity produced at depth still remains, and far enough back to permit development of the observed depth profiles. The uncertainty in the half-life of Mn^{53} (KAYE and CRESSY, 1965), 1–2 $\times$ 10^6 yr, also enters the picture. The problem is easier if the half-life is shorter. For Cl^{36} the depth effect is not marked, and the apparent excess could be ascribed mainly to target chemistry.

Sample calculations indicate that times for the most recent ejection of the order of 10^5–10^6 yr ago can give acceptable solutions for Mn^{53} and Al^{26}. Any low R_0, in the range 50–150 MV, is acceptable for the observed surface excess. The required flux is of the same order as that required for Co^{56} and Fe^{55}.

Thus Model B, at the price of extra assumptions, appears to fit the data fairly well. In considering the complexities of both models, however, we hope the reader will not lose sight of the basic fact that the existence of solar proton effects over a long time period has been demonstrated.

The lunar dust data (Table 4) are consistent with the rock observations. After allowance for chemical differences in some cases, the results are intermediate between the 0–4 mm and deeper samples of rock 10017. These data suggest that allowing for the lower density, the bulk fines come from an average depth of 2–3 cm, which is fairly consistent with the description of the sampling (SHOEMAKER *et al.*, 1970).

Consistency of short- and long-lived activities shows that gardening has not altered the profiles much on a 10^6 yr scale.

The Ni^{59} result in the soil is especially interesting. This nuclide, like Co^{56}, is expected to show a very low activity in lunar material by galactic secondaries and neutrons, because of the low content of target Co and Ni (Table 1). $Ni^{59}(n, \gamma)$, $Ni^{60}(n, 2n)$ and $Co^{59}(p, n)$ all give very low production. The activity seen must be ascribed to production by alpha particles on Fe, nearly all by the reaction $Fe^{56}(\alpha, n)Ni^{59}$. This result is independent of model. The half-life of 8 $\times$ 10^4 yr falls in an otherwise empty region.

Taking into account the limit set on sample T4-DFU, we estimate roughly that a mean flux of order 10^1 alphas cm^{-2} sec^{-1} over this mean life is required. This result is not very sensitive to R_0.

A firm decision on models, fluxes, and rigidities for solar protons must await analysis of Apollo 12 rocks.

Acknowledgments—NORMAN FONG and FLORENCE KIRCHNER provided support and assistance throughout this work. Both are unlisted members of SHRELLDALFF. Shop and electronic support were provided by JACK HOLLON, RONALD WORLEY and DONALD SULLIVAN; general laboratory support was provided by BARRY NALL. The success of E. SCHONFELD in orienting rock 10017 by non-destructive counting laid the foundation for this work. We have profited by discussion with dozens of colleagues, and cannot list them all. J. HSIEH, J. N. KINSEY, F. B. MCDONALD and JOHN SIMPSON have provided much information and advice about solar particles. Discussions with G. ARRHENIUS, K. MARTI, G. D. O'KELLEY, H. C. UREY and R. WALKER have been especially useful in formulating our ideas. This research was supported by NASA Contract NAS 9-7891.

REFERENCES

ALBEE A. L., BURNETT D. S., CHODOS A. A., EUGSTER O. J., HUNEKE J. C., PAPANASTASSIOU D. A., PODOSEK F. A., RUSS G. PRICE II, SANZ H. G., TERA F. and WASSERBURG G. J. (1970) Ages, irradiation history, and chemical composition of lunar rocks from the Sea of Tranquillity. *Science* **167,** 463–466.

AMIN B. S., LAL D., LORIN J. C., PELLAS P., RAJAN R. S., TAMHANE A. S. and VENKATAVARADAN V. S. (1969) On the flux of low energy particles in the solar system. In *Meteorite Research*, (editor P. Millman), pp. 316–327. Reidel.

ARNOLD J. R., HONDA M. and LAL D. (1961) Record of cosmic-ray intensity in the meteorites. *J. Geophys. Res.* **66,** 3519–3531.

BHANDARI N. G. (1969) A selective and versatile low level beta-, X- and gamma-ray detector assembly. *Nucl. Instrum. Methods* **67,** 251–256.

CROZAZ G., HAACK U., HAIR M., HOYT H., KARDOS J., MAURETTE M., MIYAJIMA M., SEITZ M., SUN S., WALKER R., WITTELS M. and WOOLUM D. (1970) Solid state studies of the radiation history of lunar samples. *Science* **167,** 563–566.

EVANS J. C., JR., DELANY A. C. and LOOSLI H. H. (1970) High-sensitivity counting systems used for lunar samples. In preparation.

FAN C. Y., PICK M., PYLE R., SIMPSON J. A. and SMITH D. R. (1968) Protons associated with centers of solar activity and their propagation in interplanetary magnetic field regions corotating with the sum. *J. Geophys. Res.* **73,** 1555.

FIREMAN E. L., D'AMICO J. and DEFELICE J. C. (1970) Tritium and argon radioactivities in lunar material. *Science* **167,** 566–568.

FLEISCHER R. L., HAINES E. L., HANNEMAN R. E., HART H. R., JR., KASPER J. S., LIFSHIN E., WOODS R. T. and PRICE P. B. (1970) Particle track, X-ray, thermal, and mass spectrometric studies of lunar material. *Science* **167,** 568–571.

FUSE K. and ANDERS E. (1969) Aluminum-26 in meteorites. VI Achondrites. *Geochim. Cosmochim. Acta* **33,** 653–670.

HAHN P. F. (1945) Radioactive iron procedures: Purification, electroplating and analysis. *Ind. Eng. Chem. Anal. Ed.* **17,** 45–46.

HERR W., HERPERS U., HESS B., SKERRA B. and WOELFLE R. (1970) Determination of manganese-53 by neutron activation and other miscellaneous studies on lunar dust. *Science* **167,** 747–749.

HESS W. N., PETERSON H. W., WALLACE R. and CHUPP E. L. (1959) Cosmic-ray neutron energy spectrum. *Phys. Rev.* **116,** 445–457.

HILLEBRAND W. F., LUNDELL G. E. F., BRIGHT H. A. and HOFFMAN J. I. (1953) *Applied Inorganic Analysis*, 2nd edition, Chap. 31, pp. 499–500. John Wiley.

HONDA M. (1962) Spallation products distributed in a thick iron target bombarded by 3-BeV protons *J. Geophys. Res.* **67,** 4847–4858.

HONDA M. and ARNOLD J. R. (1967) Effects of cosmic rays on meteorites. In *Handbuch der Physik XLVI/2*, pp. 613–632. Springer-Verlag.

HONDA M., UMEMOTO S. and ARNOLD J. R. (1961) Radioactive species produced by cosmic rays in Bruderheim and other stone meteorites. *J. Geophys. Res.* **66,** 3541–3546.

HSIEH J. and SIMPSON J. A. (1970) Private communication.

KAYE J. H. and CRESSY P. J. (1965) Half-life of manganese-53 from meteorite observations. *J. Inorg. Nucl. Chem.* **27,** 1889–1892.

KEAYS R. R., GANAPATHY R., LAUL J. C., ANDERS E., HERZOG G. F. and JEFFERY P. M. (1970) Trace elements and radioactivity in lunar rocks: Implications for meteorite infall, solar-wind flux, and formation conditions of moon. *Science* **167,** 490–493.

KINSEY J. H. and MCDONALD F. B. (1970) Private communication.

KIRBY L. J. (1961) The radiochemistry of Ni. *Nucl. Sci. Ser. NAS-NS*-3501, 32–33.

KIRSTEN T. A. and SCHAEFFER O. A. (1970) High energy interactions in space. In *Interactions of Elementary Particle Research in Science and Technology*, (editor L. C. L. Yuan). Academic Press, to be published.

KOHMAN T. P. and BENDER M. L. (1967) Nuclide production by cosmic rays in meteorites and on the moon. In *High Energy Nuclear Reactions in Astrophysics*, (editor B. S. P. Shen). Benjamin.

LAL D., MACDOUGAL D., WILKENING L. and ARRHENIUS G. (1970) Cosmic ray spectra, transuramic elements and mixing of the lunar regolith: New evidence from fossil particle-track studies. *Geochim. Cosmochim. Acta*, Supplement I.

LAL D. and PETERS B. (1967) Cosmic-ray produced radioactivity on the earth. In *Handbuch der Physik XLVI/2*, pp. 551–612. Springer-Verlag.

LAL D., RAJAGOPALAN G. and RAMA (1967) Sensitive and descript β and β–γ counting assemblies. In *Radioactive Dating and Methods of Low-Level Counting*, pp. 615–627. I.A.E.A., Vienna.

LAL D. and VENKATAVARADAN V. S. (1967) Activation of cosmic dust by cosmic ray particles. *Earth Planet. Sci. Lett.* **3,** 299–310.

LAL D. and VENKATAVARADAN V. S. (1970) Private communication.

LIBBY W. F. (1956) Relations between energy and half-thickness for absorption of beta radiation. *Phys. Rev.* **103,** 1900–1901.

MARTI K., LUGMAIR G. W. and UREY H. C. (1970) Solar wind gases, cosmic ray spallation products, and the irradiation history. *Geochim. Cosmochim. Acta*, Supplement I.

MARTI K., SHEDLOVSKY J. P., LINDSTROM R. M., ARNOLD J. R. and BHANDARI N. G. (1969) Cosmic-ray produced radionuclides and rare gases near the surface of St. Severin meteorite. In *Meteorite Research* (editor P. Millman), pp. 246–266. Reidel.

MCDONALD F. B. (1963) Solar proton manual NASA report. NASA Tr R-169, pp. 1–17.

MORRISON G. H. and FREISER H. (1962) *Solvent Extraction in Analytical Chemistry*, pp. 195, 201–202, 231. John Wiley.

O'KELLEY G. D., ELDRIDGE J. S., SCHONFELD E. and BELL P. R. (1970) Elemental compositions and ages of lunar samples by nondestructive gamma-ray spectrometry. *Science* **167,** 580–582.

PARKIN D. W. (1965) Private communication.

PERKINS R. W., RANCITELLI L. A., COOPER J. A., KAYE J. H. and WOGMAN N. A. (1970) Cosmogenic and primordial radionuclides in lunar samples by nondestructive gamma-ray spectrometry. *Science* **167,** 577–580.

ROSSI B. (1952) *High Energy Particles*, Appendix V. Prentice-Hall.

RUDSTAM G. (1966) Systematics of spallation yields. *Z. Naturforsch.* **21a,** 1027–1041.

SHOEMAKER E. M., BAILEY N. G., BATSON R. M., DAHLEM D.H., FOSS T. H., GROLIER M. J., GODDARD E. N., HAIT M. H., HOLT H. E., LARSON K. B., RENNILSON J. J., SCHABER G. G., SCHLEICHER D. L., SCHMITT H. H., SUTTON R. L., SWANN G. A., WATERS A. C. and WEST M. N. (1970) Geologic setting of the lunar samples returned by the Apollo 11 mission. In *Apollo 11 Preliminary Science Report*, *NASA SP*-214, 41–83.

SHEDLOVSKY J. P., HONDA M., REEDY R. C., EVANS J. C., LAL D., LINDSTROM R. M., DELANY A. C , ARNOLD J. R., LOOSLI H.-H., FRUCHTER J. S. and FINKEL R. C. (1970) Pattern of bombardment–produced radionuclides in lunar soil and in rock 10017. *Science* **167,** 574–576.

SIMPSON J. A. (1960) Variations of solar origin in the primary cosmic radiation. *Astrophys. J. Suppl.* **IV,** 393–401.

TANAKA E., ITOH S., MARUYAMA T., KAWAMURA S. and HIRAMOTO T. (1967) Low-level beta spectroscopy of solid samples by means of a coincidence-type scintillation spectrometer combined with a logarithmic amplifier. *Int. J. Appl. Rad. Isotop.* **18,** 161–175.

WÄNKE H., BEGEMANN F., VILCSEK E., RIEDER R., TESCHKE F., BORN W., QUIJANO-RICO M., VOSHAGE H. and WLOTZKA F. (1970) Major and trace elements and cosmic-ray produced in lunar samples. *Science* **167,** 523–525.

WEBBER W. B. (1962) Time variations of low rigidity cosmic rays during the recent sunspot cycle. *Progress in Elementary Particle and Cosmic Ray Physics* **VI,** 75–243.

WILLIAMS I. R. and FULMER C. B. (1967) Excitation functions for radioactive isotopes produced by protons below 60 MeV on Al, Fe and Cu. *Phys. Rev.* **162,** 1055–1061.

WILLIAMSON C. F., BOUJOT J. P. and PICARD J. (1966) Tables of range and stopping power of chemical elements for charged particles of energy 0·5 to 500 MeV. Report CEA-R 3042, Commissariat a l'Energie Atomique.

Proceedings of the Apollo 11 Lunar Science Conference, Vol. 2, pp. 1533 to 1574.

Uranium–thorium–lead isotopes in some Tranquillity Base samples and their implications for lunar history

LEON T. SILVER
Division of Geological Sciences, California Institute of Technology, Pasadena, California 91109

(*Received* 25 *February* 1970; *accepted in revised form* 5 *March* 1970)

Abstract—U–Th–Pb isotopic studies of four rocks, a breccia, and lunar fines yield: (1) U (ppm) in 4 rocks average 0·53; breccia, 0·66; fines, 0·55. (2) Th (ppm) in rocks average 1·90; breccia, 2·02; fines, 2·17. (3) Pb (ppm) in rocks average 1·11; breccia, 1·86; fines, 1·46. (4) Th/U for rocks is 3·6; breccia, 3·05; fines, 3·9. U^{238}/Pb^{204} ranges from 65 to 912, and implies extreme Pb^{204} depletion on the moon compared to earth and meteorites. All leads are very radiogenic, 75 per cent to more than 95 per cent, and well-defined radiogenic daughter–parent relations exist. Apparent Pb^{207}/Pb^{206} ages for all four rocks range from 4·13–4·22 billion years and Pb–U ages are nearly concordant. U–Th–Pb systematics in the breccia and fines yield nearly concordant apparent "ages" at 4·60–4·63 billion years. These ages have no real time significance but indicate the presence of old radiogenic lead components in the composite debris. Acid leaching studies have demonstrated that the leads in the fines and breccia are mixtures of extraordinarily heterogeneous components having Pb^{207}/Pb^{206} ratios which vary by more than 50 per cent. Two ancient components, tentatively identified, are (1) old radiogenic rock leads with associated parents and (2) old parentless radiogenic leads. Model calculations which remove lead isotope components of Tranquillity Base volcanic age from the regolith, indicate an excess of radiogenic lead with a composite Pb^{207}/Pb^{206} ratio of $1·96 \pm 0·14$. This yields an apparent age of $4·95 \pm 0·10$ billion years which would be a minimum age for the oldest debris in the regolith and, presumably, for the moon. Volatile transfer is believed responsible for the parentless leads and is suggested as a major lunar geological process.

INTRODUCTION

THE GREAT age of the lunar materials returned from Tranquillity Base is one of the most significant conclusions that has emerged from the initial scientific studies. Their antiquity is supported by isotopic relations involving K^{40}–Ar^{40}, Rb^{87}–Sr^{87} and the U–Th–Pb isotope systems in rocks, breccias and fines as reported by many workers. It is further apparent that the several different isotopic systems *do not* provide identical time interpretations. They suggest, instead, a complex, rather than simple, history of evolution for the materials which contain the radiogenic isotopes. Initial attempts to infer an "age" for lunar samples or for the moon from the available data must be viewed, in fact, as preliminary rather than definitive. Many assumptions involving the geological setting, sampling, initial conditions, lunar physical processes and the geologic stability of isotopic systems go into each age interpretation. In this new (lunar) world even elementary assumptions must be validated carefully.

It is appropriate, therefore, at this point in the evolution of lunar chronology to consider all isotope ages as "apparent" and subject to possible revision when the assumptions used in interpreting the measured isotope ratios can be tested adequately.

* Contribution No. 1726, Division of Geological Sciences, California Institute of Technology, Pasadena, California 91109.

This work is a modification, expansion and more extended discussion of the uranium, thorium and lead isotope systematics observed in the chemical and isotopic analysis of four rocks, a breccia sample, and a sample of lunar fines, recently reported in *Science* (SILVER, 1970). There are some slight to significant revisions of earlier concentration data. Several series of leaching experiments on the fines and on one rock have added significantly to the understanding of the distribution of uranium, thorium and lead isotopes in these materials. From this new work a preliminary model for both the timing and nature of the evolution of these isotopic systems is emerging which may place a significant minimum on the age of the moon as well as refine the age of the rock-forming events which produced the various rock fragments that have been studied. Perhaps equally of interest is information leading to a new insight into important lunar processes responsible for the observed distributions of lead, uranium and thorium on the lunar surface.

SAMPLES

Six aliquots of lunar material were received. Lunar samples 10017-34, 10045-30 and 10072-39 were fine-grained volcanic rocks, each consisting of a dozen or more small chips. Lunar sample 10047-30, a medium-grained gabbro, consisted of a single sawed fragment and some loose granules. Lunar sample 10060-15, a breccia, consisted of eight or ten small pieces, showing internal fragments of rock and glass, including glass spherules, in a fine-grained, poorly sorted, dark matrix. A sample of lunar fines, 10084-35, which had been passed through a 1 mm sieve, completed the suite.

The three finer-grained samples consisted of freshly broken fragments but in each case evidence of some external surfaces was observed. In all cases, numerous internal cavities, both spherical and irregular in form, often with handsome coarser crystal aggregates, were prominent. In the coarser sample, similar cavities, somewhat less abundant, were visible, although the sawn face covered with a greyish brown dust was the most prominent surface. The breccia had one or two fragments with a slightly lighter colored, perhaps browner, surface.

The mineralogy and texture of these rocks have been described elsewhere (AGRELL *et al.*, 1970; ANDERSON *et al.*, 1970; KEIL *et al.*, 1970; and others). Two of them, 10072 and 10045, are olivine-bearing types with average grain sizes of about 0·1 mm and are easily assigned to the Lunar Sample Preliminary Examination Team (LSPET, 1969) classification, group A. Sample 10017 is similar in average grain size to the previous two but has scattered lath-shaped feldspar crystals up to 5 mm in length, with cristobalite but little or no olivine. It is perhaps a transition type as suggested by AGRELL *et al.* (1970). Sample 10047 is a medium-grained cristobalite gabbro, clearly belonging to LSPET group B. All of the rocks share the major minerals, sub-calcic clinopyroxenes, calcic plagioclase and ilmenite. Olivine and/or cristobalite may also be present. A small and variable amount of a significant mesostatic glass, rich in silica and alkali feldspar molecules, is also present together with traces of metallic iron, troilite, apatite and other more rare minerals. The distribution of uranium, thorium and lead among these various phases bears directly on any interpretation of the chemical and isotopic abundances reported here and will be discussed elsewhere in the paper.

Analytical Procedures

Sample aliquots of the rock were taken in the range of $1\frac{1}{2}$ to 4 g by selecting an appropriate number of typical fragments. The original samples were not homogenized initially by intensive grinding in order to avoid excessive exposure to contamination and to preclude reaching an irreversible physical state for second stage experiments. Because of uncertainty concerning the mineralogy no preliminary washing of fragments was attempted. The rock fragments for a specific experiment were crushed in a stellite-faced diamond mortar, with about half a dozen blows, to reach 0·2 mm or less. The powder was transferred via glassine boat directly to a teflon reaction vessel.

Samples of the dust were simply aliquoted by dipping into the original polystyrene container with an acid-washed stainless steel spatula and transferring directly to the weighing boat and teflon dish. No attempt was made to split or handle the dust further.

The basic chemical procedure is an evolutionary one modified by our experience during this work, and utilizing the excellent $Ba(NO_3)_2$ precipitation techniques of Tatsumoto (1966, and personal communications). The sample is attacked with HF and $HClO_4$ in teflon. The residue is leached with hot concentrated HNO_3 for a prolonged period. The acid is separated by centrifugation from a final residue rich in titanium compounds and insoluble fluorides. Care must be taken to insure complete attack of the original rock materials and adequate leaching of the residue. We have found a strong tendency for thorium, in particular, to coprecipitate in the titanium residue. The acid solution is aliquoted for lead and uranium composition analysis and for concentration determinations of uranium, thorium and lead by isotope dilution.

Lead is isolated with a $Ba(NO_3)_2$ coprecipitation step followed by a dithizone–chloroform solvent extraction procedure. The uranium and thorium are isolated by organic solvent extraction from salted aqueous solutions.

Parallel blanks were run for all rock samples. Blanks for lead ranged from 0·04–0·12 μg per rock analysis and tended to decrease to about 0·05 μg as our procedural experience with the lunar samples increased. In the acid leaches described in a later section, blanks were about 0·04 μg for each leach analysis step. Uranium and thorium blanks were consistent at $0{\cdot}002 \pm 0{\cdot}001$ μg per analysis.

Checks on yields have been made by spiking aliquots of sample prior to chemical attack for comparison and by reprocessing residues. Small quantities of uranium and significant quantities of the thorium (up to 30 per cent) were observed in some residues. It was found necessary, therefore, to reprocess all rock residues to confirm our original yields. Significant secondary recoveries were made for two rocks, 10060-15 and 10017-34, included in the preliminary report (Silver, 1970).

A series of simple leaching experiments were performed on the lunar fines 10084 and on a crushed powder of rock 10017. These consisted of stirred exposures of the sample to cold and hot concentrated HNO_3 in a centrifuge bottle for various periods of time as will be described. The acid was separated by centrifugation and extracted as described for the whole rocks. The residues were treated as rocks and analysis was initiated by an appropriate HF and $HClO_4$ attack.

Isotopic analyses of lead were made in a silica gel-phosphoric acid medium on a rhenium filament (modified from the procedure of Cameron *et al.*, 1969). Uranium and thorium were analyzed isotopically as nitrate salts by a rhenium triple filament procedure. Samples were analyzed in a 12-in.-radius solid source mass spectrometer using an electron multiplier and a digital data readout system.

Mass spectrometric precision (1σ) generally was on the order of 1–3 per cent for Pb^{206}/Pb^{204} ratios and better than one per cent for Pb^{206}/Pb^{207}, Pb^{206}/Pb^{208}, U^{238}/U^{235} and Th^{232}/Th^{230}. In a few cases of poor intensity or instability, the uncertainties for the lead data increase as much as 2–3 times. The best indications of the mass spectrometric reproducibility are found in the data for the radiogenic Pb^{207}/Pb^{206} ratios obtained in a given sample in various analyses as can be seen in Table 1, for example. In general the precision on this ratio is about 1 per cent. The method was calibrated with repeated runs of Caltech Shelf Standard Lead. Average values obtained for this standard are: $\frac{206}{204} = 16{\cdot}553$; $\frac{206}{207} = 1{\cdot}0768$; $\frac{206}{208} = 0{\cdot}46112$. Comparison with National Bureau of Standards values for this lead, suggests a small fractionation correction (0·2% per mass unit), which has been applied to all of the reported lead data.

The concentration precision is on the order of $\frac{1}{2}$–1 per cent for the uranium and thorium and is essentially unaffected by the blanks. For total lead, the concentration precision varies with sample

Table 1. U–Th–Pb isotope data for some Apollo 11 materials

Sample		Pb composition						Concentrations μg/g of rock treated Corrected for blanks			
		Observed			Corrected for blank						
No. and wt.	Type	$\frac{206}{204}$	$\frac{207}{204}$	$\frac{208}{204}$	$\frac{206}{204}$	$\frac{207}{204}$	$\frac{208}{204}$	U	Th	Th/U	Pb^{total}
10084-35* 4·73 g	D-Fines	141·2	92·80	153·8	171·2	111·2	181·7	0·562	2·172	3·86	1·465
10060-15 3·00 g	C-Breccia	77·77	51·97	82·31	83·24	55·27	86·42	0·663	2·024	3·05	1·859
10017-34† 2·54 g	A-Rock	206·9	99·81	225·6	257·0	122·1	289·2	0·784	2·961	3·78	1·673
10072-39 No. 1 1·07 g SBA	A-Rock							0·884	3·348	3·79	1·586
1·43 g No. 2		242·2	115·8	266·8	563·3	259·0	594·1	0·831	2·935	3·54	1·618
10045-30 No. 1 2·81 g	A-Rock	83·2	45·46	102·0	117·8	61·24	135·9	0·259	0·869	3·36	0·450
1·72 g No. 2 SBA								0·199	0·662	3·33	0·514
10047-34 No. 1 1·00 g SBA	B-Rock							0·246	0·849	3·45	0·741
4·26 g No. 2		67·79	39·25	82·25	79·03	44·41	92·14	0·203	0·600	2·95	0·798
Average of four rocks								0·530	1·900	3·58	

Table 1. (continued)

Sample			Concentrations in units of 10^{-8} moles/g of rock treated					
			Radiogenic isotopes					
No. and wt.	Type	Pb^{total}	Pb^{206}	Pb^{207}	Pb^{208}	Pb^{204}	U	Th
10084-35	D-Fines	0·7076	‡0·2333	‡0·1457	‡0·2192	0·00152	0·2359	0·9358
4·73 g			0·2463	0·1536	0·2318			
10060-15	C-Breccia	0·8983	0·2594	0·1584	0·1845	0·00366	0·2785	0·8720
3·00 g			0·2939	0·1791	0·2268			
10017-34†	A-Rock	0·8084	0·2885	0·1288	0·3045	0·00089	0·3294	1·2758
2·54 g			0·2977	0·1329	0·3129			
10072-39 No. 1	A-Rock	0·7664	0·2950	0·1317	0·3008	0·00054	0·3712	1·4428
1·07 g SBA			0·2991	0·1335	0·3046			
1·43 g No. 2		0·7815	0·3053	0·1361	0·3128	0·00038	0·3490	1·2649
			3·3055	0·1363	0·3130			
10045-30 No. 1	A-Rrock	0·2174	0·0685	0·0314	0·0675	0·00067	0·1032	0·3660
2·81 g			0·0740	0·0366	0·0722			
1·72 g No. 2		0·2485	0·0786	0·0360	0·0772	0·00079	0·0836	0·2853
SBA			0·0847	0·0387	0·0828			
10047-34 No. 1	B-Rock	0·3582	0·1011	0·0479	0·0900	0·00166	0·1088	0·3744
1·00 g SBA			0·1139	0·0535	0·1016			
4·26 g No. 2		0·2857	0·0806	0·0382	0·0717	0·00132	0·0853	0·2587
Average of four rocks			0·0908	0·0427	0·0810			

Table 1. (continued)

Sample No. and wt.	Type	Isotope ratios Pb^{206}/U^{238}	Pb^{207}/U^{235}	Pb^{207}/Pb^{206}	Pb^{207}/Pb^{206} §	Pb^{208}/Th^{232}	Pb^{208}/Pb^{206}	U^{238}/Pb^{204}	Apparent ages (m.y.) Pb^{206}/U^{238}	Pb^{207}/U^{235}	Pb^{207}/Pb^{206}	Pb^{208}/Th^{232}
10084-35*	D-Fines	0·996	85·8	0·628	0·634	0·234	0·940	155	4500	4600	4635	4310
4·73 g		1·052	90·4	0·623	0·629	0·248	0·941		4675	4645	4630	4535
10060-15	C-Breccia	0·938	79·0	0·610	0·607	0·217	0·731	76	4305	4505	4600	4030
3·00 g		1·062	89·3	0·609	0·605	0·260	0·772		4530	4625	4595	4735
10017-34†	A-Rock	0·882	54·3	0·446	0·447	0·239	1·051	370	4115	4125	4130	4385
2·54 g		0·904	56·0	0·446	0·449	0·245	1·055		4190	4160	4130	4495
10072-39 No. 1	A-Rock	0·900	49·3	—	0·440	0·208	—	682	3825	4030	4130	3880
1·07 g SBA		0·812	50·0		0·438	0·211			3865	4045	4130	3925
1·43 g No. 2		0·881	54·2	0·446	0·445	0·247	1·019	912	4110	4125	4130	4530
		0·882	54·2	0·446	0·444	0·247	1·018		4115	4125	4130	4530
10045-30 No. 1	A-Rock	0·670	42·3	0·458	0·453	0·184	0·982	153	3340	3875	4170	3460
2·81 g		0·721	46·4	0·457	0·453	0·197	0·977		3530	3970	4170	3685
1·72 g No. 2		0·946	59·8	—	0·456	0·271	—	105	4335	4225	4165	4910
SBA		1·020	64·2		0·455	0·290			4520	4300	4160	5220
10047-34 No. 1	B-Rock	0·936	61·1	—	—	0·240	—	65	4300	4245	4220	4410
1·00 g SBA		1·054	68·3			0·271			4680	4360	4210	4950
4·26 g No. 2		0·952	62·2	0·474	0·470	0·277	0·889	65	4355	4265	4220	5015
Average of four rocks		1·073	69·5	0·470	0·468	0·313	0·892		4745	4380	4210	5580

* See also composite leaching experiment data, Table 3.
† See also composite leaching experiment data, Table 2.
SBA = Spiked before attack.
‡ Paired values: Upper concentration based upon a contamination lead correction. Lower concentration based upon a model initial lead correction.
§ $\frac{Pb^{207}}{Pb^{206}}$ ratios calculated from Pb concentration analyses, where blank and spike corrections are small.

size and the size of the procedural blanks. Samples generally involved about 2 ± 1 μg Pb and under good conditions involved an uncertainty of about 1–2 per cent. For rocks with less lead (10045, 10047), 2–3 per cent is probably a better value. In determining concentrations of radiogenic lead components an additional uncertainty is derived from the selection of the composition of the initial lead to be used in correcting the isotopic composition. This will be discussed in the section on analytical results. Tests of the overall precision of the concentration determinations may be made by comparing the results of the composite analyses of the multi-step leach experiments with the results of the direct sample analyses (see Tables 2 and 3).

The derived daughter–parent isotope ratios from which apparent ages have been calculated contain all of the uncertainties discussed. For the Pb^{206}/U^{238} and Pb^{207}/U^{235} ratios a 2–3 per cent error assignment is appropriate. A larger uncertainty exists for Pb^{208}/Th^{232} because of thorium retentivity in the residues, and an error of about 5 per cent is tentatively assigned.

The constants used in calculating ages from the measured isotope ratios are: $U^{238}\lambda = 1{\cdot}537 \times 10^{-10}$; $U^{235}\lambda = 9{\cdot}72 \times 10^{-10}$; $Th^{232}\lambda = 4{\cdot}88 \times 10^{-11}/yr^{-1}$. The U^{238}/U^{235} ratio is taken as 137·8 following the work of ROSHOLT and TATSUMOTO (1970) and FIELDS *et al.* (1970).

ANALYTICAL RESULTS

Sample analyses

Presented in Table 1 are analytical data, calculated isotope ratios, and apparent ages derived from the direct analysis of all of the samples. For three of the rocks there are duplicate concentration determinations from experiments in which spikes were added to the reaction vessels before chemical attack. Additional replication can be found for samples 10084-35 and for 10017-34 in Tables 2 and 3 where composite analyses derived from leach experiments can be compared with the direct analytical results. Concentration data is presented both in micrograms/gram and moles/g to facilitate interpretation of abundances and isotope ratios.

The observed lead isotopic compositions are shown along with the composition ratios corrected for blanks. The blank lead correction has values Pb^{206}/Pb^{204}, 18·2; Pb^{207}/Pb^{204}, 15·6; Pb^{208}/Pb^{204}, 38·0; and is the average of a number of direct measurements. The uranium, thorium and lead concentrations have all been corrected for blanks. The thorium:uranium ratio has been calculated on a weight ratio basis.

The observed lead isotopic compositions are from 75 to more than 95 per cent radiogenic. Comparable enrichments have not been reported from terrestrial basaltic materials. These enrichments reflect (a) very low concentrations of initial or original leads containing significant Pb^{204}, relative to the observed uranium and thorium concentrations, and (b) very great ages for these materials.

Pure initial leads, taken up by the rocks at the time of their crystallization or at the time of their isolation as isotopic systems, have not been extracted successfully from the Apollo 11 samples in this or other work. To estimate the quantity of original lead in each rock, it was necessary to infer a composition. Model lead compositions based upon earth and meteorite lead evolution data have been assumed. The apparent values calculated range from 0·06 to 0·4 ppm initial lead and are lower than concentrations found in the most primitive oceanic tholeites (TATSUMOTO, 1966). Yet these calculated values are probably upper limits for the true levels. The blank procedures apply only to contamination in chemistry, and do not correct for those contaminations developed during prior physical processing.

It has been suggested (TATSUMOTO and ROSHOLT, 1970) that the initial lead taken

Table 2. Apollo 11 Lunar Fines Sample 10084-35

	Pb composition						Concentrations $\mu g/g$ of rock treated Corrected for blanks			
	Observed			Corrected for blank and spike contribution*						
	206/204	207/204	208/204	206/204	207/204	208/204	U	Th	Th/U	Pb^{total}
Experiment No. 4 1·2660 g of fines leached with conc. HNO_3, 80°C for 1½ hr, centrifuged for 1 hr										
Leachate 4A*	116·79	85·67	n.a.	205·26*	151·10*	n.a.	0·262	1·201	4·58	n.a.
Residue 4B*	91·06	55·43	n.a.	128·63*	76·00*	n.a.	0·277	0·910	3·28	n.a.
Combined data No. 4							0·539	2·111	3·91	n.a.
Experiment No. 6 1·4131 g of fines leached with conc. HNO_3, sequentially										
Leachate 6A* 15 min, 25°C, stirred. 15 min centrifuge	72·09	56·22	n.a.	89·59*	69·34*	n.a.	0·080	0·591	7·36	n.a.
Leachate 6B* 30 min, 70°C, stirred 15 min centrifuge	43·75	35·16	n.a.	56·07*	44·64*	n.a.	0·037	0·201	5·44	n.a.
Leachate 6C 7 hr, 70°C, stirred. 1 hr centrifuge,	139·14	91·64	143·96	272·66	175·56	260·53	0·230	0·923	4·02	0·551
Residue 6D HF, $HClO_4$ attack	74·59	45·01	81·24	100·38	58·28	100·74	0·190	0·519	2·73	0·425
Combined data No. 6							0·537	2·234	4·16	n.a.
Total sample, experiment No. 2	141·8	92·80	153·79	171·22	111·16	181·74	0·562	2·172	3·86	1·465

Table 2. (continued)

	Isotope ratios						Apparent ages			
	Pb^{206}/U^{238}	Pb^{207}/U^{235}	Pb^{207}/Pb^{206}	Pb^{208}/Th^{232}	Pb^{208}/Pb^{206}	U^{238}†$/Pb^{204}$	Pb^{206}/U^{238}	Pb^{207}/U^{235}	Pb^{207}/Pb^{206}	Pb^{208}/Th^{232}
Experiment No. 4 1·2660 g of fines leached with cone NO_3, 80°C for 1½ hr, centifuged for 1 hr										
Leachate 4A*	1·259	120·9	0·718	n.a.	n.a.	139·4	5300	4940	4835	n.a.
	1·328	126·4	0·713				5490	4985	4825	
Residue 4B*	0·818	61·7	0·547	n.a.	n.a.	138·6	3890	4260	4435	n.a.
	0·881	67·0	0·551				4110	4340	4445	
Combined data No. 4	1·032	90·5	0·636	n.a.	n.a.	139·0	4615	4645	4655	n a.
	1·098	95·9	0·633				4820	4705	4650	
Experiment No. 6 1·4131 g of fines leached with conc. HNO_3, sequentially										
Leachate 6A* 15 min, 25°C, stirred. 15 min centrifuge	1·537	159·5	0·753	n.a.	n.a.	46·8	6055	5225	4900	n.a.
	1·723	174·2	0·736				6520	5315	4865	
Leachate 6B* 30 min, 70°C, stirred. 15 min centrifuge	1·182	125·1	0·767	n.a.	n.a.	31·7	5080	4975	4930	n.a.
	1·456	147·6	0·736				5850	5145	4870	
Leachate 6C 7 hr, 70°C, stirred. 1 hr centrifuge,	0·998	86·4	0·629	0·208	0·874	254·7	4505	4600	4640	3870
	1·031	89·3	0·628	0·216	0·877		4525	4630	4640	4000
Residue 6D HF, $HClO_4$ attack	0·819	58·8	0·521	0·222	0·763	101·2	3895	4210	4360	4105
	0·896	63·4	0·514	0·246	0·776		4165	4290	4340	4515
Combined data No. 6	1·028	90·1	0·631	n.a.	n.a.	95·2	4600	4640	4645	n.a.
	1·116	96·8	0·629				4875	4715	4640	
Total sample, experiment No. 2	0·996	85·8	0·628	0·248	0·940	155·1	4500	4600	4635	4535
	1·052	90·4	0·623	0·234	0·941		4675	4645	4630	4310

Table 2. (continued)

	Concentrations in units of 10^{-8} moles/g of rock treated						
		Radiogenic isotopes					
	Pb^{total}	Pb^{206}	Pb^{207}	Pb^{208}	Pb^{204}‡	U	Th
Experiment No. 4 1·2660 g of fines leached with conc. HNO_3, 80°C for 1½ hr, centrifuged for 1 hr							
Leachate 4A*	n.a.	0·1376 § 0·1451	0·0959 0·1002	n.a.	0·00079	0·1101	0·5173
Residue 4B*	n.a.	0·0946 0·1019	0·0518 0·0562	n.a.	0·00084	0·1164	0·3919
Combined data No. 4	n.a.	0·2322 0·2470	0·1476 0·1564	n.a.	0·00163	0·2265	0·9092
Experiment No. 6 1·4131 g of fines leached with conc. HNO_3, sequentially							
Leachate 6A* 15 min, 25°C, stirred. 15 min centrifuge	n.a.	0·0514 0·0577	0·0387 0·0423	n.a.	0·00072	0·0337	0·2547
Leachate 6B* 30 min, 70°C, stirred. 15 min centrifuge	n.a.	0·0182 0·0224	0·0140 0·0165	n.a.	0·00049	0·0155	0·0847
Leachate 6C 7 hr, 70°C, stirred. 1 hr centrifuge,	0·266	0·0955 0·0987	0·0599 0·0620	0·0835 0·0866	0·00038	0·0964	0·4016
Residue 6D HF, $HClO_4$ attack	0·205	0·0650 0·0711	0·0338 0·0365	0·0396 0·0552	0·00079	0·0799	0·2237
Combined data No. 6	n.a.	0·2302 0·2500	0·1463 0·1573	n.a.	0·00238	0·2256	0·9647
Total sample, experiment No. 2	0·7076	0·2333 0·2463	0·1457 0·1536	0·2192 0·2318	0·00152	0·2359	0·9358

n.a. Not available because of Pb^{208} spike interference.
* Pb^{208} spike added before analysis.
† Because of possible pre-chemistry contamination these are limiting minimum values.
‡ Pb^{204} remaining after blank correction.
§ Paired values: Upper concentration based upon a contamination lead correction. Lower concentration based upon a model initial lead correction.

Table 3. Apollo 11 lunar sample 10017-34 leach experiment

	Pb composition						Concentrations (ppm)			
	Observed			Corrected for blank						
	206/204	207/204	208/204	206/204	207/204	208/204	U	Th	Th/U	Pb^{total}
Experiment No. 2 1·1710 g of rock crushed <0·5 mm leached with conc. HNO_3, 80°C, for 1½ hr, centrifuged for 1 hr Leachate 2A	114·92	58·99	188·36	192·05	93·62	307·88	0·276	1·911	6·92	0·7046
Residue 2B	229·20‡	107·86‡	205·26‡	370·50	169·70	317·14	0·580	1·167	2·01	1·0832
Combined data for Experiment No. 2							0·856	3·078	3·60	1·788
Total Sample, Experiment No. 1	206·87	99·81	225·64	256·99	122·11	289·95	0·784	2·961	3·78	1·673
TATSUMOTO and ROSHOLT (1970)§				410·0	191·9	435·0	0·854	3·363	4·07	1·56

Table 3. (continued)

	Concentrations in units of 10^{-8} moles/g of rock treated						
		Radiogenic isotopes					
	Pb^{total}	Pb^{206}	Pb^{207}	Pb^{208}	Pb^{204}	U	Th
Experiment No. 2							
1·1710 g of rock							
crushed <0·5 mm							
leached with conc. HNO_3,							
80°C, for 1½ hr,							
centrifuged for 1 hr							
Leachate 2A	0·3397	0·0955†	0·0455	0·1545	0·00057	0·1159	0·8237
		0·1045	0·0466	0·1585			
Residue 2B	0·5238	0·2153	0·0941	0·1706	0·00061	0·2436	0·5027
		0·2200	0·0962	0·1749			
Combined data for Experiment No. 2	0·8635	0·3148	0·1386	0·3251	0·00118	0·3595	1·3264
		0·3245	0·1428	0·3334			
Total Sample, Experiment No. 1	0·8084	0·2885	0·1288	0·3045	0·00089	0·3294	1·2758
		0·2977	0·1329	0·3129			
TATSUMOTO and ROSHOLT (1970)§	0·7536	0·2848	0·1282	0·2885	0·00072	0·3587	1·4489
		0·2904	0·1306	0·2936			

Table 3. (continued)

	Isotope ratios						Apparent ages			
	Pb^{206}/U^{238}	Pb^{207}/U^{235}	Pb^{207}/Pb^{206}	Pb^{208}/Th^{232}	Pb^{208}/Pb^{206}	U^{238}*/Pb^{204}	Pb^{206}/U^{238}	Pb^{207}/U^{235}	Pb^{207}/Pb^{206}	Pb^{208}/Th^{232}
Experiment No. 2										
1·1710 g of rock										
crushed <0·5 mm										
leached with conc, HNO_3,										
80°C, for 1½ hr										
centrifuged for 1 hr										
Leachate 2A	0·865	53·3	0·449	0·188	1·552	203	4055	4110	4140	3520
	0·908	55·8	0·446	0·192	1·556		4205	4155	4130	3605
Residue 2B	0·890	53·7	0·437	0·339	0·792	399	4140	4115	4100	5985
	0·910	55·8	0·438	0·348	0·795		4210	4135	4100	6120
Combined data for Experiment No. 2	0·882	53·6	0·440	0·245	1·033	305	4115	4115	4110	4490
	0·909	55·6	0·440	0·251	1·027		4210	4150	4110	4595
Total Sample, Experiment No. 1	0·882	54·3	0·446	0·239	1·051	370	4115	4125	4130	4385
	0·904	56·0	0·446	0·245	1·055		4190	4160	4130	4495
Tatsumoto and Rosholt (1970)	0·800	49·6	0·450	0·199	1·013	493	3825	4035	4145	3720
	0·816	50·6	0·450	0·203	1·011		3905	4080	4165	3780

* Minimum values because of possible pre-chemistry contamination.
† Paired values: Upper value based upon a contamination lead correction. Lower value based upon a model initial lead correction.
‡ Includes a large correction (20%) for background peak at the 204 position.
§ Recalculated using the common lead corrections applied in this work.

up in the rocks may be a secondary lead with a significant radiogenic component. This suggestion is certainly worthy but evidence will be presented that argues against such a component forming a significant fraction of the presently observed lead in the rocks. If the assumed initial leads are erroneous, the data still permits calculation of values for the apparent concentrations of Pb^{204} in the rocks, as shown. These, for contamination reasons, are maximum values.

The radiogenic lead isotope concentrations are reported in pairs of values. The upper value is based on the assumption that all Pb^{204} observed is associated with terrestrial contamination. The lower value assumed an initial lead, for about 4·2 b.y. ago, based on earth lead evolution of: Pb^{206}/Pb^{204}, 10·5; Pb^{207}/Pb^{204}, 12·2; Pb^{208}/Pb^{204}, 31·0. There is no independent support for the existence of such a lead, but its effect on the concentration calculations provides a probable maximum limit to the concentrations of the radiogenic isotopes generated in these rocks. Almost all other possible initial leads will tend to reduce concentration levels relative to these two sets of values.

Comparison of the observed and blank-corrected lead compositions reported in this work for 10017 and 10084, with the more radiogenic values reported on the same materials by TATSUMOTO (1970) is instructive.

			Observed			Corrected for blank	
		206/204	207/204	208/204	206/204	207/204	208/204
LTS	10017	206·9	99·81	225·6	257·0	122·1	289·2
MT	10017	367·7	173·7	390·2	410·0	191·9	435·0
LTS	10084	141·2	92·8	153·8	171·2	111·2	181·7
MT	10084	237·4	155·3	246·7	261·9	171·0	270·1

The fine analytical work of Tatsumoto yielded observed leads with less Pb^{204} by a factor of nearly two than is reported here. That the data from the two laboratories do not converge *after* blank correction indicates that (a) the blank assignments in one laboratory or the other are incorrect, or (b) that pre-chemistry contamination of the samples is different, and greater in this work. The effects of all initial lead corrections are greatest where the Pb^{204} content is highest. At the levels observed in this work it is reasonable to infer the Pb^{204} is predominantly derived from a terrestrial common lead contamination. Therefore, in all further discussions the upper lead values and the isotope ratios derived from them are preferred.

The uranium concentrations for the four rocks range from 0·2 to 0·9 ppm and average about 0·53 ppm, a figure essentially identical to that for the fines and close to that for the breccia. Similarly the thorium concentrations in the rocks range from 0·6 to 3·35 ppm and average 1·90 ppm, a figure also similar to that found in the soil and breccia. In these concentrations these rocks are similar to oceanic alkali basalts and are enriched 5–10 times compared to oceanic tholeites. The Th/U ratio in the rocks is 3·6 and in the soil is 3·86. The Th/U ratios also resemble the values observed in alkali basalts, while contrasting with the oceanic tholeites.

Comparison of these data with the results of TATSUMOTO and ROSHOLT (1970) on other rock and soil samples confirms the similarities of the soil values to the average of nine rocks.

The values listed for the ratio U^{238}/Pb^{204} range from 65 to 912, and are minimum values for the true indigenous isotope ratios in the rocks. They are extraordinary in comparison with the same ratio in terrestrial basalts which rarely exceeds 10.

For each of the four rocks analyzed in this work, there is one duplicate analysis. For the rock 10017-34 there is a composite value derived from the leach experiment data reported in Table 3. For the other three rocks replication is in the form of aliquots of crushed rock to which spikes were added before the chemical attack. The pairs of concentration data show large internal differences between aliquots in concentrations of uranium and thorium in almost all cases. Uranium differences range from 6 to 30 per cent and thorium differences range from 4 to 40 per cent of the smaller value.

	U ppm	Th ppm
Nine rocks	0·53	1·96
Soil (ave. of 3)	0·55	2·17
Soil (T & R)	0·54	2·09

The Th/U ratios remain relatively constant (±10 per cent) indicating the sympathetic variation of the two elements. The derived Pb^{206}/U^{238} and Pb^{208}/Th^{232} isotope ratios do not reflect the concentration variations except, perhaps, in the case of 10045-30. That analytical reproducibility is generally within the error limits assigned is believed adequately demonstrated in the results of the three major analytical experiments on the lunar fines 10084, reported in Table 2. Sample inhomogeneity and inadequate sample size appear to be the origin of the observed differences in the small aliquots of rock. The distribution of uranium and thorium in these rocks is apparently too nonuniform for representative sampling at levels of 1–4½ g. Future plans for sample distributions of lunar samples should consider this problem.

Isotope ratios relevant to age and geochemical considerations are presented along with apparent ages calculated from the three radiogenic isotope systems. The ratios are reported in pairs reflecting the two assumed initial lead values and their effects on lead isotope concentrations.

The Pb^{206}/U^{238} and Pb^{207}/U^{235} ratios for the pairs of rock analyses are in excellent agreement for 10047-34 and 10017-34 (Tables 1 and 3). The values for 10072-39 differ by 10 per cent, falling somewhat outside the assigned analytical precision limits. Values for 10045-30 differ by 45 per cent of the lesser value and constitute a major discrepancy. For the breccia, 10060-15, a single set of data is available. For the soil, 10084-35, ratios by direct analysis shown in Table 1 compare very closely with two composite sets of ratios shown in Table 2.

The reasons for the two discrepant sets of rock data can only be inferred. Unrecognized analytical errors or natural heterogeneity in daughter-parent isotope distribution on the scale of the sampling may be responsible factors. Subjectively, the former is minimized and the latter is suspected, but until additional work on these samples is completed, no more definitive statement can be made. In the case of 10045-30 the smaller values, representing an apparent 30 per cent loss of daughter lead isotopes, would imply a greater isotopic disturbance than reported for any other lunar rock. Review of the analytical data in the light of our experience with the high TiO_2 residues, leads us to prefer the second, more concordant set of data for 10045, at this time, but a set of averaged values may be more correct.

The Pb^{208}/Th^{232} results present similar problems of uncertainty. Great pains were taken to insure complete yields of thorium, but Pb^{208}/Th^{232} apparent ages tend to be distinctly higher (up to 15 per cent) than the associated U–Pb ages. TATSUMOTO and ROSHOLT (1970) did not observe a similar displacement. In this work, therefore, the Pb^{208}/Th^{232} results are treated with suitable caution.

The most impressively reproducible isotopic data, and seemingly independent of most of the analytical concerns, are the radiogenic lead Pb^{207}/Pb^{206} ratios. The ratios are reproducible both from the isotopic composition runs and the better concentration determinations. They show a limited range of values for all four rocks, 0·446–0·474, implying ages of 4·13 to 4·22 b.y. They are essentially insensitive to the two types of initial lead corrections evaluated here, but would be somewhat more sensitive to an initial radiogenic lead correction. In the excellent data of TATSUMOTO and ROSHOLT (1970) where such a correction has been applied, the calculated Pb^{207}/Pb^{206} ages have a range of 350 m.y. Following the premise that essentially all observed Pb^{204} in lunar rocks is associated with terrestrial contamination leads, a recalculation of that data reduces the spread of apparent ages to 90 m.y., from $4{\cdot}14$ to $4{\cdot}23 \times 10^9$ y, in remarkable agreement with this work. Conversely, because the blank-corrected data from this work is less radiogenic, the application of a radiogenic initial lead correction would disperse the Pb^{207}/Pb^{206} ratios even more than is observed in their treatment of their data. Additional arguments will be introduced against a significant radiogenic initial lead contribution.

In summary, the basic pattern of isotope relations in the rocks analyzed here is one of close similarity in radiogenic character of the lead; concordance or near concordance of apparent ages in the U–Pb isotope systems; and near agreement to moderately higher ages in the Pb^{208}/Th^{232} apparent ages.

The isotope systematics in the fines, 10084-35, and breccia, 10060-15, are analytically internally consistent, and distinctly different isotopically from the rocks. The Pb^{206}/Pb^{027} apparent ages are 4·63–4·65 b.y. for the fines and 4·60 b.y. for the breccia. The pairs of concentration-dependent isotope values shown (Tables 1, 2) are based upon initial leads of contamination (upper) and primordial (lower) composition. The primordial lead composition is taken from MURTHY and PATTERSON (1962) as Pb^{206}/Pb^{204}, 9·6; Pb^{207}/Pb^{204}, 10·4; Pb^{208}/Pb^{204}, 29·7. The Pb^{206}/Pb^{238} and Pb^{207}/U^{235} apparent ages are close to concordant for both samples. The Pb^{208}/Th^{232} ages are slightly younger. Since both of these samples are aggregates of debris from diverse sources the implications of near concordance must be carefully considered.

The U–Pb isotope systematics of all of the analyzed samples are graphically presented in Fig. 1, a "concordia" diagram (WETHERILL, 1956) suitable for representing radiogenic Pb^{206}/U^{238}–Pb^{207}/U^{235} systems. The smooth curve, "concordia," is the locus of modern isotope ratios in ideal closed uranium systems of all ages which originated without initial leads, and which give ages in agreement over the ratio range shown in the graph. The dimensions of the sample figures are dimensions of uncertainty which incorporate various analytical error assignments and the uncertainties due to selection of initial lead corrections. The figures tend to be conservative and the preferred values are the small points. For sample 10045-30, the second analysis has been selected over the first. Data points for the first analysis would fall markedly off "concordia," about 30 per cent closer to the origin. With this exception, the

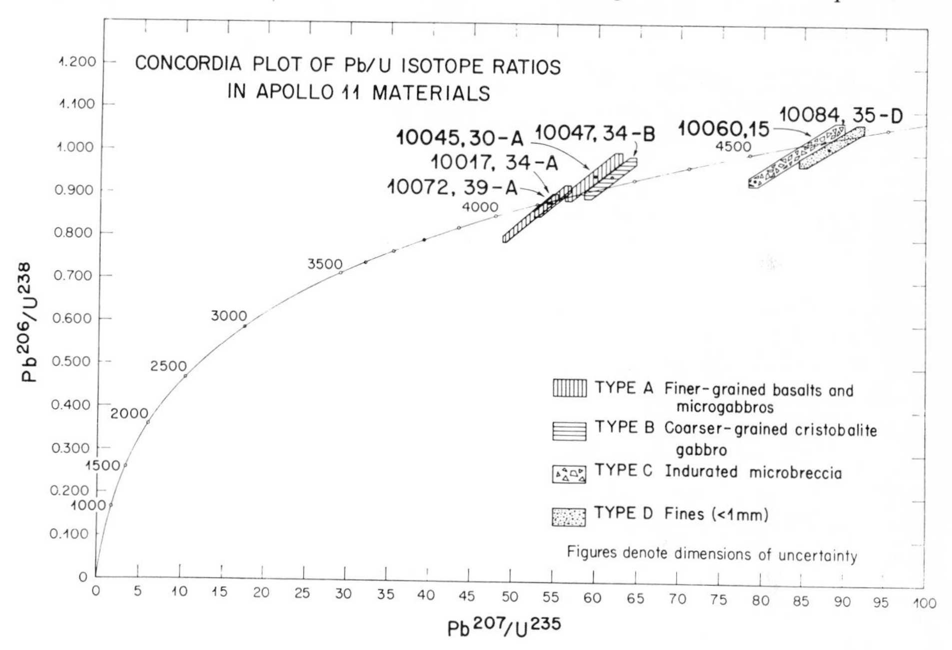

Fig. 1. "Concordia" diagram representation of Pb^{206}/U^{238} and Pb^{207}/U^{235} isotope relations in the analyzed rocks, breccia and lunar fines. Samples are represented by points and error figures whose dimensions reflect analytical and calculation uncertainties discussed in the text.

plotted values are close to "concordia" reflecting the near agreement of the Pb–U calculated ages. The difference in apparent ages between the soil and breccia and the four rocks is clearly illustrated.

Acid leach experiments

Two types of lunar materials, fines and a rock, were subjected to leaching with concentrated HNO_3 in searches for evidence of isotopic inhomogeneity. The leaches and residues have been analyzed for the elemental and isotopic abundance data in the same fashion as the whole rocks and the results are given in Tables 2 and 3.

Two aliquots of lunar fines ($<1{\cdot}0$ mm) were leached. In experiment No. 4, 1·266 g of fines were leached with concentrated HNO_3 at 80°C for $1\frac{1}{2}$ hr with intermittent stirring, then were centrifuged for 1 hr. The insoluble residue resembled the original sample in volume. The supernatant leachate (4A) and the residue (4B) were

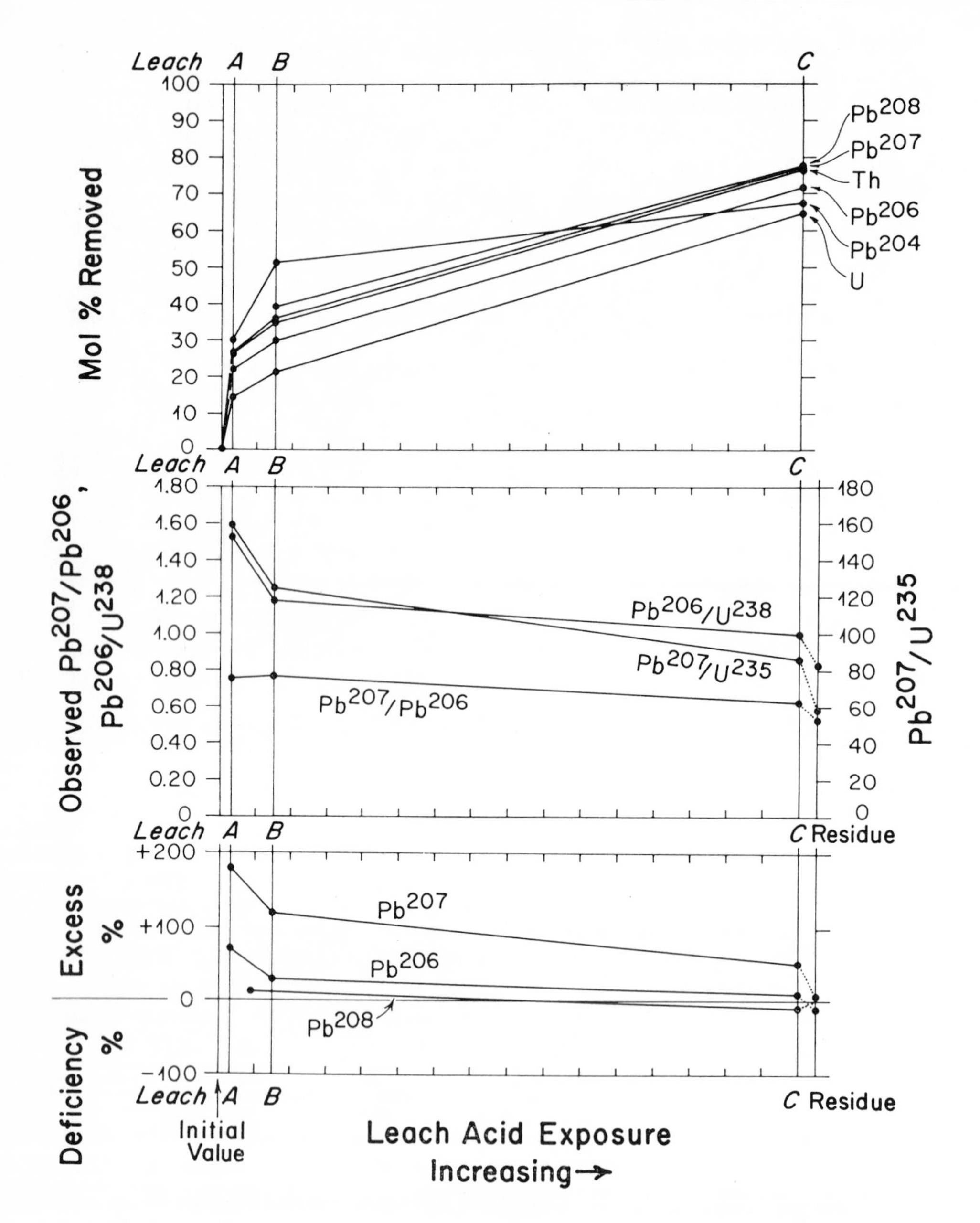

Fig. 2. Graph showing results of acid leach experiment No. 6 on lunar fines 10084-35. Upper portion shows cumulative isotope yields in mol.%. Middle portion shows variations in isotope ratios of U–Pb systems in successive leaches. Lower portion shows the excess of deficiency of lead isotopes observed over that calculated to have been produced from observed uranium and thorium in the last 4·175 b.y. Abscissa is an approximate index of leach acid exposure taking into account duration and temperature of leaches.

each spiked directly and analyzed separately and the results were combined into a composite set of values to check the material balance. The leachate contained 49% of the uranium, 58% of the thorium, 48% of the Pb^{204}, 59% of radiogenic Pb^{206} and 65% of the radiogenic Pb^{207}. Data for Pb^{208} is not directly available because of interference of the Pb^{208}-rich spike.

The Pb^{207}/Pb^{206} ratio observed in the leachate from excellent mass spectrometric data is 0·718, 13 per cent higher than that found in the whole sample. The Pb^{207}/Pb^{206} value found in the residue is 0·547, appropriately reduced. These results clearly established an extreme degree of isotopic heterogeneity among the lead components which comprise the lead of the total lunar soil. The Th/U ratio in the leachate differs by nearly 50 per cent from the residue suggesting that thorium and uranium are partitioned in a non-uniform manner in the various materials of the fines. An excellent material balance was established.

To further investigate this effect, a series of sequential HNO_3 leaches on a second aliquot of fines was attempted in experiment No. 6. The conditions are described in Table 2. They were intended to provide, through time and temperature control, three successive steps differing by roughly an order of magnitude in chemical reactivity in the presence of an excess of acid. The first two leaches were spiked directly with Pb^{208}; only Pb^{206}/Pb^{207} concentration data were obtained in the individual leaches. Composition and concentration analyses were made on both leach C and the residue. From the lead composition data and material balance considerations, the radiogenic Pb^{208} behavior could be traced for the combined leach A and B. Some of the results are presented graphically in Fig. 2. The abscissa, leach acid exposure, is not truly linear because of the problems of converting centrifugation time to exposure time. In addition the interval of the first exposure (leach A, in cold acid) is exaggerated on the scale by a factor of about 3, for purposes of resolution in the illustration.

Experiment No. 6 confirmed the great lability of these isotope systems. The first, quick leach in the series yielded for Pb^{204}, 30% of total sample content; for the radiogenic isotopes, Pb^{206}, 22%; Pb^{207}, 27%; Pb^{208}, 25–30% (estimated); for the parent isotopes U, 15%, Th, 26%. The Pb^{207}/Pb^{206} ratio corrected from the spiked lead aliquot was 0·753. The Th/U ratio of the first leach is 7·4, nearly twice that of the total sample. Although the precise value is not known, the Pb^{208}/Pb^{206} ratio clearly tends to follow the Th/U ratio. These data indicate selective rates of attack on the various isotope sites in the fines.

In the successive leaches the yield rates declined considerably but dissolution was apparently sustained throughout the entire series. By the end of the series the cumulative leach yields were Pb^{204}, 67%; radiogenic isotopes, Pb^{206}, 73%; Pb^{207}, 77%; Pb^{208}, 77%; parents, U, 64%; Th, 77%. A good material balance was established from the analysis of the residue which had retained most of the major chemical constituents.

In the second leach the Pb^{207}/Pb^{206} ratio rose slightly to 0·767 and then declined sharply. In the residue the ratio was 0·521, more than 30 per cent lower than the initial yield values. The Pb^{206}/U^{238} and the Pb^{207}/U^{235} isotope ratios were initially extraordinarily high and then declined very rapidly to values in the residue that were 53 per cent and 37 per cent respectively of the initial leached values.

The variations of the Pb^{206}/U^{238} and Pb^{207}/U^{235} in the sequence are shown in Fig. 3, a "concordia" diagram covering a larger ratio range than Fig. 1.
This expansion was necessary because the apparent "ages" of the first leach were 6055 and 5225 millions of years! Significant relations visible in the plot are:

(1) The extreme discordance and location of leach 6A.
(2) The drastic successive displacement and near concordance of leaches No. 6B and 6C and of the residue.
(3) The apparent concordance of the composite which contains such extremely heterogeneous components.
(4) The close approach of the residue to values observed in the nine rocks which have been analyzed from the Apollo 11 site.

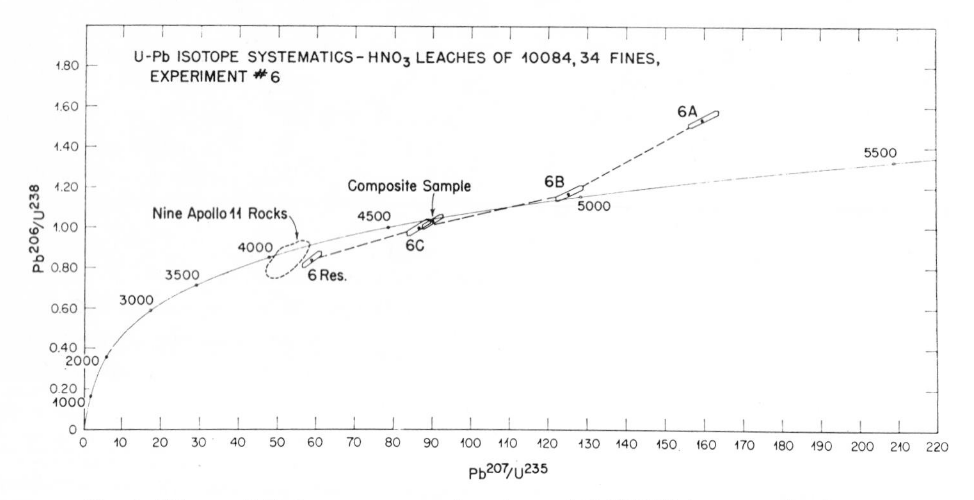

Fig. 3. "Concordia" diagram representing observed Pb^{206}/U^{235} isotope ratios in successive leaches of leach experiment No. 6 on lunar fines 10084-35. Composite value is also shown. Error figures represent dimensions of uncertainty. Dashed enclosed area is occupied by nine Apollo 11 rock analyses.

A leach experiment was performed on a crushed powder of lunar rock 10017-34 to isolate and compare the leach response of a typical rock constituent found in the lunar debris with the response observed in the lunar surface fines. The leach conditions were quite similar to the experiment No. 4 conducted on the fines. They are reported in Table 3, together with the analytical results. Complete composition and concentration data were obtained on leachate and residue.

The leachate of the single leach step yielded large fractions of the total content of parent and daughter isotopes in the rocks, although most of the original sample material was retained in the residue. The Th/U ratio in the leachate is 6·9 compared to 3·6 for the whole sample. The analyzed residue completed a satisfactory material balance for the isotope systems by comparison with the direct analysis of the rock and with an independent analysis by TATSUMOTO and ROSHOLT (1970). The leach experiment aliquot had somewhat higher uranium and thorium concentrations, another reflection of probable sample inhomogeneity at this scale of sampling.

The isotope ratios calculated for the leachate and the residue are nearly identical with the total sample for the U–Pb systems, but differ drastically in the Th–Pb behavior. The Pb^{207}/Pb^{206}, Pb^{206}/U^{238} and Pb^{207}/U^{235} ratios are constant within ± 2 per cent. The apparent ages calculated from these ratios for either fraction are essentially concordant, as they are for the composite sample. In contrast, the Pb^{208}/Th^{232} in the leachate is nearly a factor of two less than in the residue. The Pb^{208}/Pb^{206} ratio is a factor of two greater in the leachate than in the residue. It is clear (1) that thorium and Pb^{208} are not uniformly distributed with uranium and its daughters throughout the rock; (2) that the acid attacks the thorium sites preferentially and differentially, and introduces an apparent chemical discordance in the Pb^{208}/Th^{232} system; (3) that the acid-soluble uranium sites are attacked with no differentiation between parents and daughters; and (4) that since all of the sites that contributed acid-soluble Pb^{207} and Pb^{206} are uniform in their Pb^{207}/Pb^{206} ratios and since all remaining sites in the residue have similar Pb^{207}/Pb^{206} ratios, the Pb^{207}/Pb^{206} ratio is uniform throughout the rock.

A comparison of the responses of the crushed rock and the soil exposed to similar leach conditions is given in Table 4 which shows that the rock yields its thorium and

Table 4. Comparison of acid leach responses of lunar fines 10084-35 and crushed rock 10017-35 under similar conditions
Yields in percent of total content

	U	Th	Pb^{206}	Pb^{207}	Pb^{208}	Pb^{204}
10017-34 Experiment No. 2A	32	62	30	33	48	48
10084-35 Experiment No. 4A	49	58	59	65	45*	48
Yield rate ratio	0·65	1·07	0·51	0·51	1·07	1·00

* Estimated from cumulative curve for Pb^{208} from Fig. 2 for Experiment No. 6.

Pb^{208} approximately at the same percentage rate as the soil; its uranium at about two-thirds the rate of the soil; and its Pb^{206} and Pb^{207} at about half the rate of the soil. The Pb^{204} yields are very similar in the rock and soil.

It appears from this comparison that if the fine debris in the soil is derived in large part from mineral and rock fragments like 10017-34, much of the soluble U–Th–Pb isotopes observed in the soil leaches may be contributed from that material. It is also apparent that much of the Pb^{204} is very readily removed by the leach procedures from soil and rock, tending to support an attribution to contamination sources.

Discussion

Geologic context of the samples

The regolith on the lunar surface at Tranquillity Base from which Armstrong and Aldrin collected these rock and soil samples does not provide a simple geological context for the interpretation of these analytical results. Its debris-strewn and crater-pocked character has been well-described and photographed (Shoemaker *et al.*, 1969, 1970a; Aldrin *et al.*, 1969). Although nearby craters probably penetrated to bedrock, no samples were obtained that can be positively assigned to rock exposures at the site. It has been necessary to construct circumstantial arguments for the attribution of the samples to their sources.

Tranquillity Base is in the southwestern part of Mare Tranquillitatis, about 40 km north of the nearest highlands region (Kant Plateau) and 40 km or more from the nearest large craters (Moltke and Sabine). Although some distinct rays from large distant craters cross the Mare within 15 km of the site, it appears that the regolith at the site probably is comprised dominantly from the materials which form this part of Mare Tranquillitatis (SHOEMAKER *et al.*, 1970) and is a result of local impact cratering. Fragmental debris derived from the local rocks has accumulated to an estimated depth of 3–6 m, and SHOEMAKER *et al.* (1970b) have developed a model for the evolution of this debris layer in which about 5 per cent is expected to come from distances greater than 100 km. This provides for the possibility of perhaps 5 per cent of the regolith, predominantly in the finer size fractions, having lunar highlands as their source.

All of the larger rock fragments collected at Tranquillity (excluding breccias derived from regolith) have the textures, structures and bulk chemistry of basaltic igneous rocks. Many petrological and mineralogical studies converge on the conclusion that they formed as lavas at, or within a few meters of, the surface. They are distinctive, both terrestrially and from our limited lunar sampling, in their high titanium content. The bulk chemistry of the regolith fines reflects this character. Surveyor 5 landed on the Mare Tranquillitatis 25 km to the north and reported a similar high titanium composition (TURKEVICH *et al.*, 1969). In contrast, data from Sinus Medii (Surveyor 6), Tycho (Surveyor 7) and Oceanus Procellarum (Apollo 12) indicate surface compositions clearly lower in titanium.

The combined evidence and arguments support the interpretation that the Apollo 11 lunar rock samples are fragments of lavas which formed part of Mare Tranquillitatis in the vicinity of Tranquillity Base. Isotopic properties observed in the rocks may be assigned plausibly to the lava-forming events and to the sources of the lava.

The fines and breccia samples are more complex. Although they are both part of the regolith, contain similar rocks, minerals and glasses, are nearly indistinguishable from each other chemically (see AGRELL *et al.*, 1970). and are similar in bulk chemistry to the rocks, they must be considered as separate sampling situations.

The sample 10084-35 is material of the upper few centimeters of the regolith passed through a 1 mm sieve. It has as principal constituents (DUKE *et al.*, 1970; CHAO *et al.*, 1970; and others): (1) basaltic rock fragments and related mineral fragments ranging from very fine-grained to medium-grained; (2) shocked rock and mineral fragments; (3) glasses of great varieties of forms and compositions reflecting partial to complete melting and shock transformation from basalts and basaltic minerals; and (4) breccia fragments (i.e. indurated regolithic debris) up to 25 per cent or more. In general, in the finest grain size there is an increase in the proportion of plagioclase component in the glasses. This is similar to the appearance of anorthosite fragments in the coarse fines (1 mm to 1 cm) but not in the larger rocks, reported by WOOD *et al.* (1970), KING *et al.* (1970) and others. The bulk chemistry of the soil may be explained reasonably as being derived principally from the local rocks with a significant increment of plagioclase-rich material (anorthosite, anorthositic gabbro, glass) and a significant depletion in ilmenite ($FeTiO_3$), a heavy, refractory mineral (COMPSTON *et al.*, 1970). Estimates of the non-local components, by various workers, range from 2 to 10 per cent, including a trace (<2 per cent) of meteoritic material.

The breccia sample 10060-15 differs from the lunar fines in its lunar history and laboratory treatment. It has not been sorted in the laboratory and includes rock and glass fragments up to one centimeter in diameter. These coarse fragments move the bulk chemical composition very slightly toward the average composition of the larger rocks from the composition of the fines. The site from which this fragment was derived is not precisely known but its characteristics suggest a nearby source. The effect of the induration episode on its isotopic systems cannot be evaluated independently. Its depth of burial prior to its last movement is not known.

The geological significance of observed isotopic properties of such composite materials as the lunar fines and breccia, containing a majority of identified local debris constituents but with a significant fraction of non-local constituents, must be treated carefully. There have been no independent measurements of U–Th–Pb isotope systematics on such candidates for non-local materials as the anorthositic fragments reported in the coarse fines. Interpretations of these mixed systems contain some hazards which must be identified if they cannot be completely avoided.

History of the rocks

The levels of uranium and thorium, the unusually favorable U^{238}/Pb^{204} and Th^{232}/Pb^{204} ratios, the great ages, and the unique power of the paired Pb–U daughter–parent systems clearly make the U–Th–Pb isotope relationship the most sensitive of the long-lived radioactivity systems which can be applied to the interpretation of the ages of the lunar rocks. It is remarkable, then, that the nine rocks analyzed in two laboratories (TATSUMOTO and ROSHOLT, 1970) reveal such a closely-grouped set of isotope ratios, when they have been treated for the same initial lead correction. In Fig. 4, the nine rock points are shown on a "concordia" diagram which clearly emphasizes the limited range of variation of the measured isotope ratios among the rocks. If they were treated as a single population dispersed by random sampling and analytical errors, they would yield the following ratios and ages:

Pb^{207}/Pb^{206}	Pb^{206}/U^{235}	Pb^{207}/U^{206}	Pb^{208}/Th^{232}
0·863 ± 0·054	54·7 ± 4·0	0·459 ± 0·012	0·225 ± 0·021
4·050 ± 0·19 b.y.	4·130 ± 0·07 b.y.	4·175 ± 0·04 b.y.	4·155 ± 0·35 b.y.

The average value would plot just under the 10017-34 point of this work, very close to "concordia."

The principal direction of dispersion of the group of points is along a line passing close to the origin of the concordia plot suggesting quite strongly that analytical errors (including laboratory bias) may be a major contributor to the data spread. The smallest dispersion is reflected in the 2·5 per cent standard deviation for the Pb^{207}/Pb^{206} ratios, which has less than a 1 per cent effect on the age.

As impressive and as nearly concordant as this group value is, there is no geologic requirement that all of the rocks in the Apollo 11 collection be of a single age, although the mare-forming process may have occurred in a very limited interval of time. SILVER (1970) suggested an apparent greater age for 10047 than for 10017, 10045 and

10072. In Fig. 5, a frequency plot of radiogenic Pb^{207}/Pb^{206} ratios in all of the samples indicates a bimodal distribution for the rocks. The two sharply defined maxima fall at Pb^{207}/Pb^{206} equal 0·451 ± 0·005 and 0·472 ± 0·002 in the older group are 10047, 10050, 10057 and 10071; the younger group includes 10003, 10017, 10020, 10045 and 10072.

The averaged isotope ratios and apparent ages for the two groups are:

	$Pb/^{207}Pb^{206}$	Pb^{206}/U^{238}	Pb^{207}/U^{235}	Pb^{208}/Th^{232}
Younger group	0·451 ± 0·005	0·851 ± 0·056	52·9 ± 3·7	0·226 ± 0·026
age × 10^9 yr	4·145 ± 0·020	4·005 ± 0·100	4·100 ± 0·065	4·175 ± 0·440
Older group	0·472 ± 0·002	0·882 ± 0·052	57·3 ± 3·4	0·224 ± 0·024
age × 10^9 yr	4·215 ± 0·007	4·115 ± 0·080	4·180 ± 0·060	4·140 ± 0·400

The dispersion of the concentration data is too great to clearly resolve the two groups on the basis of the daughter–parent ratios, but the suggestion is there. Inspection of Fig. 4 shows that within each group there is a distinct linear dispersion toward the origin (along the dashed lines), indicating effects of analytical errors and laboratory bias. The near concordance of both groups is visible, nevertheless. The age distinction on the basis of the Pb^{207}/Pb^{206} ratios is clearly within the practical capability of the mass spectrometric method, under these favorable conditions. A difference of 1·7 per cent in relative *age* is clearly resolvable when the standard deviations for the *ratios* are between 0·5 and 1·0 per cent.

With the analytical uncertainties, it is not possible to evaluate more precisely the degree of discordance. If one attempts to eliminate dispersion due to laboratory bias by choosing the larger and coherent group of data of TATSUMOTO and ROSHOLT, a 3–5 per cent discordance appears. This is not confirmed by the work done in our laboratory but our error assignments approach the magnitude of the discrepancy between us. Since no more refined explanation is available, we cannot treat the discordance as significant, at this time. The near concordance supports, but does not uniquely confirm, the use of the Pb^{207}/Pb^{206} apparent ages as the best indices of relative age. The small possible discordance is not accompanied by a sufficient spread of data points to provide a guide to its interpretation.

These two apparent age groups do not seem to correspond to LSPET groups A and B. They do not correspond to the two consistent compositional groups characterized by (1) higher and (2) lower K–Rb–Ba–U–Th concentrations as defined by COMPSTON *et al.* (1970). The differences in concentration levels between these two groups are as much as factors of five or ten. In Fig. 5 the high U and Th rocks are distinguished by symbol from the low rocks and there is no apparent correlation between concentrations and isotope ratios.

COMPSTON *et al.* have suggested that such marked minor and trace element differentiation could not be the result of simple fractional crystallization of a basaltic magma

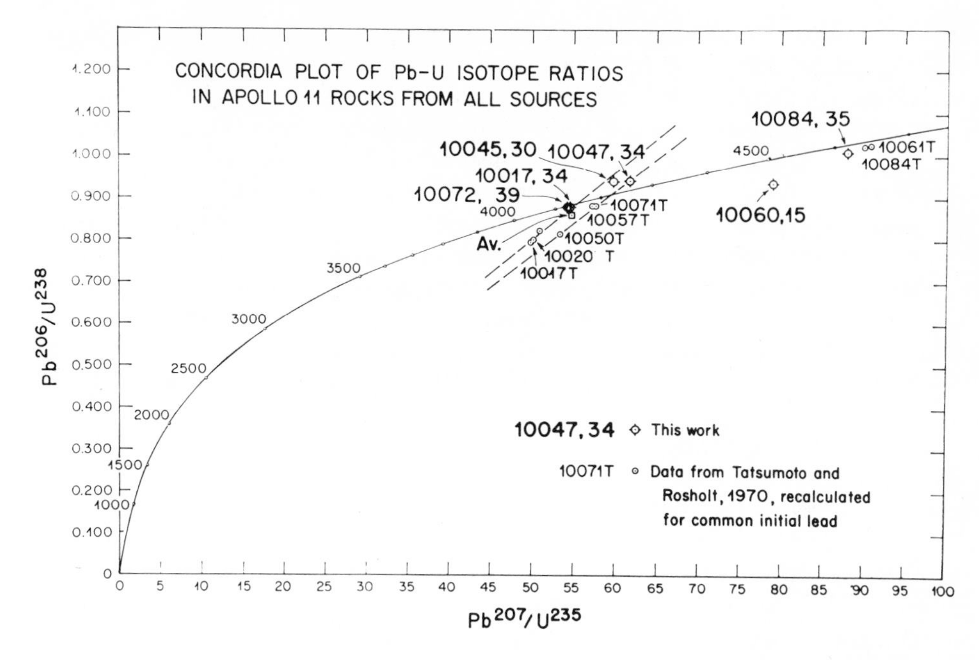

Fig. 4. "Concordia" diagram representing Pb^{206}/U^{238} and Pb^{207}/U^{235} ratios for lunar samples from this work and that of TATSUMOTO and ROSHOLT (1970). The latter points have been recalculated for a common initial lead. Dashed lines are loci of point dispersion from concentration errors.

because the two groups do not appear fractionated in bulk chemistry. They suggest instead derivation by partial melting from two chemically different source regions.

Close inspection of the available rock bulk chemical and mineral data suggests there are indeed group differences in the major constituents of the groups recognized by COMPSTON *et al.* Group (1) has more TiO_2 (and ilmenite) than group (2). If sufficient FeO is assigned to all of the TiO_2 to convert it to $FeTiO_3$, ilmenite, the remaining silicates in group (1) have a distinctly higher Mg/Mg + Fe ratio (about 10 per cent) than the silicates in group (2). There is about 20 to 25 per cent more plagioclase in group (2) than there is in group (1). The plagioclase in group (1) appears about 10 per cent more sodic than in group (2). This has been confirmed in large part by STEWART *et al.* (1970), who reports plagioclase values of An 73–75 for group (1) and An 78–85 for group (2) rocks. It is noteworthy that high Na (and K, Rb, Cs) does not parallel high Fe in these group differences.

Neither textural characteristics, nor the two age groups suggested by the Pb^{207}/Pb^{206} ratios correlate with the compositional groups. With the limited information available at this time, the genetic differences between the two age groups cannot be clearly read. However, fractional crystallization and/or partial melting of a single source would be expected to produce correlated alkali, U, and Th, and Fe enrichment. This does not appear to be the case and the argument for more than one source is supported. If two age groups are accepted, the different sources were tapped more than once to produce the observed assemblage of rocks. The interval of time represented is 70 ± 30 m.y. and this would be a minimum for the interval of mare lava accumulation in this region.

The Th/U ratio and the U^{238}/Pb^{204} ratio are non-radiogenic isotope ratios that offer important information related to the histories of these rocks. The Th/U ratio for the 4 rocks analyzed in this work averages 3·6. With the results of TATSUMOTO and ROSHOLT (1970) nine rocks average 3·7. The average for soil and breccia is nearly the same. Considering the limited sampling represented, these figures are remarkably close to the value of about 3·8 which can be calculated from evolutionary models for earth and meteorite systems. The variations in the rock values for Th/U are sufficiently large to indicate that real fractionation of thorium from uranium does occur; indeed, our leach experiments have confirmed it independently. Th/U ratios may be expected to be a useful clue to lunar igneous differentiation.

The U^{238}/Pb^{204} ratios reported in this work range from 65 to 912. These values are minimum values because they include pre-chemistry contamination Pb^{204}, which for practical purposes is probably the principal Pb^{204} source. It has already been pointed out that this ratio rarely exceeds 10 for terrestrial basalts. A good terrestrial average taken from earth lead evolution is between 8·5 and 9·0. A few unusual terrestrial rocks attain values of 30 or 35, but are numerically very rare.

These first lunar samples have been derived from a chemical system which has been extraordinarily depleted in non-radiogenic initial leads (by 10–100 times, or more) such as those found in iron meteorites and chondrites, or those from which earth leads have evolved. This depletion must have occurred prior to the separation of these rocks from their parent reservoirs. It may be a general lunar characteristic which, if confirmed, will make it possible to place important time limit and feasibility constraints on models relating the moon to the earth and meteorites (GOPALAN *et al.*, 1970). There is good reason, from the soil data discussed in a following section, to believe that extreme fractionation of lead from lunar rocks was a continuing process on the lunar surface for much of its early history. A more extensive sampling of the moon is required to ascertain whether the Pb^{204} isotope is truly depleted for the entire moon or simply depleted from the igneous rocks at this mare site, and concentrated elsewhere. Several workers (KEAYS *et al.*, 1970; KOHMAN *et al.*, 1970) have shown that other volatile elements, particularly Ag, Cd, Zn, In, Tl and Bi, are depleted by one or two orders of magnitude compared to terrestrial basalts. Lead must be responding to a fundamental process in lunar evolution which has systematically modified many geochemical abundances in the parent materials of the rocks.

The preceding tentative interpretation of the rock histories must be reconciled with interpretations offered by others for the ages of the Apollo 11 rocks. TATSUMOTO and ROSHOLT (1970) did not develop precise age interpretations for the rocks; they simply stated the data indicated the rocks solidified several hundred million years after the formation of the moon, about 4·65 b.y. ago.

Their rock age data are moderately dispersed between 3·76 and 4·17 b.y. Their calculated (not observed) correction is strongly enriched in radiogenic Pb^{207} and Pb^{206} with a high Pb^{207}/Pb^{206} ratio compared to the contamination lead correction we have used. They argue that a moon of high U^{238}/Pb^{204} would have added a large radiogenic increment to its primordial lead by the time the Tranquillity rocks were formed and that this lead would have entered significantly into these rocks. Their calculated ages show a distinct correlation with the Pb^{206}/Pb^{204} ratio of the lead isotopic compositions. This strongly suggests that their initial lead correction is manipulating the calculated isotope ratios.

In this work the remarkable depletion of Pb^{204} in the analyzed rocks is striking and there is no convincing evidence that any significant quantity of Pb^{204} is indigenous rather than contamination. One test of the suitability of any systematic correction is whether the corrected values tend to converge on those values for which the correction is smallest. The TATSUMOTO and ROSHOLT corrections provide higher ages as the correction effect becomes very small in increasingly radiogenic samples. The highest ages in their calculation converge on the complete set of ages obtained when all the data is corrected with a modern terrestrial contamination lead. Any initial lead selected other than a terrestrial model evolution lead which is applied to the Apollo rock data will produce a dispersion similar to the one they created.

Isotopic analyses by the Rb^{87}–Sr^{87} whole rock and mineral isochron techniques and the K^{40}–Ar^{40} method on minerals and rocks have been reported by various workers (ALBEE *et al.*, 1970; COMPSTON *et al.*, 1970; GOPALAN *et al.*, 1970; TURNER, 1970; and many others) who have observed younger apparent ages for the rocks. The Rb^{87}–Sr^{87} mineral isochron ages cluster near 3·7 b.y. with a range from 3·5 to 4·05 b.y. TURNER (1970) reported K^{40}–Ar^{40} ages determined by the Ar^{40}–Ar^{39} release technique averaging 3·7 billion years with a maximum value of $3{\cdot}93 \pm 0{\cdot}07$ b.y. for 10003-43.

In their summary of the Apollo 11 Lunar Science Conference, the LSAPT (1970) concluded that the ages determined by Rb^{87}–Sr^{87} and K^{40}–Ar^{40} methods showed the basaltic rocks formed at 3·7 b.y. ago. This age was alleged to be supported by U–Th–Pb data. Thoughtful reexamination of this interpretation is appropriate. Consideration of the information presented at the conference, and developed further in this work shows that major discrepancies involving 300–600 m.y. exist in the apparent ages of the rocks derived by the various methods.

An explanation that might be attractive would be one that attributed the greater ages of the U–Th–Pn systems to a multi-stage history. This would permit large fractions of the radiogenic daughter–parent isotope systems to be evolved and derived from the other lunar environments during significant intervals of time preceding the emplacement and crystallization of the Mare Tranquillitatis volcanics, something like 3·7 b.y. ago. The rocks could have inherited (a) a significant fraction of radiogenic lead unrelated in abundance to the new concentrations of uranium and thorium or (b) a correlated initial radiogenic lead and the associated parent uranium and thorium.

The first alternative can be examined easily. In Fig. 5 it is shown that the rocks vary by nearly a factor of six in uranium concentration but vary in Pb^{207}/Pb^{206} ratio within a ± 3 per cent range. This requires a closely correlated co-variation of initial lead and uranium concentrations so that, at all concentration levels, the observed modern composite of (1) lead evolved since 3·7 b.y. ago and (2) the inherited lead

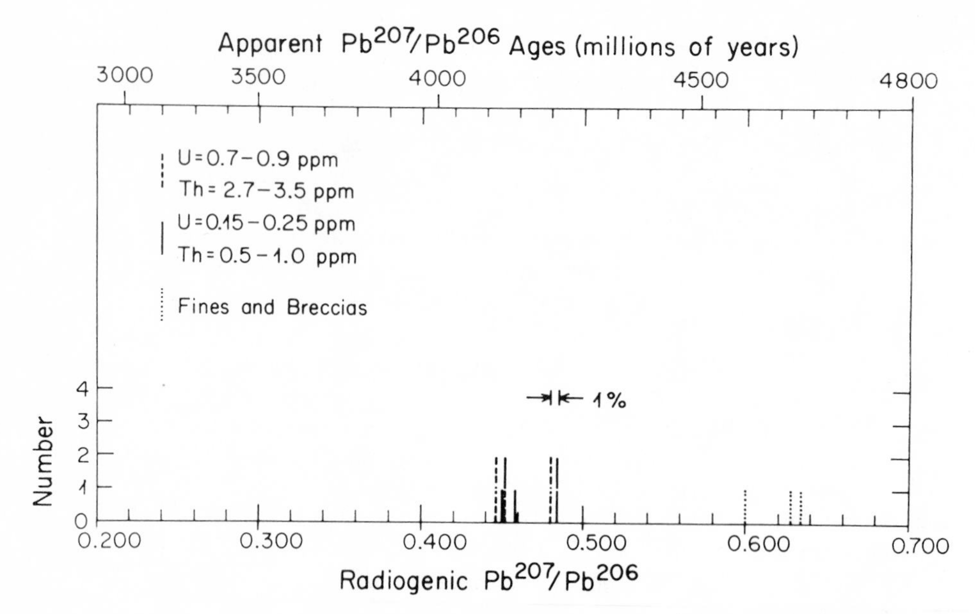

Fig. 5. Frequency diagram for radiogenic Pb^{207}/Pb^{206} ratios observed in nine rocks from this work and from TATSUMOTO and ROSHOLT (1970). Breccias and fines are also shown. Uranium and thorium concentration levels are indicated by pattern. A 1% range of variation is indicated.

would have the appropriate proportions. Since the inherited lead would have a Pb^{207}/Pb^{206} ratio greater than 1·0, it is required that the Pb and U entered the various rocks in a constant ratio to within ± 6 per cent. On geochemical grounds, this is very difficult to accept. The observed U^{238}/Pb^{204} ratios vary by a factor of 10, probably because of contamination. The Th/U ratio in these same rocks varies by ± 15 per cent. GAST and HUBBARD (1970), COMPSTON *et al.* (1970), PHILPOTTS and SCHNETZLER (1970) KEAYS *et al.* (1970) and MURTHY *et al.* (1970) provide precise relative elemental abundances which challenge the geochemical probability of such a constant Pb/U ratio. The closely related K/U ratio varies by more than a factor of two. For seven of the rocks the Ba/U ratio varies by more than ± 15 per cent. The Tl/U ratio, another close geochemical analogue, varies by more than a factor of two for five of the rocks, with precision on the order of ± 10 per cent.

If one considers the population of points on the concordia diagram in Fig. 4, it is apparent that none of them could have been produced by such a mixture if the original daughter–parent system in the reservoir had been fractionated more than about ± 15 per cent. Assuming a uniform population of rocks dispersed by analytical errors, the original fractionation is constrained to about ± 6–7 per cent of the original daughter–parent ratio. It is also required that the lead be derived from an environment of essentially identical Th/U ratio or a systematic difference in the Pb^{206}/U^{238} and Pb^{208}/Th^{232} ages would appear. In short, the observed isotopic patterns cannot be satisfied by a system of fractionated inherited lead, yet geochemical arguments derived from other elemental ratios in these same rocks make it difficult to accept magmatic derivation without fractionation of lead from uranium. Inasmuch as U^{238}/Pb^{204} is so large in these rocks, and in the dust, it would appear that extreme fractionation is a fundamental process in the lunar environment.

If we accept that two independent sources of partial melting, each with a distinctive history and with different degrees of partial melting, provided the groups (1) and (2) of COMPSTON *et al.*, then an incredibly fortuitous set of fractionations and elemental co-variations are required to explain the U–Th–Pb ages as the products of inheritance.

If one considers the possibility that in the original lava a completely unfractionated U–Th–Pb system is inherited, there are some additional tests for its existence in the rocks. Numerous microprobe studies of the various phases in the rocks show that during crystallization some pronounced fractionations of minor and trace elements occurred among the principal phases. Uranium, thorium and lead have not yet been identified in specific phases of the rocks. Observations on the distributions of the rare earths and zirconium suggest that apatite or other phosphate minerals, baddeleyite, dysanalyte and the acid glass mesostasis may be hosts for the U and Th. The plagioclase and the glassy mesostasis are the principal sites for the potassium and barium in the rocks and it is reasonable to expect the lead to tend to follow these elements during crystallization.

In the leach experiment on the rock 10017-34, it was demonstrated that thorium and uranium were fractionated in the rocks, and that the thorium sites were preferentially attacked by the acids. It is reasonable to assume at least as much Pb–U as Th–U fractionation in the rock during crystallization. The leach provided an opportunity for solution and extraction of different combinations of old inherited lead and indigenous radiogenic lead. WANKE *et al.* (1970) have shown that most of the feldspar component is readily attacked with hot HNO_3. Certainly the glassy residue would be at least as vulnerable as the feldspar to the acid. The Pb^{207}/Pb^{206} ratio of the leachate showed no significant difference from the total rock ratio, even though the Pb^{208}/Pb^{206} changed sympathetically with the Th–U ratio. There cannot be a significant fraction of inherited lead in 10017-34.

If it does not seem possible to call on pre-crystallization lead isotope memory to explain the differences between the uranium–thorium–lead ages and those derived by Rb^{87}–Sr^{87} and K^{40}–Ar^{40} techniques, a search must be turned in other directions. Perhaps, as is so commonly the case terrestrially, the Rb^{87}–Sr^{87} mineral isochrons are giving evidence of secondary episodes in the history of these rocks. It is striking that the Ar^{40}–Ar^{39} release ages for 10003 (TURNER, 1970) at $3{\cdot}93 \pm 0{\cdot}07$ million years exceeds most of the mineral isochrons. It is also noteworthy that there is a 400 m.y. spread in the observed Ar^{40}–Ar^{39} release data for various rocks. Turner prefers to take the mean value of the various apparent ages as representing a significant event. Terrestrial experience requires one to consider whether the *highest*, analytically-sound Ar^{40}-based age values should be taken as *minimum* values for the time of lava crystallization. Our geochronological experience with lunar samples is limited. If we recognize the discrepancies, we can look forward to resolving them in the future with a better understanding of the lunar processes which treat each isotopic system with such different sensitivities.

U–Th–Pb *isotopes in lunar regolith*

The uranium–thorium–lead isotope systematics of the samples of lunar regolith which have been analyzed have yielded extremely old apparent ages. Various workers reported Pb^{207}/Pb^{206} ages at 4·63 (SILVER, 1970), 4·65 (TATSUMOTO and ROSHOLT, 1970), 4·69 (KOHMAN *et al.*, 1970), 4·75 (GOPALAN *et al.*, 1970) and 4·76 billion years (WANLESS *et al.*, 1970) for the lunar fines 10084. SILVER (1970) reported a Pb^{207}/Pb^{206} apparent age of 4·60 billion years for breccia 10060-15 and TATSUMOTO and ROSHOLT (1970) reported 4·66 billion years as the apparent age of breccia 10061.

Some workers considered these ages as a basis for determining the age of the moon. TATSUMOTO and ROSHOLT argued that, inasmuch as (1) the observed data fit into the meteorite lead array, (2) the data lie close to "concordia" and (3) the age is similar to the "older" meteorite ages measured by the Rb–Sr method, the age may be a correct age for the moon. GOPALAN *et al.*, interpreted the Pb^{207}/Pb^{206} age as the time of major U–Pb fractionations on the surface of the moon and with awareness of the assumptions, an "age of the moon" analogous to ages calculated for the earth from terrestrial modern lead and meteorite primordial lead. Key assumptions include a well-mixed sample representative of all the lunar crust, and a closed system. The authors recognized the regolith did not represent a simple rock system.

SILVER (1970) felt that a major discrepancy existed between the rock ages and the regolith ages. The geological and geochemical evidence indicated that the largest portion of the debris in the lunar regolith at Apollo 11 site was probably derived from the near vicinity in Mare Tranquillitatis and probably contained isotopic systems similar to those in the rocks. Therefore an additional component of lead with higher Pb^{207}/Pb^{206} ratios than the composite value must be present in the fines, implying an even greater apparent age for some part of the lunar surface. Silver argued that near concordance is not definitive in this age range and that the available points on "concordia" did not permit use of discordance patterns to reach a unique age interpretation from the obviously heterogeneous dust samples.

In this work the study of the lead isotopes in the soil has been extended by the acid–leaching experiments. These experiments have demonstrated (1) that there is extreme isotopic heterogeneity of various lead components in the regolith, (2) that at least one component has a Pb^{207}/Pb^{206} ratio 25 per cent greater than the composite value, (3) that a substantial portion of the radiogenic lead, uranium and thorium are labile, with the lead isotopes distinctly more leachable than the parent uranium and thorium (Fig. 2), and (4) that extended leaching changed the uranium–thorium–lead isotopic composition of the more insoluble residue to values very close to those observed in the larger analyzed rocks from the same sample vicinity. The data on the Pb–U isotope pairs in leach series No. 6 on 10084-35, presented graphically in Fig. 3, illustrates many important points.

The first important point is that near-concordance as indicated by the position of the composite value for the series may be a deceptive and dangerous criterion for unique interpretations. It is a necessary condition but not sufficient to establish that a Pb–U isotope system of this great age has been a single closed system. The nearly straight form of "concordia" beyond 4000 m.y. permits many different degrees of mixing old Pb–U systems without creating analytically detectable deviations from the curve.

Significant deviations from "concordia" can occur only when there is a large fractionation of daughter from parent. In leach No. 6A, the observed isotope ratios are displaced above the curve so as to suggest a 25 per cent excess of daughter over parent, for the Pb^{207}/Pb^{206} ratio (and age). This could reflect a preferential acid extraction of lead from parent in the residual material, or it could demonstrate the existence of a significant fraction of parentless lead. In the former alternative, the residual parent should appear as a significant factor which moves later leaches or the residue below "concordia." The first leach involves about 25 per cent of the total radiogenic lead in the rock and would produce an excess of residual uranium of about 10 per cent of the remaining uranium value. There is insufficient discordance below the curve to support this alternative.

Our rock leach experiment has shown that acid leaching of rock constituents can maintain closely the original ratios between daughter and parent for the uranium system, and that for the thorium system, if anything, the thorium came out more rapidly ($\sim$15%) than the lead produced by it.

The very mobile early lead in leach No. 6A is interpreted as a mixture of two components: (1) leads of unknown ages associated with parents in rock, mineral and glass sites, and (2) parentless lead (on particle surfaces?) highly available to the acid. The rock leads probably enter the acids with their parents, as indicated by the more rapid yields of Pb^{208} and Th compared to Pb^{206}, Pb^{207} and uranium. If one assumes that the rock and mineral isotope systems are largely derived from material like the nine rocks we have analyzed in age and isotopic character, and that the associated parent–daughter systems are essentially unfractionated (as in 10017-34), it is possible to calculate the excess radiogenic components of lead in the leach. (The effects of the assumptions will be reviewed at a later point.) In Table 5, calculated quantities of leads produced by the observed uranium and thorium in 4·175 billion years in the leaches, residue, and composite sample No. 6 are compared with quantities of lead observed. The difference is treated as excess and deficiency, and is reported as percent of the calculated value. The Pb^{207}/Pb^{206} ratio of this excess component is calculated and compared with the observed Pb^{207}/Pb^{206}. There is no significant difference between the lead isotope concentrations observed for the nine rocks and these calculated values. It has further been shown that, happily, the average uranium and thorium concentration of the nine rocks is essentially the same value found in the soil samples and the breccia. As a first approximation, this simplification is not extreme.

For the composite system an excess of 57·3% Pb^{207} and 13·4% Pb^{206} exists over that which would have been produced in a closed system since 4·175 b.y. ago. No direct value for Pb^{208} is possible from this composite because leaches A and B did not have independent lead composition analysis, and the Pb^{208} is concealed by the Pb^{208}-rich spike. (A combined Pb^{208} for leach A and B is calculated from a material balance comparison with the total analysis of the fines reported in Table 1.) The Pb^{207}/Pb^{206} has a value of 1·96. In Fig. 3, it would require parentless lead of this composition in the proportions indicated in Table 5 to displace the mean of the nine Apollo 11 rocks to the observed composite sample position.

Leach No. 6A has an incredible excess of radiogenic Pb^{206} (71%) and Pb^{207} (180%) but the excess Pb^{207}/Pb^{206} ratio is reduced by nearly one-half. It is, of course, far removed from "concordia" in Fig. 3.

Table 5. Calculation of excess radiogenic components of lead found in leach fractions and residue of experiment No. 6, lunar fines, 10084-35, over calculated production by observed uranium and thorium in $4{\cdot}175 \times 10^9$ yr

	Units of 10^{-8} moles/g rock treated			Pb^{207}/Pb^{206}	Pb^{207}/Pb^{206}
	Pb^{206}	Pb^{207}	Pb^{208}	Excess	Observed
Composite					
Observed	0·2302	0·1463	n.a.		
Calculated	0·2030	0·0930			
Excess	0·0272 +13·4%	0·0533 +57·3%		1·96	0·631
Leach A					
Observed	0·05145	0·0387	n.a. (Combined A + B)		
Calculated	0·03012	0·0138			
Excess	0·02133 +70·8%	0·0249 +180.%		1·17	0·753
Leach B					
Observed	0·01822	0·0140	Combined A + B: 0·0850*		
Calculated	0·01395	0·0064	0·0767		
Excess	0·00427 +30·6%	0·0076 +120.%	0·0083 +10·8%	1·79	0·767
Leach C					
Observed	0·0955	0·0599	0·0835		
Calculated	0·0861	0·0395	0·0908		
Excess	0·0094 +10·9%	0·0204 +51·9%		2.17	0·629
Deficiency			0·0073 −8·0%		
Residue					
Observed	0·0650	0·0338	0·0496		
Calculated	0·0714	0·0327	0·0506		
Excess		0·0011 +3·2% 13·3%†			
Deficiency	0·0064 −9·0%		0·0010 −2·0%	—	0·521

* Calculated by material balance from independent analysis of total sample.
† 13·3% over Pb^{207} required for Pb^{206} found.
n.a. = not available because of Pb^{208}-rich spike.

Leach No. 6B has somewhat less excess lead but with a higher Pb^{207}/Pb^{206} ratio. It is only slightly above "concordia" and near and to the right of a line from leach No. 6A to the origin. Leach No. 6A and No. 6B show combined 10 per cent excess of Pb^{208}.

Leach No. 6C has a further reduced but significant excess of Pb^{206} and Pb^{207} and a fairly high Pb^{207}/Pb^{206} ratio. It is only slightly below "concordia" in Fig. 3 and gives apparent ages very close to that of the total sample. A 9 per cent deficiency in Pb^{208} has appeared.

Leach No. 6 residue has a modest deficiency of Pb^{206} (9%), a slight excess of Pb,^{207} and a slight deficiency in Pb^{208}. The trends of excess and deficiency observed for the various lead isotopes in the entire series based on this model calculation are shown graphically at the bottom of Fig. 2. The modest dimensions of these residual deficiencies support the argument that large separations of daughters from parent have not occurred as a result of the leaching.

The position of the leach No. 6 residue on Fig. 3 is quite close to the region occupied by the nine analyzed rocks. The trend from leach No. 6A to leach No. 6 residue has moved convincingly toward the values for these rocks suggesting that the most insoluble material (coarsest? least shocked? least vitrified?) containing about 25–30 per cent of the total system, is predominantly related to the local rocks.

Table 6 reports a similar analysis for the simpler experiment No. 4 on the dust. It is completely supportive of leach experiment No. 6.

Table 6. Calculation of excess radiogenic components of lead found in leach fractions and residue of experiment No. 4, lunar fines, 10084-35, over calculated production by observed uranium and thorium in 4·175 × 10^9 yr

	Units of 10^{-8} moles/g rock treated			Pb^{207}/Pb^{206}	Pb^{207}/Pb^{206}
	Pb^{206}	Pb^{207}	Pb^{208}	Excess	Observed
Composite					
Observed	0·2322	0·1476	n.a.		
Calculated	0·2023	0·0928			
Excess	0·0299	0·0548		1·83	0·636
	+14·8%	+59·1%			
Leach No. 4A					
Observed	0·1376	0·0959	n.a.		
Calculated	0·0983	0·0451			
Excess	0·0393	0·0508		1·293	0·718
	+40·0%	+112·6%			
Residue No. 4B					
Observed	0·0946	0·0518	n.a.		
Calculated	0·1040	0·0477			
Excess		0·0041			
		+8·6%		(>1·83)	0·547
Deficiency	0·0094				
	−0·9%				

n.a. = not available because of Pb^{208}-rich spike.

One problem that may be explained well by these leach trends is the differences in Pb^{207}/Pb^{206} ages reported by various workers. Sample homogeneity probably was more closely attained in 10084 than any other Apollo 11 aliquots. It is possible that some of the higher ages reflect incomplete reactions and/or digestions of the samples.

Leach experiments of this type are not the most incisive techniques for fractionating isotope systems. None of the leach steps can be said to have sampled only one, or a limited number, of the isotopic systems in the fines. We can only interpret qualitatively the various contributions in the light of the various experiments, the information available on mineralogy and composition of the lunar fines, and the geological setting of the collection site. Three major contributors to the soil are probably present; a fourth may be suspected. They are:

(1) Rock leads and associated parents, comprising about 90 per cent of the Pb^{206}, 65 per cent of the Pb^{207} and 90 per cent of the Pb^{208} in the soil, contained in Tranquillity rocks. This might comprise about 90–95 per cent of the volume of the regolith and would include most of the coarser material.

(2) Rock leads and associated Parents, comprising about 5 per cent of the Pb^{206},

15 per cent of the Pb^{207}, and 5 per cent of the Pb^{208}, derived from older materials on the lunar surface (e.g. the highlands as suggested by SHOEMAKER *et al.*, 1970b). This debris might be 5–10 per cent of the total volume and would be found mostly in the finer size fractions because of its greater distance from the source regions. It probably includes a much larger fraction of acid-soluble glass and feldspathic component.

(3) Parentless leads containing 5 per cent of the Pb^{206}, 20 per cent of the Pb^{207} and 5 per cent of the Pb^{208}. This occupies no volume. It is presumably volatilized lead derived from a composite of generally older rocks.

(4) A small component of younger rocks, which would tend to counter isotopically some of the older rocks, might be present; but it is probably not significant in volume.

In the leach sequence, No. 6A extracted most of the parentless lead and the easily soluble fractions of the other two groups. The volume of Tranquillity rocks probably outweighed the older rock contribution. This combination is responsible for pulling the data so far above "concordia."

Leach No. 6B is predominantly both rock leads with a last remnant of the parentless leads. In the hot acid most of the fine-grained glassy and feldspathic older rock fragments are rapidly dissolving and the coarse young debris is yielding more slowly but continuously. The yield of this leach was the smallest fraction of the series. This is largely a mixing of older and younger parent-and-daughter systems and is thus very close to "concordia." Its position above the line indicates the presence of the last contribution of parentless lead.

Leach No. 6C dissolved most of the small remainder of older material and a moderate portion of the younger rocks. It is a mixture of older and younger coherent Pb–U systems and is very close to "concordia." The deficiency of Pb is a minor indication of cumulative daughter fractionation from parent.

The residue has a little memory of the older material. It may have suffered moderately from the cumulative effects of preferential daughter loss. Its position below "concordia" might also be influenced if there was a contribution from a somewhat younger component. But most remarkably, considering its extended exposure to acid attack with a loss of 65–75 per cent of its radiogenic families, it still closely resembles a coherent isotopic system.

Calculation of excess radiogenic leads such as have been applied to the leaches may also be applied to the total fines sample. This has been done, in effect, for the composite values shown at the top of Tables 5 and 6. In Table 7, the complete direct analysis of 10084-35 in this work and the analysis by TATSUMOTO and ROSHOLT (1970) have been treated in the same way. The contribution of excess radiogenic leads above that produced by the observed uranium and thorium, averages, for all four cases, Pb^{206}, $13{\cdot}2 \pm 1{\cdot}8\%$ and Pb^{207}, $56{\cdot}4 \pm 3{\cdot}8\%$. For two cases the excess Pb^{208} is $6{\cdot}2 \pm 3{\cdot}6\%$. The Pb^{207}/Pb^{206} ratio for the four cases averages $1{\cdot}96 \pm 0{\cdot}14$.

From the model interpretation given above, it is believed that this Pb^{207}/Pb^{206} value represents a composite lead comprised of parentless old radiogenic leads in the soil and of that fraction of older rock leads associated with uranium, which was generated in those older rocks before 4·175 b.y. ago. The precise mixtures cannot be described

Table 7. Calculation of excess radiogenic components of lead found in two aliquots of lunar fines, 10084, over calculated production by observed uranium and thorium in $4 \cdot 175 \times 10^9$ yr and in $3 \cdot 600 \times 10^9$ yr

	Units of 10^{-8} moles/g rock treated			Pb^{207}/Pb^{206}	Pb^{207}/Pb^{206}
	Pb^{206}	Pb^{207}	Pb^{208}	Excess	Observed
$4 \cdot 175 \times 10^9$ yr					
10084-35 (LTS)					
Observed	0·2333	0·1457	0·2192		
Calculated	0·2107	0·0966	0·2115		
Excess	0·0226	0·0491	0·0077	2·17	0·634
	+10·7%	+50·8%	+3·6%		
10084, (TATSUMOTO and ROSHOLT, 1970)					
Observed	0·2328	0·1484	0·2217		
Calculated	0·2040	0·0936	0·2037		
Excess	0·0288	0·0548	0·0180	1·90	0·638
	+14·1%	+58·5%	+8·8%		
$3 \cdot 600 \times 10^9$ yr					
10084-35 (LTS)					
Observed	0·2333	0·1457	0·2192		
Calculated	0·1731	0·0545	0·1798		
Excess	0·0602	0·0912	0·0394	1·51	0·634
	+34·8%	+167·4%	+21·9%		
10084, (TATSUMOTO and ROSHOLT, 1970)					
Observed	0·2328	0·1484	0·2217		
Calculated	0·1676	0·0528	0·1731		
Excess	0·0652	0·0956	0·0486	1·47	0·638
	+38·9%	+181·1%	+28·1%		

but some important limits on the time of formation of such a mixture can be established. In Fig. 6, two curves* for the production of Pb^{207}/Pb^{206} ratios in leads generated in uranium systems with the accepted constants are shown. The lower curve is the cumulative integrated production curve Pb^{207}/Pb^{206} to be found now in closed systems for times back to 5 billion years ago. The upper curve is the instantaneous Pb^{207}/Pb^{206} production from U^{235}/U^{238} as a function of time before the present. These two curves limit all possible Pb^{207}/Pb^{206} ratios generated in the last 5 b.y., assuming no isotopic fractionation or intermediate daughter fractionation. The ranges of observed values in Apollo 11 rocks and fines are shown as horizontal lines. The range of Pb^{207}/Pb^{206} ages in the rocks is shown between two vertical lines.

The quantities of lead produced instantaneously are negligible and a finite interval of time is required for a contribution. For an old uranium system to generate a

---

* For instantaneous production of Pb^{207}/Pb^{206} at time T before present:

$$\frac{Pb^{207}}{Pb^{206}} = \frac{e^{\lambda 235 T} \cdot \lambda 235}{K e^{\lambda 238 T} \cdot \lambda 238}, \quad \text{where } K = 137 \cdot 8.$$

For cumulative integrated production of Pb^{207}/Pb^{206} from time T before present:

$$\frac{Pb^{207}}{Pb^{206}} = \frac{e^{\lambda 235 T} - 1}{K(e^{\lambda 238 T} - 1)}.$$

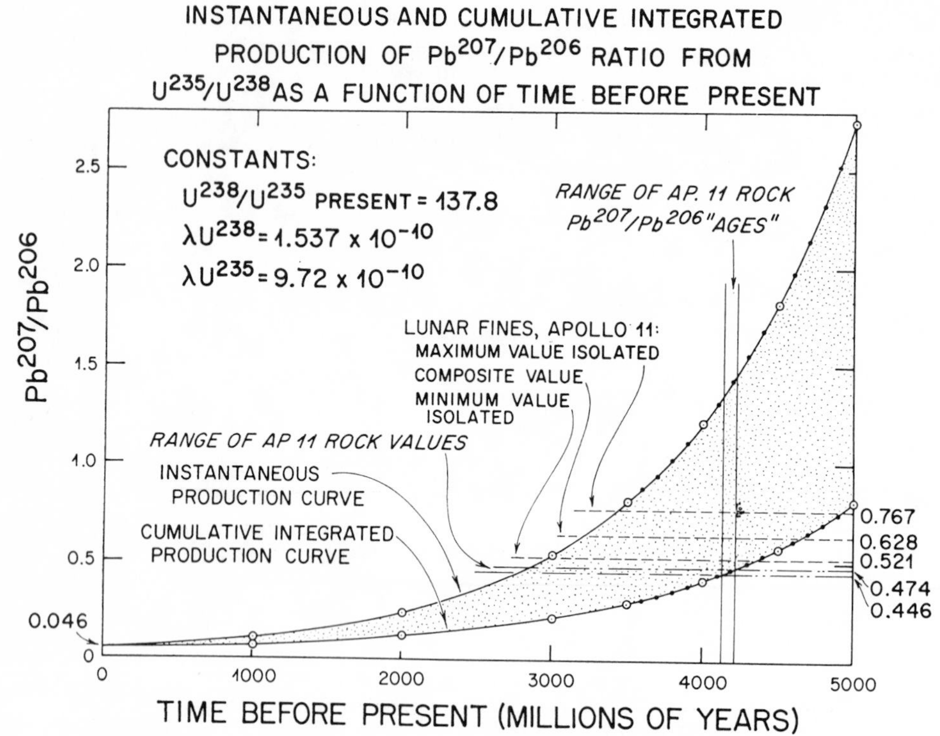

Fig. 6. Graph showing production curves for Pb^{207}/Pb^{206} ratios as a function of time before the present. Lower curve is the cumulative integrated production in ideal closed systems. Upper curve is instantaneous production. Various significant observed Pb^{207}/Pb^{206} ratios are plotted as horizontal lines. Age limits of the nine Apollo 11 rocks are shown. Constants assumed are shown.

composite lead with the Pb^{207}/Pb^{206} ratio of 1·96 and to produce it in significant quantities, it must have existed a considerable time before 4·63 b.y. which, from the upper curve, was the time when 1·96 was the instantaneous ratio. The interval of time must be roughly centered about the 1·96 point on the instantaneous production curve.

In our model we cannot define the time limits for the generation of the parentless lead, nor estimate the composite ages of the older rock debris. We do know that the regolith cannot be older than the rocks on which we believe they rest. We assign an average age to them of 4·175 b.y. and believe the regolith to have been initiated shortly thereafter. Parentless leads in that regolith were derived from other sources after the Mare Tranquillitatis rocks formed. Presumably they were associated with parent isotopes prior to reaching Tranquillity Base. It is possible to calculate the time interval, Δt, prior to 4·175 b.y. required to generate leads with the 1·96 value. It makes no difference in the model whether this excess lead was generated by uranium at some other site during this interval, followed by separation, or whether that uranium and the associated leads were carried together in rock debris to the Tranquillity Base site to continue to accumulate additional lead up to the present. The effect of the model is to treat post 4·175 b.y. old lead as Tranquillity leads. All others are excess.

In Fig. 7, the curve* for the production of integrated Pb^{207}/Pb^{206} ratios from U^{235}/U^{238} systems as a function of Δt prior to daughter–parent separation at $T_1 = 4{\cdot}175 \times 10^9$ yr are shown. The time interval, Δt, required is 0·76 b.y. Total time before the present is 4·94 ± 0·10 b.y.

If one considers the suggestion that the upper profile of the regolith formed and was accumulating lead and highland debris as recently as 3·6 or 3·7 × 10^9 yr ago, a similar model calculation can be made. An example is given in Table 7 where the

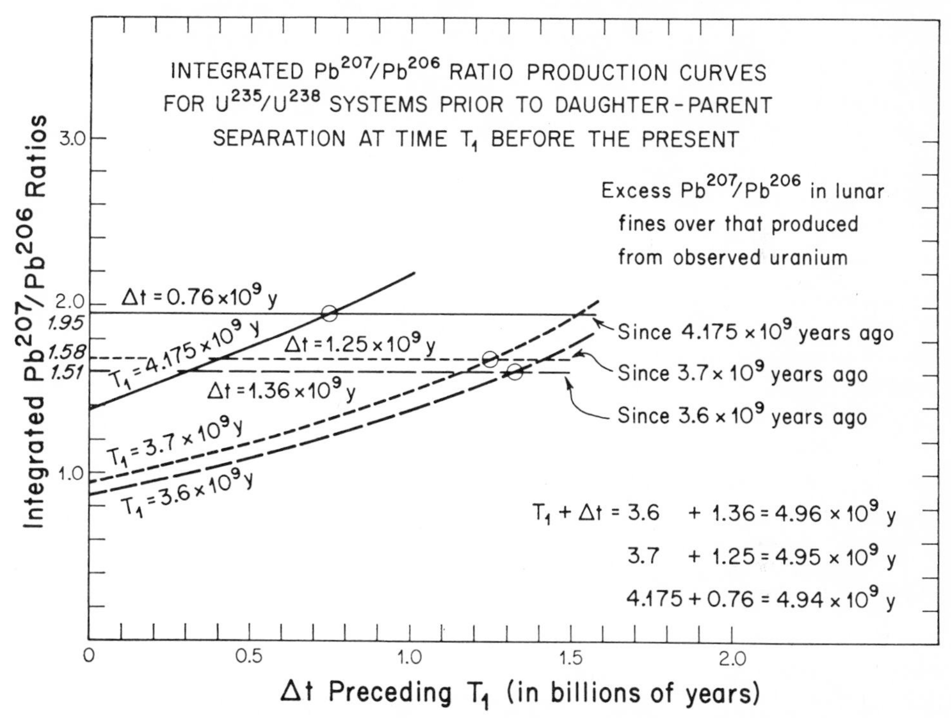

Fig. 7. Graph showing integrated Pb^{207}/Pb^{206} ratio production curves for U^{235}/U^{238} systems as a function of the interval time, Δt, before daughter–parent separations at various times, T_1. Calculated excess Pb^{207}/Pb^{206} in lunar fines over that produced by the observed uranium since time T_1, is fitted to appropriate T_1 curves to obtain Δt.

lunar fines data is treated to yield leads in excess of that produced by the observed uranium isotopes in the 3·6 b.y. before the present. The excess lead has a Pb^{207}/Pb^{206} ratio of 1·51. In Fig. 7, $T_1 = 3{\cdot}6$ b.y. ago, the value Δt is 1·36 b.y. and a composite age for the excess lead is 4·96 b.y. An uncertainty of about 100 m.y. attaches to this number from the available data.

* Curves in Fig. 7 are based on the relations

$$\frac{Pb^{207}}{Pb^{206}} = \frac{e^{\lambda_{235}\Delta t} - 1}{K'(e^{\lambda_{238}\Delta t} - 1)}, \text{ where } K' = \frac{137{\cdot}8e^{\lambda_{238}T_1}}{e^{\lambda_{235}T_1}}$$

$$= \frac{e^{\lambda_{235}T_1}}{137{\cdot}8e^{\lambda_{238}T_1}} \cdot \frac{(e^{\lambda_{235}\Delta t} - 1)}{(e^{\lambda_{238}\Delta t} - 1)}.$$

The similarity in the two answers is not coincidence. The model proceeds on the assumption that there is uranium somewhere (Tranquillity, or highlands, or?) which produced the excess fraction of leads continuously before T_1 as an integrated product. If early lead production at these sites was not continuous, it must nevertheless have extended back considerably before 4·63 b.y. ago.

The implications of such a calculated value in the light of presently accepted models for the age of meteorites, the earth and solar system, drive one right back to a re-examination of the basic assumptions on which this astonishing apparent age is based. As has been just pointed out, the Pb^{207}/Pb^{206} calculation is independent of where the lead was made. It does require that the U^{235}/U^{238} abundance as a function of time not be disturbed by any unanticipated physical process. Although this cannot be defended absolutely, no contrary data exists. It requires that the decay constants be known with good precision. The combined errors assigned to the two decay constants for U^{238} and U^{235} generate an uncertainty of about ± 160 m.y. at five billion years. Most of this uncertainty is assigned to the U^{235} decay constant (2·2%) and BANKS and SILVER (1966) have offered natural evidence that the λU^{235} error is probably less than 1 per cent. Isotopic fractionation cannot be ruled out but has not been demonstrated for lead in nature. Intermediate daughter loss and migration might be considered but since the moon is essentially a closed system for high mass atoms, such effects, whatever their direction and magnitude, should be averaged out on a long-term basis for the lunar surface.

If one wishes to search for the leach data that most closely represents the postulated older rock-contaminated leads and uranium, consider leach No. 6B. In this leach the Pb^{207}/Pb^{206} ratio *rose* slightly from leach No. 6A value. The position on concordia is nearly concordant but suggests a small contribution of parentless lead. The lead and uranium are primarily mixtures of coherent systems. It appears that the leach must obtain its high Pb^{207}/Pb^{206} ratio principally from a source independent of the parentless lead. Since it is a composite sample, it appears there are very old rock Pb–U systems incorporated in the leach which contribute to the composite Pb^{207}/Pb^{206} ratio of 0·767. In this case the composite value must also be less than the highest component in it. Yet calculated as the product of a mixture of coherent lead–uranium systems, it has a Pb^{207}/Pb^{206} apparent age of 4·93 b.y. and the Pb–U ages are within 3 per cent agreement. The existence of the ancient component seems particularly visible in this leach.

It appears that this apparent age, a minimum age for a significant fraction of leads in the lunar fines, must be given serious consideration. It cannot be considered a lunar age because it lacks the sound geological and geochemical evidence necessary to firmly establish the origins of the lead. Additional geographic sampling, vertical profiles, more carefully documented geological sampling and site studies can test the existence and general distribution of these excess leads. Apollo 13 will probably sample the highlands material to provide a check on its age. At this point, it can only be stated that 4·95 $\pm$ 0·1 b.y. appears to be a minimum age for the moon from the U–Pb isotope systematics.

KEAYS *et al.* (1970) have suggested that there may be a 2% carbonaceous chondrite contribution to the soil. Metallic iron meteorite fragments have been recognized by

MASON *et al.* (1970) and other workers. Lead isotopic studies of these meteorites (MARSHALL, 1962) have shown them to have primitive leads; no radiogenic increment has been recognized in them. In any event their abundance levels and lead concentrations preclude a significant effect on the radiogenic leads in the soil. Indeed the observed Pb^{204} in the soil probably can be used to limit the probable abundances of carbonaceous chondrites to significantly less then 2 per cent of the soil.

The case for a labile lead in the lunar fines presented from the leach experiments strongly suggests volatilization as a major transfer process on the lunar surface. In a preliminary experiment, vacuum furnace heating of lunar fines to 950°C for $1\frac{1}{2}$ hr, at 1 mm vacuum, transferred approximately 50 per cent of the lead to a cold finger. Poor Pb^{204} data from the mass spectrometer limits the precise value of the Pb^{207}/Pb^{206} radiogenic ratio to $0{\cdot}67 \pm 0{\cdot}03$.

The existence of an easily volatilized lead in the soil supports the possible significance of volatilization as a lunar process. KEAYS *et al.* (1970) have found Ag, Zn, Cd, Bi, Tl and In enrichment in the soil by factors of 2 to 100 over the values in the rocks. These volatile elements are depleted in the rocks compared to terrestrial basalts, similar to the Pb^{204} depletion. It would appear that volatile metal depletions in the rocks reflect a high temperature interval in the early history of the moon as suggested by KEAYS *et al.* But it also appears that volatile transfer was a continuing and important geochemical process on the lunar surface until after the lavas of Mare Tranquillitatis formed. It may not be necessary to invoke carbonaceous chondrite accretion to the lunar surface to explain the observed trace element enrichments in the soil.

Among several interesting consequences if significant volatile transfer has affected elemental abundances in the lunar soil, are the possible effects on other radioactive daughter–parent systems. Whereas daughter leads are more volatile than their parent uranium and thorium, parent Rb^{87} is more volatile than its radiogenic Sr^{87} daughter. Volatile transfer could introduce an effect on the apparent Rb^{87}–Sr^{87} ages for regolith samples which would be opposite in sense and hence divergent from that produced in the U–Th–Pb systems. One would expect consistently younger Rb^{87}–Sr^{87} ages than the Pb^{206}/U^{238} ages for the soil. Neither would have unique time significance. Possible rubidium enrichment in the soil over the rocks may be reflected in the K/Rb ratios which are lower in the soil and breccia than in any of the rocks (GAST *et al.*, 1970).

SUMMARY

Despite some analytical difficulties, the study of U–Th–Pb isotope systematics in four rocks, a breccia and fines, has revealed some important systematics. The data obtained fits closely with data reported by TATSUMOTO and ROSHOLT (1970), particularly when the latter observations are calculated with the same initial lead assumptions utilized in this work. From the combination of nine rock samples, two breccia analyses and several analyses of lunar fines, from the insight learned from several experiments leaching the fines and rock, from the geological and geochemical reports of many workers, it is possible to tentatively identify some important episodes and to place limits on their timing in lunar history.

(1) The oldest materials on the lunar surface apparently exceed 4·95 b.y. in age.

(2) Prior to that time, an early high temperature episode in lunar history is suggested. It produced an apparent depletion in volatile elements, including lead, as indicated by the extraordinary high U^{238}/Pb^{204} ratios of lunar materials from Tranquillity Base, compared to terrestrial and chondritic materials.

(3) The lavas in Mare Tranquillitatis, near the sample site, are very limited in apparent ages, having formed in a 70 ± 30 m.y. interval centered at 4·18 billion years ago. Two distinct episodes in the short interval are suggested by the data. The earlier is represented by four rocks at 4·22 ± 0·01; the later is found in five rocks at 4·15 ± 0·02 b.y. ago. The lavas, on geochemical grounds, appear to have been derived from more than one source region during each volcanic episode.

(4) The regolith formed on the Tranquillity Base rocks was derived principally from the local volcanic rocks. It contains a small proportion (5%?) of much older rocks, probably anorthositic in character, which may have been derived from the highlands. These materials appear to provide the old leads from which the 4·95 b.y. apparent age is derived.

(5) A parentless component of old lead is also present in the fine fraction of the upper few cm of the regolith. This component was transferred into the soil from distant sources, probably over a significant interval of time after the Tranquillity volcanism. It was accompanied by an enrichment of other volatile elements such as Ag, Cd, Bi, Tl, Zn and In. This enrichment suggests a prolonged post-4·15 b.y. thermal history for the moon during which the cumulative transfer of volatile components took place.

(6) Some time after the Tranquillity volcanism, additional geochemical events have occurred to which Rb^{87}–Sr^{87}, K^{40}–Ar^{40} and U–Th–Pb isotope systems in the same materials have responded with significant differences in behavior. The reasons for these discrepancies are not understood, but their resolution may shed some new light on lunar history.

These data appear to challenge some well-established interpretations concerning the age of the earth, meteorites and the solar system. Nothing reported in this work need be incompatible with previously reported data. It is probably in the assumptions involved in this and other models that contradictions may appear. Continuous examination of basic assumptions provides some of the greatest harvests in science. Hopefully lunar exploration will continue to challenge many old ideas and assumptions with new facts.

Acknowledgments—Mrs. GERALDINE BAENTELI, in the chemistry, and Mrs. MARIA PEARSON, in the mass spectrometry, have made major contributions to this work. E. VICTOR NENOW and CURTIS BAUMAN built and fine-tuned the mass spectrometer used in the isotopic analyses. Professor G. J. WASSERBURG generously permitted use of some basic design information in the mass spectrometer construction. K. LUDWIG and T. ANDERSON provided many kind assistances. JOSEPH MAYER maintained laboratory security against contamination. Mrs. EVELYN BROWN provided great administrative assistance in managing the NASA contract No. NAS 9-7963 under which this work was carried out. I am indebted to many colleagues, particularly E. SHOEMAKER, M. DUKE and H. H. SCHMITT, for stimulating discussions of the geology and materials of the lunar surface. Dr. MITSUNOBU TATSUMOTO U.S. Geological Survey, shared many valuable insights into analytical techniques and problems with brotherly generosity. The California Institute of Technology provided many essential facilities. P. GOLDREICH, D. S. BURNETT and H. P. TAYLOR have commented constructively on various parts of this manuscript.

REFERENCES

AGRELL S. O., SCOON J. H., MUIR I. D., LONG J. V. P., MCCONNELL J. D. C. and PECKETT A. (1970) Mineralogy and petrology of some lunar samples. *Science* **167**, 583–586.

ALBEE A. L., BURNETT D. S., CHODOS A. A., EUGSTER O. J., HUNEKE J. C., PAPANASTASSIOU D. A., PODOSEK F. A., RUSS G. P. II, SANZ H. G., TERA F. and WASSERBURG G. J. (1970) Ages, irradiation history, and chemical composition of lunar rocks from the Sea of Tranquillity. *Science* **167**, 463–466.

ALDRIN E. E., JR., ARMSTRONG N. A. and COLLINS M. (1969) Crew observations. Apollo 11 preliminary science report, NASA SP-214, 35–39.

ANDERSON A. T., CREWE A. V., GOLDSMITH J. R., MOORE P. B., NEWTON J. C., OLSEN E. J., SMITH J. V. and WYLLIE P. J. (1970) Petrologic history of moon suggested by petrography, mineralogy and crystallography. *Science* **167**, 587–590.

BANKS P. O. and SILVER L. T. (1966) Evaluation of the decay constant of uranium 235 from lead isotope ratios. *J. Geophys. Res.* **71**, 4037–4046.

CAMERON A. E., SMITH D. H. and WALKER R. L. (1969) Mass spectrometry of nanogram-size samples of lead. *Anal. Chem.* **41**, 525–526.

CHAO E. C. T., JAMES O. B., MINKIN J. A., BOREMAN J. A., JACKSON E. D. and RALEIGH C. B. (1970) Petrology of unshocked crystalline rocks and shock effects in lunar rocks and minerals. *Science* **167**, 644–647.

COMPSTON W., ARRIENS P. A., VERNON M. J. and CHAPPELL B. W. (1970) Rubidium–strontium chronology and chemistry of lunar material. *Science* **167**, 474–476.

DUKE M. B., WOO C. C., BIRD M. L., SELLERS G. A. and FINKELMAN R. B. (1970) Lunar soil: Size distribution and mineralogical constituents. *Science* **167**, 648–650.

FIELDS P. R., DIAMOND H., METTA D. N., STEVENS C. M., ROKOP D. J. and MORELAND P. E. (1970) Isotopic abundances of actinide elements in lunar material. *Science* **167**, 499–501.

GAST P. W. and HUBBARD N. J. (1970) Abundance of alkali metals, alkaline and rare earths and strontium-87/strontium-86 ratios in lunar samples. *Science* **167**, 485–487.

GOPALAN K., KAUSHAL S., LEE-HU C. and WETHERILL G. W. (1970) Rubidium–strontium, uranium and thorium–lead dating of lunar material. *Science* **167**, 471–473.

KEAYS R. R., GANAPATHY R., LAUL J. C., ANDERS E., HERZOG G. F. and JEFFERY P. M. (1970) Trace elements and radioactivity in lunar rocks: Implications for meteorite infall, solar-wind flux and formation conditions of moon. *Science* **167**, 490–493.

KEIL K., PRINZ M. and BUNCH T. E. (1970) Mineral chemistry of lunar samples. *Science* **167**, 597–599.

KING E. A., JR., CARMAN M. F. and BUTLER J. C. (1970) Mineralogy and petrology of coarse particulate material from lunar surface at Tranquillity Base. *Science* **167**, 650–652.

KOHMAN T. P., BLACK L. P., IHOCHI H. and HUEY J. M. (1970) Lead and thallium isotopes in Mare Tranquillitatis surface material. *Science* **167**, 481–483.

LSAPT (LUNAR SAMPLE ANALYSIS PLANNING TEAM) (1970) Summary of Apollo 11 lunar science conference. *Science* **167**, 449–451.

LSPET (LUNAR SAMPLE PRELIMINARY EXAMINATION TEAM) (1969) Preliminary examination of lunar samples from Apollo 11. *Science* **165**, 1211–1227.

MARSHALL R. R. (1962) Mass spectrometric study of the lead in carbonaceous chondrites. *J. Geophys. Res.* **67**, 2005–2015.

MASON B., FREDRIKSSON K., HENDERSON E. P., JAROSEWICH E., MELSON W. G., TOWE K. M. and WHITE J. S., JR. (1970) Mineralogy and petrography of lunar samples. *Science* **167**, 656–659.

MURTHY V. R. and PATTERSON C. C. (1962) Primary isochron of zero age for meteorites and the earth. *J. Geophys. Res.* **67**, 1161–1167.

MURTHY V. R., SCHMITT R. A. and REY P. (1970) Rubidium–strontium age and elemental and isotopic abundances of some trace elements in lunar samples. *Science* **167**, 476–479.

PHILPOTTS J. A. and SCHNETZLER C. C. (1970) Potassium, rubidium, strontium, barium and rare-earth concentrations in lunar rocks and separated phases. *Science* **167**, 493–495.

ROSHOLT J. N. and TATSUMOTO M. (1970) Isotopic composition of uranium and thorium in Apollo 11 samples. *Geochim. Cosmochim. Acta*, Supplement I.

Shoemaker E. M., Bailey N. G., Batson R. M., Dahlem D. H., Foss T. H., Grolier M. J., Goddard E. N., Hait M. H., Holt H. E., Larson K. B., Rennilson J. J., Schaber G. G., Schleicher D. L., Schmitt H. H., Sutton R. L., Swann G. A., Waters A. C. and West M. N. (1969) Geologic setting of the lunar samples returned by the Apollo 11 mission. Apollo 11 Preliminary Science Report NASA SP-214, 41–83.

Shoemaker E. M., Hait M. H., Swann G. A., Schleicher D. L., Dahlem D. H., Schaber G. G., and Sutton R. L. (1970a) Lunar regolith at Tranquillity Base. *Science* **167,** 452–455.

Shoemaker E. M., Hait M. H., Swann G. A., Schleicher D. L., Schaber G. G., Sutton R. L., Dahlem D. H., Goddard E. N. and Waters A. C. (1970b) Origin of the lunar regolith at Tranquillity Base. *Geochim. Cosmochim. Acta*, Supplement I.

Silver L. T. (1970) Uranium–thorium–lead isotope relations in lunar materials. *Science* **167,** 468–471.

Stewart D. B., Appleman D. E., Huebner J. S. and Clark J. R. (1970) Crystallography of some lunar plagioclases. *Science* **167,** 634–635.

Tatsumoto M. (1966) Isotopic composition of lead in volcanic rocks from Hawaii, Iwo Jima and Japan. *J. Geophys. Res.* **71,** 1721–1733.

Tatsumoto M. (1970) Age of the moon: an isotopic study of lunar samples from Apollo 11—II. *Geochim. Cosmochim. Acta*, Supplement I.

Tatsumoto M. and Rosholt J. N. (1970) Age of the Moon: An isotopic study of uranium–thorium–lead systematics of lunar samples. *Science* **167,** 461–463.

Turner G. (1970) Argon-40/argon-39 dating of lunar rock samples. *Science* **167,** 466–468.

Turkevich A. L., Franzgrote E. J. and Patterson J. H. (1969) Chemical composition of the lunar surface in Mare Tranquillitatis. *Science* **165,** 277–279.

Wanke H., Begemann F., Vilcsek E., Rieder R., Teschke F., Born W., Quijano-Rico M., Voshage H. and Wlotzka F. (1970) Major and trace elements and cosmic-ray produced radioisotopes in lunar samples. *Science* **167,** 523–525.

Wanless R. K., Loveridge W. D. and Stevens R. D. (1970) Age determinations and isotopic abundance measurements on lunar samples. *Science* **167,** 479–480.

Wetherill G. W. (1956) Discordant uranium–lead ages, I. *Trans. Amer. Geophys. Union* **37,** 320–326.

Wood J. A., Dickey J. S., Jr., Marvin U. B. and Powell B. N. (1970) Lunar anorthosites. *Science* **167,** 602–604.

Proceedings of the Apollo 11 Lunar Science Conference, Vol. 2, pp. 1575 to 1581.

Elemental composition of lunar surface material

A. A. SMALES, D. MAPPER, M. S. W. WEBB, R. K. WEBSTER and J. D. WILSON
Analytical Sciences Division, Atomic Energy Research Establishment, Harwell, England

(*Received* 16 *February* 1970; *accepted* 17 *February* 1970)

Abstract—Elemental abundances, so far obtained, derived from the analysis of Apollo 11 lunar material are reported. Similarities and differences exist between lunar material, the eucritic achondrites, and the augite achondrite Angra dos Reis, the analysis of which is also reported.

WE PRESENT here the results of our initial analytical work on (i) lunar fines (sample 10084), (ii) lunar rock (sample 10060,26), (iii) glassy spheres (sample 10085/30/5) received, respectively, on 19 September, 23 October, and 19 November 1969. We also give some data on the elemental abundances in the augite achondrite Angra dos Reis, of which we were able to get 240 mg (BM 63233). In order to give approximate results in an initial survey of as many elements as possible, we used the following methods: spark source mass spectrography (SSM), emission spectrography (E), and X-ray fluorescence spectrometry (XRF). To give more accurate results for a smaller number of elements, we used activation analysis (AA; instrumental, neutron, activation analysis, INAA; radiochemical, neutron activation analysis, RNAA; radiochemical, high-energy gamma, activation analysis, RGAA) and mass spectrometric isotope dilution analysis (MSID). Not all methods were used on the glassy spheres or the meteorite because of the limited amount of sample available.

In the SSM method we used an A.E.I. MS702 instrument. Powdered samples were mixed with graphite previously doped with rhenium as an internal standard and were pressed into electrodes. The electrode surfaces were cleaned by presparking (~100 nc), and the analysis was carried out over a wide exposure range (300–0·003 nc) at intervals of × 1·5 to × 3. The photographic plates were examined by microdensitometry for the rare earths and by inspection for other elements, results being obtained by the "just detectable line" method. Results recorded for "standard" rocks W-1 and BCR-1 were used to provide empirical corrections for ionization efficiency factors in a number of cases. The accuracy of the method is believed to be within a factor of 3 of the stated values. The cleaved glassy spherule was examined directly by mounting it in an indium support and sparking against a carbon counter electrode. Plates were analyzed visually, and the relative ion exposures were verified by densitometry for several multiisotopic elements. Values for Ti, Mn and Dy from separate neutron activation analysis were used as internal standards.

In the ES method we used the technique of WEBB and WORDINGHAM (1968); synthetic standards and "standard" rocks (G-1, W-1, AGV-1, BCR-1, DTS-1, T-1 and syenite rock-1) were used for calibration, the spectra being assessed visually and by microphotometry. Coefficient of variation for a single determination was between 5 and 10 per cent at concentrations between 20 and 100 parts per million. Accuracy is estimated as ±20 per cent.

In the XRF method we used for major elements a fusion mix consisting of lithium tetraborate, lithium carbonate and lanthanum oxide (Spectroflux 105) to which had been added 0·8 per cent lithium nitrate, the general method being that of Norrish and Hutton (1969). We measured the minor and trace elements using lithium tetraborate only. Synthetic standards were used in both cases. Comparison of results by these methods was made with published figures on "standard" rocks AGV-1, BCR-1, G-1 and DTS-1 and from this the estimated accuracy is, for major elements (>1 per cent) ±3 per cent, minor elements (0·1 to 1 per cent) ±5 per cent, trace elements (<0·1 per cent) ±15 per cent.

For activation analysis both neutrons (from DIDO flux $1 \cdot 3 \times 10^{13}$ neutron

Table 1. Elements determined by activation analysis* or mass spectrometric isotope dilution or both

		Concentration of elements			
Atomic number	Element	Lunar fines	Lunar rock	Lunar glassy spheres	Meteorite Angra dos Reis
			Percentages		
11	Na	0·32	0·38	0·14	0·03
12	Mg		6	4	
13	Al	7·0	5·2	7·0	5·7
19	K	0·12‖	0·17‖		
20	Ca	~8†	~11†	~10†	~17†
22	Ti	4·7	5·2	4·9	1·7
26	Fe	12·5	14·2	12·3	6·5
			Parts per million		
9	F	74§			
21	Sc	66	70	57	57
23	V		~90†	~120†	
24	Cr	2300	2800	2300	1700
25	Mn	1630	1620	1610	700
27	Co	~25†	~30†	~30†	~10†
29	Cu		9·1‡		
31	Ga	3·8‡	4·7‡		
32	Ge	0·5‡	0·4‡		
33	As	0·03‡	0·01‡		
37	Rb	3·0‖	4·2‖		
38	Sr	177‖	180‖		
46	Pd		0·006‡		
49	In	2·0¶, 0·5¶	0·004‡		
55	Cs	0·12‖	0·19‖		
56	Ba	174‖	212‖		
57	La	19	25	27	10
58	Ce	46‖	56‖		
60	Nd	37‖	45‖		
62	Sm	13‖, 14	16‖, 16	13	8
63	Eu	1·8‖, 2·0	1·9‖, 2·6	2·1	2·0
65	Tb	4	6	5	2
66	Dy	23	24	21	12
67	Ho	5	7	5	3·5
70	Yb	12	14	12	6
71	Lu	1·5	1·8	1·4	0·7
72	Hf	11	12	11	3

* Accuracy by INAA, ±5 per cent for all results, except where noted.
† Accuracy by INAA, ±10 per cent.
‡ Accuracy by RNAA, ±5 per cent.
§ Accuracy by RGAA, ±10 per cent.
‖ Accuracy by MSID, ±5 per cent.
¶ Contamination of samples suspected.

cm^{-2} sec^{-1}) and high-energy gamma photons (bremsstrahlung produced from the Harwell electron linear accelerator) were used. For INAA, irradiation and decay periods were 12 sec and 7 min, followed by 5 min and 14 hr, respectively, for one set of samples; we used irradiation and decay periods of $2\frac{1}{4}$ days and $3\frac{1}{2}$ days, respectively, for a different set of samples. Counting was done with two Ge(Li) detectors, a small planar, high-resolution detector (1 keV at 120 keV) for identification of photopeaks and a larger 30-cm^3 coaxial detector (resolution, 5 keV at 1·32 MeV) for general counting. The associated PDP-8 gamma spectrometer has been described by PIERCE *et al.* (1969); we processed data from this instrument on the main Harwell computer (IBM 360/65), using a specially developed least-squares fitting programme which applies corrections for both decay and dead-time. Estimated accuracy is ± 10 per cent except where stated. For RNAA we used a flux of 5×10^{12} neutron cm^{-2} sec^{-1}

Table 2. Major elements in lunar fines, expressed as oxides

Oxide	By XRF (%)	By INAA (%)	By MSID (%)
Na_2O		0·43	
MgO	7·6		
Al_2O_3	13·0	13·2	
SiO_2	43·0		
K_2O	0·15		0·14
CaO	12·2		
TiO_2	7·7	7·8	
Cr_2O_3	0·3	0·34	
MnO	0·2	0·21	
FeO	15·6	16·1	

generally, but for short-lived radionuclides the higher flux ($1 \cdot 3 \times 10^{13}$) was used as a pneumatic "rabbit" tube into that flux position. Radiochemical methods used have been described elsewhere (SMALES *et al.* 1964, 1967a, b; FOUCHÉ and SMALES 1967a, b). Estimated accuracy is ± 5 per cent. For the fluorine determination by RGAA, samples and standards were irradiated in Al-foil containers for 30 min in the bremsstrahlung produced in a tungsten converter $\frac{1}{8}$ in. thick from a 10-μA beam of 24- to 28-MeV electrons. Radiochemical separation was achieved by distillation of H_2SiF_6 counting of the 0·51MeV annihilation radiation (after gamma spectrometry to establish absence of other gamma emitters) was then used for measurement of ^{18}F. Decay of this nuclide was followed for four half-lives as further confirmation of radiochemical purity. "Standard" rocks G-1 and W-1 were also analyzed and the estimated accuracy is ± 10 per cent.

In the MSID methods we used enriched isotopes of the elements concerned, Rb, Cs, Ba, Sr, Ce, Nd and Sm being determined in the same sample, K being determined in a separate sample. Cation exchange was used to separate the alkali, alkaline earths and rare earths as three groups, for subsequent mass analysis on a laboratory-constructed instrument with a radius of 6 in. and a sector of 60° for the two former groups and on an A.E.I. MS5 instrument for the latter group. Details are given elsewhere (MURUGAIYAN *et al.*, 1968). Estimated accuracy is ± 5 per cent.

Results obtained by AA and MSID are reported in Table 1; results obtained for the major elements by XRF, MSID and AA are given in Table 2 and data from some supplementary analyses by SSM, E and XRF are presented in Table 3. Table 4 lists conservative upper limits for several other elements derived from SSM measurements; some of these, for example, P, S and Pb, were detected but not measured.

By comparison with the work of the LSPET (1969), who used mainly emission spectrography for elemental analysis, and where results for 23 elements and upper limits (for that method) for 30 more were given, we list results so far for 35 elements in Tables 1 and 2, less firm results for 11 more in Table 3 and upper limits (generally at lower levels) for 25 elements in Table 4. Of those we undertook to determine in

Table 3. Supplementary results by spark source mass spectography, emission spectrography and X-ray fluorescence spectrometry

Atomic number	Element	Lunar fines (ppm)			Lunar rock (ppm)		Lunar spheres (ppm)	Meteorite Angra dos Reis (ppm)	
		SSM	E	XRF	SSM	E	SSM	SSM	E
19	K							50	
23	V	50	50			45	40	160	150
28	Ni	170	180		100	150	80	30	50
29	Cu	10			10		10	10	
30	Zn	50			40		20	10	
31	Ga							6	
37	Rb	3					3	1	
38	Sr	200	170	170	200	170	200	100	150
39	Y	100	120	100	100	120	60	40	30
40	Zr		330	320	300	340	200	100	100
41	Nb	10			10		5	5	
56	Ba	200		180	200		150	30	30
58	Ce						40	20	
59	Pr	20			20		20	10	
60	Nd						40	20	
64	Gd	15			20		20	10	
68	Er	10			10		10	5	
69	Tm	2			2		2	1	
90	Th	1			1			1	
92	U	0·4			0·4			0·4	

our original proposal, we have not yet any result for seven elements; on the other hand, we quote results for 18 elements which we did not include in our original proposal. We have not as yet done all the cross-checking of results by our several methods that we would wish to do. (This may explain the conservative accuracy estimates quoted.)

Our results broadly confirm those of the LSPET, the only significant difference being that for ytterbium (LSPET reports 1·3–7 parts per million; our value is 12–14 ppm). We do not know at present whether our samples are taken from those analyzed by the LSPET or whether they are different. If the samples are different, then our ytterbium figures may simply indicate a larger range for that element in the lunar material rather than an analytical disagreement. The LSPET comment that "samples are apparently free from inorganic contamination ... from ... indium, which forms the seal of the rock box ..." may no longer hold on the basis of methods of higher sensitivity. Our evidence is not strong but we believe that the disagreement between duplicate indium determinations for lunar fines, coupled with the comparatively high values, indicates contamination. The value for the lunar rock of 0·004 ppm was obtained by irradiating a piece of rock, then leaching it after irradiation with boiling HNO_3–HF and thereafter proceeding with radiochemical separation on the leach liquor and on the remaining rock as separate samples. The leach liquor contained the equivalent of an additional 0·004 ppm of indium; this result possibly implies that even here some readily removable indium (that is, contamination) was present.

Table 4. Upper limits by our spark source mass spectrographic method for elements not quantified in lunar fines and rocks

Atomic number	Element	Upper limit (ppm)
5	B	2
15	P	1500
16	S	500
17	Cl	70
34	Se	20
35	Br	3
42	Mo	10
44	Ru	1
45	Rh	1
47	Ag	1
48	Cd	1
50	Sn	10
51	Sb	1
52	Te	1
53	I	1
73	Ta	3
74	W	1
76	Os	1
77	Ir	1
78	Pt	1
79	Au	1
80	Hg	1
81	Tl	1
82	Pb	10
83	Bi	1

The results for rare earths and barium on the Angra dos Reis meteorite compare quite reasonably with those of SCHNETZLER and PHILPOTTS (1969) and our approximate values for potassium, thorium and uranium are not dissimilar to published values (MÜLLER and ZÄHRINGER, 1969; MORGAN and LOVERING, 1964). Our results for the major elements are in broad agreement with those of LUDWIG and TSCHERMAK (1887) but not unexpectedly our data for sodium and potassium differ markedly from theirs [as also do the results of MÜLLER and ZÄHRINGER (1969) for potassium].

We had some 10 mg of the glassy spheres; one-half sphere we analyzed by SSM; the other whole sphere, after grinding, was analyzed by AA. On grinding the color changed from black to yellow-brown. The analysis shows the spheres to be quite similar to the lunar fines and rocks, except for the sodium content which is distinctly lower in the spheres. The result is very similar to that of FREDRIKSSON and KRAUT

(1967) for the analysis of Cachari eucrite glass. The analogy with the arguments about tektites and their probably terrestrial parent material is also obvious.

Several attempts have been made to compare the chemical composition of Apollo 11 lunar material with that of known terrestrial and meteoritic silicates. Only the suggestion of terrestrial anorthositic gabbro (OLSEN, 1969; PHINNEY *et al.*, 1969) seems to meet the requirements for the major elements, but it fails on the alkali metals and carbon. Several authors have suggested similarities with the basaltic achondrites, especially the eucrites. TURKEVICH *et al.* (1969) say, however, that in the lunar material "the high value for titanium and the relatively low value for iron are variance with the content of these elements in eucrites and other calcium-rich achondrites." The statement about titanium is true, but that about iron is incorrect. Moore County, a eucrite containing 12·4 per cent of FeO, comes well within the accuracy range of TURKEVICH's value of 12·1 per cent. Our values for FeO (15·6 and 16·1 per cent) on Apollo lunar fines are higher than the latter value and fit well into the eucrite range (NICHIPORUK *et al.*, 1967).

As a result of these considerations we felt that it would be worthwhile to examine Angra dos Reis, a unique calcium-rich achondrite, which is reported to have the highest titanium content of any meteorite so far analyzed. We analyzed a small sample by SSM and by AA and found that there are some quite interesting similarities to Apollo 11 lunar material. Concentrations of the rare earths, especially, are within a factor of 2 of those in lunar samples, closer than values for any other meteorite. The concentration of europium, which seems very low in lunar material by comparison with the other rare earths, is in this meteorite almost exactly the same as that in the lunar material. (Incidentally, the "normalized to ytterbium" figure for europium in lunar material is lower than for any meteoritic material.) The uranium and thorium abundances in the two materials are also quite similar.

Dissimilarities exist in that for Angra dos Reis, the titanium content is lower [TURKEVICH *et al.* (1969) comment that Sinus Medii material may have lower titanium content than Mare Tranquillitatis material], the sodium and potassium content is much lower and the calcium content is higher than in lunar material. Also the points on the plot of CaO content as a function of Fe/(Fe + Mg) (MASON, 1962), are in quite different positions (that for lunar material fitting well with eucrites).

We agree with TURKEVICH *et al.* (1969) that "there appears to be no common material (available) on earth that matches in all respects the chemical composition of lunar surface material at Mare Tranquillitatis." It may well be that the comment by HOYLE (1969) is significant; that "planetary material separated from the sun ... to the terrestrial distance ... would be expected to experience temperatures ... (such that) ... melting and chemical segregation could have taken place within the primitive planetary material, even though this hot phase was short-lived, $\sim 10^4$–10^5 y", and that in the moon we have a unique fraction of that segregation process.

Acknowledgments—We thank Dr. Moss of the British Museum (Natural History) for the Angra dos Reis sample; Dr. S. O. AGRELL of Cambridge University for the lunar glassy spheres; Dr. J. S. HISLOP for the fluorine determination by RGAA; and R. HALLET, T. C. HUGHES, R. P. KAY, C. A. J. MCINNES, C. R. MATTHEWS, A. G. MORTON, P. F. RALPH for experimental assistance.

Reference

Fouché K. F. and Smales A. A. (1967a) The distribution of trace elements in chondritic meteorites. 1. Gallium, germanium and indium. *Chem. Geol.* **2,** 5–33.

Fouché K. F. and Smales A. A. (1967b) The distribution of trace elements in chondritic meteorites. 2. Antimony, arsenic, gold, palladium and rhenium. *Chem. Geol.* **2,** 105–134.

Fredriksson K. and Kraut F. (1967) Impact glass in the Cachari eucrite. *Geochim. Cosmochim. Acta* **31,** 1701–1704.

Hoyle F. (1969) Planetary formation and lunar material. *Science* **166,** 401.

Ludwig E. and Tschermak G. (1887) Der Meteorit von Angra dos Reis. *Mineral. Petrogr. Mitt.* **8,** 341–350.

Mason B. (1962) *Meteorites*, p. 107. John Wiley.

Morgan J. W. and Lovering J. F. (1964) Uranium and thorium abundances in stony meteorites. 2. The achondritic meteorites. *J. Geophys. Res.* **69,** 1989–1994.

Müller H. W. and Zähringer J. (1969) Rare gases in stony meteorites. In *Meteorite Research*, (edited by P. M. Millman), pp. 845–856. Reidel.

Murugaiyan P., Verbeek A. A., Hughes T. C. and Webster R. K. (1968) Separation scheme for the determination of alkali metals (K, Rb, Cs) alkaline earths (Sr, Ba) and rare earths (Ce, Nd, Sm, Eu) in silicate materials by isotopic dilution analysis. *Talanta* **15,** 1119–1124.

Nichiporuk W., Chodos A., Helin E. and Brown H. (1967) Determination of iron, nickel cobalt, calcium, chromium and manganese in stony meteorites by X-ray fluorescence. *Geochim. Cosmochim. Acta* **31,** 1911–1930.

Norrish K. and Hutton J. T. (1969) An accurate X-ray spectrographic method for the analysis of a wide range of geological samples. *Geochim. Cosmochim. Acta* **33,** 431–453.

Olsen E. (1969) Pyroxene gabbro (anorthosite association): Similarity to Surveyor V Lunar analysis. *Science* **166,** 401–402.

LSPET (Lunar Sample Preliminary Examination Team) (1969) Preliminary examination of lunar samples from Apollo 11. *Science* **165,** 1211–1227.

Phinney R. A., O'Keefe J. A., Adams J. B., Gault D. E., Kuiper G. P., Masursky H., Collins R. J. and Shoemaker E. M. (1969) Implications of the Surveyor 7 results. *J. Geophys. Res.* **74,** 6053–6080.

Pierce T. B., Webster R. K., Hallett R. and Mapper D. (1969) Developments in the use of small digital computers in activation analysis systems. In *Modern Trends in Activation Analysis*, (editor J. R. DeVoe). *U.S. Nat. Bureau Stand. Spec. Publ. 312,* **2,** 1116–1120.

Schnetzler C. C. and Philpotts J. A. (1969) Genesis of the calcium-rich achondrites in light of rare earth and barium concentrations. In *Meteorite Research*, (editor P. M. Millman) pp. 206–216. Reidel.

Smales A. A., Hughes T. C., Mapper D., McInnes C. A. and Webster R. K. (1964) The determination of rubidium and caesium in stony meteorites by neutron activation analysis and by mass spectrometry. *Geochim. Cosmochim. Acta* **28,** 209–233.

Smales A. A., Mapper D. and Fouché K. F. (1967a) The distribution of some trace elements in iron meteorites, as determined by neutron activation. *Geochim. Cosmochim. Acta* **31,** 673–720.

Smales A. A., Mapper D. and Fouché K. F. (1967b) The distribution of some trace elements in iron meteorites, as determined by neutron activation. Part 2. Analytical Methods. U.K.A.E.A. AERE-R 5254. H.M.S.O., London.

Turkevich A. L., Franzgrote E. J. and Patterson J. H. (1969) Chemical composition of the lunar surface in Mare Tranquillitatis. *Science* **165,** 277–279.

Webb M. S. W. and Wordingham M. L. (1968) A versatile spectrographic method for the analysis of a wide-range of materials. U.K.A.E.A. Report AERE-R 5799. H.M.S.O., London.

Proceedings of the Apollo 11 Lunar Science Conference, Vol. 2, pp. 1583 to 1594.

Cosmic-ray production of rare-gas radioactivities and tritium in lunar material*

R. W. STOENNER, W. J. LYMAN and RAYMOND DAVIS, JR.
Chemistry Department, Brookhaven National Laboratory, Upton, New York 11973

(*Received* 3 *February* 1970; *accepted in revised form* 21 *February* 1970)

Abstract—The argon radioactivities ^{37}Ar (half-life 35-days) and ^{39}Ar (half-life 269-years) were obtained by vacuum melting from interior and exterior portions of rock 10057, and from a portion of the fines from the bulk sample container. The release of argon and tritium from the fine material was followed as a function of the temperature. The tritium activity for the rock and fine material was found to be 220 $\pm$ 15, and 314 $\pm$ 13 dpm/kg respectively, which is similar to that commonly found in stone meteorites. The hydrogen content of the lunar fine material was found to be 0·84 cm^3 (STP)/g, and the H/He atom ratio was 7·8. These gases presumably are trapped solar wind. The rock was found to contain 18·5 $\pm$ 0·9 dpm/kg ^{37}Ar and 11·2 $\pm$ 0·3 dpm/kg ^{39}Ar, and the fine material contained 18·7 $\pm$ 1·2 dpm/kg ^{37}Ar and 9·2 $\pm$ 0·4 dpm/kg ^{39}Ar. A comparison is made of the argon activities observed in the lunar samples with those expected from the spallation of iron, titanium and calcium. From these data and the ^{38}Ar content, the cosmic-ray exposure age of the rock is deduced to be 110 million years.

INTRODUCTION

A MEASUREMENT of the stable and radioactive argon isotopes in lunar material can give valuable information on cosmic ray interactions on the lunar surface and can be used to deduce cosmic-ray exposure ages of fragmented material resting on the lunar surface. Of particular interest are the stable isotopes, ^{36}Ar and ^{38}Ar, and the radioactive isotopes, ^{37}Ar (half-life 35 days) and ^{39}Ar (half-life 269 yr). These isotopes are produced primarily by high energy bombardment of the relatively abundant elements iron, titanium and calcium. The primary goal of the present investigation was to measure the radioactive isotopes, ^{37}Ar and ^{39}Ar, in surface rocks and in the fine material. The argon is evolved from the solid by vacuum melting and is counted in a small low-level gas proportional counter. By using pulse-height analysis the Auger-electron spectrum from the electron capture of ^{37}Ar can be easily distinguished from the beta spectrum of ^{39}Ar, and both isotopes can be measured simultaneously. Since ^{37}Ar has a half-life of only 35 days, the measurements were made at the Lunar Receiving Laboratory (LRL) during the biological quarantine.

The vacuum melting technique used to evolve and count the argon radioactivities permits a search to be made as well for radioactive isotopes of the higher rare gases, namely ^{81}Kr, ^{85}Kr, ^{127}Xe, ^{133}Xe and ^{222}Rn. The isotope ^{81}Kr (half-life 2·1 $\times$ 10^5 yr) has been observed in meteorites (MARTI, 1967; EUGSTER *et al.*, 1967) and has been attributed to high energy spallation of Sr, Y and Zr. Since the lunar rocks are relatively high in these elements, measurable amounts of ^{81}Kr and ^{85}Kr may be present. Radon-222 will arise from the decay of ^{238}U and a measurement of the ^{222}Rn content can be used as a sensitive measurement of the uranium concentration.

* This work was supported by NASA and the AEC.

Of special interest is the tritium radioactivity in lunar materials that is also released by vacuum melting. This isotope is produced by high energy cosmic-ray bombardment. The rate of tritium production is an essential quantity needed to determine the tritium-^{3}He cosmic-ray exposure age. Of far greater interest is the possibility that some of the solar-flare particles are tritons. Fireman and his associates observed tritium in the Discoverer 17 satellite and in a sounding rocket that was exposed to the solar flare of 12 November, 1960. An analysis of their measurements indicates that the tritium to hydrogen ratio in this event was about 10^{-3} (FIREMAN *et al.*, 1961; FIREMAN, 1963; TILLES *et al.*, 1963; BISWAS and FICHTEL, 1965). A search for tritium in material exposed on the surface of the moon may afford a unique opportunity to remeasure the tritium content of solar-flare particles.

EXPERIMENTAL

When the experiment to measure ^{37}Ar in lunar material was planned, it was difficult to estimate how long the biological quarantine period would last. In view of the short half-life of ^{37}Ar it was considered advisable to make the measurements at the Lunar Receiving Laboratory (LRL) during the quarantine period. The extraction, purification and counting apparatus was installed at LRL, and

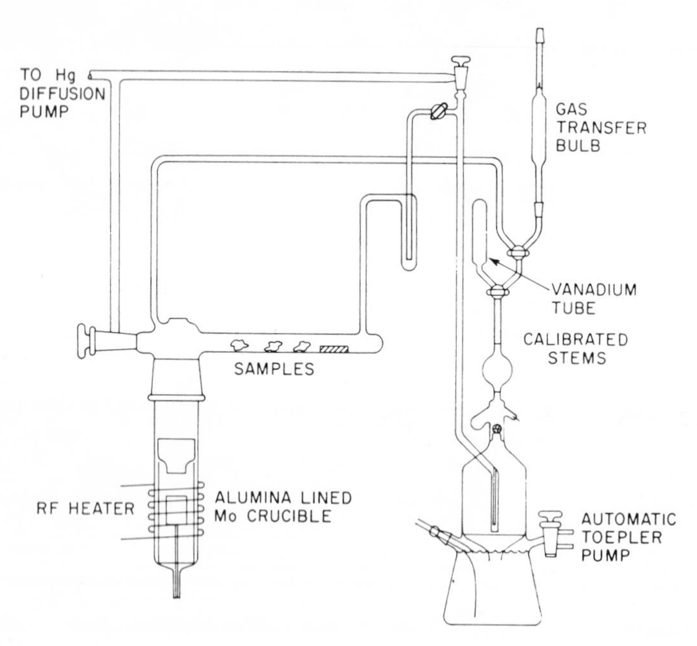

Fig. 1. The apparatus for vacuum fusion and extraction of rare gases and hydrogen.

measurements were made on two samples: fine material (10002-6) and an exterior sample of a rock (10057-3). After the quarantine period, an interior sample of this same rock (10057-27) was measured on an essentially identical apparatus at Brookhaven National Laboratory.

The samples measured at LRL were sterilized by heating for 24 hr at 120°C in a sealed stainless steel container. Loss of argon during this modest heating was rather unlikely, but to test this point a search was made for ^{37}Ar and ^{39}Ar in the container. This measurement showed that less than 0·6 dpm/kg of ^{37}Ar was lost during this treatment. It is more likely that tritium is released by this heating. However, if tritium were released it would be lost by absorption in the stainless steel of the container, so measurements were not attempted. It will be shown later that it is unlikely that tritium activity was lost by the sterilization heating.

The apparatus for vacuum melting the sample and isolating the rare gases and hydrogen is shown in Fig. 1. The samples were dropped into an alumina-lined molybdenum crucible heated by a radio-frequency induction heater. The gases evolved were collected in the Toepler pump and placed in a tube containing hot vanadium metal. The vanadium, at 850°C, serves as a getter to remove chemically active constituents. The vanadium was then cooled to room temperature to absorb hydrogen as vanadium hydride. The rare gases were removed from the vanadium tube, their volume was measured,

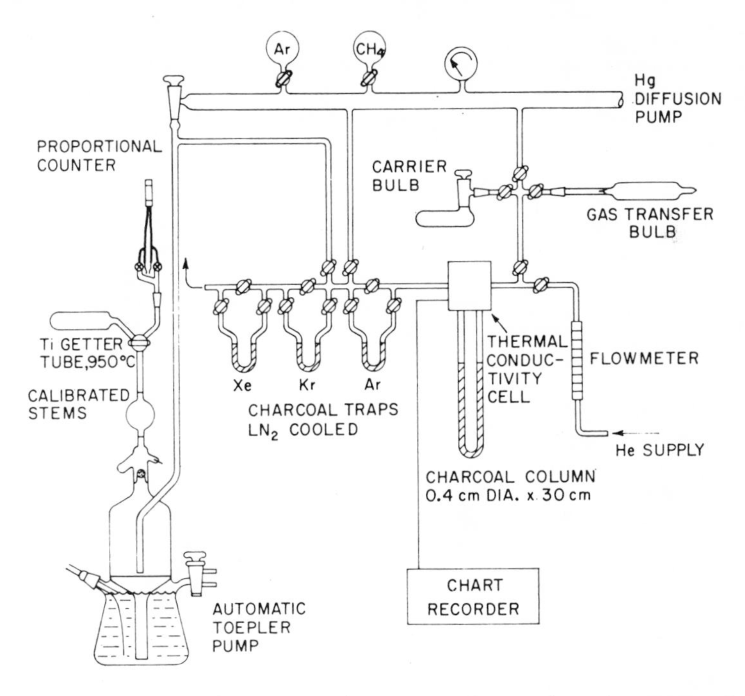

Fig. 2. Apparatus used to separate the rare gases by gas phase chromatography.

and they were placed in a sample-transfer bulb. The vanadium tube was then reheated to 850°C to dissociate the vanadium hydride. The hydrogen was collected with a Toepler pump, measured, and sealed in a second sample-transfer bulb. The apparatus for vacuum fusion and extraction of the rare gases and hydrogen from the samples was located in the gas analysis laboratory behind the secondary biological barrier. The sealed glass ampules containing the rare gases and hydrogen were then sterilized and taken through the biological barrier to a purification and counter filling system outside the barrier.

The isolated rare gases were then separated by gas-phase chromatography on a charcoal column with helium as the carrier gas. The apparatus used is shown in Fig. 2. Small measured volumes (0·3 cm^3 STP) of ^{36}Ar, Kr, and Xe carrier gases were introduced into the system prior to opening the gas-transfer bulb (see Fig. 2). These carrier gases and the rare-gas sample were adsorbed on top of the charcoal column while it was evacuated and cooled with liquid nitrogen. The column was then filled with helium and a flow of approximately 70 cm^3 (STP)/min started through the system, by-passing the individual charcoal traps for Ar, Kr and Xe. The column temperature was then warmed up to -25°C by quickly removing the liquid nitrogen Dewar and replacing it with a Dewar containing alcohol at −25°C. The emergence and passage of each rare gas from the column was observed with a thermal conductivity cell and chart recorder. In about 1 minute argon was observed to leave the column and in 4 min all of the argon was removed. During this period argon was trapped from the helium gas stream in the charcoal trap provided. Approximately 6 min later the krypton fraction appeared at the end of the column. The column temperature was then increased by replacing the alcohol Dewar with one containing water at 50°C. At this temperature krypton was eluted in 4 min, and trapped in the second charcoal trap. In about 8 min xenon began to leave the column, and at this time the column temperature was raised to 90°C to increase the rate of removal of xenon. It was trapped in the third charcoal trap. Only radon now remains on the column. The helium flow was stopped, the column was cooled to liquid nitrogen temperature, all traps and the column were evacuated to remove helium, and all stopcocks were closed.

The individual rare gases were then collected sequentially with the automatic Toepler pump and each gas was placed into the hot (950°C) titanium getter tube. Argon was easily removed from its trap at −25°C, but krypton and xenon required warming to 50°C and 100°C, respectively. Since there was no carrier gas for radon, a special procedure was used. Prior to removing radon, a small volume of argon gas (0·3 cm^3 STP) was adsorbed on the column at liquid nitrogen temperatures, and then both gases were desorbed by heating the column to 350°C. There is some difficulty with radon emanating from the natural radium in the charcoal used in the column and traps. We found that Columbia JXC charcoal prepared from petroleum residues emanates only 5·6 ± 1·1 ^{222}Rn atoms min^{-1} g^{-1}. Recovery of the initially introduced rare gases is essentially quantitative. The yields are measured and corrections for yield are made as required. The argon carrier gas is 99·5 per cent ^{36}Ar, so that in the case of contamination of the sample with air argon, the yield of carrier gas can be corrected by making a mass analysis of the recovered argon. In the experiments reported, air argon was not introduced, so this additional precaution was not necessary.

The separated argon, krypton, and xenon were placed in small gas proportional counters for the radioactivity measurements. A small amount of methane (3–5 per cent) is added to the counter and then the entire rare-gas sample is introduced into the counter. The final gas pressure in the counter is measured for each filling since the ^{39}Ar efficiency and pulse-height spectrum shape depends upon the gas pressure. In these experiments the gas pressure ranged from 0·89 to 1·71 atm. The counters had a 0·0125 mm tungsten center wire in a cylindrical cathode of zone refined iron enclosed in an envelope of silica glass. The counter had a thin window at one end to permit energy calibration of the multichannel analyzer with an external ^{55}Fe X-ray source. The active volume of these counters ranged from 0·3 to 0·6 cm^3. The counter was operated inside a ring of anti-coincidence gas proportional counters inside a lead shield 10 cm thick. The shield was located 15 m underground in the Radiation Counting Laboratory. Eight counters could be operated simultaneously, recording the spectra with 256-channel analyzers. The counter voltage was adjusted so that the 2·8-keV Auger electrons from the decay of ^{37}Ar would center at channel 90. The counters had a 30-per cent resolution (full width at half maximum) for these events. The amplifier was designed so that all events with energy greater than 6·8 keV were stored in channels 250–254 (hereafter referred to as pile-up channels). This arrangement enabled one to distinguish clearly the events arising from the electron-capture decay of ^{37}Ar, and the continuous beta spectrum (maximum energy 570 keV) arising from ^{39}Ar decay, see Fig. 3. Individual counter efficiencies for ^{37}Ar and ^{39}Ar were measured by filling the counters with argon containing known amounts of these radioactivities. The exact efficiency of the individual counters was measured over the pulse-height spectrum and the pile-up channels. Their individual distributions depended upon their specific dimensions and gas pressure. Counter efficiencies for ^{37}Ar and ^{39}Ar depend upon measuring the specific activity of argon gas containing these individual activities in a series of large proportional counters of measured volume efficiencies. The large proportional counters had cathode dimensions ranging from 9·0 cm dia. to 1·2 cm dia. by 30 cm long. This range of counter dimensions was chosen to explore the importance of wall effects. The efficiencies of the small proportional counters used in the measurements reported here are known to ±3 per cent accuracy. In general they exhibited total efficiencies in the range of 50–70 per cent for ^{37}Ar and ^{39}Ar. To measure krypton, xenon and radon radioactivities the counter voltage was adjusted so that the ^{55}Fe X-ray would center at channel 212, and only those decays in the pile-up channels were used in the analysis of the data.

Hydrogen was purified by passing the gas through a palladium metal thimble at 800°C and the gas placed in a gas proportional counter. The cathode was made of copper and the outer envelope of silica glass. The cathode had an i.d. of 1·0 cm and a length of 12·5 cm. The gas filling was a mixture of 90 per cent argon, and 10 per cent methane. The counter operated at one atmosphere and contained less than 0·2 atm of hydrogen gas. Counting was performed in the same system used in the rare-gas counting, with the voltage on the counter adjusted so that all events from an externally placed ^{60}Co source occurred in the pile-up channels. This procedure is the equivalent of operating the counter on its plateau.

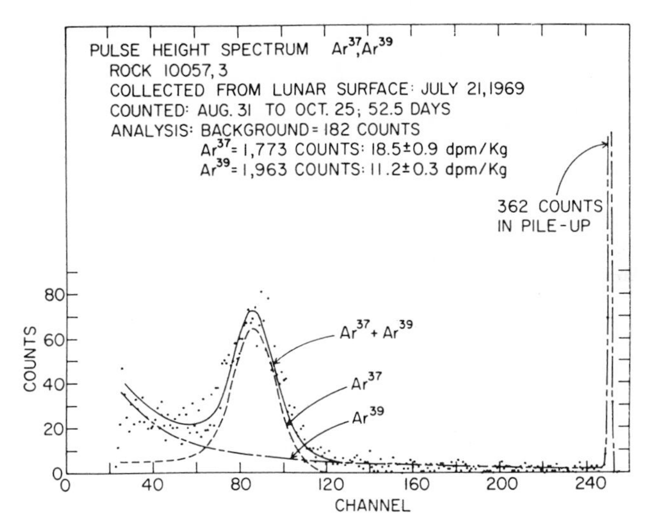

Fig. 3. Pulse height spectrum for ^{37}Ar and ^{39}Ar from lunar rock 10057-3.

Results and Discussion

Preliminary studies (LSPET, 1969) showed that the fine material contained large quantities of rare gas that probably can be attributed to solar-wind bombardment. It was therefore of interest to measure the amount of rare gases and hydrogen evolved from this finely divided material with increasing temperature. As mentioned above, this sample had been heated previously to 120°C for 24 hr; the quantity of gas released during this period was not measured. The sample was then heated in stages reaching the maximum temperatures 600°, 900°, 1200° and 1600°C; the volume of rare gas and hydrogen released at these temperatures, and the periods of time that the fines were maintained at the maximum temperature are given in Table 1. Detailed studies of the rare gas composition of the fine material show that the rare gas is 98 per cent helium (LSPET, 1969; Funkhouser *et al.*, 1970). This gas evolution experiment shows that more than 67 per cent of the helium and 80 per cent of the hydrogen is removed from this fine material at 600°C. A temperature of 900°C releases the remaining quantities of these gases. It is interesting to compare the H/He atom ratio of 7·8 measured from the fines to the H/He ratio in the sun. The recently accepted solar value (Iben, 1969; Taylor, 1967; Lambert, 1967) is 17, higher by a factor of two than the value obtained by heating the fine material. This may be explained by the

Table 1. Volumes of rare gases and hydrogen from lunar material

Sample	Extraction temperature and heating period	Rare gases (cm^3(STP)/g)	Hydrogen (cm^3(STP)/g)
Fines 10002-6	Sterilization, 120°C 24 hr	not meas.	not meas.
	120–600, 64 min	0·145	0·67
	600–900, 73 min	0·068	0·12
	900–1200, 92 min	0·003	0·04
	1200–1600, 68 min	0·0005	0·007
	Total 120–1600	0·217	0·84
Rock 10057-3 (Exterior)	120–1600	not meas.	0·1
Rock 10057-27 (Interior)	120–1600	not meas.	0·16

preferential loss of hydrogen with respect to helium in these finely divided silicates while they are on the lunar surface. On the moon the surface is heated for prolonged periods to temperatures exceeding 100°C. The volume of hydrogen observed in these measurements would correspond to a hydrogen abundance of 75 ppm. The two fragments of rock 10057 were heated to 1600°C; they yielded 0·1 and 0·16 cm^3 of hydrogen (Table 1). The larger volume of hydrogen obtained from the interior sample can probably be accounted for by the fact that this sample had been stored in air for several weeks prior to the measurement. We conclude the hydrogen abundance of the rock was less than or equal to 10 ppm.

The tritium activity released from the fine material in the various temperature intervals is given in Table 2. Sixty-three per cent of the tritium was released at 600°C,

Table 2. Rare gas and tritium radioactivities

Sample	Radioactivities in dpm/kg: Tritium	^{37}Ar*	^{39}Ar	$^{37}Ar/^{39}Ar$
Fines, No. 10002-6				
Sterilization heating	not meas.	$\leqslant$0·6	—	—
Heat I: 120–600°C	199 ± 12	$\leqslant$1	—	—
Heat II: 600–900°C	70 ± 3	6·7 ± 0·7	3·6 ± 0·2	1·84 ± 0·20
Heat III: 900–1200°C	41 ± 3	11·3 ± 0·8	5·3 ± 0·3	2·14 ± 0·18
Heat IV: 1200–1600°C	3 ± 2	$\leqslant$2	—	—
Combined argon Heats I and IV	—	0·7 ± 0·2	0·2 ± 0·2	—
Totals	314 ± 13	18·7 ± 1·2	9·2 ± 0·4	2·03 ± 0·16
Rock 10057-3, Exterior†	224 ± 15	18·5 ± 0·9	11·2 ± 0·3	1·65 ± 0·09
Rock 10057-27, Interior†	214 ± 20	23·7 ± 4·1	14·8 ± 0·4	1·60 ± 0·30

* Values have been corrected for decay to 0000 hours, c.d.t., July 21, 1969. If re-entry is taken as the true EOB time, the ^{37}Ar results should be decreased by a factor of 1·076.

† $^{81}Kr + ^{85}Kr$ radioactivities were measured for rock 10057: Exterior sample, −0·5 ± 0·3; interior sample, 0·0 ± 0·3 dpm/kg.

closely paralleling the evolution of hydrogen (80 per cent). The total tritium activity observed in the fine material was approximately the same as observed in the rock. All of these samples have approximately the same amount of tritium activity that is usually observed in stone meteorites (DAVIS *et al.*, 1963; FIREMAN *et al.*, 1963; ST. CHARALAMBUS *et al.*, 1969), namely 200 to 700 dpm/kg. It is clear that the high tritium content that one might expect from solar-flare tritons imbedded in an exposed surface is not observed. Perhaps the samples measured were not truly surface samples. The fines were well mixed, and, if there were a high tritium content on the surface, it could be highly diluted with deeper lying material. The exterior rock sample did have micro-meteorite pits on its surface, but it is not at all certain that this surface was exposed during the last 20 years or so. The interesting observation of Fireman and his associates (FIREMAN *et al.*, 1961; FIREMAN, 1963; TILLES *et al.*, 1963) of a triton component in solar-flare particles should be tested with lunar material known to have been exposed in recent years on the surface.

The tritium activities measured in the sample of fines (10002-6) agree very well with the tritium activity of 325 $\pm$ 17 dpm/kg observed on sample of fines (10084-24) by FIREMAN *et al.* (1970). These same investigators also measured the tritium activity in rocks (10017-14), (10072-11), and in the surface and interior samples of rock 10061, obtaining the values respectively: 219 $\pm$ 7, 234 $\pm$ 10, 235 $\pm$ 15 and 231 $\pm$ 10 dpm/kg. Again, the range of values is very close to those we measured for rock 10057. It is concluded from these observations that the tritium activity in surface lunar rocks is quite uniform, and one could use an average value of 226 $\pm$ 10 dpm/kg in calculating tritium–helium-3 exposure ages.

Tritium is primarily a high-energy product of cosmic-ray bombardment. It is reasonable to expect that the tritium production should be fairly insensitive to changes in the elemental composition of silicate-like target materials since the target and product masses are so far apart. Estimates of the tritium production rates in chondrites (and thus on the moon because of the not too dissimilar composition) due to galactic cosmic rays have been made by TRIVEDI and GOEL (1969) and by TRIVEDI (1969). TRIVEDI (1969) calculates that the tritium production rate should rise from about 130 dpm/kg at one cm below the lunar surface to a maximum of 230 dpm/kg at a depth of 9 cm; thereafter it falls, reaching 25 dpm/kg at a depth of 80 cm. Typical diameters for the lunar rocks in which tritium was measured range from 6 to 13 cm, thus it is quite probable that most, if not all, of the tritium found in the lunar rocks was due to production by galactic cosmic rays. The agreement between observed and calculated results is quite remarkable considering uncertainties and fluctuations in the galactic cosmic ray spectrum.

The ^{37}Ar and ^{39}Ar activities observed from the fine material and rock 10057 are listed in Table 2. The release of ^{37}Ar and ^{39}Ar radioactivities from the fine material during the step-wise heating shows that the argon evolved almost completely in the temperature interval 600°–1200°C. Measurements on the combined sample evolved below 600°C and above 1200°C (see Table 2) showed that less than 4 per cent of both of these activities was released outside the 600° to 1200°C interval. This result indicates that the argon produced throughout this silicate material with an average particle size of 100 μ (LSPET, 1969) is strongly retained. The total ^{37}Ar and ^{39}Ar activity measured in this material is approximately the same as that measured in the rock sample.

These argon activities may be compared to those observed by other investigators. FIREMAN *et al.* (1970) measured the ^{37}Ar and ^{39}Ar contents of a sample of fines (10084-24) and in two rock samples (10017-14) and (10072-11). Their values for the

^{39}Ar activity (dpm/kg) and ^{37}Ar/^{39}Ar activity ratio are for these samples: (10084-24)—12·1 ± 0·7, 2·25 ± 0·35; (10017-14)—16·4 ± 0·9, 1·28 ± 0·20; (10072-11)—15·8 ± 1·0, 1·62 ± 0·25. Their values agree very well with those reported here. Argon activities were also measured by WÄNKE *et al.* (1970) for the lunar fines (10084-18) with the following results: ^{37}Ar—33·2 ± 4·5 dpm/kg, ^{39}Ar—7·5 ± 0·5 dpm/kg and ^{37}Ar/^{39}Ar—4·43 ± 0·67. These investigators also observed 9·08 ± 0·29 dpm/kg of ^{39}Ar in rock (10057-40). Their value for the ^{37}Ar activity for the fine sample (10084-18) agrees well with the value of FIREMAN *et al.* (1970), though the ^{39}Ar content is about 40 per cent lower. The ^{39}Ar activity that WÄNKE *et al.* (1970) measured for rock 10057-40 is about 30 per cent lower than the value (averaged) we report for rock 10057. It should be noted, however, that all of these measurements were performed on samples that are not strictly comparable, and one might not expect exact agreement. In order to make detailed comparisons it would be very important to know the exact location of the measured samples in the rock, and the orientation of the sample on the surface of the moon. Although this important information could be obtained, it is not known at the present time.

It is interesting to compare the ^{37}Ar and ^{39}Ar activities produced in the fines and the rock with those expected from cosmic-ray bombardment. The magnitude of the activities is about the same as those observed in recently fallen meteorites (DAVIS *et al.*, 1963; FIREMAN *et al.*, 1963; ST. CHARALAMBUS *et al.*, 1969). The meteorite samples studied are regarded as interior pieces of objects weighing 10–100 kg, that receive an isotropic cosmic-ray exposure. The surface lunar material might be expected to contain higher activities produced by low energy solar and galactic cosmic rays, but on the other hand the lunar surface receives only about one half of the isotropic flux. Apparently the geometric and shielding effects nearly cancel.

It is of interest to derive information about the intensity of galactic and solar cosmic rays, and the time variation in their respective intensities by comparing the observed ratio ^{37}Ar/^{39}Ar to that expected from these two sources of high energy particles. The galactic cosmic-ray intensity is modulated by the solar wind. The flux of particles below a few GeV energy decreases with increasing solar activity, following the 11 year solar activity cycle. Since Apollo 11 rocks were recovered during a period of maximum solar activity, the intensity of galactic cosmic rays is lower than its average value. Since ^{37}Ar has only a 35-day half-life the observed ^{37}Ar activity would measure the present cosmic-ray intensity, whereas ^{39}Ar has a 269-yr half-life and its activity level would reflect the average particle intensity over many hundreds of years. Solar cosmic rays on the other hand occur as a sudden burst of particles following a solar flare. If the moon is in the path of a solar flare, its surface will be exposed to an intense flux of particles for a relatively short period of time, hours to a few days. Argon-37 and ^{39}Ar will both be produced, but while the activity level of ^{37}Ar will suddenly increase the ^{39}Ar activity level will not be greatly affected. The ^{39}Ar activity would reflect the long time average of the total particle flux from both solar and galactic cosmic rays. The solar-flare particle intensity on the moon was very low during the summer of 1969. A solar event starting on April 12 did, however, increase the proton flux in the earth–moon vicinity (ESSA, 1969) significantly for a period of a few days. It is estimated that no more than 10 per cent of the measured

^{37}Ar can be attributed to production by this event. One might then conclude that the ^{37}Ar activity level observed in Apollo 11 samples corresponds to the amount of activity produced by galactic cosmic rays during a period of high solar activity, and the ^{39}Ar activity level corresponds to the long-time-average particle flux.

To examine these questions quantitatively we would have to know the relative production rates of ^{37}Ar and ^{39}Ar in lunar material as a function of the energy of the bombarding particles, protons and alphas, and as a function of the depth in the lunar surface. High energy particles entering the surface interact directly with the elements K, Ca, Ti and Fe to produce ^{37}Ar and ^{39}Ar. In addition, the primary particles produce high energy cascade particles, mainly protons, neutrons and pions, and also lower energy nuclear evaporation particles, mainly protons, neutrons, and alphas. Measurements have been made of the production of various radioisotopes in iron and simulated stone targets (SHEDLOVSKY and RAYUDU, 1963, 1964; TRIVEDI, 1969; DAVIS *et al.*, 1963) and these measurements have been interpreted with theoretical internuclear cascade processes to derive the production of various isotopes on the lunar surface (EBEOGLU and WAINIO, 1966; KOHMAN and BENDER, 1967). Experiments are now in progress at Brookhaven to measure the production of ^{37}Ar and ^{39}Ar with 600-MeV protons in a thick target of simulated lunar material. Since this information is not available, we will estimate the relative production rates of ^{37}Ar and ^{39}Ar in lunar material from measured thin target cross sections of K, Ca, Ti and Fe with 600-MeV protons and neutron capture cross-sections for 14-MeV neutrons on K and Ca.

The proton-induced cross section values along with values for ^{36}Ar and ^{38}Ar obtained by other workers are given in Table 3. The values listed for the stable argon

Table 3. Production cross sections for the argon isotopes in the lunar material

Element	Per cent in sample		Thin target cross section for 600 MeV protons (mb)*				
	Fines	Rock 10057	^{36}Ar	^{37}Ar	^{38}Ar	^{39}Ar	^{42}Ar
Fe	12·4	15·5	1·4†	5·6	12†	6·3	0·075
Ti	4·2	7·5	3·8‡	11·6	30‡	15·7	0·34
Ca	8·6	7·1	16 ‡	46·7	32‡	2·0	0·039
K	0·10	0·15	—	47·4	50§	6·3	0
Mn + Cr	0·5	1·3	—	5§	12§	5§	0·1§

* Absolute errors are in the range of 5–10 per cent.
† Value of GOEBEL *et al.* (1964) normalized to the ^{37}Ar and ^{39}Ar values given above for Fe.
‡ Derived from the ^{38}Ar/^{39}Ar ratio measured with 380-MeV protons on Ti by STOENNER *et al.* (1965), and with 3-GeV protons on Ca by SCHAEFFER (1969).
§ These cross sections are rough estimates from nuclear systematics.

isotopes were normalized to our measured value for the ^{39}Ar cross section at 600 MeV. It should be noted that the values for ^{36}Ar and ^{38}Ar produced from titanium and calcium were for a proton energy of 380 MeV, and 3 GeV respectively. The cross sections for ^{37}Ar, ^{38}Ar, ^{39}Ar and ^{42}Ar represent the entire chain yield for these masses since the beta decay half-lives in these chains are short. This is not the case for ^{36}Ar since the half-life of ^{36}Cl is $3{\cdot}1 \times 10^5$ yr. Using these cross sections and the chemical composition (LSPET, 1969) listed in Table 3 the calculated $^{37}Ar/^{39}Ar$ atom production ratio is 3·2 for the fine material, and 2·2 for rock 10057. These relative production ratios calculated for 600-MeV protons may be regarded as characteristic for the primary cosmic-ray protons, and cascade protons and positive pions. The observed $^{37}Ar/^{39}Ar$ ratios for both the fine material and rock 10057 are about 30 per cent lower than these calculated proton production ratios.

The production of ^{37}Ar and ^{39}Ar in meteorites and lunar samples has been attributed to (n, p) and (n, α) reactions from nuclear-evaporation neutrons by Wänke *et al.* (1970), by Begemann *et al.* (1967) and by Begemann *et al.* (1969). The processes of importance here are the fast neutron reactions $^{39}K(n, p)^{39}Ar$, $^{40}Ca(n, \alpha)^{37}Ar$ and $^{40}Ca(n, 2p)^{39}Ar$. We may make an estimate for the relative production of ^{37}Ar and ^{39}Ar using the cross sections for these reactions measured with 14 MeV neutrons. The $^{39}K(n, p)^{39}Ar$ reaction has a cross section of 350 mb (Jessen *et al.*, 1966). We have measured the $^{40}Ca(n, \alpha)^{37}Ar$ and $^{40}Ca(n, 2p)^{39}Ar$ reaction cross sections with 14 MeV neutrons, and found 180 mb and 35 mb respectively. Using these cross sections and the K and Ca compositions of the fine material and rock 10057 given in Table 3, we calculate the $^{37}Ar/^{39}Ar$ neutron production ratios of 4·5 and 4·2 respectively. These ratios are higher than the corresponding ratios calculated from 600-MeV proton cross sections. The calculations of Ebeoglu and Wainio (1966) show that for shallow depths in the lunar surface the relative number of primary protons exceed the number of neutrons by a factor of three. However, since the effective cross sections for neutron capture are higher than for proton capture, the neutrons may be of equal importance to protons. The observed $^{37}Ar/^{39}Ar$ ratio is lower than these production ratios. This could be attributed to the fact that the galactic cosmic ray intensity is now lower than its average value, and also that the solar-flare particle intensity was low at the time of the Apollo 11 mission.

Since the lunar samples are high in Sr, Y and Zr it is possible that measurable amounts of ^{81}Kr and ^{85}Kr may be present. A search was made for krypton activities in rock 10057, but only an upper limit of less than 0·5 dpm/kg could be set. From the composition of the rock and an estimate of the production cross sections, we conclude that the level of both ^{81}Kr and ^{85}Kr activities is about 0·2 dpm/kg, about a factor of two below our upper limit.

The radon fractions from these samples have been counted. Extraction and counter efficiencies are being determined. These data are not ready for presentation at this time.

Cosmic-Ray Exposure Ages

To deduce a cosmic-ray exposure age based upon the ^{38}Ar content of rock 10057 one must evaluate the relative production rates of ^{38}Ar and ^{39}Ar. Using the cross sections listed in Table 3, an $^{38}Ar/^{39}Ar$ cross section ratio of 2·8 is calculated. The ^{38}Ar content of rock 10057 was measured by Schaeffer and his associates (Funkhouser *et al.*, 1970) to be $9{\cdot}7 \times 10^{-8}$ cm^3 (STP)/g. This value must be corrected for the primordial and atmospheric ^{38}Ar also in this sample. If this correction is made using an $^{36}Ar/^{38}Ar$ ratio of 5·35 for primordial and atmospheric argon, and a spallation $^{36}Ar/^{38}Ar$ production ratio of 0·97, one obtains a spallation ^{38}Ar content of

$7{\cdot}5 \times 10^{-8}$ cm^3 (STP)/g. Using this ^{38}Ar content, the measured ^{39}Ar production rate (average), and the calculated ^{38}Ar/^{39}Ar cross section ratio we estimate rock 10057 was exposed for 110 million years on the lunar surface. The fine material contains large quantities of solar wind ^{38}Ar, and it is therefore impossible to determine the small amount of ^{38}Ar produced by cosmic rays. For this reason an effective exposure age of the fine material cannot be determined.

The helium-3 exposure age of rock 10057 can also be estimated from the spallation ^{3}He content and estimates of the ^{3}He production rate. The ^{3}He content of rock 10057 is 44×10^{-8} cm^3 (STP)/g (FUNKHOUSER *et al.*, 1970) and this is almost entirely spallation produced. If the ^{3}He production rate is taken to be double the tritium production rate that we observe, namely $2{\cdot}3 \times 10^5$ atoms ^{3}He/yr, an exposure age of 52 m.y. is estimated. This value is about a factor of two lower than the ^{39}Ar–^{38}Ar exposure age. Possible sources of this discrepancy are: the ^{3}He production rate estimated here may be incorrect, there may be He loss in lunar samples, or the relative production rates for ^{38}Ar and ^{39}Ar used may be incorrect.

Exposure ages for rock 10057 have been estimated by several investigators (HINTENBERGER *et al.*, 1970; REYNOLDS *et al.*, 1970; MARTI *et al.*, 1970) with values ranging from 47 to 125 m.y. In general these ages were calculated using the production rates for ^{3}He, ^{21}Ne, ^{38}Ar, ^{82}Kr or ^{83}Kr from meteorites. In view of the lack of knowledge of isotope production rates in lunar rocks, one may regard exposure ages to be only approximately determined.

Acknowledgments—We acknowledge the generous assistance of the personnel of the Lunar Receiving Laboratory, in particular the staff of the Radiation Counting Laboratory and the Gas Analysis Laboratory. We also are grateful for the use of the cyclotron at the Space Radiation Effects Laboratory, NASA, Newport News, Va.

REFERENCES

BEGEMANN F., VILCSEK E. and WÄNKE H. (1967) The origin of the 'excess' argon-39 in stone meteorites. *Earth Planet. Sci. Lett.* **3,** 207–212.

BEGEMANN F., RIEDER R., VILCSEK E. and WÄNKE H. (1969) Cosmic-ray produced radionuclides in the Barwell and Saint-Severin meteorites. In *Meteorite Research*, (editor P. M. Millman), pp. 267–274. R. Riedel.

BISWAS S. and FICHTEL C. E. (1965) Composition of solar cosmic rays. *Space Sci. Rev.* **4,** 709–736.

DAVIS R., JR., STOENNER R. W. and SCHAEFFER O. A. (1963) Cosmic-ray produced ^{37}Ar and ^{39}Ar activities in recently fallen meteorites. In *Radioactive Dating*, pp. 355–365. IAEA, Vienna.

EBEOGLU D. B. and WAINIO K. M. (1966) Solar proton activation of the lunar surface. *J. Geophys. Res.* **71,** 5863–5872.

ESSA (1969) Solar protons (Explorer 34). *Solar-Geophysical Data* **303,** Part 2, pp. 119–123.

EUGSTER O., EBERHARDT P. and GEISS J. (1967) The isotopic composition of krypton in unequilibrated and gas-rich chondrites. *Earth Planet. Sci. Lett.* **2,** 385–393.

FIREMAN E. L. (1963) Solar surface nuclear reactions. In *Proc. Int. Conf. Cosmic Rays*, Jaipur **3,** 487–503.

FIREMAN E. L., D'AMICO J. C. and DEFELICE J. C. (1970) Tritium and argon radioactivities in lunar material. *Science* **167,** 566–568.

FIREMAN E. L., DEFELICE J. and TILLES D. (1961) Solar flare tritium in a recovered satellite. *Phys. Rev.* **123,** 1935–1936.

FIREMAN E. L., DEFELICE J. and TILLES D. (1963) Tritium and radioactive argon and xenon in meteorites and satellites. In *Radioactive Dating*, pp. 323–334. IAEA, Vienna.

FUNKHOUSER J. G., SCHAEFFER O. A., BOGARD D. D. and ZÄHRINGER J. (1970) Gas analysis of the lunar surface. *Science* **167**, 561–563.

GOEBEL K., SCHULTES H. and ZÄHRINGER J. (1964) Production cross sections of tritium and rare gases in various target elements. CERN-Report 64–12.

HINTENBERGER H., WEBER H. W., VOSHAGE H., WÄNKE H., BEGEMANN F., VILSCEK E. and WLOTZKA F. (1970) Rare gases, hydrogen and nitrogen: concentrations and isotopic composition in lunar material. *Science* **167**, 543–545.

IBEN I. (1969) The ^{37}Cl solar neutrino experiment and the solar helium abundance. *Ann. Phys.* **54,** 164.

JESSEN P., BORMANN M., DREYER F. and NEUERT H. (1966) Experimental excitation functions for (n, p), (n, t), (n, α), (n, 2n), (n, np) and (n, nα) reactions. *Nucl. Data* **1A,** 103–202.

KOHMAN T. P. and BENDER M. L. (1967) Nuclide production by cosmic rays in meteorites and on the moon. In *High-Energy Nuclear Reactions in Astrophysics*, (editor B. S. P. Shen), Chap. 7, pp. 169–245. W. A. Benjamin.

LAMBERT D. L. (1967) Abundance of helium in the sun. *Nature* **215,** 43–44.

LSPET (LUNAR SAMPLE PRELIMINARY EXAMINATION TEAM) (1969) Preliminary examination of lunar samples from Apollo 11. *Science* **165,** 1211–1227.

MARTI K. (1967) Mass-spectrometric detection of cosmic-ray-produced ^{81}Kr in meteorites and the possibilities of Kr–Kr dating. *Phys. Rev. Lett.* **18,** 264–266.

MARTI K., LUGMAIR G. W. and UREY H. C. (1970) Solar wind gases, cosmic ray spallation products, and the irradiation history. *Science* **167,** 548–550.

REYNOLDS J. H., HOHENBERG C. M., LEWIS R. S., DAVIS P. K. and KAISER W. A. (1970) Isotopic analysis of rare gases from stepwise heating of lunar fines and rocks. *Science* **167,** 545–548.

SCHAEFFER O. A. (1969) Private communication.

SHEDLOVSKY J. P. and RAYUDU G. V. S. (1963) Simulated cosmic-ray irradiation of a thick stone target with 0·44, 1, and 3 GeV protons. Nuclear Chemistry Research at Carnegie Institute of Technology 1962–1963, Progress Report, pp. 30–37.

SHEDLOVSKY J. P. and RAYUDU G. V. S. (1964) Radionuclide production in thick iron targets bombarded with 1- and 3-GeV protons. *J. Geophys. Res.* **69,** 2231–2242.

ST. CHARALAMBUS, GOEBEL K. and STÖTZEL-RIEZLER (1969) Tritium and argon-39 in stone and iron meteorites. *Z. Naturforsch.* **24a,** 234–244.

STOENNER R. W., SCHAEFFER O. A. and KATCOFF S. (1965) Half-lives of argon-37, argon-39, and argon-42. *Science* **148,** 1325–1328.

TAYLOR J. R. (1967) The helium problem. *Quart. J. Roy. Astro. Soc.* **8,** 313–333.

TILLES D., DEFELICE J. and FIREMAN E. L. (1963) Measurements of tritium in satellite and rocket material, 1960–1961. *Icarus* **2,** 258–279.

TRIVEDI B. M. P. (1969) Production of sodium-22 and tritium by high energy protons in thick silicate targets and cosmic-ray production of nuclides in stone meteorites. Thesis, Indian Institute of Technology, Kanpur, India.

TRIVEDI B. M. P. and GOEL P. S. (1969) Production of ^{22}Na and ^{3}H in a thick silicate target and its application to meteorites. *J. Geophys. Res.* **74,** 3909–3917.

WÄNKE H., BEGEMANN F., VILCSEK E., RIEDER R., TESCHKE F., BORN W., QUIJANO-RICO M., VOSHAGE H. and WLOTZKA F. (1970) Major and trace elements and cosmic-ray produced radioisotopes in lunar samples. *Science* **167,** 523–525.

Proceedings of the Apollo 11 Lunar Science Conference, Vol. 2, pp. 1595 to 1612.

Age of the moon: An isotopic study of U–Th–Pb systematics of Apollo 11 lunar samples—II*

MITSUNOBU TATSUMOTO
U.S. Geological Survey, Denver, Colorado 80225

(*Received* 10 *February* 1970; *accepted in revised form* 28 *February* 1970)

Abstract—Concentrations of U, Th and Pb in the Apollo 11 samples studied are low (U, 0·16–0·87; Th, 0·53–3·4; Pb, 0·29–1·7, in ppm) but the extremely radiogenic nature of the lead allows radiometric dating of whole rock samples. The U^{238}/Pb^{204} ratios are very large and indicate depletion of volatile elements in primordial lunar material. Nearly concordant ages of the dust and breccia are interpreted to indicate that the initial age of the moon is about 4·6 b.y. The crystalline rocks at Tranquillity Base are 3·4–3·8 b.y. old. Reasons for an old age of the dust and breccia on the younger underlying crystalline rocks are given by an interpretation of U–Pb evolution diagram. If the moon was captured by, or derived from the earth, it must have occurred 4·6 b.y. ago or during the very initial stages of formation of the solar system.

INTRODUCTION

THIS paper reports the results of a lead isotope study of Apollo 11 lunar samples to determine the age and postcrystallization history of lunar surface material. PATTERSON (1955, 1956), using isotopic lead data, determined the age of meteorites and the age of the earth as 4·55 ± 0·07 b.y., assuming that lead in stony meteorites and modern terrestrial lead have evolved from a primordial lead compositionally like that in iron meteorites. This lead–lead method has the advantage that it is necessary to measure only the relative abundances of lead isotopes, but not uranium and thorium abundances; however, the initial lead isotopic composition must be known. PATTERSON calculated the uranium and thorium concentrations required to produce the lead concentration and isotopic composition. These calculated values were an order of magnitude higher than those actually determined later in stony meteorites (HAMAGUCHI *et al.*, 1957; BATE *et al.*, 1957; REED *et al.*, 1960). Thus one might suspect the lead–lead age of meteorites. However, these uranium, thorium and lead data were obtained on different specimens of the meteorite by different experimental techniques. In this paper on lunar samples, the concentrations of uranium, thorium and lead, as well as the lead isotopic composition, were determined on each of the samples.

Attempts to obtain U–Pb and Th–Pb internal isochrons from different density fractions were made in order to find the initial lead composition. In a previous paper (TATSUMOTO and ROSHOLT, 1970), the age of the breccia and dust was reported to be $4{\cdot}66^{+0{\cdot}07}_{-0{\cdot}16}$ b.y., whereas the age of the crystalline rocks from Tranquillity Base was reported to be about 4 b.y. In this paper, by refined data treatment and results from additional analyses, more closely defined ages will be given. Reasons for an old age of the dust and breccia but younger ages of the underlying crystalline rocks will also be discussed.

---

* Publication authorized by the Director, U.S. Geological Survey.

EXPERIMENTAL

Sampling

The crystalline rock samples were washed with double-distilled acetone by ultrasonic vibration and crushed in a boron-carbide mortar. The breccia sample was washed quickly with acetone, but the fine dust sample was not washed. Duplicate analyses (indicated by *a*'s and *b*'s on sample number in the tables) of a sample were made from separate sampling from different vials.

Samples 10017, 10050 and 10084 were separated into density fractions as follows: A rock fragment (~3 g) was gently crushed, and the majority of metal and troilite was separated with a covered hand magnet. The crushed fragments were sized with bolting cloth (silk) into −50 +100, −100 +200, −200 +325 and −325 +400 meshes. Each size fraction was separated into density fractions using bromoform and diiodomethane. The heavy fractions in diiodomethane was further divided into two fractions with a magnetic separator. All heavy liquids and acetone were distilled before use. The heavy liquid separations were made in a centrifuge using polyethylene test tubes and a Teflon filter. The equal-density fractions of different sizes were recombined in order to obtain enough material for analysis.

Analytical procedure

All chemistry was carried out in Teflon-ware or quartz-ware. Lead was first separated by barium coprecipitation (TATSUMOTO, 1966) and further purified by conventional dithizone extraction (TILTON *et al.*, 1955) by a micro-technique using Teflon centrifuge tubes and quartz pipettes.

For the determination of the lead isotopic composition about 2 grams of a sample were dissolved in 2 ml of triple-distilled $HClO_4$ and 20 ml of high-purity HF (TATSUMOTO, 1969). The residue from the acid decomposition was dissolved in 100 ml of double-distilled concentrated HNO_3. After the residue was centrifuged out, about 3 ml of $Ba(NO_3)_2$ saturated water solution were added to the supernatant liquid until $Ba(NO_3)_2$ precipitate appeared. The precipitate was centrifuged out in a Teflon tube, dissolved into water, and extracted directly into dithizone-chloroform solution in the centrifuge tube in the presence of ammonium citrate and KCN. The chloroform layer was taken out by a quartz pipette. Lead was back-extracted into 1% HNO_3 solution in a small test tube. The lead extraction by dithizone was repeated. The residue remaining after the concentrated HNO_3 dissolution and the supernatant liquid from barium precipitation were used for determination of uranium and thorium isotopic compositions (ROSHOLT and TATSUMOTO, 1970).

The concentrations of lead, uranium, and thorium were determined by isotope dilution (TATSUMOTO, 1966) with about 0·5 g of sample. Because of the appearance of meta-titanic acid in the acid dissolution, the HF treatment was repeated in order to facilitate isotopic equilibrium. Due to the importance of blank corrections, each set of analyses for the determination of both isotopic composition and concentrations was accompanied by a blank analysis. The lead blanks for concentration and composition determinations are shown in Table 1. The blanks yielded 0·015–0·032 μg lead, and except for

Table 1. Lead blank run

Date (1969)	Pb (μg)	For sample†
9/20	0·018	84c
9/28	0·032	84p
10/10	0·029	20c, 57c, 71c
10/23	0·019	20p, 57p, 61p, 71p, 84p
11/9	0·024	3c, 17c, 50c, 61c
11/11	0·016	3p, 17p, 50p, 57p
11/20	0·017	50m
12/3	0·023	17m
12/11	0·018	84m
12/19	0·020	3c, 3p, 20c, 20p, 61c, 71c, 71p

† Sample numbers are final digits. Concentration and composition runs of whole rock are indicated by c and p; those for mineral separates are indicated by m.

sample 10050 these were less than 3 per cent of the total lead used for the composition analyses of the whole-rock samples. The blanks for uranium and thorium concentration determinations were in the range of 0·0001–0·0002 μg.

Lead was analyzed on a 12-in. mass spectrometer (SHIELDS, 1966) using the phosphoric acid–silica gel technique of CAMERON *et al.* (1969) with some modification (TATSUMOTO and DELEVAUX, unpublished data). The standard deviations of the lead isotope ratios are: Pb^{206}/Pb^{204} ($\pm$1 per cent, except sample 10050 which has $\pm$3 per cent uncertainty); Pb^{206}/Pb^{207} ($\pm$0·2 per cent); Pb^{206}/Pb^{208} ($\pm$0·3 per cent). A fractionation correction factor determined from replicate analyses of NBS and USGS lead standards (0·15 per cent per mass unit) is applied to convert the observed data to absolute values. Due to the low concentrations of lead in the samples, the procedure in making blank corrections introduces relatively large overall uncertainties, as shown in Table 2. The blank correction has the most effect on Pb^{204} abundance, but because these corrections only change points nearly parallel to the isochron slopes (Fig. 1), no serious difficulty arises in the age measurements. The isotopic composition of blank lead used in this study is assumed to be: Pb^{206}/Pb^{204} = 18·5, Pb^{207}/Pb^{204} = 15·6, Pb^{208}/Pb^{204} = 38·4.

RESULTS

Results of the whole-rock analyses are shown in Table 2. Lead, uranium and thorium concentrations range from 0·29 to 1·74 ppm, 0·16 to 0·87 ppm and 0·53 to 3·43 ppm, respectively. These uranium and thorium concentrations are in good agreement with those reported by LSPET (1969).

Table 2. Isotopic composition of lead and concentrations of lead, uranium and thorium

		Concentration (ppm)†			Atomic ratio: Raw data			Atomic ratio: Corrected for blank					
Sample No.	Type	Pb	U	Th	Pb^{206}/Pb^{204}	Pb^{207}/Pb^{204}	Pb^{208}/Pb^{204}	Pb^{206}/Pb^{204}	Pb^{207}/Pb^{204}	Pb^{208}/Pb^{204}	Error‡ %	U^{238}/Pb^{204}	Th^{232}/U^{238}
10003a§	crystalline	0·51	0·268	$1{\cdot}02_9$	311·9	147·6	333·3	423·9	198·0	448·9	6	491·3	3·96
b		0·48	0·239	0·922	282·3	136·1	305·1	377·5	179·6	401·4	3	417·0	3·99
10017	crystalline	1·56	0·854	$3{\cdot}36_3$	367·7	173·7	390·2	410·0	191·9	435·0	3	492·8	4·07
10020a	vesicular	0·37	0·202	0·694	238·9	116·3	243·5	288·7	139·0	289·7	6	338·3	3·55
b		0·35	0·182	0·662	188·5	93·3	199·0	283·1	136·6	288·4	6	315·0	3·77
10050	crystalline	0·29	0·156	0·531	187·7	95·8	190·2	295·8	147·0	287·3	10	351·4	3·53
10057a	vesicular	1·68	0·865	$3{\cdot}41_5$	938·6	447·9	973·6	1241·5	590·1	1281·3	2	1392·2	4·08
b					623·7	298·8	646·3	742·9	354·5	766·1	2	833·2	—
10071a	vesicular	1·71	0·873	$3{\cdot}43_4$	637·1	306·6	662·3	753·4	361·2	779·5	2	839·2	4·06
b		1·67	0·845	$3{\cdot}29_9$	520·2	255·9	546·3	742·6	362·5	771·6	2	822·6	4·03
10061a	breccia	1·74	0·674	$2{\cdot}57_4$	238·8	152·8	232·7	249·1	163·2	258·2	2	225·6	3·94
b					264·6	173·0	272·7	274·5	179·3	282·1	2	248·6	—
10084	fine material	1·39	0·544	$2{\cdot}09_2$	237·4	155·3	246·7	261·9	171·0	270·1	2	239·6	3·97

† Estimated uncertainty for the concentrations are smaller than 2% for Pb and 1% for U and Th.
‡ Estimated error (2σ in percentage) for isotopic composition of lead.
§ The a and b indicate separate sampling from different vials.

The concentrations of lead, uranium and thorium in the fine-grained rocks (10017, 10057 and 10071) are three to five times higher than those in the coarse-grained rocks (10003, 10020 and 10050). The average uranium and thorium contents are higher than those of stone meteorites (BATE *et al.*, 1957; REED *et al.*, 1960); however, the lowest values of uranium and thorium observed for sample 10050 are similar to those of the Nuevo Laredo achondrite. Lead contents of the lunar samples are in the range of stone meteorites (REED *et al.*, 1960). The uranium and thorium contents are two to four times higher than those of terrestrial primitive oceanic basalts and Hawaiian tholeiites, and are similar to those of oceanic alkali basalts (TATSUMOTO,

1966; TATSUMOTO and KNIGHT, 1968). By inference from the present quiescent nature of the moon, these high uranium and thorium concentrations in the lunar rock must result from slight partial melting of the lunar mantle and extensive fractional crystallization, as GAST and HUBBARD (1970) suggested to explain europium depletion of the lunar sample.

The Th/U ratios range from 3·5 to 4·1 and average 3·9. The ratios in the fine-grained rocks (4·03–4·08) are greater than in the coarse-grained rocks (3·55–3·99). The Th/U ratios of breccia (10061) and the dust (10084) are close to the average.

The lead isotopic composition in all Apollo 11 samples examined is extremely radiogenic, and this allows radiometric dating. The analytically determined present day values of U^{238}/Pb^{204} range from 200 to 1400. Here too, the ratios of the fine-grained rocks are higher than those of the coarse-grained rocks. The ratios of the dust and breccia are smaller, although their uranium and lead contents are close to those for fine-grained rocks. The U^{238}/Pb^{204} values are much greater than comparable values for terrestrial igneous rocks and meteorites; this indicates that the volatile element lead was depleted, relative to uranium and thorium, on the lunar surface as compared to that on the earth and in the meteorites, as LSPET (1969) suggested. It is reasonable to consider that depletion of volatile elements is an intrinsic characteristic of primordial lunar material, as discussed later. This characteristic is fundamentally important for the interpretation of the present lead data. That is, the great values of U^{238}/Pb^{204} significantly affect the common lead (initial lead) correction for the lunar rock age measurements even though the abundance of Pb^{204} is small. Thus, use of terrestrial model lead for the common lead correction in an age calculation of the lunar rock is basically incorrect for lunar rock significantly younger than 4·6 b.y.

Data for the whole rocks are plotted in three ways: Pb^{206}/Pb^{204} vs. U^{238}/Pb^{204}, Pb^{207}/Pb^{204} vs. U^{235}/Pb^{204} and Pb^{208}/Pb^{204} vs. Th^{232}/Pb^{204} (Fig. 1). Using this technique enables us to obtain the initial Pb^{206}/Pb^{204}, Pb^{207}/Pb^{204} and Pb^{208}/Pb^{204} values, as well as isochron ages, on the assumption that some samples are coeval. The dust (10084) and breccia (10061) clearly show ages older than the crystalline rocks. The isochrons through these data points and primordial lead correspond to 4·70 b.y., 4·67 b.y. and 4·60 b.y., respectively, in plots involving Pb^{206}–U^{238}, Pb^{207}–U^{235} and Pb^{208}–Th^{232}. The primordial lead isotopic composition and decay constants used in this paper are: $Pb^{206}/Pb^{204} = 9{\cdot}346$, $Pb^{207}/Pb^{204} = 10{\cdot}218$ and $Pb^{208}/Pb^{204} = 28{\cdot}96$ (OVERSBY, 1970); $\lambda_{238} = 0{\cdot}15369 \times 10^{-9}\ yr^{-1}$, $\lambda_{235} = 0{\cdot}97216 \times 10^{-9}\ yr^{-1}$ (FLEMING *et al.*, 1952) and $\lambda_{232} = 0{\cdot}048813 \times 10^{-9}\ yr^{-1}$ (SENFTLE *et al.*, 1956). U^{238}/U^{235} atomic ratio $= 137{\cdot}8$ (ROSHOLT and TATSUMOTO, 1970). The data points of the crystalline rocks (except samples 10057 and 10071) fall in a linear grouping and indicate, respectively, isochrons of 3·78 b.y., 4·01 b.y. and 3·83 b.y. with more radiogenic initial lead given in Table 3. The fine-grained rocks 10057 and 10071 appear to be somewhat older than the other four crystalline rocks. The initial lead values shown in Table 3 were obtained by a least-squares method (YORK, 1966) based on these four crystalline rocks. The intercept values are greater than those of the terrestrial model lead but seem reasonable if lead is depleted compared to uranium and thorium on the moon surface since its formation. The initial U^{238}/Pb^{204} ratio is calculated to be about 80 based on the observed intercept values. In this paper, the U^{238}/Pb^{204} ratio is always calculated to

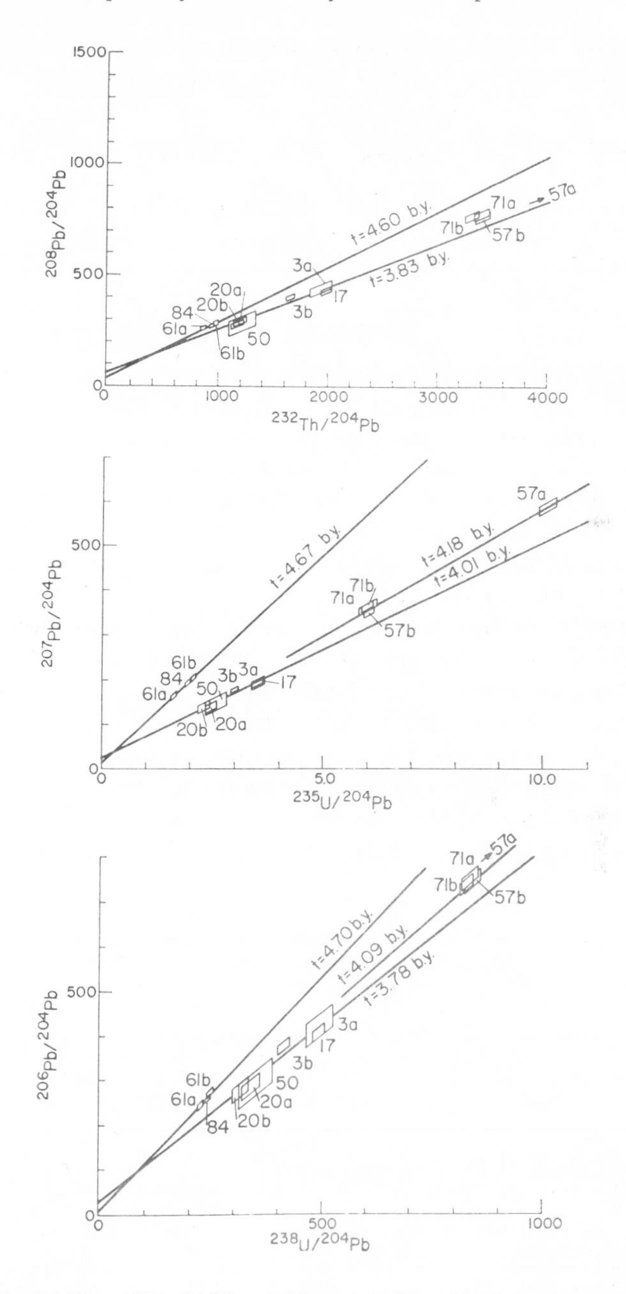

Fig. 1. The $Pb^{206} - U^{238}$, $Pb^{207} - U^{235}$ and $Pb^{208} - Th^{232}$ isochrons for whole rocks of Apollo 11 samples. Indicated errors (boxes) include blank corrections. Number on data points are final digits of sample number, and appended *a*'s and *b*'s indicate duplicate determinations.

Table 3. The initial lead and isochron ages obtained by whole rock analysis of four crystalline rocks (10003, 10017, 10020, 10050) and by mineral fractions of sample 10017.

Samples	Initial lead			Isochron age (in b.y.)		
	Pb^{206}/Pb^{204}	Pb^{207}/Pb^{204}	Pb^{208}/Pb^{204}	Pb^{206}/U^{238}	Pb^{207}/U^{235}	Pb^{208}/Th^{232}
Four crystalline rocks	28·7 ± 22·5	23·1 ± 10·8	54·6 ± 18·9	3·78 ± 0·4	4·01 ± 0·4	3·66 ± 0·5
Fractions 2, 3 and 4 of 10017	25·9 ± 5·5	22·3 ± 0·5	132·2 ± 73·7	3·54 ± 0·1	3·76 ± 0·4	2·01 ± 1·5
All fractions and whole rock of 10017	20·3 ± 21·6	100 ± 19·5	147·9 ± 56·6	3·66 ± 0·2	3·95 ± 0·2	2·39 ± 0·7

its present value, as is customarily done in lead isotope studies. The initial values for lead and U/Pb ratio, however, are highly uncertain. The isochron ages of the crystalline rocks obtained by these three plots are discordant but the ages of the dust and the breccia are nearly concordant; and the Pb^{206}/U^{238} age of crystalline rocks is younger than their Pb^{207}/U^{235} age, but the Pb^{206}/U^{238} age of the dust and the breccia is older than their Pb^{207}/U^{235} age.

These intercepts are further tested by internal isochrons. Crystalline samples 10017 and 10050 and the dust 10084 were separated into "mineral composites". Sample 10050 fractions were analyzed only for lead, uranium and thorium concentrations. Fractions of crystalline rock 10017 and the dust 10084 were analyzed for lead isotopes and for the concentrations of lead, uranium and thorium. Due to the small amount of material available for analysis compared to the lead blank, the obtained data have about twice the uncertainty of the whole-rock data. The results are listed in Table 4. Uranium, thorium and lead concentrations are high in the light (2·9–3·3 density range) fraction in the three samples and low in the heavy nonmagnetic fraction ($\rho > 3{\cdot}3$) which consists predominantly of clinopyroxene. The light fraction consists

Table 4. Isotopic composition of lead and concentration of lead, uranium and thorium in mineral concentrate

Sample No.	Fraction	Concentration (ppm)			Atomic ratios Corrected for blank					Error %
		Pb	U	Th	U^{238}/Pb^{204}	Th^{232}/U^{238}	Pb^{206}/Pb^{204}	Pb^{207}/Pb^{204}	Pb^{208}/Pb^{204}	
10017	1 $\rho < 2{\cdot}9 \sim 80\%$ plag.	1·00	0·696	$2{\cdot}32_6$	558·3	3·45	365·1	146·2	381·7	10
	2 $2{\cdot}9 < \rho < 3{\cdot}3$ plag. ilm. py.	2·25	$1{\cdot}25_3$	$2{\cdot}49_3$	322·1	2·06	264·3	109·5	272·9	10
	3 $3{\cdot}3 < \rho$, nonmagnet. $\sim 70\%$ py.	0·59	0·289	$1{\cdot}20_2$	119·0	4·30	111·8	54·7	136·3	5
	4 $3{\cdot}3 < \rho$, magnetic $\sim 80\%$ ilm.	1·41	0·828	$4{\cdot}06_5$	364·8	5·07	284·3	122·1	313·3	5
	Σ sum of fractions	1·41	0·820	$2{\cdot}92_9$		3·67				
	whole rock	1·56	0·854	$3{\cdot}36_3$	492·8	4·07	410·0	191·9	435·0	3
10050	1 $\rho < 2{\cdot}9 \sim 95\%$ plag.	0·21	0·113	0·404		3·68				
	2 $2{\cdot}9 < \rho < 3{\cdot}3$ mixture	0·47	0·285	0·937		3·40				
	3 $3{\cdot}3 < \rho$, nonmagnet. $> 95\%$ py.	0·01	0·008	0·024		3·20				
	4 $3{\cdot}3 < \rho$, mix. $\sim 60\%$ ilm. $\sim 40\%$ py.	0·17	0·101	0·286		2·94				
	5 $3{\cdot}3 < \rho$, magnetic $\sim 90\%$ ilm.	0·22	0·150	0·463		3·20				
	Σ sum of fractions	0·14	0·129	0·409		3·28				
	whole rock	0·29	0·156	0·531	351·4	3·53	295·8	147·0	287·3	10
10084	1 $\rho < 2{\cdot}9$	1·23	0·526	1·984	124·2	3·90	125·1	85·9	139·5	3
	2 $2{\cdot}9 < \rho < 3{\cdot}3$	1·45	0·625	$2{\cdot}40_5$	199·1	3·98	198·4	121·9	209·5	3
	3 $3{\cdot}3 < \rho$	0·84	0·384	$1{\cdot}42_6$	139·1	3·83	132·0	83·5	171·7	3
	Σ sum of fractions	1·22	0·534	$2{\cdot}02_7$		3·93				
	whole rock	1·39	0·544	$2{\cdot}09_2$	239·6	3·97	237·5	155·3	246·7	

mainly of mixture of plagioclase with adhered ilmenite and plagioclase with adhered pyroxene; greater elemental concentrations in that fraction might be due to the presence of phosphate minerals (apatite) (c.f. AGRELL *et al.*, 1970).

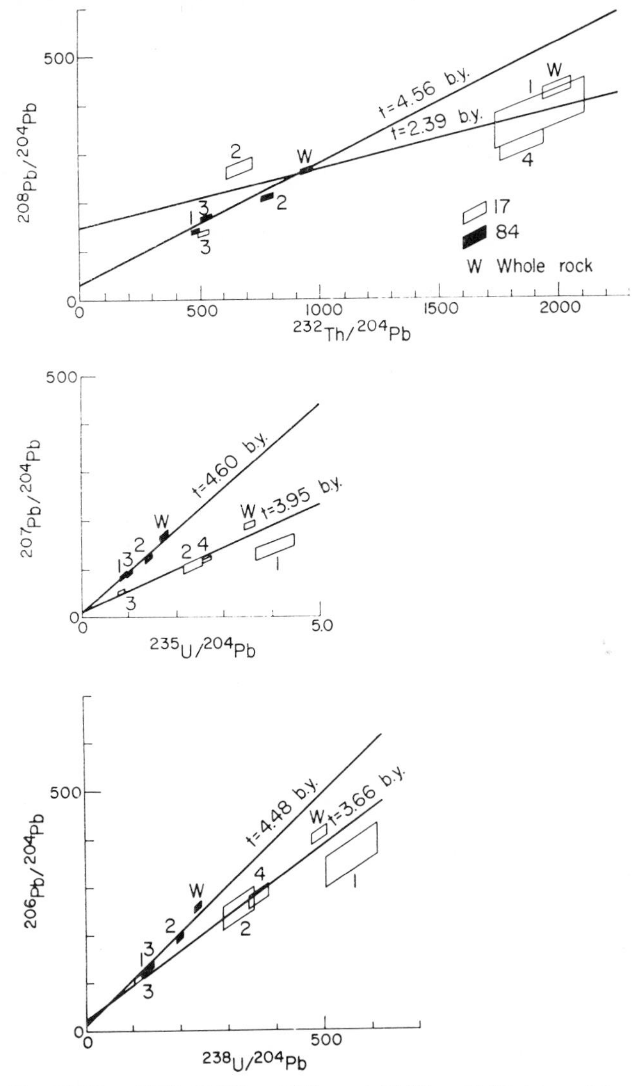

Fig. 2. The $Pb^{206} - U^{238}$, $Pb^{207} - U^{235}$, and $Pb^{208} - Th^{232}$ internal isochrons for whole-rock (W) and fractions (1–4) of samples 10017 and 10084. Open boxes for 10017 and closed boxes for 10084. Size of boxes indicates the experimental error.

The isotope data are plotted in Fig. 2. The internal isochrons of 10017 and 10084 indicate a slightly younger age than those of the whole rocks. The reason for the difference is not well understood, but it may be due to a complex history of the sample such as its having been subjected to shock metamorphism or leaching of lead during mineral separation. Alternatively, the four crystalline rocks, from which the whole-rock isochrons are obtained, may not be coeval. However, the intercepts for the initial Pb^{206}/Pb^{204} and Pb^{207}/Pb^{204} obtained from sample 10017 fractions are the same, within the experimental error, as the intercept of the four crystalline rocks (Table 3). The intercepts obtained from fractions of sample 10084 are the same within experimental error as the values obtained from meteorite lead. Therefore, the following values were chosen, as a first approximation, to correct common lead from the crystalline rocks: $Pb^{206}/Pb^{204} = 29$, $Pb^{207}/Pb^{204} = 39$ and $Pb^{208}/Pb^{204} = 45$. These values were calculated from the Pb^{206}/Pb^{204} intercept assuming that the lead had evolved from 4·6 b.y. to 3·8 b.y. prior to crystalline rock formation. (This choice for the age interval is discussed later in the report.) The apparent ages are shown in Table 5. The initial lead used for whole rock samples 10061 (breccia) and 10084 (dust) is primordial meteorite lead (OVERSBY, 1970).

Table 5. Apparent ages (in million years) and first stage parameters

Sample No.	Run	Corrected for common lead† Pb^{206}/U^{238}	Pb^{207}/U^{235}	Pb^{207}/Pb^{206}	Pb^{208}/Th^{232}	Calculated parameters for concordant ages‡ $(U^{238}/Pb^{204})_{t_0 \sim t_1}$	t_1	$(Pb^{206}/Pb^{204})_{t_1}$	$(Pb^{207}/Pb^{204})_{t_1}$
10003	a	3838	3929	3976	3861	197·7	3528	70·24	91·50
	b	3953	3970	3979	3978	114·7	3924	32·30	45·31
10017		3727	3887	3970	3643	221·4	3210	95·71	114·44
10020	a	3706	3830	3908	3799	157·9	3254	69·18	83·41
	b	3849	3885	3902	3824	118·3	3728	39·44	53·17
10050		3675	3877	3984	3654	185·5	2965	92·93	104·00
10057	a	4075	4132	4160	4035	510·6	3833	124·48	180·67
	b	4027	4088	4117	3942	318·5	3785	85·39	120·92
10071	a	4049	4102	4215	4144	308·4	3841	78·21	112·44
	b	4064	4125	4242	4056	389·8	3639	117·88	160·53
10061		4710	4678	4663	4696				
10084		4685	4668	4657	4626				

† Assumed common lead obtained from isochron plot are: $Pb^{206}/Pb^{204} = 29$, $Pb^{207}/Pb^{204} = 39$, $Pb^{208}/Pb^{204} = 45$. The initial lead for 10061 and 10084 is troilite lead (OVERSBY, 1970); $Pb^{206}/Pb^{204} = 9{\cdot}346$, $Pb^{207}/Pb^{204} = 10{\cdot}218$, $Pb^{208}/Pb^{204} = 28{\cdot}96$.

‡ Using observed U^{238}/Pb^{204} and determined isotopic composition (Table 2), the first stage U^{238}/Pb^{204} ratio and lead ratios at t_1 were calculated for concordant age t_1 by Pb^{206}/U^{238}, Pb^{207}/U^{235}, and Pb^{207}/Pb^{206} methods. [Refer to equations (3) and (4) in text.]

DISCUSSION

Ages

The preceding section considered the initial lead isotopic composition of the lunar samples by either internal or whole-rock isochron, from which one can then calculate the lead age of the samples. The apparent Pb^{207}/Pb^{206} age of the dust and breccia is 4·66 b.y.—as old as the meteorites and the earth (PATTERSON, 1955, 1956). The lead is extremely radiogenic and the age obtained is not very dependent on the initial lead. Thus, use of the primordial lead of meteorites as the primordial lead of the moon seems reasonable and, within broad limits, supported by the data. Mineral fractions of the dust (10084) also intersect the *Y* axis close to the meteorite primordial value and support its use. The Pb^{206}/U^{238} and Pb^{207}/U^{235} ages of the dust and breccia, as well as the age of Pb^{208}/Th^{232}, are nearly in agreement with Pb^{207}/Pb^{206} age, that is, nearly concordant (Table 5) when the meteorite initial lead is used. However, the Pb/U ages are slightly higher than the Pb–Pb ages. This is important in estimating the initial age of the moon. Use of $\lambda_{Th^{232}} = 4{\cdot}8813 \times 10^{-11}\ yr^{-1}$ rather than $4{\cdot}99 \times 10^{-11}\ yr^{-1}$ in Pb^{208}/Th^{232} age calculation results in better agreement with the U–Pb and the Pb–Pb ages.

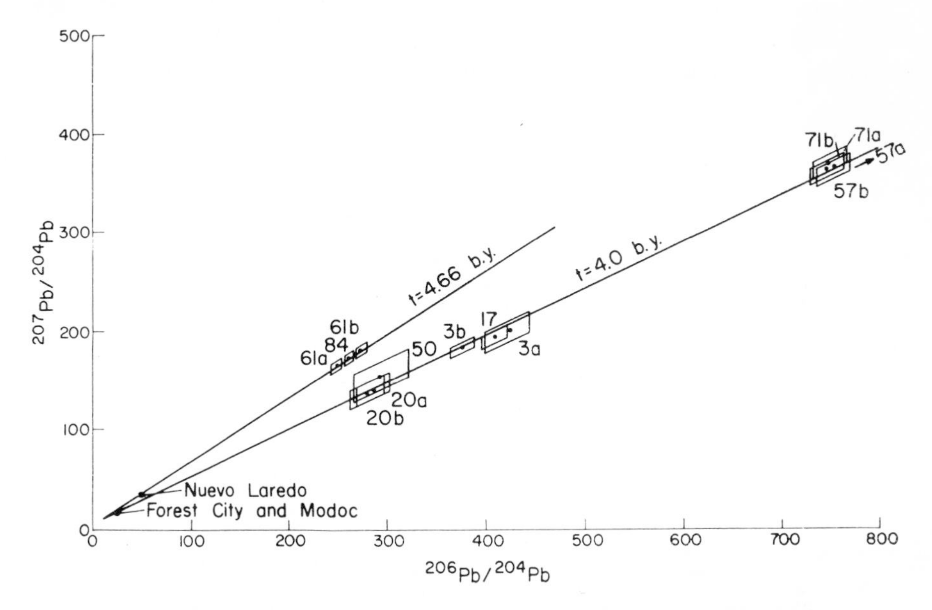

Fig. 3. The Pb^{206}/Pb^{204} – Pb^{207}/Pb^{204} evolution diagram for lunar samples and meteorites. Indicated Pb^{206}/Pb^{207} isochrons are obtained by use of primordial lead of meteorites for samples 10061 and 10084. The initial lead used for other samples is $Pb^{206}/Pb^{204} = 29$ and $Pb^{207}/Pb^{204} = 39$. Numbers on data points are final digits of sample number, and appended a's and b's indicate duplicate determinations. Size of boxes indicates the experimental error.

In Fig. 3, Pb^{206}/Pb^{204} and Pb^{207}/Pb^{204} for lunar dust and breccia are plotted, along with PATTERSON's data (1955) for stony meteorites. In a previous paper (TATSUMOTO and ROSHOLT, 1970), we interpreted these plots to indicate that the system, of which the earth, moon and meteorites are parts, probably was formed $4{\cdot}66^{+0{\cdot}07}_{-0{\cdot}16}$ b.y. ago, for the following reasons: (i) observed data fit into the extension of the meteorite lead array, (ii) the data lie near the U–Pb evolution curve as discussed later, and (iii) this age is very close to the "older" meteorite ages obtained by Rb–Sr method (BURNETT and WASSERBURG, 1967; WASSERBURG and BURNETT, 1968; KAUSHAL and WETHERILL, 1969). Further sample analyses and refined treatment of the data disclosed, first, that the slight discordance among the U–Pb, Th–Pb and Pb–Pb ages for the dust and the breccia (Table 5) is significant and, second, that the initial age for the moon can be reduced to 4·60–4·66 b.y. (even though this reduction is in the error range of the assigned moon age in the previous paper).

The apparent ages of the crystalline rocks (Table 5) definitely appear to be younger than the dust and breccia. These ages, furthermore, are discordant, and the Pb–Pb ages are older than U–Pb and Th–Pb ages. Thus, let us now plot these data corrected for primordial meteorite lead in a U–Pb evolution diagram (Fig. 4). The diagram was first formally interpreted by WETHERILL (1956), and an application for oceanic basalt leads was discussed by ULRYCH (1967) and OVERSBY and GAST (1968). The points for the dust and breccia are *slightly* above the concordia line, whereas the points for the crystalline rocks are below. The chord through the data points of the dust, breccia and samples 10057 and 10071 intersects the concordia at 3·8 b.y. and 4·60 b.y. Re-

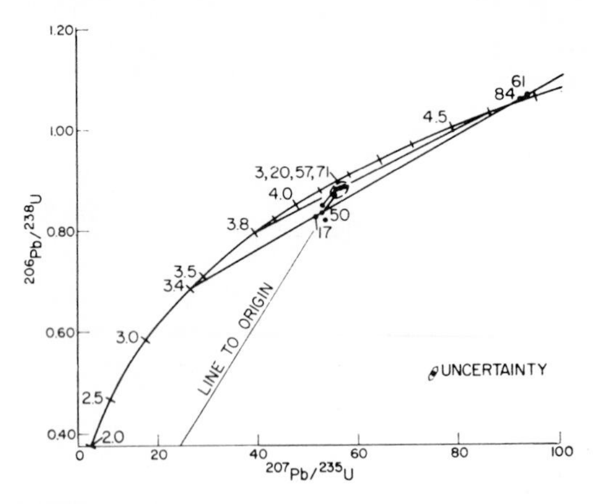

Fig. 4. The U–Pb evolution diagram. Plotted points are $(Pb^{207}_{observed}-Pb^{207}_{primordial}/U^{235}$ against $(Pb^{206}_{observed}-Pb^{206}_{primordial})/U^{238}$. Numbers on data points are final digits of sample numbers, and numbers on the concordia indicate billions of years. Duplicate runs of samples 10003 and 10020 are tied. The second values of 10003 and 10020 are close to data points of 10057 and 10071.

analyses of samples 10003 and 10020 have smaller uncertainties than the original determinations (indicated by tieline in Fig. 4), and fall essentially on the same chord with 10057 and 10071. The chord for data point 10017, the dust, and the breccia intersects the concordia at 3·4 b.y. and 4·63 b.y. Sample 10050, which has a significantly greater experimental uncertainty, may belong to the latter chord.

Now, suppose that by a simple two-stage model the moon originated t_0 b.y. ago (the time at which the gross structure of the moon developed), and then, t_1 b.y. ago, the moon's surface material was melted and was differentiated (Fig. 5). The U/Pb ratios in the lunar rocks differentiated (or fractionated) and increased at this time, as follows:

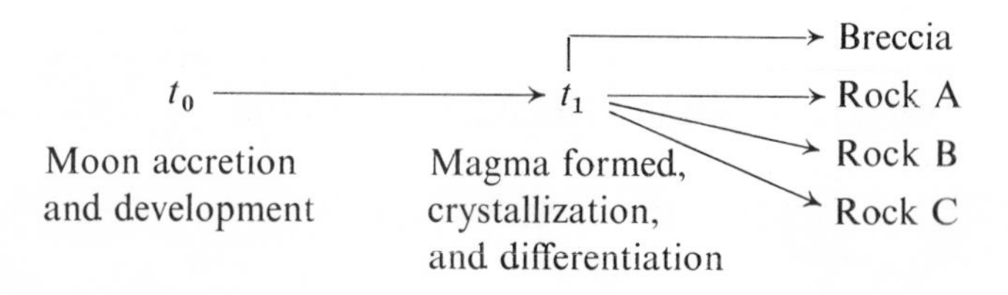

As WETHERILL (1956) has shown, the radiogenic components of the present-day Pb^{206}/U^{238} and Pb^{207}/U^{235} ratios of this two-stage system fall on a straight line given by

$$\left(\frac{Pb^{206}}{U^{238}}\right)^* = \frac{(Pb^{206})_{t=0} - (Pb^{206})_{t_0}}{(U^{238})_{t=0}} = R_1\,(e^{\lambda_8 t_0} - e^{\lambda_8 t_1}) + (e^{\lambda_8 t_1} - 1) \tag{1}$$

$$\left(\frac{Pb^{207}}{U^{235}}\right)^* = \frac{(Pb^{207})_{t=0} - (Pb^{207})_{t_0}}{(U^{235})_{t=0}} = R_1\,(e^{\lambda_5 t_0} - e^{\lambda_5 t_1}) + (e^{\lambda_5 t_1} - 1) \tag{2}$$

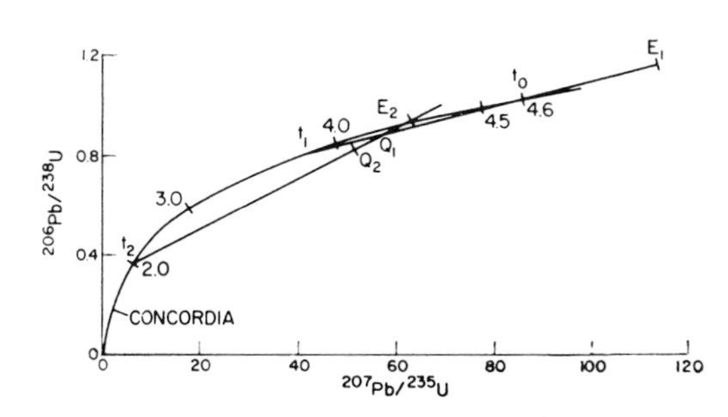

Fig. 5. U–Pb evolution diagram indicating U–Pb ages.

where λ_8 and λ_5 are decay constants of U^{238} and U^{235} and subscripts t_0 and $t = 0$ indicate the value t_0 b.y. ago and at present, respectively; $R_1 = [(Pb^{206}/U^{238})^*$ after $t_1]/[(Pb^{206}/U^{238})^*$ before $t_1]$. R_1 can be expressed in similar fashion by Pb^{207}/U^{235}. This straight line passes through t_0 and t_1 on the concordia when $R_1 = 1$ and 0. On the other hand,

$$\left(\frac{Pb^{206}}{Pb^{204}}\right)^* = \left(\frac{U^{238}}{Pb^{204}}\right)_{t_0\sim t_1}(e^{\lambda_8 t_0} - e^{\lambda_8 t_1}) + \left(\frac{U^{238}}{Pb^{204}}\right)_{t=0}(e^{\lambda_8 t_1} - 1) \tag{3}$$

$$\left(\frac{Pb^{207}}{Pb^{204}}\right)^* = \left(\frac{U^{235}}{Pb^{204}}\right)_{t_0\sim t_1}(e^{\lambda_5 t_0} - e^{\lambda_5 t_1}) + \left(\frac{U^{235}}{Pb^{204}}\right)_{t=0}(e^{\lambda_5 t_1} - 1). \tag{4}$$

Where $(Pb^{206}/Pb^{204})^*$ and $(Pb^{207}/Pb^{204})^*$ indicate the present lead ratios minus the primordial ratio; $(U^{238}/Pb^{204})_{t_0\sim t_1}$ is the U^{238}/Pb^{204} ratio before t_1 (calculated to the present value). If we know the common lead isotopic composition at t_1 and subtract it from (3) and (4), we calculate the radiogenic components:

$$\left(\frac{Pb^{206}}{U^{238}}\right)_{Rad} = e^{\lambda_8 t_1} - 1; \quad \left(\frac{Pb^{207}}{U^{235}}\right)_{Rad} = e^{\lambda_5 t_1} - 1.$$

These are the same as can be derived from (1) and (2) when $R_1 = 0$, and they are the coordinates of t_1 on concordia. In this simple two-stage model we only apply a catastrophic change of parent and daughter ratio and do not consider any other mechanism such as diffusion loss of daughter.

If this simple two-stage model can be applied to the lunar lead data, then the upper intercept 4·60–4·63 b.y. indicates the original time of the moon's surface development, and the lower intercept 3·8–3·4 b.y. indicates the crystallization age of rocks from Tranquillity Base. For convenience of discussion first we consider that the breccia and dust came mainly by a physical breakdown of primary (4·6 b.y. old) surface material from the 3·8-b.y. event, probably by intensive meteorite bombardments. These bombardments might have triggered magma formation in Mare Tranquillitatis. The data points in Fig. 4 indicate that the magma (or primary material) lost about two-thirds of the lead (or gained uranium) at the 3·8 b.y. event, that is, $R_1 = \frac{1}{3}$ (Q_1 in Fig. 5). If there has been loss of lead by either volatilization from magma or by the impact shock, then the other component must have been enriched in the lead. Conceivably, this other component is in the breccia and dust. If this argument is true, extremely fine particles (which contain the most volatile component) in the dust must be enriched in radiogenic lead and show higher apparent ages up to about 5 b.y. (E_1 in Fig. 5). SILVER (1970), indeed, found that the acid-leached lead from the dust is more radiogenic and shows somewhat higher ages than the whole-rock samples. This simple two-stage model enables us to explain why the dust and breccia of Apollo 11 appear to be older than the crystalline rocks. On the other hand, if the data point ($R_1 = \frac{1}{3}$) was produced by uranium gain (compared to lead) by magmatic differentiation, which is most likely by analogy with terrestrial differentiation, then the uranium depleted component must be in the lower lunar crust (or upper lunar mantle). In this case, the dust must be a good mixture of the previously existed dust, uranium-gain component, and uranium-loss component in order to maintain original age in the dust, as discussed later.

Now let us calculate the common lead for these six crystalline rocks, following this simple two-stage evaluation and assuming that these rocks were formed at 3·8 b.y. ago, and that the primary moon age is 4·6 b.y. If these crystalline rocks were formed from a single, homogeneous magma, then they presumably should have a unique common lead at time of formation. No such unique common lead existed, as is shown in Table 5. The parameters for the first stage in the table were calculated, using determined lead isotopic composition and observed U^{238}/Pb^{204} ratio, so that each sample has a concordant age by the Pb^{206}/U^{238}, Pb^{207}/U^{235} and Pb^{207}/Pb^{206} methods. Samples 10003 and 10020, which have measured $(U^{238}/Pb^{204})_{t=0}$ = about 450, must have had a common lead which evolved with $(U^{238}/Pb^{204})_{t_0 \sim t_1}$ = about 120 in order to move the observed points onto the concordia. Samples 10057 and 10071, for which $(U^{238}/Pb^{204})_{t=0}$ = about 830, should have a common lead evolved in $(U^{238}/Pb^{204})_{t_0 \sim t_1}$ = 300 system (Fig. 6). Samples 10017 and 10050 need an intermediate U^{238}/Pb^{204} value to generate a suitable common lead. No unique common lead can move the data points of the six crystalline rocks to the concordia, even though all data points fall in a narrow range in Fig. 4. The position of the data points indicates that the lead losses (or uranium enrichment) for all crystalline rocks are two-thirds of the original amounts. The reason for two-thirds lead loss (or uranium gain) for every sample is not known. Even though the Pb^{204} amount is small in all the samples, the different common lead composition required for age calculation cannot be attributed to experimental error: the difference in the necessary U^{238}/Pb^{204} ratio far exceeds the experimental error, and estimated primary U^{238}/Pb^{204} ratios are about one-third of the secondary ones. A unique common lead may be lacking because, most likely, rock was formed from several magma generations during the period of 3·8 ~ 3·4 b.y. Alternatively, isotopic homogenization was not reached within the magma reservoir (or lava lake), or magma may have been contaminated by upper-layer lead or dust lead which covered the lava lake or migrated in the lava.

Had these rocks been formed 4·6 b.y. ago and been metamorphosed 3·8 b.y. ago, instead of crystallizing directly from the molten magma, the observed points in Fig. 4 would be scattered along the chord depending upon the degree of metamorphism. However, the observed data points are clustered essentially at one place on the chord. Thus, these crystalline rocks may reasonably be assumed to have formed from a magma with fractionation.

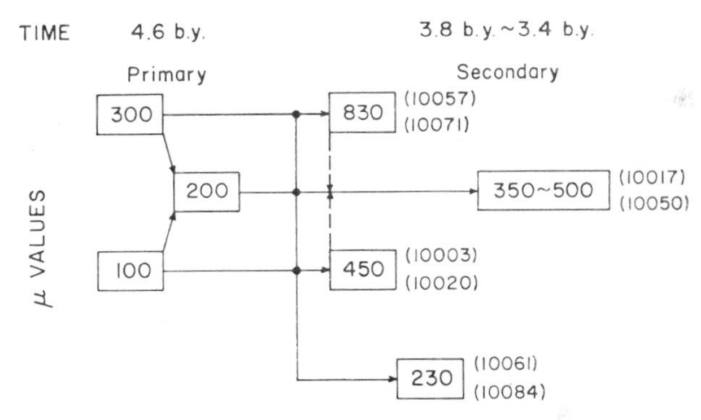

Fig. 6. Schematic diagram for $U^{238}/Pb^{204}(\mu)$ change. Numbers in rectangles are μ values and those in parentheses are sample numbers.

Though the concordant ages in Table 5 are somewhat younger than ages obtained from common lead correction estimated from isochron plots, the concordant ages are, nonetheless, in good agreement with the concordia interpretation, and with Rb–Sr ages [ALBEE *et al.*, 1970 (3·65 b.y. mineral ages); COMPSTON *et al.*, 1970 (3·8 b.y. from whole-rock isochron); GAST and HUBBARD, 1970 (3·5 b.y. from whole-rock isochron); GOPALAN *et al.*, 1970 (3·4 and 4·0 b.y. of two mineral ages)].

Samples 10017 and 10050 appear to be younger than other crystalline rocks in Table 5. The internal isochron age of 10017 is about 3·5 ~ 3·7 b.y. and is also younger than the other whole-rock isochron age (Table 3 and Fig. 2). Using only the mineral fractions (not including 10017 whole-rock data) produces a younger age. The high-temperature Ar^{40}/Ar^{39} age of 10017 is 3·2 b.y. whereas that of other crystalline rocks is in the range of 3·5 ~ 3·9 b.y. (TURNER, 1970). The K–Ar age of 10017 seems also to be somewhat younger (2·6 b.y.) while the K–Ar age of other crystalline rocks is about 3·2–3·3 b.y. (FUNKHOUSER *et al.*, 1970; BOGARD, personal communication, even though the difference is within their experimental error). If 10017 formed from the primary lunar material, the intercepts of the chord through 10017 and the dust and breccia on the concordia indicate that the original moon formed about 4·63 b.y.

ago and sample 10017 formed about 3·4 b.y. ago. However, the data points of the 10017 fractions on the U–Pb evolution diagram (Fig. 7) are not distributed along with the chord and might indicate that 10017 was not formed directly from the original lunar material. Sample 10017 also exhibits a hornfelsic character (CHAO *et al.*, 1970). If 10017 is formed from crystalline rocks 3·8 b.y. old, the low intercept of the chord through data points of the crystalline rocks and 10017 (t_2 in Fig. 5) should indicate an age of the secondary episode. The chord intersects the concordia at about 4·2 and

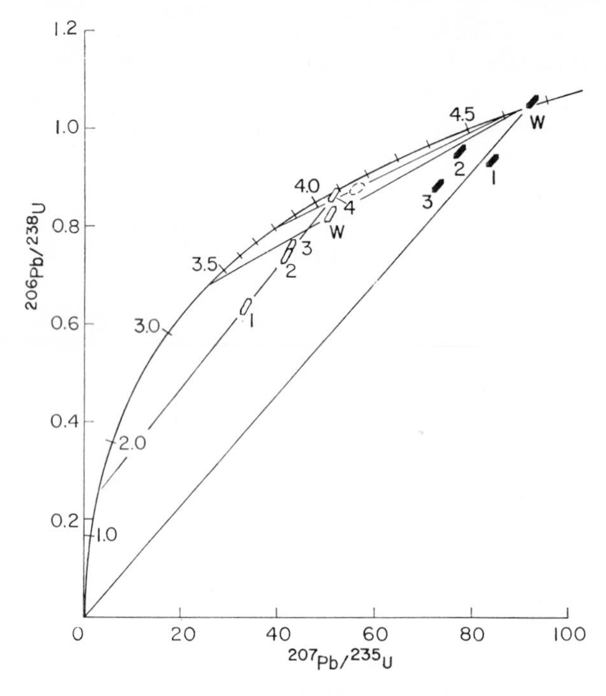

Fig. 7. The U–Pb evolution diagram for whole-rock (W) and fractions (1–4) of 10017 and 10084. Plotted points are $(Pb^{207}_{observed}-Pb^{207}_{primordial})/U^{235}$ against $(Pb^{206}_{observed}-Pb^{206}_{primordial})/U^{238}$. Open polygons for 10017 and closed polygons for 10084. Size of polygons indicates the experimental error. Best fit line for 10017 fractions intersects the concordia at 4·1 b.y. and 1·4 b.y.

2·0 b.y.; however, the best fit line on the data points of 10017 fractions intersects the concordia at 4·1 and 1·4 b.y. (Fig. 7). However, the lower intercepts (2·0 and 1·4 b.y.) are not in agreement with Ar/Ar age (TURNER, 1970), K–Ar age (FUNKHOUSER *et al.*, 1970) and Rb–Sr ages (ALBEE *et al.*, 1970). Thus the lower intercepts may not have physical meaning but be caused by common lead correction.

The earlier discussion of the two-stage model assumed that the dust consists solely of fragments, formed about 3·8 b.y. ago, from the original 4·6 b.y. old material. Actually, the dust and breccia are fragmental mixtures of lunar surface material, not only from Mare Tranquillitatis but also perhaps from other areas, in various ages from many meteorite bombardments and also they contain a small amount of meteorite fragments. Meteorites are neglected in the following discussion because of the small contribution in uranium, thorium and lead contents of the dust.

Consider three cases. (Case A) Suppose the dust and breccia to be mainly a mixture of fragments which were formed from the 4·6 b.y. old original material and bombarded continually since then; lead loss from the original material had occurred in these numerous events. The chord which ties between the data point of the breccia and the dust and concordia curve up to recent will intersect the upper concordia at between 4·60 and 4·66 b.y. The fractions of the dust (10084) may indicate this case (Fig. 7). This is also the circumstance if the dust had formed at the Highland area (assuming the rocks of the Highland area to be 4·6 b.y. old). (Case B) Suppose the dust consists solely of material 3·4–3·8 b.y. old. Then the data point of the dust must come near the upper extension of the secondary chord which extends through the crystalline rocks and t_2 in Fig. 5. In this case, the older age of the dust is difficult to explain. (Case C) Suppose Q_1 had a secondary geological event and differentiated to Q_2 and E_2 (Fig. 5), then Q_2 was added into the existing dust by mechanical crushing of meteorite impact. The point of the resultant dust should be inside the triangle E_1 E_2 Q_2. The position of the point depends on the mixing rate of the components E_1, E_2, Q_1 and Q_2. When the components are mixed well, the point for the dust should be close to t_0. When Q_2 further differentiates in the following stage and then forms dust by meteorite impacts, the above explanation can still apply, provided that no components are lost from the dust or added from other systems. Thus, dust and breccia must be a good mixture compositionally approximating the lunar crust in order to produce a 4·6 b.y. age on the U–Pb concordia (Fig. 4) and Pb–Pb method (Fig. 3).

I favor Case C and do not believe the dust age is accidental. If it is accidential, it is a remarkable coincidence that the indicated age is the same as found in meteorites by Pb–Pb and Rb–Sr methods. The agreement of Rb–Sr age of the dust (ALBEE *et al.*, 1970) with U–Pb ages is also a strong indication that the dust age dates a real event.

Restrictions on lunar origin

Of course, this initial age of the moon cannot define uniquely the origin of the moon; and I shall merely state here some of the restrictions imposed on the prevailing theories by the age determination. The theories can be broadly divided as to whether the moon (1) was captured by the earth (GERSTENKORN, 1955; UREY, 1962; ALFVÉN and ARRHENIUS, 1969; BAILEY, 1969); (2) was derived from the earth by fission (WISE, 1969) or volatilization-sedimentation (RINGWOOD, 1966); or (3) was formed by direct accretion close to the earth (KUIPER, 1959).

It is evident from the age of the moon that the capturing of the moon must have occurred early in the development of the solar system, if capture is catastrophic. The theory of moon delivery from the earth at a later stage (MACDONALD, 1964) also is untenable. If the moon was captured by or derived from the earth, it must have occurred at 4·6 b.y. ago or during the very initial stages of formation of the earth. However, later capturing by means of spin-orbit resonance is still possible (ALFVÉN and ARRHENIUS, 1969).

RINGWOOD and ESSENE (1970) argued that RINGWOOD's chemical evolution model

(1966) of the terrestrial planets and the moon agrees well with the results obtained on the Apollo 11 samples. His hypothesis is that the moon accumulated from a sediment-ring of cold planetismals formed as a massive primitive atmosphere of the earth. This model explains some features of moon chemistry such as volatile element depletion (predicted by RINGWOOD, 1966, p. 85) and lower density of the moon. RINGWOOD's theory also requires depletion of radioactive and refractory elements such as uranium and thorium (*op. cit.*, p. 85). GAST and HUBBARD (1970) inferred that the uranium, thorium and potassium contents of the lunar mantle are $<0{\cdot}08$, $<0{\cdot}3$ and <100 ppm, respectively. The highest uranium and thorium concentrations of the lunar mantle (which they inferred) are about ten times higher than the concentrations in chondritic material, even though the potassium concentrations in both are similar. The high concentrations seem to be not consistent with RINGWOOD's prediction. Conversely, if the inferred uranium and thorium values are much greater than the true values of sediment-ring and if potassium, uranium and thorium all were depleted in the original lunar material, then a longer time is needed to melt lunar mantle, and one might have difficulty explaining the old ages on lunar surface. On this point, RINGWOOD's hypothesis, it seems to me, is not particularly better than WISE's (1969) fission hypothesis.

We have determined (ROSHOLT and TATSUMOTO, 1970) that the isotopic composition of uranium (U^{238}/U^{235} ratio) is exactly the same as that of terrestrial material, within the experimental error (0·3 per cent). This suggests that a close time-genetic relation once existed between the earth and the moon. WASSERBURG *et al.* (1969) discussed nucleosynthesis chronologies for the galaxy and estimated time span (Δt) between the termination of the nucleosynthesis and xenon retention. It seems reasonable to infer that isotopic homogenization continued until the formation of the solar system by the collapse of the solar nebula (HOHENBERG, 1969) and lunar material was separated from a high-temperature dispersed nebula, as indicated by high original U/Pb ratio.

ADAMS *et al.* (1969) stated that evolution of lead (Pb isotopic growth) on the moon and on the earth should not be very different, if the moon fissioned in bulk from the earth after earth-differentiation. This is an oversimplification, because if a large amount of volatile elements has been lost in fission, the U/Pb ratios in the moon surface material would be high from the outset and lead of the moon surface material today should be much more radiogenic than that of the earth.

Lead in tektites (TILTON, 1958; WAMPLER *et al.*, 1969) is quite different from the moon lead; this difference confirms TILTON's conclusion that tektites were not derived from the moon but are of terrestrial origin. By the same reasoning, the hypothesis that stone meteorites originate from the moon (UREY, 1960; see also RINGWOOD, 1966, p. 92) might be untenable unless unradiogenic lead is found on the moon's surface.

Acknowledgments—I wish to thank R. E. WILCOX for petrographic examinations of the lunar samples and mineral fractions analyzed in this study. I thank R. J. KNIGHT and D. M. UNRUH for invaluable laboratory assistance, and M. DELEVAUX for mass spectrometer maintenance. I was privileged to have the benefit of discussions with R. E. ZARTMAN and criticism of B. R. DOE, Z. E. PETERMAN and G. R. TILTON. This study was financed by NASA Contract T-75445.

REFERENCES

ADAMS J. B., CONEL J. E., DUNNE J. A., FANALE F., HOLSTROM G. B. and LOOMIS A. A. (1969) Strategy for scientific exploration of the terrestrial planets. *Rev. Geophys.* **7**, 623–661.

AGRELL S. O., SCOON J. H., MUIR I. D., LONG J. V. P., MCCONNELL J. D. C. and PECKETT A. (1970) Mineralogy and petrology of some lunar samples. *Science* **167**, 583–586.

ALBEE A. L., BURNETT D. S., CHODOS A. A., EUGSTER O. J., HUNEKE J. C., PAPANASTASSIOU D. A., PODOSEK F. A., RUSS G. PRICE II, SANZ H. G., TERA F. and WASSERBURG G. J. (1970) Ages, irradiation history and chemical composition of lunar rocks from the Sea of Tranquillity. *Science* **167**, 463–466.

ALFVÉN H. and ARRHENIUS G. (1969) Two alternatives for the history of the moon. *Science* **165**, 11–17.

BAILEY J. M. (1969) The moon may be a former planet. *Nature* **223**, 251–253.

BATE G. L., HUIZENGA J. R. and POTRATZ H. A. (1957) Thorium content of stone meteorites. *Science* **126**, 612–614.

BURNETT D. S. and WASSERBURG G. J. (1967) Sr^{87}–Rb^{87} ages of silicate inclusions in iron meteorites. *Earth Planet. Sci. Lett.* **2**, 397–408.

CAMERON A. E., SMITH D. H. and WALKER R. L. (1969) Mass spectrometry of nanogram-size samples of lead. *Anal. Chem.* **41**, 525–526.

CHAO E. C. T., JAMES O. B., MINKIN J. A., BOREMAN J. A., JACKSON E. D. and RALEIGH C. B. (1970) Petrology of unshocked crystalline rocks and shock effects in lunar rocks and minerals. *Science* **167**, 644–647.

COMPSTON W., ARRIENS P. A., VERNON M. J. and CHAPPELL B. W. (1970) Rubidium–strontium chronology and chemistry of lunar material. *Science* **167**, 474–476.

FLEMING E. H., JR., GHIORSO A. and CUNNINGHAM B. B. (1952) The specific alpha-activities and half-lives of U^{234}, U^{235} and U^{236}. *Phys. Rev.* **88**, 642–652.

FUNKHOUSER J. G., SCHAEFFER O. A., BOGARD D. D. and ZÄHRINGER J. (1970) Gas analysis of the lunar surface. *Science* **167**, 561–563.

GAST P. W. and HUBBARD N. J. (1970) Abundance of alkali metals, alkaline and rare earths, and strontium-87/strontium-86 ratios in lunar samples. *Science* **167**, 485–487.

GERSTENKORN H. (1955) Über Gezeitenreibung beim Zweikörpenproblem. *Z. Astrophys.* **26**, 245–274.

GOPALAN K., KAUSHAL S., LEE-HU C. and WETHERILL G. W. (1970) Rubidium–strontium, uranium and thorium–lead dating of lunar material. *Science* **167**, 471–473.

HAMAGUCHI H., REED G. W., JR. and TURKEVICH A. L. (1957) Uranium and barium in stone meteorites. *Geochim. Cosmochim. Acta* **12**, 337–347.

HOHENBERG C. M. (1969) Radio isotopes and the history of nucleosynthesis in the galaxy. *Science* **166**, 212–215.

KAUSHAL S. K. and WETHERILL G. W. (1969) Rb^{87}–Sr^{87} age of bronzite (H group) chondrites. *J. Geophys. Res.* **74**, 2717–2726.

KUIPER G. P. (1959) The moon. *J. Geophys. Res.* **64**, 1713–1719.

LSPET (LUNAR SAMPLE PRELIMINARY EXAMINATION TEAM) (1969) Preliminary examination of lunar samples from Apollo 11. *Science* **165**, 1211–1227.

MACDONALD G. J. F. (1964) Tidal friction. *Rev. Geophys.* **2**, 467–541.

OVERSBY V. M. (1970) The isotopic composition of lead in iron meteorites. *Geochim. Cosmochim. Acta* **34**, 65–75.

OVERSBY V. M. and GAST P. W. (1968) Oceanic basalt leads and the age of the earth. *Science* **162**, 925–927.

PATTERSON C. C. (1955) The Pb^{207}/Pb^{206} ages of some stone meteorites. *Geochim. Cosmochim. Acta* **7**, 151–153.

PATTERSON C. C. (1956) Age of meteorites and the earth. *Geochim. Cosmochim. Acta* **10**, 230–237.

REED G. W., KIGOSHI K. and TURKEVICH A. (1960) Determinations of concentrations of heavy elements in meteorites by activation analysis. *Geochim. Cosmochim. Acta* **20**, 122–140.

RINGWOOD A. E. (1966) Chemical evolution of the terrestrial planets. *Geochim. Cosmochim. Acta* **30**, 41–104.

RINGWOOD A. E. and ESSENE E. (1970) Petrogenesis of lunar basalts and the internal constitution and origin of the moon. *Science* **167,** 607–610.

ROSHOLT J. N. and TATSUMOTO M. (1970) Isotopic composition of uranium and thorium in Apollo 11 samples. *Geochim. Cosmochim. Acta*, Supplement I.

SENFTLE F. E., FARLEY T. A. and LAZAR N. (1956) Half-life of Th^{232} and the branching ratio of Bi^{212}. *Phys. Rev.* **104**, 1629.

SHIELDS W. R., (editor) (1966) Analytical mass spectrometry section: Instrumentation and procedures for isotopic analysis. *NBS Technical Note* **277**, 99.

SILVER L. T. (1970) Personal communication; and companion paper in this publication.

TATSUMOTO M. (1966) Isotopic composition of lead in volcanic rocks from Hawaii, Iwo Jima and Japan. *J. Geophys. Res.* **71,** 1721–1733.

TATSUMOTO M. (1969) New method for preparing ultrapure hydrofluoric acid. *Anal. Chem.* **41,** 2088–2089.

TATSUMOTO M. and KNIGHT R. J. (1968) Isotopic composition of lead in volcanic rocks from Hawaii. In *Abstracts for* 1966. *Geol. Soc. Amer. Spec. Paper* **101,** 218–219.

TATSUMOTO M. and ROSHOLT J. N. (1970) Age of the moon: an isotopic study of uranium–thorium–lead systematics of lunar samples. *Science* **167,** 461–463.

TILTON G. R. (1958) Isotopic composition of lead from tektites. *Geochim. Cosmochim. Acta* **14,** 323–330.

TILTON G. R., PATTERSON C. C., BROWN H. S., INGHRAM M. G., HAYDEN R., JR., HESS D. C. and LARSEN E. S., JR. (1955) Isotopic composition and distribution of lead, uranium, and thorium in a Precambrian granite. *Bull. Geol. Soc. Amer.* **66,** 1131–1148.

TURNER G. (1970) Argon-40/argon-39 dating of lunar rock samples. *Science* **167,** 466–468.

ULRYCH T. J. (1967) Oceanic basalt leads—A new interpretation and an independent age for the Earth. *Science* **158,** 252–256.

UREY H. C. (1960) On the chemical evolution and densities of the planets. *Geochim. Cosmochim. Acta* **18,** 151–153.

UREY H. C. (1962) Origin and history of the moon. In *Physics and Astronomy of the Moon*, (editor Z. Kopal), Chapter 13, pp. 481–523. Academic Press.

WAMPLER J. M., SMITH D. H. and CAMERON A. E. (1969) Isotopic composition of lead in tektites with lead in earth materials. *Geochim. Cosmochim. Acta* **33,** 1045–1055.

WASSERBURG G. J. and BURNETT D. S. (1968) The status of isotopic age determinations on iron and stone meteorites. In *Meteorite Research*, (editor Millman), pp. 467–479. D. Reidel.

WASSERBURG G. J., SCHRAMM D. N. and HUNEKE J. C. (1969) Nuclear chronologies for the galaxy. *Astrophys. J.* **157,** L91–L96.

WETHERILL G. W. (1956) Discordant uranium–lead ages, 1. *Trans. Amer. Geophys. Union* **37,** 320–326.

WISE D. U. (1969) Origin of the moon from the earth: some new mechanisms and comparisons. *J. Geophys. Res.* **74,** 6034–6045.

YORK D. (1966) Least-squares fitting of a straight line. *Can. J. Geophys. Res.* **44,** 1079–1086.

Proceedings of the Apollo 11 Lunar Science Conference, Vol. 2, pp. 1613 to 1626.

O^{18}/O^{16} ratios of Apollo 11 lunar rocks and minerals*

HUGH P. TAYLOR, JR. and SAMUEL EPSTEIN
Division of Geological Sciences, California Institute of Technology, Pasadena, California 91109

(*Received* 16 *February* 1970; *accepted in revised form* 25 *February* 1970)

Abstract—Oxygen isotope analyses were obtained on coexisting minerals in 5 lunar microgabbros and basalts. The δO^{18} values are: cristobalite (+7·16), plagioclase (+6·06 to +6·33), clinopyroxene (+5·67 to +5·95), and ilmenite (+3·85 to +4·12). The uniformity of δO^{18} suggests a close approach to isotopic equilibrium, with a calculated temperature of formation of about 1200°C, or about the same as terrestrial basalts. These rocks must have crystallized rapidly at low P_{H_2O}, as they show no evidence of further oxygen isotope exchange during cooling.

The lunar fines and breccias are 0·5 per mil richer in O^{18} than the above rocks, indicating addition of an O^{18}-rich constituent (perhaps carbonaceous chondrites), and/or that O^{16} was depleted by vapor fractionation during formation of the abundant glass in these materials. The lunar rocks are very similar in δO^{18} to terrestrial mafic and ultramafic igneous rocks and to the chondritic meteorites, suggesting a genetic relationship. However, they are significantly richer in O^{18} than several other meteorite groups, such as the basaltic achondrites. The O^{18} data also indicate that tektites are unlikely to have originated on the moon.

INTRODUCTION

OXYGEN isotope analyses have been obtained on mineral separates and whole-rock samples of several types of material returned from the Sea of Tranquillity by the Apollo 11 expedition. A preliminary report on these experiments is given by EPSTEIN and TAYLOR (1970).

O^{18}/O^{16} analyses of lunar samples are useful in two principal ways: (1) They can provide an estimate of the temperature of formation of a given rock if it can be shown that oxygen isotope equilibrium was attained during crystallization of the mineral assemblage. Actually, the calculated "isotopic temperature" will indicate the temperature of final "freezing in" of the O^{18}/O^{16} ratios in a set of coexisting minerals during their cooling history. In a very rapidly quenched rock this temperature will be essentially the formation temperature of the rock. (2) They can provide information on possible genetic relationships among the various moon samples, between the moon and the different classes of meteorites, and between the moon and the earth.

SAMPLE PREPARATION

Mineral separates were prepared by crushing the 0·8–1·0 g rock specimens in a stainless steel diamond mortar, sieving into various size-fractions, and performing magnetic separations with hand magnet and Frantz separator. For the ilmenite microgabbros (10044, 10058 and 10003) further purification was done solely by hand-picking. However, for the finer-grained ilmenite basalts (10017, 10057) the concentrates were additionally purified in heavy liquids.

All plagioclase separates analyzed were at least 96–98 per cent pure, as were all the clinopyroxene separates except that of 10057 (where it was 82 per cent). However, even though most of the ilmenite separates were prepared by careful hand-picking, the purity of these separates was only 70–90 per

* Contribution No. 1713, Division of Geological Sciences, California Institute of Technology, Pasadena, California 91109.

cent, and in one case (10057) was only 35 per cent. As will be described below, the amount of impurity (mainly clinopyroxene) in these ilmenite separates could be calculated very accurately through the measurement of the quantity of SiF_4 gas produced during reaction of the mineral separates with fluorine gas. As there is essentially no Si in the lunar ilmenites (<0·1 per cent on the basis of electron microprobe measurements, WARE and LOVERING, 1970; ADLER *et al.*, 1970) this measurement of SiF_4 gives an unequivocal indication of the amount of impurity phase. This is particularly important in the case of the ilmenites, because they are intergrown with clinopyroxene on a fine scale and as they are opaque it is very difficult to estimate the amount of impurity visually. In fact, our own visual estimates of the amount of pyroxene impurity were consistently much lower than was indicated by the SiF_4 analyses.

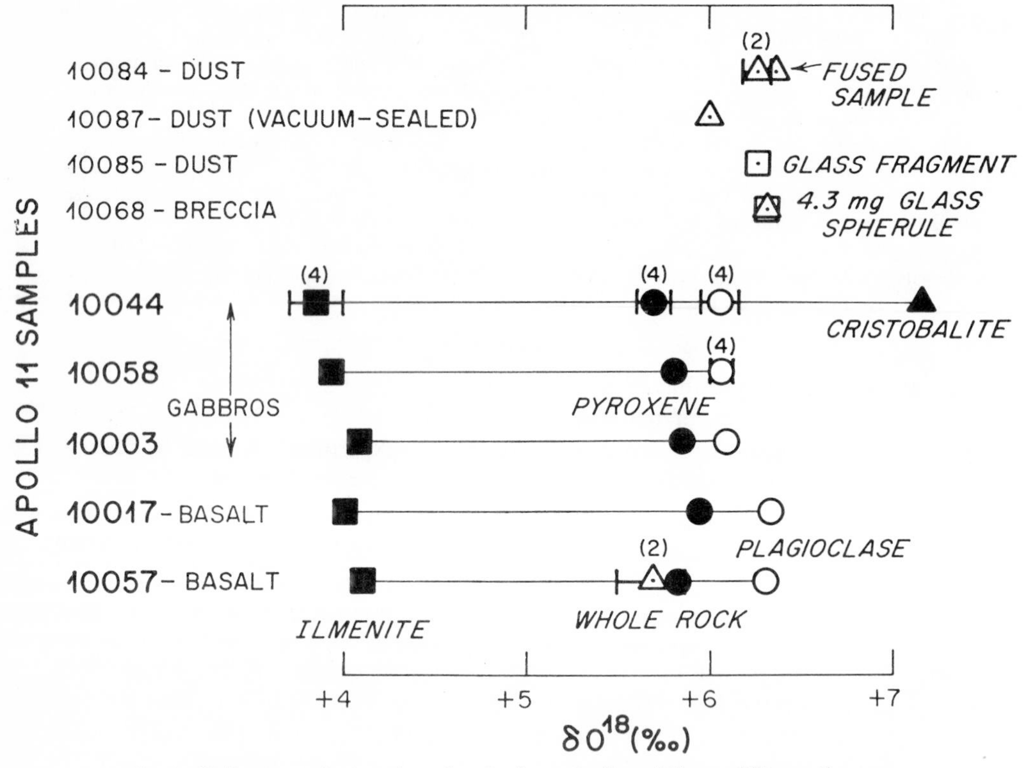

Fig. 1. δO^{18} values of minerals and rocks from the Sea of Tranquillity on the moon.

All of the O^{18} analyses of ilmenites in this work were therefore corrected for the amount and O^{18}/O^{16} ratio of the impurity. An analogous correction was made for the pyroxene of 10057. Finally an O^{18} correction was also made on the cristobalite separate from 10044, as it was only 75 per cent pure, based on both visual estimates and on the oxygen yield obtained from this separate during reaction with fluorine.

ANALYTICAL RESULTS

The oxygen isotope data are given in Table 1 and Fig. 1. In general analyses were made on 10–20 mg aliquots, except for the 4.3 mg brown glass spherule from 10068 and the 2·3 mg cristobalite sample from 10044. Oxygen was extracted from the minerals and rocks by reaction with fluorine gas at 550°–600°C; the oxygen was subsequently converted quantitatively to CO_2 by reaction with a hot carbon rod. The CO_2

Table 1. O^{18}/O^{16} ratios of Apollo 11 rocks and minerals

Sample	δO^{18}(‰)	Avg. dev.	No. of runs
10084-130, fines (Type D)			
Whole rock	+6·26	±0·07	2
Whole rock (fused sample)	+6·36		1
10068-21, breccia (Type C)			
Whole rock	+6·31		1
Brown glass spherule (4·3 mg)	+6·32		1
10085, coarse fines (Type D)			
Brown glass fragment	+6·27		1
10087-1, fines (Type D)			
Whole rock	+6·00		1
10058-71, microgabbro (Type B)			
Plagioclase	+6·07	±0·06	4
Clinopyroxene	+5·75		1
Ilmenite	+3·94		1
Whole rock (calc.)	+5·70		
10044-31, microgabbro (Type B)			
Cristobalite	+7·16		1
Plagioclase	+6·06	±0·15	4
Clinopyroxene	+5·67	±0·09	4
Ilmenite	+3·85	±0·14	4
Whole rock (calc.)	+5·62		
10003-10, microgabbro (Type B)			
Plagioclase	+6·09		1
Clinopyroxene	+5·86		1
Ilmenite	+4·10		1
Whole rock (calc.)	+5·71		
10017-32, basalt (Type A)			
Plagioclase	+6·33		1
Clinopyroxene	+5·95		1
Ilmenite	+4·02		1
Whole rock (calc.)	+5·76		
10057-39, basalt (Type A)			
Plagioclase	+6·30		1
Clinopyroxene	+5·83		1
Ilmenite	+4·12		1
Whole rock	+5·68	±0·19	2
Whole rock (calc.)	+5·71		

samples were then each analyzed twice on a sensitive dual inlet, double-collector mass spectrometer. These procedures are the same as those used by TAYLOR and EPSTEIN (1962), except that (1) no HF was added to the samples, (2) the nickel reaction vessels were only $\frac{1}{2}$ the length (and hence $\frac{1}{2}$ the volume) of the vessels used previously, and (3) a new cylinder of fluorine gas was utilized that contains only $\frac{1}{3}$ the amount of oxygen contaminant of the cylinder used by TAYLOR and EPSTEIN (1962). This oxygen contaminant has a δO^{18} of +4·9 per mil, very similar to most of the analyses in Table 1. Therefore the correction of each analysis for this effect was typically only 0·00–0·02 per mil.

All samples were loaded into reaction vessels in a specially-designed vacuum air-lock dry box in an atmosphere of P_2O_5-dried nitrogen. The inside of this dry box was *never* exposed to atmospheric moisture either before or after the sample loadings. One of the analyzed samples (10087) had been packed in vacuum and sealed in an aluminum cylinder. This sample of fine dust from the lunar surface was opened in the dry box and loaded directly into a nickel reaction vessel. All other samples listed in Table 1 were exposed to the normal laboratory atmosphere, but one in which the dust and humidity levels were controlled. Except as noted previously, no water or other reagents were allowed to come into contact with the samples.

The SiF_4 gas released from all samples was collected, and its volume measured in a Hg manometer. These samples were then analyzed for Si^{30}/Si^{28} in another mass spectrometer (see EPSTEIN and TAYLOR, 1970).

The δO^{18} values listed in Table 1 have been corrected in the normal way for background and valve-mixing and for change of standard from our working mass spectrometer standard to SMOW (standard mean ocean water), as well as for the aforementioned minor O_2 contaminant in the F_2. In addition, however, all the ilmenite analyses have been corrected for the presence of silicate impurities on the basis of the amount of SiF_4 obtained during fluorination of these mineral separates. These data are shown in Table 2. Various tests were made which show that this is an exceedingly

Table 2. Oxygen isotope corrections for silicate impurities in analyzed ilmenite concentrates

Sample	μm O_2*	μm SiF_4†	Wt. % ilmenite‡	$\delta O^{18}_{Meas.}$	$\delta O^{18}_{Corr.}$§
10044	156	18	86·4	+3·97	+3·61
10044	147	21	83·0	+4·47	+4·14
10044	149	17	86·6	+4·15	+3·83
10044	173	30	79·2	+4·29	+3·80
10058	259	56	73·5	+4·53	+3·94
10003	117	23	76·1	+4·62	+4·10
10017	146	54	69·5	+4·73	+4·02
10057	194	93	34·3	+5·35	+4·12

* Micromoles of oxygen formed during fluorination reaction.
† Micromoles of SiF_4 formed during fluorination reaction.
‡ Calculated weight per cent of ilmenite in sample, assuming that all silicate impurities are clinopyroxene.
§ Corrected δO^{18} value of ilmenite.

accurate method of correcting the ilmenite analyses. Note that even though the corrections are in some cases quite large, the corrected ilmenite analyses given in Table 1 vary from only +3·85 to +4·12 per mil. Accurate knowledge of the extent of silicate impurity in each sample is particularly important because the clinopyroxene and plagioclase contain 1·3–1·5 times as much oxygen per unit weight as does ilmenite. In order not to over-correct for these impurity effects, a conservative procedure was used. We assumed that the ratio of oxygen atoms to silicon in the pyroxenes is exactly 3 to 1 (e.g. like $MgSiO_3$) and therefore that 1·5 μmoles of O_2 were released for each μ mole of SiF_4 formed. The actual ratio is a little higher than this, and in addition some plagioclase (in which the ratio is about 4 to 1) was also present in each concentrate, and the plagioclase is even higher in O^{18} than the pyroxene. Hence by assuming that the impurity is *all* pyroxene and correcting on this basis, our corrected ilmenite analyses listed in Table 1 are if anything too high in δO^{18}. The best-determined ilmenite δ-value in terms of number of replicate analyses and purity of mineral separates is that given for 10044, and it is interesting that this also has the lowest O^{18}/O^{16} ratio.

Whole-rock δO^{18} values were calculated for the 3 microgabbros and 2 basalts, based upon the mineral δ-values and on modal abundance data for the various rocks given by HAGGERTY *et al.* (1970) and BAILEY *et al.* (1970). In one case where direct replicate whole-rock analyses were carried out (10057) the calculated value agrees very well with the direct analysis.

As far as isotope geothermometry is concerned, the important relationships are the oxygen isotope fractionations among coexisting minerals, ordinarily expressed as $1000 \ln \alpha_{A-B}$ where α is O^{18}/O^{16} in mineral A divided by O^{18}/O^{16} in mineral B (e.g. see O'NEIL and TAYLOR, 1967). The term $1000 \ln \alpha$ (hereafter abbreviated as Δ) is to a very close approximation equal to $\delta A - \delta B$. Note that the plagioclase-ilmenite Δ-values of the lunar rock vary from 1·98–2·20 in the microgabbros and 2·17–2·30 in the basalts. This uniformity suggests a close approach to isotopic equilibrium, and it is also interesting that slightly bigger fractionations are observed in the basalts, as this would be predicted because of the more sodic composition of the plagioclase in the finer-grained rocks (typically An_{78} vs. An_{90}). O'NEIL and TAYLOR (1967) demonstrated that at isotopic equilibrium, the more Na-rich plagioclase is enriched in O^{18}.

COMPARISON WITH DATA OF OTHER WORKERS

The O^{18} analyses given in this paper and in EPSTEIN and TAYLOR (1970) agree exceedingly well with data presented by ONUMA *et al.* (1970) and FRIEDMAN *et al.* (1970), except that all of the ilmenites analyzed by ONUMA *et al.* are 0·3–0·5 per mil higher in O^{18} than our analyses. We have discussed in some detail the δO^{18} corrections made on our ilmenite analyses, and we may note here that ONUMA *et al.* made no such corrections because the purity of their mineral separates "was $>95\%$ in all cases." FRIEDMAN *et al.* (1970) corrected their two ilmenite analyses on the basis of excess yields of oxygen obtained during reaction with BrF_5 (this excess was attributed to pyroxene impurities).

Our mineral separates were prepared differently than those of ONUMA *et al.* (1970) in the sense that our microgabbro ilmenite separates were prepared by hand-picking. Therefore, our analyses are actually on the coarser-grained plates of ilmenite in the rock. If the ilmenites exhibit isotopic variations as a function of grain size in each rock, with the coarser ilmenites being lower in δO^{18}, this might explain the discrepancy; however, it is doubtful that such an effect could produce such a large discrepancy.

The published electron microprobe analyses of ilmenites from various lunar rocks indicate that the extent of $MgTiO_3$ (geikielite) substitution varies from an average of about 5–9 mole per cent in the Type A basalts to about 1–3 mole per cent in the Type B microgabbros. If we utilize such analyses in connection with the reported oxygen contents of ilmenites by ONUMA *et al.* (1970), it is apparent that excess yields of oxygen are being obtained. Making the assumption that the excess oxygen is due to silicate impurities, we calculate that the amount of true ilmenite-derived oxygen from each of the separates analyzed by ONUMA *et al.* (1970) is no more than 84–91 per cent for the fine-grained Type A rocks (including 10017) and 94–99 per cent for the coarse-grained Type B rocks. These relationships are exactly those expected if the separates are impure because it is much more difficult to get perfectly clean separates from the finer-grained rocks.

If we assume, therefore, that the excess yields reported by ONUMA *et al.* are a result of the same effects that were corrected for by FRIEDMAN *et al.* (1970) and by ourselves, the discrepancies in the various O^{18} analyses of ilmenites are practically eliminated.

For example, if we assume that typical oxygen yields of 97–99 per cent are obtained during fluorination of pure ilmenite concentrates, which is in line with our own experience, we calculate that the plagioclase-ilmenite fractionations reported by ONUMA *et al.* become 1·90–2·10 instead of their reported 1·64–1·90. This is in much better agreement with our results of 1·98–2·20 for samples 10044, 10058, 10003 and 10057.

TEMPERATURES OF CRYSTALLIZATION OF THE LUNAR MAGMAS

The uniformity of mineral O^{18}/O^{16} ratios in all the lunar basalts and gabbros argues strongly that oxygen isotope equilibrium was essentially attained in these rocks. The order of enrichment in O^{18} in a single mineral assemblage, namely ilmenite–clinopyroxene–plagioclase–free silica, is the same as that observed in numerous

Table 3. Calculated oxygen isotope temperatures of lunar microgabbros and basalts

Sample	Plagioclase composition	$\Delta_{\text{Plagioclase–Ilmenite}}$	Temperature (°C)
10044	An_{90}	2·20	1202°
10058	An_{90}	2·12	1243°
10003	An_{90}	1·98	1325°
10017	An_{78}	2·30	1208°*
10057	An_{78}	2·17	1277°*

* Calculated on basis of equation.

$$1000 \ln \alpha_{\text{Plagioclase}(An_{78})\text{-Ilmenite}} = 0{\cdot}81 + 3{\cdot}27(10^6 T^{-2}).$$

terrestrial igneous rocks and meteorites (TAYLOR *et al.*, 1965; TAYLOR, 1968). However, the O^{18} fractionations between plagioclase and ilmenite in the lunar gabbros are smaller than in terrestrial gabbros such as the Skaergaard and Kiglapait intrusions (2·0 to 2·3 vs. 3·3 to 4·0). This difference can in part be accounted for by the higher Ca content of the lunar plagioclase (e.g. see O'NEIL and TAYLOR, 1967), but it also unquestionably indicates that the O^{18}/O^{16} ratios of coexisting minerals in the lunar microgabbros were "frozen in" at a considerably higher temperature than in the terrestrial gabbros.

Although an isotopic temperature calibration curve involving ilmenite has not yet been determined, an estimate of the quartz–ilmenite and plagioclase–ilmenite geothermometer curves can be made based upon (a) the plagioclase–magnetite curve utilized by TAYLOR (1968) in estimating isotopic "temperatures" of various igneous rocks, (b) the plagioclase–H_2O calibration curves of O'NEIL and TAYLOR (1967), (c) the quartz–H_2O curve of CLAYTON *et al.* (1967), and (d) the observation that in natural mineral assemblages, 1000 ln α for quartz–ilmenite is equal to 0·92 times 1000 ln α for quartz–magnetite (GARLICK and EPSTEIN, 1967). The constructed curves are:

$$1000 \ln \alpha_{\text{Quartz–Ilmenite}} = 1{\cdot}35 + 4{\cdot}56(10^6 T^{-2})$$

$$1000 \ln \alpha_{\text{Plagioclase}(An_{90})\text{–Ilmenite}} = 0{\cdot}72 + 3{\cdot}22(10^6 T^{-2}).$$

The plagioclase–ilmenite "temperatures" calculated for the lunar rocks are given in Table 3 and Fig. 2.

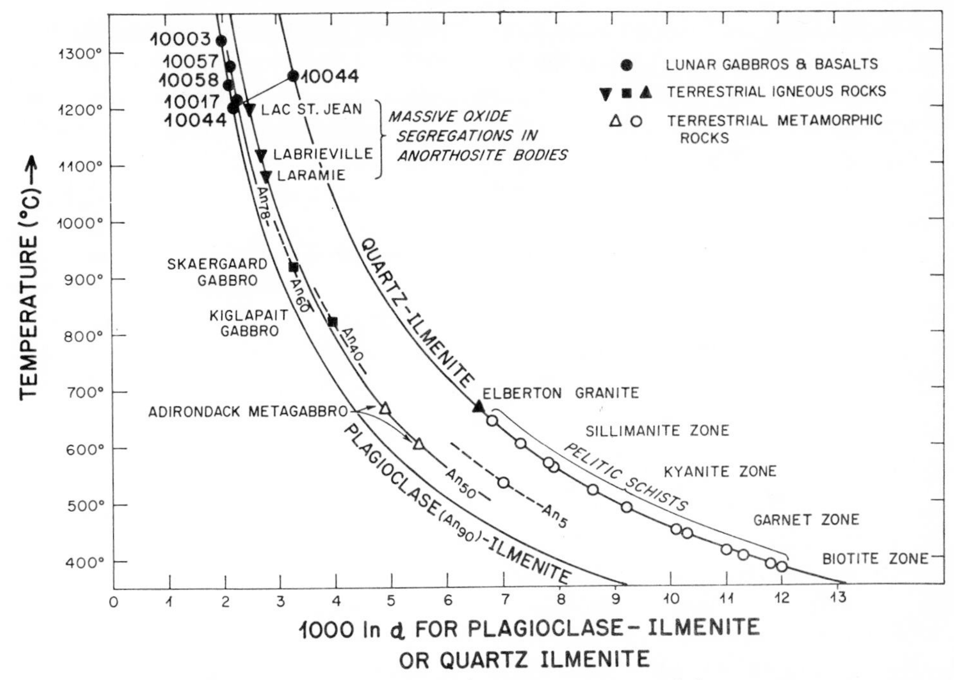

Fig. 2. Plot of equilibrium temperature (°C) vs. $\Delta_{\text{Plagioclase-Ilmenite}}$ and $\Delta_{\text{Quartz(or Cristobalite)-Ilmenite}}$ for various terrestrial igneous and metamorphic rocks and for the lunar microgabbros and basalts. The data on terrestrial rocks are from GARLICK and EPSTEIN (1967), TAYLOR (1968), and ANDERSON (1968). The derivation of the various isotope geothermometer curves is discussed in the text.

Utilizing the quartz–ilmenite curve, 10044 gives a cristobalite–ilmenite "temperature" of 1260°C, almost concordant with the 1202°C plagioclase–ilmenite "temperature" of this rock. Inasmuch as the plagioclase–ilmenite fractionation of 10044 is known with a higher degree of certainty than other values given in Table 3, this temperature of 1200°C represents our best isotopic temperature estimate for the lunar rocks.

The above temperatures may be compared with "temperatures" of 920°C for a Skaergaard gabbro and 820°C for a Kiglapait gabbro, both of which are compatible with estimates made on these rocks based on the independent Fe–Ti oxide geothermometer of BUDDINGTON and LINDSLEY (1964). The inference may be drawn that the lunar gabbros crystallized and cooled very rapidly under exceptionally "dry" conditions as compared with many terrestrial gabbros; if P_{H_2O} had been appreciable we should expect to observe more evidence of the type of post-crystallization oxygen isotopic exchange observed in terrestrial gabbros, particularly those containing hornblende (TAYLOR, 1968, Table 4).

It must be emphasized that the isotopic temperature of 1200°C quoted here for the lunar microgabbros is based on such a large number of assumptions that it should only be considered accurate to ± 100°C at the present time. However, this temperature can be quite reliably utilized within the framework of the isotopic temperatures for other igneous rocks quoted by TAYLOR (1968), as well as those given on Fig. 2, because all are based on the same assumptions. It is very possible that these assumptions result in temperature estimates that are as much as 100°C too high, because they also imply temperatures of 1200°–1300°C for terrestrial basalts analyzed by GARLICK (1966) and ANDERSON and CLAYTON (1966). Hence, irrespective of any laboratory calibrations, it may be stated with certainty that the isotopic "temperatures of formation" of the ilmenite microgabbros and basalts from the moon are essentially identical to those of terrestrial basalts.

Oxygen isotope temperatures of various terrestrial igneous and metamorphic rocks are shown on Fig. 2, to emphasize the point that O^{18}/O^{16} ratios of coexisting minerals in the Type A and B lunar rocks were "frozen in" at much higher temperatures than most terrestrial plagioclase–ilmenite and quartz–ilmenite pairs. Note that the isotopic temperatures of massive ilmenite–magnetite segregations in Precambrian anorthosite bodies are slightly lower than those of the lunar microgabbros and basalts. The anorthosite "temperatures" are based upon δO^{18} analyses of essentially monomineralic masses of plagioclase adjoining the massive oxide segregations. By utilizing such rock types, one can eliminate most problems of post-formation O^{18} exchange (see ANDERSON, 1968; TAYLOR, 1968). Hence, we can also say with some certainty that the primary temperatures of formation of the Type A and B lunar rocks were at least as high or somewhat higher than those of terrestrial anorthosite bodies.

Another problem that deserves mention is the total lack of hematite solid solution in the lunar ilmenites (ANDERSON *et al.*, 1970). Inasmuch as ilmenites from terrestrial gabbros and basalts generally contain appreciable hematite in solid solution, for purposes of comparison with the terrestrial plagioclase–ilmenite fractionations, those from lunar rocks should probably be increased by as much as 0·05; this results from the fact that at isotopic equilibrium hematite is lower in O^{18} than coexisting ilmenite. This effect alone could reduce the previous temperature estimate from 1200°C to 1170°C.

In view of the assumptions and analytical errors involved in the isotopic temperature estimates, they must be considered to be in excellent agreement with other estimates of the temperature range of crystallization of the lunar basalts and microgabbros based upon melting relationships at 1 atm pressure on materials of similar chemical composition: 1075–1155°C (ANDERSON *et al.*, 1970); 1075–1210°C (ROEDDER and WEIBLEN, 1970); and 1075–1210°C (O'HARA *et al.*, 1970). In addition, SKINNER (1970) has calculated that a large portion of the lunar basaltic magma must have crystallized above 1140°C on the basis of intergrowths of metallic Fe in FeS.

The oxygen isotope data are also in good agreement with other evidence that the lunar microgabbros and basalts crystallized very rapidly. The sectorally zoned pyroxenes (HARGRAVES *et al.*, 1970), the presence of an apparently metastable Fe-rich pyroxenoid not reported in terrestrial basalts and gabbros (ANDERSON *et al.*, 1970), the occurrence of cristobalite, and the strong crystallization zoning of all the minerals, particularly the pyroxenes, all are indicative of exceedingly rapid crystallization. Inasmuch as melts of the composition of the lunar rocks have considerably lower viscosities than terrestrial basalts (WEILL *et al.*, 1970), one can expect that even rocks as coarse-grained as the lunar microgabbros might have been "quenched" as rapidly as fine-grained basalts at the earth's surface. Although the composition of the vapor phase that coexisted with the lunar magmas is unknown, it might also have helped to promote a coarse grain size.

The question arises as to whether oxygen isotope equilibrium could have been attained in the lunar rocks, considering the rapid crystallization and lack of chemical homogeneity from grain to grain (or even within the same crystal) in a single rock. The answer to this question is that isotopic equilibrium was very likely attained only at each crystal–melt interface, but that as soon as a subsequent crystalline layer formed over this interface, it acted as an "armoring" layer that prevented further communication of oxygen between the melt and the interior of the crystal. Had the melt steadily changed in O^{18} during this process, the crystals would be isotopically zoned. The latter is unlikely because of the very small O^{18} fractionations between minerals and silicate melts at these high temperatures, and because almost no change in O^{18} occurs during differentiation from basalt to rhyolite obsidian or from gabbro to ferrosyenite in terrestrial igneous rocks (TAYLOR, 1968). To sum up, the apparent oxygen isotope equilibrium observed in the lunar microgabbros and basalts is probably a result of (1) surface O^{18} equilibrium at each melt–crystal interface and (2) the lack of any significant change in O^{18} in the melt phase during its crystallization history (except perhaps in the very end-stages where the effects would be quantitatively unimportant).

LUNAR BRECCIAS AND DUST

As shown in Fig. 1 and Table 1, the whole-rock O^{18}/O^{16} ratios of the breccia (10068, Type C) and the lunar fines (10084 and 10087, Type D) are essentially identical at +6·00 to +6·36 per mil. One of the analyses of 10084 was done on a sample previously fused in vacuum to a clear, brown glass by induction heating. The fusion had no apparent effect on the O^{18}/O^{16} ratio of the dust sample. The glass fragment from 10085 and the 4·3 mg glass spherule from 10068 also have δO^{18}-values that are indistinguishable from the whole-rock analyses of the breccia and fines

The whole-rock δO^{18} values of Type C and D lunar rocks are 0·3–0·7 per mil higher than those of the Type A and B rocks. Therefore, either (1) an O^{18}-rich constituent has been added to the breccias and fines in addition to Type A and B material, or (2) during formation of both the breccias and the fines, an enrichment in O^{18} takes place.

Note that the Type C and D materials are largely composed of glass fragments and spherules, so the enrichment in O^{18} may have occurred during formation of the glass (presumably by impact melting). In this connection it is interesting that WALTER and CLAYTON (1967) demonstrated that fractional vaporization of a tektite-like glass at 2800°C produces an enrichment of O^{18} in the residual material. However, in their experiments a drastic lowering of SiO_2 content from 82 per cent to 45 per cent was accompanied by at most a 1·2 per mil increase of δO^{18} in the residual glass. The lunar fines, although about 3 wt.% higher in Al_2O_3 content, are not radically different in chemical composition from the Type A and B rocks. Without vaporization experiments on analogous materials it is therefore premature to ascribe the average 0·5 per mil difference between the Type A–B and C–D groups to a vapor fractionation process.

Possible O^{18}-rich components that may have been added to the lunar dust and breccia are material from the lunar highlands (plagioclase?) or debris derived from meteorite infall. If the latter process is to explain the enrichment in O^{18} of the glass and dust, it must be due to infall of Type I or II carbonaceous chondrites (or ureilites), because these are the only meteorite groups which are O^{18}-rich (Taylor *et al.*, 1965). Carbonaceous chondrites are probably abundant among the meteorites that strike the earth and moon, but Keays *et al.* (1970) have estimated that they make up no more than 2–3 per cent of the lunar dust. Assuming that the analyses of Taylor *et al.* (1965) give typical values for all Type I and II carbonaceous chondrites, a 0·5 per mil enrichment requires 10 per cent of the dust and breccia be made up of this type of meteorite. This assumes that no appreciable O^{18} fractionations occur in the carbonaceous chondrite material during its impact fusion; inasmuch as large amounts of H_2O must be driven off during this process, some O^{18} changes are likely, and they would probably involve enrichment in O^{18} (see Taylor and Epstein, 1964, Fig. 5). It is practically inconceivable, however, that an enrichment of 15 per mil would be produced, and this is the amount required if carbonaceous chondrites make up only 2·5 per cent or less of the breccias and dust.

Addition of plagioclase or plagioclase glass to the lunar fines from elsewhere on the moon could account for the higher Al_2O_3 content of the fines and might also be expected to produce a small O^{18}-enrichment (also, see Wood *et al.*, 1970). Appreciable amounts of solar wind hydrogen and rare gases are present in the lunar breccias and dust, and the isotopic composition of this hydrogen is very peculiar in that it is almost deuterium-free (Epstein and Taylor, 1970). If the solar wind should happen to be very O^{18}-rich this could also help to account for the observed differences.

Whatever the origin of the O^{18}-rich constituent in these Type C and D rocks, it is also present in the glass spherules and glass fragments. Thus, any explanation involving simple mechanical mixing of impact debris from other parts of the moon's surface would imply that this mixing occurred prior to or during the formation of the glass. It seems likely that a combination of the different mechanisms described above is responsible for the O^{18} enrichment observed in the lunar fines; no one mechanism seems adequate by itself.

Isotopic Relationships with Meteorites and the Earth

In Fig. 3, δO^{18}-values of pyroxenes from various types of meteorites and terrestrial igneous rocks are compared with data on the lunar pyroxenes. A plot of whole-rock δO^{18}-values for the rock types shown in Fig. 3 would show identical relationships, because the δO^{18} values of the pyroxenes are essentially determined by variations in the "oxygen reservoir" of the rocks themselves.

Figure 3 shows clearly that there are no significant differences in O^{18}/O^{16} between the lunar rocks on the one hand and the chondritic meteorites and terrestrial mafic and ultramafic igneous rocks on the other. Other meteorites which have similar O^{18}/O^{16} ratios are those designated as Group II by Taylor *et al.* (1965). The oxygen isotope data thus support a genetic relationship among the Group II meteorites (the bulk of which are ordinary chondrites), the moon and the earth. This genetic relationship, however, could readily be placed as far back in time as the formation of the solar system itself.

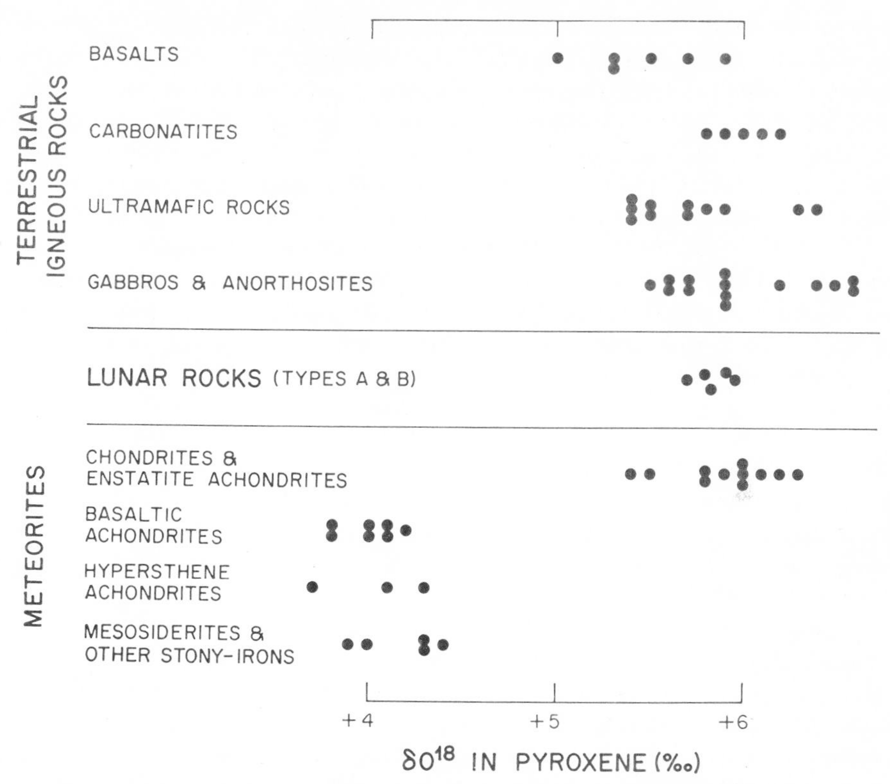

Fig. 3. Comparison of δO^{18} values of pyroxenes from the Apollo 11 lunar rocks with those from terrestrial mafic and ultramafic igneous rocks (TAYLOR, 1968; GARLICK, 1966) and from meteorites (TAYLOR *et al.*, 1965).

The Group I meteorites of TAYLOR *et al.* (1965) are consistently 1·5–2·0 per mil lower in O^{18} than the lunar rocks. This is significant because one class of Group I meteorites, the basaltic achondrites, have several chemical and mineralogical features in common with the lunar rocks. Also, DUKE and SILVER (1967) previously suggested that basaltic achondrites may have come from the moon.

More O^{18} analyses of other parts of the lunar surface must be obtained before definite conclusions can be reached, but it is considered unlikely that igneous rocks with O^{18}/O^{16} ratios similar to the basaltic achondrites will be found on the moon. This reasoning is as follows: (1) There is a remarkable uniformity of O^{18}/O^{16} in basalts, gabbros and ultramafic rocks on earth, irrespective of age or geographic location; by analogy it is logical to suppose that such rock types on the moon might also show such uniformity in δO^{18}. (2) The Group I meteorites themselves show a remarkable uniformity of δO^{18} (Fig. 3) over a wide range of rock types from iron meteorites with silicate inclusions, to stony-irons (the mesosiderites), to ultramafic rocks (the hypersthene achondrites), to the basaltic achondrites which themselves show a large variation in Fe/(Fe + Mg) from about 0·3 to 0·7 (TAYLOR *et al.*, 1965).

The oxygen isotope data therefore suggest that the basaltic achondrites and other meteorites of Oxygen Isotope Group I perhaps originated from a completely different oxygen isotope reservoir (another part of the solar system?) than did the earth, moon and chondritic meteorites. However, the arguments of DUKE and SILVER (1967) remain valid that these meteorites probably formed on a body having an appreciable gravitational field, perhaps similar in size to the moon. The basaltic achondrites are also similar in alkali abundances and basic mineralogy to the lunar rocks, and both sets of rocks crystallized under low-pressure volcanic conditions under very low fugacities of O_2 and H_2O. We may note, however, that the cristobalite and tridymite of the basaltic achondrites is much richer in O^{18} than the cristobalite of the lunar microgabbro (TAYLOR *et al.*, 1965). Thus, whereas the free silica phase appears to have equilibrated with coexisting plagioclase and pyroxene in the lunar rocks it is drastically out of oxygen isotope equilibrium with these minerals in the basaltic achondrites, and therefore must have had a different origin.

ORIGIN OF TEKTITES

With analyses of the first set of samples from the moon, it is possible to place new restrictions on the origin of tektites. Although it will be necessary to obtain samples from a site in the lunar highlands before anything definitive can be said, the present geochemical evidence argues strongly against the hypothesis that tektites are derived from the moon.

Recently, TAYLOR and EPSTEIN (1969) demonstrated that several groupings of tektites display a clear-cut correlation between SiO_2 content and O^{18}/O^{16} ratio. A plausible explanation for these features is that they represent mixing curves involving a high-SiO_2 igneous component and a low SiO_2 component formed in the presence of aqueous solutions at much lower temperatures. Weathered or hydrothermally altered igneous rocks on earth satisfy these requirements, but we can now say with some certainty that such processes are unlikely to have occurred on the moon. Certainly, there has been no atmosphere on the moon in the last 3·7 b.y. that could have been involved in such a low-temperature hydrous alteration event; otherwise we should have seen some evidence of it in the Apollo 11 rocks.

The other plausible explanation for the observed SiO_2–δO^{18} correlations in tektites is that they are the result of impact melting and vapor fractionation of parent material that initially had a SiO_2 content of 80–84 wt.%. Such a parent material could not be widespread on the lunar surface, because the lunar mare are now known to be essentially basaltic in composition, and the lunar highlands may be largely anorthosite (WOOD *et al.*, 1970).

Therefore, neither of the plausible explanations of the O^{18}/O^{16} variations in tektites make much sense in terms of a lunar origin. Other geochemical data on lunar materials, in particular the Pb isotope data, are also impossible to reconcile with existing data on tektites. In terms of the oxygen isotope data alone, the last hope is that reasonably abundant SiO_2-rich igneous rocks exist on unsampled parts of the lunar surface, and these rocks have undergone localized hydrothermal alteration of some type. Inasmuch as at least 3 separate tektite-forming events have occurred in the last 33 m.y. producing glassy objects that are all very similar in their chemical and isotopic properties, it is not very reasonable to assume that localized parent materials having just these characteristics either (1) exist on the lunar surface, or (2) if they do exist, that they are the only materials melted and ejected during meteorite impacts on the moon.

Note that it is not enough that SiO_2-rich igneous rocks (perhaps similar in composition to the late-stage, immiscible globules of silicate melt observed by ROEDDER and WEIBLEN, 1970) simply exist on the lunar surface. Unless they have been hydrothermally altered, such SiO_2-rich igneous rocks would probably have δO^{18} values very similar to the Type A and B lunar rocks because (1) differentiation of basalt to rhyolite or rhyolite obsidian on earth generally involves little or no change in O^{18}/O^{16} ratios, particularly if differentiation proceeds at low fugacities of H_2O and O_2 (TAYLOR, 1968), and (2) the cristobalite in microgabbro 10044 has a δ-value of $+7{\cdot}16$, indicating that any rhyolitic melt that might have been differentiated from such a magma would have a δ-value even lower than 7·16. Thus the O^{18} data on lunar rocks, meager though they are at present, suggest that even though high-SiO_2 igneous rocks may exist on the moon (and may even be abundant in certain localities) it is very doubtful that they would have δO^{18}-values in the $+9{\cdot}0$ to $+11{\cdot}5$ per mil range typical of most tektites. Prior to our analyses of the Apollo 11 rocks, it remained conceivable that the moon might be 4–5 per mil richer in O^{18} than terrestrial basalts and gabbros. We now know that it is not, and reasonable inference tells us that high-O^{18} rocks that are suitable tektite parent materials are rare or non-existent on the lunar surface.

Acknowledgments—Sample 10017 was provided by D. S. BURNETT and G. J. WASSERBURG. Some of the laboratory work was done by P. YANAGISAWA. This research was supported by National Aeronautics and Space Administration Contract No. NAS 9-7944.

REFERENCES

ADLER I., WALTER L. S., LOWMAN P. D., GLASS B. P., FRENCH B. M., PHILPOTTS J. A., HEINRICH K. J. F. and GOLDSTEIN J. I. (1970) Electron microprobe analysis of lunar samples. *Science* **167,** 590–592.

ANDERSON A. T. (1968) Oxidation of the La Blache Lake titaniferous magnetite deposit, Quebec. *J. Geol.* **76,** 528–546.

ANDERSON A. T. and CLAYTON R. N. (1966) Equilibrium oxygen isotope fractionation between minerals in igneous, meta-igneous and meta-sedimentary rocks. *Trans. Amer. Geophys. Union* **47,** 212.

ANDERSON A. T., CREWE A. V., GOLDSMITH J. R., MOORE P. B., NEWTON R. C,. OLSEN E. J., SMITH J. V. and WYLLIE P. J. (1970) Petrologic history of moon suggested by petrography, mineralogy and crystallography. *Science* **167,** 587–590.

BAILEY J. C., CHAMPNESS P. E., DUNHAM A. C., ESSON J., FYFE W. S., MACKENZIE W. S., STUMPFL E. F. and ZUSSMAN J. (1970) Mineralogical and petrological investigations of lunar samples. *Science* **167,** 592–594.

BUDDINGTON A. F. and LINDSLEY D. H. (1964) Iron–titanium minerals and synthetic equivalents. *J. Petrol.* **5,** 310–357.

CLAYTON R. N., O'NEIL J. R. and MAYEDA T. (1967) Unpublished data.

DUKE M. B. and SILVER L. T. (1967) Petrology of eucrites, howardites and mesosiderites. *Geochim. Cosmochim. Acta* **31,** 1637–1665.

EPSTEIN S. and TAYLOR H. P. (1970) $^{18}O/^{16}O$, $^{30}Si/^{28}Si$, D/H, and $^{13}C/^{12}C$ studies of lunar rocks and minerals. *Science* **167,** 533–535.

FRIEDMAN I., O'NEIL J. R., ADAMI L. H., GLEASON J. D. and HARDCASTLE K. (1970) Water, hydrogen, deuterium, carbon, carbon-13, and oxygen-18 content of selected lunar material. *Science* **167,** 538–540.

GARLICK G. D. (1966) Oxygen isotope fractionation in igneous rocks. *Earth Planet. Sci. Lett.* **1,** 361–368.

GARLICK G. D. and EPSTEIN S. (1967) Oxygen isotope ratios of coexisting minerals in regionally metamorphosed rocks. *Geochim. Cosmochim. Acta* **31**, 181–214.

HAGGERTY S. E., BOYD F. R., BELL P. M., FINGER L. W. and BRYAN W. B. (1970) Iron–titanium oxides and olivine from 10020 and 10071. *Science* **167**, 613–615.

HARGRAVES R. B., HOLLISTER L. S. and OTALORA G. (1970) Compositional zoning and its significance in pyroxenes from three coarse-grained lunar samples. *Science* **167**, 631–633.

KEAYS R. R., GANAPATHY R., LAUL J. C., ANDERS E., HERZOG G. F. and JEFFERY P. M. (1970) Trace elements and radioactivity in lunar rocks: Implications for meteorite infall, solar-wind flux, and formation conditions of Moon. *Science* **167**, 490–493.

O'HARA M. J., BIGGAR G. M. and RICHARDSON S. W. (1970) Experimental petrology of lunar material: The nature of mascons, seas and the lunar interior. *Science* **167**, 605–607

O'NEIL J. R. and TAYLOR H. P. (1967) The oxygen isotope and cation exchange chemistry of feldspars. *Amer. Mineral.* **52**, 1414–1437.

ONUMA N., CLAYTON R. N. and MAYEDA T. K. (1970) Oxygen isotope fractionation between minerals and an estimate of the temperature of formation. *Science* **167**, 536–538.

ROEDDER E. and WEIBLEN P. W. (1970) Silicate liquid immiscibility in lunar magmas, evidenced by melt inclusions in lunar rocks. *Science* **167**, 641–644.

SKINNER B. J. (1970) High crystallization temperatures indicated for igneous rocks from Tranquillity Base. *Science* **167**, 652–654.

TAYLOR H. P. (1968) The oxygen isotope geochemistry of igneous rocks. *Contrib. Mineral. Petrol.* **19**, 1–71.

TAYLOR H. P. and EPSTEIN S. (1962) Relationship between O^{18}/O^{16} ratios in coexisting minerals of igneous and metamorphic rocks. Part 1: Principles and experimental results. *Bull. Geol. Soc. Amer.* **73**, 461–480.

TAYLOR H. P. and EPSTEIN S. (1964) Comparison of oxygen isotope analyses of tektites, soils and impactite glasses. In *Isotopic and Cosmic Chemistry*, (editors H. Craig, S. L. Miller and G. J. Wasserburg), North-Holland. pp. 181–199.

TAYLOR H. P., DUKE M. B., SILVER L. T. and EPSTEIN S. (1965) Oxygen isotope studies of minerals in stony meteorites. *Geochim. Cosmochim. Acta* **29**, 489–512.

TAYLOR H. P. and EPSTEIN S. (1969) Correlations between O^{18}/O^{16} ratios and chemical compositions of tektites. *J. Geophys. Res.* **74**, 6834–6844.

WALTER L. S. and CLAYTON R. N. (1967) Oxygen isotopes: Experimental vapor fractionation and variations in tektites. *Science* **156**, 1357–1358.

WARE N. G. and LOVERING J. F. (1970) Electron-microprobe analyses of phases in lunar samples. *Science* **167**, 517–520.

WEILL D. F., MCCALLUM I. S., BOTTINGA Y., DRAKE M. J. and MCKAY G. A. (1970) Petrology of a fine-grained igneous rock from the Sea of Tranquillity. *Science* **167**, 635–638.

WOOD J. A., DICKEY J. S., MARVIN U. B. and POWELL B. N. (1970) Lunar anorthosites. *Science* **167**, 602–604.

Proceedings of the Apollo 11 Lunar Science Conference, Vol. 2, pp. 1627 to 1635.

Preliminary chemical analyses of Apollo 11 lunar samples

S. R. TAYLOR
Lunar Sample Preliminary Examination Team, NASA Manned Spacecraft Center, Houston Texas 77058; Lunar Science Institute, Houston Texas 77058; and Australian National University, Canberra, Australia

and

P. H. JOHNSON, R. MARTIN, D. BENNETT, J. ALLEN and W. NANCE
Brown and Root-Northrop, Lunar Receiving Laboratory, NASA Manned Spacecraft Center, Houston, Texas 77058

(*Received* 2 *February* 1970; *accepted* 4 *February* 1970)

Abstract—A description is given of the analytical methods employed in the preliminary examination of the Apollo 11 lunar samples. The reasons for the selection of the techniques are discussed. Values used for calibration standards are listed, and a table of spectral lines examined for those elements not detected is given. The analytical data are briefly discussed and compared with the Apollo 12 data.

INTRODUCTION

THE preliminary examination of the lunar rocks (LSPET, 1969) necessitated a rapid survey of their chemical composition and emission spectroscopy was selected early in the design stages of the Lunar Receiving Laboratory since it fulfilled the requirements of

(1) rapid analysis
(2) wide element coverage
(3) sensitivity down to ppm levels
(4) milligram size sample.

Various guidelines were adopted for developing methods to deal with an essentially unknown sample, in the knowledge that there would be no time to modify techniques following sample return. The Surveyor V analysis, 25 km from the Apollo 11 landing site, indicated a basaltic-like material. Shortly before the Apollo 11 sample return TURKEVICH *et al.* (1969) reported a high Ti content. Other suggestions (e.g. DUKE and SILVER, 1967) that basaltic achondrites or chondrites might be present led to the collection of several of these meteorites for comparison. They subsequently proved of great value for the alkali elements since K and Rb contents were close to chondritic values, and the Na concentration was similar to that of the eucrites. Basalts from Hawaii, Galapagos, and the East Pacific Rise were also acquired on the assumption that the lunar material might be analogous. The USGS standard rocks, covering the range in terrestrial igneous rocks from ultrabasic to granitic were available, and a late and fortunate addition to the bank of comparative material was the Canadian Syenite standard, high in Zr, Y, and rare earth elements.

The problem of carrying out all procedures behind the biological barrier led to modifications in the standard techniques for sample preparation.

Based on the need for wide element coverage and geochemical considerations of elements expected to be present, and on the limitations of various spectrographic techniques, the following analytical scheme was set up.

(1) A method to determine the major elements which were known from the Surveyor V analysis to include Si, Al, Fe, Mg, Ca (and Ti);
(2) A sensitive method to determine the involatile trace elements (including Sr) to ppm levels; to search for all possible detectable elements;
(3) Methods to determine Rb to ppm levels, to establish Rb/Sr ratios;
(4) Methods to detect Pb at ppm levels.

The procedures adopted to solve these problems are given in the following pages. Generally they follow the principles of spectrographic analysis given by TAYLOR and AHRENS (1960) and AHRENS and TAYLOR (1961) based on the behaviour of elements of differing volatility during d.c. arc excitation of the samples.

The following conditions were constant for all methods:

(1) Jarrell-Ash Mark IV Ebert Spectrograph; 1500 line per inch grating (1st order) and constant current D.C. arc;
(2) 80% Argon–20% oxygen atmosphere;
(3) 8 step sector ratio 1/1·585;
(4) 4 mm arc gap;
(5) Model 23·100 Jarrell-Ash densitometer with drive rate of 1 mm/min, slit height of 5 mm and slit width of 10 μ was used for all plate reading;
(6) All plates developed for 4 min in Kodak D19 developer and fixed for 4 min.

$SrCO_3$ METHOD

This method was originally designed by KVALHEIM (1947) as a general quantitative method. It has been used in modified form by many subsequent workers (see AHRENS and TAYLOR, 1961, Section 26-2) and was further modified in the LRL.

The technique was developed to utilize a minimum of sample (3 mg) for three exposures to give fair to good sensitivities for most trace elements and a general analysis for all minor and major elements. The use of the $\frac{1}{8}$ in. upper and small lower electrodes resulted in a constant spherical arc column which burned homogeneously resulting in good precision. An analysis of the volatilization curves showed that the internal standard and the sample elements were evolved simultaneously during the entire burn.

The conditions adopted were as follows:

Sample preparation: 1 part sample mixed with 1 part $SrCO_3$ and 4 parts carbon. Lower electrode: National SPK L, 4258. Upper electrode: Ultra carbon 1992. Preburn: 0.5 sec. Current: 8 amp D.C. Arcing time: 30 sec. burn to completion. Wavelength range: 2450–4950 Å. Slit: 20 μ. Plate: Kodak SA3.

The following lines were read:

Sr 2932 (int. std.) Si 2532, Al 2575, 2652, Fe 2929, 2941, Ca 3181, Mg 2781, 2783, Ti 2646, Mn 2801, 2939, Cu 3274, Ni 3515 (background correction) Co 3453 (background correction) Zr 3438 (background correction at Fe 3439), V 4379, Ba 4554, 4934, Sc 4246 (background correction).

To determine Na and K, a separate exposure was made, using the same conditions except that the wavelength range was 5800–8300 Å, Kodak IN plates were used and a Corning 3486 filter was used to cut out u.v. second order lines. The following lines were read:

Sr 7673 (int. std.), Na 5890, 5895, K 7664, 7699.

C/Pd METHOD (INVOLATILE METHODS)

This procedure was adapted from the method described by AHRENS and TAYLOR (1961) and used for trace element determination in a wide variety of geological samples for several years at the Australian National University. The reasons for adapting this method were to determine Sr and to have a sensitive trace element method to supplement the $SrCO_3$ method. It was also necessary to provide for an unusually wide range in concentrations because of the unknown nature of the lunar material. This led to an increase of sample size relative to carbon to obtain as much sensitivity as possible.

The conditions adopted were as follows:

Sample preparation: Sample mixed with an equal weight of National Carbon Co. Sp-2 graphite containing 50 ppm Pd added as Johnson Matthey Specpure $(NH_3)_4$ Pd $(NO_3)_2$ (Australian National University C/Pd mix No. 12). Mixing was carried out in a hand agate mortar, and electrodes were loaded by inversion packing. Lower electrode: National Carbon Co. SPK L. 4258. Upper electrode:

Table 1. Values used for international rock standards

	G-1	W-1	G-2	GSP-1	AGV-1	PCC-1	DTS-1	BCR-1	SY-1
B ppm	—	12	—	—	—	—	—	—	72
Ba ppm	1000	170	1800	1250	1200	—	—	710	300
Ca%	0·99	7·8	1·42	1·45	3·56	0·38	0·11	4·97	7·2
Co ppm	2·3	48	4·5	7	15	120	140	37	19
Cr ppm	20	115	8	13	10	2800	4100	13	52
Cu ppm	13	115	10	33	60	10	7	22	22
Fe%	1·37	7·8	1·9	3·0	4·8	5·9	6·1	9·4	5·9
Ga ppm	19	16	22	20	18	—	—	21	20
Li ppm	25	10	40	22	12·5	1·9	—	13	76
Mg%	0·25	4·0	0·47	0·58	0·92	—	—	1·98	2·45
Mn ppm	230	1300	260	340	760	900	980	1350	3100
Ni ppm	1·2	75	4	8	16	2500	2440	11	37
Rb ppm	220	22	170	255	68	—	—	50	140
Sc ppm	2·8	34	3.5	6·0	12	8	3	32	15
Sr ppm	250	185	480	240	660	—	—	335	260
Ti%	0·16	0·64	0·29	0·40	0·64	—	—	1·34	0·29
V ppm	14	245	36	53	120	31	11	370	88
Y ppm	13	26	11	27	21	—	—	45	450
Yb ppm		2·1	—	—	—	—	—	—	70
Zr ppm	210	95	310	525	230	—	—	180	3000

National Carbon Co. SPK L. 3719 or ultra carbon 1992. Current: 6 amp d.c. Arcing time: 45 sec. Wavelength range 2450–4950 Å. Plate: Kodak SA3. Slit: 20μ.

The following lines were read:

Pd 3460, Pd 3516, Pd 3421, Ti 2647, Mg 2780, Mg 2783, Mn 2801, Cr 2843, Fe 2929, Ga 2943, Cu 3274, (Ti 3274 interferes if Ti 3276·8 is visible), Y 3327 (background correction), Ni 3414, Zr 3438, (corrected for Fe interference), Co 3453 (corrected for background and interference from Cr 3453·7), Sr 3464 (background), Yb 3987 (background correction), Sc 4246, V 4379 (background correction), Ba 4554 (no interference from Cr 4554), Sr 4607, Ba 4934.

Relative intensity values were obtained using the self-calibration method (AHRENS and TAYLOR, 1961). The use of the step sector with ratio 1/1·585 enabled an intensity range of about 300 to be covered. Use of an 8 step 1/2 ratio sector in Apollo 12 enabled an intensity range of 1000 to be covered and greatly extended the flexibility of the technique in covering the extremes of concentration encountered in the lunar samples.

Intensity vs concentration ("working") curves were constructed using the values for the international rock standards quoted in Table 1. The C/Pd method, as developed, used Pd 3421 as internal standard, and working curves were constructed using this line. The first lunar samples contained 2000–4000 ppm Cr and revealed serious interference from Cr 3421. This effect is only found in terrestrial ultrabasic rocks. A search for alternative Pd internal standard lines provided the following information: Pd 3460·8, Pd 3516·9 and Pd 3242·7 are clear from interference. Other possible Pd lines suffered from interferences as follows: Pd 3609 (Ni, Cr, Ti interference), Pd 3634 (Fe, Co),

Pd 3259 (Cr, Zr), Pd 3373 (Co, Ti), Pd 3404 (Fe), Pd 3433 (Co, Fe, Cr, Ni), Pd 3441 (Cr, Fe). Correction factors based on the ratios Pd 3516/3421, 3460/3421, and 3242/3421 in low Cr samples (G-1, W-1, Sy-1) were used. For the Apollo 12 mission, all working curves were redrawn, based on Pd 3460, which was free from interference.

Alkali Method

An additional LSAPT requirement for rubidium determination led to the development of a method to determine Rb down to 0·2 ppm. This method was based on the principle of selective volatilibility enhanced by adding a volatile compound. The procedure adopted is based on techniques developed by Ahrens and Taylor (1961) and Taylor (1960), and was based on the premise that the alkali metal content of the samples would be low.

The conditions adopted are as follows:

Sample preparation: Samples mixed with an equal weight of Johnson Matthey Specpure Na_2CO_3 Lab. No. 57183. Lower electrode: Ultra gold label 1678 or National SPK L. 4260. Upper electrode: National SPK L. 3719 or Ultra 1992.

Current 8 amp d.c. Arcing time: 3 min. Wavelength range: 6100–8600. Plates: Kodak IN. Slit: 30 μ Filter: Corning No. 3486 with cutoff at 5100 Å to exclude 2nd order u.v. lines.

The following lines were used:

Na 6154 (int. std.), Li 6707, Rb 7800, Rb 7947. Cs 8521 was sought but not detected in any sample.

Volatile Elements (Cu, Pb, B, T1)

An adaptation of the Alkali method, using the same principles was used in the ultraviolet to look for the volatile elements. The presence of Na served to suppress the spectra of the refractory elements by lowering the arc temperature (Ahrens and Taylor, 1961). The conditions were the same as for the alkali method except as follows: Slit 20 μ—no filter. Wavelength range 2450–4950. Kodak SA-3 plates. This technique was used to determine copper since Ti 3274 interference on Cu 3274 was minimal. Likewise Mn 4057 and Ti 4057 interference on Pb 4057, the sensitive lead line, was minimal. Pb was not detected in any Apollo 11 samples, although it was found in some Apollo 12 samples. A detection limit of about 2 ppm was reached by this technique.

Discussion

A number of qualitative analyses were carried out on individual mineral grains, and a quantitative analysis was made of a small metal fragment separated from the fine material. It contained 88% Fe, 11% Ni and 0·4% Co and thus has the composition of an iron meteorite.

There is no sign of contamination from the rocket exhaust pintle or skirt which contained 88% Nb, 11% Ta and 1% W. These elements were not detected. Indium is a very sensitive element with a detection limit of about one part per million. It is not detectable and there is no apparent contamination from the indium vacuum seal of the rock box.

One spectra of the Biopool sample showed detectable silver lines at 3280 and 3382 Å. These lines were not detected in other spectra of the same sample, nor in other samples, and are due to contamination either from a spot electrode impurity, or from within the Class III biological cabinets. Selenium has no lines in the spectral region accessible with the spectrograph and cannot be determined.

The high Fe, Ti, Zr and Cr contents relative to terrestrial rocks did not produce undue interference from weak lines and in a majority of cases the spectrum was clear of interfering lines at the wavelengths of the strong lines of the elements being looked for.

Table 2. Elements not detected. These elements listed in parentheses in column 3 e.g. (Zr) are possible interfering elements which were not observed

Element	Spectral lines wave-lengths (Å)	Line interferences and comments	Detection limit (ppm)
Cesium	8521	not visible	1
Cs			
Beryllium	3130	not visible	3
Be	3131	Ti3131 present but clear. Be not visible	
	2348	not visible	
Lanthanum	4333	not visible	30
La			
Neodymium	4303	not visible	30
Nd			
Uranium	4050	not visible	100
	4244	Fe interference	
	4241	Zr interference	
	3859	Fe interference	
Thorium	4019	not visible	100
Th			
Hafnium	3399	not visible	50
Hf	2866	not visible	
	3072	not visible	
Niobium	4058	(Fe) not visible	50
Nb	4079	(Ti) not visible	
	4100	Fe interference	
Tantalum	3311	(Cr) not visible	100
Ta	2714	(Fe) not visible	
Molybdenum	3170	not visible	5
Mo	3798	Fe interference	
	3864	Zr CN	
Tungsten	4302	not visible	20
W		clear from Fe 4302	
	4074	not visible	
	4008	Ti interference	
Rhenium	3464	not visible	100
Re	3460	Cr interference	
Ruthenium	3436	not visible	10
Ru	3728	Fe interference	
	3498	Ti interference	
Rhodium	3434	not visible	10
Rh	3528	Ni interferes	
	3692	V, Ti interferes	
Palladium	3421	Cr interferes	
Pd	3242	not visible	10
Iridium	3220	not visible	50
Ir	3133	(Zr) not visible	
Platinum	2659	not visible	50
Pt	3064	(Zr) not visible	
Gold	2675	(Ti) not visible	10
Au	3122	not visible	
	2427	not visible	
Zinc	3345	not visible	30
Zn			
Cadmium	2288	not visible	10
Cd	3261	not visible	
	3610	Fe interferes	
Mercury	4358	not visible	100
Hg	3125	not visible	
	2536	Fe 2536	
Boron	2497	Fe interferes	5
B			
Gallium	2943	not visible	10
Ga			
Indium	4511	not visible	1
In	4101	not visible	
Thallium	3775	(Ni) not visible	1
Tl			
Germanium	2651	not visible	5
Ge	2709	not visible	
	3039	not visible	
Tin	3175	not visible	10
Sn			
Lead	4057	Ti, Mn inter-ference	2
Pb	2843	not visible	
Arsenic	2860	not visible	100
As	2780	not visible clear from Mg 2780	
Antimony	3267	not visible	20
Sb	2598	Fe interferes	
	2877	Cr interferes	
	3232	Ti interferes	
Bismuth	3067	not visible	20
Bi	2877	not visible	
Tellurium	2385	(Cr) not visible	100
Te			

The elements given in Table 2 were not detected. The strongest spectral lines have been sought. Where these have suffered interference or blending by lines of other elements, observations have been made on the strongest lines which are free from detectable interference.

A list of average values, and ranges in composition is given in Tables 3 and 4 for the four types of material encountered. Individual sample analyses are given by LSPET (1969).

The major constituents of the samples are Si, Al, Ti, Fe, Ca and Mg. The major mineral phases present are silicates and oxides, allowing the deduction that oxygen

Table 3. Element abundances

	Type A rocks (vesicular)		Type B rocks (crystalline)		Type C rocks (breccias)		Type D (Fine Material)	
	Average	Range	Average	Range	Average	Range	Average	Overall average
Rb ppm	4·8	1·5–6·5	2·7	0·8–6·0	3·1	—	2·2	3·3
Ba ppm	120	50–180	100	60–140	100	90–105	70	100
K%	0·13	0·053–0·17	0·11	0·053–0·18	0·13	0·12–0·15	0·10	0·12
Sr ppm	120	55–230	110	55–190	105	60–150	115	110
Ca%	7·5	6·4–8·6	7·6	7·1–9·3	7·4	6·4–7·9	8·4	7·6
Na%	0·37	0·30–0·44	0·41	0·36–0·48	0·30	0·15–0·37	0·39	0·37
Yb ppm	4·4	2–7	3	1·3–5	3·2	1·8–4·5	2·5	3·5
Y ppm	230	185–310	210	100–310	210	115–300	130	210
Zr ppm	1100	870–2000	870	250–2300	950	400–1500	400	930
Cr ppm	4000	2100–6500	3500	1500–4800	2950	2500–3400	2500	3500
V ppm	32	20–40	41	15–80	27	22–32	42	36
Sc ppm	94	45–110	98	36–170	62	55–68	55	88
Ti%	6·5	4·7–7·5	5·5	4·2–7·0	5·3	5·2–5·4	4·2	5·8
Ni ppm	—		—		225	215–235	250	—
Co ppm	13	3–22	9	2·5–19	12	12–13	18	11
Cu ppm	5	—	6	—	—	—	—	—
Fe%	14·2	12–16	14	12·4–15·5	13·7	12·4–14·8	12·4	13·9
Mn ppm	2560	1940–3800	2900	2100–4300	2450	1700–3300	1750	2650
Mg%	4·6	3·9–5·7	4·8	3·6–6·6	5·4	5·0–6·6	4·8	4·8
Li ppm	18	11–27	16	10–25	12·5	—	15	16
Ga ppm	5	—	6	—	—	—	—	5
Al%	5·2	4·1–6·9	6·2	4·5–7·4	6·2	5·8–7·1	6·9	5·9
Si%	19·3	16·8–21	19·0	17·8–21	19·6	18·7–20	20	19·3
No. of samples	7		8		3		2	

comprises the major anion. Sodium, Cr, Mn, K and Zr are minor constituents with concentrations ranging between a few hundred parts per million and 0·5% by weight. Most of the other constituents are present at concentrations below 200 ppm (0·02%). All the samples show a general similarity in composition but there are some interesting and significant differences in detail. The vesicular (Type A) and crystalline (Type B) rocks are quite similar. Within each group, there are significant differences in the

Table 4. Abundances expressed as weight per cent oxides

	Type A rocks (vesicular)		Type B rocks (crystalline)		Type C rocks (breccias)		Type D (Fine Material)	
	Average	Range	Average	Range	Average	Range	Average	Overall average
SiO_2	41	36–45	41	37–45	42	37–43	43	41
TiO_2	11	7·8–12·5	9	7–11·6	9	8·6–10	7·0	9·6
Al_2O_3	9·8	7·7–13	11·7	8·5–14	11·7	11–13·5	13	11·2
FeO	18·3	15·5–21	18	16–20	17·6	17·7–19	16	17·9
MgO	7 6	6·5–9·5	8	6–10	9	7·4–12	8	8
CaO	10·5	9–12	10·6	10–13	10·4	9–11	11·8	10·6
Na_2O	0·50	0·40–0·60	0·55	0·48–0·65	0·40	0·20–0·50	0·53	0·50
K_2O	0·16	0·064–0·21	0·13	0·064–0·22	0·16	0·15–0·17	0·12	0·14
MnO	0·40	0·25–0·49	0·46	0·27–0·55	0·39	0·22–0·43	0·28	0·42
Cr_2O_3	0·58	0·31–0·95	0·51	0·22–0·69	0·43	0·37–0·69	0·36	0·51
ZrO_2	0·15	0·11–0·27	0·11	0·03–0·17	0·13	0·04–0·20	0·05	0·13
NiO	—	—	—	—	—	0·02–0·04	0·03	—
Σ	100·0		100·1		101·2		100·2	100·0
No. of samples	7		8		3		2	

abundances of K and Rb by a factor of about 5. The breccias and the fine material resemble one another quite closely, but there are marked differences between them and the Type A and B rocks for some elements. Ti, Zr and Y are lower in the breccias, and Ni is much more abundant. The high concentration of nickel in the fines is attributed to a meteoritic contribution. Assuming an average value of 1·5% Ni in meteorites, the contribution is of the order of one to two per cent. The similar chemistry of the fines and breccias particularly for nickel, indicates a common origin.

There are notable differences in composition in comparison with terrestrial rocks, meteorites or estimates of cosmic abundances. The most striking feature is the high concentration of refractory elements Ti, Zr and Y. In comparison with the composition of chondritic meteorites, Fe and Mg are lower, Ca and Al are higher, and Zr, Sr, Ba, Y and Yb are enriched by one to two orders of magnitude. K and Rb are present at similar concentration levels and Ni and Co are depleted by large factors. The Fe/Ni ratio is very high in the lunar samples. The chemistry shows some analogies to that of the eucrites (basaltic achondrites). The contents of Al, Fe, Mg, Na and Cr are similar. Although the abundances of Si, K, Rb, Ba, U and Th are much higher in the Apollo 11 samples, the possibility of finding material of eucrite chemistry on the moon remains open. Rb/Sr ratios are very low, reminiscent of oceanic tholeiites. Barium is abundant, with concentration levels similar to terrestrial continental basalts, as is lithium. Chromium is particularly abundant, and reaches levels only equalled by ultrabasic rocks terrestrially. The overall chemistry of the Apollo 11 rocks is distinct from terrestrial basalts. If they are volcanic rocks derived by partial melting, then the composition of the lunar interior does not resemble that of the earth's mantle.

Preliminary chemical data has been published for the Apollo 12 samples (LSPET, 1970) and a comparison of these samples from Oceanus Procellarum with the Apollo 11 samples from Mare Tranquillitatas shows that the chemistry at the two Maria sites is clearly related. Both suites show the distinctive features of high concentrations of "refractory" elements and low contents of "volatile" elements, which most clearly distinguish lunar from other materials. This overall similarity indicates that the Apollo 11 sample composition was not unique. Taken in conjunction with the Surveyor V and VI chemical data, it is suggestive of a similar chemistry for the Maria basin fill.

Figure 1 shows a comparison between the composition of the Apollo 11 and Apollo 12 crystalline rocks. Points lying along the 45° diagonal line indicate equality in composition. Increasing distance from the diagonal line indicates increasing disparity in composition.

In detail, there are numerous and interesting differences between the Apollo 12 and 11 rocks. These include

(1) Lower concentration of Ti both for rocks and fine material of Apollo 12. The range in composition is 0·72–3·4% Ti (1·2–5·1% TiO_2) compared with the range in the Apollo 11 rocks of 4·7–7·5% Ti (7–12% TiO_2).

(2) Lower concentrations of K, Rb, Zr, Y, Li and Ba in Apollo 12.

(3) Higher concentrations of Fe, Mg, Ni, Co, V and Sc in the crystalline rocks

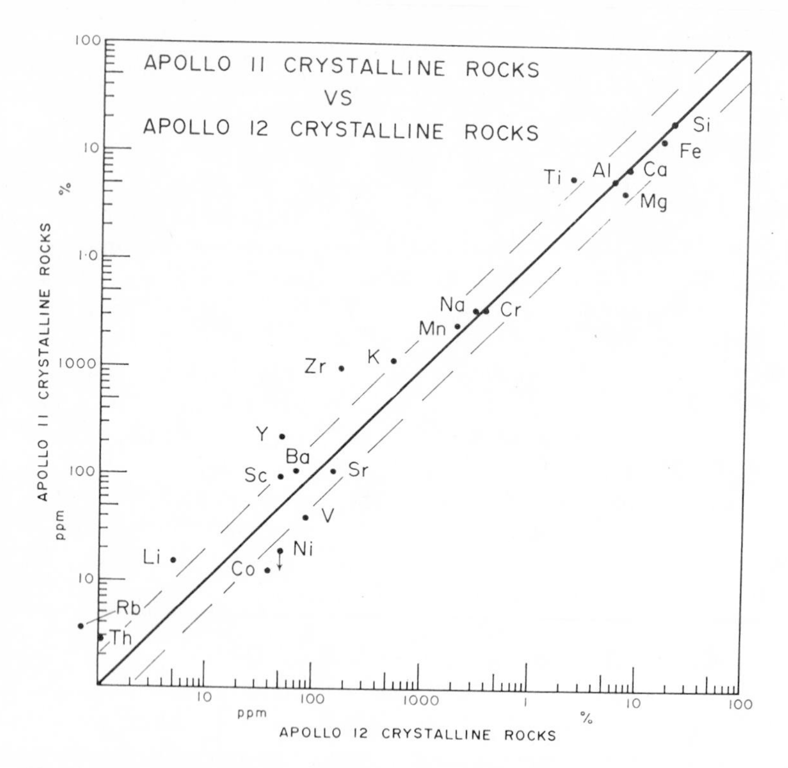

Fig. 1. Comparison of average composition of Apollo 11 and Apollo 12 crystalline rocks. Points lying along the 45° diagonal line indicate equal concentrations. Increasing distance from the diagonal line indicates increasing disparity in composition between the two sites. The dashed lines indicate the limits for differences by a factor of two. Elements lying to the right of the diagonal line are enriched in Apollo 12 rocks relative to Apollo 11. Conversely, elements plotting to the left of the line are enriched in Apollo 11 rocks. Note the higher concentrations of Ti, K, Zr, Y, Ba, Sc, Li and Th in Apollo 11 rocks, and the lower contents of Mg, V, Ni and Co.

from Apollo 12. These data are consistent with the more "basic" character of the Apollo 12 rocks.

(4) The significant variation in the Apollo 12 rocks is in the variation among the ferromagnesian elements, entering the principal mineral phases. In the Apollo 11 suite there was a much wider variation in the concentration of elements such as K, Rb concentrated in residual melts indicating a wider range of crystallization in history. However, the unique sample 12013 represents a much more extreme composition (and probably later crystal fraction) than any of the Apollo 11 rocks.

Figure 2 shows a comparison of the Apollo 11 and 12 fine material.

The fine material at the Apollo 12 sites differs from that at Apollo 11 in containing about half the titanium content, more Mg and possibly higher amounts of Ba, K, Rb, Zr and Li. Thus the chemistry of the fine material is not uniform in the different

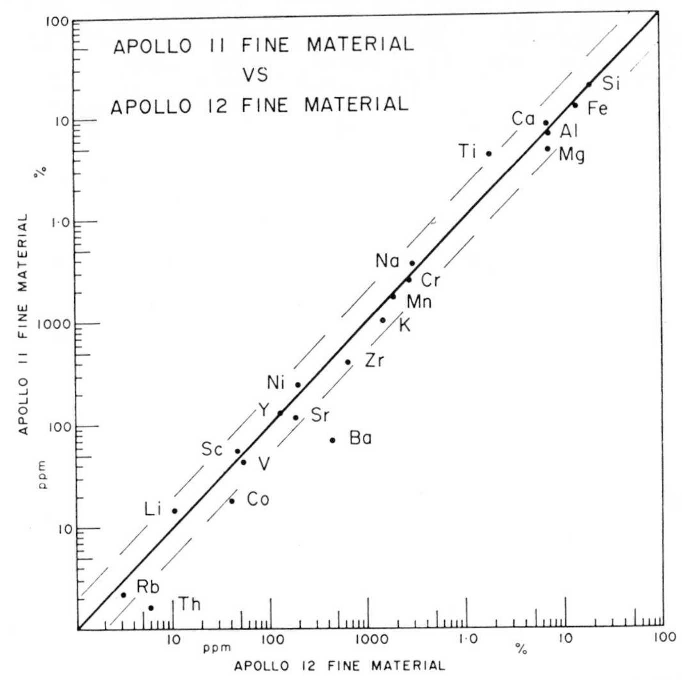

Fig. 2. Comparison of average composition of Apollo 11 and Apollo 12 fine material. Construction of diagram similar to Fig. 1. Note the higher content of Ti in Apollo 11 and the lower amounts of Mg, Ba, Co and Th.

maria. It appears to be an exceedingly complex sample influenced both by near and distant contributions.

Acknowledgements—Carleton Moore contributed the meteorite samples and A. R. McBirney, Dale Jackson and Lee Peck supplied analysed basaltic rocks. Mrs. Maureen Kaye, Australian National University, provided the carbon–palladium mix, and much assistance with standard samples and computer programs.

References

Ahrens L. H. and Taylor S. R. (1961) *Spectrochemical Analysis*. Addison-Wesley.

Duke M. B. and Silver L. T. (1967) Petrology of eucrites, howardites and mesosiderites. *Geochim. Cosmochim. Acta* **31,** 1637–1665.

Kvalheim A. (1947) Spectrochemical determination of the major constituents in minerals and rocks. *J. Opt. Soc. Amer.* **37,** 585.

LSPET (Lunar Sample Preliminary Examination Team) (1969) Preliminary examination of lunar samples from Apollo 11. *Science* **165,** 1211–1227.

LSPET (Lunar Sample Preliminary Examination Team) (1970) Preliminary examination of lunar samples from Apollo 12. *Science* **167,** 1325–1339.

Taylor S. R. (1960) Abundance and distribution of alkali elements in australites. *Geochim. Cosmochim. Acta* **20,** 85–100.

Taylor S. R. and Ahrens L. H. (1960) Spectrochemical analysis *Instrumental Methods of Analysis,* (editors A. A. Smales and L. R. Wager), Chap. 3. Interscience.

Turkevich A. L., Franzgrote E. J. and Patterson J. H (1969) Chemical composition of the lunar surface in Mare Tranquillitatis. *Science* **165,** 277–279.

Proceedings of the Apollo 11 Lunar Science Conference, Vol. 2, pp. 1637 to 1657.

Comparative study of Li, Na, K, Rb, Cs, Ca, Sr and Ba abundances in achondrites and in Apollo 11 lunar samples*

F. TERA, O. EUGSTER, D. S. BURNETT and G. J. WASSERBURG
The Charles Arms Laboratory, Division of Geological Sciences,
California Institute of Technology, Pasadena, California 91109

(*Received* 10 *March* 1970; *accepted* 10 *March* 1970)

Abstract—This study provides analytically precise abundance data for Li, Na, K, Rb, Cs, Ca, Sr and Ba in five lunar rocks, a breccia and the lunar soil. For comparison, analyses have also been made for eight eucrites, Angra dos Reis, and a terrestrial basalt.

The chemical compositions of lunar rocks fall into two distinct groups which can be clearly distinguished by their K concentrations. A parallel grouping is also found for concentrations of Li, Rb, Cs and Ba which is the same as that observed for K. These elements plus many others can be shown to reside in interstitial phases. Gross abundance features, e.g. the Ba enhancement relative to carbonaceous chondrites, reflect the composition of the lunar crust. It is not possible to precisely form the soil by a mixture of the two rock types. A component with high Rb, Cs and Ba relative to K is required. Breccias form a well-defined chemical group whose composition can be described as a mixture of soil and high K rocks. There is no evidence for differential volatilization of alkalis between the rocks and the soil.

Compared to carbonaceous chondrites, Ca, Sr, Ba and Li are strongly enhanced and Rb and Cs are systematically depleted relative to K in a similar manner in both lunar samples and achondrites. Similar abundance patterns are found in terrestrial basalts. The concentrations of K, Rb, Cs and Ba in the lunar samples are distinctly higher than the achondrites. The Na (and Ca) concentrations of achondrites and lunar samples are essentially equal.

We conclude that the composition of the moon as a whole is characterized by relative abundances for these elements that are distinct from those found in chondrites.

INTRODUCTION

THE PURPOSE of this investigation was to measure the abundances of Li, Na, K, Rb, Cs, Ca, Sr and Ba in Apollo 11 lunar samples and to compare these with the analogous abundance patterns for related terrestrial and meteoritic material. The remote chemical analyses of three different points on the lunar surface during the Surveyor missions (TURKEVICH *et al.*, 1969; FRANTZGROTE *et al.*, 1970) were consistent (except for high Ti in the Mare Tranquillitatis site) with a composition resembling either terrestrial basalts or Ca-rich achondrites. The preliminary investigation of the Apollo 11 samples (LSPET, 1969) confirmed these results. Consequently, for comparison with the lunar sample data, we have analyzed a series of achondrites and a high TiO_2 terrestrial alkali basalt.

The alkali and alkaline earth elements have traditionally assumed an important role in considerations of the chemical evolution of the earth, the meteorites, and the solar system as a whole. The distribution of these elements has been extensively studied in a wide variety of rocks and meteorites; consequently, from an observational point of view, more is known about the geochemical behavior of these elements than any others (see TAYLOR, 1966 for a recent review).

* Division of Geological Sciences Contribution No. 1717.

Despite the fact that a detailed theoretical basis for the interpretation of the abundance patterns does not exist, a comparison of the relative abundances of the alkalis and alkaline earths in lunar rocks with those for meteorites and terrestrial samples should provide clues to:

(a) the formation of the local rocks and soil at Tranquillity Base; (b) the composition of the lunar crust; (c) the composition of the moon as a whole.

Some data on the concentrations of the alkalis and the alkaline earths in achondrites were previously available, particularly from GAST (1965). In contrast to previous studies we have analyzed all elements in the same sample. This should produce more reliable relative abundances because problems due to sample heterogeneity will be minimized. The analyses reported here are also more accurate than any reported previously.

EXPERIMENTAL

Sampling

Although they are destructive, our analytical techniques permit an accurate analysis of lunar samples less than 10 mg. The following procedure was adopted in order to minimize sampling errors but, at the same time, exploit our analytical capabilities and hence avoid excessive consumption

Table 1. Alkali and alkaline earths in lunar samples*

Sample	Weight sampled (g)	Weight dissolved (g)	Li (ppm)	Na (%)	K (ppm)	Rb (ppm)	Cs (ppb)	Ca (%)	Sr (ppm)	Ba (ppm)
Soil (10084-12)	4·43	0·074	12·5	0·314	1110	2·68	101	8·52†	162·8	169
Breccia (10059-29)	1·09	0·060	13·9	0·344	1459	3·66	126	8·28	163·1	203
High K rocks										
Rock (10017-32)	0·39	0·041	18·1	0·348	2388	5·57	159	7·35	156·5	280
Rock (10069-26)	0·37	0·037	17·2	0·340	2438	5·70	163	7·20	155·6	288
Low K rocks										
Rock (10044-30)	0·10	0·051	11·8	0·334	781	0·886	34	8·34		95
Rock (10050-24)	0·76	0·083	11·0	0·273	664	0·750	29	8·17	171·3	92
Rock (10058-20)	0·79	0·046	11·4	0·327	837	1·02	37	8·58		113

* Unless specifically indicated, the maximum analytical uncertainties in the tabulated concentrations are: Li, ±1%; Na, ±1·5%; K, ±2%; Rb, ±2%; Cs, ±3%; Ca, ±0·7%; Sr, ±1·5%; Ba, ±1·5%. See text for discussion.

† Analysis performed on a separate split.

of lunar and meteorite samples. Typically, chips of 0·4–1 g were crushed to at least −300 μ and carefully split. Typically, an aliquot of 40–80 mg was taken for dissolution, although larger meteorite samples were used. Tables 1–3 give the amounts sampled and amounts dissolved for individual lunar and meteorite samples. All lunar sample handling was performed in a clean room. As judged by the lack of surface erosional features, all lunar samples analyzed were interior fragments. The samples were ultrasonically cleaned in acetone to remove any adhering surface particles, e.g. lunar dust. The soil was sieved to −420 μ before splitting; however, comparison with other workers indicates that no chemical differences were introduced by this sizing.

All meteorite samples were inspected and usually ground with a dental burr before analysis in order to remove any surface contamination.

Six dissolutions of four chondrite samples (Table 3) were also performed. Single chips of Bruderheim and Forest City were used. The Abee and Leedey samples were crushed in a manner similar to that described above except that coarse metallic particles were removed. This will make the measured concentrations for these samples slightly higher than those which would be obtained from a large

Table 2. Alkali and alkaline earth concentrations in achondrites*

Meteorite	Weight sampled (g)	Weight dissolved (g)	Li (ppm)	Na (%)	K (ppm)	Rb (ppm)	Cs (ppb)	Ca (%)	Sr (ppm)	Ba (ppm)
Nuevo Laredo	0·738	0·162	12·4	0·399	414	0·324	13·8	7·31	80·5	39.3
Nuevo Laredo (Gast, 1965)					367	0·37	18		84·4	44
Stannern	1·401	0·090[S]	12·7	0·431	657	0·696	14·2	7·40	87·7	53·0
Juvinas	4·354	0·249[S]	10·4	0·335	322	0·167	5·7	7·30	77·1	30·2
Sioux County	0·984	0·243		0·330		0·201	9·0	7·39	75·9	27·2
			8·84[i]	0·330	306	0·203	9·2	7·37	76·0	27·2
				0·332		0·207	9·3			
					307	0·204				
		0·239[S]	8·91	0·327	305	0·209		7·35	75·3	27·2
Sioux County (Gast, 1965)					335	0·24	12		68·8	25
Pasamonte Spec. 197g	0·35	0·094[S]	9·67	0·347	322	0·269	11·2	7·23	75·0	29·1
		0·071[S]		0·351	333	0·267	11·7	7·27	77·5	29·8
		0·092	9·86[i]	0·341	327	0·267	11·2		74·9	29·6
Pasamonte Spec. 297y	1·08	1·08	11·1	0·313	275	0·194	7·6	7·32	78·0	28·6
					277	0·193	7·4			28·6
						0·188	7·6			
							7·7			
Pasamonte (Gast, 1965)					425	0·23	11		82·7	38
Moore County	0·182	0·182	2·95[i]	0·275	159	0·0487 ±0·0014	0·71 ±0·02		64·3	
			2·96[i]	0·278		0·0507 ±0·0012		6·75	63·9	18·6
Moore County (Gast, 1965)					187	0·16	5		79·5	22
Bereba	1·0	0·250	9·94	0·331	258	0·178	5·77	7·15	74·7	28·6
Jonzac	5·11	0·348	9·65	0·332	329	0·405	13·9	7·16	74·6	29·0
		0·301	9·59[i]	0·334	324	0·405	14·6	7·15	74·1	29·1
			9·43[i]		327	0·404	14·6		74·2	29·3
Angra dos Reis	0·655	0·043	2·02[i]	0·0264 ±0·0005	12·9 ±1·6	0·0311 ±0·0022	0·435 ±0·065	16·48	133·0	21·5

* Multiple concentration entries for a given "weight dissolved" figure refer to replicate analyses of the same dissolved sample solution.
See footnote, Table 1, for analytical errors.
[S] Sample spiked during dissolution.
[i] Ion current integrated.

single unseparated specimen. However, the *relative* alkali and alkaline earth abundances, which are the primary concern of this study, should not be affected.

Dissolution and spiking

All samples were dissolved in a mixture of $HClO_4$ and HF, typically one ml of each. Evaporation to dryness was carried out in teflon containers under a stream of filtered nitrogen (Sanz and Wasserburg, 1969). Based on visual inspection, all samples appeared to completely dissolve.

Table 3. Alkali and alkaline earth concentrations in chondrites*

Meteorite	Weight sampled (g)	Weight dissolved (g)	Li† (ppm)	Na (%)	K (ppm)	Rb (ppm)	Cs (ppb)	Ca (%)	Sr (ppm)	Ba (ppm)
Abee	3·2	0·826[S]	2·28	0·795	867·8	3·45	237	0·92	6·9	2·25
Bruderheim	4·79	4·79	1·87		882·7	2·62	2·96	1·32	10·8	3·37
Forest City	0·92	0·92[S]	1·72	0·662	786·0	2·81	98·7‡	1·22	10·0	3·37
	0·50	0·50	1·97	0·655	787·9	2·79	107	1·20	10·2	3·30
Leedey	42·46	0·331	2·14	0·742	905	2·86	2·27	1·38	11·4	3·64
		0·306		0·750		2·90	2·5‡	1·36	11·3	3·75
			2·03[i]		874		2·6‡			3·76
Knippa basalt spec. K47	0·729	0·014	31·9[i]	2·15	9466	27·84	779	8·89	1012	645

* See footnote, Table 1, for analytical errors.
† Ion current not integrated.
‡ Poor run.
[i] Ion current integrated.
[S] Sample spiked during dissolution.

The analyses of Li, K, Rb, Cs, Ca, Sr and Ba were done by isotope dilution. For this we used the following enriched isotopes: Li^6, K^{40}–K^{41} mixture, Rb^{87}, Cs^{135}–Cs^{137} mixture, Ca^{42}–Ca^{48} mixture, Sr^{84} and Ba^{136}. During the course of the measurements each spike solution has been calibrated with at least two normal solutions prepared from high purity salts. We have observed a well-documented *decrease* in the concentration of the Ba spike solution of about 1½ per cent over a period of about one year. Although spike solutions are always highly acidic (2–3 N HCl), we may be observing precipitation or adsorption on the walls of the polyethylene bottle. The weights of all normal solution bottles and spike solutions which are not in active use have been followed for periods of two years. Evaporation of transpiration losses appear to be negligible (~0·1 per cent or less). Similarly, a 2 per cent decrease over a period of about 1½ years was observed in the concentration of the Cs spike solution; however, this shift is not as well documented as that for Ba. All other spike solution concentrations have shown adequate reproducibility. These variations are covered in our overall error estimates given below.

Most meteorites and all lunar samples were spiked simultaneously for Li, K, Rb, Cs, Sr and Ba. Five achondrite samples were also spiked simultaneously for Ca. Thus, the cross-contamination correction from each spike on the other elements had to be determined All such contributions were negligible except for the Ba and Sr corrections from the Ca spike and the Rb and Cs corrections from the K spike. The isotopic compositions of the contaminants in the spike solutions were found to be normal except for an anomalous abundance of ^{84}Sr in the Ca spike. This presumably results from the presence of $^{84}Sr^{2+}$ ions at mass 42 during the calutron separation of ^{42}Ca. The maximum Sr and Ba corrections due to the Ca spike were 2 per cent and 7 per cent, respectively, for some achondrites; however, errors introduced by these corrections should be small. Ca analyses were performed on separate aliquots of the dissolved lunar and chondrite samples. Due to a combination of high K/Rb and K/Cs and overspiking of K, the Rb and Cs cross-contamination corrections from the K spike in one of the Moore County analyses were 30 per cent and 25 per cent, respectively. The corresponding Rb and Cs corrections were less than 8 per cent for the other eucrites, less than 2 per cent for chondrites, and less than 1½ per cent for the lunar samples. In order to perform replicate analyses, most meteorite samples were not spiked during dissolution. Those that were are indicated on Table 1. However, because of the unusual bulk chemistry of the lunar samples, spikes (except for Ca) were always added during the sample dissolution in order to avoid errors due to partial dissolution, coprecipitation on the walls of vessels, etc.

Chemical separations

An aliquot of a sample solution corresponding to 50–80 mg solid was evaporated to dryness, dissolved in 2 ml 1·5 N HCl and loaded onto a 1 cm i.d. column made from SiO_2 glass and packed with Dowex 50W-X8 resin. The resin height was 17 cm in 4 N HCl. Elution was carried out with 1·5 N, 2·5 N and then 4 N HCl to affect the elution sequence shown in Fig. 1. Calibrations were carried out several times during the course of this study and were always reproducible within 5 ml.

In the above separation K, Rb and Cs are eluted together with Fe and Mg. In order to improve the Rb and Cs mass spectrometer runs, we have applied the procedure of RUCH *et al.* (1964) to remove Fe and Mg, as well as any traces of Ca and Al. The separations were carried out with an anion exchange (Dowex 1-X8) column using a mixture of 80% *p*-dioxane and 20% 2·5 N HCl (by vol.) as the eluent. The alkali fraction from the cation exchange column was evaporated to dryness, the residue dissolved in 2 ml of mixed solvent and loaded onto the anion exchange column which was preconditioned by 30 ml of mixed solvent. The alkaline earths, Fe and Al, are all held on the resin as negatively charged complexes, which presumably form because of the low dielectric constant of *p*-dioxane. Using a 23 × 1 cm resin bed (in H_2O), the alkalis were eluted in the 15–35 ml fraction. Experiments have shown that the alkali elements can be separated as a group from all other major constituents directly from a dissolved sample with this procedure. On evaporation the mixed solvent leaves behind a considerable quantity of organic residue which is not readily eliminated by HNO_3, $HClO_4$ or mixtures of both. In addition, treatment of the residue from the evaporation of *p*-dioxane, which is an ether, with $HClO_4$ constitutes an explosion hazard (P. GAST, personal experience). To avoid this problem, the alkali fraction from the anion column was passed through a small (0·5 × 10 cm) cation column in order to strip the alkalis directly from the mixed solvent without additional dilution or evaporation. No breakthrough was observed when 20 ml of mixed solvent containing Cs^{137} tracer was passed through the column. It is necessary to precondition the column with the mixed solvent prior to separation in order to avoid bubble formation. After passing the sample solution, the residual mixed solvent was washed out with 30 ml of HO_2 and the alkalis eluted with 7 ml of 4 N HCl.

After each separation the large cation column was purged by ~300 ml 4 N HCl. Fe, Mg, Ca, Al and other elements were removed from the anion column by washing it with ~50 ml 0·5 N HCl followed by ~50 ml H_2O. The H_2O washing destroys the complexes of the above-mentioned elements and causes their desorption from the resin phase.

Sometimes, when the large cation column is bypassed and the sample introduced directly into the anion column, the upper portion of the resin bed acquires a faint coloration which cannot be

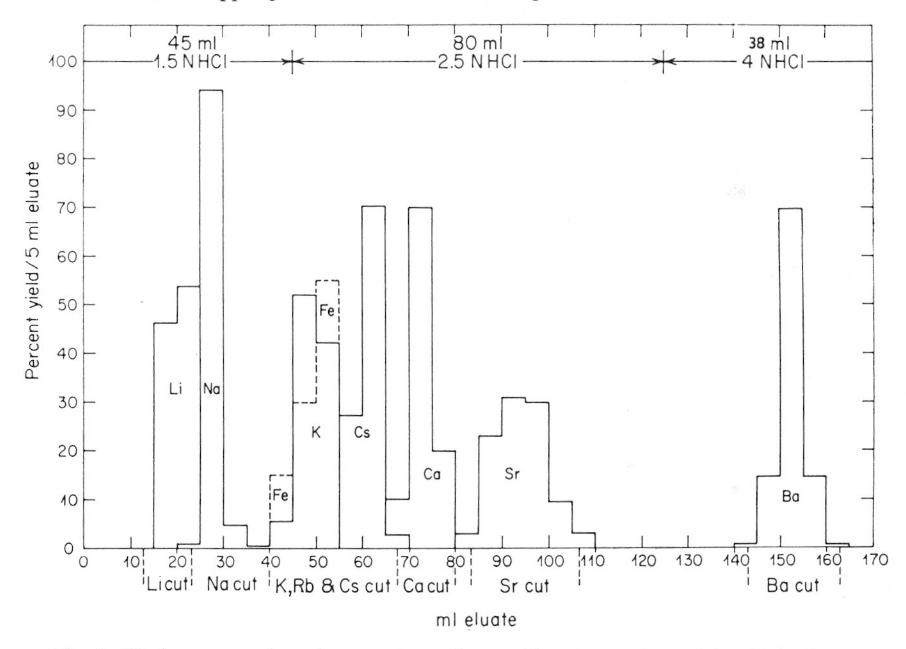

Fig. 1. Elution curves from large cation column. Fractions collected for the various elements are indicated.

eluted with H_2O or HCl. This may be due to an exceptional case of irreversible sorption of the chloro-complexes. These complexes can be destroyed by washing the column with ~50 ml of 2 N NHO_3 The resin must be converted back to the chloride form before use.

In some cases we collected separate K + Rb and Cs fractions from the large cation column. Each fraction was then purified independently by the anion and small cation columns. Cs prepared this way gave a higher and more stable signal in the mass spectrometer. It may be that the presence of excess K on the filament suppresses the Cs ionization and enhances the production of hydrocarbons in the Cs mass region.

Determination of Na

Although a few early measurements were made by flame photometry, most Na analyses were carried out by atomic abosrbtion (Perkin–Elmer Model 303). Because Na has been separated from all other major components, we have neglected any interference effects in the determination. For this determination we depend upon quantitative recovery of Na from the large cation columns. Trial elutions with Na^{22} tracer showed that the amount of Na in the adjoining column fractions (see Fig. 1) was less than 0·1 per cent and the amount of Na^{22} added to the column was recovered to better than 99 per cent. We also routinely checked for Na in the preceding column fraction (Li fraction). No amount greater than 0·2 per cent of the total Na in the sample was ever found.

Mass spectrometry

Ca, Sr, Ba and Li were measured separately on the Lunatic I digital mass spectrometer (Wasserburg *et al.*, 1969). Single Ta filaments were used for Ca, Sr and Ba; single W filaments for Li. K, Rb and Cs were usually determined simultaneously on a conventional mass spectrometer using single Ta filaments. All filaments were outgassed under controlled conditions prior to use. The outgassing procedure was determined by a series of test experiments. Filament background problems are particularly severe for K and Ba. For these elements the measured isotopic ratios during sample analyses were monitored over a wide range of intensity in order to check for contamination.

Table 4. Column blanks in units of 10^{-9} g

Column type	Li	Na	K	Rb	Cs	Ca	Sr	Ba
	Range of blank in the period between February to August 1969							
Big cation	0·2–0·6	200–300	40–102	0·07–0·14	0·001–0·002	43–102	0·5–1	0·05–0·07
Anion + small cation			49–128	0·01–0·07	0·0006–0·006			
Anion column alone			10*	0·011*	0·0006*			

* Single value.

Table 5. Reagent blanks

Reagent	Li	K	Rb	Cs	Sr	Ba
	Element concentration in ng/ml					
$HClO_4$		2	0·0025	0·0002	0·14	0·97
HF		2·25	0·002	<0·0002	0·02	0·03
6 N HCl Batch 1	0·03	7·2	0·002	0·00003	0·02	0·0084
6 N HCl Batch 2		3·1	0·001	<0·0001	0·036	0·03

During the mass spectrometric analysis of Li the Li^7/Li^6 ratio changed by up to 30 per cent over the course of the run due to mass fractionation. In order to correct for this effect we integrated the total sample ion current for all lunar and some meteorite samples using basically the procedure described by Krankowsky and Mueller (1964). Four analyses of reagent Li gave the following integrated values for Li^6/Li^7: 12·15, 12·12, 12·11 and 12·25. Because of the high yields of Li ions for surface ionization from a W filament, only about 5×10^{-11} g Li was loaded on the filament for each analysis. Polyethylene tubing was used for loading the sample. The blank introduced by the loading procedure was less than 5×10^{-15} g Li.

Analytical errors

The error limits quoted in this section are meant to be the maximum errors arising from the chemical and mass spectrometric analysis of a given sample. They do not include any estimate of the extent to which the results of our analyses deviate from the true average concentrations because of inhomogeneities. The magnitude of such sample variations in the *relative* concentrations of the elements determined will be the limiting factor in the interpretation of the differences in relative abundances among various samples.

The amounts of contamination introduced by the reagents and the ion exchange columns are shown on Tables 4 and 5. The cation exchange columns are the dominant source of contamination. A blank correction based on these data has been subtracted from the results in Tables 1–3. For lunar samples, chondrites, and most achondrites the blank levels do not represent a serious source of error, even allowing for fluctuations of ±50 per cent in the measured values for each column. For these samples, errors due to uncertainties in the blank corrections are covered in our overall error estimates given below. For Angra dos Reis and Moore County the uncertainties in the blank corrections for K, Rb and Cs were not negligible compared to our overall error estimates; consequently, specific error estimates are given for these samples which allow for a ± 50 per cent uncertainty in the blank levels. Overall errors have been assessed in two ways: (1) replicate analyses of dissolved sample solutions and (2) replicate analyses of splits of homogenized meteorite samples. Item (2) checks for effects such as incomplete dissolution which would not be revealed in analyses of the first type; however, it will also include variations due to incomplete homogenization and splitting of meteorite samples. Thus, the type (2) analyses will set comparatively firm upper limits on the error in the analysis of a given sample. Variations in the type (1) analyses represent a closer estimate of our actual analytical error, particularly for lunar samples which were spiked during dissolution. The results of type (1) analyses for Sioux County, Pasamonte (297 y), Moore County, Jonzac and Leedey are tabulated in Tables 2 and 3. The maximum per centage spreads for Li, Na, K, Rb, Cs, Ca, Sr and Ba are 2, 1, 1, 3, 4, 0·3, 0·8 and 0·7 per cent, respectively. The results of type (2) replicate analyses for Sioux County, Leedey, Pasamonte (197 g), and Jonzac show a spread which is slightly larger than those of the type (1) replicate analyses, as expected. Thus, as a maximum estimate, we have adopted $\pm$ error limits as given in Table 1 which encompass the total spread in the type (2) replicate analyses. For Ba and Cs an additional error equal to the variations in the spike calibrations has been added quadratically. The quoted Li errors apply only to those concentrations calculated using the beam integrating technique. Based on the Leedey and Pasamonte 197 g type (2) analyses, concentrations obtained without beam integration agree with those obtained by integration to within 5 per cent.

Results

Our meteorite results are in good accord with typical literature values for chondrites (see, e.g. the review by UREY, 1964) and Ca-rich achondrites (see, e.g. GAST, 1965; DUKE and SILVER, 1967). The data of GAST (1965) for four of the eucrites studied in this work are included in Table 2. There are distinct differences, far outside the quoted errors, for the two sets of values. With the exception of K for Nuevo Laredo, alkali contents by GAST are consistently higher than ours. These variations may be due to sample heterogeneity; however, this cannot be proved at the present time. Two different Pasamonte specimens (197 g and 297 y) differ in K concentrations by about 20 per cent, hence variations of at least this amount exist for samples $\gtrsim 0{\cdot}4$ g. Less understandably, the relative as well as the absolute concentrations show definite variations. The K/Rb ratio varies from 1200 to 1400 in our two Pasamonte samples, whereas GAST's value is 1850. An even larger variation exists for the K/Rb in Moore Co. Our value is 3200; GAST's 1200. This results mainly from the fact that our Rb value is about $\frac{1}{3}$ that of Gast's. Thus, care should be taken in drawing conclusions based on subtle variations in the *relative* alkali abundances of eucrites.

A more modern comparison of analytical results is possible using the analyses of lunar soil from the Apollo 11 mission. The soil samples distributed to various laboratories were not representative splits. However, based on our own experience, this is a fairly homogeneous material which should provide a good basis for comparison. Such a comparison is given in Table 6 for K, Rb, Cs, Sr and Ba based on data reported at the First Lunar Science Conference. We restrict the comparison to include only isotopic dilution measurements because of the higher inherent accuracy of this method for these elements and to those papers where at least three of the above elements were analyzed. Although the total spread in the results exceeds the quoted errors for individual analyses and probably reflects sampling problems, it appears that the average concentrations of these elements in the lunar soil are well-defined to well within

Table 6. Lunar soil analyses*

	K (ppm)	Rb (ppm)	Cs (ppb)	Sr (ppm)	Ba (ppm)
This work and ALBEE *et al.* (1970)	1110	2·68	101	163	169
GAST and HUBBARD (1970)	1200	2·79	104	171	176
SMALES *et al.* (1970)	1200	3·0	120	177	174
PHILPOTTS and SCHNETZLER (1970)	1120	2·78	—	162	170
MURTHY *et al.* (1970)	1020	2·88	—	168	200
GOPALAN *et al.* (1970)	1100	2·83	—	165	—
Maximum spread	16%	11%	20%	9%	17%

* Isotopic dilution data only.

20 per cent. This will be sufficiently accuracte for the purposes of this paper. Larger variations are reported in the concentrations of these elements in the same lunar rock analyzed by these laboratories. These are also probably due to sampling. It is unfortunate that the only modern interlaboratory comparison must be made on lunar samples rather than a well-homogenized terrestrial standard.

Despite sampling problems, there is a clear chemical grouping of the Apollo 11 lunar rocks into two groups depending on their K content. This was noted by several workers at the First Lunar Science Conference (see e.g. COMPSTON *et al.*, 1970; GAST and HUBBARD, 1970). The high K rocks (10017 and 10069 in Table 1) have K contents of about 2400 ppm; the low K rocks (10050, 10058, 10044 in Table 1) are characterized by 600–800 ppm K; the K concentrations of the soil and breccia are intermediate. This grouping shows very clearly in the rock analyses reported in the papers listed on Table 6 and in most other chemical data reported at the First Lunar Science Conference. The high K and low K rocks appear to be about equal in abundance among the samples distributed for investigation and correspond approximately to the Type A and Type B rocks of LSPET (1969). However, there are several differences, e.g. 10017 and 10020 are high and low K rocks, respectively, but are classified as B and A rocks. There is a relationship between K content and texture (in particular, grain size) with the high K rocks being fine-grained and viceversa. However, the precise definition of what constitutes a "coarse-grained rock" is somewhat ambiguous and an Apollo 11 rock classification based on K content seems preferable. Based on data from the First Lunar Science Conference, nine high K rocks were distributed for analysis: 10017, 10022, 10024, 10049, 10056, 10057, 10069, 10071, 10072. Also, nine low K rocks were distributed: 10003, 10020, 10044, 10045, 10047, 10050, 10058, 10062, 10070. Some low K rocks contain olivine; others do not.

In Figs. 2a and 2b we have plotted the concentrations of the elements investigated along with Gd in representative lunar samples and achondrites relative to carbonaceous chondrite abundances. The plot is analogous to that used to display rare earth concentration data (see e.g. HASKIN *et al.*, 1966). To the extent that carbonaceous chondrite abundances for these elements approximate primitive material, these plots display the degree to which these materials have chemically evolved from the average composition of the solar system. The type of chondritic abundances chosen for normalization is not critical except for Cs. The ordinary chondrites (see Table 3) appear to be depleted in Cs relative to enstatite or carbonaceous chondrites by factors

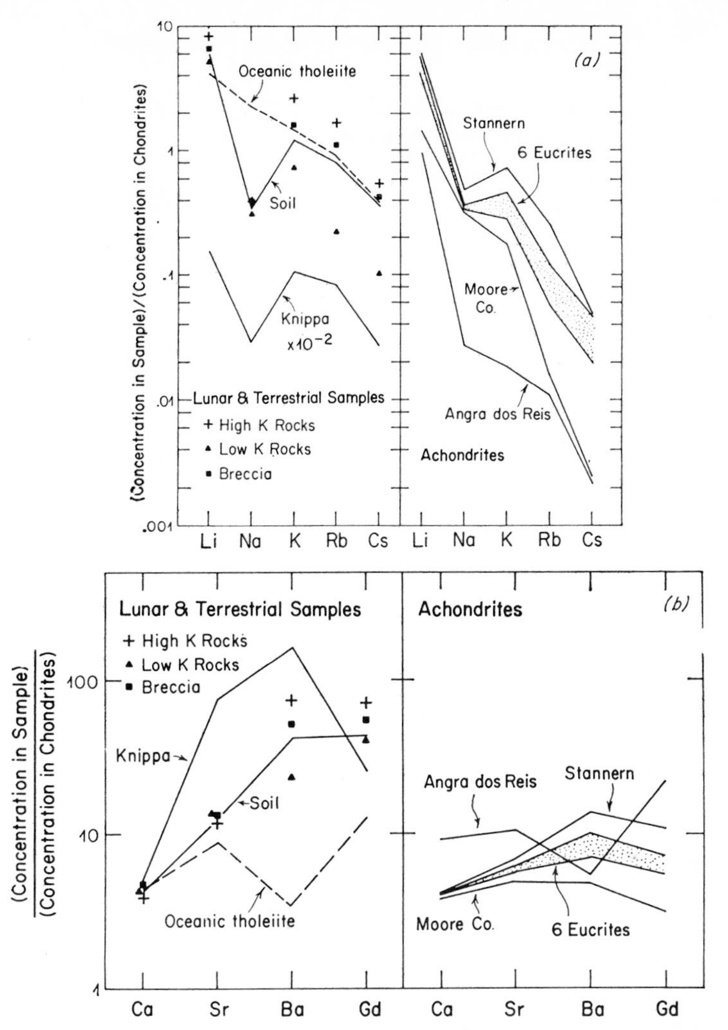

Fig. 2. Relative concentrations of alkalis (Fig. 2a) and alkaline earths (Fig. 2b) in lunar, terrestrial and achondritic samples compared to those for carbonaceous chondrites. Gd is included as a representative rare earth. The reference carbonaceous chondrite concentrations are effective values claculated from abundances for Type I carbonaceous chondrites tabulated by UREY (1967). These differ from measured values by a factor of 1·64 which approximately corrects for the volatile constituents in Type I carbonaceous chondrites. The assumed carbonaceous chondrite concentrations in ppm are Li = 2·1, Na = 9000, K = 910, Rb = 3·40, Cs = 0·29, Ca = 18,000, Sr = 13, Ba = 4·0, Gd = 0·39. Lunar Gd concentrations are the mean values of GAST and HUBBARD (1970) and SCHNETZLER and PHILPOTTS (1970). The other lunar sample concentrations are from this work (average of 10017 and 10069 for high K rocks; 10050 for low K rock). Achondritic Gd concentrations are from SCHNETZLER and PHILPOTTS (1968). Terrestrial Gd concentrations from GAST and HUBBARD (1970, and private communication). The Knippa basalt (this work) and an average oceanic tholeiite (ENGEL *et al.*, 1965; GAST, 1965) are shown as typical terrestrial samples. The stippled area contains the abundance patterns of Nuevo Laredo, Pasamonte, Jonzac, Juvinas, Bereba and Sioux County. See text for discussion of the abundance patterns.

ranging up to 100 (compare Larimer and Anders, 1967). Consequently, we have used type (I) carbonaceous chondrite concentrations calculated from abundances tabulated by Urey (1967). The adopted carbonaceous chondrite concentrations are given in the Fig. 2 caption. Any sample whose concentrations differ from carbonaceous chondritic abundances by a constant enrichment or depletion factor will plot as a flat horizontal line shifted up or down from unity by this factor. An abundance pattern with a high inclination on this plot indicates that relative elemental abundances are highly fractionated compared to carbonaceous chondrites. Abundance patterns for two samples which may be superimposed by a vertical displacement have identical relative abundances.

For comparison, abundance patterns for oceanic tholeiites and the Knippa basalt (Table 3; a high Ti, continental alkali basalt) are shown in Figs. 2a and 2b.

There is a gross similarity in the abundance patterns for terrestrial, lunar and meteoritic basaltic material in that Ca, Sr, Ba, RE and Li are all systematically enriched relative to carbonaceous chondrites by about a factor of 6–100, and Rb and Cs appear to be systematically depleted relative to K.

For clarity we show the lunar samples and the Knippa basalt on a separate diagram from the basaltic achondrites. The total range in the compositions of the basaltic achondrites studies is defined by the curves for Stannern and Moore County. The remaining six eucrites define a narrower band lying between these two curves. Angra dos Reis, an unusual, Ti-rich meteorite, is also shown for comparison.

The basic pattern for the lunar samples is one of a decrease in the relative abundances of the alkalis with increasing atomic weight. The only exception to this is sodium which is depleted relative to the general trend, producing a notch in the curve. This general pattern is well-defined and is also seen for the Knippa basalt and for all of the eucrites. In contrast to this pattern, there is no notch in the representative curve for an oceanic tholeiite which reflects its high sodium content and high Na/K ratio. For the alkaline earths the general pattern for the lunar samples is one of increasing relative enrichment with increasing atomic weight. The Knippa basalt is again similar and this pattern is distinct from that of the oceanic tholeiite which has a comparatively low Ba abundance. The general trend may also be seen for the eucrites but is not nearly as pronounced.

The above similarities in terms of abundance levels and relative abundance patterns strongly suggest that all of these samples have been produced by a qualitatively similar chemical evolution. This is compatible with the same inference drawn from their petrographic and mineralogic properties and major element chemistry. The processes which generate basalts thus seem to produce an associated characteristic pattern for the alkali and alkaline earth elements.

From these data the lunar and terrestrial basalts and eucrites would appear to be more closely related to each other than to the chondrites. In contrast to this correlation, the $^{16}O/^{18}O$ ratios show that terrestrial and lunar basalts are similar to chondrites whereas the eucrites are distinct (Epstein and Taylor, 1970; Onuma *et al.*, 1970; Taylor *et al.*, 1965). Because there are not large variations in $^{18}O/^{16}O$ for terrestrial basalts, it appears, following Taylor *et al.* (1965), that major oxygen fractionation is not involved in the generation of basalts and hence that eucrites must have formed from a different parent material, although along chemically similar paths in order to account for the similarities in the alkali, alkaline earth, and rare earth abundance patterns.

It is striking that the terrestrial Ti-rich alkali basalt has an abundance pattern very similar to that of the lunar samples. The resemblence is particularly strong for the alkalis. However, in contrast, the rare earth abundance patterns in achondrites and lunar samples are very similar to the tholeiites, whereas those for terrestrial alkali basalts are highly enriched in the lighter rare earths relative to the heavies (compare HASKIN *et al.*, 1966). Thus, there is not a consistent similarity in the trace

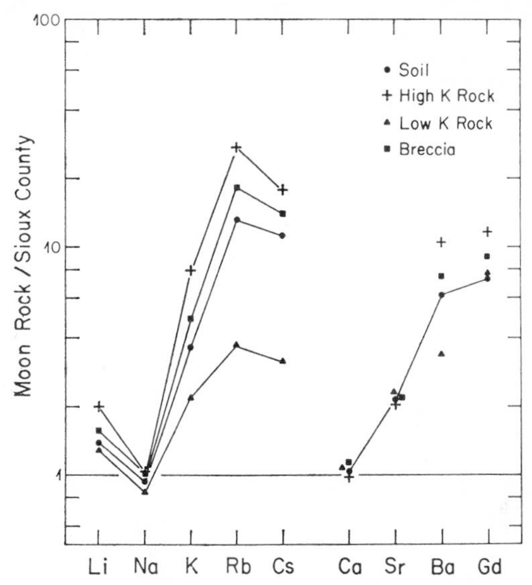

Fig. 3. Concentrations of alkalis, alkaline earths, and Gd in lunar rocks relative to Sioux County, a representative eucrite. Refer to caption for Fig. 2 for detailed information on samples. Note the systematic enrichment in the heavier alkalis and alkaline earths with respect to the lighter ones.

element abundance patterns for the lunar basalts and those for a specific type of terrestrial basalt.

The achondrites show a higher degree of depletion in Rb and Cs than either the terrestrial or lunar samples. This may be seen by the steeper slope for the achondrites in Fig. 2a. The lunar samples (and the terrestrial alkali basalt) also show greater enrichments in Sr, Ba and REE than the achondrites. These results are brought out more explicitly in Fig. 3 where we have plotted the abundances of lunar samples relative to Sioux County (a typical eucrite). The distinct difference in the relative abundances is quite evident

The variations in composition between the lunar samples themselves and between the achondrite samples themselves can be seen in Figs. 4–7. Correlation plots of Rb, Cs, Ba and Li vs. K are shown for the lunar samples and achondrites separately. All the figures are on linear scales. One point for each lunar rock is plotted. Using only the data from the references in Table 6, an average was taken for duplicate

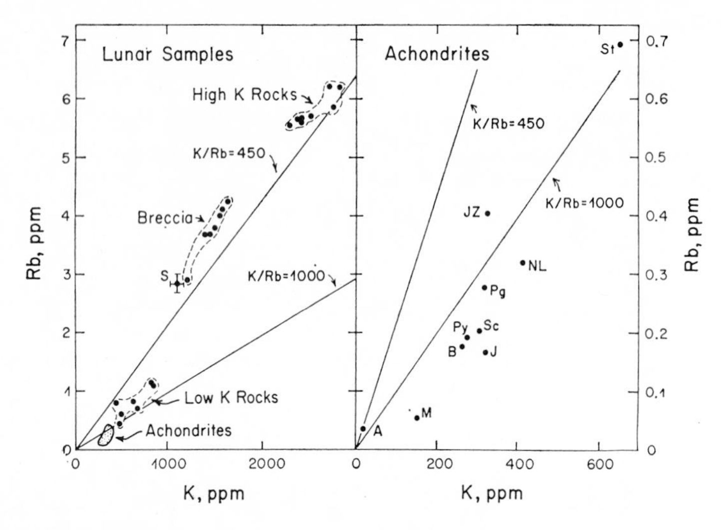

Fig. 4. Correlation plots of K vs. Rb for lunar samples and achondrites. Reference lines of constant K/Rb are shown. Note differences in scales for the lunar sample and achondrite graph. Dotted lines delineate the regions of the high K and low K lunar rocks and the breccias. S = soil analyses. Data points are average of analyses for a single lunar rock from references in Table 6. The stippled area on the lunar graph shows the region in which the achondrite analyses lie. A = Angra dos Reis; M = Moore County; B = Bereba; J = Juvinas; Py = Pasamonte, specimen 197 y; Pg = Pasamonte, specimen 197 g; SC = Sioux County; NL = Nuevo Laredo; St = Stannern. All achondrite analyses from this work.

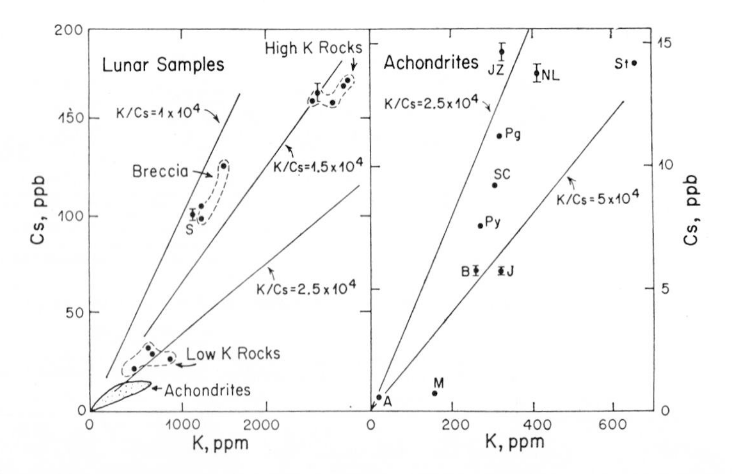

Fig. 5. Correlation plots of Cs vs. K for lunar samples and achondrites. Note the differences in scales for the lunar sample and achondrite graphs. Lines of constant K/Cs are shown for reference. See caption for Fig. 4 for detailed information.

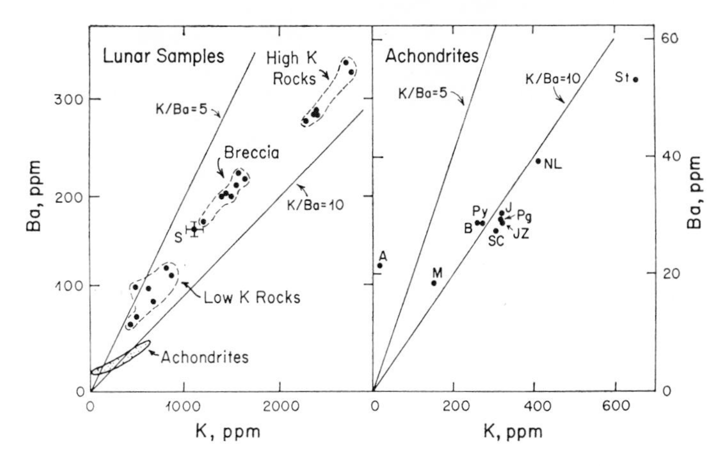

Fig. 6. Correlation plot of Ba vs. K for lunar samples and achondrites. Note the differences in scales for the lunar sample and achondrite graphs. Lines of constant K/Ba are shown for reference. See caption for Fig. 4 for detailed information.

measurements on the same rock. Samples having a fixed element ratio, e.g. K/Rb, will lie on a line passing through the origin. Examples of lines of constant K/Rb, K/Cs, K/Ba and K/Li are shown for reference on Figs. 4–7. Except for Li, the achondrite scales are considerably expanded over those for lunar samples. As indicated by the stippled regions the achondrites form a distinct compositional group lying below the low K rocks. The closest compositional affinity is with the low K rocks; but, as shown in Figs. 4–7, they are distinct.

The two groups formed by high K and low K lunar rocks are clearly distinguished on these plots. Moreover, Rb, Cs, Ba and Li show enrichments correlated with K

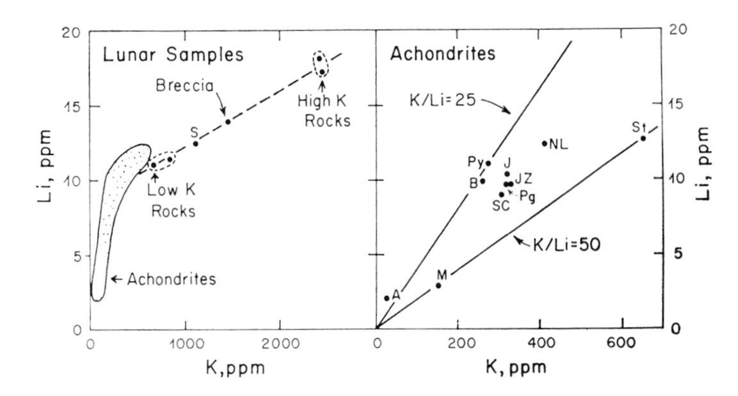

Fig. 7. Correlation plot of Li vs. K for lunar samples and achondrites. All Li analyses from this work. Lines of constant K/Li are shown on the achondrite plot. The dotted line on the lunar sample plot is a mixing line drawn between the high K and low K rocks. The soil and breccia points lie very close to this line. See caption for Fig. 4 for detailed information.

between the two types of lunar rocks. Ca, Sr or Na do not vary much in concentration between the high K and low K rocks; however, there are small systematic differences in the Na and Ca contents with Na being slightly enriched and Ca slightly depleted in the high K rocks relative to the low K rocks. This can be seen in our data and also very clearly in the data of COMPSTON *et al.* (1970).

Reasonably well-defined inter-element correlations exist for the achondrites among themselves for Rb, Li and Ba relative to K. In general Cs and K do not correlate well, although the large compositional difference between Moore County and Stannern is correlated. A better correlation exists for Rb vs. Cs, as noted previously by GAST (1960a). Except for Moore County and Stannern, the compositional range of the eucrites is comparatively small (see also Figs. 2a and 2b). To a large extent the existence of inter-element correlations among the achondrites is established solely by Moore County and Stannern. The Li–K and Ba–K correlations in eucrites have not been noted previously.

The element correlations mentioned above refer to both the trends between the lunar rock groups and within each rock group. Correlations also may be seen for the soil and the breccias taken together as a group. This will be discussed later. The compositional variations for the high and low K rocks are limited and possibly just reflect variations from the sampling of two individual rock fragments. While the spread in cosmic ray exposure ages proves the distinct nature of many of the fragments, it is nonetheless plausible to suppose that they represent samples of only two rock bodies. The possibility must also be considered that the exposure age variations reflect differences in depth and hence shielding of the samples within these rock bodies. If a detailed analysis of the relative abundances of spallation products shows this to be possible, then the Tranquillity Base rocks need only be ejecta from one or, at most, two impacts.

Petrological studies on lunar rocks have shown that K is concentrated in fine-grained or glassy interstitial areas in both the high K and low K rocks (see e.g. ROEDDER and WEIBLEN, 1970; ALBEE *et al.*, 1970; ALBEE and CHODOS, 1970). These are probably quenched residual magmatic liquids. This phase contains most of the K in the high K rocks: for rocks 10017 and 10057 (high K rocks) plagioclase obtained by magnetic separation gave 1660 and 2820 ppm, respectively. The presence of some of the interstitial phase in the separates may mean that these values are upper limits for the K contents of the plagioclase. The total rock K concentrations are 2390 (this work) and 2590 ppm (WANLESS *et al.*, 1970), respectively. It is unreasonable to believe that the ilmenite and clinopyroxene contain appreciable K. Also, some pyroxenes have been shown to have low K (74 ppm measured in rock 10044). Assuming 25% plagioclase (by weight) in these rocks, then no more than 16 per cent (rock 10017) and 27 per cent (rock 10057) of the K can be accounted for by the major minerals. Rock 10017 is required to contain ~10 per cent interstitial phase (by weight) if the interstitial phase has a K content of 2 per cent. PHILPOTTS and SCHNETZLER (1970) have demonstrated and discussed the lack of a material balance for K, Rb, Ba and RE in the major mineral phases. From the fact that Rb, Cs, Ba and Li are correlated with K and the K is to a large extent in the interstitial phases, it is almost certain that Rb, Cs, Ba and Li are also highly concentrated in this residual phase. COMPSTON *et al.* (1970) have pointed out that there are distinct differences in the concentrations of many other elements between the high K and low K rocks, viz. Ti, P, S, Zr, RE, Co, Na and Th. The data of other workers support this list (GOLES *et al.*, 1970; WÄNKE *et al.*, 1970; MORRISON *et al.*, 1970). The Ti, S and Na variations are small because these elements occur primarily in major phases. From abundance measurements on fractions of different grain size, GOLES *et al.* (1970) also inferred that the RE, as well as Hf, were concentrated in a residual phase. Hf appears to be correlated with K. The distribution of U and Th is of particular importance. Many workers (see e.g. LSPET, 1969; O'KELLEY *et al.*, 1970) have noted the constancy of K/U and Th/U in lunar samples. The correlation of U and K is demonstrated on Fig. 8 using data from the First Lunar Science Conference. As K

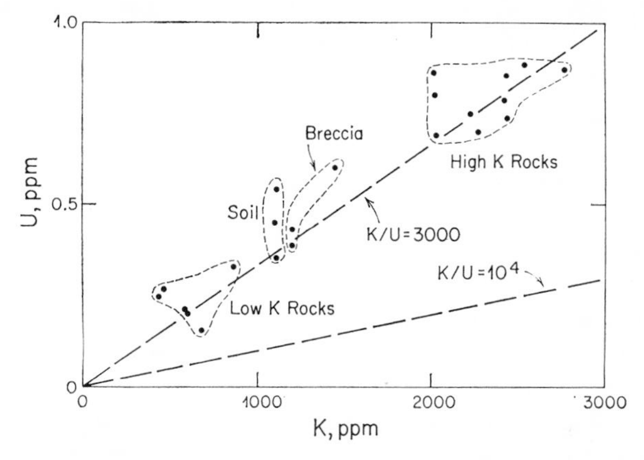

Fig. 8. Correlation plot of K vs. U. K data from sources given in Table 6; U data from SILVER (1970), TATSUMOTO and ROSHOLT (1970), LSPET (1969), and WÄNKE *et al.* (1970). A reference line of $K/U = 10^4$ (a typical terrestrial value) is shown. The line for chondritic $K/U = 8 \times 10^4$ has an even lower slope. The achondrites would fall close but slightly below the correlation line for lunar samples.

is concentrated in the interstitial phase, this implies that U, Th (and hence, radiogenic Pb) are also concentrated in the residual phase. The "interstitial phase" consists in reality of several phases including glasses and fine-grained crystals including apatite, whitlockite and baddeleyite which undoubtedly contain large amounts of U and Th.

DISCUSSION

The elemental abundance patterns reflect two distinct stages of chemical fractionation. The first stage corresponds to fractionation processes in the solar nebula which produced the bulk composition of the moon and the second corresponds to the fractionation processes on the moon itself which eventually led to the formation of the Tranquillity Base rocks and soil. The discussion will proceed on two levels. In Part I we will consider what information can be obtained on the second phase, viz. on the processes which formed lunar rocks and soil. This discussion will primarily involve the inter-element correlations shown on Figs. 4–8. In Part II we consider the "gross abundance features" which are primarily displayed on Figs. 2–3. From these it is possible to make some reasonable speculations on the composition of the moon as a whole.

I. The alkalis, alkaline earths, rare earths, Hf, Zr, U and Th have been shown to be concentrated in interstitial phases for both rock types. Thus, the differences in the concentrations of these elements between the high and low K rocks (see Figs 4–8) reflect a greater proportion of interstitial phases in the high K rocks. The interstitial phases are interpreted from petrographic observations to be quenched residual liquids. The high concentration of these elements in residual liquid is what would be expected, based on terrestrial geochemical experience.

As can be seen on Fig. 3, the Na (and Ca) contents of the various lunar samples are very similar. The constant and low Na/Ca ratios compared with the comparatively large and variable K, Rb, Cs, Li and Ba contents deserve serious consideration (see also, GAST and HUBBARD, 1970). Despite the complex melting relationships displayed by the lunar rocks and their possible source materials (O'HARA *et al.*, 1970; RINGWOOD and ESSENE, 1970), it is reasonable to believe that Na/Ca would be increased in the liquid phase of a melt which is undergoing fractional crystallization or in the liquid produced by various degrees of partial melting of the source material. In order to explain the low Na/Ca ratio either (a) the lunar interior, as a whole, is very depleted in Na or (b) the lunar rocks were crystallized from a magma produced by melting material which was a solid residue from some previous stage of partial fusion and which had been depleted in Na. However, alternative (b) alone is not compatible with the comparatively high K, Rb, Cs and, particularly, Ba contents if we assume that these elements would be concentrated in the liquid phase during magmatic processes. Alternative (b) alone is also not compatible with the wide range in K, Rb, Cs and Ba concentrations because we would expect to observe sympathetic variations in Na which are not present.

Although alternative (a) is simpler, it is not clear how approximately constant Na contents and highly variable K, Rb and Cs would be produced by magmatic processes acting on a single source material.

The alkali (including Na) and Ba abundances can be understood by assuming that the high K and low K rocks were produced by contaminating a single alkali-depleted parent magma with an alkali-rich "country rock" (except for Na). This hypothesis is compatible with either alternative (a) or (b) above. The initial $^{87}Sr/^{86}Sr$ has been shown (ALBEE *et al.*, 1970; PAPANASTASSIOU *et al.*, 1970) to be higher for the high K rocks as compared to the low K rocks. While this clearly proves that the "magmatic reservoirs" are different, it is also compatible with contamination of a single magma by older "country rock" or dust.

It should be noted that there is insufficient variety within the two major crystalline rock types to demonstrate that we are dealing with samples of a differentiation *series* rather than a single rock type. Certainly, the mechanical mobility of the high K interstitial phase when it was liquid in these two very similar rocks must be taken into consideration in ascribing a more profound character to the variation in alkali abundance.

The existence of a "missing component" in the soil is required in order to explain the $4 \cdot 6 \times 10^9$ yr model Rb–Sr and Pb–U–Th ages for the soil relative to the $3 \cdot 6 \times 10^9$

yr age from internal Rb–Sr isochrons on the rocks (ALBEE *et al.*, 1970; PAPANASTASSIOU *et al.*, 1970; TATSUMOTO and ROSHOLT, 1970; SILVER, 1970). The chemical composition of this missing component can be investigated using Figs. 4–7. If the soil can be accounted for by a mixture of the high K and low K rocks, its composition should plot on a line joining the high K rock and low K rock regions on Figs. 4–7. These figures show that the extra component in the soil does not necessarily contribute greatly to the abundances of the elements in the soil because the soil composition lies *close* to the mixing line between the high K and low K rocks. However, except for Li, the soil composition is distinctly off the mixing line, indicating the existence of a third component. The composition of the third component lies in a region above and to the left of lines joining the soil composition with that of the rocks. The missing component must have lower K/Rb, K/Cs and K/Ba than the soil with Cs $\geqslant$ 50 ppb, Rb $\geqslant$ 0·7 ppm and Ba $\geqslant$ 50 ppm.

Figures 4–8 indicate that the breccias comprise a well-defined chemical group (see also the data of ANNELL and HELZ, 1970). Inspection of the figures shows the breccias to lie very close to the line between the soil and the high K rocks. No breccia lies on the line between soil and the low K rocks indicating that no breccias studied can be considered to be a mixture dominated by the low K rocks and soil. This is suggestive that the breccias are made solely of a mixture of "soil" and high K rocks. However, petrographic study of the breccias clearly shows the presence of coarse-grained rocks with gabbroic texture and the data on Rb and Cs would permit as much as 20 per cent of the breccias to be made up of low K rock fragments. The trend shown in the individual breccia analysis for Rb and Cs (Figs. 4 and 5) suggest that the breccias contain an additional component which has higher Rb/K and Cs/K and perhaps higher Rb and Cs concentrations than the high K rocks.

The well-defined chemical composition of the breccias means that there is no evidence that the breccias returned by Apollo 11 were formed in more than one impact event.

The lunar soil is an aggregate of many types of material, much of which has been highly shocked and some vitrified. It is conceivable that extensive volatile loss of Cs with respect to Rb and Rb with respect to K would accompany the formation of the soil; however, the soil appears to have a similar K–Rb–Cs abundance pattern (Fig. 2) as the rocks. Thus, there is no evidence for differential volatilization losses of Rb and Cs. This question is of considerable importance because it would be difficult to regard the $4{\cdot}6 \times 10^9$ model ages for the soil as being significant if extensive Rb volatile loss occurred during the formation of the soil.

II. For this discussion we assume that the alkali and alkaline earth abundances of Tranquillity Base samples are typical of the lunar surface, at least for the mare regions. Similarities in the major element composition between various mare sites are supported by the Surveyor chemical analyses (TURKEVICH *et al.*, 1968; FRANTZGROTE *et al.*, 1969). We also assume that the geochemical behavior of these elements in magmatic processes on the moon is the same as on the earth and not totally different because of the dry, reducing conditions (compare GOLES *et al.*, 1970).

Evidence for chemical and physical differentiation in the moon comes from: (1) the high U and Th contents of lunar rocks and soil. Unless an almost totally molten moon can be accepted, it is necessary to concentrate U and Th to a high degree at the surface. (2) Consideration of phase equilibria (WETHERILL, 1968; RINGWOOD and ESSENE, 1970; O'HARA *et al.*, 1970) shows that the low density of the moon cannot be explained if the whole moon has a similar composition to that of the surface rocks or soil.

By analogy with the crust of the earth, the enrichment of U and Th in the outer crust should be accompanied by enrichment in the alkalis (except perhaps Na), Sr, Ba and RE. Consequently, the uniformly high concentrations of Li, Sr, Ba in all lunar samples probably reflects the composition of a lunar crust rather than the moon as a whole (compare, however, GAST and HUBBARD, 1970). Relative to Li, Ba, Sr or RE, the K, Rb, Cs abundances do not appear to be strongly enriched, whereas, following the terrestrial analogy, one would expect them to be enriched by comparable factors. Furthermore, the heavier alkalis appear to be depleted relative to K. The expected effect of the differentiation processes which formed a lunar crust would be either to preserve or to enhance the Rb/K and Cs/K ratios over that which occurred in the planet as a whole. Further, the low initial Sr^{87}/Sr^{86} values observed for Tranquillity Base rocks (ALBEE *et al.*, 1970; PAPANASTASSIOU *et al.*, 1970) indicate that these materials have never spent appreciable time in an environment with Rb/Sr significantly higher than that found in the soil. Thus, it appears that at least the source material of the Tranquillity Base samples and probably the moon as a whole is depleted in alkalis relative to the alkaline earths or rare earths and is especially depleted in Rb and Cs.

The above arguments are essentially identical to those given by GAST (1960b, 1965) for the earth and for basaltic achondrites. GAST pointed out that these materials cannot represent a fraction produced by differentiation of a chondritic source material but must represent material derived from a parent body already highly depleted in K, Rb, Cs and enriched in Ca, Sr, Ba, U and Th. From these considerations we conclude that the earth, the moon and the basaltic achondrites were derived from parent materials which were quite distinct from chondritic (or carbonaceous chondritic) abundances for the above elements.

It is evident that processes were operative in the early solar system which produced extensive and variable chemical fractionation of primordial material prior to or during the accretion of the various planetary bodies. Amongst these mechanisms, it is necessary to hypothesize a particular one which produced the chemical patterns of the earth and moon. These cannot have been a local characteristic, e.g. at 1 a.u., if basaltic achondrites do not come from the moon. These mechanisms must have been more widespread in the solar system and may possibly have been responsible for the chemistry of all of the terrestrial planets.

In postulating parent materials for the earth and moon, it would be most convincing if examples of such stuff were present among the meteorites which represent some of the oldest undisturbed objects in the solar system. The simplest possible choice would be the eucrites themselves; however, as pointed out by WETHERILL (1968) and later by RINGWOOD and ESSENE (1970) and O'HARA *et al.* (1970), a moon whose total composition is basaltic would yield an excessive density because of the high pressure phase transitions. It is conceivable that some comparatively small variation from a basaltic composition could get around this difficulty (e.g. lower normative feldspar).

From the point of view of the elements considered in this paper, a parent material with Li, Na, K, Rb, Cs, Sr, Ba and RE abundances similar to some eucrites would be a plausible composition for the total moon. This would also be satisfactory for the mantle and crust of the earth but could not explain the earth's Fe core. No obvious examples of possible parent materials are known to us at present which could serve as an appropriate source for the total moon or earth. The fission hypothesis for the origin of the moon would provide a mechanism for producing the moon and earth from one parent material by assuming that scission occurred after the formation of the earth's core.

Another feature of eucrites which is strikingly similar to the Apollo 11 basalts is their low nickel content (DUKE, 1965). This may be due to the low abundance of olivine in both these materials. If the very low abundance of Ni in lunar materials found for Apollo 11 proves to be characteristic, then some special mechanism is required. This would certainly provide a difficulty for the fission hypothesis. The formation and separation of a metallic iron phase is an exceedingly efficient mechanism for stripping Ni from silicates. If this took place in conjunction with the separation of a FeS phase, then it would be possible to strip both chalcophile elements (including Pb and Tl) and siderophile elements from the silicates. This mechanism would provide a plausible alternative to the hypothesis that the chalcophile elements were depleted by volatilization which has been inferred for the lunar basalts by some workers (see, for example Keays, *et al.* 1970).

The separation of chalcophile elements could have taken place locally on the moon itself, but the element abundance and fractionation pattern for the alkali elements must have been produced prior to or during the formation of the moon.

Acknowledgments—The authors acknowledge the careful work of H. DERKSEN, particularly for the sodium analyses. A. L. ALBEE gave freely of his time for discussion and argumentation. D. A. PAPANASTASSIOU frequently aided and advised on instrumental problems concerning the Lunatic. I. V. DEBELAK prepared this manuscript with great care and patience under the usual impossible circumstances. Meteorite samples were generously provided by W. CURVELLO (Museo Nacional, Rio de Janeiro); C. B. MOORE (Nininger Meteorite Collection); G. KURAT (Vienna Natural History Museum); P. Pellas, J. ORCEL, and F. KRAUT (Paris National History Museum); and L. T. SILVER. This work was supported by NASA Contract NAS 9-8074.

REFERENCES

ALBEE A. L., BURNETT D. S., CHODOS A. A., EUGSTER O. J., HUNEKE J. C., PAPANASTASSIOU D. A., PODOSEK F. A., RUSS G. PRICE II, SANZ H. G., TERA F. and WASSERBURG G. J. (1970) Ages, irradiation history, and chemical composition of lunar rocks from the Sea of Tranquillity. *Science* **167,** 463–466.

ALBEE A. L. and CHODOS A. A. (1970) Microprobe investigations on Apollo samples. *Geochim. Cosmochim. Acta*, Supplement I.

ANNELL C. and HELZ A. (1970) Emission spectrographic determination of trace elements in lunar samples. *Science* **167,** 521–523.

COMPSTON W., ARRIENS P. A., VERNON M. J. and CHAPPELL B. W. (1970); Rubidium–strontium chronology and chemistry of lunar material. *Science* **167,** 474–476.

DUKE M. B. (1965) Metallic iron in basaltic achondrites. *J. Geophys. Res.* **70,** 1523–1527.

DUKE M. B. and SILVER L. T. (1967) Petrology of eucrites, howardites and mesosiderites. *Geochim. Cosmochim. Acta* **31**, 1637–1665.

ENGEL A. E. J., ENGEL C. G. and HAVENS R. G. (1965) Chemical characteristics of oceanic basalts and the upper mantle. *Bull. Geol. Soc. Amer.* **76**, 719–734.

EPSTEIN S. and TAYLOR H. P. (1970) $^{18}O/^{16}O$, $^{30}Si/^{28}Si$, D/H and $^{13}C/^{12}C$ studies of lunar rocks and minerals. *Science* **167**, 533–536.

FRANTZGROTE E. J., PATTERSON J. H., TURKEVICH A. L., ECONOMOV T. E. and SOWINSKI K. P. (1970) Chemical composition of the lunar surface in Sinus Medii. *Science* **167**, 376–379.

GAST P. W. (1968) Implications of the Surveyor V chemical analysis. *Science* **159**, 896.

GAST P. W. (1965) Terrestrial ratio of K to Rb and the composition of earth's mantle. *Science* **147**, 858–860.

GAST P. W. (1960a) Alkali metals in stone meteorites. *Geochim. Cosmochim Acta* **19**, 1–4.

GAST P. W. (1960b) Limitations on the composition of the upper mantle. *J. Geophys. Res.* **65**, 1287–1297.

GAST P. W. and HUBBARD N. J. (1970) Abundance of alkali metals, alkaline and rare earths, and strontium-87/strontium-86 ratios in lunar samples. *Science* **167**, 485–487.

GOLES G. G., OSAWA M., RANDLE K., BEYER R. L., JEROME D. Y., LINDSTROM D. J., MARTIN M. R., MCKAY S. M. and STEINBORN T. L. (1970) Instrumental neutron activation analyses of lunar specimens. *Science* **167**, 497–499.

GOPALAN K., KAUSHAL S., LEE-HU C. and WETHERILL G. W. (1970) Rb–Sr, U, and Th–Pb dating of lunar material. *Science* **167**, 471–473.

HASKIN L. A., FREY F. A., SCHMITT R. A. and SMITH R. H. (1966) Meteoritic, solar, and terrestrial rare-earth distributions. *Phys. Chem. Earth* **7**, 167–321.

HASKIN L. A., HELMKE P. A. and ALLEN R. O. (1970) Rare earth elements in returned lunar samples. *Science* **167**, 487–490.

KEAYS R. R., GANAPATHY R., LAUL J. C., ANDERS E., HERZOG G. F. and JEFFERY P. M. (1970) Trace elements and radioactivity in lunar rocks: implications for meteorite in fall. *Science* **167**, 490–493.

KRANKOWSKY D. and MÜLLER O. (1964) Isotopenhäufigkeit und Konzentration des Lithiums in Steinmeteoriten. *Geochim. Cosmochim. Acta* **28**, 1625–1635.

LARIMER J. W. and ANDERS E. (1967) Chemical fractionations in meteorites II. abundance patterns and their interpretation. *Geochim. Cosmochim. Acta* **31**, 1239–70.

LSPET (LUNAR SAMPLE PRELIMINARY EXAMINATION TEAM) (1969) Preliminary examination of lunar samples from Apollo 11. *Science* **165**, 1211–1227.

MORRISON G. H., GERARD J. T., KASHUBA A. T., GANGADHARAM E. V., ROTHENBERG A. M., POTTER N. M. and MILLER G. B. (1970) Multielement analysis of lunar soil and rocks. *Science* **167**, 505–507.

MURTHY V. R., SCHMITT R. A. and REY P. (1970) Rb–Sr age and elemental and isotopic abundances of some trace elements in lunar samples. *Science* **167**, 476–479.

O'HARA M. G., BIGGAR G. M. and RICHARDSON S. W. (1970) Experimental petrology of lunar material: the nature of mascons, seas, and the lunar interior. *Science* **167**, 605–607.

O'KELLEY G. D., ELDRIDGE J. S., SCHONFELD E. and BELL P. R. (1970) Elemental compositions and ages of lunar samples by non destructive gamma-ray spectroscopy. *Science* **167**, 580–583.

ONUMA N., CLAYTON R. N. and MAYEDA T. K. (1970) Oxygen isotope fractionation between minerals and an estimate of the temperature. *Science* **167**, 536–538.

PAPANASTASSIOU D. A., BURNETT D. S. and WASSERBURG G. J. (1970) Rb–Sr ages of lunar rocks from the Sea of Tranquillity. To be published in *Earth. Planet. Sci. Lett.*

PHILPOTTS J. A. and SCHNETZLER C. C. (1970); Potassium, rubidium, strontium, barium and rare-earth concentrations in lunar rocks and separated phases. *Science* **167**, 493–495.

RINGWOOD A. E. and ESSENE E. (1970) Petrogenesis of lunar basalts and the internal constitution and origin of the moon. *Science* **167**, 607–610.

ROEDDER E. and WEIBLEN P. W. (1970) Silicate liquid immiscibility in lunar magma, evidenced by melt inclusions in lunar rocks. *Science* **167**, 641–644.

RUCH R. R., TERA F. and MORRISON G. H. (1964) Anion exchange behaviour of alkali and alkaline earth elements in dioxane–mineral acid media. *Anal. Chem.* **36**, 12–15.

SANZ H. G. and WASSERBURG G. J. (1969) Determination of an internal ^{87}Rb–^{87}Sr isochron for the Olivenza chondrite. *Earth Planet. Sci. Lett.* **6,** 335–345.

SCHNETZLER C. C. and PHILPOTTS J. A. (1968) Genesis of Ca-rich achondrites in light of rare earth and Ba concentrations. In *Meteorite Research*, (editor P. Millman), pp. 206–217. Reidel.

SILVER L. T. (1970) Uranium–thorium–lead isotope relations in lunar materials. *Science* **167,** 468–470.

SMALES A. A., MAPPER D., WEBB M. S. W., WEBSTER R. K. and WILSON J. D. (1970) Elemental composition of lunar surface material. *Science* **167,** 509–512.

TATSUMOTO M. and ROSHOLT J. N. (1970) Age of the moon: An isotopic study of uranium–thorium–lead systematics of lunar samples. *Science* **167,** 461–463.

TAYLOR H. P., DUKE M. B., SILVER L. T. and EPSTEIN S. (1965) Oxygen isotope studies of minerals in stony meteorites. *Geochim. Cosmochim. Acta* **29,** 489–512.

TAYLOR S. R. (1966) A review of trace element fractionation. *Phys. Chem. Earth* **7,** 133–213.

TURKEVICH A. L., FRANTZGROTE E. J. and PATTERSON J. H. (1969) Chemical composition of the lunar surface in Mare Tranquillitatis. *Science* **165,** 277–279.

UREY H. C. (1964) A review of atomic abundances in chondrites and the origin of meteorites. *Rev. Geophys.* **2,** 1–34.

UREY H. C. (1967) The abundance of the elements with special reference to the problem of the iron abundance. *Quart. J. Roy. Astron. Soc.* **8,** 23–47.

WÄNKE H., BEGEMANN F., VILCSEK E., RIEDER R., TESCHKE F., BORN W., QUIJANO-RICO M., VOSHAGE H. and WLOTZKA F. (1970) Major and trace elements and cosmic-ray produced radioisotopes in lunar samples. *Science* **167,** 523–527.

WANLESS R. K., LOVERIDGE W. D. and STEVENS R. D. (1970) Age determinations and isotopic abundance measurements on lunar samples. *Science* **167,** 479–481.

WASSERBURG G. J., PAPANASTASSIOU D. A., NENOW E. V. and BAUMAN C. A. (1969) A programmable magnetic field mass spectrometer with on-line data processing. *Rev. Sci. Instrum.* **40,** 288–295.

WETHERILL G. W. (1968) Lunar interior: constraint on basaltic composition. *Science* **160,** 1256.

Proceedings of the Apollo 11 Lunar Science Conference, Vol. 2, pp. 1659 to 1664.

Neutron activation analysis of milligram quantities of Apollo 11 lunar rocks and soil

KARL K. TUREKIAN and D. P. KHARKAR*
Department of Geology and Geophysics, Yale University, New Haven, Connecticut 06520

(*Received* 20 *February* 1970; *accepted in revised form* 2 *March* 1970)

Abstract—A neutron activation scheme for determining 26 elements in lunar samples weighing 20 mg is described and applied to a suite of Apollo 11 lunar materials. Concentrations of Ti, Cr, Sc, Ta, Hf and rare earths are higher than in average basalt while Co, Ni and Cu are lower. Chemical variations show groupings of elements possibly associated with the major phases, pyroxene, plagioclase and ilmenite. The high concentration of "refractory oxides" and low volatile content implies that the raw material for the Apollo 11 samples was condensed from the primitive solar nebula at high temperatures.

INTRODUCTION

THE AIM of our program of investigation of lunar material was to develop and employ a neutron activation procedure for the analysis of 20 mg of sample for 26 elements. The method is particularly useful when only small quantities of material are available for analysis. This situation may arise either from a paucity of sample or more likely in the analysis of mineral separate fractions from lunar rock and soil. The procedure was adapted from the one described by PERLMAN and ASARO (1969). This paper is based on the first report of our data (TUREKIAN and KHARKAR, 1970) with the inclusion of some additional analyses and some corrections.

PROCEDURE

The general analytical scheme starts with a short neutron irradiation immediately followed by counting with a Ge(Li) diode detector then proceeding to a longer irradiation and counting with the same system after longer waiting periods. Finally chemical separations of specific fractions are performed on the sample and counting is done by either Ge(Li) detector or NaI(Tl) detector depending on efficiency and purity requirements.

Approximately 250 mg of a lunar rock or soil sample is crushed and ground in an alundum mortar. Accurately weighed 20 mg aliquots are mixed in a small agate mortar with 80 mg of chromatographic grade cellulose and pelletized with a standard KBr-pelletizer under 4000 psi pressure. The diameter of the pellet is 1·3 cm and its thickness is less than a millimeter.

Initially our procedure involved the transfer of the pellet to an accurately weighed high purity (99·9999%) aluminum foil planchet just large enough to receive the pellet. The planchet was capped and crimped to form a thin disc. This planchet we shall designate as the Type 1 planchet. A standard silicate-matrix material (the "standard pot" made and analyzed by PERLMAN and ASARO and analyzed by us independently) was treated in identical fashion. In our initial irradiation sequence (called "irradiation 1") the samples and standard were placed radially on edge around a cylinder and irradiated. During irradiation the cylinder is rotated around the axis to guarantee a homogeneous flux for all samples and standards.

---

* On leave of absence from the Tata Institute of Fundamental Research, Bombay, India.

In irradiation 1 using Type 1 planchets the samples and the aluminum planchets were counted as a whole and subtraction for the contaminants in the aluminum made using a blank for both irradiation sequences A and B (see below). The high amounts of ^{24}Na produced in the aluminum, the presence of a large amount of copper and a variable gold contamination in the aluminum foil as well as the desirability of determining uranium by fission track counting (BERTINE *et al.*, 1970) led us to develop a new removable container for the pellets. The new planchet is constructed so that an aluminum ring clamps a sheet of lexan plastic (to detect induced uranium fission tracks) over the pellet seated in a machined slot. This planchet, hereafter designated the Type 2 planchet, was used in part A of a second irradiation sequence (irradiation 2A).

Throughout all the irradiations the cellulose binder, although blackened, maintains its cohesive role and the pellet can be transferred without loss.

The detailed program of irradiations and countings is given below. All irradiations were performed at the Union Carbide "swimming pool" reactor at Sterling Forest, New York, a two hour drive from New Haven, Connecticut and the counting was done at Yale.

(A) A short irradiation of 30 min duration in Type 2 planchets in a highly thermalized portion of the reactor at a flux of about $4{\cdot}5 \times 10^{12}$ neutrons cm^{-2} sec^{-1} was made (Irradiation 2A). Three hours after removal from the pile and after separation of the lexan, which was etched and spark counted for induced fission tracks from uranium (BERTINE *et al.* 1970), the pellet was transferred to a Type 1 planchet and counted for Mn and Dy. Because of excess interference of Ga with the Mn peak in the standard pot, the USGS standard diabase W-1 was used as the standard. Decay for 18 hr is then followed by counting for Na, K, Cu, Sm and Eu. The Ge(Li) diode detector is used for this part.

(B) A long irradiation of the samples in Type 1 planchets was next done for three days in the core of the reactor with a thermal neutron flux of 3×10^{13} neutrons cm^{-2} sec^{-1} and a large fast neutron component. Two separate irradiations were made this way. The first we shall call Irradiation 1B and the second, Irradiation 2B. In the case of 1B, the total planchet was counted for short-lived nuclides (2–4 day half-lives) 8 days after removal from the pile and recounted 22 days later to obtain results for the long-lived nuclides. In 2B counting was done after 3 days cooling and the sample was immediately processed chemically as described below. The Ge(Li) diode detector was used for this part.

(C) After the Ge(Li) detector counting sequence was over the samples were removed from the Type 1 planchet into a nickel crucible, mixed with 500 mg of a solid carrier mixture charge containing 50 mg each of oxides of the elements sought and fused with 2 g of Na_2O_2 for half an hour in a muffle furnace at 650°–700°C. [For A(1) 10 g of NaOH and 10 g Na_2O_2 were used.] The fused mass was disaggregated in boiling water and centrifuged several times to separate the water-soluble components from the insoluble residue. The water-soluble portion contained Se, As and Mo. The residue contained the remainder of the elements of interest.

In the 1B irradiation sequence the residue was dissolved in hydrochloric acid and passed through a column (15 cm long × 3 cm dia.) containing Dowex-1-resin (50–100 mesh) in chloride form. Two column volumes each of the following acid solutions were passed and collected sequentially: 12N HCl carries the rare-earths, Ag; 8N HCl carries Ti, Zr, Hf; 4N HCl carries Co. Other elements were, of course, also removed from the column with the sought elements. From the appropriate fraction the following elements are separated, purified where necessary, and counted: silver was separated and counted with the NaI detector, chemical yield determined and compared with a silver standard; the remaining solution containing Sc and the rare-earths was evaporated to dryness on a planchet and counted on the Ge(Li) detector—the Ce/Sc count ratio combined with the previously-determined Sc concentration gave the Ce concentration when compared to the standard pot without the necessity of yield determination or further purification; the cobalt fraction containing ^{60}Co and ^{58}Co from (n, p) on ^{58}Ni was purified and counted on the NaI detector and the count ratio $^{58}Co/^{60}Co$ compared to the standard pot and the previously determined cobalt concentration gave the nickel concentration. However, because of sample mix-up during fusing these results are not reported for the Apollo 11 sample (see note 8, Table 1).

The Zr- and Hf-bearing fraction was evaporated to dryness and counted on the Ge(Li) detector. The Zr/Hf count ratio could be used with the previously determined Hf concentration provided a standard with a known Hf concentration and Zr/Hf was available. In the absence of a suitable standard rock, Zr was determined directly using a standard solution pipetted on aluminum foil and irradiated with the samples. Yields were determined (commonly 80%) and corrections made.

In the 2B sequence short-lived As, Mo and Au were sought particularly and the procedure using the water-soluble portion was as follows. Selenium was separated from the acidified water soluble portion by passing a strong stream of SO_2. Selenium was later distilled in HBr–HCl–H_2SO_4 mixture, reduced with SO_2, weighed and counted on the NaI detector. The solution after Se removal is made strongly acidic with HCl, As was precipitated by ammonium phosphite, and dissolved in conc. HCl containing $KClO_3$, reduced by SO_2 and extracted as AsI_3 in benzene. Arsenic is then back extracted from the benzene layer in 2% Na_2CO_3 solution, precipitated as As_2S_3, weighed and counted on the NaI detector.

The solution after As separation is evaporated to dryness; Mo was precipitated as sulfide from dilute acidified solution, extracted in isopropyl ether from 6 N HCl, and finally precipitated with α-benzoin oxime. The precipitate is heated at 550°C for 1 hr, weighed and counted on the NaI detector. The U^{235} (n, f) Mo^{99} contribution to the Mo^{99} peak was measured to be equal to that from Mo^{98} (n, γ) Mo^{99} and was subtracted from the Mo peak with the use of the uranium values.

Silver and gold are separated from the fusion residue after disolving it in HCl. Silver was purified by ammonium hydroxide, hydrogen-sulfide precipitation in ammoniacal medium, $Fe(OH)_3$ scavenging and finally was precipitated as AgCl, weighed, and counted on the NaI detector. Gold was then separated by SO_2, solvent-extracted in ethyl acetate, and weighed and counted as elemental Au on the NaI detector. A significant gold blank was found, presumably from the cellulose. On the assumption (perhaps not completely justified) that the gold contamination was in the cellulose and distributed homogeneously, a correction was made.

W-1 diabase and the standard pot were similarly processed and used as the standards for these determinations.

Results and Discussion

The results of our analyses to date are listed in Table 1. Because of probable weighing problems in the irradiation 1B suite between samples and standards all the numbers obtained from this irradiation sequence were corrected by the use of data from irradiation 2A. All values from irradiation 1B are normalized to the europium determined carefully together with the other short-lived nuclides in irradiation 2A.

Duplicate results indicate an error of about 5 per cent–10 per cent (coefficients of variation determined from the ranges of a pooled sample) except where noted. We think that most of this error is due to heterogeneities or sample handling since the counting errors in most cases are lower.

Many of the distinctive chemical properties of the Apollo 11 materials were outlined in the published preliminary results by LSPET. We confirm the high rare-earth, Ti, Zr, and Cr contents and the low Mn, Co, and Ni contents. In addition the results of Table 1 show a deficiency of europium relative to other rare-earths when compared to chondrites. Titanium, Cr, Sc, Ta and Hf are higher in Apollo 11 site materials than in terrestrial basalts (Turekian and Wedepohl, 1961).

The elemental abundances vary among the several Apollo 11 site materials we have examined. This in part is due to the small size of our samples, which allows for heterogeneities based on the relative proportions of key minerals and in part due to real bulk differences among major groups of covarying elements. Considering only our analyses, for interval consistency, we see that one group varying together is Mn, Sc, Ta, Hf, and the rare earths; a second group appears to be Fe, Ti, Co, and Cr; and a third group is Na and U. If we conceive of the Apollo 11 material to be composites of different proportions of three dominant phases, pyroxene, ilmenite

Table 1. Chemical composition of Apollo 11 lunar rocks and soil. Values in ppm except Fe and Ti in %

		Type A "vesicular basalt"			Type B "gabbro"			Type C "breccia"		Type D "soil"
Element	Note	10044-27	10049-33	10057-79	10020-27	10062-25	10056-21	10021-33	10046-24	10084-57
Na	3	3630*	3800	4050	2680*	2990*		3470	3720*	3000*
K	3	1090*	2350	3610	1260*	230*		1720	1200*	1000*
Mn	1	2290*	1790	1930	2130*	1510*	2140*	1580	1760*	1250*
Cu	8	15*			20*	4*	15*		25*	15*
Dy	1	36*	31	37	18*	22*	34*	23	20*	17*
Eu	3	3·0*	2·1	2·5	1·5*	1·8*	(2·9)	1·8	1·9*	2·0*
Sm	3	18*	18	17	7·6*	10*	(17)	12	13*	8·7*
Sm	7	15*	15	15	6·4*	9·5*	12*	10	11*	8·1*
La	4	14			7·7	12	12		18	20
La	7	14	25	25	7·7*	13*	10*	18	19*	13*
Yb	4	9·5			6·6	6·3	8·3		8·7	9·5
Lu	4	1·3			0·81	0·87	1·2		1·2	1·3
Ce	6	86			47	48	75		78	79
Fe%	5	13			14	12	13		12	14
Cr	5	1600			2400	1310	1410		1970	2220
Co	5	13			21	13	14		30	33
Ni	8	15*			15*	15*	15*		100*	150*
Sc	5	100			98	76	109		71	80
Ti%	4	3·7			5·1	3·9	3·6		3·9	4·4
Ti%	7	4·0*	4·3	4·9	4·6*	4·2*	3·4*	3·5	3·9*	2·8*
Hf	5	13			6·6	10	11		11	11
Zr	6	660*			260*				370*	380*
Ta	5	2·7			2·2	1·8	2·6		1·8	1·8
U	2	0·41*	0·74	0·47	0·22*	0·28*	0·35*	0·39*	0·69*	0·29*
Ag	6	0·20*			0·65	0·08	0·08*		0·65	0·12
Ag	9	0·20*	0·064	0·052	0·32*	0·062*	0·13*	0·36	0·36*	0·15
Mo	9	<0·03	0·055	0·10	0·24*	0·16	<0·03	0·20	<0·03	0·35
As	9	<0·05	<0·05	<0·05	0·06*	<0·05	<0·05	0·05*	<0·05	0·1 0*
Au	9	0·0009*	0·0047	0·0064	0·0032*	0·0058*	<0·0009	0·0041	0·0048*	0·0085*
Se	9	0·23*	0·20	0·12	0·25*	0·23*	0·28*	0·17*	0·28*	0·39*

Notes

1. Irradiation 2A in highly thermalized neutron flux of $4{\cdot}5 \times 10^{12}$ neutrons cm^{-1} sec^{-1} for 30 min. Counting began 3 hr after removal from pile. Irradiation in Type 2 planchet, counting in Type 1 planchet.
2. Irradiation 2A, uranium determined by spark counting of induced fission tracks in lexan clamped on Type 2 planchet. Slightly corrected downward in some cases from preliminary report.
3. Irradiation 2A, counting began 18 hr after removal from pile. Irradiation in Type 2 planchet, counting in Type 1 planchet.
4. Irradiation 1B in core of reactor at thermal neutron flux of 3×10^{13} neutrons cm^{-2} sec^{-1} and large fast neutron component for 3 days. Counting began 8 days after removal from pile. Irradiation and counting in Type 1 planchet. Normalization to europium values from Irradiation 2A.
5. Irradiation 1B, counting began 30 days after removal from pile. Normalization to europium values from Irradiation 2A.
6. Chemical separation and counting as described in text from Irradiation 1B. Normalization to europium values from Irradiation 2A. Silver values half of those in preliminary report because of error in calculation.
7. Irradiation 2B in core of reactor at thermal flux of 3×10^{13} neutron cm^{-2} sec^{-1} for 3 days. Counting began 3 days after removal from pile. Irradiation and counting in Type 1 planchet.
8. Semi-quantitative emission spectrographic analysis using W-1 (110 ppm Cu, 78 ppm Ni) and BCR-1 (25 ppm Cu, 15 ppm Ni) as standards. Replaces values reported in preliminary report because: (1) Error in determination of sodium contribution to 0·511 MeV annihilation peak due to very low Cu content. (2) Mix-up suspected during chemical processing and confirmed by spectrographic analysis. Spectrographic values are used only in this report and have an estimated error of $\pm 50\%$.
9. Chemical separation and counting as described in text from Irradiation 2B. Estimated reproducibility error from duplicates: Ag, 10%; Mo, 5%; As, 30%; Au, 30%; Se, 20%.

* Indicates single determination. All others average of duplicate determinations.
A blank indicates that no determination was made.

and plagioclase, the trace element groupings may be crudely explained as strong associations with these phases. Separated portions must be analyzed before this point can be made explicit. Overall bulk composition variations may be expected.

The silver and gold analyses are of particular interest in that of the elements we have determined the greatest disparity exists among the investigators reporting data in the January 30, 1970 issue of *Science*, (Vol. 167, No. 3918).

Our silver determinations were made twice at two separate times, the second time including some additional samples received at a later time than the first group. In the first irradiation, 1B, silver nitrate pipetted on aluminum foil was used as the standard while the second irradiation, 2B, used W-1 diabase as the standard. The agreement between the silver concentrations on common samples for the two irradiations was good (Table 1). Our range of silver concentration is from 0·052 ppm (for 10057-a Type A rock) to 0·36 ppm (for 10021 and 10046, both Type C rocks) but with no consistent pattern. In comparing our results with those of KEAYS *et al.* (1970) our values are generally higher although we seem to fall in the same range of values determined by MORRISON *et al.* (1970). We would tend to want to believe the lower values of KEAYS *et al.* because of fears of contamination but cannot understand how our particular samples could have become contaminated with silver.

If we accept only the last three rocks received by us (10021, 10049 and 10057) as least subject to contamination because of our being particularly alerted to the possibility of such dangers, the Type C rock is 5–6 times higher in silver concentration than the two Type A rocks but are not quantitatively the same values reported by KEAYS *et al.* (1970). For example we report 52 ppb for 10057 while KEAYS *et al.* report 0·69 ppb—a factor of 70 difference.

We found variability for gold which did not show any correlation with rock type. It is possible that the gold blank in the cellulose was not homogeneously distributed thus giving our results their spread and obscuring low values but this was not obvious to us in reviewing our data.

We expect that with further work and cross-checking these difficulties will be resolved but at the present time we have no explanation for the disparity.

We have resolved our future procedure for the analysis of lunar materials through our Apollo 11 experience into the following steps, all of which have been now performed separately:

(1) A 5 mg aliquot is mixed with 5 mg of Specpure $CaCO_3$ and examined by emission spectrography. This sets the limits for many elements we are determining by neutron activation analysis. It also monitors the risks of later mixups during radiochemical separations. We determine Ni and Cu in particular this way and survey the Ag and Co lines as well as any others of interest.

(2) The planchet (Type 2) that is to be irradiated in the short irradiation is first measured by X-ray fluorescence to obtain preliminary determinations of Fe and Ti.

(3) The instrumental and radiochemical neutron activation techniques described above are then used.

Although it is premature to present a quantitative model for the chemical make-up of Apollo 11 site materials, several constraints are imposed by the data. The low cobalt and nickel concentrations relative to chromium and scandium, and the low abundance of metallic iron implies that these components were either not accumulated or were separated out by a secondary process. The high abundances of Ti, Cr, Sc, Ta, Hf and rare earths imply that the raw material for Apollo 11 material could be a condensate from the solar nebula at temperatures between the higher temperature for the condensation of metallic iron and nickel and the lower temperature for the condensation of sulfides and proton-rich materials, a general model suggested by several people (WOOD, 1962; LARIMER, 1967; ANDERS, 1968; TUREKIAN and CLARK, 1969). The phases and their variations now observed may be due to the redistribution of the elements at the time of impact or as the result of some other magma-producing process.

Acknowledgments—This research was supported by NASA grant NAS-9-8032. At various times, in various ways we have benefited from the advice and help of a group of our close associates including K. K. BERTINE, G. BRASS, L. CHAN, L. GROSSMAN and E. PERRY.

REFERENCES

ANDERS E. (1968) Chemical processes in the early solar system as inferred from meteorites. *Acc. Chem. Res.* **1,** 289–298.

BERTINE K. K., CHAN L. and TUREKIAN K. K. (1970) *Geochim. Cosmochim. Acta* in press.

KEAYS R. R., GANAPATHY R., LAUL J. C., ANDERS E., HERZOG G. F. and JEFFERY P. M. (1970) Trace elements and radioactivity in lunar rocks: Implications for meteorite infall, solar-wind flux and formation conditions of moon. *Science* **167,** 490–493.

LARIMER J. W. (1967) Chemical fractionations in meteorites—I. Condensation of the elements. *Geochim. Cosmochim. Acta* **31,** 1215–1235.

MORRISON G. H., GERARD J. T., KASHUBA A. T., GANGADHARAM E. V., ROTHENBERG A. M., POTTER N. M. and MILLER G. B. (1970) Multi-element analysis of lunar soil and rocks. *Science* **167,** 505–507.

PERLMAN I. and ASARO F. (1969) Pottery analysis by neutron activation. *Archaeometry* **11,** 21–52.

TUREKIAN K. K. and CLARK S. P. (1969) Inhomogeneous accumulation of the earth from the primitive solar nebula. *Earth Planet. Sci. Lett.* **6,** 346–348.

TUREKIAN K. K. and KHARKAR D. P. (1970) Neutron activation analysis of milligram quantities of lunar rocks and soils. *Science* **167,** 507–509.

TUREKIAN K. K. and WEDEPOHL K. H. (1961) Distribution of the elements in some major units of the Earth's crust. *Geol. Soc. Amer. Bull.* **72,** 175–192.

WOOD J. A. (1962) Chondrites and the origin of the terrestrial planets. *Nature* **194,** 127–130.

Proceedings of the Apollo 11 Lunar Science Conference, Vol. 2, pp. 1665 to 1684.

Argon-40/argon-39 dating of lunar rock samples

G. TURNER
Department of Physics, University of Sheffield

(*Received* 6 *February* 1970; *accepted in revised form* 26 *February* 1970)

Abstract—Seven crystalline rock samples (10003-43, 10017-49, 10022-46, 100024-26, 10044-40, 10062-30, 10072-45) have been analysed in detail using the ^{40}Ar–^{39}Ar dating technique. The extent of radiogenic argon loss varies from 7 per cent to $\geqslant$ 48 per cent. Potassium argon ages corrected for the effects of this loss lie between $3{\cdot}52 \times 10^9$ yr and $3{\cdot}92 \times 10^9$ yr and presumably indicate the time when much of the vulcanism associated with the formation of the Mare Tranquillitatis occurred. The rocks appear to fall into two chemical groups with distinct ages.

A major cause of the escape of gas from lunar rock is probably the impact event which ejected it from its place of origin to its place of discovery. Variations in the $^{40}Ar/^{39}Ar$ ratio indicate that some loss of radiogenic ^{40}Ar occurred relatively recently, within the last $(0{\cdot}3–1{\cdot}0) \times 10^9$ yr. Cosmic ray exposure ages have been measured and range from 70 to 440 million years.

INTRODUCTION

THE MEASUREMENT of accurate and meaningful ages of rocks and soil from carefully selected regions of the moon will be of prime importance in unravelling the sequence of events which have occurred there since the moon formed as an independent object in space. A comparison with relative ages determined on the basis of morphological features, such as cratering density and depth of the regolith layer (OBERBECK and QUAIDE, 1968), should ultimately lead to the establishment of an absolute time scale for the whole of the lunar surface. This paper is concerned with the measurement and interpretation of potassium–argon ages of crystalline rocks returned by Apollo 11.

In its most simple form the determination of a potassium–argon age involves the measurement of the total amount of potassium and the total amount of radiogenic ^{40}Ar in the sample. The assumptions are made that the rock was free of argon when formed and has quantitatively retained ^{40}Ar, from the decay of ^{40}K, since that time. The assumption of quantitative argon retention is particularly *inappropriate* for the lunar rocks. The rocks returned to earth have been picked up loose from the surface of the moon, presumably at some distance from their place of origin. It is likely that high-energy events, possibly meteorite impacts, may have transported them from their place of origin to their place of discovery and it is very probable that argon loss occurred at that time. In an attempt to estimate the extent of any gas loss and to obtain an age which is more representative of the original formation time of the rock than that based on total argon content, the ^{40}Ar–^{39}Ar method of correlated gas release (MERRIHUE and TURNER, 1966; TURNER *et al.*, 1966; TURNER, 1968, 1969, 1970a) has been applied to seven of the crystalline lunar rocks.

The ^{40}Ar–^{39}Ar method (MERRIHUE and TURNER, 1966) permits some indication to be obtained of the relative distribution of potassium and radiogenic argon within the rock. Briefly the technique consists of converting a measured fraction of ^{39}K in the rock to ^{39}Ar by neutron activation (WÄNKE and KÖNIG, 1959) and then heating the sample in stages to release this ^{39}Ar, together with radiogenic ^{40}Ar, by thermal diffusion. The argon is subsequently analysed in a mass spectrometer. In a sample which has quantitatively retained argon the two isotopes are expected to reside in fixed proportions in equivalent crystal sites, since they have both been produced from potassium. Having nearly identical diffusion coefficients the two isotopes will be released in proportion during a thermal diffusion experiment and the (constant) $^{40}Ar/^{39}Ar$ ratio may be converted directly to a meaningful potassium–argon age.

In contrast, a rock which has lost some of its radiogenic argon or alternatively has acquired argon by any process other than the *in situ* decay of potassium will invariably contain mineral sites which have different potassium to ^{40}Ar ratios. During an ^{40}Ar–^{39}Ar thermal diffusion experiment these differences in distribution of potassium and argon will be reflected by variations in the measured $^{40}Ar/^{39}Ar$ ratio.

The simplest interpretation of such a variable $^{40}Ar/^{39}Ar$ ratio is the direct implication that the age based on total potassium and total argon is unreliable. However, if the variations can be related to a reasonable model of the samples past history it may still be possible to extract a meaningful age. For example, a rock which has lost some of its argon at an intermediate stage of its history, due to heating, will contain easily outgassed crystal sites which are low in ^{40}Ar relative to ^{40}K. These sites may be minerals of high diffusion coefficient, small mineral grains, surface regions of large mineral grains or regions surrounding crystal defects. When such a sample is analysed by the ^{40}Ar–^{39}Ar method, the gas released in the early stages of the diffusion experiment is expected to come from the more easily outgassed sites and therefore be low in ^{40}Ar relative to ^{39}Ar. As the experiment proceeds the more retentive sites will release their argon and the $^{40}Ar/^{39}Ar$ ratio increase. The $^{40}Ar/^{39}Ar$ ratio of the gas released at high temperatures may be converted to an age which will be closer, though not necessarily equal, to the 'true' age of the rock, while the initial $^{40}Ar/^{39}Ar$ ratio may, in suitable circumstances, indicate the most recent time at which argon loss occurred (MERRIHUE and TURNER, 1966; TURNER, 1969). The likelihood of loose rocks on the surface of the moon having had a history of argon loss similar to that described above formed the basis for the author's proposal to apply the ^{40}Ar–^{39}Ar method to the returned lunar samples.

EXPERIMENTAL PROCEDURE

(a) *Neutron Irradiation*

The rocks investigated were 10003-43, 10017-49, 10022-46, 10024-26, 10044-40, 10062-30 and 10072-45, all of which were crystalline rocks of textural types A or B (LSPET, 1969). Approximately 100 mg of each rock was analysed (see Tables 2–8). Each sample was broken into small chips, 1–3 mm in size using a stainless steel mortar. They came into contact with the atmosphere during this operation and also briefly later, when being loaded into the gas extraction system, but during neutron irradiation they were vacuum encapsulated in quartz ampules to minimize contamination with atmospheric argon. Sample 10022–46 was from the exterior of rock number 10022 but the fragments selected for irradiation were chosen to exclude the exterior surface. All other samples appeared to be interior fragments.

The samples were irradiated for two days in the core of the Herald reactor A.W.R.E., Aldermaston and received an integrated fast neutron flux of $6{\cdot}0 \times 10^{18}$ neutrons cm^{-2}. The irradiation was monitored by the inclusion of two terrestrial hornblende samples of known age. This monitor (reference no. 341A) was chosen for its high age and low potassium content. D. C. Rex, of Leeds

University Department of Earth Sciences, who provided the monitors, has determined the radiogenic ^{40}Ar content ($1{\cdot}636 \times 10^{-4}$ cm^3 STP/g) and the potassium content (0·61%) which correspond to a K–Ar age of $2{\cdot}87 \times 10^9$ yr. ($\lambda_\beta = 4{\cdot}72 \times 10^{-10}$ yr^{-1}; $\lambda_c = 0{\cdot}584 \times 10^{-10}$ yr^{-1}; $^{40}K/K = 0{\cdot}0119$ atm %). At the time of writing this report an ^{40}Ar–^{39}Ar comparison of the monitor with several interlaboratory standards has not been completed, and an uncertainty of 4 per cent has been assigned to the $^{40}Ar/K$ ratio of the monitor for the time being.

The quartz ampules (6 mm dia.) containing samples 10003-43 to 10022-46 were placed radially around one monitor ampule (341A-SH4) and packed in a watertight cylindrical A1 container for irradiation. The remaining samples, 10024-26 to 10072-45, were packaged in a similar fashion around the second monitor ampule (341A-SH5) in a second A1 container. The two containers were irradiated side by side in a region of relatively uniform flux. The horizontal separation of the two monitors was 3 cm. Ni wires were included in each can to measure the fast neutron flux and also to provide an additional check on its uniformity.

Table 1. Measurements on neutron flux monitors

Monitor	Integrated fast neutron flux* (neutrons/cm^2)	$^{40}Ar/^{39}Ar$†
341A–SH4		$104{\cdot}8 \pm 0{\cdot}6$
		$103{\cdot}7 \pm 0{\cdot}6$
Ni–SH4	$6{\cdot}1 \times 10^{18}$	
341A–SH5		$118{\cdot}6 \pm 0{\cdot}6$
Ni–SH5	$5{\cdot}9 \times 10^{18}$	

* Trotman (1969).
† $^{40}Ar/^{39}Ar$ is *inversely* proportional to the flux.

The results of measurements on both sets of flux monitor are summarized in Table 1. The hornblende monitors indicated a flux variation of 7 per cent over a distance of 3 cm. (The Ni monitors indicate a variation of only half this amount but the figures may not be directly comparable on account of differences in excitation function for the reactions concerned.) An uncertainty of 2 per cent in the flux measurement for the individual lunar samples has been assigned on the basis of the observed flux gradient.

(b) *Argon analysis*

Following the irradiation, samples were heated in a tantalum crucible for periods from fifteen minutes to one hour, at a series of successively higher temperatures, from 400°C to 1600°C. The argon evolved at each stage was adsorbed on activated charcoal at liquid nitrogen temperatures, later released and cleaned by means of heated titanium foil and analysed using a uhv stainless steel, sector focussing mass spectrometer. The temperature of the sample furnace, which has a self-supporting tantalum mesh heating element and is enclosed in a stainless steel envelope, was not recorded directly but is estimated to within 50°C using a current vs. temperature calibration curve obtained previously through a viewing port by means of an optical pyrometer. The temperatures quoted therefore have only semiquantitative significance but this in no way affects the conclusions of the experiment.

Corrections for atmospheric argon released from the gas extraction system were applied by measuring system blanks before and after each series of extractions. In general blanks were measured at 400–500°C and at 1200–1600°C since at these temperatures gas release is small and the blank correction most critical. Between these temperatures, where the correction is less critical, interpolated values were used. Typical blanks ranged from 10^{-9} cm^3 STP at low temperatures to 10^{-8} cm^3 STP at higher temperatures. In the tables of results presented later the ^{40}Ar air correction can be seen at a glance. The 'error' figure quoted for this, the reference isotope, is in fact just 50 per cent of the blank correction. (An absolute upper limit for the radiogenic ^{40}Ar release at any temperature can thus be obtained by adding on twice the stated 'error'.)

The spectrometer was calibrated between samples by the introduction of a measured volume of atmospheric argon from a gas pipette.

RESULTS

The release patterns of all the argon isotopes from the seven samples analysed are presented in Tables 2–8. The form in which the experimental errors are presented requires some explanation. The ^{40}Ar 'error' figures have just been referred to.

Table 2. Argon release pattern (sample 10003-43)

Temperature (°C)	Time (min)	^{40}Ar‡	^{39}Ar	^{38}Ar	^{37}Ar	^{36}Ar	(^{40}Ar*/^{39}Ar)§	(^{37}Ar/^{39}Ar)	(^{40}Ar*/^{39}Ar$_K$)‖	Apparent K–Ar age¶ ($\times 10^9$yr)
		($\times 10^{-8}$ cm^{3}STP/g)								
500	75	5·94 ±2·80	0·275 ±0·007	0·021 ±0·005	0·773 ±0·011	0·037 ±0·007	21·4 ±10·0	2·81 ±0·08	21·5 ±10·0	1·0 ±0·4
610	60	94·2 ±3·5	1·156 ±0·007	0·217 ±0·006	6·81 ±0·02	0·562 ±0·005	81·0 ±3·0	5·89 ±0·04	81·4 ±3·0	2·51 ±0·05
720	60	271·3 ±3·5	1·613 ±0·011	0·894 ±0·006	27·77 ±0·09	2·736 ±0·009	166·5 ±2·4	17·22 ±0·13	168·8 ±2·4	3·61 ±0·02
810	60	405·8 ±3·4	2·041 ±0·011	1·918 ±0·014	70·23 ±0·35	5·326 ±0·020	196·2 ±2·0	34·42 ±0·25	201·7 ±2·1	3·90 ±0·02
900	60	568·3 ±3·6	2·803 ±0·017	2·469 ±0·010	138·4 ±0·5	3·464 ±0·015	201·5 ±1·7	49·36 ±0·31	200·7 ±1·8	3·96 ±0·02
990	60	495·3 ±3·6	2·443 ±0·008	1·917 ±0·012	120·0 ±0·3	1·607 ±0·011	202·0 ±1·6	49·11 ±0·22	210·2 ±1·6	3·97 ±0·02
1060	60	230·7 ±4·6	1·206 ±0·011	1·363 ±0·007	65·9 ±0·3	1·360 ±0·007	190·2 ±4·4	54·69 ±0·56	198·8 ±4·6	3·88 ±0·04
1140	30	91·4 ±4·8	0·604 ±0·007	3·589 ±0·009	192·7 ±0·4	2·574 ±0·011	147·1 ±8·0	319·2 ±3·7	196·7 ±10·9	3·86 ±0·09
1220	30	78·0 ±4·9	0·699 ±0·007	7·84 ±0·04	409·3 ±0·7	5·145 ±0·008	109·0 ±6·9	612·1 ±6·4	210·9 ±13·8	3·97 ±0·10
1290	30	13·5 ±6·9	0·120 ±0·010	1·261 ±0·010	66·1 ±0·2	0·824 ±0·007	105 ±54	548 ±44	186 ±98	3·8 ±0·9
1400	20	3·73 ±6·90	0·021 ±0·007	0·056 ±0·007	2·35 ±0·07	0·034 ±0·007	—	113 ±37	—	—
Total		2258	12·95	21·54	1100	23·67	172·6	85·0	185·0	3·76

† Sample weight 0·0903 g, flux monitor 341A, SH4. $J = 0·0341 \pm 0·0007$.
‡ The 'error' figure for ^{40}Ar is 50 per cent of the blank correction.
The 'error' figures for the other isotopes are statistical errors *relative* to ^{40}Ar, the reference isotope.
§ ^{40}Ar* = ^{40}Ar − ^{36}Ar (see text).
‖ ^{39}Ar$_K$ = ^{39}Ar − 0·00079 ^{37}Ar (see text).
¶ The 'error' figures in the apparent ages do *not* include uncertainties in the measurement of neutron flux or uncertainties inherent in the correction for trapped ^{40}Ar.

Uncertainties given for the other isotope abundances are measured statistical errors in the appropriate isotope *relative* to the ^{40}Ar abundance, that is to say they are indicative of uncertainties in *isotope ratios*. Absolute abundances on the other hand are probably accurate to little better than ±20 per cent.

The abundances of ^{37}Ar and ^{39}Ar have been corrected for decay during and after the irradiation using the expression.

$$A_0 = A(t)\left[1 + \frac{k}{2}(t_2 - t_1)\right] \exp k(t - t_2)$$

where A_0 is the corrected abundance, $A(t)$ is the measured abundance at time t, t_1 and t_2 the times of the start and end of the irradiation respectively, k is the appropriate decay constant ($k_{37} = 0·0203$ day^{-1}, $k_{39} = 7·2 \times 10^{-6}$ day^{-1}).

Before calculating ^{40}Ar–^{39}Ar ages from the figures presented it is necessary first to consider the possibility of interference from trapped ^{40}Ar and from argon isotopes produced by reactions on elements other than potassium.

Table 3. Argon release pattern (sample 10017-49†)

Temperature (°C)	Time (mins)	^{40}Ar‡	^{39}Ar	^{38}Ar	^{37}Ar	^{36}Ar	($^{40}Ar^*/^{39}Ar$)§	($^{37}Ar/^{39}Ar$)	($^{40}Ar^*/^{39}Ar)_K$‖	Apparent K–Ar age¶ (10^9 yr)
			($\times 10^{-8}$ cm^{3}STP/g)							
400	320	12·5	2·596	0·057	0·880	0·068	4·2	0·30	4·2	0·25
		±6·7	±0·016	±0·007	±0·018	±0·005	±2·3	±0·01	±2·3	±0·13
500	60	48·4	4·310	0·139	1·431	0·168	11·2	0·33	11·2	0·61
		±2·0	±0·013	±0·007	±0·009	±0·007	±0·5	±0·00	±0·5	±0·02
610	60	487·4	13·98	0·760	6·875	1·229	34·8	0·49	34·8	1·48
		±2·3	±0·09	±0·017	±0·045	±0·009	±0·3	±0·00	±0·3	±0·01
720	60	1074	14·85	2·000	18·65	4·299	72·0	1·26	72·1	2·35
		±2	±0·03	±0·012	±0·11	±0·015	±0·2	±0·01	±0·2	±0·01
790	40	236·5	3·232	0·440	4·166	0·938	72·9	1·29	73·0	2·36
		±2·0	±0·015	±0·008	±0·016	±0·007	±0·7	±0·01	±0·7	±0·01
850	40	512·1	4·963	3·059	46·17	5·123	102·2	9·30	102·9	2·85
		±2·3	±0·020	±0·020	±0·26	±0·027	±0·6	±0·07	±0·6	±0·01
900	60	547·4	4·987	3·570	69·14	3·545	109·1	13·86	110·3	2·95
		±2·7	±0·013	±0·010	±0·20	±0·009	±0·6	±0·04	±0·6	±0·01
990	60	658·3	5·717	4·386	82·53	3·562	114·5	14·43	115·8	3·02
		±4·0	±0·020	±0·015	±0·17	±0·009	±0·8	±0·06	±0·8	±0·01
1140	30	380·7	3·209	18·40	271·9	12·81	114·7	84·76	122·9	3·11
		±5·4	±0·009	±0·07	±0·7	±0·03	±1·7	±0·34	±1·8	±0·02
1280	30	245·5	2·004	22·78	329·8	14·22	115·4	164·57	132·7	3·23
		±5·4	±0·011	±0·07	±0·8	±0·07	±2·6	±0·98	±3·0	±0·04
1400	30	52·9	0·393	±4·33	62·6	2·742	128	159·5	146	3·38
		±8·7	±0·004	±0·01	±0·1	±0·013	±21	±1·7	±24	±0·26
Total		4256	60·6	59·9	894	48·7	69·4	14·75	70·2	2·31

† Sample weight 0·093 g. Flux monitor 341A, SH4. $J = 0{\cdot}0341 \pm 0{\cdot}0007$.
‡ The 'error' figure for ^{40}Ar is 50 per cent of the blank correction.
The 'error' figures for the other isotopes are statistical errors *relative* to ^{40}Ar, the reference isotope.
§ $^{40}Ar^* = {}^{40}Ar - {}^{36}Ar$ (see text).
‖ $^{39}Ar_K = {}^{39}Ar - 0{\cdot}00079\ {}^{37}Ar$ (see text).
¶ The 'error' figures in the apparent ages do *not* include uncertainties in the measurement of neutron flux or uncertainties inherent in the correction for trapped ^{40}Ar.

(a) *Trapped and cosmogenic argon*

The $^{40}Ar/^{36}Ar$ ratio of trapped argon in the Apollo 11 lunar soil is of the order of 1 whereas in the breccia it is roughly 2 (LSPET, 1969). The origin of this difference is not clear at the moment though it seems to be generally accepted that the explanation lies in the fact that the trapped ^{40}Ar is not of solar origin. Whatever the cause of the variation it renders uncertain any attempt to apply a correction for trapped ^{40}Ar. Nevertheless on the approximate assumption that the $^{40}Ar/^{36}Ar$ ratio of the trapped gas is unity the abundance of radiogenic argon, $^{40}Ar^*$, has been calculated as:

$$^{40}Ar^* = {}^{40}Ar - {}^{36}Ar \qquad (1)$$

Inspection of Tables 2–8 will reveal that this correction is in general small and the uncertainty introduced by ignoring it completely would be less than that arising from the uncertainty in the atmospheric ^{40}Ar correction in most cases.

Table 4. Argon release pattern (sample 10022-46†)

Temperature (°C)	Time (min)	^{40}Ar‡	^{39}Ar	^{38}Ar	^{37}Ar	^{36}Ar	(^{40}Ar*/^{39}Ar)§	(^{37}Ar/^{39}Ar)	(^{40}Ar*/^{39}Ar)$_K$‖	Apparent K–Ar age¶ ($\times 10^9$yr)
		($\times 10^{-8}$ cm^{3}STP/g)								
400	75	6·24	0·445	0·015	0·175	0·000	14·0	0·39	14·0	0·74
		±0·43	±0·006	±0·004	±0·006	±0·003	±1·0	±0·01	±1·0	±0·04
500	75	21·4	0·946	0·038	0·356	0·007	22·6	0·38	22·6	1·08
		±1·0	±0·005	±0·005	±0·008	±0·003	±1·1	±0·01	±1·1	±0·04
610	60	209·0	3·883	0·194	1·734	0·102	53·8	0·45	53·8	2·00
		±1·0	±0·017	±0·007	±0·013	±0·008	±0·4	±0·00	±0·4	±0·01
720	60	981·7	8·447	0·863	7·229	0·597	116·2	0·86	116·2	3·03
		±1·0	±0·040	±0·012	±0·058	±0·007	±0·6	±0·01	±0·6	±0·01
790	70	1332·8	9·36	1·938	21·59	1·405	142·3	2·31	142·6	3·34
		±1·3	±0·12	±0·029	±0·19	±0·015	±1·8	±0·04	±1·8	±0·02
850	60	629·3	3·877	1·732	23·29	1·191	162·0	6·01	162·8	3·55
		±1·5	±0·022	±0·008	±0·09	±0·005	±1·0	±0·04	±1·0	±0·01
900	60	647·7	3·807	2·485	35·93	1·657	169·7	9·44	171·0	3·63
		±1·9	±0·030	±0·009	±0·09	±0·007	±1·4	±0·08	±1·5	±0·01
1020	60	861·0	5·202	3·768	53·95	2·460	165·1	10·37	166·4	3·59
		±2·5	±0·032	±0·010	±0·20	±0·013	±1·1	±0·07	±1·1	±0·01
1140	60	824·8	5·093	11·48	159·1	7·447	160·5	31·23	164·5	3·57
		±3·0	±0·036	±0·04	±0·2	±0·015	±1·3	±0·22	±1·3	±0·01
1280	60	152·5	1·059	24·24	320·3	15·19	129·6	302·3	170·3	3·63
		±3·9	±0·005	±0·05	±1·6	±0·06	±3·4	±2·0	±4·6	±0·04
1400	60	18·58	0·125	2·420	32·19	1·513	136	256	170	3·63
		±4·50	±0·005	±0·012	±0·16	±0·006	±33	±10	±42	±0·40
Total		5685	42·2	49·2	656	31·5	133·8	15·52	135·5	3·26

† Sample weight 0·0904 g. Flux monitor 341A, SH4. $J = 0{\cdot}0341 \pm 0{\cdot}0007$.
‡ The 'error' figure for ^{40}Ar is 50 per cent of the blank correction.
The 'error' figures for other isotopes are statistical errors *relative* to ^{40}Ar, the reference isotope.
§ ^{40}Ar* = ^{40}Ar–^{36}Ar (see text).
‖ ^{39}Ar$_K$ = ^{39}Ar–0·00079^{37}Ar (see text).
¶ The 'error' figures in the apparent ages do *not* include uncertainties in the measurement of neutron flux or uncertainties inherent in the correction for trapped ^{40}Ar.

Note that in applying the correction no attempt has been made to distinguish that part of the ^{36}Ar which is cosmogenic in origin. Cosmogenic ^{36}Ar will be accompanied by cosmogenic ^{40}Ar but the production ratio for these two isotopes is not known. The error introduced into the ^{40}Ar* abundance due to cosmogenic ^{36}Ar will be proportional to the amount by which the cosmogenic (^{40}Ar/^{36}Ar) ratio differs from 1.†

(b) *Neutron interactions*

Argon isotopes are produced by a number of neutron induced reactions. The principle target nuclei involved are potassium, calcium and chlorine. The only argon isotope which might possibly be produced in significant amounts from chlorine is ^{38}Ar, which is produced by the reaction ^{37}Cl (n, γ)^{38}Cl (β^-)^{38}Ar. This reaction has no bearing on ^{40}Ar–^{39}Ar dating and the ^{38}Ar measurements will be considered separately in a later section.

Note added in proof.

† It is possible to show from correlations between the ratios (^{40}Ar/^{39}Ar), (^{38}Ar/^{39}Ar) and (^{36}Ar/^{39}Ar) that the cosmogenic (^{40}Ar/^{36}Ar) ratio is $1{\cdot}0 \pm 0{\cdot}5$.

MITCHELL (1968) has carried out argon measurements on potassium and calcium salts irradiated in the core of the Herald reactor, under conditions very similar to those of the present experiment. The isotope ratios presented by MITCHELL are affected by the presence of atmospheric argon and cannot be converted unambiguously to relative production rates except for ^{37}Ar and ^{39}Ar. Nevertheless with certain realistic assumption it is possible to estimate approximate production rates for the stable argon isotopes. On the assumption that there is negligible production

Table 5. Argon release pattern (sample 10024-26†)

Temperature (°C)	Time (min)	^{40}Ar‡	^{39}Ar	^{38}Ar	^{37}Ar	^{36}Ar	($^{40}Ar^*/^{39}Ar$)§	($^{37}Ar/^{39}Ar$)	($^{40}Ar^*/^{39}Ar)_K$‖	Apparent K–Ar age¶ ($\times 10^9$yr)
			($\times 10^{-8}$ cm^{3}STP/g)							
400	80	1·67	0·455	0·008	0·164	0·005	3·7	0·36	3·7	0·21
		±0·46	±0·005	±0·005	±0·009	±0·003	±1·0	±0·02	±1·0	±0·06
500	60	7·75	1·573	0·035	0·456	0·024	4·9	0·29	4·9	0·28
		±0·70	±0·005	±0·005	±0·005	±0·005	±0·4	±0·00	±0·4	±0·02
610	60	72·6	3·775	0·119	1·414	0·076	19·2	0·37	19·2	0·90
		±0·7	±0·007	±0·005	±0·008	±0·005	±0·2	±0·00	±0·2	±0·01
720	60	663·7	12·54	0·731	8·83	0·682	52·9	0·70	52·9	1·87
		±0 7	±0·03	±0·005	±0·04	±0·005	±0·1	±0·00	±0·1	±0·01
810	60	780·6	8·273	1·488	23·07	1·820	94·1	2·79	94·3	2·63
		±0·9	±0·033	±0·012	±0·08	±0·006	±0·4	±0·01	±0·4	±0·01
900	60	817·9	6·822	2·768	47·83	2·506	119·5	7·01	120·2	2·98
		±1·4	±0·026	±0·012	±0·10	±0·010	±0·5	±0·03	±0·5	±0·01
1020	60	772·0	5·293	3·372	54·27	2·321	145·5	10·25	146·6	3·28
		±2·0	±0·028	±0·010	±0·15	±0·010	±0·9	±0·05	±0·9	±0·01
1140	60	535·9	3·634	4·348	61·08	3·204	146·6	16·81	148·6	3·30
		±2·7	±0·022	±0·021	±0·43	±0·026	±1·2	0·16	±1·2	±0·01
1280	60	500·8	3·274	34·03	485·4	21·11	146·5	148·2	166·0	3·48
		±3·5	±0·013	±0·08	±1·4	±0·06	±1·2	±0·7	±1·5	±0·01
1400	60	9·1	0·035	0·180	2·031	0·107	255	57·4	268	4·3
		±4·6	±0·003	±0·003	±0·040	±0·005	±131	±4·6	±138	±0·8
Total		4162	45·7	47·1	685	31·85	90·4	14·99	91·5	2·58

† Sample weight 0·0968 g. Flux monitor 341A, SH5. $J = 0{\cdot}0319 \pm 0{\cdot}0007$.
‡ The 'error' figure for ^{40}Ar is 50 per cent of the blank correction.
The 'error' figures for the other isotopes are statistical errors *relative* to ^{40}Ar, the reference isotope.
§ $^{40}Ar^* = {}^{40}Ar - {}^{36}Ar$ (see text).
‖ $^{39}Ar_K - {}^{39}Ar - 0{\cdot}00079\,{}^{37}Ar$ (see text).
¶ The 'error' figures in the apparent ages do *not* include uncertainties in the measurement of neutron flux or uncertainties inherent in the correction for trapped ^{40}Ar.

of Ar^{36} from neutron reactions on potassium it is a simple matter to calculate the composition of the potassium derived argon as: $^{36}Ar/^{37}Ar/^{38}Ar/^{39}Ar/^{40}Ar = 0/10^{-5}/0{\cdot}0114/1{\cdot}00/0{\cdot}0063$. ^{39}Ar is the dominant product. In a similar way, if one assumes a negligible contribution to ^{40}Ar from neutron reactions on Ca, the composition of the calcium derived argon is calculated as: $^{36}Ar/^{37}Ar/^{38}Ar/^{39}Ar/^{40}Ar = 0{\cdot}0009/1{\cdot}00/0{\cdot}00010/0{\cdot}00304/0$. ^{37}Ar is the dominant product from calcium. ^{40}Ar will be produced to some extent from (n, α) reactions on ^{43}Ca (natural abundance 0·13 per cent). If one assumes that ^{40}Ar is produced in comparable amounts to ^{39}Ar, which is the product of (n, α) reactions on ^{42}Ca (natural abundance 0·64 per cent), the above ratios, apart from the ^{40}Ar figure itself, are virtually unchanged.

Table 6. Argon release pattern (sample 10044-40†)

Temperature (°C)	Time (min)	^{40}Ar‡	^{39}Ar	^{38}Ar	^{37}Ar	^{36}Ar	($^{40}Ar^*/^{39}Ar$)§	($^{37}Ar/^{39}Ar$)	($^{40}Ar^*/^{39}Ar)_K$‖	Apparent K–Ar age¶ ($\times 10^9$yr)
			($\times 10^{-8}$ cm^{3}STP/g)							
500	60	3·50	0·216	0·011	0·463	0·003	16·2	2·15	16·3	0·79
		±0·60	±0·003	±0·003	±0·004	±0·003	±2·8	±0·04	±2·8	±0·11
570	60	6·87	0·257	0·010	0·593	0·009	26·7	2·31	26·7	1·17
		±0·60	±0·004	±0·003	±0·021	±0·003	±2·4	±0·09	±2·4	±0·08
640	60	28·8	0·571	0·030	1·976	0·020	50·4	3·46	50·6	1·82
		±0·8	±0·003	±0·003	±0·021	±0·003	±1·4	±0·04	±1·5	±0·03
720	60	145·8	1·299	0·109	7·541	0·124	112·2	5·81	112·7	2·88
		±0·8	±0·006	±0·003	±0·026	±0·003	±0·8	±0·03	±0·8	±0·01
810	60	462·4	2·595	0·272	21·53	0·377	178·0	8·30	179·2	3·60
		±1·0	±0·015	±0·010	±0·07	±0·008	±1·1	±0·05	±1·1	±0·01
900	60	854·2	4·455	0·685	64·14	0·634	191·6	14·40	193·8	3·73
		±1·2	±0·015	±0·008	±0·14	±0·006	±0·7	±0·06	±0·7	±0·01
990	60	1060	5·430	1·082	111·2	0·756	195·2	20·48	198·4	3·77
		±1	±0·022	±0·006	±0·5	±0·010	±0·8	±0·12	±0·8	±0·01
1060	60	777·4	4·044	0·784	73·08	0·500	192·1	18·07	124·9	3·74
		±1·8	±0·039	±0·006	±0·15	±0·004	±1·9	±0·18	±2·0	±0·02
1180	60	553·0	3·022	2·315	207·4	1·475	182·5	68·64	193·0	3·72
		±2·2	±0·019	±0·006	±0·6	±0·007	±1·3	±0·47	±1·5	±0·01
1280	60	221·8	1·333	2·910	278·6	1·904	165·0	209·0	197·6	3·76
		±3·0	±0·006	±0·007	±0·7	±0·007	±2·4	±1·0	±2·9	±0·02
1400	60	64·3	0·360	0·736	74·12	0·495	177	205	212	3·87
		±3·6	±0·006	±0·006	±0·14	±0·007	±10	±3	±12	±0·10
Total		4179	23·58	8·94	841	6·30	177·0	35·7	182·1	3·62

† Sample weight 0·1113 g. Flux monitor 341A, SH5. $J = 0{\cdot}0319 \pm 0{\cdot}0007$.
‡ The 'error' figure for ^{40}Ar is 50 per cent of the blank correction.
The 'error' figures for the other isotopes are statistical errors *relative* to ^{43}Ar, the reference isotope.
§ $^{40}Ar^* = {}^{40}Ar – {}^{36}Ar$ (see text).
‖ $^{39}Ar_K = {}^{39}Ar – 0{\cdot}00079\, {}^{37}Ar$ (see text).
¶ The 'error' figures in the apparent ages do *not* include uncertainties in the measurement of neutron flux or uncertainties inherent in the correction for trapped ^{40}Ar.

These measurements indicate that interference with the $^{40}Ar–^{39}Ar$ dating technique may occur from ^{40}Ar produced from neutron interactions on potassium and from ^{39}Ar (and possibly ^{40}Ar) produced by (n, α) reactions on calcium. The $^{40}Ar/^{39}Ar$ ratio for the potassium derived argon is small (0·0063) and the necessity for any correction on this account may be avoided by choosing a neutron dose which yields a sufficiently large $^{40}Ar^*/^{39}Ar$ ratio, say $\geqslant 1$.

The possibility of interference from calcium derived ^{39}Ar is more serious, particularly in samples with a high Ca/K ratio. However, since ^{37}Ar is produced almost entirely from calcium, a straightforward correction can be applied by way of the ^{37}Ar measurement, if the ($^{39}Ar/^{37}Ar$) ratio of the calcium derived argon is known. This ratio may be measured by irradiating a calcium salt along with the samples. Alternatively, and this is the method adopted in the present experiment, the ratio may be deduced from measurements on the samples themselves.

Applying the $^{40}Ar–^{39}Ar$ technique to meteorites MERRIHUE and TURNER (1966) noted that the release patterns of ^{39}Ar and ^{37}Ar were quite different, the ^{39}Ar being released at comparatively low temperature while the ^{37}Ar was released predominantly at the highest temperatures. They used this observation to show that calcium derived ^{39}Ar made a negligible contribution over most of the release of radiogenic

Table 7. Argon release pattern (sample 10062-30†)

Temperature (°C)	Time (min)	^{40}Ar‡	^{39}Ar	^{38}Ar	^{37}Ar	^{36}Ar	(^{40}Ar*/^{39}Ar)§	(^{37}Ar/^{39}Ar)	(^{40}Ar*/^{39}Ar)$_K$ ‖	Apparent K–Ar age¶ ($\times 10^9$yr)
			($\times 10^{-8}$ cm^{3}STP/g)							
500	60	8·3	0·378	0·008	0·524	0·007	22	1·39	22	1·0
		±3·4	±0·003	±0·003	±0·019	±0·003	±9	±0·05	±9	±0·3
610	60	33·3	1·077	0·048	2·033	0·032	30·9	1·89	31·0	1·30
		±4·3	±0.007	±0·005	±0·019	±0·003	±4·0	±0·02	±4·1	±0·12
720	60	173·6	2·082	0·212	12·48	0·251	83·3	5·99	83·7	2·46
		±4·3	±0·012	±0·003	±0·04	±0·006	±2·1	±0·04	±2·2	±0·04
810	60	371·2	2·210	0·625	45·13	0·530	167·7	20·42	170·5	3·52
		±4·4	±0·005	±0·004	±0·14	±0·004	±2·0	±0·08	±2·0	±0·02
900	60	420·8	2·082	1·137	89·27	0·829	201·7	42·87	208·8	3·85
		±4·4	±0·011	±0·005	±0·13	±0·005	±2·3	±0·23	±2·4	±0·02
1020	60	487·5	2·468	1·518	116·8	1·013	197·1	47·34	204·8	3·82
		±4·4	±0·009	±0·009	±0·3	±0·009	±1·9	±0·22	±2·0	±0·02
1140	60	250·6	1·302	1·337	90·7	0·895	191·7	69·65	202·9	3·80
		±4·4	±0·005	±0·005	±0·3	±0·007	±3·4	±0·22	±3·6	±0·03
1280	60	162·5	1·060	5·84	384·2	3·756	150	362·6	210	3·86
		±9·6	±0·005	±0·02	±1·1	±0·010	±9	±1·4	±12	±0·10
1400	60	51·2	0·146	0·824	52·93	0·616	346	362	485	5·3††
		±12·1	±0·005	±0·005	±0·15	±0·007	±82	±12	±116	±0·4
Total		1959	12·8	11·55	794	7·93	152·4	62·0	160·2	3·42

† Sample weight 0·0982 g. Flux monitor 341A SH5. $J = 0·0319 \pm 0·0007$.
‡ The 'error' figure for ^{40}Ar is 50 per cent of the blank correction.
The 'error' figures for the other isotopes are statistical errors *relative* to ^{40}Ar, the reference isotope.
§ ^{40}Ar* = ^{40}Ar–^{36}Ar (see text).
‖ ^{39}Ar$_K$ = ^{39}Ar–0·00079 ^{37}Ar (see text).
¶ The 'error' figures in the apparent ages do *not* include uncertainties in the measurement of neutron flux or uncertainties inherent in the correction for trapped ^{40}Ar.
†† The atmospheric ^{40}Ar correction appears to have been underestimated in this extraction.

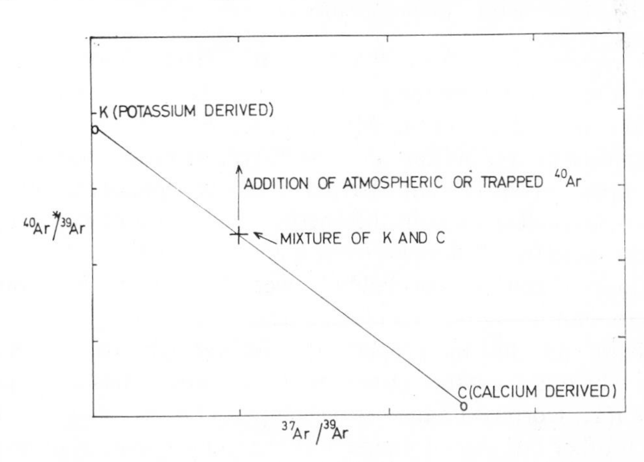

Fig. 1. Correlation plot (schematic) of ^{40}Ar*/^{39}Ar against ^{37}Ar/^{39}Ar showing the effect of mixing argon derived from potassium (K) with argon derived from calcium (C).

argon. A similar situation arises with the lunar samples and in fact the ^{39}Ar/^{37}Ar ratio of the calcium derived argon may be estimated directly from an isotope correlation plot with ordinate (^{40}Ar*/^{39}Ar) and abscissa (^{37}Ar/^{39}Ar).

Figure 1 illustrates the method schematically. It is applicable only to the gas

Table 8. Argon release pattern (sample 10072-45†)

Temperature (°C)	Time (min)	^{40}Ar‡	^{39}Ar	^{38}Ar (× 10^{-8} cm^{3}STP/g)	^{37}Ar	^{36}Ar	(^{40}Ar*/ ^{39}Ar)§	(^{37}Ar/ ^{39}Ar)	(^{40}Ar*/ ^{39}Ar)$_K$‖	Apparent K–Ar age¶ (× 10^{9}yr)
400	60	6·5 ±1·8	0·522 ±0·003	0·006 ±0·003	0·221 ±0·022	0·012 ±0·003	12·5 ±3·5	0·42 ±0·04	12·5 ±3·5	0·63 ±0·15
500	60	23·8 ±1·6	1·011 ±0·004	0·028 ±0·004	0·353 ±0·037	0·049 ±0·004	23·5 ±1·6	0·35 ±0·04	23·5 ±1·6	1·06 ±0·05
610	60	370·9 ±1·6	5·891 ±0·014	0·298 ±0·007	2·450 ±0·066	0·593 ±0·005	62·9 ±0·3	0·42 ±0·01	62·9 ±0·3	2·08 ±0·01
720	60	1453 ±2	10·13 ±0·04	1·063 ±0·013	9·39 ±0·12	2·565 ±0·026	143·2 ±0·7	0·93 ±0·01	143·3 ±0·7	3·25 ±0·01
810	60	1549 ±3	9·28 ±0·04	1·950 ±0·020	28·45 ±0·09	4·290 ±0·010	166·4 ±0·8	3·06 ±0·8	166·8 ±0·8	3·48 ±0·01
900	60	1433 ±3	8·27 ±0·03	2·455 ±0·026	56·38 ±0·18	3·182 ±0·018	172·9 ±0·7	6·82 ±0·03	173·8 ±0·7	3·55 ±0·01
1020	60	1152 ±4	6·75 ±0·04	2·473 ±0·021	65·0 ±0·2	2·032 ±0·013	170·3 ±1·2	9·62 ±0·07	171·6 ±1·3	3·53 ±0·01
1140	60	731 ±4	4·367 ±0·013	3·901 ±0·009	79·5 ±0·2	2·471 ±0·010	166·9 ±1·1	18·21 ±0·06	169·3 ±1·1	3·51 ±0·01
1280	60	619 ±6	3·834 ±0·028	16·74 ±0·03	449·8 ±1·3	10·47 ±0·04	158·9 ±1·9	117·3 ±0·9	175·2 ±2·2	3·56 ±0·02
1400	60	97·1 ±7·9	0·541 ±0·008	2·322 ±0·007	61·83 ±0·14	1·476 ±0·008	176 ±14	114·3 ±1·8	194 ±16	3·73 ±0·14
Total		7438	50·6	30·43	753	27·14	146·4	14·89	148·1	3·30

† Sample weight 0·0903 g. Flux monitor 341A SH5. $J = 0·0319 \pm 0·0007$.
‡ The 'error' figure for ^{40}Ar is 59 per cent of the blank correction.
The 'error' figures for the other isotopes are statistical errors *relative* to ^{40}Ar, the reference isotope.
§ ^{40}Ar* = ^{40}Ar–^{36}A (see text).
‖ ^{39}Ar$_K$ = ^{39}Ar–0·00079 ^{37}Ar (see text).
¶ The 'error' figures in the apparent ages do *not* include uncertainties in the measurement of neutron flux or uncertainties inherent in the correction for trapped ^{40}Ar.

released from regions of uniform ^{40}Ar*/K ratio. The composition of potassium derived argon is represented by the point *K*. (^{40}Ar*/^{39}Ar)$_K$ is related to the age of the sample and is to be determined. MITCHELL'S (1968) measurements on potassium sulphate, referred to above, indicate that the ^{37}Ar/^{39}Ar ratio of point *K* is essentially zero. The composition of calcium derived argon is represented by point *C*. The position of *C* is dependent only on the energy distribution of the neutrons and will therefore be the same for all samples from a given irradiation. As stated previously the ^{40}Ar/^{39}Ar ratio of point *C* is probably between 0 and 1. In what follows it will be taken to be zero with negligible loss of accuracy.

The correlation plot has the property (see for example, REYNOLDS and TURNER, 1964) that gas which is a mixture of components K and C has a composition which lies on the line *KC*. Correlation lines from samples of different ages will be displaced relative to each other but should intersect at the point *C* and in principle permit an estimate to be made of (^{40}Ar/^{39}Ar)$_C$. Note in passing that the presence of atmospheric ^{40}Ar will displace points vertically above the line.

Figure 2 is a correlation plot of argon isotopes released from sample 10003-43. As with meteorite samples the major ^{39}Ar release occurs before that of ^{37}Ar and consequently most of the ^{39}Ar observed up to and including the 1060°C point is potassium derived. The ^{40}Ar*/^{39}Ar ratio increases regularly in the first three extractions (500°C,

610°C, 720°C) in a way which would be expected for a specimen which had suffered a partial loss of radiogenic argon. These points represent 23 per cent of the total ^{39}Ar release. From 810 to 1060°C the $^{40}Ar^*/^{39}Ar$ ratio remains relatively constant indicating that the release is occurring from more retentive sites which have a uniform $^{40}Ar^*/K$ ratio. By 1060°C, 89 per cent of the ^{39}Ar has been released but only 39 per cent of the ^{37}Ar. The major release of ^{37}Ar occurs between 1140°C and 1300°C and the $^{40}Ar^*/^{39}Ar$ ratio decreases in these fractions in a way which can be understood in terms of the presence of calcium derived ^{39}Ar. The correlation line in Fig. 2 is the best fit to the 810–1220°C points and by extrapolation to $(^{40}Ar^*/^{39}Ar) = 0$ indicates that $(^{37}Ar/^{39}Ar)_C = 1270 \pm 80$. MITCHELL'S (1968) value for $(^{37}Ar/^{39}Ar)_C$ is 330, in

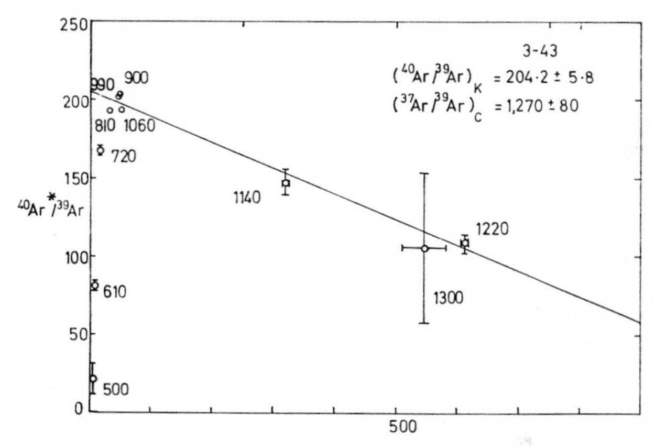

Fig. 2. Correlation plot of $^{40}Ar^*/^{39}Ar$ against $^{37}Ar/^{39}Ar$ for lunar rock 10003-43. Numbers indicate the extraction temperature in degrees Celsius.

complete disagreement with the present value and indeed is only half the $(^{37}Ar/^{39}Ar)$ ratio of the 1220°C extraction. It is not clear to what extent the value of 330 has been corrected for the decay of ^{37}Ar.

On the basis of Fig. 2 abundances of potassium derived ^{39}Ar have been calculated from the ^{39}Ar abundances in Tables 2–8 by subtracting 1/1270 of the ^{37}Ar abundance. When this is done the 'high temperature' $(^{40}Ar^*/^{39}Ar)_K$ ratios of samples 10022-46, 10044-40, 10062-30 and 10072-45 are found to be essentially constant for a given sample. When correlation plots are made (Fig. 3) the 'high-temperature' argon from these rocks yield intercepts, $(^{37}Ar/^{39}Ar)_C$, in substantial agreement with the value obtained from sample 10003-43. However the correlation plots for these rocks do not extend to such high values of $(^{37}Ar/^{39}Ar)$ and the estimates of $(^{37}Ar/^{39}Ar)_C$ are consequently less precise.

(c) *Age calculations*

The corrected $(^{40}Ar^*/^{39}Ar)$ ratios may be converted to 'apparent' potassium–argon ages by means of the following relationship (MERRIHUE and TURNER, 1966):

$$\frac{(^{40}Ar^*/^{39}Ar)}{(^{40}Ar^*/^{39}Ar)_m} = \frac{(\exp(t/T) - 1)}{(\exp(t_m/T) - 1} \tag{3}$$

t is the potassium argon age of the sample, T is the mean life of ^{40}K, and the subscript m refers to the monitor of known age, t_m.

In order to calculate t explicitly (3) may be rewritten as:

$$t = T \ln [1 + J (^{40}Ar^*/^{39}Ar)] \tag{4}$$

where

$$J = (\exp (t_m/T) - 1)/(^{40}Ar^*/^{39}Ar)_m. \tag{5}$$

(Note—the expression for J in TURNER (1970b) is incorrect. However, the ages presented in that paper were calculated using the correct expression.)

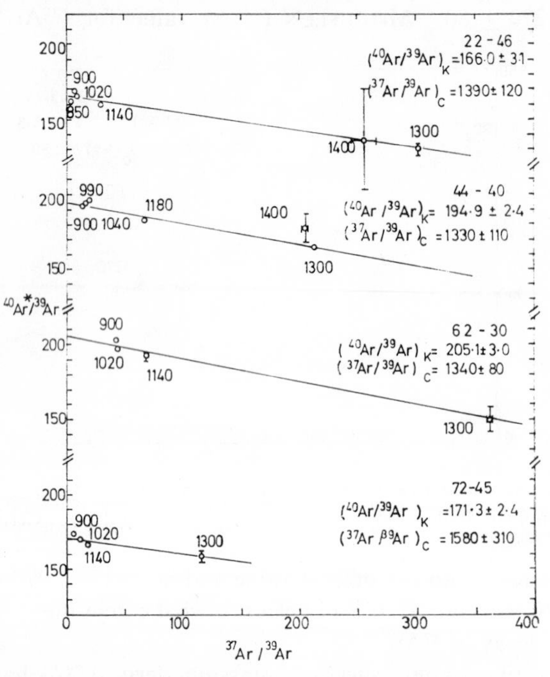

Fig. 3. Correlation plot of $^{40}Ar^*/^{39}Ar$ against $^{37}Ar/^{39}Ar$ for "high temperature" argon from lunar rocks, 10022-46, 10044-40, 10062-30 and 10072-45. Numbers indicate the extraction temperature in degrees Celsius.

The apparent age indicated by the argon from each extraction has been calculated and included in Tables 2–8.

(d) ^{38}Ar *measurements*

^{38}Ar abundances in a virgin specimen of lunar rock may be used to calculate cosmic rays exposure ages. In an irradiated specimen ^{38}Ar may also be present as a result of neutron interactions on chlorine, potassium and calcium, and the importance of these effects must be considered before attempting to draw any conclusions from the measured ^{38}Ar abundances.

MITCHELL's (1968) measurements indicate an $^{38}Ar/^{39}Ar$ ratio of 0·0114 for the potassium derived argon and an $^{38}Ar/^{37}Ar$ ratio of 10^{-4} for the calcium derived argon.

Assuming these ratios to be applicable to the present experiment it is apparent, on inspecting Tables 2–8, that neither source of ^{38}Ar can significantly affect the *overall* ^{38}Ar abundance. Interactions on potassium may however be responsible for an appreciable proportion of the ^{38}Ar released in some low temperature fractions but interactions in calcium are responsible for less than 1 per cent of the ^{38}Ar in all extractions.

The effects of (n, γ) reactions on ^{37}Cl should be recognizable by high ^{38}Ar/^{36}Ar ratios, at least in some individual extractions if not overall. No samples, regardless of absolute ^{38}Ar abundances, show excessively high ^{38}Ar/^{36}Ar ratios and the presence of chlorine derived ^{38}Ar has not been detected. The gas released at high temperatures

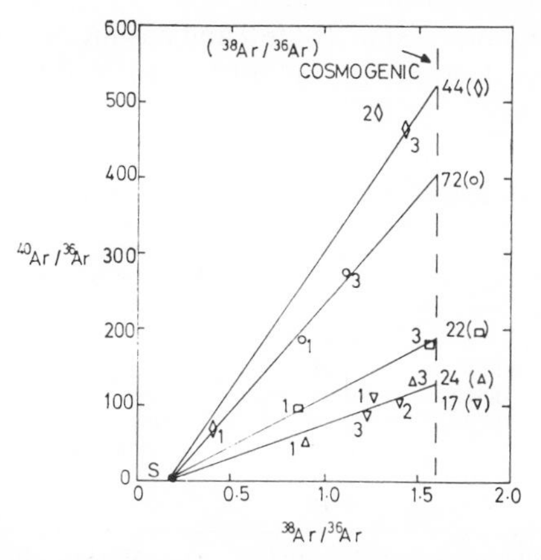

Fig. 4. Correlation plot of ^{40}Ar/^{36}Ar against ^{38}Ar/^{36}Ar for the total argon released from lunar rock samples. The lines indicate the variation to be expected if the samples from different regions of a particular rock differ only in the amount of trapped argon (S) present. The numbers refer to; 1—FUNKHOUSER *et al.*, 1970; 2—HINTENBERGER, 1970; 3—this work.

is of fairly constant composition ^{38}Ar/^{36}Ar = 1·55–1·60, and is presumably mainly cosmogenic. The ratio at low temperatures is generally lower but variable due to the additional presence of trapped argon, ^{38}Ar/^{36}Ar = 0·19, LSPET (1970), and small amounts of potassium derived ^{38}Ar.

Some of the samples analysed have also been investigated elsewhere (FUNKHOUSER *et al.*, 1970; HINTENBERGER, 1970). A comparison with these measurements reveals appreciable differences in isotope ratios. However, these differences can largely be understood in terms of variable concentrations of trapped gas in the different samples.

The effect is illustrated in Fig. 4 which is a correlation diagram of ^{40}Ar/^{36}Ar against ^{38}Ar/^{36}Ar. These isotopes constitute a three component system (trapped, cosmogenic and radiogenic argon) and as such would not normally be expected to yield a linear correlation. However, if for a given rock and for pieces of sufficient size to avoid effects of inhomogeneity, the concentration of cosmogenic ^{38}Ar and ^{36}Ar and radiogenic ^{40}Ar are constant, then the total argon will behave effectively as a two component system. Samples with different amounts of trapped argon will define a correlation line which passes through the points: $^{40}Ar/^{36}Ar \simeq 1{\cdot}0$, $^{38}Ar/^{36}Ar = 0{\cdot}19$. To a first approximation samples from different laboratories do show the expected trend. Depth dependence in the amount and composition of the cosmogenic argon probably accounts for the departures observed. The presence in the irradiated samples of ^{36}Ar from neutron interactions on Ca will deflect the points slightly towards the origin.

It appears that most of the ^{38}Ar in all the samples analysed is either cosmogenic or trapped in origin and cosmic ray exposure ages can therefore be calculated. The amount of cosmogenic ^{38}Ar present in each sample has been determined assuming $^{38}Ar/^{36}Ar = 1{\cdot}60$ for the cosmogenic argon and $^{38}Ar/^{36}Ar = 0{\cdot}19$ for the trapped argon. The results are summarized in Table 9 and cosmic ray exposure ages calculated for an assumed ^{38}Ar production rate of $0{\cdot}13 \times 10^{-8}$ cm^3 STP/g m.y. (EBERHARDT *et al.*, 1970).

Table 9. ^{38}Ar cosmic ray exposure ages

Sample	($^{36}Ar/^{38}Ar$)	$^{38}Ar_C/^{38}Ar$	$^{38}Ar_C$‡ ($\times 10^{-8}$ cm^3 STP/g)	Exposure age† ($\times 10^6$ yr)
10003-43	1·10	0·90	19	150
10017-49	0·81	0·96	57	440
10022-46	0·64	0·99	49	380
10024-26	0·68	0·99	46	360
10044-40	0·70	0·98	8·8	70
10062-30	0·69	0·98	11	90
10072-45	0·89	0·94	29	220

† Calculated from an assumed ^{38}Ar production rate of $0{\cdot}13 \times 10^{-8}$ cm^3 STP/g m.y. (EBERHARDT *et al.*, 1970).
‡ Absolute abundances accurate to $\pm 20\%$ (see text).

DISCUSSION

(a) ^{40}Ar–^{39}Ar *ages*

Figures 5 and 6 summarize the ^{40}Ar–^{39}Ar age determinations on each of the samples analysed. The apparent potassium argon age, and the corresponding $^{40}Ar^*/^{39}Ar$ ratio, are plotted as a function of the fraction of ^{39}Ar released. This method of presenting the results has the advantage of indicating the significance of each extraction, relative to the total gas released. Without exception the release patterns indicate that the more easily outgassed mineral sites are low in radiogenic ^{40}Ar relative to potassium.

In all but two samples (10017-49 and 10024-26) the $^{40}Ar^*/^{39}Ar$ ratio attains a constant value in the last five or so high temperature extractions. Note that the calcium derived ^{39}Ar abundance is only significant over a small portion of the release pattern in every case. The solid circles indicate those extractions where this correction amounted to more than 3 per cent of the apparent age. The author has argued previously that an age calculated on the basis of the more retentively held 'high-temperature' argon is a truer indication of the time the rock formed than an age based on total gas release. A 'high-temperature' age of each sample has been calculated

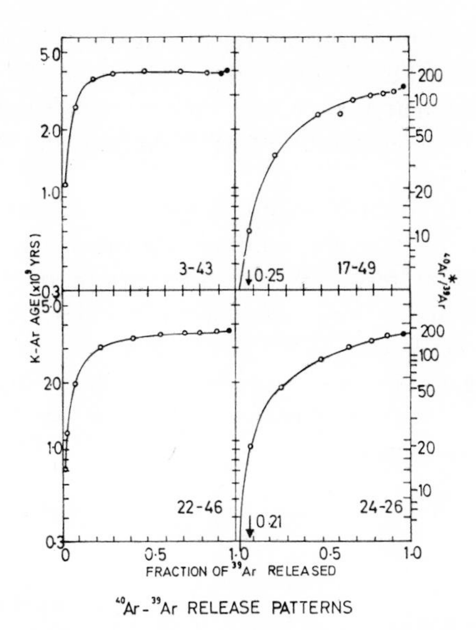

Fig. 5. $^{40}Ar^*$–^{39}Ar release patterns of lunar rock samples showing the increase in apparent K–Ar age as the more retentive sites release argon. The closed circles indicate those samples for which the correction for ^{39}Ar derived from calcium accounts for more than 3 per cent of the age.

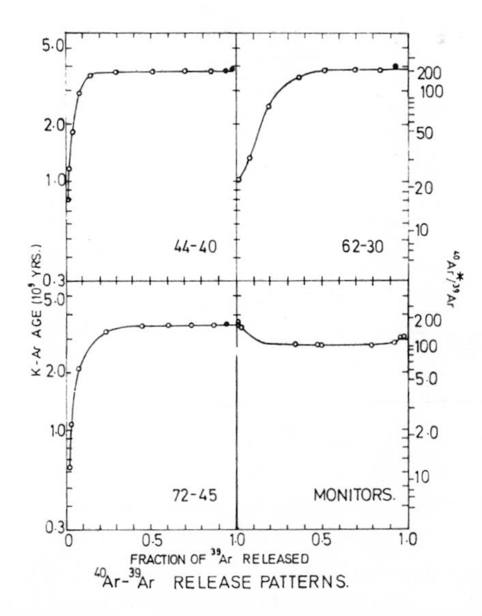

Fig. 6. $^{40}Ar^*$–^{39}Ar release patterns of lunar rock samples showing the increase in apparent K–Ar age as the more retentive sites release argon. The closed circles indicate those samples for which the correction for ^{39}Ar derived from calcium accounts for more than 3 per cent of the age. The monitor release pattern is a composite for the three samples analysed and is characteristic of the horneblende used (TURNER, 1968).

using the intercepts $(^{40}Ar^*/^{39}Ar)_K$ from Figs. 2 and 3. The ages are summarized in Table 10. The ages based on total argon release are included for comparison and also an estimate of the extent of argon loss from each sample. This is simply the fractional difference between the high temperature $^{40}Ar^*/^{39}Ar$ ratio and the total gas $^{40}Ar^*/^{39}Ar$ ratio.

Samples 10017-49 and 10024-26 do not show plateaux in the argon release pattern and it is probably that the 'high temperature' age calculated on the basis of the last significant argon extraction, represents only a lower limit on the true age of the rock. Likewise the fractional argon losses calculated (48 and 45 per cent respectively) can only be regarded as lower limits.

The effect of taking account of argon loss is impressive in that the spread of ages in the seven samples is considerably reduced. The ages based on total argon cover a range of $1{\cdot}4 \times 10^9$ years, from $2{\cdot}30 \times 10^9$ yr to $3{\cdot}76 \times 10^9$ yr. Taking account of

Table 10. $^{40}Ar^{39}Ar$ ages of lunar samples

Sample No.	Chemical group[d]	Fractional ^{40}Ar loss	Total argon age ($\times 10^9$ yr)	Ref.	High temperature argon age ($\times 10^9$ yr)	Rb-Sr age ($\times 10^9$ yr)	Ref.
10003	2	0·10	$3{\cdot}76 \pm 0{\cdot}05$		$3{\cdot}92 \pm 0{\cdot}07$		
10017	1	$\geqslant 0{\cdot}48$	$2{\cdot}31 \pm 0{\cdot}05$		$\geqslant 3{\cdot}23 \pm 0{\cdot}06$	$3{\cdot}40 \pm 0{\cdot}60$	(a)
			2·45	(b)		$3{\cdot}59 \pm 0{\cdot}08$	(b)
			2·60	(c)		$3{\cdot}80 \pm 0{\cdot}11$	(d)
10022	1	0·19	$3{\cdot}26 \pm 0{\cdot}05$		$3{\cdot}59 \pm 0{\cdot}06$		
10024	1	$\geqslant 0{\cdot}45$	$2{\cdot}58 \pm 0{\cdot}05$		$\geqslant 3{\cdot}48 \pm 0{\cdot}05$	$4{\cdot}05 \pm 0{\cdot}70$	(a)
10044	2	0·07	$3{\cdot}62 \pm 0{\cdot}05$		$3{\cdot}73 \pm 0{\cdot}05$	$3{\cdot}70 \pm 0{\cdot}07$	(b)
			3·90	(c)			
10062	2	0·22	$3{\cdot}42 \pm 0{\cdot}05$		$3{\cdot}82 \pm 0{\cdot}06$		
10072	1	0·14	$3{\cdot}30 \pm 0{\cdot}05$		$3{\cdot}52 \pm 0{\cdot}05$	$3{\cdot}80 \pm 0{\cdot}30$	(d)

Ages are calculated on the basis of, $\lambda_\beta = 4{\cdot}72 \times 10^{-10}$ yr^{-1}; $\lambda_e = 0{\cdot}584 \times 10^{-10}$ yr^{-1}; $^{40}K/K = 0{\cdot}0119$ atm %.

The quoted errors in the ages determined in the present experiment includes the effects of uncertainty arising from non-uniformity of the measured neutron flux. Absolute uncertainties in the monitor age affect all samples equally and introduce an additional uncertainty of $\pm 0{\cdot}06. \times 10^9$ yr in the above ages.

(a) GOPALAN *et al.* (1970)
(b) ALBEE *et al.* (1970)
(c) HINTENBERGER (1970)
(d) COMPSTON *et al.* (1970)

the fact that the high temperature ages of 10017-49 and 10024-26 are now to be regarded only as lower limits the significant spread in age is reduced to $0{\cdot}4 \times 10^9$ yr; from $3{\cdot}52 \times 10^9$ yr to $3{\cdot}92 \times 10^9$ yr.

Ages determined by conventional K–Ar dating and by Rb–Sr dating using mineral separates have been included in Table 10 for comparison. The K–Ar ages compare well with the 'total gas' $^{40}Ar^*/^{39}Ar$ ages indicating that the argon loss observed in the present experiment occurred on the moon and not in the Herald reactor. Previous measurements on meteorites with high ages (TURNER, 1969) and calculations based on estimated sample temperatures ($\sim$200°C) in the reactor also indicate that gas loss during irradiation is negligible.

Agreement, within the experimental uncertainties, of the Rb–Sr ages and the 'high-temperature' ^{40}Ar–^{39}Ar ages gives added confidence that the ages are meaningful and represent the original cooling time of the rock.

Having largely eliminated the effects of argon loss the problem of interpretation of these ages is considerably eased provided some estimate can be made of the area of the lunar surface represented by the samples. It is probable that much of the regolith is of comparatively local origin. This viewpoint is supported by the relative uniformity of composition of many of the Apollo 11 rocks (LSPET, 1969), the existence of well defined colour boundaries on the lunar surface, associated with differences in underlying rock type (WHITAKER, 1966) and the relationship of regolith thickness to local cratering density (OBERBECK and QUAIDE, 1968). Despite the rather limited age determinations available on the crystalline rocks we can therefore say with a fair degree of confidence that much of the vulcanism associated with this region of Mare Tranquillitatis occurred around $3{\cdot}7 \times 10^9$ yr ago.

A note of caution is raised by the observation of a more varied selection of rock types in the lunar fines (WOOD *et al.*, 1970) and differences in U, Th–Pb ages between fines and rocks (TATSUMOTO, 1970). Evidently the sampling of the lunar surface by the fine material is different from that of the crystalline rocks analysed so far. Perhaps the fine material is representative of a wider region of the surface.

Although the correction for $^{40}Ar^*$ loss has removed much of the spread in K–Ar ages there remain differences which are difficult to ignore. The observed spread (400 million years) in 'high-temperature' ages corresponds to a range of $^{40}Ar^*/K$ ratios of around 20 per cent. It is difficult to account for this difference either in terms of the uncertain contribution of cosmogenic ^{40}Ar or in terms of any failure of the ^{40}Ar–^{39}Ar method to correct fully for the $^{40}Ar^*$ loss. Samples 10003-43 and 10072-45 show the greatest divergence in age yet the correction for $^{40}Ar^*$ loss in each is quite small (10 and 14 per cent respectively). It is difficult to escape the conclusion that the spread is real and represents different rock forming events which have occurred in the vicinity of the Apollo 11 site.

COMPSTON (1970) has divided the crystalline rocks into two distinct chemical groups on the basis of the abundances of 13 elements. Compston's group 1 for example is much higher in the elements Rb, K, Ba and Th than is group 2. ALBEE *et al.* (1970) find that rock 10044 (group 2) has a $(^{87}Sr/^{86}Sr)_I$ ratio distinctly lower (i.e. more primitive) than $(^{87}Sr/^{86}Sr)_I$ from rocks 10017, 10057, 10069 and 10071 (group 1) indicating that the samples represent at least two different rock bodies.

The chemical groups of the samples analysed have been included in Table 10. Although the sampling at the moment is too restricted to draw any firm conclusions the ages of the group 2 rocks measured are distinctly higher than the ages of the group 1 rocks. Group 1 rocks, 10022 and 10072 have a mean age of $(3{\cdot}56 \pm 0{\cdot}06) \times 10^9$ yr while the group 2 rocks 10003, 10044 and 10062 have a mean age of $(3{\cdot}83 \pm 0{\cdot}09) \times 10^9$ yr. The distribution of ^{40}Ar–^{39}Ar ages therefore tends to substantiate the division of the samples made on chemical grounds and indicates, albeit tentatively, that the group 2 samples are older than the group 1 samples, possibly by as much as 200–300 million years.

(b) *Cosmic ray exposure ages and low temperature* ^{40}Ar–^{39}Ar *ages*

The cosmic ray exposure ages presented in Table 9 are comparable with values obtained by other experimenters on unirradiated specimens of the same rock

(FUNKHOUSER *et al.*, 1970). From a comparison of the ^{38}Ar and ^{40}Ar–^{39}Ar measurements it appears unlikely that the cosmic ray 'clock' has been appreciably reset by heating in any of the samples since exposure started. In the laboratory experiments ^{39}Ar release occurs at much lower temperatures than the ^{38}Ar release. Assuming that the outgassing *sequence* is not critically dependent on the thermal history of the sample one may conclude that a heating episode sufficient to release significant quantities of cosmogenic ^{38}Ar would have resulted in almost complete loss of radiogenic ^{40}Ar. The heating episodes which led to the loss of around 50 per cent of the radiogenic argon in samples 10017 and 10024, even if they had occurred recently would probably not have released more than 10 per cent of the cosmogenic ^{38}Ar.

The extent to which the exposure age can be equated to the time when the rocks were first brought to the surface depends on the extent of overturning of the regolith layer due to meteorite impact. The present experiments do not convey much information on this aspect. The author has argued (TURNER, 1969) that it is possible to date the times of radiogenic argon loss, in suitable circumstances, by measuring the apparent K–Ar age indicated by the 'low temperature' gas release. The assumption is made that the ^{40}Ar/K ratio in the easily outgassed sites associated with the low temperature release, would fall to zero at the time of any major gas loss. The technique has been applied to chondritic meteorites with low K–Ar ages, where the release patterns are sufficiently flat in the low temperature portion to give a well defined initial $^{40}Ar^*/^{39}Ar$ ratio. The release patterns for the lunar samples, rise so sharply in the initial stages that it is only possible to infer from the initial $^{40}Ar^*/^{39}Ar$ ratios (see Tables 2–8) upper limits for the times of outgassing. A possible exception is sample 10062-30 for which the initial $^{40}Ar^*/^{39}Ar$ ratio is relatively well defined at $(1{\cdot}0 \pm 0{\cdot}3) \times 10^9$ yr (the error figure takes account of all possible variation in the atmospheric ^{40}Ar correction). In contrast the cosmic ray exposure age of 10062-30 is only 90 million years. One might explain the observations in terms of the rock having been produced as an individual fragment in an impact 10^9 yr ago, at which time ^{40}Ar loss occurred, and having lain for most of the intervening time under several metres of regolith. This interpretation is speculative in the absence of corroborating evidence.

The ^{40}Ar–^{39}Ar ages indicated by the low temperature argon from samples 10017-49 and 10024-26 are in each case lower than the ^{38}Ar cosmic ray exposure age. Either the atmospheric ^{40}Ar blank correction has been overestimated in each case or the samples have lost a small amount of radiogenic ^{40}Ar during the heating associated with a recent impact.

SUMMARY

In conclusion the information obtained in the present experiment may be summarized as follows:

(1) All of the lunar crystalline rock samples analysed have lost radiogenic argon. The amounts involved range from 7 per cent to $\geq$48 per cent and render K–Ar ages based on total ^{40}Ar measurements inaccurate.

(2) When account is taken of this argon loss the ages obtained lie in the range $3{\cdot}5 \times 10^9$ yr to $3{\cdot}9 \times 10^9$ yr. The ages are in good agreement with those obtained by Rb–Sr dating.

(3) Two groups of rocks identified by chemical differences appear to have significantly different ages, one group being $(0{\cdot}2 - 0{\cdot}3) \times 10^9$ yr older than the other.

(4) The rock ages probably indicate the time of the major vulcanism associated with the formation of Mare Tranquillitatis.

(5) ^{38}Ar exposure ages in the rocks analysed range from 70 to 440 million years and have not been seriously affected by loss of ^{38}Ar. They represent a lower limit on the time when the rocks were first ejected, as a result of meteorite impact, from their place of origin.

(6) Loss of radiogenic argon in the samples analysed has occurred relatively recently (within the last $(0{\cdot}3\text{–}1{\cdot}0) \times 10^9$ yr) but no precise information can be obtained which would help to indicate the extent of mixing in the regolith.

(7) From the point of view of the ^{40}Ar–^{39}Ar dating technique a method has been presented for estimating the extent of interference from calcium derived ^{39}Ar. Corrections for calcium derived ^{39}Ar are found to be small for most argon fractions from irradiated lunar samples despite the high Ca/K ratio.

Acknowledgments—I am grateful to the Science Research Council and the United Kingdom Atomic Energy Authority for making the facilities of the Herald reactor available to me, and to DAVID TROTMAN and the staff of the Herald reactor for performing the neutron irradiation. I also wish to thank MARTIN ASHWORTH for invaluable help in writing the computer program for the reduction of mass spectrometer data and COLIN GOLDSBOROUGH for constructing the spectrometer and extraction system which functioned flawlessly during the analysis.

REFERENCES

ALBEE A. L., BURNETT D. S., CHODOS A. A., EUGSTER O. J., HUNEKE J. C., PAPANASTASSIOU D. A., PODOSEK F. A., RUSS G. PRICE II, SANZ H. G., TERA F. and WASSERBURG G. J. (1970) Ages, irradiation history and chemical composition of lunar rocks from the Sea of Tranquillity. *Science* **167,** 643–466.

COMPSTON W., ARRIENS P. A., VERNON M. J. and CHAPPELL B. W. (1970) Rubidium-strontium chronology and chemistry of lunar material. *Science* **167,** 474–476.

EBERHARDT P., GEISS J., GRAF H., GRÖGLER N., KRÄHENBÜHL U., SCHWALLER H., SCHWARZMÜLLER J. and STETTLER A. (1970) Trapped solar wind noble gases, Kr^{81}/Kr exposure ages and K/Ar ages in Apollo 11 lunar material. *Science* **167,** 558–560.

FUNKHOUSER J. G., SCHAEFFER O. A., BOGARD D. D. and ZÄHRINGER J. (1970) Gas analysis of the lunar surface. *Science* **167,** 561–563.

GOPALAN K., KAUSHAL S., LEE-HU C. and WETHERILL G. W. (1970) Rubidium-strontium, uranium and thorium-lead dating of lunar material. *Science* **167,** 471–473.

HINTENBERGER H., WEBER H. W., VOSHAGE H., WÄNKE H., BEGEMANN F., VILCSEK and WLOTZKA F. (1970) Rare gases, hydrogen and nitrogen: concentrations and isotopic composition in lunar material. *Science* **167,** 543–545.

LSPET (LUNAR SCIENCE PRELIMINARY EXAMINATION TEAM) (1969) Preliminary Examination of Lunar Samples from Apollo 11. *Science* **165,** 1211–1227.

MERRIHUE C. M. and TURNER G. (1966) Potassium-argon dating by activation with fast neutrons. *J. Geophys. Res.* **71,** 2852–2857.

MITCHELL J. G. (1968) The Argon-40/Argon-39 method for potassium-argon age determination. *Geochim. Cosmochim. Acta* **32,** 781–790.

OBERBECK V. R. and QUAIDE W. L. (1968) Genetic implications of lunar regolith thickness variations. *Icarus* **9,** 446–465.

REYNOLDS J. H. and TURNER G. (1964) Rare gases in the chondrite renazzo. *J. Geophys. Res.* **69,** 3263–3281.

TROTMAN, D. (1969) Private Communication.

TATSUMOTO M. and ROSHOLT J. N. (1970) Age of the moon. An isotopic study of uranium–thorium–lead systematics of lunar samples. *Science* **167**, 461–463.

TURNER G., MILLER J. A. and GRASTY R. L. (1966) The thermal history of the Bruderheim meteorite. *Earth Planet. Sci. Lett.* **1**, 155–157.

TURNER G. (1968) The distribution of potassium and argon in chondrites. In *Origin and Distribution of the Elements* (editor L. H. Ahrens), pp. 387–389. Pergamon.

TURNER G. (1969) Thermal histories of meteorites by the ^{39}Ar–^{40}Ar method. In *Meteorite Research* (editor P. M. Millman), pp. 407–417. D. Reidel.

TURNER G. (1970a) Thermal histories of meteorites. In *Palaeogeophysics* (editor S. K. Runcorn). Academic Press. In press.

TURNER G. (1970b) Argon–40/Argon–39 dating of lunar rock samples. *Science* **167**, 466–468.

WÄNKE H. and KÖNIG H. (1959) Eine Neue Method zur Kalium–Argon–Alterbestimmung und ihre Anwendung auf Steinmeteorite. *Z. Naturforsch.* **14a**, 860–866.

WHITAKER E. A. (1966) The surface of the moon. In *The Nature of the Lunar Surface* (editors W. N. Hess, D. H. Menzel and J. A. O'Keefe), Chapter 3, pp. 79–98. Johns Hopkins.

WOOD J. A., MARVIN U. B., POWELL B. N. and DICKEY J. S., JR. (1970) Mineralogy and petrology of the Apollo 11 lunar sample. Smithsonian Astrophysical Observatory Special Report 307.

Proceedings of the Apollo 11 Lunar Science Conference, Vol. 2, pp. 1685 to 1717.

Elemental abundances of major, minor and trace elements in Apollo 11 lunar rocks, soil and core samples

H. WAKITA, R. A. SCHMITT and P. REY
Department of Chemistry and The Radiation Center, Oregon State University, Corvallis, Oregon 97331

(*Received* 2 *February* 1970; *accepted in revised form* 25 *February* 1970)

Abstract—Elemental abundances of ten major and minor elements (i.e. Si, Ti, Al, Fe, Mg, Ca, Na, K, Cr and Mn) and twenty-six minor and trace elements (i.e. fourteen rare earth elements (REE), Y, Sc, Cd, In, Ba, V, Co, Zr, Hf, Th, Rb and Cs) have been determined via both techniques of instrumental and radiochemical neutron activation analysis (INAA and RNAA) in one Type A fine grained rock (10017-31), four Type B coarser grained rocks (10044-29, 10045-28, 10047-33, 10050-28), two Type C breccia rocks (10019-24 and 10059-26) and Type D soil (10084-56) returned by the Apollo 11 mission. Abundances of seven major (i.e. Ti, Al, Fe, Ca, Na, Cr and Mn) and twelve minor and trace elements were also determined via INAA in five samples from core tube No. 1 (10005-29). Total REE + Y abundances found in Types B and C rocks and soil are about ~60 and ~3 times the abundances found in chondrites and abyssal subalkaline basalt. All lunar REE distribution patterns, relative to chondrites, resemble those found in terrestrial abyssal subalkaline basalt, but with Eu depleted by 71 $\pm$ 2 per cent in Type A rocks and 58 $\pm$ 4 per cent in Type B rocks. Sm/Eu ratios of 6·5 $\pm$ 0·7 of this and other studies are very similar for Types B and C rocks and soil and markedly different from 9·5 $\pm$ 0·6 for Type A rocks. This suggests that Types B and C rocks and lunar soil had some common chemical processes which were distinct from Type A rocks. Separation of plagioclase and hypersthene minerals from a melt could account for Eu depletion and the observed REE distribution pattern. From major, minor and trace elemental abundances of this and other studies, we conclude that (1) Type A rocks are different from Type B rocks, especially in their alkali elements, REE, Th, Co, Ba, Zr, Hf and to a lesser degree in their Al, Ca and Ti abundances; (2) since Types A and B rocks are compositionally different from Type C rocks and soil, especially for Ti, Al and Fe, it seems implausible that soil and breccia rocks at Tranquillity Base (T.B.) have originated from impact fragmentation and subsequent compaction (breccia rocks from soil) of Types A and B rocks returned from T.B.; (3) abundances of ten bulk elements overlap for Type C rocks and soil; however, from elemental abundance differences in seventeen lithophilic trace elements, Type C rocks at T.B. may have been derived from impaction processes on soil other than that at T.B. Average elemental abundances in five core tube soil samples agree with the corresponding elemental abundances in soil. In general, elemental abundances for all core samples were quite uniform with possible 10 per cent differences in REE abundances for the 0-cm compared to the 10·5-cm sample. Possibly higher (by 8–11 per cent) abundances of Ca, Ti and Fe were observed in the 0-cm sample compared to the 10·5-cm sample.

INTRODUCTION

THE ABUNDANCES of ten bulk elements and twenty-six trace elements in lunar samples (one Type A fine grained igneous rock, four Type B coarser grained igneous rocks, two Type C breccia rocks and Type D fine soil matter) returned by the Apollo 11 lunar mission from the Sea of Tranquillity have been determined via both techniques of INAA and RNAA. Abundances of seven bulk and 12 trace elements were also determined in five core tube soil samples via INAA. Knowledge of the chemical composition of lunar rocks is considered basic per se and also extremely valuable in the interpretation of data gathered by other scientific disciplines. A preliminary report of this work has been previously published (SCHMITT *et al.*, 1970a).

EXPERIMENTAL

The technique of sequential INAA was used to determine the elemental abundances of ten bulk (major and minor) elements, i.e. Si, Ti, Al, Fe, Mg, Ca, Na, K, Cr and Mn in ~0·3–1 g specimens of chips, powders or fine matter from one Type A rock (10017-31), four Type B rocks (10044-29, 10045-28, 10047-33 and 10050-28), two Type C rocks (10019-24 and 10059-26), Type D soil (10084-56). Abundances of the major and minor elements, Ti, Al, Fe, Ca, Na, Mn and Cr were determined via INAA in 70–140 mg each of five core tube samples (10005-29) taken at 2·6-cm interval depths.

The method of sequential INAA has been previously described in detail by SCHMITT *et al.* (1970b). For lunar sample analysis, some modifications have been adopted. Nuclear reactions and γ-ray energies used for abundance measurements are listed in Table 1. Briefly, the method consists of six

Table 1. Sequential activation analysis for major, minor and trace elements in lunar rocks—I

Element	Nuclear reaction	Gamma-ray measured (keV)
Si	^{28}Si(14-MeV n, p) 2·3-m ^{28}Al	1780
Al	^{27}Al(n, γ) 2·3-m ^{28}Al	1780
Ca	^{48}Ca(n, γ) 8·8-m ^{49}Ca	3100
Ti	^{50}Ti(n, γ) 5·8-m ^{51}Ti	320
V	^{51}V(n, γ) 3·8-m ^{52}V	1434
Na	^{23}Na(n, γ) 15-h ^{24}Na	2754
Mn	^{55}Mn(n, γ) 2·56-h ^{56}Mn	847
K	^{41}K(n, γ) 12·4-h ^{42}K	1524
Mg	^{25}Mg(23-MeV bremss. γ, p) 15-h ^{24}Na	2754
Sc	^{45}Sc(n, γ) 84-d ^{46}Sc	889
Cr	^{50}Cr(n, γ) 28-d ^{51}Cr	320
Fe	^{58}Fe(n, γ) 45-d ^{59}Fe	1100
Co	^{59}Co(n, γ) 5·2-y ^{60}Co	1173
Ba	^{130}Ba(n, γ) 11·6-d ^{131}Ba	496

sequential activations; (1) 14-MeV neutron activation for Si determination; (2) INAA for Al at a low thermal-neutron flux of $\sim 2 \times 10^7$ cm^{-2} sec^{-1} for 1 min; (3) INAA for Ca, Ti, V, Na and Mn at $\sim 8 \times 10^{10}$ cm^{-2} sec^{-1} for 5 min; (4) INAA for K at $\sim 2 \times 10^{11}$ cm^{-2} sec^{-1} for 30 min; (5) IPAA (instrumental 23-MeV bremsstrahlung photonuclear activation analysis) for Mg at $\sim 10^5$ roentgens min^{-1} for 1 min; (6) INAA for Cr and Fe at $\sim 2 \times 10^{12}$ cm^{-2} sec^{-1} for 2 hr. For small amounts of specimen, especially for the core samples, the total neutron flux was increased by factors of 5–10. After each activation, γ-ray spectra were taken by 3 in. × 3 in. NaI(Tl) and 30 cm^3 Ge(Li) detectors coupled to either 400, 2048 or 4096 multichannel analyzers. During activation for the above bulk elements, additional minor and trace elements were obtained via INAA; i.e., with the Cr and Fe group, Sc, Co, Ba, La, Sm, Eu, Yb, Lu, Zr, Hf and Th were analyzed. For the above group of elements, the general INAA technique of GORDON *et al.* (1968) was applied. After INAA, chips of Type A rock (10017-31a), Type B rock (10050-28b), Type C rock (10019-24a), and fine soil matter (10084-56a), were further subjected to RNAA for determination of the 14 REE; Y, Cd, In, K, Rb and Cs. Nuclear reactions and γ-ray energies for minor and trace elemental determinations are summarized in Table 2. The radiochemical separation procedures, described by MOSEN *et al.* (1961), SCHMITT *et al.* (1963a) and REY *et al.* (1970) were used.

Chips from rocks 10017-31, 10045-28, 10050-28, 10019-24 and 10059-28 were split into two smaller chips. Powders from rocks 10044-29 and 10047-33 were sieved through 100 nylon mesh size into a coarse fraction, >100 mesh, and a fine fraction, <100 mesh. For 10044-29, the relative coarse and fine fraction weights were 0·85 and 1·03 g and for 10047-33, 0·35 and 0·56 g, respectively. From a

Table 2. Sequential activation analysis for minor and trace elements in lunar rocks—II

Element	Nuclear reaction	Gamma-ray measured (keV)
Rb*	^{85}Rb(n, γ) 18·7-d ^{86}Rb	1077
Cs*	^{133}Cs(n, γ) 2·1-y ^{134}Cs	605, 796
Zr	^{94}Zr(n, γ) 65-d ^{95}Zr	757
Hf	^{180}Hf(n, γ) 43-d ^{181}Hf	133
Cd*	^{114}Cd(n, γ) 2·3-d ^{115}Cd	beta count
In*	^{115}In(n, γ) 54-m ^{116}In	1290
La†	^{139}La(n, γ) 40-h ^{140}La	1595
Ce*	^{142}Ce(n, γ) 33-h ^{143}Ce	293
Pr*	^{141}Pr(n, γ) 19-h ^{142}Pr	beta count
Nd*	^{146}Nd(n, γ) 11·1-d ^{147}Nd	91, 531
Sm†	^{152}Sm(n, γ) 47-h ^{153}Sm	103
Eu†	^{151}Eu(n, γ) 9·3-h ^{152m} Eu	842, 964
Gd*	^{158}Gd(n, γ) 18-h ^{159}Gd	364
Tb*	^{159}Tb(n, γ) 72-d ^{160}Tb	879, 966
Dy*	^{164}Dy(n, γ) 2·35-h ^{165}Dy	95
Ho*	^{165}Ho(n, γ) 27-h ^{166}Ho	81
Er*	^{170}Er(n, γ) 7·5-h ^{171}Er	296, 308
Tm*	^{169}Tm(n, γ) 125-d ^{170}Tm	beta count
Yb†	^{168}Yb(n, γ) 32-d ^{169}Yb	63, 177, 198
	^{174}Yb(n, γ) 4·2-d ^{175}Yb	396
Lu†	^{176}Lu(n, γ) 6·7-d ^{177}Lu	208
Y*	^{89}Y(n, γ) 64-h ^{90}Y	beta count
Th	^{232}Th(n, γ) 22-m ^{233}Th → 27-d ^{233}Pa	312

* Using radiochemical method after neutron activation.
† Determined by both radiochemical and instrumental activation analysis.

total of 5·3 g of lunar soil, three aliquants were split into two unsieved 1 g samples and the third sample of 3·1 was sieved through 100 and 200 nylon mesh. Approximate weights of the third sample in sieved quantities were: $>$100 mesh, 0·73 g, 100–200 mesh, 0·47 g, and $<$200 mesh, 1·90 g. To check the overall INAA technique and data reduction, the U.S.G.S. standard rocks W-1, BCR-1 and DTS-1 were simultaneously analyzed with the lunar samples.

Results and Discussion

Major elements

The abundances of ten major and minor elements in Types A, B and C rocks and Type D soil are given in Table 3. In Table 4, average abundances of the major and minor elements in Types A, B and C rocks and Type D soil of this work have been compared with other IAA (instrumental activation analysis) values obtained by Goles *et al.* (1970a) and CH (classical chemical and IAA analysis) values by A-W (1970) (Agrell *et al.*, 1970; Compston *et al.*, 1970; Engel and Engel, 1970; Frondel *et al.*, 1970; Rose *et al.* 1970; Wänke *et al.*, 1970 and Wiik and Ojanperä 1970). In general, abundances of Ti, Al, Fe, Mg and Ca of this work agree with those reported by LSPET (1969), A-W (1970), Taylor *et al.* (1970) and others.

Si. The average Si abundances of this work for one Type A, four Type B and two Type C rocks of 20·3 ± 1·0, 20·8 ± 0·8 and 18·7 ± 1·1 per cent, respectively, agree with Si abundances reported by Ehmann and Morgan (1970), Taylor *et al.* (1970) and A-W (1970). Abundances of Si in the soil of 17·5 ± 1·0 per cent are just

Table 3. Abundances of major and minor elements in Types A, B and C rocks and Type D lunar soil by IAA (instrumental activation analysis)*

	Type A rock		Type B rocks								
Element	10017-33 a chip 0·637 g	10017-33 b chip 0·343 g	10044-29 a powder >100 mesh 0·849 g	10044-29 b powder <100 mesh 1·027 g	10045-28 a chip 0·496 g	10045-28 b chip 0·544 g	10047-33 a powder >100 mesh 0·351 g	10047-33 b powder <100 mesh 0·563 g	10050-28 a chip 0·600 g	10050-28 b chip 0·575 g	Av.
Si(%)	20·0	20·8	24·3	19·4	19·9	20·6	20·8	21·5	19·7	20·4	20·8 ±0·8
Ti(%)	—	7·8	5·9	5·7	6·9	7·2	6·5	5·7	7·7	—	6·7 ±0.9
Al(%)	4·2	4·0	4·6	6·0	5·3	5·4	3·9	5·9	5·5	5·7	5·4 ±0·2
Fe†(%)	17·0	18·1	15·9	14·4	16·0	16·5	17·0	15·4	15·6	14·8	15·7 ±0·6
Mg(%)	4·8	4·6	4·1	3·5	5·1	5·9	4·1	2·9	4·4	4·2	4·3 ±0·9
Ca(%)	—	7·6	8·4	8·9	8·3	8·3	9·7	9·0	8·0	—	8·6 ±0·6
Na(ppm)	3660	3730	2730	3660	2590	2580	2590	3630	2930	3040	3020 ±290
K‡(ppm)	1430(R) 1800	2200	910	1060	460	570	680	500	400	490	630 ±250
Cr(ppm)	2450	2710	1570	1190	2730	3200	1810	1380	2410	2230	2020 ±760
Mn(ppm)	1830	1920	2360	2060	1940	2090	2730	2250	2360	2150	2230 ±180

Type C rocks					Type D, soil							
10019-24 a chip 0·532 g	10019-24 b chip 0·497 g	10059-26 a chip 0·600 g	10059-26 b chip 0·353 g	Av.	10084-56 a 1·125 g	10084-56 b 1·068 g	10084-56 d >100 mesh 0·731 g	10084-56 e 100– 200 mesh 0·468 g	10084-56 f <200 mesh 0·957 g	10084-56 g <200 mesh 0·937 g	Av. 84a, b (2·19 g)	mass wted. av. (5·29 g
19·2	15·7	19·4	20·2	18·7 ±1·1	18·5	16·6	16·2	17·4	18·6	17·5	17·5 ±1·4	17·5 ±1·0
—	5·3	4·9	5·2	5·2 ±0·2	—	4·7	5·2	4·9	4·5	4·6	4·7 ±0·2	4·7 ±0·3
6·7	7·0	6·8	6·9	6·9 ±0·1	7·2	7·1	6·8	6·8	7·5	7·4	7·1 ±0·1	7·2 ±0·3
13·3	12·9	14·1	14·3	13·7 ±0·6	12·0	11·8	12·4	13·0	11·6	12·4	11·9 ±0·6	12·1 ±0·5
5·3	4·3	4·4	4·4	4·6 ±0·3	4·8	4·8	5·1	5·1	4·6	5·5	4·8 ±0·2	5·0 ±0·3
—	8·9	8·8	9·6	9·0 ±0·7	—	8·3	9·3	9·2	8·5	9·1	8·3 ±0·6	8·8 ±0·5
3470	3410	3360	3550	3440 ±70	3010	3140	3350	3140	3100	3270	3070 ±100	3170 ±140
890(R) 870	1080	1230	1450	1200 ±300	910(R) 790	840	1170	1130	940	770	850 ±100	910 ±170
2110	1990	2260	2340	2180 ±170	1720	1850	1950	2180	1910	1960	1780 ±100	1900 ±150
1720	1910	1650	1690	1750 ±130	1600	1600	1750	1750	1600	1600	1600 ±10	1630 ±80

* Large chips from rocks 10017, 10045, 10050, 10019 and 10059 were split into two smaller chips, e.g. 10017 into 17a and 17b. Powders from rocks 10044 and 10047 were sieved into two fractions: coarse fraction, >100 mesh and fine fraction, <100 mesh. Lunar soil, 10084, was split into three portions, two, 10084a and 10084b were roughly of equal mass and the third portion was sieved into three sized, >100 mesh, 100–200 mesh, and <200 mesh. One standard deviation due to counting statistics and other errors for single determinations are approximately ±5% for Si, Ti, Fe and Mg; ±2% for Al, Na and Mn; ±3–5% for Cr; ±8% for Ca and ±10–20% for K.

† Fe abundances are relative to the Fe concentration in the U.S.G.S. standard rocks DTS-1 and W-1.

‡ Some abundances were determined by radiochemical activation analysis, designated by (R). These (R) values are considered far superior to IAA values.

outside the 2σ limit of most other published values. For Si determinations, samples and standards were activated simultaneously and counted for the same time intervals but with separate counting systems. In order to achieve the same neutron activation geometry, nearly identical volumes of the U.S.G.S. standard rock AGV-1 were used as standards for analyses of Types A, B and C rocks. A possible explanation for the

Table 4. Summary of average abundances for major and minor elements in Types A, B and C rocks and in Type D soil via instrumental activation and chemical analyses

	Type A			Type B			Type C			Type D		
Element	1* IAA	4† IAA	6‡ CH	4* IAA	5† IAA	7‡ CH	2* IAA	18† IA A	5‡ CH	IAA*	IAA†	CH‡
Si(%)	20·3 ±1·0	18·9 ±0·8	18·9 ±0·1	20·8 ±0·8	18·7 ±0·8	18·9 ±0·9	18·7 ±1·1	19·7 ±0·7	19·5 ±0·2	17·5 ±1·4	20·2 ±0·2	19·6 ±0·1
O(%)	—	38·5 ±1·2	—	—	39·4 ±1·0	—	—	41·1 ±1·0	—	—	40·8 ±1·2	—
Al(%)	4·1 ±0·1	4·3 ±0·6	4·6 ±0·3	5·4 ±0·2	5·5 ±0·8	5·5 ±0·6	6·9 ±0·1	7·4 ±0·9	6·7 ±0·4	7·1 ±0·1		7·3 ±0·3
Fe(%)	17·4 ±0·8	15·0 ±0·7	14·8 ±0·2	15·7 ±0·6	14·8 ±0·7	14·4 ±0·7	13·7 ±0·6	13·0 ±0·6	12·4 ±0·4	11·9 ±0·6		12·1 ±0·2
Mg(%)	4·7 ±0·3	3·9 ±0·3	4·6 ±0·2	4·3 ±0·9	4·6 ±1·1	4·2 ±0·5	4·6 ±0·3	4·6 ±0·6	4·7 ±0·1	4·8 ±0·2		4·8 ±0·1
Ca(%)	7·6 ±0·7	7·2 ±0·1	7·7 ±0·2	8·6 ±0·4	8·5 ±0·5	8·2 ±0·4	9·0 ±0·7	8·8 ±0·6	8·4 ±0·2	8·3 ±0·6		8·5 ±0·2
Na(ppm)	3680 ±70	3600 ±70	3700(4) ±200	3020 ±290	3180 ±440	3100(5) ±300	3440 ±70	3520 ±160	3600(4) ±100	3070 ±100		3300(5) ±200
K(ppm)	1820 ±100	—	2500 ±300	630 ±240	—	600 ±200	1200 ±300	—	1400 ±100	850 ±100		1200 ±100
Ti(%)	7·8 ±0·2	7·2 ±0·2	7·1 ±0·3	6·7 ±0·9	6·5 ±0·5	6·3 ±0·7	5·2 ±0·2	5·0 ±0·4	5·1 ±0·3	4·7 ±0·2		4·5 ±0·1

* This work.

† Goles *et al.* (1970a).

‡ CH, chemical analyses by Agrell *et al.* (1970), Compston *et al.* (1970), Engel and Engel (1970), Frondel *et al.* (1970), Rose *et al.* (1970), Wänke *et al.* (1970) and Wiik and Ojanperä (1970). Numbers in parenthesis after some values indicate number of 'reliable' analyses for Na. Type A rocks were 10017, 10022, 10049, 10057, 10072 and 10024. (Rock 10024 had been designated a Type B rock (LSPET, 1969); however, abundances of Na, K, Rb and Th justify a Type A reclassification.) Type B rocks were 10003, 10044, 10045, 10047, 10050, 10058 and 10062 and Type C rocks, 10018, 10019, 10048, 10060 and 10061. 'Type C rock' 10056, analyzed by Rose *et al.* (1970), was not included because abundances of Al, Fe, Mg, Ti, K, Rb and Th are more in line with Type B rock abundances.

discrepancy in low Si abundances for the soil (10084-56) may be attributed to differences in volumes between soil samples and AGV-1 standards, so that appreciable activity differences resulted in errors in relative live time correction between the two counting systems. From the summation of total elemental concentrations in soil, the real value of soil should be closer to 20 per cent. In any case, Si abundances in lunar material are very low compared to the 22–23% Si value found in terrestrial basalts and in eucrite achondrites. From an evaluation of all Si data (Table 4), it seems evident that Si does not differentiate the four types of lunar rocks and soil. However, the 18·9 ± 0·1% Si for six type A rocks via chemical (CH) methods (column 4) are statistically lower than Si of five C rocks and soil, all analyzed by CH methods (column 10). The statistical criterion used in this paper is that two means are different if their difference is greater than the square root of the sums of squares of their respective standard deviations (Parratt, 1961). Of course, both IAA and CH data for Si result in identical average Si abundances in Types A and B rocks, but with a considerably larger dispersion for Type B rocks, obtained by IAA and CH methods. We concur with LSPET (1969) that Si abundances are uniform within experimental error for all Apollo 11 rocks and soil at T.B.

Ti. Turkevich *et al.* (1969) and Patterson *et al.* (1969) have reported markedly high Ti enrichments in lunar soil and rocks in the Sea of Tranquillity (~25 km from T.B.) and Sinus Medii (both Mares) and lesser amounts of Ti in the lunar highlands near the Tycho crater. Average abundances of Ti, 7·8 ± 0·2 and 6·7 ± 0·9 per cent in one Type A and four Type B rocks, respectively, are more abundant than the 5·2 ± 0·2% Ti found in breccia rocks and the 4·7 ± 0·2% Ti in soil. Preliminary Ti and Ca values reported by SCHMITT *et al.* (1970a) were systematically lower by 10–15 per cent than the other published values. In the sequential INAA procedure for these two elements during the same activation and counting steps, the radioactivity levels of the specimens were very high and, therefore, resulted in appreciable analyzer dead times of ~15–25 per cent. The low Ti and Ca values have been attributed to pulses losses at high analyzer dead times. In a redetermination of Ti and Ca in these samples, at reduced analyzer dead times (<10 per cent) and with the addition of a base-line restorer circuit, we have obtained the revised Ti and Ca abundances (Table 3), which agree well with those obtained by classical techniques. Within the 1σ limit, the average Ti values of this work agree with the results of other independent determinations of Ti in four Type A, five Type B and eighteen Type C rocks (GOLES *et al.* 1970a) and in six Type A, seven Type B and five Type C rocks (A-W, 1970). The value of 4·7 ± 0·2% Ti in lunar soil of this work agrees well with the average Ti abundance of 4·5 ± 0·1% in eight independent determinations by A-W (1970) and TAYLOR *et al.* (1970) and with the average value of 4·6% Ti, reported by TURKEVICH *et al.* (1969). The value 7·8 ± 0·2% Ti in rock 10017 compares with the 7·0% Ti, reported by WÄNKE *et al.* (1970), the 7·1% Ti by WIIK and OJANPERÄ (1970) and the 7·2% Ti by COMPSTON *et al.* (1970), respectively. For a comparison with the more homogenized sample 10044-29, our value of 5·8 ± 0·2% Ti agrees with that of 5·3% Ti by ENGEL and ENGEL (1970) and 5·5% Ti by AGRELL *et al.* (1970), and is less than WÄNKE *et al.* (1970) value of 6·3%; for rock 10045, the 7·1% Ti of this work agrees with 6·8 and 6·7% Ti reported by COMPSTON *et al.* (1970) and AGRELL *et al.* (1970), respectively. For rock 10047, the 6·0 ± 0·5% Ti of this work agrees with the 5·7 and 6·1% Ti reported by COMPSTON *et al.* (1970) and ROSE *et al.* (1970), respectively; for rock 10050, the 7·7 ± 0·5% Ti of this work compares with 7·6% Ti obtained by ROSE *et al.* (1970); for rock 10019, the 5·3 ± 0·3% Ti of this work compares with 5·0% by ROSE *et al.* (1970).

The general agreement of average elemental abundances and their respective dispersions by various investigators, using different analytical techniques, is quite remarkable and emphasizes the homogeneous composition of these igneous rocks. This, of course, indicates a well mixed molten magma which cooled quite rapidly with only minor degrees of chemical differentiation, as noted by many investigators (e.g. see LSPET, 1969 and MASON *et al.*, 1970). As previously noted by TURKEVICH *et al.* (1969), the absolute abundances of Ti are about five and ten times more in lunar samples than those of average oceanic tholeiites (ENGEL *et al.*, 1965) and eucrite achondrites (DUKE and SILVER, 1967), respectively. Compared with high-titania alkali–olivine basalts from the John Day formation, Oregon (ROBINSON, 1969), lunar rocks and soil are enriched by about a factor of 2–3. OLSEN (1969) reported that the chemical composition of pyroxene gabbros (anorthositic gabbros), high in titanium, from the Adirondack Mountains, most closely resemble the composition of the lunar rocks and soil. It is clear from either the IAA and/or the CH data in Table 4 that average Ti abundances for Type A and B igneous rock are statistically higher than those of Type C and soil. This evidence indicates that breccia rocks and soil at T.B. can not be derived directly from fragmentation impacts of Type A rocks, and perhaps also not from degradation of Type B rocks. Note that the average Ti abundances via the CH data for Types B and C rocks coincide within the statistical criterion, and that the average Ti contents via CH data are just different for Type C rocks and soil. Although the average Ti abundances in Type A is about 10 per cent higher than in Type B rocks, the two Ti mean values just barely coincide on a statistical basis (see Al, Ca and Fe discussions below for additional correlative studies).

Al. The average values of 4·1 ± 0·1 and 5·4 ± 0·2% Al found in Types A and B rocks of this work are considerably lower than those of 6·9 ± 0·1% Al found in breccia and 7·1 ± 0·1% Al in soil. The agreement between our Al values and many other published values is excellent (Table 4).

The value of 7·1 ± 0·1% Al in lunar soil of this work overlaps the Al abundance of 7·6% obtained by TURKEVICH *et al.* (1969) and the average Al abundances of 7·2 ± 0·2% in nine independent determinations by EHMANN and MORGAN (1970), A-W (1970), and TAYLOR *et al.* (1970). For rock 10044, the 5·4 ± 0·1% Al agrees with the 5·4% Al obtained by AGRELL *et al.* (1970), and these values are less than the 6·2 and 6·3% Al reported by ENGEL and ENGEL (1970) and WÄNKE *et al.* (1970), respectively. For rock 10017, the 4·1 ± 0·1% Al of this work agrees with the 4·1, 4·2 and 4·4% Al reported by COMPSTON *et al.* (1970), WIIK and OJANPERÄ (1970), and WÄNKE *et al.* (1970), respectively. The 5·4 ± 0·1% Al in rock 10044 agrees with the 5·2 and 5·4% Al obtained by ENGEL and ENGEL (1970) and by AGRELL *et al.* (1970), respectively. These values are less than the 6·3% Al found by WÄNKE *et al.* (1970). For rock 10045, and 5·4 ± 0·1% Al of this work agrees with the 5·0% Al reported by AGRELL *et al.* (1970) and COMPSTON *et al.* (1970). For rock 10047, the 5·1 ± 0·1% Al of this work agrees with the 5·2% abundance reported by ROSE *et al.* (1970) and COMPSTON *et al.* (1970). For rock 10050, the 5·6 ± 0·1% Al is higher than the 4·7% Al reported by ROSE *et al.* (1970). For rock 10019, 6·9 ± 0·1% Al agrees with the 7·3% obtained by ROSE *et al.* (1970). In general, the agreements of Al abundances are very satisfactory.

For Types A and B *and* Type C rocks and Type D soil, a strong anticorrelation is apparent between average Al and Ti abundances (Table 4); i.e. it suggests different proportions of more ilmenite and less plagioclase minerals in Type A and B rocks compared to Type C breccia rocks and soil. The average Al abundances via IAA of Type C rocks and soil are almost identical, and both higher than those of Type A and B rocks. Again, it seems implausible that soil and breccia rocks at T.B. have originated from impact fragmentation and subsequent compaction (breccia rocks from soil) of Types A and B rocks at T.B.

The average Al (or equivalent plagioclase) abundance in Type A rocks is about 20 per cent less than that found in Type B rocks. Comparisons of IAA data or CH data shows that Al mean values are statistically different and anticorrelate with the observed Ti average values for these two groups. The range of Al abundances in lunar rocks and soil material resembles the average value of 6·5 ± 0·8% Al in eucrite achondrites (LOVELAND *et al.*, 1969) and of the pyroxene gabbro 7·4% Al (OLSEN, 1969). The Al values of 8–9% found in oceanic tholeiites and high-titania alkali-olivine basalts are more abundant than that found in lunar material.

Fe. The average Fe abundances in Types A and B rocks of 17·4 ± 0·8 and 15·7 ± 0·6% are higher than the Fe contents in breccia rocks and soil of 13·7 ± 0·6 and 11·9 ± 0·6%, respectively. The value of 11·9 ± 0·6% Fe in lunar soil agrees with the average Fe abundances of 12·2 ± 0·2% in eight independent determinations by A-W (1970), and TAYLOR *et al.* (1970). A lower Fe abundance of 9·4% for the Sea of Tranquillity, reported by TURKEVICH *et al.* (1969), probably reflects experimental errors inherent in the alpha backscattering measurements. The following comparison of Fe abundances have been made with other published values: (1) for rock 10017, 17·4 ± 0·8% Fe of this work with 15·4% Fe by WIIK and OJANPERÄ (1970), 14·6% Fe by WÄNKE *et al.* (1970), and 15·1% Fe by COMPSTON *et al.* (1970); (2) for rock 10044, 15·1 ± 0·7% Fe of this work with 14·0% Fe by ENGEL and ENGEL (1970), 13·7% Fe by AGRELL *et al.* (1970) and 13·3% Fe by WÄNKE *et al.* (1970); (3) for rock 10045, 16·3 ± 0·6% Fe of this work with 14·9% Fe by AGRELL *et al.* (1970) and 15·0% Fe by COMPSTON *et al.* (1970); (4) for rock 10047, 16·0 ± 0·7% Fe of this work with 14·8% Fe by ROSE *et al.* (1970) and COMPSTON *et al.* (1970); (5) for rock 10050, 15·2 ± 0·6% Fe of this work with 13·4% Fe by ROSE *et al.* (1970); and (6) for rock 10019, 13·1 ± 0·5% Fe of this work with 12·2% Fe by ROSE *et al.* (1970), respectively. TAYLOR *et al.* (1970) reported 14% Fe, with values ranging from 12·4–15·5% Fe in eight Type B rocks, and also 13·7% Fe (12·4–14·8% Fe) in three Type C rocks and 12·4% Fe in soil. The agreement between our data and theirs is satisfactory, but our value for Type A rock 10017-31 is just outside of their 12–16 per cent range for Type A rocks. In summary, the Fe values for soil of this work agree exactly with those of eight other investigators; for five Apollo 11 rocks the Fe abundances of this work are 8 per cent relatively higher on the average than those reported by other investigators. The source of the discrepancy in Fe abundances in rocks is unknown since all samples (rocks and soil) and the Fe standards (U.S.G.S. rocks DTS-1 and W-1) were activated simultaneously and counted with the same detector and multichannel analyzer. As seen in Table 4, for average Fe abundances there is a positive correlation for average Ti abundances and a negative correlation for average Al abundances between Types A and B rocks *and* Type C rock and soil, i.e. both Fe and Ti are statistically higher in Types A and B rocks compared to Type C and soil. No significant Fe differences are observed between Types A and B rocks. The average Fe content in lunar igneous rocks is higher than the 11 and 12% Fe contents found in high-titania alkali–olivine basalts and pyroxene gabbros, respectively. The Fe abundances of oceanic tholeiites (6·7 ± 1·6% Fe) by ENGEL *et al.* (1965) are lower than this value. Eucrite achondrites with a value of 14·3 ± 0·7% Fe (DUKE and SILVER, 1967) are close to the lunar material. Significant differences exist between the lunar and terrestrial samples, such as the anhydrous character and the absence of ferric iron in lunar rocks, as observed by many investigators (for instance see ENGEL and ENGEL, 1970).

Mg. The magnesium abundances are distributed very uniformly in all Types A, B, C and D lunar samples with values of 4·7 ± 0·3, 4·3 ± 0·9, 4·6 ± 0·3 and 4·8 ±

0·2%, respectively. Excellent concordance exists between the 4·8 ± 0·2% Mg in lunar soil of this work and the 4·8 ± 0·1% Mg for the average of 8 independent determinations by A-W (1970) and TAYLOR *et al.* (1970). The Mg content of 2·7%, reported by TURKEVICH *et al.* (1969) for Sea of Tranquillity soil is considerably less than the average Mg content of 4·8% for T.B. soil and rocks.

Comparing our Mg abundances in rocks 10017, 10044, 10045, 10047, 10050 and 10019 with other published values, satisfactory agreements are noted: (1) for rock 10017, 4·7 ± 0·3% Mg of this work agrees with 4·6% Mg by COMPSTON *et al.* (1970), 4·7% Mg by WIIK and OJANPERÄ (1970) and 4·8% by Mg WÄNKE *et al.* (1970); (2) for rock 10044, 3·8 ± 0·2% Mg of this work agrees with 3·8, 3·9 and 3·6% Mg obtained by ENGEL and ENGEL (1970), WÄNKE *et al.* (1970) and AGRELL *et al.* (1970), respectively; (3) for rock 10045, 5·5 ± 0·3% Mg agrees within 2σ of the 4·9% Mg by AGRELL *et al.* (1970) and 4·7% Mg by COMPSTON *et al.* (1970); (4) for rock 10047, 3·4 ± 0·2% Mg agrees with 3·4% Mg by ROSE *et al.* (1970); (5) for rock 10050, 4·3 ± 0·3% Mg agrees with 4·8% Mg by ROSE *et al.* (1970); and (6) for rock 10019, 4·8 ± 0·3% Mg agrees with 4·7% Mg by ROSE *et al.* (1970).

Magnesium abundances in Apollo 11 rocks and soil overlap those observed in oceanic tholeiites, eucrite achondrites or pyroxene gabbro. On the other hand, the Mg content of high-titania alkali–olivine basalt is depleted by a factor of 0·6, relative to the above rock samples.

Ca. Calcium concentrations in Types A, B and C rocks and soil are 7·6 ± 0·7, 8·6 ± 0·6, 9·0 ± 0·1 and 8·3 ± 0·6%, respectively. As already stated in the Ti discussion, preliminary Ca values reported SCHMITT *et al.* (1970a) were systematically lower by 10 ∼ 15 per cent. Comparison of the revised Ca values of this work with other published values is now quite satisfactory. The following comparison of Ca abundances have been made with the other published values: (1) for rock 10017, 7·6 ± 0·7% Ca of this work agrees with 7·6% by WIIK and OJANPERÄ (1970), 7·7% by COMPSTON *et al.* (1970), and 8·2% by WÄNKE *et al.* (1970); (2) for rock 10044, 8·7 ± 0·6% Ca agrees well with 8·7% by ENGEL and ENGEL (1970) and 8·8% by AGRELL *et al.* (1970), and is higher than 5·1 per cent by WÄNKE *et al.* (1970); (3) for rock 10045, 8·3 ± 0·7% Ca agrees with 7·9% by AGRELL *et al.* (1970) and 8·1% by COMPSTON *et al.* (1970); (4) for rock 10047, 9·3 ± 0·8% Ca agrees with 8·7% by ROSE *et al.* (1970) and COMPSTON *et al.* (1970); (5) for rock 10050, 8·0 ± 0·6% Ca agrees with 8·6% by ROSE *et al.* (1970); and (6) for rock 10019, 8·9 ± 9·7% Ca agrees with 8·5% by ROSE *et al.* (1970). Our value of 8·3 ± 0·6% Ca in soil agrees well with the average Ca abundance of 8·4 ± 0·2% from eight independent determinations by A-W (1970), and TAYLOR *et al.* (1970).

Evaluation of the more accurate CH data for Ca (Table 4) shows that the average Ca abundance in Type A rocks is statistically less than that observed for Type C rocks and Type D soil. This observation is consistent with similar Al variations (plagioclase mineral depletion) in these same rock groups. For the Type A and B rock groups, the average Ca content is less in Type A rock, but the means of Ca in these two groups just barely coincide on a statistical basis. This average Ca content of 8·4 ± 0·2% for soil at Tranquillity Base is ∼20 per cent less than the Ca abundance of 10·4%, reported by TURKEVICH *et al.* (1969) for the Sea of Tranquillity. The average

Ca abundances in lunar samples fall within the range of terrestrial basalts and eucrite achondrites.

Na. The average Na abundances are 3680 ± 70, 3020 ± 290, 3440 ± 70 and 3070 ± 100 ppm in Types A, B, C rocks and soil. The 4450 ppm Na for soil at the Sea of Tranquillity, reported by TURKEVICH *et al.* (1970), is ~40 per cent higher than the 3070 ± 100 ppm Na for soil T.B. In general, values by LSPET (1969) and by TAYLOR *et al.* (1970) are considerably higher. For rock 10017, 3680 ± 70 ppm Na of this work agrees with 3780 ppm obtained by WIIK and OJANPERÄ (1970) and COMPSTON *et al.* (1970) and 3470 ppm by WÄNKE *et al.* (1970). For rock 10044, 3240 ± 70 ppm Na of this work agrees with 3560 ppm reported by ENGEL and ENGEL (1970) and AGRELL *et al.* (1970) and 3580 ppm obtained by WÄNKE *et al.* (1970) and 3630 ppm by TUREKIAN and KHARKAR (1970), respectively. For rock 10045, 2590 ± 60 ppm Na agrees well with the 2670 ppm found by AGRELL *et al.* (1970) and COMPSTON *et al.* (1970), respectively. For rock 10047, 3230 ± 70 ppm Na of this work agrees with 3340 ppm obtained by COMPSTON *et al.* (1970), and is 30 per cent less than the 4820 ppm reported by ROSE *et al.* (1970). For rock 10050, 3000 ± 60 ppm Na of this work is 40 per cent less than the 4900 ppm obtained by ROSE *et al.* (1970). For rock 10019, the 3440 ± 70 ppm Na of this work is ~50 per cent less than the 6900 ppm reported by ROSE *et al.* (1970). In general, the values obtained by ROSE *et al.* (1970) are much higher than the values obtained via INAA.

The agreement of Na (and Cr and Mn) abundances of this work is excellent with the average values obtained via NAA by MORRISON *et al.* (1970), WÄNKE *et al.* (1970), SMALES *et al.* (1970), TUREKIAN and KHARKAR (1970), WIIK and OJANPERÄ (1970) and others.

Type B rock 10045-28 has a lower Na abundance by ~28 per cent in comparison to the average abundance which is calculated for three other Type B rocks (see Table 3).

A real statistical difference exists for the mean Na abundances between Type A and B rocks (Table 4), that is, the average Na abundances of Type A rock via IAA or CH data are statistically higher than in Type B rock by ~15 per cent. An average Na abundance of 3500 ± 160 ppm in 20 Type C rocks exceeds the average Na content of 3070 ± 100 ppm in soil, all determined via IAA; for the less reliable Na values obtained by CH methods, the average Na abundance just coincides.

Average Na concentration in Apollo 11 lunar rocks is approximately half of that found in chondrites (SCHMITT *et al.*, 1970c) and about 0·2 times that observed in terrestrial basalts.

Cr. The range of Cr in Apollo 11 lunar rocks and soil ranges from 1360 to 3000 ppm, largely representative of the variation observed in Type B rocks. The value of 1780 ± 100 ppm Cr in soil agrees with Cr abundances of 1880 and 1830 ppm, reported by WIIK and OJANPERÄ (1970) and WÄNKE *et al.* (1970), respectively. Higher Cr abundances of 2000, 2300 and 2500 ppm have been found by MORRISON *et al.* (1970), SMALES *et al.* (1970) and TAYLOR *et al.* (1970), respectively.

For rock 10017, the 2540 ± 100 ppm Cr of this work agrees with 2310 and 2260 ppm Cr, obtained by WÄNKE *et al.* (1970) and WIIK and OJANPERÄ (1970), respectively, within the 2σ limit. For rock 10044, the 1360 ± 60 ppm Cr of this work agrees with the 1300, 1440 and 1600 ppm Cr reported by WÄNKE *et al.* (1970) and AGRELL *et al.* (1970) and TUREKIAN and KHARKAR (1970), respectively. For rock 10045, the 3000 ± 130 ppm Cr agrees with the 3220 ppm by AGRELL *et al.* (1970) and is higher than 2400 ppm reported by COMPSTON *et al.* (1970). For rock 10047, the 1540 ± 80 ppm Cr of this work agrees with 1510 ppm by ROSE *et al.* (1970) and is higher than 1220 ppm by COMPSTON *et al.* (1970). For rock 10050, the 2200 ± 100 ppm Cr of this work is lower than the 2400 ppm reported by ROSE *et al.* (1970). For rock 10019, the 2050 ± 100 ppm Cr in this work agrees with the 2190 ppm by ROSE *et al.* (1970). Such differences may reflect heterogeneity of Cr in 0·5–1 g specimens from the same rock (see below for a more detailed discussion). Within experimental error, the average Cr abundances in Types B and C rocks and soil are essentially the same and are significantly lower by ~20 per cent than the Cr abundance in one Type A rock.

The average Cr abundances in lunar rocks and soil are ~0·6 times those found in ordinary chondrites and almost the same in eucrite achondrites (NICHIPORUK *et al.*, 1967; and SCHMITT *et al.*, 1970c).

Mn. The average Mn abundance of 2230 ± 180 ppm in four Type B rocks may be higher than the average Mn abundances of 1750 ± 130 and 1600 ± 10 ppm, observed in Type C rocks and soil, respectively. The average abundance of 1600 ± 10 ppm Mn for lunar soil overlaps the average Mn abundances of 1640 ± 110 ppm in eight independent determinations by A-W (1970), and TAYLOR *et al.* (1970). Comparisons of Mn values found in rocks 10017, 10044, 10045, 10047, 10050 and 10019 with those obtained by other investigators are listed as follows: (1) in rock 10017, the 1860 ± 40 ppm Mn of this work agrees with 1700 ppm by WIIK and OJANPERÄ (1970) and is higher than 1480 ppm reported by WÄNKE *et al.* (1970) and is lower than 2170 ppm found by COMPSTON *et al.* (1970); (2) in rock 10044, 2200 ± 40 ppm Mn of this work agrees with 2170 and 2290 ppm obtained by AGRELL *et al.* (1970) and TUREKIAN and KHARKAR (1970), respectively, and is higher than 2000 ppm reported by WÄNKE *et al.* (1970); (3) in rock 10045, 2020 ± 40 ppm Mn agrees with 2090 and 2170 ppm obtained by COMPSTON *et al.* (1970) and AGRELL *et al.* (1970), respectively; (4) in rock 10047, the 2430 ± 50 ppm Mn of this work is higher than 2170 and 2250 ppm obtained by COMPSTON *et al.* (1970) and ROSE *et al.* (1970), respectively; (5) in rock 10050, 2260 ± 30 ppm Mn of this work is higher than 2090 ppm reported by ROSE *et al.* (1970); (6) in rock 10019, the 1810 ± 30 ppm Mn of this work is compared to 1700 ppm obtained by ROSE *et al.* (1970).

The average Mn values in lunar samples lie between those of 1300 ± 200 ppm Mn in oceanic tholeiites, (ENGEL *et al.*, 1965), 740–1550 ppm in pyroxene gabbros (OLSEN, 1969) and 3900 ± 600 ppm in eucrite achondrites (SCHMITT *et al.*, 1970c).

K. Potassium abundances in Apollo 11 lunar rocks and soil vary from 500 to 2000 ppm. This wide K concentration range encompasses K abundances of eucrite achondrites (200–700 ppm, CLARK *et al.*, 1966), chondritic meteorites (850 ppm, UREY, 1964), and oceanic tholeiites (500–2000 ppm, ENGEL *et al.*, 1965). The K content of pyroxene gabbros and high-titania alkali–olivine basalts are significantly higher than this range. The average K concentration of 1820 ppm in the Type A rock 10017-31 is less than the 2610 ppm and 2210 ppm abundances reported by GAST and HUBBARD (1970) and MURTHY *et al.* (1970), respectively, for the same rock. The average K abundance of 630 ± 250 ppm in four Type B rocks agrees with the average K content of ~620 ± 190 ppm in four Type B rocks, reported by GAST and HUBBARD (1970) and ~700 ± 240 ppm in three Type B rocks, obtained by MURTHY *et al.* (1970). For rock 10044, our K value of 980 ± 120 ppm agrees with those of 816 and 1090 ppm reported by MURTHY *et al.* (1970) and TUREKIAN and KHARKAR (1970), respectively. For rock 10045, the 510 ± 180 ppm K of this work agrees with 424 ppm obtained by MURTHY *et al.* (1970). As observed by many other investigators, e.g. GAST and HUBBARD (1970), Type A rocks are markedly enriched in K by a factor of ~4, compared to Type B rocks (Table 4). Also, average abundances in Type C rocks and soil are quite distinct from Types A and B crystalline rocks. The value of 910 ± 50 ppm K via RNAA in soil seems 10–20 per cent lower than those of many other published values (for example, see WÄNKE *et al.*, 1970, GAST and HUBBARD, 1970). A possible explanation for our lower K abundances involves some loss of K in the three precipitations of K-tetraphenylborate. Since ^{137}Cs tracer was used as a chemical yield indicator for K, Rb and Cs, it seems possible that small quantities (1–2 mg) of Rb and Cs are totally co-precipitated within completely precipitated K-tetraphenylborate.

For soil sample (10084-56), if a Si abundance of 20 per cent is adopted, the summation of total elemental concentration approaches ~100% within the 1σ uncertainties of the individual measurements.

The elements Na and Al are enriched and Ti, Fe and Mg are depleted in the fine fractions relative to the coarse fractions of powders 10044-29 and 10047-33; Cr and Mn are significantly depleted in the fines relative to the coarse fractions. These measurements are consistent with visual inspection of both fraction; i.e. the fine fractions were enriched in light colored minerals indicative of an enrichment of plagioclase and a relative depletion of pyroxene and ilmenite minerals.

In soil, we do not observe, in general, any significant differences in elemental abundances as a function of minerals sizes, i.e. between the original unsieved samples and the various sieved samples. Elemental abundance differences for Cr in the 10084-56e sieved (100–200 mesh) fraction, compared to the lower abundances in the unsieved and other sieved fractions, might be attributed to either preferential concentration of Cr in that particular mineral size or, more likely, to a proportionately larger fraction of Cr containing minerals of that grain size. The latter suggestion is compatible with enrichments of both Fe and Ti in the 10084-56e sieved fraction; this suggests a larger fraction of ilmenite grains is present in the 100–200 mesh size, compared to other mineral sizes.

Average values of the major and minor elements of this work for the three types of rocks agree within 1σ limit with the average abundances of the same elements observed in chips from twenty-seven different rocks, i.e. in two Type A, seven Type B and eighteen Type C rocks, respectively (GOLES *et al.*, 1970a). Our Type B rock 10045-28 has been classified as Type A by AGRELL *et al.* (1970). The elemental abundances of Ti, Mg and Cr in rock 10045 are markedly higher from three other Type B rocks; i.e. Ti, Mg and Cr are ~30 per cent more enriched in rock 10045 than in the other three rocks, and are quite similar to those of Type A rock 10017-31. However, since the ranges of abundances of these elements in two other Type A rocks and seven Type B rocks (GOLES *et al.*, 1970a) overlap each other, and the relative REE abundance pattern of rock 10045-28 closely resembles the Type B REE pattern, at the present stage rock 10045 should either be classified as a Type B rock or a transitional A–B rock.

These major and minor elemental abundance data support the conclusion that the closest meteoritic or terrestrial analogs to the composition of lunar rocks and soil at T.B. are eucrite achondrites and terrestrial pyroxene gabbro (TURKEVICH *et al.*, 1969, FRANZGROTE *et al.*, 1970, OLSTEN, 1969, and others); but as previously pointed out, differences between Ti (lunar soil—eucrite achondrites) and Na (lunar soil—pyroxene gabbro) rule our strict compositional identity.

From a comparison of ratios of Type C abundance/Type D abundance for Ti ($1{\cdot}10 \pm 0{\cdot}06$), Al ($0{\cdot}97 \pm 0{\cdot}02$), Fe ($1{\cdot}15 \pm 0{\cdot}10$), Mg ($0{\cdot}96 \pm 0{\cdot}07$), Ca ($1{\cdot}08 \pm 0{\cdot}08$), Na ($1{\cdot}12 \pm 0{\cdot}05$), K ($1{\cdot}41 \pm 0{\cdot}36$), Cr ($1{\cdot}22 \pm 0{\cdot}12$) and Mn ($1{\cdot}09 \pm 0{\cdot}08$), it appears that breccia rocks are closely related to soil in chemical composition (LSPET, 1969, SCHMITT *et al.*, 1970a, and others). The results of oxygen isotope measurements also support the above conclusion (ONUMA *et al.*, 1970). From rare-gas and shock measurements (LSPET, 1969), it had been suggested previously that breccia rocks may be compacted from lunar soil. However, since the abundances of seventeen lithophilic trace elements, including the fourteen REE of this and other work (e.g. HASKIN *et al.*, 1970), are 5–25 per cent higher in breccia rocks, it is unlikely that breccia rocks were compacted from soil whose composition is comparable to that collected at T.B. However, a 10–25 per cent variation in average abundances of seventeen lithophilic trace elements in 1 g chips from breccia rocks suggests that a similar variation may exist in Sea of Tranquillity soil; therefore, it is possible that breccia rocks may be derived from soil similar to, but not identical to, that at T.B. FUNKHOUSER *et al.* (1970) have observed significant differences in the $^{40}Ar/^{36}Ar$ ratios for breccia rocks and soil and they also conclude that breccia rocks at T.B. were derived from impaction processes on soil other than at T.B.

In summary, from an evaluation of elemental abundances of the bulk major and minor elements of this and many other works, (alluded to in our major and minor elemental abundance comparisons), we conclude that: (1) Types A and B igneous rocks are quite different in chemical composition. ONUMA *et al.*, (1970), measuring $^{18}O/^{16}O$ ratios, concluded that both Types A and B rocks originated from magma of the same oxygen isotopic ratios and perhaps from the same chemical composition. MASON *et al.* (1970) suggest that all crystalline rocks were blown to their present sites by meteoritic impact. However, Type A rocks differ markedly from Type B, especially in their alkali element, and to a lesser degree in their Al, Ca and Ti abundances. (2) Types A and B rocks differ significantly from Type C rocks and soil, especially in their Ti, Al and Fe abundances. (3) Supported by considerable differences in their elemental abundances of trace elements, such as REE and Th (discussed in the latter part), the Type C rock and soil at T.B. are not derived from fragmentation of Types A and B rocks. Age differences between the igneous rocks and the soil and breccia rocks of $\sim 10^9$ yr (e.g. ALBEE *et al.*, 1970, COMPSTON *et al.*, 1970, GAST and HUBBARD, 1970, TATSUMOTO and ROSHOLT, 1970, and others) also precludes any simple derivation of soil and breccia at T.B. from the rock found at T.B.

TRACE ELEMENTS

Rare earth and yttrium

The abundances of twenty-six minor and trace elements, i.e. fourteen rare earth elements, Y, Sc, Cd, In, Ba, V, Co, Zr, Hf, Th, Rb and Cs in one Type A, four Type B, two Type C rocks and Type D soil are listed in Table 5. Total abundances of all fourteen REE and Y included are 525 ppm in Type A rock 10017-31, about 296 ppm in Type B rock 10050-28, 307 ppm in Type C rock 10019-24, and 300 ppm in Type D soil 10084-56, respectively. These total REE + Y abundances are near the upper limit of total REE + Y concentrations observed in terrestrial continental basalts and gabbros, which average ~190 ppm REE + Y content, with a range from 20 to 530 ppm (HASKIN *et al.*, 1966, and FREY *et al.*, 1968). The total REE + Y of 300 ppm found in Types B and C rocks and soil, is about ~60 and ~5 times the total REE + Y found in chondrites and eucrite achondrites, respectively (SCHMITT *et al.*, 1963a, 1964). However, relative to the REE distribution in chondrites or eucrite achondrites, continental basalts are significantly enriched in the light REE (La–Sm). The REE pattern in soils and breccia rocks, relative to the average pattern in twenty-nine chondritic meteorites of all classes, assumed as the distribution in the primordial solar nebula [these assumed best values of SCHMITT *et al.* (1963a, 1964) and HASKIN *et al.* (1968), have been added to the legend of Fig. 1], resembles that found in eucrite achondrites (SCHMITT *et al.*, 1963a, 1964; PHILPOTTS *et al.*, 1967; PHILPOTTS and SCHNETZLER, 1970), abyssal subalkaline basalts (FREY *et al.*, 1968; HASKIN *et al.*, 1970; KAY *et al.*, 1970) and Red Sea submarine tholeiitic basalts (SCHILLING, 1969). However, the total REE abundances in lunar rocks at T.B. are about three and five times higher compared to the total REE abundances in twelve abyssal basalts (Fig. 1) and of five Red Sea tholeiites, which have average total REE + Y concentrations of 88 and 55 ppm, respectively, with a range from 40 to 126 ppm.

Europium is severely depleted in lunar rocks and soils by 71 ± 2 per cent and 58 ± 4 per cent in the average Types A and B rocks, respectively, relative to chondritic meteorites (see Fig. 1 and 2). It seems plausible that plagioclase, which readily incorporates divalent Eu into its structure and also accept trivalent REE in a monotonic selection of increasing abundance with increasing ionic radius (TOWELL *et al.*, 1965, PHILPOTTS *et al.*, 1967; and PHILPOTTS and SCHNETZLER, 1968), may have separated out of a lunar magma and markedly depleted Eu and partially depleted the light REE (GAST and HUBBARD, 1970; GOLES *et al.*, 1970b; HASKIN *et al.*, 1970; MURTHY *et al.*, 1970; PHILPOTTS and SCHNETZLER, 1970; SCHMITT *et al.*, 1970a; and others). However, since the very heavy REE are also depleted relative to a chondritic or achondritic REE distribution pattern (Fig. 1), separation of plagioclase mineral from the melt alone will not account for the observed REE distribution pattern in lunar rocks and soil. The relation between K content and K/Sr ratio also suggest fractional crystallization of plagioclase from a melt (MURTHY *et al.*, 1970). The results of low oxygen fugacity (ANDERSON *et al.*, 1970; and ARRHENIUS *et al.*, 1970), the presence of metallic Fe (CAMERON, 1970; HAGGERTY *et al.*, 1970; and others), of possible divalent Cr (HAGGERTY *et al.*, 1970), and of trivalent Ti (ROSE *et al.*, 1970), indicates

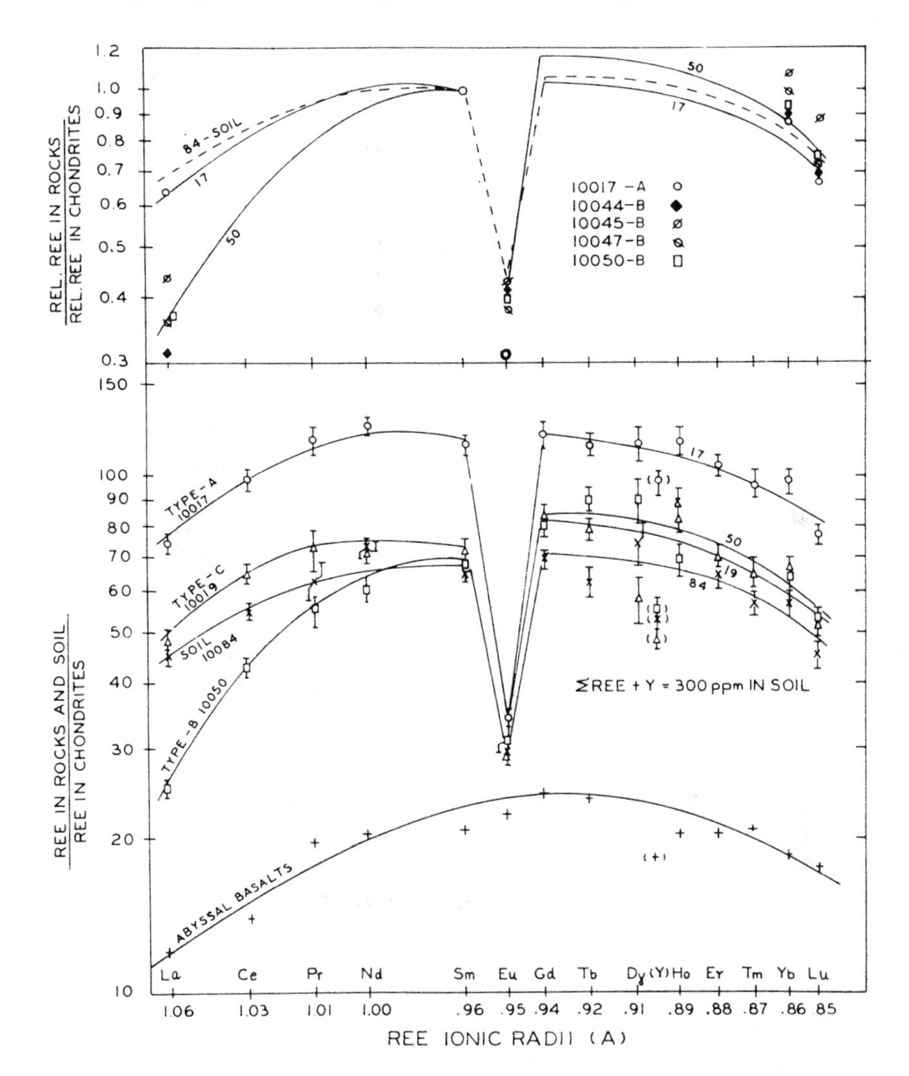

Fig. 1. Upper half: Ratios of relative REE (La, Sm, Eu, Yb and Lu) abundances (Sm ≡ 1·00) in Types A and B rocks to the same relative REE (Sm ≡ 1·00) abundances in twenty-nine chondrites. Lower half: Ratios of absolute REE and Y abundances in Type A rock 10017-31, Type B rock 10050-28, Type C rock 10019-24, lunar soil 10084-56 and twelve terrestrial subalkaline abyssal basalts (Frey *et al.*, 1968) to absolute average REE and Y abundances in twenty-nine chondrites (La 0·32 ppm, Ce 0·86, Pr 0·11, Nd 0·59, Sm 0·196, Eu 0·071, Gd 0·25, Tb 0·048, Dy 0·31, Ho 0·071, Er 0·20, Tm 0·031, Yb 0·19, Lu 0·033 and Y 1·9; ΣREE + Y = 5·2 ppm by Schmitt *et al.*, 1963a, 1964; and Haskin *et al.*, 1968); REE radii, by Templeton and Dauben (1954).

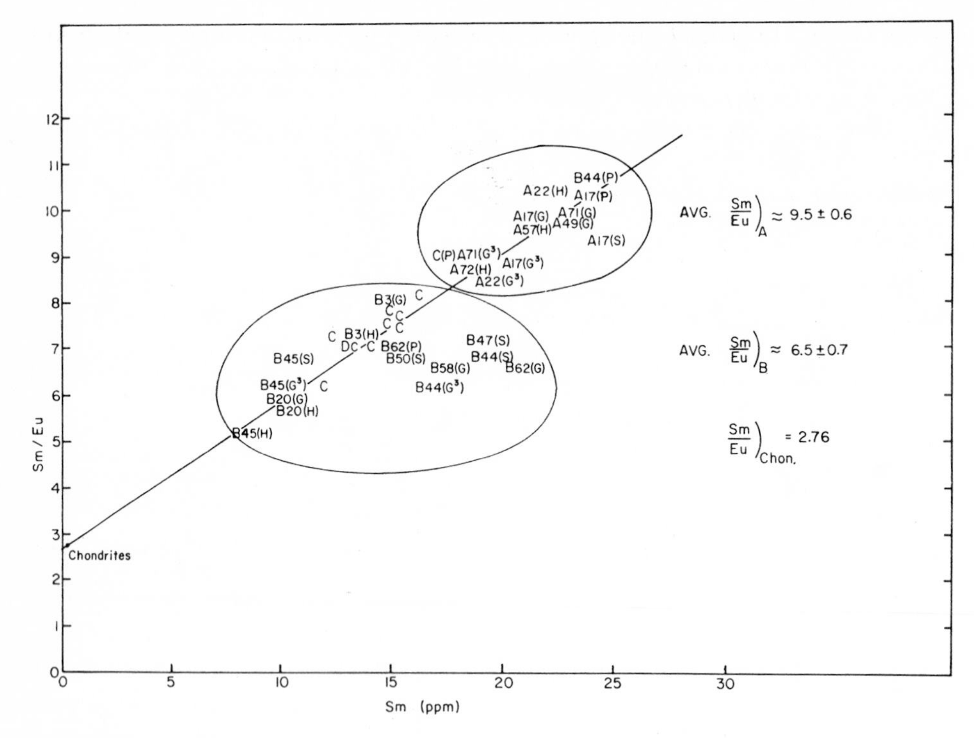

Fig. 2. Sm/Eu ratios versus Sm (ppm) are plotted for Types A, B and C rocks and Type D soil. Center of each letter A, B, C and D indicates exact position of rock Types A, B, C and D. Following each letter, the rock number is listed and the names of investigators are shown in parentheses. G: GAST and HUBBARD (1970), G³: GOLES *et al.* (1970b), H: HASKIN *et al.* (1970), P: PHILPOTTS and SCHNETZLER (1970), and S: SCHMITT *et al.* (1970a).

that mineral crystallization from the melt occurred under highly reducing conditions very conducive for Eu reduction to the divalent state with subsequent Eu depletion by separation of plagioclase from such a melt. The normal Ce abundances relative to the adjacent REE, La, and Pr, and the highly reducing conditions suggest a trivalent state for Ce in the magma.

Yttrium is similarly depleted by ~20–30 per cent on a relative basis in all rocks and soil samples and ~15 per cent in subalkaline basalts (Fig. 1). Scandium, with trivalent properties that approximate those of the 14 REE + Y, is enriched by factors of ~7 and ~2 relative to Sc abundances in chondrites and eucrite achondrites (SCHMITT *et al.*, 1963a, 1964, 1970c).

Before discussing in detail elemental abundances of trace elements in lunar sample analyses, it seems worthwhile to compare values of this study with other published values. In Table 6 the results of this work are compared with other data for lunar soil (10084). Also results of GAST and HUBBARD (1970), GOLES *et al.* (1970b), HASKIN *et al.* (1970), PHILPOTTS and SCHNETZLER (1970) and of this work are compared in Fig. 2, using a linear correlation between Sm/Eu vs. Sm as suggested by HASKIN *et al.* (1970). In general, the agreement between this work and others is quite satisfactory. It is also noted that the La, Sm, Eu, Yb and Lu abundances obtained via RNAA agree with those via INAA.

Lanthanum, Sm and Eu, and Yb and Lu, represent the light, middle and heavy REE, respectively. The selective summation of La, Sm, Eu, Yb and Lu abundances of 74 ppm was found in Type A rock 10017-31; 50, 30, 52 and 40 ppm (averaged 43 ppm) were found in Type B rocks 10044-29, 10045-28, 10047-33 and 10050-28, respectively. (Total abundances of all 14 REE + Y will be about 7·5 times greater than the selective abundances of these five REE, La, Sm, Eu, Yb and Lu.) Corresponding abundance summations of 45 ppm and 52 ppm were obtained in Type C rocks 10019-24, and 10059-26, respectively, and the value of 41 ppm was found in lunar soil. The selective REE abundance total of 74 ppm in Type A rock 10017-31 is 1·7 times higher than the average selective REE in Type B rocks and 1·5 times higher than the average selective REE abundances in Type C rocks and soil.

Abundances of Sm in six Type A rocks, one rock analyzed in this work and by five other investigators (see legend of Fig. 2 for references), and higher than in Types B and C rocks and soil, i.e. 21 $\pm$ 2 ppm Sm in Type A rocks with abundances ranging from 18 to 24 ppm Sm for 11 individual determinations. Abundances of Sm in eight Type B rocks, four rocks from this work and from four others (see Fig. 2 for references) average 13·8 $\pm$ 4·0 ppm Sm, with a range from 8 to 20 ppm Sm. It seems evident from Fig. 2 that the Sm/Eu ratios fall into two distinct groups, naturally divided into Types A and B rock groups. The higher average Sm/Eu ratio of 9·5 $\pm$ 0·6 for the Type A rock group implies a significant degree of Eu depletion, 71 $\pm$ 2 per cent, compared to the average in twenty-nine chondrites. The lower Sm/Eu ratio of 6·5 $\pm$ 0·7 for Type B rocks (average of eight rocks) implies 58 $\pm$ 4% Eu depletion, relative to twenty-nine chondrites.

For rock powder sample 10044, PHILPOTTS and SCHNETZLER (1970) reported markedly higher Sm abundance and Sm/Eu ratio that the average of the three determinations for rock 10044 powder by GOLES *et al.* (1970b), SCHMITT *et al.* (1970a), and TUREKIAN and KHARKAR (1970), whose values agree within experimental errors. All four Sm and Sm/Eu ratios for Type C rocks reported by PHILPOTTS and SCHNETZLER (1970) are consistently higher than other 9 determinations (see Fig. 2 for reference). For rock 10060, the 17·5 ppm Sm obtained by PHILPOTTS and SCHNETZLER (1970) is also higher than 15·4 ppm Sm reported by HASKIN *et al.* (1970) and 14·8 ppm Sm found by GOLES *et al.* (1970b). In calculations of the average Sm/Eu ratio for the Type B rock group of Fig. 2, the PHILPOTTS and SCHNETZLER (1970) rock 10044 value has been excluded.

Abundance differences of some trace elements, viz. REE and Th, are observed for two rock chips split from larger fragments for rocks 10017-31, 10050-28 and 10059-26 (Table 5). In addition, variations of ~10 per cent observed in four Sm values for rock 10017, two Sm values for rock 10022 and two Sm values for rock 10071 (Fig. 2), may imply heterogeneous distribution of Sm in 0·5–1 g specimens from the same rock. For Type A rock 10017, the following Sm abundances are reported, i.e. 21·0 ppm (GAST and HUBBARD, 1970), 20·0 $\pm$ 0·2 ppm (GOLES *et al.*, 1970b), 23·4 ppm (PHILPOTTS and SCHNETZLER, 1970), 22·5 $\pm$ 1 and 27 $\pm$ 1 ppm in two different

Table 5. Abundances (ppm) of the fourteen rare-earth elements and Y, Sc, Co, Cd, In, V, Zr, Hf, Th, Rb, activation

	Type A rock		Type B rocks								
Element	10017-33 a chip 0·637 g	10017-33 b chip 0·343 g	10044-29 a powder >100 mesh 0·849 g	10044-29 b powder <100 mesh 1·027 g	10045-28 a chip 0·496 g	10045-28 b chip 0·544 g	10047-33 a powder >100 mesh 0·351 g	10047-33 b powder <100 mesh 0·563 g	10050-28 a chip 0·600 g	10050-28 b chip 0·575 g	Av.
Rb	6·0*										
Cs	0·26*										
Sc	87	93	105	84	88	92	120	90	96	89	94 ±5
V	49	43	50	27	93	113	73	57	99	97	75 ±31
Co	35	37	15	16	23	26	15	19	19	19	19 ±3
Zr	1200	1570	1140	420	270	430	70	800	650	390	520 ±140
Hf	23	26	15	18	9	9	15	16	15	12	14 ±3
Cd	0·044*										
In	0·0010*										
Ba	350	410	260	230	370	360	250	290	70	68	240 ±120
La	23*									8·0*	8·9 ±1·6
	25	28	8·7	10·0	7·3	6·7	8·5	12·2	8·9	7·5	
Ce	84*									37*	
Pr	12·9*									6·2*	
Nd	74*									36*	
Sm	23*									12·9*	15·5 ±4·2
	22	27	17·7	19·7	10·1	9·6	16·1	20·2	16·2	14·0	
Eu‡	2·3*									2·2*	2·3 ±0·6
	2·6	2·9	2·0	3·4	1·5	1·4	2·2	2·9	2·2	2·1	
Gd	30*									19·9*	
Tb	5·5*									4·3*	
Dy	36*									28*	
Ho	8·3*									4·9*	
Er	21*									—	
Tm	3·0*									—	
Yb	17·8*										14·2 ±3·4
	19	23	16	16	10·5	9·8	18	18	13·6	12·2	
Lu	2·5*										2·0 ±0·4
	2·6	3·0	2·1	2·2	1·4	1·5	2·3	2·3	1·9	1·7	
Y	184*									104*	
Th‡	4·3	6·4	0·6	1·4	1·8	1·9	2·4	1·6	2·0	1·6	1·6 ±0·4

Cs and Ba in Types A, B and C lunar rocks and in Type D lunar soil by instrumental and radiochemical analysis†

Type C rocks					Type D, soil								
10019-24 a chip 0·532 g	10019-24 b chip 0·497 g	10059-26 a chip 0·600 g	10059-26 b chip 0·353 g	Av.	0084-56 a 1·125 g	10084-56 b 1·068 g	10084-56 d >100 mesh 0·731 g	10084-56 e 100–200 mesh 0·468 g	10084-56 f <200 mesh 0·957 g	10084-56 g <200 mesh 0·937 g	avg. 84a, b (2·19 g)	mass wt'd. av. (5·29 g)	Anal. Error %
3·4*					3·0*								5–10
0·41*					0·15*								10
64	64	68	67	66 ±3	58	58	63	70	56	55	58 ±1	59 ±6	5
86	98	59	101	86 ±15	60	67	52	98	51	38	63 ±4	58 ±21	15–20
36	34	42	40	38 ±4	33	29	36	38	35	39	31 ±3	34 ±4	10–15
490	220	330	360	350 ±110	720	200	570	300	270	400	460 ±360	420 ±200	20–35
13	14	16	13	14 ±1	9	9	10	13	12	10	9 ±1	10 ±2	10
—					0·056*						0·056* ±0·005		10
0·0052*					0·58*								5
340	130	210	180	210 ±90	160	240	60	130	230	120	200 ±60	170 ±70	20–70
15·1*				16·4 ±2·2	14·3*						14·3 ±0·7		5
15·6	14·1	18·4	17·3		14·4	14·0	14·4	13·5	14·2	13·9	14·2 ±0·2	14·1 ±0·3	5
56*					47*						47 ±2		5
7·9					6·9*						6·9 ±0·7		10
42*					43*						43 ±2		5
13·8*				15·2 ±1·2	14·0*								5
14·3	14·2	16·0	16·5		12·3	12·2	13·5	13·2	12·0	12·8	12·6 ±0·6	12·8 ±0·6	5
1·9*				2·0 ±0·1	2·2*								5
2·2	2·0	2·1	1·9		2·0	1·8	1·7	1·6	1·5	2·0	2·0 ±0·2	1·8 ±0·2	10
20·5*					17·2*						17·2 ±0·8		5
3·8*					3·0*						3·0 ±0·2		5
18*					23*						23 ±2		10
5·9*					6·3*						6·3 ±1·3		20
14·1*					13*						13 ±1		5
2·0*					1·8*						1·8 ±0·1		5
12·4*				13·0 ±0·8	11·1*								5
13·0	12·0	13·7	13·6		10·6	10·9	11·7	11·9	10·5	11·6	10·8 ±0·2	11·1 ±0·6	5
1·6*				1·8 ±0·1	1·5*								5
1·7	1·6	1·9	1·9		1·5	1·4	1·6	1·6	1·4	1·5	1·5 ±0·1	1·5 ±0·1	5
91*					96*						96 ±5		5
2·7	3·2	4·6	3·8	3·6 ±0·8	3·1	2·1	2·3	2·8	2·3	3·2	2·6 ±0·7	2·6 ±0·5	10–20

* Abundance obtained via radiochemical neutron activation analysis.
† Sample identifications are identical to those described in Table 3. Estimated analytical errors of single determinations (see last column) include counting statistics and other experimental errors for counting geometry, pipetting, standard solution analyses, etc.
‡ Instrumental neutron activation analysis results of Eu and Th are relative to Eu and Th abundances of 1·1 and 2·6 ppm in the U.S.G.S. standard rock W-1.

0·5 g chips (SCHMITT *et al.*, 1970c) and 25 ppm (WIIK and OJANPERÄ, 1970). An average Sm abundance of 23·2 $\pm$ 2·6 ppm indicates a homogeneity of $\pm$10 per cent in 0·5–1 g specimen for this rock. As seen in Fig. 1, the relative REE distribution pattern for Type A rock 10017 is very similar to those of breccia rocks and soil. On the other hand, the light REE, especially La, have been highly fractionated in four Type B rocks (Fig. 1 upper half). The average La/Sm ratio of 0·57 with a range from 0·5 to 0·7, found in Type B rocks is markedly different from those of 1·0, 1·1 and 1·1 found in Types A and C rocks and soil, respectively. For the heavier REE, we can not observe any significant difference in all rock types and soil samples, i.e. average Lu/Sm ratios for Types A, B and C and soil are 0·11, 0·13, 0·12 and 0·12 respectively, and average Yb/Lu ratios are 7·4, 7·1, 7·2 and 7·1, respectively.

The Sm/Eu ratios of 9·5 $\pm$ 0·6 and 6·5 $\pm$ 0·7 for Types A and B rock groups respectively, may be compared to ratios of 7·4 $\pm$ 0·4 and 7·1 $\pm$ 0·2 for Type C rocks and Type D soil, respectively. Within experimental error, the ratios of Type C breccia rocks and soil coincide and that for Type B rocks may also overlap. From this Sm/Eu ratio criterion, it appears that any derivation of soil from fragmentation of rocks must be linked to rocks with similar Sm/Eu ratios like in Type B rocks. Since there are significant differences in major and minor elemental abundances (Table 4) and also different La/Sm ratios (discussed above) between Type B rocks and soil, it is implausible that the soil at T.B. has been derived from Type B rocks. Similar comparison precludes Type A rocks as precursors for soil or breccia rocks.

From the results of REE abundances of this study and other work (see Fig. 2 references), we conclude that: (1) Type A rocks are distinguished from Types B and C rocks and soil with higher abundances in total REE and higher depletion of Eu in Type A rocks; (2) For Type B rocks total REE + Y abundances (average about 300 ppm) range within a factor of two (Sm abundances range from 8 to 20 ppm); (3) In Type B rocks, La is fractionated, relative to the chondritic or achondritic pattern, to a greater degree than are the heavier light REE, i.e. Ce, Pr and Nd: (4) Despite the general similarities in the REE distribution patterns between lunar rocks and soil and terrestrial subalkaline abyssal basalts (Fig. 1), the average Sm/Eu ratios of 9·5 $\pm$ 0·6 and 6·5 $\pm$ 0·7 in Types A and B rocks and 7·4 and 7·1 ratios in Type C rocks and soil, respectively, differ strikingly from the average Sm/Eu ratio of 2·6 in twelve abyssal basalt (FREY *et al.*, 1968) and the 4·1 ratio observed in one olivine tholeiite D-3 basalt from the East Pacific Rise (FREY *et al.*, 1968). Clearly, different chemistries are involved in the generation of lunar and terrestrial basalts; (5) Ce abundances are normal, thereby indicating a trivalent oxidation state and suggesting a reducing environment; (6) Yttrium, relative to the chondritic REE + Y distribution pattern, is depleted to about some degree in lunar rocks and soils and in twelve abyssal basalts (FREY *et al.*, 1968); and (7) in general, only the light REE (La–Eu) are enriched in the fine fractions, i.e. $<$100 mesh and rich in plagioclase mineral, of Type B rock 10044-29 (15% La enriched) and 10047-33 (48% La enriched). As expected, Eu has been enriched to a greater percentage in both fine fractions compared to Sm. This again indicates a divalent Eu oxidation state and enrichment of Eu in plagioclase minerals as reported by GOLES *et al.* (1970b) and PHILPOTTS and SCHNETZLER (1970). (8) The average REE and other trace elemental abundances in

Type B rocks are about half the average abundances in Type A rocks; this observation is consistent with compositional differences between Types A and B rocks (see Table 4) for major and minor elements. This suggests either different cooling rates for these two types of rocks, probably related to different depths of formation (CAMERON, 1970; and MASON *et al.*, 1970) or dissimilar parental magma were responsible for generation of these two types of rocks (FREDRIKSSON *et al.*, 1970).

If a chondritic-like composition is assumed for the original moon and the REE distribution pattern observed in Apollo 11 rocks is the result of melting and fractional recrystallization of the moon with almost complete concentration (>90 per cent) of the REE in the lunar surface, then a 9-km thick crust may be calculated. Since it is probably that the Ca-rich rocks and soil of the Sea of Tranquillity are abnormally enriched in REE and other lithophilic trace elements, the thickness of the crystal layer will increase inversely with the average enrichment factor. Since complete melting of the moon does not seem possible (UREY, 1968; and UREY and MACDONALD, 1969) either overall surface heating (e.g. a high temperature sun) and accretion heating or localized surface melting resulting from meteoritic impact (UREY, 1968) may have fractionated only the outer surface of the moon into two or more layers. In any case, for every volume of lunar surface rock that has a composition similar to the Apollo 11 rocks about sixty volumes of the subsurface layer probably are severely depleted in REE and other heavy lithophilic trace elements relative to the average of chondritic meteorites. The average REE content in this subsurface layer may approximate that found in Ca-poor hypersthene achondrites, such as Johnstown and Shalka (SCHMITT *et al.*, 1963a, 1964) in which the total REE contents are considerably depleted relative to chondrites, i.e. about 5–40 per cent of the REE content in chondritic meteorites. For this type of achondrite, the light REE (La–Gd) are normal while the heavy REE (Tb–Lu) are progressively enriched from Tb–Lu, all relative to the chondritic pattern. Material balance arguments suggest that separation of plagioclase and hypersthene achondritic like minerals (specifically like Shalka) from an original melt, depleted in high temperature minerals, would yield the REE distribution pattern observed in Apollo 11 rocks and soil (SCHMITT *et al.*, 1970a).

Other trace elements

In Table 7, average abundances of Sc, Co, Zr, Hf, Ba, V, Th, Rb and Cs in Types A, B and C rocks and Type D soil of this work have been compared with other reliable values (see Table 7 for references).

Sc. For all types of rocks and soil, average Sc abundances of this work agree well with those obtained by seven other independent investigators (see Table 7 for references) and by TAYLOR *et al.* (1970). The trivalent chemical properties of Sc and the REE are similar to each other in some aspects. However, there is no Sc abundances difference between Types A and B rocks and a significant difference of a factor of $\sim$2 in total REE abundances between these two rock types. The average value of 86 ± 6 ppm Sc for six Type A rocks and the 87 ± 6 ppm for five Type B rocks are identical. The average Sc abundances for these two types of rocks are $\sim$30 per cent higher than the 67 ± 3 ppm Sc for 10 Type C rocks and are $\sim$40 per cent higher than the 60 ± 4 ppm for soil. The dispersion range of Sc abundances of $\sim$7 per cent is

Table 6. Comparison of 25 trace

Element	10084-56 (this work)	HASKIN *et al.* (1970)	PHILPOTTS and SCHNETZLER (1970)[1]	GAST and HUBBARD (1970)	WÄNKE *et al.* (1970)	GOLES *et al.* (1970b)	SMALES *et al.* (1970)
La	14·2 ± 0·2	16·6 ± 0·4		16·3 ± 0·5	15 ± 1	14·3 ± 0·4	19
Ce	47 ± 2	47 ± 1	46·1	47·6 ± 1·4	63 ± 10	51·6 ± 1·8	46 ± 2
Pr	6·9 ± 0·7				5·3 ± 0·5		
Nd	43 ± 2	43 ± 6	40·5	36·8 ± 1·1	47 ± 5		37 ± 2
Sm	12·6 ± 0·2	13·7 ± 0·1	13·9	13·1 ± 0·4	7·6 ± 1·9	12·6 ± 0·4	13, 14
Eu	2·0 ± 0·2	1·77 ± 0·02	1·77	1·77 ± 0·05	1·67	1·78 ± 0·07	1·8, 2·0
Gd	17·2 ± 0·8	15 ± 2		17·2 ± 0·9	18 ± 2		
Tb	3·0 ± 0·2	3·2 ± 0·2			2·8	2·6 ± 0·2	4
Dy	23 ± 2	20·1 ± 0·1	19·5	19·7 ± 1·0	17 ± 2		23
Ho	6·3 ± 1·3	4·2 — 4·7			4·6 ± 0·5		5
Er	13 ± 1	12 ± 2	11·7	12·1 ± 0·6	9·5 ± 0·1		
Tm	1·8 ± 0·1						
Yb	10·8 ± 0·2	11·3 ± 0·3	10·6	11·5 ± 0·6	8·3	10·9 ± 0·6	12
Lu	1·5 ± 0·1	1·66 ± 0·01		1·58 ± 0·11	1·30	1·58 ± 0·06	1·5
Y	96 ± 5						
Sc	58 ± 1				61	59·5 ± 0·7	66
Ba	200 ± 60		170	176 ± 9		155 ± 30	174 ± 9
In	0·58 ± 0·03				0·75 ± 0·07		0·5, 2
Cd	0·056 ± 0·005						
Rb	3·0 ± 0·2		2·78	2·79 ± 0·08	3·0 ± 0·3		3·0 ± 0·2
Cs	(0·18)			0·104 ± 0·01	0·12 ± 0·01		0·12 ± 0·01
Co	31 ± 3				27·2	30·5 ± 0·5	25 ± 3
V	63 ± 4						
Zr	460 ± 360					240 ± 50	
Hf	9 ± 1				15·6	10·6 ± 0·3	11
Th	2·6 ± 0·7				1·25 ± 0·13		
Method	INAA + RNAA	RNAA	MS[b]	MS	INAA + RNAA	INAA	NAA

very tight for all types of rocks. The mean Sc abundances in ten breccia rocks is statistically higher than the mean abundance found in soil, this observation reinforces the concept that breccia rocks at T.B. are not derived from the analyzed soil at T.B. The striking uniformity of Sc, together with Si, O, Fe and Mg, in Types A and B rocks, suggest that these groups were derived by chemical differentiation from a common source or magma. Scandium is enriched in soil by factors of $\sim$7 and $\sim$2 relative to Sc in chondrites and achondrites (SCHMITT *et al.*, 1963a, 1964 and 1970c); this contrasts with REE enrichments in lunar soil of $\sim$60 and $\sim$5 relative to the stone meteoritic types. Clearly, chemical characteristics such as different solubilities and complexing properties, must be responsible for the large observed variations in enrichments. Possibly a large fraction of Sc coprecipitated with crystallization and settling of the high temperature olivine and pyroxene minerals. Almost the same amounts of average Sc abundances in lunar rocks and soil are found in oceanic tholeiite (ENGEL *et al.*, 1965).

Co. Average Co abundances of this and other studies (see Table 7 for references) agree within $\sim 1\sigma$ limit and are about two times higher than those of reported by LSPET (1969) and TAYLOR *et al.* (1970). A consistent correlation is observed between elements in Type D soil (10084)

Element	Wiik and Ojanperä (1970)	Morrison *et al.* (1970)[2]	Turekian and Kharkar (1970)[3]	Taylor *et al.* (1970)	Compston *et al.* (1970)	Others
La		22[a,d]	20		21	16[4]
Ce		50[a,d]	79		58	
Pr		9[d]			10	
Nd		46[a,d]			33	
Sm	18[a]	18[a,d]	8·7			
Eu	1·8[a]	1·9[a,d]	2·0			
Gd		20[a,d]				
Tb	3[a]	3·8[a,d]				
Dy		25[d]	17			
Ho		6[a,d]				
Er		15[d]				
Tm		1·2[a,d]	1·3			
Yb	12[a]	12[a,d]	9·5	2·5		
Lu	3[a]	1·4[a,d]				
Y	120[c]	150[d]		130	99	81[4]
Sc	59[c]	60[a]	80	55		56[4]
Ba		220[a,d]		70	134	210[4]
In		0·5[d]				0·77,[5] 0·68[12]
Cd		0·3[d]				0·053, 0·035[5]
Rb		4·4[a,d]			2·96	3·21,[5] 2·9,[11] 2·83,[7] 2·7[4]
Cs		0·2[a,d]				0·094, 0·098[5]
Co	32[a,c]	40[a,d]	33	18	34	26·8, 28·1,[5] 24[4]
V	71[c]	78[a,d]		42	36	50[4]
Zr	380[c]	390[a]	380	400	318	273[4]
Hf	8[a]	9[a,d]	11			
Th	3[a]	2·3[a,d]			2·5	2·09,[6] 2·27, 2·08,[8] 2·17,[9] 2·26[10]
Method	RNAA[a] + ES[c]	NAA[a] + SSMS[d]	INAA + RNAA	S[e]	X[f]	

(a) INAA, RNAA, Neutron activation analysis.
(b) MS, Mass spectroscopy.
(c) ES, Emission spectroscopy.
(d) SSMS, spark source mass spectrography.
(e) S, optical spectrography.
(f) X, X-ray fluorescence spectrometry.
1. The standard deviation (σ) is about ± 2 per cent.
2. σ by NAA and by SSMS are 10 per cent and 5–25 per cent, respectively.
3. 20 milligram samples were used. σ is 5–10 per cent.
4. ES by Annell and Helz (1970).
5. RNAA by Keays *et al.* (1970).
6. MS by Tatsumoto and Rosholt (1970).
7. γ-ray spectrometry by Perkins *et al.* (1970).
8. MS by Gopalan *et al.* (1970).
9. MS by Silver (1970).
10. MS by Wanless *et al.* (1970).
11. MS by Murthy *et al.* (1970).
12. NAA by Baedecker and Wasson (1970).

average Co and REE abundances in Types A and B rocks. The average abundances of 29 $\pm$ 2 ppm Co for six Type A rocks, 30 $\pm$ 3 ppm Co for eleven Type B rocks and 32 $\pm$ 5 ppm Co for soil are identical and two times higher than those of 15 $\pm$ 3 ppm Co for eight Type B rocks. Cobalt also serves to distinguish Type A from Type B rocks.

Table 7. Comparison of trace elemental abundances in Apollo 11 lunar rocks and soil

	Type A			Type B			Type C			Type D	
Element	This work (1)*	Others	TAYLOR *et al.* (1970) (7)	This work (4)	Others	TAYLOR *et al.* (1970) (8)	This work (2)	Others	TAYLOR *et al.* (1970) (3)	This work	Others
Sc	89 ±3	86[1](6) ±6	94	94 ±5	87[2](5) ±5	98	66 ±3	67[3](10) ±3	62	58 ±1	60[4] ±4
Co	36 ±3	29[5](6) ±2	13	19 ±3	15[6](8) ±3	9	38 ±4	30[7](11) ±3	12	31 ±3	32[8] ±5
Ba	370 ±80	320[9](7) ±40	120	240 ±120	170[9](8) ±80	100	160 ±30	230[10](10) ±20	100	200 ±60	180[11] ±30
V	47 ±17	70[12](5) ±10	32	75 ±30	60[13](3) ±20	41	86 ±15	62[14](7) ±6	27	63 ±4	66[15] ±15
Zr	1330 ±300	560[16](6) ±50	1100	540 ±140	360[17](8) ±80	870	720 ±150	410[17](11) ±90	950	460 ±360	360[18] ±50
Hf	24 ±3	19[19](5) ±2	—	14 ±3	11[20](5) ±3	—	14 ±1	11[21](5) ±1	—	12 ±1	11[22] ±2
Th	5·0 ±0·6	3·5[23](3) ±0·1	—	1·6 ±0·4	1·0[23](6) ±0·1	—	3·6 ±0·8	2·5[24](5) ±0·3	—	2·6 ±0·7	2·3[25] ±0·3
Rb	6·0 ±0·3	5·6[26](8) ±0·4	4·8	—	1·0[27](8) ±0·3	2·7	3·7(1) ±0·4	3·6[28](10) ±0·5	3·1	3·0 ±0·2	2·9[29] ±0·2
Cs	0·26 ±0·03	0·17[30](4) ±0·02	—	—	0·03[31](2) ±0·01	—	0·41(1) ±0·05	0·13[31](4) ±0·03	—	0·15 ±0·02	0·11[32] ±0·01
Sample	17	17, 22, 44, 49, 57, 69, 71, 72		44, 45, 57, 50	3, 20, 44, 45, 47, 50, 58, 62		19, 59	18, 19, 21, 46, 48, 59, 60, 61, 65, 68, 73		84	84

* Number of analyzed rocks are given in parentheses.

1. AN, GOL, MO, WÄ and WI (1970); first two or three letters of each full name reference are given.
2. AN, GOL, MO and WÄ.
3. AN, GOL, MO, SM and WÄ.
4. AN, GOL, MO, SM, TAY, WÄ and WI.
5. AN, GOL, KE, MO, WÄ and WI.
6. AN, GOL, KE, MO, TU and WÄ.
7. AN, GOL, MO, SM, TU and WÄ.
8. AN, CO, GOL, KE, MO, SM, TU WÄ and WI.
9. AN, CO, GA, GOL, MO, MU and PH.
10. AN, CO, GA, GOL, MO, MU, PH and SM.
11. AN, CO, GA, GOL, MO, PH and SM.
12. AN, MO, WI.
13. AN, MO.
14. AN, MO and SM.
15. AN, MO and WI.
16. AN, CO, MO and WI.
17. AN, CO, GOL, MO and TU.
18. CO, MO, TAY, TU and WI.
19. GO, MO and WÄ.
20. GO, MO, TU and WÄ.
21. GO, MO, SM and TU.
22. GO, MO, SM, TU and WÄ.
23. CO, MO, O, PER, SIL, TAT and WÄ.
24. CO, MO, O, SIL and TAT.
25. CO, GOP, MO, PER, SIL, TAT, WAN and WI.
26. AN, CO, GA, GOP, HUR, KE, MU, PH and WÄ.
27. AN, CO, GOP, HUR, KE, MO, MU and PH.
28. AN, CO, GA, KE, MO, MU, PH and SM.
29. AN, CO, GA, GOP, KE, MU, PH, SM and WA.
30. GA, KE and WÄ.
31. GA and KE.
32. GA, KE, SM and WÄ.

Assuming an average Co content of 700 ppm in ordinary chondritic meteorites (UREY, 1964), the calculated <4 per cent admixture of chondritic meteorites to lunar rocks and soil is consistent with about 3 per cent meteoritic imput on lunar surface calculated by ÖPIK (1969). If the primitive moon had a chondritic composition (700 ppm Co abundance) and was subjected to melting in a reducing environment, it is plausible that Fe metal, Co and other siderophilic elements would have precipitated out of the magma (UREY, 1964). Such a simple melting and precipitation mechanism is not compatible with Ni/Co ratios of about 24 (UREY, 1964), 7 and <0·3 (WIIK and OJANPERÄ, 1970 and WÄNKE *et al.*, 1970) in chondritic meteorites, lunar soil and

Types A and B rocks, respectively. KEAYS *et al.* (1970) suggest that an admixture of 1·5–2 per cent carbonaceous chondrite-like material to Types A and B rocks will yield the observed trace elemental abundances in Type C breccia and Type D lunar soil. Assuming 1·5 per cent addition of 480 ppm Co in Cc_1 (carbonaceous) chondrites (SCHMITT *et al.*, 1970c) to averages of 30 ppm and 16 ppm Co in Types A and B rocks, respectively, lunar dust and breccia should contain either 37 or 23 ppm Co, with an average value of 30 ppm. This calculated Co average overlaps the observed 31 ppm in breccia rocks and soil. Such an elementary analogy suggests that lunar soil and breccia are the average composition of roughly equal amounts of fragmented A *and* B rocks. As previously emphasized, significant differences in major and minor elemental abundances preclude a direct derivation of breccia and lunar soil from Types A and B rocks at T.B.

Ba. Because of the relatively low Ba abundances in lunar samples, Ba determinations via INAA have rather large standard deviations due to counting statistics. However, agreements between average Ba abundances of this and other works (see Table 7 for references) are quite satisfactory. These average Ba abundances, considered very reliable, are two to three times higher than those of LSPET (1969) and TAYLOR (1970). In general, Ba abundances in Type A rocks are markedly higher than Ba in Type B and C rocks and soil; Ba abundances in Type B and C rocks and soil are identical within 1σ errors. The 320 ± 40 ppm Ba average abundance for seven Type A rocks is ~60 per cent higher than the 170 ± 80 ppm Ba for eight Type B rock, 230 ± 20 ppm Ba for 10 Type C rocks and the 180 ± 30 ppm for soil, respectively. We note that the trace elements Ba, the 14 REE + Y, Zr and Hf have been enriched to about the same degree in Type A rocks relative to Type B rocks. Average Ba abundances of ~4 ppm and ~14 ppm are found in chondrites (UREY, 1964) and in oceanic tholeiites (ENGEL *et al.*, 1965), respectively. Relative to chondritic meteorites and oceanic tholeiites, enrichment factors of Ba in Apollo 11 lunar matter are 50 ~ 80 (the REE are enriched by ~60) and 10 ~ 20 (the REE are enriched by ~3), respectively.

It has been postulated by many investigators (see REE discussion) that plagioclase separation will deplete a magma in Eu (as Eu^{2+}) relative to the other REE. PHILPOTTS and SCHNETZLER (1970) have also observed an enrichment of Ba in plagioclase relative to pyroxene and opaque minerals all separated from two Type B rocks. If the original lunar material were chondritic-like, followed by melting and plagioclase segregation with the observed partition coefficients, then Eu^{2+} and Ba^{2+} must have been decidedly depleted in the residue. This does not seem compatible with the experimental observations that the REE and Ba have been enriched to about the same degree in both Types A and B rocks relative to chondrites. Either Ba was enriched with respect to the REE in the primary magma or the presently observed partition coefficients do not apply to plagioclase precipitation from a primary magma. The same impasse is reached if the original magma composition were like the eucrite achondrites.

V. For Types B and C rocks and soil, average V abundances of this work agree with other values within ~1σ limit (see Table 7). The average V abundances obtained by LSPET (1969) and TAYLOR *et al.* (1970) are ~40 per cent lower than these average values.

In general, V abundances (60–70 ppm) in Apollo 11 lunar rocks and soil are uniformly distributed. These values fall within the range for chondrite and eucrite achondrites (NICHIPORUK, 1970). On the other hand, for oceanic tholeiites, considerably higher V abundances of ~300 ppm are reported (ENGEL *et al.*, 1965).

Hf. Due to the large variation of reported Hf values and the lack of enough Hf determinations, it is rather difficult to compare our Hf values with other published values. For example, for Type A rock 10017, the following Hf abundances are reported: 17·9, 12, 23·2 and 24 ± 3 ppm by GOLES *et al.* (1970), WIIK and OJANPERÄ (1970), WÄNKE *et al.* (1970) and this work, respectively. Significant variations in Hf abundances for other rocks exist for those reported by the above investigators with values obtained by MORRISON *et al.* (1970), TUREKIAN and KHARKAR (1970) and SMALES *et al.* (1970). In general, the values obtained by WIIK and OJANPERÄ (1970) are lower. Despite the apparent experimental bias in Hf analyses of this and all other studies average Hf abundances of this work are compared with other published values (Table 7).

A real difference in average Hf abundances exists between Type A rocks *and* Types B and C rocks and soil. The average value of 19 ± 2 ppm Hf for five Type A rocks is about two times higher than those of 11 ppm for five Type B rocks, five Type C rocks and soil. There is a positive correlation between average Hf, Ba and the REE abundances in these rock groups. As expected, this correlation holds for average Zr abundances.

Zr. Our Zr values have large errors mainly derived from the high Compton continuum subtraction. However since all samples are activated at the same time and counted under the same conditions, our Zr abundance values, systematically higher by ~50 per cent, are meaningful on a relative basis.

Ratios of the average Zr abundance (of other studies)/average Hf abundance (of this and other studies) for Types A, B and C rocks and soil are 28 ± 4, 29 ± 12, 34 ± 9 and 31 ± 6, respectively. GOLES *et al.* (1970b) also using INAA for Zr and Hf analyses, reported Zr/Hf ratios that were 10–40 per cent lower for the rock and soil groups. An average Zr/Hf ratio of ~30 in lunar rocks and soil is compared with ~36 in chondrites and ~60 in eucrite achondrites (EHMANN and REBAGAY, 1970) and 50, 50, 44, 33 in selected terrestrial granites, basalts, subalkaline tholeiite basalt, ecologite (REBAGAY, 1969), all determined by radiochemical neutron activation analysis. In conclusion, the Zr/Hf ratio in lunar rocks and soil approximates that found in basic terrestrial rocks and chondritic meteorites.

Th. In general, average Th abundances of this work agree with those obtained by other more accurate methods within 1·5σ (Table 7). Because of a very high Compton continuum subtraction in the γ-ray spectra, a systematic bias of 50 per cent may have been introduced. The rather divergent Th abundances of 0·53 and 3·4 ppm, for Type B rocks 10050 and 10071, respectively by TATSUMOTO and ROSHOLT (1970), have been excluded in calculation of the average Th content of 1·0 ± 0·1 ppm for six Type B rocks. Comparing the average abundances of Th obtained by others (see Table 7) we note the average Th content in Type A rocks is enriched by a factor of ~3·5 relative to Type B rocks which are also different from Type C rocks and soil. Corresponding ratios for other heavy trace elements, viz. REE and Ba that also have large ionic radii and are present in the lithphilic phase, are ~1·8. The average values of the 2·5 ± 0·3 ppm Th for five Type-C rocks and 2·3 ± 0·3 ppm Th for soil are identical within the error.

Rb and *Cs*. The values of our Rb and Cs abundances for each Type A and C rocks and soil have been compared with average abundances obtained by others (Table 7). For Rb the agreement between other values and ours are excellent. For lunar soil, 0·15 ± 0·02 ppm Cs of this work is just outside of the average Cs abundances of 0·11 ± 0·01 ppm obtained by four other investigators (see Table 7). Our Cs abundances for two Types A and C rocks differ considerably from the average Cs values obtained by others. No explanation for these apparent discrepancies is offered since the standard Cs solution used in our work agreed within ±10 per cent of that used by GAST and HUBBARD (1970). As observed by the previous authors, Rb and Cs abundances distinguish Type A rocks from Types B and C rocks and soil.

Cd. Cadmium abundances of 0·044 ± 0·004 and 0·056 ± 0·005 ppm were obtained in one Type A rock 10017 and in soil, respectively (Table 5). KEAYS *et al.* (1970) reported lower Cd abundances of 0·003 and 0·006 ppm for two Type A rocks (10057-31 and 10072-23) and 0·044 ± 0·009 ppm for soil. The higher Cd abundance, by a factor of ~8 in rock 10017 of this work may be attributed to either Cd contamination of the sample chip or to real Cd heterogeneity at the 0·5 g sampling size. Our Cd content of soil overlaps those reported by KEAYS *et al.* (1970). Abundances of Cd by MORRISON *et al.* (1970) seem anomalously high in Types A, B and C rocks by factors of 10–300 in comparison to this work and KEAYS *et al.* (1970). Cadmium abundance in lunar rocks and soil correspond to the general Cd level of ~0·057 ppm found in ordinary chondrites, 0·042 ppm in eucrite achondrites, and 0·3–0·5 ppm Cd in terrestrial basalts (SCHMITT *et al.*, 1963b).

Following the KEAYS *et al.* (1970) model for addition of 1·5–2% Cc_1 chondrites to pulverized rocks, such as Types A and B to yield the present soil and subsequent breccia rocks, we have calculated that a 1·5 per cent admixture of 1·0 ± 0·1 ppm Cd found in Cc_1 chondrites (SCHMITT *et al.*, 1963b) to 0·044 ppm Cd obtained in Type A rock of this work would result in 0·059 ppm Cd in the soil; this calculated value agrees with the reported Cd abundance in soil of this work.

In. In abundances of 1·0, 5·2 and 580 ppb were found in Type A rock 10017-31, Type C rock 10019-24 and soil, respectively. For Type A rock 10072, 52 ppb In was reported by BAEDECKER and WASSON (1970); for Type A rock 10057, 2·7 and 3·0 ppb In are reported by WÄNKE *et al.* (1970) and BAEDECKER and WASSON (1970), respectively. For Type C rock 10060, 4 ppb In by SMALES *et al.* (1970) and for three Type C rocks 10021, 10046 and 10048, BAEDECKER and WASSON (1970) reported 22 ± 8, 16 ± 9 and 60 ppb In, respectively. An In abundance of 580 ppb in soil, about seven times that found in carbonaceous chondrites, has been attributed to contamination at the Lunar Receiving Laboratory. Also for Type C rock 10060, 1030 ppb In is reported by BAEDECKER and WASSON (1970).

The values of 1–5 ppb In abundances found in lunar rocks are compared to 0·05–2 ppb (average 0·3 ppb) in H5, H6, L5 and L6 equilibrated chondrites and 1–55 ppb in L3 unequilibrated chondrites (TANDON and WASSON 1968, SCHMITT and SMITH, 1968) and to 60–90 ppb in terrestrial basalts (HAMAGUCHI *et al.*, 1967; ONUMA *et al.*,

1968; and SCHMITT and SMITH, 1968). Assuming 1·5 per cent addition (after KEAYS *et al.*, 1970) of 80 ppb In for Cc_1 carbonaceous chondrites (AKAIWA, 1966; FOUCHÉ and SMALES, 1967, and SCHMITT and SMITH, 1968) to 1–3 ppb In found in Types A and B rocks, we calculate an expected mean 2·2–4·2 ppb In in lunar breccias and soil.

CORE SAMPLE

Abundances of four major elements (Ti, Al, Fe and Ca), three minor elements (Na, Cr and Mn) and twelve trace elements (Sc, V, Co, Zr, Hf, Ba, La, Sm, Eu, Yb,

Table 8. Elemental abundances in Apollo 11 core tube soil (10005-29)*

Element	0 (82 mg)	Depth of samples (cm) 2·6 (73 mg)	5·2 (78 mg)	7·8 (131 mg)	10·5 (146 mg)	Average	Abund. in 2·2 g lunar soil†
Ti(%)‡	5·0	4·7	5·0	4·5	4·6	4·8 ±0·2	4·7 ±0·2
Al(%)	7·3	7·3	7·4	7·6	7·4	7·4 ±0·1	7·1 ±0·1
Fe(%)	14·1	13·1	13·3	12·7	12·9	13·2 ±0·6	11·9 ±0·6
Ca(%)‡	9·9	8·2	8·4	8·8	8·8	8·8 ±0·7	8·3 ±0·6
Na(ppm)‡	3480	3320	3080	3260	3190	3270 ±150	3070 ±100
Cr(ppm)	2140	2020	2010	1930	2050	2030 ±80	1780 ±100
Mn(ppm)	1650	1630	1620	1580	1630	1620 ±30	1600 ±10
Sc(ppm)	66	63	62	60	60	62 ±3	58 ±1
V(ppm)	69	58	80	67	55	66 ±10	63 ±4
Co(ppm)	36	31	31	30	30	32 ±3	31 ±3
Zr(ppm)	310	350	280	190	590	340 ±150	460 ±360
Hf(ppm)	8	8	8	8	7	8 ±1	9 ±1
In(ppm)				0·057			0·56
Ba(ppm)	120	160	120	150	140	140 ±20	200 ±60
La(ppm)	16·6	15·3	15·6	15·3	14·6	15·5 ±0·7	14·2 ±0·2
Sm(ppm)	12·4	12·0	12·4	11·4	11·2	11·9 ±0·5	12·6 ±0·6
Eu(ppm)	2·0	2·3	2·2	1·8	2·0	2·1 ±0·2	2·0 ±0·2
Yb(ppm)	11·6	11·6	11·5	10·4	10·5	11·1 ±0·6	10·8 ±0·2
Lu(ppm)	1·7	1·7	1·6	1·5	1·6	1·6 ±0·1	1·5 ±0·1
Th(ppm)	0·9	1·0	0·8	0·8	0·7	0·8 ±0·1	2·6 ±0·7

* One standard deviation errors due to counting statistics and other errors are approximately ±2% for Na and Mn; ±3–4% for Ti, Al, Ca and Sc; ±10% for V, Co, Hf and Th (on a relative basis); ±15% for Ba; and ±20–30% for Zr.

† Averages were taken from Tables 3 and 5.

‡ Average abundances and standard deviations for duplicate analyses of each sample are given. Lower Ti and Ca abundances for these same specimens were first reported by SCHMITT *et al.* (1970a). Their early values were in error by ~12–20% due to counting losses attributable to high analyzer dead times.

Lu and Th) in drive-tube Core No. 1 (10005–29) have been determined via INAA. The results of the analyses of five specimens (70–140 mg each), taken at 2·6 cm intervals, are listed in Table 8. The descriptions of core samples have been published by LSPET (1969) and FRYXELL *et al.* (1970).

In general, average elemental abundances for five core samples agree with the corresponding elemental abundances in the soil samples that were scooped randomly for the bulk sample (LSPET, 1969). Generally, elemental abundances for all core samples were quite uniform and indicate a well mixed regolith, at least down to 10-cm. This agrees with visual inspection of the core samples by FRYXELL *et al.* (1970), who state that the primary characteristics were essentially uniform in Core No. 1.

In detail, no differences are observed for Al, Cr, Mn, and Hf abundances in all core samples. However, for Ti and Ca abundances in core samples (least accurately determined of all the major and minor elements) considerable variations are observed; i.e. $5{\cdot}0 \pm 0{\cdot}1\%$ Ti and $9{\cdot}9 \pm 0{\cdot}2\%$ Ca for 0-cm depth sample are higher by $\sim$10 per cent than $4{\cdot}6 \pm 0{\cdot}1\%$ Ti and $8{\cdot}8 \pm 0{\cdot}2\%$ Ca obtained for the 10·5-cm depth sample. In addition, a comparison of average elemental abundance values of two near surface samples (0 and 2·6-cm) and the deepest two samples (7·8–10·5-cm) indicates slight 5–10 per cent variations for Fe, Na, Sc, the five REE (La, Sm, Eu, Yb and Lu) and Th; that is, all of these elements seem to be more enriched in the near surface samples. This observation may suggest a relative $\sim$10 per cent enrichment of ilmenite minerals near the lunar surface compared to the 8–10 cm depths. The constant Al abundance (most accurately determined element) indicate a uniform distribution of plagioclase at all depths. From cosmic-ray track studies, FLEISCHER *et al.* (1970) reported that lunar surface soil is stirred often by incoming meteorites and that the residence time of the soil in the top 10·5-cm of the surface is approximately 15×10^6 yr. From a statistical model, they estimate that the top layer has been re-mixed every 10^4–10^5 yr. In general, the data of this work are consistent with these observations.

There is considerable disagreement between average Th abundances in core and soil samples. The difficulties of Th determination has already been stated (in the trace element discussion). Five core samples and the same volumes of U.S.G.S. standard rocks, BCR-1, W-1 and GSP-1 were activated under identical activation conditions and then counted on the same geometry. Abundance values for BCR-1, W-1 and GSP-1 are 7·0, 2·2 and 122 ppm Th, respectively, all in line with literature values (FLANAGAN, 1969; and FLEISCHER, 1969). We have no explanation for the factor of three difference; in any case, Th abundances in the core tube should be considered only reliable on a relative basis.

In order to obtain the uncontaminated In abundance in lunar soil, one of the core samples (7·8 cm) has been analyzed by RNAA (Table 8). The value of $0{\cdot}057 \pm 0{\cdot}003$ ppm In is ten times lower than that found in soil. These high values of In in soil have been attributed to contamination at the L.R.L. Note that this In abundance in the core sample is still $\sim$10 times higher than In abundances found in Types A and B igneous rocks. It is not clear whether this In abundance of 0·057 ppm represents the true In value in lunar soil or has been contaminated during sample preparation at the L.R.L.

Acknowledgments—This study was supported by NASA grant NGR 38-002-020 and by NASA contract 9-8097. We acknowledge the assistance of the Oregon State University TRIGA reactor group for neutron activations, and T. Cooper, D. G. Coles, T. W. Osborn, D. B. Curtis and R. G. Warren for help in analysis. Dr. V. P. Guinn and J. Mackenzie of Gulf General Atomic provided assistance in Mg analyses.

References

Agrell S. O., Scoon J. H., Muir I. D., Long J. V. P., McConnell J. D. C. and Peckett A. (1970) Mineralogy and petrology of some lunar samples. *Science* **167,** 583–586.

Akaiwa H. (1966) Abundances of selenium, tellurium, and indium in meteorites. *J. Geophys. Res.* **71,** 1917–1923.

Albee A. L., Burnett D. S., Chodos A. A., Eugster O. J., Huneke J. C., Papanastassiou D. A., Podosek F. A., Russ G. Price, II, Sanz H. G., Tera F. and Wasserburg G. J. (1970) Ages, irradiation history, and chemical composition of lunar rocks from the Sea of Tranquillity. *Science* **167,** 463–466.

Anderson A. T., Jr., Crewe A. V., Goldsmith J. R., Moore P. B., Newton J. C., Olsen E. J., Smith J. V. and Wyllie P. J. (1970) Petrologic history of Moon suggested by petrography, mineralogy, and crystallography. *Science* **167,** 587–590.

Annell C. and Helz A. (1970) Emission spectrographic determination of trace elements in lunar samples. *Science* **167,** 521–523.

Arrhenius G., Asunmaa S., Drever J. I., Everson J., Fitzgerald R. W., Frazer J. Z., Fujita H., Hanor J. S., Lal D., Liang S. S., MacDougall D., Reid A. M., Sinkankas J. and Wilkening L. (1970) Phase chemistry, structure, and radiation effects in lunar samples. *Science* **167,** 659–661.

Baedecker P. A. and Wasson J. T. (1970) Gallium, germanium, indium, and iridium in lunar samples. *Science* **167,** 503–505.

Cameron E. N. (1970) Opaque mineral in lunar samples. *Science* **167,** 623–625.

Clark S. P., Jr., Peterman Z. E. and Heier K. S. (1966) Abundances of uranium, thorium, and potassium. In *Handbook of Physical Constants*, (editor S. P. Clark, Jr.), pp. 521–541. Geol. Soc. Amer.

Compston W., Arriens P. A., Vernon M. J. and Chappell B. W. (1970) Rubidium–strontium chronology and chemistry of lunar material. *Science* **167,** 474–476.

Duke M. B. and Silver L. T. (1967) Petrology of eucrites, howardites and mesosiderites. *Geochim. Cosmochim. Acta* **31,** 1637–1655.

Ehmann W. D. and Morgan J. W. (1970) Oxygen, silicon, and aluminum in lunar samples by 14 MeV neutron activation. *Science* **167,** 528–530.

Ehmann W. D. and Rebagay T. V. (1970) Zirconium and hafnium in meteorites by activation analysis *Geochim. Cosmochim. Acta* (in press).

Engel A. E. J. and Engel C. G. (1970) Lunar rock compositions and some interpretations. *Science* **167,** 527–528.

Engel A. E. J., Engel C. G. and Havens R. G. (1965) Chemical characteristics of oceanic basalts and the upper mantle. *Geol. Soc. Amer. Bull.* **76,** 719–734.

Flanagan F. J. (1969) U.S. Geological Survey standard—II. First compilation of data for the new U.S.G.A. rocks. *Geochim. Cosmochim. Acta* **33,** 81–120.

Fleischer M. (1969) U.S. Geological Survey standards—I. Additional data on rocks G-1 and W-1, 1965–1967. *Geochim. Cosmochim. Acta* **33,** 65–79.

Fleischer R. L., Haines E. L., Hanneman R. E., Hart H. R., Jr., Kasper J. S., Lifshin E., Woods R. T. and Price P. B. (1970) Particle track, X-ray, thermal, and mass spectrometric studies of lunar material. *Science* **167,** 568–571.

Fouché K. F. and Smales A. A. (1967) The distribution of trace elements in chondritic meteorites. 1. Gallium, germanium and indium. *Chem. Geol.* **2,** 5–33.

Franzgrote E. J., Patterson J. H., Turkevich A. L. and Economou T. E. (1970) Chemical composition of the lunar surface in Sinus Medii. *Science* **167,** 376–379.

Fredriksson K., Nelen J., Melson W. G., Henderson E. P. and Andersen C. A. (1970) Lunar glasses and micro-breccias: properties and origin. *Science* **167,** 664–666.

FREY F. A., HASKIN M. A., POETZ J. A. and HASKIN L. A. (1968) Rare earth abundances in some basic rocks. *J. Geophy. Res.* **73,** 6085–6098.

FRONDEL C., KLEIN C., JR., ITO J. and DRAKE J. C. (1970) Mineralogy and composition of lunar fines and selected rocks. *Science* **167,** 681–683.

FRYXELL R., ANDERSON D., CARRIER D., GREENWOOD W. and HEIKEN G. (1970) Apollo 11 drive-tube core samples: An initial physical analysis of lunar surface sediment. *Science* **167,** 734–737.

FUNKHOUSER J. G., SCHAEFFER O. A., BOGARD D. D. and ZÄHRINGER J. (1970) Gas analysis of the lunar surface. *Science* **167,** 561–563.

GAST P. W. and HUBBARD N. V. (1970) Abundance of alkali metals, alkaline and rare earths, and strontium-87/strontium-86 ratios in lunar samples. *Science* **167,** 485–487.

GOLES G. G., SCHMITT R. A., EHMANN W. D., RANDLE K., OSAWA M., WAKITA H. and MORGAN J. W. (1970a) Elemental abundances by instrumental activation analyses in chips from twenty-seven lunar rocks. *Geochim. Cosmochim. Acta*, Supplement I.

GOLEŠ G. G., OSAWA M., RANDLE K., BEYER R. L., JEROME D. Y., LINDSTROM D. J., MARTIN M. R., MCKAY S. M. and STEINBORN T. L. (1970b) Instrumental neutron activation analyses of lunar specimens. *Science* **167,** 497–499.

GOPALAN K., KAUSHAL S., LEE-HU C. and WETHERILL G. W. (1970) Rubidium–strontium, uranium, and thorium-lead dating of lunar material. *Science* **167,** 471–473.

GORDON G. E., RANDLE K., GOLES G. G., CORLISS J. B., BEESON M. H. and OXLEY S. S. (1968) Instrumental activation analysis of standard rocks with high-resolution γ-ray detectors. *Geochim. Cosmochim. Acta* **32,** 369–396.

HAGGERTY S. E., BOYD F. R., BELL P. M., FINGER L. W. and BRYAN W. B. (1970) Iron–titanium oxides and olivine from 10020 and 10071. *Science* **167,** 613–615.

HAMAGUCHI H., TOMURA K., ONUMA N., HIGUCHI H. and SUDA K. (1967) Determination of indium in rocks by neutron activation analysis. *Bunseki Kagaku* **16,** 1233–1238.

HASKIN L. A., FREY F. A., SCHMITT R. A. and SMITH R. H. (1966) Meteoritic, solar and terrestrial rare-earth distributions. In *Phys. Chem. Earth*, (editors L. H. Ahrens, F. Press, S. K. Runcorn and H. C. Urey) Vol. 7, pp. 167–321. Pergamon Press.

HASKIN L. A., HASKIN M. A., FREY F. A. and WILDEMAN T. R. (1968) Relative and absolute terrestrial abundances of the rare earths. In *Proc. Symp. Origins and Distribution of the Elements*, (editor L. H. Ahrens) pp. 889–912. Pergamon Press.

HASKIN L. A., HELMKE P. A. and ALLEN R. O. (1970) Rare earth elements in returned lunar samples. *Science* **167,** 487–490.

HURLEY P. M. and PINSON W. H., JR. (1970) Rubidium–strontium relations in Tranquillity Base samples. *Science* **167,** 473–474.

KAY R., HUBBARD N. J. and GAST P. W. (1970) *J. Geophys Res.* (in press).

KEAYS R. R., GANAPATHY R., LAUL J. C., ANDERS E., HERZOG G. F. and JEFFERY P. M. (1970) Trace elements and radioactivity in lunar rocks: Implications for meteorite infall, solar-wind flux, and formation conditions of moon. *Science* **167,** 490–493.

LOVELAND W., SCHMITT R. A. and FISHER D. E. (1969) Aluminum abundances in stony meteorites. *Geochim. Cosmochim. Acta* **33,** 375–385.

LSPET (LUNAR SAMPLE PRELIMINARY EXAMINATION TEAM) (1969) Preliminary examination of lunar samples from Apollo 11. *Science* **165,** 1211–1227.

MASON B., FREDRIKSSON K., HENDERSON E. P., JAROSEWICH E., MELSON W. G., TOWE K. M. and WHITE J. S., JR., (1970) Mineralogy and petrography of lunar samples. *Science* **167,** 656–659.

MORRISON G. H., GERARD J. T., KASHUBA A. T., GANGADHARAM E. V., ROTHENBERG A. M., POTTER N. M. and MILLER G. B. (1970) Multielement analysis of lunar soil and rocks. *Science* **167,** 505–507.

MOSEN A. W., SCHMITT R. A. and VASILEVSKIS J. (1961) A procedure for the determination of the rare earth elements, lanthanum through lutetium, in chondritic, achondritic, and iron meteorites by neutron-activation analysis. *Anal. Chim. Acta* **25,** 10–24.

MURTHY V. R., SCHMITT R. A. and REY P. (1970) Rubidium–strontium age and elemental and isotopic abundances of some trace elements in lunar samples. *Science* **167,** 476–479.

NICHIPORUK W. (1970) Vanadium (23). In *Elemental Abundances in Meteoritic Matter*, (editor B. Mason). To be published.

NICHIPORUK W., CHODOS A., HELIN E. and BROWN H. (1967) Determination of iron, nickel, cobalt, calcium, chromium and manganese in stony meteorites by X-ray fluorescence. *Geochim. Cosmochim. Acta* **31,** 1911–1930.

O'KELLEY G. D., ELDRIDGE J. S., SCHONFELD E. and BELL P. R. (1970) Elemental compositions and ages of lunar samples by nondestructive gamma-ray spectrometry. *Science* **167,** 580–582.

OLSEN E. (1969) Pyroxene gabbro (anorthosite association): Similarity of Surveyor V lunar analysis. *Science* **166,** 401–402.

ONUMA N., CLAYTON R. N. and MAYEDA T. K. (1970) Oxygen isotope fractionation between minerals and an estimate of the temperature of formation. *Science* **167,** 536–538.

ONUMA N., HIGUCHI H., WAKITA H. and NAGASAWA H. (1968) Trace element partition between two pyroxenes and the host lava. *Earth Planet. Sci. Lett.* **5,** 47–51.

ÖPIK E. J. (1969) The moon's surface. *Ann. Rev. Astron. Astrophys.* **7,** 473–526.

PARRATT L. G. (1961) *Probability and Experimental Errors in Science.* p. 121. John Wiley.

PATTERSON J. H., FRANZGROTE E. J., TURKEVICH A. L., ANDERSON W. A., ECONOMOU T. E., GRIFFIN H. E., GROTCH S. L. and SOWINSKI K. P. (1969) Alpha-scattering experiment on Surveyor 7: Comparison with Surveyors 5 and 6. *J. Geophys. Res.* **74,** 6120–6148.

PERKINS R. W., RANCITELLI L. A., COOPER J. A., KAYE J. H. and WOGMAN N. A. (1970) Cosmogenic and primordial radionuclides in lunar samples by nondestructive gamma-ray spectrometry. *Science* **167,** 577–580.

PHILPOTTS J. A., SCHNETZLER C. C. and THOMAS H. H. (1967) Rare earth and barium abundances in the Bununu howardite. *Earth Planet. Sci. Lett.* **2,** 19–22.

PHILPOTTS J. A. and SCHNETZLER C. C. (1968) Europium anomalies and the genesis of basalt. *Chem. Geol.* **3,** 5–13.

PHILPOTTS J. A. and SCHNETZLER C. C. (1970) Potassium, rubidium, strontium, barium, and rare earth concentrations in lunar rocks and separated phases. *Science* **167,** 493–495.

REBAGAY T. V. (1969) The determination of zirconium and hafnium in meteorites and terrestrial materials by activation analysis. Ph.D. thesis, University of Kentucky.

REY P., WAKITA H. and SCHMITT R. A. (1970) Radiochemical activation analysis of In, Cd and the 14 rare earth elements and Y in rocks. In preparation.

ROBINSON P. T. (1969) High-titania alkali-olivine basalts of north-central Oregon, U.S.A. *Contrib. Mineral. Petrol.* **22,** 349–360.

ROSE H. J., JR., CUTTITTA F., DWORNIK E. J., CARRON M. K., CHRISTIAN R. P., LINDSAY J. R., LIGON D. T. and LARSON R. R. (1970) Semimicro chemical and X-ray fluorescence analysis of lunar samples. *Science* **167,** 520–521.

SCHILLING J. G. (1969) Red Sea floor origin: Rare-earth evidence. *Science* **165,** 1357–1360.

SCHMITT R. A., SMITH R. H., LASCH J. E., MOSEN A. W., OLEHY D. A. and VASILEVSKIS J. (1963a) Abundances of the fourteen rare earth-elements, scandium, and yttrium in meteoritic and terrestrial matter. *Geochim. Cosmochim. Acta* **27,** 577–622.

SCHMITT R. A., SMITH R. H. and OLEHY D. A. (1963b) Cadmium abundances in meteoritic and terrestrial matter. *Geochim. Cosmochim. Acta* **27,** 1077–1088.

SCHMITT R. A., SMITH R. H. and OLEHY D. A. (1964) Rare-earth yttrium and scandium abundances in meteoritic and terrestrial matter—II. *Geochim. Cosmochim. Acta* **28,** 67–86.

SCHMITT R. A. and SMITH R. H. (1968) Indium abundances in chondritic and achondritic meteorites and in terrestrial rocks. Origin and distribution of the elements, (editor L. H. Ahrens), pp. 283–300. Pergamon Press.

SCHMITT R. A., WAKITA H. and REY P. (1970a) Abundances of 30 elements in lunar rocks, soil, and core samples. *Science* **167,** 512–515.

SCHMITT R. A., LINN T. A., JR. and WAKITA H. (1970b) The determination of fourteen common elements in rocks via sequential instrumental activation analysis. *Radiochim. Acta.* In press.

SCHMITT R. A., GOLES G. G. and SMITH R. H. (1970c) Elemental abundances in stony meteorites. In preparation.

SILVER L. T. (1970) Uranium–thorium–lead isotope relations in lunar materials. *Science* **167,** 468–471.

SMALES A. A., MAPPER D., WEBB M. S. W., WEBSTER R. K. and WILSON J. D. (1970) Elemental composition of lunar surface material. *Science* **167,** 509–512.

TANDON S. N. and WASSON J. T. (1968) Gallium, germanium, indium and iridium variations in a suite of L-group chondrites. *Geochim. Cosmochim. Acta* **32,** 1087–1109.

TATSUMOTO M. and ROSHOLT J. N. (1970) Age of the moon: An isotopic study of uranium–thorium–lead systematics of lunar samples. *Science* **167,** 461–463.

TAYLOR S. R., JOHNSON P. H., MARTIN R., BENNETT D., ALLEN J. and NANCE W. (1970) Preliminary chemical analyses of 20 Apollo 11 lunar samples. Preprint.

TEMPLETON D. H. and DAUBEN C. H. (1954) Lattice parameters of some rare earth compounds and a set of crystal radii. *J. Amer. Chem. Soc.* **76,** 5237–5239.

TOWELL D. G., WINCHESTER J. W. and SPIRN R. V. (1965) Rare-earth distributions in some rocks and associated minerals of the batholith of southern California. *J. Geophys. Res.* **70,** 3485–3496.

TUREKIAN K. K. and KHARKAR K. D. (1970) Neutron activation analysis of milligram quantities of lunar rocks and soil. *Science* **167,** 507–509.

TURKEVICH A. L., FRANZGROTE E. J. and PATTERSON J. H. (1969) Chemical composition of the lunar surface in the Mare Tranquillitatis. *Science* **165,** 277–279.

UREY H. C. (1964) A review of atomic abundances in chondrites and the origin of meteorites. *Rev. Geophys.* **2,** 1–34.

UREY H. C. (1968) Mascons and the history of the moon. *Science* **162,** 1408–1410.

UREY H. C. and MACDONALD G. J. F. (1969) Geophysics of the moon. *Science J.* **5,** 60–65.

WÄNKE H., BEGEMANN F., VILCSEK E., RIEDER R., TESCHKE F., BORN W., QUIJANO-RICO M., VOSHAGE H. and WLOTZKA F. (1970) Major and trace elements and cosmic-ray produced radioisotopes in lunar samples. *Science* **167,** 523–525.

WANLESS R. K., LOVERIDGE W. D. and STEVENS R. D. (1970) Age determinations and isotopic abundance measurements on lunar samples. *Science* **167,** 479–480.

WIIK H. B. and OJANPERA P. (1970) Chemical analyses of lunar samples 10017, 10072 and 10084. *Science* **167,** 531–532.

Proceedings of the Apollo 11 Lunar Science Conference, Vol. 2, pp. 1719 to 1727.

Major and trace elements in lunar material

H. WÄNKE, R. RIEDER, H. BADDENHAUSEN, B. SPETTEL,
F. TESCHKE, M. QUIJANO-RICO and A. BALACESCU
Max-Planck-Institut für Chemie (Otto-Hahn-Institut), 65 Mainz, Germany

(*Received* 3 *February* 1970; *accepted in revised form* 23 *February* 1970)

Abstract—Analytical data for forty-five major and trace elements in lunar fines and rock-types A, B and C were obtained by combined instrumental activation analysis with fast and thermal neutrons and radiochemical neutron activation analysis. The observed abundance pattern of lunar fines is compared with patterns in carbonaceous chondrites, the average earth crust and eucrites, for which the smallest deviations are found. The rare-earth abundances show a distinct negative Eu-anomaly relative to average chondritic composition.

Besides the well-known rare gases, hydrogen, and nitrogen, also chlorine and perhaps indium are found to be present in considerable excess in the lunar fines, compared with rock samples, the excess amounts appearing to be correlated with solar wind phenomena.

An excess of siderophile elements in lunar fines, in particular of nickel, was also observed. These elements are apparently seated in small metal grains, the amount of which has been determined as 0·6 per cent by weight.

EXPERIMENTAL PROCEDURE

Samples of between 0·2 and 1 g were pulverized in a steel mortar to pass a 60 μ sieve and irradiated subsequently under 4 different conditions: Irradiations 1 and 2 were performed with fast neutrons (14 MeV, $^{3}H(d, n)^{4}He$) from a neutron generator, irradiations 3 and 4 with thermal neutrons from a TRIGA-research reactor.

Irradiation 1 extended over a period of 10 sec and the activities of ^{16}N and ^{28}Al were counted with a 5 $\times$ 5 in. NaI(TL) detector for the determination of oxygen and silicon. To minimize statistical errors this step was repeated at least five times. High purity quartz was used as a standard and irradiation and counting geometry were carefully adjusted to reduce systematic errors from different sample and standard geometries.

Irradiation 2 extended over a period of 5 min and the activities of ^{28}Al, ^{27}Mg and ^{24}Na were counted with a large volume Ge(Li) detector for the determination of Mg and Al. Silicon was used as an internal standard and the activity-ratios were converted into absolute figures by comparison with standards made up of mixtures of metal-oxides and irradiated under identical conditions.

Irradiation 3 extended over a period of 6 hr at a flux of 7×10^{11} n/cm^2 sec and during a cooling period of 3 weeks activities of ^{24}Na, ^{42}K, ^{46}Sc, ^{47}Sc (from Ti and Ca), ^{48}Sc (from Ti), ^{51}Cr, ^{56}Mn, ^{59}Fe, ^{60}Co, ^{140}La, ^{141}Ce, ^{147}Nd, ^{153}Sm, ^{152}Eu, ^{160}Tb, ^{169}Yb, ^{177}Lu, ^{181}Hf, and ^{182}Ta could be counted with a large volume Ge(Li) detector for the determination of these elements.

Irradiation 4. Prior to the irradiation the samples were packed in a high purity quartz ampoule and evacuated. After a degassing period of several hours at 180°C the quartz ampoule was sealed under vacuum. The irradiation lasted for 6 hr at a flux of 7×10^{11} n/cm^2 sec. The samples were then radiochemically processed to separate and determine the following elements Li, F, Cl, Cu, Ga, Ge, Ni (via ^{58}Co), Rb, Sr, Pd, In, Cs, Ba, Pr, Gd, Dy, Ho, Er, W, Ir, Au, Th (via ^{233}pa), U (via ^{135}Xe).

RADIOCHEMICAL SEPARATION

Because of the different chemical composition of the lunar samples and in order to make use of the advantage of large volume high resolution Ge(Li) detectors we had to change our standard procedure (RIEDER and WÄNKE, 1969) developed for the chemical analysis of meteoritic samples considerably.

The quartz ampoule was opened and the sample mixed with the eight-fold amount of Na_2O_2 and transferred to a Zr-crucible. About 10 mg of carrier of each element to be determined had been added previously and the crucible put into vacuum. After several minutes of pumping 0·2 scc Ar and 0·2 scc Xe were added as carriers for the extracted rare gases and the sample fused for 30 min at 850°C by means of high-frequency induction heating. After fusion Ar and Xe were adsorbed on charcoal at liquid nitrogen temperature and further purified (KÖNIG and WÄNKE, 1959) and the Zr-crucible removed from the vacuum system. The melt was treated with H_2O, the unsoluble part centrifuged off and dissolved in HNO_3. The two solutions were combined and the hydroxides precipitated with NaOH at pH 8. The supernatant was adjusted to pH 11 by addition of NaOH and the precipitating hydroxides centrifuged off (1. pH 11 fraction). To the supernatant Na_2CO_3 was added, the precipitating carbonates filtrated (separation Cu, Sr, Ba), the solution adjusted to pH 2 with HCl and K, Rb, and Cs precipitated with tetraphenylboron ('kalignost') (Separation K, Rb, Cs).

Gold separation: The first pH 8-precipitate was dissolved in HNO_3 dil. and reprecipitated at pH 8 and subsequently at pH 11 (2. pH 11 fraction). The second pH 8 precipitate was dissolved in HBr + Br_2, adjusted to 5·5 n in HBr by addition of HBr conc. and extracted with diethylether. The organic phase was washed three times with 5·5 n HBr and stripped with small volumina of H_2O and again with 1 n HCl. The gold containing organic phase was further purified by two washing extractions with 1 n HCl and the ether evaporated.

Indium separation: The aqueous phases were combined, adjusted to 7 n in HCl by addition of HCl fum. and extracted twice with di-iso-propylether. The aqueous phase was evaporated to dryness.

Gallium separation: The first di-iso-propylether was washed three times with equal amounts of 7 n HCl + $TiCl_3$. After addition of a few ml HNO_3 dil. the di-iso-propylether was evaporated.

Germanium and Arsenic separation: The aqueous phase of the diethylether extraction was combined with the first wash-HBr and extracted with benzene to remove the dissolved ether from the acid. After addition of the 0·6-fold amount of HCl fum. the aqueous phase was extracted two times with 100 ml benzene each. The combined organic phases were washed with 50 ml 10 n HCl and stripped with two times 10 ml of H_2O.

Protactinium separation: The aqueous phase was extracted with a mixture of 80% iso-butylmethylketone (IBMK) and 20% benzene. After two wash extractions with 7 n HCl the organic phase was stripped with 30 ml 3% HF.

Rare-earth separation: The aqueous phase was washed with 100 ml benzene, crystalline NaOH was added and the solution adjusted to pH 8. The precipitate was centrifuged off and the supernatant adjusted to pH 11 (3. pH 11 fraction). The pH 8 precipitate was dissolved in HCl dil., centrifuged to separate from eventually precipitating SiO_2 and pH 8 and pH 11 precipitation repeated in the supernatant (4. pH 11 fraction). The pH 8 precipitate was again dissolved in HCl dil. All the eventually precipitated SiO_2 was put into this solution and 40% HF added dropwise until the SiO_2 dissolved and precipitation occurred. After centrifugation the fluoride precipitation was repeated in the supernatant, centrifuged, the precipitates combined and the rare earth fluorides dissolved in 6 ml saturated H_3BO_3 + HNO_3 conc. After addition of 14 ml H_2O the solution was centrifuged and further purified by alternate hydroxide-(NH_4OH) and fluoride precipitation.

Palladium separation: The supernatant of the fluoride precipitation was combined with the four pH 11 fractions (dissolved in HCl dil.), the solution adjusted to 1–2 n in acid, heated to 60°C and H_2S introduced. After cooling the precipitate was dissolved in aqua regia, evaporated to almost dryness, taken up with 2 n HCl, 4–6 ml 1% di-methyl-glyoxime added and extracted several times with $CHCl_3$. The precipitation and extraction was repeated in the aqueous phase, the organic phases were combined and washed with 100 ml 2 n HCl. Then 1 ml di-methyl-glyoxime and 40 ml 1 n HCl were added to the organic phase, $CHCl_3$ evaporated, the solution diluted with H_2O and filtrated.

Copper separation: The aqueous phase of the $CHCl_3$ extraction contains copper.

Tungsten separation: The supernatant of the H_2S precipitation was transferred to a teflon dish, H_2SO_4 added and the HF fumed off. This step is repeated twice and the residue redissolved in HNO_3 conc., diluted with H_2O and heated to hydrolyse WO_3, the precipitate centrifuged off, dissolved in NH_4OH and reprecipitated with HNO_3. This purification procedure was repeated 2–3 times.

Cobalt and iridium separation: The supernatant of the HNO_3-hydrolysis was further processed for the separation of iridium and cobalt by repeated hydroxide precipitation in the presence of H_2O_2.

Chemical yields were determined by reactivation, except for Ar, Xe, Pd and Pa. In these cases the yields were determined volumetrically, gravimetrically and—in case of Pa—by a tracer technique (^{231}Pa).

Solutions of the separated elements were filled into 20 ml standard containers and the activities counted in a 3 × 3 in. NaI well-type crystal. In a few cases, e.g. Pa, a large volume high-resolution Ge(Li) detector was also used. Pd was β-counted with a GM-end window counter; Ar and Xe by β-γ-coincidence counting.

Before radiochemical processing aliquants were set aside for the measurement of Li, Cl and B. Li and Cl were also determined by neutron activation via the reactions ^{7}Li(n, α)^{3}H, ^{16}O(^{3}H, n)^{18}F and ^{37}Cl(n, γ)^{38}Cl, resp., while for B an especially developed fluorimetric method was applied. The procedures have been described in detail previously (Quijano-Rico and Wänke, 1969; Quijano-Rico, 1970).

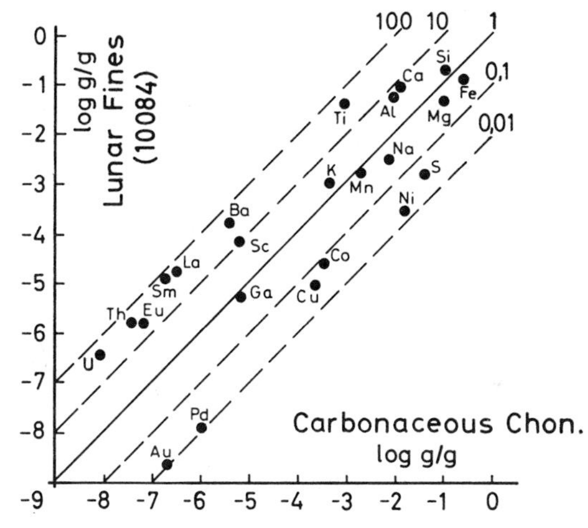

Fig. 1. Comparison of the concentrations of various elements in lunar fines and carbonaceous chondrites.

Results and Discussion

The results are listed in Table 1. In general our results were found to be in good agreement with the results of other workers. To demonstrate this close agreement, not only of our own data with those of particular other groups, but of the results of practically all investigators with each others, we have collected in Table 2 all data on lunar fines available at this time and calculated mean values and standard deviations for single data. Only few data have been excluded from the calculation of the mean value (figures in brackets).

The concentrations found in the lunar fines for a number of characteristic elements were compared with the concentrations of these elements in other samples of different origin:

Figure 1 shows the comparison with material of carbonaceous chondrites. As can be seen, the enrichment for a number of highly fractionated elements in the lunar fines amounts to a factor of up to 100, while other elements, especially the siderophile ones are depleted by a factor of up to 100. Yet, as has been pointed out by KEAYS *et al.* (1970), the major portion of these elements must be ascribed to extralunar, i.e. meteoritic material admixed to the lunar surface layers. Such an origin of these elements is indicated by their even lower abundance in some of the lunar rocks.

Table 1. Major and trace elements in lunar fines and rock-types A, B and C.

	Type A	Type B		Type C		Type D
	10057-40	10017-33	10044-32	10018-22	10060-16	10084-18
%						
O	40·4	40·7	41·5	40·7	41·4	41·5
Mg	4·2	4·8	3·9	5·0	4·6	4·8
Al	4·0	4·4	6·3	6·1	6·2	6·9
Si	18·9	19·6	20·1	19·6	19·8	19·7
Ca	8·4	8·2	5·1	8·3	8·3	8·1
Ti	6·5	7·0	6·3	5·5	4·6	4·3
Fe	14·0	14·6	13·3	13·8	13·8	12·0
ppm						
B	0·8	0·7	1·2			1·03
Li	14·0	23		13·3	10·4	10·4
Na	3000	3470	3580	3920	3600	3150
Cl	12·0	12·2		16·5	15·5	27·1
K	2010	2060	860	1320	1330	1090
Sc	87	86	92	69	70	61
Cr	2160	2310	1300	1900	1820	1830
Mn	1800	1480	2000	1050	1340	1560
Co	25·4	24·5	11·0	33·8	30·1	27·2
Ni	<10			370		280
Cu	4·3	7·7	4·2			8·2
Ga	5·2	4·2	5·1			4·9
Ge	<1	<1	<1			1·4
Rb	5·2	4·2	5·1			3·0
Sr	100	151		195		176
Pd	0·01	0·001				0·012
In	0·0027	0·138	0·0032	0·36		0·75
Cs	0·20	0·12				0·12
Ba	208	256	234			176
La	25	21	12	18	18	15
Ce	79	64	44	72	60	47
Pr	9·0	7·3				5·0
Nd	60	58	50	60		47
Sm	20·0	18·9	17·9	13·5	13·8	12·1
Eu	1·80	1·89	2·69	1·68	1·61	1·67
Gd	30	19	24			18
Tb	5·0	4·6	4·5	4·1	4·2	2·8
Dy	24	19				17
Ho	5·5	3·8				4·6
Er	16	13				9·5
Yb	16·8	15·6	15·0	11·1	10·9	8·3
Lu	2·15	2·12	1·96	1·56	1·57	1·30
Hf	16·9	16·5	12·0	13·4	14·0	10·2
Ta	2·0	2·2	2·0	1·7	1·9	1·3
W	0·43	0·40	0·24			0·22
Au	0·0016	0·0081	0·0019	0·0050		0·0021
Th	3·94	3·94	0·98	3·72		1·61
U	0·80	0·69	0·28	0·61		0·35

Table 2. Major and trace elements in lunar fines, comparison of data from different authors.

Element	This work	Mean	Standard deviation	Other authors
	%		%	
O	41·5			41·94[b]
Mg	4·8	4·73	1·8	4·81[b] 4·6[e] 4·62[f] 4·68[j] 4·75[k] 4·8[m] 4·8[n]
Al	6·9	7·13	2·3	7·32[b] 7·3[e] 7·11[f] 7·2[j] 7·3[k] 6·9[m] 7·1[n] 7·0[p]
Si	19·7	19·53	4·3	19·74[b] 20·2[e] 19·7[j] 19·8[k] 20[m] 17·5[n] 19·6[f]
Ca	8·1	8·40	6·8	8·55[b] 8·14[c] 9·6[e] 8·68[f] 8·55[j] 8·6[k] 8·4[m] 7·4[n] 8[p]
Ti	4·3	4·36	5·7	4·52[b] 4·61[c] 4·4[d] 4·1[e] 4·25[f] 4·65[j] 4·3[k] 4·2[m] 3·9[n] 4·7[p]
Fe	12·0	12·34	6·2	11·2, 11·4[a] 12·28[b] 14[d] 12·5[e] 12·4[f] 12·5[j] 13·1[k] 12·4[m]
	ppm			
B	1·03			2[e]
Li	10·4	9·8	42	6[e](1·9) 7·2[h] 15[m] 11[q]
Na	3150	3300	8·2	3210 3260[a] 3200[b] 3200[c] 3000[d] 3300[e] 3800[f] 3500[j] 3120[k] 3900[m] 3070[n] 3200[p]
Cl	27·1			350[e] 2·9, 11·7[h]
K	1090	1107	9·7	1080[b] 1200[c] 1000[d] 1100[e] 1200[f] 1300[j] 1080[k] 1120[l] 1000[m] 910[n] 1200[p]
Sc	61	61·4	12	58·9, 60·0[a] 59[b] 80[d] 60[e] 55[m] 58[n] 56[q] 66[p]
Cr	1830	2005	11	1900, 1940[a] 1880[b] 2200[d] 2000[e] 1850[f] 2000[j] 2250[k] 2400[m] 1780[n] 1740[q] 2300[p]
Mn	1560	1603	9·6	1590, 1610[a] 1550[b] 1250[c] 1600[e] 1600[f] 1550[j] 1600[k] 1750[m] 1600[n] 1750[m] 1600[n] 1630[p] 1960[q]
Co	27·2	29·2	18	30·3, 30·6[a] 32[b] 33[c] 40[e] 34[f] 26·8, 28·1[i] 18[m] 31[n] 24[q] 25[p]
Ni	280	229	22	200[b] 198[c] 320[d] 170[e] 230[f] 250[m] 185[q]
Cu	8·2	9·49	21	13[b] (38[d]) 9·9[e] (33[f]) 8·07, 7·75[i] 10[q]
Ga	4·9	4·54	14	4·6[e] 4[f] 5·41, 5·24[i] 4·6[o] 3·8[p] 3·8[q]
Ge	1·4	0·735	64	0·7[e] 0·34[o] 0·5[p]
Rb	3·0	2·99	18	2·79[c] 4·4[e] 2·96, 2·65[f] 3·33, 3·09[i] 2·78[l] 2·2[m] 3·0[p] 2·7[q]
Sr	176	162	16	171·4[c] 200[e] 164·8, 164·4[f] 162[l] 115[m] 130[q] 177[p]
Pd	0·012	0·018	81	0·04[e] 0·011, 0·0094[i]
In	0·75	0·788	64	0·5[e] 0·524, 0·768[i] 0·58[n] 0·679[o] 2·0, 0·5[p]
Cs	0·12	0·122	32	0·104[c] 0·2[e] 0·098, 0·094[i] 0·12[p]
Ba	176	167	25	140, 170[a] 176[c] 220[e] 134[f] 170[l] 70[m] 200[n] 210[q] 174[p]
La	15	17·2	17	14·5, 14·2[a] 16·3[c] 20[d] 22[e] 21[f] 16·6[g] 14·2[n] 16[q] 19[p]
Ce	47	49·4	8	50·4, 52·8[a] 47·6[c] (79[d]) 50[e] 58[f] 47[g] 46·1[l] 46[p]
Pr	5·0	8	33	9[e] 10[f]
Nd	47	40·5	13	36·8[c] 43[e] 33[f] 43[g] 40·5[l] 37[p]
Sm	12·1	13·5	19	12·6, 12·5[a] 18[b] 13·1[c] 8·7[d] 18[e] 13·7[g] 13·9[l] 12·6[n] 13·14[p]
Eu	1·67	1·84	6	1·76, 180[a] 1·8[b] 1·77[c] 2·0[d] 1·9[e] 1·77[g] 1·77[l] 2·0[n] 1·8, 22·0[p]
Gd	18	17·6	12	17·2[c] 20[e] 15[g]
Tb	2·8	3·14	18	2·47, 2·7[a] 3[b] 3·8[e] 3·2[g] 4[p]
Dy	17	20·1	15	19·7[c] 17[d] 25[e] 20[g] 19·5[l] 23[p]
Ho	4·6	5·03	14	6[e] 4·5[g] 5[p]
Er	9·5	12·1	16	12·1[c] 15[e] 12[g] 11·7[l]
Yb	8·3	10·9	11	10·5, 11·3[a] 12[b] 11·5[c] 9·5[c] 9·5[d] 12[e] 11·3[g] 10·6[l] (2·5[m]) 10·8[n] 12[p]
Lu	1·30	1·49	8·6	1·57, 1·58[a] (3[b]) 1·58[c] 1·3[d] 1·4[e] 1·66[g] 1·5[n] 1·5[p]
Hf	10·2	10·0	11	10·2, 10·8[a] 8[b] 11[d] 9[e] 11[p]
Ta	1·3	1·40	17	1·2, 1·4[a] (3[b]) 1·8[d] 1·3[e]
W	0·22	0·235	9	0·25[e]
Au	0·0021	0·00275	44	0·00415, 0·00201[i]
Th	1·61	2·35	24	3[b] 2·3[e] 2·5[f]
U	0·35	0·37	28	0·45, 0·42[a] 0·33[d] 0·48[e] 0·19[h]

(a) Goles *et al.* (1970)
(b) Wiik and Ojauperä (1970)
(c) Gast and Hubbard
(d) Turekian and Kharkar (1970)
(e) Morrison *et al.* (1970)
(f) Compston *et al.* (1970)
(g) Haskin *et al.* (1970)
(h) Reed *et al.* (1970)
(i) Keays *et al.* (1970)
(j) Agrell *et al.* (1970)
(k) Maxwell *et al.* (1970)
(l) Philpotts and Schnetzler (1970)
(m) Taylor *et al.* (1970)
(n) Schmitt *et al.* (1970)
(o) Baedecker and Wasson (1970)
(p) Smales *et al.* (1970)
(q) Annell and Helz (1970)

In case of the earth (Fig. 2) comparison was not made with some special type of basalt, but with the average concentration in the rocks of the upper earth crust (CORRENS, 1968). In this case the correlation is somewhat better than for the carbonaceous chondrites, but nevertheless deviations of about a factor of ten in both directions persist.

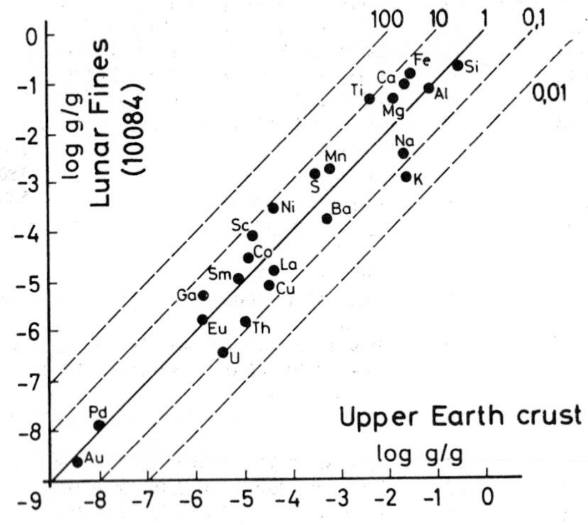

Fig. 2. Comparison of the concentrations of various elements in lunar fines and the average concentration in magmatic rocks of the upper earth crust.

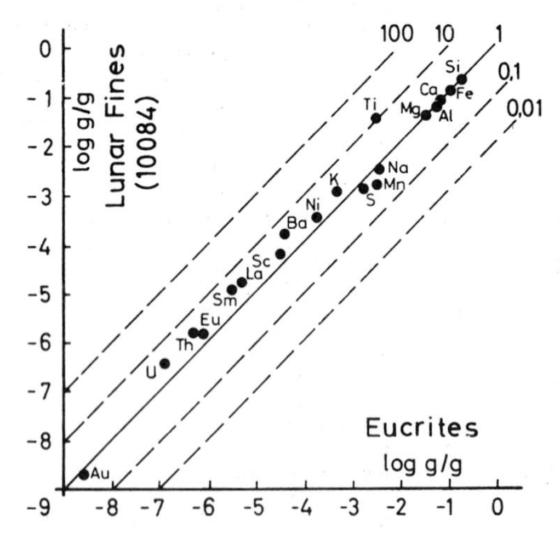

Fig. 3. Comparison of the concentrations of various elements in lunar fines and eucrites.

By far the best agreement for nearly all elements is found for the eucrites, titanium being the element with the highest deviation. It is interesting to note that the K/U-ratio of about 3000 found in lunar material agrees best with a ratio of 4000 for eucrites and other achondrites. Furthermore the close similarity in the abundance pattern of trace elements in the metal particles from lunar fines and that in the eucrite Juvinas is striking (WÄNKE *et al.*, 1970a). The question of trace elements in these metal particles will be discussed in a separate paper (WÄNKE *et al.*, 1970b).

As compared with chondrites the REE are enriched by a factor of up to 100 in the lunar samples, the Eu showing a negative anomaly. This Eu-anomaly has been reported by various authors (GAST and HUBBARD, 1970; HASKIN *et al.*, 1970; PHILPOTTS and SCHNETZLER, 1970; SCHMITT *et al.*, 1970). Among the lunar samples given to us the Eu-concentration was found highest in rock 10044-32 (2·69 ppm). The rare earth pattern of this rock differs from those of other lunar rocks insofar as the abundance curve shows a much steeper decrease towards cerium and lanthanum.

An excess of siderophile elements, in particular of Ni, has been observed in lunar fines, compared with lunar rocks. These elements are apparently seated in small metal grains, the amount of which has been determined from the amount of H_2 evolved during the treatment of a sample of lunar fines with diluted sulfuric acid under vacuum: $Fe_{met} = 0{\cdot}6\%$ by weight in sample 10084-18.

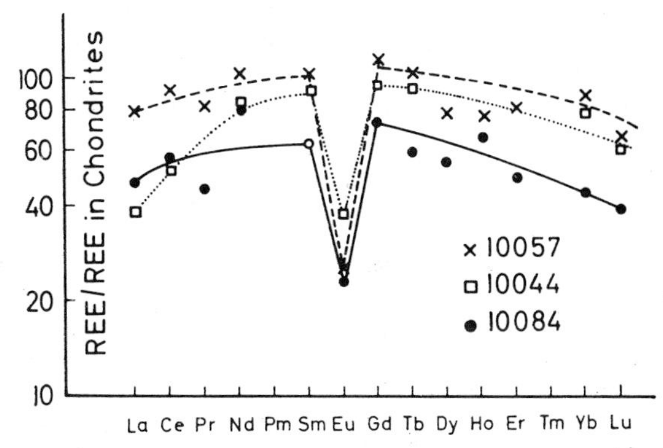

Fig. 4. Rare earth concentration in lunar samples, normalized to chondritic values (rare earth data in chondrites taken from SCHMITT *et al.*, 1963, 1964).

Besides the siderophile elements chlorine and indium are found in lunar fines in excess amounts compared to rocks. In the case of In contamination prior to the distribution of the samples is a serious problem, but we nevertheless wish to point out that the value of 0·75 ppm In in lunar fines is only five times higher than the In-concentration found in the dark portion of the gas-rich bronzite chondrite Leighton (RIEDER and WÄNKE, 1969). The In-concentration in the lunar fines is only a factor of 2 higher than that in the breccia 10018-22, a chip of rock where the problem of contamination should be considerably less important. The In-excess in Leighton-dark and in the dark portions of other gas-rich meteorites is still an unsolved problem. Clean—will say: definitely uncontaminated—samples of lunar fines and breccias would be highly desirable in order to enlighten this very puzzling mystery. It might be of importance that the major fraction of the elements nitrogen (HINTENBERGER *et al.*, 1970), chlorine and indium are liberated together with the solar wind implanted rare gases by dissolution of the surface layers of the individual grains of lunar fines.

Acknowledgments—We are grateful to NASA for making available such a generous supply of lunar material for our investigation. We wish to thank the staff of the TRIGA research reactor of the Institut für Anorganische Chemie und Kernchemie der Universität Mainz and the technical staff of our Institute, in particular Mr. E. JAGOUTZ and Miss H. PRAGER.

The financial support by the Bundesministerium für Wissenschaftliche Forschung is gratefully acknowledged.

REFERENCES

AGRELL S. O., SCOON J. H., MUIR I. D., LONG J. V. P., MCCONNELL J. D. C. and PECKETT A. (1970) Mineralogy and petrology of some lunar samples. *Science* **167**, 583–586.

ANNELL C. and HELZ A. (1970) Emission spectrographic determination of trace elements in lunar samples. *Science* **167**, 521–523.

BAEDECKER P. A. and WASSON J. T. (1970) Gallium, germanium, indium, and iridium in lunar samples. *Science* **167**, 503–505.

COMPSTON W., ARRIENS P. A., VERNON M. J., CHAPPELL B. W. (1970) Rubidium–strontium chronology and chemistry of lunar material. *Science* **167**, 474–476.

CORRENS C. W. (1968) *Einführung in die Mineralogie*. Springer-Verlag.

GAST P. W. and HUBBARD N. J. (1970) Abundance of alkali metals, alkaline and rare earths and strontium-87/strontium-86 ratios in lunar samples. *Science* **167**, 485–487.

GOLES G. G., OSAWA M., RANDLE K., BEYER R. L., JEROME D. Y., LINDSTROM D. J., MARTIN M. R., MCKAY S. M. and STEINBORN T. L. (1970) Instrumental neutron activation analyses of lunar specimens. *Science* **167**, 497–499.

HASKIN L. A., HELMKE P. A. and ALLEN R. O. (1970) Rare earth elements in returned lunar samples. *Science* **167**, 487–490.

HINTENBERGER H., WEBER H. W., VOSHAGE H., WÄNKE H., BEGEMANN F., VILCSEK E. and WLOTZKA F. (1970) Rare gases, hydrogen and nitrogen: concentrations and isotopic composition of lunar material. *Science* **167**, 543–545.

KEAYS R. R., GANAPATHY R., LAUL J. C. ANDERS E., HERZOG G. F. and JEFFERY P. M. (1970) Trace elements and radioactivity in lunar rocks: Implications for meteorite infall, solar-wind flux, and formation conditions of moon. *Science* **167**, 490–493.

KÖNIG H. and WÄNKE H. (1959) Uranbestimmungen an Steinmeteoriten mittels Neutronenaktivierung über die Xenon-Isotope 133 und 135. *Z. Naturforsch.* **14a**, 866–869.

MAXWELL J. A., ABBEY S. and CHAMP W. H. (1970) Chemical composition of lunar material. *Science* **167**, 530–531.

MORRISON G. H., GERARD J. T., KASHUBA A. T., GANGADHARAM E. V., ROTHENBERG A. M., POTTER N. M. and MILLER G. B. (1970) Multielement analysis of lunar soil and rocks. *Science* **167**, 505–507.

PHILPOTTS J. A. and SCHNETZLER C. C. (1970) Potassium, rubidium, strontium, barium and rare earth concentrations in lunar rocks and separated phases. *Science* **167**, 493–405.

QUIJANO-RICO M. and WÄNKE H. (1969) Determination of boron, lithium and chlorine in meteorites. In *Meteorite Research* (editor P. Millman), pp. 132–145. D. Reidel.

QUIJANO-RICO M. (1970) Méthodes rapides a tres haute sensibilité pour la determination du bore et du lithium dans les meteorites pierreuses et dans les roches. In *Dosage des Elements en Trace dans les Roches et dans d'Autres Matiéres Naturelles*, edition du CNRS. In press.

REED G. W. JR., JOVANOVIC S. and FUCHS L. H. (1970) Trace elements and accessory minerals in lunar samples. *Science* **167**, 501–503.

RIEDER R. and WÄNKE H. (1969) Study of trace element abundance in meteorites by neutron activation. In *Meteorite Research*, (editor P. Millman), pp. 75–86. D. Reidel.

SCHMITT R. A., SMITH R. H., LASCH J. E., MOSEN A. W., OLEHY D. A. and VASILEVSKIS J. (1963) Abundance of fourteen rare-earth elements, scandium and yttrium in meteorites and terrestrial matter. *Geochim. Cosmochim. Acta* **27**, 577–622.

SCHMITT R. A., SMITH R. H. and OLEHY D. A. (1964) Rare-earth, yttrium and scandium abundances in meteoritic and terrestrial matter—II. *Geochim. Cosmochim. Acta* **28**, 67–86.

SCHMITT R. A., WAKITA H. and REY P. (1970) Abundances of 30 elements in lunar rocks, soil, and core samples. *Science* **167**, 512–515.

SMALES A. A., MAPPER D., WEBB M. S. W., WEBSTER R. K. and WILSON J. D. (1970) Elemental composition of lunar surface material. *Science* **167**, 509–512.

TAYLOR S. R., JOHNSON P. H., MARTIN R., BENNETT D., ALLEN J. and NANCE W. (1970) Preliminary chemical analyses of 20 Apollo 11 lunar samples. *Science*. In press.

TUREKIAN K. K. and KHARKAR D. P. (1970) Neutron activation analysis of milligram quantities of lunar rocks and soils. *Science* **167**, 507–509.

WÄNKE H., BEGEMANN F., VILCSEK E., RIEDER R., TESCHKE F., BORN W., QUIJANO-RICO M., VOSHAGE H. and WLOTZKA F. (1970a) Major and trace elements and cosmic-ray produced radioisotopes in lunar samples. *Science* **167**, 523–525.

WÄNKE H., WLOTZKA F., JAGOUTZ E. and BEGEMANN F. (1970b) Compositions and structure of metallic iron particles in lunar 'fines'. *Geochim. Cosmochim. Acta.* Supplement I.

WIIK H. B. and OJANPERÄ P. (1970) Chemical analyses of lunar samples 10017, 10072 and 10084. *Science* **167**, 531–532.

Proceedings of the Apollo 11 Lunar Science Conference, Vol. 2, pp. 1729 to 1739.

Age determinations and isotopic abundance measurements on lunar samples (Apollo 11)

R. K. WANLESS, W. D. LOVERIDGE and R. D. STEVENS
Geological Survey of Canada, Ottawa, Canada

(*Received* 2 *February* 1970; *accepted* 18 *February* 1970)

Abstract—Three samples of crystalline rock yield K–Ar whole-rock ages of 2270, 2875 and 3370 m.y., while Rb–Sr results are compatible with a 4600 m.y. reference isochron having an initial $^{87}Sr/^{86}Sr$ ratio of 0·6989. The $^{207}Pb/^{206}Pb$ age determined for a sample of type D fines is 4670 m.y. The K/Rb ratios do not differ greatly from those found in chondritic meteorites and certain types of terrestrial basic rocks. The primordial ^{36}Ar content of two samples of type A crystalline rock are similar to the concentrations found for carbonaceous chondrites.

Isotopic compositions of Li, K, Rb, Sr, U and Th agree with terrestrial and meteoritic values.

INTRODUCTION

OUR OBJECTIVE in undertaking this investigation of lunar materials was three fold: (1) To determine the isotopic composition of certain geochronologically important elements; (2) to measure precisely their concentrations using isotope dilution techniques, and (3) to use the data in calculating apparent ages.

The samples received for study were classified as: (1) Type A, fine-grained vesicular crystalline rock numbers 10020-32 and 10057-34; (2) type B, medium-grained vuggy crystalline rock number 10071-26; (3) type C, breccia number 10065-35; and (4) type D, fines, number 10084-33.

The techniques employed were based on those in current use for terrestrial materials in our geochronological laboratories, which differ in some respects from those generally applied in the study of meteorites. This is particularly so in the case of the rare gas analyses.

In order to ensure that the results obtained were representative of the lunar samples, and since we assumed that great care was taken in handling and packaging the materials in the Lunar Receiving Laboratory, we decided against any form of pretreatment. All sample containers were opened in a box flushed with dry nitrogen and the contents were not exposed to the atmosphere prior to the start of the dissolution procedures. The handling of our samples of fines (10084-33) constituted the sole deviation from this procedure since in this instance a crude separation based on grain size was made. The samples received no acid or ultrasonic cleaning prior to analysis and therefore the results reported represent the total-rock analyses of materials as received, and may include contributions attributable to lunar dust adhering to the surfaces.

EXPERIMENTAL TECHNIQUES

Sample vials were opened in an atmosphere of dry nitrogen and, when necessary, larger pieces were broken in a clean mullite mortar in this atmosphere. Selected portions were weighed in sealed nitrogen flushed weighing bottles. Relatively large chips (3 mm × 5 mm) were taken for the argon extractions and several smaller fragments of these were used for the corresponding potassium determinations and also for the Rb and Sr analyses. This technique was used in order to minimize the effects of possible elemental inhomogeneity in the specimen. Since no argon analyses were planned for our sample of fines (10084-33) the precaution of handling it in a nitrogen atmosphere was not taken. A separation was made into a fine and coarse component using a new 100 mesh nylon sieve and a 0·5 g portion of each was withdrawn for Rb and Sr analyses. The remainder was recombined and used for the Pb, U, and Th extractions. Attempts were made to hand pick the white feldspar grains appearing in the coarse component but we were unable to obtain sufficient material for separate Rb and Sr analyses. Consequently this portion (0·2 g) was returned to the coarse fraction.

Argon extraction

In order to ensure a low terrestrial argon background a new molybdenum crucible was outgassed at 1600°C for one hour prior to each fusion. The sample was then transferred from the capped weighing bottle to the crucible entirely in a nitrogen atmosphere by effecting the transfer inside a large plastic bag which was continually flushed with nitrogen. The fusion chamber and extraction line were then baked at 150°C and 400°C respectively for 12–16 hr. A pressure of 1×10^{-7} mm of Hg was routinely realized in the isolated extraction apparatus prior to the commencement of the sample fusion.

Samples were melted by radio frequency induction heating and the released gases were purified by passage through hot CuO, liquid nitrogen cooled traps, and over a titanium sponge getter. The purified gases were handled in a variety of ways. For the first fusion of sample 10057-34 the gas was collected in two calibrated tubes (having volumes in the ratio of about 2:1) for transfer to the sample inlet line of the mass spectrometer. The larger portion was admitted directly to the analyser tube for isotope ratio determination, whereas the smaller volume of gas was equilibrated with an essentially pure ^{38}Ar tracer prior to isotope dilution analysis. Two sample loadings were used for the second fusion of sample 10057-34 and for each of the other three samples. The gas released from one portion of each sample was equilibrated with ^{38}Ar tracer during the course of the fusion, while the other was processed without the addition of the tracer to permit isotope ratio determinations.

Argon blanks were determined both by re-heating outgassed crucibles for periods in excess of the time required to extract the gas from a sample, and by re-fusing the sample residue of a previously melted lunar sample. In both instances the blank was found to be 6×10^{-8} cm^3 STP ^{40}Ar. The magnitude of the correction applied is given for each extraction listed in Table 1.

From inspection of the molybdenum crucibles following gas extraction it was clear that the samples had been completely molten as they had cooled to form a small fillet around the bottom edge of the crucible. It is concluded that no additional gas was released on re-fusion since the blank determined in this manner was identical with that measured when an empty preheated crucible was taken through the normal fusion procedure.

Chemical extractions

(1) *Lithium, potassium, rubidium and strontium.* Rock chips were weighed in platinum dishes and dissolved in $HClO_4$ and HF. After excess $HClO_4$ was expelled, the residue was taken up in 6 N HCl and quantitatively transferred to a polyethylene bottle from which separate aliquots were taken for isotope ratio and isotope dilution analyses. Enriched ^{87}Rb, ^{84}Sr and ^{40}K–^{41}K tracers were added to the aliquots destined for isotope dilution determinations. Li, K and Rb were precipitated as perchlorates and converted to the sulphate form for isotopic analysis. Strontium was extracted from the supernatant liquid using standard cation exchange techniques.

The extracted samples were dissolved in a drop of distilled, demineralized water and placed on the Ta side filaments of the triple-filament mass spectrometer source assembly using a clean micropipette (polyethylene for K and pyrex for Rb and Sr). Excess moisture was driven off under a heat lamp and/or by passing a current through the filaments.

Blank extractions accompanied the preparations and the corrections applied in each instance are given in Tables 1 and 2.

(2) *Lead, uranium and thorium.* Approximately 10 g of sample 10084-33 (type D fines) were dissolved in platinum and transferred quantitatively to a polyethylene bottle as outlined in the extraction procedure for Li, K, Rb and Sr. Enriched ^{208}Pb, ^{235}U and ^{232}Th tracers were added to an

aliquot amounting to about 30% of the total solution for isotope dilution analysis. The remaining 70 per cent was set aside for Pb, U and Th isotopic abundance measurements.

Lead was extracted from both aliquots using anion exchange techniques and was further purified by passage through a second anion column (CATANZARO and KULP, 1964). H_2S was bubbled through the solution to precipitate PbS which was placed on the single Re filament of the mass spectrometer source assembly. A drop of ammonium nitrate and boric acid was added and evaporated to dryness on the filament.

Uranium was extracted from the eluate of the first Pb anion exchange column by passage through another anion exchange column (TERA and KORBISCH, 1961). It was further purified, again by anion exchange and mounted, as the nitrate, on a single Re filament for isotopic analysis.

The eluate containing the Th was purified using a cation exchange column. The Th was converted to the chloride form and transferred to a single Re filament assembly.

All extractions were carried out in a contamination-free laboratory and only high purity reagents were used. Pb, U and Th blanks amounting respectively to 0·2, 0·003 and 0·005 μg/g of sample were observed and appropriate corrections were applied.

Isotopic determinations

All argon isotopic analyses were carried out with a modified A.E.I. MS-10 mass spectrometer operated in the static mode, and having a sensitivity for argon of 1×10^{-8} cm^3 STP/mV. Mass discrimination factors were determined for analyses of atmospheric argon. Ion currents were selected by rapidly switching the source voltage, and the vibrating reed amplifier output was measured with an integrating digital voltmeter (Hewlett-Packard model 2401C) and recorded on paper tape. For about 30 per cent of the analyses it was necessary to apply zero-time corrections to the observed ratios. The standard deviation of a typical analysis was 0·04 per cent and in no instance did this exceed 0·1 per cent.

Two Nier type, 90 degree deflection solid source mass spectrometers were used for the isotopic analysis of the remaining elements studied. Potassium determinations were generally carried out on the 6 in. model, whereas Li, Rb, Sr, Pb, U and Th analyses were carried out on the larger 10 in. radius instrument. Both instruments are fitted with identical triple-filament source assemblies (INGHRAM and CHUPKA, 1953) comprising Ta side filaments mounted with a Re center filament. Only the center filament was used for the Pb, U and Th analyses. All filament assemblies were baked overnight in a high temperature vacuum oven to reduce contamination levels. Assemblies treated in this manner yield no residual or background Rb or Sr ion currents and only very small K ion currents which rapidly decay to zero when the center filament current is momentarily raised to a high level. Ion currents were detected with a nine-stage electron multiplier, the output of which was amplified by a Cary vibrating reed electrometer. Ion currents were selected by switching the magnet current between peaks and the electrometer output was fed to an integrating digital voltmeter and printed on paper tape.

Rubidium, Sr and Li isotopic abundance measurements were corrected for electron multiplier discrimination by the use of factors proportional to the square-root of the isotopic masses, whereas factors proportional to the isotopic masses were required to correct K, Pb, U and Th measurements. The use of tracers enriched in ^{84}Sr and ^{40}K–^{41}K in the isotope dilution analyses permitted the determination of the amount of fractionation in each K and Sr analysis. No fractionation corrections were applied to our Rb, Pb, U and Th analyses.

RESULTS

Potassium and argon

The K concentrations (Table 1) range from a low of 474 ppm to a maximum value in excess of 2600 ppm. Surprisingly, this total range was found for our two examples of type A crystalline rock (LSPET, 1969), samples 10057-34 and 10020-32. Our single example of type B material (10071-26) is also found at the top of the range with a concentration essentially identical to rock 10057-34. The type C breccia sample

10065-35 and the type D fines 10084-33 occupy a mid-position with a range from 1000 to 1500 ppm. The results of three individual K determinations for sample 10057-34 vary by about 6 per cent, and since this difference is greater than the estimated error of the separate determinations it is attributed to a heterogeneous distribution within the specimen. In order to minimize the effects of possible K heterogeneity on the calculated K–Ar ages, two of the K determinations for this sample and each of the determinations for the other four samples were carried out on small chips and fragments broken from the same chip used for the argon extraction.

Argon isotope ratios measured for both spiked and unspiked samples are also given in Table 1. Breccia sample 10065-35 contains very large quantities of gas with a $^{36}Ar/^{38}Ar$ ratio of 5·38 which indicates a primordial source (ANDERS, 1964; HEYMANN, 1965; HEYMANN and MAZOR, 1968; ZÄHRINGER, 1968). The ^{40}Ar

Table 1. Potassium and argon analyses

Sample		Potassium			Argon						K–Ar
No.	Type	wt. (g)	ppm	Blank (%)	wt. (g)	$^{40}Ar/^{36}Ar$	$^{36}Ar/^{38}Ar$	^{40}Ar ($\times 10^{-8}$cm³stp/g)	^{36}Ar ($\times 10^{-8}$cm³stp/g)	Blank ^{40}Ar%	Age* (m.y.)
10071-26	B	0·135	2574	1·1	0·288	204·4†	—	6947	34	0·3	2875
		—	—	—	0·511	205·4	0·740	—	—	0·5	—
10057-34	A	0·154	2520	0·7	—	—	—	—	—	—	—
		0·161	2588	0·9	0·144	83·38†	—	4513	54	0·9	2260
		—	—	—	0·382	83·82	2·769	—	—	—	—
		0·092	2647	1·6	0·500	160·2†	—	4646	29	0·2	2270
10020-32	A	0·141	474	5·7	0·182	40·7†	—	1762	43	1·8	3370
		—	—	—	0·508	33·8	2·136	—	—	0·6	—
10065-35	C	0·129	1469	2·0	0·575	2·36†	—	140,880	59,670	0	—
		—	—	—	0·224	2·09	5·383	—	—	—	—
10084-33‡	D	0·121	1090	3·0	—	—	—	—	—	—	—
10084-33§	D	0·136	1110	2·6	—	—	—	—	—	—	—

^{40}K: $\lambda_\beta = 4·72 \times 10^{-10}$ yr^{-1}, $\lambda_\varepsilon = 0·585 \times 10^{-10}$ yr^{-1}, atomic abundance $= 1·19 \times 10^{-4}$.
† Values calculated from spiked analyses.
* No primordial correction applied to ^{40}Ar (see text).
‡ Fine fraction of sample of lunar fines.
§ Coarse fraction of sample of lunar fines.

content is much too large to be attributed to the *in situ* decay of the K present (1500 ppm) and must be designated as excess primordial gas that may have been derived from the solar wind (LSPET, 1969). No meaningful age can be calculated since only about 1 per cent of the ^{40}Ar could be accounted for on the basis of a K–Ar age of 4500 m.y. Since the $^{36}Ar/^{38}Ar$ ratios for the other samples are all lower than that found for the breccia sample their ^{38}Ar components are mixtures of primordial and cosmogenic argon. The largest cosmogenic component was found in sample 10071-26 which has a $^{36}Ar/^{38}Ar$ ratio of 0·74. The ^{40}Ar concentration was determined from the known amount of ^{38}Ar after adjustment based on the measured $^{36}Ar/^{38}Ar$ ratios obtained for the unspiked samples. In addition it was necessary to distinguish the cosmogenic and radiogenic components of ^{40}Ar. The ratio of the spallation production rates for cosmogenic $^{36}Ar/^{38}Ar$ is 0·6 (HEYMANN, 1965) and from this we may determine that the cosmogenic ^{36}Ar accounts for 40 per cent, 60 per cent and 100 per cent of the total ^{36}Ar for samples 10057-34, 10020-32, and 10071-26 respectively.

The spallation produced ^{40}Ar is 31 per cent (KAISER and ZÄHRINGER, 1968) of this cosmogenic ^{36}Ar component, thus the cosmogenic ^{40}Ar contribution to the total measured ^{40}Ar is only 0·1 per cent, 0·5 per cent and 0·1 per cent for samples 10057-34, 10020-32, 10071-26 and consequently no correction was applied to the calculated K–Ar ages.

The ^{40}Ar contents of the two portions of sample 10057-34 are in essential agreement but the ^{36}Ar contents differ by almost a factor of 2, indicating a heterogeneous distribution of ^{36}Ar. This variation may be a function of the position of our sample with respect to the outer surface of the rock. If the chip used for the second extraction was more effectively shielded by the outer layers of the rock, cosmogenic production rates may have been reduced thereby resulting in a lower observed ^{36}Ar concentration. Under these circumstances the ^{38}Ar content would also be expected to vary, and consequently we may not be justified in employing the observed ^{36}Ar/^{38}Ar value derived for the first extraction in determining the ^{40}Ar concentration for the second analysis.

The measured ^{40}Ar values have been corrected as indicated in Table 1 to account for the blank determined for our extraction apparatus, and it has been shown that the cosmogenic component is of negligible proportions. We therefore consider that the ^{40}Ar concentrations listed in Table 1 have been derived solely from the *in situ* decay of ^{40}K and have used them in the derivation of the K–Ar ages given in the last column of the table. For sample 10057-34 the two results are in excellent agreement at 2270 m.y. whereas sample 10071-26 yields an age of 2875 m.y. and sample 10020-32 an age of 3370 m.y. The results are in agreement with the range of ages averaging 3·0 $\pm$ 0·7 b.y. for seven lunar samples as reported by LSPET (1969). They are gas retention ages that provide an indication of the time since the rocks cooled below the temperature at which radiogenic argon was retained within the crystal lattices. From the information available it is not possible to ascertain if the ages approximate times of extrusion or provide an indication of elevated temperatures subsequent to initial crystallization as a consequence of lunar thermal activity or resulting from collisions of extra-lunar material. The ages do not correlate with the rock type since our two samples of type A crystalline rock yield highly divergent results.

The total ^{36}Ar contents are also listed in Table 1. For breccia sample 10065-35 the very large concentration is assumed to be totally primordial, whereas it has been shown that sample 10071-26 contains only cosmogenically produced ^{36}Ar and that samples 10057-34 and 10020-32 have mixtures of cosmogenic and primordial components. The primordial content for 10057-34 is 32×10^{-8} cm^3 STP/g and for sample 10020-32 it is 17×10^{-8} cm^3 STP/g. ZÄHRINGER (1966) has shown that the primordial ^{36}Ar content of chondritic meteorites is inversely related to the increasing degree of their mineralogical metamorphic texture, and that carbonaceous chondrites have the highest primordial ^{36}Ar contents. The calculated values reported above for 10057-34 and 10020-32, while slightly lower than those reported by ZÄHRINGER, are in the general range of those found for carbonaceous chondrites. The correlation is markedly improved if one considers the total ^{36}Ar concentrations as given in Table 1.

Table 2. Rubidium and strontium analyses

Sample		Wt	Rubidium		Strontium				
No.	Type	(g)	(ppm)	Blank (%)	(ppm)	Blank (%)	$^{87}Sr/^{86}Sr$*	$^{87}Rb/^{86}Sr$	K/Rb
10071-26	B	0·135	5·88	0·2	170·6	0·05	0·7059 ± 0·0004	0·0999 ± 0·0030	438
10057-34(1)	A	0·154	5·85	0·16	171·0	0·08	0·7053 ± 0·0002	0·0991 ± 0·0030	431
10057-34(2)		0·161	5·96	0·27	175·5	0·06	0·7058 ± 0·0004	0·0983 ± 0·0029	434
10020-32	A	0·141	0·754	1·7	146·7	0·05	0·6999 ± 0·0002	0·0149 ± 0·0004	628
10065-35	C	0·129	3·74	0·5	165·5	0·07	0·7035 ± 0·0002	0·0654 ± 0·0020	393
10084-33†	D	0·121	2·73	0·6	165·1	0·08	0·7033 ± 0·0002	0·0479 ± 0·0014	399
10084-33‡	D	0·136	2·63	0·5	148·9	0·08	0·7028 ± 0·0004	0·0510 ± 0·0015	422

† Fine fraction of sample of lunar fines.
‡ Coarse fraction of sample of lunar fines.
* Average of spiked and unspiked analyses.

Rubidium and strontium

The rubidium and strontium results obtained are presented in Table 2 and are plotted on a strontium evolution diagram Fig. 1. The rocks possess distinctly different $^{87}Rb/^{86}Sr$ ratios as also noted by COMPSTON *et al.* (1970) and are in accord with the range of potassium values as reported here. The highest $^{87}Rb/^{86}Sr$ values have been found in samples 10071-26 and 10057-34 whereas sample 10020-32 yields the lowest Rb concentration of 0·754 ppm. Breccia sample 10065-35, and both the fine and coarse components of our type D fines sample 10084-33 again occupy an intermediate range. While there is no reason to assume that the three crystalline rock samples represent different phases of the same total rock system and that their Rb–Sr isotopic concentrations should define an isochron, the experimentally determined points do fall close to a 4·6 b.y. ($\lambda_{Rb}{}^{87} = 1{\cdot}39 \times 10^{-11}\ yr^{-1}$) reference isochron (Fig. 1) with a $^{87}Sr/^{86}Sr$ intercept of 0·6989. Many more whole rock sample analyses

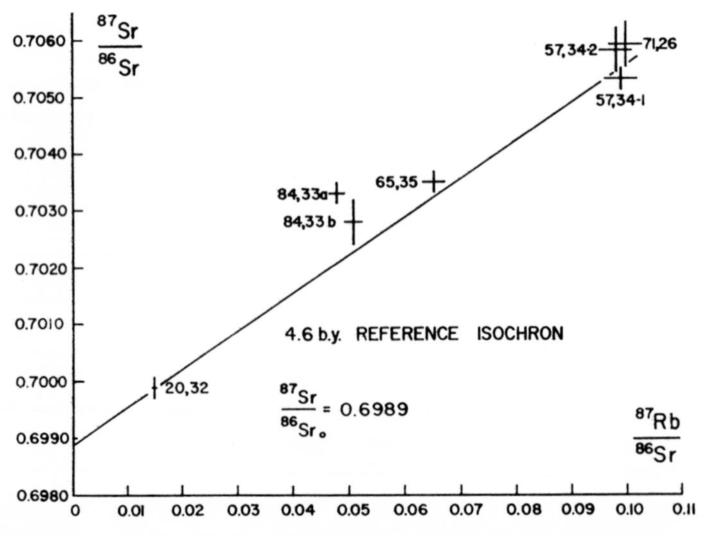

Fig. 1. Rb–Sr evolution diagram, Apollo 11 samples.

would be required to establish a meaningful age relationship but the preliminary indications are that the result would not be incompatible with the age established for the earth and meteorites. The results for breccia and lunar fines are displaced to the left of this line in Fig. 1 indicating that if they are of the same age as the crystalline rocks they have subsequently lost a portion of their rubidium while retaining essentially all of their initial strontium.

Potassium–rubidium ratios

K/Rb ratios calculated for our five samples are listed in Table 2. With the exception of sample 10020-32 all ratios fall between 393 and 438 and are within the range of recently obtained values for terrestrial basic rocks (ERLANK *et al.*, 1968; HART, 1969), although a little higher than the proposed total earth ratio of 350 (HURLEY, 1968).

Table 3. Lead, uranium and thorium analyses, sample 10084-33 type D fines

	Lead							$^{207}Pb/^{206}Pb$
	ppm		Relative Abundances			Uranium	Thorium	Age
	Primordial	Radiogenic	$^{206}Pb/^{204}Pb$	$^{207}Pb/^{204}Pb$	$^{208}Pb/^{204}Pb$	(ppm)	(ppm)	(m.y.)
Observed			45·81	33·61	64·77			
Corrected for for reagent blank	0·71	1·55	51·90	37·59	70·64	0·53	1·31	4670
Common lead correction: 86% of ^{204}Pb (see text)	0·10	1·27	257·0	171·9	268·4			

A relatively high ratio of 628 was obtained for sample 10020-32 which, however, is still within the range of reported terrestrial values.

There is an additional similarity between our results and values reported for chondritic meteorites (GAST, 1965), but the K content of our lunar samples is higher (again with the exception of sample 10020-32). The lunar values are distinctly different than values reported by GAST (1965) for basaltic achondrites.

Lead, uranium and thorium

The isotopic results obtained for a single sample of type D fines number 10084-33 are given in Table 3. If, having corrected for the reagent lead blank, one assumes that all of the nonradiogenic lead is of primordial composition (MURTHY and PATTERSON, 1962), an extremely discordant age pattern ranging from 4670 m.y. for the $^{207}Pb/^{206}Pb$ ratio to 8200 m.y. for the $^{208}Pb/^{232}Th$ ratio is found*. When the $^{206}Pb/^{238}U$, $^{207}Pb/^{235}U$ data point is plotted on a concordia diagram it falls above the curve indicating either a loss of uranium or an addition of modern, or zero-age lead to the

* ^{238}U: $\lambda = 1·537 \times 10^{-10}$ yr^{-1}; ^{235}U: $\lambda = 9·722 \times 10^{-10}$ yr^{-1}; ^{232}Th: $\lambda = 4·881 \times 10^{-11}$ yr^{-1}; Atomic ratio $^{238}U/^{235}U = 137·7$.

system. Since the uranium concentration is in agreement with that reported by LSPET (1969), TATSUMOTO and ROSHOLT (1970) and SILVER (1970) for type D fine material and since the lead isotopic ratios reported by both TATSUMOTO and SILVER are distinctly more radiogenic than those measured in our sample, we believe the cause of the age discordance is excess modern lead. Accordingly we subtracted sufficient zero-age lead to produce concordant $^{207}Pb/^{206}Pb$, $^{206}Pb/^{238}U$ and $^{207}Pb/^{235}U$ ages at 4690 m.y. In order to accomplish this, it was necessary to remove 86 per cent of the ^{204}Pb content and this resulted in the reduction of the lead concentration by 39 per cent. It is interesting to note that the resultant lead relative isotopic abundances are now nearly identical to those reported by TATSUMOTO and ROSHOLT (1970) and they are decidedly more radiogenic than those measured by SILVER (1970).

As noted, our results were corrected to account for the blank determined with the extraction and hence any modern lead component must in fact represent a contaminant in our sample. TATSUMOTO and ROSHOLT (1970) pre-treated their samples by washing in double-distilled acetone but this was not done in our laboratory since, as stated at the outset, we wished to obtain data for the essentially unaltered lunar material.

While the modification to the lead isotope values served to bring the Pb–Pb and Pb–U ages into general accord the $^{208}Pb/^{232}Th$ age remained discordantly high near 6900 m.y. Our Th value is in general agreement with that reported by LSPET (1969), but decidedly lower than reported by TATSUMOTO and ROSHOLT (1970) and SILVER (1970) and we are at a loss to explain this fact. At this time we feel that the number of analyses available for the lunar fines is too small to rule out the possibility that large elemental concentration variations may exist in the lunar regolith, especially since the discrepancy is so large–it would require a Th content of 2·03 ppm to realize a concordant age of 4700 m.y. Consequently, for the moment we prefer to consider the anomaly unexplained.

Isotopic abundances of elements in lunar samples

The isotopic abundances of Li, K, Rb and Sr were measured in our four rock samples and the fine and coarse size fractions of the lunar fines. The results obtained are compared with terrestrial and meteoritic values in Table 4. In all instances the average lunar values agree with accepted terrestrial and meteoritic ratios to within less than 1 per cent. $^{85}Rb/^{87}Rb$ ratios for the lunar rocks are generally somewhat lower than those obtained for terrestrial rocks in this laboratory. We believe this is an instrumental effect attributable to the low rubidium concentration levels in these samples. This hypothesis finds support from results carried out on two low rubidium meteorite samples which were found to be in good agreement with the lunar results. This effect is particularly evident in the analysis of sample 10020-32 which has a much lower Rb concentration than the other samples and yields an anomalously low $^{85}Rb/^{87}Rb$ ratio. Our experience with Li is limited, but we believe a similar effect to be present in our Li analyses.

The isotopic compositions of U and Th in the sample of lunar fines were also determined. We obtained a value of 137·3 $\pm$ 1·3 for the $^{238}U/^{235}U$ ratio, in good agreement with the terrestrial figure of 137·7. ^{234}U amounting to about 0·005 per cent of the ^{238}U abundance was detected, and this also is in good agreement with the terrestrial

Table 4. Isotopic abundance ratios, lunar rocks

Sample		Lithium	Potassium	Rubidium	Strontium	
No.	Type	$^7Li/^6Li$	$^{39}K/^{41}K$	$^{85}Rb/^{87}Rb$	$^{84}Sr/^{88}Sr$	$^{86}Sr/^{88}Sr$
10071-26	B	12·094	(13·274)*	2·5775	0·006748	0·11936
10057-34	A	12·118†	13·504†	2·5810†	0·006718†	0·11901†
10020-32	A	12·232	13·491	(2·5484)*	0·006759	0·11933
10065-35	C	12·077	13·447	2·5787	0·006766	0·11928
10084-33‡	D	12·132	13·377	2·5693	0·006721	0·11911
10084-33 §	D	12·178	(13·263)*	2·5673	(0·006685)*	(0·11843)*
Average		12·139	13·455	2·5747	0·006742	0·11922
Terrestrial		12·222‖	13·461	2·5926 ± 0·006¶	0·006733**	0·11914**
Meteoritic		12·047††	—	2·5787‡‡	—	

* Not included in average.
† Average of 2 extractions.
‡ Fine fraction of sample of lunar fines.
§ Coarse fraction of sample of lunar fines.
‖ SVEC and ANDERSON (1965).
¶ Average of 10 terrestrial rocks—unpublished data this laboratory.
** Average of 12 analyses of M.I.T. E. & A. Sr CO_3 lot No. 492327.
†† KRANKOWSKY and MÜLLER (1967). Correction applied to published data to account for electron multiplier discrimination.
‡‡ Average of 2 analyses of Bruderheim sample Bru-7-8–unpublished data this laboratory.

abundance of 0·0056 per cent. No ^{233}U was detected (less than 0·003 per cent) and no isotopes of Th with abundances greater than 0·01 per cent of the ^{232}Th abundance were detected.

CONCLUSIONS

While the samples studied represent a suite of rocks with limited variability, they may nevertheless be clearly distinguished on the basis of their K, Ar and Rb elemental concentrations. The K content, while relatively constant in a single sample varies from a low near 500 ppm to a high of more than 2600 ppm. The Rb variation range is also large (0·754–5·96 ppm) and parallels that of the K, whereas all Sr values fall within ±10 per cent of the mean value 163 ppm. Our two examples of type A crystalline rock appear to represent the extremes of the range with respect to both K and Rb concentration and our single sample of type B crystalline material falls precisely at the high end of the type A range. The breccia and fines occupy an intermediate position. Of the two type A rocks, it is sample 10020-34 which is most widely at variance with the results obtained for our other lunar samples. This is most apparent in its K, Ar and Rb concentrations and the data derived from these values. These parameters also serve to place the type B sample near the upper part of the range found for the type A rocks. The breccia sample, however, contains extremely large quantities of argon presumably of primordial composition and presumably derived from the solar wind. The primordial ^{36}Ar content of two of the crystalline rocks is in the range of values reported for carbonaceous chondrites. It is also noted that the K/Rb ratios resemble those found for chondritic meteorites and certain terrestrial basic rocks.

K–Ar ages are variable, ranging over a span of 1000 m.y. from 2270 to 3370 m.y. indicating a complex lunar history. Whole-rock Rb–Sr results for three crystalline rocks are not incompatible with an age of 4·6 b.y. in essential agreement with the $^{207}Pb/^{206}Pb$ age of 4·67 b.y. obtained for the type D fines.

Isotopic abundance ratios for lithium, potassium, rubidium, strontium, uranium and thorium are indistinguishable from those found for terrestrial and meteoritic materials.

Acknowledgments—The authors wish to express their gratitude to R. W. SULLIVAN for his valued contribution to the chemical procedures involved in all sample preparations, and to the entire technical staff of the Geochronology Section for their enthusiastic assistance throughout the course of this investigation.

REFERENCES

ANDERS E. (1964) Origin, age and composition of meteorites. *Space Sci. Rev.* **3,** 583–714.

CATANZARO E. J. and KULP J. L. (1964) Discordant zircons from the Little Belt (Montana), Beartooth (Montana) and Santa Catalina (Arizona) mountains. *Geochim. Cosmochim. Acta* **28,** 87–124.

COMPSTON W., ARRIENS P. A., VERNON M. J. and CHAPPELL B. W. (1970) Rubidium–Strontium chronology and chemistry of lunar material. *Science* **167,** 474–476.

ERLANK A. J., DANCHIN R. V. and FULLARD C. C. (1968) High K/Rb ratios in rocks from the Bushveld Complex, South Africa. *Earth Planet. Sci. Lett.* **4,** 22–29.

GAST P. W. (1965) Terrestrial ratio of potassium to rubidium and the composition of the earth's mantle. *Science* **147,** 858–860.

HART S. R. (1969) K, Rb, Cs contents and K/Rb, K/Cs ratios of fresh and altered submarine basalts. *Earth Planet. Sci. Lett.* **6,** 295–303.

HEYMANN D. (1965) Cosmogenic and radiogenic helium, neon, and argon in amphoteric chondrites. *J. Geophys. Res.* **70,** 3735–3743.

HEYMANN D. and MAZOR E. (1968) Noble gases in unequilibrated ordinary chondrites. *Geochim. Cosmochim. Acta* **32,** 1–20.

HURLEY P. M. (1968) Absolute abundance and distribution of Rb, K and Sr in the earth. *Geochim. Cosmochim. Acta* **32,** 273–283.

INGHRAM M. G. and CHUPKA W. A. (1953) Surface ionization source using multiple filaments. *Rev. Sci. Instrum.* **24,** 518–520.

KAISER W. and ZÄHRINGER J. (1968) K/Ar age determinations of iron meteorites. IV. New results with refined experimental procedures. *Earth Planet. Sci. Lett.* **4,** 84–88.

KRANKOWSKY D. and MÜLLER O. (1967) Isotopic composition and abundance of lithium in meteoritic matter. *Geochim. Cosmochim. Acta* **31,** 1833–1842.

LSPET (LUNAR SAMPLE PRELIMINARY EXAMINATION TEAM) (1969) Preliminary examination of lunar samples from Apollo 11. *Science* **165,** 1211–1227.

MURTHY V. R. and PATTERSON C. C. (1962) Primary isochron of zero age for meteorites and the earth. *J. Geophys. Res.* **67,** 1161–1167.

SILVER L. T. (1970) Uranium–thorium–lead isotope relations in lunar materials. *Science* **167,** 468–471.

SVEC H. J. and ANDERSON A. R., JR. (1965) The abundance of the lithium isotopes in natural sources. *Geochim. Cosmochim. Acta* **29,** 633–641.

TATSUMOTO M. and ROSHOLT J. N. (1970) The age of the moon: An isotopic study of uranium–thorium–lead systematics of lunar samples. *Science* **167,** 461–463.

TERA F. and KORKISCH J. (1961) Separation of uranium by anion exchange. *Analyt. Chim. Acta* **25,** 222–225.

ZÄHRINGER J. (1966) Primordial argon and the metamorphism of chrondrites. *Earth Planet. Sci. Lett.* **1,** 379–382.

ZÄHRINGER J. (1968) Rare gases in stoney meteorites. *Geochim. Cosmochim. Acta* **32,** 209–237.

APPENDIX 1

Results obtained on the normal Rb and Sr solutions prepared at Cal Tech and distributed with the lunar samples are listed in the table below.

Bottle No.	Concentration of Rb solution $\times 10^{-9}$ moles $^{87}Rb/g$		Bottle No.	Concentration of Sr solution $\times 10^{-9}$ moles $^{88}Sr/g$	
	This paper	Cal Tech		This paper	Cal Tech
3	0·4389	0·43986	15	14·62	14·6092
9	0·4393	0·43986	21	14·64	14·6092

Four analyses on the MIT isotopic standard Sr CO_3 (Eimer and Amend Lot 492327) were made during the course of the analyses of the lunar rocks. The average $^{87}Sr/^{86}Sr$ result, normalized to $^{86}Sr/^{88}Sr = 0{\cdot}1194$ is $0{\cdot}7090 \pm 0{\cdot}0003$.

Proceedings of the Apollo 11 Lunar Science Conference, Vol. 2, pp. 1741 to 1750.

Ga, Ge, In, Ir and Au in lunar, terrestrial and meteoritic basalts

JOHN T. WASSON and PHILIP A. BAEDECKER
Department of Chemistry, and Institute of Geophysics and Planetary Physics,
University of California, Los Angeles, California 90024

(*Received* 5 *February* 1970; *accepted in revised form* 25 *February* 1970)

Abstract—Neutron-activation analyses of Ga, Ge, In, Ir and Au in nine Apollo 11 samples, five basaltic achondrites, and a number of terrestrial basalts and standard rocks are reported. The lunar samples were contaminated with In during prior handling. The concentrations of Ge, Ir and Au are somewhat higher in lunar breccias and fines than in crystalline rocks, and are consistent with the formation of the former types by the addition of about 1 per cent of a Cl chondrite-like substance to material similar to the crystalline rocks. The Al data of other investigators indicates that the fines and breccias include about 20 per cent of material as rich in Al as the anorthositic (and possibly of highlands derivation) fragments in the coarse fines. Gallium is slightly enriched in the breccias as compared to the fines and crystalline rocks. The Ga content of the highlands rocks appears to be the same as or slightly higher than the Tranquillitatis samples. Trends established in a suite of Hawaiian basalts indicate that Ga, In and Ti are enriched during the crystallization of a basaltic magma, and suggest that these elements were present in lower concentrations in the parent magma of the lunar basalts. Germanium remained nearly constant during the crystallization of the Hawaiian magma.

INTRODUCTION

THE LUNAR Sample Preliminary Examination Team (LSPET, 1969) has shown that the Apollo 11 lunar samples are of basaltic composition, but with unusually large amounts of refractory elements, such as Ti, and unusually low amounts of relatively volatile elements, such as Na. We report here investigations of Ga, Ge, In, Ir and Au in nine lunar samples and in several terrestrial and meteoritic rocks of basaltic composition.

The lunar samples have been classified as fine-grained crystalline rocks (type A); medium-grained crystalline rocks (type B); microbreccias (type C); and fines (type D). Chemical, petrological and rare-gas evidence favors the interpretation that the breccias are compacted fines. The A- and B-type rocks appear to be of igneous origin, the former having cooled somewhat more rapidly than the latter.

The preliminary report (LSPET, 1969) indicated the presence of native iron and troilite in the crystalline rocks, and this has since been confirmed. This indicates an environment somewhat more reducing than that in the crust of the Earth; if metal or troilite were lost due to gravitational phase separation during the period that the crystalline rocks were molten, one would expect to see a depletion of Ir, Au and Ge, all of which are easily reduced to the free metals. Indium is a rather volatile element, and might be expected to have been depleted relative to terrestrial basalts by the same process(es) which resulted in the low Na and Pb contents of the lunar rocks. Gallium is not as easily reduced as the former elements, and is not especially volatile; little could be said in advance with regard to its behavior in the lunar rocks, other than that it, like the other elements, might be expected to show concentrations similar to those in terrestrial basalts or those in basaltic achondrites.

Rather few determinations have been made of these five trace elements in Ca-rich achondrites. The available data which are reported in the literature are discussed in the Results section. With the exception of In (WAGER *et al.*, 1958), we know of no studies of the behavior of these elements during the crystallization of a large intrusion, such as the Skaergaard, nor during the course of an extended period of extrusive igneous activity, such as is present in some of the Hawaiian volcanos. We report here in preliminary form what is intended to be the first of a series of studies of these and other elements in differentiation suites of basic terrestrial rocks.

Some questions regarding the moon are the following, roughly in order of increasing difficulty. What are the compositions of lunar surface rocks? What is the composition of the interior? (Assuming the moon did not originate by geofission), is the mean lunar composition chondritic, or was there substantial fractionation during the processes of condensation and agglomeration from the solar nebula? What were the major geochemical processes which have occurred in the moon? Our data have bearing on all of these.

EXPERIMENTAL

Samples and sample preparation

Approximately one gram each of eight rock samples was provided for our study, plus approximately five grams of the lunar fines. Rock sample 10060 was ground in the Lunar Receiving Laboratory, for the purpose of interlaboratory comparison. The other samples were received as chips, which were ground in a Diamond percussion mortar. Two splits were packaged for replicate analysis. The USGS standard rocks were received as fine powder. The other terrestrial rocks were prepared by grinding approximately 5–10 g, and the achondrites by grinding 1–2 g in a Diamond mortar. Approximately 500 mg aliquots were taken for each analysis. Sources of the samples are given in Tables 2 and 3, which will be discussed later.

The samples were packaged in sealed quartz vials which had been cleaned in boiling aqua regia. Duplicate flux monitors were prepared by evaporating approximately 12 mg Ge, 2·5 mg Ga, 0·25 mg In, 10 ng Au, and 4 ng Ir onto separate high-purity aluminum foils. The samples and flux monitors were irradiated for two days at a flux of about 2×10^{13} cm^{-2} sec^{-1} at the Ames Laboratory Research Reactor, Ames, Iowa, to induce sufficient Ir^{192} activity for our analyses. The samples were reirradiated in the UCLA reactor for three hours at a flux of about 2×10^{12} cm^{-2} sec^{-1} to produce the short lived Ga, Ge, In and Au activities.

Analytical and radiometric procedures

The five elements Ga, Ge, In, Ir and Au were determined simultaneously in each rock sample by a procedure which was similar to that employed by TANDON and WASSON (1968), modified to include Au. Aliquots of carrier solutions of the investigated elements as well as As and Mn were placed in Zr crucibles, made basic with NH_4OH, and evaporated to dryness. The irradiated samples were fused with about 5 g of Na_2O_2, and after cooling the fusion cake dissolved in H_2O. The resulting solution was made acid with HBr, and then brought to a pH of 8 with NaOH. The resulting precipitate of silica and the hydroxides of Ga, Ge, In and Ir was collected by centrifugation and the aqueous layer retained for gold chemistry. The hydroxide precipitate was dissolved in concentrated HBr and the solution made 5·5 M. The bromide complexes of Ga and In were extracted into diisopropyl ether. The ether layer was washed with 5 M HBr, and In was back extracted with 7 N HCl, precipitated with NH_4OH, redissolved with HBr, and the extraction procedure repeated following the addition of Fe and Mn holdback carriers. The In was finally precipitated again with NH_4OH, redissolved and counted. The chemical yield was determined by atomic absorption spectroscopy.

Gallium (and Fe) were back extracted from the ether with H_2O, the Fe (III) reduced with $TiCl_3$, and the solution made 6 N in HCl. Gallium was extracted into diisopropyl ether, the ether phase washed with 7 N HCl, and Ga back extracted with H_2O. The extraction procedure was repeated following the addition of Fe holdback carrier. The final aqueous Ga solution was counted and the chemical yield determined by atomic absorption.

The aqueous solution containing Ge and Ir was mixed with an equal volume of HCl, and Ge was extracted into CCl_4, back extracted with 4 N HCl, and the solution made 6 N in HCl. After distillation of the Ge in a stream of Cl_2, the distillate was counted, and the carrier recovery determined colorimetrically by the phenylfluorone method.

The aqueous solution containing Ir was concentrated by evaporation and the Ir and bromide oxidized with HNO_3. The solution was filtered to remove silica, and introduced to a Dowex-1 anion-exchange column. The column was washed with 1 N HCl, and Ir was batch eluted by first reducing the Ir (IV) with a warm mixture of hydrazine and dilute HCl, and then washing the resin with 8 N HCl. The Ir-containing solution was treated with HNO_3 to bring Ir back to the +4 state, Sc holdback carrier added, and the solution evaporated to a few ml. The solution was made up to 200 ml with H_2O and passed through an Amberlite IR-120 cation-exchange column. The volume of the aqueous solution was reduced and the anion-exchange procedure was repeated. The final aqueous Ir solution was counted and the chemical yield determined by reactivation.

The aqueous solution containing Au was processed by the procedure employed by BAEDECKER and EHMANN (1965). The solution was made acid with HBr and the Au bromide complex was extracted into diisopropyl ether. The Au was back extracted with H_2O, and the Au was adsorbed onto a Dowex-1 anion-exchange column. The Au was eluted with an 0·1 M thiourea solution, which was 0·1 M in HCl. The eluate was heated to boiling and Au was precipitated as the sulfide by the addition of NH_4OH. The precipitate was collected by centrifugation and dissolved in aqua regia. Gold was extracted into ethyl acetate, back extracted with a 5% NH_4OH solution, the solution made acid with HCl, and the Au precipitated with SO_2. The Au metal was filtered, washed with water and ethanol, dried, weighed to determine chemical yield, and counted.

Gamma counting was used exclusively to measure the induced activity in our samples. Gallium was determined using the 834-keV photopeak of ^{72}Ga, Ge the 266-keV peak of ^{75}Ge, In the 406-keV photopeak of ^{116}In and Au the 411-keV peak of ^{198}Au. Counting was normally carried out with 3 × 3 in. NaI well-detector coupled to an RIDL 400-channel analyzer. Iridium was determined by measuring the induced ^{192}Ir activity. However, our samples were frequently contaminated with Cr^{51} and Pa^{232} which have gamma energies near the 317-keV photopeak of ^{192}Ir, and ^{181}Hf which has a line which would interfere with the 468-keV peak of ^{192}Ir, if the samples were counted using NaI (TI) spectroscopy. Therefore, most Ir samples were counted on a high-resolution (2·5-keV for the 1332 keV ^{60}Co photopeak) Ge (Li) detector, which has a 6 per cent efficiency relative to NaI. The 468-keV line was generally employed in such analyses.

Precision and accuracy

TANDON and WASSON (1968) have determined relative standard deviations for our methods when applied to the analysis of L-group chondrites, of ±3·5%, 30%, 18% and 8% for the elements Ga, Ge, In and Ir. The precision during our current set of analyses has not been adequately retested, but we estimate that our Ga (all values) limits are about the same as those previously reported. Our current Ge (values above 1 ppm) and In (values above 1 ppb) errors appear to be somewhat lower, amounting to relative standard deviations of 5 and 8 per cent respectively. The improvement in Ge may result from the fact that the chemical yield was generally much higher (by a factor of 2) and thus determined more precisely using our current procedure. The improvement in In is perhaps a reflection of generally greater sample homogeneity in the terrestrial rocks. Concentrations less than those listed are less precise, and some of our lowest results are listed only as upper limits.

We have too few replicate analyses to provide precision limits on our analyses for Au and Ir at the low concentrations at which these elements are observed in terrestrial and lunar basalts. A summation of all accountable errors would suggest relative standard deviations of ±10 per cent for these elements at concentrations greater than 1 ppb.

Ninety-five per cent confidence limits on the means of duplicate determinations should be about 1·6 times greater than the standard deviations listed above.

The Ge results reported in this paper are higher by a factor of about 1·08 than those previously reported (BAEDECKER and WASSON, 1970) due to a recalibration of our monitor solution. Our Ir results may be systematically incorrect due to the fact that on 2 of our 3 irradiations our flux monitors were contaminated. We were able to obtain relative Ir values, which we then normalized to previous and subsequent analyses of duplicate rock samples. The problem of post-irradiation contamination of our samples with ^{192}Ir is important in our laboratory, where Ir is routinely determined at levels above 1 ppm. Cases of obvious contamination are indicated in the tables and the values discarded in the determination of the means. In two of our recent runs, the In results appear to be systematically low by factors of approximately 1·2–1·5, suggesting problems in our flux monitor preparation. These corrected values are also indicated in the tables. Additional experiments are planned to investigate this problem.

The accuracy of our results can best be judged by comparison of our analyses of USGS standard rocks with those of other research groups, and, where possible, between the results of various research groups on specific lunar rocks. Table 1 provides a comparison of data on three USGS standard rocks

Table 1. Comparison of our data for USGS standard rocks and lunar fines with that of other research groups

Rock: Element	W–1	G–2	BCR–1	10084
Ga (ppm)	*17·3*, 20[1], 15·5[2], 16[3]	*22·0*, 24[1]	*20·6*, 24[1], 19·7[4]	*4·8*, 5·3[12], 4·9[13], 4·6[14], 3·8[15]
Ge (ppm)	*2·0*, 1·7[5], 1·7[2], 1·6[4]	*1·2*	*1·6*, 1·7[4]	*0·38*, 1·4[13], 0·7[14] 0·5[15]
In (ppm)	*66*, 67[6], 66[2], 68[3]	*32*, 28[6]	*95*, 95[4], 109[6]	*656*, 920[12], 750[13], 500[14], 1250[15]
Ir (ppb)	(*0·32*), 0·26[9]	(*0·12*)	(*1·3*), ($\leq$0·2),1·1[7]	*10·7*, 7·14[12]
Au (ppb)	(*2·0*), 3·5[8], 4·9[10], 5·0[11]	(*0·86*), 1·1[7], 0·8[8]	*0·73*, 0·9[7], 0·6[8]	*1·4*, 2·85[12], 2·1[13]

Our data are in italics. Single analyses are enclosed in parentheses.

[1] BRUNFELT *et al.* (1967). [2] FOUCHÉ and SMALES (1967). [3] NICHOLLS *et al.* (1967). [4] TANDON and WASSON (1968). [5] MORRISS and BATCHELOR (1966). [6] JOHANSEN and STEINNES (1966). [7] BAEDECKER (1967). [8] MILLARD, ROWE and BROWN as reported in FLANAGAN (1969). [9] CROCKET as reported in BARKER and ANDERS (1968). [10] BAEDECKER and EHMANN (1965). [11] SHCHERBAKOV and PEREZHOGIN (1964). [12] KEAYS *et al.* (1970). [13] WÄNKE *et al.* (1970). [14] MORRISON *et al.* (1970). [15] SMALES *et al.* (1970).

and the lunar fines. In general the agreement is satisfactory. The greatest discrepancies in the lunar data are for Ge. The Ge values of MORRISON *et al.* (1970) are near 1·2 ppm for the lunar crystalline rocks, approximately 20 times higher than ours. The WÄNKE *et al.* (1970) value of 1·4 ppm for the lunar fines is about 3·7 times higher than our own mean value. No evidence of systematic differences greater than about 5 per cent can be discerned from comparison of our data with those of other researchers.

RESULTS

The results of our analyses of the Apollo 11 returned lunar samples are presented in Table 2. Data for the Ca-rich achondrites, terrestrial basalts and USGS standard rock samples are presented in Table 3. The data for Ga are presented only as means, since in most cases the replicates differed by less than 8 per cent. For the other elements the agreement between replicate analyses was less satisfactory, and the individual determinations are listed.

As was indicated in previous studies on In in the lunar rocks, most specimens are contaminated, probably by In seals in the rock box or in containers used for samples transferred out of the vacuum system at the Manned Spacecraft Center. Perhaps the values of 3 ppb found in duplicate samples of rocks 10057 and 10056 are near the true levels. For now, all In values must be considered suspect. As discussed above, some high Ir values have resulted from contamination during processing. These are indicated in Tables 2 and 3, and have not been included in the calculations of the means. Some Ir and Au "means" are from one determination only, and are subject to larger errors than those based on replicate determinations.

As pointed out by BAEDECKER and WASSON (1970), Ga is constant at about 5 ppm (to within about 2σ) for all lunar samples. The concentrations of the easily reduced elements Ge, Ir and Au are considerably greater in the fines and breccias than in

Table 2. Concentrations of Ga, Ge, In, Ir and Au in lunar surface materials returned by Apollo 11 mission

Sample	Petrologic class	Ga (ppm) mean	Ge (ppm) replicates	Ge (ppm) mean	In (ppb) replicates	In (ppb) mean	Ir (ppb) replicates	Ir (ppb) mean	Au (ppb) replicates	Au (ppb) mean
10057-27	Crys A	**4·9**	≤0·04, ≤0·09	**≤0·07**	3·0, 2·9	**3·0**	0·18, ≤0·02	**0·10**	0·15, ≤0·47	**≤0·31**
10072-28	Crys A	**4·9**	0·03, 0·09	**0·06**	49, 53	**52**	24*, 0·46	**0·46**	0·07, 0·25	**0·16**
10058-30	Crys B	**4·8**	0·03, 0·09	**0·06**	158†, 211‡	**185**			0·53, 0·92	**0·72**
10056-20	Brec§	**5·0**	0·02, ≤0·05	**≤0·04**	2·4, 3·5	**3·0**	0·05, 0·21	**0·13**	0·07, 1·15	**0·61**
10021-25	Brec	**5·5**	0·49, 0·33	**0·41**	14, 30	**22 ± 8**	7·7, 7·2	**7·5**	3·4, 1·4	**2·4**
10046-28	Brec	**5·4**	0·41, 0·36	**0·39**	25, 6	**16 ± 9**	15·9, 7·2	**11·6**	2·2, 3·4	**2·8**
10048-41	Brec	**5·2**	0·38, 0·31	**0·35**	56, 65	**60**	12·6, 8·1	**10·4**	1·6, 2·1	**1·8**
10060-27‖	Brec	**5·2**	0·21, 0·26	**0·24**	1150, 911	**1030 ± 150**	5·1, 5·6	**5·4**	1·6, 1·2	**1·4**
10084-26	Fines	**4·8**	0·36, 0·38, 0·39, 0·41	**0·39**	681, 677, 600†, 664‡	**656**	11·5, 9·8	**10·7**	1·3, 1·9 1·8, 0·7	**1·4**

* Contamination suspected.
† The In values from this irradiation have been increased by a factor of 1·5.
‡ The In values from this irradiation have been increased by a factor of 1·2.
§ Based on chemical composition this is a B-type crystalline rock. See text for details.
‖ Ground and split at NASA-Manned Spacecraft Center.

Table 3. Ga, Ge, In, Ir and Au concentrations in Ca-rich achondrites, terrestrial basalts and other terrestrial rocks

Sample	Type	Source*	No.	Ga (ppm) mean	Ge (ppm) replicates	Ge (ppm) mean	In (ppb) replicates	In (ppb) mean	Ir (ppb) replicates	Ir (ppb) mean	Au (ppb) replicates	Au (ppb) mean
Juvinas	eucrite	UTü	—	**1·7**	0·06, ≤0·10	**≤0·08**	1·4, 1·4	**1·4**	3·0†, 0·09, 0·22	**0·16**	25, 12, 14	**17**
Pasamonte	eucrite	UCLA	308	**1·7**	≤0·04, ≤0·01	**≤0·03**	3·4, 3·0	**3·2**	0·61, 0·56	**0·59**	0·52	**0·52**
Stannern	eucrite	UTü	—	**1·7**	0·15, ≤0·02, 0·07	**0·08**	0·27, 0·45, 0·25‡	**0·32**	1·2†, 0·17, 0·05	**0·11**	2·3, 0·99, 2·5	**1·9**
Le Teilleul	howardite	MHNP	711	**0·89**	0·05, 0·10	**0·08**	0·88, 0·72	**0·80**	2·1, 2·4	**2·3**	15, 16	**16**
Kapoeta	howardite	HS	—	**1·1**	1·0	**1·0**	2·7§	**2·7**			11	**11**
Waianae, Haw.	thol. basalt	GAM	C–6	**17·2**	1·5, 1·6	**1·6**	85, 69	**77**	1·3, 0·69	**1·0**	2·4, 2·0	**2·2**
Waianae, Haw.	thol. basalt	GAM	C–11	**20·6**	1·6, 1·7	**1·7**	95, 78	**87**	0·49, 0·66	**0·58**	1·2, 1·1	**1·2**
Waianae, Haw.	thol. basalt	GAM	C–20	**22·0**	1·3, 1·7	**1·5**	117, 81	**99**	16†, 0·39	**0·39**	3·2, 2·1	**2·7**
Waianae, Haw.	alk. basalt	GAM	C–31	**25·1**	1·5, 1·7	**1·6**	87, 78	**82,**	0·14, 0·12	**0·13**	1·3, 1·4	**1·4**
Waianae, Haw.	hawaiite	GAM	C–178	**27·6**	1·6	**1·6**	108, 115	**112**	0·25, 0·18	**0·22**	0·78, 0·49	**0·64**
Mid-Atlantic Ridge	thol. basalt	GOSA	AD–3	**16·8**	1·8, 1·8	**1·8**	80‡, 82§	**81**			1·9, 2·9	**2·4**
Bridal Veil, Ore.	basalt	USGS	BCR–1	**20·6**	1·5, 1·6, 1·6	**1·6**	98, 94, 92	**95**	1·3, ≤0·19		0·46, 0·91	**0·73**
Westerly, R.I.	granite	USGS	G–2	**22·0**	1·4, 1·2	**1·3**	30, 35	**33**	13†, 0·12	**0·12**	0·86	**0·86**
Centerville, Va.	diabase	USGS	W–1	**17·3**	1·8, 2·0	**1·9**	61, 71	**66**	0·32	**0·32**	2·0	**2·0**

* The sources are abbreviated as follows: GOSA, G.O.S. Arrhenius, University of California, San Diego; GAM, G.A. MacDonald, University of Hawaii; HS, Hans Suess, University of California, San Diego; MHNP, Muséum d'Histoire Naturelle, Paris; UCLA, University of California, Los Angeles; USGS, United States Geological Survey; and UTü, Mineralogisches Institut, Universität Tübingen.
† Contamination suspected.
‡ The In values from this irradiation have been increased by a factor of 1·5.
§ The In values from this irradiation have been increased by a factor of 1·2.

the crystalline rocks.* As discussed in the next section, these data are interpreted in terms of the production of breccias and fines from crystalline rocks by the addition of material resembling Cl chondrites.

The data for the basaltic achondrites (three eucrites and two howardites) are presented in Table 3. The eucrites show uniform Ga contents of 1·7 ppm, and the howardites, somewhat lower Ga contents, of approximately 1 ppm. These are well below the 5·1 ppm Ga concentrations of the Tranquillitatis samples. Germanium in the achondrites was less than or equal to 0·15 ppm with the exception of a single analysis of Kapoeta. Ir was also found to be low. Au was found to be low in Stannern and Pasamonte, and significantly higher (about 10 ppm) in the other achondrites analyzed. All In concentrations were observed to be low, near or below the lowest In concentrations observed for the lunar rocks. No systematic differences could be observed between the howardites and eucrites for Ge, In, Ir or Au.The concentrations of Ge in achondrites were about the same as those observed in lunar rocks, while the Au and Ir contents were in general higher. Our data clearly indicate that typical basaltic achondrites have not originated at the Apollo 11 site. It is too early to discuss whether they might derive from some other location on the moon.

In terrestrial mafic and felsic rocks, Ga concentrations were about 20 ppm (four times higher than in the mafic lunar rocks). Germanium was found to be remarkably uniform in terrestrial mafic rocks, and averaged about 1·6 ppm. This is more than an order of magnitude higher than the values in basaltic achondrites, with the exception of the unreplicated Kapoeta value, and about 20 times higher than the Ge concentrations in crystalline lunar rocks. Terrestrial mafic rocks had In concentrations in the range of 77–112 ppm, which was considerably higher than the In content of the basaltic achondrites or the lowest values in lunar rocks. The Ir results were generally below 1 ppb for terrestrial basalts, and the Au contents ranged from 0·5 to 3 ppb. The data seem to indicate comparable levels of Ir in crystalline lunar rocks and terrestrial basalts, while the Au concentrations were generally higher in terrestrial basalts than in the crystalline rocks from Mare Tranquillitatis. The terrestrial basalts are from Hawaii, the Columbia River and the Mid-Atlantic Ridge. It is remarkable that the Ga, Ge and In concentrations of basalts from such diverse locations should be so similar.

The five Hawaiian rocks are all from the Waianae Volcano. The first three are from the lower member of this volcano (MACDONALD and KATSURA, 1964), and may represent material emplaced during a single period of activity. The latter two rocks are from the upper member, and are less closely related. We have treated the five

* One rock, 10056, which is classified as a breccia by workers at MSC, has contents of Ga, Ge, Ir and Au which are more consistent with its classification as a crystalline rock. Data on Ti and Ni are also indicative of such a classification. ROSE *et al.* (1970) have shown that TiO_2 in crystalline rocks is always greater than 9·5 per cent, whereas those of breccias and fines are always less than 9·2 per cent. They (and others) find Ni contents which are less than 8 ppm in crystalline rocks, and concentrations of up to about 160 ppm in breccias. The TiO_2 and Ni contents reported for 10056 are 11·1 per cent and <8 ppm respectively. On the basis of its K and Al contents, 10056 appears to be a B-type rock.

rocks as a group, and in Fig. 1 are plotted our Ga, In and Ge data against the crystallization index of POLDERVAART and PARKER (1964). A lower value of this index presumably corresponds to a later differentiate. The index is calculated from the bulk chemical analyses for these rocks reported by MACDONALD and KATSURA (1964) and MACDONALD (1968). Also plotted in Fig. 1 are the TiO_2 and Na_2O values reported by these authors. Our Ir and Au values are not plotted, as no trend is evident in the data listed in Table 3, perhaps as a result of sampling errors. Figure 1 shows that Ga, In, Ti and Na are increased in the residual or late magmas during the course of crystallization of the Hawaiian rocks, whereas Ge remains remarkably constant.

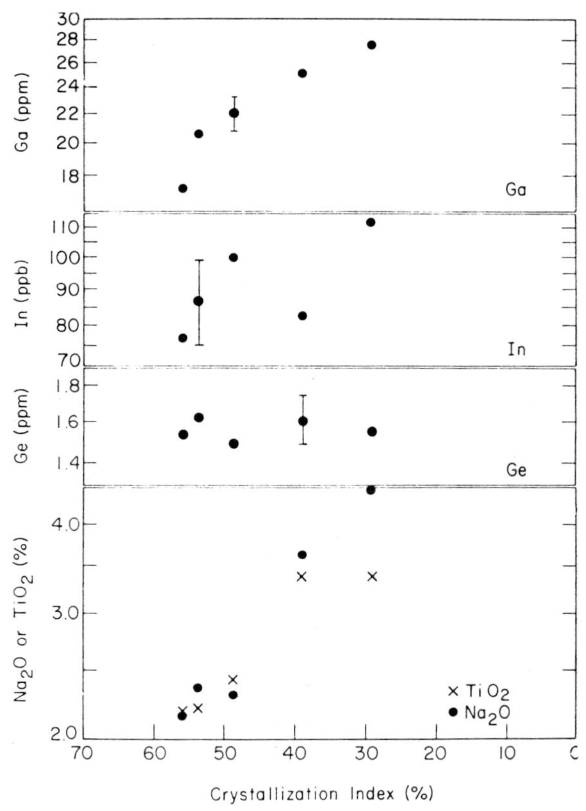

Fig. 1. Plots of our Ga, Ge and In data and TiO_2 and Na_2O data of MACDONALD and KATSURA vs. crystallization index for a suite of Waianae, Hawaii, rocks. The crystallization indices decrease with increasing sample Nos. listed in Table 3. The error bars correspond to 95 per cent confidence limits on the means of duplicate determinations.

DISCUSSION

Initial studies of Apollo 11 samples have yielded the following facts which are applicable to our study. The composition of the lunar fines and breccias is roughly intermediate to those of the A- and B-type rocks. However, detailed differences exist. KEAYS *et al.* (1970) have shown that nine trace elements are more abundant in the fines and breccias, and that this excess is best explained by the addition of 1–2 per cent of material having a composition similar to type-1 carbonaceous chondrites. There is substantial evidence for an anorthositic composition of the lunar highlands, best summarized by WOOD *et al.* (1970). The rare earths show high concentrations in the Tranquillitatis samples, and are fractionated similarly to abyssal tholeiitic basalts, except that Eu is anomalously low (e.g. HASKIN *et al.*, 1970; SCHMITT *et al.*, 1970; PHILPOTTS and SCHNETZLER, 1970). Europium tends to concentrate in plagioclase in a reducing environment, and the missing Eu may be in the lunar highlands (GAST and HUBBARD, 1970). The Preliminary Examination Team (LSPET, 1970) has found the Apollo 12 samples to be basaltic, with less Ti and more ferrous Fe than the Apollo 11 samples. Despite the prevalence of basaltic material on the lunar surface, the interior cannot be basaltic without making the bulk density considerably greater than that observed (WETHERILL, 1968; O'HARA *et al.*, 1970; RINGWOOD and ESSENE, 1970).

Our Ge, Ir and Au data are also indicative of a Cl chondrite-like component* which has been added to the crystalline rocks to produce the breccias and fines. On a water-free basis, the Cl-chondrite concentrations of Ge (FOUCHÉ and SMALES, 1967) and Ir and Au (EHMANN *et al.*, 1970) are 43, 0·61 and 0·18 ppm, respectively. Combining these with our average lunar data, we calculate that the Cl-chondrite component amounts to 0·7, 1·4 and 0·9 per cent respectively. KEAYS *et al.* (1970) have pointed out that the rate of influx of cosmic matter (based on an assumed regolith thickness of 4·6 m for Mare Tranquillitatis) is in reasonable agreement with present day estimates of the influx of cosmic dust to the earth, but is considerably higher than that expected for a present day solar-wind flux.

Gallium is enriched in the breccias but not in the fines with respect to the crystalline rocks, the values being 5·34, 4·78 and 4·92 ppm, respectively. The Ga content of the fines rises to 4·92 ppm if our one very low result is discarded. Despite their similarities, the fines and breccias have clearly had different histories. They both clearly consist of material derived from local rocks as well as from more distant locations, in addition to the 1% Cl-chondrite-like material. Presumably a portion of the fines and breccias are derived from the lunar highlands, which WOOD *et al.* (1970) propose to consist of anorthositic material containing about 14·5–17·9% Al. EHMANN and MORGAN (1970) and ROSE *et al.* (1970) show that the Al content of the breccias and fines are higher than those of the crystalline rocks, and the Al contents of the B-type rocks is somewhat higher than that of the A-type objects (especially if rock 10024 is reclassified from B to A on the basis of its low Al and high K contents and 10056 is reclassified from breccia to B-type rock, as indicated earlier). EHMANN and MORGAN report Al values of 7·1, 6·7, 4·0 and 5·2 per cent and ROSE *et al.* values of 6·8, 7·5, 4·9 and 5·7 per cent in the fines, breccias, A- and B-type rocks, respectively. In order to account for the differences in Al between the Ehmann and Morgan CD and AB samples, about 24 $\pm$ 8 per cent of the breccias and fines must be of highlands origin if the remainder is to resemble the crystalline rocks. The corresponding value based on ROSE's data is 17 $\pm$ 6 per cent. The large uncertainties arise from the use

* Although our data could also be accounted for by iron-meteorite component, the Ag, Zn, Cd, etc. data of KEAYS *et al.* (1970) clearly indicate a volatile-rich material such as C1, C2, E3 or E4 chondrites.

of different assumptions regarding the proper mixture of the two anorthosite analyses given by WOOD *et al.* (1970) and of the two types of crystalline rocks. We will use a value of 20 per cent, which is considerably higher than the value of 3·6 per cent which WOOD *et al.* estimate on the basis of 1–5 mm diameter fines, but is consistent with the fact that the highest velocity ejecta in cratering events is in the finest size range. A calculation based on the assumption that the excess Ga content of the breccia samples is the result of the addition of 20 per cent highlands material yields a Ga content of the highlands of 7 ppm. This is in contrast to the situation with the fines, where the Ga concentration is the same as that of the crystalline rocks, and suggests that the highlands concentration is about 5 ppm. In fact, the crystalline-rock component of either or both the breccias and fines may differ from the Apollo 11 rocks, but the magnitudes calculated are still likely to be reasonably correct limits. This belief is reinforced by the remarkable constancy of Ga in terrestrial basalts. It is of interest to note that the concentrations of Eu in the breccias and fines are lower than those in most crystalline rocks. Thus, although the Eu missing from the crystalline rocks may be in the highlands, the absolute Eu concentrations in the highlands appear to be less than those in Apollo 11 crystalline rocks.

Insofar as the data shown in Fig. 1 can be applied to the fractionation of lunar magmas, we would predict that the parent magma of the Apollo 11 basalts was lower in Ga, In, Ti and Na concentrations, whereas Ge was about the same. These trends should be checked by further determinations on carefully selected suites of lunar crystalline rocks. It will also be of interest to determine the concentrations of these and other trace elements in the low-Ti rocks returned from Oceanus Procellarum.

Acknowledgments—We are greatly indebted to J. KAUFMAN, J. KIMBERLIN, R. SCHAUDY, S. MOISSIDIS, J. ELZIE, S. DORRANCE, D. GONDAR, W. SIMPSON and C. PAWLAK for assistance. Terrestrial samples were kindly provided by G. ARRHENIUS, F. J. FLANAGAN and G. A. MACDONALD. Meteoritic samples were provided by W. VON ENGELHARDT, P. PELLAS and H. E. SUESS. Neutron irradiations at the UCLA and Ames Laboratory reactors were capably handled by J. HORNBUCKLE, A. F. VOIGT and their associates. This research has been supported in part by NASA contract NAS 9-8096 and NSF grant GA-1347.

REFERENCES

BAEDECKER P. A. (1967) The distribution of gold and iridium in meteoritic and terrestrial materials. Ph.D. Thesis, Univ. of Kentucky, 110 pp. USAEC Technical Report ORO-2670-17.

BAEDECKER P. A. and EHMANN W. D. (1965) The distribution of some noble metals in meteorites and natural materials. *Geochim. Cosmochim. Acta* **29,** 329–342.

BAEDECKER P. A. and WASSON J. T. (1970) Gallium, germanium, indium, and iridium in lunar samples. *Science* **167,** 503–505.

BARKER J. L. and ANDERS E. (1968) Accretion rate of cosmic matter from iridium and osmium contents of deep-sea sediments. *Geochim. Cosmochim. Acta* **32,** 627–645.

BRUNFELT A. O., JOHANSEN O. and STEINNES E. (1967) Determination of copper, gallium and zinc in "standard rocks" by neutron activation. *Anal. Chem. Acta* **37,** 172–178.

EHMANN W. D., BAEDECKER P. A. and MCKOWN D. M. (1970) Gold and iridium in meteorites and some selected rocks. *Geochim. Cosmochim. Acta* in press.

EHMANN W. D. and MORGAN J. W. (1970) Oxygen, silicon, and aluminum in lunar samples by 14 MeV neutron activation. *Science* **167,** 528–530.

FLANAGAN F. J. (1969) U.S. Geological Survey Standards—II. First compilation of data for the new USGS rocks. *Geochim. Cosmochim. Acta* **33,** 81–120.

FOUCHÉ K. F. and SMALES A. A. (1967) The distribution of trace elements in chondritic meteorites. 1. Gallium, germanium and indium. *Chem. Geol.* **2,** 5–33.

GAST P. W. and HUBBARD N. J. (1970) Abundance of alkali metals, alkaline and rare earths, and strontium-87/strontium-86 ratios in lunar samples. *Science* **167,** 485–487.

HASKIN L. A., HELMKE P. A. and ALLEN R. O. (1970) Rare-earth elements in returned lunar samples. *Science* **167,** 487–490.

JOHANSEN O. and STEINNES E. (1966) Determination of In in standard rocks by neutron activation. *Talanta* **13,** 1177–1181.

KEAYS R. R., GANAPATHY R., LAUL J. C., ANDERS E., HERZOG G. F. and JEFFERY P. M. (1970) Trace elements and radioactivity in lunar rocks: Implications for meteorite infall, solar-wind flux, and formation conditions of moon. *Science* **167,** 490–493.

LSPET (LUNAR SAMPLE PRELIMINARY EXAMINATION TEAM) (1969) Preliminary examination of lunar samples from Apollo 11. *Science* **165,** 1211–1227.

LSPET (LUNAR SAMPLE PRELIMINARY EXAMINATION TEAM) (1970) Preliminary examination of lunar samples from Apollo 12. *Science* **167,** 1325–1339.

MACDONALD G. A. (1968) Composition and origin of Hawaiian lavas. In *Studies in Volcanology*, (editors R. R. Coats, R. L. Hay and C. A. Anderson). *Geol. Soc. Amer. Mem.* **116,** 477–522.

MACDONALD G. A. and KATSURA T. (1964) Chemical composition of Hawaiian lavas. *J. Petrol.* **5,** 82–133.

MORRIS D. F. C. and BATCHELOR J. S. P. (1966) Germanium in the rocks G-1 and W-1 determined by neutron activation analysis. *Geochim. Cosmochim. Acta* **30,** 737–738.

MORRISON G. H., GERARD J. T., KASHUBA A. T., GANGADHARAM E. V., ROTHENBERG A. M., POTTER N. M. and MILLER G. B. (1970) Multi-element analysis of lunar soil and rocks. *Science* **167,** 505–507.

NICHOLLS G. D., GRAHAM A. L., WILLIAMS E. and WOOD M. (1967) Precision and accuracy in trace element analysis of geological materials, using solid source mass spectrography. *Anal. Chem.* **39,** 584–590.

O'HARA M. J., BIGGAR G. M. and RICHARDSON S. W. (1970) Experimental petrology of lunar material: The nature of mascons, seas, and the lunar interior. *Science* **167,** 605–607.

PHILPOTTS J. A. and SCHNETZLER C. C. (1970) Potassium, rubidium, strontium, barium, and rare-earth concentrations in lunar rocks and separated phases. *Science* **167,** 493–495.

POLDERVAART A. and PARKER A. B. (1964) The crystallization index as a parameter of igneous differentiation in binary variation diagrams. *Amer. J. Sci.* **262,** 281–289.

RINGWOOD A. E. and ESSENE E. (1970) Petrogenesis of lunar basalts and the internal constitution and origin of the moon. *Science* **167,** 607–610.

ROSE H. J., JR., CUTTITTA F., DWORNIK E. J., CARRON M. K., CHRISTIAN R. P., LINDSAY J. R., LIGON D. T. and LARSON R. R. (1970) Semimicro chemical and X-ray fluorescence analysis of lunar samples. *Science* **167,** 520–521.

SCHMITT R. A., WAKITA H. and REY P. (1970) Abundances of 30 elements in lunar rocks, soil, and core samples. *Science* **167,** 512–515.

SHCHERBAKOV Y. G. and PEREZHOGIN G. A. (1964) Geochemistry of gold. *Geochemistry* **6,** 489–496. (English translation).

SMALES A. A., MAPPER D., WEBB M. S. W., WEBSTER R. K. and WILSON J. D. (1970) Elemental composition of lunar surface material. *Science* **167,** 509–512.

TANDON S. N. and WASSON J. T. (1968) Gallium, germanium, indium and iridium variations in a suite of L-group chondrites. *Geochim. Cosmochim. Acta* **32,** 1087–1110.

WÄNKE H., BEGEMANN F., VILCSEK E., RIEDER R., TESCHKE F., BORN W., QUIJANO-RICO M., VOSHAGE H. and WLOTZKA F. (1970) Major and trace elements and cosmic-ray produced radioisotopes in lunar samples. *Science* **167,** 523–525.

WAGER L. R., SMIT J. R. and IRVING H. (1958) Indium content of rocks and minerals from the Skaergaard intrusion, East Greenland. *Geochim. Cosmochim. Acta* **13,** 81–86.

WETHERILL G. W. (1968) Lunar interior: Constraint on basaltic composition. *Science* **160,** 1256–1257.

WOOD J. A., DICKEY J. S., JR., MARVIN U. B. and POWELL B. N. (1970) Lunar anorthosites. *Science* **167,** 602–604.

Proceedings of the Apollo 11 Lunar Science Conference, Vol. 2, pp. 1751 to 1755.

Al^{26} and Na^{22} in lunar surface materials: Implications for depth distribution studies

ROBERT C. WRIGLEY and WILLIAM L. QUAIDE
Ames Research Center, NASA, Moffett Field, California 94035

(*Received* 2 *February* 1970; *accepted in revised form* 20 *February* 1970)

Abstract—Concentrations of the cosmogenic radionuclides Al^{26} and Na^{22} in two aliquots of lunar surface fine material and fragments of three lunar surface rocks returned by Apollo 11 were measured by nondestructive γ-ray spectrometry. The concentrations of Al^{26}, 138 $\pm$ 5 and 130 $\pm$ 5 dpm/kg in the fines and 75 $\pm$ 6, 83 $\pm$ 7 and 74 $\pm$ 7 dpm/kg in the rocks are very high compared to expected values based on meteorite concentrations; the Na^{22}:Al^{26} ratio is inverted in comparison with meteorites. These differences have been interpreted as due to chemical differences and to production by solar particle bombardment of a thin surface layer. The combination of both effects make future studies of the depth distributions of Al^{26} and Na^{22} promising; it may be possible to determine turnover rates in the regolith with a special 10 cm dia. core of a few centimeters depth.

INTRODUCTION

CURRENT theories of the generation of the lunar regolith involve repeated fragmentation of an originally hard, rock surface by meteoroid impact (OBERBECK and QUAIDE, 1968). Extensive reworking with concurrent increase in thickness of the regolith with time is predicted. With knowledge of the flux of impacting meteoroids and the scaling laws for crater size and morphology, an estimate of an average turnover rate as a function of depth in the regolith could be made. Neither of these quantities is well known so that an experimental estimate is necessary. Measured solidification ages (TATSUMOTO and ROSHOLT, 1970) of large rock fragments of assumed local origin at the Apollo 11 site in Mare Tranquillitatis, a relatively young surface, are very old ($\sim$4 b.y.) and therefore suggest that the average turnover rate is very low. Long-lived radionuclides produced by cosmic and solar particle bombardment in the uppermost layers of the lunar regolith may provide an appropriate time scale for estimating this rate: their depth distributions may show evidence of mixing when compared to the depth distributions of short-lived radionuclides. The candidate radionuclides with the longest half-lives are: Be^{10} (2·5 m.y.), Mn^{53} (2 m.y.) and Al^{26} (0·74 m.y.). Both Be^{10} and Mn^{53} require chemical separation for measurement, but Al^{26} can be determined nondestructively by γ-ray spectrometry. Hence, nondestructive measurements of Al^{26} activities appeared to be the best choice. In addition, two short-lived radionuclides, Na^{22} (2·6 yr.) and Mn^{54} (303 days) were thought to be useful as comparisons since they should show no evidence of mixing or turnover.

The lunar samples studied were two aliquots of fine material ($<$1 mm) from Apollo 11 sample 10084-113 and three rock fragments, samples 10003-25, 10017-37 and 10057-30; they were received in special stainless steel and plastic containers filled with dry nitrogen. The samples of fine material represent part of the bulk sample collected within a few centimeters of the surface, but are not further segregated on the basis of depth. Consequently, the depth dependence of Al^{26} in the fine material could not be measured, but the unexpectedly high amounts of Al^{26} showed its measurement could be made in smaller samples than estimated from meteoritic concentrations.

PROCEDURE

The equipment used to measure nondestructively the radionuclides of interest consists of an anti-coincidence shielded, two-parameter γ-ray spectrometer. A pair of NaI(Tl) scintillator crystals (12·5 cm × 10 cm) with 8·5 per cent resolution are placed 180° apart. The anti-coincidence mantle completely surrounding the NaI(Tl) crystals consists of NE-102 scintillator plastic with walls 10 cm thick. Events in the mantle cause rejection of any events in NaI(Tl) crystals. The rejection threshold is set high so as not to act as an anti-Compton device which would have reduced coincidence counting efficiencies. A large iron shield with 20 cm thick walls of conventional whole-body counter design surrounds the scintillation detectors and is itself contained in a clean room atmosphere with controlled temperature ($\pm\frac{1}{2}$°C). Accepted events in the NaI(Tl) crystals are recorded in either singles or coincident modes by a 4096 channel, two-parameter pulse height analyzer set up in a 32 × 128 channel array with equal energy ranges, 0–2·8 Mev.

Mock-ups of each Apollo 11 sample were made for each of the radionuclides of interest. For the samples of fine material, identical stainless steel containers were filled with crushed dunite ($\rho \sim 1{\cdot}7$) for background determinations; known amounts of each radionuclide were added to the dunite.

Table 1. Al^{26}, Na^{22}, thorium and uranium in lunar surface materials

Sample No.	10084-113-1	10084-113-2	10057-30	10017-37	10003-25
Sample type	Fines < 1 mm	Fines < 1 mm	Rock (A)	Rock (B)	Rock (B)
Sample mass (g)	270·8	327·0	230·0	115·5	117·2
Al^{26} (dpm/kg)	138 ± 5	130 ± 5	75 ± 6	83 ± 7	74 ± 7
Na^{22} (dpm/kg)	78 ± 7	73 ± 7	49 ± 8	47 ± 7	56 ± 7
Thorium (ppm)	2·30 ± 0·06	2·19 ± 0·06	3·27 ± 0·08	2·90 ± 0·10	0·90 ± 0·10
Uranium (ppm)	0·64 ± 0·02	0·64 ± 0·02	0·97 ± 0·04	0·90 ± 0·06	0·25 ± 0·04

Notes: (1) Quoted standard deviations include only those due to all counting statistics. Systematic errors are discussed in the text.
(2) Values for Na^{22} have been corrected for decay to July 21, 1969.
(3) Values for Th and U assume terrestrial isotopic abundances and radioactive equilibrium in their decay chains.

In the cases of Al^{26}, uranium and thorium, approximately 30 small, sealed pellets containing the radioactivity were distributed in a layered grid in the dunite. For Na^{22}, a solution was pipetted into the dunite, dried and mixed thoroughly. The thorium and uranium standards were analyzed ores from the New Brunswick Laboratory of the AEC, numbers 79 and 42 respectively and are known to ±1 per cent. The Na^{22} standard was an NBS solution, number 4922-E, known to ±2 per cent. Al^{26} was calibrated by direct comparison of net 4π counting rates with NBS Na^{22} standards. The error in the Al^{26} calibration may approach 5 per cent. A similar procedure provided mock-ups for the rock fragments except that replicas (produced by pressing aluminum foil around approximate wax models of the fragments and covering the foil with epoxy resin for strength) were filled with iron powder ($\rho \sim 2{\cdot}8$) instead of dunite. The replicas were placed in a plastic container identical to those used for the rock fragments and oriented similarly. Data were reduced by subtracting background and interfering interactions caused by nuclides previously determined and then comparing full energy peaks with those of the appropriate mock-up. The full energy peaks used were Al^{26}: 0·51 × 2·32 MeV, Na^{22}: 0·51 × 1·79 MeV, Th: 0·58 × 2·61 MeV, U: 0·61 × 1·12 MeV. All samples were analyzed for 12,000 min except sample 10003-25 which was available for only 10,500 min of analysis.

RESULTS

The concentrations for Al^{26} and Na^{22} are shown in Table 1 along with the concentrations of thorium and uranium for the two aliquots of the fine material, Apollo 11 sample 10084-113, and for three fragments of rock, samples 10003-25, 10017-37 and 10057-30. (Values for thorium and uranium are shown since they made sizeable contributions to the full energy peaks of Al^{26} and Na^{22}.) The values for Al^{26}, thorium and uranium agree reasonably well with other reported results (PERKINS *et al.*, 1970; O'KELLEY *et al.*, 1970) but those for Na^{22} are consistently 14 per cent high by comparison with the results of PERKINS *et al.* (1970) for identical samples. Since our Na^{22} and Al^{26} standards agree within the standard deviations of 3 per cent, a possible explanation for the discrepancy may be the different methods used to make mock-ups. A major correction must be made for the Al^{26} present to determine Na^{22} since its full energy peak coincides with a secondary peak of Al^{26}; that correction is sensitive to subtle differences. A completely uniform distribution of Al^{26} and not a pellet distribution seems necessary. HERZOG (1970) offered an alternate explanation for the discrepancy in Na^{22} results: Co^{56} contributions to the Na^{22} full energy peak were not subtracted and could account for some of the difference. The magnitudes of the systematic errors due to the two effects are not clear at this time, but the total may approach 20 per cent. Systematic errors due to improper mock-ups for the other nuclides must be considerably less ($\sim$5 per cent) because their full energy peaks do not coincide with a secondary peak. In fact, Na^{22} mock-ups provided by PERKINS *et al.* (1970) and made of a mixture of iron powder, plaster of paris and water agreed with the mock-ups used in this study within $\sim$5 per cent. The relatively large amounts of thorium and uranium shown in Table 1 are comparable to amounts in terrestrial basalts and they effectively mask the Mn^{54} activity as seen by the NaI(Tl) crystals. Brief use of a borrowed high resolution detector, a 45 cm^3 Ge(Li) diode, allowed a very approximate determination of the Mn^{54} in one of the aliquots of fine material, sample 10084-113-1, as 44 $\pm$ 30 dpm/kg.

DISCUSSION

The Al^{26} concentrations in the rock fragments and particularly in the fine material are high by any measure of comparison with meteoritic concentrations. A simplistic estimate of lunar surface concentrations of Al^{26} would merely halve the stony meteoritic average to account for the flux reduction due to 2π geometry at the lunar surface: this ignores differences of chemistry or flux. The result would be $\sim$30 dpm/kg. Detailed galactic production calculations by SHEDLOVSKY *et al.* (1970) and by PERKINS *et al.* (1970) yielded 31 and 42 dpm/kg, respectively, for Apollo 11 sample 10017. In contrast, the measured values are 83 $\pm$ 7 dpm/kg for sample 10017-37 and 138 $\pm$ 5 for the sample of fine material 10084-113-1. A similar, but less dramatic, situation exists for Na^{22}. SHEDLOVSKY *et al.* (1970) and O'KELLEY *et al.* (1970) observed concentration gradients with depth for Al^{26} and Na^{22}, among others, in the top centimeter of sample 10017. They attribute the gradients to surface production due to energetic solar particle bombardment. This excess production would account for much but perhaps not all of the difference between observed and calculated concentrations: SHEDLOVSKY *et al.* (1970) find a discrepancy even in their "deep" sample which they attribute to other effects. In stony meteorites, the Na^{22}:Al^{26} ratio is somewhat greater than unity (for example, see FIREMAN, 1967) whereas that ratio for the Apollo 11 rock fragment 10057-30 is 0·65 and for the fine material 10084-113 is 0·56. This inversion of the Na^{22}:Al^{26} ratio compared to meteorites has been interpreted as due to chemical differences and to surface production (HEYDEGGER and TURKEVICH, 1970; WÄNKE *et al.*, 1970; PERKINS *et al.*, 1970).

The surface gradients of Al^{26} and Na^{22} concentrations reported by SHEDLOVSKY *et al.* (1970) and O'KELLEY *et al.* (1970) clearly indicate a similar effect must occur for the fine material; the lighter density of the fines would extend the effect to larger depths (~2 cm). Excess surface production of Na^{22} would seem to occur by a somewhat different reaction than Al^{26} since the results of WÄNKE *et al.* (1970) would seem to indicate the major target nuclide for both radionuclides is Al^{27}; that is, Na^{22} would appear to be a product of higher energy reactions. If that were the case, use of Na^{22} as a short-lived comparison with Al^{26} will have to be done with some care. It is interesting to note there is little apparent excess surface production of Mn^{54} (SHEDLOVSKY *et al.*, 1970) making its use as a short-lived comparison questionable.

The indicated excess surface production of Al^{26} and Na^{22} in the fine material makes the future study of turnover rates promising for two reasons: (a) the excess production occurs only in the top centimeters so that any noted excess Al^{26} below that depth would be due to mixing, and (b) the high concentrations of Al^{26} make its measurement in small samples easier. Given a counting container with near optimum geometry, 50 g samples would be adequate to determine the Al^{26} and Na^{22} specific activities with depth nondestructively. Assuming the sample split at 0·5 cm depth intervals and with 50 g of sample per interval, a special 10 cm dia. core sampling to a depth of several centimeters is required. Such a sampling device requires a reliable bottom closing mechanism although the bottom of the core contains the least significant information. A slotted inner core tube would enable personnel to split the core accurately. Sampling at the lunar surface should include sufficient photography of the core site to allow accurate identification of nearby craters greater than a few centimeters in diameter. Eventually, it may be possible with several such cores from a single area to establish the effects of individual small craters and thus to date their formation.

Acknowledgments—We wish to thank Mr. PIERO PIANETTA for preparing the thorium and uranium standards and Robert Thompson Associates for the loan of the Ge(Li) detector. We also thank Dr. LOUIS A. RANCITELLI for valuable discussions on data reduction and one of the referees, Dr. GREGORY HERZOG, for pointing out the probable contribution of Co^{56} to the 1·8 MeV peak of Na^{22}.

REFERENCES

FIREMAN E. L. (1967) Radioactivities in meteorites and cosmic ray variations. *Geochim. Cosmochim. Acta* **31**, 1691–1700.

HERZOG G. (1970) Personal communication.

HEYDEGGER H. R. and TURKEVICH A. (1970) Evidence for solar flare induced radioactivity in lunar surface material returned by Apollo 11. *Science* in press.

OBERBECK V. R. and QUAIDE W. L. (1968) Genetic implications of lunar regolith thickness variations. *Icarus* **9**, 446–465.

O'KELLEY G. D., ELDRIDGE J. S., SCHONFELD E. and BELL P. R. (1970) Elemental compositions and ages of lunar samples by nondestructive gamma-ray spectrometry. *Science* **167**, 580–582.

PERKINS R. W., RANCITELLI L. A., COOPER J. A., KAYE J. H. and WOGMAN N. A. (1970) Cosmogenic and primordial radionuclides in lunar samples by nondestructive gamma-ray spectrometry. *Science* **167**, 577–580.

SHEDLOVSKY J. P., HONDA M., REEDY R. C., EVANS J. C., JR., LAL D., LINDSTROM R. M., DELANEY A. C., ARNOLD J. R., LOOSLI H.-H., FRUCHTER J. S. and FINKEL R. C. (1970) Pattern of bombardment-produced radionuclides in rock 10017 and in lunar soil. *Science* **167**, 574–576.

TATSUMOTO M. and ROSHOLT J. N. (1970) The age of the moon: An isotopic study of Uranium–thorium–lead systematics of lunar samples. *Science* **167**, 461–463.
WÄNKE H., BEGEMANN F., VILCSEK E., RIEDER R., TESCHKE F., BORN W., QUIJANO-RICO M., VOSHAGE H. and WLOTZKA F. (1970) Major and trace elements and cosmic-ray produced radioisotopes in lunar samples. *Science* **167**, 523–525.

Organic Geochemistry

Proceedings of the Apollo 11 Lunar Science Conference, Vol. 2, pp. 1757 to 1773.

Organic analysis of the returned Apollo 11 lunar sample

P. I. ABELL, C. H. DRAFFAN,* G. EGLINTON, J. M. HAYES,† J. R. MAXWELL and C. T. PILLINGER

School of Chemistry, University of Bristol, Bristol BS8 ITS, England

(*Received* 9 *February* 1970; *accepted in revised form* 21 *February* 1970)

Abstract—Apollo 11 lunar fines have been examined for organic compounds by solvent extraction, vacuum crushing, programmed heating and hydrofluoric acid etching. Products were examined by low resolution mass spectrometry. Solvent extraction indicated that no single volatile compound was present in excess of 5 ppb. A variety of small organic molecules including methane and other hydrocarbons accompanied the release of carbon monoxide, carbon dioxide and the rare gases when the sample was heated in a stepwise fashion to 900°C under vacuum. Methane is more abundant (ca. 2 ppm) than argon in the gases liberated by hydrofluoric acid etch of the fines. Methane is also liberated from a dark portion of the gas-rich meteorite, Kapoeta but smaller quantities are evolved from the carbonaceous chondrite, Pueblito de Allende.

INTRODUCTION

WE HAVE undertaken analytical procedures for the detection and identification of carbon compounds over a wide molecular weight range.

The preliminary examination of the Apollo 11 rocks and fines (LSPET, 1969) pointed to an origin from high temperature melts subsequently exposed on the surface for some millions of years. This hostile environment makes biogenic products extremely improbable. Sources of carbon apart from indigeneous carbon, are likely to include the solar wind and comet and meteorite impact (DRAFFAN *et al.*, 1969).

Preliminary data on the organic carbon content of the fines were available from two experiments carried out at Houston (LSPET, 1969). The results indicated an organic carbon content of between 1 and 10 ppm. A measure of total carbon was subsequently made available by MOORE *et al.* (1970). Weighted mean analyses of two samples (10086 A and 10086 B) of the bulk fines indicated levels of 142 and 226 ppm.

We decided to examine the sample in four main ways:

(1) Extract lipids and polymeric materials soluble in organic solvents and subject these to combined gas chromatography–mass spectrometry (gc–ms) with and without prior derivatisation. Techniques modified to minimize terrestrial contamination were devised.

Although indigenous carbon compounds of biological origin were not expected in the lunar fines (DRAFFAN *et al.*, 1969) a search for such compounds had to be an essential part of the analytical study. The preliminary results from the Lunar Receiving Laboratory (LRL) indicated that a minimum of 10 g of sample would be necessary.

---

* Present address: Department of Clinical Pharmacology, Royal Postgraduate Medical School, Hammersmith Hospital, London W.12.

† Present address: Department of Chemistry, Indiana University, Bloomington, Indiana.

(2) Pyrolyse a sample under vacuum in a carefully controlled stepwise manner to 900°C and analyse the products by mass spectrometry. A trapping system would be incorporated to permit fractionation. In this way we planned to study temperature and time dependence of the evolution of individual organic compounds, whether occluded or the products of pyrolysis of polymeric materials.

(3) Crush the lunar fines *in vacuo*, trap the head-space gases and examine them by mass spectrometry (GOGUEL, 1963). We hoped to detect occluded low molecular weight organic compounds, the progress of the crushing procedure being monitored by the release of rare gases. This experiment, like that listed below, would allow the detection of compounds such as methane which might conceivably have been formed as a result of solar wind bombardment.

(4) Etch the sample with hydrofluoric acid and again analyse the released gases by mass spectrometry, using a trap system.

The organic content of meteorites that have fallen on the earth would be expected to furnish additional information about extra-terrestrial organic matter. Accordingly, the gas-rich achondritic howardite, Kapoeta and the carbonaceous chondrite, Allende, were selected for comparative analyses.

The low levels of organic compounds reported in the preliminary examination of lunar material (LSPET, 1969) together with the large surface area and presumed highly adsorptive nature of the fines, demanded that stringent measures be taken to avoid organic contamination at all stages of our analyses. These arrangements permitted the classical geochemical procedures of solvent extraction, separation, evaporation, derivatisation and chromatography using only small volumes of solvent and with only brief exposure of samples, extracts and fractions to the atmosphere and the laboratory environment.

EXPERIMENTAL

General

The method of contamination control consisted of a secondary organic barrier in the form of clean filtered air supplied to a clean working station, and a primary organic barrier composed of all-glass and Teflon systems pressurised with inert gas. The gases used were nitrogen (British Oxygen Ltd., white spot grade) and helium (Air Products Ltd. C. P. 99·99%). Normal clean room equipment, apparatus, tools, clothing and procedures were used. Glass apparatus, precleaned with hot chromic acid, was baked at 500°C (12–16 hr) when required for direct contact with lunar fines. Equipment cleaned in this way was stored in glass containers under nitrogen passed through 5 Å molecular sieve. Other glassware was stored in closed glass containers and washed with solvent immediately prior to use. Stainless steel, aluminium and Teflon items were degreased by extraction with solvent in an ultrasonic tank.

Solvents used were either undistilled Nanograde (Mallinkrodt) or distilled Analar grade (Hopkins and Williams).

Establishment of a clean area–secondary organic barrier

The ventilation system of a small (150 ft^2) laboratory, reserved for the lunar sample analysis, was modified so that the pressurising supply of fresh air to the room was ducted through the filters of a "class 100" laminar flow bench (Microflow Ltd.). Materials such as lubricants, various plastics, rubber, paper and other substances likely to shed organic particles were, where practical, excluded from the laboratory.

The ambient background was monitored in two ways.

(i) Aliquots (10–20 g) of alumina (Woelm, Grade 1 neutral), previously soxhlet-extracted with benzene–methanol (3:1) and then baked at 500°C for 16 hr in air, were exposed in open vessels for 24 hr. The recovered alumina was eluted with benzene–methanol (30 ml) and the eluate evaporated to dryness under closed conditions (Fig. 1) in a tapered micro-vial (capacity 3 ml). Solvent (1 μl.) was added and an aliquot (0·5 μl.) immediately removed for glc. Concentration of solutions of standard mixtures by this method indicates that components more volatile than n-C_{14} alkane are substantially lost.

A Perkin–Elmer F-11 mark 2 gas chromatograph, containing a 10 ft $\times \frac{1}{16}$ in. column (3% OV-17 on Chromosorb W, 100–120 mesh) programmed from 90° to 280° at 5°/min, was used for contamination assays. A major contaminant was consistently recognised in the extracts from the exposed alumina and shown to be absent from control extracts of unexposed alumina. A mass spectrum obtained by gc–ms was consistent with an unresolved mixture. An abundant ion at m/e 149 strongly suggested the presence of phthalate esters. The FID response to the major unidentified

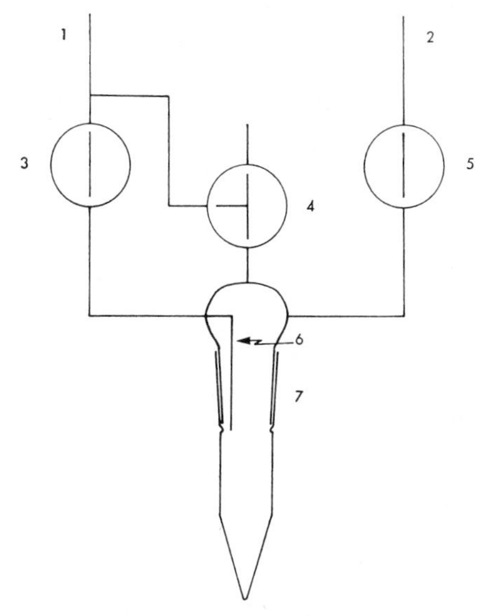

Fig. 1. Diagrammatic representation of a portion of derivatisation closed system for evaporation and derivatisation at ambient temperature. 1. Connection to nitrogen supply *via* pressure-regulated manifold (1–2 psi above atmospheric pressure); 2. Connection to vent manifold; 3, 4, 5. Teflon taps (solvent or reagent may be introduced by syringe needle through the bore of tap 4); 6. Capillary providing a jet of nitrogen; 7. Standard 10 mm ground glass joint on a tapered vessel of 3 ml capacity.

peak (carbon number 22·7) was arbitrarily compared with that of n-C_{22} alkane. Crude quantitation afforded a spread of results for the principal contaminant in the range 0·3–2·5 $\times$ 10^{-9} g/hr per g of exposed adsorptive material. Thin layers of dispersed alumina gave the highest results. An additional unresolved envelope of peaks from carbon numbers 15–30 was also present at an estimated combined level of 20 $\pm$ 10 $\times$ 10^{-9} g/hr per g of adsorptive material.

(ii) Solvent (ether or hexane, 200 ml) was allowed to evaporate to dryness in an open vessel. Concentration of the residue, followed by glc, afforded qualitatively similar results to the alumina exposure assay. Silylation of residues with bis-(trimethylsilyl)-trifluoroacetamide gave no shift in retention time of the major contaminant.

Soxhlet extraction of the following materials with benzene/methanol followed by gas chromatographic analysis of the residues established the main airborne contaminant to be absent: PVC tubing, polythene gloves, laminar flow bench and central ventilation system filter materials. No contamination attributable to the above materials was detectable in any of the analyses.

Closed analytical systems–Primary organic barrier

Extraction closed system. The major portion of this system is shown schematically in Fig. 2. The complete apparatus permits extraction, evaporation, separation on columns of adsorbents, and collection of fractions—all in a closed apparatus held under a positive pressure of pure inert gas. The relatively large reserve volumes (1 l.) of solvent ensure that a considerable number of experiments, including the necessary blank and control studies, may be performed with a single batch of solvent. Solvent contamination levels therefore remain constant. Calibrated volumes of different solvents can be run under gas pressure into the extraction vessel. Samples are extracted by placing an

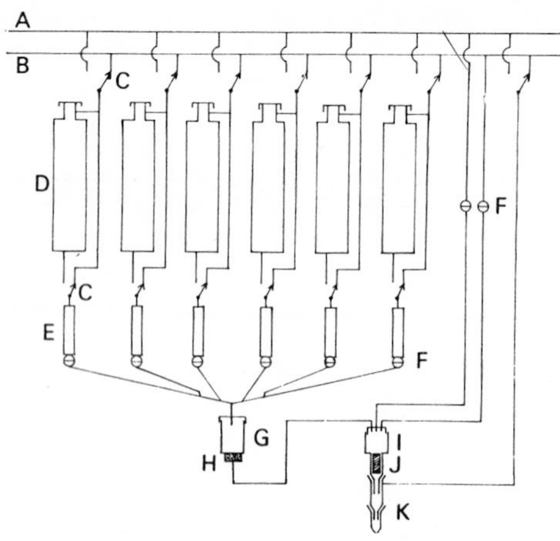

Fig. 2. Diagrammatic representation of extraction closed system. The apparatus as shown allows extraction, evaporation and collection. When required, a chromatographic column and fraction collector replace the collection unit shown. A. Vent manifold; B. Pressure manifold; C. Two way taps (upper Teflon, lower glass); D. Solvent reservoirs (1 l. capacity); E. Calibrated delivery volumes; F. One way taps (upper Teflon, lower glass), G. Extraction vessel; H. Glass sinter; I. Evaporation vessel; J. Magnetic valve; K. Collection vessel.

ultrasonic tank, containing water, around the extraction vessel. Opening and closing of appropriate taps provides a sufficient pressure of gas to transfer solvents and solutions from the extraction vessel to the evaporation vessel, collection vessel or chromatographic column and fraction collector (if present). The magnetic valve permits solutions to be contained in the evaporation vessel while volumes are reduced for collection or for chromatography. The unit for extraction, evaporation and collection is detachable but the remainder of the apparatus, once assembled, is not exposed to air. Teflon taps do not come in contact with liquid solvents and only briefly with solvent vapour; all other taps are greaseless glass taps. Cross contamination between solvents is avoided by flushing taps and lines with gas.

Derivatisation closed system. The possibility of polar material being present in a lunar sample extract required that a chemical derivatisation step be included to increase volatility before gc-ms analysis. A diagrammatic representation of a simple glass unit designed to allow several operations to be carried out under a purified nitrogen stream is shown in Fig. 1. A number of these units are incorporated in the pressure and vent manifolds of the extraction closed system. Dilute solutions containing the material to be derivatized may be concentrated and reagent or solvent introduced to the reaction vessel by syringe while an atmosphere of nitrogen is maintained in the system. Reactions which will go to completion at room temperature and which employ reagent readily removed under a stream of nitrogen can be conveniently handled in a unit of this type. Silylation with bis-(trimethylsilyl)-trifluoroacetamide is one such procedure. The efficiency of this procedure was determined by treating a mixture of β-naphthol, α-naphthylamine, n-C_{16} and n-C_{18} alcohols and n-C_{22} fatty acid (100 μg/component) with bis-(trimethylsilyl)-trifluoroacetamide (30 μl; Applied Science) at ambient temperature for 18 hr under a positive pressure of nitrogen. Gas chromatographic analysis (column 10 ft $\times \frac{1}{16}$ in., 3% OV-17 on Chromosorb W, programmed from 150° to 280°C at 5°/min) showed only peaks arising from the appropriate trimethylsilyl derivatives. No unreacted starting materials were observed. In another experiment to investigate the efficiency of the transfer procedure the same mixture (100 ng/component) containing n-hexadecanoic acid 1-^{14}C (4 ng) was similarly treated with bis-(trimethylsilyl)-trifluoroacetamide (10 μl.). A portion (25%) of the radiolabelled material remained in the reaction vessel after evaporation and removal of the products for glc.

The contamination level in the system as operated in purified inert gas was determined by attaching vessels containing aliquots (6 g) of alumina and leaving them for four hours under a pressure of nitrogen. Gas chromatographic analysis of the extracts of the alumina revealed no detectable contamination at the FID setting used (detection level equivalent to less than 5 ng/component). However, if the nitrogen pressure was shut off and the alumina left in the system for a further 16 hr to allow equilibration via joints and taps with the air in the laminar flow bench the ubiquitous phthalate esters contamination appeared at a level of ca. $0{\cdot}5 \times 10^{-9}$ g/hr per g alumina. This indicated that the two closed systems were indeed necessary as primary organic barriers.

Sample handling and storage

The lunar sample container (sample No. 10,086.19, bulk fines D, 105 g) was opened and aliquots of the sample transferred, to glass tubes which were then attached to a unit of the derivatisation closed system and sealed off under nitrogen with a hydrogen/oxygen torch. Exposure time was kept to a minimum. The soxhlet extracted (benzene/methanol) samples of alumina (heated to 550°C for 16 hr) needed for blank experiments were exposed, handled and stored in the same way as the lunar fines.

The sample of Pueblito de Allende meteorite, which had been received in a Teflon bag, was stored in a desiccator until required. The exterior surface (ca. 2 mm) was removed and the sample (5 g) pulverised in a porcelain mortar and pestle to pass 22 mesh. The Kapoeta meteorite sample (4·3 g after removal of outer surface), was handled in the same way. All operations, prior to the pyrolysis and the hydrofluoric acid etching procedure, were carried out in the clean area.

Mass spectrometry

Mass spectra were determined on a Varian-MAT CH-7 single focussing mass spectrometer. Gas mixtures were admitted from the appropriate reaction system, by means of a variable leak (valve: Hoke Engineering, 413-HT). The sensitivity of the mass spectrometer was sufficient to detect a flow of 0·1 ng/sec of methane.

The response was calibrated with aliquots of a gas mixture (British Oxygen Ltd, Grade X purity) comprising neon 80·7%, argon 7·5%, krypton 1·9%, methane 7·8% and ethane 2·1%. The ion current in coulombs/g was measured for an individual calibrant gas by calculating the percentage contribution to the total ionisation produced by the mixture using the ten most abundant peaks in its mass spectrum. Peak heights were converted to absolute peak heights using the relationship between the height of the base peak and that of the ion m/e 43 of n-butane (CORNU and MASSOT, 1966). The sensitivity of detection and the accuracy of the isotopic abundance measurements for rare gases are inferior to those obtainable by the static method normally used in rare gas analysis. The amounts of krypton and xenon in the lunar fines are insufficient for detection at the sample sizes and sensitivities employed. Quantitative data for low molecular weight species (H_2, ^{3}He, ^{4}He) were not feasible under the ion-focussing conditions used.

Solvent extraction of lunar fines

An aliquot (14·8 g) of the lunar sample was added to the extraction chamber (G) of the detachable unit (Fig. 2). This unit was then re-attached to the extraction closed system with the minimum exposure of the sample to the air (ca. 2 min). Benzene/methanol (3:1, 9 ml) was added from the appropriate reservoir and the sample extracted (15 min.) in an ultrasonic tank. The extract was passed under helium pressure (2 psi) to the evaporation chamber (I) and the solvent evaporated at ambient temperature under a flow of helium. The extraction was repeated (4·5 ml benzene/methanol) and the extract again evaporated. The magnetic valve (J) was lowered and benzene/methanol (3 ml) was passed through the system under helium pressure and the washings collected. This washing procedure was repeated (2 ml benzene/methanol) and the washings collected separately.

The extracts were centrifuged in stoppered tubes, to remove any lunar fines which had passed through the sinter (H) in the extraction chamber, and the supernatant removed. Evaporation of the supernatant and washings to dryness in the derivatisation closed system (Fig. 1) afforded the extract as a yellow gum (ca. 2 μl). The extract did not dissolve appreciably in benzene/methanol (6 μl) but dissolved in methanol (2 μl). An aliquot (0·6 μl) of the yellow methanolic solution was injected into the gc–ms instrument. No peaks were detected in the total ion current gas chromatogram at the minimum level of detection (equivalent to 25 ng of n-C_{20} alkane). The remainder of the extract was silylated by heating (45 min, 60°C) with 5 μl of reagent drawn from a freshly-opened ampoule of bis-(trimethylsilyl)-trifluoroacetamide. The gas chromatogram of the total product, when injected into the gc–ms, showed only two distinct peaks, which were also present in the product of a blank

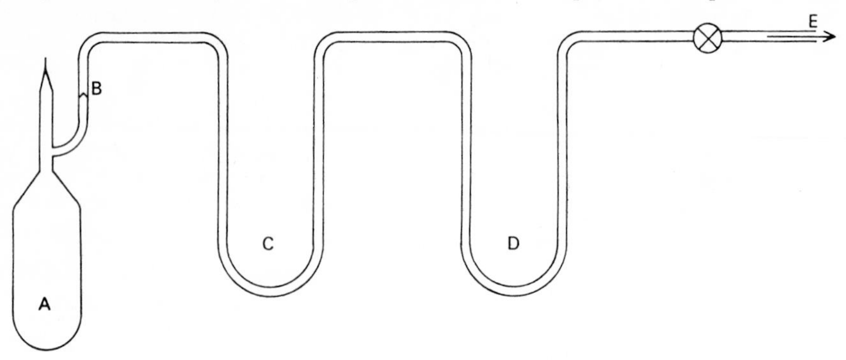

Fig. 3. Lunar sample pyrolysis sytem. A. Pyrolysis vessel; B. Break seal; C. Trap cooled in solid carbon dioxide/acetone; D. Trap cooled in liquid nitrogen; E. Entry valve to mass spectrometer ion source.

silylation procedure. The efficiency of the extraction and isolation procedures was determined by adding a solution of n-octadecane I-^{14}C (100 ng) in hexane (100 μl) to activated alumina (10 g Grade I neutral) contained in the extraction closed system. Extraction of this material and subsequent evaporation and collection in a manner similar to that used for the lunar sample gave a 70 per cent recovery of n-octadecane 1-^{14}C. Immediately prior to each extraction and isolation procedure a blank experiment was carried out in the apparatus with the same volumes of solvent alone. Alumina (15 g) which had been exposed, handled and stored with the lunar fines was also used as a blank in the analytical system. No peaks were detectable in the gas chromatograms of the products from the blank experiments.

Vacuum crushing

A stainless steel (type EN58BM; C, 0·1%) capsule (volume ca. 35 ml), equipped with a stainless steel ball and two glass break-seals, and previously baked out (36 hr, 550°C) under vacuum (10^{-6} torr) was used in this experiment. The capsule containing an aliquot (2–5 g) of lunar sample was baked out (30 min, 150°C) under the mass spectrometer vacuum (10^{-6} torr) to remove adsorbed terrestrial gases, as far as possible, from the sample. Heating was continued until the background had diminished to the level observed prior to break-seal opening. After cooling, the capsule, sealed under vacuum, was removed from the trap system and the sample crushed by subjecting the capsule to 4 min of intermittent shaking (Glen Creston Pulverizer, catalogue No. 8000). The capsule was then re-connected to the trap system, the system evacuated, the second break-seal broken and the gases allowed to flow into a molecular sieve (5 Å)/liquid nitrogen trap. A mass spectrum of the non-condensed gases was recorded. Cooling of an unpacked trap to liquid nitrogen temperature and warming of the molecular sieve trap to ambient allowed a spectrum of sieve-trapped gases to be recorded. Finally, a spectrum of the remaining gases was obtained when the unpacked trap was warmed to ambient temperature.

Blank experiments were carried out with the capsule and ball alone.

Vacuum pyrolysis

The system used is shown in Fig. 3. The pyrolysis vessel (ca. 30 ml) was baked out before use (12 hr 550°C) and was constructed of pyrex or quartz, depending on the maximum temperature required (500° or 900°C, respectively). Baking (130°C) of the entire system, containing an aliquot (ca. 2 g) of lunar fines, was carried out under the mass spectrometer vacuum until the background spectrum reached a constant low intensity. The temperature in the pyrolysis vessel was raised in

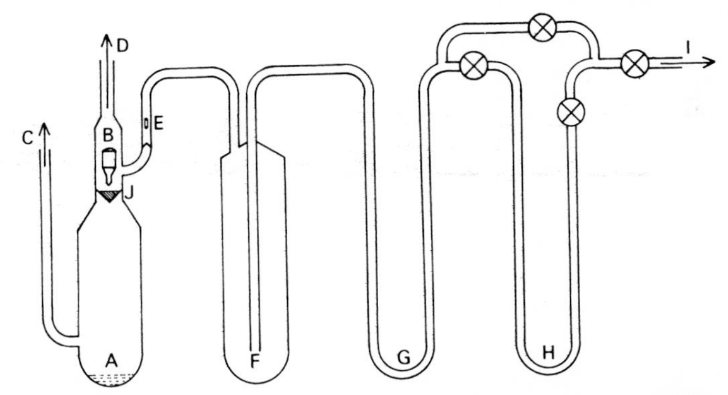

Fig. 4. Hydrofluoric acid etching apparatus. A. Reaction vessel containing hydrofluoric acid; B. Sample vessel; C. Hydrofluoric acid degas exit; D. Sample degas exit; E. Break seal; F, G. Liquid nitrogen traps; H. Molecular sieve/liquid nitrogen trap; I. Entry valve to mass spectrometer ion source; J. Dump seal.

100° steps (heating time ca. 5 min) and the baked-out gases examined after the temperature had been held steady (ca. 15 min) at the end of each temperature increment. The untrapped, non-condensible gases were metered into the mass spectrometer. The total ion current was recorded continuously and spectra recorded repeatedly over a period of up to 30 min. Similarly, mass spectra of the trapped gases were subsequently recorded by raising the liquid nitrogen trap to ambient temperature and metering the gases into the mass spectrometer over a further period of time (up to 10 min). A blank experiment was also carried out with empty pyrolysis vessels.

Hydrofluoric acid etching

The apparatus shown in Fig. 4 was used for all etching experiments. The molecular sieve trap was baked (220°C) under vacuum (10^{-6} torr) prior to each experiment. After removal of adsorbed gases from the sample vessel and the reaction vessel under vacuum (150°C, 10^{-6} torr), an aliquot (ca. 0·2 g) of the sample was added to the sample vessel and the adsorbed gases again removed. The sample vessel was then sealed under vacuum. Hydrofluoric acid (ca. 1 ml, 40%, Hopkin and Williams, "Analar" grade) was added to the reaction vessel and this then sealed on to a vacuum line. Gases dissolved in the hydrofluoric acid were removed by several freeze (liquid nitrogen)/pump/thaw cycles. Reaction of the hydrofluoric acid with the pyrex glass of the reaction vessel was evidently slow since no gas evolution was visible, either before or after degassing. The reaction vessel was next cooled in liquid nitrogen, sealed off under vacuum, removed from the vacuum line and the dump-seal broken by shaking the vessel. When all of the sample was in contact with the hydrofluoric acid the reaction vessel was warmed to ambient and the reaction allowed to proceed (30 min). Copious evolution of gases occurred. The mixture was then cooled to liquid nitrogen temperature, the reaction vessel sealed on to the trap system, and the system pumped down. Only when the background in the mass spectrometer was at an acceptable low level was the trapping system cooled to liquid nitrogen temperature and the break-seal broken. The reaction vessel was allowed to warm to ambient and the mass spectra of untrapped gases (H_2, He, Ne) recorded. Hydrofluoric acid gas, silicon tetrafluoride and water were retained in the two preliminary traps, along with any high molecular weight organic compounds. Gases trapped in the molecular sieve/liquid nitrogen trap (Ar, CH_4, O_2 and N_2) were analysed by warming the trap to ambient.

In another experiment the molecular sieve trap was by-passed and the gases examined directly by the mass spectrometer. Blank experiments with hydrofluoric acid alone and with alumina were also carried out.

RESULTS

Contamination control

Calibration of the FID response allowed background contamination to be expressed in semi-quantitative terms. The results indicated that the system of contamination control is adequate for measurement of low levels of organic compounds. The level of contamination of the principal contaminant within the secondary barrier, that is within the laminar flow bench in the clean area, ranged from 0·3 to $2{\cdot}5 \times 10^{-9}$ g/hr per g of exposed adsorptive material. Over a period of three months the qualitative reproducibility of these results was satisfactory.

Within the main primary barrier, that is the extraction and derivatisation closed systems, no contaminant was observed using the experimental procedures outlined above.

The double barrier approach described herein should have wide applicability in organic geochemistry; difficulties and limitations associated with glove box working are avoided.

Solvent extraction

Gc–ms examination of an aliquot of the benzene methanol extract and the silylated residue from 15 g of fines afforded no indication of any peaks assignable to the lunar sample. From these data we infer that individual compounds in the approximate volatility range corresponding to n-C_{14} to n-C_{32} alkanes are not present in excess of ca. 5 ppb. A wide variety of common classes of carbon compounds would be revealed by these procedures, including hydrocarbons, esters, alcohols, amines, phenols and carboxylic acids. Some handling losses are inherent at the nanogram level and have been studied in the extraction closed system. They do not greatly affect these qualitative estimates. Low molecular weight compounds, highly polar compounds, polymeric materials and matrix-bound or trapped compounds would not be detected by this experiment. It is noteworthy that there was a yellow gum remaining after evaporation of the extraction solvent. This gum gave no volatile materials either by direct injection, or after silylation on gc–ms but cannot be ascribed to elemental sulphur (KAPLAN and SMITH, 1970).

Vacuum crushing

Rare gases, including ^{20}Ne and ^{22}Ne, ^{36}Ar, ^{38}Ar and ^{40}Ar, in the isotope abundances observed by LSPET (1969), were liberated in this experiment, indicating that the crushing was at least partially effective. However, electron microscope studies, both scanning and transmission, carried out before and after the crushing showed little change in the overall appearance of the sample (at magnifications of 500 to 12,000 times), but some fractured bubble-filled fragments were seen. Other gases including CO, CH_4, C_2H_6 and C_3H_8 were observed in this experiment but comparable amounts of the same gases were observed in blank experiments when the capsule and ball were subjected to the shaking procedure. Metal carbides and other carbon inclusions in the steel are the likely sources of these gases (ROBINSON, 1960).

Vacuum pyrolysis

Vacuum pyrolysis (Fig. 3) of the lunar fines indicated that approximately 150 ppm of carbon (based on ion monitor response) can be released as low molecular weight

Table 1. Gas content of pyrolysed lunar sample

	130–200°C	200–300°C	300–400°C	400–500°C	500–600°C	600–700°C	700–800°C	800–900°C
H_2	+	+	++	++	++	++	+	+
^4He	+	+	+	+	+	+	+	+
^{20}Ne				+	+	+	+	+
^{22}Ne				+	+	+	+	+
^{36}Ar				+	+	+	+	+
^{38}Ar				+	+	+	+	+
^{40}Ar				+	+	+	+	+
N_2	+	+	+	+	+	+	+	+
O_2	+	+						
CO	+++	+++	+++	+++	+++	+++	+++	+++
CO_2	+++	+++	+++	+++	+++	+++	+++	+++
NH_3		+	++	++	+			
CH_4	+	+	++	++	++	+		
C_2H_4	+	+	+	++	+			
C_4H_8	++	+	++	+++	+			
C_6H_6	+	+	++	+++	+++	++	+	+
C_7H_8	+	+	+	++	++	+	+	+
C_8H_8				+	+	+		
C_8H_{10}				+	+	+	+	+
C_9H_8					+	+	+	+
C_9H_{10}					+	+	+	+
C_9H_{12}						+	+	+
C_4H_4S		+	+	+	+			

pyrolysis products. The major components of the total pyrolysate were CO and CO_2, especially at the higher temperatures. The untrapped gases proved to be H_2, He, CH_4, Ne, CO and Ar (Table 1), all of which were present in the mass spectrum of the 500°C pyrolysate (Fig. 5A, 5B). The rare gases first appear near 300°C and continue to be evolved thereafter, but with some variation in observed isotope ratios, presumably due to selective baking out of different sites and the presence of terrestrial gases at the lower temperatures. Methane appears predominantly between 300 and 600°C. Analysis of the condensed gases liberated from the traps indicated the presence of (Table 1) NH_3, CO_2 and unsaturated and aromatic hydrocarbons, including benzene and alkylbenzenes (Fig. 5C, 5D). The compounds identified from their principal ions were NH_3, CO_2, C_2H_4, C_4H_8, C_6H_6, C_7H_8, C_8H_{10}, C_9H_8, C_9H_{10}, C_4H_4S. A pyrolysis to 500°C in a pyrex vessel without the dry ice/acetone trap, also indicated the presence

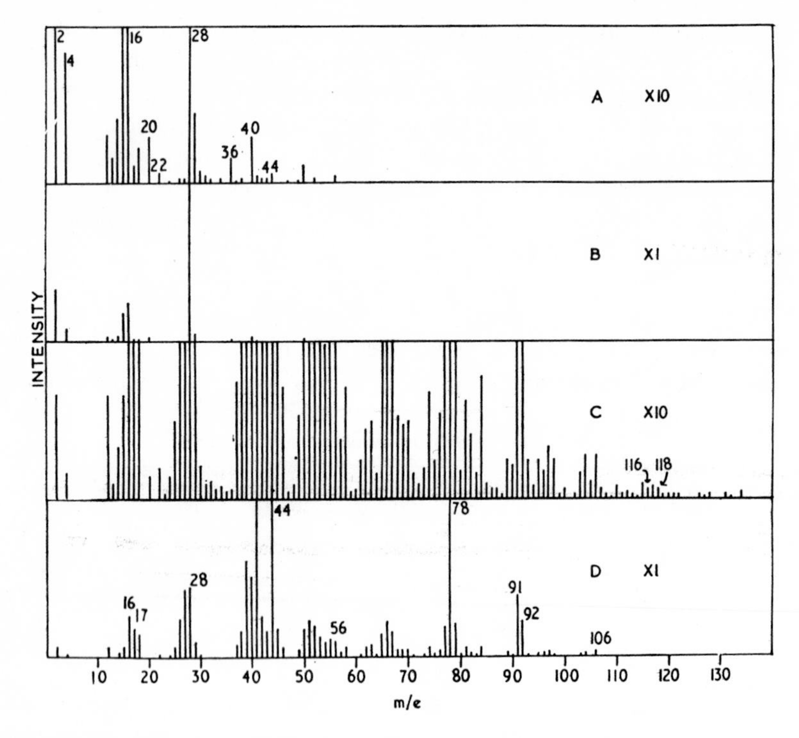

Fig. 5. Mass spectra (sensitivities × 1 and × 10) of the non-condensible (A + B) and the condensable (C + D) gases released during the 400°–500°C pyrolysis step.

of $C_{10}H_8$ and $C_{11}H_{10}$ compounds. Otherwise, the results of this pyrolysis were comparable with most from the earlier stages of the 900°C pyrolysis.

It should be noted that the identifications are based on low resolution mass spectra and some heteroatomic species may contribute to the observed ion abundances. The relative amounts indicated by +, ++ and +++ in Table 1 refer to each temperature increment only and do not allow comparisons between temperature increments. The only carbon-containing compound present in a blank pyrolysis (900°C) conducted on the empty quartz capsule was carbon dioxide. This component was present in very small quantities and probably arose from desorption of atmospheric gases from the capsule. Nitrogen, oxygen and nitric oxide were also identified.

Hydrofluoric acid etching

Utilization of the molecular sieve trap in the system (Fig. 4) permitted a fractionation step, Ar, CH_4, O_2 and N_2 being retained at liquid nitrogen temperatures, but not when the sieve was warmed to 20°C. Methane was the most abundant non-atmospheric gas (Fig. 6) and was present at a level of ca. 2 ppm by weight ($3–4 \times 10^{-3}$ cm^3/g) of the fines (Table 2). The CH_4 was identified by its fragmentation pattern and was distinguished from ^{16}O at a resolvlng power of 1000 (10% valley) in the mass spectrometer. The etching procedure also released H_2 and the rare gases, He, Ne and Ar.

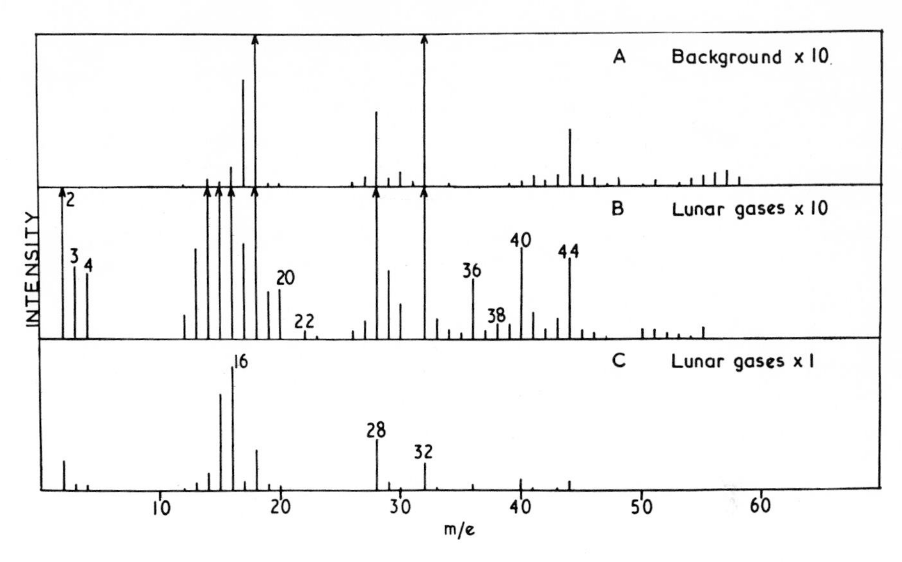

Fig. 6. Mass spectra (sensitivities × 1 and × 10) of total lunar gases (B + C) released by hydrofluoric acid etching. Mass spectrum (sensitivity × 10) of the analysis system residual gas background (A).

The amount (Table 2) of ^{36}Ar (7–9 × 10^{-4} cm^3/g) is similar to that measured for the lunar fines at the LRL (LSPET, 1969). Precise quantitation was not feasible with the techniques used: variables include the extent and rate of dissolution of the samples.

Methane and ^{36}Ar were released at the levels of 6 × 10^{-5} cm^3/g (0·04 ppm) and 7 × 10^{-6} cm^3/g, respectively, from the gas-rich achondrite Kapoeta which has a low total carbon content (0·07 per cent). (MASON and WIIK, 1966). On the contrary, the type C-3 chondrite Pueblito de Allende, which has a total carbon content of 0·35 per cent HAN *et al.* (1969), released 2 × 10^{-5} cm^3/g (0·02 ppm) methane under the hydrofluoric acid etching conditions (Table 2), but no ^{36}Ar was detected.

The quantities of CH_4 and ^{36}Ar released from an aliquot of the exposed alumina

Table 2. Argon-36 and methane content by 40% aqueous HF etch

Sample	Wt. of sample (g)	Etch	Yield in 10^{-8} cm^3/g Argon-36	Methane
Lunar fines	0·270	- -	90,000	300,000
Lunar fines	0·201	- -	70,000	400,000
Lunar fines after pyrolysis at 900°C	0·213	- -	5000	100,000
Packaging blank (Al_2O_3)	0·200	-	Below detection limit	
System blank	0	0	Below detection limit	
Allende	0·234	- - -	Below detection level	2000
Kapoeta	0·312	- -	700	6000

Degree of etch indicated by number of dashes.
Minimum detectable: 100 × 10^{-8} cm^3/g.

blank and from empty reaction vessels were below the detection limits of the experiment. Methane was not detected when a fresh sample of fines was heated to 150°C at 10^{-6} torr: we presume that measurable quantities of this gas are not present on the surface of the sample. The LM exhaust studies make no mention of CH_4 as a component of the exhaust gases (SIMONEIT *et al.*, 1969).

The isotope ratios of Ne and Ar observed in the HF etch experiments (and the pyrolysis experiments) of the lunar material and Kapoeta meteorite are approximately those observed in the LSPET experiment (LSPET, 1969) and the data on Kapoeta reported by ZAHRINGER (1963). The argon isotope ratios we observe for the Apollo 11 lunar fines ($^{36}Ar:^{38}Ar:^{40}Ar \simeq 3:1:5$) are not terrestrial (1:0·2:300); terrestrial argon is highly depleted in the lighter isotopes. The results we find for Kapoeta ($^{36}Ar:^{40}Ar = 1:10$) are comparable to ZAHRINGER's data (1:15). ^{38}Ar was not detected. These ratios are obviously different from those obtained for terrestrial argon and for argon from the Apollo 11 lunar fines.

DISCUSSION

Carbon content of lunar fines and rocks

The results reported by MOORE *et al.* (1970) show that the largest quantities of total carbon (500 ppm) occur in the finest fractions of the fines and the smallest (64 ppm) in the interior of the igneous rocks. Similarly, the lunar fines have the highest rare gas content; therefore, the content of total carbon broadly parallels that of the solar wind rare gases. It is unlikely that these large differences are to be explained by contamination. The elemental composition of the solar corona is known from spectroscopic evidence and consideration of solar abundance tables (UNSÖLD, 1969) leads to the conclusion that carbon, nitrogen and other important organogenic elements must accompany hydrogen, helium, argon, etc. in the solar wind flux. Although it is possible that rare gases could be primordial, cosmogenic or radiogenic it is widely accepted that the hydrogen and the rare gases in the lunar fines derive predominantly from the solar wind. If so certainly some of the carbon in the fines must also be derived from that source.

The question is, how much? The rare gas values for lunar material indicate saturation of the surface layers of the particles (HINTENBERGER *et al.*, 1970; HEYMANN *et al.*, 1970). Losses by diffusion and shock events are not known but the observed depletion of light rare gases and the difference from solar isotope composition indicate that these must have been considerable. Comparing solar and lunar abundances of nuclides, the observed depletion factors relative to ^{84}Kr are of the order of 10, 50 and 300 for ^{36}Ar, ^{20}Ne and ^{4}He, respectively. But the loss of carbon compared to that of the lower molecular weight rare gases should be less owing to retention by reaction with the matrix. Making some rough allowance for this factor by assuming a carbon depletion factor more like those of ^{36}Ar or ^{84}Kr than of ^{20}Ne or ^{4}He, the figure for carbon implanted by the solar wind in the Apollo 11 surface fines should be around 50 ppm. Similar arguments have been advanced by MOORE *et al.* (1970). Precise calculations are not as yet possible for little is known of the diffusion losses, the surface exposure ages, lunar gardening processes, etc.

Two other sources for the lunar carbon merit discussion: carbonaceous chondrites and the igneous rocks themselves. KEAYS *et al.* (1970), using the proportions of certain elements as an indication, have suggested that the contribution of carbonaceous chondrites to the regolith approximates to about 2 per cent. The amount of carbon provided from this source must therefore be very small, bearing in mind both the low carbon content (approximately 2 per cent) of such meteorites and the extremely high temperatures following impact which should vapourize most of the carbon, partially as carbon monoxide. The parent igneous rocks of the moon, like their terrestrial counterparts (BARKER,1965), could well contain a small abundance of truly indigenous primordial carbon, but the results of our own and of other investigations (BURLINGAME *et al.*, 1970; ORÓ *et al.*, 1970) show that much of the carbon present in the Apollo 11 rocks and fines is very rapidly lost, mainly as CO, on heating to high temperatures of 700°C and over. The vesicular nature of some of the rocks may have resulted from just such outgassing at the time of crystallization.

Since the bulk of the "lunar soil" has been derived from the crystalline rocks, it seems difficult to escape the conclusion that the higher carbon content of the fines is to be ascribed to additional quantities provided from an outside source, namely the solar wind.

Major organic contaminants which we had anticipated might be present in the sample from collection, handling in the biopreparation facility at the LRL, and packaging, include butyl rubber, plasticizers, hydrocarbon oils, Teflon, cellulose fibres, contingency bag contaminants, peracetic acid and ethylene oxide and their reaction products. Further, the bulk fines had been collected over an area within about 30 feet of the LM engine, where they must certainly have received considerable effluent during the final few seconds of the descent. The oxidizer, N_2O_4, was also vented on that side of the craft. We were unable to positively identify any of these contaminants in our sample of fines, but high resolution mass spectrometric studies of the products of controlled heating by MURPHY *et al.* (1970), BURLINGAME *et al.* (1970), revealed hydrocarbons and heteroatomic species known to be present in the LM exhaust, and the presence of phthalate contaminants. MURPHY *et al.* (1970), are of the opinion that most of the ions seen in the M.S. analysis of the products of pyrolysis of the fines are derived from contaminants, such as ethylene oxide and peracetic acid. However, the high positive $\delta^{13}C$ values found by KAPLAN and SMITH (1970), ORÓ *et al.* (1970) and EPSTEIN and TAYLOR (1970) preclude a terrestrial origin for most of the carbon in the lunar samples.

In agreement with most other bioscience investigators, we were unable to detect any of the conventional biolipids. Our levels of detection, arrived at by solvent extraction of the 15-g sample of fines, were equivalent to putting an upper limit of ca. 5 ppb on any single component in the volatility range studied. Thus the anhydrous state of the lunar fines and rocks, their high temperature history, the low abundance of carbon and the non-detectability of biolipids, the extremes of the lunar environment in terms of radiation, temperature and lack of atmosphere, all parallel the anticipated and confirmed absence of viable lunar organisms (LSPET, 1969; OYAMA *et al.*, 1970).

However, pyrolysis of the lunar sample (150–900°) did give evidence of a variety of aromatic compounds (Table 1). These compounds are probably not present as such in the lunar fines and they most likely have been formed by pyrolysis of polymeric material. Thermal equilibration in a C, H, O ternary system produces a variety of aliphatic and aromatic compounds (DAYHOFF *et al.*, 1964; ECK *et al.*, 1966). Calculations of thermal equilibrium concentrations for a variety of compounds at selected regions of the CHO ternary diagrams have been largely verified by experimental studies of equilibria in plasma (WEIFFENBACH *et al.*, 1969; GRIFFITHS *et al.*, 1969). Aromatics may constitute as much as 70% volatiles in an equilibrium mixture. Aromatic products similar to those reported herein have been recognized in the pyrolyses products of carbonaceous chondrites (HAYES and BIEMANN, 1968), and the Onverwacht chert (SCOTT, 1970). In the case of the lunar fines it is difficult at this stage to rule out pyrolysis of contaminants as at least a partial source. Pyrolysis of the methane in the sample, even if present as such, is insufficient to produce the quantity of aromatics detected (HAN and ORÓ 1967).

Carbon monoxide is certainly the major carbon containing molecule evolved on heating the lunar fines. Our figures, and those of other workers (BURLINGAME *et al.*, 1970; ORÓ *et al.*, 1970), show that most of the carbon in the sample is eventually released in the form of carbon monoxide as the temperature is raised to 900°C and beyond. We do not as yet know the form, bound or unbound, of most of this carbon—that is prior to heating of the fines or the dissolution in acid. We have been unable to confirm the release of large quantities, equivalent to 66 ppm, of carbon monoxide on HF treatment of the fines (BURLINGAME *et al.*, 1970). Some of the carbon is evidently present as graphite, for one fragment has been reported in the fines (ARRHENIUS *et al.*, 1970). The presence of small amounts of calcium carbonate (calcite and aragonite) (AGRELL, 1970) has been reported. Iron nickel carbide $(Fe, Ni)_3C$ (cohenite) has been recognized in meteorites (BRETT, 1967) and meteoritic fragments from the lunar soil (FRONDEL *et al.*, 1970). Carbon-containing materials may also be present in the opaque portion of the rocks and fines, since these are difficult to characterize.

Methane and lunar rare gases

The impressive quantities of the lunar rare gases, first reported in the preliminary analyses (LSPET, 1969), prompted our search for methane and other likely products of solar wind implantation, for the vast excess of hydrogen (as H^+ and $H^\bullet$) over carbon reaching the surface of the moon might reasonably be expected to give rise to hydrogenated products. Thus, proton bombardment of glasses in the laboratory produces hydroxyl groups (ZELLER *et al.*, 1966). Model solar wind experiments involving bombardment of glass and of other silicate materials with mixtures of nuclei (hydrogen, carbon, etc.) are clearly desirable. The situation on the lunar surface is further complicated by the existence of the intense u.v. radiation, the thermal regime of the lunar day and night, and of the cosmic ray bombardment. High energies released spallogenically provide further means for alteration of the carbon content of the lunar fines through 'hot atom' chemistry (MACKAY and WOLFGANG, 1965). Hence, we believe our most significant finding to be the release of approximately

2 ppm of methane when the lunar sample is etched with hydrofluoric acid. The lunar rare gases are concomitantly released. Again, methane and rare gases are liberated when the sample alone is heated above 300°C, but here the methane might have a pyrolitic origin. Methane and other hydrocarbons have been reported by PONNAMPERUMA *et al.* (1970), in the gaseous products evolved on dissolution of the sample in 6 N-hydrochloric acid. They ascribe the formation of these hydrocarbons to hydrolysis of indigenous metal carbides. Our present trapping techniques unfortunately preclude study of hydrocarbons other than methane, but acidic hydrolysis of carbides could conceivably account for some of the methane evolved in our hydrofluoric etching experiments. A modification of the crushing experiment may well provide a non-chemical method which would test our belief that the methane is present as such in the fines.

If methane is indeed present as such in the lunar fines, then it could well have one or more of the following origins: (1) it could be primordial; (2) it could have been entrapped during formation of the fines; (3) it could be a primary product of the solar wind; or (4) a secondary product of indigenous carbon or of solar wind carbon. We hope to distinguish between these possibilities in due course, but we have sought information from two meteorites. Methane has been reported in carbonaceous chondrites by STUDIER *et al.* (1965), but analyses of Allende by us revealed only small quantities of this component. The gas-rich achondrite Kapoeta which consists of aggregated interplanetary dust bombarded by the solar wind (MULLER and ZAHRINGER, 1966) and has a very low carbon content, contains approximately three times as much methane. The cosmic ray exposure ages are presumably similar for both meteorites. Thus, our limited data indicate that the quantity of methane is approximately proportional to the rare gas content, but not to the carbon content. We have no information on the carbide content (if any) of these meteorites. There appears to be no previous report of the presence of methane in gas-rich achondrites or in ordinary chondrites.

We anticipate that new information on the solar wind implantation and cosmic ray irradiation of the moon and of meteorites will become available by the kind of approach herein described. The compounds of carbon and other elements liberated from the sample simultaneously with the rare gases should be identified and quantitatively measured. In particular, it will be important to do this by careful etching studies designed to reveal precise locations and state of the carbon compounds trapped and bound to the matrix.

Acknowledgments—We thank the Science Research Council and the Natural Environment Research Council for financial assistance, and the National Aeronautics and Space Administration for a sub-contract covering organic geochemical studies (NGL-05-003-003), made through the University of California, Berkeley. The award of Fellowships from the following bodies is gratefully acknowledged: The Science Research Council (G. H. D., C. T. P.). Shell International Petroleum (J. R. M.), the National Science Foundation (J. M. H.) and the Petroleum Research Fund of the American Chemical Society (P. I. A.). The work was conducted during leave of absence from the University of Indiana (J. M. H.) and the University of Rhode Island (P. I. A.). We also thank Dr. R. HUTCHESON of the British Museum for the sample of Kapoeta, Mr. B. SIMONEIT of the Space Sciences Laboratory of the University of California, Berkeley, for that of Allende, Dr. K. H. G. ASHBEE, Mr. NAYLOR FIRTH and Dr. T. E. THOMPSON, all of the University of Bristol, for light and electron microscope studies, and Mr. D. MORGAN and Mrs. PAT MUIR for technical assistance. One of us (G. E.) acknowledges helpful comment and suggestions relating to the solar wind made to him by Drs. R. WALKER, J. ARNOLD and R. O. PEPIN.

REFERENCES

AGRELL S. O. (1970) The moon at Houston; Mineralogy. *Nature* **225**, 324–325.

ARRHENIUS G., ASUNMAA S., DREVER J. I., EVERSON J., FITZGERALD R. W., FRAZER J. Z., FUJITA H., HANOR J. S., LAL D., LIANG S. S., MACDOUGALL D., REID A. M., SINKANKAS J. and WILKENING L. (1970) Phase chemistry, structure and radiation effects in lunar samples. *Science* **167,** 659–661.

BARKER C. G. (1965) Mass spectrometric analysis of the gas evolved from some heated natural minerals. *Nature* **205,** 1001–1002.

BRETT R. (1967) Cohenite: its occurrence and a proposed origin. *Geochim. Cosmochim. Acta* **31,** 143–159.

BURLINGAME A. L., CALVIN M., HAN J., HENDERSON W., REED W. and SIMONEIT B. R. (1970) Lunar organic compounds: Search and characterization. *Science* **167,** 751–752.

CORNU A. and MASSOT R. (1966) *Compilation of Mass Spectral Data.* Heyden.

DAYHOFF M. O., LIPPINCOTT E. R. and ECK R. V. (1964) Thermodynamic equilibria in prebiological atmospheres. *Science* **146,** 1461–1464.

DRAFFAN G. H., EGLINTON G., HAYES J. M., MAXWELL J. R. and PILLINGER C. T. (1969) Organic analysis of the returned lunar sample. *Chem. Brit.* **5,** 296–307.

ECK R. V., LIPPINCOTT E. R., DAYHOFF M. O. and PRATT Y. T. (1966) Thermodynamic equilibrium and the inorganic origin of organic compounds. *Science* **153,** 628–633.

EPSTEIN S. and TAYLOR H. P., JR. (1970) $^{18}O/^{16}O$, $^{30}Si/^{28}Si$, D/H, and $^{13}C/^{12}C$ studies of lunar rocks and minerals. *Science* **167,** 533–535.

FRONDEL C. JR., KLEIN C., ITO J. and DRAKE J. C. (1970) Mineralogy and composition of lunar fines and selected rocks. *Science* **167,** 681–683.

GOGUEL R. (1963) Die chemische zusammensetzung der in den mineralien einiger granite und ihrer pegmatite engeschlossenen gase und flussigkeiten. *Geochim. Cosmochim. Acta* **27,** 155–181.

GRIFFITHS P. R., SCHUMANN P. J. and LIPPINCOTT E. R. (1969) High temperature equilibria from plasma sources. II. Hydrocarbon systems. *J. Phys. Chem.* **73,** 2532–2537.

HAN J. and ORÓ J. (1967) Application of combined gas chromatography–mass spectrometry to the analysis of aromatic hydrocarbons formed by pyrolysis of methane. *J. Gas Chromatogr.* **5,** 480–485.

HAN J., SIMONEIT B., BURLINGAME A. and CALVIN M. (1969) Organic analysis of Pueblo de Allende meteorite. *Nature* **222,** 264–265.

HAYES J. M. and BIEMANN K. (1968) High resolution mass spectrometric investigations of the organic constituents of the Murray and Holbrook chondrites. *Geochim. Cosmochim. Acta* **32,** 239–267.

HEYMANN D., YANIV A., ADAMS J. A. S. and FRYER G. E. (1970) Inert gases in lunar samples. *Science* **167,** 555–558.

HINTENBERGER H., WEBER H. W., VOSHAGE H., WÄNKE H., BEGEMANN F., VILCSEK E. and WLOTZKA F. (1970) Rare gases, hydrogen and nitrogen: Concentrations and isotopic compositions in lunar material. *Science* **167,** 543–545.

KAPLAN I. R. and SMITH J. W. (1970) Concentration and isotopic composition of carbon and sulfur in Apollo 11 lunar samples. *Science* **167,** 541–543.

KEAYS R. R., GANAPATHY R., LAUL J. C., ANDERS E., HERZOG G. F. and JEFFERY P. M. (1970) Trace elements and radioactivity in lunar rocks: Implications for meteorite infall, solar-wind flux and formation conditions of moon. *Science* **167,** 490–492.

KRANZ R. (1969) Organic compounds in the gas inclusions of feldspars. In *Organic Geochemistry. Methods and Results,* (editors G. Eglinton and M. E. Murphy), Chap. 21, pp. 521–533. Springer-Verlag.

LSPET (LUNAR SAMPLE PRELIMINARY EXAMINATION TEAM) (1969) Preliminary examination of lunar samples of Apollo 11. *Science* **165,** 1211–1227.

MACKAY C. and WOLFGANG R. (1965) Free carbon atom chemistry. *Science* **148,** 899–907.

MASON B. and WIIK H. E. (1966) The composition of the Barratta, Carraweena, Kapoeta, Mooresfort and Ngawi meteorites. *Amer. Mus. Novitates* No. 2273, 1–25.

MOORE C. B. (1969) Personal communication.

MOORE C. B., LEWIS C. F., GIBSON E. K. and NICHIPORUK W. (1970) Total carbon and nitrogen abundances in lunar samples. *Science* **167,** 495–497.

MÜLLER O. and ZÄHRINGER J. (1966) Chemische unterschiede bei uredelgashaltigen steinmeteoriten. *Earth Planet. Sci. Lett.* **1,** 25–29.

MURPHY R. C., PRETI G., NAFISSI-V M. M. and BIEMANN K. (1970) Search for organic material in lunar fines by mass spectrometry. *Science* **167,** 755–757.

NAGY B., DREW C. M., HAMILTON P. B., MODZELESKI V. E., MURPHY M. E., SCOTT W. M., UREY H. C. and YOUNG M. (1970) Organic compounds in lunar samples: pyrolysis products, hydrocarbons, amino acids. *Science* **167,** 770–773.

ORÓ J., UPDEGROVE W. S., GIBERT J., MCREYNOLDS J., GIL-AV E., IBANEZ J., ZLATKIS A., FLORY D. A., LEVY R. L. and WOLF C. (1970) Organogenic elements and compounds in surface samples from the Sea of Tranquility. *Science* **167,** 765–767.

OYAMA V. I., MEREK E. L. and SILVERMAN M. P. (1970) A search for viable organisms in a lunar sample. *Science* **167,** 773–775.

PONNAMPERUMA C., KVENVOLDEN K. A., CHANG S., JOHNSON R., POLLOCK G., PHILPOTT D., KAPLAN I., SMITH J., SCHOPF J. W., GEHRKE C., HODGSON G., BREGER I. A., HALPERN B., DUFFIELD A., KRAUSKOPF K., BARGHOORN E., HOLLAND H. and KIEL K. (1970) Search for organic compounds in the lunar dust from the Sea of Tranquility. *Science* **167,** 760–762.

ROBINSON N. W. (1960) The action of molybdenum, tungsten, tantalum and nickel on residual gases in a vacuum system. *Vacuum* **10,** 75–80.

SCOTT W. M. (1970) Personal communication.

SIMONEIT B. R., BURLINGAME A. L., FLORY D. A. and SMITH I. D. (1969) Apollo 11 Lunar Module engine exhaust products. *Science* **166,** 733–738.

STUDIER M. H., HAYATSU R. and ANDERS E. (1965) Organic compounds in carbonaceous chondrites. *Science* **149,** 1455–1459.

UNSÖLD A. O. J. (1969) Stellar abundances and the origin of the elements. *Science* **163,** 1015–1025.

WEIFFENBACH C. K., GRIFFITHS P. R., SCHUHMANN P. J. and LIPPINCOTT E. R. (1969) High-temperature equilibria from plasma sources. I. Carbon–hydrogen–oxygen systems. *J. Phys. Chem.* **73,** 2526–2531.

ZÄHRINGER J. (1963) Uber die erdelgase in den achondriten Kapoeta und Staroe Pesjanoe. *Geochim. Cosmochim. Acta* **26,** 665–680.

ZELLER E. J., RONCA L. B. and LEVY P. W. (1966) Proton induced hydroxyl formation on the lunar surface. *J. Geophys. Res.* **71,** 4855–4860.

Proceedings of the Apollo 11 Lunar Science Conference, Vol. 2, pp. 1775 to 1777.

Micropaleontological study of lunar material from Apollo 11

ELSO S. BARGHOORN
Department of Biology and The Botanical Museum, Harvard University,
Cambridge, Massachusetts 02138

(*Received* 29 *January* 1970; *accepted in revised form* 27 *February* 1970)

Abstract—Samples of lunar dust, rock chips (microbreccia) and thin sections of microbreccia from Tranquillity Base have been examined by light, transmission and scanning electron optics. The material reveals no indication of biological morphology although certain of the minute mineral constituents in the form of glassy polymorphic spherules simulate the form and structure of microorganisms. It is concluded from these studies that the surficial rocks and sediments in this area of the moon have been devoid of life since their postulated time of origin, 4·6 billion years ago. It may be inferred that the lunar regolith has never possessed life and is now inimical to life.

THE POSSIBILITY that some form of life might conceivably exist on the surface of the moon was seriously considered in planning the Apollo missions and in bringing back to earth fragments of the lunar surface. For this reason elaborate precautions were devised for prevention of possible contamination of the earth by extra-terrestrial life. Although the likelihood of extant life on the moon was regarded in the scientific community as infinitely remote, the possibility of pre-existent life, even in an evanescent biosphere in the early history of the moon, seemed intriguing and within the realm of plausibility. Detailed examination of the material made available for this study, however, indicates no morphological evidence of pre-existent life on the moon. In terms of paleontological analysis, the results reported here are entirely negative with respect to the former existence of a biosphere on the moon. The techniques and methodologies which were employed, however, are pertinent to similarly oriented studies in the future directed to determining possible evidence of previous life in both terrestrial and extra-terrestrial rocks and sediments and are herewith noted.

The lunar samples investigated in this work consisted of the following: (1) lunar fines (10086-8 and 10086-3) from the bulk sample container, sieved to various size fractions; (2) rock chips of microbreccia each approximately 1·0 g in weight (10091-6 and 10091-7); (3) thin sections of microbreccia (10059-32; 10065-25; 10046-56 and 10021-29).

Samples of the lunar fines were examined as free powder on the surface of glass microscope slides by normal and polarized transmitted light, and by dark-field and light-field reflected light. In addition, slurries of dust which had been sieved to ⟨100 mesh/inch were prepared with various mounting media, including microscope immersion oil, gun damar and diaphane. In white light microscopy diaphane yielded the highest contrast of structural features in the lunar dust and was superior to other mounting media employed. Index of refraction of constituent dust particles was not determined. The dust dispersed freely in the mounting media yielding remarkably clear preparations for white light microscopy and obviated the need for standard maceration procedures routinely employed in freeing microfossil residues such as pollen and spores from their mineral matrix.

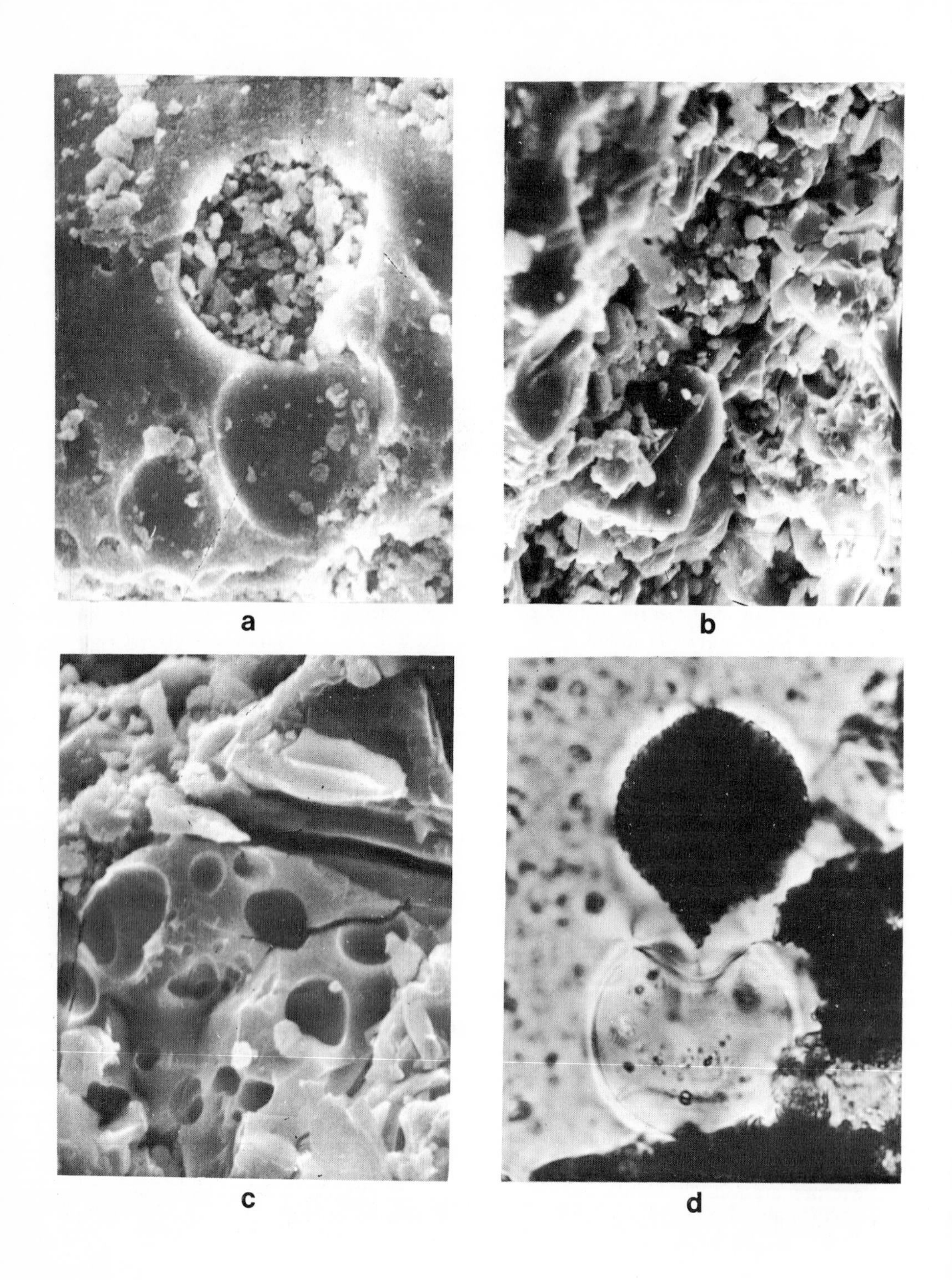
a
b
c
d

The two chips made available for study by the Manned Spacecraft Center were from a microbreccia from the bulk sample box (sample 10091-6 from an outside weathered surface and sample 10091-7 from the inside fractured surface of the same rock). The chips were prepared for study by scanning electron microscopy by securing in appropriate mounts and coated on a rotating stage *in vacuo* with *ca.* 500 Å of gold without precoating with carbon. It was found that despite this thickness of gold the material developed an electron charge sufficient to slightly distort the image (Fig. 1(a), upper centre) of the back scattered electrons It is probable that the excessive electron charging is a consequence of the minuteness of particle size of material in both the weathered and fresh fractured surfaces of the microbreccia or possibly due to the absence of a carbon film below the gold coating. The existence of unconsolidated and non-fused particles of such minute size are enigmatic in terms of the postulated thermal history of the rock since much of the fine dust is free of cohesion or any evidence of melting.

Thin sections supplied by the Manned Spacecraft Center were studied by optical microscopy utilizing transmitted and light field and dark field reflected light. Sections covered by microscope cover glasses were found to be unusable for other than transmitted light microscopy.

All samples examined appear to be totally devoid of any biological morphology, although pseudomorphs of biological organization may not uncommonly be found in the epoxy resin preparations of thin sections as is shown in Fig. 1(d), as viewed in white light.

The results of optical and electron microscopy in demonstrating the absence of microstructures of biological origin are consistent with the results of chemical analyses showing an extremely low content of carbon or carbonaceous compounds which would result from pre-existent life.

Acknowledgments—This contribution is a part of the Ames (NASA) Consortium group effort on the Apollo 11 project and was financed by NASA Contract NAS 9-8060. The author wishes to express his appreciation to Dr. SHELDON MOLL of American Metals Research Inc. of Burlington, Mass. for assistance in the SEM studies, and to Mr. UMESH BANERJEE, Harvard University for his assistance in photography and sample preparation.

Fig. 1. (a) Surface features of "weathered" surface of outside face of chip from sample 10091-6. Note particles contained within cavity upper center. ×3300 SEM (scanning electron micrograph).

(b) Surface features of fractured (inside) surface of microbreccia. Sample 10091-7. Note what appear to be flow lines of fusion or melting of various minerals. ×1850 SEM.

(c) Surface features revealed on glassy remnants within the microbreccia upon fracturing. Inside chip of sample 10091-7. The term "effet frommage" is suggested. It is postulated that this morphology was produced by degassing of the glass prior to cooling of the multi-brecciated matrix. ×3300 SEM.

(d) Structure produced by condensation and polymerization of epoxy resins used in preparation of thin sections. Epoxy structures are most commonly found in preparations which are mounted under a cover glass. Photograph taken with partially polarized light. Note the strain anisotropy of the pellicle and its extruded contents. Slide 10046-56 ×670.

Proceedings of the Apollo 11 Lunar Science Conference, Vol. 2, pp. 1779 to 1791.

Study of carbon compounds in Apollo 11 lunar samples*

A. L. BURLINGAME, MELVIN CALVIN, J. HAN, W. HENDERSON,
W. REED and B. R. SIMONEIT
Space Sciences Laboratory, Department of Chemistry and Department of Geology,
University of California, Berkeley, California 94720

(*Received* 2 *February* 1970; *accepted in revised form* 4 *March* 1970)

Abstract—Several methods of analyses of the Apollo 11 lunar fine material have confirmed that the carbon level is of the order of 200 ppm. By far the largest single component of this carbon appears to be carbon monoxide. Some of this appears to occur as gas bubbles in the glass microspheres, but most of it appears to be in some complex form other than gaseous. This result would be consistent with the idea that most of the fines passed through high temperatures during which the carbon was oxidized by mineral oxides.

INTRODUCTION

IN THE search for clues as to the various stages of chemical evolution and the origin of life on earth, no terrestrial material has yet been found or examined which is free from the biologically derived organic compounds found everywhere on earth (CALVIN, 1969). At present, our knowledge of the early stages of chemical evolution and indeed of the origin of life is limited to knowledge of hypothetical reaction sequences. The opportunity offered by the return to earth of the Apollo 11 lunar materials represents the first possibility of examining extra-terrestrial material which has not been contaminated by terrestrial biota or biologically derived organic matter. This examination, and succeeding ones from future lunar landing missions, should result in the acquisition of knowledge concerning not only the origin of the moon, but also perhaps further insight into chemical evolution of the solar system and the origin of life on earth. The presence, or absence, of indigenous organic molecules in the lunar surface or regolith rocks and their molecular architecture would indeed be a most significant finding.

The cosmic abundance of carbon, next only to hydrogen, helium and oxygen, would indicate that there should be a reasonable probability of finding lunar carbon containing molecules. In addition, the contribution from meteoritic carbonaceous materials collected by the moon's gravitational field should be significant (KEAYS *et al.*, 1970). It has been estimated (MASON, 1962) that 10^5–10^6 tons per annum of meteorites and cosmic dust are collected by the earth; of this, 4 per cent is carbonaceous chondrites, probably a very low value due to the friable nature of carbonaceous chondrites and the small probability of their being preserved on the surface of the earth. Organic analyses of meteorites (HAYES, 1967) indicate an organic content generally within the range of 1–5 per cent by weight (siderites 1200 ppm C and chondrites 1600 ppm C). The question of endogenosity of carbon containing molecules in

* This represents Part XXXIII in the series High Resolution Mass Spectrometry in Molecular Structure Studies. For Part XXXII, see B. R. SIMONEIT, H. K. SCHNOES, P. HAUG and A. L. BURLINGAME, *Nature*, in press.

extra-terrestrial materials is still being examined (HAYES, 1967; SMITH and KAPLAN, 1970). If the moon only collects a small fraction of 10^5–10^6 tons per annum, over the 4×10^9 yr of its known history, this must surely represent a significant contribution to the lunar surface.

The preliminary analysis of the lunar materials (LSPET, 1969) indicated that the total carbon content was in the range 100–350 ppm as determined by combustion and that the volatile organic compound content was of the order of 5 ppm or less (as determined by mass spectrometry). The total carbon level is at least one order of magnitude less than predicted. The volatile organic carbon level is extremely low and in fact might be mostly contamination (LSPET, 1969). The low level of carbon is mystifying even after allowing for thermal effects, radiation, proton flux and loss by escape from the lunar gravitational field.

To confirm and extend these preliminary findings, we proceeded with three experimental approaches to examine volatilizable, non-volatile and gaseous carbon containing molecules in the lunar fines. First, we conducted direct pyrolysis under vacuum into a high resolution mass spectrometer at temperatures of 150°, 500° and 1150°C. This vacuum pyrolysis at 1150°C depends on the blast furnace reaction, where the carbon is oxidized to CO by minerals such as iron oxides. Secondly, solvent extraction of the surface of the fines was followed by high resolution mass spectrometric analysis. Subsequent solution of the fines with 20% HF was followed by solvent extraction and mass spectrometric analysis of the extract. Finally, the fines were dissolved in cold 20% HF in a closed system with a sequence of cold traps so as to condense any volatile constituents or products which might be released by the action of the HF. Pyrolysis at 150°, 500° and 1150°C of the demineralized residue thus obtained was then carried out as previously indicated.

EXPERIMENTAL

A. *Optical microscopy*

The general mineralogic composition of Bulk Fines C sample No. 10086 (Monopole Can) was examined using a Zeiss W-Pol petrographic microscope, with the mineral grains mounted in calibrated refractive index oils. Objectives used were flat field 50×, 38·6× and 13·6× in conjunction with a 12× ocular.

Mineral identifications were accomplished by successive immersions of individual grains in the refractive index oils, in addition to the usual tests for optical properties.

B. *Pyrolyses*

All pyrolysis experiments were carried out in the apparatus described below. It consisted of an Alundum tube (Coors Alumina, Denver, Colo.) heatable under vacuum to at least 1200°C, connected to an expansion volume inlet system on the mass spectrometer ion source. The expansion volume was made of two bulbs with a volume ratio of 1:10, equipped with three Westef stopcocks (West-Glass Corporation, El Monte, California). This system was used for pressure reduction in both calibration and sampling. The inlet manifold was comprised of five stainless steel needle valves (Nuclear Products Company, Cleveland, Ohio) of low internal volume, connected by 5 mm (outside diameter) tubing. This manifold was connected to the roughing and diffusion pump system of the mass spectrometer, to the expansion volumes, and via a gold leak directly into the ion source. A mercury manometer was also attached. The total volume of the inlet system to the gold leak and including the large expansion bulb was 550 ml. The mass spectrometer system consisted of a modified G.E.C.–A.E.I. MS-902 high resolution mass spectrometer online to an XDS Sigma 7 computer (BURLINGAME,

1968; BURLINGAME *et al.*, 1969, 1970). Multiple spectra of each sample were recorded online under the following spectrometer conditions: resolution 5000, ionizing current 500 μA, ionizing voltage 55 eV, ion source temperature 200°C, and mass range 12–300 amu at a scan rate of 16 sec/decade. The high resolution mass spectral data are presented as heteroatomic plots (BURLINGAME and SMITH, 1968). The Alundum sample tube was connected to the inlet manifold by a Cajon fitting with a nickel gasket. A Fisher microcombustion furnace was used to heat the tube. A temperature calibration, both inside and outside the tube, indicated no temperature differential after thermal equilibration.

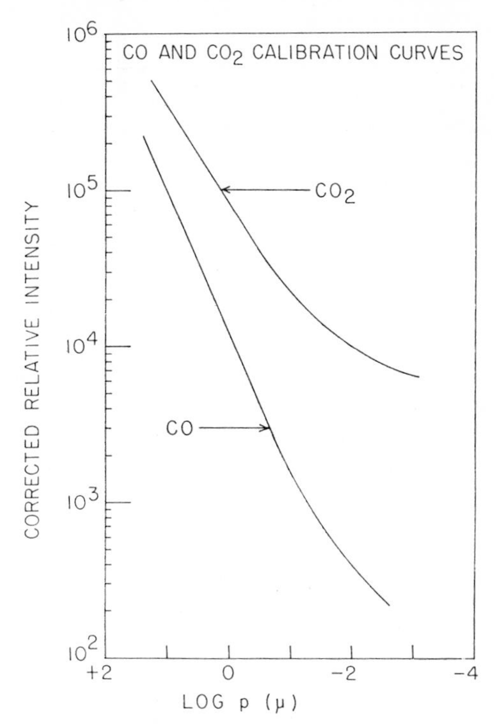

Fig. 1. CO and CO_2 calibration curves.

The system was calibrated for carbon monoxide (CO) and carbon dioxide (CO_2). A known pressure of CO or CO_2 (Matheson Company) was introduced into the expansion system and the instrument response was measured by multiple scans after successive tenfold pressure reductions. Before and after each pressure reduction multiple scans of the background were also taken. The calibration curves are shown in Fig. 1, where the corrected relative intensity is plotted versus the pressure range of 10^2–10^{-1} μ.

The system background was determined in each experiment by heating the evacuated (10^{-3} μ) tube at 1150°C for 60 min. The sample size was usually 100 mg. Untreated lunar fines and HF demineralization residue samples were examined by the following operation sequence: evacuate, background analysis, heat at 150°C for 30 min, product analysis and so on at 500°, 1150°C for 60 min and 1150°C for a further 15 min. Using the same sequence of operations a Hawaiian basalt (collected by I. Kaplan, UCLA) was examined at 150° and 1150°C. Similarly, a sample of baked graphite (1150°C in air) was fused into the pyrolyzed lunar fines at 1150°C and the products analyzed.

C. *Pyrolysis–gas chromatography–mass spectrometry*

The lunar fines (12 mg) were also subjected to a pyrolysis–gas chromatography–mass spectrometry analysis (HAN *et al.*, 1969) where the sample was heated to 500°C for 30 min (after having been baked

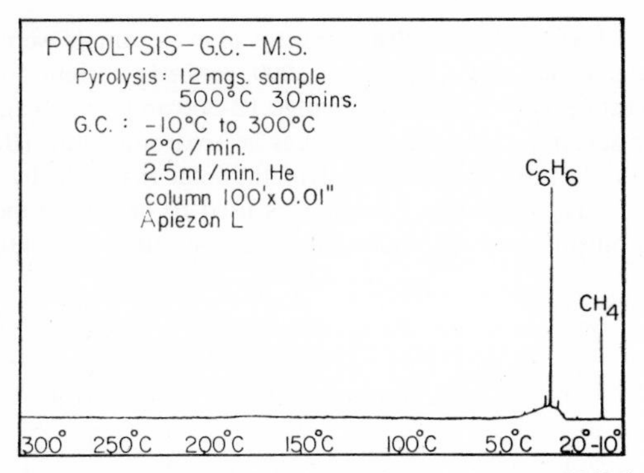

Fig. 2. Pyrolysis–GC-MS-12 of lunar fines at 500°C.

at 150°C for 15 min to remove adsorbed material). The sample was pyrolyzed onto a 100 ft × 0·01 in. stainless steel capillary column (Apiezon L, Varian Aerograph 204, programmed from −10 to 300°C at 2°/min with a He flow rate of 2·5 ml/minute). The resultant GC trace is shown in Fig. 2. The effluent was analyzed by a G.E.C.–A.E.I. MS-12 single focusing mass spectrometer.

D. *Solvent extraction*

A sample (19·8 g) of fines was extracted with benzene–methanol (3:1) and analyzed according to the procedure outlined in Fig. 3. The solvents used were redistilled nanograde (Mallinckrodt) quality. The extract (100 ml) was monitored by capillary GC–MS between successive ten-fold volume reductions. The final residue (3 μl) was analyzed also by direct introduction into the MS-902 ion source, the data being taken in the multiple online scan model (BURLINGAME, 1968; BURLINGAME *et al.*, 1969, 1970).

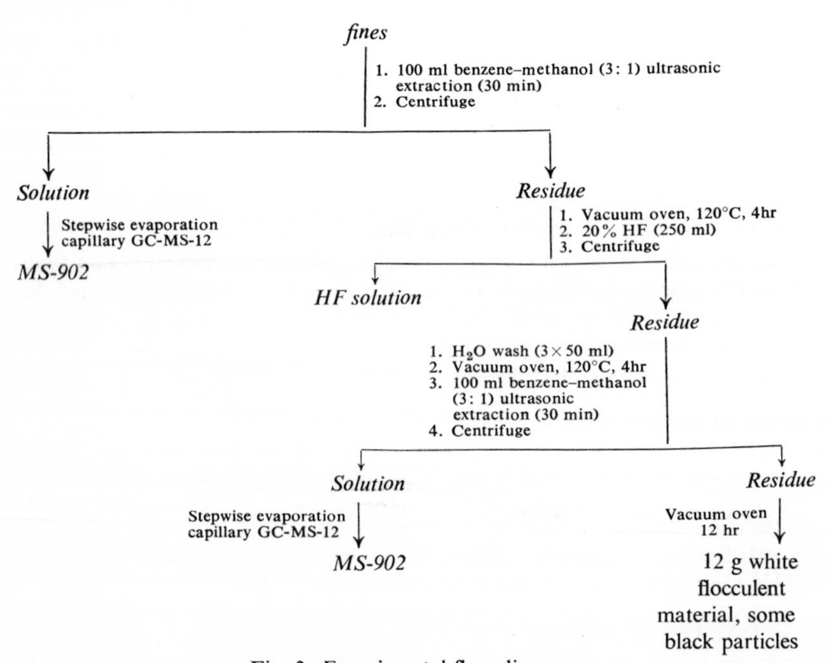

Fig. 3. Experimental flow diagram.

E. *HF demineralization and solvent extraction*

The exhaustively extracted fines were demineralized as indicated in Fig. 3 with 20% HF, prepared as described by HAN *et al.* (1969), followed by drying and benzene–methanol (3:1) extraction. A sample of the extracted demineralized residue was pyrolyzed as described in Section B.

F. *Gases liberated by HF treatment*

The schematic of the apparatus used is shown in Fig. 4. The system was made of glass with Westef stop-cocks and Cajon trap disconnectors. It was He leak checked using a C.E.C. 24-120B leak detector. The 20% HF (HAN *et al.*, 1969) was degassed with He before use. In order to check the recovery efficiency of the system a silica gel blank was run. The whole system was alternately purged with He (three times) and pumped down (two times) before the flask was doped with 1 ml (STP) of CO. The flask was valved off and stirred during the step-wise addition of 150 ml HF. The traps were under vacuum and at the respective temperatures of the coolants. After purging all the volatiles from the flask into the Dry Ice-acetone trap with He, the flask was again isolated. By

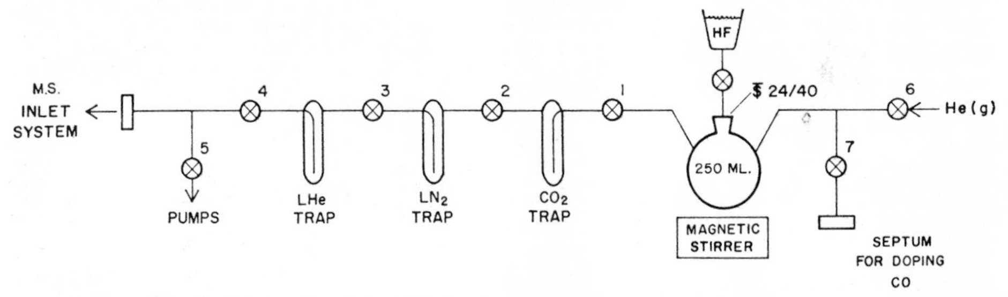

Fig. 4. Schematic of the HF demineralization gas sampling apparatus.

a series of heating and freezing cycles, the volatiles were transferred to the liquid He trap and valved off. The He trap pressure was measured and the contents analyzed by high resolution mass spectrometry.

The fines (17·8 g) were degassed at 150°C for 15 min in the same system under a He atmosphere. The same procedure was followed as described for the blank run.

RESULTS

A. *Optical microscopy*

Examination of the lunar fines from Bulk Fines C No 10086 (Monopole Can) by optical microscopy revealed particle sizes within the range from nearly 1 mm down to <1 μm, with one particle in the bottom of the can being approx. 0·2 cm $\times$ 1 cm. Among the more intriguing components to be found in the dust were the glassy, sphere-like objects, comprising from approximately 1 per cent to as much as 10 per cent of the various aliquots; and the quantity of spherules increases as the average particle size decreases. These spherules grade smoothly from completely clear, low index ($n = 1{\cdot}46$) objects to those which are deeply colored and have a refractive index ($n > 1{\cdot}75$), apparently in response to a similar smooth gradation in composition. While the particle size of the spheres ranges from <1 μm to >200 μm, the deeply colored, high index type seems confined to the larger size range (>50 μm).

The surfaces of the spherules often show pitting or etching on an exceedingly fine scale, although most show a roughness apparently created by concentric flaking, as though by exfoliation (Fig. 5). Few of the larger grains show a glassy or polished surface, whereas most of the small ones do.

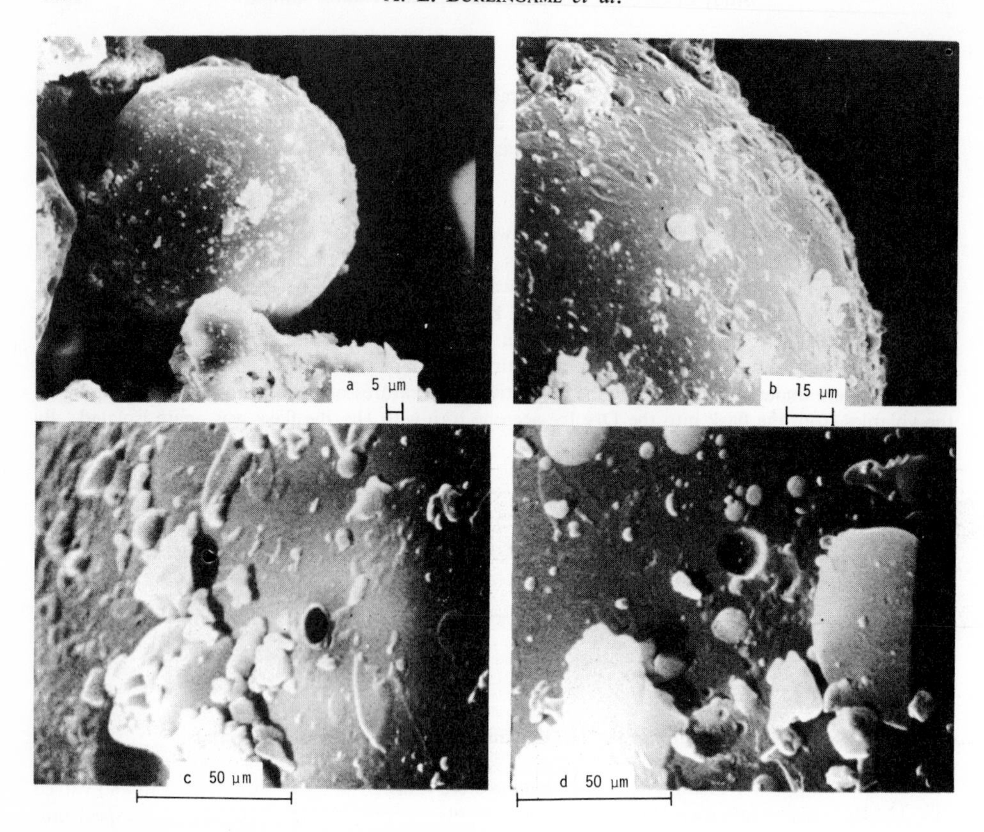

Fig. 5. Scanning electron micrographs of lunar dust. (a) Glass spherule from lunar dust; (b) Detail of *a* showing concentric "exfoliation" fractures; (c) Detail of glass surface showing "frozen" eruption of vesicle at once molten surface; (d) Detail of glass surface showing "zap" craters.

Inclusions within the glassy objects are of 3 general types (Fig. 6): (1) discrete mineral grains, (2) fine granular, optically dense material, (3) gas (?). Little need be said of the first two types in that they represent a simple physical incorporation. Examination of some of this glassy material with the dark granular inclusions by 650 kV electron microscopy showed that quite often these amorphous materials contain very small crystallites. From the electron diffraction patterns we have determined that at least a portion of this opaque material is native α-iron. The more intriguing "inclusion" type is the "gas bubble" (Fig. 6b). It should be pointed out here that the glass in the lunar fines is by no means confined to the spherules. Indeed, there are more angular fragments than spherules. These angular fragments show similar variation in refractive index, and in composition, as do the spheres. One systematic difference is that the larger angular glass fragments are often highly vesicular, whereas even the largest spherules rarely have more than one "bubble." Figure 5c shows what can only be such a vesicle which was "frozen" as it erupted at the surface of the silicate melt. Particles containing such "bubbles" are quite rare, comprising from 0·1 per cent to 0·5 per cent of the fines.

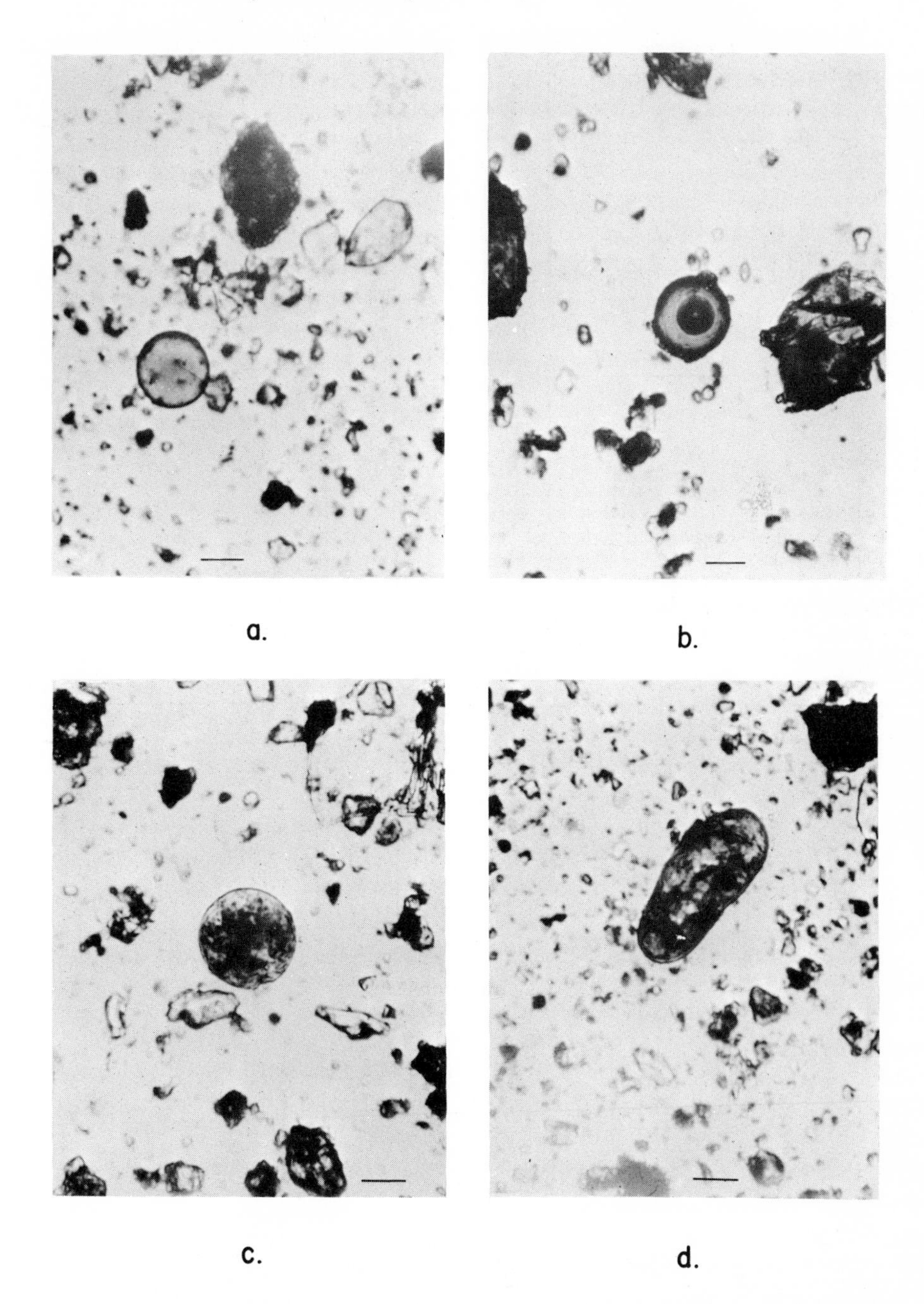

Fig. 6. Apollo 11 fines. (a) Clear glass sphere with minor dark inclusions; (b) Clear glass sphere containing vesicle; (c) Cloudy glass sphere showing granular dark inclusions; (d) Cloudy elongate glass "spherule" showing granular dark inclusions, and vesicle. Length of line on all photographs is 20 μm. Oil mounts ($n = 1{\cdot}569$) of Apollo 11 fines.

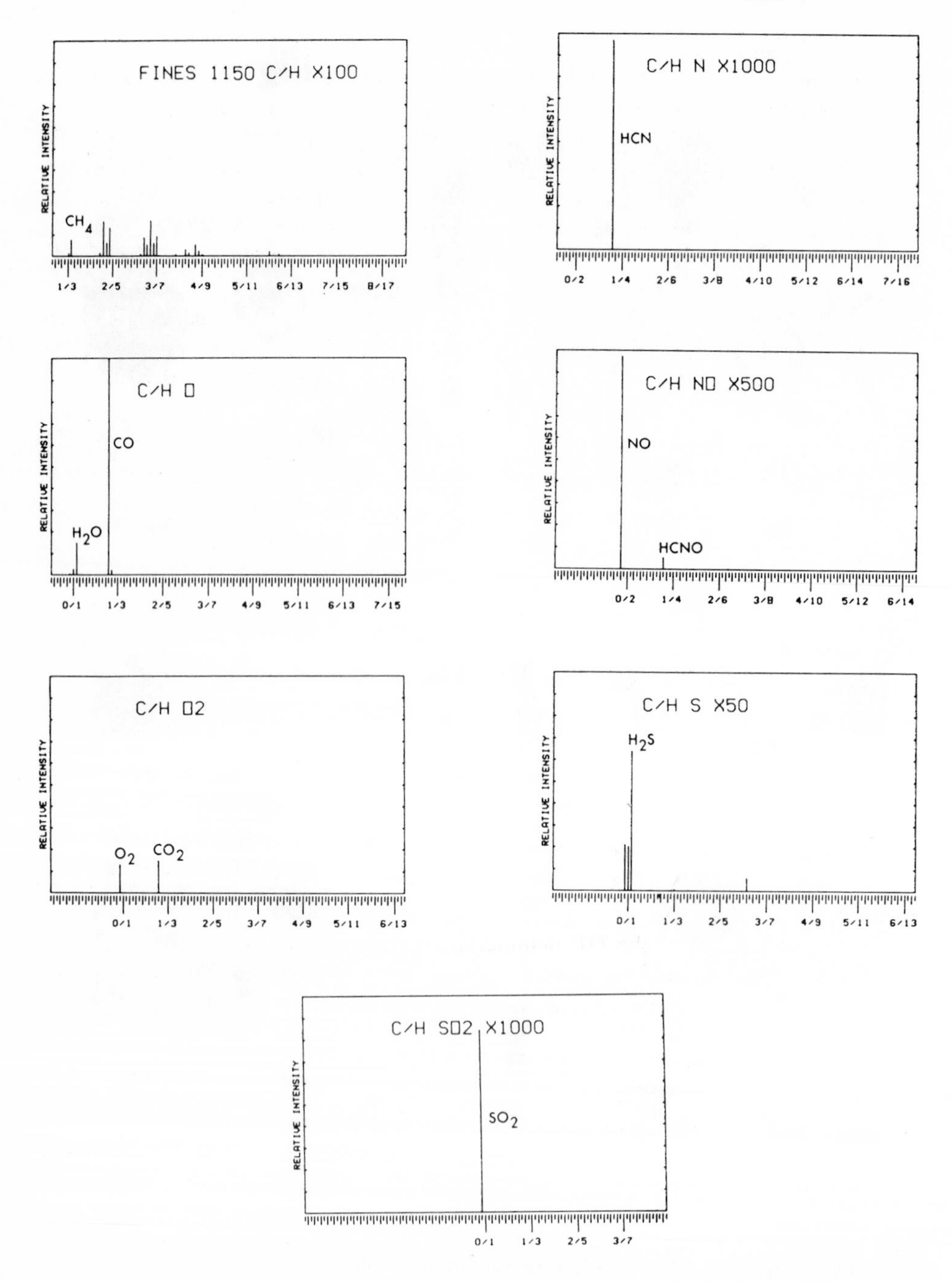

Fig. 7. High resolution mass spectral data of a typical pyrolysis (1150°C) scan. (Note: the odd electron molecular ions, e.g. H_2O and H_2S, occur one amu above the saturated hydrocarbon, C_nH_{2n+1}, fragment).

B. *Pyrolyses*

Two types of direct pyrolysis experiments on lunar fines were performed. In the first, the sample was initially heated at 150°C and the volatile products analyzed by mass spectrometry. The presence of the following gases was established: N_2, O_2, Ar, H_2O, CO, CO_2 and CH_4. Their respective concentrations were at the level of instrument background. The sample was then heated directly to fusion at 1150°C for one hour. The gases so evolved were introduced directly into the MS-902 and consisted of CO (156 ppm C), CO_2 (12 ppm C), traces of HCN, H_2S, SO_2, Ne, Ar, C_2H_2, N_2, O_2 (nitrogen was 20 times more abundant than oxygen).

A typical scan is shown as heteroatomic plots in Fig. 7. The pressures of CO and CO_2 equivalent to the respective signal intensities were obtained from the calibration curves in Fig. 1, and then converted to their respective elemental carbon concentration in ppm.

In the second experiment the sample was again heated at 150°C and the volatiles analyzed. The constituents as listed above were found to be background levels.

Table 1. Summary of total carbon analysis by pyrolysis and gases liberated by HF treatment

	Temperature			
	Ambient	150°C	500°C	1150°C
Lunar fines	—	air, H_2O, CO CO_2, CH_4	—	CO 156 ppm C*
Lunar fines	—	air, H_2O, CO CO_2, CH_4	CO 9 ppm C CO_2 <1 ppm C	—
HF(20%) residue	—	CO 18 ppm C	CO 13 ppm C	CO 88 ppm C
HF(20%) liberated CO–LHe trap	CO 66 ppm C trace CH_4, Ne and Ar	—	—	—

* Traces of HCN, H_2S, SO_2, Ne, Ar, C_2H_2, N_2, O_2.

Subsequent heating at 500°C evolved CO (9 ppm C), CO_2 (<1 ppm C) and traces of the same series as discussed above, but including some CH_4.

A sample (22 mg) of the HF demineralized residue was subjected to sequential pyrolyses at 150, 500 and 1150°C and analyzed by high resolution mass spectrometry after each increment. The principal products in each case were H_2O, CO, CO_2 and SiF_4. In the volatiles from the 500°C increment the presence of benzene was indicated. The carbon concentrations for the three temperature steps were 18, 13 and 88 ppm C, respectively, (weight of C per weight of the original dust), making a total of 119 ppm.

In order to check that this vacuum pyrolysis involving iron oxides, silicates and carbon does indeed result in accurate total carbon concentration data a sample of a Hawaiian basalt was analyzed. This basalt had previously been shown to contain 120–150 ppm C by combustion (KAPLAN, 1969). Our technique corroborated this result with 140 ppm C.

Finally, the addition of baked graphite (1 mg) to fused fines, followed by heating at 1150°C resulted in less than 3 per cent conversion of this graphite to CO.

The results of all the pyrolyses are summarized in Table 1.

C. *Pyrolysis–gas chromatography–mass spectrometry*

Pyrolysis of fines at 500°C, followed by subsequent online capillary GC and low resolution mass spectrometric analysis indicated the presence of methane and benzene at a concentration of the order of 1 ppm (cf. Fig. 2). No higher molecular weight material was found under conditions which would have detected organic constituents up to at least C_{30} at a sensitivity level of 10 ppb.

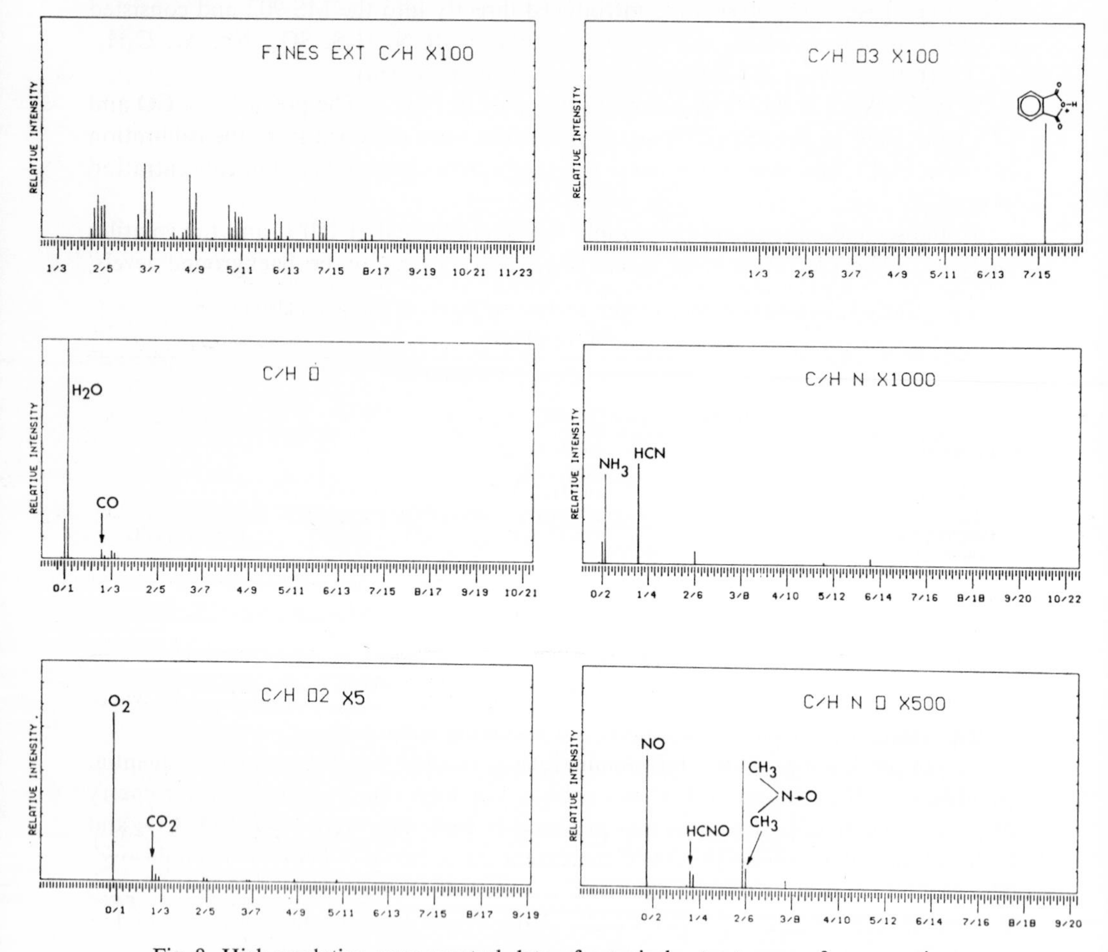

Fig. 8. High resolution mass spectral data of a typical extract scan, after correction for background.

D. *Solvent extraction*

The extract residue, solvent blank residue and the direct introduction probe background were analyzed on the MS-902 mass spectrometer by multiple high resolution scans. Heteroatomic plots of a typical scan after background corrections are shown in Figure 8. The important features are the presence of a number of contaminants derived from sample handling procedures, which have been previously documented

(LSPET, 1969; FLORY *et al.*, 1969; SIMONEIT and FLORY, 1969) and retro-rocket exhaust products (SIMONEIT *et al.*, 1969). The sampling contaminants are hydrocarbons ranging to C_{10} (cf. C/H plot of Fig. 8) and phthalates as indicated by the ion composition $C_8H_5O_3$ (cf. C/H O_3 plot of Fig. 8). The ions attributable to the LM retro-rocket exhaust are NH_3, HCN, NO, HCNO, CH_2NO, C_2H_6N and C_2H_6NO. The individual concentrations of hydrocarbons and phthalates alone is less than 10 ppb as determined by capillary GC.

E. *HF demineralization and solvent extraction*

The benzene–methanol (3:1) extract of this residue yielded no detectable material above solvent background, either by capillary GC or by direct introduction into the MS-902 ion source.

F. *Gases liberated by HF treatment*

The recovery efficiency of the apparatus (cf. Fig. 4), as estimated using a silica gel blank with a known quantity of CO added, was found to be 58 per cent.

After solution of the fines with HF at room temperature followed by condensation of the released volatiles in the liquid He trap, the contents were analyzed by high resolution mass spectrometry. A total of 66 ppm carbon (cf. Table 1) was recovered as CO from the liquid He trap. The presence of methane, neon and argon was also indicated.

CONCLUSIONS

In addition to a number of contaminants which are derived from sample handling procedures, the presence of material resulting from retro-rocket exhaust is evident in the extract of the fines. The benzene/methanol extract of the demineralized residue yielded no detectable material indigenous to the fines. This demineralized residue was subjected to sequential pyrolyses yielding CO at the level of 18 ppm C, 13 ppm C and 88 ppm C, respectively (wt./wt. of the original dust), adding up to a total of 119 ppm C.

The relatively low value of carbon, together with the presence of vesicles in the glass spheres, suggested the dust might contain volatile material which would be liberated upon solution in HF. When untreated moon dust was dissolved in HF, a total of 66 ppm C was recovered as CO from the liquid He trap. This, when added to the 119 ppm C as CO obtained by pyrolysis of residual silica after HF digestion, added up to 185 ppm C, and may be compared with the value of 168 ppm C found in the dust by direct pyrolysis at 1150°C.

It is of interest to estimate the amount of gaseous CO which might be present in the vesicles of the glass spheres. A rough statistical analysis of their distribution would require that they contain 10^4 atm partial pressure CO in order to be equivalent to the 66 ppm C liberated upon solution. This seems excessive and we therefore feel a considerable amount of HF-liberated CO must be in other than gaseous form in the original moon dust, e.g. carbonyls.

It appears that the largest chemical component form of the carbon which is present in the lunar fines which we have so far analyzed is carbon monoxide. This could be the result of the oxidation of any reduced carbon by the heat of meteorite impact or other sources of energy available to the lunar surface, using the mineral oxides as the oxidizing agent. Most of the carbon monoxide so formed would have escaped the lunar gravitational field, thus accounting for the extremely low levels of total carbon in any of the samples yet analyzed (<0.02 per cent). We might expect a much higher level of carbon in those lunar materials which have never been molten. These would be either the primordial aggregates from which the moon might have been formed, or the collection of meteorite infall at low energy levels.

Considerable mission planning and significant effort was devoted towards assuring that the first samples returned to earth in the Apollo mission would be free of terrestrial organic and biological contamination. However, it was clearly established during the mission simulations in the spring of 1969, and also in the preliminary organic investigations at the Lunar Receiving Laboratory which took place after return of the first lunar samples, that the levels of system contamination were in the 5 ppm range, except in the organic reserve samples. While organic contamination has virtually precluded definitive characterization of possible indigenous organic matter in the Apollo 11 samples, our results would indicate certain things clearly. These include the liberation of relatively large quantities of CO upon dissolution of the sample in HF, positive identification of methane, positive identification of certain products previously identified in studies of the LM descent engine retrorocket exhaust. In addition, other organic materials which have been recognized previously to be contaminants of the sample packaging and processing facility, both in terms of the ALSRC and the Lunar Receiving Laboratory prior to sample distribution to the outside laboratories were also identified. It would, of course, be feasible to characterize in detail organic matter which is present at the ppb and ppm level in a geologic sample were there sufficient lunar sample available for a single laboratory to study. However, the paucity of material available at this particular moment from the Apollo 11 mission in this laboratory precludes definitive characterization of those organic materials which are indicated from our investigations. It may be noted that the preliminary examination of Apollo 12 returned lunar material would indicate that the probable levels of organic contamination in the Apollo 11 samples has been reduced by at least one, possibly two, orders of magnitude in the Apollo 12 samples (LSPET, 1970). It would certainly be premature to draw further conclusions which might be regarded as definitive on the organic content either from contamination or of a possible indigenous nature on the Apollo 11 samples. It would be reasonable to expect that were there sufficient Apollo 12 material made available to the organic investigators, upon completion of detailed studies on these samples, the interpretative ambiguities of the data which we are reporting in this manuscript would be alleviated.

Acknowledgments—We acknowledge the contributions of Professors GARETH THOMAS (Metallurgy), T. EVERHART (Electrical Engineering), THOMAS HAYES (Medical Physics) for electron microscopy; Professor GABOR SOMORJAI (Chemistry) for auger spectroscopy; Dr. D. H. SMITH, Miss M. PETRIE, Mr. F. C. WALLS, Mr. P. HARSANYI, Mr. C. SPROWLS, Mr. J. WILDER (Space Sciences Laboratory) for technical assistance in high resolution mass spectrometry and computation; Mr. P. M. HAYES (Laboratory of Chemical Biodynamics) and Safety Services/Health Chemistry Engineering Group (Lawrence Radiation Laboratory, Berkeley) for technical assistance. This work was supported by NASA MSC Contract No. NAS 9-7889.

REFERENCES

BURLINGAME A. L. (1968) Data acquisition, processing and interpretation via coupled high speed real-time digital computer and high resolution mass spectrometer systems. *Advances in Mass Spectrometry*, Vol. 4 (editor E. Kendrick), pp. 15–35. Institute of Petroleum, London.

BURLINGAME A. L. and SMITH D. H. (1968) High resolution mass spectrometry in molecular structure studies II. Automated heteroatomic plotting as an aid to the presentation and interpretation of high resolution mass spectral data. *Tetrahedron* **24**, 5749–5761.

BURLINGAME A. L., SMITH D. H., WALLS F. C. and OLSEN R. W. (1969) Performance of a computer-coupled, double-focussing mass spectrometer system at medium resolution. *Proc. 17th Annual Conf. Mass Spectrometry and Allied Topics*, Dallas, Texas, May 18–23, pp. 28–30.

BURLINGAME A. L., SMITH D. H., MERREN T. O. and OLSEN R. W. (1970) Real-time high resolution mass spectrometry. In *Computers in Analytical Chemistry*. (*Progress in Analytical Chemistry Series*, Vol. 4, (editors C. H. Orr and J. Norris), pp. 17–38. Plenum Press.

BURLINGAME A. L., CALVIN M., HAN J., HENDERSON W., REED W. and SIMONEIT B. R. (1970) Lunar organic compounds: Search and characterization. *Science* **167**, 751–752.

CALVIN M. (1969) *Chemical Evolution*. Oxford University Press.

FLORY D. A., SIMONEIT B. R. and SMITH D. H. (August 1969) Apollo 11 organic contamination history. NASA-MSC report for internal use only.

HAN J., SIMONEIT B. R., BURLINGAME A. L. and CALVIN M. (1969) Organic analysis on the Pueblito de Allende meteorite. *Nature* **222**, 364–365.

HAYES J. M. (1967) Organic constituents of meteorites—a review. *Geochim. Cosmochim. Acta* **31**, 1395–1440.

KAPLAN I. R. (1969) Personal communication.

KEAYS R. R., GANAPATHY R., LAUL J. C., ANDERS E., HERZOG G. F. and JEFFERY P. M. (1970) Trace elements and radioactivity in lunar rocks: Implications for meteorite infall, solar-wind flux, and formation conditions of moon. *Science* **167**, 490–493.

LSPET (LUNAR SAMPLE PRELIMINARY EXAMINATION TEAM) (1969) Preliminary examination of lunar samples from Apollo 11. *Science* **165**, 1211–1227.

LSPET (LUNAR SAMPLE PRELIMINARY EXAMINATION TEAM) (1970) Preliminary examination of lunar samples from Apollo 12. *Science* **167**, 1325–1339.

MASON B. (1962) *Meteorites*. John Wiley.

SIMONEIT B. R., BURLINGAME A. L., FLORY D. A. and SMITH I. D. (1969) Apollo lunar module engine exhaust products. *Science* **166**, 733–738.

SIMONEIT B. R. and FLORY D. A. (1969) Apollo 12 organic contamination history and supplementary report to: Apollo 11 organic contamination history. NASA-MSC report for internal use only.

SMITH J. W. and KAPLAN I. R. (1970) Endogenous carbon in carbonaceous meteorites. *Science* in press.

Proceedings of the Apollo 11 Lunar Science Conference, Vol. 2, pp. 1793 to 1798.

Micromorphology and surface characteristics of lunar dust and breccia*

PRESTON CLOUD, STANLEY V. MARGOLIS, MARY MOORMAN,
J. M. BARKER and G. R. LICARI
Department of Geology, University of California, Santa Barbara, California 93106
DAVID KRINSLEY
Queens College, City University of New York, Flushing, New York 11367
and
V. E. BARNES
Bureau of Economic Geology, University of Texas, Austin, Texas 78712

(*Received* 3 *February* 1970; *accepted* 10 *February* 1970)

Abstract—Although nothing of direct biologic interest was observed in the sample studied, small shaped glass particles and glazed pits resemble objects which elsewhere have been described as fossils. These features, although nonbiological, do bear on processes of lunar weathering and outgassing. The glazed pits are impact features. Fusion of their surfaces released gases. Electron microscopy of the glasses, pits, and angular microfractured mineral grains indicates a prevalence of destructive weathering processes—thermal expansion and contraction, abrasion by by-passing particles, and, of course, impact.

GENERAL FEATURES OF SAMPLE STUDIED

OUR SAMPLE consisted of 7·4 g of fine black surface dust (sample 10086-15), 0·92 g of fragments from the interior of a breccia block (10091-4), a single 1·22 g fragment showing an external surface of the breccia (10091-5), four thin sections (10019-15, 10046-56, 10061-27, 10065-25), and three probe mounts (10059-32, 10059-37, 10067-28).

Both breccia and fines consist of a matrix of mostly very fine grained, mainly angular particles of mineral, rock, and glass, within which are scattered larger mineral and rock fragments as well as spheroids and fragments of glass. The main difference is the induration of the breccia, which shows a strong (shock?) coherence between grains and probably some welding between glass particles. Point counts show the order of abundance to be glass, rock fragments, plagioclase, pyroxene, ilmenite, olivine; but identification of minerals and estimates of their abundance and chemistry was not the goal of this study.

GOALS, PROCEDURES, INSTRUMENTATION

Our mission was to search for evidences of present or former life on the moon and, failing that, to consider aspects of the lunar sample that might bear indirectly on the question of life processes.

Toward this end a variety of sample preparations was employed, resulting in end-products that were examined under varied modes of light microscopy, by scanning

* Contribution No. 1 of Biogeology Clean Laboratory, University of California, Santa Barbara, California.

and transmission electron microscopy, and in the mass spectrometer. In addition to the study of optical thin sections prepared in Houston, our procedures involved the preparation of strewn slides of the dust, both before and after HCl–HF maceration; the preparation and vacuum-coating of standard plug mounts, grids, and replications of dust and rock fragments for electron microscopy; and the fusion (under vacuum) of dust at temperatures above 1200°C followed by studies similar to the above and by mass spectrometry of volatiles obtained. Techniques employed are familiar or obvious ones.

As it became evident that our sample contained neither life nor fossils, interest focused on the abundant tiny glass spheroids and ovoids which, if carbonaceous, could be mistaken for microfossils. We also studied features indicative of outgassing and weathering processes (using techniques described by KRINSLEY and MARGOLIS, 1969) which might suggest the former presence of a lunar atmosphere in which biogenic or prebiogenic processes could have occurred.

SHAPED PARTICLES OF LUNAR GLASS

Although glass appears to be the most abundant component of our sample, it is hard to estimate just how much there is. The minute size of many of the matrix grains, the degree of overlap in thin section, and the abundance of opaque minerals make it difficult to employ standard point-count procedures with confidence. Nevertheless, it seems that from 20 to 50 per cent or more of the total dust and breccia is glass. This glass has an index of refraction mostly greater than 1·64, implying a density greater than 2·94 (BARNES, 1940). Its color ranges from dark amber through yellow to clear, presumably varying with iron content. Small glass particles adhere to a weakly magnetized knife blade. Some are embayed, fractured, crazed, cracked, and locally devitrified, and they adhere to surrounding mineral particles as if hot at the time of deposition. One dumbell-shaped fragment consists of two fused glass spheres, implying the possible significance of fusion of hot glass as a cementation mechanism.

What is of interest here, however, is that between 1 and 5 per cent of the total rock consists of mostly tiny spheroids, ovoids, and other shaped particles of glass (Figs. 1–3). The abundant spheroids and ovoids are similar in shape to some algal and bacterial unicells, and the smaller ones are comparable in size. Indeed, were such glassy particles to be encased in carbon and then dissolved or altered (as might happen in a parent body such as that of the carbonaceous chondrites), they would make impressive pseudomicrofossils.

The size distribution of the glass spheroids in fact resembles the mortality curves of some microorganisms, ranging from diameters of well under 1 μm to 392 μm in populations counted and skewing so sharply toward smaller sizes (Fig. 1) as to suggest that, with better statistics in the smaller grain sizes the curve would approach both ordinate and abscissa asymptotically. Indeed, examination of the lunar dust under the scanning and transmission electron microscopes shows a multitude of submicron sized spheroids going right down to a few tens of angstroms. Such a size distribution, however, is not like that usual for the morphologically simple procaryotic unicells which the glass spheres resemble in a crude way, but rather like the mortality curves of sexually reproducing microorganisms. This is *not* to propose that there are or were

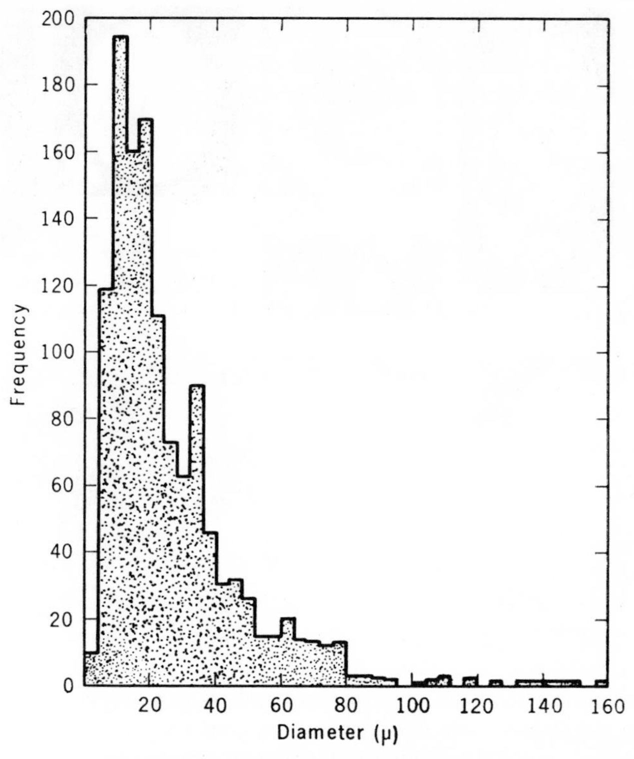

Fig. 1. Size distribution of glass spheroids from Apollo 11 thin sections 10059-37; 10061-15; 10061-27; and 10067-28.

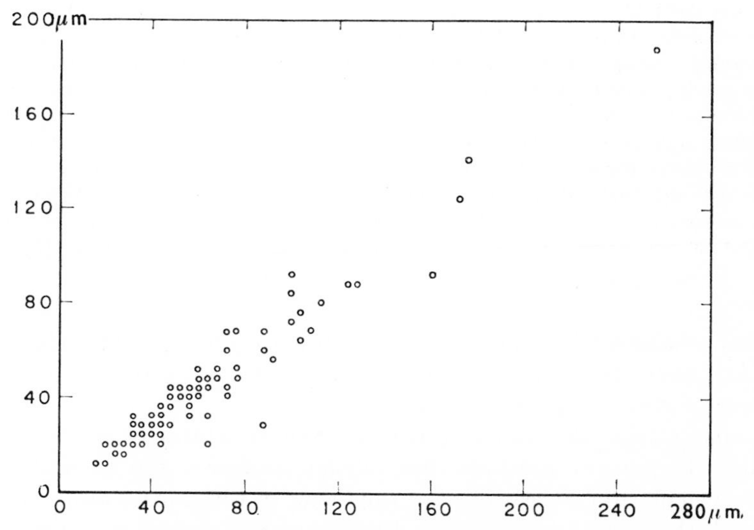

Fig. 2. Size distribution of glass ovoids from Apollo 11 thin sections 10059-37; 10061-15; 10061-27; and 10067-28.

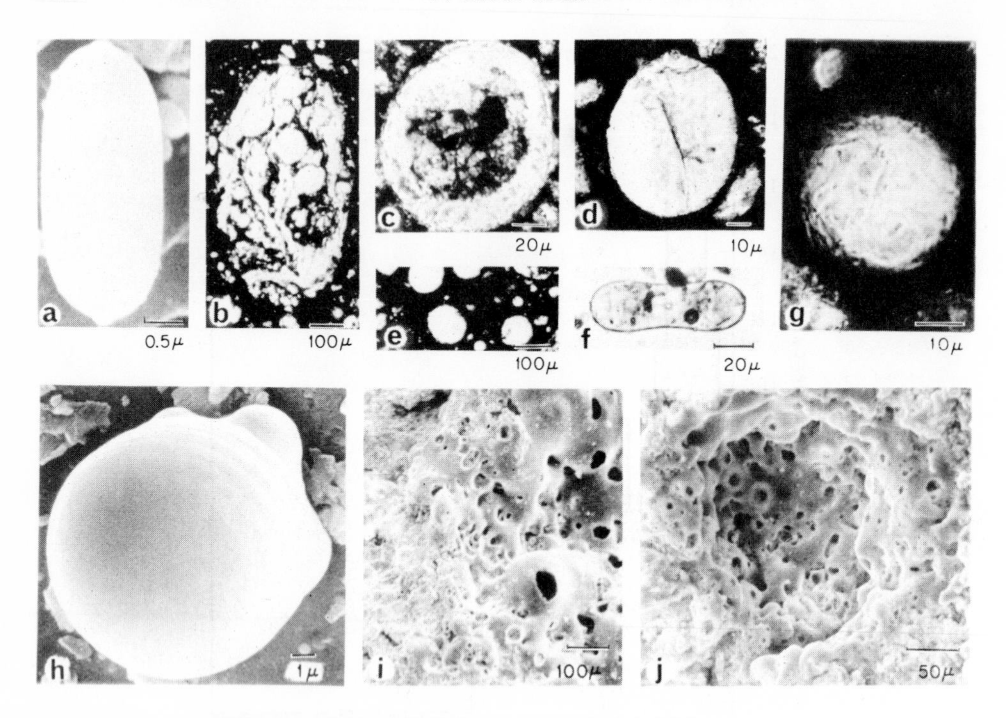

Fig. 3. Shaped glass particles and glazed microcraters from Apollo 11 dust and breccia. (Bar scales give magnifications as indicated. Photographs b–g with optical microscope, a and h–j with scanning electron microscope.)

(a) Glass ellipsoid with terminal nipples from dust sample 10086-15. (b) Large composite glassy ovoid including matrix material, circular voids (where spheres ripped out during preparation) and glassy spheroids within a framework of wipsy glass. Thin section 10019-15. (c) Section of a spheroid showing glassy rind around matrix and mineral core. Thin section 10019-15. (d) Slightly oblate and fractured glass spheroid with nipple on upper end. Thin section 10019-15. (e) Glass spheroids surrounded by fine grained matrix. Thin section 10019-15. (f) Dumbell-shaped glass object with bubbles and mineral inclusions. From strewn slide of dust sample 10086-15b. (g) Glass spheroid with textured surface. Slide 10019-15. (h) Glass spheroid with protruberances and concentric flanges. Grain in dust sample 10086-15. (i) Margin between glazed pit (at right) and unglazed portion of microbreccia (at left) showing burst bubbles with overhanging rims. Surface of chip picked from dust sample 10086-15. (j) Crater-like glazed pit showing terracing and bubbles. External surface of breccia sample 10091-5.

solid glass Protozoa on the moon, but to add one more warning to the many that have already been given about a too-ready interpretation of exotic objects as of vital origin on the basis of gross morphology alone.

The lunar glasses also contain many internal structures of interest—bubbles, "organelle"-like mineral inclusions, fine tubular structures, and microlites such as are well known in natural glasses on earth and some of which make rather spectacular pseudofossils (ROSS, 1962; BRAMLETTE, 1968). Some of the glasses involve complex composite structures.

The warning deserves emphasis. Elsewhere on the lunar or martian surface may be lifelike artifacts that will be harder to discriminate from the real thing.

Surface Characteristics of Lunar Breccia and Glazed Aggregates of Dust

In addition to the glass observed as shaped particles and fragments, crusts of glass partly envelop rock or mineral particles, completely envelop dust aggregates of different shapes (including spheres), and glaze the surfaces of small rimmed pits or microcraters (Fig. 3 (i–j)). Such glazed surfaces, moreover, show bubble aggregates indicative of outgassing as well as local microfracture patterns consistent with origin of the microcraters by impact.

Although less than 1 cm^2 of external surface (10091-5) was available for study, examination of this and of smaller fragments from a dust sample (10086-15) revealed a number of roughly circular depressions with raised rims and glazed surfaces (Fig. 3 (i–j)). Pits observed range from 5 mm to 50 μm in dia., average 0·5 mm, and occur with a frequency of about $10/cm^2$. Curiously, no overlapping of such pits was observed. Examination by scanning electron microscope shows these glazed surfaces to be pitted with innumerable smaller pits and bubble-like features, many of which contain still smaller "bubbles." The bubbles, both large and small, usually also have slightly raised margins, but they also commonly show their largest diameter beneath an overhang, are preferentially located on elevated surfaces, and are clearly burst bubbles and evidence of outgassing.

The glazed surfaces of the larger pits within which the bubbles are located are only a few microns thick, and the glazing extends the full width of the rim, sometimes bubbling over slightly onto the unglazed surrounding rock. Subconcentric microfractures may parallel the margins between glazed pit and unglazed rock, and microfractures within the glazed areas run between the bubbles with no apparent displacement. The sides of the pits are occasionally terraced (Fig. 3 (j)). Depth is roughly proportional to pit diameter. The bubbles within the pits tend to cluster and have rounded bottoms as well as (commonly) overhanging edges. They are comparable to the bubbles of scoriaceous terrestrial basalts.

We interpret the glazed pits as impact craterlets. The bubbles are just that—the product of outgassing resulting from high-velocity impact and local fusion of the lunar surface. The gas may have come from the projectile itself (for example, cometary particles) or from fusion of the lunar breccia on impact. The subconcentric fractures at the margins, as well as the terraces within some of the glazed pits, were probably caused by the initial impact.

On the "biological" side these glazed pits are comparable to some impressions that have been described as pre-Paleozoic fossils. Of more interest, however, are their implications for a lunar atmosphere.

We wanted to know the source of the gas evaded during surface fusion and whether the glass spheroids and other shaped particles observed could be derived from fusion of local materials. With the help of Mark Stein, therefore, a small quantity of the lunar dust was fused under vacuum in a tantalum crucible after outgassing to 10^{-7} torr. Fusion took place at 1200°–1400°C, producing glass spheroids having identical appearance and physical properties to those of the lunar dust and breccia. Vigorous bubbling during the fusion process indicates a sufficient quantity of indigenous volatilizable materials to account for the observed vesiculation of the glassy crusts, although very little of this remained gaseous at room temperature.

Acknowledgments—Supported by NASA Contract No. 9-7882 for major equipment and travel, NASA Grant No. NGR 05-010-035 for operations, University of California for facilities and partial services, and National Science Foundation Grant No. GB-7851 for equipment and general support of the Biogeology Clean Laboratory.

REFERENCES

BARNES V. E. (1940) North American tektites. *University of Texas Publ.* 3945, p. 521.

BRAMLETTE M. N. (1968) Primitive microfossils or not? *Science* **158,** 673.

KRINSLEY D. H. and MARGOLIS S. V. (1969) A study of quartz sand grain surface textures with the scanning electron microscope. *Trans. N.Y. Acad. Sci. Ser.* II **31,** 457–477.

ROSS C. S. (1962) Microlites in glassy volcanic rocks. *Amer. Mineral.* **47,** 723–740.

Proceedings of the Apollo 11 Lunar Science Conference, Vol. 2, pp. 1799 to 1803.

Analyses for amino acids in lunar fines

P. E. HARE
Geophysical Laboratory, Carnegie Institution, Washington, D.C. 20008

and

K. HARADA and S. W. FOX
Institute of Molecular Evolution, University of Miami, Coral Gables, Florida 33134

(*Received* 2 *February* 1970; *accepted in revised form* 20 *February* 1970)

Abstract—Amino acids were sought by aqueous extraction of lunar fines 10086 and by hydrolysis of the aqueous extracts, with an ultrasensitive amino acid analyzer. RTs (retention times) of glycine and alanine were observed in the unhydrolyzed samples, and RTs of these plus those of glutamic acid, aspartic acid, serine, and threonine were identified in the hydrolyzate. While analyses are necessarily so few as to make quantitation tenuous, the amino acids recovered on hydrolysis represent several times as much as do the free amino acids recovered. Controls of water extracts showed virtually no amino acids. Total recovery approximates 50 ppb. Although the results may well be explained by contamination, the amino acid pattern is not one usually found in other materials when contamination has occurred.

SAMPLES of lunar fines collected during the Apollo 11 mission were examined in order to determine if the material contained (a) amino acids or polymers thereof and/or (b) microparticles with resemblance to terrestrial cells, proteinoid microspheres, microfossils (FOX, 1965, 1969), organic particles from hydrothermal veins (MUELLER, 1968, 1969), or microparticles from carbonaceous meteorites (FOX, 1965, 1969; MUELLER, 1968, 1969). Other naturally occurring rounded particles of similar morphology and more certain inorganic nature are tectites and microtectites (GLASS, 1967) and the products of terrestrial volcanism (MUELLER, 1968, 1969) and of meteoritic impact (WRIGHT and HODGE, 1969). Comparisons of these were also sought. The planned procedure, if substances and regular structures were found, involved identification of the chemical composition of the particles.

Other objectives that developed during the investigation included determining if the oxidation of 1,1-dimethylhydrazine (rocket fuel) could produce amino acids, and a quantitative assessment of terrestrial contaminants at the subnanomole level.

For assay of amino acids, 10–15 g of lunar fines was extracted with 30 ml of especially purified water by gentle refluxing for 17 hr. The water had been purified by fresh distillation followed by filtration through washed millipore filters (GSWPO 4700, pore size 0·22 μ). The cooled extract was filtered through a millipore filter and the filtrate was concentrated to dryness in a vacuum desiccator. Some of the residues were hydrolyzed for 24 hr in evacuated sealed tubes in an oil bath at 110°C with 6 N HCl purified by redistillation. Others were dissolved in 0·01 N HCl and analyzed directly. All operations in the open were conducted in an Edcraft laminar flow hood in a clean room. Glassware was cleaned with dichromate-sulfuric acid, washed copiously with water and baked at 500°C overnight. Controls were run in the same manner with the lunar material omitted.

The lunar fines were first extracted with water and the aqueous extracts were concentrated and then hydrolyzed. This procedure was chosen in preference to direct hydrolysis of the lunar fines, because of the separate experiences of each of us in independent laboratory studies. We have observed that amino acids can be recovered from some geological samples by extracting with water first and then hydrolyzing, whereas they are not found when the sample is hydrolyzed directly, presumably due to decomposition catalyzed by the high proportions of minerals in the presence of hot HCl. Direct hydrolysis of lunar fines with HCl was also found to produce unmanageable proportions of salt. The extraction with water was continued for 17 hr at 100°C in order to favor more complete solution, although some hydrolysis could be anticipated under these conditions. The extracts were hydrolyzed because the experience of one of us (S.W.F.) with volcanic samples is that amino acids in such materials are predominantly or entirely in polymerized form.

Suitable aliquots were first analyzed on a Phoenix K-5000 Amino Acid Analyzer but mainly on the ultrasensitive modified amino acid analyzer of one of the authors (HARE, 1966), an apparatus which employs a 1·5 mm bore ion-exchange column. This apparatus permits quantitation of amino acids in the range of 1 ppb.

In order to determine if amino acids could be formed by oxidation of the fuel, 0·05 ml of 1,1-dimethylhydrazine was treated under nitrogen with 0·04 ml of conc. HNO_3. The yellow liquid changed to red and back to yellow over 1 hr. The mixture was then concentrated in a vacuum desiccator, dissolved in 0·01 N HCl and analyzed.

RESULTS

Analysis of hydrolyzates of lunar extracts on the Phoenix instrument revealed a peak with the RT (retention time) of ammonia, another in the region of basic amino acids, and a minimal third peak at about the RT of glycine.

Other samples were analyzed on the ultrasensitive apparatus (HARE, 1966). Seven peaks corresponding to amino acids were observed in two chromatograms. The one peak for which the identity was most uncertain corresponded to the RT of α,β-diaminopropionic acid. Six others in one chromatogram (Fig. 2) had RTs in common with proteinogenous amino acids. The amino acids tentatively identified and their amounts in nanomoles in another analysis are: aspartic acid, 0·05; threonine, 0·02; serine, 0·09; glutamic acid, 0·09; glycine, 0·48; alanine, 0·23, corresponding to an extract of 1·7 g of lunar fines.

The fine-free controls (Fig. 3), which were carried through all steps of the analytical sequence, contained significantly lower levels of amino acids (0·01 nmol), and different patterns. Samples of the unhydrolyzed extracts (Fig. 1) showed relatively low levels of amino acids. Analyses of other aliquots of these samples after four weeks of storage showed slightly increased amino acid contents in all cases. These latter results indicate the necessity of thoroughly evaluating the effect of storage on determinations at the subnanomole level.

Analyses of fuel exhaust products supplied by Dr. A. L. Burlingame and by Dr. D. Flory, showed only trace levels of amino acids when hydrolyzed. Samples of 1,1-dimethylhydrazine oxidized by nitric acid showed a number of as yet unidentified ninhydrin-positive peaks.

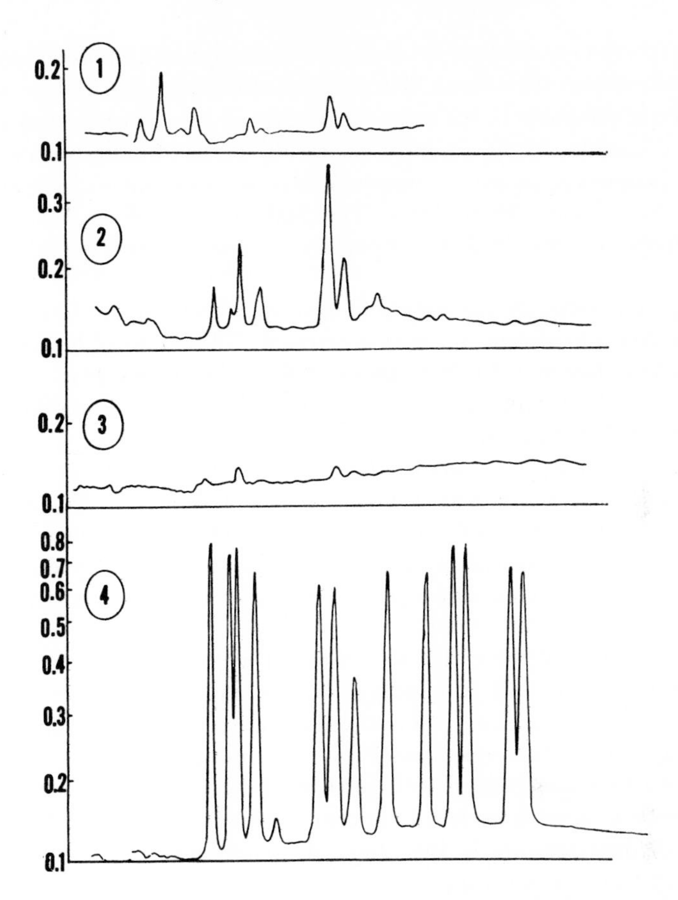

Figure 1. Amino acid analyzer chromatogram of unhydrolyzed aqueous extract of lunar fines no. 10086. Rightmost peaks correspond to RT's of glycine and alanine. Three main peaks on left are probably neither amino acids nor peptides. Figure 2. RT's of principal peaks (left to right) correspond to asp, thr (a shoulder), ser, glu, gly and ala for amino acids following hydrolysis of aqueous extract. Figure 3. Control of water sample carried through entire process without lunar fines. Figure 4. Chromatogram of standard amino acids, peaks corresponding to (left to right) asp, thr, ser, glu, pro, gly, ala, $cys/_2$, val, met, ile, leu, tyr, phe. (1.0 nmole each).

Discussion

The amino acids found in the chemical analysis of bulk fines could be present as the result of terrestrial contamination, as the result of chemical conversion of fuel, or as the result of indigenous chemical processes. Stronger likelihood of contamination may be provided by tests for dominant L isomers, as are under way. Stronger likelihood of indigenous synthesis on the moon would be suggested by concordant analyses of subsurface samples. Resolution of these alternative possibilities requires further samples and analyses.

The amino acids appear to be predominantly in a polymerized structure. Finding of amino acids in the quantities recorded required recognition of the possibility that the conditions on the moon could be conducive to formation and preservation of their polymers (FOX, 1966). This result is consistent also with terrestrial contamination.

The two principal peaks in the amino acid range of Fig. 1 correspond to the RT's of glycine and alanine. If the amino acids are the result of indigenous synthesis the much higher proportion in the polymerized form is consistent with observations on synthesis of amino acids (FOX, 1966). The peaks on the left in Fig. 1 are almost certainly not peptides since the R.T.'s of peptides tend to be longer than that of aspartic acid.

The results are consistent, at least qualitatively, with those of NAGY *et al.* (1970), who found glycine and alanine in aqueous extracts. The results of NAGY *et al.* and of our group at first glance differ from those of PONNAMPERUMA *et al.* (1970) and of ORÓ *et al.* (1970) who found no amino acids in extracts. Their extraction procedures were, however, different and their methods of analysis required recognition of amino acid derivatives, whereas the NAGY *et al.* group and ours used direct analysis of free or liberated amino acids. PONNAMPERUMA *et al.* did analyze hydrolyzates but they report analyzing hydrolyzates of lunar fines extracted by water rather than hydrolyzates of extracts. We used hydrolyzates of extracts, for reasons given earlier. The patterns observed in our chromatograms do not resemble closely patterns of known contamination.

Other explanations of the varied results might include (a) that the samples were inhomogeneous and (b) that two samples were contaminated whereas two others were not. These uncertainties may be resolved by further tests of Apollo 11 samples or of deeper samples from subsequent missions, or from other locales.

The microparticles observed in the lunar fines (FOX *et al.*, 1970) are predominantly glass and represent approximately 0·1 per cent of the material. Since the amino acids found in the hydrolyzate are in the range of 50 ppb they can hardly constitute a significant fraction of the formed particles. Accordingly, the lunar micromorphology and the amino acid contents appear to bear no significant relationship one to the other, even if the organic material is not contaminant.

REFERENCES

FOX S. W. (1965) A theory of macromolecular and cellular origins. *Nature* **205**, 328–340.

FOX S. W. (1966) The development of rigorous tests for extraterrestrial life. In *Biology and the Exploration of Mars*, (editors C. S. Pittendrigh, W. Vishniac, and J. P. T. Pearman), pp. 213–228. National Academy of Sciences National Research Council Publication 1296.

FOX S. W. (1969) Self-ordered polymers and propagative cell-like systems. *Naturwiss.* **56**, 1–9.

FOX S. W., HARADA K., HARE P. E., HINSCH G. and MUELLER G. (1970) Bio-organic compounds and glassy microparticles in lunar fines and other materials. *Science* **167**, 767–770.

GLASS B. (1967) Microtectites in deep-sea sediments. *Nature* **214**, 372–374.

HARE P. (1966) Automatic multiple column amino acid analysis—the use of pressure elution in small bore ion-exchange columns. *Fed. Proc.* **25**, 709.

MUELLER G. (1968) El origen de la vida (Spanish). *Bol. Soc. Biol. Concepcion* **40**, 161–173.

MUELLER G. (1969) The cosmo-petrological significance of the coalescence of microchondrules with the decrease of volatiles in chondrites. *Astrophys. Space Sci.* **4**, 3–43.

NAGY B., DREW C. M., HAMILTON P. B., MODZELESKI V. E., MURPHY M. E., SCOTT W. M., UREY H. C. and YOUNG M. (1970) Organic compounds in lunar samples: pyrolysis products, hydrocarbons, amino acids. *Science* **167**, 770–773.

ORÓ J., UPDEGROVE W. S., GIBERT J., MCREYNOLDS J., GIL-AV E., IBANEZ J., ZLATKIS A., FLORY D. A., LEVY R. L. and WOLF C. (1970) Organogenic elements and compounds in surface samples from the Sea of Tranquillity. *Science* **167**, 765–767.

PONNAMPERUMA C., KVENVOLDEN K., CHANG S., JOHNSON R., POLLOCK G., PHILPOTT D., KAPLAN I., SMITH J., SCHOPF J. W., GEHRKE C., HODGSON G., BREGER I. A., HALPERN B., DUFFIELD A., KRAUSKOPF K., BARGHOORN E., HOLLAND H. and KEIL K. (1970) Search for organic compounds in the lunar dust from the Sea of Tranquillity. *Science* **167**, 760–762.

WRIGHT F. W. and HODGE P. W. (1969) *32nd Annual Meeting Meteoritical Soc.*, Houston, 29–31 October.

Proceedings of the Apollo 11 Lunar Science Conference, Vol. 2, pp. 1805 to 1812.

Total organic carbon in the Apollo 11 lunar samples

RICHARD D. JOHNSON and CATHERINE C. DAVIS
National Aeronautics and Space Administration, Ames Research Center,
Moffett Field, California 94035

(*Received* 2 *February* 1970; *accepted in revised form* 28 *February* 1970)

Abstract—Analysis for total organic carbon content by pyrolysis hydrogen flame ionization detection is reported. The analytical technique, its scope and its limitations are described in detail. The apparent organic carbon content of fifteen different lunar samples ranged from 10 to 126 ppm. The amounts correlated with the sample handling history. Comparison of these results with those reported by other investigators indicates that the higher levels reported here are due to fewer limitations in this technique of liberating and analyzing organic matter. Lacking qualitative information about the nature and source of this organic matter, an upper limit of 10 ppm is estimated for any possible indigenous lunar organic material.

INTRODUCTION

THE EXAMINATION of extraterrestrial material for molecular species which can form the structural backbone for biochemistry leads almost inevitably to the choice of organic compounds, that is, those with carbon—carbon and carbon—hydrogen bonds. Organic rather than exotic chemistries are favored for this biochemical role on the basis of cosmic elemental abundances, bond strengths, polymer formation, thermal stability, multiplicity of functional groups, hydrolytic stability, and conformational utility (YOUNG *et al.*, 1965; SAGAN, 1966; LEDERBERG, 1966; PIMENTEL *et al.*, 1966). Although current theories of the origin of life and the actual analysis of meteoritic material indicate a strong possibility for the presence of organic matter on the surface of other planets, the actual confirmation of the ubiquity of organic matter still remains. Prior to the landed Apollo missions, the only data on carbon on the surface of an extraterrestrial body was provided by TURKEVITCH *et al.* (1969) from the Surveyor alpha scattering experiment which placed an upper limit of two to three per cent carbon without differentiation into organic and inorganic species.

This analysis for organic carbon in the lunar samples was performed in the Lunar Receiving Laboratory as part of the preliminary examination (LSPET, 1969; APOLLO 11 PRELIMINARY SCIENCE REPORT, 1969; JOHNSON and DAVIS, 1970). During this examination fourteen different specimens of lunar material were analyzed during the quarantine period behind both the primary and secondary biological barrier systems. One additional related sample was also analyzed as part of the subsequent lunar sample distribution to PONNAMPERUMA *et al.* (1970).

Although we have reported the presence of organic matter in the Apollo 11 samples, the technique employed is incapable of distinguishing indigenous material from trace contamination. Through examination of the degree of exposure to possible sources of contamination and by comparison of our results with those reported by other investigators, an upper limit of 10 ppm is placed upon possible indigenous organic matter.

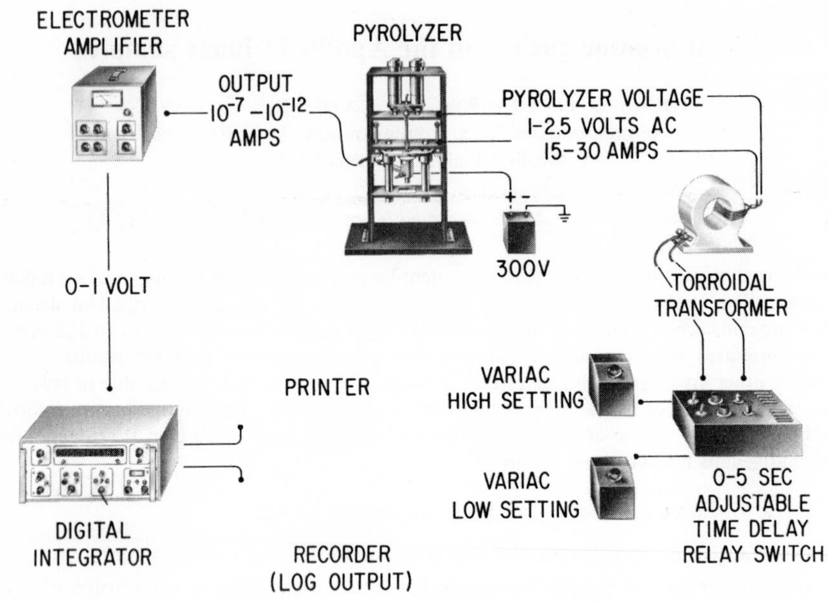

Fig. 1. Organic carbon electronic schematic.

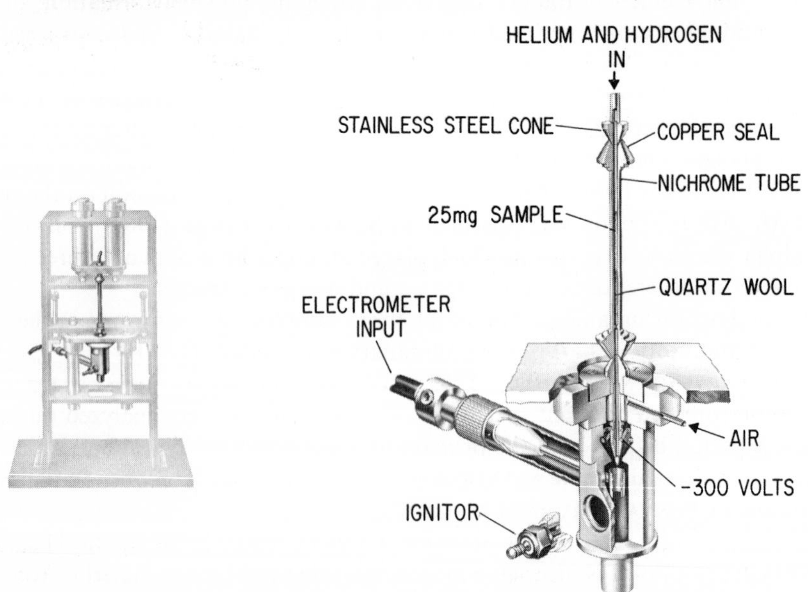

Fig. 2. Organic carbon pyrolyzer–hydrogen flame ionization detector.

Experimental Technique

The analytical method is based upon the response of the hydrogen flame ionization detector to the volatile products arising from pyrolysis of the sample. Approximately 25 mg of the particulate matter is placed in a nichrome tube and heated resistively in a flowing stream of helium and hydrogen to approximately 800°C. Any organic compounds in the sample are pyrolyzed with substantial conversion into volatile fragments which are swept by the flowing gas into the flame detector. The volatilized sample in the hydrogen and helium is burned at a jet in an atmosphere of air.

The chemical processes which take place in the flame are such that the organic compounds are thermally cracked in the inner flame zone to fragments which oxidize exothermally in the outer flame zone to excited species which undergo chemi-ionization to ionic species as postulated by Sternberg *et al.* (1962). In the polarized electric field above the flame these ions are collected, and the resulting ion current is amplified. The single peak produced by the ionized organic matter is integrated electronically to give an area response directly proportional to the original organic content of the sample.

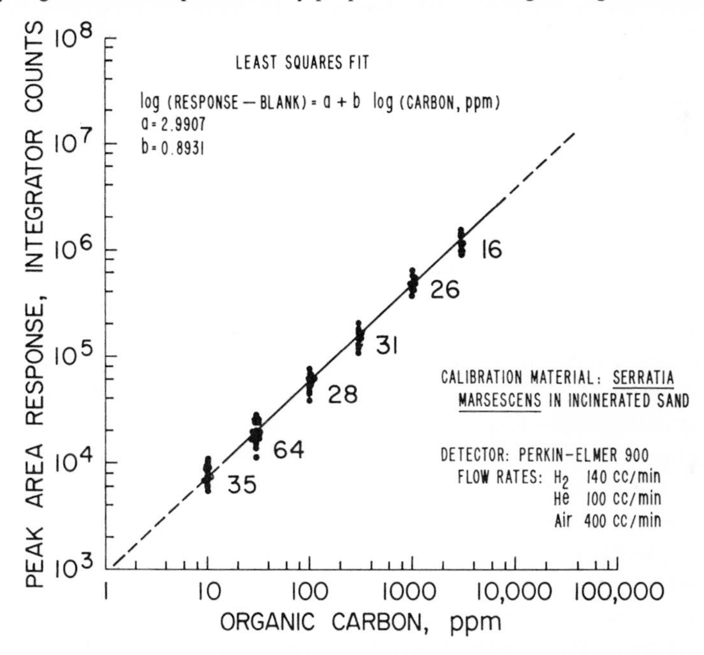

Fig. 3. Organic carbon calibration curve.

The quantitation of unknown samples is based upon retrofitting of the response to the calibration curve obtained from known samples (*Serratia marsescens* with 43·77 ± 0·67 per cent carbon freeze dried into incinerated sand at concentrations from 10 to 1000 ppm of carbon) as described by Mandel (1964) with the exception that the data is treated as the log transform to cover the dynamic range of the analysis. Schematic illustrations of the apparatus are shown in Figs. 1 and 2. A typical calibration curve is shown in Fig. 3.

This analytical method has been applied to terrestrial soils, meteorites, ocean sediments, volcanic debris and ancient sediments. A majority of these samples have also been analyzed with good agreement by classical wet combustion techniques. The results from the hydrogen flame method have never been too high, indicating no response is obtained from inorganic substances. This indication has been confirmed by subjecting the detector to a wide variety of inorganic gases and by pyrolysis of inorganic salts. In no instance has any significant response been obtained. In particular, inorganic forms of carbon such as carbon dioxide, carbon monoxide, carbonates, graphite and some carbides (zirconium and silicon) do not respond. Results from very ancient sediments where most of the organic matter is quite graphitic are often only 10 per cent of that obtained by classical combustion methods.

During the Apollo 11 mission the samples were analyzed as part of the normal sample flow in the Physical Chemistry Test Laboratory in the Lunar Receiving Laboratory. This facility consists of a bank of glove cabinets swept with dry nitrogen at slightly reduced pressure. The previous handling of the samples in this sequence of examinations, which included geology, physical properties, and elemental analyses, was not always under our control with respect to contamination precautions. Wherever possible, the sources of known contamination were documented.

The lunar material allocated for this analysis was treated upon receipt with clean steel implements and stored in clean glass vials. The small rock chips were always sufficiently friable to be crushed with tweezers in a glass vial. The amounts of sample were determined by weighing on an analytical balance or by a calibrated volumetric scoop.

Incinerated sand blanks and calibration samples were analyzed daily prior to the analysis of the lunar samples to insure proper operation of the entire system and to continuously refine and update the calibration curve. Exposure of the incinerated sand to the cabinet atmosphere did not result in noticeable increase in organic content from gaseous contamination.

RESULTS AND DISCUSSION

From the first organic analysis it was apparent that the organic content of the lunar samples was extremely low and easily obscured by trace contamination. The results for all of the samples are listed in Table 1.

The contingency sample was extracted from its pouch with relatively simple handling procedures and clean implements under the dry nitrogen atmosphere of the Physical Chemistry Test Laboratory. Although exposed to the spacecraft and terrestrial atmospheres during the return trip, these samples were relatively uncontaminated. As indicated in Table 1, some of the rocks were handled in the course of other examinations which resulted in a measureable increase in organic levels.

The documented sample box was opened in the vacuum chamber where the complexities of operation caused considerably higher degrees of contamination. The pump oil, the cumbersome gloves, and the heat sterilization of the samples and chamber all contributed markedly to the higher levels of organic compounds.

The bulk sample was processed in a dry nitrogen cabinet system in a manner that permitted minimum contamination of some samples. All material from the bulk box was inadvertently exposed to ethylene oxide and Freon which are normally quite volatile. The extent of this exposure and the involatile residues remaining are unknown but probably not great.

The fines distributed to the investigators (sample 10086-A) is closely related to sample 1002-73. Both were removed from the bulk box at the same time and stored in similar clean cans. Sample 1002-73 was sieved, and some portions were periodically removed prior to our analysis. The handling, however, was minimal. Sample 10086-A was subjected to at least one extraneous examination (the monopole incident) during which it was transferred to another container prior to packaging and distribution to investigators. From the higher level of organic matter in the distributed sample (40 ppm) as compared with this related sample (18 ppm) it must be assumed that some of these intervening operations caused traces of contamination to be introduced. It is sample 86 upon which most of the reported organic investigations were made.

The data on the organic content correlate directly with the degree and type of exposure in sample processing. The major sources of contamination appear to be

Table 1. Organic carbon content of Apollo 11 lunar samples

Sample No.	Type of material	Organic carbon (ppm)*	No. of determinations	Blank (ppm)†	Comments
Contingency Sample‡					
10010-5	fines	16 ± 7	2	4	minimal handling; one day after pouch opened
10010-26	fines	48 ± 18	3	1	same as 10010-5 eleven days later after extensive handling; fibers
10021-21	aggregate	44 ± 31	3	0	extensive handling for geology; stored in plastic box
10024-4	crystalline	17 ± 6	2	−1	moderate handling; stored in plastic box
10032-1	crystalline	10 ± 9	1	3	minimal previous handling
Documented box§					
10051-10	fines	126 ± 64	1	0	from gas reaction cell, very heavy handling; fibers noted
10061-4	aggregate	62 ± 19	2	1	vacuum chamber, routine handling sterilized; processed on clean bench
10064-4	aggregate	96 ± 30	2	1	as 10061-4
10068-3	aggregate	39 ± 18	1	1	as 10061-4 except with special handling in vacuum chamber
Bulk Box‖					
10002-73	fines	18 ± 8	2	0	sieved; minimal handling, minimal contamination
10086-A	fines	40 ± 8	2	1	similar to 10002-73; distributed to investigators
10046-3	aggregate	80 ± 26	2	1	routinely chipped for geological examination
10048-3	aggregate	88 ± 43	2	1	as 10046-3
10049-7	crystalline	54 ± 23	1	−2	as 10046-3.
10050-3	crystalline	28 ± 19	3	−2	as 10046-3
10056-12	aggregate	10 ± 4	2	1	specially prepared chip for organic analysis; also analyzed by organic mass spectrometer

* Errors are derived from 90% confidence limits.
† The organic carbon values have been corrected for this blank value.
‡ Contingency samples were processed under dry nitrogen in Physical Chemistry Test Laboratory.
§ Documented box was opened in vacuum chamber; samples and chamber were heat sterilized.
‖ Bulk box was opened under dry nitrogen in Biological Preparation Laboratory; samples were exposed to ethylene exode and Freon.

pump oil, tissue fibers, teflon, rubber glove fragments and residues from sterilizing agents. Contamination from the cabinet systems, the tools and the storage containers is generally smaller but also more uncertain.

Five of the samples listed in Table 1 had organic contents less than 20 ppm. All were processed under dry nitrogen with minimal handling. All of the other samples reported here were treated less carefully. On the basis of these results, we have placed our estimate of the indigenous organic matter at less than 10 ppm, realizing that even much of this may be due to traces of contamination.

The fundamental questions about organic matter in the lunar sample concern the amounts of such material from all sources and then its nature and source. The pyrolysis–hydrogen flame ionization detector method is capable of measuring the amount of organic matter but incapable of determining the type of sources of the compounds. In general the amount of organic matter reported here is greater than that reported by other investigators. This factor is probably due to the direct connection of the pyrolyzer to the flame detector and the subsequent ability to sweep the pyrolysis products efficiently at high temperature into the detector.

ORÓ *et al.* (1970) reported 0·6 ppm of organic matter detectable with a hydrogen flame ionization detector after sample pyrolysis at 600°C and gas chromatography on a silicone oil column at 130°C. Under these conditions only a fraction of the volatilized material would pass through the column. In addition, pyrolysis at this temperature is incomplete. ORÓ points out that when the products were analyzed on the mass spectrometer, methane appeared only after pyrolysis at temperatures above 600°C. CHANG *et al.* (1970) report a similarly low amount of C_1 to C_4 hydrocarbons upon pyrolysis to 750°C, separation on a Poropak Q column programmed to 140°C, and detection on a flame ionization detector. In this instance there was an intervening trapping system which would lower the efficiency of transfer of material to the gas chromatograph. Low yields detectable by gas chromatography are easily explained by retention of the barely volatile pyrolysis products on the chromatographic column.

The effect of pyrolysis temperature noted above by ORÓ, by CHANG, by NAGY *et al.* (1970) and by ABELL *et al.* (1970) is that increased yields of volatile pyrolysis products are obtained at the higher pyrolysis temperatures, usually above 600°C. The direct vaporization of organic matter into the inlet of a mass spectrometer as reported by MURPHY *et al.* (1970) and by LSPET (1969) involved temperatures of only 400 and 500°C respectively with reported maximum organic values of 2·5 and 5 ppm. Both of these analyses were quantitated on the basis of volatile hydrocarbon standards. The low pyrolysis temperatures and volatile calibration standards would not suffice for analysis of normally involatile organic compounds, particularly those that have withstood the thermal and vacuum environment of the moon.

This problem was investigated by heating natural and artificial low organic soils (sand from Death Valley and our calibration material) to 500 ± 50°C at 7×10^{-6} torr for thirty minutes. Analysis of this heated material, as compared with the unheated samples and with appropriate blanks, resulted in as much as 90 per cent of the organic matter remaining unvolatilized after such treatment. Thus the results from the mass spectrometer could be considerably low.

The investigations based upon solvent extract of the lunar material generally reported very little or no detectable organic matter. However GEHRKE *et al.* (1970) report organosiloxanes at levels of 35–50 ppm after extraction and hydrolysis with hydrochloric acid and derivatization with n-butanol and trifluroacetic anhydride. These amounts conveniently agree with the levels of organic matter reported here. It remains, however, to be shown more conclusively that most of the organic portion of these compounds did not originate with the derivatizing agents. It is quite possible, for example, that the lunar sample contains a derivatizable inorganic silicon compound with little or no indigenous carbon.

CHANG *et al.* (1970) report evidence for carbides. Although the pyrolysis–hydrogen flame ionization detector method was tested and found not to respond to silicon and zirconium carbides, it remains possible, although unlikely, that other carbides would react with the hydrogen at 800°C to produce volatile hydrocarbons. Further work in this area is indicated.

CHANG also indicates that it is theoretically possible for hydrogen and carbon monoxide to react in the heated inorganic matrix to form hydrocarbons in a Fisher–Tropsch synthesis. This possible reaction was tested by CHANG with essentially negative results. It is unlikely that the response reported here from pyrolysis at 800°C could be due to this factor.

CONCLUSIONS

The techniques utilized by the investigators were designed to extract maximum information about the nature of the organic compounds by use of mild techniques such as heating to moderate temperatures or extraction with solvents coupled to high resolution techniques such as chromatography. In these efforts to obtain maximum qualitative data without extensive degradation of the organic compounds, the quantitation of the total organic content suffers. The results reported here, void of qualitative detail, do more fully quantitate the amount of organic matter.

Because of the very high temperatures required to liberate significant quantities of organic matter from the lunar soil, it is tempting to assign at least some of this material as indigenous to the moon. As reported here, however, liberation of terrestrial organic matter from soils is inefficient even at temperatures up to 500°C. Therefore characterization of the organic matter in the lunar samples on this basis is tenuous.

While we have been unable to do more than place a 10 ppm upper limit on the possible indigenous organic compounds, the levels of organics reported by other investigators may be pessimistic, and that the true levels of indigenous organic matter may lie between these two extremes.

REFERENCES

ABELL P. I., DRAFFAN G. H., EGLINGTON G., HAYES J. M., MAXWELL J. R. and PILLINGER C. T. (1970) Organic analysis of the returned lunar sample. *Science* **167,** 757–759.

APOLLO 11 PRELIMINARY SCIENCE REPORT (1969) Preliminary examination of lunar samples. NASA SP-214 pp. 136–138.

CHANG S., SMITH J. W., KAPLAN I., LAWLESS J., KVENVOLDEN K. A. and PONNAMPERUMA C. (1970) Carbon compounds in lunar fines from Mare Tranquillitatis—IV. Evidence for oxides and carbides *Geochim. Cosmochim. Acta*, Supplement I.

GEHRKE C. W., ZUMWALT R. W., AUE W. A., STALLING D. L., DUFFIELD A., KVENVOLDEN K. and PONNAMPERUMA C. (1970) Carbon compounds in lunar fines from Mare Tranquillitatis—III. Organosiloxanes in the hydrochloric acid hydrolysates. *Geochim. Cosmochim. Acta*, Supplement I.

JOHNSON R. D. and DAVIS C. C. (1970) Pyrolysis-hydrogen flame ionization detection of organic carbon in a lunar sample. *Science* **167,** 759–760.

LEDERBERG J. (1966) Signs of Life. In *Biology and the Exploration of Mars*, (editors C. S. Pittendrigh, W. Vishniac and J. P. T. Pearman), pp. 127–140. *Nat. Acad. Sci. Nat. Res. Council Publ.* 1296.

LSPET (LUNAR SAMPLE PRELIMINARY EXAMINATION TEAM) (1969) Preliminary examination of lunar samples from Apollo 11. *Science* **165,** 1211–1227.

MANDEL J. (1964) *The Statistical Analysis of Experimental Data*, 1st edition, pp. 272–282. Interscience.

MURPHY R. C., PRETI G., NAFISSI V. M. and BIEMANN K. (1970) Search for organic material in lunar fines by mass spectrometry. *Science* **167,** 755–757.

NAGY B., DREW C. M., HAMILTON P. B., MODZELESKI V. E., MURPHY M. E., SCOTT W. M., UREY H. C. and YOUNG M. (1970) Organic compounds in lunar samples: Pyrolysis products, hydrocarbons, amino acids. *Science* **167,** 770–773.

ORÓ J., UPDEGROVE W. S., GIBERT J., MCREYNOLDS J., GIL-AV E., IBANEZ J., ZLATKIS A. FLORY D. A., LEVY R. L. and WOLF C. (1970) Organogenic elements and compounds in surface samples from the Sea of Tranquillity. *Science* **167,** 765–767.

PIMENTEL G. C., ATWOOD K. C., GAFFRON H., HARTLINE H. K., JUKES T. H., POLLARD E. C. and SAGAN C. (1966) Exotic biochemistries in Exobiology. In *Biology and the Exploration of Mars*, (editors C. S. Pittendrigh, W. Vishniac, and J. P. T. Pearman), pp. 243–251. *Nat. Acad. Sci. Nat. Res. Council, Publ.* 1296.

PONNAMPERUMA C., KVENVOLDEN K., CHANG S., JOHNSON R., POLLOCK G., PHILPOTT D., KAPLAN I., SMITH J., SCHOPF J. W., GEHRKE C., HODGSON G., BREGER I. A., HALPERN B., DUFFIELD A., KRAUSKOPF K., BARGHOORN E., HOLLAND H. and KEIL K. (1970) Search for organic compounds in the lunar dust from the Sea of Tranquillity. *Science* **167**, 760–762.

SAGAN C. (1966) The cosmic setting. In *Biology and the Exploration of Mars*, (editors C. S. Pittendrigh, W. Vishniac and J. P. T. Pearman), pp. 73–113. *Nat. Acad. Sci. Res. Council Publ.* 1296.

STERNBERG J. C., GALLAWAY W. S. and JONES D. T. L. (1962) The mechanism of response of flame ionization detectors. In *Gas Chromatography*, (editors N. Brenner, J. E. Callen and M. D. Weiss), pp. 321–267. Academic Press.

TURKEVICH A. L., ANDERSON W. A., ECONOMOU T. E., FRANZGROTE E. J., GRIFFIN H. E., GROTCH S. L., PATTERSON J. H. and SOWINSKI K. P. (1969) The alpha-scattering chemical analysis experiment on the Surveyor lunar missions. In *Surveyor Program Results*, pp. 271–350. NASA, SP-184.

YOUNG R. S., PAINTER R. B. and JOHNSON R. D. (1965) An analysis of the extraterrestrial life detection problem. NASA SP-75.

Proceedings of the Apollo 11 Lunar Science Conference, Vol. 2, pp. 1813 to 1828.

Carbon compounds in lunar fines from Mare Tranquillitatis—I. Search for molecules of biological significance

K. A. KVENVOLDEN, S. CHANG, J. W. SMITH*, J. FLORES, K. PERING, C. SAXINGER, F. WOELLER, K. KEIL†, I. BREGER‡ and C. PONNAMPERUMA
Exobiology Division, Ames Research Center, NASA, Moffett Field, California 94035

(*Received* 3 *February* 1970; *accepted in revised form* 2 *March* 1970)

Abstract—This four-part paper describes an extensive analysis for carbon compounds in lunar fines labelled "10086-3 Bulk A" from the Sea of Tranquillity. The purpose of the work has been to define the chemical state of carbon in this sample. A search was made for the presence of the following organic compounds: high molecular weight alkanes and aromatic hydrocarbons, fatty acids, porphyrins, amino acids, purines, pyrimidines, and carbohydrates. The search utilized a sequential procedure in which the sample always remained in the container into which it was originally placed. With the exception of trace quantities of pigment having porphyrin-like spectral characteristics, there was no evidence at the nanogram level for the presence of the other classes of organic compounds. Acid hydrolysis of the lunar fines produced (1) low molecular weight hydrocarbons, suggesting the presence of carbides, (2) hydrogen sulfide, the isotopic composition of the sulfide being enriched in ^{34}S by +8·2 per mil relative to the Canyon Diablo meteorite standard, and (3) organosiloxanes which were found when the hydrolysates were acylated and esterified.

INTRODUCTION

IN THE study of chemical evolution we are interested in the path by which molecules of biological significance may have arisen in the universe, in the absence of life (PONNAMPERUMA and GABEL, 1968). Studies of the organic matter in very ancient sediments of the earth have provided us with information about the earliest evolutionary stages of life and given us some clues to the nature of prebiotic processes. Microfossils suggestive of such life-forms as bacteria and blue-green algae have been preserved in rocks at least as old as 3·1 b.y. (BARGHOORN and SCHOPF, 1966; SCHOPF and BARGHOORN, 1967). In these same rocks have been found organic compounds (HOERING, 1965; ORO and NOONER, 1967) which are believed to be molecular evidence for early life. Apparently life was well established on earth 3 b.y. ago. Older rocks exist on earth, but none has been discovered which is older than about 3·6 b.y. (DONN *et al.*, 1965).

For samples of carbon older than this, one must look outside the earth to meteorites and now to lunar samples. For the first time extraterrestrial samples are available which have not been exposed for extended periods of time to the earth's biosphere. From such samples new understandings of the role and distribution of carbon in

* Present address: Institute of Geophysics and Planetary Physics, University of California, Los Angeles, California 90024. Permanently at Division of Mineral Chemistry, C.S.I.R.O., Chatswood, N.S.W., Australia.

† Present address: Department of Geology and Institute of Meteoritics, The University of New Mexico, Albuquerque, New Mexico 87106.

‡ Present address: U.S. Geological Survey, Washington, D.C. 20242.

the universe may be obtained. Perhaps sequestered in some lunar sample are carbon-containing substances that will give clues to prebiotic organic synthesis (SAGAN, 1961). Finding and identifying such components is the purpose of the analysis reported here. Such knowledge is important in understanding the cosmological history of carbon and in providing new data points in studies related to the evolutionary history of the solar system.

Preliminary examinations of lunar samples (LSPET, 1969) indicated a total concentration of carbon compounds in the order of about 10 $\mu g/g$ or less. A scheme of analysis (Fig. 1) was designed, therefore, to obtain maximum information from samples

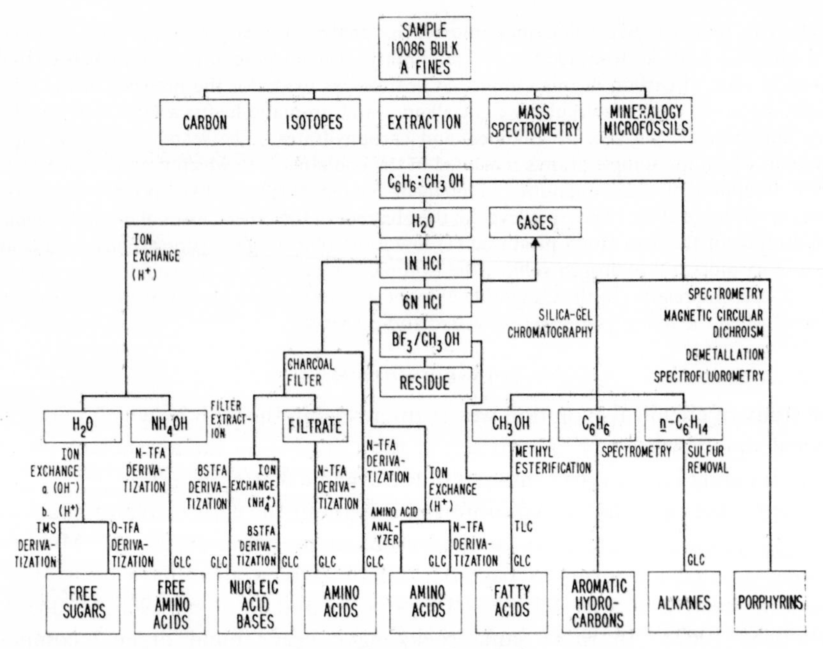

Fig. 1. Scheme of analysis.

containing minimal concentrations of carbon compounds. The scheme follows generally accepted organic geochemical practices but is unique in its application to a single sample. The analysis included an examination of lunar material for the chemical state of carbon in the sample, isotopic fractionation, microfossils and mineralogy. Sequential treatment of the sample with benzene–methanol, water, hydrochloric acid and boron trifluoride in methanol provided extracts for examination by chromatographic and spectrometric techniques.

The sample "10086-3 Bulk A Fines" was analyzed according to the scheme by a consortium of investigators who pooled their efforts in order to determine as much as possible about the organic chemistry of a single sample of lunar dust. The sample was a fine black powder of less than 1 mm mesh size. Because of its fine grain size, the sample was not pulverized but rather was analyzed as received from the Lunar Receiving Laboratory, Houston, Texas.

A portion of the sample was divided into several aliquots of unequal weights, the weights being determined by the nature of the experiments to be performed. After a complete analysis of the sample, those analyses providing the greatest amount of information were repeated on additional aliquots of the sample. Table 1 shows the total weights of sample utilized in this analysis.

Table 1. Weights of lunar material used for analysis

Solvent extraction (Includes confirmatory analyses)	68·8 g
Carbon and isotopes (Includes analyses for organic carbon, carbides and pyrolysis products)	23·3 g
Mass spectrometry	0·02 g
Mineralogy and microfossils	2·25 g

This report is organized into four parts. This first section (I) describes the general scheme of analysis and summarizes the results. Certain portions of the analysis were more rewarding than others, and those portions where positive results were obtained are described in detail in parts II, III and IV: II describes the finding of pigments which may be porphyrins; III describes the discovery of organosiloxane compounds; and IV describes the chemical state of the carbon in the sample.

Prior to analyzing the lunar sample the scheme of analysis was applied to several different kinds of geological materials:

(1) A modern sediment from Saanich Inlet, British Columbia.
(2) A basalt bomb from near Aloi Crater, Island of Hawaii.
(3) Pueblito de Allende meteorite from Mexico described by KING *et al.* (1969).
(4) Ottawa sand (Matheson, Coleman and Bell) prepared by heating at 1000°C for 48 hr.

Samples 1 and 2 were provided by I. Kaplan, University of California, Los Angeles. The results obtained on these samples confirmed that the scheme could be successfully applied in the analysis of carbon-containing materials and that the background contamination was sufficiently low to permit the determination of specific organic compounds, if present, in the nanogram range.

During the analysis of the lunar sample parallel experiments were conducted on Ottawa sand. Results for this blank were compared with those obtained from the lunar sample in order to monitor the background. Experiments were also conducted on an interior sample from a 6-kg piece of the Pueblito de Allende meteorite, and the results were compared.. To minimize contamination, the analyses were carried out in a clean laboratory designed specifically for lunar sample work.

EXTRACTION

A major portion of the lunar sample was treated with a sequence of solvents [benzene–methanol (9:1), water, 1 N and 6 N hydrochloric acid, and 7% boron trifluoride in methanol] according to the scheme shown in Fig. 1. The procedures were designed to minimize the handling of both solids and liquids. In this sequential scheme, which involved a minimum of glassware, the solid lunar sample was retained in all steps in the vessel in which it was originally placed. Figure 2 shows the specially modified glassware. Addition of extracting solvents to the sample was calculated by weight to avoid the use of volumetric measuring glassware. Extraction was facilitated by means of a sonic bath and stirring. Solutions were recovered from the solid by centrifugation and decantation. Particulate material in solutions was removed by filtering through clean glass wool.

All solvents and reagents were carefully evaluated prior to use. Pesticide quality benzene, methanol, hexane and carbon tetrachloride were distilled in $1{\cdot}25 \times 50$ cm

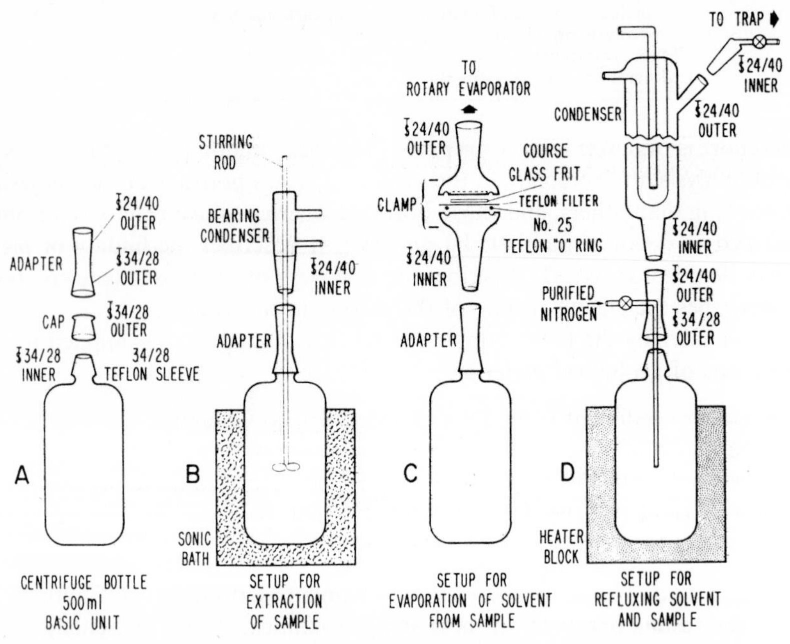

Fig. 2. Glassware used in sequential extraction of lunar sample.

vacuum jacketed glass columns packed with 6 mm glass helices. Take off rate for these solvents was about 1 ml/min. Hydrochloric acid was prepared by bubbling HCl gas through triple distilled water until a 12 N solution resulted. Acids of required normality were obtained by dilution with triple distilled water. A 14 per cent solution of BF_3 in methanol was prepared by bubbling BF_3 gas into distilled methanol (METCALF and SCHMIDT, 1961). Ammonium hydroxide was also prepared by bubbling ammonia gas into triple distilled water.

All glassware was carefully washed in detergent (Alconox) and hot water, rinsed and placed in a bath of sulfuric acid and sodium dichromate at 90–100°C for at least 1 hr. Each piece of glassware was rinsed with single distilled water and finally with triple distilled water and dried in an oven at 150°C. The glassware was stored in covered stainless steel trays. Wherever possible, each piece of glassware was rinsed with the solvent to be used in the analysis.

Procedure

The following is a brief outline of the sample treatment.

1. *C_6H_6–MeOH*

The sample (54·6 g) was extracted with a mixture of benzene-methanol (9:1) by sonication with stirring for 15 min. (Fig. 2B). The extract was recovered by centrifugation, decantation and filtration. This procedure was repeated twice. Of the solution recovered 151 g was used for hydrocarbon and fatty acid analysis and 45 g was used for porphyrin analysis. Residual solvent which remained with the sample was removed by rotary evaporation against an aspirator vacuum (Fig. 2C). The evaporator was connected to the aspirator through two 500 ml cold finger condensers cooled with liquid nitrogen.

2. *H_2O*

The dried sample was extracted with triple distilled water in the same manner as with benzene–methanol. About 230 g of water extract were recovered. Free amino acids and sugars were sought in this extract.

3. *1 N HCl*

1 N HCl was added to the residue left after the water extraction. This mixture was placed in an aluminum heating block at about 125°C, and the sample was refluxed for one hour (Fig. 2D). During this period, a stream of purified nitrogen was passed through the reaction vessel and vented through a series of gas bubblers (Fig. 3). After hydrolysis the mixture was cooled, centrifuged, and the

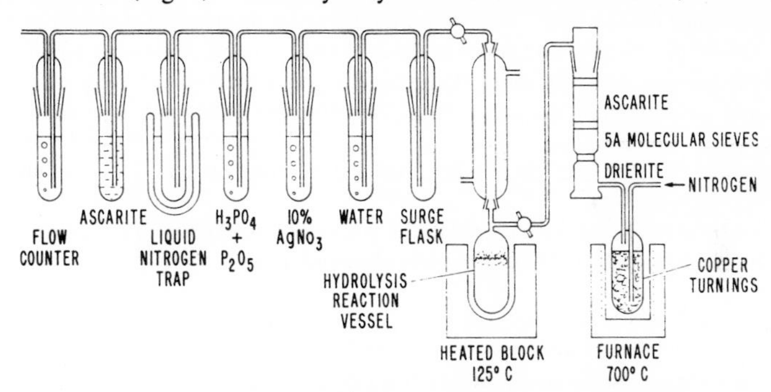

Fig. 3. System for trapping gases evolved during hydrolysis.

hydrolysate decanted through a glasswool filter. The sample was rinsed twice with 1 N HCl, and the rinse was recovered by centrifugation and decantation. About 219 g of acid extract were recovered. Nucleic acid bases and amino acids were sought in this extract.

4. *6 N HCl*

To the residue of the 1 N HCl hydrolysis 6 N HCl was added. The sample was then hydrolyzed under reflux conditions at about 125°C for 19 hr (Fig. 2D). During the first 16 hr of hydrolysis, purified nitrogen was passed through the reaction vessel and vented through a series of traps as described earlier. After hydrolysis the mixture was cooled and centrifuged. The hydrolysate was decanted through a glass wool filter. The sample was rinsed two additional times with 6 N HCl, and the rinse was recovered by centrifugation and decantation. The weight of 6 N HCl hydrolysate recovered was about 207 g. This extract was examined for amino acids. The residual sample was dried by rotary evaporation (Fig. 2C).

5. *BF_3/CH_3OH*

To the dried sample was added a 7 per cent solution of BF_3 in methanol. The mixture was refluxed at 90°C for 4 hr (Fig. 2D, w/o nitrogen bleed). This technique for obtaining fatty acids was based on the method of LAWLOR and ROBINSON (1967). The extract was recovered by centrifugation, decantation and fitration. The sample was washed twice more with methanol, centrifuged and the solution decanted. About 153 g of extract was recovered for fatty acid analysis. The sample was dried by rotary evaporation (Fig. 2C). The residue recovered weighed 37·6 g.

Summary of Analyses for Organic Compounds

The extracts and hydrolysates were examined for specific organic compounds—porphyrins, alkanes, aromatic hydrocarbons, fatty acids, amino acids, nucleic acid bases and sugars. The presence of these compounds would be significant in the context of chemical evolution. Of these classes of compounds, only porphyrins appear to be present on the basis of spectrofluorometric analysis.

1. C_6H_6–CH_3OH *Extract*

(a) *Porphyrins.* This extract was divided into two portions by weight: 23 per cent was allotted for porphyrins and 77 per cent for hydrocarbons and fatty acids. Porphyrins were sought by absorption and magnetic circular dichroism spectrometry and by spectrofluorometric techniques involving analytical demetallation of porphyrin complexes and remetallation of free-base porphyrins (HODGSON *et al.* 1969). Fluorescence excitation and emission spectra of demetallated extracts suggest that porphyrin-like pigments are present in this lunar sample. However, similar spectral responses were also observed in exhaust products from tests of lunar descent rocket engines. Approximately 10^{-4} μg/g of porphyrin-like pigments were detected in the lunar sample. A detailed account of this work appears in part II.

(b) *Alkanes.* The extract was examined for both straight chain and isoprenoid hydrocarbons. The benzene–methanol (9:1) extract was evaporated just to dryness on a rotary evaporator. The residue was dissolved in *n*-hexane and was fractionated on a column containing 9 g of activated silica gel (Davison 923) pre-wet with *n*-hexane. The column was eluted successively with 15 ml each of *n*-hexane, benzene and methanol according to a procedure modified from MEINSCHEIN and KENNY (1957). The fractions eluted by each eluant were collected separately. The *n*-hexane fraction was examined for alkanes, the benzene fraction for aromatic hydrocarbons and the methanol fraction for fatty acids.

Activated copper metal strips were placed in the *n*-hexane fraction for 24 hr in order to remove any dissolved sulfur. No tarnishing of the copper was observed, indicating that no appreciable free sulfur was likely to be present.

The *n*-hexane fraction was concentrated by rotary evaporation and finally in a stream of purified nitrogen. The concentrate was analyzed by gas–liquid chromatography (Table 2). No peaks corresponding to *n*-alkanes or isoprenoid hydrocarbons in the range from about C_{15}–C_{32} were detected. Partially resolved peaks with retention times corresponding to about n-C_{21} and equivalent to approximately 20 ng were found, but these peaks can be explained as contamination since similar chromatographic peak patterns had been observed on a random basis previously; the source of this contamination could not be traced. Minimum level of detection for individual alkanes at the conditions of chromatography used was about 10 ng.

(c) *Aromatic hydrocarbons.* Aromatic hydrocarbons were sought in the benzene eluate from chromatography. The fraction was evaporated to dryness and dissolved in methanol. Absorption spectra were obtained in the ultraviolet region from 200 to 350 nm and in the visible region from 350 to 600 nm (Cary 14 Spectrophotometer). Absorption was observed at 224, 274 and 280 nm. Because these absorption bands were present in both the lunar sample and the sand blank, the presence of aromatic hydrocarbons indigenous to the lunar sample cannot be inferred.

(d) *Fatty acids.* Fatty acids were sought in the methanol eluate from chromatography as well as in the boron trifluoride–methanol extract which was obtained after the sample had been extracted with water and hydrolyzed with hydrochloric acid. When the comprehensive scheme of analysis (Fig. 1) was applied earlier to a modern sediment sample from Saanich Inlet, slightly different distributional patterns of fatty acids were obtained in the methanol eluate and the boron trifluoride–methanol extract. This finding set a precedent for applying the same analytical scheme to the lunar sample, because fatty acids are bound in different ways in sediments, and therefore, can be recovered by different techniques.

(1) The methanol eluate from chromatography of the benzene–methanol extract of the lunar sample was concentrated on a rotary evaporator and taken to dryness on a 40°C bath with a purified nitrogen stream. The extract was esterified with a solution of 14 per cent BF_3 in CH_3OH according to the procedure of METCALF and SCHMITZ (1961). Any methyl esters of fatty acids which might be present were recovered in benzene.

(2) The boron trifluoride–methanol extract obtained as the last step in the sequential analysis was concentrated on a rotary evaporator. Water was added, and the mixture was extracted with CCl_4. This CCl_4 solution was concentrated by rotary evaporation and taken to dryness as described above. The residue was dissolved in benzene.

Table 2. Summary of gas–liquid chromatography

Instrument	Compounds sought	Liquid phase and support	Dimensions	Helium flow rates and pressures	Attenuation	Temperature program	Injector temperature	Detector temperature	Minimum detectable concentration of individual compound
Perkin Elmer 880	Alkanes	OV-17	Support coated open tubular 50 ft × 0·02 in.	3 ml/min 10 psi	×50	70°C for 10 min 70°–250°C at 2°/min	190°C	Same as oven	~10 ng
Hewlett Packard 5750	Fatty acids (as methyl esters)	3% SE 30 on 100–120 mesh gas chrom Z	5 ft × 1/8 in. stainless steel	20 ml/min 60 psi	×20 ×40 ×80	118° to 300°C at 4°/min	220°C	280°C	~10 ng
	Amino acids (as N-TFA derivatives)	0·325% EGA on 80–100 mesh AW chromasorb G	1·5 m × 4 mm glass	60 ml/min 60 psi	×2 ×4	60°C for 4 min 60–190°C at 4°/min	160°C	240°C	~1 ng
	Amino acids (as N-TFA derivatives)	3% OV-17 on 80–100 mesh HP chromasorb W	1·5 m × 4 mm glass	80 ml/min 60 psi	×10 ×20	65°C for 6 min 65–240°C at 4°/min	185°C	240°C	~1 ng
Hewlett Packard 400	Fatty acids (as methyl esters)	3% OV-17 on 60–80 mesh gas chrom Q	6 ft × 1/4 in. glass	50 ml/min 40 psi	×40 ×80	100–320°C at 5°/min	200°C	280°C	~20 ng
	Nucleic acid bases (as TMS derivatives)	3% OV-17 on 60–80 mesh gas chrom Q	6 ft × 1/4 in. glass	50 ml/min 40 psi	×40 ×80	100°C for 6 min 100–300°C at 7·5°/min	200°C	280°C	~20 ng
Varian 1520	Sugars (as O-TFA derivatives)	Trifluoromethyl propyl silicone	Support coated open tubular 50 ft × 0·02 in.	5 ml/min 60 psi	×0·4	130°C for 8 min 130–175°C at 4°/min	180°C	200°C	~5 ng
	Sugars (as TMS derivatives)	Neopentyl glycol adipate	Support coated open tubular 50 ft × 0·02 in.	5 ml/min 60 psi	×0·8	140°C for 8 min 140–175° at 4°/min	180°C	200°C	~5 ng

The benzene solutions from (1) and (2) were analyzed by thin layer chromatography on silica gel plates (LAWLOR and ROBINSON, 1967). Bands with R_f values equivalent to those of standard methyl esters were recovered from the plates and extracted with benzene. Benzene solutions were concentrated and analyzed by gas–liquid chromatography (Table 2).

Fatty acids n-$C_{16:0}$, n-$C_{16:1}$, n-$C_{18:0}$ and n-$C_{18:1}$ were detected in both the lunar sample extract and the sand blank extract at concentrations for each acid ranging from 8×10^{-3} to 1×10^{-3} μg/g. Because fatty acids are present in about equal concentrations in extracts from the sample and the blank, there is no evidence for the presence of fatty acids indigenous to the lunar sample at concentrations exceeding 10^{-2} μg/g.

2. *H_2O Extract*

(a) *Free sugars.* The extract was examined for its content of free sugars and free amino acids. The H_2O extract was reduced in volume by rotary evaporation. About 1·3 per cent of the extract was set aside for analysis of free amino acids. Particulate material was removed by centrifugation and the concentrated extract was treated by microscale ion-exchange chromatography modeled after DEGENS and REUTER (1964). The extract was washed on to a 9 mm glass column packed with 1·5 ml AG-50 W-X8, 200–400 mesh ion-exchange resin (H^+ form). The column was eluted with water and the eluate passed through a column containing 1·5 ml AG-3-X4, 200–400 mesh (OH^- form) and a second column containing 1·5 ml AG-50 W-X8, 200–400 mesh (H^+ form). The final neutral eluate was reduced in volume by evaporation, divided and freeze-dried. In one-sixth of the neutral fraction, any monosaccharides and other compounds containing hydroxyl or mercapto groups which might be present were converted to trimethylsilyl ether derivatives (SWEELEY *et al.*, 1963) for gas–liquid chromatographic analysis (Table 2). One-third of the neutral extract was reduced with sodium borohydride (ABDEL-AKHER *et al.*, 1951). Excess borate ion was removed as methyl borate (ALBERSHEIM *et al.*, 1967). The resulting reduction products were converted to trifluoroacetates (SHAPIRA, 1969) and analyzed by gas–liquid chromatography (Table 2). The limit of detection for a single monosaccharide as the trimethylsilyl derivative was established at 30 ng for the bulk sample. If sugars are in the lunar sample they are present in concentrations less than 6×10^{-4} μg/g for individual pentoses and hexoses and less than 1×10^{-3} μg/g of any other directly derivatizable compound. Four unidentified trifluoroacetylated reduction products were detected. Each was present in concentrations less than about 2×10^{-3} μg/g. There was insufficient material for mass spectral identification.

(b) *Free amino acids.* After the neutral fraction had been eluted from the first ion exchange column used in the treatment of free sugars, the column was disconnected from the system and eluted with 2 N ammonium hydroxide. This basic solution was evaporated to dryness. At the same time the 1·3 per cent of the water extract which had been set aside was also evaporated to dryness. These two samples were derivatized according to a scheme modified from ROACH and GEHRKE (1969) to form N-trifluoroacetyl (N-TFA) *n*-butyl esters of any amino acids which might be present. Gas–liquid chromatography (Table 2) of the water extract which had not passed through the ion exchange column revealed the presence of three minor chromatographic peaks which did not correspond in chromatographic retention times to any common amino acids. Gas–liquid chromatography (Table 2) of the derivatized ammonium hydroxide eluate showed no significant chromatographic peaks. Therefore, the three compounds present in the untreated water extract were not amino acids. With this technique 1 ng of any amino acid would have produced a significant chromatographic peak of about 10 per cent of a full scale deflection.

3. 1 *N HCl Hydrolysate*

The 1 N HCl hydrolysate was prepared in order (1) to obtain nucleic acid bases without greatly altering their structure by deamination reactions, (2) to recover any amino acids that did not dissolve during water extraction, and (3) to obtain any amino acids which can be easily hydrolyzed from large molecules.

(a) *Nucleic acid bases.* About 90 per cent of the 1 N HCl hydrolysate was filtered through a 6 mm column containing about 500 mg of a 1 : 1 mixture of Norit A charcoal (Nutritional Biochemical Corp.) and Celite 545 (Johns-Manville). This mixture had been previously prepared by refluxing with a sequence of solvents: 2 N HCl, 6N NH_4OH, pyridine: water (1 : 1), pyridine, water, methanol, a mixture of benzene–ethanol–water, benzene, formic acid and water. The filtration removed purines, pyrimidines and other non-polar, hydrophobic and aromatic molecules. The filter was washed with 50 ml of 1 N HCl and 50 ml of water to remove loosely adsorbed material, e.g. amino acids, sugars and salts. The charcoal mixture was extracted with formic acid for 5 hr at 40°C and the extract evaporated to dryness in a 40°C bath. The last traces of formic acid were removed by NaOH pellets in a desiccator. The charcoal elution procedure must be followed closely to ensure quantitative elution without extraction of background contaminants from charcoal. Half of the residue was reacted with bis (trimethysilyl) trifluoracetamide (GEHRKE and RUYLE, 1968), in order to make trimethylsilyl derivatives of any purines and pyrimidines which might have been present. The derivatized mixture was analyzed by gas–liquid chromatography (Table 2). None of the common nucleic acid bases was detected. The detection limit for the procedure is about 20 ng. The remaining half of the residue was chromatographed using Dowex 50 (NH_4^+). Gas–liquid chromatography of the trimethylsilyl derivatives of the product revealed no nucleic acid bases. The presence of heterocyclic base ion fragments could not be determined by direct probe analytical mass spectrometry (CEC 21-110) of a sample representing 8 per cent of the 1 N HCl hydrolysate.

(b) *Amino acids*. About 2 per cent of the 1 N HCl hydrolysate and 3 per cent of the filtrate from the charcoal–celite column was analyzed for the presence of amino acids by taking the hydrolysate and filtrate to dryness and reacting the residues to form N-TFA *n*-butyl esters (ROACH and GEHRKE, 1969) of any derivatizable material present. Gas chromatography (Table 2) of the reaction products produced almost identical chromatograms for the hydrolysate and filtrate, but none of the chromatographic peaks present corresponded to the common amino acids. No chromatographic peaks were found in the sand blanks which had been treated in the same manner as the samples. Later it was discovered that the unknown compounds needed both esterification and acylation in order to be successfully chromatographed by gas–liquid chromatography. Mass spectrometric analysis of these compounds showed they are organosiloxanes. Further discussion of these compounds appears in part III.

4. 6 *N HCl Hydrolysate*

The scheme of analysis was designed so that, if tightly bound amino acids should be present in the sample, they would be concentrated mainly in the 6 N HCl hydrolysate. A portion of hydrolysate equivalent to 1 g of lunar sample was evaporated to dryness, redissolved in H_2O and charged on a 10 ml Dowex 50 (H^+) ion exchange column. The column was eluted with H_2O and 2 N NH_4OH according to procedures modified from DEGENS and REUTER (1964). The NH_4OH eluate was evaporated to dryness. The residue was dissolved in H_2O, the pH of the solution adjusted to 2, and charged on an automated amino acid analyzer (Beckman 120 C). About 0·1 nm each of glycine, alanine and serine was detected. In a repeat analysis of the sample the same amount of glycine and alanine was found, but serine was absent. Analysis of the sand blank showed about 0·1 nm of glycine. A portion of the NH_4OH eluate equivalent to 1 g of lunar sample was examined for amino acids by gas chromatographic techniques as described before (ROACH and GEHRKE, 1969). No common amino acids were detected, therefore, the presence in the lunar sample of glycine and alanine as determined by ion exchange chromatography could not be confirmed by gas–liquid chromatography. Two gas chromatographic peaks were found, however, which did not correspond in retention times to any common amino acids but did have retention times similar to peaks observed during the gas–liquid chromatography of the 1 N HCl hyrolysate.

5. *Gases evolved during hydrolysis*

Figure 3 shows the system in which gases evolved during hydrolysis were entrained in a stream of nitrogen and passed through a series of traps. Carbon dioxide was to be frozen out in the liquid nitrogen trap. A small quantity of nitrogen was condensed in the liquid nitrogen trap during hydrolysis and was subsequently removed at liquid nitrogen temperatures. The volume of the residual condensed gases was measured manometrically at ambient temperatures. No carbon dioxide was detected. The condensation of nitrogen within this trap and its subsequent evaporation may have resulted in the loss of small volumes of carbon dioxide, if present. This method of trapping, therefore, does not appear to be useful in situations where only traces of carbon dioxide may be released during acidification. Further experiments indicated that if carbon dioxide from carbonate is present, it must be in concentrations less than 5 μg/g.

Hydrogen sulfide, evolved during hydrolysis, was removed by precipitation in a 10 per cent silver nitrate solution. The precipitated silver sulfide was collected on a pre-weighed Millipore filter and thoroughly washed with 1:1 ammonium hydroxide solution and finally with water. The sulfide was dried at 105°C, weighed and quantitatively converted to sulfur dioxide for mass spectrometric analysis (KAPLAN and SMITH, 1970). The concentration of sulfide in the sample was about 700 μg/g with an average $\delta^{34}S$ (relative to Canyon Diablo meteorite) of +8·2. Sulfur isotopic compositions will be discussed in detail later in this report.

6. *Residue*

The residual of the sample remaining after treatment with 7% BF_3 in methanol weighed 37·6 g. An aliquot of this residue was dried overnight in an air oven at 105°C and was combusted in oxygen at 1050°C. The resulting CO_2 was measured manometrically. The total carbon concentration of the residual sample was greater than 3000 μg/g. Another sample was dried under vacuum for 50 hr at 150°C and combusted in a similar manner. Its carbon content was determined to be 1300 μg/g. Thus it appears that treatment of the sample with BF_3 in methanol caused an enrichment of carbon which could not be removed by vacuum at 150°C.

CARBON

The determination of the total carbon and the chemical state of the majority of the carbon in this sample of lunar fines are described in detail in part IV. A summary of this information follows:

The total carbon concentration in this lunar sample was 143 and 170 μg/g in two determinations. These values compare favorably with those reported by MOORE *et al.* (1970). Carbon monoxide and carbon dioxide are major products resulting from pyrolysis under vacuum at temperatures ranging from 150 to 750°C.

The lunar sample was also examined by R. D. JOHNSON and C. C. DAVIS, Ames Research Center, by pyrolysis in a stream of hydrogen and helium. This experiment measured the presence of carbon compounds which volatilized at 800°C and burned in a hydrogen flame causing an ionization current response in a flame ionization detector. The resulting single-peak area was compared with that from calibrated standards, corrected for the blank and converted to carbon concentrations (JOHNSON and DAVIS, 1970). The average results obtained for the lunar sample were compared with results from sand blanks and the Pueblito de Allende Meteorite:

Sand blank control	4·9 ± 6·5 μg/g
Pueblito de Allende	14·4 ± 3·2 μg/g
Lunar sample fines	39·5 ± 8·3 μg/g

The amount of pyrolyzable organic compounds in this lunar sample is higher than the reported preliminary value estimated by the same technique to be less than 10 μg/g (LSPET, 1969).

When the silicate matrix of the lunar sample was destroyed by treatment with hydrofluoric and hydrochloric acids the amount of carbon in the residue, which may consist of highly polymeric organic matter, acid resistant carbides and elemental carbon, was between 30 and 42 μg/g.

Acid hydrolysis of the lunar sample resulted in the evolution of low molecular weight C_1 through C_4 hydrocarbons. This discovery plus mineralogic evidence (ANDERSON *et al.*, 1970) suggests that carbides are present in the sample. In parallel experiments with a sample of Mighei carbonaceous chondrite and cohenite (Fe_3C) from Canyon Diablo iron meteorite similar results were obtained. Hydrocarbons were identified by gas chromatography and mass spectrometry. The amount of carbon accounted for as carbides was about 20 μg/g.

Although it is not possible at this time to account for all of the carbon in the lunar fines, it appears that a significant portion of the carbon is present in a form which on pyrolysis gives oxides of carbon and as carbides.

Isotopes

Both carbon and sulfur isotopic compositions were determined for this sample of lunar fines.

Carbon Isotopic Composition

The $\delta^{13}C$ composition of the total carbon of this sample was determined to be $+20$ per mil relative to the PDB standard. This $\delta^{13}C$ value is anomalously heavy compared to total meteoritic carbon which ranges from -4 to -25 per mil and terrestrial carbon with a range of $+5$ to -60 per mil. The carbon of this lunar sample is not as enriched in ^{13}C as the carbonate phase of meteorites which seems to have a range from $+40$ to $+70$ per mil. For a complete discussion of this and other isotopic determinations on lunar sample see Kaplan and Smith (1970).

Sulfur Isotopes

Gases evolved during the hydrolysis steps of the general scheme of analysis (Fig. 1) contained sulfides with unusually heavy isotopic compositions. This finding prompted additional experiments. Total sulfur was determined on a separate 3 g sample of lunar fines by oxidation of sulfides to sulfates with bromine and aqua regia. After the metal (iron) hydroxide was precipitated with ammonia, the solution was filtered, re-acidified and brought to boiling. Barium chloride solution was added, and the precipitated barium sulfate was removed by filtration after standing overnight. The barium sulfate was quantitatively converted to barium sulfide by reduction with graphite powder at 1050°C for 1 hr in a quartz crucible under nitrogen. After the contents of the crucible cooled, they were dispersed in de-oxygenated water and rapidly filtered into 5 per cent silver nitrate solution to convert the sulfide to silver sulfide which was oxidized to sulfur dioxide for mass spectrometric analysis.

Sulfides in samples for the Canyon Diablo, Murray and Pueblito de Allende meteorites were reacted with either 85 per cent phosphoric acid or 2 N hydrochloric acid under vacuum (Kaplan and Smith, 1970). The sulfide gas generated was precipitated as silver sulfide which was converted to sulfur dioxide for mass spectrometric analysis. The concentration and isotopic composition of sulfur in the lunar fines and in three meteorites are given in Table 3. The sulfur content of the lunar fines, whether determined by acid hydrolysis or oxidation, varied little, with contents from 640, 670 and 680 μg/g being found by three separate experiments.

Values for $\delta^{34}S$ show very high enrichment of the heavy isotope in the fines compared to meteoritic sulfur. In general, values for meteoritic troilite produced by HCl treatment fall within ± 1 per mil of a Canyon Diablo standard (Kaplan and Hulston, 1966). The data here show reasonably consistent results for the lunar material analyzed by acid treatment and by aqua regia oxidation, but a somewhat lower $\delta^{34}S$ for the sample treated by phosphoric acid. To determine if this difference was due to some experimental artifact, two intact meteorites (Murray and Pueblito de Allende) and troilite from Canyon Diablo meteorite were reacted with HCl or H_3PO_4. The results in Table 3 show that variation can occur, both in amounts of sulfide released and in isotope ratios. This finding is interpreted as partly due to inhomogeneity and partly to multiple forms of metal sulfides being differentially soluble in the two acids. In this set of experiments, the greatest variations for $\delta^{34}S$ is +1·1 per mil relative to the standard. Therefore, the values falling around +8 per mil are considered to be real and the value of +5·4 per mil obtained by H_3PO_4 treatment represents heterogeneity, e.g. possible inclusion of some larger grain fragments. Examination of more samples of lunar fines is necessary before an explanation of these observations can be offered. Certainly, the fines are heavily enriched in ^{34}S. The degree of enrichment apparently is not constant.

Table 3. Concentration and isotopic composition of sulfur in lunar fines, Pueblito de Allende and Murray Meteorites and troilite from Canyon Diablo meteorite

Sample	Sample Wt (g)	Acid	$\delta^{34}S$ (relative to Canyon Diablo Meteorite standard)	S content μg/g
Lunar fines	54·6	1 N HCl	+8·6	260
	54·6	*6 N HCl	+7·4	412
	1·79	85% H_3PO_4	+5·4	680
	3·01	Aqua regia	+8·2	640
Pueblito de Allende	107·6	1 N HCl	+1·1	410
	107·6	*6 N HCl	+1·1	5900
	0·50	2 N HCl	+0·9	19,900
	0·48	85% H_3PO_4	+0·7	10,200
Murray	0·28	2 N HCl	−0·1	2600
	0·36	85% H_3PO_4	+0·8	4600
Canyon Diablo Troilite	0·06	2 N HCl	+0·5	105,900
	0·04	85% H_3PO_4	+0·2	82,600

* Hydrolysis with 6 N HCl followed initial treatment with 1 N HCl as described in text.

Mass Spectrometry

This section refers only to the high resolution mass spectrometric investigation of the bulk lunar sample before any chemical processing took place. The experiment was designed to determine the presence and structure of volatile organic material which could be analyzed by directly inserting a 7 mg sample on a probe into the ion source of a CEC 21-110 high resolution mass spectrometer and pyrolyzing at temperatures ranging from 80 to 300°C. Ion fragments containing C, H, N, O and S were observed, but their presence was only marginal over background providing little information for interpretational purposes.

Mineralogy and Microfossils

Mineralogical studies were conducted primarily to determine the mineral matrix in which the carbon compounds occur. An aliquot weighing about 0·5 g of lunar fines was examined. The sample was sieved through a set of nylon sieves and the +100 mesh fraction was studied under a stereo microscope. Six fractions consisting largely of feldspar (and anorthositic rock fragments), pyroxene, olivine, rock fragments, glass spherules and glass fragments were handpicked for preparation of polished thin sections. These sections were studied in detail with an optical microscope and an electron microprobe X-ray analyzer. These studies reveal that the sample consists of fragments ranging from fine-grained to coarse-grained basaltic type rocks to anorthositic rocks containing largely anorthite and augite. The sample also contains the minerals plagioclase ranging in composition from labradorite to anorthite, and pyroxene ranging in composition from augite to subcalcic augite to pigeonite and pyroxmangite. Olivine occurs in accessory amounts. Ilmenite is the dominant opaque phase, but minor amounts of metallic nickel–iron of meteoritic origin and of troilite were also observed. The presence of metallic nickel–iron of typical meteoritic composition suggests that at least part of the loose surface material is of impact origin. However, the amount of meteoritic material present in the sample is small (probably <1% by weight) and, hence, it is concluded that contributions from common meteorites have not appreciably altered the bulk composition of the loose surface fines. Glass spherules and irregularly shaped glass fragments are common. Composition of the glass spherules varies considerably (Table 4) and is probably in large part due

Table 4. Electron microprobe analyses of glass spherules separated from loose surface fines (wt.%)

	Reddish brown	Reddish brown	Light brown	Brown	Opaque	Opaque	Dark brown
SiO_2	38·4	40·0	43·5	43·0	39·8	37·0	39·4
Al_2O_3	10·0	12·7	12·9	16·7	12·0	5·1	14·5
Cr_2O_3	0·45	0·35	0·32	0·20	0·25	0·65	0·29
TiO_2	6·9	7·2	4·3	4·8	8·3	10·7	7·3
FeO	19·8	16·9	13·6	12·1	17·6	25·0	15·4
MgO	13·9	10·4	14·7	8·9	9·3	14·4	8·6
MnO	0·20	0·16	0·16	0·11	0·17	0·24	0·14
CaO	9·7	11·4	9·9	11·8	12·5	7·3	12·3
Na_2O	0·03	0·07	0·11	0·58	0·56	0·25	0·37
K_2O	<0·02	0·04	<0·02	<0·02	0·08	0·07	0·08
ZrO_2	<0·02	0·04	<0·02	<0·02	0·06	0·05	0·05
P_2O_5	0·03	0·03	0·03	0·03	0·09	0·04	0·07
Total	99·41	99·29	99·52	98·22	100·71	100·80	98·50

to differences in composition of the parent rocks and to different mixing ratios of the dominant minerals plagioclase, pyroxene and ilmenite. The minerals observed in the sample do not appear to be altered by chemical weathering. Furthermore, no evidence of significant amounts of hydrous minerals was found in any of the sections studied. The textures and compositions of the rock fragments studied are typical of comparatively rapidly cooled rocks and are analogous to terrestrial volcanic or hypabyssal rocks.

Optical and electron microscope studies of the sample also showed a total absence of structures that can be interpreted as biological in origin (Barghoorn *et al.*, 1970; Schopf, 1970).

Discussion and Summary

From this comprehensive organic analysis of fines from the Sea of Tranquillity no evidence was found for the presence of the common classes of organic compounds which have been sought previously in many organic geochemical studies and chemical evolution experiments. At the nanogram level of detection no high molecular weight alkanes, aromatic hydrocarbons or fatty acids could be found. At the same time there was no evidence for sugars, purines and pyrimides. Although two amino acids were suggested by chromatograms from an amino acid analyzer, these compounds could not be verified by gas chromatographic techniques. Only a trace (10^{-4} μg/g) of a pigment behaving chemically like a porphyrin was found, and this material may be a product from rocket exhaust.

Compounds which appear to contain carbon bound to silicon were recovered during hydrolysis with hydrochloric acid. These compounds could not be made sufficiently volatile for gas–liquid chromatography unless they were both acylated and esterified as N-trifluoroacetyl *n*-butyl derivatives. Numerous experiments were performed (part III) in order to trace the source of these organosiloxanes. That the compounds resulted from reaction of inorganic lunar silicates with the derivatizing reagents is possible, but the data collected thus far suggest that the organosiloxanes may be indigenous to the lunar fines. Their significance, however, is not clear. These compounds may have resulted from bombardment of the silicates at the lunar surface by carbon in the solar wind.

Much of the carbon in the lunar sample appears to be present in a form which on pyrolysis yields oxides of carbon. Chemical evidence strongly suggests that a portion of the carbon in the lunar sample is there as carbides; mineralogical evidence supports this contention (ANDERSON *et al.*, 1970). Because carbides are found in meteorites, their discovery on the moon is not completely unexpected. LATIMER (1950) noted earlier the importance of carbides in the formation of the earth. From the point of view of chemical evolution, carbides may have played an important role in the formation of simple organic compounds under natural conditions of hydrolysis on the primitive earth.

The total carbon in the lunar sample of about 160 μg/g can be the carbon in the primordial dust cloud (MASON, 1966), or carbon from the solar wind or from meteoritic impact (MOORE *et al.*, 1970). Therefore, the carbon concentration of this lunar sample is not unexpectedly high or low. The carbon isotopic composition of this carbon is unusual, however. The $\delta^{13}C$ values of +20 relative to the PDB standard for this sample lies between terrestrial carbon and meteoritic carbonate carbon. It will be of interest in the future to measure the isotopic composition of the individual major carbon species in the lunar sample.

The hydrogen sulfide generated during the hydrolysis of this lunar sample was trapped, and the sulfide examined. Its concentration was measured at about 700 μg/g. The source of much of the sulfide was troilite which was identified in the mineralogical study of the sample. The isotopic composition of the sulfide was unusually heavy with the $\delta^{34}S$ measuring as high as +8·2 per mil relative to the Canyon Diablo meteoritic standard. Sulfides in meteorites have $\delta^{34}S$ values falling within ± 1 per mil. Thus, both the carbon and sulfur isotopic abundances are heavy relative to most other cosmological occurrences of these elements. Explanation of these observations must await further studies.

Although no significant amounts of common organic compounds were found in this exercise, the concept of performing chemical analyses in a single system of glassware without transfer of sample may have useful application in other geochemical studies. Certainly contamination due to sample handling can be reduced with the system that was described. Nevertheless, inadvertent contamination at the nanogram level was noticed in analyses for hydrocarbons, fatty acids and amino acids. The sources of these contaminations have not as yet been ascertained. The use of blanks analyzed in parallel with the lunar sample provided a measure of the background contamination level.

This lunar sample analysis has succeeded in defining the chemical state of some of the carbon in a sample from the lunar surface. This study, we hope, represents only a beginning. Future work will have different priorities concerning which organic analyses will be most meaningful. These future studies will be directed toward understanding the relationships among the carbon species and in ascertaining the source and possible cosmologic history of carbon.

Acknowledgments—We are grateful to the other members of the NASA Ames Research Center Consortium: R. Johnson, G. Pollock, D. Philpott, I. Kaplan, J. Schopf, C. Gehrke, G. Hodgson, B. Halpern, A. Duffield, K. Krauskopf, E. Barghoorn and H. Holland; to Jerri Mazzurco and JoAnn Williams for their technical assistance; to Don Moody, Dan Dempsey, Harry Horn and Bob Bettendorff for constructing the glassware; to John Reitman, Ray Einberger and Joe Ribera for machine fabrication; to F. Busche for electron microprobe analysis done under NASA contract NAS9-9365 (K. Keil, Principal Investigator).

References

Abdel-Akher M., Hamilton J. K. and Smith F. (1951) Reduction of sugars with sodium borohydride. *J. Amer. Chem. Soc.* **73,** 4691–4692.

Albersheim P., Nevins D. J., English P. D. and Carr A. (1967) A method for analysis of sugars in plant cell wall polysaccharides by gas–liquid chromatography. *Carbohyd. Res.* **5,** 340–345.

Anderson A. T., Crewe A. V., Goldsmith J. R., Moore P. B., Newton J. C., Olsen E. J., Smith J. V. and Wyllie P. J. (1970) Petrologic history of moon suggested by petrography, mineralogy, and crystallography. *Science* **167,** 587–590.

Barghoorn E. S., Philpott D. and Turnbill C. (1970) Micropaleontological study of lunar material. *Science* **167,** 775.

Barghoorn E. S. and Schopf J. W. (1966) Microorganisms three billion years old from the Precambrian of South Africa. *Science* **152,** 758–763.

Degens E. T. and Reuter J. H. (1964) Analytical techniques in the field of organic geochemistry. In *Advances in Organic Geochemistry. Proc. Int. Meeting in Milan* 1962, (editors U. Columbo and G. D. Hobson), pp. 389–390. Macmillan.

Donn W. L., Donn B. D. and Valentine W. G. (1965) On the early history of the earth. *Geol. Soc. Amer. Bull.* **76,** 287–306.

Gehrke C. W. and Ruyle C. D. (1968) Gas–liquid chromatographic analysis of nucleic acid components. *J. Chromatogr.* **38,** 473–491.

Hodgson G. W., Peterson E. and Baker B. L. (1969) Trace porphyrin complexes: Fluorescence detection by demetallation with methanesulfonic acid. *Mikrochim. Acta* (*Wein*) 805–814.

Hoering T. C. (1965) The extractable organic matter in Precambrian rocks and the problem of contamination. *Carnegie Inst. Wash. Yearb.* **64,** 215–218.

Johnson R. D. and Davis C. C. (1970) Pyrolysis-hydrogen flame ionization detection of organic carbon in a lunar sample. *Science* **167,** 759–760.

Kaplan I. R. and Hulston J. R. (1966) The isotopic abundance and content of sulfur in meteorites. *Geochim. Cosmochim. Acta* **30,** 479–496.

Kaplan I. R. and Smith J. W. (1970) Concentration and isotopic composition of carbon and sulfur in Apollo 11 lunar samples. *Science* **167,** 541–543.

KING E. A., SCHONFELD E., RICHARDSON K. A. and ELDRIDGE J. S. (1969) Meteorite fall at Pueblito de Allende, Chihuahua, Mexico: Preliminary information. *Science* **163,** 928–929.

LATIMER W. M. (1950) Astrochemical problems in the formation of the earth. *Science* **112,** 101–104.

LAWLOR D. L. and ROBINSON W. E. (1967) Fatty acids and *n*-alkanes in Green River Oil Shale: Changes with depth. *Fuel Division, Amer. Chem. Soc.* 153rd National Meeting, Miami, Florida, pp. 480–486, Preprint.

LSPET (LUNAR SAMPLE PRELIMINARY EXAMINATION TEAM) (1969) Preliminary examination of lunar samples from Apollo 11. *Science* **165,** 1211–1227.

MASON B. (1966) *Principles of Geochemistry*, 3rd edition. John Wiley.

MEINSCHEIN W. G. and KENNY G. S. (1957) Analysis of a chromatographic fraction of organic extracts of soils. *Anal. Chem.* **29,** 1153–1161.

METCALF L. D. and SCHMITZ A. A. (1961) The rapid preparation of fatty acid esters for gas chromatographic analysis. *Anal. Chem.* **33,** 363–364.

MOORE C. B., LEWIS C. F., GIBSON E. K. and NICHIPORUK W. (1970) Total carbon and nitrogen abundances in lunar samples. *Science* **167,** 495–497.

ORO J. and NOONER J. W. (1967) Aliphatic hydrocarbons in Precambrian rocks. *Nature* **213,** 1082–1085.

PONNAMPERUMA C. and GABEL N. W. (1968) Current status of chemical studies on the origin of life. *Space Life Sci.* **1,** 64–96.

ROACH D. and GEHRKE C. W. (1969) Direct esterification of the protein amino acids gas–liquid chromatography of N-TFA *n*-butyl esters. *J. Chromatogr.* **44,** 269–278.

SAGAN, C. (1961) Organic matter on the moon. Nat. Acad. Sci., Nat. Res. Council, Publ. 757, 49p.

SCHOPF J. W. (1970) Micropaleontological studies of lunar samples. *Science* **167,** 779–780.

SCHOPF J. W. and BARGHOORN E. S. (1967) Alga-like fossils from the early Precambrian of South Africa. *Science* **156,** 508–512.

SHAPIRA J. (1969) Identification of sugars as their trifluoroacetyl polyol derivatives. *Nature* **222,** 792–793.

SWEELEY C. C., BENTLEY R., MAKITA M. and WELLS W. W. (1963) Gas–liquid chromatography of trimethylsilyl derivatives of sugars and related substances. *J. Amer. Chem. Soc.* **85,** 2497–2507.

Proceedings of the Apollo 11 Lunar Science Conference, Vol. 2, pp. 1829 to 1844.

Carbon compounds in lunar fines from Mare Tranquillitatis—II. Search for porphyrins

GORDON W. HODGSON
Exobiology Research Group, University of Calgary, Calgary 44. Alberta, Canada

EDWARD BUNNENBERG and BERTHOLD HALPERN
Stanford University, Stanford, California 94305

and

ETTA PETERSON, KEITH A. KVENVOLDEN and CYRIL PONNAMPERUMA
Chemical Evolution Branch, NASA Ames Research Center, Moffett Field, California 94035

(*Received* 28 *January* 1970; *accepted in revised form* 20 *February* 1970)

Abstract—Lunar fines contain porphyrin-like pigments as indicated by fluorescence spectrometry and analytical demetallation. Major fluorescence excitation at 390 nm was obtained for 600–690 nm emission. The abundance of porphyrin-like material was estimated to be about 10^{-4} μg/g. Similar pigments were found in exhaust products from a lunar descent engine. Although the infall of meteoritic dust to the lunar surface is appreciable and may be expected to contain considerable carbon and associated organic compounds including porphyrins, the data suggest that most if not all of the content of porphyrin-like pigments of the lunar samples was probably introduced during landing of the lunar module. If the pigments in the lunar fines are indeed the product of rocket combustion, a novel synthesis has taken place. Analogous types of syntheses perhaps occur in the cosmos where simple compounds of carbon, hydrogen and nitrogen interact at high temperatures.

INTRODUCTION

PORPHYRINS in meteorites were once regarded as a strong criterion for the existence of extraterrestrial life (BERNAL, 1961), but more recent work has shown that such pigments can arise in experiments simulating prebiotic planetary conditions (HODGSON and PONNAMPERUMA, 1968), and the existence of prophyrins in extraterrestrial samples now has a much broader implication. The appearance of simple organic compounds in interstellar space (SNYDER *et al.*, 1969; CHEUNG *et al.*, 1969), the condensation of graphite from cool stars (DONN *et al.*, 1968) and evidence for molecular absorption of organic compounds in interstellar space (JOHNSON, 1967) suggest that astrochemical synthesis of organic matter of considerable complexity is within reason. The current study based on analysis of lunar fines produced evidence that prophyrin-like pigments were present. These pigments may have been synthesized from unsymmetrical dimethylhydrazine at the temperatures of rocket engines—2000 to 3000°C. Analogous types of syntheses may occur naturally in stellar atmospheres and interstellar space.

This report presents a part of the results obtained by a group of investigators established as the NASA Ames Research Center Consortium to analyse the lunar sample labelled "10086-3 Bulk A fines". Results on the search for porphyrins by the Consortium were recently published in abbreviated form (HODGSON *et al.*, 1970), and the object of this report is to provide more details of the data and to discuss more fully the implications of the findings. Related analyses for porphyrins in lunar dust were

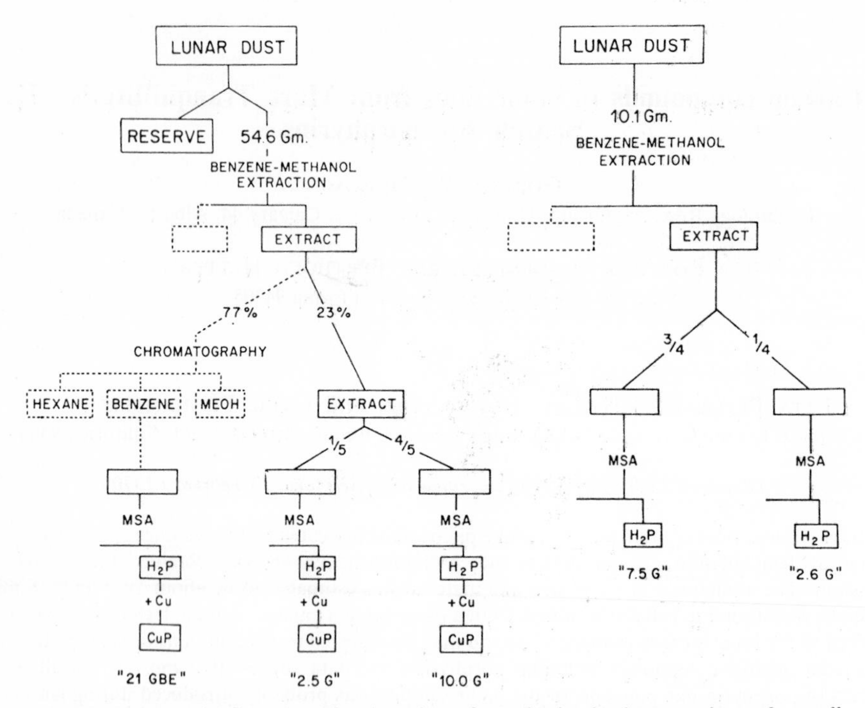

Fig. 1. Left, partial flow sheet for detection of porphyrins in three portions of overall extract obtained for comprehensive organic analysis of lunar sample; right, flow sheet for confirmatory analysis of porphyrins in second sample of lunar fines.

carried out by RHO *et al.* (1970) who used a sequential extraction procedure and gross demetallation by methanesulfonic acid, but no porphyrins were observed.

ANALYTICAL SCHEME

Porphyrins were sought in lunar fines by absorption spectrometry, magnetic circular dichroism (MCD) and fluorescence spectrometry. In addition to passive observations of such spectra, several chemical and physical transformations were carried out on extracts of lunar fines to test the identification of lunar pigments as porphyrins. The analytical approach was briefly outlined by PONNAMPERUMA *et al.* (1970), and in Part I of this series (KVENVOLDEN *et al.*, 1970).

A flow sheet relating the analyses for porphyrins to the overall analytical scheme is given in Fig. 1. In the course of examining the lunar samples for porphyrins, five specific portions of lunar extracts were examined as outlined in Table 1. In addition to these, numerous blank runs were made on solvents, reagents and glassware; parallel analyses were done on two sand blanks and on a sample of

Table 1. Description of extracts analysed

Item	Extraction	Portion for porphyrin analysis	Equivalent lunar fines (g)	Sample designation
1	First: 54·6 g	1/5 aliquot of 23% of extract	2·5	2·5 G
2		4/5 aliquot of 23% of extract	10·0	10·0 G
3		1/2 of benzene eluate of 77% of extract	21·0	21 GBE
4	Second: 10·1 g	3/4 aliquot of total extract	7·5	7·5 G
5		1/4 aliquot of total extract	2·6	2·6 G

the Pueblito de Allende carbonaceous chondrite. Similar analyses were performed on substances pertaining to the landing of the lunar module: unsymmetrical dimethylhydrazine, and rocket exhaust products collected in traps A and E described by SIMONEIT *et al.* (1969). With the exception of some of the control analyses and the MCD measurements, all of the work was done in a "clean" laboratory which was established specifically for analyses of lunar samples.

Absorption and fluorescence spectrometry have been extensively used in previous studies of porphyrins in numerous terrestrial and extraterrestrial samples (HODGSON *et al.*, 1968; KVENVOLDEN and HODGSON, 1969; HODGSON and BAKER, 1969). The fluorescence technique has been especially useful in detecting nanogram concentrations of porphyrins in geochemical substances (HODGSON *et al.*, 1969). The technique involves incremental demetallation of the extracted porphyrin complexes with methanesulfonic acid and the measurement of excitation and emission spectra at each demetallating step. The use of magnetic circular dichroism is novel in its application to the identification of porphyrins. This type of spectrometry was applied to lunar extracts in the hope that confirmation could be obtained for the presence of the porphyrin structure suggested from the fluorescence spectra. Because of this novel application of MCD it will be described in some detail.

Absorption spectrometry

Absorption spectra were obtained directly on lunar extracts using a Cary 14 spectrophotometer with semi-micro cells and a 0–0·1 optical density slidewire.

Magnetic circular dichroism

Magnetic circular dichroism is a sensitive nondestructive spectroscopic method for the detection of metalloporphyrins, which show three prominent absorption bands in the 350–700 nm region. The positions of these three bands depends on the metal and on the ring substituents. Adopting the nomenclature of PLATT (1956), the transition of lowest energy, at about 570 nm, is designated Q_{0-0}; the transition at about 535 nm is a vibrational overtone of the 0–0 band; and the much more intense transition around 400 nm is designated B. The analytically important features in the MCD spectra of metalloporphyrins are the shapes of the MCD bands associated with the Q_{0-0} and B transitions. The effective symmetry of metalloporphyrins is D_{4h} and all bands are degenerate. In a magnetic field the degeneracy is lifted, and one observes the S-shaped MCD bands characteristic of A terms (BUCKINGHAM and STEPHENS, 1966). Previous studies (STEPHENS *et al.*, 1966) as well as the reference spectra collected for this project (HODGSON and BUNNENBERG, unpublished) show that the magnitudes of the A terms of the B and Q_{0-0} bands are comparable even though the absorption coefficients of these bands differ by an order of magnitude. As a representative example, the MCD curve for Mg(II) deuteroporphyrin IX dimethyl ester is shown in Fig. 2. Some metalloporphyrins [e.g. Cu(II) porphine], however, exhibit considerably more intense A terms in the B band than in the Q band. The utility of MCD for detecting small amounts of metalloporphyrins derives from: (1) the intensity of the two prominent MCD bands, (2) their characteristic S-shape, and (3) the observation of two such bands in particular regions of the spectrum. It should be noted that MCD could also be used for the detection of metal-free porphyrins. However, since the MCD bands are much weaker a considerably larger sample would be required.

MCD data were displayed directly in analog form, and in addition were processed by computer in order to gain increased sensitivity. The most secure and conservative sampling mode was used for the lunar sample measurements. The visible portion (335–650 nm) of the spectrum was scanned. The raw data obtained from the lunar sample and the sand blank measurements were corrected for the reference blank (cell plus solvent). The resulting curves were smoothed and are presented in Fig. 3. Measurements were made on a specially constructed MCD instrument (to be described elsewhere) at ambient temperature in a magnetic field of 49,500 G. The light-path length of the cells was 1 cm.

Spectrofluorometry and demetallation

Spectrofluorometry was done with an unmodified Turner 210 instrument using standard stoppered 1 × 1 cm cells. For demetallation analysis, the procedures previously described were used (HODGSON *et al.*, 1969).

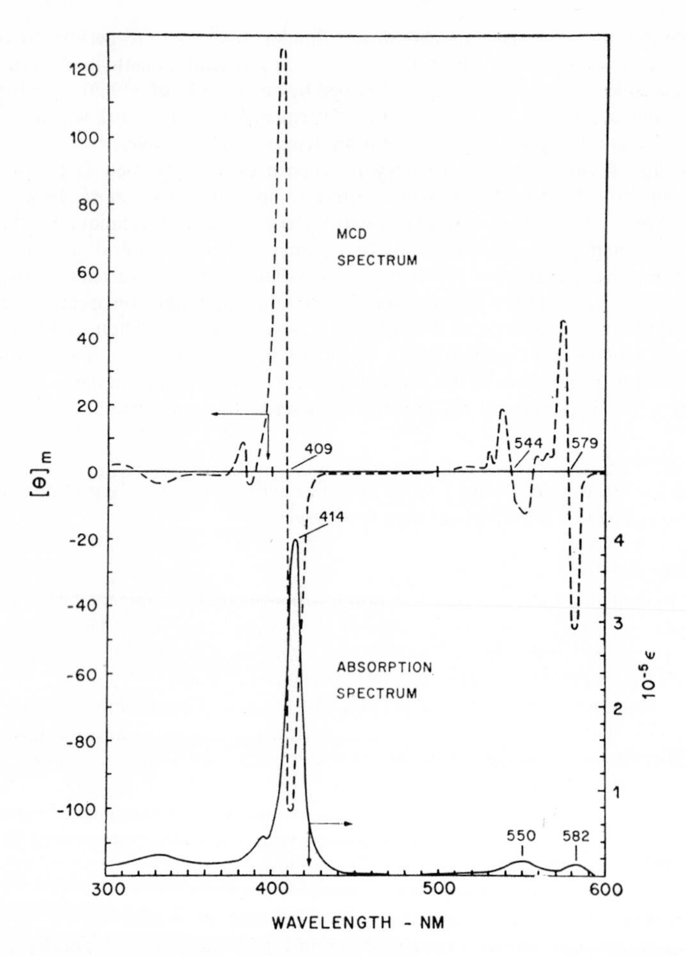

Fig. 2. Magnetic circular dichroism (MCD) and absorption spectra of Mg(II) deuteroporphyrin IX dimethyl ester (7×10^{-6} M) in benzene. A cell of 1-mm light-path was used. The ordinate, $[\theta]_M$, is molar magnetic ellipticity (deg cm^2 G^{-1} $decimole^{-1}$).

RESULTS

Abundance of specific pigmented compounds in extracts of lunar samples was so low that absorption spectrophotometry failed to reveal any spectral features in the range of porphyrin compounds, i.e. from 350 nm to 700 nm, from which it is concluded that porphyrins, if present, were less than 100 ng per sample. For 10 g samples, the indicated content was less than 10^{-2} μg/g. While magnetic circular dichroism and spectrofluorometry are substantially more sensitive than absorption spectrophotometry, it is clear that any data obtained by the latter methods were not far above their limits of detection, and for that reason were not of customary precision nor entirely free of spectral and chemical noise. In the light of this, a certain subjectivity in treating the data is to be expected, but the major findings are above such questions, with only auxiliary data open to variable interpretation. In some instances, extracts were

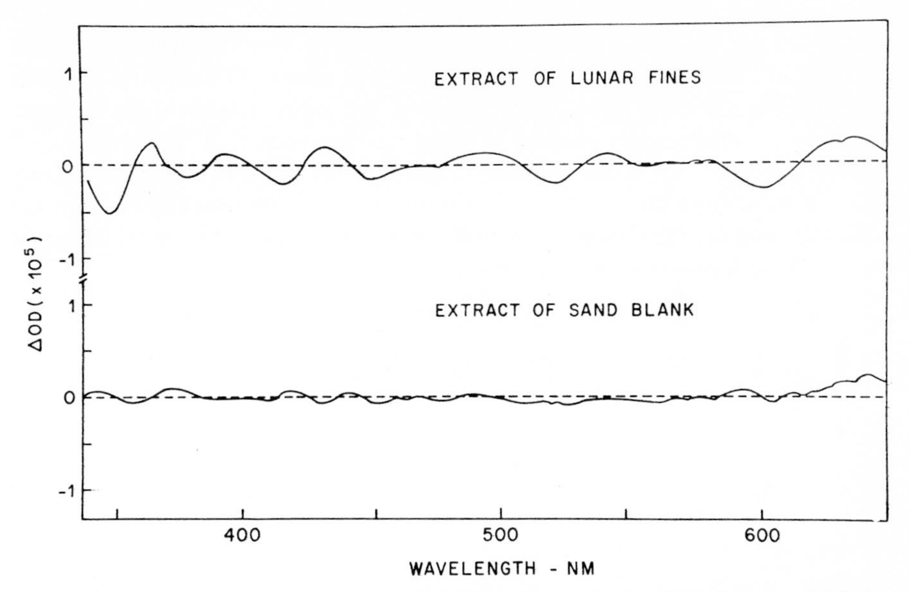

Fig. 3. Plot of magnetic circular dichroism data obtained from lunar and blank extracts as measured in a magnetic field of 49,500 G. The curves have been corrected for the cell plus solvent reference blank. A cell of 1-cm light-path length was used; solvent, methanol.

identified to the analysts only by a code designation, in an attempt to promote complete objectivity.

Absorption spectrophotometry

No evidence of Soret nor non-Soret bands was obtained.

Magnetic circular dichroism

A critical examination of the MCD curve of the lunar extract (in methanol) shown in Fig. 3 reveals that:

1. The *A* term in the vicinity of 570 nm is definitely not present.
2. The approximately equal spacing between the positive and negative signals in the 335–475 nm region strongly suggests that these signals are actually instrumental artifacts, notwithstanding their relatively greater intensities as compared to the MCD curve of the sand blank.

On the basis of the criteria established, we conclude that the amount of metalloporphyrins, if any were present in the lunar extract, must have been less than 7 ng. Thus, examination of the total of samples 2·5 G and 10·0 G before they were divided, i.e. the extract corresponding to 12·5 g of lunar soil, failed to show any spectral responses that could be attributed to porphyrins and their concentration in the lunar dust was therefore less than 6×10^{-4} μg/g.

Fluorescence spectrometry

Two extracts of lunar dust corresponding respectively to 12·5- and 10·1-g aliquots of the lunar fines showed considerable fluorescence, as illustrated in Fig. 4. For emission at 625 nm, excitation scans through the Soret region of the spectrum showed a major peak at about 460 nm, with a distinct peak at about 387 nm and a shoulder at about 482 nm. Similar scans were obtained for other emission wavelengths, and in general the practice was adopted of doing multiple excitation "cross scanning" through regions of interest in the emission range, in order to discern more sensitively any emission maxima which on direct observation were badly obscured by background noise. In this manner, families of excitation curves were obtained from which could be established the location of emission maxima.

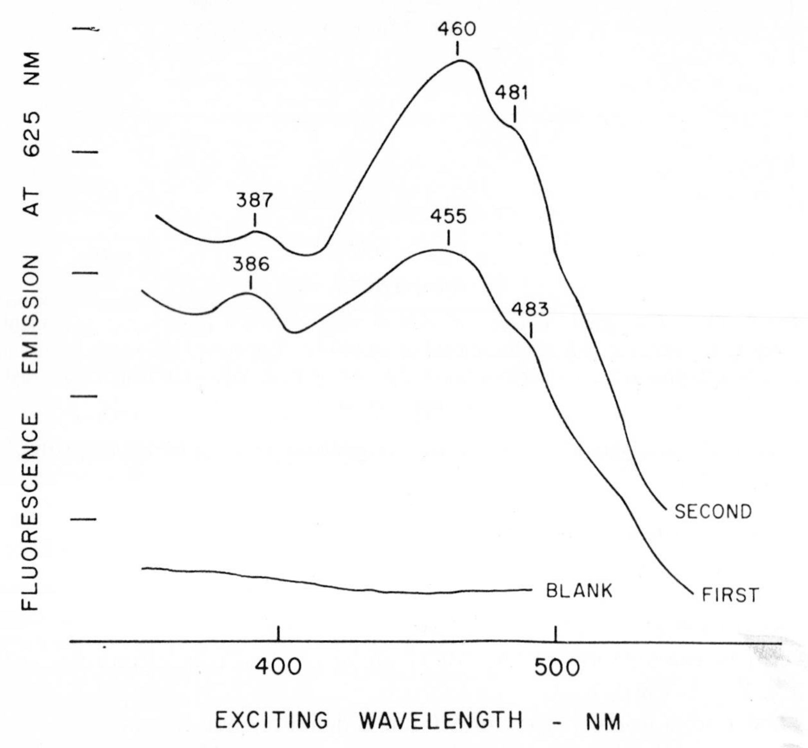

Fig. 4. Excitation spectra of lunar extracts. Solvent, methanol.

Porphyrins do not commonly have Soret absorption above 430 nm, and the excitation features at 460 and 481 nm were not believed to be due to porphyrins. The smaller peaks at 387 nm were tentatively regarded as possible Soret excitations of porphyrins, either free-base or complexed with diamagnetic metals. Complexes with paramagnetic metals were unlikely, since, if present, their Soret bands would have been detected in absorption spectrometry at the levels of fluorescence sensitivity involved.

While information on the chromatographic behavior of the observed pigments would have been helpful, paucity of material precluded such an approach. Some information, however, was available from the examination of solution 21GBE, equivalent to a 21-g sample, which had been chromatographed on silica gel (Part I of this series). Sample 21GBE was the benzene eluate, and it contained some of the same pigments but in substantially lower quantities, from which it is concluded that the pigments of Fig. 4 would have eluted with either n-hexane or methanol. Complexes of porphyrins in terrestrial geochemical samples are generally nickel, vanadyl or iron, and these compounds elute respectively with n-hexane–benzene (9/1 v/v), benzene and methanol (HODGSON *et al.*, 1968).

Analytical demetallation

Methanesulfonic acid was added incrementally to the extracted pigments in diethyl ether with fluorescence monitoring for the onset of demetallating reactions. Figure 5

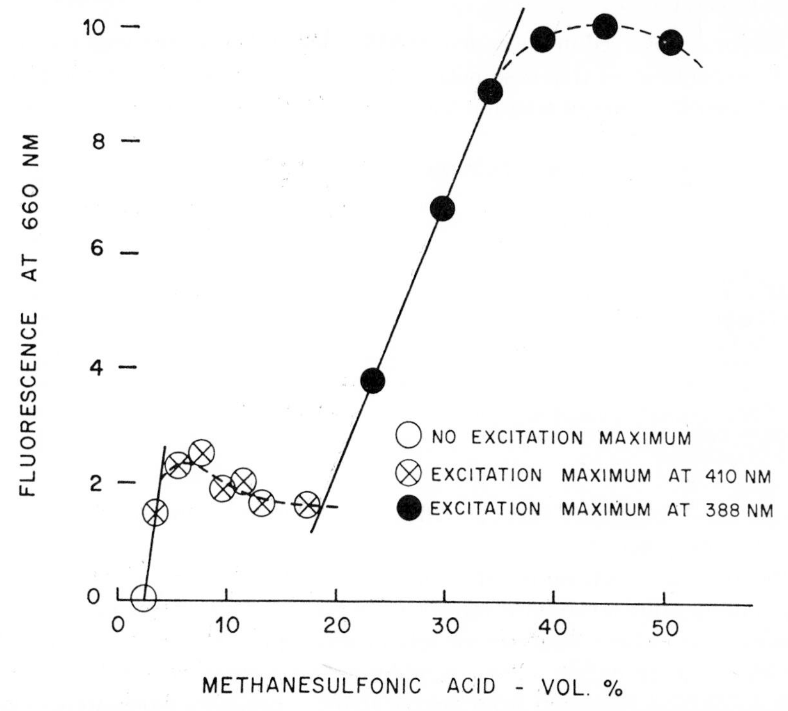

Fig. 5. Demetallation analysis of lunar extract "2·5 G" showing emergence of very weak fluorescence early in the reaction, followed by the major reaction at about 20% MSA.

shows data typical of this approach for the lunar samples. For sample 2·5 G, it shows a short initial period in which no change in fluorescence is noted, after which 2% methanesulfonic acid (MSA) produced a very small excitation feature at 410 nm for emission at 660 nm. This weak band remained unchanged in intensity until about 20% MSA had been added, at which point it was replaced by a much stronger excitation peak emerging at 390 nm. This increased in intensity until 30% MSA was reached, after which it remained nearly constant. Similar behavior was observed for the demetallation of authentic porphyrins of terrestrial rocks (HODGSON *et al.*, 1969) and for porphyrins of carbonaceous chondrites (HODGSON and BAKER, 1969). Of particular interest was the observation that the process involved two steps. The other four samples listed in Table 1 showed similar demetallation patterns with the same two-stage increase in fluorescence. Summary data for these samples are given in Table 2.

Demetallation at about 25% MSA for the emergence of a Soret band at 388 nm for 660 nm emission is not commonly observed for naturally occurring terrestrial complexes of prophyrins. Threshold values are normally higher, generally in the range above 40% MSA. Soret bands generally appear at longer wavelengths, commonly at 410 nm.

Free-base recovery

The response observed in the course of MSA demetallation analyses was probably due to the generation of free-base porphyrins, and an attempt was made to recover them for further spectral examination and chemical processing. Partition of the products of the MSA treatment between ether and saturated aqueous sodium acetate was used for this purpose.

Table 2. Demetallation of lunar pigments

Sample	Demetallation threshold (%MSA) Initial	Major	Soret position (nm) Early	Late
2·5 G	2	19	410	388
10·0 G	3	31	410	388
21 GBE	3	25	384	386
7·5 G	5	26	388	387
2·6 G	—	25	(400)	391

Note: "Initial" threshold in all instances was very weak, and estimation of the position of an emerging Soret band was difficult to make.
( ) uncertain.

In the ether layer, excitation attributable to porphyrins was barely discernable. Extraction of the ether layer, however, with 6 N HCl transferred free-base pigments to the aqueous layer where fluorescence was clearly detectable. For example, Fig. 6 shows a family of excitation curves in addition to an emission curve for presumed free-base porphyrins recovered from sample 10·0 G. Emission bands were evident at 600, 630 and 680 nm. Three-band emission spectra are not common for ordinary terrestrial geochemical porphyrins in acid solution, and it remains to be seen whether the observed pigments represent a single class of compounds or a group of two or more with different fluorescent spectra. Terrestrial geochemical porphyrins generally show a single intense emission band at about 660 nm in aqueous acid solution.

Fluorescence characteristics of free-base pigments recovered from the four other samples are shown in Table 3. These data again are different from those for terrestrial geochemical porphyrins in detail, viz. 389 vs. 410 nm for Soret excitation. Porphyrins of carbonaceous chondrites in HCl show a range of positions with a number of samples exciting at 390 nm, closely resembling the pigments of the lunar fines.

Transfer of the lunar pigments to glacial acetic acid for subsequent complexing with copper caused a shift of a few nanometers toward longer wavelengths, as shown in Table 3. This kind of behavior is not unexpected for free-base porphyrins.

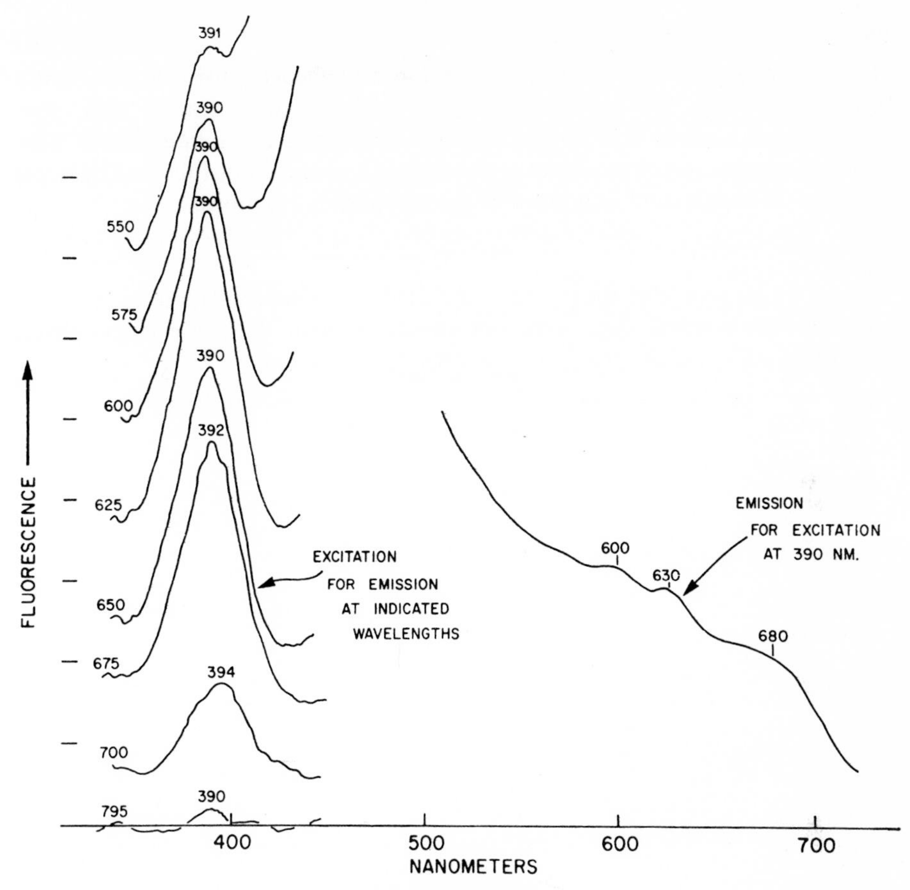

Fig. 6. Emission spectrum of lunar sample "10·0 G" showing fluorescence cross-plotted to confirm relationship between excitation at 390 nm and weak emission bands in the red. Solvent, 6 N HCl.

Table 3. Fluorescence of free-base porphyrins after demetallation

Sample	Ether	Solvent 6 N HCl	HOAc
2·5 G	385/660*	389/660	395/660
10·0 G	390/625	390/660 389/630	390/630
21 GBE	388/625 387/690	389/600	392/660
7·5 G	387/625	n/a†	n/a†
2·6 G	n/a	388/620	n/a

* 385/660: Excitation maximum at 385 nm for emission at 660 nm.

† Pigment apparently destroyed by trace of H_2S in HCl solution prepared from HCl gas.

Complexing with copper

To further demonstrate behavior of the lunar pigments consistent with that of authentic porphyrins, the recovered free-base pigments were reacted with cupric acetate in glacial acetic acid to produce non-fluorescing metal complexes. Thus, Fig. 7 shows excitation in the Soret region before the reactions, followed by apparently complete suppression as a result of the complexing reactions.

Abundance

On the assumption that the pigments detected in the demetallation analyses were porphyrins—the spectral data were not consistent with those for azaporphyrins

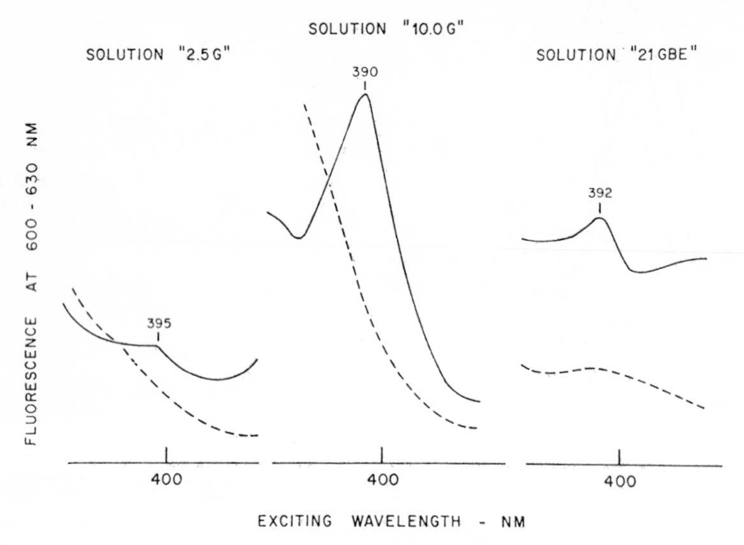

Fig. 7. Fluorescence spectra before (solid lines) and after (dotted lines) complexing with copper, in glacial acetic acid.

(ELVIDGE, 1956), nor phthalocyanines (MOSER and THOMAS, 1963)—the amount present in the larger samples was estimated to be 1 ng. The smaller samples showed smaller amounts, and the general concentration was therefore indicated to be 10^{-4} $\mu g/g$ in the lunar fines.

Blanks

(i) Solvents and reagents: diethyl ether was purified by distillation; MSA was used as received. Considerable variation was noted, however, in the quality of the MSA as received from the supplier (Eastman #6320) and occasionally freshly-opened bottles could not be used because of the appearance of strong excitation at 380–390 nm in the course of blank runs involving only ether and MSA. In the best blank runs, a small excitation was always observed, and this could not be prevented by ordinary purification procedures involving adsorption or distillation. No evidence of spurious fluorescence was noted for blank analyses involving the other reagents: hydrochloric acid, sodium acetate, cupric acetate and distilled water.

(ii) Complete procedural blanks were run in parallel with analyses of lunar extracts. For lunar samples 2·5 G and 10·0 G a single sand blank was used. It consisted of Ottawa sand fired at 1000°C for 48 hr in the atmosphere. Another aliquot of the same sand blank solution was used with sample 21GBE. A further sand blank prepared in the same manner was used in connection with lunar samples 2·6 G and 7·5 G. Although, as noted above, minor excitation at 380–390 nm was always present in demetallation analyses at high concentrations of MSA, the data obtained for the lunar extracts were always significantly stronger. This is clearly illustrated in Fig. 8

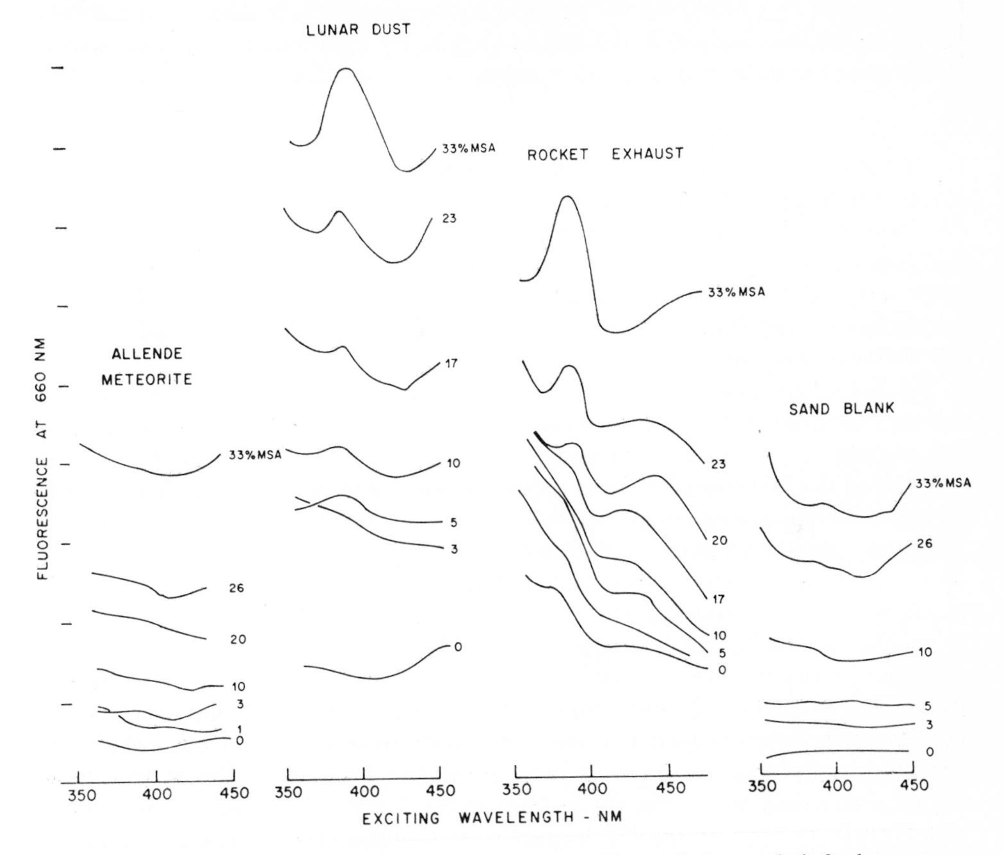

Fig. 8. Comparison of Allende and sand blanks with demetallation analysis for lunar and rocket exhaust determinations.

by a direct comparison of the lunar results with those for a sand blank, which shows that the peak height for the lunar extract was more than 12 times greater than that for the blank.

(iii) Figure 8 also illustrates the difference between the lunar results and those obtained for the demetallation of an extract of a sample of the Pueblito de Allende carbonaceous chondrite included in the same overall analytical program (Part I of this series). Although some samples of this meteorite showed detectable amounts of porphyrins (HODGSON *et al.*, unpublished), the particular sample included in the present study gave demetallation results in which no evidence of the pigments was observed. Thus, the Allende results provided a further check on the control of spurious substances in the course of the analysis for porphyrins.

While the behavior of procedural blanks as outlined above showed that confidence could be placed in the observations obtained for the lunar substances, the 380–390 excitation for emission in the red was unfortunate. The unknown substance which exhibited this fluorescence (which typically amounted to 1–3 per cent of the baseline fluorescence at 33% MSA) did not however, follow porphyrins in subsequent recovery operations for free-base porphyrins and as a result, the likelihood of misinterpretation of demetallation data because of trace impurities in the MSA was very small.

Rocket fuel and exhaust

(i) *Rocket fuel* (*unsymmetrical dimethylhydrazine*). The lunar descent rocket engine was fueled with unsymmetrical dimethylhydrazine, and while the possibility that this substance contained porphyrins was very slight, it was nevertheless examined for fluorescence responses similar to those observed for the lunar fines. To 1 ml of ether was added 100 μl of unsymmetrical dimethylhydrazine and the customary demetallation analysis carried out. No evidence of porphyrins was obtained, nor was any similarity with the lunar analyses noted.

(ii) *Trap A Exhaust*. A sample of exhaust products from a test firing of a rocket descent engine described by SIMONEIT *et al.* (1969) was obtained from Trap A, through the courtesy of Dr. A. L. Burlingame, University of California, Berkeley. It was collected in the first trap of a series cooled by liquid nitrogen, and contained a wide variety of decomposition and polymerization products involving carbon, nitrogen and hydrogen. As shown earlier (HODGSON *et al.*, 1970), on direct examination by fluorescence spectrometry it exhibited broad excitation at 450 nm and minor excitation at 385 nm, with a shoulder at 510 nm for emission measured at 600 nm. These data are similar to those for the lunar fines illustrated in Fig. 4 measured under essentially the same conditions. Direct extraction of Trap A material with 6 N HCl gave a strong peak at 390 nm; smaller peaks were observed at 470 and 527 nm, as shown in Fig. 9. These results suggest that the compounds tentatively identified in the lunar extracts as porphyrin-like pigments were present in the rocket exhaust to a large degree as free-base porphyrins (or perhaps as metal complexes readily demetallated by 6 N HCl). In addition, the 470 nm peak resembled that of the immediate precursors of porphyrins generated from ammonia, methane and water (HODGSON and PONNAMPERUMA, 1968).

MSA demetallation analysis of Trap A material was obscured by background substances, but initial excitation at 412 nm for emission at 630 nm was detected at 5% MSA. This increased in intensity and persisted until a concentration of about 20% MSA was reached. At approximately 30% MSA the customary 390 nm excitation band appeared and intensified with further additions of MSA. The data obtained on this sample suggest that the pigments detected in extracts of lunar fines were derived directly from the exhaust of the lunar module rocket engine.

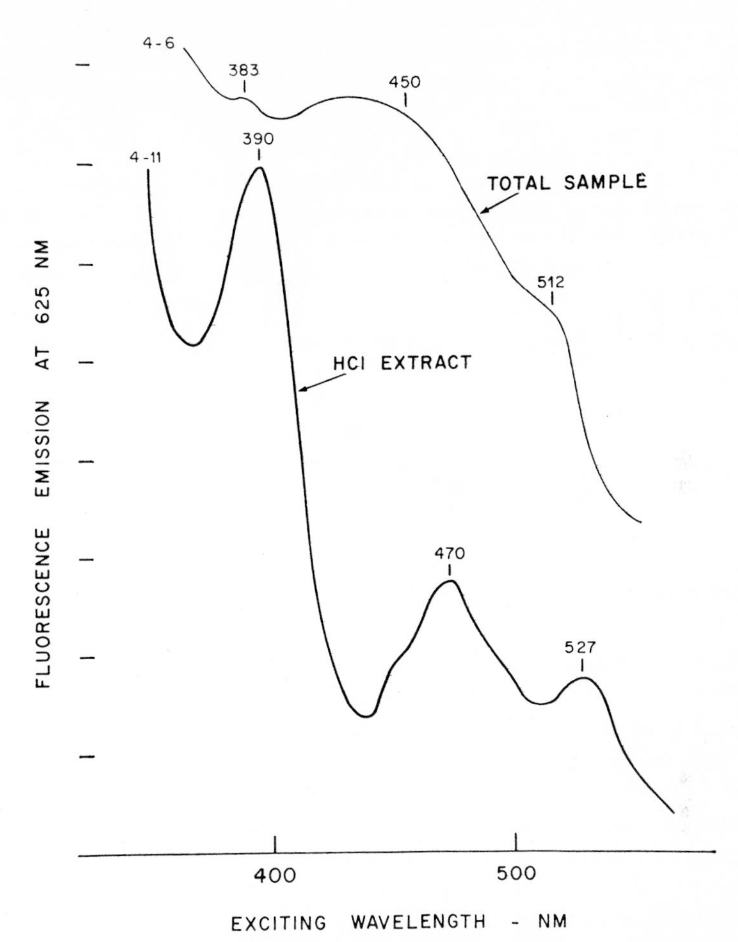

Fig. 9. Fluorescence of trap A rocket exhaust showing evidence for free-base porphyrins in portion soluble in hydrochloric acid.

(iii) *Trap E.* Exhaust products collected in Trap E (SIMONEIT *et al.*, 1969) supplied by Dr. D. A. Flory, Manned Spacecraft Center, Houston, were extracted from the dunite contained in the trap and treated in the same manner as the lunar fines. Spectrophotometry revealed strong absorption bands at 385 and 365 nm. Figure 10 shows the progress of demetallation analysis after initial examination. It gave essentially the same excitation pattern as for Trap A, as summarized in Table 4. Comparison of the demetallation analysis for this sample with that of a lunar sample is afforded

Table 4. Demetallation of rocket-exhaust pigments

Sample	Demetallation threshold (%MSA) Initial	Major	Soret position (nm) Early	Late	in HCl
Trap A	5	30	412	390	n/a
Trap E	3	29	430	387	388

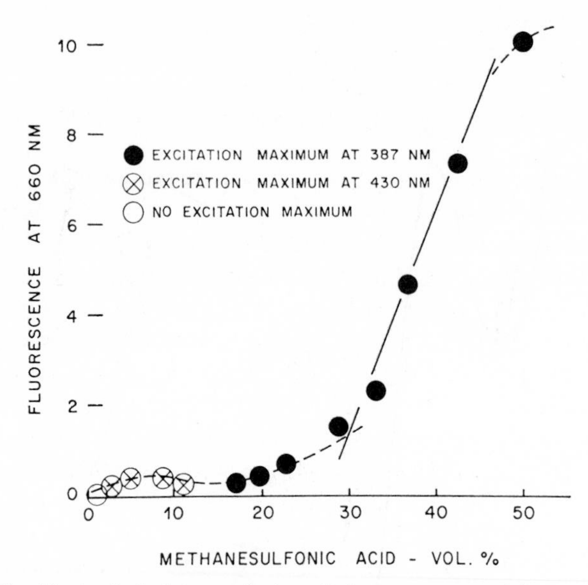

Fig. 10. Analytical demetallation of pigments extracted from trap E.

in Fig. 8 and again, there can be little doubt about the similarity between the behavior of the two samples. The amount of fluorescing pigment in the dunite from Trap E was estimated to be about 50 times that present in the lunar fines.

MCD analyses on rocket exhaust substances failed to confirm the presence of porphyrins. A large number of compounds were indicated to be present, however, and an S-shaped spectral feature was observed at 380 nm. In detail, the band was reversed in sign and rather sharper than is customary for metal complexes of porphyrins. Chromatography on silica gel of rocket products extracted from the Trap E dunite resolved the mixture to some degree with the 380 nm material appearing in the n-hexane eluate. Subsequent eluates (n-hexane–benzene and benzene) were nearly free of this substance and showed much broader bands, but these could not be confidently attributed to porphyrins. Absorption spectrophotometry showed strong absorption at 383 and 361 nm for the hexane eluate, in good agreement with the foregoing data.

CONCLUSIONS

The analytical data collected in the analyses of the lunar fines for porphyrins may be summarized as follows:

1. Fluorescent pigments were present in the lunar fines returned by the crew of Apollo 11.
2. A portion of the pigments appeared to be porphyrin-like.
3. The porphyrin-like pigments of the lunar fines resembled porphyrins detected in carbonaceous chondrites.
4. Rocket exhaust products contained pigments which were similar to those in the lunar fines.

The conclusions to be drawn from the analyses are:

(a) The lunar porphyrin-like pigments were probably derived from exhaust products of the lunar landing module; and

(b) Porphyrin-like pigments appeared to be synthesized from simple compounds containing carbon, nitrogen, hydrogen and oxygen under conditions of very high temperatures.

It is possible that indigenous porphyrins exist on the surface of the moon. These may result from (a) incorporation of carbonaceous primordial matter during the accumulation of the moon (UREY, 1968), and/or (b) the infall of extra-lunar materials during the lifetime of the moon. Little can be evaluated with respect to (a), but (b) can be briefly tested on the basis of indicated infall (KEAYS *et al.*, 1970) and the porphyrin content of carbonaceous chondrites (HODGSON *et al.*, 1969), from which a porphyrin content of about 10^{-5} μg/g of lunar fines is estimated. This is an order of magnitude lower than the measured level, and is therefore in keeping with attributing the bulk of the indicated porphyrins to the combustion products of the rocket.

If a synthesis of porphyrin-like compounds at the temperatures of rocket engines (2000–3000°C) takes place from simple precursors of carbon, nitrogen, hydrogen and oxygen, such a synthesis appears to be a novel process. It is not clear whether the carbon–nitrogen–carbon bond of unsymmetrical dimethylhydrazine remains intact in the process; if it were to do so, it would give rise to a major part of the pyrrole subunits making up the structure of the porphyrins. While the synthesis itself is of interest, it has interesting implications in the context of the generation of moderately complex organic matter wherever simple compounds of carbon, nitrogen, hydrogen and oxygen exist together at very high temperatures. Such conditions may prevail in the cosmos (VARDYA, 1966; FIELD *et al.*, 1969).

Acknowledgement—This work was supported by NASA grant NGR-05-020-004.

REFERENCES

BERNAL J. D. (1961) Significance of carbonaceous meteorites in theories on the origin of life. *Nature* **190,** 129–131.

BUCKINGHAM A. D. and STEPHENS P. J. (1966) Magnetic optical activity. *Ann. Rev. Phys. Chem.* **17,** 399–432.

CHEUNG A. C., RANK D. M., TOWNES C. H., THORNTON D. D. and WELCH W. J. (1969) Detection of NH_3 molecules in the interstellar medium by their microwave emission. *Phys. Rev. Lett.* **21,** 1701–1705.

DONN B., WICKRAMASINGHE N. C., HUDSON J. P. and STECHER T. P. (1968) On the formation of graphite grains in cool stars. *Astrophys. J.* **153,** 451–464.

ELVIDGE J. A. (1956) The synthesis of azaporphins and related macrocycles. *J. Chem. Soc.* **1956,** 28–41.

FIELD G. B., GOLDSMITH D. W. and HABING H. J. (1969) Cosmic-ray heating of the interstellar gas. *Astrophys. J.* **155,** L149–L154.

HODGSON G. W. and BAKER B. L. (1969) Porphyrins in meteorites: Metal complexes in Orgueil, Murray, Cold Bokkeveld and Mokoia carbonaceous chondrites. *Geochim. Cosmochim. Acta* **33,** 943–958.

HODGSON G. W., HITCHON B., TAGUCHI K., BAKER B. L. and PEAKE E. (1968) Geochemistry of porphyrins, chlorins and polycyclic aromatics in soils, sediments and sedimentary rocks. *Geochim. Cosmochim. Acta* **32,** 737–772.

HODGSON G. W., PETERSON E. and BAKER B. L. (1969) Trace porphyrin complexes: Fluorescence detection by demetallation with methanesulfonic acid. *Mikrochim. Acta* **1969,** 805–814.

HODGSON G. W., PETERSON E., KVENVOLDEN K. A., BUNNENBERG E., HALPERN B. and PONNAMPERUMA C. (1970) Search for porphyrins in lunar dust. *Science* **167**, 763–765.

HODGSON G. W. and PONNAMPERUMA C. (1968) Prebiotic porphyrin genesis: Porphyrins from electric discharge in methane, ammonia, and water vapor. *Proc. Nat. Acad. Sci.* **59**, 22–28.

JOHNSON F. M. (1967) Diffuse interstellar lines and chemical characterization of interstellar dust, pp. 229–240. In *Interstellar Grains NASA SP*-140, (editor J. M. Greenberg and T. P. Roark), 269 pp.

KEAYS R. R., GANAPATHY R., LAUL J. C., ANDERS E., HERZOG G. F. and JEFFERY P. M. (1970) Trace elements and radioactivity in lunar rocks: implications for meteorite infall, solar-wind flux, and formation conditions of moon. *Science* **167**, 490–493.

KVENVOLDEN K. A. and HODGSON G. W. (1969) Evidence for porphyrins in early Precambrian Swaziland System sediments. *Geochim. Cosmochim. Acta* **33**, 1195–1202.

KVENVOLDEN K. A., CHANG S., SCHOPF J. W., FLORES J., PERING K., SAXINGER C., WOELLER F., KEIL K., BREGER I. and PONNAMPERUMA C. (1970) Carbon compounds in lunar fines from Mare Tranquillitatis. I. Search for molecules of biological significance. *Geochim. Cosmochim. Acta* Supplement I.

MOSER F. H. and THOMAS A. L. (1963) *Phthalocyanine Compounds*, 365 pp. Reinhold.

PLATT J. R. (1956) Electronic structure and excitation of polyenes and porphyrins. In *Radiation Biology*, (editor A. Hollaender), Vol. 3, pp. 71–123. McGraw-Hill.

PONNAMPERUMA C., KVENVOLDEN K., CHANG S., JOHNSON R., POLLOCK G., PHILPOTT D., KAPLAN I., SMITH J., SCHOPF J.W., GEHRKE C., HODGSON G., BREGER I., HALPERN B., DUFFIELD A., KRAUSKOPF K., BARGHOORN E., HOLLAND H. and KEIL K. (1970) Search for organic compounds in the lunar dust from the Sea of Tranquillity. *Science* **167**, 760–762.

RHO J. H., BAUMAN A. J., YEN T. F. and BONNER J. (1970) Fluorometric examination of a lunar sample. *Science* **167**, 754–755.

SIMONEIT B. R., BURLINGAME A. L., FLORY D. A. and SMITH I. D. (1969) Apollo lunar module engine exhaust products. *Science* **166**, 733–738.

SNYDER L. E., BUHL D., ZUCKERMAN B. and PALMER P. (1969) Microwave detection of interstellar formaldehyde. *Phys. Rev. Lett.* **22**, 679–681.

STEPHENS P. J., SUETAAK W. and SCHATZ P. N. (1966) Magneto-optical rotatory dispersion of porphyrins and phthalocyanines, *J. Chem. Phys.* **44**, 4592–4602.

UREY H. C. (1968) Primary objects. In *International Dictionary of Geophysics*. Pergamon Press.

VARDYA M. S. (1966) March with depth of molecular abundances in the outer layers of K and M stars. *Mon. Not. Roy. Astron. Soc.* **134**, 347–370.

Proceedings of the Apollo 11 Lunar Science Conference, Vol. 2, pp. 1845 to 1856.

Carbon compounds in lunar fines from Mare Tranquillitatis—III. Organosiloxanes in hydrochloric acid hydrolysates

CHARLES W. GEHRKE,* ROBERT W. ZUMWALT,† WALTER A. AUE,*
DAVID L. STALLING,‡ ALAN DUFFIELD, §
KEITH A. KVENVOLDEN and CYRIL PONNAMPERUMA
Exobiology Division, Ames Research Center, NASA, Moffett Field, California 94035

(*Received* 3 *February* 1970; *accepted in revised form* 2 *March* 1970)

Abstract—The hydrochloric acid hydrolysates of the lunar fines were examined by gas–liquid chromatography for compounds which could be esterified with *n*-butanol·3 N HCl and/or trifluoroacetylated with trifluoroacetic anhydride. Gas–liquid chromatography combined with mass spectrometry indicated the presence of siloxanes in the acid hydrolysates at levels of 35–50 ppm based on the hydrogen flame response to standard amino acid N-trifluoroacetyl *n*-butyl ester derivatives. None of the common amino acids were detected in the lunar fines.

INTRODUCTION

THIS investigation was initially dedicated to the determination of amino acids in lunar sample "10086-3 bulk A fines", as part of the analyses done by a group of investigators established as the NASA–Ames Research Center Consortium. As the study progressed, compounds other than amino acids were found, and the planned experiments were consequently modified according to results obtained in the course of the investigations. This introduction, therefore, describes in chronological order the logical sequence for initiating new experiments and the broad objectives behind the choice of experimental design.

Originally, the 1 normal (N) and 6 N HCl hydrolysates (Part I) were analyzed by the gas–liquid chromatographic (GLC) method developed by GEHRKE *et al.* (1968) and ROACH and GEHRKE (1969), in which amino acids are converted to N-trifluoroacetyl-*n*-butyl esters prior to chromatography. Other organic compounds amenable to acylation or esterification are also derivatized. When the method was used to analyze hydrolysates of the lunar sample, no amino acids could be found. However, a number of gas-chromatographic peaks appeared which did not originate from amino acids. Low levels of an amino acid and other derivatizable compounds were added to the lunar sample and recovered and derivatized according to the analytical scheme. These experiments proved that had amino acids been present above 2×10^{-3} μg/g, the limit of detection, they would have been found.

Following this conclusion, experiments were designed to elucidate the nature of the major unknown GLC peaks. First, elaborate precautions were taken to exclude

* Present address: University of Missouri, Columbia, Missouri.

† Present address: University of Missouri, Columbia, Missouri, National Science Foundation Predoctoral Fellow Professors.

‡ Present address: Fish Pesticide Research Laboratory, U.S. Bureau of Sport Fisheries and Wildlife, Columbia, Missouri, Chief Chemist.

§ Present address: Department of Chemistry, Stanford University, Stanford, California.

the possibility of contamination introduced by the techniques. Parallel with the lunar sample, a sand blank was run through exactly the same experiments. Later, and again parallel with a second portion of the lunar sample, a basalt sample from Hawaii and quartz crystal samples from South Africa were analyzed by the same procedures. In none of these three materials could peaks similar to those observed from the lunar sample be found; in fact, there appeared no peaks at all of significant size.

Another possible source of contamination could have been the rocket exhaust of the descending lunar module. To exclude this possibility, olivine chips were analyzed which previously had been exposed to lunar descent module rocket exhaust by NASA at White Sands, New Mexico. Again, no chromatographic peaks corresponding to those in the lunar sample were observed. Having thus established that the observed peaks did indeed originate with the lunar sample as received at Ames Research Center, combined gas–liquid chromatography—mass spectrometry (GLC–MS) was employed to elucidate the structure of the unknown compounds. Mass spectrometry clearly showed the presence of silicon-containing ions, similar to those in spectra of the dimethylpolysiloxanes commonly used in silicone oils, silicone rubber, etc. This indicated the obvious possibility of contamination. Consequently, several experiments were designed to clarify whether silicones were present in the lunar sample and to demonstrate how silicone-based materials would behave in the analytical scheme.

First, a virgin lunar sample was exhaustively extracted with benzene to remove any possible silicones present as contaminants. The extract was analyzed by the same procedures as used previously. Second, a mixture of silicone oils and rubbers was taken through these procedures. Third, silicone monomer, both neat and partially hydrolyzed was subjected to the above procedures. In none of the three cases could peaks corresponding to those from the lunar sample be found.

One study was initiated to define the particular reactions in the analytical scheme necessary to obtain GLC peaks. Another study involved the extraction of the derivatized compounds by pentane. This eliminated some of the background and chromatographic interferences caused by the inorganic salts formed during acid hydrolysis of the lunar fines. In a third study, derivatization of the lunar material was again reproduced, the peaks analyzed by GLC–MS and a thorough study of the obtained mass spectra performed. Furthermore, samples from various states in the analytical procedure were analyzed by direct-probe mass spectrometry. At the relatively low level of derivatizable compounds observed, these were the techniques most likely to yield a maximum of information. To add another material to the ones already studied, pulverized Pueblito de Allende (PDA) meteorite was analyzed by the same methods. Surprisingly, the same major GLC peaks were found.

Experimental

Apparatus and reagents

Three gas chromatographs were used in this investigation. One was a Varian Aerograph Model 1520 gas chromatograph with a two column oven bath, equipped with flame ionization detectors, a dual channel electrometer, a linear temperature programmer and a Westronics strip chart recorder. The other two instruments were Hewlett-Packard Model 5750 research gas chromatographs with dual column oven baths and detectors and Honeywell Electronik 16 strip chart recorders.

All of the amino acids used in this study were obtained from Mann Research Laboratories, Inc. or Nutritional Biochemicals Corporation and were chromatographically pure. Standard amino acid N-trifluoroacetyl-*n*-butyl ester derivatives were obtained from Regis Chemical Company, Chicago, Illinois. *n*-Butanol was a "Baker Analyzed Reagent." The trifluoroacetic anhydride obtained from Distillation Industries was an 'Eastman Grade' chemical. Anhydrous HCl, 99·0% minimum purity, was obtained from Matheson Company.

The *n*-butanol and methylene chloride were redistilled from all-glass systems and protected from atmospheric moisture by storage in all-glass inverted top bottles. The methylene chloride and *n*-butanol were first refluxed over calcium chloride before distillation. The anhydrous HCl gas was passed through a H_2SO_4 drying tower before bubbling through the *n*-butanol.

GLC instrument settings and conditions

The chromatograms presented were obtained from a strip chart recorder with a one millivolt full scale response. The amplification of the signal from the hydrogen flame detector was accomplished by use of an electrometer with a maximum sensitivity of 1×10^{-12} amp/mV. The designation XI appearing on the figures denotes this sensitivity as maximum. Other sensitivity designations (e.g. X16, X32, etc.) represent reductions in sensitivity by that particular factor. The changes in sensitivity during the analyses were made to obtain the optimum amount of information, although they do not add to the convenience of interpretation. The symbols ADJ appearing on the chromatograms signify changes in the suppression amperage only and do not alter the sensitivity setting of the analysis.

Sand blank

The sand blank was prepared by heating a sample of Ottawa sand for 48 hr at 1000°C.

Hydrolysis—1 N HCl

The lunar sample, PDA and the sand blank were hydrolyzed for one hour with 1 N HCl under reflux. The samples were then centrifuged at 2700 rev/min for 4 min and the 1 N HCl decanted off for analysis. Although the 1 N HCl fraction contained the highest concentration of the components of interest, the 6 N HCl hydrolysate of the residue remaining after the 1 N HCl hydrolysis, was also analyzed. In addition, the 6 N hydrolysate of a separate one gram lunar sample which had been used in investigations for the presence of carbides (Part IV) was studied.

Derivatization conditions

Aliquots of the acidic hydrolysates corresponding to 0·25 to 0·50 g of the samples were placed in small pear-shaped glass reaction vials. The solutions were then evaporated just to dryness using an infrared heat lamp.

Esterification of carboxyl groups in the residue was performed by adding *n*-butanol·3 N HCl to the vial (200 μl/0·25 g of sample), capping the vial with a teflon lined cap, ultrasonic mixing and then placing the vial in a 100° sand bath for 30 min. The *n*-butanol was then removed by evaporation just to dryness under an i.r. lamp. Dichloromethane was used for azeotropic removal of any moisture present at the end of the evaporation step. The samples were then trifluoroacetylated by the addition of 100 μl of an acylating reagent consisting of trifluoroacetic anhydride (TFAA) in the presence of dichloromethane (0·5 μl TFAA/100 μl CH_2Cl_2) for each 0·25 g of lunar sample, PDA, or sand blank. The vials were again tightly capped and placed in the 100° sand bath for 10 min. After removal from the bath, they were allowed to cool and then analyzed by GLC, GLC–MS and/or direct probe MS.

In subsequent studies, it was found that the derivatized compounds of interest could be extracted with pentane from the derivatization mixture. This pentane extraction eliminated most of the soluble salts from the derivatized sample, thus minimizing interferences in GLC analysis due to salt deposits in the injection port of the GC.

Columns

The stabilized grade of EGA was obtained from Analabs, Inc., Hamden, Conn. The support material, 80/100 mesh acid washed (A.W.) Chromosorb W, was obtained from Supelco, Inc.,

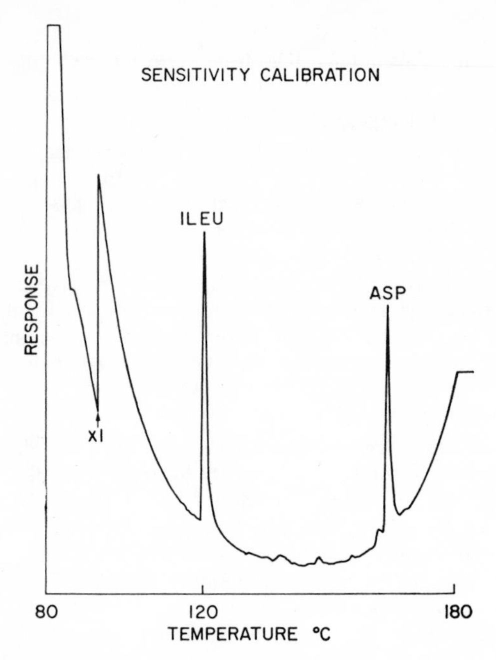

Fig. 1. Instrumental sensitivity calibration. 10 ng of each amino acid N-TFA *n*-butyl ester injected. Column: 1·5 w/w% OV-17 on 80/100 mesh H.P. Chromosorb G, 1·5 m × 4 mm i.d. glass. Initial temp. 80°C, program rate 4°C/min.

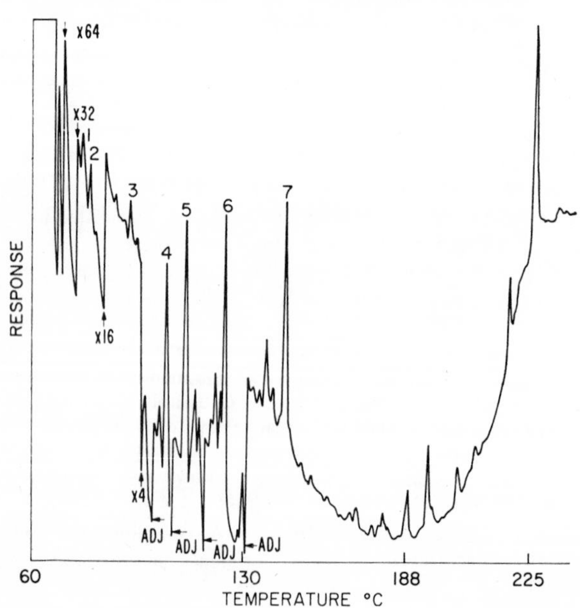

Fig. 2. 1 N HCl Hydrolysate of lunar fines—esterified and acylated. Sample equivalent to 25 mg of lunar material injected. Column: 1·5 w/w% OV-17 on 80/100 mesh, H.P. chromosorb G. 1·5 m × 4 mm i.d. glass. Initial temp. 60°C, program rate 4°C/min.

Bellefonte, Pa. Columns were prepared using the 80/100 mesh A.W. Chromosorb W after drying in an oven for 12 hr at 140°. The packing consisted of 0·325 w/w% EGA coated on the support material and the glass column dimensions were 1·5 m × 4 mm i.d. The OV-17 siloxane substrate was obtained from Supelco, Inc. and glass columns 1·5 m × 4 mm i.d. containing 1·5 w/w% OV-17 on 80/100 mesh high performance (H.P.) Chromosorb G were prepared.

Gas–liquid chromatography—mass spectrometry

A Varian Aerograph gas chromatograph was used with the EGA and OV-17 columns as described in the previous section. The *total* column effluent was introduced into a modified Ryhage type molecular separator. The two separator acceptor jets were of larger inner diameter than commonly employed, and improved pumping by a mechanical and an oil diffusion pump was used to cope with the large amounts of helium carrier gas. The mass spectrometer was a high-resolution CEC-21-110 Model, equipped with photoplate detection. The ion beam current was monitored on a strip chart recorder to ensure a permanent record of the peaks entering the mass spectrometer and the corresponding elution temperatures in the gas chromatograph. The photoplates were scanned by a high resolution densitometer and the signals digitized and their exact mass determined by computer curve fitting, using the fragments from PFA (perfluoroalkane) as a standard. PFA was introduced via the heated inlet system during all runs. With available computer software, element mapping was achieved. The results were confirmed, where applicable, on a Perkin-Elmer Model 270 B low-resolution GLC–MS system. On both high- and low-resolution mass spectrometers, several direct probe experiments were run and interpreted through photoplate-densitometer-computer from the CEC and through oscillographic recorder traces from the Perkin-Elmer model. Correction for background was performed manually for both the GLC–MS and the direct probe-MS runs.

Results and Discussion

The initial experiments using the above derivatization procedures were conducted on the 1 N and 6 N hydrolysates of bulk lunar sample fines, 10086-3. These samples were chromatographed on columns containing ethylene glycol adipate and OV-17 (a phenyl methyl polysiloxane) as liquid phases. These studies revealed the presence of derivatizable compounds which did not correspond to any of the common amino acids. These compounds were present at a total concentration of 35–50 ppm, estimated from the hydrogen flame detector response for standard amino acid derivatives. A larger number of chromatographic peaks were observed when the chromatography was performed on the OV-17 column, indicating that some of the derivatives were decomposed when chromatographed on the more polar EGA polyester column. Chromatography of the derivatized lunar sample on the EGA column showed only two major peaks with retention temperatures of 122 and 141°, using an initial temperature of 60°C, an 8 min initial hold, and a program rate of 4°C per min. Apparently, some kind of substrate-derivative interaction occurred. This may be due to adsorption or decomposition as a larger number of major chromatographic peaks were obtained on the OV-17 column. These peaks are identified in Fig. 2 with numbers of 1, 2, 3, 4, 5, 6 and 7.

The sensitivity of the instrumentation in all experiments was carefully monitored. A typical chromatogram resulting from injection of 10 ng of derivatized isoleucine and aspartic acid is presented in Fig. 1.

Derivatization studies

To gain information on the organic functional groups associated with the derivatizable compounds of the lunar sample that were observed, an underivatized 1 N hydrolysate of the lunar sample was analyzed, as well as two 1 N lunar hydrolysate samples which had been esterified *only* and acylated *only*. No significant chromatographic peaks were observed in any of these three experiments, indicating the reactions associated with *both* esterification and acylation were required to achieve GLC analysis of these compounds.

Hydrolysis and chromatography of lunar sample, PDA meteorite and sand blank

In order to exclude the possibility of contamination of the lunar sample during the water and benzene-methanol extraction steps, (which were made in the first experiments described in Part I on 54·6 g of the lunar fines) a separate sample of 2·1 g

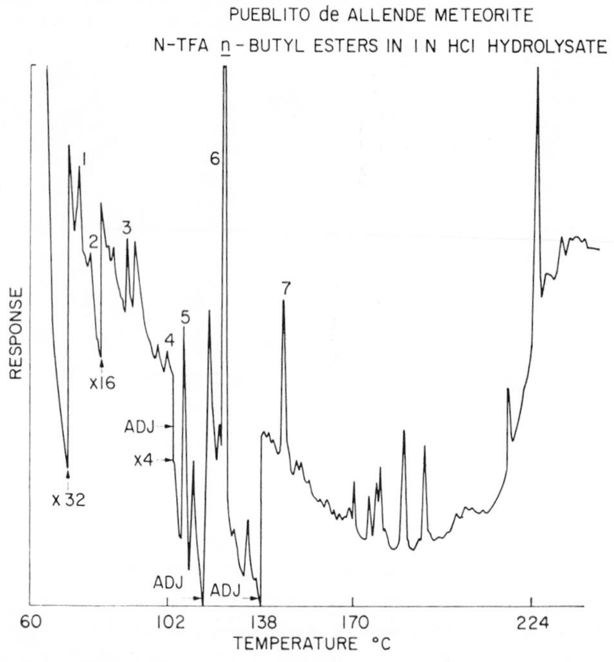

Fig. 3. 1 N HCl of Pueblito de Allende meteorite—esterified and acylated. Sample equivalent to 25 mg of meteorite injected. Chromatographic conditions on Fig. 2.

of the lunar fines (10086-3) was hydrolyzed under reflux with 1 N HCl for one hour. This experiment was performed to obtain sufficient quantities of the derivatized compounds for mass spectral studies. The previously mentioned benzene–methanol and water extraction steps (which were made on the 54·6 g sample) were omitted. A sand blank and 2·1 g of the PDA meteorite were also hydrolyzed and analyzed under identical conditions. The hydrolysates were centrifuged and the 1 N HCl was decanted. Solution aliquots corresponding to 0·5 g of the lunar sample, PDA meteorite and sand blank were placed in reaction vials, the HCl evaporated and the residue esterified and acylated, yielding a final acylation mixture of 100 μl volume.

Aliquots of the derivatized samples were then chromatographed on a 1·5 m $\times$ 4 mm i.d. glass chromatographic column containing 1·5 w/w% OV-17 on 80/100 mesh H.P. Chromosorb G. A typical chromatogram obtained on analysis of the equivalent of 25 mg of the lunar sample is shown in Fig. 2. Note the occurrence of the numbered major peaks. Figure 3 presents the chromatographic analysis of the meteorite sample and a comparison of this chromatogram with Fig. 2 (lunar) reveals striking similarities. Several of the major peaks coincide, as do some of the minor peaks throughout these two chromatograms. Complete sand blanks which were analyzed as parallel samples for these experiments did not give chromatographic peaks (Fig. 4) even though the GLC analysis was conducted with a fourfold increase in instrument sensitivity.

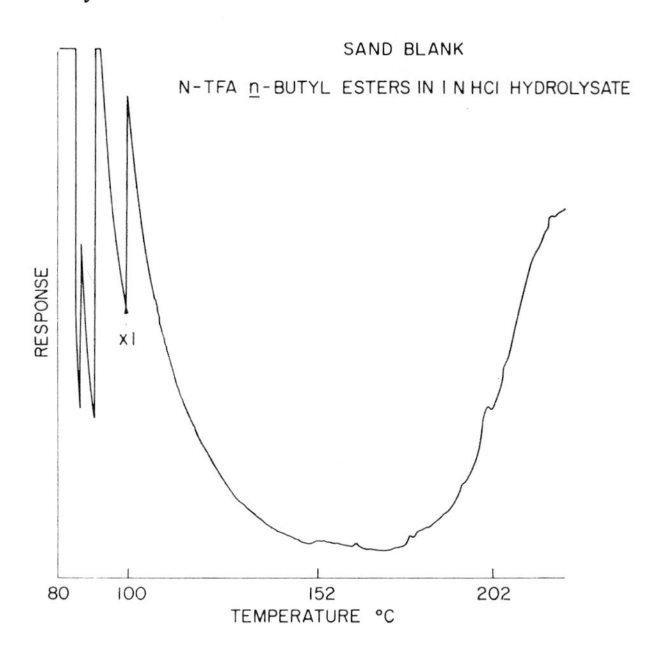

Fig. 4. 1 N HCl Hydrolysate of sand blank—esterified and acylated. Sample equivalent to 50 mg of sand injected. Chromatographic conditions on Fig. 2 except initial temp. 80°C.

These experiments were followed by the analysis of another 6 N HCl hydrolysate from one gram of lunar sample which resulted in the chromatogram presented in Fig. 5, thus confirming the presence of these derivatizable compounds in the lunar fines and again the appearance of the same major peaks (1–7). It is apparent that the 6 N HCl removes different relative amounts of the derivatizable compounds than does the 1 N HCl. Also, a greater amount and number of peaks were observed on analysis of the 6 N HCl hydrolysate than for the 1 N HCl hydrolysate. This may be due to the more rigorous hydrolysis method, or perhaps the larger amount of soluble salts in the 6 N HCl hydrolysate causes some structural changes in the major compounds during extraction, derivatization, or GLC analysis. In Fig. 5, an initial temperature of 70°C with an initial isothermal hold, plus an increased amount of sample injected were responsible for the non-resolution of peaks numbered 1 and 2.

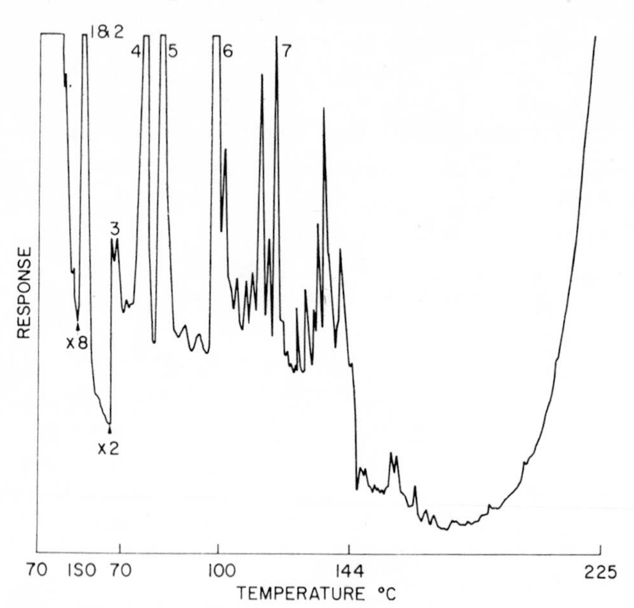

Fig. 5. 6N HCl Hydrolysate of lunar fines—esterified and acylated. Sample equivalent to 50 mg of lunar material injected. Chromatographic conditions on Fig. 2 except initial temp. 70°C.

The 6 N HCl lunar hydrolysate was also derivatized and chromatographed on an ethylene glycol adipate column. The two major peaks and the minor peaks observed again did not correspond to any of the common amino acids. Further, in earlier experiments for amino acids, a water extract, equivalent to 50 g of lunar sample, was analyzed and amino acids were not found at a 1 ng/g level. Also, amino acids were not found in any of the benzene–methanol extracts.

Salt interference studies

Experiments were performed to determine whether the presence of the salts formed during acid hydrolysis of lunar fines interfered with the derivatization reactions and subsequent GLC chromatographic analysis. Leucine was successfully derivatized and chromatographed in the presence of the lunar salts. In this experiment, 250 ng of leucine were added to 0·25 g of a lunar sample 1 N hydrolysate, derivatized and analyzed. These results further support that if amino acids were present in a 1 N hydrolysate of the lunar sample, they would have been observed on GLC analysis. Additional information on possible interferences by lunar salts was gained by the successful analysis for 1 μg each of valine, 1-octanol, 1-amino hexane and lauric acid added to the lunar sample. Recovery levels were better than 80 per cent. Adipic acid, which was also used in this study, was the only compound not recovered.

Experiments on silicones

Initial mass spectrometric data revealed the presence of silicon-containing species in the GLC peaks. Also, octoil (di-*n*-octyl phthalate) was identified as being present in the lunar sample by direct probe high-resolution MS. Therefore, numerous laboratory materials were analyzed, using the same reaction and chromatographic conditions as those used in the analysis of the lunar sample, PDA meteorite and sand blank.

Octoil and a mixture of organosiloxanes were first hydrolyzed, then derivatized and chromatographed. The mixture of silicones contained the commonly used oils SE30, SF96, DC200 (all dimethyl polysiloxanes) and DC 550 (phenyl methyl polysiloxane, similar to OV-17 used in the GLC column). The hydrolysis was conducted by adding 20 mg of the silicone mixture, plus a silicone rubber O-ring and a silicone rubber septum (containing methyl silicones) to 0·5 g of Chromosorb W and heating the mixture at 100° for one hour under reflux in the presence of 5 ml of 1 N HCl. Octoil (20 mg) was hydrolyzed in the same manner as was a sample of Chromosorb W—1 N HCl. The HCl was then decanted from each of the samples, evaporated and the residue subjected to esterification and acylation using the same experimental conditions as for the previous samples. The samples were then chromatographed at the sensitivity level used for the analysis of the lunar samples. No corresponding chromatographic peaks were observed in these experiments, (in fact, there appeared only one peak of comparable size), suggesting that the GLC peaks from the lunar sample and meteorite are not due to laboratory contamination from silicones, or octoil.

Extraction of lunar sample with benzene

In an effort to further investigate the source and nature of the organosiloxanes present in the lunar sample, a separate 1 g sample of the lunar fines (10086-3) was extracted for 8 hr with benzene under reflux, the liquid decanted and the procedure repeated twice. Under these conditions, surface organic silicone greases present on the surface as contamination would have been extracted (AUE, 1969).

The combined benzene extracts were evaporated and the residue hydrolyzed, derivatized and chromatographed under the same experimental conditions used in the previous analysis of the lunar sample. No chromatographic peaks were observed in the benzene extract of the lunar sample, or in a similarly treated benzene blank, thus excluding silicone contamination incurred in the history of the lunar sample from collection on the moon to analysis.

Rocket exhaust analysis

In order to determine if the derivatizable compounds present in the lunar sample could be due to the rocket exhaust of the descending lunar module, olivine chips obtained from D. A. Flory which had been exposed to a similar rocket exhaust in New Mexico were analyzed (SIMONEIT, 1969). No chromatographic peaks were observed on analysis of the olivine. Also, rocket exhaust materials, which had been condensed with liquid nitrogen traps, were analyzed. Only one chromatographic peak had a retention time similar to a chromatographic peak observed in the lunar and PDA samples.

Analysis of basalt and quartz crystals

Although the possibility seemed remote, further experiments were made to show that the organosiloxane peaks observed on analysis of the lunar samples and PDA meteorite could not have arisen from the reaction of inorganic silicon compounds with the employed reagents, namely aqueous HCl, n-C_4H_9OH, HCl, TFAA and CH_2Cl_2. Approximately three ppm of organic material were found by GLC in the basalt sample and 0·3 ppm in the quartz crystals. These samples were processed in the same manner as were the lunar fines. Neither of these materials gave peaks which were chromatographically similar to those from the lunar sample or the PDA meteorite. Silicone contaminants were not found by exhaustive extraction of lunar material with benzene and chromatographic peaks could not be obtained through analysis of terrestrial samples such as basalt, olivine, Chromosorb and quartz crystals. These results provide evidence that peaks observed in chromatograms of derivatized acid hydrolysates of lunar and meteorite material could be attributed to compounds actually extracted from these samples.

Mass spectrometric determinations

GLC–MS of derivatized lunar sample. The major peaks in the chromatograms of derivatized lunar sample were examined by high resolution mass spectrometry. The major chromatographic peaks obtained from the lunar fines, seen both on the OV-17 and EGA column, are due to organosiloxanes. These are the peaks numbered 4, 5, 6 and 7 in Fig. 2 and the peaks eluting at 122° and 141° on EGA.

It was not possible to deduce a particular organic structure from the various mass spectra. All the peaks showed some of the ion fragments typical of methylpolysiloxanes with the characteristic mass 74·0186 a.m.u. between members in ion series. Three typical series were identified from the major GLC peaks as indicated by the beam monitor and the GLC trace. All had the general composition $[R[OSi(CH_3)_2]_n]^+$, with the R groups and "n" as follows:

(a) $R = C_3H_9O_2Si_2$ (n = 0, 1, 2, 3, 4) (MCLAFFERTY, 1957)
(b) C_3H_9Si [n = 0 (also in background), 1, 2, 5?] (MCLAFFERTY, 1957)
(c) $CH_3Si_2O_2$ (n = 0, 1, 2?, 3?)

(The question mark denotes a slight discrepancy between the theoretical and the computer-calculated masses, which, we assume, is caused by the data acquisition system.)

Some other silicon-containing fragments such as $C_4H_{13}Si_3O_3^+$ (MCLAFFERTY, 1957) were also found, which are known to occur in dimethylpolysiloxane. The analyses were repeated at the University of Missouri and the Fish Pesticide Research Laboratory, Columbia, Missouri, on a low-resolution GLC–MS and the presence of silicon-containing ions was confirmed by their typical isotope pattern.

No definite conclusion could be reached regarding the type(s) of derivative formed. C_4H_9O and its various fragments are profusely present in the spectra and obviously originate from several peaks entering the MS during the GLC run. These peaks, however, do not coincide in all cases with the dimethylpolysiloxane peaks as seen in the rise and fall of alkoxy species versus silicon species in successively recorded spectra. This behavior could be caused by decomposition in the GC column or the molecular separator. To assume occurrence of some decomposition appears reasonable, since the EGA column showed fewer GLC peaks than the OV-17 column. EGA decomposes certain compounds as was found in the analysis of some amino acids (GEHRKE *et al.*, 1968) derivatized by the same methods as the lunar sample. Methylpolysiloxanes may have perhaps been derivatized by butoxy groups on a terminal SiOH. The chemistry of this type of derivatization is under further investigation. The CF_3CO fragment, on the other hand, was conspicuously absent from the mass spectra. It is not clear what acylation contributes toward forming volatile derivatives since their mass spectra do not contain trifluoroacetyl ions.

It is known from the literature that compounds amenable to gas chromatography have been obtained in several instances from inorganic silicates by trimethylsilylation of silicic acids (e.g. LENTZ, 1964). However, in those experiments it was necessary to introduce organically bound silicon, such as in hexamethyldisiloxane or trimethylchlorosilane. In our investigations, however, no such compounds were introduced during the described analytical scheme and their formation during the derivatization reactions is hardly conceivable. Specifically, while it is easy to introduce $-\overset{|}{\underset{|}{Si}}-O-\overset{|}{\underset{|}{C}}-$ and $-\overset{|}{\underset{|}{Si}}-O-CO-\overset{|}{\underset{|}{C}}-$ units, the formation of Si—C bonds such as present in $[(CH_3)_2SiO]_n$ requires a different chemistry.

The lunar fines, when hydrolyzed, derivatized and introduced via heated direct probe into the mass spectrometer, gave several additional fragments of importance. Thus, we found a strong series $C_2H_nN_3$ ($n = 2$ to 5) suggestive of the lunar descent module's propellant, unsymmetrical dimethylhydrazine and N_2O_4; several species containing N_3; some weak fragments containing C, N and Si; and C, O and Si; some evidence of octoil; and, of course, an abundance of alkyl and alkyloxy fragments arising from the derivatization reagents.

SUMMARY AND CONCLUSIONS

Derivatizable organic compounds were found in the 1 N and 6 N HCl hydrolysates of the lunar sample and PDA meteorite; however, they were not the common amino acids. These peaks could not be observed without conducting both esterification and acylation of the samples. The total concentration of these organic compounds is estimated at 35–50 ppm in the lunar sample and 45–60 ppm in the PDA meteorite based on the hydrogen flame detector response to standard amino acid derivatives. These derivatizable compounds contain organosiloxanes. There is a striking similarity between the GLC chromatograms of the derivatizable matter in the acid hydrolysates of the lunar sample and Pueblito de Allende meteorite showing the same major chromatographic peaks and a number of minor peaks.

Neither the sand blank nor the olivine exposed to rocket exhaust, the basalt, the quartz crystals, the mixture of silicones, or Chromosorb W, showed any of the peaks found in the lunar or PDA samples. Further, the peaks observed in the analysis of the rocket exhaust trappings did not coincide, with the exception of one, with those for the lunar sample or PDA meteorite.

High and low resolution mass spectrometric data (GLC-MS) obtained from the major gas chromatographic peaks in the derivatized lunar sample identified its major organic constituents as polydimethylsiloxanes through several fragment series of the general formula $[R[OSi(CH_3)_2]_n]^+$. These siloxanes are possibly in the form of *n*-butoxy derivatives.

Other MS evidence points to octoil (contamination) and organic compounds with two or three nitrogen atoms (rocket exhaust?) in the lunar sample.

Although it is possible that the polydimethylsiloxanes found originated from laboratory contamination, the many different kind of blanks analyzed and the failure to obtain GLC peaks (similar to those derived from lunar materials) from various types of common silicones, provide a strong argument for ruling out the possibility of contamination.

Addendum—That the organosiloxanes could be artifacts from the analytical procedures at first seemed unlikely for a number of reasons. Work done since this paper was accepted in revised form for publication, however, suggests that the formation of some organosiloxane artifacts has occurred. Re-examination of basalt and other ultramafic rocks shows that these materials, when acid hydrolyzed for 15 hours (instead of one hour as done previously), produce the same type of organosiloxane compounds as were found in the lunar sample and PDA meteorite. A detailed report concerning our recent work is in preparation.

Acknowledgments—The authors are grateful to Dr. SHERWOOD CHANG, DON RAMEY and DON QUINSLAND of the Ames Research Center for their many suggestions and technical assistance throughout these investigations and to SAI CHANG, ROY RICE, STEVE HARRIS and ROSE SWEENEY of the Experiment Station Chemical Laboratories, University of Missouri, Columbia, for their most helpful assistance in chromatographic analysis and in the preparation of the mass spectrometric data graphs.

The authors also want to recognize the helpful suggestions of Messrs. EGLINTON, PILLINGER and the reviewers of this paper. Contribution from the Missouri Agricultural Experiment Station, Journal Series No. 5881. Approved by the Director. Supported in part by grants from the National Aeronautics and Space Administration (NGR 26-004-011), the National Science Foundation (GB-7182), the Missouri Agricultural Experiment Station (Analytical Services 132), and the United States Bureau of Sport Fisheries and Wildlife, Fish Pesticide Research Laboratory.

REFERENCES

AUE W. A. and HASTINGS C. R. (1969) Preparation and chromatographic uses of surface-bonded silicones. *J. Chromatogr.* **42,** 319–335.

GEHRKE C. W., ZUMWALT R. W. and WALL L. L. (1968) Gas–liquid chromatography of protein amino acids—separation factors. *J. Chromatogr.* **37,** 398–413.

LENTZ C. W. (1964) Silicate minerals as sources of trimethylsilyl silicates and silicate structure analysis of sodium silicate solutions. *Inorg. Chem.* **3,** 574–579.

MCLAFFERTY F. W. (1957) *Appl. Spectrosc.* **11,** 148; Cited by K. BIEMANN (1962) *Mass Spectrometry: Applications to Organic Chemistry*, pp. 171–172. McGraw-Hill.

ROACH D. and GEHRKE C. W. (1969) Direct esterification of protein amino acids—GLC of N-TFA *n*-butyl esters. *J. Chromatogr.* **44,** 269–278.

SIMONEIT B. R., BURLINGAME A. L., FLORY, D. A. and SMITH I. D. (1969) Apollo lunar module engine exhaust products. *Science* **166,** 733–738.

Proceedings of the Apollo 11 Lunar Science Conference Vol. 2, pp. 1857 to 1869.

Carbon compounds in lunar fines from Mare Tranquillitatis—IV. Evidence for oxides and carbides

S. CHANG, J. W. SMITH,* I. KAPLAN,† J. LAWLESS,
K. A. KVENVOLDEN and C. PONNAMPERUMA
Exobiology Division, Ames Research Center, NASA, Moffett Field, California 94035

(*Received* 3 *February* 1970; *accepted in revised form* 2 *March* 1970)

Abstract—Identification was made of types of nonextractable and volatilizable carbon compounds in 10086-3 Bulk A lunar fines. The total carbon concentration was 157 ± 14 μg/g and appears to vary with particle size. Products obtained by pyrolysis and acid hydrolysis were examined by mass spectroscopy and gas chromatography. Volatile products obtained in pyrolysis experiments were primarily carbon monoxide and carbon dioxide with traces of C_1–C_4 hydrocarbons. Hydrolysis experiments consistently yielded hydrocarbons indicating that carbon was in the form of metallic carbides in concentrations varying from 5–20 μg/g.

INTRODUCTION

EVIDENCE supporting the hypothesis of abiotic primordial synthesis of organic matter on the Earth has accumulated since MILLER (1953) first demonstrated the synthesis of organic compounds from methane, ammonia and water (FOX, 1965; PONNAMPERUMA and GABEL, 1968). The occurrence of extraterrestrial synthesis is indicated by carbonaceous material isolated from meteorites (HAYES, 1967), and the possibility of cosmochemical organic synthesis has been raised by the identification of simple compounds containing carbon, hydrogen, nitrogen and oxygen in comets (SWINGS and HASER, 1956), interstellar dust clouds (CHEUNG *et al.*, 1969; SNYDER *et al.*, 1969) and in stellar atmospheres (VARDYA, 1966).

It was hoped that analysis of the organic substances in lunar material could confirm or shed new light on the mechanisms for synthesis of carbon compounds in the solar system (SAGAN, 1961; DAYOFF *et al.*, 1964; ECK *et al.*, 1966; STUDIER *et al*, 1968). However, with the possible exceptions of porphyrins (Part II) and organosiloxanes (Part III), no significant amounts of extractable carbon–hydrogen compounds were detected (Part I). This report describes another part of the results obtained by the NASA–Ames Research Center Consortium during analysis of the lunar sample labelled "10086-3 Bulk A fines". Attention was focused on the determination of total carbon and the identification of some types of nonextractable and volatile carbon compounds in the lunar sample. This knowledge could provide clues to processes that may possibly be, or have been, involved in lunar organic synthesis. Therefore, products derived from a variety of experiments involving pyrolysis and acid treatment of the sample have been examined.

---

* Present address: Institute of Geophysics and Planetary Physics, University of California, Los Angeles, California 90024, Permanently at Division of Mineral Chemistry, C.S.I.R.O., Chatswood, N.S.W., Australia.

† Present address: Institute of Geophysics and Planetary Physics, University of California, Los Angeles, California 90024.

EXPERIMENTAL

Preliminary treatment

All lunar samples prior to analysis were degassed by heating at 150°C at 10^{-3} torr pressure for 48 hr. Condensable gases were collected in a sampling bulb immersed in liquid nitrogen. Gas chromatography of the bulb contents showed the presence of traces of C_2, C_3 and C_4 hydrocarbon which may represent contamination in the samples and/or diffusion pump vapors arising from back-streaming. δC^{13} ratios (KAPLAN and SMITH, 1970) have clearly demonstrated the need for removal of possible volatile contaminants in this manner.

Total carbon

Total carbon was measured by combustion of one gram samples in an atmosphere of oxygen at about 700 mm Hg pressure at 1050°C. The gaseous products were passed over copper oxide catalyst, silver wire and lead oxide (to remove halogens and oxides of nitrogen and sulfur, respectively). The resulting CO_2 was purified from water by distillation through a dry ice-acetone trap and its volume measured on a manometer. Details of the combustion apparatus are given elsewhere (KAPLAN *et al.*, 1970).

Volatile carbon compounds obtained by acid treatment

Estimations of the concentration of carbide carbon in samples of lunar fines, the Mighei carbonaceous chondrite and cohenite, containing Fe_3C, (removed from Canyon Diablo iron meteorite) were obtained from analysis of gaseous hydrolysis products generated by the following procedure:

A reaction tube, sealed at one end with a breakable glass diaphragm, was cooled to −120°C in a bath of isohexane–liquid nitrogen, and 3 ml of 4·8 N hydrochloric acid was added. When the acid had frozen, the weighed, powdered sample was introduced, and the reaction tube, still at a temperature of −120°C, was connected to a vacuum system. Entrapped air was exhausted from the tube as quickly as possible to a pressure of less than 10^{-2} torr, and the tube was sealed with a flame. After shaking to thoroughly mix the contents, the tube was heated at 98°C for 16 hr. The tube was cooled and reconnected to the evacuated system, and the vacuum pumps isolated from the collection system. The gas-retaining seal was fractured with an externally controlled iron rod. The gas evolved from the Mighei chondrite was collected directly at −195°C in a tube containing about 100 mg freshly activated charcoal, sealed at one end with a breakable glass diaphragm. Those gases from the lunar fines and cohenite, however, were held over outgassed, concentrated solutions of ether zinc sulfate or potassium hydroxide to remove hydrogen sulfide or hydrogen sulfide and carbon dioxide, respectively, prior to their final collection. Noncondensable gases were present in every experiment and could not be collected quantitatively. The gas sample tubes were flame sealed when the pressure in the system had fallen to a constant minimum.

The residue from hydrochloric acid hydrolysis of the lunar fines was extracted with boiling benzene. The extract was exaporated almost to dryness and examined by gas chromatography.

During analysis for sulfur (Part I and KAPLAN and SMITH, 1970) other lunar materials (10002-54, breccia; 10049, hard rock; 10057-40, hard rock; 10060-22, breccia; 10084, fines) were warmed to about 100°C in 85% phosphoric acid and left standing overnight at 60°C Those gases not trapped in 5% silver nitrate were collected at −195°C over freshly activated charcoal and preserved for analysis.

In an experiment modeled after those of ABELL *et al.* (1970) and BURLINGAME *et al.* (1970), the lunar sample was treated with hydrofluoric acid. Estimations of the concentration of carbon monoxide, carbon dioxide and other volatile compounds were obtained by analysis of the gaseous products. The reaction was carried out in a teflon-lined tube and the mixture was stirred with a teflon-coated magnet for 18 hr at 25°C. Evolved gases were collected at −195°C on 5Å molecular sieves freshly activated at 400°C and evacuated to 10^{-3} torr pressure. Otherwise the procedure used was essentially that described for hydrolysis of carbides. In all hydrolysis experiments, gas collection methods and reagents were monitored by parallel control experiments involving identical reaction and sample collection conditions.

Residues from repeated treatment of lunar fines with hydrochloric and hydrofloric acid were combusted in oxygen as described above and the total residual carbon determined as carbon dioxide.

Volatile carbon compounds obtained by pyrolysis

The simple pyrolysis system shown in Fig. 1 was connected to the combustion apparatus (KAPLAN *et al.*, 1970). A 1-g sample was again outgassed at 150°C (for 2 hours) at 10^{-3} torr pressure. The vacuum pumps were isolated from the pyrolysis and gas collection system and the sample was pyrolyzed over three temperature ranges: 150–250°C, 250–500°C and 500–750°C. The volatile products released were frozen into a sampling bulb immersed in liquid nitrogen, the products from each pyrolysis step being condensed directly into separate sample bulbs. The contents of these bulbs were reserved for analysis by gas chromatography and mass spectroscopy. The total quantity of carbon remaining in the final pyrolysis residue was determined by measurement of the carbon dioxide produced on combustion at 1050°C.

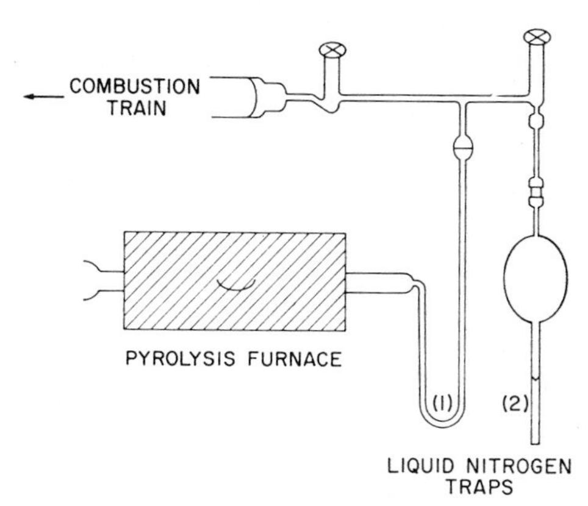

Fig. 1. Pyrolysis apparatus.

In a separate experiment, after completion of pyrolysis at 750°C, the pyrolysis chamber was evacuated and cooled to 25°C. It was then filled to 76 torr pressure with a mixture of carbon monoxide and hydrogen in a 1 to 4 ratio. Products obtained by reheating at 750°C for one hour were collected in the usual fashion.

Methods of analysis

High and low resolution mass spectra were obtained with CEC 21-110 and 21-491 mass spectrometers, respectively, at an ionizing voltage of 70 eV. Quantitation on the 21–110 was accomplished by comparison of the abundances of characteristic ions in sample spectra with those of corresponding ions in spectra of known amounts of reference compounds. Gases, in collection bulbs at measured pressures, were admitted to the 21–110 expansion reservoir. Analyses were performed on gases passing through the gold leak into the mass spectrometer source compartment. Under these circumstances the detectable limits of carbon monoxide and carbon dioxide were about 5 ng. For the bulk sample this corresponds to a minimum detectable concentration of 8 μg per gram of lunar fines.

Gas chromatographic analyses for hydrocarbons were conducted on two instruments equipped with single flame ionization detectors: Varian Aerograph Models 1520B and 200B. The model 1520B was operated in the dual-column, thermal conductivity mode for detection of carbon dioxide. All analyses of material with molecular weight less than 75 were performed on a 6 ft × 0·125 in stainless steel column packed with 100–120 mesh Poropak Q. For higher molecular weight material a 5 ft × 0·125 in stainless steel column packed with 3% SE-30 on 100–120 mesh Gaschrom Z was used.

Gas sample tubes were fitted with ground-glass vacuum stopcocks and evacuated to 10^{-6} torr pressure before the gas-retaining seals were broken. Prior to gas chromatography, the end of the stopcock was sealed with a silicone rubber septum and evacuated through a syringe needle. The tubes were heated to 200°C for 10 min, and sampling was conducted through the septum with a gas-tight Hamilton syringe. Since sample tubes were usually at reduced pressure, the syringe always was partially filled with laboratory air prior to insertion in the injection port. When appropriate, corrections were made for possible atmospheric contamination. The carbon dioxide determination is considered primarily of qualitative significance because repeated gas chromatographic measurements varied as much as ± 25 per cent. Methane was present in laboratory air, but only in barely detectable amounts. Quantitation was based on detector responses to known volumes of standard gases. Hydrocarbon determinations were routinely reproducible to ± 10 per cent. A detection limit could be set at 4×10^{-4} μg of any C_1 to C_4 hydrocarbon. For the bulk sample this corresponds to a detectable concentration limit of 4×10^{-3} μg per gram of lunar fines.

RESULTS AND DISCUSSION

Total carbon

The average total carbon concentration in duplicate analyses of 10086-3 Bulk A lunar fines was 157 ± 14 μg/g (Part 1). In good agreement with this figure were values of 142 ± 10 μg/g (MOORE *et al.*, 1970) and 168 μg/g BURLINGAME *et al.*, 1970), obtained by different methods of analysis. Examination of sieved material showed that carbon was more abundant in fine than coarse particles. In a sample of 10084 lunar fines, particles >60–140 mesh contained 92 μg/g total carbon; from 140 to 300 mesh, 183 μg/g; and finer than 300 mesh, 261 μg/g (KAPLAN *et al.*, 1970). This trend is consistent with that reported by MOORE *et al.* (1970).

Of the 157 ± 14 μg/g total carbon, JOHNSON and DAVIS (1970) have estimated that 40 ± 8 μg/g could be attributed to pyrolyzed organic matter. Results in Part I indicate that virtually no organic compounds were extractable from the lunar sample by water or organic solvents. Extraction with hydrochloric acid, however, yielded organosiloxanes (Part II) which might account for some of the organic carbon.

Volatile carbon compounds released by acids

In experiments designed to detect metallic carbides, gases derived from lunar samples by acid hydrolysis were initially examined with a low resolution mass spectrometer. Ions which were considered significant were at least an order of magnitude more abundant than corresponding ones in spectra of instrument background and control blanks. Ions of particular importance appeared with m^+/e 15, 16, 25, 26, 27, 29, 30, 41, 42 and 43. These are ions expected to be prominent in a spectrum of a mixture of C_1, C_2 and C_3 hydrocarbons. An ion with m^+/e 28 was also observed, possibly arising from nitrogen and carbon monoxide as well as hydrocarbon; however, the abundance of the corresponding ion in the control blank was comparable. Similar results were obtained with the meteorite samples.

Gas chromatography confirmed the presence of hydrocarbons. Typical chromatograms are depicted in Fig. 2. The excellent agreement in elution times between peaks in the samples and those in the standard gas mixture, coupled with the nonexistence of substances having similar chromatographic properties and flame-ionization detector responses, leaves no doubt regarding the identity of the C_1, C_2 and C_3 hydrocarbons.

The amounts of carbon represented in the samples by the hydrocarbons are summarized in Table 1. Because of their finite vapor pressure at -195°C and their solubility in the reagent solutions, all evolved gases were never completely recovered. Therefore, the data in Table 1 represent minimum values. Methane, ethene, acetylene, ethane, propene, propane and small amounts of unidentified C_3 and C_4 hydrocarbons

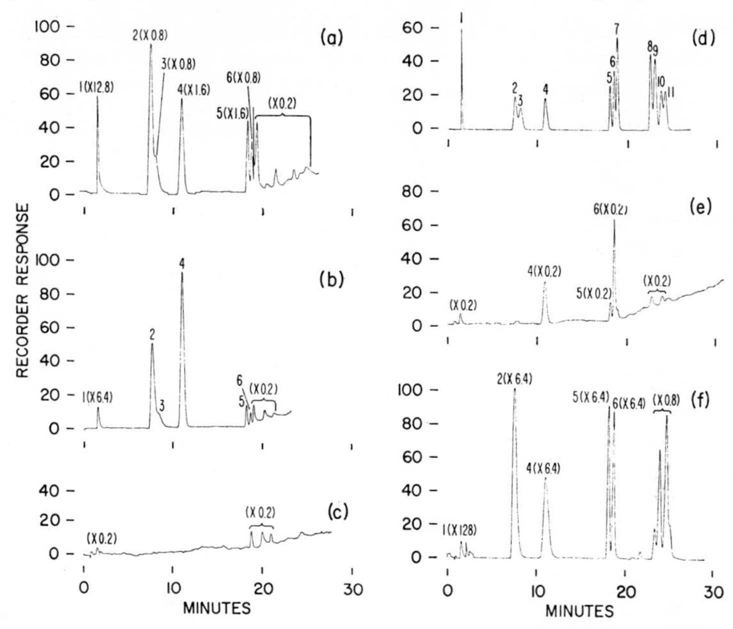

Fig. 2. Gas chromatograms of hydrocarbons obtained from: (a) and (b) HCl treatment of 10086-3 Bulk A fines; (c) control blank; (d) standard mixture containing about 20 μl each of (1) methane, (2) ethene, (3) acetylene, (4) ethene, (5) propene, (6) propane, (7) allene, (8) isobutane, (9) butene-1, (10) butane, (11) *cis*-2-butene (attenuation $\times$ 160 throughout run); (e) and (f) HCl treatment of Mighei meteorite and cohenite, Fe_3C, respectively. A 6 ft $\times$ 0·125 in. stainless steel column packed with 100–120 mesh Poropak Q was used on a Varian-Aerograph Model 1520 B instrument equipped with flame detector. Flow rate was 20 ml/min. Oven temperature initially held at 35°C for 10 min then programmed at 10°/min to 140°C where it was held. Unless otherwise indicated, peak attenuations were $\times$ 0·4 (attenuation 4, range 0·1). Each injection represented 7 per cent of the total sample.

were obtained from the lunar fines representing up to 21 μg/g in one case and 4·9 μg/g in another (all values designated by "μg/g" are calculated to signify μg of carbon per gram of lunar sample). As expected considerably more hydrocarbons were produced from the sample of cohenite. On the other hand, yields from Mighei meteorite were lower, and methane and ethene were conspicuously absent.

Table 1. Amounts of carbon (μg/g) produced as hydrocarbons during hydrochloric acid hydrolysis of lunar and meteorite samples

Experiment	CH_4	C_2H_4	C_2H_2	C_2H_6	C_3H_6	C_3H_8	Other C_3	C_4	Total
1. 10086-3 fines	6·3	3·8	1·4	5·0	2·9	1·3	0·58	0·20	21·3
2. 10086-3 fines	0·86	1·10	0·25	2·0	0·20	0·19	0·29	0	4·9
3. Mighei	0	0	0	0·46	0·18	1·34	0·12	0	2·1
4. Cohenite (Fe_3C)	240	750	0	370	510	690	0	150	2710
5. Blank	0	0	0	0	0	0	0·24	0·05	0·29

When H_2S in evolved gases was removed with KOH, (Experiment 1, Table 1) the total hydrocarbon yield was about four times greater than when $ZnSO_4$ was used (Experiment 2, Table 1). This difference could have been a result of sample inhomogeneity, since preferential solution of hydrocarbons in aqueous zinc sulfate does not seem likely.

The residue from hydrochloric acid treatment of the lunar fines was extracted with boiling benzene. Gas chromatography of the concentrated extract on a general purpose SE-30 column revealed no compounds above C_{10} in concentration greater than 10^{-3} $\mu g/g$. This was confirmed by a separate experiment in which hexane was

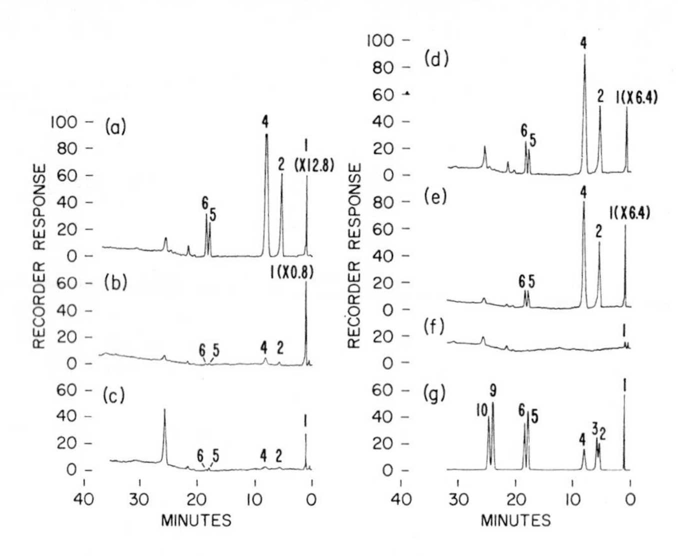

Fig. 3. Gas chromatograms of hydrocarbons obtained from: H_3PO_4 treatment of lunar samples (a) 10002-54; (b) 10049; (c) 10057-40; (d) 10060-22; (e) 10084; (f) control blank; (g) standard mixture containing 2 μl of each gas (identified in Fig. 2), attenuation $\times$ 64. Conditions were the same as Fig. 2 except that helium flow was 30 ml/min.

sealed in the hydrolysis tube with hydrochloric acid and lunar sample. Analysis of the hexane, in the same manner as the benzene extract, revealed no detectable hydrocarbons above C_{10}. Hydrocarbons in the C_5 to C_{10} range could not be determined conveniently under the concentration and chromatographic conditions employed. Apparently hydrocarbons above C_{10} were not generated during acid treatment.

Chromatograms of hydrocarbons released during mild phosphoric acid treatment of other lunar samples (KAPLAN and SMITH, 1970; KAPLAN *et al.*, 1970) are shown in Fig. 3. The products were essentially the same, in the C_1 to C_3 region, as those derived from acid treatment of the cohenite sample and, except for the absence of acetylene, from the 10086-3 lunar fines. The amount of carbon represented by the hydrocarbons are shown in Table 2. Values for total carbon concentration determined by KAPLAN and SMITH (1970) are included. In no instance did hydrocarbon yields attain the values found in the samples of 10086-3 fines (Table 1). Probably, the

Table 2. Amounts of carbon (μg/g) produced as hydrocarbons during mild phosphoric acid hydrolysis of various lunar samples

Experiment		CH_4	C_2H_4	C_2H_2	C_2H_6	C_3H_6	C_3H_8	Other C_3	C_4	Total
1. 10002-54, breccia	(190)*	0·66	0·11	0	0·24	0·034	0·052	0	0·036	1·13
2. 10049, hard rock	(70)	0·029	0·007	0	0·009	tr†	tr	0	tr	0·045
3. 10057, hard rock	(16)	0·01	tr	0	tr	tr	tr	0	tr	0·01
4. 10060-22, breccia	(134)	0·30	0·10	0	0·27	tr	0·05	0	0·06	0·78
5. 10084, fines	(132)	0·36	0·10	0	0·20	tr	tr	0	tr	0·76
6. Blank		0	0	0	0	0	0	0	0·39	0·39

* Total carbon content in μg/g presented in parentheses taken from data in KAPLAN and SMITH (1970)
† Trace amounts; about 4×10^{-3} μg/g.

conditions of phosphoric acid hydrolysis were not as favorable as those with hydrochloric acid. Interestingly, total hydrocarbon yields paralleled total carbon concentrations. Control blanks for all hydrolysis experiments contained no detectable amounts of C_1, C_2 and C_3 hydrocarbons. Evidently, the corresponding hydrocarbons found in lunar samples did not arise from contamination in the reaction system and handling.

Possibly, the hydrocarbons were not derived from carbides, but rather were freed from mineral enclosures by HCl treatment. Data obtained from experiments described elsewhere have some bearing on this question. During pyrolysis experiments most of the total carbon was detected as oxides. Hydrocarbons should be as easily released from mineral enclosures as oxides of carbon, therefore, a major portion of the hydrocarbons should have been released by pyrolysis. In fact the C_1, C_2 and C_3 hydrocarbons amounted to less than 0·6 μg/g, (see below) as compared to the much higher values (Table 1) generated by HCl treatment. Evidently, if hydrocarbons were trapped in mineral enclosures, and subsequently freed by HCl, they could have contributed only a small fraction of the hydrocarbons in the hydrolysis products.

Heating lunar materials of different types in aqueous acids consistently yielded low molecular weight hydrocarbons. Carbides (SCHMIDT, 1934) appear to be the only substances which would yield hydrocarbons under such conditions. Therefore our results, combined with mineralogic observations made by ANDERSON *et al.* (1970), constitute convincing evidence for the existence of carbide carbon in lunar samples.

In an experiment modeled after those reported by ABELL *et al.* (1970) and BURLINGAME *et al.* (1970), gases evolved during hydrofluoric acid digestion of the lunar sample were examined by high resolution mass spectroscopy. No evidence for hydrocarbons, carbon dioxide, carbon monoxide, or other low molecular weight products was found. The minimum detectable concentration of a compound in the sample corresponded to 8 μg/g. Gas chromatography, however, revealed the presence of a hydrocarbon mixture, similar in composition to that obtained by hydrochloric acid hydrolysis (Table 1), in amount corresponding to 7·0 μg/g.

The absence of carbon dioxide was consistent with the observation noted earlier in Part I that hydrochloric acid treatment produced only a trace concentration (less than 5 μg/g) of carbon as carbon dioxide. These results indicate that a gram of lunar fines contained, at most, 5 μg of carbon in the form of carbonates. Furthermore, acid treatment did not appear to free carbon dioxide or carbon monoxide from mineral enclosures. ABELL *et al.* (1970) also did not report finding carbon monoxide during hydrofluoric acid treatment. BURLINGAME *et al.* (1970), however, reported 66 μg/g. Although carbon monoxide amounting to 8 μg/g could have been undetected in our experiments, it is possible that the disparate results reflect inhomogeneity in the 10086-3 Bulk A fines.

Volatile carbon compounds produced by pyrolysis

Products obtained during stepwise heating of lunar fines at 150°–250°C, 250°–500°C and 500°–750°C were examined by gas chromatography. Carbon dioxide amounting to about 50 μg/g was detected, most appearing in the lower temperature ranges. Since large variations in the measurements occurred ($\pm$25%) this result was primarily of qualitative value. Carbon monoxide could not be determined under the gas chromatographic conditions used.

After ABELL *et al.* (1970), BURLINGAME *et al.* (1970), EPSTEIN and TAYLOR (1970) and ORÓ *et al.* (1970) reported detection of carbon monoxide during pyrolysis of lunar samples, efforts were made in this laboratory to determine its presence by means other than gas chromatography. Thus in a second experiment, pyrolysis products were examined by high resolution mass spectroscopy. Carbon monoxide was identified only in the 500°–750°C range and amounted to 121 μg/g. Up to 8 μg/g could have been produced in each of the other temperature ranges without being detected. Although no carbon was found as carbon dioxide at any pyrolysis stage by this method, a total of 24 μg/g (3 $\times$ 8 μg/g) could have gone undetected. The discrepancy in the carbon dioxide determinations does not appear so serious when the detection limits in the second experiment and the qualitative nature of the first are taken into consideration.

Apparently, carbon monoxide was the major oxide of carbon evolved during pyrolysis. Release of carbon dioxide below 500°C has been reported by LIPSKY *et al.* (1970), ABELL *et al.* (1970) and ORO *et al.* (1970). The last group found more carbon dioxide than carbon monoxide between 300°—500°C, but from 600°–750°C, the carbon monoxide exceeded carbon dioxide by a factor of ten. EPSTEIN and TAYLOR (1970) reported a similar trend. In the range of 150°–1150° BURLINGAME *et al.* (1970) detected 156 μg/g of carbon monoxide and 12 μg/g of carbon dioxide. FRIEDMAN *et al.* (1970) reported 199 carbon dioxide between 300°–950°C, but no carbon monoxide. Apparently, carbon monoxide and carbon dioxide determinations were reproducible between some laboratories, but not others. Inhomogeneity in lunar samples plus different laboratory procedures may have been important factors in the varying results.

BURLINGAME *et al.* (1970) have suggested that the major part of the carbon monoxide detected was formed on the moon by oxidation of reduced carbon with metal oxides under the influence of meteorite impact or other energy sources and entrapped in mineral enclosures. It is also conceivable that some of the carbon monoxide generated in experiments at 1100°C or higher resulted from reduction of silicate or metal oxides by elemental carbon during pyrolysis of the sample. Some possibilities are shown below. (All thermodynamic data taken from DOW CHEMICAL COMPANY, THERMAL RESEARCH LABORATORY, 1965.) In the first two cases, the free energies are positive, but the equilibrium pressures of carbon monoxide for the systems are appreciable. The thermodynamic calculations indicate that the reactions could occur to a

		ΔF° (kcal)	P_{CO} (atm)
$SiO_2 + 3C \rightarrow SiC + 2CO$	(1100°C)	28·5	$5·4 \times 10^{-3}$
$TiO_2 + C \rightarrow TiO + CO$	(1100°C)	12·0	$1·26 \times 10^{-2}$
$FeO + C \rightarrow Fe + CO$	(1100°C)	−11·8	74

significant extent around 1100°C. Confirmation of this point was provided in an experiment modeled after one performed by BURLINGAME *et al.* (1970). One hundred mg each of lunar fines and graphite were intimately mixed and pyrolyzed at 1100°C for one hour. During this time the mixture fused. When the gases were combusted in oxygen, 6·0 cm³ (25°C, 1 atm) of carbon dioxide was obtained, an amount 200 times that expected from pyrolysis of all the carbon in the lunar sample. Pyrolysis of 100 mg of graphite alone yielded less than 0·2 cm³. It can be calculated that 3·1 mg of graphite was converted to carbon monoxide by 100 mg of lunar material. Clearly, such a process should be seriously considered as a source of carbon monoxide both in pyrolysis experiments and on the moon. On the other hand, high temperature conversion of carbon to carbon monoxide by way of the water–gas reaction ($C + H_2O \rightarrow CO + H_2$) at 750°, although thermodynamically favorable ($\Delta F = -2·13$ kcal), seems unlikely since most of the water evolved during pyrolysis of lunar material is expected to be driven off below 650°C (EPSTEIN *et al.*, 1970). This reaction, however, may have been significant in the genesis of carbon monoxide on the lunar surface.

In the experiment where carbon dioxide was detected by gas chromatography, low resolution mass spectra of the products at each pyrolysis stage revealed the absence of hydrocarbon ions in abundances above instrument background. Gas chromatography failed to reveal any hydrocarbons at 150°–250°C. However, between 250° and 500°C methane and an unknown C_4 hydrocarbon appeared in concentrations of 0·03 μg/g and 0·58 μg/g respectively. At 500°–750°C, 0·05 μg/g methane, 0·27 μg/g ethene, 0·23 μg/g propene, and 0·44 μg/g unidentified C_4 hydrocarbon were produced. Thus, from 150°–750°C less than 0·6 μg/g of carbon identified as C_1 to C_3 hydrocarbons was generated, most of it appearing at the upper end of the temperature range. The chromatograms are reproduced in Fig. 4. The absence of hydrocarbons from 150° to 250°C indicates that the C_1 to C_4 hydrocarbons found at higher temperatures were notdue to contamination from the pyrolysis system. Had detectable amounts of residual hydrocarbons (from vacuum pump vapor, for instance) been present in the system, they would have been collected during the low temperature pyrolysis stage. These results are quite similar to those reported by others. Pyrolysis of lunar samples up to 400°C by MURPHY *et al.* (1970) and above 400°C by ORO *et al.* (1970) afforded traces of methane. From 130° to 700°C, ABELL *et al.* (1970) obtained small amounts ($<$1 μg/g) of methane. BURLINGAME *et al.* (1970), from 500°–1150°C, detected traces of hydrocarbons. At 510°C, NAGY *et al.* (1970) also report traces, but at 700° larger unstated amounts, predominantly methane, were found. In all instances except the last, only trace amounts ($\leq$1 μg/g) of hydrocarbons were produced by pyrolysis. Notably, comparison of the yields of hydrocarbons and oxides of carbon produced by pyrolysis show that gaseous carbon from lunar material was primarily in oxidized states.

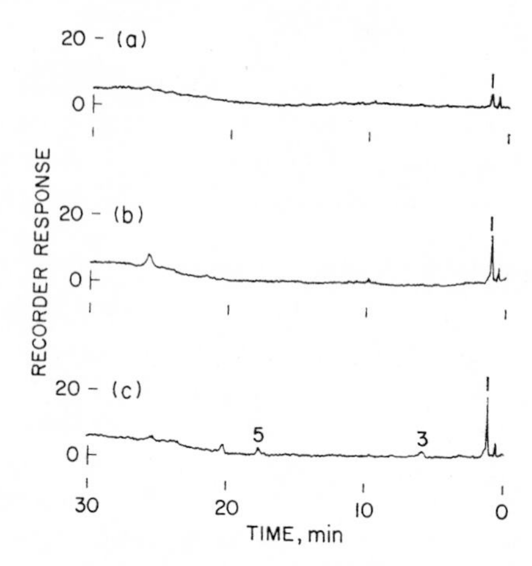

Fig. 4. Gas chromatograms of hydrocarbons produced by pyrolysis of 10086-3 lunar fines at (a) 150°–250°C; (b) 250°–500°C; and (c) 500°–750°C. Conditions were the same as Fig. 3, except that 0·4 per cent of each sample was injected.

Two obvious processes for pyrolytic production of hydrocarbons are thermal cracking of more complex organic compounds and release from mineral enclosures in the inorganic matrix. Trapped gaseous hydrocarbons have been freed from meteorites by STUDIER *et al.* (1965) and BELSKY and KAPLAN (1970). Another possibility was suggested by STUDIER *et al.* (1968) who demonstrated that in the presence of iron meteorite powder, hydrocarbons could be synthesized from carbon monoxide and hydrogen at temperature above 150°C. Lunar samples contain small amounts of native iron (LSPET, 1969), and carbon monoxide and hydrogen (EPSTEIN and TAYLOR 1970) were produced by pyrolysis up to 750°C. Therefore, all the ingredients were available for the Fischer–Tropsch type synthesis. When a previously pyrolyzed lunar sample was reheated to 750° in the presence of carbon monoxide and hydrogen, only traces of methane, ethane and propene, which were absent in a control experiment, were detected (less than 0·05 μg/g) by gas chromatography. Apparently, a Fischer–Tropsch type reaction was not favorable. Although the reaction could account for a small fraction of the hydrocarbons generated during pyrolysis, most were probably produced by cracking of organic compounds or liberation from the inorganic matrix.

Residual carbon

In the experiment where gas chromatography indicated generation of carbon dioxide during pyrolysis of lunar fines to 750°C, the residue remaining contained 63 μg/g of carbon. When a pyrolysis residue from a similar experiment was heated to 1050°C and the resulting gases combusted to carbon dioxide, 100 per cent of the initial total carbon was accounted for. Direct combustion of the residual material produced no additional carbon dioxide. Apparently, no carbon remained after pyrolysis at 1050°C. In light of the experiment involving heating of lunar fines and graphite described previously, the results of pyrolysis and combustion of 1050°C suggest that some residual carbon may have been in the form of elemental carbon.

BURLINGAME *et al.* (1970) reported that, after digesting a sample of lunar fines in hydrofluoric acid, pyrolysis of the residue to 1100°C yielded 119 μg/g of carbon in the form of carbon monoxide. Although, as they suggest, the carbon monoxide could have been freed from the residual silica matrix by pyrolysis, it seems equally possible that some was produced by reduction of residual silica, metal oxides or salts by elemental carbon.

Material remaining after repeated treatment of the lunar sample with hydrochloric and hydrofluoric acids contained 36 $\pm$ 6 μg/g of carbon in duplicate experiments. Since carbides would have been removed by this treatment, the residual carbon should be largely composed of elemental carbon (graphite). The presence of carbides in the pyrolysis, but not in the hydrolysis, residues would account for part of the 27 μg/g difference in residual carbon content.

Summary and conclusions

Pyrolysis and hydrolysis experiments indicated that the carbon in the 10086-3 Bulk A lunar fines is present primarily in the form of carbide, carbon monoxide and carbon dioxide (or substances which yield oxides of carbon during pyrolysis or hydrolysis), and possibly elemental carbon. A crude, partial estimate of the carbon distribution in a 1-g sample of fines is as follows: Carbon dioxide accounted for about 50 μg of carbon in a pyrolysis experiment. Traces (<2 μg) of C_1 to C_4 hydrocarbons were also found. The pyrolysis residue contained 63 μg of carbon. Of that figure some was probably elemental carbon, and from 5 to 21 μg was carbide. Thus, of the total carbon content of 157 $\pm$ 14 μg, 113 μg, could be accounted for. Although carbon monoxide was not determined in this instance, the results of another pyrolysis experiment, in which it was identified by mass spectroscopy, suggested that carbon monoxide could account for the difference between 113 μg and the total carbon content. Control experiments indicate that the carbon compounds obtained from the lunar sample were indigenous to the lunar surface and not introduced as contamination from our laboratory.

The cosmological significance of lunar carbon is not understood at present. Since high temperatures may have been involved in the formation of materials in the Sea of Tranquillity (LSPET, 1969), products of primordial organic synthesis are not likely to have survived. This is consistent with the low concentration of organic compounds found in the samples.

Exposure of carbides, carbon monoxide, carbon dioxide, elemental carbon, hydrogen, metals and metal oxides, to sufficiently intense thermal and/or radiation energy on the moon's surface could convert some of these substances to organic molecules. Entrapment in the inorganic matrix could possibly preserve some of the products from further loss by volatilization or radiation-induced destruction. Results in Part I, however, indicate that if organic synthesis had occurred in the area of the Apollo 11 landing, very little evidence remains.

Acknowledgments—We thank Mr. FRITZ WOELLER and Mr. EDWARD RUTH for their valuable assistance in part of this work.

REFERENCES

ABELL P. I., DRAFFAN G. H., EGLINTON G., HAYES J. M., MAXWELL J. R. and PILLINGER C. T. (1970) Organic analysis of the returned lunar sample. *Science* **167**, 757–759.

ANDERSON A. T., JR., CREWE A. V., GOLDSMITH J. R., MOORE P. B., NEWTON J. C., OLSEN E. J., SMITH J. V. and WYLLIE P. J. (1970) Petrologic history of moon suggested by petrography, mineralogy and crystallography. *Science* **167**, 587–590.

BELSKY T. and KAPLAN I. R. (1970) Light hydrocarbon gases, C^{13}, and origin of organic matter in carbonaceous chondrites. Accepted for publication in *Geochim. Cosmochim. Acta* **34,** 257–278.

BURLINGAME A. L., CALVIN M., HAN J., HENDERSON W., REED W. and SIMONEIT B. R. (1970) Lunar organic compounds: search and characterization. *Science* **167**, 751–752.

CHEUNG A. C., RANK D. M., TOWNES C. H., THOMSON D. D. and WELCH W. J. (1968) Detection of NH_3 molecules in the interstellar medium by their microwave emission. *Phys. Rev. Lett.* **21**, 1701–1705.

DAYHOFF M. O., LIPPINCOTT E. R. and ECK R. V. (1964) Thermodynamic equilibria in prebiological atmospheres. *Science* **146,** 1461–1464.

DOW CHEMICAL COMPANY, THERMAL RESEARCH LABORATORY (1965) JANAF Thermochemical tables, National Bureau of Standards, Institute for Applied Technology.

ECK R. V., LIPPINCOTT E. R., DAYHOFF M. O. and PRATT Y. T. (1966) Thermodynamic equilibrium and the inorganic origin of organic compounds. *Science* **153,** 628–633.

EPSTEIN S. and TAYLOR H. P., JR. (1970) $^{18}O/^{16}O$, $^{30}Si/^{28}Si$, D/H, and $^{13}C/^{12}C$ Studies of lunar rocks and minerals. *Science* **167,** 533–535.

FOX S. W. (editor) (1965) The origins of prebiological systems and of their molecular matrices. Proceedings of a Conference held at Wakulla Springs, Oct. 1963, Academic Press, N.Y.

FRIEDMAN I., O'NEIL J. R., ADAMI L. H., GLEASON J. D. and HARDCASTLE K. Water, hydrogen, deuterium, carbon, C-13, and O-18 content of selected lunar material. *Science* **167,** 538–540.

HAYES J. M. (1967) Organic constituents of meteorites—a review. *Geochim. Cosmochim. Acta* **31,** 1395–1440.

JOHNSON R. D. and DAVIS C. C. (1970) Pyrolysis–hydrogen flame ionization detection of organic carbon in a lunar sample. *Science* **167,** 759–760.

KAPLAN I. R. and SMITH J. W. (1970) Concentration and isotopic composition of carbon and sulfur in Apollo 11 lunar samples. *Science* **167,** 541–543.

KAPLAN I. R., SMITH J. W. and RUTH E. (1970) Carbon and sulfur concentration and isotopic composition in Apollo 11 lunar samples. *Geochim. Cosmochim. Acta*, Supplement I.

LIPSKY S. R., CUSHLEY R. J., HORVATH C. G. and MCMURRAY W. J. (1970) Analysis of lunar material for organic compounds. *Science* **167,** 778–779.

LSPET (LUNAR SAMPLE PRELIMINARY EXAMINATION TEAM) (1969) Preliminary examination of lunar samples from Apollo 11. *Science* **165,** 1211–1227.

MILLER S. L. (1953) A production of amino acids under possible primitive earth conditions. *Science* **117,** 528–529.

MOORE C. B., LEWIS C. F., GIBSON E. K. and NICHIPORUK W. (1970) Total carbon and nitrogen abundances in lunar samples. *Science* **167,** 495–497.

MURPHY R. C., PRETI G., NAFISSI-V M. M. and BIEMANN K. (1970) Search for organic materials in the lunar fines by mass spectroscopy. *Science* **167,** 755–756.

NAGY B., DREW C. M., HAMILTON P. B., MODZELESKI V. E., MURPHY M. E., SCOTT W. M., UREY H. C. and YOUNG M. (1970) Organic compounds in lunar samples: Pyrolysis products, hydrocarbons, amino acids. *Science* **167,** 770–773.

ORÓ J., UPDEGROVE W. S., GILBERT J., MCREYNOLDS J., GIL-AV E., IBANEZ J., ZLATKIS A., FLORY D. A., LEVY R. L. and WOLF C. (1970) Organogenic elements and compounds in surface samples from the Sea of Tranquillity. *Science* **167,** 765–767.

PONNAMPERIUM C. and GABEL N. (1968) Current status of studies on the origin of life. *Space Life Sci.* **1,** 64–96.

SAGAN C. (1961) Organic matter on the moon. *Nat. Acad. Sci., Nat. Res. Council, Publ.* **757,** 49 pp.

SCHMIDT J. (1934) Decomposition of carbides by water or dilute acids. *Z. Elecktrochem.* **40,** 170–174.

SNYDER L. E., BUHL D., ZUCKERMAN B. and PALMER P. (1969) Microwave detection of interstellar formaldehyde. *Phys. Rev. Lett.* **22,** 679–681.

STUDIER M. H., HAYATSU R. and ANDERS E. (1965) Organic compounds in carbonaceous chondrites. *Science* **149,** 1455–1459.

STUDIER M. H., HAYATSU R. and ANDERS E. (1968) Origin of organic matter in early solar system—I Hydrocarbons. *Geochim. Cosmochim. Acta* **32,** 151–173.

SWINGS P. and HASER L. (1956) Atlas of representative cometary spectra. University of Liege Astrophysical Institute, Louvain.

VARDYA M. S. (1966) March with depth of molecular abundances in the outer layers of K and M stars. *Mon. Notc. Roy. Astron. Soc.* **134,** 347–370.

Proceedings of the Apollo 11 Lunar Science Conference, Vol. 2, pp. 1871 to 1873.

Analysis of lunar material for organic compounds*

S. R. LIPSKY, R. J. CUSHLEY, C. G. HORVATH and W. J. MCMURRAY
Section of Physical Sciences, Yale University School of Medicine, New Haven, Connecticut 06520

(*Received* 11 *February* 1970; *accepted* 11 *February* 1970)

Abstract—A sample of lunar material from Apollo 11 was subjected to analysis by several techniques, which included mass spectrometry, gas chromatography, liquid chromatography and nuclear magnetic resonance and their variations, in an effort to detect the presence of organic compounds. None were found. On the basis of the sensitivity ascribed to certain of the methods employed, it is assumed that if organic matter were present it would exist in concentrations less than 1 part per million.

THREE sample-handling procedures were utilized during efforts to determine the existence of organic compounds in the Apollo 11 lunar sample 10086.5 (class D). First, an aliquot of untreated sample was directly analyzed by gas chromatography and mass spectrometry. Second, the sample was subjected to extraction by certain organic solvents MEINSCHEIN (1965) the solvents were removed, and portions of the residue, if any, were analyzed by gas chromatography, mass spectrometry, and nuclear magnetic resonance. Third, an aliquot of the sample was acid-hydrolyzed, and the hydrolyzate was extracted and subjected to analysis by high-performance liquid chromatography on pellicular resins, by gas chromatography, and by mass spectrometry.

All solvents were of spectroscopic grade and were redistilled when necessary. A series of blanks was run prior to assay of the material in all instances.

For the nuclear-magnetic-resonance analysis a 23-g sample of lunar material was placed in an extraction thimble and extracted with a solution containing 80 ml of a 4:1 benzene–methanol (spectrograde, redistilled solvents) mixture in a Soxhlet extraction apparatus for 24 hr and then was stirred as a slurry for 24 hr. The contents were decanted into four test tubes and centrifuged for 2 hr. The supernatants were combined, and an aliquot was evaporated to dryness in a stream of pure nitrogen. The contents were triturated 18 hr with 0·8 ml of spectrograde carbon tetrachloride, and the carbon tetrachloride solution was added to a Wilmad Imperial 507PP sample tube (5 mm). A small amount of Matheson tetramethylsilane was added as a reference signal for the field-frequency lock. A blank was determined with identical materials and apparatus.

The nuclear-magnetic-resonance study was conducted with a Bruker HFX-3 spectrometer operating at 90 Mhz (H^1). The sensitivity was increased by use of a 4096 channel computer of average transients (CAT). The spectrum was determined over a sweep range from 30 to 1166 hz downfield with tetramethylsilane. The sweep speed, determined by the CAT, was 80 msec per channel, and the signal was filtered with an input integrator (time constant = 80 msec per channel).

* This article was reprinted from Science (Vol. 167, pp. 778–779, 1970) without further refereeing or revision by the author.

After 27 hr, the spectrum showed peaks at 133 hz and a multiplet at 305 hz. The multiplet was also present in the blank and is a solvent impurity. No other peaks were present. The peak at 133 hz was probably due to the intense tetramethylsilane peak due to repeated scans by the signal-averaging device as evidenced by the spectrum of the blank.

The sensitivity of the Bruker HFX-3 spectrometer for a 1-per cent (by volume) ethylbenzene solution, using a 5-mm sample tube, is given as a 50-to-1 signal-to-noise ratio. The CAT further increases the signal-to-noise ratio by a factor of 40. Therefore the sensitivity of a sample with the same molecular weight and number of hydrogen atoms as ethylbenzene will be approximately 5×10^{-4}. Any material present in a concentration of less than 1×10^{-4} will be difficult to detect.

An analysis of the sample extract was made with a high-performance liquid chromatography system (HORVATH *et al.*, 1967; HORVATH and LIPSKY, 1969a,b) equipped with an ultraviolet detector. With this technique, minute quantities of substances, such as nucleic acid constituents, which have absorptivity at 250 nm, can be separated and determined.

Ten grams of the finely ground lunar sample were digested with 50 ml of 2·5N hydrochloric acid at 50°C for 2 days. After centrifugation the extract was lyophilized. Then 1 ml of distilled water was added to the vessel in order to prepare the sample solution for the chromatographic investigation. Another portion of this sample solution was then processed for analysis by mass spectrometry and gas chromatography.

The LCS 1000 liquid chromatograph (Varian Aerograph) with 1-mm i.d. columns packed with pellicular cation exchange or anion exchange resins (particle dia. 50 μm) was used in a fashion described earlier (2). The length of both columns was 150 cm. Dilute potassium hydrogen phosphate solutions having different pH values were used as eluents. In some experiments gradient elution (concentrated KH_2PO_4 served as strong eluent) was employed. The flow rate was 10 ml/hr, and the column temperature was 60°C in all experiments. A 10- or 20-μl sample solution was injected onto the chromatographic columns with a Hamilton No. 701 microsyringe.

Chromatographic runs were made on both cation exchanger and anion exchanger columns with eluents of various pH values in the range 2–7. No peaks attributable to retarded solutes have been found. In gradient elution with both columns small peaks appeared regularly at a certain point of the chromatogram, but these peaks were also obtained when a solvent blank was analyzed. Thus it is concluded that under the experimental conditions described no nucleic acid constituents or u.v.-absorbing materials of similar chromatographic behavior have been extracted from the lunar sample in an appreciable quantity. Considering the sensitivity of the technique, therefore, the amount of individual nucleic acid constituents, if such were present, must be less than 10^{-10} mole/g of sample.

For the gas-chromatography analysis a 25-mg sample of the lunar material without prior treatment was placed in a small quartz tube containing quartz wool and attached to a support coated open tubular (SCOT) column, 15·1 m long with a 0·058-cm inside diameter, containing silicone fluid (SF96, 1000 cs) and Igepal CO-880 (nonylphenoxy-polyethyleneoxy-ethanol). The column was connected to a flame ionization detector (sensitivity 10^{-12} g/sec), and the entire system was flushed with carrier gas (nitrogen) for 30 min. The sample was heated to 150°C, then to 250°C, and finally to 500°C in the flash heater zone of the gas chromatograph. The temperature of the sample tubes was maintained at each temperature level while the column was temperature-programmed from 50° to 180°C at the rate of 3°C/min. A control containing solvent-washed glass beads was substituted for the lunar sample and analyzed in a similar fashion. There was no evidence of organic compounds' being eluted from the sample. Similar results were obtained when these analyses were repeated on a 1·5-m by 0·058-cm SCOT column coated with Apiezon L and programmed to 210°C.

Aliquots from both the acid hydrolyzate and the organic solvent extracts of the sample and reagent blanks were concentrated and then subjected to the gas chromatographic procedures described above. Again, the findings were negative.

In all mass-spectrometry experiments, the spectrometer (M-S 9, Associated Electronics Industries, Manchester, England) was operated at 70 eV and at 100 or 300 μA trap current. The resolution was approximately 1:1000.

All samples, extracts, and reagent blanks were introduced via the direct insertion probe. Prior to sample introduction, the operating parameters of the instrument were optimized with perfluorotributylamine as a reference compound. After this material was pumped off, the ion source was baked at 250°C for 4 days.

A 10-mg portion of the lunar sample was loaded into a special quartz capillary tube attached to the direct-insertion probe. The probe was inserted into the mass spectrometer and spectra were recorded at various temperatures as the temperature of the ion source was increased from 50° to 250°C. The various spectra were dominated by peaks due to H_2O, N_2, O_2 and CO_2. The mass spectral patterns of the remaining peaks were indicative of very small quantities of low-molecular-weight hydrocarbons consistent with that noted as background. A repeat analysis gave similar results.

Studies of the extracts and reagent blanks by mass spectrometry did not reveal the presence of any components which could be considered to be indicative of endogenous material derived from the lunar sample.

In conlusion it is assumed that the principal components of the various spectra were contaminant gases and traces of low-molecular-weight hydrocarbons which had adsorbed onto the surface of the lunar sample during handling and exposure to the atmosphere.

Acknowledgments—This work was supported by NASA grants NAS 9-8072 and NGL 07-004-008.

References

Horvath C. and Lipsky S. R. (1969a) Rapid analysis of ribonucleosides and bases at the picomole level using pellicular cation exchange resin in narrow base columns. *Anal. Chem.* **41,** 1227-1234.

Horvath C. and Lipsky S. R. (1969b) Column Design in High Pressure Liquid Chromatography. *J. Chrom. Sci.* **7,** 109–116.

Horvath C., Preiss B. and Lipsky S. R. (1967) Fast liquid chromatography: An investigation of operating parameters and separation of nucleotides on pellicular ion exchangers. *Anal. Chem.* **39,** 1422–1428.

Meinschein W. G. (1965) Soudan formation: Organic extracts of early Precambrian rock. *Science* **150,** 601–605.

Proceedings of the Apollo 11 Lunar Science Conference, Vol. 2, pp. 1875 to 1877.

Search for alkanes of 15–30 carbon atom length in lunar fines

W. G. MEINSCHEIN, T. J. JACKSON and J. M. MITCHELL
Department of Geology
and
EUGENE CORDES and V. J. SHINER, JR.
Department of Chemistry, Indiana University, Bloomington, Indiana 47401

(*Received* 22 *January* 1970; *accepted* 22 *January* 1970)

Abstract—A 50-g sample of lunar fines was subjected to stepwise extraction in a mixture of benzene and methanol while intact, after being pulverized, and after being digested in hydrofluoric acid. None of these three extracts contained detectable quantities of C_{15}–C_{30} alkanes. No C_{15} to C_{30} alkane was present in this lunar sample at a concentration exceeding 1 part per billion by weight.

ALKANES are saturated organic compounds that are composed solely of carbon and hydrogen atoms. These hydrocarbons are ubiquitous but minor constituents of the waxes, fats, and oils of plants or animals as reported by GERARDE and GERARDE (1962). Because alkanes are less readily assimilated than other food substances and are chemically less reactive than most biological compounds, alkanes are preferentially preserved relative to other organic materials in sedimentary environments. Natural gas and petroleum deposits commonly appear to be concentrates of alkanes partially derived from preexistent organisms (MEINSCHEIN, 1961; MEINSCHEIN *et al.*, 1968). Many alkanes from petroleum have molecular structures that are either identical or similar to alkanes and related compounds in plants, animals, soils, marine sediments and sedimentary rocks (MEINSCHEIN, 1959; BURLINGAME *et al.*, 1965). The marked structural and distributional resemblances between certain alkanes in rocks and compounds in organisms have led to the widespread use of alkanes as molecular or chemical fossils (MEINSCHEIN, 1959; BARGHOORN *et al.*, 1965; EGLINTON and CALVIN, 1967) and to the recommendation that C_{15}–C_{30} alkanes be employed in exobiological research (MEINSCHEIN, 1962). This report deals with a search for C_{15}–C_{30} alkanes in a lunar sample.

All of the lunar fines allocated for organic analyses came from the 15 kg bulk sample. This sample was collected near the side of the landing module that faced to the lunar northeast (see KING, 1969). A 50 g portion of lunar fines with grain sizes in the 20 μm to 1 mm range was used. Repeated tests established that all solvents, reagents, and apparatuses were free of detectable amounts of organic contaminants with vapor pressures equivalent to C_{15}–C_{30} alkanes. The gas–liquid chromatograph at the sensitivity setting used to monitor contamination gave 5–12 per cent of full scale deflections with 10^{-7} g of pristane (2, 6, 10, 14-tetramethyl pentadecane), a C_{19} alkane. Sensitivity checks were run on standard solutions of pristane before and after each gas–liquid chromatographic analysis of a lunar extract sample.

The lunar fines were initially extracted with an azeotropic mixture of benzene and methanol in a specially designed combination Soxhlet extractor and ballmill as described in MEINSCHEIN *et al.* (1968). Throughout these extractions, a positive pressure of nitrogen was maintained in the system. Nitrogen flowed through a drying tower filled with silica gel into the extractor, and the nitrogen stream was exhausted through a bubbler tube with a 1 cm head of mercury. After approximately 500 ml/hr of solvent had been refluxed through the lunar fines for a period of 170 hr, a 200 ml sample of solvent was collected as it flowed from the extractor. The volume of this sample was reduced to several microliters by removal of solvent in a stream of filtered nitrogen at 40°C, and all of the residual sample that could be taken into a 10 μl syringe was injected into a Apeizon L coated capillary gas–liquid chromatographic column. No organic material except solvent was detected in this sample. The extractor was then disconnected from the reflux condenser and distillation flask, sealed with two glass stoppers, and rolled for 48 hr on a ballmill. All the extraction solvent was transferred from the distillation flask, and the volume of the extract was reduced by removal of solvent in a stream of filtered nitrogen at 40°C to approximately 100 μl. After a 2 μl portion of the remaining extract failed to yield detectable C_{15}–C_{30} alkane peaks in the gas chromatograph, the volume of the extract was further reduced to a few microliters which were injected into the gas chromatograph. Only a solvent peak was observed in this chromatogram.

The extractor, containing the ball-milled lunar sample, was reconnected to the reflux condenser and distillation flask, and azeotropic benzene and methanol was refluxed through the pulverized sample at a rate of approximately 500 ml/hr for 72 hr. At the end of this period, the solvent bumped causing the mercury within the bubbler to be drawn into the extractor. The extraction was discontinued at the time of this mishap, and gas chromatographic analyses of the extract of the pulverized lunar sample indicated an absence of detectable quantities of organic materials with vapor pressures equivalent to C_{15}–C_{30} alkanes.

All contents of the extractor, except for the stainless steel grinding balls, were transformed to a 2 l. Teflon beaker. Mercury was removed from the pulverized lunar minerals with an acid cleaned gold wire, and 500 g of 48% hydrofluoric acid was poured slowly with stirring into the beaker. The hydrofluoric acid layer in the beaker was covered with a 5 cm layer of benzene, and the contents of the beaker was stirred periodically until reactions ceased. The beaker was then warmed for a period of approximately 2 hr to 80°C. After most of the benzene layer evaporated, the reaction mixture was transferred to a 2 l. separatory funnel containing 500 ml of distilled water. This mixture was shaken thoroughly with three successive 200 ml portions of benzene. These benzene extracts were composited and concentrated. These extracts did not contain detectable quantities of volatile organic materials.

The lunar fines analyzed contained no C_{15}–C_{30} alkane at a concentration exceeding 1 ppb by weight, whereas the concentrations of C_{15}–C_{30} alkanes in rocks from the surface of the earth commonly exceed 100 ppm by weight.

Acknowledgements—The authors express their sincere thanks to the National Aeronautics and Space Administration, who sponsored this analysis, under contract NAS 9-9974, and supplied the lunar sample. We also deeply appreciate the funds provided by the National Science Foundation, under grant GB-6583, which made possible the establishing of our laboratory for biogeochemical research. We thank Mr. R. P. NEWLIN and Mr. S. M. SCHERER for technical assistance.

References

BARGHOORN E. S., MEINSCHEIN W. G. and SCHOPF J. W. (1965) Paleobiology of a Precambrian shale. *Science* **148,** 461–472.

BURLINGAME A. L., HAUG P., BELSKY T. and CALVIN M. (1965) Occurrence of biogenic steranes and pentacyclic triterpanes in an Eocene shale (52 million years) and in an early Precambrian shale (2·7 billion years): A preliminary report. *Proc. Nat. Acad. Sci.* **54,** 1406–1412.

EGLINTON G. and CALVIN M. (1967) Chemical fossils. *Sci. Amer.* **216,** 32–43.

GERARDE H. W. and GERARDE D. F. (1962) The ubiquitous hydrocarbons. *Quart. Bull. Assoc. Food Drug Officials U.S.* **26,** 65–89.

KING E. A. Jr. (1969) *Lunar Sample Catalog Apollo 11.* p 24. Lunar Receiving Laboratory, Science and applications Directorate, Manned Spacecraft Center, Houston, Texas.

MEINSCHEIN W. G. (1959) Origin of petroleum. *Bull. Amer. Assoc. Petrol. Geol.* **43,** 925–943.

MEINSCHEIN W. G. (1961) Significance of hydrocarbons in sediments and petroleum. *Geochim. Cosmochim. Acta* **22,** 58–64.

MEINSCHEIN W. G. (1962) Preliminary proposal for government contract research on development of hydrocarbon analyses as a means of detecting life in space. Esso Research and Engineering Company, 17 pp. Submitted to NASA January 31.

MEINSCHEIN W. G., STERNBERG Y. M. and KLUSMAN R. W. (1968) Origins of natural gas and petroleum. *Nature* **220,** 1185–1189.

Proceedings of the Apollo 11 Lunar Science Conference, Vol. 2, pp. 1879 to 1890.

Analysis of Apollo 11 lunar samples by chromatography and mass spectrometry: Pyrolysis products, hydrocarbons, sulfur, amino acids

SISTER MARY E. MURPHY,* VINCENT E. MODZELESKI, BARTHOLOMEW NAGY, WARD M. SCOTT and MARIA YOUNG
The University of Arizona, Tucson, Arizona 85721

CHARLES M. DREW
Naval Weapons Center, China Lake, California 93555

PAUL B. HAMILTON
Alfred I. du Pont Institute, Wilmington, Delaware 19899
and
HAROLD C. UREY
University of California at San Diego, La Jolla, California 92037

(*Received* 2 *February* 1970; *accepted in revised form* 19 *February* 1970)

Abstract—Lunar fines and rocks from the Apollo 11 mission were analyzed by chromatography, mass spectrometry and light and scanning electron microscopy. Extraction of fines with benzene–methanol yielded trace quantities of hydrocarbons and sulfur; extraction with water led to isolation of what appear to be simple amino acids, alanine, glycine, as well as urea and ethanolamine. Hot water also liberated small amounts of hydrogen sulfide, and possibly other organic sulfur compounds. Pyrolysis at 510°C and 700°C yielded hydrogen, methane, carbon dioxide and higher molecular weight aromatic hydrocarbons. Microscopic studies, especially with the scanning electron microscope, revealed cavities in fine particles, rocks and surfaces of impact craters. Interpretation of results is difficult because of the low carbon content of the Apollo 11 samples and possible contamination prior to analysis.

INTRODUCTION

LUNAR fines and rocks returned by the Apollo 11 astronauts are under investigation by a number of scientists in an attempt to elucidate the composition, origin and environment of the moon. As part of the NASA planned program, these lunar samples were analyzed for total and organic carbon content (LSPET, 1969; JOHNSON and DAVIS, 1970; MOORE *et al.*, 1970). Of particular interest to the organic geochemist are the amount and distribution of carbonaceous matter on the lunar surface. In the present investigation, analysis consisted of extraction for lipids with organic solvents, aqueous extraction for free amino acids, pyrolysis for the non-extractable carbon, and microscopy with transmitted light and scanning electron microscopy. The experiments were conducted with lunar fines (10086) received in an aluminum screw-cap container, and outside and inside chips of a lunar breccia (10002, 54) obtained from the Lunar Receiving Laboratory, Manned Spacecraft Center, Houston, Texas.

---

* On leave from St. Joseph College, West Hartford, Connecticut 06117.

EXPERIMENTAL

Solvent Extraction

Lunar fines were extracted with benzene–methanol (75 ml 4:1; v/v) for 1 hr by sonication in a Mettler Ultrasonic tank (28·1 kc). To ensure optimum contact between solvent and sample, the fines were stirred every 15 min in the flasks placed in a water bath maintained at room temperature. Duplicate procedure blanks in which no sample was used preceded each extraction to determine possible laboratory contamination; a single parallel procedure control was always analyzed at the same time as lunar samples.

The lunar fines were separated from the solvent in a funnel with a glass frit of medium porosity. It was possible to hold back all fines in the funnel by initial gravity flow followed by partial vacuum from a water aspirator. The fines were then carefully washed three times with small amounts of benzene to remove traces of extracted material. The extracted matter was isolated by removing excess solvent in a Buchler rotary evaporator and transferred to weighed vials with minimum amounts of benzene. Weights of all extracts were obtained prior to subsequent tests by removing traces of solvents under a stream of nitrogen purified by flowing through 5 Å molecular sieve. Identical procedure blanks showed no contamination.

All spectral grade solvents were further purified by distillation using a 300 × 29 mm column of 6 × 6 mm raschig rings. All glassware was rinsed with solvents followed by cleaning with hot acid (H_2SO_4–HNO_3; 85:15; v/v).

Thin-layer chromatography

One-tenth of each extract was chromatographed on 0·25 mm thin layers of Silica Gel G (Brinkmann) using both *n*-hexane and *n*-hexane–ether–acetic acid (95:5:1, v/v/v) as developing solvent systems to detect compound classes present in the lunar extracts (MANGOLD, 1964). The thin layer plates were prewashed in hexane; chloroform prewashing was used for the more polar system. Samples, procedure controls, and standards were chromatographed on the same thin layer plate. After visualization with Rhodamine 6G (0·0005%) (ROUSER *et al.*, 1963), the developed chromatograms were examined under u.v. light (254 and 365 mμ) and photographed with a Polaroid 110B camera using Pola Pan Type 42 film (MURPHY *et al.*, 1967).

Gas chromatography–mass spectrometry

The remaining nine-tenths of the extract was analyzed in duplicate by mass spectrometry. One-half of the first extract (1·7 mg) was injected into a Perkin-Elmer 226 gas chromatograph (polyphenyl ether column, 50 ft × 0·02 in.) coupled to a Hitachi RMU-6E single focusing mass spectrometer (MODZELESKI *et al.*, 1968). All other extracts were introduced into the mass spectrometer on the direct inlet probe which was heated in a step-wise manner from 100°C to 300°C. Mass spectra were recorded at approximately 25°C intervals.

Aqueous extraction: Amino acids

Lunar fines (32·66 g) were extracted in a Soxhlet apparatus with triply distilled water for 2 hr to remove free amino acids. The liquids in the flask and thimble were analyzed separately because of the slow percolation of the liquid through the sample. A second extraction of the same lunar fines with water followed immediately. A parallel procedure blank was also analyzed for free amino acids. The contents of the flask and thimble for this blank were combined. Excess water was removed in a rotary evaporator and residues were stored at −68°C until analysis.

A bacterial control was performed prior to chromatographic analysis on the contents of all sample flasks to determine any contribution of amino acids from microorganisms. The contents of each bacterial control flask were dissolved in 2·0 ml of water, 0·1 ml of which was plated on nutrient agar and colonies were counted. The water was evaporated and the residue was redissolved in 1·0 ml of 0·1 N HCl and one-third was next chromatographed on an ion exchange column as described by HAMILTON (1963). Three flasks showed no bacterial growth and three flasks had very low colony counts (80 colonies/ml). This level of contamination was far below the amino acid detection limits. The source of this contamination was proven to be the water added.

Several test procedures preceded the lunar sample analysis: (a) a test at the 10^{-9} mole level was completed to determine column recovery and integration values; (b) two tests were completed on unopened columns to ensure that the column contained no free amino acids; (c) a sensitivity test was performed at the 1×10^{-10} mole level to determine the lower limits of detectability; and (d) 1200 ml of water (10 times the volume used in the lunar sample analysis) was evaporated to dryness and analyzed in a similar manner.

A standard mixture of amino acids (10^{-9} mole level) was co-injected with one-fifth of the lunar extract to compare the retention times of the standards and lunar sample components. Finally, a dunite rock standard, subjected by NASA to the lunar module descent engine exhaust in an initial vacuum of 10^{-4} torr (SIMONEIT *et al.*, 1969; FLORY, 1969) and the corresponding procedure blank were also analyzed for free amino acids.

Aqueous extraction: Volatile components

An experiment was designed to detect certain volatile components from the aqueous treatment of lunar fines. The apparatus consisted of a flask in which reactor grade helium was introduced below the water level to direct any released material through consecutive ice–NaCl and liquid N_2 traps to the atmosphere. The experiments were conducted for various time periods from $\frac{1}{2}$–2 hr. In two experiments, a prewashed thin layer plate was placed under the effluent He stream to collect any material which may have passed through both traps in the form of an aerosol. These chromatograms were developed as described above.

After heating the contents of the reaction flask, both traps and corresponding procedure blanks were extracted with freshly distilled ether to isolate organic components. The ether extract was first dried over anhydrous $MgSO_4$ previously washed with distilled ether. Excess ether was removed under vacuum from a water aspirator. One-third of each extract was examined by thin layer chromatography: the remaining two-thirds were analyzed by mass spectrometry using the direct inlet probe.

In another experiment, the lunar fines were (a) heated *without* water, (b) treated with *cold* water, and (c) heated in water, each for $1\frac{1}{2}$ hr. The condensate in both traps was extracted after each step; the supernatant liquid from the flask was extracted at the end of the experiment. To test for the presence of hydrogen sulfide, the effluent gas not retained in the traps was bubbled into an aqueous silver nitrate solution.

Pyrolysis

A modified Hamilton Multi-Purpose Sampling System used for pyrolysis consisted of a quartz tube (43 cm $\times$ 4 mm i.d.) heated externally by an electric furnace, connected by a heated transfer line through a liquid N_2 trap to the gas chromatograph–mass spectrometer described above. Helium carrier gas passed through the system continuously prior to and during pyrolysis, at a flow rate of 3–5 ml/min. The volatilized material was first condensed in the liquid N_2 trap which was then rapidly heated to 250–300°C. This effected instantaneous injection of the trapped volatiles onto the gas chromatographic column. Individual components were identified by mass spectrometry.

Not all volatile compounds, e.g. H_2, Ne, etc., were trapped under these conditions. However, those compounds which do not condense in the trap but give a positive signal with the flame ionization detector (FID) may still be identified. It should be noted that components which do not give a FID signal may also be identified from the mass spectrum of the mixture if they are eluted from the column coincident with a component which does give a FID signal.

RESULTS AND DISCUSSIONS

Solvent extraction

All extracted material (Table 1) had a red-brown color. A grey precipitate formed as the volume of solvent decreased; this material was partly soluble in hexane and carbon tetrachloride. It is probable that this latter fraction contained elemental sulfur

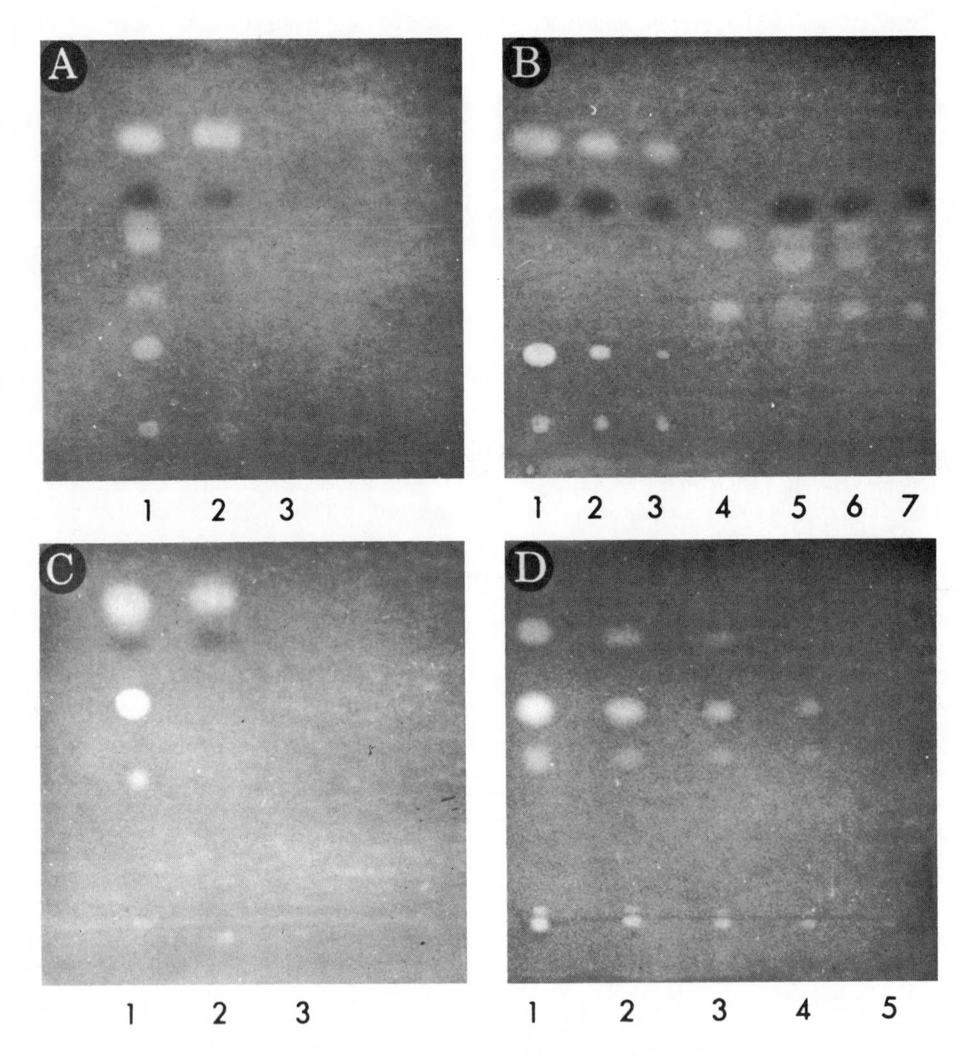

Fig. 1. Thin layer chromatograms of solvent extract of lunar fines, Silica Gel G, ascending development, Rhodamine 6G (0·0005%) visualizer, viewed under u.v. light, 254 and 365 mμ. Plate A: Lunar sample extract, hexane development. Applications: (1) standards: octadecane, sulfur, octadecylthiol, phenanthrene, methyl octadecanoate, 4 μg each, except 2 μg sulfur; (2) Lunar sample extract, 1/10 aliquot; (3) procedure blank, 1/10 aliquot; Plate B: standards, concentration effect, hexane development. Applications: (1, 2, 3) standards as in A1, 5, 2, 1 μg, each; (4) octadecylthiol, 5 μg (upper spot R_f 0·5, octadecyl chloride impurity); (5, 6, 7) elemental sulfur and octadecylthiol, 5, 2, 1 μg, note product R_f 0·4. Plate C: Lunar sample extract; hexane–ether–acetic acid (95:5:1; v/v/v). Applications: (1) standards, same as A1; (2) lunar sample extract, 1/10 aliquot; (3) procedure blank, 1/10 aliquot. Plate D: Standards, concentration effect, hexane–ether–acetic acid. Applications: (1, 2, 3, 4, 5) standards: octadecane, phenanthrene, methyl octadecanoate, octadecanoic acid, and octadecanol, 5, 2, 1, 0·5, 0·2 μg each.

(MURPHY *et al.*, 1967). Similar extracts were reported by other laboratories analyzing the Apollo 11 samples (ABELL *et al.*, 1970).

The thin-layer chromatograms are shown in Fig. 1. The two major components in the small amount of extract from the fines were saturated hydrocarbons and elemental sulfur, both of which were confirmed by mass spectrometry (NAGY *et al.*, 1970; MURPHY *et al.*, 1970). Two other components, a compound less polar than methyl octadecanoate but more polar than phenanthrene, and a yellow fluorescent material which remained at the origin even in the more polar solvent system were seen in trace quantities on the chromatograms but not evident on the photographs. It should be noted that these components were absent in the parallel procedure blank. Known amounts of standards determined the sensitivity of the method to be 0·25 μg for

Table 1. Solvent extraction of lunar fines

Sample	Weight (g)	Extract (mg)	Hydrocarbon (ppm)*	Sulfur (ppm)*†
1. (top of can, fine grained portion)	22·81	1·7	2·0	0·2
2. (middle of can)	22·16	0·3	0·5	0·9
3. (bottom of can, coarse grained portion)	16·44	0·2	0·3	1·8

* Estimated from thin layer chromatograms.
† Values represent sulfur in solution of the extract, but not total extractable sulfur.

octadecane and 0·5 μg for elemental sulfur (ROUSER *et al.*, 1968). In terms of the weight of lunar sample taken for analysis, the detection limits of thin layer chromatography were approximately 10 and 20 ppb for saturated hydrocarbons and sulfur, respectively.

Analysis of the extract by combined gas chromatography–mass spectrometry showed the presence of nanogram quantities of toluene, C_2 alkyl benzene, phenol, diphenyl and a methyl ester, which are probably contaminations (MURPHY *et al.*, 1970). These compounds, however, did not appear in the procedure blank.

When the lunar samples were being weighed for analysis, it appeared that the material in the bottom of the container was coarser than that on top. In fact, an uncatalogued rock chip was found in the lower half inch of fines. The quantities of hydrocarbons from the three samples used for solvent extraction experiments markedly decreased with depth in the container. MOORE *et al.* (1970) has also observed that lunar fines <300 mesh gave 500 ppm total carbon; 140–300 mesh, 210 ppm; and 60–140 mesh, 115 ppm total carbon. From the fact that sulfur tended to precipitate out of solution in the experimental procedure the variations in amounts of sulfur extracted may not be real.

Aqueous extraction: Amino acids

The aqueous extract of lunar fines contained what appeared to be glycine, alanine, ethanolamine and urea (32, 36, 22 and 65 ppb, respectively) based on absorbance at 5700 Å of these ninhydrin positive species. The one-fifth aliquot of the extract co-injected with known amounts of standards resulted in overlapping symmetrical peaks on the chromatogram. The parallel procedure blank gave a different distribution of amino acids present in trace quantities. Analysis of the second extract showed that the first extract had been essentially quantitative. The sample of dunite subjected to a lunar module descent engine exhaust by NASA and a parallel procedure blank did not show the same amino acid distribution as the lunar sample.

The results are presented in Fig. 2, along with the rocket exhaust control, the corresponding procedure blanks and a typical laboratory, mainly hand, contamination control (HAMILTON, 1965). The amino acid distributions in the bacterial controls were completely different. The other procedure controls completed prior to the lunar

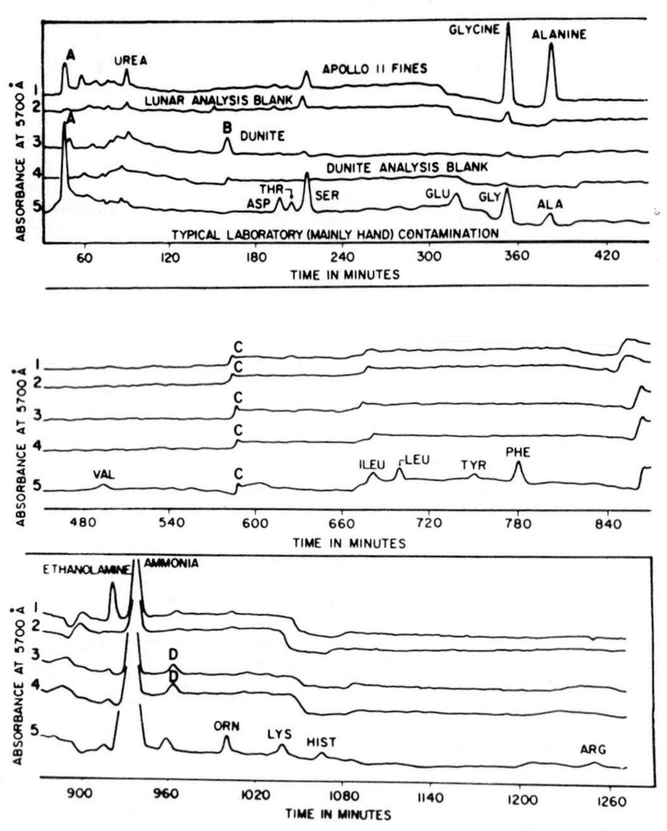

Fig. 2. Ion-exchange chromatograms of amino acids in lunar fines and controls. (1) Apollo 11 lunar fines; (2) Lunar fines procedure control; (3) Terrestrial rock, dunite, exposed to the lunar module descent engine exhaust at White Sands, N. Mex.; (4) Dunite procedure control; (5) Typical laboratory, mainly hand, contamination. A, B, D, unknown ninhydrin positive substances; C, one of five buffer changes. Note the amino acid distributions for the lunar fines and dunite blank are different from the laboratory contamination control.

sample analysis showed: (a) quantitative data, subject to a 10–20 per cent error, could be obtained for all amino acids at the 10^{-9} mole level; (b) amino acids did not arise, *sui generis* from the columns; (c) the detection limit for many of the amino acids in this procedure was at the 10^{-10}–10^{-11} mole level.

What appeared to be glycine, alanine, and ethanolamine were greater than background by a factor of 10; urea was greater by a factor of approximately 3. On the basis of these amounts and the overlap of peaks on co-injection with appropriate standards, tentative identification of free amino acids has been made. Absolute identification of a peak in the chromatograms is, of course, contingent upon isolation and characterization by gas chromatography–mass spectrometry or other methods. Fox *et al.* (1970) also reported traces of amino acids in lunar fines. Further studies by other laboratories will still be necessary to confirm the present results.

Abiotic syntheses from methane, ammonia and water in simulated primitive earth environments do produce glycine and alanine as major components (MILLER, 1955; MILLER and UREY, 1959; ORO, 1965). Synthesis of these compounds by the lunar module descent engine exhaust may possibly account for the free amino acids; however, no amino acid has been identified in such engine exhaust controls. However, it is difficult to account for free amino acids under lunar surface conditions.

Aqueous extraction: Volatile components

When hot water reacted with lunar fines, hydrogen sulfide was detected by its odor and gave a black precipitate with Ag^+ in aqueous solution. At the same time (but not necessarily related) the pH of the water, boiled immediately prior to the experiments, changed slightly during the reaction from pH 4 to 5 in 2 hr, whereas the water in the procedure blank changed from pH 4 to 4·3. The nature of the compound(s) which will react with hot water under these conditions to produce H_2S is not known. Some of the hydrogen sulfide may be a reaction product of troilite, but there are other possible sources of sulfur in the lunar rocks. Perhaps more interesting is the appearance of what may be organic sulfur compounds on the thin layer chromatograms (see Fig. 3); this substance was detected in the nitrogen trap and on the thin layer plate placed under the effluent gas. MURPHY *et al.* (1970) reported a number of sulfur compounds by high resolution mass spectrometry. The experiment designed to test the effects on the sample of (a) dry heat and (b) cold water, did not give the same results. Therefore, hot water appears necessary for the reaction. Only H_2S was detected when the experiment was repeated. It is possible that the sulfur-containing substances vary in distribution within the sample container. It should be noted that the samples were subjected to many handling procedures prior to analysis in this laboratory.

In addition to procedure controls, a sample of the polysulfide resin (FLORY, 1969) used to seal the walls of the glovebox at the LRL where samples were packaged for distribution to analysts, was extracted with hot water in an identical experiment, but gave different results. Procedure controls which accompanied each aqueous extraction did not yield any elemental sulfur or organic sulfur compounds.

Pyrolysis

Pyrolysis at 510°C of 148·6 mg of untreated lunar fines yielded the products shown in chromatogram A (Fig. 4). Peak 1 corresponds to those compounds which did not condense in the liquid N_2 trap, but gave a positive FID response, i.e. hydrogen and methane. The compounds corresponding to the remaining peaks are listed in Table 2.

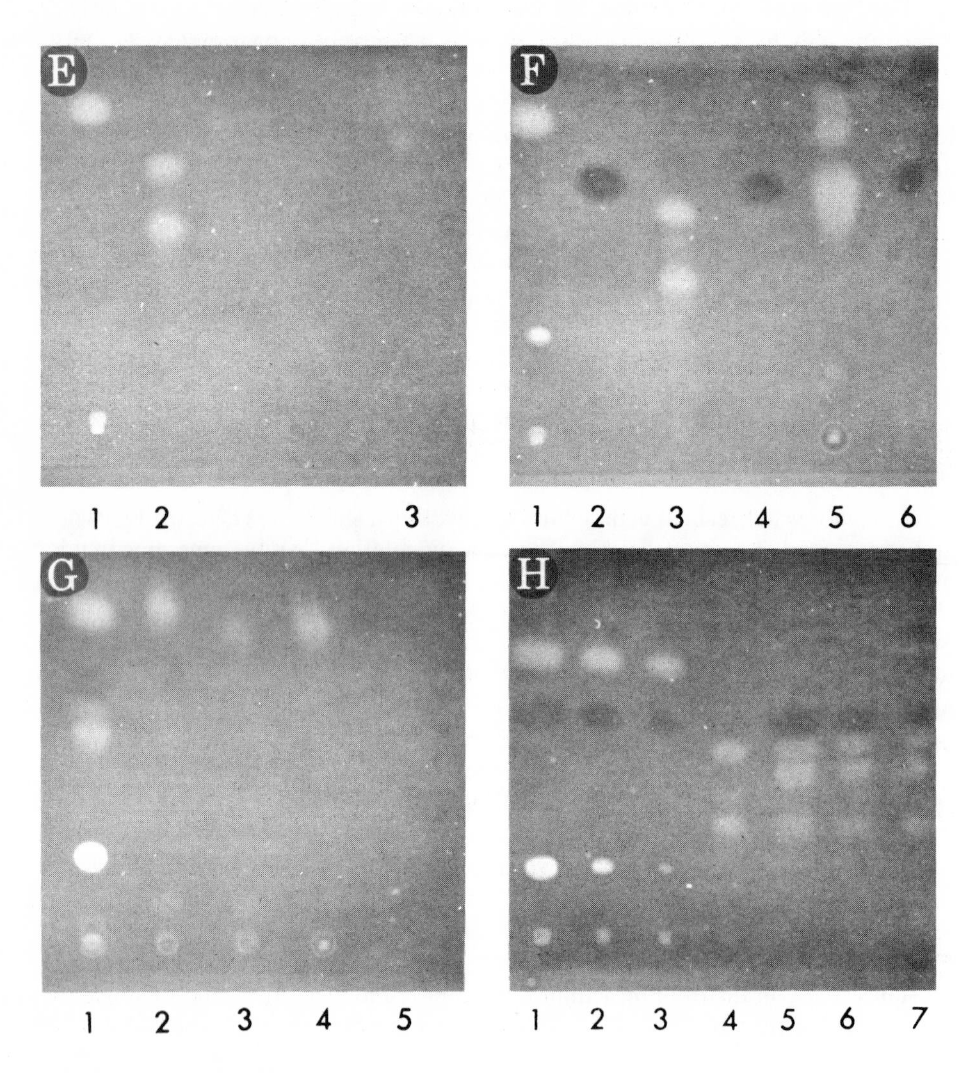

Fig. 3. Thin layer chromatograms of volatiles from lunar fines treated with hot water Silica Gel G, hexane ascending development, Rhodamine 6G (0·0005%) visualizer, viewed under u.v. light, 254 and 365 mμ. Plate E: Gas effluent from hot water treatment of lunar fines. Applications: (1) standards: octadecane, phenanthrene (R_f 0·4), methyl octadecanoate, octadecanoic acid, 4 μg each; (2) octadecylthiol, 4 μg; (3) volatiles from lunar sample gas effluent. Plate F: Ice–NaCl trap condensate. Applications: (1) standards, same as E1; (2, 4, 6) sulfur, 10, 5, 2 μg; (3) octadecylthiol, 4 μg; (5) Ice–NaCl trap condensate, 2/3 aliquot. Plate G: Traps and supernatant. Applications: (1) standards, same as A1; (2) Ice–NaCl trap contents, 1/3 aliquot; (3) liquid N_2 trap contents, 1/3 aliquot; (4) supernatant, 1/3 aliquot; (5) procedure blank, 1/3 aliquot. Plate H: Standards, concentration effect. Applications: (1, 2, 3) same as B1, B2, B3; (4) octadecylthiol, 5 μg; (5, 6, 7) elemental sulfur and octadecylthiol, 5, 2, 1 μg each.

These compounds did not appear in the corresponding procedure blank (chromatogram B); the benzene and toluene confirm preliminary experiments at 500°C (LSPET, 1969).

Pyrolysis of larger samples (~700 mg) at 700°C resulted in more readily detectable products. The product distributions from pyrolysis at 700°C of (a) untreated lunar fines, 623·4 mg, (b) solvent extracted lunar fines, 717·2 mg (from which remaining traces of solvent were removed by heating for 6 hr at 134°C at 35 mm Hg pressure) and

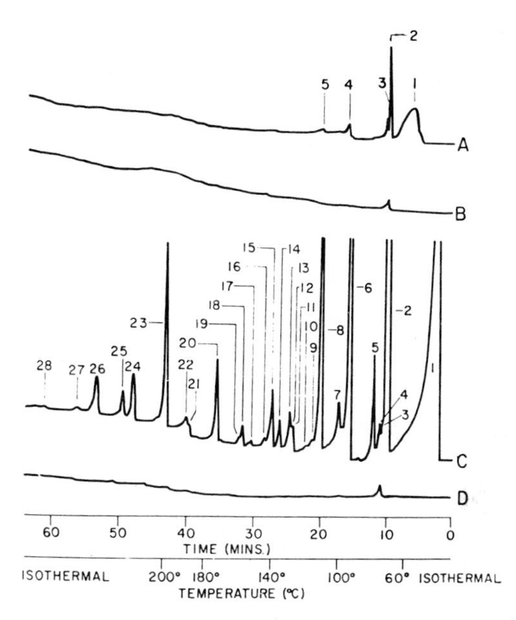

Fig. 4. FID Gas chromatograms of pyrolysis products

Chromatogram	Sample	Pyrolysis temperature (°C)
A	148·6 mg lunar fines (10086)	510
B	Procedure blank	510
C	675·8 mg powdered rock chip from lunar breccia (10002, 54)	700
D	Procedure blank	700

Column: 50 ft × 0·02 in i.d. capillary, polyphenyl ether liquid phase.

(c) a powdered rock chip, 675·8 mg, from a lunar breccia were very similar. Several investigators have reported detecting CH_4, CO_2 and aromatic hydrocarbons by pyrolysis (ABELL *et al.*, 1970; BURLINGAME *et al.*, 1970; ORO *et al.*, 1970). PONNAMPERUMA *et al.* (1970) have reported C_1–C_4 hydrocarbons by acid treatment of lunar fines.

The products from the powdered rock chip are shown in chromatogram C. The compounds, identified on the basis of low resolution mass spectral data, are listed in Table 2.

The chromatographic column was unable to resolve the components of certain peaks, but individual compounds could be identified from spectra of mixtures (scan time, 4 sec) obtained at intervals under the peak.

While results shown in chromatograms A and C are not quantitative, it is evident that at 700°C, more organic material is released than can be accounted for by using approximately five times the sample size. Many of the peaks in chromatogram C correspond to pyrolysis products of methane. It has been shown by ORO and HAN (1967) that methane, when passed over silica gel at 1000°C, at a flow rate of 0·2 ml/min, undergoes approximately 10% conversion to many of the products listed in Table 2.

Table 2

Chromatogram A		Chromatogram C			
Peak	Identity	Peak	Identity	Peak	Identity
1	H_2, CH_4	1	H_2, CH_4	12, 13, 14	*p*-, *m*-, and *o*-Xylene
2	C_2H_6, C_2H_4 CO_2	2	C_2H_6, C_2H_4, CO_2 C_3H_8, C_3H_6, C_4H_8 C_4H_6	15	Styrene
3	C_3H_8, C_3H_6			16, 17, 19	C_3-Alkylbenzenes
4	Benzene			18	Methylstyrene
5	Toluene	3	C_5H_{10}	20	Indene
		4	C_5H_8	21, 22	Methylindenes
		5	C_5H_6	23	Naphthalene
		6	Benzene	24	2-Methylnaphthalene
		7	Thiophene	25	1-Methylnaphthalene
		8	Toluene	26	Diphenyl
		9, 10	Methylthiophenes	27, 28	C_2-Alkylnaphthalenes
		11	Ethylbenzene		

It is probable that the other aromatic compounds, of lower volatility than those identified, were also released in the pyrolysis experiments, but the upper temperature limit of 200°C of the chromatographic column precluded their detection.

There are several possible sources of these carbon compounds, e.g. polymeric material (which could not be removed by solvent extraction), lattice entrapped material or material located in mineral enclosures. Figure 5 shows scanning electron micrographs (taken with a Stereoscan Mark II instrument) of untreated particles of lunar fines. These electron micrographs show small cavities which could possibly have contained carbonaceous substances before they were broken open. Other micrographs of glass beads, broken beads and the interior of an impact crater have been shown previously (NAGY *et al.*, 1970). The material located in mineral enclosures could represent organic carbon truly indigenous to the moon, i.e. not accumulated from any galactic processes occurring later than the formation of the moon.

CONCLUSIONS

In interpreting these results on the organic carbon in the lunar samples, the possibility of contamination, acquired during sample collection and processing, either terrestrial or lunar, must be considered. Although some of the results have been reported by other investigators, confirmation of analytical data by other scientists is necessary before conclusions can be made. It should be noted that laboratories attempting to duplicate results should receive similar samples, if not homogeneous specimens. Mesh size of fine samples may be a determining factor in distributing lunar samples for organic geochemical studies because of reported variations in total carbon content.

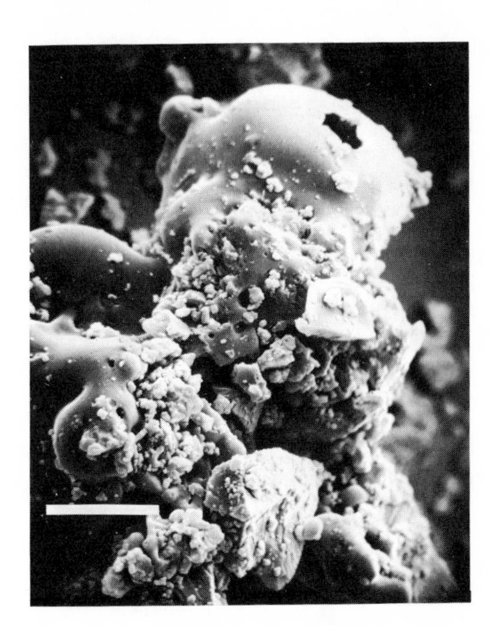

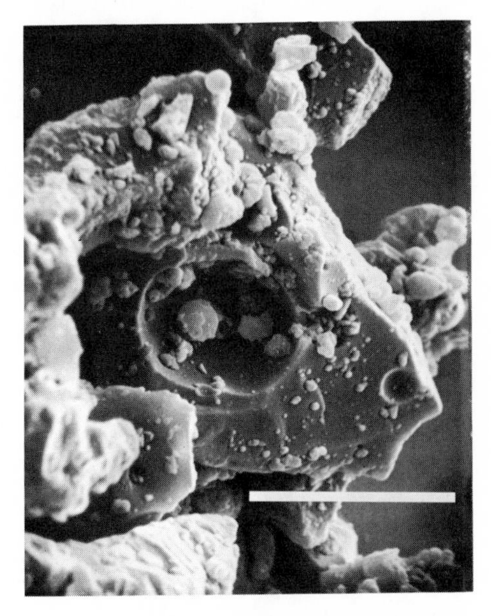

Fig. 5. Scanning electron micrographs of (left) a hollow bubble-like glassy particle in the lunar fines; (right) fine particle broken open showing internal cavities. The main cavity apparently contains secondary particles. Bubbles and cavities may have originally contained carbonaceous and/or gaseous substances. White lines indicate 25 μ.

Data from Apollo 11 and future returned samples may be useful in determining the role of carbon in any theory on the origin and history of the moon. The results of this investigation suggest that some of the carbon compounds, such as methane, are present in trace quantities on the moon.

Acknowledgments—The authors are grateful for the assistance of Mrs. ALICE LUMPKINS, JUDITH MODZELESKI and LOIS ANNE NAGY of The University of Arizona; Mrs. ROBERTA DE FIORE, JOHANNE C. DICKINSON, JANET GREENQUIST, Drs. MARJORIE LOU and T. TIMOTHY MYODA of the Alfred I. du Pont Institute; JOSEPH E. THOMAS and Mrs. REBA WARD of the Naval Weapons Center, China Lake. This investigation was supported by NASA at The University of Arizona, Contract NAS-9-8747 and Grant NGR-03-002-171, with additional support from St. Joseph College (NSF), Naval Weapons Center, and the Alfred I. du Pont Institute. Contribution No. 199 from the Department of Geochronology, The University of Arizona.

REFERENCES

ABELL P. I., DRAFFAN G. H., EGLINTON G., HAYES J. M., MAXWELL J. R. and PILLINGER C. T. (1970) Organic analysis of the returned lunar sample. *Science* **167**, 757–759.

BURLINGAME A. L., CALVIN M., HAN J., HENDERSON W., REED W. and SIMONEIT B. R. (1970) Lunar organic compounds: search and characterization. *Science* **167**, 751–752.

FLORY D. (1969) Personal communication.

FOX S. W., HARADA K., HARE P. E., HINSCH G. and MUELLER G. (1970) Bio-organic compounds and glassy microparticles in lunar fines and other materials. *Science* **167**, 767–770.

HAMILTON P. B. (1963) Ion exchange chromatography of amino acids: a single column, high resolving, fully automatic procedure. *Anal. Chem.* **35**, 2055–2064.

HAMILTON P. B. (1965) Amino acids on hands. *Nature* **205**, 284–285.

JOHNSON R. D. and DAVIS C. C. (1970) Pyrolysis–hydrogen flame ionization detection of organic carbon in a lunar sample. *Science* **167**, 759–760.

LSPET (LUNAR SAMPLE PRELIMINARY EXAMINATION TEAM) (1969) Preliminary examination of lunar samples from Apollo 11. *Science* **165**, 1211–1227.

MANGOLD H. (1964) Thin layer chromatography of lipids. *J. Amer. Oil Chem. Soc.* **41**, 762–777.

MILLER S. L. (1955) Production of some organic compounds under possible primitive earth conditions. *J. Amer. Chem. Soc.* **77**, 2351–2361.

MILLER S. L. and UREY H. C. (1959) Organic compound synthesis on the primitive earth. *Science* **130**, 245–251.

MODZELESKI V. E., MACLEOD W. D., JR. and NAGY B. (1968) A combined gas chromatographic-mass spectrometric method for identifying *n*- and branched-chain alkanes in sedimentary rocks. *Anal. Chem.* **40**, 987–989.

MOORE C. B., LEWIS C. F., GIBSON E. K. and NICHIPORUK W. (1970) Total carbon and nitrogen abundances in lunar samples. *Science* **167**, 495–497.

MURPHY SR. M. T. J., NAGY B., ROUSER G. and KRITCHEVSKY G. (1965) Identification of elemental sulfur and sulfur compounds in lipid extracts by thin-layer chromatography. *J. Amer. Oil Chem. Soc.* **42**, 475–480.

MURPHY R. C., PRETI G., NAFISSI-V M. M. and BIEMANN K. (1970) Search for organic material in lunar fines by mass spectrometry. *Science* **167**, 755–757.

NAGY B., DREW C. M., HAMILTON P. B., MODZELESKI V. E., MURPHY SR. M. E., SCOTT W. M., UREY H. C. and YOUNG M. (1970) Organic compounds in lunar samples: Pyrolysis products, hydrocarbons, amino acids. *Science* **167**, 770–773.

ORO J. (1965) Stages and mechanisms of prebiological organic synthesis. In *The Origins of Prebiological Systems*, (editor S. Fox), Part II, p. 146. Academic Press.

ORO J. and HAN J. (1967) Application of combined chromatography–mass spectrometry to the analysis of aromatic hydrocarbons formed by pyrolysis of methane. *J. Gas Chromatogr.* **5**, 480–485.

ORO J., UPDEGROVE W. S., GIBERT J., MCREYNOLDS J., GIL-AV E., IBANEZ J., ZLATKIS A., FLORY D. A., LEVY R. L. and WOLF C. (1970) Organogenic elements and compounds in surface samples from the Sea of Tranquillity. *Science* **167**, 765–767.

PONNAMPERUMA C., KVENVOLDEN K., CHANG S., JOHNSON R., POLLOCK G., PHILPOTT D., KAPLAN I., SMITH J., SCHOPF J. W., GEHRKE C., HODGSON G., BREGER I. A., HALPERN B., DUFFIELD A., KRAUSKOPF K., BARGHOORN E., HOLLAND H. and KEIL K. (1970) Search for organic compounds in the lunar dust from the Sea of Tranquillity. *Science* **167**, 760–762.

ROUSER G., KRITCHEVSKY G., HELLER D. and LIEBER E. (1963) Lipid composition of beef brain, beef liver, and the sea anemone: two approaches to quantitative fractionation of complex lipid mixtures. *J. Amer. Oil Chem. Soc.* **40**, 425–454.

SIMONEIT B. R., BURLINGAME A. L., FLORY D. A. and SMITH I. D. (1969) Apollo lunar module engine exhaust products. *Science* **166**, 733–738.

Proceedings of the Apollo 11 Lunar Science Conference, Vol. 2, pp. 1891 to 1900.

Search for organic material in lunar fines by mass spectrometry

R. C. MURPHY, GEORGE PRETI, M. M. NAFISSI-V and K. BIEMANN
Department of Chemistry, Massachusetts Institute of Technology, Cambridge, Massachusetts 02319

(*Received* 2 *February* 1970; *accepted in revised form* 20 *February* 1970)

Abstract—Three kinds of experiments were performed in an effort to detect and identify organic compounds present in the lunar material (sample 10086): vaporization of the volatilizable components directly into the ion source of a high resolution mass spectrometer, and extraction of the material with organic solvents before and after dissolving most of the inorganic substrate in hydrochloric and hydrofluoric acid. The extracts were investigated by a combination of gas chromatography and mass spectrometry. Although a number of organic compounds or compound types have been detected, none appears to be indigenous to the lunar surface.

INTRODUCTION

THE SEARCH for organic compounds on other celestial bodies is a complex problem which has many facets. Basically, they center around three aspects: are there any organic compounds at all, and if so, what is their nature or structure, and most importantly, how were they produced. The last problem is of course intimately related to the hope of finding somewhere a system that reflects the period just prior to or at the beginning of the generation of living systems. This would thus give us an insight into the origin of life, even though it may not necessarily be identical to that on earth.

Quantitative elemental analysis, i.e. the relative abundance of the various elements (and their isotopes) is of importance for mineralogical or cosmological considerations. On the other hand a quantitative analysis of the various elements of which most organic compounds are composed, namely carbon, hydrogen, and to a lesser extent nitrogen and oxygen, is rather useless information. It is the structure of the individual compounds, or at least their structural type, which has to be determined if one wishes to obtain any insight into their origin. An analysis for total carbon is therefore of no direct value. Even a differentiation of total carbon from 'organic carbon' is not a very informative number except that it might indicate the ease with which organic compounds may be isolated and identified.

After all the controversy that has been raging over the nature and origin of organic compounds isolated from meteorites, the material returned from the lunar surface by the Apollo 11 mission provided for the first time authenticated extraterrestrial specimens for investigation. Unfortunately the moon's surface represents by no means a very likely depository of organic compounds, since the absence of an atmosphere or water eliminates gas phase or solution reactions capable of producing organic compounds. Furthermore, the rather high temperature prevailing at the surface during the lunar day creates conditions under which most organic compounds would evaporate. There is however, the finite possibility that organic compounds were produced if there ever was an atmosphere and water, or volcanic activity. These could be retained at subsurface regions and brought to the surface much more recently by the 'gardening effect' of meteorite impact.

The deposition of organic compounds by meteorites would be another possible source. In either case, only highly condensed, polymeric substances (like 'kerogen' on earth) would have sufficiently low vapor pressure to survive under the conditions prevailing at or near the surface of the moon. Furthermore, the heat generated upon meteor impact would most likely completely vaporize even such components and at best leave a deposit of carbon.

It is therefore highly unlikely that there could be appreciable concentrations of organic compounds at the surface of the moon. However, it was felt that an effort had to be made to detect the presence of organic material and, if possible to determine the structure of these compounds. It should be kept in mind that the proof of the absence of organic compounds, or at least certain types is of considerable significance and by no means a futile effort. Consequently the detection and identification of organic compounds calls for a technique that combines high sensitivity with a high degree of structural information. This can be achieved in two quite different ways. Either using a very sensitive method which is specific for a particular compound or one that produces structural information for any type of compound (i.e. a general survey method). Of the latter type mass spectrometry is unquestionably the method of choice because the mass spectrum of a compound is directly related to its structure. While it may not reach the sensitivity of light absorption or emission measurements for compounds with a very high molar extinction coefficient or very efficient fluorescence, the qualitative information content of even a single mass spectrum is vastly higher.

Our own investigation was aimed at the detection of organic compounds that would indicate organic synthesis either random or biogenic. Of particular interest is the detection of compounds containing heteroatoms, such as nitrogen, oxygen, sulfur, and phosphorous because these are the types of compounds that would have to be formed at the beginning of any pathway ultimately leading to systems that could be called 'living'.

In an effort to characterize the organic compounds present in the surface covering the Apollo 11 landing site, a portion of fines (sample 10086) has been subjected to mass spectrometric analysis. Three different types of experiments were undertaken: (i) Vaporization of organic compounds present in lunar material; (ii) solvent extraction of lunar material; (iii) dissolution of lunar material in acid followed by extraction of the aqueous phase.

Vaporization of the organic material by heating of the sample placed into the ion source of a mass spectrometer combines the least risk of contamination of the sample with a maximum of sensitivity: no solvents or reagents are employed and volatile components are not lost during the evaporation of the solvent. Performing the experiment in a spectrometer of the Mattauch–Herzog type combines the advantages of high resolution, which allows the direct and, in most cases, unambiguous assignment of the elemental composition of each ion, with continuous recording of all ions and the integrating property of the photographic plate placed in the focal plane of the instrument. Thus, a series of successive exposures taken without interruption while heating the sample from ambient to the upper temperature limit represents a complete and permanent record of the ions produced from all the material that vaporized under these conditions. Substances not sufficiently volatile at this temperature range may pyrolyze and lead to the mass spectrum of the pyrolysis products.

PROCEDURE

The sample (number 10086 0·4 g) was placed in a small glass bulb mounted on the ion source chamber of the high resolution mass spectrometer (CEC 21-110B) as shown in Fig. 1. The circular oven permits heating the sample to 400°C. In contrast to the procedure described previously by HAYES and BIEMANN (1968) the sample was placed into the spectrometer by venting the ion source housing with dry, purified nitrogen each time a sample was run to eliminate any chance of picking up a trace of organic material while the rather bulky sample container slides through the vacuum

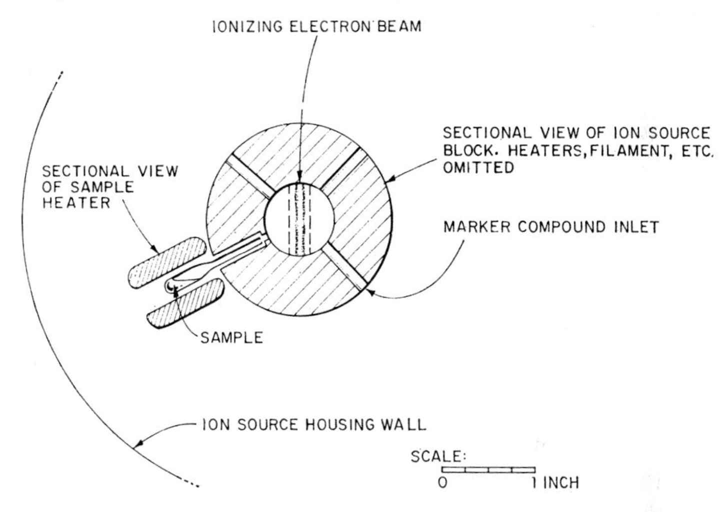

Fig. 1. Schematic of sample holder and oven attached to the ion source of a high resolution mass spectrometer.

lock of the spectrometer. Seven 3-min exposures were recorded while the sample was heated from ambient to 400°C. An evaporated silver bromide plate (provided by Technical Operations, Inc., Burlington, Mass.) was used because of its relatively high sensitivity attributable to the lack of grain and fog. The line positions were measured with an automatic comparator (D.A. Mann Corp., Burlington, Mass.) operated on line with a IBM-1802 computer which also converts the line positions to masses and finally elemental compositions. Perfluoroalkane (high boiling) was used as a mass standard.

RESULTS

The majority of the material vaporized at approximately 150°C. The composition of the ions produced is given in Table 1. Most of the species are hydrocarbon ions of various degrees of unsaturation, but there are also some that contain one to three oxygen atoms or one nitrogen atom. Furthermore, there are a few ions that contain nitrogen and oxygen and some rather abundant ions which contain sulfur, either alone or in combination with oxygen. It should be noted that these sulfur-containing ions are either free of carbon (hydrogen sulfide and sulfur dioxide) or have only one carbon atom (carbon disulfide).

A few of the ions may represent compositions with three nitrogen atoms but these differ by only 1·3 millimass units from those having C_2H_2O instead of N_3 and can therefore not be unambiguously assigned. The absence of species with two nitrogen atoms (for which no such ambiguity exists) would argue in favor of the oxygen containing composition and against the corresponding one with three nitrogens which are therefore listed in brackets in Table 1.

Table 1. Ions observed upon heating lunar sample (0·4 g) to 400°C into the ion source of a high resolution mass spectrometer

C, H	
++C_2H_4*†	++C_5H_9†
++C_2H_5*	++C_5H_{10}
+C_2H_6	++C_5H_{11}
++C_3H	+C_6H_2
++C_3H_2*†	+C_6H_3
+++C_3H_3*†	+C_6H_4†
++C_3H_4†	+C_6H_5
++C_3H_5*	+C_6H_6*
++C_3H_6*	+C_6H_7
+++C_3H_7	+C_6H_9
+C_3H_8	+C_6H_{10}
+C_4H	++C_6H_{11}
++C_4H_2†	+C_6H_{12}
++C_4H_3	+C_6H_{13}
+C_4H_4†	++C_7H_7
++C_4H_5	+C_7H_8
+C_4H_6	+C_7H_9
+++C_4H_7	+C_7H_{11}
+++C_4H_8*	+C_7H_{12}
+++C_4H_9	+C_7H_{13}
+C_5H_2	+C_8H_9
+C_5H_3	+C_8H_{13}
+C_5H_4	+C_8H_{15}
+C_5H_5	+C_8H_{16}
+C_5H_6	+C_9H_7
++C_5H_7	+C_9H_{11}
+C_5H_8	

C, H, O	
+++CO_2†	+C_3H_5O
++CH_2O	++C_3H_6O
++CH_3O	+C_3H_7O
++CH_4O†	+C_4H_6O
+C_2O	+C_4H_8O
+C_2HO	+C_6H_6O
++C_2H_2O†	+C_7H_4O
+++C_2H_3O*†	+C_7H_5O
+++C_2H_4O†	+CH_2O_2
+++C_2H_5O	++$C_2H_4O_2$
+C_3HO	+$C_3H_5O_2$
+C_3H_2O	+$C_8H_3O_2$
++C_3H_3O*	+$C_4H_2O_3$
+C_3H_4O	+$C_8H_5O_3$

C, H, N, O	
+CH_2N	+C_3H_3N
+CH_4N	+CNO†
+C_2N	+++HCNO†
++C_2HN†	+HNO
++C_2H_2N	+++NO_2†
+C_2H_3N†	[++HN_3†]
+C_2H_4N†	[+HCN_3†]
+C_2H_6N	[+CH_3N_3†]

C, H, S, O	
+++S	+COS
+HS	+++SO
++H_2S	+++SO_2
++CS	
+CS_2	

* Also present in background.
† Found in rocket exhaust (SIMONEIT *et al.*, 1969).
+ Relative intensity.

It is important to keep in mind that these combinations of elements represent ions formed under electron impact and do not necessarily indicate molecules present as such on the lunar material. They do however indicate beyond doubt that upon heating of the sample, compounds vaporize which do contain at least these atoms within their molecular structure. For example, a unipositively charged ion of mass 57·0704 must consist of four carbon atoms and nine hydrogen atoms which in turn requires the presence of a compound containing a butyl group. Whether it is butane, a higher hydrocarbon or a butyl substituted compound can not be stated at this point, particularly since one most likely deals with a complex mixture. On the other hand, it is certainly not due to butyl thiophene, for example, because there are no highly unsaturated sulfur containing ions present.

Such an evaluation of the data listed in Table 1 indicates that there seems to be present a series of hydrocarbon ions having up to nine carbon atoms and ranging from almost completely saturated species (C_8H_{16} corresponds to a hydrocarbon with one double bond or ring) to more unsaturated ones. The aromatic ions (C_9H_7, C_8H_9, C_7H_8 and C_6H_6, which would correspond to an indenyl ion, phenylethyl ion, toluene and benzene) appeared to be present in much lower abundance. Because of the high relative abundance of these ions in the mass spectra of the corresponding pure compounds, these aromatic compounds are thus present in very low concentration.

Aside from the large quantities of CO_2, the most abundant oxygen-containing ions consisted of one or two carbons and one or two oxygens, as they may be produced upon ionization of small alcohols or glycols. Such ions might have come from either ethylene oxide or its hydrolysis product, ethylene glycol. (Ethylene oxide was used as a sterilizing agent at the Lunar Receiving Laboratory.) Similarly, the presence of an ion whose elemental composition ($C_2H_4O_2$) would indicate the presence of acetic acid, which may arise from the reduction of peracetic acid, also used as a sterilizing agent at L.R.L. The ions containing three carbon atoms and one oxygen atom may have been derived from acetone which in turn may have been produced by the well known thermal decomposition of salts of acetic acid. The $C_8H_5O_3$ ions probably arose from traces of dialkyl phthalates which are common laboratory contaminants.

The nitrogen-containing ions all represent small molecules or fragments thereof, and some of them (indicated in Table 1 by an asterisk) had been previously found in the products of the lunar retro-rocket exhaust (SIMONEIT *et al.*, 1969). Last but not least, there appears to be a considerable amount of sulfur or sulfur containing substances (hydrogen sulfide, carbon disulfide, and sulfur dioxide) produced.

The organic material which can be volatilized out of the sample represents small molecules which are probably of terrestrial origin, with the exception of CO_2, SO_2, H_2S, COS and CS. These are probably produced from indigenous material in the sample but it should be noted that a sulfur-base caulking compound was used to seal various partitions within the cabinet systems at L.R.L. It should also be pointed out that most of the organic material began to appear in the spectra when the sample had reached temperatures of 150–200°C and disappeared in later exposures taken at a sample temperature of 400°C. This makes it unlikely that they are pyrolysis products of otherwise nonvolatile polymeric organic material. Equally significant is the absence of a series of hetero-aromatic systems such as pyridines, furanes and thiophenes. These hetero-aromatic compounds would be produced from the pyrolysis of more complex heteroatom-containing molecules which would be of greater interest to the organic and biochemist. In view of the high relative intensity of the molecular ion of these types of compounds, it should have been possible to detect them in rather small amounts (a few nanograms).

It is expected that the above described direct vaporization technique is an efficient method for investigating the materials deposited on or near the surface of the individual matrix particles but is probably less so for organic compounds occluded within the particles. Solvent extraction or removal of the inorganic material by dissolution in acid followed by extraction of the liberated organic substances is often more efficient and has the added advantage that one can extract a much larger sample and thus increase the detection limit of the organic substances. As indicated earlier, the risk of contamination is increased because of the use of solvents and necessary manipulations (even though the solvents are highly purified and the work is done in a clean-room with the usual precautions). More importantly, only substances with a boiling point appreciably above that of the solvent used will be detectable, while the more volatile ones are lost when the large (relative to the extracted material) amount of solvent is evaporated.

As mentioned earlier, it was our major objective in this first investigation of lunar material, to detect and identify compounds more complex than simple hydrocarbons. For the same reason we used a general purpose gas chromatographic column (OV-17) rather than one specifically suited for a certain compound type. Under the conditions employed, toluene is just barely resolved from the solvent peak and compounds of shorter retention times escape detection. The gas chromatograph (Varian Aerograph) is operated in conjunction with a low resolution mass spectrometer (Hitachi RMU-6D) coupled to a computer (IBM 1802) to enable one to permanently record all data in a form that can be analyzed with the help of the computer (HITES, 1968) as outlined below.

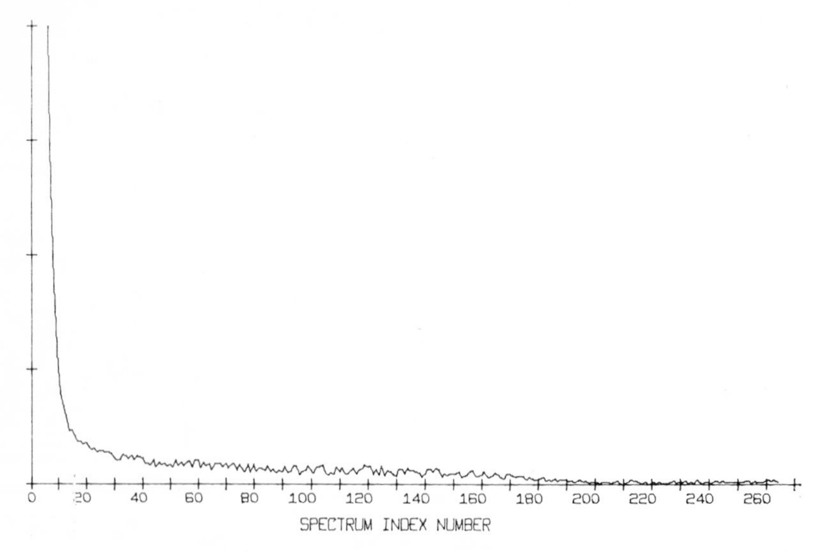

Fig. 2. Gas chromatogram of dichloromethane extract of hydrochloric acid solution.

PROCEDURE

The fines (50 g of sample 10086) were extracted with 50 ml of benzene-methanol (1:1) in a sealed ampoule by shaking in an ultrasonic bath for 48 hr at 60°C. After settling, the solvent was decanted and the residue was stirred with 25 ml more solvent, the combined extracts were evaporated in a stream of purified helium to a small fraction of a milliliter.

The insoluble residue was dissolved in hydrochloric acid (50 ml of 18 per cent), the acid solution decanted followed by treatment of the residue with hydrofluoric acid (100 ml of 48 per cent). The respective acid solutions were diluted, extracted with dichloromethane, made basic and extracted again. The combined extracts were dried with anhydrous magnesium sulfate (previously extracted

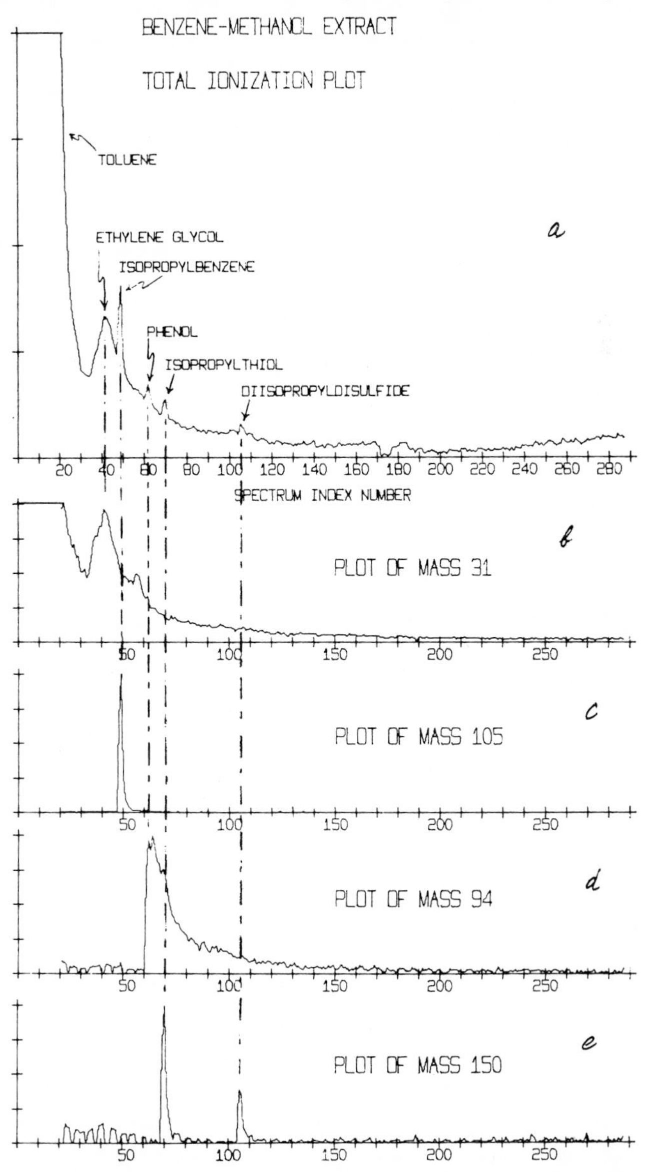

Fig. 3(a). Gas chromatogram of benzene–methanol extract; (b)–(e) Mass chromatogram for mass 31, 105, 94 and 150, respectively (for explanation see text).

with solvent) and concentrated in a stream of helium to 0·1–0·2 ml each. Solvents used were Mallinckrodt 'Nanograde' and J. T. Baker hydrofluoric and 'Ultrex' hydrochloric acids. Doubly distilled water was used in all cases. Another aliquot (30 g) of sample number 10086 was treated with hydrochloric acid without prior extraction with benzene–methanol. The acidic solution was extracted and concentrated in the same manner as described above. It was further concentrated (using a helium stream) to 10–20 μ and a major portion thereof injected into the gas chromatograph—mass spectrometer—computer system at a column temperature of 50°C which was then heated to 200°C at a rate of 10°/min and a carrier gas flow rate of 25 ml/min. The mass spectrometer scans continuously (every 4 sec from mass 10 to 600) and the computer records, digitizes, and stores the mass spectra, which are then converted to mass and intensity data.

The gas chromatograms are recorded not only from a flame ionization detector but also plotted by the computer. This is accomplished by adding all 8000 data points recorded during a mass spectral scan and plotting this sum versus scan index number (HITES and BIEMANN, 1970). Figures 2 and 3 represent such plots. Division of the scan index number by fifteen gives the time, in minutes, elapsed after the solvent peak, because the repetitive scanning of the mass spectrometer is started at that point.

The mass spectra (a few hundred for each gas chromatogram) are evaluated (a) by plotting the ion intensity at certain characteristic mass-to-charge values in each of the consecutive spectra ('mass chromatograms', see Fig. 3b–e); (b) by automatic comparison of selected spectra (those at maxima in the gas chromatogram or mass chromatogram) with a collection of mass spectra (approximately 7000) stored in the secondary memory of the computer; and (c) conventional interpretation of the individual mass spectra. High resolution mass spectra were obtained by introducing an aliquot of the residue of the extract directly into the ion source of the mass spectrometer.

RESULTS

None of the gas chromatograms revealed noticeable amounts of organic compounds that could be definitely identified as indigenous to the moon. As an example the record obtained from the dichloromethane extract of the hydrochloric acid solution is shown in Fig. 2. The detection limit was about 20–50 ng per component. Beyond mere inspection of this gas chromatogram, the data were searched for the presence of characteristic peaks in their spectra; this was unsuccessful. The utility of this technique is outlined in Fig. 3, which represents the gas chromatogram of the benzene–methanol extract of the 50 g sample (see above). A computer evaluation of all 287 spectra recorded during this gas chromatogram using a program that searches for the peaks present in some of the mass spectra but not in others, yielded a series of mass numbers which were then plotted as a mass chromatogram (see above). The maxima of these plots coincided with individual gas chromatographic peaks as shown in Fig. 3 (b–e) for m/e 31, 94, 105 and 150 as examples. The plot of m/e 150 demonstrates the capability of this technique to 'dig out' components not clearly visible in the gas chromatogram.

The mass spectra recorded at the maximum of these peaks revealed the various components to be toluene, ethylene glycol, isopropylbenzene, phenol and diisopropyl disulfide (the spectra centered around scan 70 as well as 105). The first disulfide is, in fact due to dehydrogenation of isopropylthiol, probably on the metal surfaces of the gas chromatograph exit line). All these identifications were confirmed using authentic samples; however all of these compounds are artifacts or contaminants. The toluene, isopropylbenzene and phenol are present in traces in the benzene used for the extraction and also appear in the blank. Ethylene glycol is most probably a hydrolysis product of the ethylene oxide that had leaked into the Bulk Sample Box (ALSRC). The presence of the two sulfur compounds which were definitely not detectable in the direct vaporization experiment is puzzling. They are thus not an artifact present in the sample but must have been produced during the extraction. Reaction of a metal sulfide (FeS perhaps) with isopropanol or acetone which may have been present in the solvent is a possible source. Model experiments to test this possibility are presently underway.

High resolution mass spectra of the extract without gas chromatographic separation should be capable of detecting substances which give characteristic and abundant ions at lower levels of concentration than possible with the gas chromatograph—mass spectrometer system. However the compounds must be of a volatility low enough not to be lost when placed into the vacuum lock of the spectrometer and high enough to be vaporized upon heating at 10^{-6} mm Hg. In this manner, ions of the composition C_8H_8, C_9H_{10}, $C_{10}H_8$, $C_{10}H_{10}$ and $C_{14}H_{14}$ were detected in the high resolution mass spectrum of the dichloromethane extract of the hydrochloric acid solution. These could be due to the presence of styrene, indane, naphthalene, dihydronaphthalene and anthracene at the nanogram level, i.e. less than a part per billion with respect to the lunar sample. At that level it is difficult to exclude the possibility that they are indeed indigenous.

It may be noted that all high resolution spectra of the extracts obtained after acid treatment showed the presence of elemental sulfur. Since free sulfur is not present in the original sample it must be produced from inorganic sulfides. The mass spectrum of the gas produced upon treatment of a portion of sample 10086 with concentrated phosphoric acid also shows the evolution of hydrogen sulfide.

Conclusion

The examination of material from the Sea of Tranquillity (fines sample 10086) by direct vaporization into the ion source of a mass spectrometer at temperatures up to 400°C or by solvent extraction followed by gas chromatographic separation monitored with a mass spectrometer did not reveal the presence of organic compounds that could conclusively be considered indigenous to the moon. The substances which were identified in the course of this work are at concentrations below the part per million range (ethylene glycol, isopropylthiol, diisopropyl disulfide and dialkyl phthalates) can be traced to terrestrial contamination or artifacts. There is some indication of the presence of a few aromatic compounds (biphenyl, styrene, naphthalene, dihydronaphthalene, indane and anthracene) at the part per billion level and these may possibly be indigenous. The experiments performed to date do not necessarily cover very polar or polymeric substances present below the parts per million range and no effort was made to detect compounds of molecular weights below mass 28·0000.

It appears that in order to search for organic compounds in specimens from the surface layer of the moon, one has to concentrate on those which can be produced from carbon and other elements originating from the solar wind (Moore *et al.*, 1970; Abell *et al.*, 1970; Oró *et al.*, 1970). This necessitates the search for small and rather uncomplicated organic molecules at extremely high sensitivity.

Acknowledgments—We are indebted to James Biller and Harry Hertz for assistance with the computer evaluation of the data. This work was supported in part by NASA research contract NAS 9-8099.

References

Abell P. I., Draffan G. H., Eglinton G., Hayes J. M., Maxwell J. R. and Pillinger C. T. (1970) Organic analysis of the returned lunar sample. *Science* **167,** 757–759.

Hayes J. M. and Biemann K. (1968) High resolution mass spectrometric investigation of the organic constituents of the Murray and Holbrook condrites. *Geochim. Cosmochim. Acta* **32,** 239–267.

Hites R. A. (1968) Computer recording and processing of mass spectra. Ph.D. Thesis, Massachusetts Institute of Technology, Cambridge, Mass.

Hites R. A. and Biemann K. (1970) Computer evaluation of continuously scanned mass spectra of gas chromatographic effluents. *Anal. Chem.* In press.

Moore C. B., Lewis C. F., Gibson E. K. and Nichiporuk W. (1970) Total carbon and nitrogen abundances in lunar samples. *Science* **167,** 495–497.

Oró J., Updegrove W. S., Gibert J., McReynolds J., Gil-av E., Ibancz J., Zlatkis A., Flory D. A., Levy R. L. and Wolf C. (1970) Organogenic elements and compounds in surface samples from the Sea of Tranquillity. *Science* **167,** 765–767.

Simoneit B. R., Burlingame A. L., Flory D. A. and Smith I. D. (1969) Apollo Lunar Module engine exhaust products. *Science* **166,** 733–738.

Proceedings of the Apollo 11 Lunar Science Conference, Vol. 2, pp. 1901 to 1920.

Organogenic elements and compounds in type C and D lunar samples from Apollo 11

J. ORÓ, W. S. UPDEGROVE, J. GIBERT,
J. MCREYNOLDS, E. GIL-AV,* J. IBANEZ, A. ZLATKIS
Departments of Biophysical Sciences and Chemistry, University of Houston, Houston, Texas 77004

and

D. A. FLORY
NASA, Manned Spacecraft Center, Houston, Texas 77058

and

R. L. LEVY, C. J. WOLF
McDonnell Research Laboratories, McDonnell Douglas Corp., St. Louis, Missouri 63166

(*Received* 2 *February* 1970; *accepted in revised form* 3 *March* 1970)

Abstract—Ten different analytical methods have been used for a comprehensive study of carbonaceous, organic and organogenic matter in lunar materials retrieved by the crew of Apollo 11 from the surface of the Sea of Tranquillity. Organogenic elements have been found in the fine particulate material and in a breccia rock in the following amounts (upper values in ppm): S (4, 300), P (270), C (41), N (20) and H (1). Other elements essential to life, Fe (12%), Mg (4·2%), Cu (12 ppm), Co (8·7 ppm), etc., are also present in the lunar samples. These measurements were obtained by spark ionization mass spectrography and reflect the amounts of these elements present as solid compounds. Other mass spectrometric methods were used for the characterization of gases and organogenic compounds. In addition to noble gases and nitrogen the major compounds of carbon released by heating are carbon monoxide and carbon dioxide, with very small amounts of methane ($CO > CO_2 \gg CH_4$). Hydrogen sulfide, CO_2, CH_4 and traces of C_2H_2, C_2H_4 and C_2H_6 were also released from the fines by acid treatment. The carbon isotopic ratio values obtained by combustion of the fines were found to be definitely non-terrestrial (+2·6 to +18·5 per mil vs. PDB). The concentration of total carbon was found to be surface-correlated. Amounts of the order of 200 ppm were measured in the sieved fines.

In conclusion, if all the above gases, together with water, were generated in a planetoidal body of sufficient gravitational field (e.g. the earth) the resulting atmosphere and hydrosphere would have eventually produced organic molecules of biological significance.

INTRODUCTION

IN THE light of the results from Surveyor V (TURKEVICH *et al.*, 1967, 1969), an igneous origin has been considered likely for the Sea of Tranquillity (GAULT *et al.*, 1967) and has since been confirmed by an examination of its surface rocks (LSPET, 1969; WOOD *et al.*, 1970). This precludes the existence of large amounts of organic matter on this part of the lunar surface. Therefore, it was considered pertinent to analyze for organogenic elements and compounds since these chemical species are necessary

* On leave of absence from the Weizmann Institute, Rehovoth, Israel.

for a possible abiological or biological synthesis of organic molecules (ORÓ, 1965). Thus the composition of the fine particulate material and of a small piece of a breccia sample was studied by different methods of analysis in accordance with the general objectives of our proposed work: "Comprehensive study of carbonaceous, organic and organogenic matter in lunar materials." Our study included analyses of organogenic (H, C, N, P, S) and other elements essential to life (Fe, Mg, Cu, etc.); CO, CO_2, N_2 and other gases; organic and organogenic compounds released by heat,

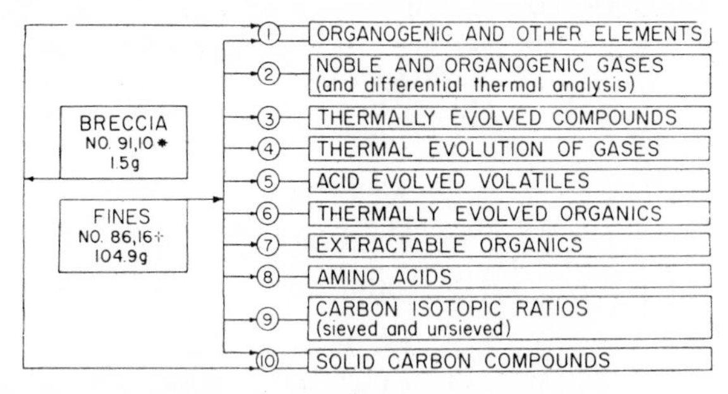

Fig. 1. Schematic flow diagram of sample distribution for the analysis of organogenic elements and compounds in lunar fines and breccia from Apollo 11. These are the same samples studied by ORÓ et al. (1970).

* The number 91-10 (or 10091-10) is a new series number corresponding to an old series number (10002-54) used previously.

† The number 86-16 (or 10086-16) corresponds to the number 10086-6 used previously, where the sixth digit (1) was inadvertently omitted.

acid or other treatment; carbon isotopic ratios; and solid carbonaceous compounds. In order to avoid contamination and the introduction of artifacts, the samples were analyzed as directly as possible and with a minimum of handling. Thus, the fines were not pulverized, being sieved in just one case, and the breccia was cleaved only once.

ANALYTICAL PROCEDURES AND RESULTS

Figure 1 shows a simplified schematic diagram of the different substances analyzed (or measurements made) in the two samples received from MSC (breccia and fines) by the methods briefly described below. The weight and code numbers of the samples are also listed in Fig. 1. Two different techniques were applied to the breccia and ten to the fines.

1. *Spark ionization mass spectrography*

This method was utilized to determine the abundances of hydrogen, carbon, nitrogen, phosphorus and sulfur, following established techniques. A survey analysis of thirty-six other elements was also carried out for comparative purposes. We used a Mattauch-Herzog type double focusing (CEC X21-110) mass spectrograph that operated at a resolution of approximately 3000 and at a relative sensitivity that ranged from 0·005–0·03 ppm. This was achieved by the use of ion pumping, gold sample holders, tantalum slits and photoplate exposures of 1×10^{-7} C. Data reduction was accomplished by a precalibrated hybrid digital–analog computer system.

Each sample, about 250 mg, was sparked against a pure gold electrode. Up to 20 mg of each sample was actually consumed in the analysis. All the samples were isolated from the laboratory environment by means of glove boxes up to and including their insertion into the mass spectrometer vacuum system. Cryogenic rough pumping, ion pumping and preliminary sparking assured a minimum of contamination present during actual measurements. The fine material was vacuum encapsulated, hydrostatically pressed to 4570 kg/cm^2 and then cleaved for analysis. The piece of breccia was cleaved and immediately analyzed at selected surfaces. Previous calibration analyses with six National Bureau of Standards specimens of Special Ingot Iron having carbon contents of from 370–4000 ppm showed agreement to within 20 per cent of the NBS certified values. It should be noted that if the sample is not homogenous, variations in the analytical values will be observed depending on the composition of the area sparked. Also, preliminary sparking causes degassing of the sample so that the measured values correspond to solid or chemically bound species of the elements analyzed.

The results obtained from three analyses of the fines and the breccia are shown in Table 1. With the exception of the six major elements, the concentrations are given in ppm. Very small amounts of hydrogen, small amounts of carbon and nitrogen and larger amounts of phosphorus and sulfur were observed. Since the samples were neither pulverized nor homogenized, it was not surprising to see variations of less than one order of magnitude for C, N, P and S. As could be expected, if only solid or chemically bound species were measured, the values for C and N were found lower than those obtained for total C and N (Moore *et al.*, 1970). Medium to small amounts of solid compounds such as sulfides, phosphates, graphite, carbides and nitrides could be responsible for the values of these organogenic elements. The low values for hydrogen are significant because they demonstrate the absence of any but trace amounts of solid organic compounds, such as aromatic or aliphatic hydrocarbons or other hydrogen-bearing compounds. Nitrogen and phosphorus are two organogenic elements that were also determined along with other elements in other laboratories (e.g. Morrison *et al.*, 1970). The values for these elements and for most of the other elements, particularly those essential to life, Fe, Mg, Cu, Co, etc. (with possibly the exception of the halogens, which are lower in our case) agree fairly well with the measurements obtained by Morrison *et al.* (1970) using spark ionization mass spectrography, neutron activation analysis and other methods.

Additional experiments were carried out by pyrolysis of samples and characterization of the gases or volatiles evolved using the following three mass spectrometric methods.

2. *Volatilization—high resolution mass spectrography and differential thermal analysis*

Measurement of the evolution of helium, neon, argon, CO and CO_2 with increasing temperature was accomplished by programmed heating (100°–1400°C) of small aliquots of fines (usually about 35 mg) in a high vacuum DTA cell attached to a CEC 21-110B high resolution mass spectrograph.

As shown in Fig. 2, the noble gases first appeared at temperatures of from 200–400°C and reached maxima from 500°–800°C depending on the atomic mass of each gas and the heating rate. The appearance of CO, CO_2 and N_2 varied according to time and temperature. Carbon monoxide prevailed over CO_2 at high temperatures as illustrated in Fig. 2 ($CO/CO_2 \simeq 10$). Nitrogen was also present in substantial amounts as shown by the high resolution mass spectrum of Fig. 3 where typically $CO/N_2 > 2$. This gas also appeared intermittently in sharp bursts ($N_2/CO > 3$) as the temperature of the fines was increased. This indicated sudden escape from vesicles in the lunar material.

Differential thermal analysis was carried out simultaneously with the mass spectrographic analysis of the gases released. The curve obtained is depicted in Fig. 4. The melting temperature of the bulk fines was 1134°C after which an exotherm, followed by another significant transition at 1211°C were observed. In agreement with these observations, examination by other workers of phase melt inclusions has revealed evidence for the occurrence of major phase-transition phenomena at these two temperatures. Thus, on cooling the original magma, ilmenite (and/or ferropseudo brooklite) appears at 1210°C; and pyroxene at 1140°C (Roeder and Weiblen, 1970); solid metallic iron and iron sulfide (troilite) intergrowths are also formed from an initially homogenous liquid phase at the latter temperature (Skinner, 1970).

Plotting the product of the intensity of the carbon monoxide ion and the temperature against the inverse of the temperature (Fig. 5) and, making use of a modified Clausius–Clapeyron equation

Table 1. Spark ionization mass spectrographic analysis of organogenic and other elements in Apollo 11 samples

		Rock			Fines		
Element	Detection limit	Cleaved surface	Outer surface	White spot	Unsieved (1st run)	Unsieved (2nd run)	Sieved
(%)							
Si	3	20	18	20	23	21	20
Fe	0·3	12	15	15	12	10.4	12
Ca	0·3	5·8	9·2	8·3	6·3	6·4	8·0
Mg	0·3	3·9	4·6	3·3	4·2	5·0	6·3
Al	1	4·4	7·9	6·2	4·0	5·2	6·2
Ti	2	5·3	3·8	5·2	3·8	5·6	2·4
(ppm)							
S	5	2800	2100	1000	4200	590	960
Na	0·07	2600	340	780	2000	1660	1300
Cr	0·1	1800	2100	1300	1900	1700	1640
Mn	0·1	1000	2000	1300	1400	1650	1500
K	0·07	1200	2300	1400	600	1200	3600
P	0·3	110	250	59	270	187	110
Ni	0·3	160	420	<1	63	56	72
Ba	0·3	110	18	16	41	170	230
Sr	0·3	40	42	31	29	42	64
V	0·1	31	25	24	27	11	23
Zr	0·5	32	15	17	23	140	100
N	2	15	<10	<10	14	20	7
Zn	5·0	N.D.	N.D.	N.D.	14	<5	<5
Rb	0·07	13	7·0	9·2	14	13	7·4
Cu	3·0	N.D.	N.D.	N.D.	12	<3	<3
Li	0·03	4·1	3·7	2·6	9·2	4·9	3·5
C	1		6·0	5·5	<5–9*	24	41
Co	0·3	14	9·6	3·3	8·7	12	20
F	0·2	3·7	2·4	7·0	6·3	3·5	5·0
Ga	0·3	1·6	3·0	N.D.	4·6	3·9	2·9
As	0·3	0·39	0·28	0·42	4·0	0·57	0·31
Ce	0·3	2·8	1·0	0·9	3·8	7·3	15
Cl	0·3	3·3	2·0	1·4	3·5	0·91	1·0
Nb	0·3	2·3	1·8	0·69	3·1	10	16
B	0·03	0·63	0·11	0·37	1·4	0·71	0·74
Y	0·3	1·4	1·3	0·74	1·4	11	9·3
H	0·03	0·13	0·11	0·08	1·2	1·2	0·41
Ge	0·3	0·45	1·3	0·83	1·0	1·3	0·38
Ag	0·5	1·0	3·0	3·0	0·91	3·9	0·44
Sn	1·0	N.D.	N.D.	N.D.	0·5	<1	<1
La	0·3	0·73	0·34	0·9	0·3	0·67	1·0
Cs	0·1	0·22	0·89	1·2	0·19	0·24	0·16
Se	0·3	N.D.	N.D.	N.D.	N.D.	N.D.	N.D.
Br	0·3	N.D.	N.D.	N.D.	N.D.	N.D.	N.D.
Be	0·07	0·42	N.D.	0·1	N.D.	1·3	2·1

N.D. = not detected.
* Two separate analyses of the same sample.

(CHUPKA and INGHRAM, 1953) energies of 12 kcals/mole (between 480° and 590°C) and 27 kcals/mole (between 610° and 760°C) were obtained for the release of carbon monoxide. These two values correspond to two regions of a semicontinuous curve which measures the carbon monoxide binding energies. It is likely, although not certain, that chemidesorption and decomposition processes may be involved in the release of this gas.

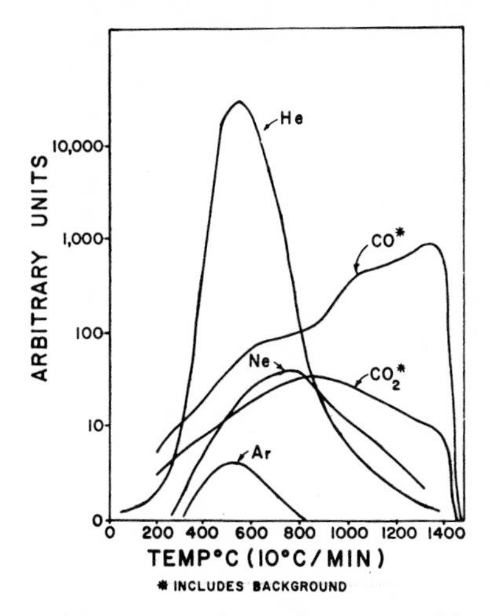

Fig. 2. Temperature–release curves for carbon monoxide, carbon dioxide, helium, neon and argon obtained in a high vacuum DTA cell. The conversion factor for the helium curve is $\times 10^{10}$ cm^3/sec (STP). For the other gases the arbitrary units correspond to photoplate densities. The range from 0 to 1000 corresponds to a density of 0–80 for Ne and Ar, and 40–100 for both CO and CO_2.

3. *Volatilization—quadrupole mass spectrometry*

Several experiments were carried out by direct pyrolysis–mass spectrometry of small amounts of fines. A specially designed T-type stainless steel oven, which could be heated quickly to preselected temperatures (up to 750°C) by means of a molten metal bath, was attached directly to a vacuum chamber-quadrupole mass spectrometric system, which included a monopole mass filter for total pressure measurements (UPDEGROVE and ORÓ, 1969). Samples of about 100–450 mg were used. The pressures developed by the samples were at least two orders of magnitude above the pressures at ambient temperature. The empty sample tube was cleaned, baked out at 750°C for 20 min and exposed to the atmosphere for a time comparable to an actual sample loading. The tube was replaced on the instrument, evacuated and heated to 450°C to obtain a background spectrum which was subtracted from the sample spectrum. Figure 6 shows a mass spectrum of lunar sample, minus background, at 2×10^{-7} torr, after heating for 1 min at 450°C. Evidence for the presence of parent ion masses 18 (H_2O), 28 (CO and N_2) and 44 (CO_2) and traces of other compounds including C_6H_6

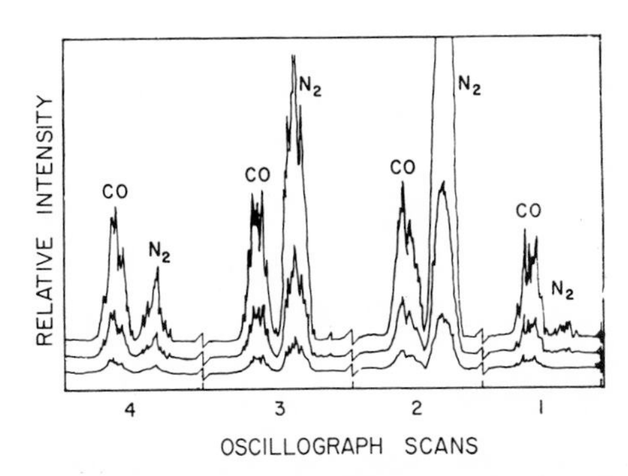

Fig. 3. High resolution mass spectrometric ion intensities for CO and N_2 recorded every 2 sec., at 914°C (1 and 4, typical).

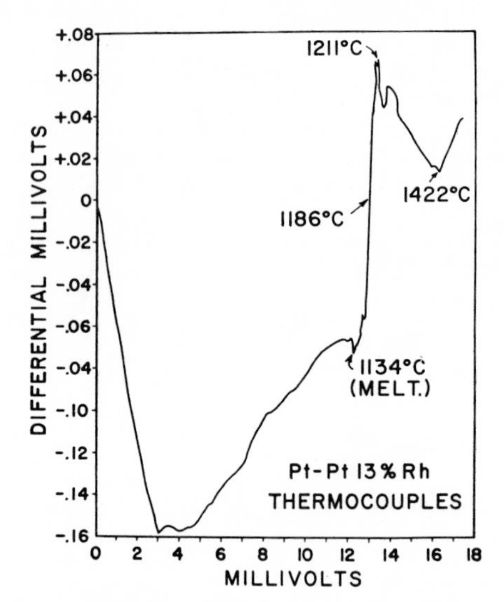

Fig. 4. Differential thermogram for the temperature range 200–1500°C.

and $C_7H_7^+$ (tropillium ion) was obtained by this method. The 18 peak is assumed to be water adsorbed on the surface of the fines due to exposure of the sample to the atmosphere. It is believed that the surface area of the lunar fines is adequate to account for the intensity of this peak.

4. *Volatilization—gas chromatography–mass spectrometry*

The combination of gas chromatography and mass spectrometry has proved, in the last few years, to be one of the most effective techniques for the analysis of complex mixtures of compounds (ORÓ

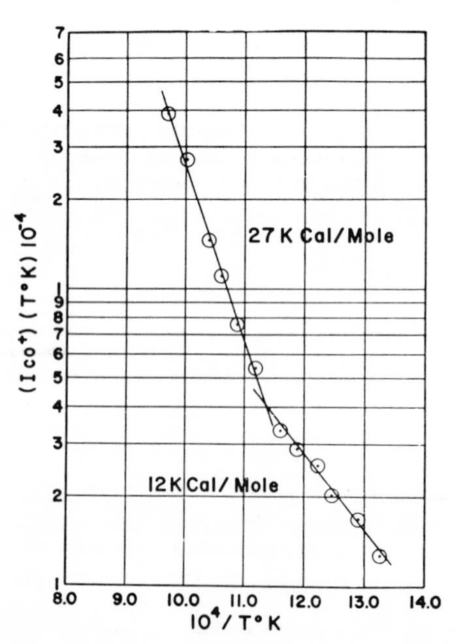

Fig. 5. Energies of release for carbon monoxide from DTA data.

et al., 1968). It was employed here to identify volatile products obtained from the fines by either stepwise heating or by acid treatment of appropriate samples.

In order to analyze the gases evolved from stepwise pyrolysis, the samples were placed in a specially designed stopped-flow pyrolyzer which consisted of a quartz tube (8 cm long by 0·8 cm i.d.) with metal ends. This was connected to a stainless steel column (120 cm long by 0·2 cm i.d.) packed with 200 mesh Porapak Q and operated at room temperatures and 0·84 kg/cm^2 of helium as carrier gas. This column was connected in turn to an LKB 9000 gas chromatograph mass-spectrometer.

A brass mantle with four cartridge type heating elements surrounding the quartz tube provided temperatures up to 750°C. A valve system allowed flow of the carrier gas for short intervals (e.g. 10–100 sec) through the pyrolyzer tube.

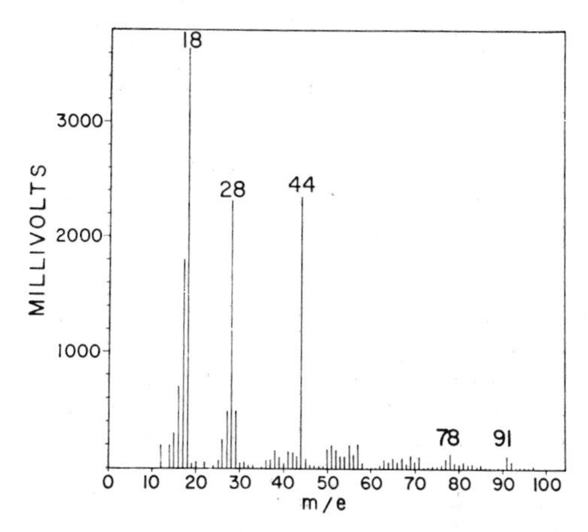

Fig. 6. Quadrupole mass spectrum of lunar material volatilized at 450°C.

After each pre-selected temperature was reached and maintained for 5 min to allow sufficient heat transfer, the carrier gas was allowed to pass through the pyrolyzer for 30 sec. The products evolved at each temperature were swept into and separated in the chromatographic column before their introduction into the mass spectrometer. The chromatographic peaks tended to spread because of the relatively long time during which the gas transfer into the column took place. Therefore, mass spectra of the effluent gases were obtained by semi-continuous mass scannings.

Figure 7 shows the gas chromatograms obtained from 1·5 g of fines sequentially heated at several temperatures. No significant amounts of gases were detected below 200°C, and only a gradual increase in the spectral masses corresponding to CO_2 was observed. From 200 to 500°C the main gas evolved was CO_2. Above 500°C the predominant gas released was CO (and some CO_2). Even though CO_2 was detected from the first heating steps, it reached a maximum about 400°C and then appeared to decrease with increasing temperature, being accompanied by an increase in CO. Traces of methane were first seen at 400°C and increased more or less linearly up to 750°C. These results are in general agreement with and complement those obtained by direct pyrolysis-mass spectrometry.

5. *Acid treatment—gas chromatography–mass spectrometry*

Several treatments with acids were performed on the lunar fines in order to gain some insight into the mineralogy of organogenic elements, especially carbon and sulfur.

A specially designed glass reaction vessel was attached to the system described above which was operated under similar chromatographic conditions. The sample was placed in the bottom of the reactor and the acid in the side arm. Before starting the reaction the system was outgased by displacement with helium at room temperature and a blank was run to insure that no gases other than helium were present in the vessel. The side arm was tipped so that the acid could be mixed with the

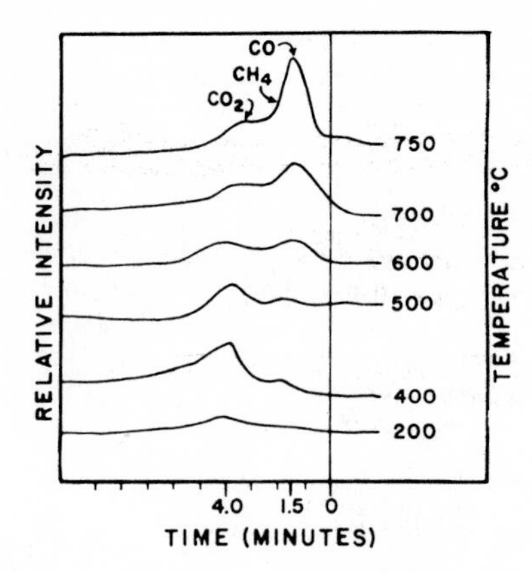

Fig. 7. Gas chromatograms of CO and CO_2 obtained by sequential heating of a 1·5 g sample of fines. Semicontinuous mass scans of the gases were obtained. CH_4 appears in the tail of the CO peak.

sample. After the reaction was complete the generated volatile products were swept into the chromatographic column, for short intervals, as described in the previous section.

Phosphoric, sulfuric and hydrochloric acids were used in three separate experiments. In each case the acidolysis was carried out at room temperature for 1 hr. In the first case, 2 ml of 60% H_3PO_4 was poured over 0·19 g of fines. This treatment yielded mainly H_2S and traces of CO_2 (Fig. 8). A determination of the ion intensities at m/e 36 ($H_2{}^{34}S$) and 34 ($H_2{}^{32}S$) gave an approximate value of 4·2 per cent for isotopic ^{34}S fully in line with its natural abundance, thus corroborating the mass spectrometric identification of this sulfur compound.

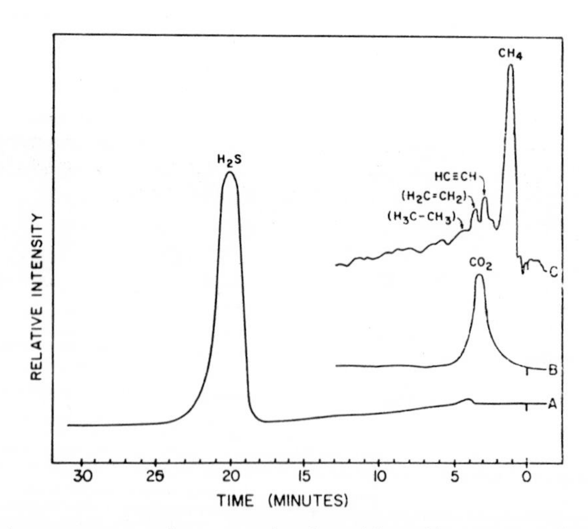

Fig. 8. Gas chromatograms of compounds released by acid treatment. Semicontinuous mass scans of the gases were obtained. (A) Phosphoric acid. GC column (1·2 m long) of 200 mesh Porapak Q operated at 0·82 kg/cm² of helium. (B) Sulfuric acid. GC column (0·6 m long) of 200 mesh Porapak Q operated at 0·35 kg/cm² of helium. (C) Hydrochloric acid. GC column (0·6 m long) of 200 mesh Porapak Q operated at 0·35 kg/cm² of helium.

In the second experiment, 2 ml of 98% H_2SO_4 was employed to treat 0·82 g of sample. In this case a small filter made of a tube filled with a cartride of copper powder in a copper mesh, was interposed between the reaction vessel and the chromatographic column. The use of this filter insured the retention of the hydrogen sulfide generated by the H_2SO_4. The main product thus measured was CO_2 (Fig. 8).

Concentrated H_3PO_4 and H_2SO_4 are not particularly good hydrolytic reagents. Therefore, in order to get more effective hydrolytic conditions, 6 N HCl was employed. Thus, in the third experiment, 2 ml of 6 N HCl was used to hydrolyze 0·66 g of lunar fines. In this case a filter of copper mesh, copper powder and sodium oxide-hydroxide beads was utilized to retain not only the H_2S but also the CO_2 (and the HCl), thus allowing only non-polar compounds such as hydrocarbons to reach the column. In this experiment the main gas identified was methane. Traces of acetylene, ethene and ethane were also detected (Fig. 8). These results on the release of light hydrocarbons by HCl are in agreement with the work of ABELL *et al.* (1970) and PONNAMPERUMA *et al.* (1970).

6. *Volatilization—flame ionization gas chromatography*

Non-graphitic compounds detectable by hydrogen flame ionization were released from the fines by heating from ambient temperature to 480°C in a flow-through pyrolyzer. The pyrolyzer consisted of a silver core heater and a quartz tube (12·7 cm long by 0·22 cm o.d.) which contained the sample and was directly connected to the GC column via a low internal volume feedthrough. Two open tubular columns (Perkin–Elmer Corp.) connected in series were used: A 15·2 m long by 0·51 mm i.d. SCOT column coated with SE-30 and a 39·6 m long by 0·51 mm i.d. wall coated column containing SE-30. Provisions were made to permit introduction of liquid nitrogen to the column oven in order to maintain the column at sub-ambient temperatures.

Liquid nitrogen traps were introduced in the carrier gas and the make-up gas lines in order to remove organic contaminants. Precautions were taken to avoid back-flow of stationary phase volatiles when the traps were immersed in liquid nitrogen. The quartz pyrolyzer tube containing a quartz wool plug was thoroughly decontaminated from any organic material by flaming it with a hydrogen–oxygen flame to glowing temperatures (800–900°C) while surgical oxygen was flowing through the tube. Other details of this method are given elsewhere (LEVY *et al.*, 1970).

Quantities of 35–55 mg of lunar fines were introduced into the pyrolyzer tube. Immediately after filling, the tube was connected to the inlet of the gas chromatograph and flow of carrier gas (between 3·0 and 3·5 cm^3/min) established. The column was cooled to approximately −60°C for 10 min and then programmed to 130°C without heating the sample in order to establish a blank. The procedure was repeated by heating the sample at a desired temperature and a pyrogram recorded at the highest obtainable sensitivity of the instrument which, in terms of minimum detectable quantity, corresponded to approximately 10^{-11} g/sec. The total amount of volatile organic compounds produced in several of these experiments was approximately 0·5 ppm. A few peaks, one of which corresponded to benzene by its elution time, were observed. Blank chromatograms obtained while the sample was kept at room temperature for 5 min prior to the pyrolysis experiments showed no measurable amounts of flame ionization detectable compounds.

7. *Extraction and flame ionization gas chromatography*

Analysis of extractable organic substances was carried out by procedures similar to the ones used extensively in this laboratory (NOONER and ORÓ, 1967) but specifically adapted for the detection of very small amounts of compounds.

Small aliquots of lunar fines were extracted in specially designed all glass Soxhlet apparatus which had an interchangeable sintered glass cup as sample holder. All the glassware used was thoroughly cleaned in the same way immediately preceeding the analysis. The solvents used were originally of nanograde quality but were triply distilled in our laboratory in a distillation apparatus which had a column efficiency of approximately ten theoretical plates.

The extractions and controls were carried out for 4 hr with a 16 cm^3 mixture 3:1 of benzene and methanol. Each extract was concentrated to about 5 μl under a stream of purified N_2 and a 1 μl aliquot was injected into a high resolution capillary column (120 m long by 0·76 mm i.d.) coated with Polysev operated at 1·2 kg/cm² of helium and programmed from 100 to 200°C at 2°C/min. An F and M Model 810 gas chromatograph with an Infotronics integrator was used for these analyses. The sequence for each analysis was as follows: programming of the column, injection of the standard, programming of the column, injection of the blank, programming of the column and injection of the sample extract.

Two series of extractions were carried out independently by two different investigators. In the first set, 0·123 g and 0·148 g were taken from the top of the original lunar sample container which had an aluminum screw cap with a transparent Teflon liner interposed. A single component with a molecular weight of about 150 was detected in both analyses, amounting in each case to approximately 1 ppm. This compound, is no doubt, a contaminant. It is not yet known whether it is derived from the Teflon liner, the York mesh used in the Apollo 11 boxes, or from other sources. Interestingly enough, a compound with similar molecular weight and retention time was also detected in the York mesh used for the Apollo 12 boxes (GIBERT and ORÓ, 1969). With the exception of this single compound the total integration of the base lines of the sample extraction chromatograms gave a similar value to the integration obtained from the base line of the controls, demonstrating the absence of any other extractable organic components.

A second set of extractions carried out with two samples of 1·105 g and 1·492 g, respectively, from the bottom of the same container yielded less than 0·001 ppm. The procedure for this second set was identically the same as for the first set and was carried out only 24 hr later. These results demonstrate the absence of significant amounts of extractable substances in the bulk of sample 10086-16.

8. *Hydrolysis, derivatization and flame ionization gas chromatography*

In order to determine the minimum amounts of amino acids that could be detected by means of this analytical technique, an amino acid standard (isoleucine) was passed through the ion-exchange column and then derivatized for gas chromatographic analysis (see below). The results indicated that amounts as small as 0·1 μg or less of this amino acid could be detected.

A 1·44 g portion of lunar fines was hydrolyzed in 6 N HCl in a sealed tube for 24 hr at 100°C. The hydrolyzate was then filtered; the residue was allowed to dry and was stored. The supernatant liquid was evaporated to dryness under reduced pressure. The remaining residue was taken up in 2·0 ml of triply distilled water and passed through 20 ml of ion exchange resin, Dowex 50(H^+). The resin was eluted with 200 ml of distilled water and then with 200 ml of 2 N NH_4OH. The latter eluate was evaporated to dryness under reduced pressure. This fraction should have contained amino acids if they were present.

Derivatization of the material was carried out with 1·25 N HCl-isopropanol followed by treatment with trifluoroacetic anhydride. The derivatized material (if any) was dissolved in chloroform and injected into a gas chromatograph (Varian Aerograph Model 1200 equipped with a flame ionization detector). The injection was made into a 122 m × 0·05 cm stainless steel capillary column coated with N-TFA-L-valyl-valine-O-cyclohexyl ester and chromatography was carried out isothermally at 110°C (GIL-AV *et al.*, 1966). The analysis indicated the absence of any amino acid N-trifluoroacetyl isopropyl ester.

9. *Stable carbon isotope ratios*

Stable carbon isotope ratio measurements were made on several samples of fines using two different methods of combustion. Two samples of approximately 10 g each were burned according to a method developed in this and other laboratories (FLORY, 1969) which is a modification of Craig's (CRAIG, 1953) original method. Carbon dioxide was formed by heating at high temperatures in the presence of oxygen and cupric oxide. One sample was heated at 800°C for one hour and another sample at 900°C for 2 hr. The amounts of CO_2 obtained from these two samples corresponded to 40 and 110 ppm of carbon, respectively. These amounts are within the range of those reported for total carbon for the Apollo 11 lunar samples (MOORE *et al.*, 1970). The $\delta^{13}C$ values of these two samples were +13·0 and +18·5 per mil vs. PDB, respectively, in essential agreement with data from other laboratories (KAPLAN and SMITH, 1970; EPSTEIN and TAYLOR, 1970). The second sample was then burned an extra 6 hr at 900°C and supplementary CO_2 corresponding to 10 ppm carbon was obtained. This additional CO_2, probably derived from difficultly combustible carbon, had a $\delta^{13}C$ value of $+14{\cdot}1$ per mil vs. PDB.

In another series of experiments a third sample of 10 g was burned using an entirely different technique. The combustion was carried out in a Leco RF induction furnace. Iron and cupric oxide catalyst were added to the sample, which was heated in a flowing oxygen stream to a temperature sufficient to completely fuse it (greater than 1400°C). The effluent oxygen and combustion products were trapped in liquid nitrogen and the CO_2 isolated and purified by fractional distillation. The small size of the ceramic combustion boats made it necessary to burn the sample in five consecutive combustions using 2 g each, and the CO_2 was combined to give sufficient gas for an isotope ratio measurement. The total CO_2 obtained corresponded to 60 ppm total carbon with a $\delta^{13}C$ value of $+2{\cdot}6$ per mil vs. PDB.

The fact that better than 90 per cent of the total CO_2 collected was produced in the last two burns, however, gave us reason to doubt the validity of the $\delta^{13}C$ value. Two additional samples were then burned using in each case sufficient material (2·1 and 2·9 g) to yield enough CO_2 for an isotope ratio measurement. The lunar fines used in these last two combustions had been passed through a 200 mesh sieve in order to increase the bulk carbon concentration, since it is known that this concentration increases with decreasing particle size (MOORE *et al.*, 1970). The CO_2 obtained corresponded to 210 and 190 ppm total carbon and the $\delta^{13}C$ values measured were $+2{\cdot}6$ per mil and $+4{\cdot}6$ per mil vs. PDB, respectively.

The carbon obtained as CO_2 and $\delta^{13}C$ values discussed in the above two series of experiments are summarized in Table 2. Standards run concurrently with the lunar samples indicate these values are accurate to within $\pm 0{\cdot}2$ per mil.

10. *Ion microanalysis*

A small chip of the breccia specimen was also investigated by ion microanalysis with oxygen and argon positive ions as the primary ionization beam. A CEC Type 27-201 ion microanalyzer was used for these studies. The erosion rate of the instrument was approximately 100 Å/sec over an area of 0·3 mm in diameter. Ionic distribution images of about 150 μm in diameter were obtained from the center of the ionized area with a spatial resolution of about 5 μm. Mass resolution of 1/300 (10 per cent valley) was achieved up to m/e 300. Several micro-images for carbon and mass spectra of C_2 negative ions (m/e = 24) were obtained showing that the distribution of this element is at times diffuse but can also be observed in more or less well differentiated distribution patterns. Some of the patterns could be followed in depth by allowing the primary beam to erode continuously for several minutes. In this way increases as well as decreases in intensity of different areas from the carbon micro-image could be observed. It is not known at present whether the observed micro-images correspond to graphite, carbides, or other forms of carbon. However, these results are consistent with the presence of graphite (e.g. ARRHENIUS *et al.*, 1970); and carbides (ADLER *et al.*, 1970; FRONDEL *et al.*, 1970), indicated by electron microprobe analysis and with other data (see experiment 5) which indirectly suggest the presence of carbides.

DISCUSSION AND CONCLUSIONS

It should be said at the outset that analyses of the samples of Apollo 11 can only give data corresponding to Mare Tranquillitatis which obviously cannot be considered representative for the whole moon. Indeed, as indicated in the Introduction, it was known well in advance of the Apollo 11 mission that the landing site, because of the presumed igneous nature of the Sea of Tranquillity, would be one of the less likely places to find organic compounds. In spite of this, the results obtained from these samples during the past few months have turned out to be quite informative in relation to the early stages of chemical evolution which may have preceded the formation of biochemical compounds in more appropriate environments of the solar system.

Elements (Experiment 1)

One of the major conclusions which can be derived from the results is that the elements necessary for the synthesis of organic molecules of biological significance are present in different forms on the surface of the moon.

The work on spark ionization mass spectrometry shows that the organogenic elements are present in small amounts in the lunar fines and breccia. Listed in order of decreasing abundance, the upper values obtained for these elements are (in ppm): S (4200) > P (270) > C (41) > N (20) $\gg$ H (1). When compared with other analytical and mineralogical data these are found to be conservative values. These elements enter into the composition of a number of minerals and solid compounds which have been identified in the lunar samples, namely troilite, apatite, calcium phosphate, schreibersite, aragonite, graphite, cohenite, carbides, nitrides, and possibly nitrates (LSPET, 1969; ADLER *et al.*, 1970; ARRHENIUS *et al.*, 1970; FRONDEL *et al.*, 1970; GAY *et al.*, 1970; HINTENBERGER *et al.*, 1970; MAXWELL *et al.*, 1970; and WOOD *et al.*, 1970).

Other elements, which are considered essential to life, such as Fe, Mg, Cu, Co, etc., are present in the fines in substantial amounts. Representative values obtained for these four elements in the unsieved lunar fines are 12 per cent, 4·2 per cent, 12 ppm and 8·7 ppm, respectively, in line with values reported by others (MORRISON *et al.*, 1970).

Elements and compounds thermally evolved (Experiments 2, 3 *and* 4)

The effluent gas analyses of the lunar fines from 200° to about 1500°C) (Experiment 2) provide data, not available so far, for either lunar samples or meteorites, on the release curves for the noble gases He, Ne and Ar, and on the evolution of organogenic gases CO, CO_2 and N_2, which are listed in the order of their abundance. Carbon monoxide predominates over CO_2 particularly at high temperatures. This is in agreement with some of our other data (Experiment 4) and with the work of BURLINGAME *et al.* (1970) and HINTENBERGER *et al.* (1970). Even though some relationship between the evolution of the carbon oxides and the noble gases was observed (see Fig. 2), the composition and the state of the carbon oxides does not necessarily reflect their condition in the solid material. The values of 12 and 27 kcal/mole obtained for the energies of CO evolution are compatible with different ligands and chemical interactions. This includes entrapping of the gas in crystalline lattices, chemisorption, binding with metals in the form of carbonyl derivatives, and reaction of finely divided carbon with oxides. The fact that the concentration of carbon increases with decreasing particle size (MOORE *et al.*, 1970; Table 2 of this work), suggests that the presence of this element in any form which is capable of evolving as CO is a surface-correlated phenomenon, as has been shown for the inert gases (HEYMANN *et al.*, 1970).

The CO_2 may exist as such or combined in the form of carbonates. Indeed, the formation of CO_2 by the action of sulfuric acid (see fifth experiment; also HINTENBERGER *et al.*, 1970, and MAXWELL *et al.*, 1970) indicates that at least part of the CO_2 is liberated from carbonates, such as aragonite, (GAY *et al.*, 1970). Moreover, the temperatures of CO_2 evolution are compatible with those necessary for the decomposition of calcium carbonates at different partial pressures of this gas.

The typical amounts of nitrogen observed at 914°C (see Fig. 3) are lower than those of CO. However as shown by Fig. 3, also sporadic bursts of N_2 were observed indicating rupture of a vesicle containing much larger quantities of this gas than CO. It is possible therefore that we are dealing with more than one source of nitrogen. Indeed other work (HINTENBERGER *et al.*, 1970) has indicated that in addition to gaseous nitrogen, nitrides and nitrates probably exist in the lunar fines.

Table 2. Total carbon content and stable isotope distribution in Apollo 11 bulk fines sample

Run No.	Sample wt.	Combustion method	Combustion temp.	Total C (ppm)	$\delta^{13}C$ (per mil vs PDB)
1	11·0 g	Craig	800°C	40	+13·0
2a	10·9 g	Craig	900°C	110	+18·5
2b		Craig	900°C	11*	+14·1*
3	9·6 g	Fusion	>1400°C	60	+2·6
4	2·1 g	Fusion	>1400°C	210†	+2·6
5	2·9 g	Fusion	>1400°C	190†	+4·6

* Additional CO_2 obtained from re-combustion of the same sample (10·9 g) after Run No. 2a.
† Samples 4 and 5 were fines passed through a 200 mesh sieve.

In general, the experiments conducted with the other two mass spectrometric techniques upon (Experiments 3 and 4) thermal treatment of the fines confirm and extend the above observations. In particular the fourth experiment (V-GC-MS) shows clearly the relative evolution of CO_2 and CO, in this sequential order, with increasing temperature. Also the appearance of small amounts of methane at higher temperatures (400–700°C) is of major significance. Whether this compound exists in the fines as such, or is formed by the action of hydrogen (which is known to be present in substantial amounts) on other compounds of carbon is not known.

Compounds evolved by acid treatment (*Experiment* 5)

The results on the evolution of methane and other light hydrocarbons (acetylene, ethene and ethane) by the action of HCl on the fines is in essential agreement with the work of ABELL *et al.* (1970), and PONNAMPERUMA *et al.* (1970). Iron and nickel carbides are considered likely candidates as precursors for these organic gases. In fact cohenite and other carbides have been detected by electron microprobe (FRONDEL *et al.*, 1970; ADLER *et al.*, 1970). The other two major gases generated by acid, CO_2 and H_2S, are most likely derived from carbonates such as aragonite (see above) and sulfides such as troilite (e.g. LSEPT, 1969).

Organic compounds from pyrolysis (*Experiment* 6)

As shown by flame ionization gas chromatography of the fines heated at 480°C, the amount of organic compounds obtained by pyrolysis of the lunar samples is very small (about 0·5 ppm). Therefore, on the basis of the available evidence, it is not possible to state with certainty whether these small amounts of organic compounds were present in the lunar fines before they were collected by the Apollo 11 astronauts, or are terrestrial contaminants. The same can be said with regard to the benzene, toluene, and other organics detected in traces by quadrupole mass spectrometry (see Fig. 6), and to the organic compounds detected in other laboratories by similar pyrolytic methods (ABELL *et al.*, 1970; BURLINGAME *et al.*, 1970; MURPHY *et al.*, 1970; NAGY *et al.*, 1970). In fact MURPHY *et al.* (1970) state that none of the organic compounds detected in their laboratory appear to be indigenous to the lunar surface.

Extractable organic compounds (*Experiments* 7 *and* 8)

It has been shown that no measurable amounts of extractable organic compounds (less than 0·001 ppm) and amino acids (less than 0·1 ppm) were found in the bulk of sample 86·16. Similar findings have been made with other samples by others (e.g. LIPSKY *et al.*, 1970; MEINSCHEIN and CORDES, 1970). On the whole, the results of these studies demonstrate that at least some of the samples have not been grossly contaminated by terrestrial organic matter. From the results of our analyses we can say with certainty that the contamination by organic compounds of sample 10086-16 was at most 1 ppm and generally much lower. The amount of 1 ppm refers to the single extractable organic component found in the top layer of our sample (10086-16) which is considered to be a contaminant. These results are particularly important for an interpretation of the carbon isotopic ratio values.

Carbon isotopic ratios (*Experiment* 9)

The salient observation in the determination of carbon isotope ratios is that positive values ($\delta^{13}C = +2{\cdot}6$ to $+18{\cdot}5‰$ vs. PDB) were obtained for all the six aliquots of sample 10086-16 analyzed in our laboratory. Similar and more positive values were obtained by other authors on lunar fines. Thus KAPLAN and SMITH (1970) have measured carbon isotope ratios on bulk fines of the same sample 10086 and obtained values of +17·2 and +20·2 per mil vs. PDB. EPSTEIN and TAYLOR (1970) obtained values from +16·1 to +18·6 per mil vs. PDB for two aliquots of sample 10084. These results are in good agreement with our values of +13·0 to +18·5 per mil for three aliquots of sample 10086-16. Therefore, the carbon found in these samples is definitely nonterrestrial, since with the exception of some carbonates from Pennsylvania (HOERING, 1970), such positive values have never been found on earth. The carbonates from carbonaceous chondrites (CLAYTON, 1963; FLORY, 1969) are the only compounds of carbon which have given similar or more positive $\delta^{13}C$ values. Terrestrial contamination cannot be an explanation for these results since any organic product would have significantly negative $\delta^{13}C$ values. Furthermore, as shown earlier, the contamination level of sample 10086-16 is only 1 ppm or less.

On the other hand, negative $\delta^{13}C$ values have been obtained by FRIEDMAN *et al.* (1970) for other samples of breccia and fines and by KAPLAN and SMITH (1970) for fine grained rocks. It is not known at present whether these variations are the result of heterogeneity or contamination of certain samples.

It should be pointed out, however, that since the presence or evolution of different compounds of carbon (graphite, carbides, carbonates, CO, CO_2, CH_4) has been observed in these samples, there is the possibility that actual variations will be shown, particularly when different types of rocks are measured and the measurements are made on samples with a very small content of carbon. Further discussion of this matter is not warranted at this time. It is obvious that measurements of the same samples from Apollo 11 by the four laboratories concerned will be necessary before interpretations of the significance of these results can be made. This strategy should also be followed in any future analysis of lunar carbon isotopes.

Direct imaging mass analysis (*Experiment* 10)

In addition to the direct characterization of carbon ion images indicated by this work and that of graphite, carbides, phosphates, etc. reported by other investigators (see discussion on elements) more work with electron microprobe and ion bombardment mass spectrometric techniques will be required in order to determine the exact nature of some of the solid phases containing carbon and other organogenic elements.

Origin of carbon

With the information presently available it is not possible to assign an unequivocal origin to the carbon found in the lunar samples. In addition to lunar igneous sources this should include to some extent extralunar sources, such as the solar wind, comets and meteorites. The data on the radiation history of the moon (CROZAZ *et al.*, 1970), on the solar abundances of carbon, and on the presence of noble gases of solar composition in lunar samples (LSPET, 1969; HEYMANN *et al.*, 1970) suggest that the solar wind is a possible contributor to the lunar carbon. This suggestion appears to be in line with the larger amounts of this element found in the finer particulate material (MOORE *et al.*, 1970). Whereas some solar carbon isotopic ratios (GREENSTEIN *et al.*, 1950; HERZBERG *et al.*, 1967) are consistent with such an interpretation, these are not yet known with sufficient accuracy. Determination of the abundances, isotopic ratio, and capturing characteristics of the carbon in the solar wind reaching the lunar surface may help to answer whether the solar contribution is significant or is small.

Prebiochemical significance

Finally, it should be said that the results discussed above point towards mixtures of gases (H_2, CO, CO_2, N_2, H_2S, CH_4, and traces of other hydrocarbons), present in or evolved by the action of heat or acid on the lunar fines which, regardless of their source, resemble in composition gas mixtures of igneous or magmatic origin. These gases and some of the solid compounds found in the lunar samples by other workers, such as calcium phosphate, are of particular interest for prebiotic syntheses. Therefore, it may be concluded that if such a mixture of gases had been evolved on a planetoidal body of higher gravitational field (e.g. the earth) it would have produced an atmosphere and hydrosphere out of which organic molecules of biological significance could have been synthesized (ORÓ, 1965, 1968).

SUMMARY

Elements essential to life, and elements and compounds which are precursors of organic molecules, were found in Apollo 11 lunar samples (fines and breccia) by the application of ten different methods of analysis (Table 3).

1. Spark ionization mass spectrometry demonstrated the presence of five organogenic elements. Listed in order of decreasing abundance, their upper values (in

Table 3. Summary of Data obtained by application of several techniques to the analysis of type D (fines) and type C (breccia) lunar samples from Apollo 11

	Analysis	Description of method	Measurements	Principal analysts
1	Elemental	Spark ionization mass spectrography	41 elements including the organogenic H, C, N, S and P	W. S. U., J. S.
2	EGA and DTA	Effluent gas analysis from 100–1400°C by high resolution mass spectrography and differential thermal analysis	Noble gases CO, CO_2 and N_2 ($CO > CO_2$)	W. S. U., F. W., W. K. S., D. E. G.
3	Volatile and/or pyrolytic products	Stepwise heating to 750°C and quadrupole mass spectrometry	H_2O, CO (and N_2), CO_2, C_6H_6, $C_7H_7^+$ and traces of other organics	J. McR., W. S. U.
4	Volatile and/or pyrolytic products	Stepwise heating to 750°C and gas chromatography (Porapak column)—mass spectrometry	CO/CO_2 release vs. temperature. CH_4 above 400°C ($CO > CO_2$)	J. G.
5	Volatile products from acidolysis	Acid treatment and gas chromatography (Porapak column)—mass spectrometry	H_2S, CO_2. CH_4 and other light hydrocarbons	J. G.
6	Trace volatile and/or pyrolytic organic products	Stepwise heating to 480°C and flame ionization gas chromatography (SE-30 column)	$<0{\cdot}5$ ppm	R. L. L., C. J. W.
7	Trace organic solvent-extractable organic compounds	Flame ionization-high resolution–gas chromatography (Polysev column)	$<0{\cdot}001$ ppm	J. G., D. W. N.
8	Amino acids	Flame ionization-gas chromatography	$<0{\cdot}1$ ppm	E. G., J. I.
9	Carbon-13	Carbon isotopic ratio mass spectrometry	$\delta^{13}C = +2{\cdot}6$ to $+18{\cdot}5$ PDB	D. A. F., P. P., A. N. F.
10	Solid carbon compounds	Ion microanalysis	Carbon image from C_2^- ions	W. S. U., R. L

ppm) are: S (4300) > P (270) > C (41) > N (20) $\gg$ H (1). These values reflect the amounts of solid compounds of these elements and do not include significant amounts of gaseous, or readily volatilizable species. Values for other elements, some of biological interest, were also measured and agreed in general with analyses performed by similar or different methods in other laboratories.

2. Thermal volatilization high resolution mass spectrography and differential thermal analysis from 200°C to about 1500°C provided data for the evolution of He, Ne and Ar (500–800°C maxima) and of CO, CO_2 and N_2. The release of these gases varied with time and temperatures. Carbon monoxide was found to predominate ($CO/CO_2 \simeq 10$) at high temperatures. Nitrogen was usually lower than carbon monoxide ($CO/N_2 > 2$) but occasionally was higher ($N_2/CO > 3$) when short-lived bursts of N_2 and CO gases appeared, indicating rupture of vesicular cavities and suggesting more than one source for these gases. Values of 12 and 27 kcal/mole for the energies of CO evolution were obtained which are compatible with different physical and chemical interpretations. The melting temperature of the bulk fines was found to be 1134°C. Another important transition was found at 1211°C. The latter temperature corresponds approximately to the separation of ilmenite (1210°C)

and the former to the separation of pyroxene, troilite and metallic iron (1140°C) observed by other investigators.

3. Low resolution direct volatilization quadrupole mass spectrometry at 450°C gave evidence for ion masses 18, 28, 44, 78, 91 and others, corresponding to H_2O, N_2, CO_2 and traces of benzene, toluene and other compounds. Some of these compounds may not necessarily be indigenous to the moon.

4. Volatilization—gas chromatography–mass spectrometry of fines measured the relative evolution of CO, CO_2 and methane ($CO > CO_2 \gg CH_4$) with increasing temperature (100–750°C). CO_2 appeared first at about 200°C, passed through a maximum at about 500°C and then decreased. CO started to appear at about 400°C and then increased rapidly and continously up to 750°C. The evolution of traces of methane increased from 400–700°C but never reached substantial levels. In general the results of CO and CO_2 evolution are in agreement with the data of the second experiment.

5. By the action of acids on the lunar fines in three separate experiments, the following compounds were generated: Large amounts of H_2S, smaller amounts of CO_2 and CH_4 and traces of C_2H_2, C_2H_4 and C_2H_6. These compounds were identified by combined gas chromatography–mass spectrometry. These results give evidence for the presence of sulfides, carbonates and carbides.

6. Flame ionization gas chromatography was utilized to measure the total amount of volatilizable or pyrolyzable organic compounds. It was found to be very small, i.e. about 0·5 ppm. Only a few compounds responsive to flame ionization were observed in the gas chromatograms, in agreement with the results obtained by quadrupole mass spectrometry. Measurement of the carbon isotope ratios of these compounds will be necessary before anything can be said about their indigenousness to the moon. At this time the data for organic compounds which are present only in total amounts less than 1 ppm does not warrant extensive discussion.

7. Extraction of the fines with a benzene:methanol (3:1) mixture and flame ionization gas chromatography of the extract indicated no detectable organic compounds (less than 0·001 ppm) in the bulk of sample 10086-16. Two small aliquots of fines from the top layer of the sample container were found to contain about 1 ppm of a single component with a molecular weight of 150. This compound is obviously a contaminant.

8. Hydrolysis of the fines with 6 N HCl, derivatization of the hydrolyzate to form N-tri-fluoroacetyl isopropyl esters of any amino acid present and flame ionization gas chromatography on optically active stationary phases, showed no detectable amounts of any amino acid derivatives where less than 0·1 ppm could have been measured.

9. Determination of carbon isotopic ratios of the fines (sample 10086-16) by Craig's, and by a modified fusion method, gave $\delta^{13}C$ per mil values of $+13·0$ to $+18·5$ and $+2·6$ to $+4·6$, respectively, in relation to the PDB standard. The former values are in agreement with those obtained by the same method on the same sample (10086) by another laboratory. These positive values are definitely non-terrestrial. This ^{13}C enrichment may be of lunar origin or it may have been partially contributed by the solar wind or other extralunar sources. Some of the variations observed by us

and other workers may be due in part to real differences in the isotopic ratios of different carbon phases (carbonates, carbides, elemental carbon, etc) detected in the samples.

10. Ion microanalysis has provided preliminary evidence for the existence of carbon distribution patterns and mass spectra of C_2 negative ions in the breccia sample.

In conclusion, H_2, CO, CO_2, N_2, H_2O, H_2S, CH_4 and traces of other hydrocarbons were evolved by the action of heat or acid on fine particulate material from the moon. It is proposed that if these gases, together with other compounds, such as calcium phosphate, were present on a body of the solar system (e.g. on the primitive earth) gravitationally capable of retaining an atmosphere and hydrosphere of this composition, the synthesis of organic molecules of biological significance would have been an inevitability. Whether these are present in some region of our satellite can only be answered by the continued study of returned samples from diverse locations of the moon.

Acknowledgments—We thank A. J. SOCHA, Electronics Materials Div., Bell and Howell, Inc., Pasadena, Calif; F. M. WACHI, W. K. STUCKEY, and D. E. GILMARTIN, Aerospace Corp., El Segundo, Calif.; R. LEWIS, Analytical Instruments Div., Bell and Howell, Inc., Monrovia, Calif.; D. W. NOONER, Texaco Inc., Bellaire, Texas; P. PARKER, Marine Institute, University of Texas, Port Aransas, Texas; and A. N. FEUX, Rice University, Houston, Texas, for their cooperation in different phases of this work, some of which will be published in more detail elsewhere. This work was supported by NASA Contract NAS-9-8012.

REFERENCES

ABELL P. I., DRAFFAN G. H., EGLINTON G., HAYES J. M., MAXWELL J. R. and PILLINGER C. T. (1970) Organic analysis of the returned lunar sample. *Science* **167**, 757–759.

ADLER I., WALTER L. S., LOWMAN P. D., GLASS B. P., FRENCH B. M., PHILPOTTS J. A., HEINRICH K. J. F. and GOLDSTEIN J. I. (1970) Electron microprobe analysis of lunar samples. *Science* **167**, 590–592.

ARRHENIUS G., ASUNMAA S., DREVER J. I., EVERSON J., FITZGERALD R. W., FRAZER J. Z., FUJITA H., HANOR J. S., LAL D., LIANG S. S., MACDOUGALL D., REID A. M., SINKANKAS J. and WILKENING L. (1970) Phase chemistry, structure, and radiation effects in lunar samples. *Science* **167**, 659–661.

BROWN G. M., EMELEUS C. H., HOLLAND J. G. and PHILLIPS R. (1970) Petrographic, mineralogic, and X-ray fluorescence analysis of lunar igneous-type rocks and spherules. *Science* **167**, 599–601.

BURLINGAME A. L., CALVIN M., HAN J., HENDERSON W., REED W. and SIMONEIT B. R. (1970) Lunar organic compounds: Search and characterization. *Science* **167**, 751–752.

CHUPKA W. A. and INGHRAM M. G. (1953) Investigation of the heat of vaporation of carbon. *J. Chem. Phys.* **21**, 371–372.

CLAYTON R. N. (1963) Carbon isotope abundance in meteoritic carbonates. *Science* **140**, 192–198.

CRAIG H. (1953) The geochemistry of the stable carbon isotopes. *Geochim. Cosmochim. Acta* **3**, 53–92.

CROZAZ G., HAACK U., HAIR M., HOYT H., KARDOS J., MAURETTE M., MIYAJIMA M., SEITZ M., SUN S., WALKER R., WITTELS M. and WOOLUM D. (1970) Solid state studies of the radiation history of lunar samples. *Science* **167**, 563–566.

EPSTEIN S. and TAYLOR H. P., JR. (1970) $^{18}O/^{16}O$, $^{30}Si/^{28}Si$, D/H, and $^{13}C/^{12}C$ studies of lunar rocks and minerals. *Science* **167**, 533–535.

FLORY D. A. (1969) Stable carbon isotope ratio measurements of meteoritic carbonaceous material. Ph.D. Dissertation, University of Houston, Houston, Texas.

FRIEDMAN I., O'NEIL J. R., ADAMI L. H., GLEASON J. D. and HARDCASTLE K. (1970) Water, hydrogen, deuterium, carbon, carbon-13, and oxygen-18 content of selected lunar material. *Science* **167**, 538–540.

FRONDEL C., KLEIN C., JR., ITO J. and DRAKE J. C. (1970) Mineralogy and composition of lunar fines and selected rocks. *Science* **167**, 681–683.
GAULT D. E., ADAMA J. B., COLLINS R. J., GREEN J., KUIPER G. P., MAZURSKY H., O'KEEFE J. A., PHINNEY R. A. and SHOEMAKER E. M. (1967) Surveyor V: Discussion of chemical analysis. *Science* **158**, 641–642.
GAY P., BANCROFT G. M. and BOWN M. G. (1970) Diffraction and Mössbauer studies of minerals from lunar soils and rocks. *Science* **167**, 626–628.
GIBERT J. and ORÓ J. (1969) Unpublished work.
GIL-AV E., FEIBUSH B. and CHARLES-SIGLER R. (1966) Separation of enantiomers by gas–liquid chromatography with an optically active stationary phase. In *Gas Chromatography*, (editor A. B. Littlewood), pp. 227–239. Institute of Petroleum.
GREENSTEIN J. L., RICHARDSON R. S. and SCHWARZCHILD M. (1950) On the abundance of C^{13} in the solar wind. *Publ. Astron. Soc. Pacific* **62**, 15–18.
HERZBERG L., DELBOUILLE L. and ROLAND G. (1967) A new determination of the C^{12}/C^{13} isotope abundance ratio in the solar photosphere. *Astrophys. J.* **147**, 697–702.
HEYMANN D., YANIV A., ADAMS J. A. S. and FRYER G. E. (1970) Inert gases in lunar samples. *Science* **167**, 555–558.
HOERING T. (1970) Private communication.
HINTENBERGER H., WEBER H. W., VOSHAGE H., WÄNKE H., BEGEMANN F., VILCSEK E. and WLOTZKA F. (1970) Rare gases, hydrogen, and nitrogen: Concentrations and isotopic composition in lunar material. *Science* **167**, 543–545.
KAPLAN I. R. and SMITH J. W. (1970) Concentration and isotopic composition of carbon and sulfur in Apollo 11 lunar samples. *Science* **167**, 541–543.
LEVY R. L., WOLF C. J. and ORÓ J. (1970) A gas chromatographic method for the characterization of the organic content present in an organic matrix. *J. Chromatog. Sci.* Submitted for publication.
LIPSKY S. R., CUSHLEY R. J., HORVATH C. G. and MCMURRAY W. J. (1970) Analysis of lunar material for organic compounds. *Science* **167**, 778–779.
LSPET (LUNAR SAMPLE PRELIMINARY EXAMINATION TEAM) (1969) Preliminary examination of lunar samples from Apollo 11. *Science* **165**, 1211–1227.
MAXWELL J. A., ABBEY S. and CHAMP W. H. (1970) Chemical composition of lunar material. *Science* **167**, 530–531.
MOORE C. B., LEWIS C. F., GIBSON E. K. and NICHIPORUK W. (1970) Total carbon and nitrogen abundances in lunar samples. *Science* **167**, 495–497.
MORRISON G. H., GERARD J. T., KASHUBA A. T., GANGADHARAM E. V., ROTHENBERG A. M., POTTER N. M. and MILLER G. B. (1970) Multielement analysis of lunar soil and rocks. *Science* **167**, 505–507.
MURPHY R. C., PRETI G., NAFISSI-V M. M. and BIEMANN K. (1970) Search for organic material in lunar fines by mass spectrometry. *Science* **167**, 755–757.
NAGY B., DREW C. M., HAMILTON P. B., MODZELESKI V. E., MURPHY M. E., SCOTT W. M., UREY H. C. and YOUNG M. (1970) Organic compounds in lunar samples: Pyrolysis products, hydrocarbons, amino acids. *Science* **167**, 770–773.
NOONER D. W. and ORÓ J. (1967) Organic compounds in meteorites—I. Aliphatic hydrocarbons. *Geochim. Cosmochim. Acta* **31**, 1359–1394.
ORÓ J. (1965) Investigation of organo-chemical evolution. In *Current Aspects of Exobiology*, (editors G. Mamikunian and M. H. Briggs), pp. 13–76. Pergamon.
ORÓ J. (1968) Synthesis of organic molecules by physical agencies. *J. Brit. Interplanet. Soc.* **21**, 12–25.
ORÓ J., NOONER D. W. and OLSON R. J. (1968) Chromatography of hydrocarbons. In *Chromatographic Analysis of Lipids* (editor G. V. Marinetti), Vol. 2, pp. 479–521. Dekker.
ORÓ J., UPDEGROVE W. S., GIBERT J., MCREYNOLDS J., GIL-AV E., IBANEZ J., ZLATKIS A., FLORY D. A., LEVY R. L. and WOLF C. (1970) Organogenic elements and compounds in surface samples from the Sea of Tranquillity. *Science* **167**, 765–767.
PONNAMPERUMA C., KVENVOLDEN K., CHANG S., JOHNSON R., POLLOCK G., PHILLPOTT D., KAPLAN I., SMITH J., SCHOPF J. W., GEHRKE C., HODGSON G., BREGER I. A., HALPERN B., DUFFIELD A., KRAUSKOPF K., BARGHOORN E., HOLLAND H. and KEIL K. (1970) Search for organic compounds in the lunar dust from the Sea of Tranquillity. *Science* **167**, 760–762.

ROEDDER E. and WEIBLEN P. W. (1970) Silicate liquid immiscibility in lunar magmas, evidenced by melt inclusions in lunar rocks. *Science* **167**, 641–644.

SKINNER B. J. (1970) High crystallization temperatures indicated for igneous rocks from Tranquillity Base. *Science* **167**, 652–654.

TURKEVICH A. L., FRANZGROTE E. J. and PATTERSON J. H. (1969) Chemical composition of the lunar surface in Mare Tranquillitatis. *Science* **165**, 227–279.

TURKEVICH A. L., FRANZGROTE E. J. and PATTERSON J. H. (1967) Chemical analysis of the Moon at the Surveyor V landing site. *Science* **158**, 635–637.

UPDEGROVE W. S. and ORÓ J. (1969) Analysis of the organic matter on the Moon by gas chromatography-mass spectrometry: A feasibility study. In *Research in Physics and Chemistry* (editor F. J. Malina), pp. 53–74. Pergamon.

WOOD J. A., MARVIN V. B., POWELL B. N. and DICKEY J. S. (1970) Mineralogy and petrology of the Apollo 11 lunar sample. Smithsonian Astrophysical Observatory Special Report 307.

Proceedings of the Apollo 11 Lunar Science Conference, Vol. 2, pp. 1921 to 1927.

A search for viable organisms in a lunar sample

VANCE I. OYAMA, EDWARD L. MEREK and MELVIN P. SILVERMAN
NASA, Ames Research Center, Moffett Field, California 94035

(*Received* 28 *January* 1970; *accepted in revised form* 2 *March* 1970)

Abstract—The hypothesis that the moon harbored viable life forms was not verified on analysis of the first samples from the Apollo 11 mission. A biological examination of 50 g of the bulk fines confirmed the negative results obtained by the Manned Spacecraft Center quarantine team. No viable life forms, including terrestrial contaminants, were found when the sample was tested in 300 environments. Only colored inorganic artifacts, resembling microbial clones, appeared around some particles. The lunar biological laboratory and sterile biological barrier system in which the work was performed are described in detail.

THIS study to elicit the growth of organisms from lunar matter is the first in a series to attempt to show the existence of extraterrestrial life. The discovery of extraterrestrial life and its characteristics would be important in constructing theories regarding the origin, evolution, distribution and frequency of life in the universe, when such findings are related to cosmology and the physical and chemical nature of the particular body from which the samples are derived.

The moon is a testing ground for the concept of panspermia, that is, the theory reviewed by OPARIN (1966), that life can be disseminated through space without the intervention of man. The moon is hardly capable of the kind of chemical evolution, proposed for the earth, that might result in the synthesis of organic matter sufficient in amount and residence time to allow life to evolve; because of its mass, the moon cannot hold an atmosphere for long (SAGAN, 1961). At the time of its formation the moon must have been a sterile body, and it appears from its impact history that it must subsequently have served as a repository for the collection of debris from space.

At the NASA-Ames Research Center, we designed and supervised the construction of a laboratory (Fig. 1), which included a biological barrier system, to perform microbiological experiments on extraterrestrial samples. The system was maintained at a positive pressure (0·2 in. of water) to protect the sample from terrestrial contamination and was housed in a Class-100 room [particle ($>0·5\ \mu$m) count less than 100 per cubic foot of air] to protect personnel from exposure to potential hazard from the sample.

The biological barrier system contained a preparation area and an incubation area. Twelve interconnected glove boxes served as the preparation area. In this area molten agar media were dispensed and hardened; the sample was weighed, sieved and distributed onto the surface of the agar media. Aseptic means of removing material from, or introducing it to, the preparation area were provided by a pass-through sterilizer equipped for steam or gaseous sterilization; a transfer box equipped for gaseous or liquid sterilization; a dunk tank containing a solution of sterilant; and a sterile filter assembly for passing thermolabile solutions into the barrier system. The preparation area was equipped with gas lines for introducing ethylene oxide for

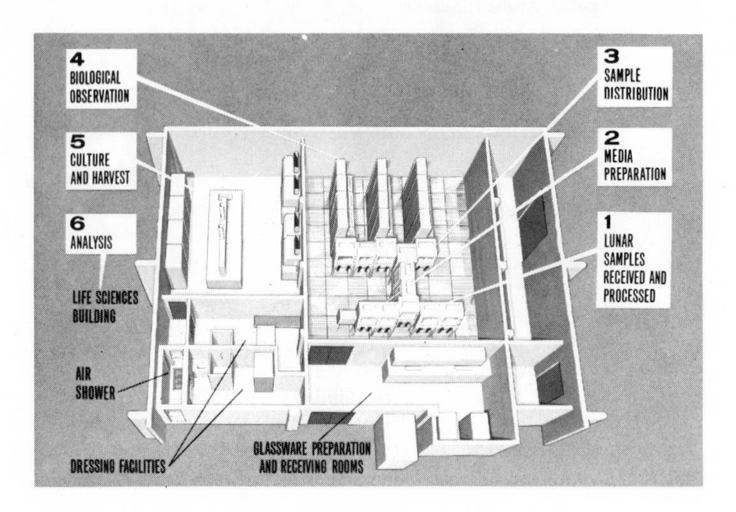

Fig. 1. Lunar sample test laboratory.

sterilizing, and dry-filtered nitrogen for purging and maintaining an oxygen-free working environment.

The incubation area consisted of three wings of incubator glove boxes arranged in banks of four. Each wing had an incubator set at 10°, 20°, 35° and 55°C. Every incubator was divided vertically into three compartments for maintaining humidity. Each compartment had nine removable trays, each of which contained ten Petri dishes, and a Teflon-lined bottom tray with a wick assembly for humidity control using water or saturated brine solutions. Fluorescent lamps, mounted behind each incubator, illuminated the Petri dishes. The compartments were moved by an elevator; each tray could be removed or inserted through an opening into a glove-box working area. This working area had built-in objective lenses, to which a stereoscope body could be mounted. Magnifications up to $\times 32$ were obtainable, and photographs could be taken. Each incubator had valved gas lines for the introduction of ethylene oxide, nitrogen or mixed gases.

The gases provided passed through a pair of biological filters and through sterile tubing before entering the biological barrier. All gases which left the barrier system, with the exception of ethylene oxide, were exhausted to the atmosphere through a furnace.

Prior to receipt of the sample, the barrier system was treated for 24 hr with a mixture of ethylene oxide and Freon, and purged with dry sterile nitrogen. The composition of the gases was monitored with a gas chromatograph during the purging cycles. The barrier system was then tested for sterility. Representative 16 in.^{2} portions of all interior surfaces of each cabinet were swabbed, and the swabs incubated in trypticase soy broth and fluid thioglycollate broth for seven days. No growth was observed in these test media.

Nine media (listed in Table 1) containing 2% agar were sterilized in the connecting autoclave, and 30 ml were dispensed into sterile glass Petri dishes.

Table 1. Media for the attempted isolation of microorganisms from the lunar sample

Designation	General Composition*
1. Water agar	
2. Salts	K_2HPO_4, KNO_3 and $(NH_4)_2SO_4$
3. Sulfur oxidation	Agar salts of 2 with surface layer of freshly precipitated sulfur
4. Formose sugar	Formaldehyde condensation products, mainly racemic mixture of sugars (SHAPIRA, 1967)
5. Complex dilute	Dilute mixture of 6
6. Complex concentrated pH 3	Glycerol, lactate, pyruvate, acetate, methanol, ethanol, K_2HPO_4, KNO_3, $(NH_4)_2SO_4$, 18 DL-amino acids, glycine, 4-OH-L-proline, β-alanine, 7 purines and pyrimidines, 16 vitamins
7. Complex concentrated pH 7	Same as 6, but adjusted to pH 7
8. Complex concentrated pH 11	Same as 6, but adjusted to pH 11
9. Spark discharge	Organics generated by spark discharge. (RABINOWITZ *et al.* (1969).

* All media combined with 0·1%, 3·0% and 20·0% synthetic sea salt formulation.

The 3000 Petri dishes containing media were placed on trays and incubated at 10°, 20°, 35° or 55°C under one of three different gas mixtures: 78% nitrogen, 20% oxygen, 2% carbon dioxide; 97% nitrogen, 1% oxygen, 2% carbon dioxide; and 98% nitrogen, 2% carbon dioxide. After a minimum 72-hr incubation, all the Petri dishes were examined. No clonal growth from contaminating organisms was observed. The barrier system was then purged with dry sterile nitrogen.

Upon delivery to Ames Research Center, the sample container was wiped with 5 per cent peracetic acid solution and treated with ultraviolet light before insertion into the barrier system. While under nitrogen gas the sample container was opened, the sample (10089,1) weighed and sieved, and the bulk density determined (Table 2). During the week of October 6, 1969, while still under nitrogen gas, 12 g of the sample were distributed uniformly in equal portions of 4 mg each on the 3000 Petri dishes,

Table 2. Physical characteristics* of lunar sample 10089,1

Sieve fractions (Tyler)	Mass (wt.%)	Bulk density
16/35	6·7	1·12
35/60	6·3	1·25
60/115	13·1	1·33
115/250	15·8	1·45
250/325	12·4	1·47
<325	45·7	1·42

* Color of sample approximated 5YR in Munsell Color Index.

SOIL DISTRIBUTION SYSTEM

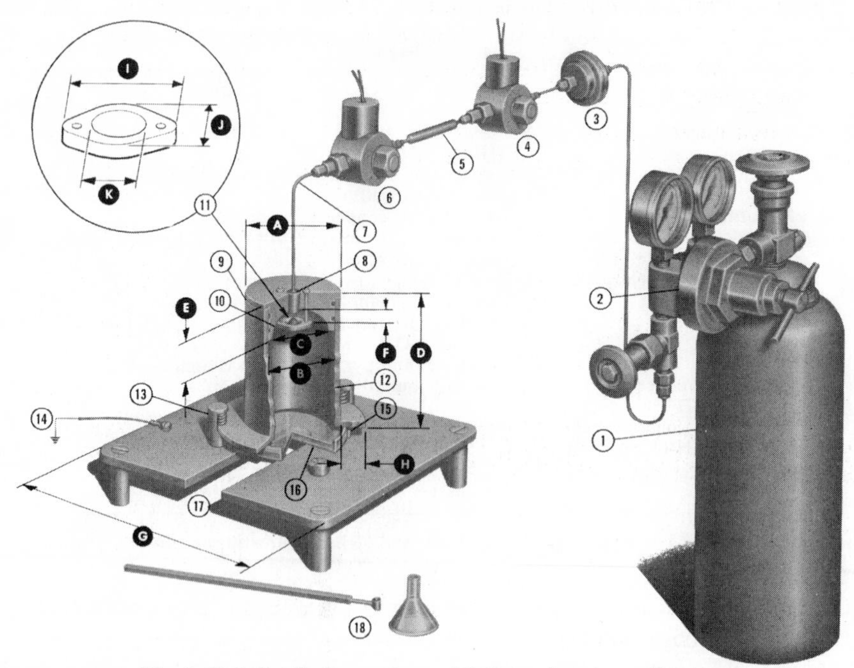

Fig. 2. Soil distribution system. All dimensions in centimeters.

A 7·6 o.d.	G 20·0 sq.
B 7 i.d.	H 2·4 flange
C 5·4	I 2·5
D 13·2	J 1·9
E 3·8	K 1·4
F 0·8 ht.	

1 High pressure gas tank	10 Curved impact surface
2 Pressure regulator—100 psi output	11 Sample holder
3 Millipore filter	12 Interior gold plated—0·01 in.
4 Solenoid valve-chamber fill	13 Spring loaded hold-down
5 Volumetric chamber—9 ml	14 Ground
6 Solenoid valve-chamber exhaust	15 Locating pin (1 of 2)
7 Gas delivery tube	16 Petri dish
8 S loading hole and plug	17 Slot—to aid Petri dish removal
9 Drop tower	18 Loading tools.

using a small settling tower (Fig. 2) within the preparation area. The inoculated Petri dishes were returned to the incubators, which were again flushed with one of the three gas mixtures. These gas mixtures have been maintained for the duration of the incubation period. The plates were observed for evidence of colony formation twice each week until the week of December 8, 1969, then discarded.

By the week of December 1, 1969, no micro-organisms had been detected in the 12 g of lunar sample. An enrichment procedure was next used with 27 g of the sample. A disk of sterile filter paper was placed upon the surface of agar media 2, 7, and 4 and 9 combined containing 0·1 % salt (Table 1), and a 1-g sample was placed on the center of the filter paper. Each representative medium was incubated at 20°, 35° and 55°C in each of the three gas mixtures. After 10 days the sample and filter paper were moved to a plate containing fresh medium. No micro-organisms were detected to date, but incubation is continuing.

Although no micro-organisms were detected in the 12 g of sample distributed on 3000 Petri dishes, numerous artifacts resembling micro-colonies were seen after 4 days

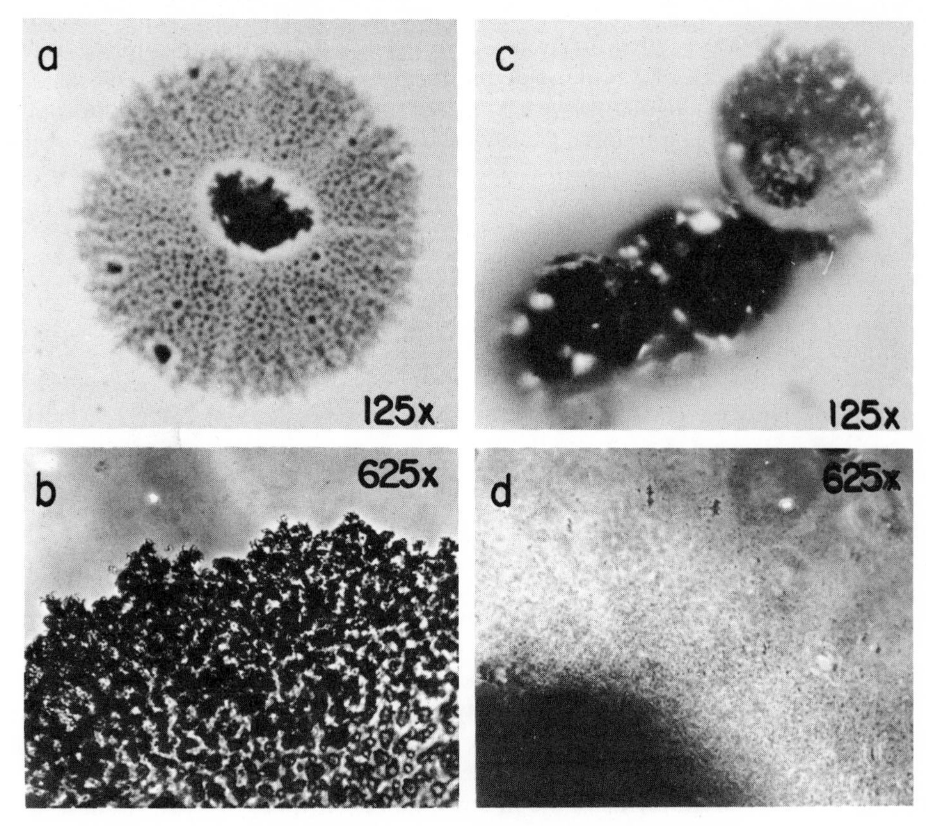

Fig. 3. Typical artifacts resembling microcolonies.
(a) Type I, ×125; (b) Type I, ×625;
(c) Type II, ×125; (d) Type II, ×625.

of incubation. Two types were distinguished. Type I was circular, flat, granular and yellow-to-brown, surrounding a central lunar particle (Fig. 3a). Type II was irregularly lobed, raised, moist and orange-to-dark-brown, and often completely covered a lunar particle (Fig. 3c). Observations at higher magnifications revealed the type I artifacts to be composed of a more coarsely crystalline material than the type II artifacts (Fig. 3b and d). After 6 weeks of incubation, transplants were made to identical media and environmental conditions, but they did not give rise to similar artifacts.

A series of spot tests [FEIGL, 1958; Fe (prussian blue reaction, p. 161; α,α'-dipyridyl reaction, p. 162), Cr (diphenylcarbazide reaction, p. 168), Mn (benzidine blue reaction, p. 175), Ti (chromotropic acid reaction, p. 197)] was performed on both types of artifacts for those elements present in the lunar material (LSPET, 1969) which might give rise to colored inorganic precipitates. Type I artifacts gave negative results when tested for Ti, Mn and Cr, but equivocal results for Fe. The artifacts were insoluble in 3 N HCl, 6 N HCl, 6 N H_2SO_4, 0·05 N NaOH and concentrated NH_4OH. They dissolved partially in NaOBr, but complete dissolution was achieved when NaOBr treatment was followed by treatment with 6 N H_2SO_4. Type II artifacts gave a negative result when tested for Ti but gave strong positive reactions for Fe when tested with $K_4Fe(CN)_6$ or α,α'-dipyridyl in thioglycollate.

Extracts of the lunar sample were prepared and analyzed by atomic absorption spectrophotometry. The results (Table 3) showed that minerals containing Fe, the

Table 3. Atomic absorption analysis of extracts of lunar sample

	Total in extract (mg/g of sample)						
Extract*	Li	Na	K	Fe	Mn	Ti	Cr
Water	<0·0001	1·76	0·062	<0·001	<0·0006	<0·002	<0·0003
1 N HCl				8·75	0·112	1·50	0·256
6 N HCl				8·00	0·106	1·99	0·120
1 M sodium citrate				2·07	0·0115	0·077	0·0115

* Thirty minute extractions of a 1-g portion of sample were made by stirring with 10 ml of water at 80°C followed by 1 N HCl at 22°C then by 6 N HCl at 22°C. A fresh 1-g sample was extracted by stirring with 10 ml of 1 M sodium citrate for 2 hr at 80–90°C. The citrate and 6 N HCl extracts were yellow; the others were colorless.

most abundant of the ions analyzed, were not solubilized by water alone, but required H^+ or Na^+. The results are consistent with cation exchange phenomena. The spot test results obtained on the type II artifacts support the cation exchange results. The reason for the equivocal response by the type I artifacts to the iron reagents is not known, but it may be that the brown color arising around the particle is a mineral containing iron at concentrations barely detectable by the two reagents. We believe that the higher concentrations of Fe in some lunar particles may well produce the lobular structures of the type II artifacts.

We conclude for this sample of the moon that there was no life present. Our conclusion agrees with that of the biological quarantine team at the Manned Spacecraft Center, Houston.

We regard, however, life on the moon to be unlikely since several factors are arrayed against the presence of viable organisms: (i) no spontaneous event is known that can launch a fragment from a source such as the earth; (ii) if a particle containing organisms is small enough to escape its source, it may be too small to protect the organisms from solar radiation; (iii) if an escaped particle is large enough to protect an organism within it, the thermal energy released upon impact might well be lethal; and (iv) even if the organism survived the rigors of launch, space travel and impact, the survival time of that organism in the lunar environment might well have been exceeded by the time the sample was examined. But regardless of the foregoing arguments, the importance of discovering extraterrestrial life justifies continuing experimentation with lunar samples.

Until more samples from various sites and depths of the moon are analyzed using a greater diversity of biological methods, the concept of panspermia with respect to the moon remains to be settled.

Acknowledgments—We thank B. J. BERDAHL, C. W. BOYLEN, G. C. CARLE, J. O. COLEMAN, C. C. DAVIS, G. HAMILTON, D. A. JERMANY, P. J. KIRK, M. E. LEHWALT, A. K. MIYAMOTO, E. F. MUNOZ, G. E. POLLOCK, B. J. TYSON and O. WHITFIELD for assistance; E. L. CRAIG and C. W. BOYLEN for photography and G. E. PETERSON and his staff for design and construction of the facility.

REFERENCES

FEIGL F. (1958) *Spot Tests in Inorganic Analysis*, 5th English edition. Elsevier.

LSPET (LUNAR SAMPLE PRELIMINARY EXAMINATION TEAM) (1969) Preliminary examination of lunar samples from Apollo 11. *Science* **165,** 1211–1227.

OPARIN A. I. (1966) The origin and initial development of life. NASA Technical Translation, NASA TTF-488, pp. 20–22.

RABINOWITZ J., WOELLER F., FLORES J. and KREBSBACH R. (1969) Electric discharge reactions in mixtures of phosphine, methane, ammonia and water. *Nature* **224,** 796–798.

SAGAN C. (1961) Organic matter and the moon. *Nat. Acad. Sci. Nat. Res. Counc. Publ.* **757,** 13, 14.

SHAPIRA J. (1967) Closed life support system. NASA SP-134.

Proceedings of the Apollo 11 Lunar Science Conference, Vol. 2, pp. 1929 to 1932.

Fluorometric examination of the returned lunar fines from Apollo 11

JOON H. RHO and A. J. BAUMAN
Jet Propulsion Laboratory, California Institute of Technology, Pasadena, California 91103
TEH FU YEN
University of Southern California, Los Angeles, California 90033
and
JAMES BONNER
California Institute of Technology, Pasadena, California 91109

(*Received* 5 *February* 1970; *accepted in revised form* 9 *February* 1970)

Abstract—We have examined organic solvent extracts of the bulk fine lunar sample for the presence of porphyrins by spectrofluorometry. Porphyrins were not found in any of the extracts even under conditions in which less than 10^{-13} moles per gram of sample could have been detected. We found, however, a material which exhibits absorption peaks at 310 and 350 mμ and fluorescence peaks at 405 and 690 mμ, the latter with excitation maxima of 360 and 410 mμ (uncorrected) respectively.

INTRODUCTION

SINCE porphyrins are essential components of all life, their presence in lunar samples would suggest that life once existed on the moon. Porphyrins have characteristic absorption spectra, which we have measured by energy-corrected fluorescence activation on nanogram quantities in organic solvents. The linear-energy spectrophotofluorometry (LEF) (SLAVIN *et al.*, 1961) in general permits measurement of the absorption spectra of fluorescent compounds at a sensitivity of the order of 100–1000 times that of absorption spectrophotometry.

We have examined several organic solvent extracts of the returned lunar surface sample from Apollo 11 for the presence of porphyrins and metallo-porphyrins by spectrophotometry and the highly sensitive LEF method.

INSTRUMENTATION

Our Perkin-Elmer LEF apparatus automatically compensates for variations of energy with wavelength of both the excitation light source and the detector sensitivity in such a way that the spectrum is free from distortion. This is accomplished by internal standardization against the output of a thermocouple detector. The instrument uses a high-pressure xenon arc source, an EMI 9558 photomultiplier (trialkali photocathode tube), and a Solid State Radiation, Inc. photoncounter. The photon counting system uses a single photoelectron pulse counting photometer to make precise measurements of low light intensities. The photometer system permits measurements at light levels limited only by photocathode quantum efficiencies and dark emission over a dynamic range greater than a million to one. We have used this system to record the complete absorption spectrum of 10^{-12} moles of porphin in 0·2 ml of methylene chloride, and similarly spectra of the dimethyl ester of nickel mesoporphyrin IX, at 10^{-11} moles and vanadyl tetra-*meso* beta-naphthyl porphin at

10^{-12} moles in the same volume. The fluorescence yield increases linearly with increasing porphyrin concentration over a wide dynamic concentration range. One can easily detect porphyrins even in the extremely low concentration range of 10^{-13} to 10^{-14} moles per 0·2 ml sample solution. Nickel porphyrin fluoresces only when nickel is removed by treatment with a strong acid such as methane sulfonic acid. The detectability of the demetallated nickel porphyrin is somewhat limited by the background light scattering due to the viscous nature of this acid solvent.

Sample Extraction

The extraction scheme is based on the use of sequential Soxhlet extractions with methanol and benzonitrile (Yen *et al.*, 1969) for 24 hr to remove loosely held material, followed by a 10-hr period of ball-milling with a melt of methane sulfonic acid and naphthalene (1 : 1, by weight) at 50°C in order to remove any material bound in solid carbon phases (Yen, 1967, unpublished; Erdman, 1965). Each extract was worked up separately for fluorescent determinations.

Fluorometric grade methanol (Harleco) and scintillation grade naphthalene (Packard) were used without treatment, but reagent grade benzonitrile and methane sulfonic acid were distilled freshly before use. Glassware was cleaned with hot aqua regia and rinsed with water and fluorometric grade methanol before use. Teflon sleeves were used on all ground joints. All operations on the sample were carried out in a dry box under an argon atmosphere.

Tests for possible contaminants in the extraction solvents were run as follows: identical volumes of the solvent systems used on the sample were used in the same way to treat a 13-g sample of clean 200-mesh optical quartz. The test extracts were treated and examined exactly as those from the sample were.

Sample 10086 (13·0 g) of lunar surface fines was layered in 5-mm thick layers separated by equally thick layers of borosilicate glass wool in a microporous fluorocarbon–polymer Soxhlet thimble. The layering of the powdery sample was necessary in order to promote free access and percolation of the extraction solvent. The sample was placed in a double-jacketed Soxhlet apparatus so designed that the thimble and contents were bathed in the vapor of the refluxing solvent and thus extracted efficiently by solvent at its boiling point. The sample was extracted first with 200 ml of fluorometric grade methanol for 24 hr. The methanol extract was then filtered through an ultrafine porosity glass filter to remove trace solids and concentrated to near dryness in a Buchler evaporator. The extract was taken to near dryness under a stream of nitrogen, then taken up in 2 ml of methylene chloride for examination of the methanol extract. The sample was next extracted with 200 ml of benzonitrile under reflux at about 67°C and a pressure of 4–5 mm-Hg, filtered as before, and finally concentrated to dryness by vacuum distillation and taken up in 2 ml of methylene chloride for examination of the benzonitrile extract. The sample was freed of benzonitrile by brief Soxhlet-extraction with methanol and transferred to a 500-ml glass-stoppered borosilicate reagent bottle. Alundum balls were added, together with methane sulfonic acid (28·9 g), and naphthalene (28·9 g), and the mixture was ball-milled as a melt at 54 rev/min for 10 hr at a temperature kept at 40°–50°C by means of a heat lamp. The methane sulfonic acid–naphthalene sample slurry was transferred by water washes into a round-bottom flask, and excess naphthalene and water were removed on a Buchler evaporator at a pot temperature of 42°C and pressure of 0·5 mm-Hg. Water (200 ml) and methylene chloride (100 ml) were used, with scraping as needed to transfer the black mineral slurry to a separatory funnel equipped with a Teflon stopcock. The glass wool used in layering the sample in the thimble was removed into an ultrafine porosity sintered glass filter apparatus and washed with water until the washes were approximately neutral; these washes were added to the main body of the acidic aqueous mineral slurry in the funnel. The aqueous phase was washed three times with 100 ml of methylene chloride, the yellow fluorescent organic phase was discarded, and the aqueous phase was centrifuged in 150-ml glass centrifuge bottles. The clear acidic aqueous supernatant was returned to the separatory funnel. The precipitate was washed twice with 50 ml of water, and these washes were added to the contents of

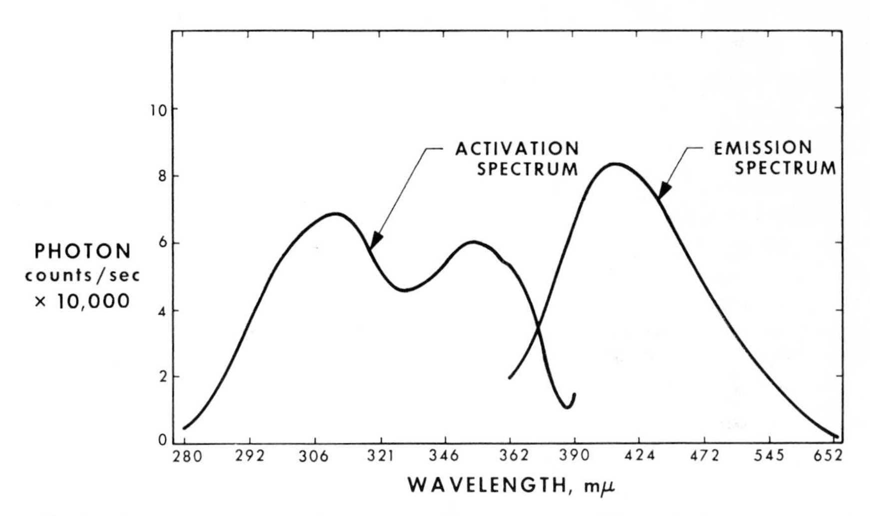

Fig. 1. Fluorescence spectra of the benzonitrile extract of the lunar sample in methylene chloride. The activation spectrum was obtained at the fluorescence maximum peak at 410 mμ by energy-corrected fluorescence activation. The emission spectrum was obtained without compensation for variations in the response of the photodetector with wavelength of the emitted light.

the separatory funnel. The clear aqueous phase, with its presumed metal-free porphyrins, was brought to pH 8 by the addition of 1 N potassium hydroxide and extracted four times with 100 ml of methylene chloride. The methylene chloride phase was then pooled and concentrated to 2 ml under a stream of nitrogen for examination as the methane sulfonic acid–naphthalene extract.

Results and Discussion

Spectral study of all extracts and solvent blanks before and after the addition of methane sulfonic acid to demetallate any metallic porphyrins present (Erdman, 1965) showed that the benzonitrile extract contained a material which exhibited a fluorescence peak at 690 mμ when excited at 410 mμ. However, on the basis of its behavior in acid and its spectral properties, we do not believe that this compound is a porphyrin. We therefore believe that porphyrins were not present in the lunar sample above the detectability limits of our method, that is, above 10^{-12} mole/g of sample (linear energy spectrofluorometer) or 10^{-13} mole/g of sample (conventional spectrofluorometer).

We have, however, found other fluorescent materials in the lunar sample but not in the solvent control blanks. Thus the benzonitrile and methane sulfonic acid–naphthalene extracts exhibit absorption maxima at 310 and 350 mμ with a fluorescence maximum at 410 mμ (Fig. 1). This material is present in the lunar sample at concentrations of the order of 10 parts per billion (quinine sulfate equivalent). Because of the low concentration of the fluorescent material in the sample we have been unable to identify it by high resolution mass spectrometry. Extraction of an ancient terrestrial sample (Gunflint Chert) by the methods outlined above yields a fluorescent material similar in its properties to those of the material present in the lunar sample.

Acknowledgments—We thank J. R. THOMPSON for technical assistance. This paper presents the results of one phase of research carried out at the Jet Propulsion Laboratory under NASA contract NAS7-100.

REFERENCES

ERDMAN J. G. (1965) Process for removing metals from a mineral oil with an alkyl sulfonic acid. U.S. Pat. 3, 190, 829.

SLAVIN W., MOONEY R. W. and PALUMBO D. T. (1961) Energy recording spectrofluorimeter. *J. Opt. Soc. Amer.* **51,** 93–97.

YEN T. F., BOUCHER L. J., DICKIE J. P., TYNAN E. C. and VAUGHAN G. B. (1969) Vanadium complexes and porphyrins in asphaltene. *J. Inst. Petrol.* **55,** 87–99.

YEN T. F. (1967) Unpublished results.

Proceedings of the Apollo 11 Lunar Science Conference, Vol. 2, pp. 1933 to 1934.

Micropaleontological studies of Apollo 11 lunar samples

J. WILLIAM SCHOPF
Department of Geology, University of California, Los Angeles, California 90024

(*Received* 19 *January* 1970; *accepted* 20 *January* 1970)

Abstract—Optical and electron microscopic studies of rock chips and dust from the Bulk Sample Box returned by Apollo 11, and of petrographic thin sections and acid-resistant residues of lunar material, have yielded no evidence of indigenous biological activity.

ALTHOUGH the present lunar environment is inimical to known biological systems, more favorable conditions may have existed in the geologic past. UREY (1966) has suggested that the moon may have become "contaminated" with terrestrial organic matter early in the evolution of the earth–moon system. If this concept is correct, and if life became established, evidence of fossil organisms might be detectable in lunar rocks. It is even conceivable that such organisms might have been the progenitors of an extant biota, adapted to the harsh conditions of the lunar surface; such organisms probably could not survive in the terrestrial environment and therefore would not be recognized in studies designed to detect vital processes (e.g. metabolism, growth, pathogenicity). The approach and techniques successfully used in Precambrian paleobiology (SCHOPF, 1970), and the criteria developed to establish the indigenous and biogenic nature of Precambrian microfossils, seem well-suited to detect, characterize and interpret any fossil or recently dead microorganisms that might occur in lunar materials (SCHOPF, 1969).

In an effort to detect evidence of lunar organisms in the Apollo 11 samples, studies were made with a light microscope (L) at magnifications ranging from 4× to 1500× and, after coating specimens with a thin gold–palladium film, with a scanning electron microscope (SEM) at magnifications ranging from 30× to 30,000×. I examined the following samples: (i) lunar dust (sample 10086,18 from the Bulk Sample Box), divided into four size-fractions by sieving (>246 μ, 246–124 μ, 124–74 μ, <74 μ), L and SEM; (ii) residue resulting from dissolution of lunar dust in hydrofluoric and hydrochloric acids, L; (iii) surfaces of rock chips from the exterior and interior of a microbreccia (sample 10002,54 from the Bulk Sample Box), and fragments of these chips, L and SEM; (iv) petrographic thin sections of microbreccias (samples 10019,15, 10046,56, 10059,32, 10059,37, 10061,27, 10061,28, and 10065,25), L. (v) As a member of the Ames Lunar Sample Consortium, I studied (L) samples being investigated by C. PONNAMPERUMA *et al.* (1970) at the Ames Research Center (sample 10086, Bulk A Fines). (vi) As a member of the Lunar Sample Preliminary Examination Team, during the Apollo 11 mission I studied (L) rocks, chips dust and bio-quarantine samples (including portions of both cores) (LSPET, 1969).

Several thin sections (e.g. 10046,56, 10059,32 and 10061,27) contain elongate, spheroidal, spinose or actinomorphic structures (Fig. 1) that superficially resemble terrestrial microfossils; many of these mineralogic "pseudofossils" are the result of partial devitrification of glassy inclusions. During preliminary studies at the Lunar

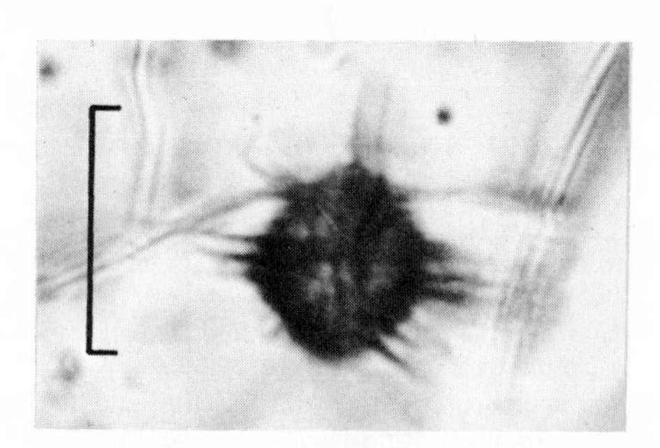

Fig. 1. Optical photomicrograph showing actinomorphic pseudofossil, apparently produced by partial devitrification of the surrounding glassy matrix, in a petrographic thin section of a microbreccia (Apollo 11 sample 10046,56); line for scale represents **10** μ.

Receiving Laboratory I detected birefringent organic fibers, a few microns in diameter, in the lunar dust and bio-quarantine samples; a few similar fibers were noted in samples (i) and (ii), above, and in the mounting medium (but *not* within mineral grains) of several petrographic thin sections. With the exception of these terrestrial contaminants, apparently derived from lens tissue or similar substances, no biogenic materials were detected in these examinations.

Acknowledgements—I thank Mrs. CAROL LEWIS for assistance, and G. OERTEL and J. CHRISTIE for suggestions. Supported by NASA Contract NAS 9-9941.

REFERENCES

LSPET (LUNAR SAMPLE PRELIMINARY EXAMINATION TEAM) (1969) Preliminary examination of lunar samples from Apollo 11. *Science* **165,** 1211–1227.

PONNAMPERUMA C., BARGHOORN E., BREGER I.A., CHANG S., DUFFIELD A., GEHRKE C., HALPERN B., HODGSON G., HOLLAND H., JOHNSON R., KAPLAN I., KEIL K., KRAUSKOPF K., KVENVOLDEN K., PHILPOTT D., POLLOCK G., SCHOPF J.W. and SMITH J. (1970) Search for organic compounds in the lunar dust from the Sea of Tranquillity. *Science* **167,** 760–762.

SCHOPF J. W. (1969) Micropaleontology and extraterrestrial sample studies. *Program. Annual Meeting, Geol. Soc. Amer.*, Atlantic City, 200 (abstr.).

SCHOPF J. W. (1970) Precambrian microorganisms and evolutionary events prior to the origin of vascular plants. *Biol. Rev. Cambridge Phil. Soc.* vol. **45**.

UREY H. C. (1966) Biological material in meteorites: A review. *Science* **151,** 157–166.

Proceedings of the Apollo 11 Lunar Science Conference, Vol. 2, pp. 1935 to 1936.

Author Index

ABELL P. I. 1757
ADAMI L. H. 1425
ALLEN J. 1627
ALLEN R. O. 1213
ANDERS E. 1117
ANDERSON M. R. 1213
ANNELL C. S. 991
ARNOLD J. R. 1503
ARRIENS P. A. 1007
AVE W. A. 1845

BADDENHAUSEN H. 1719
BAEDECKER P. A. 1741
BALACESCU A. 1719
BARGHOORN E. S. 1775
BARKER J. M. 1793
BARNES V. E. 1793
BAUMAN A. J. 1929
BEGEMANN F. 995, 1269
BELL P. R. 1407
BENNETT D. 1627
BEYER R. L. 1177
BIEMANN K. 1891
BLACK D. C. 1435
BLACK L. P. 1345
BOGARD D. D. 1111
BONNER J. 1929
BORN W. 995
BREGER I. 1813
BUNNENBERG E. 1829
BURNETT D. S. 1637
BURLINGAME A. L. 1779
BUTTERFIELD D. 1351

CALVIN M. 1779
CARRON M. K. 1493
CHANG S. 1813, 1857
CHAPPELL B. W. 1007
CHRISTIAN R. P. 1493
CLAYTON R. N. 1429
CLOUD P. 1793
COMPSTON W. 1007
COOPER J. A. 1455
CORDES E. 1875
COSCIO M. R., JR. 1393
CUSHLEY R. J. 1871
CUTTITTA F. 1493

D'AMICO J. 1029
DAVIS C. C. 1805
DAVIS P. K. 1283
DAVIS R., JR. 1583
DE FELICE J. 1029
DELANY A. C. 1503
DIAMOND H. 1097
DRAFFAN G. H. 1757
DREW C. M. 1879
DUFFIELD A. 1845
DWORNIK E. J. 1493

EBERHARDT P. 1037
EGLINTON G. 1757
EHMANN W. D. 1071, 1165
ELDRIDGE J. S. 1407
ENGEL A. E. J. 1081
ENGEL C. G. 1081
EPSTEIN S. 1085, 1613
EUGSTER O. 1637
EVANS J. C., JR. 1503
EVENSEN N. M. 1393
FIELDS P. R. 1097

FINKEL R. C. 1503
FIREMAN E. L. 1029
FLORES J. 1813
FLORY D. A. 1901
FOX S. W. 1799
FRIEDMAN I. 1103
FRUCHTER J. S. 1503
FUNKHOUSER J. G. 1111

GANAPATHY R. 1117
GANGADHARAM E. V. 1383
GAST P. W. 1143
GEHRKE C. W. 1845
GEISS J. 1037
GERARD J. T. 1383
GIBERT J. 1901
GIBSON E. K. 1375
GIL-AV E. 1901
GLEASON J. D. 1103
GOPALAN K. 1195
GOLEŠ G. G. 1165, 1177
GRÖGLER N. 1037

HALPERN B. 1829
HAMILTON P. B. 1879
HAN J. 1779
HANNEMAN R. E. 1207
HARADA K. 1799
HARDCASTLE K. 1103
HARE P. E. 1799
HASKIN L. A. 1213
HAYES J. M. 1757
HELMKE P. A. 1213
HELZ A. W. 991
HENDERSON W. 1779
HERMAN G. F. 1239
HERPERS U. 1233
HERR W. 1233
HERZOG G. F. 1239
HESS B. 1233
HEYMANN D. 1247, 1261
HINTENBERGER H. 1269
HODGSON G. W. 1829
HOHENBERG C. M. 1283
HONDA M. 1503
HORVATH G. G. 1871
HUBBARD N. J. 1143
HUEY J. M. 1345
HURLEY P. M. 1311

IBANEZ J. 1901
IHOCHI H. 1345

JACKSON T. J. 1875
JEROME D. Y. 1177
JOHNSON P. H. 1627
JOHNSON R. D. 1805
JOVANOVIC S. 1487

KAISER W. A. 1283
KAPLAN I. 1857
KAPLAN I. R. 1317
KASHUBA A. T. 1383
KAUSHAL S. 1195
KAYE J. H. 1455
KEAYS R. R. 1117
KEIL K. 1813
KHARKAR D. P. 1659
KIRSTEN T. 1331
KOHMAN T. P. 1345
KOROTEV R. L. 1213
KRÄHENÜHL U. 1037
KRINSLEY D. 1793
KVENVOLDEN K. A. 1813, 1829, 1845, 1857

LARIMER J. W. 1375
LARSON R. R. 1493
LAVL J. C. 1117
LAWLESS J. 1857
LEE-HU C. 1195
LEVY R. L. 1901
LEWIS C. F. 1375
LEWIS R. S. 1283
LICARI G. R. 1793
LIGON D. T. 1493
LINDSAY J. R. 1493
LINDSTROM D. J. 1177
LINDSTROM R. M. 1503
LIPSKY S. R. 1871
LOOSLI H. H. 1503

Loveridge W. D. 1729
Lovering J. F. 1351
Lugmair G. W. 1357
Lyman W. J. 1583

Mapper D. 1575
Margolis S. V. 1793
Marti K. 1357
Martin M. R. 1177
Martin R. 1627
Maxwell J. A. 1369
Maxwell J. R. 1757
Mayeda T. K. 1429
McKay S. M. 1177
McMurray W. J. 1871
McReynolds J. 1901
Meinschein W. G. 1875
Merek E. L. 1921
Metta D. N. 1097
Miller G. B. 1383
Mitchell J. M. 1875
Modzeleski V. E. 1879
Moore C. B. 1375
Moorman M. 1793
Moreland P. E. 1097
Morgan J. W. 1071
Morgan J. W. 1165
Morrison G. H. 1383
Murphy Sister M. E. 1879
Murphy R. C. 1891
Murthy V. R. 1393

Nafissi-v M. N. 1891
Nagy B. 1879
Nance W. 1627
Nichiporuk W. 1375
Nyquist L. E. 1435

O'Kelley G. D. 1407
O'Neil J. R. 1425
Onuma N. 1429
Oró J. 1901
Osawa M. 1165, 1177
Oyama V. I. 1921

Paster T. P. 1213
Peck L. C. 1369
Pepin R. O. 1435
Pering K. 1813
Perkins R. W. 1455
Philpotts J. A. 1471
Phinney D. 1435
Pillinger C. T. 1757
Pinson W. H., Jr. 1311
Ponnamperuma C. 1813, 1829, 1845, 1857
Potter N. M. 1383
Preti G. 1891

Quaide W. L. 1751
Quijano-Rico M. 1719

Rancitelli L. A. 1455
Randle K. 1165, 1177
Reed G. W., Jr. 1487
Reed W. 1779
Reedy R. C. 1503
Rey P. 1685
Reynolds J. H. 1283
Rho J. H. 1929
Rieder R. 995, 1719
Rokop D. J. 1097
Rose H. J. 1493
Rosholt J. N. 1499
Rothenberg A. M. 1383
Ruth E. 1317

Saxinger C. 1813
Schaeffer O. A. 1111
Schmitt R. A. 1165, 1685
Schnetzler C. C. 1471
Schonfeld E. 1407
Schopf J. W. 1933
Schwaller H. 1037
Schwarzmüller J. 1037
Scott W. M. 1879
Shedlovsky J. P. 1503
Shiner V. J., Jr. 1875
Silver L. T. 1533
Silverman M. P. 1921
Simoneit B. R. 1779
Skerra B. 1233
Smales A. A. 1575
Spettel B. 1719
Stalling D. L. 1845
Steinborn T. L. 1177
Steinbrunn F. 1331
Smith J. W. 1317, 1813, 1857
Stettler A. 1037
Stevens C. M. 1097
Stevens R. D. 1729
Stoenner R. W. 1583

Tatsumoto M. 1499, 1595
Taylor H. P., Jr. 1085, 1613
Taylor S. R. 1627
Tera F. 1637
Teschke F. 1719
Turekian K. K. 1659
Turner G. 1665

Updegrove W. S. 1901
Urey H. C. 1357, 1879

Vernon M. J. 1007
Vilcsek E. 995
Voshage H. 1269

Wakita H. 1165, 1685
Wänke H. 995, 1269, 1719
Wanless R. K. 1729
Wasserburgh G. J. 1637
Wasson J. T. 1741
Webb M. S. W. 1575
Weber H. W. 1269
Webster R. K. 1575
Wetherill G. W. 1195
Wiesmann H. 1143
Wiik H. B. 1369
Wilson J. D. 1575
Wlotzka F. 1283
Woelfle R. 1233
Woeller F. 1813
Wogman N. A. 1455
Wolf C. J. 1901
Wrigley R. C. 1751

Yaniv A. 1247, 1261
Yen Teh Fu 1929
Young M. 1879

Zähringer J. 1111, 1331
Zlatkis A. 1901
Zumwalt R. W. 1845
Zweifel K. A. 1213